Third edition

PRESTRESSED CONCRETE

PRENTICE HALL INTERNATIONAL SERIES
IN CIVIL ENGINEERING AND ENGINEERING MECHANICS

William J. Hall, Editor

AU AND CHRISTIANO, *Structural Analysis*
BARSOM AND ROLFE, *Fracture and Fatigue Control in Structures, 2/E*
BATHE, *Finite Element Procedures in Engineering Analysis*
BERG, *Elements of Structural Dynamics*
BIGGS, *Introduction to Structural Engineering*
CHAJES, *Structural Analysis, 2/E*
CHOPRA, *Dynamics of Structures*
COOPER AND CHEN, *Designing Steel Structures*
CORDING, ET AL., *The Art and Science of Geotechnical Engineering*
GALLAGHER, *Finite Element Analysis*
HENDRICKSON AND AU, *Project Management for Construction*
HIGDON ET AL., *Engineering Mechanics, 2nd Vector Edition*
HOLTZ AND KOVACS, *Introduction to Geotechnical Engineering*
HUMAR, *Dynamics of Structures*
JOHNSTON, LIN AND GALAMBOS, *Basic Steel Design, 3/E*
KELKAR AND SEWELL, *Fundamentals of the Analysis and Design of Shell Structures*
MACGREGOR, *Reinforced Concrete: Mechanics and Design, 2/E*
MEHTA, *Concrete: Structure, Properties and Materials*
MELOSH, *Structural Engineering Analysis by Finite Elements*
MEREDITH ET AL., *Design and Planning of Engineering Systems*
MINDNESS AND YOUNG, *Concrete*
NAWY, *Prestressed Concrete, 3/E*
NAWY, *Reinforced Concrete: A Fundamental Approach, 4/E*
POPOV, *Engineering Mechanics of Solids*
POPOV, *Introduction to the Mechanics of Solids*
POPOV, *Mechanics of Materials, 2/E*
SENNETT, *Matrix Analysis of Structures*
SCHNEIDER AND DICKEY, *Reinforced Masonry Design, 2/E*
WANG AND SALMON, *Introductory Structural Analysis*
WEAVER AND JOHNSON, *Finite Elements for Structural Analysis*
WEAVER AND JOHNSON, *Structural Dynamics by Finite Elements*
WOLF, *Dynamic Soil-Structure Interaction*
WRAY, *Measuring Engineering Properties of Soils*
YANG, *Finite Element Structural Analysis*

Third edition

PRESTRESSED CONCRETE

A Fundamental Approach

Dr. Edward G. Nawy, P.E.
Distinguished Professor
Department of Civil and Environmental Engineering
Rutgers University
The State University of New Jersey

PRENTICE HALL, *Upper Saddle River,* New Jersey 07458

Library of Congress Cataloging-in-Publication Data

Nawy, Edward G.
Prestressed concrete : a fundamental approach / Edward G. Nawy. --
3rd ed.
p. cm.
Includes bibliographical references (p.
ISBN 0-13-020593-1
1. Prestressed concrete. I. Title.
TA683.9.N39 1999
624. 1′83412—dc21 99-38322
CIP

Editor-in-chief: *Marcia Horton*
Editorial assistant: *Dolores Mars*
Marketing manager: *Danny Hoyt*
Production editor: *Pine Tree Composition*
Executive managing editor: *Vince O'Brien*
Managing editor: *David George*
Art director: *Jayne Conte*
Cover design: *Bruce Kenselaar*
Manufacturing manager: *Trudy Pisciotti*
Manufacturing buyer: *Beth Sturla*
Assistant vice president of production and manufacturing: *David W. Riccardi*

COVER PHOTO CREDIT:
Natchez Parkway Arches, Nashville Tennessee, America's first Segmental Arch Bridge, principal arch span is 582-ft. long and has a vertical clearance of 137 ft (Courtesy Figg Engineering Group, Tallahassee, Florida)

Printed in the United States of America
10 9 8 7 6 5 4 3 2 1

ISBN 0-13-020593-1

Prentice-Hall International (UK) Limited, *London*
Prentice-Hall of Australia Pty. Limited, *Sydney*
Prentice-Hall Canada Inc., *Toronto*
Prentice-Hall Hispanoamericana, S.A., *Mexico*
Prentice-Hall of India Private Limited, *New Delhi*
Prentice-Hall of Japan, Inc., *Tokyo*
Pearson Education Asia Pte. Ltd., *Singapore*
Editora Prentice-Hall do Brasil, Ltda., *Rio de Janeiro*

To
RACHEL E. NAWY

For her high limit state of stress endurance over the years,
which made the writing of this book in its several editions a reality.

CONTENTS

5 SHEAR AND TORSIONAL STRENGTH DESIGN 221

12 LRFD AND STANDARD AASHTO DESIGN OF CONCRETE BRIDGES **724**

13 SEISMIC DESIGN OF PRESTRESSED CONCRETE STRUCTURES 804

PREFACE

Prestressed concrete is a widely used material in construction. Hence, graduates of every civil engineering program must have, as a minimum requirement, a basic understanding of the fundamentals of linear and circular prestressed concrete. The high technology advancements in the science of materials have made it possible to construct and assemble large-span systems such as cable-stayed bridges, segmental bridges, nuclear reactor vessels, and offshore oil drilling platforms—work hitherto impossible to undertake.

Reinforced concrete's tensile strength is limited, while its compressive strength is extensive. Consequently, prestressing becomes essential in many applications in order to fully utilize the compressive strength and, through proper design, to eliminate or control cracking and deflection. Additionally, design of the members of a total structure is best achieved only by trial and adjustment: assuming a section and then analyzing it. Hence, design and analysis are combined in this work in order to make it simpler for the student first introduced to the subject of prestressed concrete design.

This third edition of the book extensively revises the previous text so as to conform to the new ACI 318-99 Code and the International Building Code, IBC 2000, for seismic design. The text is the outgrowth of the author's lecture notes developed in teaching the subject at Rutgers University over the past forty years and the experience accumulated over the years in teaching and research in the areas of reinforced and prestressed concrete inclusive of the Ph.D. level. The material is presented in such a manner that the student can become familiarized with the properties of plain concrete, both normal and high strength, and its components prior to embarking on the study of structural behavior. The book is uniquely different from other textbooks on the subject in that the major topics of material behavior, prestress loss, flexure, shear, and torsion are self-contained and can be covered in one semester at the senior level and the graduate level. The in-depth discussions of these topics permit the advanced undergraduate and graduate student, as well as the design engineer to develop with minimum effort a profound understanding of fundamentals of prestressed concrete structural behavior and performance.

The concise discussion presented in Chapters 1 through 3 on basic principles, the historical development of prestressed concrete, the properties of constituent materials, the long-term basic behavior of such materials, and the evaluation of prestress losses should give an adequate introduction to the subject of prestressed concrete. They should also aid in developing fundamental knowledge regarding the reliability of performance of prestressed structures, a concept to which every engineering student should be exposed today.

Chapters 4 and 5 on flexure, shear, and torsion, with the step-by-step logic of trial and adjustment as well as the flowcharts shown, give the student and the engineer a basic understanding of both the service load and the limit state of load at failure, thereby producing a good feel for the reserve strength and safety factors inherent in the design expressions. Chapter 4 in this edition contains the latest design procedure with numerical examples for the design of end anchorages of post-tensioned members as required by the latest ACI and AASHTO codes, inclusive of the "strut and tie" method of end-anchorage design. All examples using single-tees were replaced by double-tees since use of single-tees is no longer current. Chapter 5 presents, with design examples, the provisions on torsion combined with shear and bending, which include a unified approach to the topic of torsion in reinforced and prestressed concrete members. SI units examples included in the text in addition to having equivalent SI conversions for the major steps of examples throughout the book. Additionally, a detailed theoretical discussion is presented on the mechanisms of shear and torsion, the various approaches to the torsional problem and the plastic concepts of the shear equilibrium and torsional equilibrium theories and their interaction.

Furthermore, inclusion in this edition of new design examples is SI Units and a listing of the relevant equations in SI format extends the scope of the text to cover wider applications by the profession. In this manner, the student as well as the practicing engineer can avail themselves with the tools for using either the lb-in. (PI) system or the international (SI) system.

Chapter 6 on indeterminate prestressed concrete structures covers in detail continuous prestressed beams as well as portal frames. Numerous detailed examples illustrate the use of the basic concepts method, the C-line method and the load-balancing method presented in Chapter 1. Chapter 7 was revamped and all the examples were changed using double-tees for deflection computation for both noncomposite and composite members. The chapter discusses in detail the design for camber, deflection, and crack control considering both short- and long-term effects using three different approaches: the PCI multipliers method, the detailed incremental time steps method, and the approximate time steps method. A state-of-the-art discussion is presented, based on the author's work, of the evaluation and control of flexural cracking in partially prestressed beams. Several design examples are included in the discussion. Chapter 8 covers the proportioning of prestressed compression and tension members, including the buckling behavior and design of prestressed columns and piles and the P-Δ effect in the design of slender columns. A new section was added presenting a modified easier to use reciprocal method for biaxial bending design of columns.

Chapter 9 presents a thorough analysis of the service load behavior and yield-line behavior of two-way action prestressed slabs and plates. The service load behavior utilizes, with extensive examples, the equivalent frame method of flexural design (analysis) and deflection evaluation. Detailed discussion is given on shear-moment transfer and on deflection of two-way plates with computational examples. Extensive coverage is presented of the yield-line failure mechanisms of all the usual combinations of loads on floor slabs and boundary conditions, including the design expressions for these various conditions. Chapter 10 on connections for prestressed concrete elements covers the design of connections for dapped-end beams, ledge beams, and bearing, in addition to the design of the beams and corbels presented in Chapter 5 on shear and torsion.

This book is also unique in that Chapter 11 gives a detailed account of the analysis and design of prestressed concrete tanks and their shell roofs. Presented are the basics of the membrane and bending theories of cylindrical shells for use in the design of prestressed tanks for the various wall boundary conditions of fixed, semi-fixed, hinged, and sliding wall bases, as well as the incorporation of vertical prestressing. Chapter 11 also discusses the theory of axisymmetrical shells and domes that are used in the design of domed roofs for circular tanks.

A new and extensive Chapter 12 was added using the latest LRFD and Standard AASHTO specifications for the design of prestressed bridge deck girders for flexure, shear, torsion and serviceability, including the design of end-anchorage blocks. Several extensive examples are given using bulb-tees and box girder sections. The chapter also includes the AASHTO requirements for truck and lane loadings and load combinations as stipulated both by the LRFD and the Standard specifications.

A new and extensive Chapter 13 was added dealing with the seismic design of prestressed precast structures in high seismicity zones based on the latest ACI 318-99 and the International Building Code, IBC 2000, on sesimic design of reinforced and prestressed concrete structures. It contains several design examples and a detailed discussion of ductile moment-resistant connections in high-rise buildings and parking garages in high seismicity zones and a unique approach for the design of such ductile connections in precast beam-column joints. It also contains examples of the design of shear walls and hybrid connections—all based on the state of the art in this field.

It is important to emphasize that in this field, the use of computers is essential. Access to personal and handheld computers has made it possible for almost every student and engineer to be equipped with such a tool. Accordingly, Appendix A-1 presents a typical computer program in Q-BASIC for personal computers for the evaluation of time-dependent losses in prestress. Other programs as described in the appendix can be purchased from N.C.SOFTWARE, Box 161, East Brunswick, New Jersey, 08816. The inclusion of extensive flowcharts throughout the book and the discussion of the logic involved in them makes it possible for the reader to develop or use such programs without difficulty.

Selected photographs involving various areas of the structural behavior of concrete elements at failure are included in all the chapters. They are taken from research work published by the author with many of his MS and Ph.D. students at Rutgers University over the past four decades. Additionally, photographs of some major prestressed concrete "landmark" structures, are included throughout the book to illustrate the versatility of design in pretensioned and post-tensioned prestressed concrete. Appendices have also been included, with monograms and tables on standard properties, beam sections and charts of flexural and shear evaluation of sections, as well as representative tables for selecting sections such as PCI double-tees, PCI/AASHTO bulb tees, box girder, and AASHTO standard sections for bridge decks. Conversion to SI metric units are included in the examples throughout most chapters of the book.

The topics of the book have been presented in as concise a manner as possible without sacrificing the need for instructional details. The major portions of the text can be used without difficulty in an advanced senior-level course as well as at the graduate level for any student who has had a prior course in reinforced concrete. The contents should also serves as a valuable guideline for the practicing engineer who has to keep abreast of the state-of-the-art in prestressed concrete and the latest provisions of the ACI 318-99 Building Code and the International Building Code (IBC 2000), as well as the designer who seeks a concise treatment of the fundamentals of linear and circular prestressing.

ACKNOWLEDGMENTS

Grateful acknowledgement is due to the American Concrete Institute, the Prestressed Concrete Institute, and the Post-Tensioning Institute for their gracious support in permitting generous quotations from the ACI 318 and other relevant Codes and Reports and the numerous illustrations and tables from so many PCI and PTI publications. Special mention has to be made of the author's original mentor, the late Professor A. L. L. Baker of London University's Imperial College of Science, Technology, and Medicine, who inspired him with the affection that he has developed for systems constructed of reinforced and prestressed concrete. Grateful acknowledgement is also made to the author's countless students, both undergraduate and graduate, who have had much to do with generating the writing of this book and to the many who assisted in his research activities over the past forty years.

Acknowledgement is also made to the many who reviewed the manuscript of the first edition including Professors Carl E. Ekberg, Thomas T. C. Hsu, A. Fattah Shaikh, P. N. Balaguru, Daniel P. Jenny of the PCI, Clifford L. Freyrmuth of the PTI, and Ib Falk Jorgensen, President, Jorgensen, Hendrikson and Close, Denver. Particular thanks are due to Professor Thomas Hsu for again reviewing the revised portions on torsional theory and examples for the second edition and the shear LRFD section for the current edition. Thanks are also due to Professor Alex Aswad of Pennsylvania State University at Harrisburg for his valuable input on precast shear walls in seismic regions, to George Nasser, Editor-in-Chief, and Paul Johal, Research Director, both of the Precast/Prestressed Concrete Institute, and to Dr. Basile Rabbat, Director of Codes and Standards, Portland Cement Association, for their continuous advice and support. Thanks are due also to Mr. Khalid Shawwaf, Vice president-Engineering, Dywidag Systems International, for his cooperation and advice. Special thanks are due to Dr. Robert E. Englekirk, President, Englekirk Consulting Engineers, and Visiting Professor at the University of California, Los Angeles and San Diego, for his extensive input, discussions and advice on the subject of ductile moment-resisting frame connections in high sesimicity zones.

Thanks are also due to Professor A. Samer Ezeldin of Stevens Institute of Technology and to Robert M. Nawy, BS, BA, MBA, Rutgers Engineering Class of 1983 who assisted in work on the first edition of this book. Grateful acknowledgement is also made to the Prentice Hall officers and staff, to Marcia Horton, editor-in-chief and vice president, associate editor Alice Dworkin, Vincent O'Brien, Senior Executive Production Editor, who has rendered continuous support throughout, and Dolores Mars, technical assistant, and to Patty Donovan, Senior Project Coordinator, Pine Tree Composition, for their commendable efforts in producing this third enlarged edition of the book. Last but not least, the author is deeply grateful to his former students, Ms. Moria Treacy, MS, Princeton University, Ryan Laub, and Anand Bhatt, both MS, Rutgers University, for their diligence in processing and reviewing the extensive changes and the new additions incorporated in this third edition of the book.

Edward G. Nawy
Rutgers University
The State University of New Jersey
New Brunswick, New Jersey

1

BASIC CONCEPTS

1.1 INTRODUCTION

Concrete is strong in compression, but weak in tension: its tensile strength varies from 8 to 14 percent of its compressive strength. Due to such a low tensile capacity, flexural cracks develop at early stages of loading. In order to reduce or prevent such cracks from developing, a concentric or eccentric force is imposed in the longitudinal direction of the structural element. This force prevents the cracks from developing by eliminating or considerably reducing the tensile stresses at the critical midspan and support sections at service load, thereby raising the bending, shear, and torsional capacities of the sections. The sections are then able to behave elastically, and almost the full capacity of the concrete in compression can be efficiently utilized across the entire depth of the concrete sections when all loads act on the structure.

Such an imposed longitudinal force is called a *prestressing force,* i.e., a compressive force that prestresses the sections along the span of the structural element prior to the application of the transverse gravity dead and live loads or transient horizontal live loads. The type of prestressing force involved, together with its magnitude, are determined mainly on the basis of the type of system to be constructed and the span length and slenderness desired. Since the prestressing force is applied longitudinally along or parallel to

The Diamond Baseball Stadium, Richmond, Virginia. Situ cast and precast post-tensioned prestressed structure. (*Courtesy,* Prestressed Concrete Institute.)

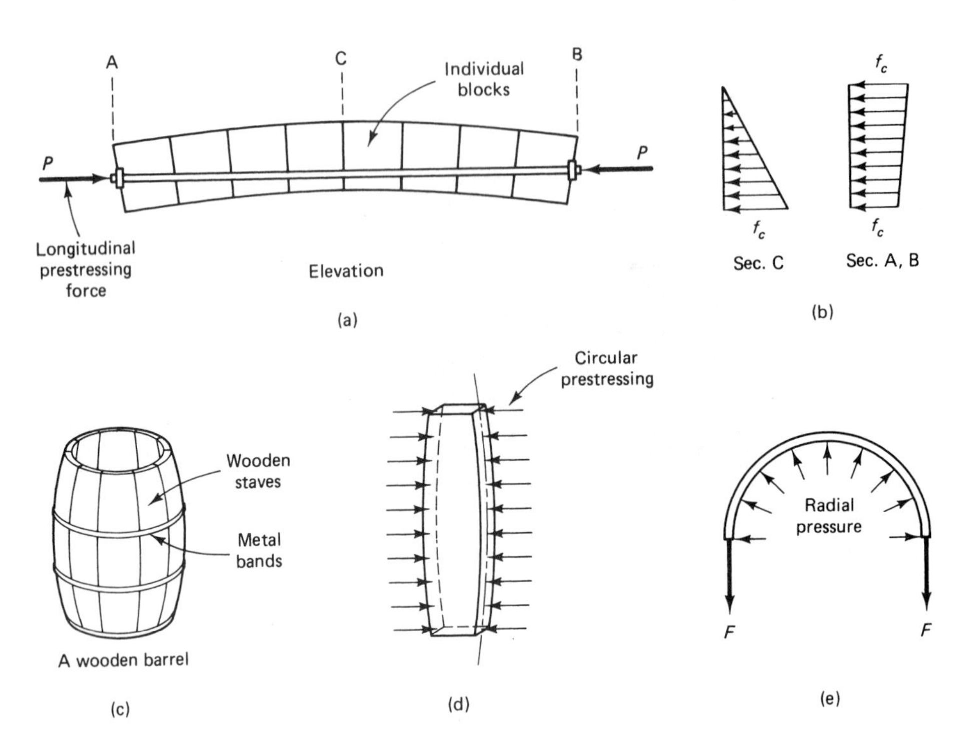

Figure 1.1 Prestressing principle in linear and circular prestressing. (a) Linear prestressing of a series of blocks to form a beam. (b) Compressive stress on midspan section C and end section A or B. (c) Circular prestressing of a wooden barrel by tensioning the metal bands. (d) Circular hoop prestress on one wooden stave. (e) Tensile force F on half of metal band due to internal pressure, to be balanced by circular hoop prestress.

the axis of the member, the prestressing principle involved is commonly known as *linear* prestressing.

Circular prestressing, used in liquid containment tanks, pipes, and pressure reactor vessels, essentially follows the same basic principles as does linear prestressing. The circumferential hoop, or "hugging" stress on the cylindrical or spherical structure, neutralizes the tensile stresses at the outer fibers of the curvilinear surface caused by the internal contained pressure.

Figure 1.1 illustrates, in a basic fashion, the prestressing action in both types of structural systems and the resulting stress response. In (a), the individual concrete blocks act together as a beam due to the large compressive prestressing force *P*. Although it might appear that the blocks will slip and vertically simulate shear slip failure, in fact they will not because of the longitudinal force *P*. Similarly, the wooden staves in (c) might appear to be capable of separating as a result of the high internal radial pressure exerted on them. But again, because of the compressive prestress imposed by the metal bands as a form of circular prestressing, they will remain in place.

1.1.1 Comparison with Reinforced Concrete

From the preceding discussion, it is plain that permanent stresses in the prestressed structural member are created before the full dead and live loads are applied, in order to eliminate or considerably reduce the net tensile stresses caused by these loads. With reinforced concrete, it is assumed that the tensile strength of the concrete is negligible

Photo 1.1 Bay Area Rapid Transit (BART), San Francisco and Oakland, California. Guideways consist of prestressed precast simple-span box girders 70 ft long and 11 ft wide. (*Courtesy,* Bay Area Rapid Transit District, Oakland, California.)

and disregarded. This is because the tensile forces resulting from the bending moments are resisted by the bond created in the reinforcement process. Cracking and deflection are therefore essentially irrecoverable in reinforced concrete once the member has reached its limit state at service load.

The reinforcement in the reinforced concrete member does not exert any force of its own on the member, contrary to the action of prestressing steel. The steel required to produce the prestressing force in the prestressed member actively preloads the member, permitting a relatively high controlled recovery of cracking and deflection. Once the flex-

Photo 1.2 Chaco-Corrientes Bridge, Argentina. The longest precast prestressed concrete cable-stayed box girder bridge in South America. (*Courtesy,* Ammann & Whitney.)

Photo 1.3 Park Towers, Tulsa, Oklahoma. (*Courtesy,* Prestressed Concrete Institute.)

ural tensile strength of the concrete is exceeded, the prestressed member starts to act like a reinforced concrete element.

By controlling the amount of prestress, a structural system can be made either flexible or rigid without influencing its strength. In reinforced concrete, such a flexibility in behavior is considerably more difficult to achieve if considerations of economy are to be observed in the design. Flexible structures such as fender piles in wharves have to be highly energy absorbent, and prestressed concrete can provide the required resiliency. Structures designed to withstand heavy vibrations, such as machine foundations, can easily be made rigid through the contribution of the prestressing force to the reduction of their otherwise flexible deformation behavior.

1.1.2 Economics of Prestressed Concrete

Prestressed members are shallower in depth than their reinforced concrete counterparts for the same span and loading conditions. In general, the depth of a prestressed concrete member is usually about 65 to 80 percent of the depth of the equivalent reinforced concrete member. Hence, the prestressed member requires less concrete, and about 20 to 35 percent of the amount of reinforcement. Unfortunately, this saving in material weight is balanced by the higher cost of the higher quality materials needed in prestressing. Also, regardless of the system used, prestressing operations themselves result in an added cost: Formwork is more complex, since the geometry of prestressed sections is usually composed of flanged sections with thin webs.

In spite of these additional costs, if a large enough number of precast units are manufactured, the difference between at least the initial costs of prestressed and reinforced concrete systems is usually not very large. And the indirect long-term savings are quite substantial, because less maintenance is needed, a longer working life is possible due to better quality control of the concrete, and lighter foundations are achieved due to the smaller cumulative weight of the superstructure.

Once the beam span of reinforced concrete exceeds 70 to 90 feet, the dead weight of the beam becomes excessive, resulting in heavier members and, consequently, greater long-term deflection and cracking. Thus, for larger spans, prestressed concrete becomes mandatory since arches are expensive to construct and do not perform as well due to the severe long-term shrinkage and creep they undergo. Very large spans such as segmental bridges or cable-stayed bridges can *only* be constructed through the use of prestressing.

1.2 HISTORICAL DEVELOPMENT OF PRESTRESSING

Prestressed concrete is not a new concept, dating back to 1872, when P. H. Jackson, an engineer from California, patented a prestressing system that used a tie rod to construct beams or arches from individual blocks. [See Figure 1.1(a).] In 1888, C. W. Doehring of Germany obtained a patent for prestressing slabs with metal wires. But these early attempts at prestressing were not really successful because of the loss of the prestress with time. J. Lund of Norway and G. R. Steiner of the United States tried early in the twentieth century to solve this problem, but to no avail.

After a long lapse of time during which little progress was made because of the unavailability of high-strength steel to overcome prestress losses, R. E. Dill of Alexandria, Nebraska, recognized the effect of the shrinkage and creep (transverse material flow) of concrete on the loss of prestress. He subsequently developed the idea that successive post-tensioning of *unbonded* rods would compensate for the time-dependent loss of stress in the rods due to the decrease in the length of the member because of creep and shrinkage. In the early 1920s, W. H. Hewett of Minneapolis developed the principles of circular prestressing. He hoop-stressed horizontal reinforcement around walls of concrete tanks through the use of turnbuckles to prevent cracking due to internal liquid pres-

Photo 1.4 Wiscasset Bridge, Maine. (*Courtesy,* Post-Tensioning Institute.)

Photo 1.5 Executive Center, Honolulu, Hawaii. (*Courtesy,* Post-Tensioning Institute.)

sure, thereby achieving watertighteness. Thereafter, prestressing of tanks and pipes developed at an accelerated pace in the United States, with thousands of tanks of water, liquid, and gas storage built and much mileage of prestressed pressure pipe laid in the two to three decades that followed.

Linear prestressing continued to develop in Europe and in France, in particular through the ingenuity of Eugene Freyssinet, who proposed in 1926 through 1928 methods to overcome prestress losses through the use of high-strength and high-ductility steels. In 1940, he introduced the now well-known and well-accepted Freyssinet system comprising the conical wedge anchor for 12-wire tendons.

During World War II and thereafter, it became necessary to reconstruct in a prompt manner many of the main bridges that were destroyed by war activities. G. Magnel of Ghent, Belgium, and Y. Guyon of Paris extensively developed and used the concept of prestressing for the design and construction of numerous bridges in western and central Europe. The Magnel system also used wedges to anchor the prestressing wires. They differed from the original Freyssinet wedges in that they were flat in shape, accommodating the prestressing of two wires at a time.

P. W. Abeles of England introduced and developed the concept of partial prestressing between the 1930s and 1960s. F. Leonhardt of Germany, V. Mikhailov of Russia, and T. Y. Lin of the United States also contributed a great deal to the art and science of the design of prestressed concrete. Lin's load-balancing method deserves particular mention in this regard, as it considerably simplified the design process, particularly in continuous structures. These twentieth-century developments have led to the extensive use of prestressing throughout the world, and in the United States in particular.

Photo 1.6 Stratford "B" Condeep offshore oil drilling platform, Norway. (*Courtesy,* Ben C. Gerwick.)

Today, prestressed concrete is used in buildings, underground structures, TV towers, floating storage and offshore structures, power stations, nuclear reactor vessels, and numerous types of bridge systems including segmental and cable-stayed bridges. Note the variety of prestressed structures in the photos throughout the book; they demonstrate the versatility of the prestressing concept and its all-encompassing applications. The success in the development and construction of all these landmark structures has been due in no small measure to the advances in the technology of materials, particularly prestressing steel, and the accumulated knowledge in estimating the short- and long-term losses in the prestressing forces.

1.3 BASIC CONCEPTS OF PRESTRESSING

1.3.1 Introduction

The prestressing force P that satisfies the particular conditions of geometry and loading of a given element (see Figure 1.2) is determined from the principles of mechanics and of stress-strain relationships. Sometimes simplification is necessary, as when a prestressed beam is assumed to be homogeneous and elastic.

Consider, then, a simply supported rectangular beam subjected to a *concentric* prestressing force P as shown in Figure 1.2(a). The compressive stress on the beam cross section is uniform and has an intensity

Photo 1.7 Sunshine Skyway Bridge, Tampa Bay, Florida. Designed by Figg and Muller Engineers, Inc., the bridge has a 1,200-ft cable-stayed main span with a single pylon, 175-ft vertical clearance, and total length of 21,878 ft. It has twin 40-ft roadways and has 135-ft spans in precast segmental sections and high approaches to elevation + 130 ft. (*Courtesy,* Figg and Muller Engineers, Inc.)

$$f = -\frac{P}{A_c} \tag{1.1}$$

where $A_c = bh$ is the cross-sectional area of a beam section of width b and total depth h. A *minus* sign is used for compression and a *plus* sign for tension throughout the text. Also, bending moments are drawn on the tensile side of the member.

If external transverse loads are applied to the beam, causing a maximum moment M at midspan, the resulting stress becomes

$$f^t = -\frac{P}{A} - \frac{Mc}{I_g} \tag{1.2a}$$

and

$$f_b = -\frac{P}{A} + \frac{Mc}{I_g} \tag{1.2b}$$

where f^t = stress at the top fibers
f_b = stress at the bottom fibers
$c = \frac{1}{2}h$ for the rectangular section
I_g = gross moment of inertia of the section ($bh^3/12$ in this case)

Equation 1.2b indicates that the presence of prestressing-compressive stress $-P/A$ is reducing the tensile flexural stress Mc/I to the extent intended in the design, either elimi-

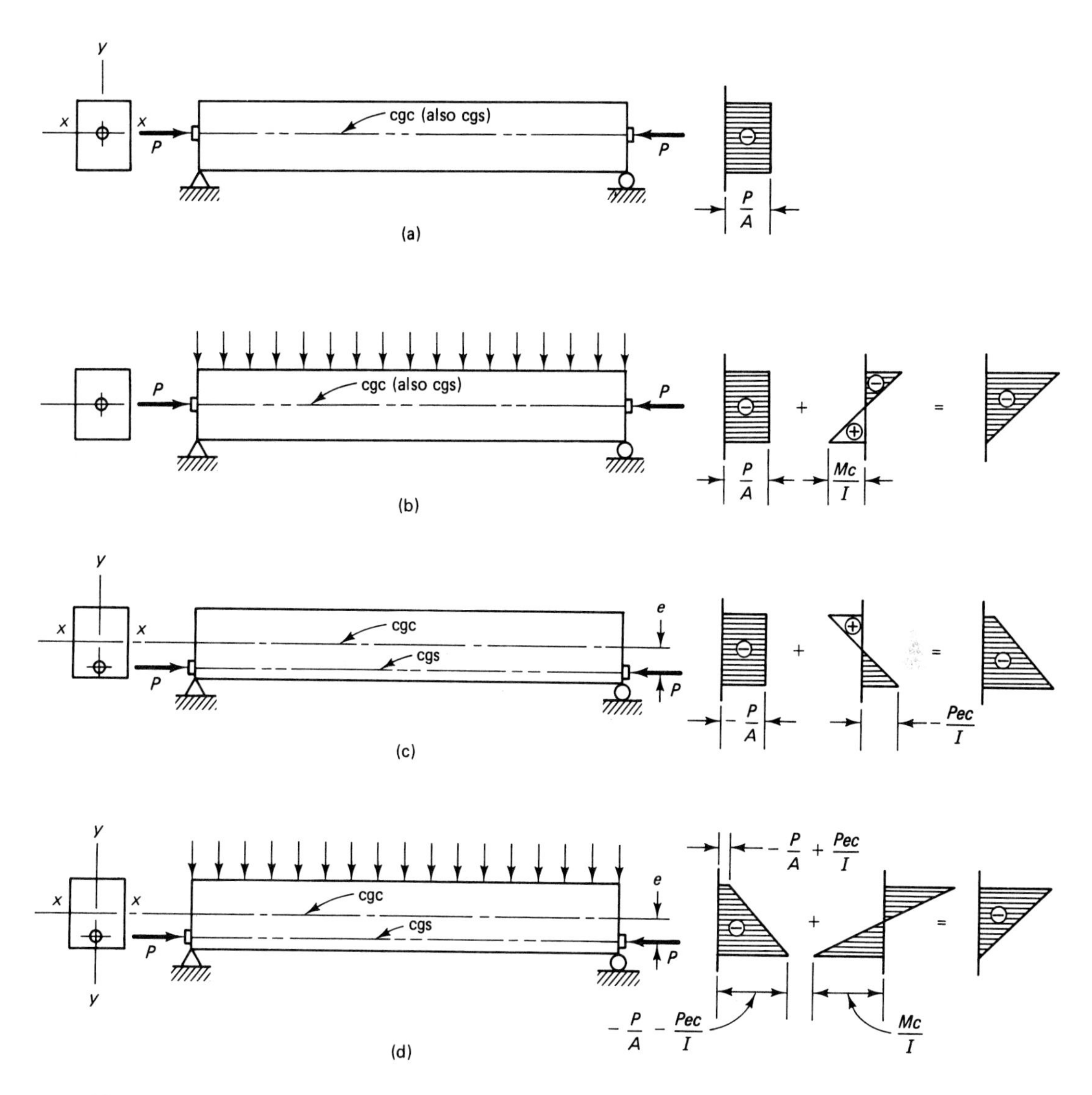

Figure 1.2 Concrete fiber stress distribution in a rectangular beam with straight tendon. (a) Concentric tendon, prestress only. (b) Concentric tendon, self-weight added. (c) Eccentric tendon, prestress only. (d) Eccentric tendon, self-weight added.

nating tension totally (even inducing compression), or permitting a level of tensile stress within allowable code limits. The section is then considered uncracked and behaves elastically: the concrete's inability to withstand tensile stresses is effectively compensated for by the compressive force of the prestressing tendon.

The compressive stresses in Equation 1.2a at the top fibers of the beam due to prestressing are compounded by the application of the loading stress $-Mc/I$, as seen in Figure 1.2(b). Hence, the compressive stress capacity of the beam to take a substantial external load is reduced by the *concentric* prestressing force. In order to avoid this limitation, the prestressing tendon is placed *eccentrically* below the neutral axis at midspan, to induce tensile stresses at the top fibers due to prestressing. [See Figure 1.2(c), (d).] If the tendon is placed at eccentricity e from the center of gravity of the concrete, termed the *cgc line,* it creates a moment Pe, and the ensuing stresses at midspan become

Photo 1.8 Douglas Bridge Crossing, Gastineau Channel, Juneau and Douglas, Alaska. (*Courtesy,* Prestressed Concrete Institute.)

$$f^t = -\frac{P}{A_c} + \frac{Pec}{I_g} - \frac{Mc}{I_g} \tag{1.3a}$$

$$f_b = -\frac{P}{A_c} - \frac{Pec}{I_g} + \frac{Mc}{I_g} \tag{1.3b}$$

Since the support section of a simply supported beam carries no moment from the external transverse load, high tensile fiber stresses at the top fibers are caused by the eccentric prestressing force. To limit such stresses, the eccentricity of the prestressing tendon profile, the *cgs line,* is made less at the support section than at the midspan section, or eliminated altogether, or else a negative eccentricity above the cgc line is used.

1.3.2 Basic Concept Method

In the basic concept method of designing prestressed concrete elements, the concrete fiber stresses are *directly* computed from the external forces applied to the concrete by longitudinal prestressing and the external transverse load. Equations 1.3a and b can be modified and simplified for use in calculating stresses at the initial prestressing stage and at service load levels. If P_i is the initial prestressing force before stress losses, and P_e is the effective prestressing force after losses, then

$$\gamma = \frac{P_e}{P_i} \tag{1.3a}$$

can be defined as the residual prestress factor. Substituting r^2 for I_g/A_c in Equations 1.3, where r is the radius of gyration of the gross section, the expressions for stress can be rewritten as follows:

(a) Prestressing Force Only

$$f^t = -\frac{P_i}{A_c}\left(1 - \frac{ec_t}{r^2}\right) \tag{1.4a}$$

$$f_b = -\frac{P_i}{A_c}\left(1 + \frac{ec_b}{r^2}\right) \tag{1.4b}$$

Photo 1.9 Tianjin Yong-He cable-stayed prestressed concrete bridge, Tianjin, China, the largest span bridge in Asia, with a total length of 1,673 ft and a suspended length of 1,535 ft, was completed in 1988. (*Credits* Owner: Tianjin Municipal Engineering Bureau. General contractor: Major Bridge Engineering Bureau of Ministry of Railways of China. Engineer for project design and construction control guidance: Tianjin Municipal Engineering Survey and Design Institute, Chief Bridge Engineer, Bang-yan Yu.)

where c_t and c_b are the distances from the center of gravity of the section (the cgc line) to the extreme top and bottom fibers, respectively.

(b) Prestressing Plus Self-weight

If the beam self-weight causes a moment M_b at the section under consideration, Equations 1.4a and b, respectively, become

$$f^t = -\frac{P_i}{A_c}\left(1 - \frac{ec_t}{r^2}\right) - \frac{M_D}{S^t} \tag{1.5a}$$

and

$$f_b = -\frac{P_i}{A_c}\left(1 + \frac{ec_b}{r^2}\right) + \frac{M_D}{S_b} \tag{1.5b}$$

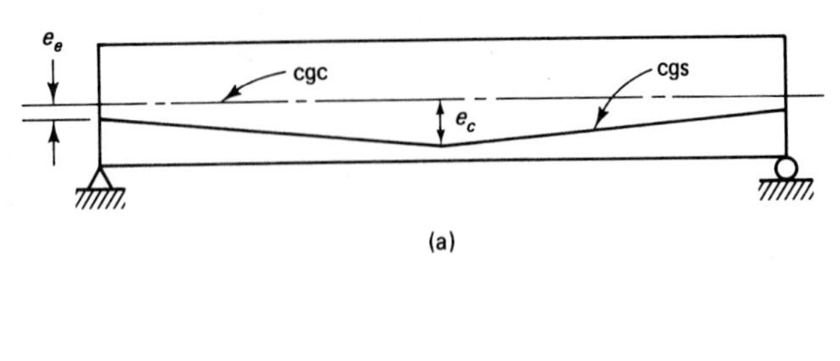

(a)

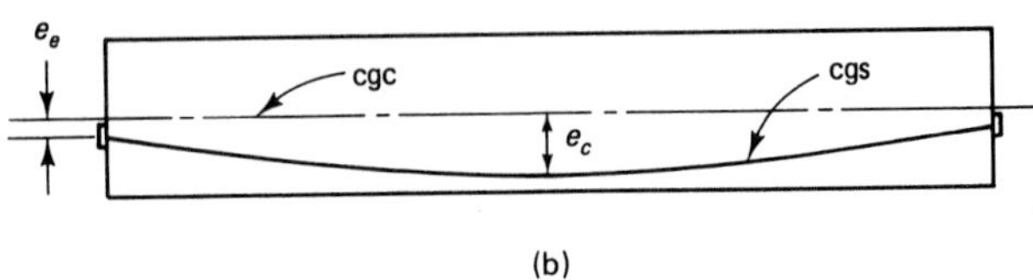

(b)

Figure 1.3 Prestressing tendon profile. (a) Harped tendon. (b) Draped tendon.

where S^t and S_b are the moduli of the sections for the top and bottom fibers, respectively.

The change in eccentricity from the midspan to the support section is obtained by raising the prestressing tendon either abruptly from the midspan to the support, a process called harping, or gradually in a parabolic form, a process called draping. Figure 1.3(a) shows a harped profile usually used for pretensioned beams and for concentrated transverse loads. Figure 1.3(b) shows a draped tendon usually used in post-tensioning.

Subsequent to erection and installation of the floor or deck, live loads act on the structure, causing a superimposed moment M_s. The full intensity of such loads normally occurs after the building is completed and some time-dependent losses in prestress have already taken place. Hence, the prestressing force used in the stress equations would have to be the effective prestressing force P_e. If the total moment due to gravity loads is M_T, then

$$M_T = M_D + M_{SD} + M_L \tag{1.6}$$

where M_D = moment due to self-weight
M_{SD} = moment due to superimposed dead load, such as flooring
M_L = moment due to live load, including impact and seismic loads if any

Equations 1.5 then become

$$f^t = -\frac{P_e}{A_c}\left(1 - \frac{ec_t}{r^2}\right) - \frac{M_T}{S^t} \tag{1.7a}$$

$$f_b = -\frac{P_e}{A_c}\left(1 + \frac{ec_b}{r^2}\right) + \frac{M_T}{S_b} \tag{1.7b}$$

Some typical elastic concrete stress distributions at the critical section of a prestressed flanged section are shown in Figure 1.4. The tensile stress in the concrete in part (c) permitted at the extreme fibers of the section cannot exceed the maximum permissible in the code, e.g., $f_t = 6\sqrt{f'_c}$ in the ACI code. If it is exceeded, bonded non-prestressed reinforcement proportioned to resist the total tensile force has to be provided to control cracking at service loads.

1.3.3 C-Line Method

In this line-of-pressure or thrust concept, the beam is analyzed as if it were a plain concrete elastic beam using the basic principles of statics. The prestressing force is considered an ex-

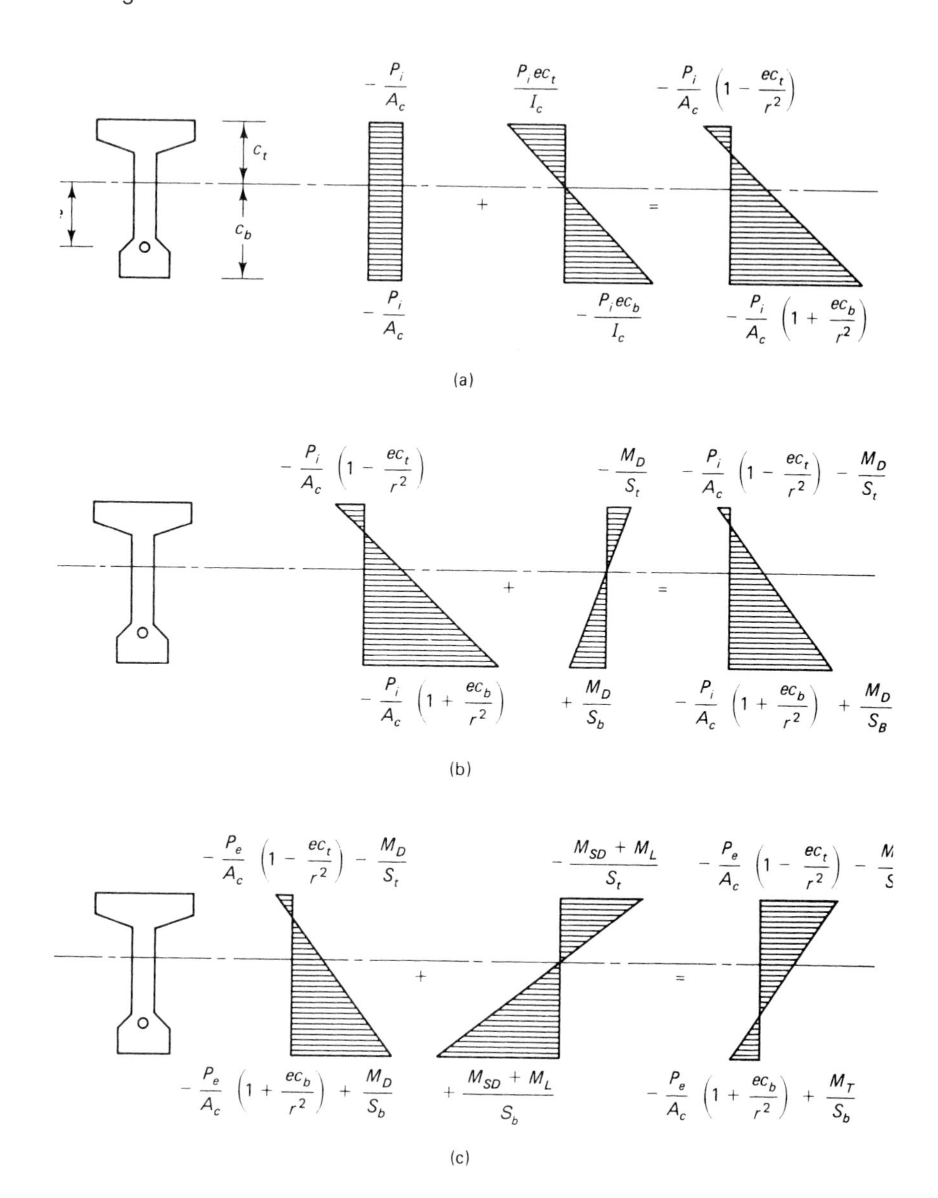

Figure 1.4 Elastic fiber stresses due to the various loads in a prestressed beam. (a) Initial prestress before losses. (b) Addition of self-weight. (c) Service load at effective prestress.

ternal compressive force, with a constant tensile force T in the tendon throughout the span. In this manner, the effects of external gravity loads are disregarded. Equilibrium equations $\Sigma H = 0$ and $\Sigma M = 0$ are applied to maintain equilibrium in the section.

Figure 1.5 shows the relative line of action of the compressive force C and the tensile force T in a reinforced concrete beam as compared to that in a prestressed concrete beam. It is plain that in a reinforced concrete beam, T can have a finite value only when transverse and other external loads act. The moment arm a remains basically constant throughout the elastic loading history of the reinforced concrete beam while it changes from a value $a = 0$ at prestressing to a maximum at full superimposed load.

Taking a free-body diagram of a segment of a beam as in Figure 1.6, it is evident that the C-line, or center-of-pressure line, is at a varying distance a from the T-line. The moment is given by

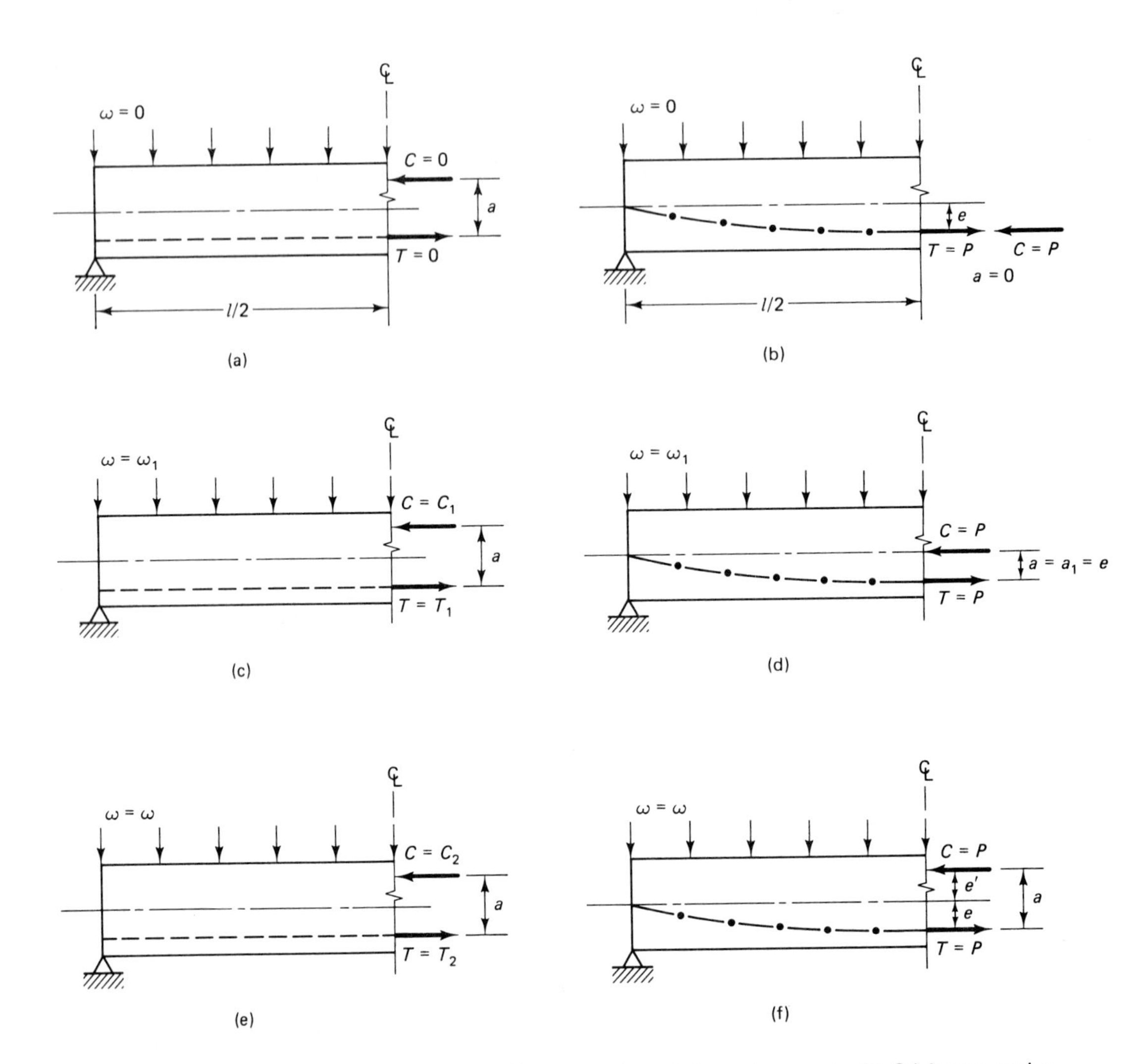

Figure 1.5 Comparative free-body diagrams of a reinforced concrete (R.C.) beam and a prestressed concrete (P.C.) beam. (a) R.C. beam with no load. (b) P.C. beam with no load. (c) R.C. beam with load w_1. (d) P.C. beam with load w_1. (e) R.C. beam with typical load w. (f) P.C. beam with typical load w.

$$M = Ca = Ta \tag{1.8}$$

and the eccentricity e is known or predetermined, so that in Figure 1.6,

$$e' = a - e \tag{1.9a}$$

Since $C = T$, $a = M/T$, giving

$$e' = \frac{M}{T} - e \tag{1.9b}$$

From the figure,

$$f^t = -\frac{C}{A_c} - \frac{Ce'c_t}{I_c} \tag{1.10a}$$

$$f_b = -\frac{C}{A_c} + \frac{Ce'c_b}{I_c} \tag{1.10b}$$

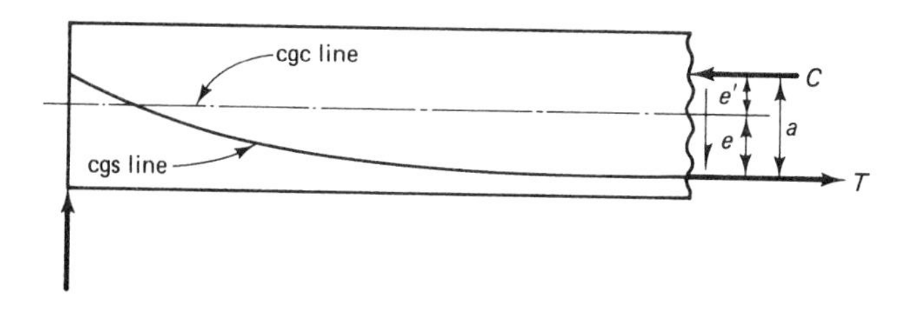

Figure 1.6 Free-body diagram for the C-line (center of pressure).

But in the tendon the force T equals the prestressing force P_e; so

$$f^t = -\frac{P_e}{A_c} - \frac{P_e e' c_t}{I_c} \tag{1.11a}$$

$$f_b = -\frac{P_e}{A_c} + \frac{P_e e' c_b}{I_c} \tag{1.11b}$$

Since $I_c = A_c r^2$, Equations 1.11a and b can be rewritten as

$$f^t = -\frac{P_e}{A_c}\left(1 + \frac{e' c_t}{r^2}\right) \tag{1.12a}$$

$$f_b = -\frac{P_e}{A_c}\left(1 - \frac{e' c_b}{r^2}\right) \tag{1.12b}$$

Equations 1.12a and b and Equations 1.7a and b should yield identical values for the fiber stresses.

1.3.4 Load-Balancing Method

A third useful approach in the design (analysis) of continuous prestressed beams is the load-balancing method developed by Lin and mentioned earlier. This technique is based on utilizing the vertical force of the draped or harped prestressing tendon to counteract or balance the imposed gravity loading to which a beam is subjected. Hence, it is applicable to nonstraight prestressing tendons.

Photo 1.10 East Huntington Bridge over Ohio River. A segmentally assembled precast prestressed concrete cable-stayed bridge spanning 200–900–608 ft. (*Courtesy,* Arvid Grant and Associates, Inc.)

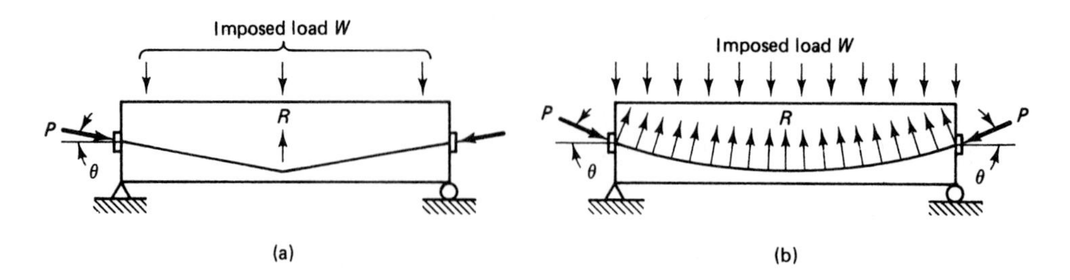

Figure 1.7 Load-balancing forces. (a) Harped tendon. (b) Draped tendon.

Figure 1.7 demonstrates the balancing forces for both harped- and draped-tendon prestressed beams. The load balancing reaction R is equal to the vertical component of the prestressing force P. The horizontal component of P, as an approximation in long-span beams, is taken to be equal to the full force P in computing the concrete fiber stresses *at midspan* of the simply supported beam. At other sections, the actual horizontal component of P is used.

1.3.4.1 Load-Balancing Distributed Loads and Parabolic Tendon Profile. Consider a parabolic tendon as shown in Figure 1.8. Let the parabolic function

$$Ax^2 + Bx + C = y \tag{1.13}$$

represent the tendon drape; the force T denotes the pull to which the tendon is subjected. Then for $x = 0$, we have

$$y = 0 \qquad C = 0$$

$$\frac{dy}{dx} = 0 \qquad B = 0$$

and for $x = l/2$,

$$y = a \qquad A = \frac{4a}{l^2}$$

But from calculus, the load intensity is

$$q = T\frac{\partial^2 y}{\partial x^2} \tag{1.14}$$

Finding $\partial^2 y/\partial x^2$ in Equation 1.13 and substituting into Equation 1.14 yields

$$q = T\frac{4a}{l^2} \times 2 = \frac{8Ta}{l^2} \tag{1.15a}$$

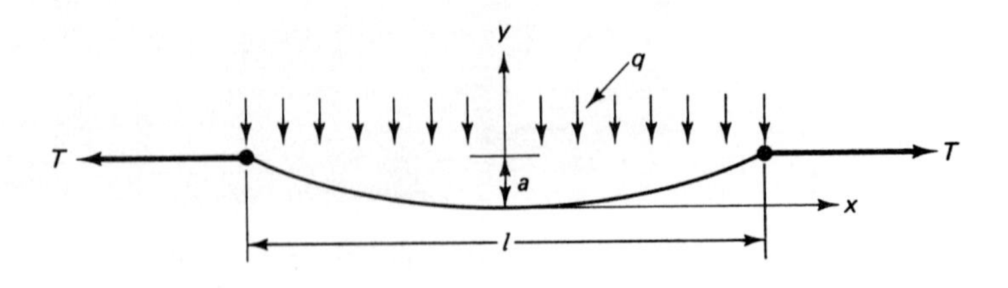

Figure 1.8 Sketched tendon subjected to transverse load intensity q.

or

$$T = \frac{ql^2}{8a} \tag{1.15b}$$

$$Ta = \frac{ql^2}{8} \tag{1.15c}$$

Hence, if the tendon has a parabolic profile in the prestressed beam and the prestressing force is denoted by P, the balanced-load intensity, from Equation 1.15a, is

$$w_b = \frac{8Pa}{l^2} \tag{1.16}$$

Figure 1.9 gives a free-body diagram of the forces acting on a prestressed beam with a parabolic tendon profile. Clearly, the two sets of equal and opposite transverse loads w_b *cancel* each other, and no bending stress is produced. This is reasonable to expect in the load-balancing method, since it is always the case that $T = C$, and C has to cancel T to satisfy the equilibrium requirement that $\Sigma H = 0$. As there is no bending, the beam remains straight, without having a convex shape, or camber, at the top face.

The concrete fiber stress across the depth of the section at midspan becomes

$$f_b^t = -\frac{P'}{A} = -\frac{C}{A} \tag{1.17}$$

This stress, which is constant, is due to the force $P' = P \cos \theta$. Figure 1.10 shows the superposition of stresses to yield the net stress. Note that the prestressing force in the load-balancing method has to act at the center of gravity (cgc) of the support section in simply supported beams and at the cgc of the free end in the case of a cantilever beam. This condition is necessary in order to prevent any eccentric unbalanced moments.

When the imposed load exceeds the balancing load w_b such that an additional *unbalanced* load w_{ub} is applied, a moment $M_{ub} = w_{ub}l^2/8$ results at midspan. The corresponding fiber stresses at midspan become

$$f_b^t = -\frac{P'}{A_c} \mp \frac{M_{ub}c}{I_c} \tag{1.18}$$

Equation 1.18 can be rewritten as the two equations

$$f^t = -\frac{P'}{A_c} - \frac{M_{ub}}{S^t} \tag{1.19a}$$

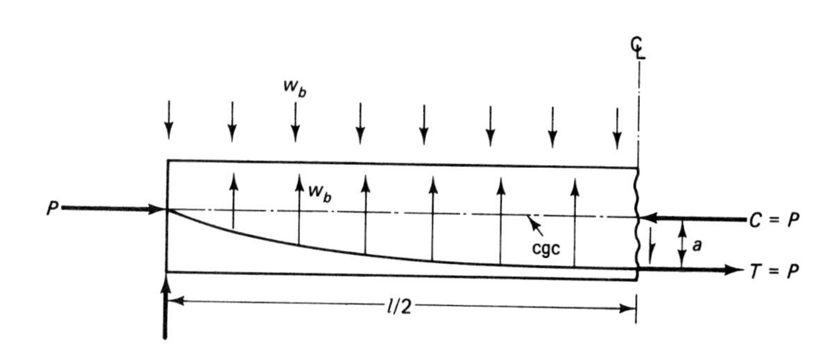

Figure 1.9 Load-balancing force on free-body diagram.

Photo 1.11 Walt Disney World Monorail, Orlando, Florida. A series of hollow precast prestressed concrete 100-foot box girders individually post-tensioned to provide six-span continuous structure. Designed by ABAM Engineers, Tacoma, Washington. (*Courtesy,* Walt Disney World Corporation.)

and

$$f_b = -\frac{P'}{A_c} + \frac{M_{ub}}{S_b} \tag{1.19b}$$

Equations 1.19 will yield the same values of fiber stresses as Equations 1.7 and 1.12. Keep in mind that P' is taken to be equal to P at the midspan section because the prestressing force is horizontal at this section, i.e., $\theta = 0$.

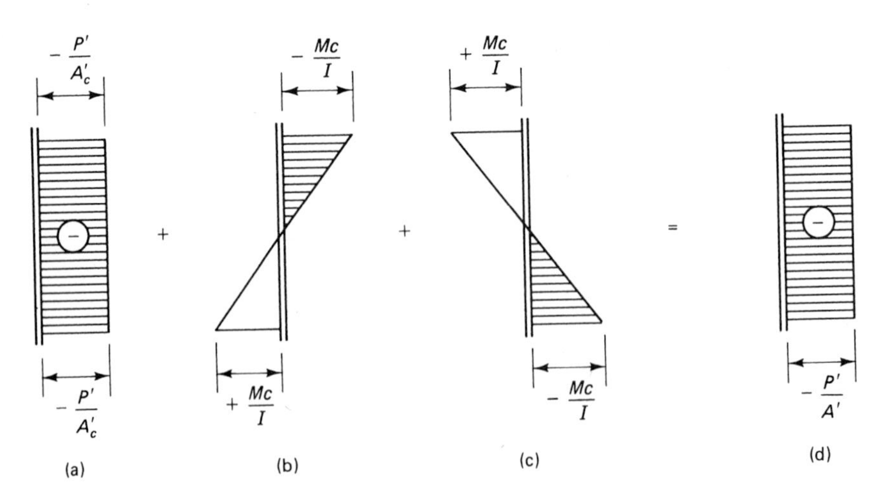

Figure 1.10 Load-balancing stresses. (a) Prestress stresses. (b) Imposed-load stresses. (c) Balanced-load stresses. (d) Net stress.

1.4 COMPUTATION OF FIBER STRESSES IN A PRESTRESSED BEAM BY THE BASIC METHOD

Example 1.1

A pretensioned simply supported 10LDT24 double T-beam without topping has a span of 64 ft (19.51 m) and the geometry shown in Figure 1.11. It is subjected to a uniform superimposed gravity dead-load intensity W_{SD} and live-load intensity W_L summing to 420 plf (6.13 kN/m). The initial prestress before losses is $f_{pi} \cong 0.70 f_{pu} = 189{,}000$ psi (1,303 MPa), and the effective prestress after losses is $f_{pe} = 150{,}000$ psi (1,034 MPa). Compute the extreme fiber stresses at the midspan due to

(a) the initial full prestress and no external gravity load
(b) the final service load conditions when prestress losses have taken place.

Allowable stress data are as follows:

$f'_c = 6{,}000$ psi, lightweight (41.4 MPa)

$f_{pu} = 270{,}000$ psi, stress relieved (1.862 MPa) = specified tensile strength of the tendons

$f_{py} = 220{,}000$ psi (1.517 MPa) = specified yield strength of the tendons

$f_{pe} = 150{,}000$ psi (1,034 MPa)

$f_t = 12\sqrt{f'_c} = 930$ psi (6.4 MPa) = maximum allowable tensile stress in concrete

$f'_{ci} = 4{,}800$ psi (33.1 MPa) = concrete compressive strength at time of initial prestress

$f_{ci} = 0.6 f'_{ci} = 2{,}880$ psi (19.9 MPa) = maximum allowable stress in concrete at initial prestress

$f_c = 0.45 f'_c$ = maximum allowable compressive stress in concrete at service

Assume that ten $\frac{1}{2}$in.-dia. Seven-wire-strand (ten 12.7-mm-dia strand) tendons with a 108-D1 strand pattern are used to prestress the beam.

$$A_c = 449 \text{ in.}^2 \ (2{,}915 \text{ cm}^2)$$

$$I_c = 22{,}469 \text{ in.}^4 \ (935{,}347 \text{ cm}^4)$$

$$r^2 = I_c/A_c = 50.04 \text{ in.}^2$$

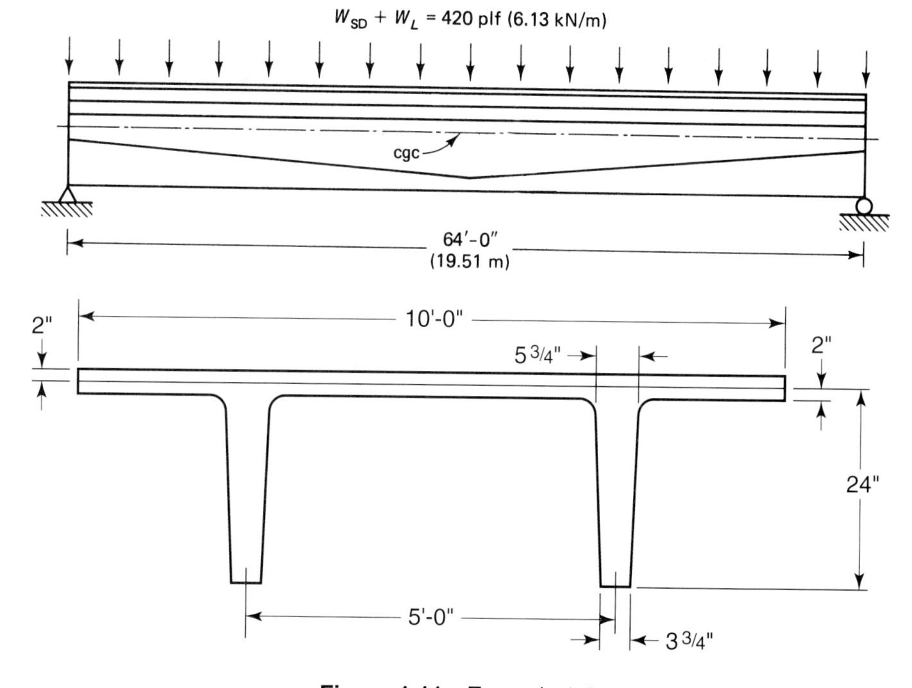

Figure 1.11 Example 1.1

Photo 1.12 Prestressed lightweight concrete mid-body for Arctic offshore drilling platform. Global Marine Development. (*Courtesy,* Ben C. Gerwick.)

$$c_b = 17.77 \text{ in. } (452 \text{ mm})$$
$$c_t = 6.23 \text{ in. } (158 \text{ mm})$$
$$e_c = 14.77 \text{ in. } (375 \text{ mm})$$
$$e_e = 7.77 \text{ in. } (197 \text{ mm})$$
$$S_b = 1{,}264 \text{ in.}^3 \ (20{,}714 \text{ cm}^3)$$
$$S^t = 3{,}607 \text{ in.}^3 \ (59{,}108 \text{ cm}^3)$$
$$W_D = 359 \text{ plf } (4.45 \text{ kN/m})$$

Solution:

(i) *Initial Conditions at Prestressing*

$$A_{ps} = 10 \times 0.153 = 1.53 \text{ in.}^2$$
$$P_i = A_{ps} f_{pi} = 1.53 \times 189{,}000 = 289{,}170 \text{ lb } (1{,}287 \text{ kN})$$
$$P_e = 1.53 \times 150{,}000 = 229{,}500 \text{ lb } (1{,}020 \text{ kN})$$

The midpan self-weight dead-load moment is

$$M_D = \frac{wl^2}{8} = \frac{359\,(64)^2}{8} \times 12 = 2{,}205{,}696 \text{ in.-lb. } (249 \text{ kN-m})$$

From Equations 1.5 and 1.7,

$$f^t = -\frac{P_i}{A_c}\left(1 - \frac{ec_t}{r^2}\right) - \frac{M_D}{S^t}$$
$$= -\frac{289{,}170}{449}\left(1 - \frac{14.77 \times 6.23}{50.04}\right) - \frac{2{,}205{,}696}{3{,}607}$$
$$= +54.03 - 611.5 \cong -70 \text{ psi } (C)$$

$$f_b = -\frac{P_i}{A_c}\left(1 + \frac{ec_b}{r^2}\right) + \frac{M_D}{S_b}$$
$$= -\frac{289{,}170}{449}\left(1 + \frac{14.77 \times 17.77}{50.04}\right) + \frac{2{,}205{,}696}{1{,}264}$$

$$= -4{,}022.1 + 1{,}745.0 \cong -2{,}277 \text{ psi } (C)$$

$$\leq f_{ci} = -2{,}880 \text{ psi allowed, O.K.}$$

(ii) ***Final Condition at Service Load*** Midspan moment due to superimposed dead and live load is

$$M_{SD} + M_L = \frac{420\,(64)^2}{8} \times 12 = 2{,}580{,}480 \text{ in.-lb}$$

$$\text{Total Moment } M_T = 2{,}205{,}696 + 2{,}580{,}480$$

$$= 4{,}786{,}176 \text{ in.-lb. (541 kN-m)}$$

$$f^t = -\frac{P_e}{A_c}\left(1 - \frac{ec_t}{r^2}\right) - \frac{M_T}{S^t}$$

$$= -\frac{229{,}500}{449}\left(1 - \frac{14.77 \times 6.23}{50.04}\right) - \frac{4{,}786{,}176}{3{,}607}$$

$$= +429 - 1{,}327 \cong -898 \text{ psi } (C) \text{ (7 MPa)}$$

$$< f_c = 0.45 \times 6{,}000 = 2{,}700 \text{ psi, O.K.}$$

$$f_b = -\frac{P_e}{A_c}\left(1 + \frac{ec_b}{r^2}\right) + \frac{M_T}{S_b}$$

$$= -\frac{229{,}500}{449}\left(1 + \frac{14.77 \times 17.77}{50.04}\right) + \frac{4{,}786{,}176}{1{,}264}$$

$$= -3{,}192 + 3{,}786 \cong +594 \text{ psi } (T) \text{ (5.2 MPa)}$$

$$< f_t = 12\sqrt{f'_c} = 930 \text{ psi, O.K.}$$

1.5 C-LINE COMPUTATION OF FIBER STRESSES

Example 1.2

Solve example 1.1 for the final service-load condition by the line-of-thrust, C-line method.

Solution:

$$P_e = 229{,}500 \text{ lb}$$

$$M_T = 4{,}786{,}176 \text{ in.-lb}$$

$$a = \frac{M_T}{P_e} = \frac{4{,}786{,}176}{229{,}500} = 20.85 \text{ in.}$$

$$e' = a - e = 20.85 - 14.77 = 6.08 \text{ in.}$$

From Equations 1.12,

$$f^t = -\frac{P_e}{A_c}\left(1 + \frac{e'c_t}{r^2}\right)$$

$$= -\frac{229{,}500}{449}\left(1 + \frac{6.08 \times 6.23}{50.04}\right) \cong -898 \text{ psi } (C)$$

$$f_b = -\frac{P_e}{A_c}\left(1 - \frac{e'c_b}{r^2}\right)$$

$$= -\frac{229{,}500}{449}\left(1 - \frac{6.08 \times 17.77}{50.04}\right) \cong +594 \text{ psi } (T)$$

Notice how the C-line method is shorter than the basic method used in Example 1.1.

Photo 1.13 Dauphin Island Bridge, Mobile County, Alabama. (*Courtesy,* Post-Tensioning Institute.)

1.6 LOAD-BALANCING COMPUTATION OF FIBER STRESSES

Example 1.3

Solve Example 1.1 for the final service load condition after losses using the load-balancing method.

Solution:

$$P' = P_e = 229{,}500 \text{ lb at midspan}$$

$$\text{At midspan, } a = e_c = 14.77'' = 1.231 \text{ ft}$$

For the balancing load, we have

$$W_b = 8\frac{Pa}{l^2} = \frac{8 \times 229{,}500 \times 1.231}{(64)^2}$$

$$= 552 \text{ plf } (8.1 \text{ kN/m})$$

Thus, if the total gravity load would have been 552 plf, only the axial load P' / A would act if the beam had a parabolically draped tendon with no eccentricity at the supports. This is because the gravity load is balanced by the tendon at the midpan. Hence,

$$\text{Total load to which the beam is subjected} = W_D + W_{SD} + W_L$$
$$= 359 + 420 = 779 \text{ plf}$$

$$\text{Unbalanced load } W_{ub} = 779 - 552 = 227 \text{ plf}$$

$$\text{Unbalanced moment } M_{ub} = \frac{W_{ub}(l)^2}{8} = \frac{227(64)^2}{8} \times 12$$
$$= 1{,}394{,}688 \text{ in.-lb}$$

Photo 1.14 Heidrun Offshore Oil Drilling Platform in the North Sea weighing 2.9 million Kg. It measures 110 m on each side; has four slip-form constructed hulls and module-support 50-ft. span beams (*Courtesy,* C. E. Morrison, CONOCO Inc., Houston, Texas.)

From Equations 1.19,

$$f^t = -\frac{P'}{A_c} - \frac{M_{ub}}{S^t} = -\frac{229{,}500}{449} - \frac{1{,}394{,}688}{3{,}607}$$

$$= -511 - 387 \cong -898 \text{ psi } (C)$$

$$f_b = -\frac{P'}{A_c} + \frac{M_{ub}}{S_b} = -\frac{229{,}500}{449} + \frac{1{,}394{,}688}{1{,}264}$$

$$= -511 + 1{,}104 \cong 594 \text{ psi } (T)$$

$$\leq f_t = 930 \text{ psi allowed, O.K.}$$

1.7 SI WORKING LOAD STRESS CONCEPTS

Example 1.4

Solve Example 1.1 using SI units

Photo 1.15 Hibernia Platform, Grand Banks, Newfoundland, 1997: 80 m water depth, 165,000 m^3, 70 MPa strength concrete and 8,000 tons of prestressing tendons; first platform designed for iceberg impacts (*Courtesy,* Dr. George C. Hoff, Mobile Research and Development Corp.)

Given

Stress Data

$$f'_c = 41.4\text{MPa}$$

$$f_{pu} = 1{,}860 \text{ MPa}$$

$$f_{pi} = 0.70\, f_{pu} = 0.70 \times 1860 = 1300 \text{ MPa}$$

$$f_{py} = 1{,}520 \text{ MPa}$$

$$f_{pe} = 1{,}034 \text{ MPa}$$

$$f_t = \sqrt{f'_c} \text{ is deflection check O.K.} = 6.4 \text{ MPa, use}$$

$$f_t = \frac{1}{2}\sqrt{f'_c} \text{ if no check made on deflection}$$

$$f'_{ci} \cong 0.8\, f'_c = 33.1 \text{ MPa}$$

Photo 1.16 Pier 37 Rebuild, Seattle Washington: a 1200-ft-long by 40-ft wide pier consisting of 20-ft-long prestressed concrete deck panel supported by situ-cast concrete pile caps and prestressed concrete piles (Designed by BERGER/ABAM Engineers, Federal Way, Washington, courtesy Robert Mast, Senior Principal)

$$f_{ci} = 0.6 f'_{ci} = 19.9 \text{ MPa}$$
$$f_c = 0.45 f'_c = 18.6 \text{ MPa}$$
$$A_{ps} = 10 \text{ tendons } 17.7 \text{ mm diameter} = 10 \times 99\text{mm}^2 = 990 \text{ mm}^2$$

Section Geometry

Try 108-D1 Strand Pattern

$$A_c = 2897 \text{ cm}^2$$
$$I_c = 935{,}346 \text{ cm}^2$$
$$r^2 = I_c/A_c = 323 \text{ cm}^2$$
$$c_b = 45.1 \text{ cm} \qquad c_t = 15.8 \text{ cm}$$
$$e_c = 37.5 \text{ cm} \qquad e_e = 19.74 \text{ cm}$$
$$S^t = 59{,}108 \text{ cm}^3$$
$$S_b = 20{,}713 \text{ cm}^3$$
$$W_D = 5.24 \text{ kN/m}$$
$$W_{SD} + W_L = 6.13 \text{ kN/m}$$
$$l = 19.51 \text{ m}$$

Solution:

1. Initial conditions at prestressing

$$A_{ps} = 990 \text{ mm}^2$$
$$P_i = A_{ps}f_{pi} = 990 \times 1{,}303 = 1{,}290 \text{ kN}$$
$$P_e = A_{ps}f_{pe} = 900 \times 1034 = 1{,}024 \text{ kN}$$

Midspan self-weight dead-load moment

$$M_D = \frac{wl^2}{8} = \frac{5.24\,(19.51)^2}{8} = 249 \text{ kN-m}$$

From Equation 1.1a,

$$f^t = -\frac{P_i}{A_c}\left(1 - \frac{ec^t}{r^2}\right) - \frac{M_d}{S^t}$$
$$= -\frac{1290}{2897}\left(1 - \frac{37.5 \times 15.8}{323}\right) - \frac{249 \times 10^2 \text{ kN-cm}}{59{,}108 \text{ cm}^2}$$
$$= +0.37 - 0.42 \text{ kN/cm}^2 = -0.5 \times 10^6 \text{ N/m}^2$$
$$= 0.5 \text{ MPa } (C) < f_{ti} \text{ in tension } < f_{ci}, \text{ O.K.}$$

From Equation 1.1b,

$$f_b = -\frac{P_i}{A_c}\left(1 + \frac{ec_b}{r^2}\right) + \frac{M_D}{S_b}$$
$$= -\frac{1290}{2897}\left(1 + \frac{37.5 \times 45.1}{323}\right) + \frac{249 \times 10^2 \text{ kN-cm}}{20{,}713 \text{ cm}^2}$$
$$= -2.33 + 1.23 \text{ kN/cm}^2 = -11.0 \times 10^6 \text{ N/m}^2$$
$$= 11.0 \text{ MPa } (C) < \text{ allowable } f_{ci} = 19.9 \text{ MPa, O.K.}$$

2. Final condition at service load

$$W_{SD+L} = 6.13 \text{ kN/m}$$
$$M_{SD+L} = \frac{6.13(19.51)^2}{8} = 292 \text{ kN-m}$$

Total Moment $M_T = 343 + 292 = 635$ kN – m. From Eq. 1.7a,

$$f^t = -\frac{P_e}{A_c}\left(1 - \frac{ec^t}{r^2}\right) - \frac{M_T}{S^t}$$
$$= -\frac{1024}{2897}\left(1 - \frac{37.5 \times 15.8}{323}\right) - \frac{541 \times 10^2 \text{ kN-cm}}{59{,}108 \text{ cm}^3}$$
$$= +0.29 - 0.92 \text{ kN/cm}^2 = -6.3 \times 10^6 \text{ N/m}^2$$
$$= -6.3 \text{ MPa } (C) < \text{ allow } f_c = 18.6 \text{ MPa, O.K.}$$

From Equation 1.7b,

$$f_b = -\frac{P_e}{A_c}\left(1 + \frac{ec_b}{r^2}\right) + \frac{M_T}{S_b}$$

$$= -\frac{1024}{2897}\left(1 + \frac{37.5 \times 45.1}{323}\right) + \frac{541 \times 10^2 \text{ kN-cm}}{20{,}713 \text{ cm}^3}$$

$$= (-2.20 + 2.16) \times 10^7 \text{ N/m}^2 = +4.1 \times 10^6 \text{ N/m}^2$$

$$= 4.1 \text{ MPa } (T) < \text{allow. } f_t = \sqrt{f'_c} = 6.4 \text{ MPa, O.K.}$$

REFERENCES

1.1 Freyssinet, E. *The Birth of Prestressing.* London: Public Translation, Cement and Concrete Association, 1954.

1.2 Guyon, Y. *Limit State Design of Prestressed Concrete,* vol. 1. Halsted-Wiley, New York, 1972.

1.3 Gerwick, B. C., Jr. *Construction of Prestressed Concrete Structures.* Wiley-Interscience, New York, 1993, 591 p.

1.4 Lin, T. Y., and Burns, N. H. *Design of Prestressed Concrete Structures.* 3d ed. John Wiley & Sons, New York, 1981.

1.5 Nawy, E. G. *Reinforced Concrete—A Fundamental Approach.* 4th ed., Prentice-Hall, Upper Saddle River, N.J. 2000, 776 p.

1.6 Dobell, C. "Patents and Code Relating to Prestressed Concrete." *Journal of the American Concrete Institute* 46, 1950, 713–724.

1.7 Naaman, A. E. *Prestressed Concrete Analysis and Design.* McGraw Hill, New York, 1982.

1.8 Dill, R. E. "Some Experience with Prestressed Steel in Small Concrete Units." *Journal of the American Concrete Institute* 38, 1942, 165–168.

1.9 Institution of Structural Engineers. "First Report on Prestressed Concrete." *Journal of the Institution of Structural Engineers,* September 1951.

1.10 Magnel, G. *Prestressed Concrete.* London: Cement and Concrete Association, 1948.

1.11 Abeles, P. W., and Bardhan-Roy, B. K. *Prestressed Concrete Designer's Handbook.* 3d ed. Viewpoint Publications, London, 1981.

1.12 Nawy, E. G., *Fundamentals of High Performance Concrete,* 2nd ed. John Wiley & Sons, New York, 2000.

1.13 Nawy, E. G., editor-in-chief, *Concrete Construction Engineering Handbook,* CRC Press, Boca Raton, FL, 1998, 1250 p.

PROBLEMS

1.1 An AASHTO prestressed simply supported I-beam has a span of 34 ft (10.4 m) and is 36 in. (91.4 cm) deep. Its cross section is shown in Figure P1.1. It is subjected to a live-load intensity $W_L = 3{,}600$ plf (52.6 kN/m). Determine the required $\frac{1}{2}$-in.-dia stress-relieved seven-wire strands to resist the applied gravity load and the self-weight of the beam, assuming that the tendon eccentricity at midspan is $e_c = 13.12$ in. (333 mm). Maximum permissible stresses are as follows:

$$f'_c = 6{,}000 \text{ psi } (41.4 \text{ MPa})$$

$$f_c = 0.45 f'_c$$

$$= 2{,}700 \text{ psi } (19.7 \text{ MPa})$$

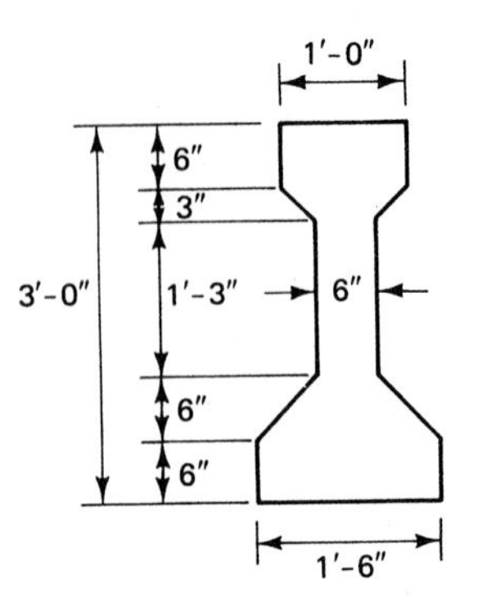

Figure P1.1.

$$f_t = 12\sqrt{f'_c} = 390 \text{ psi } (6.4 \text{ MPa})$$
$$f_{pu} = 270{,}000 \text{ psi } (1{,}862 \text{ MPa})$$
$$f_{pi} = 189{,}000 \text{ psi } (1{,}303 \text{ MPa})$$
$$f_{pe} = 145{,}000 \text{ psi } (1{,}000 \text{ MPa})$$

The section properties, are:

$$A_c = 369 \text{ in}^2$$
$$I_g = 50{,}979 \text{ in}^4$$
$$r^2 = I_c/A_c = 138 \text{ in}^2$$
$$c_b = 15.83 \text{ in.}$$
$$S_b = 3{,}220 \text{ in}^3$$
$$S^t = 2{,}527 \text{ in}^3$$
$$W_D = 384 \text{ plf}$$
$$W_L = 3{,}600 \text{ plf}$$

Solve the problem by each of the following methods:
(a) Basic concept
(b) C-line
(c) Load balancing

1.2 Solve problem 1.1 for a 45 ft (13.7 m) span and a superimposed live load W_L = 2,000 plf (29.2 kN/m).

1.3 A simply supported pretensioned pretopped double T-beam for a floor has a span of 70 ft (21.3 m) and the geometrical dimensions shown in Figure P1.3. It is subjected to a gravity live-load intensity W_L = 480 plf (7 kN/m), and the prestressing tendon has an eccentricity at midspan of e_c = 19.96 in. (494 mm). Compute the concrete extreme fiber stresses in this beam at transfer and at service load, and verify whether they are within the permissible limits. Assume that all permissible stresses and materials used are the same as in example 1.1. The section properties are:

Section Properties
Untopped

$$A_c = 1185 \text{ in.}^2$$
$$I_g = 109{,}621 \text{ in.}^4$$
$$C_b = 25.65 \text{ in.}$$
$$C_t = 8.35 \text{ in.}$$
$$S_b = 4274 \text{ in.}^3$$

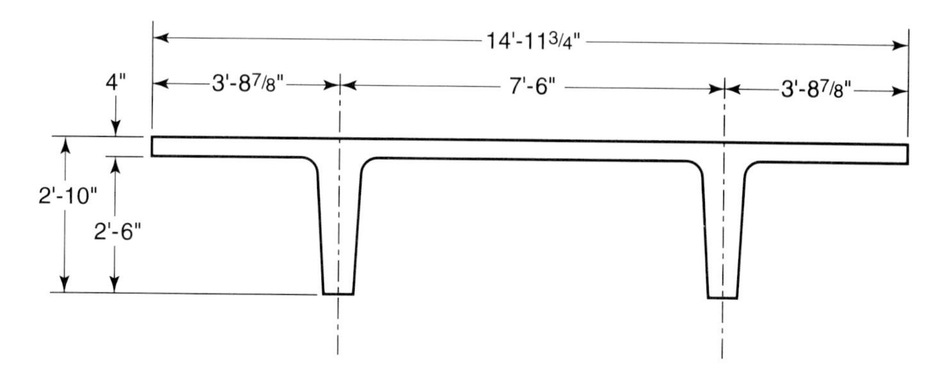

Figure P1.3.

$$S^t = 13{,}128 \text{ in.}^3$$
$$W_D = 1234 \text{ plf}$$
$$82 \text{ psf}$$
$$V/S = 2.45 \text{ in.}$$

Design the prestressing steel needed using $\frac{1}{2}$-in.-dia stress-relieved seven-wire strands. Use the three methods of analysis discussed in this chapter in your solution.

1.4 A T-shaped simply supported beam has the cross section shown in Figure P1.4. It has a span of 36 ft (11 m), is loaded with a gravity live-load unit intensity $W_L = 2{,}500$ plf (36.5 kN/), and is prestressed with twelve $\frac{1}{2}$-in.-dia (twelve 12.7-mm-dia) seven-wire stress-relieved strands. Compute the concrete fiber stresses at service load by each of the following methods:

(a) Basic concept
(b) C-line
(c) Load balancing

Assume that the tendon eccentricity at midspan is $e_c = 9.6$ in. (244 mm). Then given that

$$f'_c = 5{,}000 \text{ psi (34.5 MPa)}$$

$$f_t = 12\sqrt{f'_c} = 849 \text{ psi (5.9 MPa)}$$

$$f_{pe} = 165{,}000 \text{ psi (1,138 MPa)}$$

the section properties are as follows:

$$A_c = 504 \text{ in}^2$$

$$I_c = 37{,}059 \text{ in.}^4$$

$$r^2 = I_c/A_c = 73.5 \text{ in.}^2$$

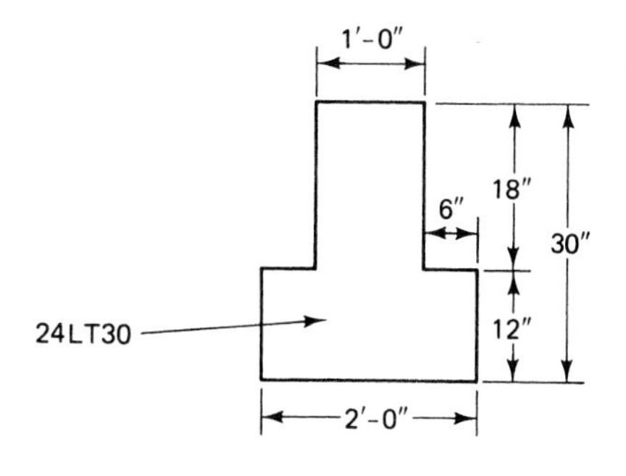

Figure P1.4.

$$c_b = 12.43 \text{ in.}$$
$$S_b = 2{,}981 \text{ in.}^3$$
$$S^t = 2{,}109 \text{ in.}^3$$
$$W_D = 525 \text{ plf}$$
$$e_c = 9.6 \text{ in.}$$
$$A_{ps} = \text{twelve } \frac{1}{2}\text{-in.-dia, seven-wire stress-relieved strands}$$

1.5 Solve problem 1.4 if $f'_c = 7{,}000$ psi (48.3 MPa) and $f_{pe} = 160{,}000$ psi (1,103 MPa).

2

MATERIALS AND SYSTEMS FOR PRESTRESSING

2.1 CONCRETE

2.1.1 Introduction

Concrete, particularly high-strength concrete, is a major constituent of all prestressed concrete elements. Hence, its strength and long-term endurance have to be achieved through proper quality control and quality assurance at the production stage. Numerous texts are available on concrete production, quality control, and code requirements. The following discussion is intended to highlight the topics directly related to concrete in prestressed elements and systems; it is assumed that the reader is already familiar with the fundamentals of concrete and reinforced concrete.

2.1.2 Parameters Affecting the Quality of Concrete

Strength and endurance are two major qualities that are particularly important in prestressed concrete structures. Long-term detrimental effects can rapidly reduce the prestressing forces and could result in unexpected failure. Hence, measures have to be taken to ensure strict quality control and quality assurance at the various stages of production

Pasco-Kennewick Intercity Bridge. Segmentally assembled prestressed concrete cable-stayed bridge, spans 407–981–407 ft. (*Courtesy,* Arvid Grant and Associates, Inc.)

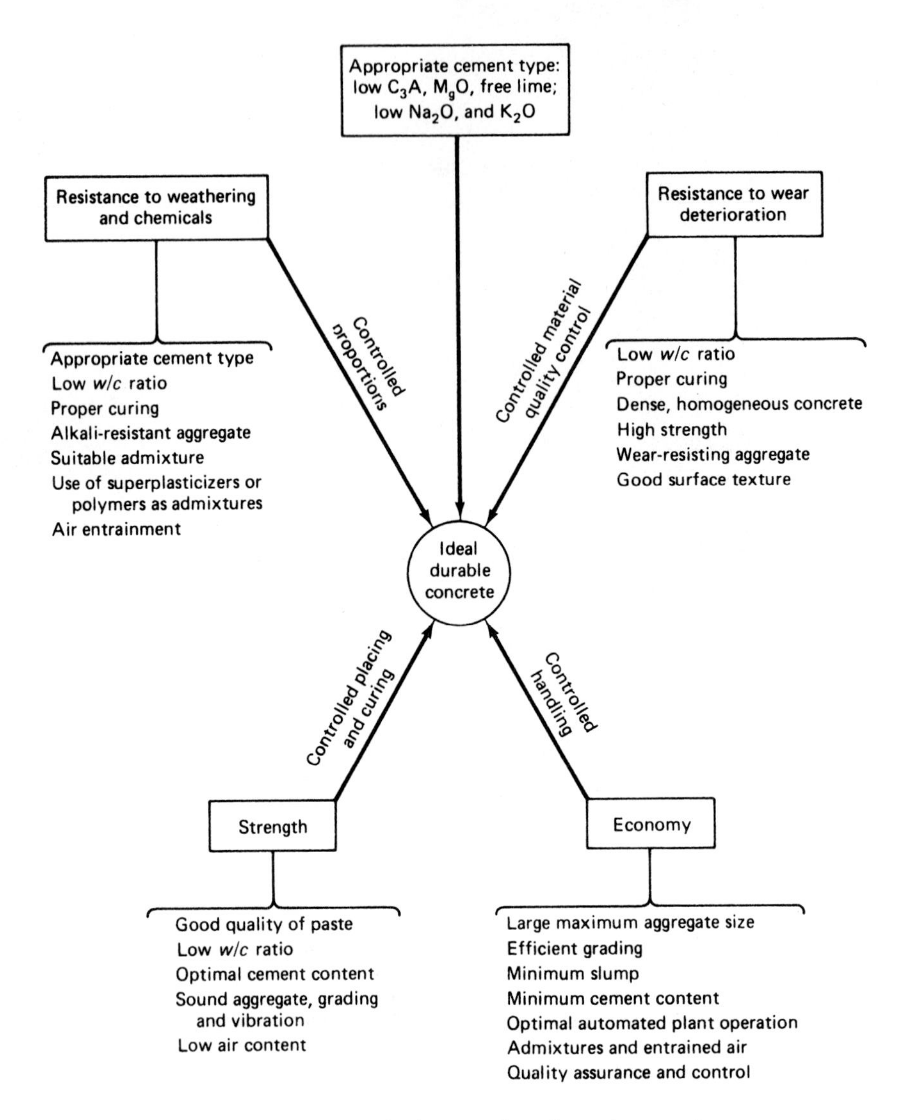

Figure 2.1 Principal properties of good concrete.

and construction as well as maintenance. Figure 2.1 shows the various factors that result in good-quality concrete.

2.1.3 Properties of Hardened Concrete

The mechanical properties of hardened concrete can be classified into two categories: short-term or instantaneous properties, and long-term properties. The short-term properties are strength in compression, tension, and shear; and stiffness, as measured by the modulus of elasticity. The long-term properties can be classified in terms of creep and shrinkage. The following subsections present some details on these properties.

2.1.3.1 Compressive Strength. Depending on the type of mix, the properties of aggregate, and the time and quality of the curing, compressive strengths of concrete can be obtained up to 20,000 psi or more. Commercial production of concrete with ordinary ag-

Photo 2.1 Concrete cylinders tested to failure in compression. Specimen A, low-epoxy-cement content; specimen B, high-epoxy-cement content. (Tests by Nawy, Sun, and Sauer.)

gregate is usually in the range 4,000 to 12,000 psi, with the most common concrete strengths being in the 6,000 psi level.

The compressive strength f'_c is based on standard 6 in. by 12 in. cylinders cured under standard laboratory conditions and tested at a specified rate of loading at 28 days of age. The standard specifications used in the United States are usually taken from ASTM C-39. The strength of concrete in the actual structure may not be the same as that of the cylinder because of the difference in compaction and curing conditions.

For a strength test, the ACI code specifies using the average of two cylinders from the same sample tested at the same age, which is usually 28 days. As for the frequency of testing, the code specifies that the strength of an individual class of concrete can be considered satisfactory if (1) the average of all sets of three consecutive strength tests equals or exceeds the required f'_c, and (2) no individual strength test (average of two cylinders) falls below the required f'_c by more than 500 psi. The average concrete strength for which a concrete mixture must be designed should exceed f'_c by an amount that depends on the uniformity of plant production.

Note that the design f'_c should not be the average cylinder strength, but rather the minimum conceivable cylinder strength.

2.1.3.2 Tensile Strength. The tensile strength of concrete is relatively low. A good approximation for the tensile strength f_{ct} is $0.10f'_c < f_{ct} < 0.20f'_c$. It is more difficult to measure tensile strength than compressive strength because of the gripping problems with testing machines. A number of methods are available for tension testing, the most commonly used method being the cylinder splitting, or Brazilian, test.

For members subjected to bending, the value of the modulus of rupture f_r rather than the tensile splitting strength f'_t is used in design. The modulus of rupture is measured by testing to failure plain concrete beams 6 in. square in cross section, having a

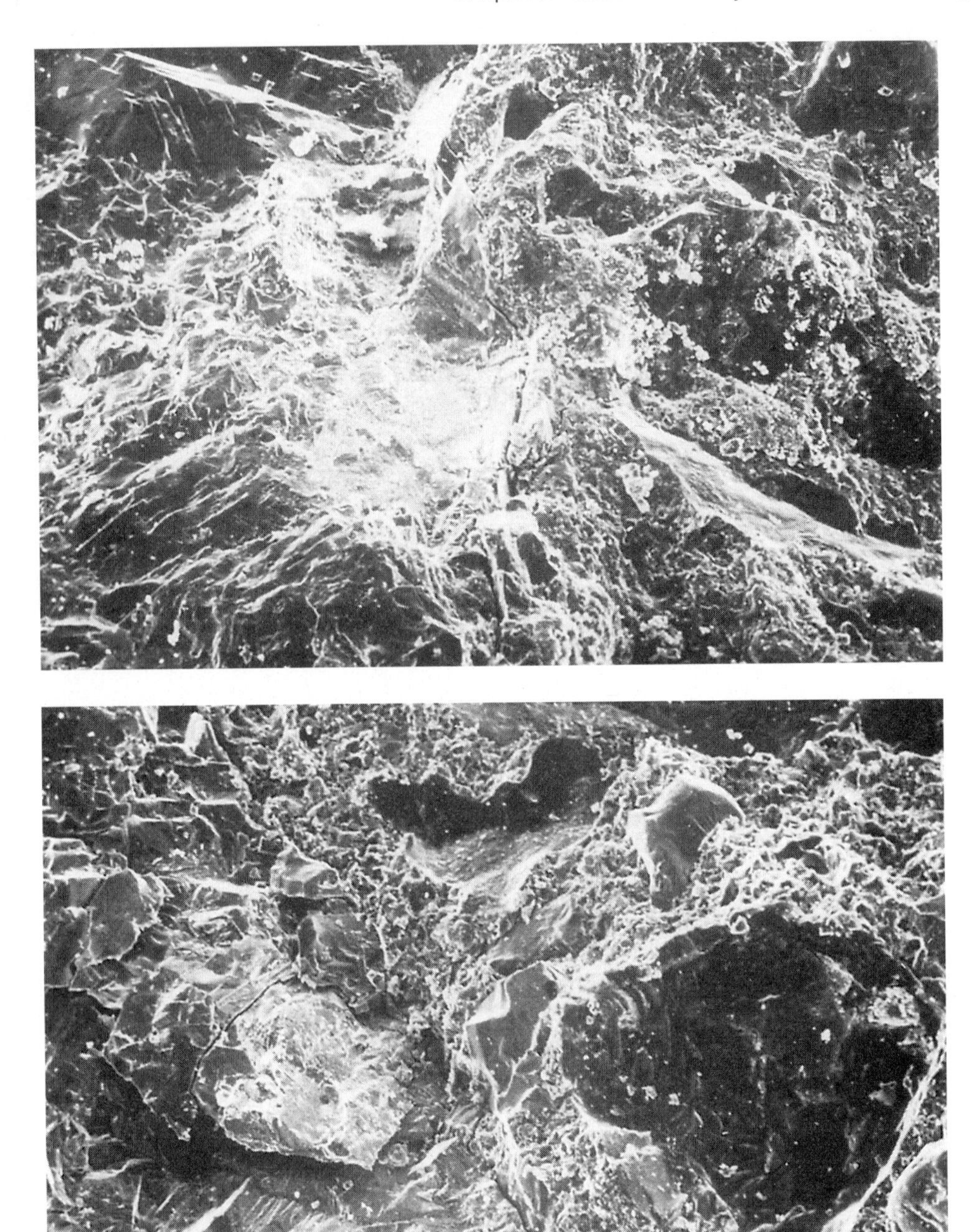

Photo 2.2 Electron microscope photographs of concrete from specimens A and B in the preceding photograph. (Tests by Nawy et al.)

span of 18 in., and loaded at their third points (ASTM C-78). The modulus of rupture has a higher value than the tensile splitting strength. The ACI specifies a value of $7.5\sqrt{f'_c}$ for the modulus of rupture of normal-weight concrete.

In most cases, lightweight concrete has a lower tensile strength than does normal-weight concrete. The following are the code stipulations for lightweight concrete:

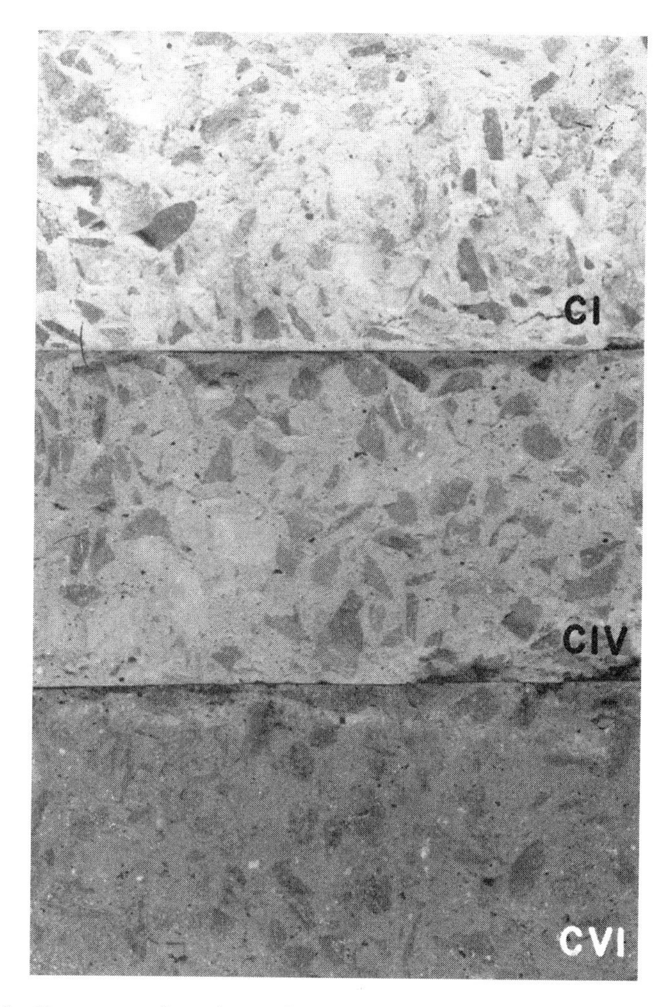

Photo 2.3 Fracture surfaces in tensile splitting tests of concretes with different *w/c* contents. Specimens CI and CIV have higher *w/c* content, hence more bond failures than specimen CVI. (Tests by Nawy et al.)

1. If the splitting tensile strength f_{ct} is specified,

$$f_r = 1.09 f_{ct} \leq 7.5\sqrt{f_c'} \tag{2.1}$$

2. If f_{ct} is not specified, use a factor of 0.75 for all-lightweight concrete and 0.85 for sand-lightweight concrete. Linear interpolation may be used for mixtures of natural sand and lightweight fine aggregate.

2.1.3.3 Shear Strength. Shear strength is more difficult to determine experimentally than the tests discussed previously because of the difficulty in isolating shear from other stresses. This is one of the reasons for the large variation in shear-strength values reported in the literature, varying from 20 percent of the compressive strength in normal loading to a considerably higher percentage of up to 85 percent of the compressive strength in cases where direct shear exists in combination with compression. Control of a structural design by shear strength is significant only in rare cases, since shear stresses must ordinarily be limited to continually lower values in order to protect the concrete from failure in diagonal tension.

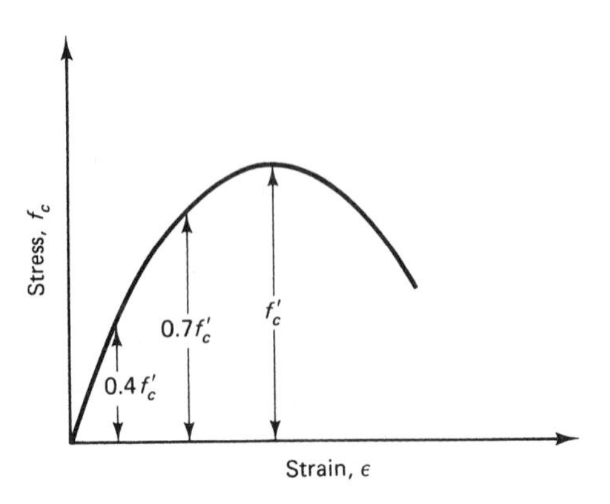

Figure 2.2 Typical stress-strain curve of concrete.

2.2 STRESS-STRAIN CURVE OF CONCRETE

Knowledge of the stress-strain relationship of concrete is essential for developing all the analysis and design terms and procedures in concrete structures. Figure 2.2 shows a typical stress-strain curve obtained from tests using cylindrical concrete specimens loaded in uniaxial compression over several minutes. The first portion of the curve, to about 40 percent of the ultimate strength f'_c, can essentially be considered linear for all practical purposes. After approximately 70 percent of the failure stress, the material loses a large portion of its stiffness, thereby increasing the curvilinearity of the diagram. At ultimate load, cracks parallel to the direction of loading become distinctly visible, and most concrete cylinders (except those with very low strengths) suddenly fail shortly thereafter. Figure 2.3 shows the stress-strain curves of concrete of various strengths reported by the Portland Cement Association. It can be observed that (1) the lower the strength of concrete, the higher the failure strain; (2) the length of the initial relatively linear portion increases with the increase in the compressive strength of concrete; and (3) there is an apparent reduction in ductility with increased strength.

2.3 MODULUS OF ELASTICITY AND CHANGE IN COMPRESSIVE STRENGTH WITH TIME

Since the stress-strain curve shown in Figure 2.4 is curvilinear at a very early stage of its loading history, Young's modulus of elasticity can be applied only to the tangent of the curve at the origin. The initial slope of the tangent to the curve is defined as the initial tangent modulus, and it is also possible to construct a tangent modulus at any point of the curve. The slope of the straight line that connects the origin to a given stress (about $0.4\,f'_c$) determines the secant modulus of elasticity of concrete. This value, termed in design calculation the *modulus of elasticity,* satisfies the practical assumption that strains occurring during loading can be considered basically elastic (completely recoverable on unloading), and that any subsequent strain due to the load is regarded as creep.

The ACI building code gives the following expressions for calculating the secant modulus of elasticity of concrete, E_c.

$$E_c = 33w_c^{1.5}\sqrt{f'_c} \qquad \text{for } 90 < w_c < 155 \text{ lb/ft}^3 \tag{2.2a}$$

where w_c is the density of concrete in pounds per cubic foot ($1 \text{ lb/ft}^3 = 16.02 \text{ kg/m}^3$) and f'_c is the compressive cylinder strength in psi. For normal-weight concrete,

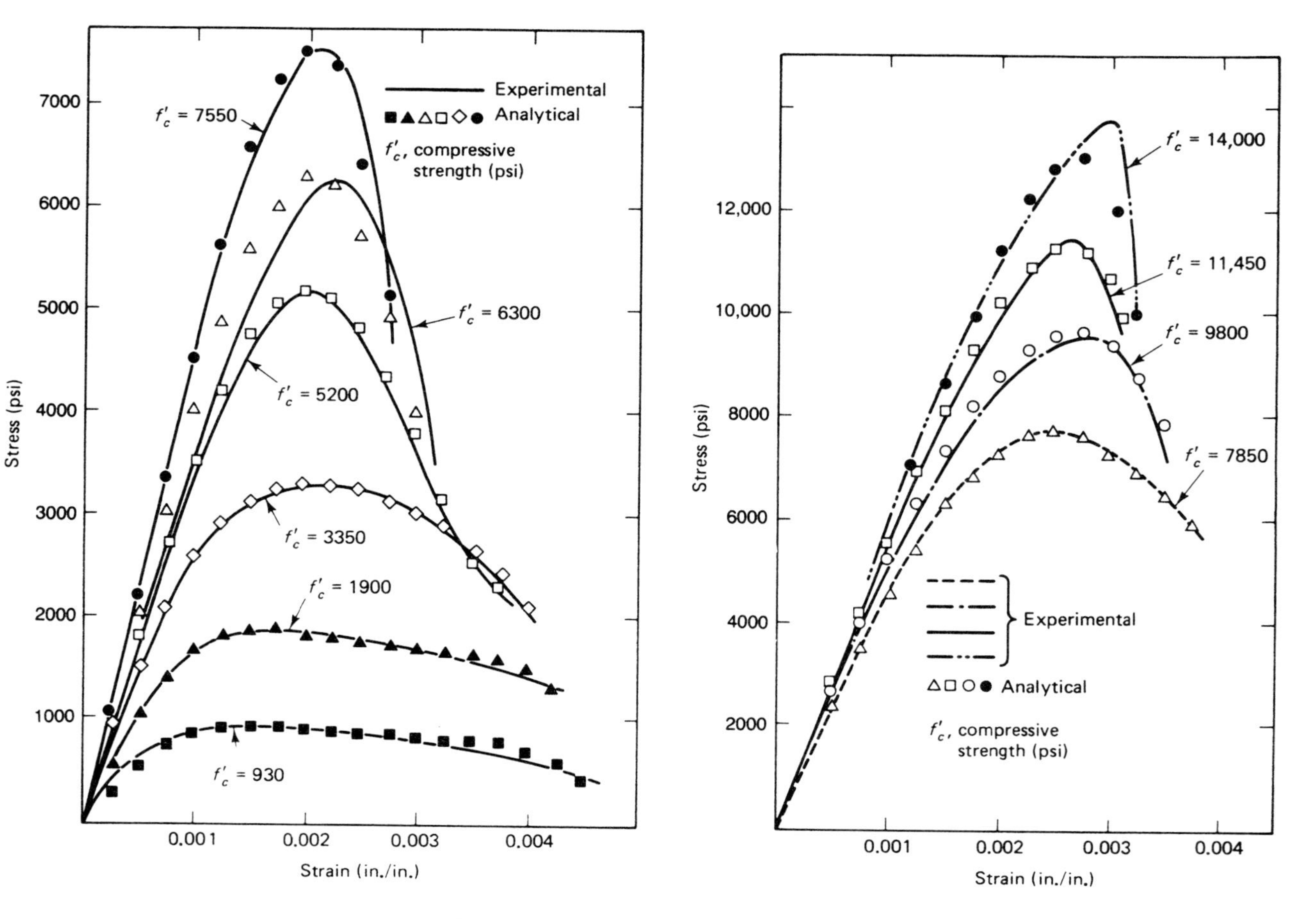

Figure 2.3 Stress-strain curves for various concrete strengths.

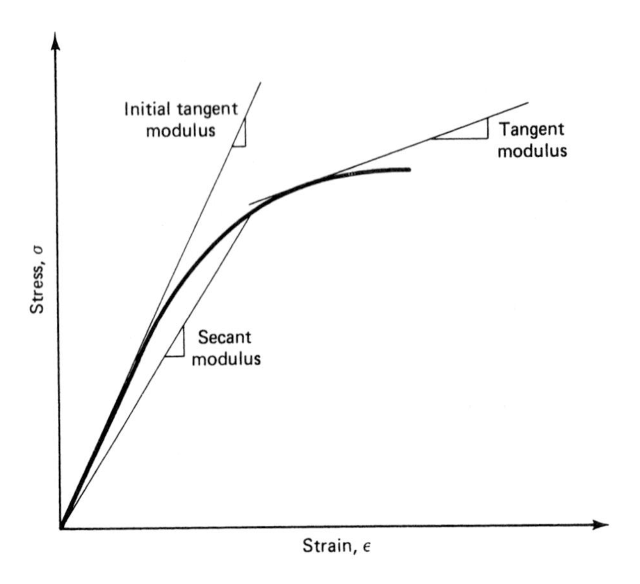

Figure 2.4 Tangent and secant moduli of concrete.

$$E_c = 57{,}000\sqrt{f'_c}\text{ psi } (4{,}700\sqrt{f'_c}\text{ MPa})$$

or

$$E_c = 0.043w^{1.5}\sqrt{f'_c}\text{ MPa} \tag{2.2.b}$$

2.3.1 High-Strength Concrete

High-strength concrete is termed as such by the ACI 318 Code when the cylinder compressive strength exceeds 6,000 psi (41.4 MPa). For concrete having compressive strengths 6,000 to 12,000 psi (42–84 MPa), the expressions for the modulus of concrete (Refs. 2.11, 2.35, 2.38)

$$E_c\text{ (psi)} = [40{,}000\sqrt{f'_c} + 10^6]\left(\frac{w_c}{145}\right)^{1.5} \tag{2.3a}$$

where f'_c = psi and w_c = lb/ft^3
or

$$E_c\text{ (MPa)} = [3.32\sqrt{f'_c} + 6{,}895]\left(\frac{w_c}{2320}\right)^{1.5} \tag{2.3b}$$

where f'_c = MPa and w_c = Kg/m^3.

Today, concrete strength up to 20,000 psi (138 MPa) is easily achieved using a maximum stone aggregate size of $\frac{3}{8}$ in. (9.5 mm) and pozzolamic cementitious partial replacements for the cement such as silica fume. Such strengths can be obtained in the field under strict quality control and quality assurance conditions. For strengths in the range of 20,000 to 30,000 (138–206 MPa), other constituents such as steel or carbon fibers have to be added to the mixture. In all these cases, mixture design has to be made by several field trial batches (five or more), modifying the mixture components for the workability needed in concrete placement. Steel cylinder molds size 4 in. (diameter) × 8 in. length have to be used, applying the appropriate dimensional correction. It is also necessary to grind the cylinder ends, then cap them with high strength capping compound for load testing, or to apply the load directly to the ground ends of the cylinder or through a removable steel cap with a hard neoprene pad bearing directly on the ground specimen ends. Preparation of the cylinders should resemble as closely as possible the field conditions of concrete placement. Mock-up placement of the high-strength concrete is advisable in order to evaluate the construction procedures and performance of the concrete in field conditions and to identify potential problems with batching, placement, and testing of the concrete at early ages. Corrective measures should be taken immediately.

A good example of the use of high-strength concrete in the range 20,000 psi (138 MPa) at 56 days and a concrete modulus $E_c = 7.8 \times 106$ psi (53.8×103 MPa) is the Two Union Square Building, Seattle, Washington (Ref. 2.11, 2.38). Actual typical mixture obtained is listed in Table 2.1, with the design mixture values in parentheses.

A slump of 8 in. with w/c $\simeq 0.22$ resulted from the mix proportions indicated. A typical compressive vs. age plot for the indicated mixture based on 4 in. × 8 in. cylinder tests is shown in Figure 2.5.

Recent work at Rutgers (Refs. 2.36, 2.37) on high-strength composite construction has resulted in considerable enhancement of the ductility of high-strength reinforced concrete beams. Prestressed concrete prisms of high-strength concrete were used in lieu of the normal mild steel bar reinforcement. The mixture proportions in lb/yd^3 were as shown in Table 2.2. The mixture was designed for a seven-day compressive strength of 12,000 psi (84 MPa). The ratio of the cementations/fine/coarse aggregate was 1:1.22:2.06 and the slump varied between 4 to 6 in. (100–150 mm). The prestressing strands were stress-relieved 270K (1900 MPa) 7 wire $\frac{3}{8}$in. (9.5 mm) diameter strands. Figure 2.6 shows the cross section of the composite beams. Concrete achieved in some of the mixtures a 7 day strength of 13,250 psi (91.4 MPa). The tested specimens were instrumented with a fiber-optic system developed by the author using Bragg Grating sensors both internally and externally.

2.3.2 Initial Compressive Strength and Modulus

Since prestressing is performed in most cases prior to concrete's achieving its 28 days' strength, it is important to determine the concrete compressive strength f'_{ci} at the prestressing stage as well as the concrete modulus E_c at the various stages in the loading history of the element. The general expression for the compressive strength as a function of time (Ref. 2.18) is

$$f'_{ci} = \frac{t}{\alpha + \beta t} f'_c \tag{2.4a}$$

Table 2.1 Mixture Proportions for $f'_c > 18{,}000$ PSI

Coarse aggregate ($\frac{3}{8}$in.) (lb)	Fine aggregate (paving sand) (lb)	Cement (lb)	Water (lb)	Silica fume (gal)	Superplasticizer W. R. Grace Dartard 40 (oz/100 lb cement)	Superplasticizer W. R. Grace Mighty 150 (oz/100 lb cement)
1872	1165	957	217	13	2.1	9.8
1894	1165	956	217	13	2.1	16.4
(1805)	(1100)	(950)	(w/c = 0.22)	(70 lb)[a]	(6.0)	(Up to 24)

[a]Weight of solid silica fume only. Water contained as part of the emulsion must be subtracted from the total water allowed.

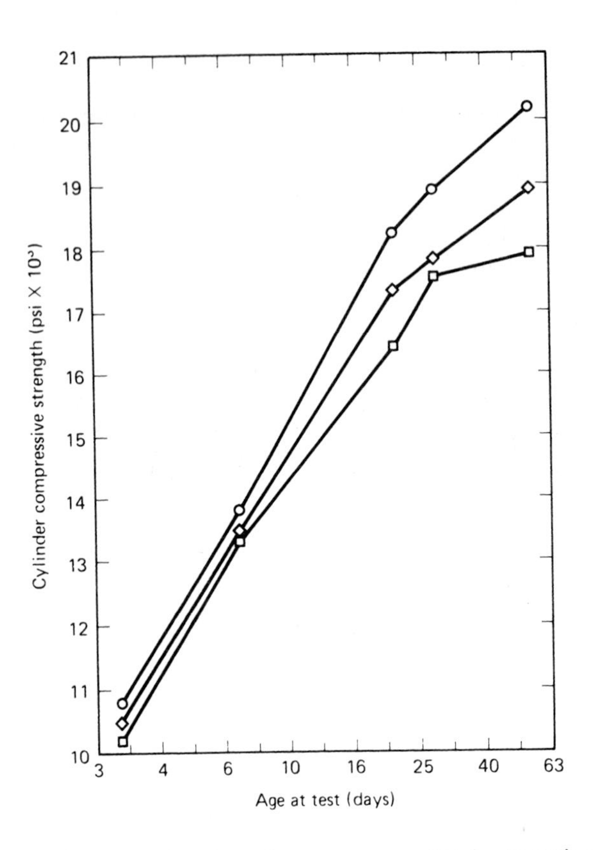

Figure 2.5 Compressive strength versus age of high-strength concrete.

where f'_c = 28 days' compressive strength
t = time in days
α = factor depending on type of cement and curing conditions
= 4.00 for moist-cured type-I cement and 2.30 for moist-cured type-III cement
= 1.00 for steam-cured type-I cement and 0.70 for steam-cured type-III cement
β = factor depending on the same parameters for α giving corresponding values of 0.85, 0.92, 0.95, and 0.98, respectively

Hence, for a typical moist-cured type-I cement concrete,

$$f'_{ci} = \frac{t}{4.00 + 0.85t} f'_c \tag{2.4b}$$

Table 2.2 Mixture Proportions in lb/yd³ For Composite Beams $f'_c > 13,000$ PSI

Coarse aggregate $\frac{3}{8}$ in.	Fine aggregate (natural sand)	Portland cement type III	Water	Powder silica fume force—10,000	Liquid super plasticizer (Grace)
(1)	(2)	(3)	(4)	(5)	(6)
1851	1100	720	288	180	54

1 lb/yd³ = 0.59 Kg/m³

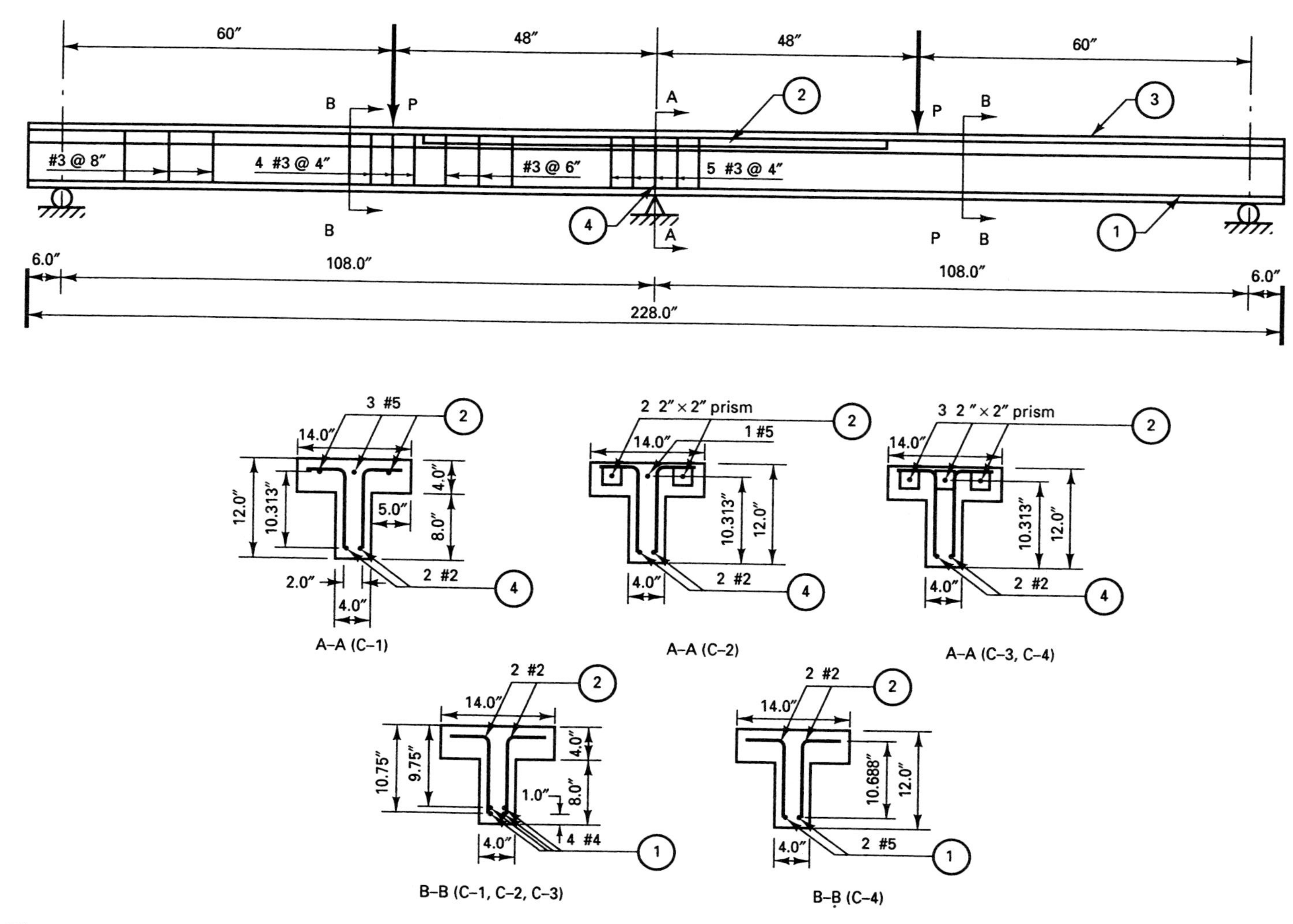

Figure 2.6 High-strength flanged sections reinforced with prestressed concrete prisms instrumented with fiber-optic sensors (Ref. 2.37).

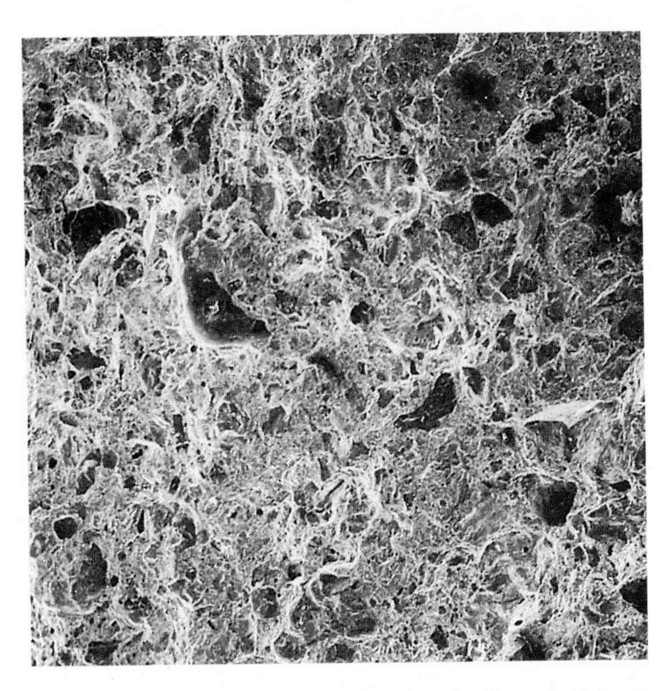

Photo 2.4 Scanning electron microscope photograph of concrete fracture surface. (Tests by Nawy, Sun, and Sauer.)

The effective modulus of concrete, E'_c, is

$$E'_c = \frac{\text{stress}}{\text{elastic strain} + \text{creep strain}} \tag{2.5}$$

and the ultimate effective modulus is given by

$$E_{cn} = \frac{E_c}{1 + \gamma_t} \tag{2.6a}$$

where γ_t is the creep ratio defined as

$$\gamma_t = \frac{\text{ultimate creep strain}}{\text{elastic strain}}$$

The creep ratio γ_t has upper and lower limits as follows for prestressed quality concrete:

$$\text{Upper:} \qquad \gamma_t = 1.75 + 2.25\left(\frac{100 - H}{65}\right) \tag{2.6b}$$

$$\text{Lower:} \qquad \gamma_t = 0.75 + 0.75\left(\frac{100 - H}{50}\right) \tag{2.6c}$$

where H is the mean humidity in percent.

It has to be pointed out that these expressions are valid only in general terms, since the value of the modulus of elasticity is affected by factors other than loads, such as moisture in the concrete specimen, the water/cement ratio, the age of the concrete, and temperature. Therefore, for special structures such as arches, tunnels, and tanks, the modulus of elasticity needs to be determined from test results.

Limited work exists on the determination of the modulus of elasticity in tension because the low-tensile strength of concrete is normally disregarded in calculations. It is, however, valid to assume within those limitations that the value of the modulus in tension is equal to that in compression.

2.4 CREEP

Creep, or lateral material flow, is the increase in strain with time due to a sustained load. The initial deformation due to load is the *elastic strain,* while the additional strain due to the same sustained load is the *creep strain.* This practical assumption is quite acceptable, since the initial recorded deformation includes few time-dependent effects.

Figure 2.7 illustrates the increase in creep strain with time, and as in the case of shrinkage, it can be seen that creep rate decreases with time. Creep cannot be observed directly and can be determined only by deducting elastic strain and shrinkage strain from the total deformation. Although shrinkage and creep are not independent phenomena, it can be assumed that superposition of strains is valid; hence,

$$\text{Total strain } (\epsilon_t) = \text{elastic strain } (\epsilon_e) + \text{creep } (\epsilon_c) + \text{shrinkage } (\epsilon_{sh})$$

An example of the relative numerical values of strain due to the foregoing three factors for a normal concrete specimen subjected to 900 psi in compression is as follows:

$$\begin{aligned}
\text{Immediate elastic strain, } \epsilon_e &= 250 \times 10^{-6} \text{ in./in.} \\
\text{Shrinkage strain after 1 year, } \epsilon_{sh} &= 500 \times 10^{-6} \text{ in./in.} \\
\text{Creep strain after 1 year, } \epsilon_c &= 750 \times 10^{-6} \text{ in./in.} \\
\epsilon_t &= 1{,}500 \times 10^{-6} \text{ in./in.}
\end{aligned}$$

These relative values illustrate that stress-strain relationships for short-term loading lose their significance and long-term loadings become dominant in their effect on the behavior of a structure.

Figure 2.8 qualitatively shows, in a three-dimensional model, the three types of strain discussed that result from sustained compressive stress and shrinkage. Since creep is time dependent, this model has to be such that its orthogonal axes are deformation, stress, and time.

Numerous tests have indicated that creep deformation is proportional to applied stress, but the proportionality is valid only for low-stress levels. The upper limit of the relationship cannot be determined accurately, but can vary between 0.2 and 0.5 of the ulti-

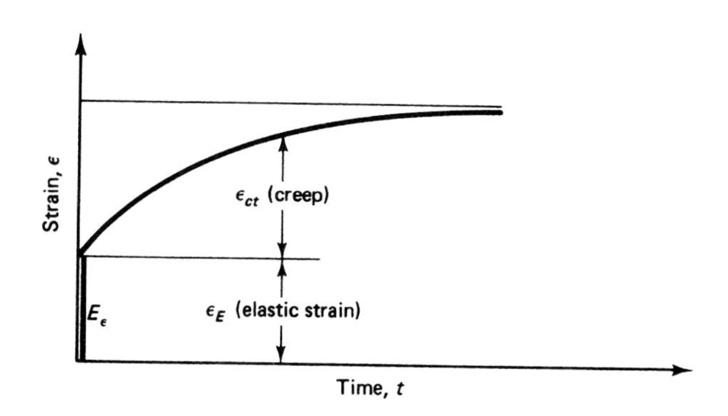

Figure 2.7 Strain-time curve.

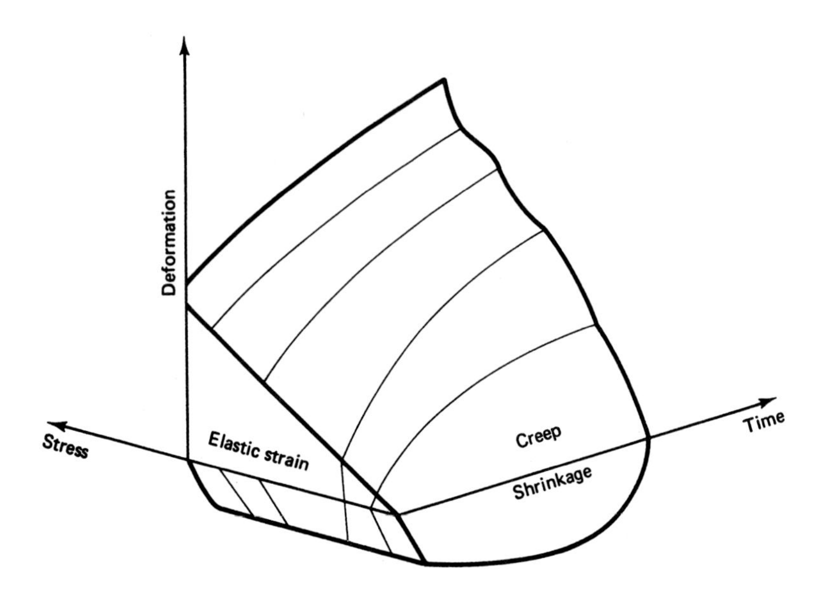

Figure 2.8 Three-dimensional model of time-dependent structural behavior.

mate strength f'_c. This range in the limit of the proportionality is due to the large extent of microcracks at about 40 percent of the ultimate load.

Figure 2.9a shows a section of the three-dimensional model in Figure 2.8 parallel to the plane containing the stress and deformation axes at time t_1. It indicates that both elastic and creep strains are linearly proportional to the applied stress. In a similar manner, Figure 2.9b illustrates a section parallel to the plane containing the time and strain axes at a stress f_1; hence, it shows the familiar relationships of creep with time and shrinkage with time.

As in the case of shrinkage, creep is not completely reversible. If a specimen is unloaded after a period under a sustained load, an immediate elastic recovery is obtained which is less than the strain precipitated on loading. The instantaneous recovery is followed by a gradual decrease in strain, called *creep recovery.* The extent of the recovery

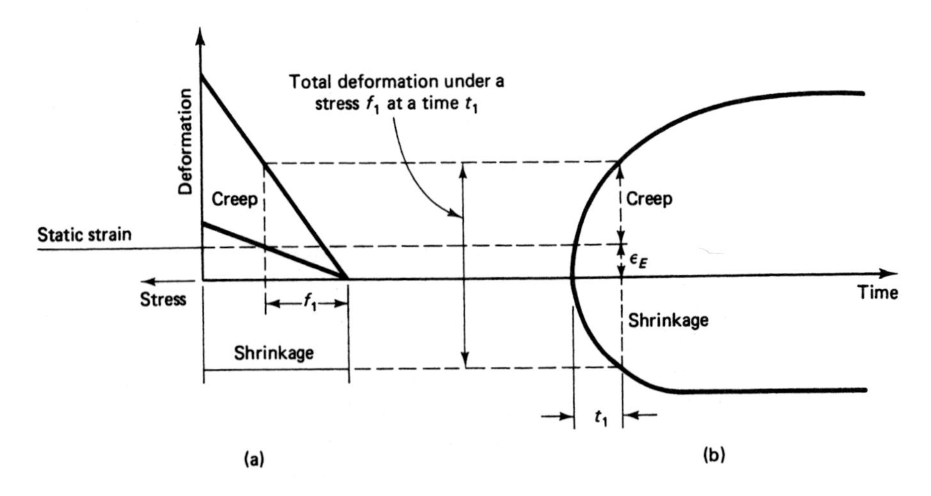

Figure 2.9 (a) Section parallel to the stress-deformation plane. (b) Section parallel to the deformation-time plane.

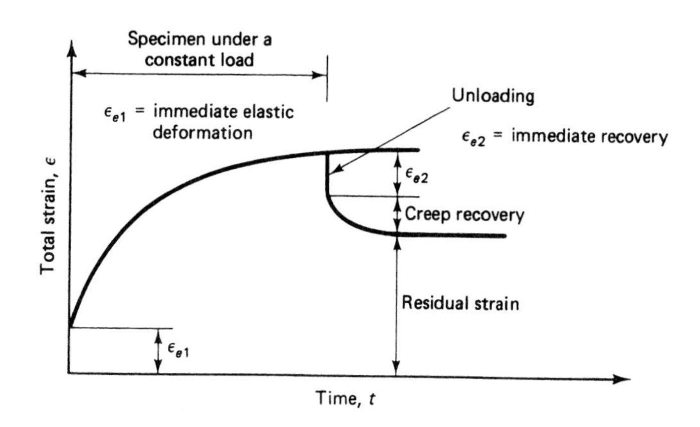

Figure 2.10 Creep recovery versus time.

depends on the age of the concrete when loaded, with older concretes presenting higher creep recoveries, while residual strains or deformations become frozen in the structural element (see Figure 2.10).

Creep is closely related to shrinkage, and as a general rule, a concrete that resists shrinkage also presents a low creep tendency, as both phenomena are related to the hydrated cement paste. Hence, creep is influenced by the composition of the concrete, the environmental conditions, and the size of the specimen, but principally creep depends on loading as a function of time.

The composition of a concrete specimen can be essentially defined by the water/cement ratio and water/cementitious ratio when admixtures are used, aggregate and cement types, and aggregate and cement contents. Therefore, like shrinkage, an increase in the water/cement ratio and in the cement content increases creep. Also, as in shrinkage, the aggregate induces a restraining effect such that an increase in aggregate content reduces creep.

2.4.1 Effects of Creep

As in shrinkage, creep increases the deflection of beams and slabs and causes loss of prestress. In addition, the initial eccentricity of a reinforced concrete column increases with time due to creep, resulting in the transfer of the compressive load from the concrete to the steel in the section.

Once the steel yields, additional load has to be carried by the concrete. Consequently, the resisting capacity of the column is reduced and the curvature of the column increases further, resulting in overstress in the concrete, leading to failure.

2.4.2 Rheological Models

Rheological models are mechanical devices that portray the general deformation behavior and flow of materials under stress. A model is basically composed of elastic springs and ideal dashpots denoting stress, elastic strain, delayed elastic strain, irrecoverable strain, and time. The springs represent the proportionality between stress and strain, and the dashpots represent the proportionality of stress to the rate of strain. A spring and a dashpot in parallel form a Kelvin unit, and in series they form a Maxwell unit.

Two rheological models will be discussed: the Burgers model and the Ross model. The Burgers model in Figure 2.11 is shown since it can approximately simulate the stress-strain-time behavior of concrete at the limit of proportionality with some limitations. This model simulates the instantaneous recoverable strain, *a;* the delayed recoverable

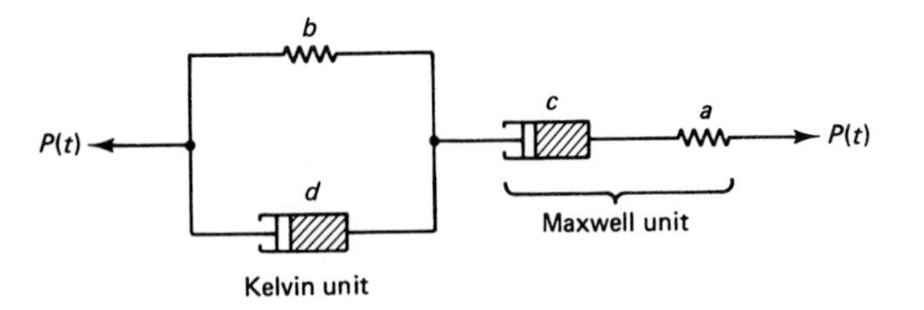

Figure 2.11 Burgers model.

elastic strain in the spring, *b;* and the irricoverable time-dependent strains in the dashpots, *c* and *d.* The weakness in the model is that it continues to deform at a uniform rate as long as the load is sustained by the Maxwell dashpot—a behavior not similar to concrete, where creep reaches a limiting value with time, as shown in Figure 2.7.

A modification in the form of the Ross rheological model in Figure 2.12 can eliminate this deficiency. *A* in this model represents the Hookian direct proportionality of stress-to-strain element, *D* is the dashpot, and *B* and *C* are the elastic springs that can transmit the applied load $P(t)$ to the enclosing cylinder walls by direct friction. Since each coil has a defined frictional resistance, only those coils whose resistance equals the applied load $P(t)$ are displaced; the others remain unstressed, symbolizing the irrecoverable deformation in concrete. As the load continues to increase, it overcomes the spring resistance of unit *B,* pulling out the spring from the dashpot and signifying failure in a concrete element. More rigorous models, such as Roll's, have been used to assist in predicting the creep strains. Mathematical expressions for such predictions can be very rigorous. One convenient expression due to Ross defines the creep *C* under load after a time interval *t* as

$$C = \frac{t}{a + bt} \tag{2.7}$$

where *a* and *b* are constants determinable from tests.

Work by Branson (Refs. 2.18 and 2.19) has simplified creep evaluation. The additional strain ϵ_{cu} due to creep can be defined as

$$\epsilon_{cu} = \rho_u f_{ci} \tag{2.8}$$

where ρ_u = unit creep coefficient, generally called *specific creep*
f_{ci} = stress intensity in the structural member corresponding to unit strain ϵ_{ci}

The ultimate creep coefficient, C_u, is given by

$$C_u = \rho_u E_c \tag{2.9}$$

or average $C_u \simeq 2.35$.

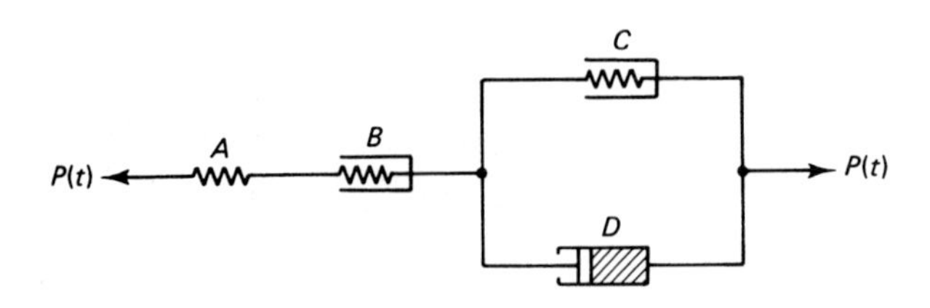

Figure 2.12 Ross model.

Branson's model, verified by extensive tests, relates the creep coefficient C_t at any time to the ultimate creep coefficient (for standard conditions) as

$$C_t = \frac{t^{0.6}}{10 + t^{0.6}} C_u \tag{2.10}$$

or, alternatively,

$$\rho_t = \frac{t^{0.6}}{10 + t^{0.6}} \tag{2.11}$$

where t is the time in days and ρ_t is the time multiplier. Standard conditions as defined by Branson pertain to concretes of slump 4 in. (10 cm) or less and a relative humidity of 40 percent.

When conditions are not standard, creep correction factors have to be applied to Equations 2.10 or 2.11 as follows:

(a) For moist-cured concrete loaded at an age of 7 days or more,

$$k_a = 1.25t^{-0.118} \tag{2.12}$$

(b) For steam-cured concrete loaded at an age of 1 to 3 days or more,

$$k_a = 1.13t^{-0.095} \tag{2.13}$$

For greater than 40 percent relative humidity, a further multiplier correction factor of

$$k_{c_1} = 1.27 - 0.0067H \tag{2.14}$$

Photo 2.5 Energy Center, New Orleans, Louisiana. (*Courtesy,* Post-Tensioning Institute.)

has to be applied in addition to those of Equations 2.12 and 2.13, where H = relative humidity value in percent.

2.5 SHRINKAGE

Basically, there are two types of shrinkage: plastic shrinkage and drying shrinkage. *Plastic shrinkage* occurs during the first few hours after placing fresh concrete in the forms. Exposed surfaces such as floor slabs are more easily affected by exposure to dry air because of their large contact surface. In such cases, moisture evaporates faster from the concrete surface than it is replaced by the bleed water from the lower layers of the concrete elements. *Drying shrinkage,* on the other hand, occurs after the concrete has already attained its final set and a good portion of the chemical hydration process in the cement gel has been accomplished.

Drying shrinkage is the decrease in the volume of a concrete element when it loses moisture by evaporation. The opposite phenomenon, that is, volume increase through water absorption, is termed *swelling.* In other words, shrinkage and swelling represent water movement out of or into the gel structure of a concrete specimen due to the difference in humidity or saturation levels between the specimen and the surroundings irrespective of the external load.

Shrinkage is not a completely reversible process. If a concrete unit is saturated with water after having fully shrunk, it will not expand to its original volume. Figure 2.13 relates the increase in shrinkage strain ϵ_{sh} with time. The rate decreases with time since older concretes are more resistant to stress and consequently undergo less shrinkage, such that the shrinkage strain becomes almost asymptotic with time.

Several factors affect the magnitude of drying shrinkage:

1. *Aggregate.* The aggregate acts to restrain the shrinkage of the cement paste; hence, concretes with high aggregate content are less vulnerable to shrinkage. In addition, the degree of restraint of a given concrete is determined by the properties of aggregates: Those with a high modulus of elasticity or with rough surfaces are more resistant to the shrinkage process.
2. *Water/cement ratio.* The higher the water/cement ratio, the higher the shrinkage effects. Figure 2.14 is a typical plot relating aggregate content to water/cement ratio.
3. *Size of the concrete element.* Both the rate and the total magnitude of shrinkage decrease with an increase in the volume of the concrete element. However, the duration of shrinkage is longer for larger members since more time is needed for drying to reach the internal regions. It is possible that 1 year may be needed for the drying

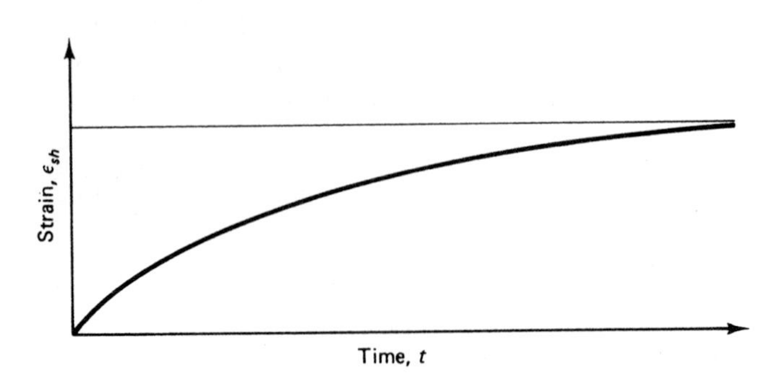

Figure 2.13 Shrinkage-time curve.

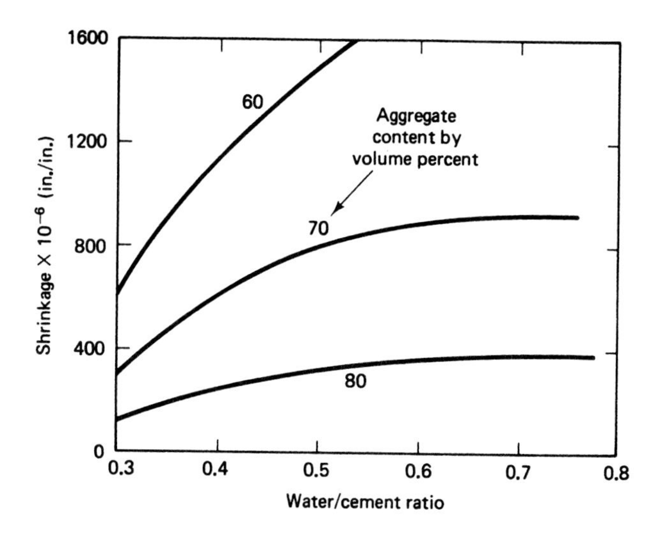

Figure 2.14 *w/c* ratio and aggregate content effect on shrinkage.

process to begin at a depth of 10 in. from the exposed surface, and 10 years to begin at 24 in. below the external surface.

4. *Medium ambient conditions.* The relative humidity of the medium greatly affects the magnitude of shrinkage; the rate of shrinkage is lower at high states of relative humidity. The environment temperature is another factor, in that shrinkage becomes stabilized at low temperatures.
5. *Amount of reinforcement.* Reinforced concrete shrinks less than plain concrete; the relative difference is a function of the reinforcement percentage.
6. *Admixtures.* This effect varies depending on the type of admixture. An accelerator such as calcium chloride, used to accelerate the hardening and setting of the concrete, increases the shrinkage. Pozzolans can also increase the drying shrinkage, whereas air-entraining agents have little effect.
7. *Type of cement.* Rapid-hardening cement shrinks somewhat more than other types, while shrinkage-compensating cement minimizes or eliminates shrinkage cracking if used with restraining reinforcement.
8. *Carbonation.* Carbonation shrinkage is caused by the reaction between the carbon dioxide (CO_2) present in the atmosphere and that present in the cement paste. The amount of the combined shrinkage varies according to the sequence of occurrence of carbonation and drying processes. If both phenomena take place simultaneously, less shrinkage develops. The process of carbonation, however, is dramatically reduced at relative humidities below 50 percent.

Branson (Ref. 2.18) recommends the following relationships for the shrinkage strain as a function of time for standard conditions of humidity ($H \cong 40$ percent):

(a) For moist-cured concrete any time t after 7 days,

$$\epsilon_{SH,t} = \frac{t}{35 + t}(\epsilon_{SH,u}) \tag{2.15}$$

where $\epsilon_{SH,u} = 800 \times 10^{-6}$ in./in. if local data are not available.

(b) For steam-cured concrete after the age of 1 to 3 days,

$$\epsilon_{SH,t} = \frac{t}{55 + t}\epsilon_{SH,u} \tag{2.16}$$

For other than standard humidity, a correction factor has to be applied to Equations 2.15 and 2.16 as follows:

(a) For $40 < H \leq 80$ percent,

$$k_{SH} = 1.40 - 0.010H \tag{2.17a}$$

(b) For $80 < H \leq 100$ percent,

$$k_{SH} = 3.00 - 0.30H \tag{2.17b}$$

2.6 NONPRESTRESSING REINFORCEMENT

Steel reinforcement for concrete consists of bars, wires, and welded wire fabric, all of which are manufactured in accordance with ASTM standards. The most important properties of reinforcing steel are:

1. Young's modulus, E_s
2. Yield strength, f_y
3. Ultimate strength, f_u
4. Steel grade designation
5. Size or diameter of the bar or wire

To increase the bond between concrete and steel, projections called *deformations* are rolled onto the bar surface as shown in Figure 2.15, in accordance with ASTM specifications. The deformations shown must satisfy ASTM Specification A616-76 for the bars to be accepted as deformed. Deformed wire has indentations pressed into the wire or bar to serve as deformations. Except for wire used in spiral reinforcement in columns, only deformed bars, deformed wires, or wire fabric made from smooth or deformed wire may be used in reinforced concrete under approved practice.

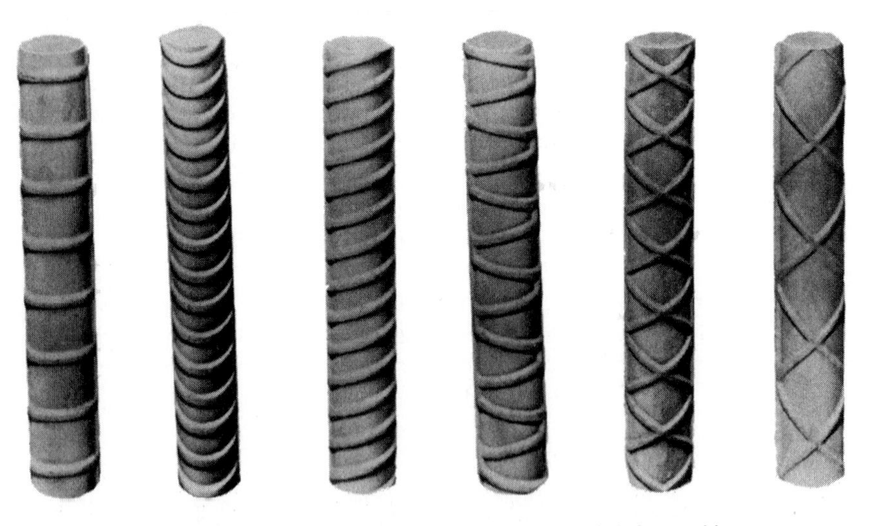

Figure 2.15 Various forms of ASTM-approved deformed bars.

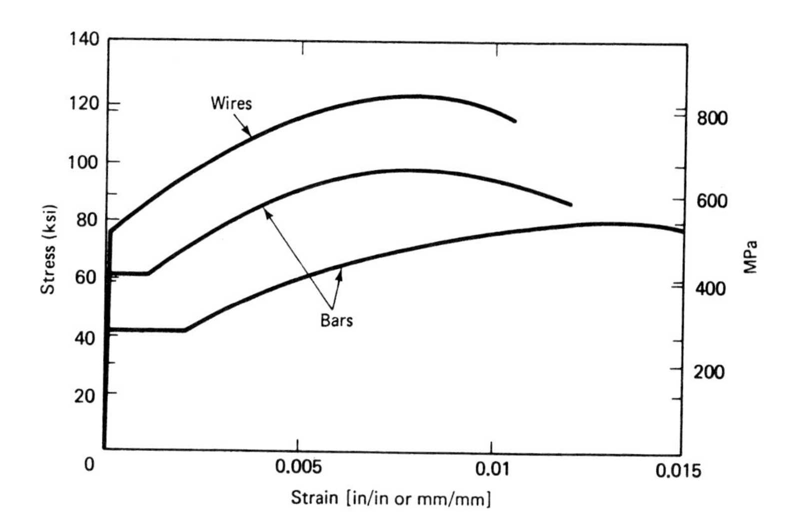

Figure 2.16 Typical stress-strain diagrams for various nonprestressing steels.

Figure 2.16 shows typical stress-strain curves for grades 40, 60, and 75 steels. These have corresponding yield strengths of 40,000, 60,000, and 75,000 psi (276, 345, and 517 N/mm^2, respectively) and generally have well-defined yield points. For steels that lack a well-defined yield point, the yield-strength value is taken as the strength corresponding to a unit strain of 0.005 for grades 40 and 60 steels, and 0.0035 for grade 80 steel. The ultimate tensile strengths corresponding to the 40, 60, and 80 grade steels are 70,000, 90,000, and 100,000 psi (483, 621, and 690 N/mm^2), respectively, and some steel types are given in Table 2.3. The percent elongation at fracture, which varies with the grade, bar diameter, and manufacturing source, ranges from 4.5 to 12 percent over an 8-in. (203.2-mm) gage length.

Welded wire fabric is increasingly used for slabs because of the ease of placing the fabric sheets, the control over reinforcement spacing, and the better bond. The fabric reinforcement is made of smooth or deformed wires which run in perpendicular directions

Table 2.3 Reinforcement Grades and Strengths

1982 Standard type	Minimum yield point or yield strength, f_y (psi)	Ultimate strength, f_u (psi)
Billet steel (A615)		
Grade 40	40,000	70,000
Grade 60	60,000	90,000
Axle steel (A617)		
Grade 40	40,000	70,000
Grade 60	60,000	90,000
Low-alloy steel (A706): Grade 60	60,000	80,000
Deformed wire		
Reinforced	75,000	85,000
Fabric	70,000	80,000
Smooth wire		
Reinforced	70,000	80,000
Fabric	65,000, 56,000	75,000, 70,000

Table 2.4 Standard Wire Reinforcement

W&D size		U.S. customary			Area (in.^{2}/ft of width for various spacings)						
					Center-to-center spacing (in.)						
Smooth	Deformed	Nominal diameter (in.)	Nominal area (in.^{2})	Nominal weight (lb/ft)	2	3	4	6	8	10	12
W31	D31	0.628	0.310	1.054	1.86	1.24	0.93	0.62	0.465	0.372	0.31
W30	D30	0.618	0.300	1.020	1.80	1.20	0.90	0.60	0.45	0.366	0.30
W28	D28	0.597	0.280	0.952	1.68	1.12	0.84	0.56	0.42	0.336	0.28
W26	D26	0.575	0.260	0.934	1.56	1.04	0.78	0.52	0.39	0.312	0.26
W24	D24	0.553	0.240	0.816	1.44	0.96	0.72	0.48	0.36	0.288	0.24
W22	D22	0.529	0.220	0.748	1.32	0.88	0.66	0.44	0.33	0.264	0.22
W20	D20	0.504	0.200	0.680	1.20	0.80	0.60	0.40	0.30	0.24	0.20
W18	D18	0.478	0.180	0.612	1.08	0.72	0.54	0.36	0.27	0.216	0.18
W16	D16	0.451	0.160	0.544	0.96	0.64	0.48	0.32	0.24	0.192	0.16
W14	D14	0.422	0.140	0.476	0.84	0.56	0.42	0.28	0.21	0.168	0.14
W12	D12	0.390	0.120	0.408	0.72	0.48	0.36	0.24	0.18	0.144	0.12
W11	D11	0.374	0.110	0.374	0.66	0.44	0.33	0.22	0.165	0.132	0.11
W10.5		0.366	0.105	0.357	0.63	0.42	0.315	0.21	0.157	0.126	0.105
W10	D10	0.356	0.100	0.340	0.60	0.40	0.30	0.20	0.15	0.12	0.10
W9.5		0.348	0.095	0.323	0.57	0.38	0.285	0.19	0.142	0.114	0.095
W9	D9	0.338	0.090	0.306	0.54	0.36	0.27	0.18	0.135	0.108	0.09
W8.5		0.329	0.085	0.289	0.51	0.34	0.255	0.17	0.127	0.102	0.085
W8	D8	0.319	0.080	0.272	0.48	0.32	0.24	0.16	0.12	0.096	0.08
W7.5		0.309	0.075	0.255	0.45	9.30	0.225	0.15	0.112	0.09	0.075
W7	D7	0.298	0.070	0.238	0.42	0.28	0.21	0.14	0.105	0.084	0.07
W6.5		0.288	0.065	0.221	0.39	0.26	0.195	0.13	0.097	0.078	0.065
W6	D6	0.276	0.060	0.204	0.36	0.24	0.18	0.12	0.09	0.072	0.06
W5.5		0.264	0.055	0.187	0.33	0.22	0.165	0.11	0.082	0.066	0.055
W5	D5	0.252	0.050	0.170	0.30	0.20	0.15	0.10	0.075	0.06	0.05
W4.5		0.240	0.045	0.153	0.27	0.18	0.135	0.09	0.067	0.054	0.045
W4	D4	0.225	0.040	0.136	0.24	0.16	0.12	0.08	0.06	0.048	0.04
W3.5		0.211	0.035	0.119	0.21	0.14	0.105	0.07	0.052	0.042	0.035
W3		0.195	0.030	0.102	0.18	0.12	0.09	0.06	0.045	0.036	0.03
W2.9		0.192	0.029	0.098	0.17	0.116	0.087	0.058	0.043	0.035	0.029
W2.5		0.178	0.025	0.085	0.15	0.10	0.075	0.05	0.037	0.03	0.025
W2		0.159	0.020	0.068	0.12	0.08	0.06	0.04	0.03	0.024	0.02
W1.4		0.135	0.014	0.049	0.084	0.056	0.042	0.028	0.021	0.017	0.014

Table 2.5 Weight, Area, and Perimeter of Individual Bars

Bar designation number	Weight per foot (lb)	Standard nominal dimensions		
		Diameter, d_b [in. (mm)]	Cross-sectional area, A_b (in.^{2})	Perimeter (in.)
3	0.376	0.375 (9)	0.11	1.178
4	0.668	0.500 (13)	0.20	1.571
5	1.043	0.625 (16)	0.31	1.963
6	1.502	0.750 (19)	0.44	2.356
7	2.044	0.875 (2)	0.60	2.749
8	2.670	1.000 (25)	0.79	3.142
9	3.400	1.128 (28)	1.00	3.544
10	4.303	1.270 (31)	1.27	3.990
11	5.313	1.410 (33)	1.56	4.430
14	7.65	1.693 (43)	2.25	5.32
18	13.60	2.257 (56)	4.00	7.09

and are welded together at intersections. Table 2.4 presents geometrical properties for some standard wire reinforcement.

For most mild steels, the behavior is assumed to be elastoplastic and Young's modulus is taken as 29×10^6 psi (200×10^6 MPa). Table 2.3 presents the reinforcement-grade strengths, and Table 2.5 presents geometrical properties of the various sizes of bars.

2.7 PRESTRESSING REINFORCEMENT

2.7.1 Types of Reinforcement

Because of the high creep and shrinkage losses in concrete, effective prestressing can be achieved by using very high-strength steels in the range of 270,000 psi or more (1,862 MPa or higher). Such high-stressed steels are able to counterbalance these losses in the

Photo 2.6 Prestressed concrete Valdez floating dock. Designed by ABAM Engineers, built in two pieces in Tacoma, Washington, then towed to Alaska by deployment. (*Courtesy,* ABAM Engineers, Tacoma, Washington.)

surrounding concrete and have adequate leftover stress levels to sustain the required prestressing force. The magnitude of normal prestress losses can be expected to be in the range of 35,000 to 60,000 psi (241 to 414 MPa). The initial prestress would thus have to be very high, on the order of 180,000 to 220,000 psi (1,241 to 1,517 MPa). From the aforementioned magnitude of prestress losses, it can be inferred that normal steels with yield strengths $f_y = 60{,}000$ psi (414 MPa) would have little prestressing stress left after losses, obviating the need for using very high-strength steels for prestressing concrete members.

Prestressing reinforcement can be in the form of single wires, strands composed of several wires twisted to form a single element, and high-strength bars. Three types commonly used in the United States are:

- Uncoated stress-relieved or low-relaxation wires.
- Uncoated stress-relieved strands and low-relaxation strands.
- Uncoated high-strength steel bars.

Wires or strands that are not stress-relieved, such as the straightened wires or oil-tempered wires often used in other countries, exhibit higher relaxation losses than stress-relieved wires or strands. Consequently, it is important to account for the appropriate magnitude of losses once a determination is made on the type of prestressing steel required.

2.7.2 Stress-Relieved and Low-Relaxation Wires and Strands

Stress-relieved wires are cold-drawn single wires conforming to ASTM standard A421; stress-relieved strands conform to ASTM standard A 416. The strands are made from seven wires by twisting six of them on a pitch of 12- to 16-wire diameter around a slightly larger, straight control wire. Stress-relieving is done after the wires are woven into the strand. The geometrical properties of the wires and strands as required by ASTM are given in Tables 2.6 and 2.7, respectively.

To maximize the steel area of the 7-wire strand for any nominal diameter, the standard wire can be drawn through a die to form a *compacted* strand as shown in Figure 2.17(b); this is opposed to the standard 7 wire strand in Figure 2.17(a). ASTM standard A 779 requires the minimum strengths and geometrical properties given in Table 2.8.

Figure 2.18(a) shows a typical stress-strain diagram for wire and strand prestressing steels, while Figure 2.18(b) shows values relative to those of mild steel.

Table 2.6 Wire for Prestressed Concrete

Nominal diameter (in.)	Min. tensile strength (psi)		Min. stress at 1% extension (psi)	
	Type BA	Type WA	Type BA	Type WA
0.192		250,000		212,500
0.196	240,000	250,000	204,000	212,500
0.250	240,000	240,000	204,000	204,000
0.276	235,000	235,000	199,750	199,750

Source: Post-Tensioning Institute

Table 2.7 Seven-Wire Standard Strand for Prestressed Concrete

Nominal diameter of strand (in.)	Breaking strength of strand (min. lb)	Nominal steel area of strand (sq in.)	Nominal weight of strands (lb per 1000 ft)*	Minimum load at 1% extension (lb)
		GRADE 250		
$\frac{1}{4}$(0.250)	9,000	0.036	122	7,650
$\frac{5}{16}$(0.313)	14,500	0.058	197	12,300
$\frac{3}{8}$(0.375)	20,000	0.080	272	17,000
$\frac{7}{16}$(0.438	27,000	0.108	367	23,000
$\frac{1}{2}$(0.500)	36,000	0.144	490	30,600
$\frac{3}{5}$(0.600)	54,000	0.216	737	45,900
		GRADE 270		
$\frac{3}{8}$(0.375)	23,000	0.085	290	19,550
$\frac{7}{16}$(0.438)	31,000	0.115	390	26,350
$\frac{1}{2}$(0.500)	41,300	0.153	520	35,100
$\frac{3}{5}$(0.600)	58,600	0.217	740	49,800

*100,000 psi = 689.5 MPa

0.1 in. = 2.54 mm; 1 in.^{2} = 645 mm^2

weight: mult. by 1.49 to obtain weight in kg per 1,000 m.

1,000 lb = 4,448 Newton

Source: Post-Tensioning Institute

2.7.3 High-Tensile-Strength Prestressing Bars

High-tensile-strength alloy steel bars for prestressing are either smooth or deformed, and are available in nominal diameters from $\frac{3}{4}$ in. (19 mm) to $1\frac{3}{8}$ in. (35 mm). They must conform to ASTM standard A 722. Cold drawn in order to raise their yield strength, these bars are stress relieved as well to increase their ductility. Stress relieving is achieved by heating the bar to an appropriate temperature, generally below 500°C. Though essentially the same stress-relieving process is employed for bars as for strands, the tensile strength of prestressing bars has to be a minimum of 150,000 psi (1,034 MPa), with a minimum yield strength of 85 percent of the ultimate strength for smooth bars and 80 percent for deformed bars.

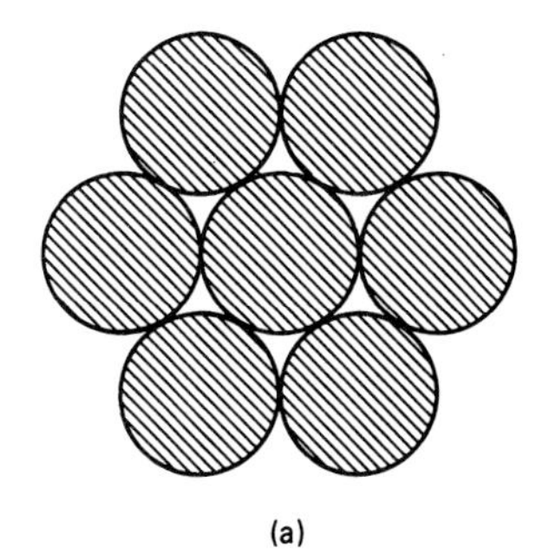

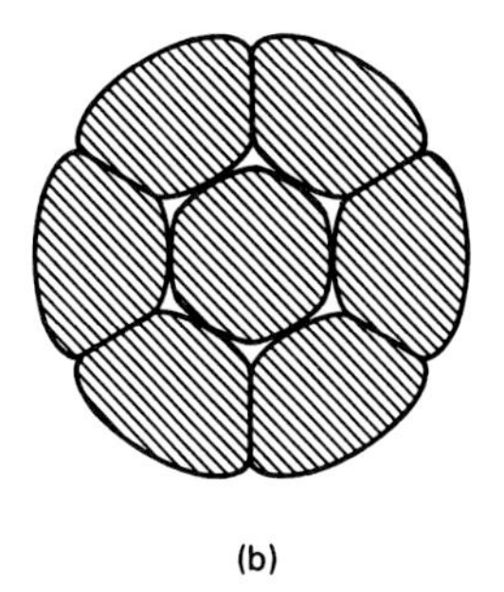

Figure 2.17 Standard and compacted 7-wire prestressing strands. (a) Standard strand section. (b) Compacted strand section.

Table 2.8 Seven-Wire Compacted Strand for Prestressed Concrete [ASTM A779]

Nominal diameter (in.)	Nominal Breaking strength of strand (min. lb)*	Nominal steel area (in.^{2})	Nominal weight of strand (per 1,000 ft-lb)
$\frac{1}{2}$	47,000	0.174	600
0.6	67,440	0.256	873
0.7	85,430	0.346	1176

*1000 lb = 4,448 Newton
Grade 270; f_{pu} = 270,000 psi ult. strength (1,862 MPa)
1 in. = 25.4 mm; 1 in.^{2} = 645 mm^2

Table 2.9 lists the geometrical properties of the prestressing bas as required by ASTM standard A 722, and Figure 2.18 shows a typical stress-strain diagram for such bars.

2.7.4 Steel Relaxation

Stress relaxation in prestressing steel is the loss of prestress when the wires or strands are subjected to essentially constant strain. It is identical to creep in concrete, except that creep is a *change* in strain whereas steel relaxation is a *loss in steel stress.* Where t = time,

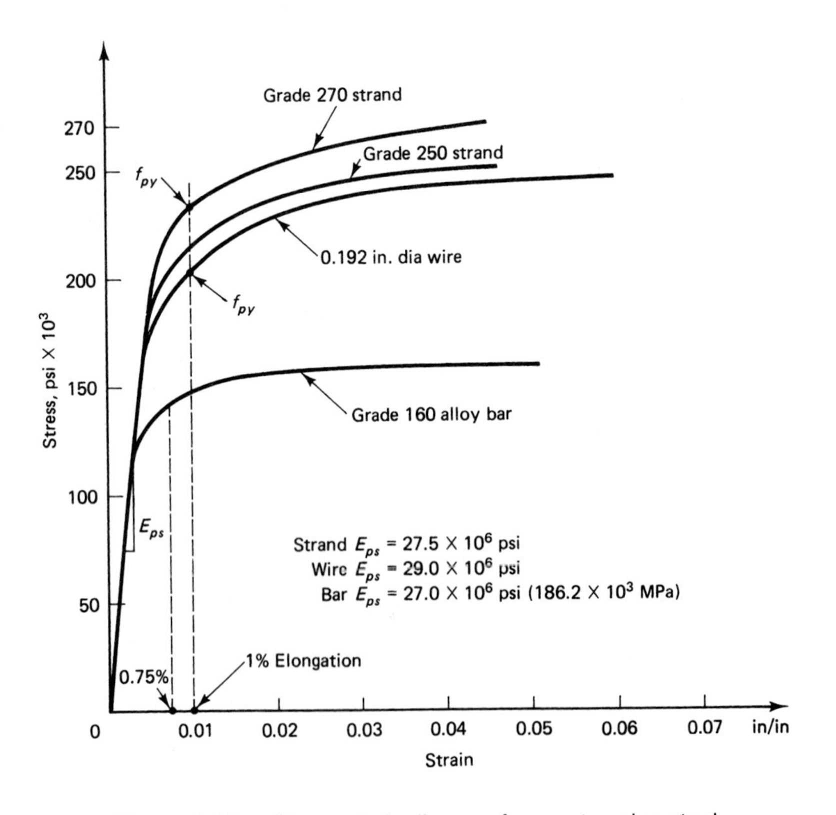

Figure 2.18a Stress-strain diagram for prestressing steel.

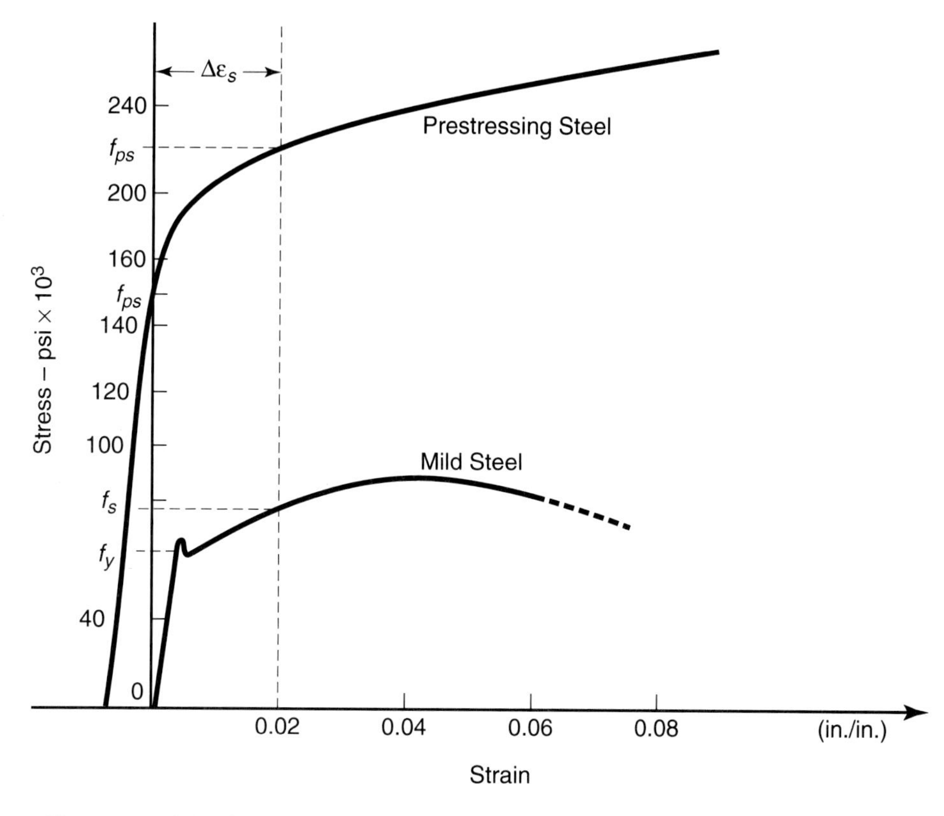

Figure 2.18(b) Stress-Strain Diagram for Prestressing Steel Strands in Comparison with Mild Steel Bar Reinforcement.

in hours, after prestressing, the loss of stress due to relaxation in stress-relieved wires and strands can be evaluated from the expression

$$\Delta f_R = f_{pi} \frac{\log t}{10} \left(\frac{f_{pi}}{f_{py}} - 0.55 \right) \tag{2.18}$$

Table 2.9 Steel Bars for Prestressed Concrete

Bar type*	Nominal diameter (in.)	Nominal steel area (in.^{2})
Smooth Alloy	0.750	0.442
Steel Grade	0.875	0.601
145 or 160	1.000	0.785
(ASTM A722)	1.125	0.994
	1.250	1.227
	1.375	1.485
Deformed	0.625	0.280
Bars	1.000	0.852
	1.250	1.295

*Grade 145; f_{pu} = 145,000 psi (1,000 MPa)
Grade 160: f_{pu} = 160,000 psi (1,103 MPa)
1 in. = 25.4 mm; 1 in.^{2} = 645 mm^2

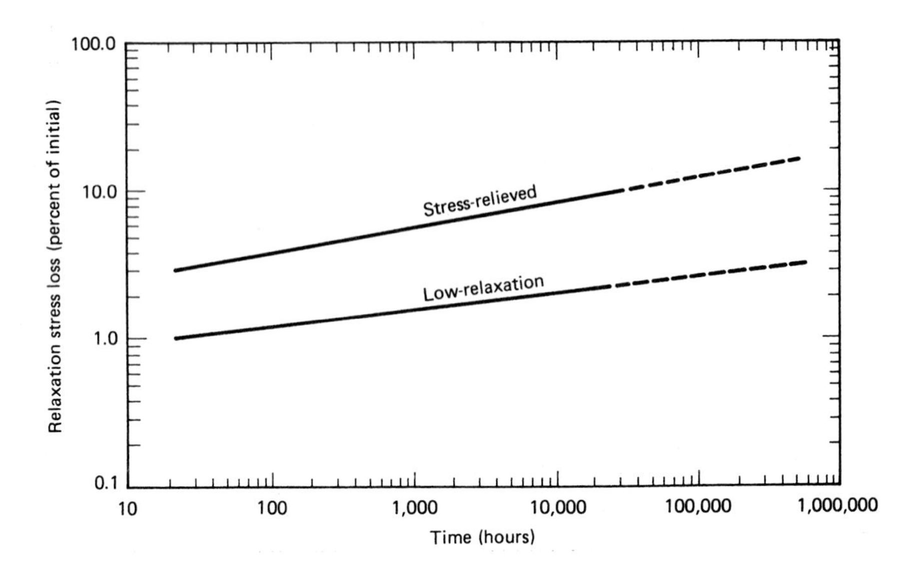

Figure 2.19 Relaxation loss vs. time for stress-relieved low-relaxation prestressing steels at 70 percent of the ultimate. (*Courtesy,* Post-Tensioning Institute.)

provided that $f_p/f_{py} \geq 0.55$ and $f_{py} \cong 0.85\, f_{pu}$ for stress-relieved strands and 0.90 for low-relaxation strands. Also, $f_{pi} = 0.82\, f_{py}$ immediately after transfer but $f_{pi} \leq 0.74\, f_{pu}$ for pre-tensioned, and $0.70\, f_{pu}$ for post-tensioned, concrete. In general, $f_{pi} \cong 0.70\, f_{pu}$.

It is possible to decrease stress relaxation loss by subjecting strands that are initially stressed to 70 percent of their ultimate strength f_{pu} to temperatures of 20°C to 100°C for an extended time in order to produce a permanent elongation—a process called *stabilization.* The prestressing steel thus produced is termed *low-relaxation steel* and has a relaxation stress loss that is 25 percent of that of normal stress-relieved steel.

The expression for stress relaxation in low-relaxation prestressing steels is

$$\Delta f_R = f_{pi}\frac{\log t}{45}\left(\frac{f_{pi}}{f_{py}} - 0.55\right) \tag{2.19}$$

Figure 2.19 shows the relative relaxation loss for stress-relieved and low-relaxation steels for 7-wire strands held at constant length at 29.5°C.

2.7.5 Corrosion and Deterioration of Strands

Protection against corrosion of prestressing steel is more critical than in the case of non-prestressed steel. Such precaution is necessary since the strength of the prestressed concrete element is a function of the prestressing force, which in turn is a function of the prestressing tendon area. Reduction of the prestressing steel area due to corrosion can drastically reduce the nominal moment strength of the prestressed section, which can lead to premature failure of the structural system. In pretensioned members, protection against corrosion is provided by the concrete surrounding the tendon, provided that adequate concrete cover is available. In post-tensioned members, protection can be obtained by full grouting of the ducts after prestressing is completed or by greasing.

Another form of wire or strand deterioration is *stress corrosion,* which is characterized by the formation of microscopic cracks in the steel which lead to brittleness and failure. This type of reduction in strength can occur only under very high stress and, though infrequent, is difficult to prevent.

2.8 ACI MAXIMUM PERMISSIBLE STRESSES IN CONCRETE AND REINFORCEMENT

Following are definitions of some important mathematical terms used in this section:

f_{py} = specified yield strength of prestressing tendons, in psi
f_y = specified yield strength of nonprestressed reinforcement, in psi
f_{pu} = specified tensile strength of prestressing tendons, in psi
f'_c = specified compressive strength of concrete, in psi
f'_{ci} = compressive strength of concrete at time of initial prestress

2.8.1 Concrete Stresses in Flexure

Stresses in concrete immediately after prestress transfer (before time-dependent prestress losses) shall not exceed the following:

(a) Extreme fiber stress in compression . . . . $0.60\,f'_{ci}$
(b) Extreme fiber stress in tension except as permitted in (c) . . . . $3\sqrt{f'_c}$
(c) Extreme fiber stress in tension at ends of simply supported members . . . . $6\sqrt{f'_{ci}}$

Where computed tensile stresses exceed these values, bonded auxiliary reinforcement (nonprestressed or prestressed) shall be provided in the tensile zone to resist the total tensile force in concrete computed under the assumption of an uncracked section.

Stresses in concrete at service loads (after allowance for all prestress losses) shall not exceed the following:

(a) Extreme fiber stress in compression due to prestress plus sustained load. $0.45\,f'_c$
(b) Extreme fiber stress in compression due to prestress plus total load. . . . . $0.60\,f'_c$
(c) Extreme fiber stress in tension in precompressed tensile zone. . . . . $6\sqrt{f'_c}$
(d) Extreme fiber stress in tension in precompressed tensile zone of members (except two-way slab systems), where analysis based on transformed cracked sections and on bilinear moment-deflection relationships shows that immediate and long-time deflections comply with the ACI definition requirements and minimum concrete cover requirements . . . . $12\sqrt{f'_c}$

2.8.2 Prestressing Steel Stresses

Tensile stress in prestressing tendons shall not exceed the following:

(a) Due to tendon jacking force . . . . $0.94\,f_{py}$
but not greater than the lesser of $0.80\,f_{pu}$ and the maximum value recommended by the manufacturer of prestressing tendons or anchorages.
(b) Immediately after prestress transfer. . . . . $0.82\,f_{py}$
but not greater than $0.74\,f_{pu}$.
(c) Post-tensioning tendons, at anchorages and couplers, immediately after tendon anchorage . . . . $0.70\,f_{pu}$

2.9 AASHTO MAXIMUM PERMISSIBLE STRESSES IN CONCRETE AND REINFORCEMENT

2.9.1 Concrete Stresses before Creep and Shrinkage Losses

Compression

Pretensioned members . . . . $0.60\,f'_{ci}$

Post-tensioned members . . . . $0.55\,f'_{ci}$

Tension

Precompressed tensile zone . . . . No temporary allowable stresses are specified.

Other Areas

In tension areas with no bonded reinforcement. . . . . 200 psi or $3\sqrt{f'_{ci}}$

Where the calculated tensile stress exceeds this value, bonded reinforcement shall be provided to resist the total tension force in the concrete computed on the assumption of an uncracked section. The maximum tensile stress shall not exceed. . . . . $7.5\sqrt{f'_{ci}}$

2.9.2 Concrete Stresses at Service Load after Losses

Compression . . . . $0.40\,f'_c$

Tension in the precompressed tensile zone

(a) For members with bonded reinforcement . . . . $6\sqrt{f'_c}$

For severe corrosive exposure conditions, such as coastal areas . . $3\sqrt{f'_c}$

(b) For members without bonded reinforcement 0

Tension in other areas is limited by the allowable temporary stresses specified in Section 2.8.1.

2.9.2.1 Cracking Stresses.

Modulus of rupture from tests or if not available.

For normal-weight concrete. . . . . $7.5\sqrt{f'_c}$

For sand-lightweight concrete . . . . $6.3\sqrt{f'_c}$

For all other lightweight concrete . . . . $5.5\sqrt{f'_c}$

2.9.2.2 Anchorage-Bearing Stresses

Post-tensioned anchorage at service load. . . . . 3,000 psi (but not to exceed $0.9\,f'_{ci}$)

2.9.3 Prestressing Steel Stresses

(a) Due to tendon jacking for . . . . $0.94\,f_{py} \leq 0.80\,f_{pu}$

(b) Immediately after prestress transfer. . . . . $0.82\,f_{py} \leq 0.74\,f_{pu}$

(c) Post-tensioning tendons at anchorage, immediately after tendon anchorage . . . . $0.70\,f_{pu}$

$f_{py} \approx 0.85\,f_{py}$ (for low-relaxation, $f_{py} = 0.90\,f_{pu}$)

Hence for 270 K tendons used in the book, f_{pi} at transfer $= 0.70 \times 270{,}000 = 189{,}000$ psi (1300 MPa) is applied for uniformity.

2.9.4 Relative Humidity Values

Figure 2.20 gives the mean annual relative humidity values for all regions in the United States in percent, to be used for evaluating shrinkage losses in concrete.

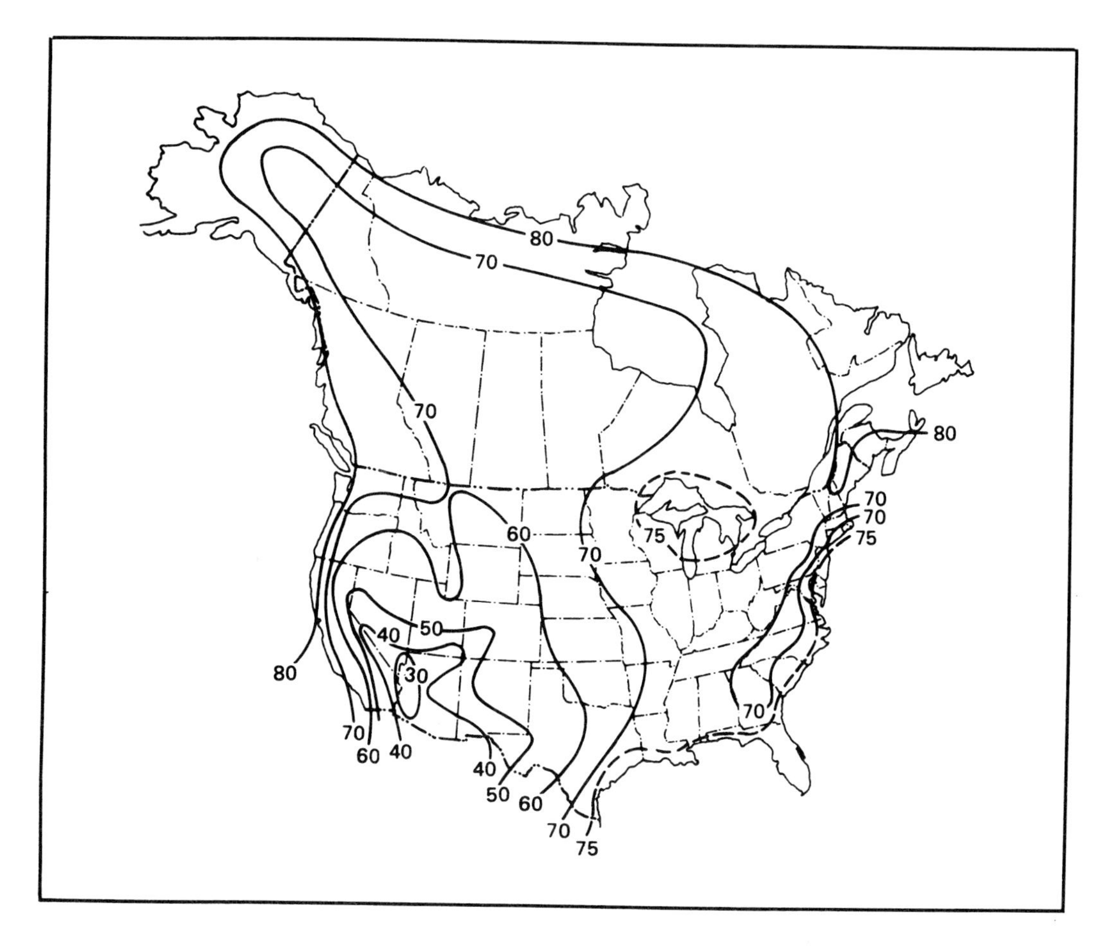

Figure 2.20 Mean annual relative humidity. (*Courtesy,* Prestressed Concrete Institute.)

2.10 PRESTRESSING SYSTEMS AND ANCHORAGES

2.10.1 Pretensioning

Prestressing steel is pretensioned against independent anchorages *prior to* the placement of concrete around it. Such anchorages are supported by large and stable bulkheads to support the exceedingly high concentrated forces applied to the individual tendons. The term "pretensioning" means pretensioning of the prestressing steel, not the beam it serves. Consequently, a *pretensioned beam* is a prestressed beam in which the prestressing tendon is tensioned prior to casting the section, while a *post-tensioned beam* is one in which the prestressing tendon is tensioned after the beam has been cast and has achieved the major portion of its concrete strength. Pretensioning is normally performed at precasting plants, where a precasting stressing bed of a long reinforced concrete slab is cast on the ground with vertical anchor bulkheads or walls at its ends. The steel strands are stretched and anchored to the vertical walls, which are designed to resist the large eccentric prestressing forces. Prestressing can be accomplished by prestressing *individual* strands, or *all* the strands at one jacking operation.

For harped tendon profiles, the prestressing bed is provided with hold-down devices as shown in Figure 2.21. Since the bed can be several hundred feet long, several precast prestressed elements can be produced in one operation, and the exposed prestressing strands between them can be cut after the concrete hardens. Pretensioning

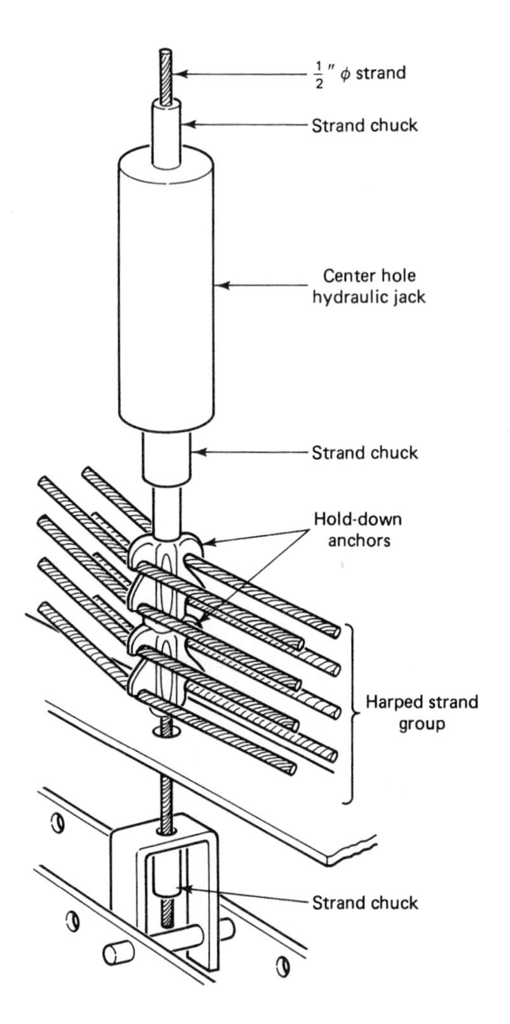

Figure 2.21 Hold-down anchor for harping pretensioning tendons. (*Courtesy,* Post-Tensioning Institute.)

several elements in a prestressing bed is represented schematically in Figure 2.22, while harping of tendons in a prestressing bed system is shown in Figure 2.23.

In pretensioning, strands and single wires are anchored by several patented systems. One of these, a chuck system by Supreme Products, is used for anchoring tendons in post-tensioning. The gripping mechanism of this system is illustrated in Figure 2.24(c). Other anchorage systems and ductile connections are shown in Figure 2.24(d), (e), and (f). A prestressing bed for moderately sized pretensioned beams up to 24 ft (7.32 m) long was developed and used by the author in Ref. 2.31 for his continuing work on the behavior of pretensioned and post-tensioned structural systems. Supreme Products anchorage chucks have been used together with the Freyssinet jack, where applicable. Figures 2.25 and 2.26 give details of the prestressing bed system also used for post-tensioning developed by Nawy and Potyondy at Rutgers University, while Figure 2.27 shows the dimensional details of the system.

2.10.2 Post-Tensioning

In post-tensioning, the strands, wires, or bars are tensioned after hardening of the concrete. The strands are placed in the longitudinal ducts within the precast concrete ele-

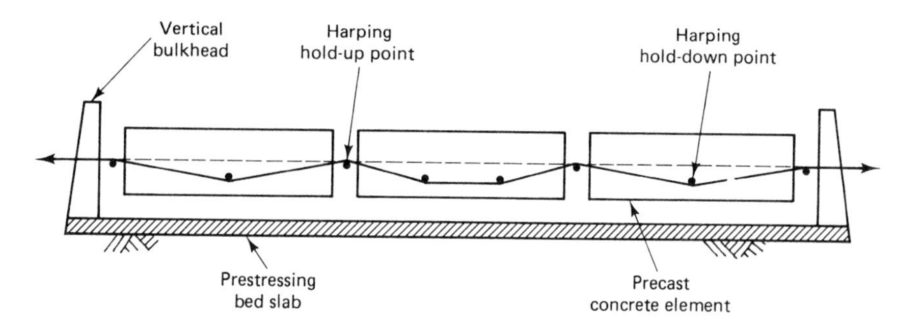

Figure 2.22 Schematic of pretensioning bed.

ment. The prestressing force is transferred through end anchorages such as the Supreme Products chucks shown in Figure 2.24. The tendons of strands should *not* be bonded or grouted prior to full prestressing.

2.10.3 Jacking Systems

One of the fundamental components of a prestressing operation is the jacking system applied, i.e., the manner in which the prestressing force is transferred to the steel tendons. Such a force is applied through the use of hydraulic jacks of capacity 10 to 20 tons and a stroke from 6 to 48 in., depending on whether pretensioning or post-tensioning is used and whether individual tendons are being prestressed or all the tendons are being stressed simultaneously. In the latter case, large-capacity jacks are needed, with a stroke of at least 30 in. (762 mm). Of course, the cost will be higher than sequential tensioning. Figure 2.28 shows a 500-tone multistrand jack for simultaneous jacking through a center hole.

Figure 2.23 Harping of tendons in a prestressing bed system.

2.10.4 Grouting of Post-Tensioned Tendons

In order to provide permanent protection for the post-tensioned steel and to develop a bond between the prestressing steel and the surrounding concrete, the prestressing ducts have to be filled under pressure with the appropriate cement grout in an injection process.

2.10.4.1 Grouting materials

1. **Portland Cement.** Portland cement should conform to one of the following specifications: ASTM C150, Type I, II, or III.

(a) Strand anchor.

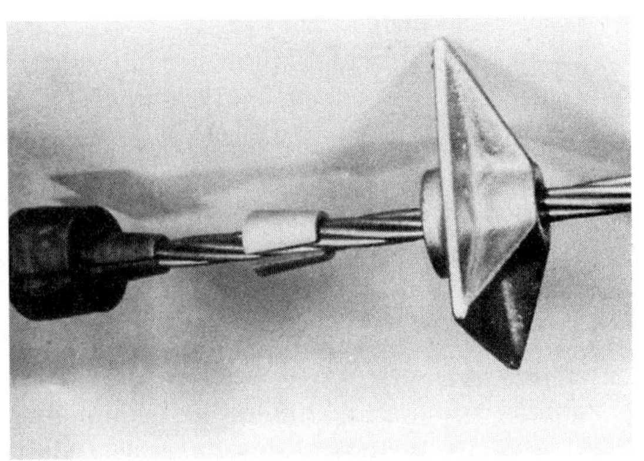

(b) Monostrand anchor.

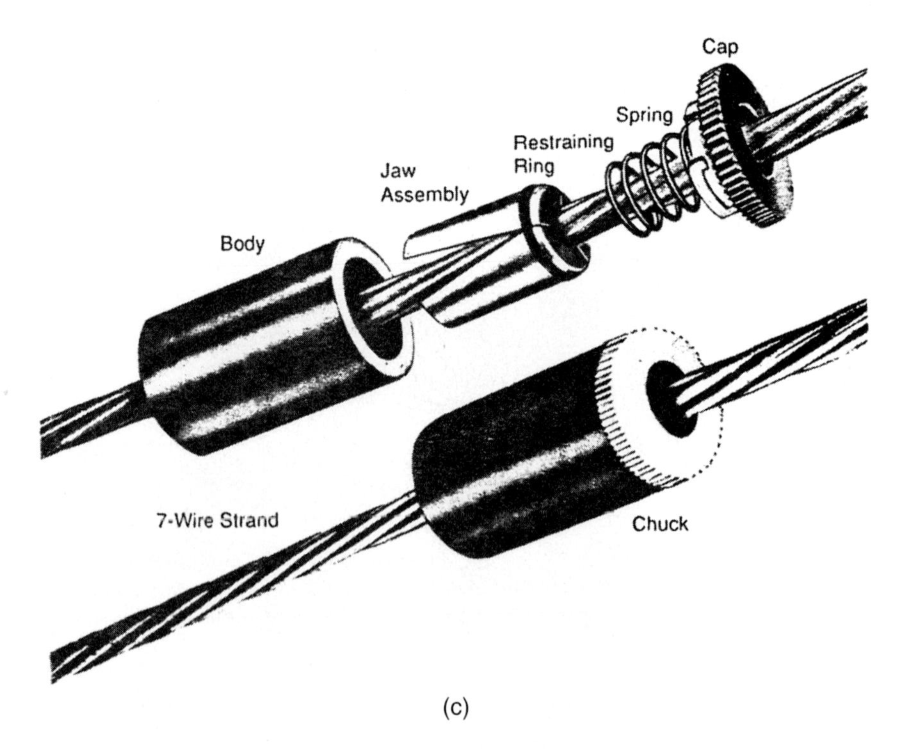

(c)

Figure 2.24 (a) Stress Strand Anchor, (b) Monostrand anchor, (c) Supreme Products anchorage chuck. (*Courtesy,* Post-Tensioning Institute.)

(d)

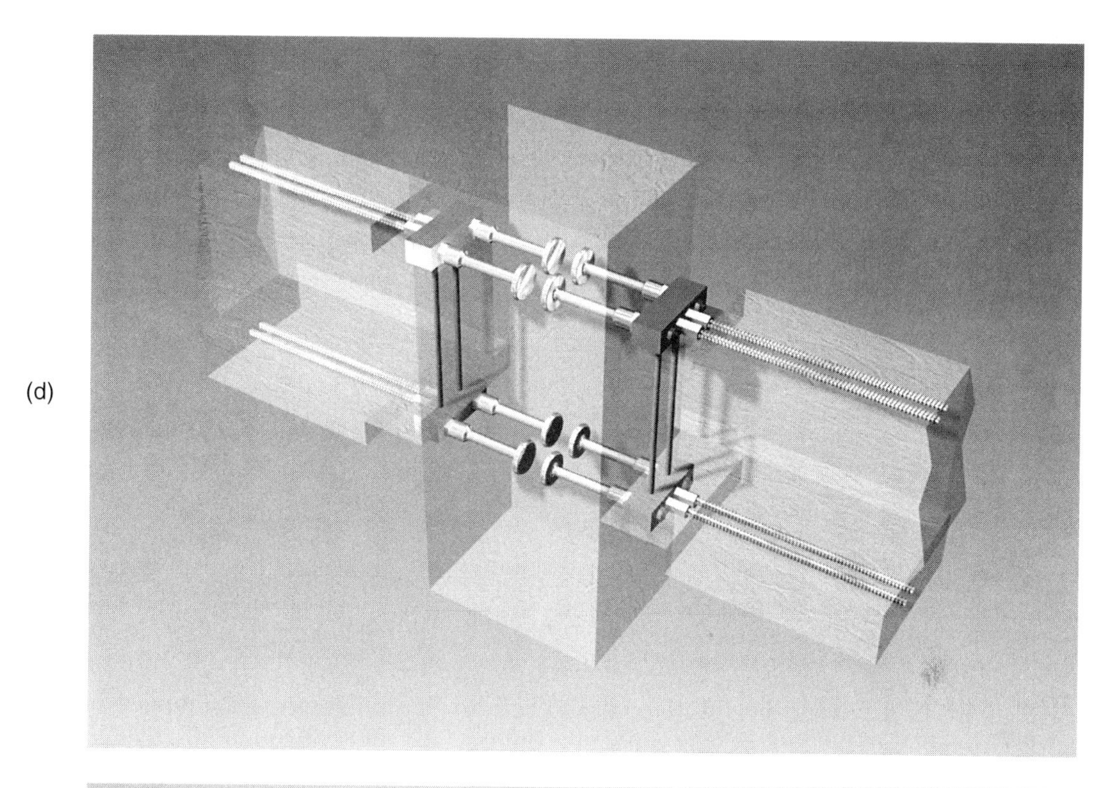

(e)

(f)

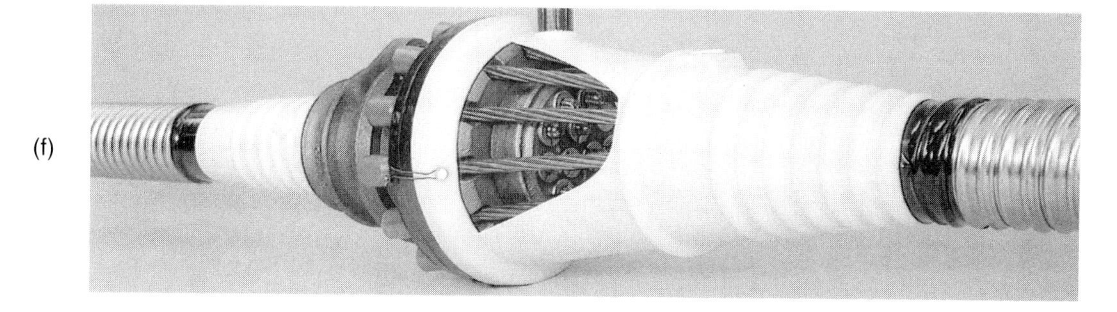

Figure 2.24 (*continued*) Multiple anchorages, couplers and ductile connectors (Courtesy Dywidag Sysems International): (d) Multiple anchorage, (e) Coupler, (f) Dywidag ductile connectors (DDC) for ductile precast beam-column connections in seismic zones. See also details in Figures 13.9 and 13.10.

Figure 2.25 Prestress tensioning arrangement (Nawy et al.).

Cement used for grouting should be fresh and should not contain any lumps or other indications of hydration or "pack set."

2. **Water.** The water used in the grout should be potable, clean, and free of injurious quantities of substances known to be harmful to portland cement or prestressing steel.
3. **Admixtures.** Admixtures, if used, should impart the properties of low water content, good flow, minimum bleed, and expansion if desired. Their formulation should contain no chemicals in quantities that may have a harmful effect on the prestressing steel or cement. Admixtures containing chlorides (as C1 in excess of 0.5 percent by weight of admixture, assuming 1 lb of admixture per sack of cement), fluorides, sulphites, or nitrates should not be used. Aluminum powder of the proper fineness and quantity, or any other approved gas-evolving material which is well dispersed through the other admixture, may be used to obtain 5 to 10 percent unrestrained expansion of the grout (see Refs. 2.11, 2.38).

2.10.4.2 Ducts

1. **Forming.**
   (a) **Formed Ducts.** Ducts formed by sheath left in place should be of a type that does not permit the entrance of cement paste. They should transfer bond

Figure 2.26 Intermediate connections between frames of the prestressing system for continuous beams (Nawy et al.).

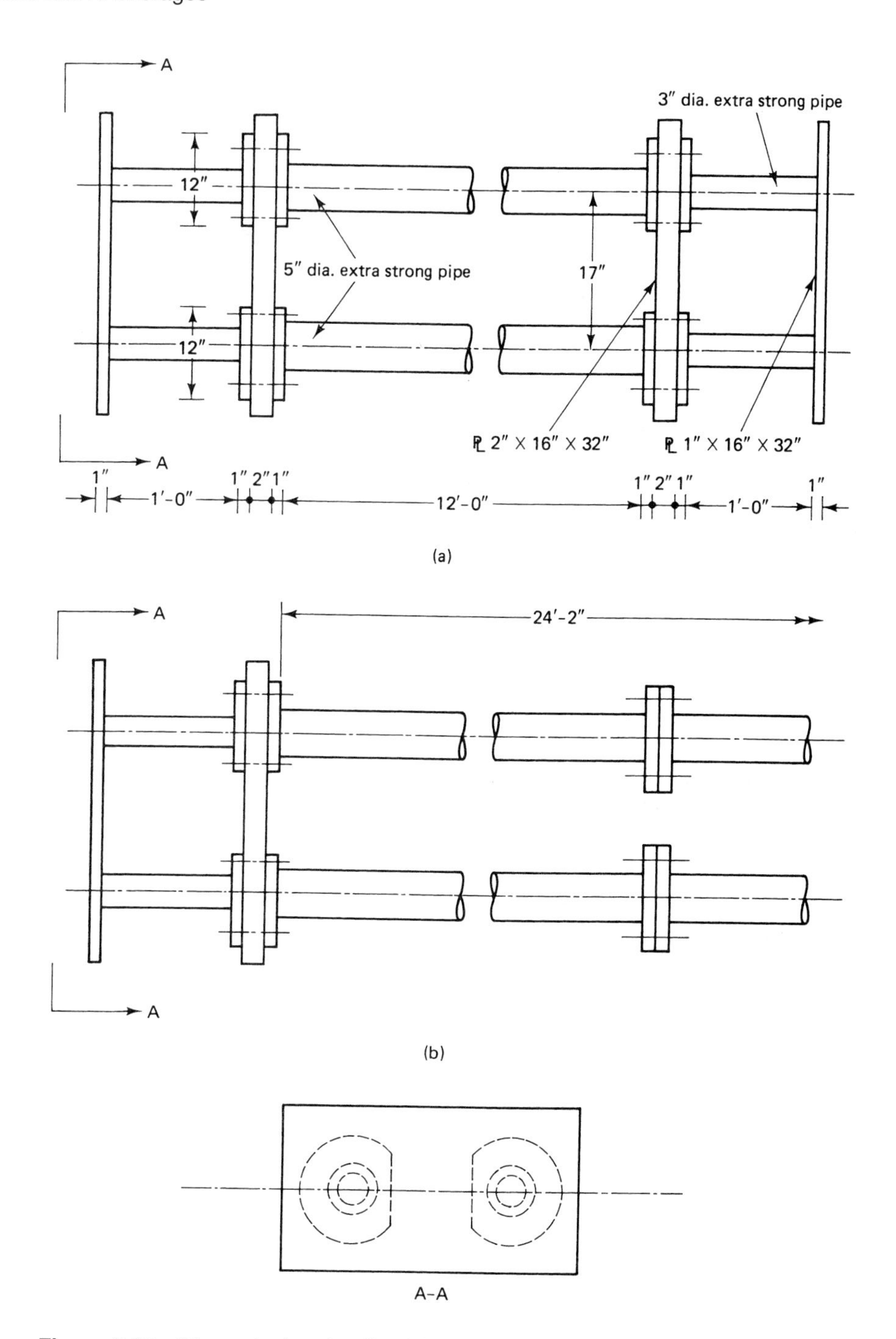

Figure 2.27 Dimensioning details of the pretensioning or post-tensioning laboratory system used for research at Rutgers (Nawy et al.).

stresses as required and should retain their shape under the weight of the concrete. Metallic sheaths should be of a ferrous metal, and they may be galvanized.

(b) Cored Ducts. Cored ducts should be formed with no constrictions which would tend to block the passage of grout. All coring material should be removed.

Figure 2.28 Stresstek Multistrand 500-ton jack. (*Courtesy,* Post-Tensioning Institute.)

2. **Grout Openings or Vents.** All ducts should have grout openings at both ends. For draped cables, all high points should have a grout vent except where the cable curvature is small, such as in continuous slabs. Grout vents or drain holes should be provided at low points if the tendon is to be placed, stressed, and grouted in a freezing climate. All grout openings or vents should include provisions for preventing grout leakage.
3. **Duct Size.** For tendons made up of a plurality of wires, bars, or strands, the duct area should be at least twice the net area of the prestressing steel. For tendons made up of a single wire, bar, or strand, the duct diameter should be at least $\frac{1}{4}$ in. larger than the nominal diameter of the wire, bar, or strand.
4. **Placement of Ducts.** After the placement of ducts, reinforcement, and forming are complete, an inspection should be made to locate possible duct damage. Ducts should be securely fastened at close enough intervals to avoid displacement during concreting. All holes or openings in the duct must be repaired prior to placement of concrete. Grout openings and vents must be securely anchored to the duct and to

Photo 2.7 Prestress conduit for a bridge deck.

either the forms or the reinforcing steel, to prevent displacement during concrete-placing operations.

2.10.4.3 Grouting Process

1. Ducts with concrete walls (cored ducts) should be flushed to ensure that the concrete is thoroughly wetted.
2. All grout and high-point vent openings should be open when grouting starts. Grout should be allowed to flow from the first vent after the inlet pipe until any residual flushing water or entrapped air has been removed, at which time the vent should be capped or otherwise closed. Remaining vents should be closed in sequence in the same manner. The pumping pressure at the tendon inlet should not exceed 250 psig.
3. Grout should be pumped through the duct and continuously wasted at the outlet pipe until no visible slugs of water or air are ejected. The efflux time of the ejected grout should not be less than the injected grout. To ensure that the tendon remains filled with grout, the outlet and/or inlet should be closed. Plugs, caps, or valves thus required should not be removed or opened until the grout has set.
4. When one-way flow of grout cannot be maintained, the grout should be immediately flushed out of the duct with water.
5. In temperatures below 32°F, ducts should be kept free of water to avoid damage due to freezing.
6. The temperature of the concrete should be 35°F or higher from the time of grouting until job-cured 2-in. cubes of grout reach a minimum compressive strength of 800 psi.
7. Grout should not be above 90°F during mixing or pumping. If necessary, the mixing water should be cooled.

Additional details and specifications on grouting are given by the Post-Tensioning Institute in Ref. 2.29.

Figure 2.29 Prestressing of preload circular tank. (*Courtesy,* N.A. Legates, Preload Technology, Inc., New York.)

2.11 CIRCULAR PRESTRESSING

Circular prestressing involves the development of hoop or hugging compressive stresses on circular or cylindrical containment vessels, including prestressed water tanks and pipes. It is usually accomplished by a wire-wound technique, in which the concrete pipe or tank is wrapped with continuous high-tensile wire tensioned to prescribed design levels. Such tension results in uniform radial compression that prestresses the concrete cylinder or core and prevents tensile stresses from developing in the concrete wall section under internal fluid pressure. Figure 2.29 shows a preload circular tank being prestressed by the wire-wrapping process along its height.

2.12 TEN PRINCIPLES

The following ten principles are taken from Abeles (Ref. 2.32) and applicable not only to prestressing concrete but to any endeavor that the engineer is called upon to undertake:

1. You cannot have everything. (Each solution has advantages and disadvantages that have to be tallied and traded off against each other.)
2. You cannot have something for nothing. (One has to pay in one way or another for something which is offered as a "free gift" into the bargain, notwithstanding a solution's being optimal for the problem.)
3. It is never too late (e.g., to alter a design, to strengthen a structure before it collapses, or to adjust or even change principles previously employed in the light of increased knowledge and experience).
4. There is no progress without considered risk. (While it is important to ensure sufficient safety, overconservatism can never lead to an understanding of novel structures.)
5. The proof of the pudding is in the eating. (This is in direct connection with the previous principle indicating the necessity of tests.)
6. Simplicity is always an advantage, but beware of oversimplification. (The latter may lead to theoretical calculations which are not always correct in practice, or to a failure to cover all conditions.)
7. Do not generalize, but rather qualify the specific circumstances. (Serious misunderstandings may be caused by unreserved generalizations.)
8. The important question is how good, not how cheap an item is. (A cheap price given by an inexperienced contractor usually results in bad work; similarly, cheap, unproved appliances may have to be replaced.)
9. We live and learn. (It is always possible to increase one's knowledge and experience.)
10. There is nothing completely new. (Nothing is achieved instantaneously, but only by step-by-step development.)

REFERENCES

2.1 American Society for Testing and Materials. *Annual book of ASTM Standards:* Part 14, *Concrete and Mineral Aggregates.* Philadelphia: ASTM, 1994.

2.2 Popovices, S. *Concrete-Making Materials.* New York: McGraw-Hill, 1979, 1997.

2.3 ACI Committee 221. "Selection and Use of Aggregate for Concrete." *Journal of the American Concrete Institute,* Farmington Hills, MI, 1992.

2.4 American Concrete Institute, *ACI Manual of Concrete Practice,* 1999. *Materials.* American Concrete Institute, Farmington Hills, MI, 2000.

2.5 Portland Cement Association. *Design and Control of Concrete Mixtures,* 13th ed. PCA, Skokie, Ill., 1994.

2.6 ACI Committee 212. "Admixtures for Concrete," in *ACI Manual of Concrete Practice 1983.* ACI, Farmington Hills, MI, 1995, ACI 212.1 R-81.

2.7 Nawy, E. G., Ukadike, M. M., and Sauer, J. A. "High Strength Field Modified Concretes," *Journal of the Structural Division, ASCE,* 103, No. ST12, December 1977, 2307–2322.

2.8 American Concrete Institute. *Super-plasticizers in Concrete,* ACI Special Publication SP-62. ACI, Farmington Hills, MI, 1979.

2.9 ACI Committee 211. "Standard Practice for Selecting Proportions for Normal, Heavyweight, and Mass Concrete," ACI 211.1–92. American Concrete Institute, Farmington Hills, MI, 1992.

2.10 ACI Committee 211. "Standard Practice for Selecting Proportions for Structural Lightweight Concrete," ACI 211.1–92. American Concrete Institute, Farmington Hills, MI.

2.11 Nawy, E. G. *Reinforced Concrete—A Fundamental Approach.* Prentice-Hall, Inc., Upper Saddle River, N.J., 4th Ed., 2000, 776 p.

2.12 ACI Committee 318. "Building Code Requirements for Structural Concrete (ACI 318–99) and Commentary (318 R–99) American Concrete Institute, Farmington Hills, MI, 2000, pp. 392.

2.13 American Society for Testing and Materials. *Significance of Tests and Properties of Concrete and Concrete Making Materials,* Special Technical Publication 169B. ASTM, Philadelphia, 1978.

2.14 Ross, A. D. "The Elasticity, Creep and Shrinkage of Concrete," in *Proceedings of the Conference on Non-metallic Brittle Materials.* Interscience Publishers, London, 1958.

2.15 Neville, A. M. *Properties of Concrete,* 4th ed. Pitman Books, London, 1996.

2.16 Freudenthal, A. M., and Roll, F. "Creep and Creep Recovery of Concrete under High Compressive Stress," *Journal of the American Concrete Institute,* Proc. 54, Farmington Hills, MI, June 1958, 1111–1142.

2.17 Ross, A. D., "Creep Concrete Data," *Proceedings, Institution of Structural Engineers,* London, 1937, 314–326.

2.18 Branson, D. E. *Deformation of Concrete Structures.* New York, McGraw-Hill, 1977.

2.19 Branson, D. E. "Compression Steel Effects on Long Term Deflections," *Journal of the American Concrete Institute,* Proc. 68, Farmington Hills, MI, August 1971, 555–559.

2.20 Mindess, S., and Young, J. F. *Concrete.* Upper Saddle River, N.J., Prentice-Hall, Inc., 1981.

2.21 Nawy, E. G., and Balaguru, P. N. "High Strength Concrete," Chapter 5 in *Handbook of Structural Concrete.* London: Pitman Books, McGraw-Hill, New York, 1983, pp. 5-1 to 5-33.

2.22 Mehta, P. Kumar. *Concrete-Structure, Properties, and Materials,* 2nd ed. Prentice Hall, Upper Saddle River, N.J., 1993.

2.23 American Society for Testing and Materials. "Standard Specification for Deformed and Plain Billet-Steel Bars for Concrete Reinforcement, A6 15–79." ASTM, Philadelphia, 1980, 588–599.

2.24 American Society for Testing and Materials. "Standard Specification for Rail-Steel Deformed and Plain Bars for Concrete Reinforcement, A6 16–79." ASTM, Philadelphia, 1980, 600–605.

2.25 American Society for Testing and Materials. "Standard Specification for Axle Steel Deformed and Plain Bars for Concrete Reinforcement, A6 17–79." ASTM, Philadelphia, 1980, 607–611.

2.26 American Society for Testing and Materials. "Standard Specification for Cold-Drawn Steel Wire for Concrete Reinforcement, A8 2-79." ASTM, Philadelphia, 1980, 154–157.

2.27 American Society for Testing and Materials. "Standard Specification for Low-Alloy Steel Deformed Bars for Concrete Reinforcement, A706–79." ASTM, Philadelphia, 1980, 755–760.

2.28 ACI-ASCE Committee 423, "Recommendation for Concrete Members Prestressed with Unbonded Tendons" (ACI 423.3R-83). *Concrete International 5* (1983): 61–76.

2.29 Post-Tensioning Institute. "Guide Specifications for Post-Tensioning Materials." In *Post-Tensioning Manual,* 5th ed. Post-Tensioning Institute, Phoenix, Ariz., 2000.

2.30 AASHTO. *Standard Specifications for Highway Bridges,* 16th ed. American Association of State Highway and Transportation Officials, Washington, D.C., 1997.

2.31 Nawy, E. G., and Potyondy, J. G. "Moment Rotation, Cracking, and Deflection of Spirally Bound Pretensioned Prestressed Concrete Beams." *Engineering Research Bulletin No. 51.* New Brunswick, N.J.: Bureau of Engineering Research, Rutgers University, 1970, pp. 1–97.

2.32. Abeles, P. W., and Bardhan-Roy, B. K. *Prestressed Concrete Designer's Handbook.* 3d ed. London: Viewpoint Publications, 1981.

2.33 Nawy, E. G. *Simplified Reinforced Concrete.* Prentice Hall, Upper Saddle River, N.J., 1986.

2.34 Nawy, E. G., "Concrete." In *Corrosion and Chemical Resistant Masonry Materials Handbook.* Park Ridge, N.J.: Noyes, 1986, pp. 57–73.

2.35 ACI Committee 435. "Control of Deflection in Concrete Structures," ACI Committee Report, E. G. Nawy, Chairman, American Concrete Institute, Farmington Hills, MI, 1995, 77 p.

2.36 Chen, B., and Nawy, E. G. "Structural Behavior Evaluation of High Strength Concrete Reinforced with Prestressed Prisms Using Fiber Optic Sensors." *Proceedings, ACI Structural Journal,* American Concrete Institute, Farmington Hills, MI, Dec. 1994, pp. 708–718.

2.37 Cen, B., Maher, M. H., and Nawy, E. G. "Fiber Optic Bragg Grating Sensor for Non-Destructive Evaluation of Composite Beams." *Proceedings, ASCE Journal of the Structural Division.* American Society of Civil Engineers, New York, Dec.1994, pp. 3456–3470.

2.38 Nawy, E. G. *Fundamentals of High Performance Concrete*, 2nd ed. John Wiley & Sons, New York, 2000.

2.39 Nawy, E. G., Editor-in-Chief, *Concrete Construction Engineering Handbook.* Boca Raton, FL: CRC Press, 1998, 1250 p.

3 PARTIAL LOSS OF PRESTRESS

3.1 INTRODUCTION

It is a well-established fact that the initial prestressing force applied to the concrete element undergoes a progressive process of reduction over a period of approximately five years. Consequently, it is important to determine the level of the prestressing force at each loading stage, from the stage of transfer of the prestressing force to the concrete, to the various stages of prestressing available at service load, up to the ultimate. Essentially, the reduction in the prestressing force can be grouped into two categories:

- Immediate elastic loss during the fabrication or construction process, including elastic shortening of the concrete, anchorage losses, and frictional losses.
- Time-dependent losses such as creep, shrinkage, and those due to temperature effects and steel relaxation, all of which are determinable at the service-load limit state of stress in the prestressed concrete element.

An exact determination of the magnitude of these losses—particularly the time-dependent ones—is not feasible, since they depend on a multiplicity of interrelated factors. Empirical methods of estimating losses differ with the different codes of practice or

Executive Center, Honolulu, Hawaii. (*Courtesy,* Post-Tensioning Institute.)

Table 3.1 AASHTO Lump-Sum Losses

Type of prestressing steel	Total loss $f'_c = 4{,}000$ psi (27.6 N/mm^2)	Total loss $f'_c = 5{,}000$ psi (34.5 N/mm^2)
Pretensioning strand		45,000 psi (310 N/mm^2)
Post-tensioning[a] wire or strand	32,000 psi (221 N/mm^2)	33,000 psi (228 N/mm^2)
Bars	22,000 psi (152 N/mm^2)	23,000 psi (159 N/mm^2)

[a]Losses due to friction are excluded. Such losses should be computed according to Section 6.5 of the AASHTO specifications.

recommendations, such as those of the Prestressed Concrete Institute, the ACI-ASCE joint committee approach, the AASHTO lump-sum approach, the Comité Eurointernationale du Béton (CEB), and the FIP (Federation Internationale de la Précontrainte). The degree of rigor of these methods depends on the approach chosen and the accepted practice of record.

A very high degree of refinement of loss estimation is neither desirable nor warranted, because of the multiplicity of factors affecting the estimate. Consequently, lump-sum estimates of losses are more realistic, particularly in routine designs and under average conditions. Such lump-sum losses can be summarized in Table 3.1 of AASHTO and Table 3.2 of PTI. They include elastic shortening, relaxation in the prestressing steel, creep, and shrinkage, and they are applicable only to routine, standard conditions of loading; normal concrete, quality control, construction procedures, and environmental conditions; and the importance and magnitude of the system. Detailed analysis has to be performed if these standard conditions are not fulfilled.

A summary of the sources of the separate prestressing losses and the stages of their occurrence is given in Table 3.3, in which the subscript i denotes "initial" and the subscript j denotes the loading stage after jacking. From this table, the total loss in prestress can be calculated for pretensioned and post-tensioned members as follows:

(i) Pretensioned Members

$$\Delta f_{pT} = \Delta f_{pES} + \Delta f_{pR} + \Delta f_{pCR} + \Delta f_{pSH} \tag{3.1a}$$

Table 3.2 Approximate Prestress Loss Values for Post-Tensioning

Post-tensioning tendon material	Prestress loss, psi: Slabs	Prestress loss, psi: Beams and joists
Stress-relieved 270-K strand and stress-relieved 240-K wire	30,000 (207 N/mm^2)	35,000 (241 N/mm^2)
Bar	20,000 (138 N/mm^2)	25,000 (172 N/mm^2)
Low-relaxation 270-K strand	15,000 (103 N/mm^2)	20,000 (138 N/mm^2)

Note: This table of approximate prestress losses was developed to provide a common post-tensioning industry basis for determining tendon requirements on projects in which the magnitude of prestress losses is not specified by the designer. These loss values are based on use of normal-weight concrete and on average values of concrete strength, prestress level, and exposure conditions. Actual values of losses may vary significantly above or below the table values where the concrete is stressed at low strengths, where the concrete is highly prestressed, or in very dry or very wet exposure conditions. The table values do not include losses due to friction.

Source: Post-Tensioning Institute.

Table 3.3 Types of Prestress Loss

Type of prestress loss	Stage of occurrence: Pretensioned members	Stage of occurrence: Post-tensioned members	Tendon stress loss: During time interval (t_i, t_j)	Tendon stress loss: Total or during life
Elastic shortening of concrete *(ES)*	At transfer	At sequential jacking	. . .	Δf_{pES}
Relaxation of tendons *(R)*	Before and after transfer	After transfer	$\Delta f_{pR}(t_i, t_j)$	Δf_{pR}
Creep of concrete *(CR)*	After transfer	After transfer	$\Delta f_{pC}(t_i, t_j)$	Δf_{pCR}
Shrinkage of concrete *(SH)*	After transfer	After transfer	$\Delta f_{pS}(t_i, t_j)$	Δf_{pSH}
Friction *(F)*	. . .	At jacking	. . .	Δf_{pF}
Anchorage seating loss *(A)*	. . .	At transfer	. . .	Δf_{pA}
Total	Life	Life	$\Delta f_{pT}(t_i, t_j)$	Δf_{pT}

where $\Delta f_{pR} = \Delta f_{pR}(t_0, t_{tr}) + \Delta f_{pR}(t_{tr}, t_s)$
t_0 = time at jacking
t_{tr} = time at transfer
t_s = time at stabilized loss

Hence, computations for steel relaxation loss have to be performed for the time interval t_1 through t_2 of the respective loading stages.

As an example, the transfer stage, say, at 18 h would result in $t_{tr} = t_2 = 18$ h and $t_0 = t_1 = 0$. If the next loading stage is between transfer and 5 years (17,520 h), when losses are considered stabilized, then $t_2 = t_s = 17{,}520$ h and $t_1 = 18$ h. Then, if f_{pi} is the initial prestressing stress that the concrete element is subjected to and f_{pJ} is the jacking stress in the tendon, then

$$f_{pi} = f_{pJ} - \Delta f_{pR}(t_0, t_{tr}) - \Delta f_{pES} \tag{3.1b}$$

(ii) Post-tensioned Members

$$\Delta f_{pT} = \Delta f_{pA} + \Delta f_{pF} + \Delta f_{pES} + \Delta f_{pR} + \Delta f_{pCR} + \Delta f_{pSH} \tag{3.1c}$$

where Δf_{pES} is applicable only when tendons are jacked sequentially, and not simultaneously.

In the post-tensioned case, computation of relaxation loss starts between the transfer time $t_1 = t_{tr}$ and the end of the time interval t_2 under consideration. Hence

$$f_{pi} = f_{pJ} - \Delta f_{pA} - \Delta f_{pF} \tag{3.1d}$$

3.2 ELASTIC SHORTENING OF CONCRETE *(ES)*

Concrete shortens when a prestressing force is applied. As the tendons that are bonded to the adjacent concrete simultaneously shorten, they lose part of the prestressing force that they carry.

3.2.1 Pretensioned Elements

For pretensioned (precast) elements, the compressive force imposed on the beam by the tendon results in the longitudinal shortening of the beam, as shown in Figure 3.1. The unit shortening in concrete is $\epsilon_{ES} = \Delta_{ES}/L$, so

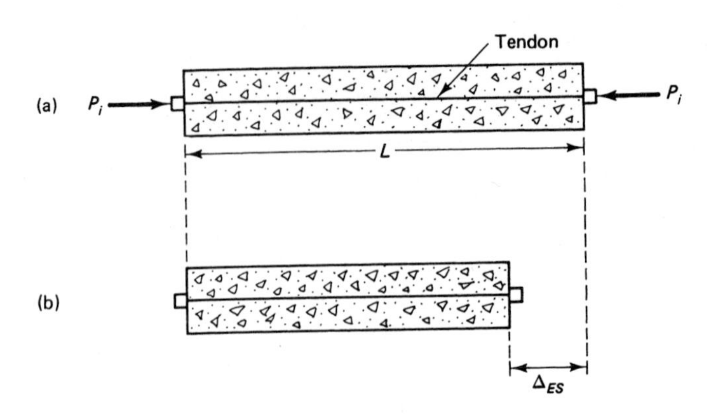

Figure 3.1 Elastic shortening. (a) Unstressed beam. (b) Longitudinally shortened beam.

$$\epsilon_{ES} = \frac{f_c}{E_c} = \frac{P_i}{A_c E_c} \tag{3.2a}$$

Since the prestressing tendon suffers the same magnitude of shortening,

$$\Delta f_{pES} = E_s\,\epsilon_{ES} = \frac{E_s P_i}{A_c E_c} = \frac{nP_i}{A_c} = nf_{cs} \tag{3.2b}$$

The stress in the concrete at the centroid of the steel due to the initial prestressing is

$$f_{cs} = -\frac{P_i}{A_c} \tag{3.3}$$

If the tendon in Figure 3.1 has an eccentricity e at the beam midspan and the self-weight moment M_D is taken into account, the stress the concrete undergoes at the midspan section at the level of the prestressing steel becomes

$$f_{cs} = -\frac{P_i}{A_c}\left(1 + \frac{e^2}{r^2}\right) + \frac{M_D e}{I_c} \tag{3.4}$$

where P_i has a lower value after transfer of prestress. The *small* reduction in the value of P_J to P_i occurs because the force in the prestressing steel immediately after transfer is less than the initial jacking prestress force P_J. However, since it is difficult to accurately determine the reduced value of P_i, and since observations indicate that the reduction is only a few percentage points, it is possible to use the initial value of P_i before transfer in Equations 3.2 through 3.4, or reduce it by about 10 percent for refinement if desired.

3.2.1.1 Elastic shortening loss in pretensioned beams

Example 3.1

A pretensioned prestressed beam has a span of 50 ft (15.2 m), as shown in Figure 3.2. For this beam,

$$f'_c = 6{,}000 \text{ psi } (41.4 \text{ MPa})$$

$$f_{pu} = 270{,}000 \text{ psi } (1{,}862 \text{ MPa})$$

$$f'_{ci} = 4{,}500 \text{ psi } (31 \text{ MPa})$$

$$A_{ps} = 10 - \frac{1}{2}\text{-in dia. seven-wire-strand tendon}$$

$$= 10 \times 0.153 = 1.53 \text{ in.}^2$$

$$E_{ps} = 27 \times 10^6 \text{ psi } (1{,}862 \text{ MPa})$$

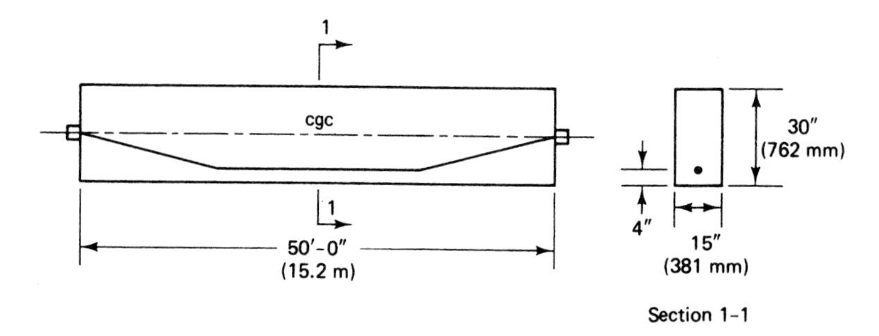

Figure 3.2 Beam in Example 3.1.

Calculate the concrete fiber stresses at transfer at the centroid of the tendon for the midspan section of the beam, and the magnitude of loss in prestress due to the effect of elastic shortening of the concrete. Assume that prior to transfer, the jacking force on the tendon was 75% f_{pu}.

Solution:

$$A_c = 15 \times 30 = 450 \text{ in.}^2$$

$$I_c = \frac{15(30)^3}{12} = 33{,}750 \text{ in.}^3$$

$$r^2 = \frac{I_c}{A_c} = 75 \text{ in.}^2$$

$$A_{ps} = 10 \times 0.153 = 1.53 \text{ in.}^2$$

$$e_c = \frac{30}{2} - 4 = 11 \text{ in.}$$

$$P_i = 0.75 f_{pu} A_{ps} = 0.75 \times 270{,}000 \times 1.53 = 309{,}825 \text{ lb}$$

$$M_D = \frac{wl^2}{8} = \frac{15 \times 30}{144} \times 150 \frac{(50)^2}{8} \times 12 = 1{,}757{,}813 \text{ in.-lb}$$

From Equation 3.4, the concrete fiber stress at the steel centroid of the beam at the moment of transfer, assuming that $P_i \cong P_J$, is

$$f_{cs} = -\frac{P_i}{A_c}\left(1 + \frac{e^2}{r^2}\right) + \frac{M_D e}{I_c}$$

$$= -\frac{309{,}825}{450}\left(1 + \frac{11^2}{75}\right) + \frac{1{,}757{,}813 \times 11}{33{,}750}$$

$$= -1{,}799.3 + 572.9 = -1{,}226.4 \text{ psi } (8.50 \text{ MPa})$$

We also have

$$\text{Initial } E_{ci} = 57{,}000\sqrt{f'_{ci}} = 57{,}000\sqrt{4{,}500} = 3.824 \times 10^6 \text{ psi}$$

$$\text{Initial modular ratio } n = \frac{E_s}{E_{ci}} = \frac{27 \times 10^6}{3.824 \times 10^6} = 7.06$$

$$\text{28 days' strength } E_c = 57{,}000\sqrt{6{,}000} = 4.415 \times 10^6 \text{ psi}$$

$$\text{28 days' modular ratio } n = \frac{27 \times 10^6}{4.415 \times 10^6} = 6.12$$

From Equation 3.2b, the loss of prestress due to elastic shortening is

$$\Delta f_{pES} = n f_{cs} = 7.06 \times 1{,}226.4 = 8{,}659.2 \text{ psi } (59.7 \text{ MPa})$$

If a reduced P_i is used with assumed 10 percent reduction,

$$\Delta f_{pES} = 0.90 \times 8{,}659.2 = 7{,}793.3 \text{ psi (53.7 MPa)}.$$

The difference of 865.9 psi in steel stress is insignificant compared to the total loss in prestress due to all factors of about 45,000 to 55,000 psi.

3.2.2 Post-tensioned Elements

In post-tensioned beams, the elastic shortening loss varies from zero if all tendons are jacked simultaneously to half the value calculated in the pretensioned case if several sequential jacking steps are used, such as jacking two tendons at a time. If n is the number of tendons or pairs of tendons sequentially tensioned, then

$$\Delta f_{pES} = \frac{1}{n}\sum_{j=1}^{n}(\Delta f_{pES})_j \tag{3.5}$$

where j denotes the number of jacking operations. Note that the tendon that was tensioned last does not suffer any losses due to elastic shortening, while the tendon that was tensioned first suffers the maximum amount of loss.

3.2.2.1 Elastic shortening loss in post-tensioned beam

Example 3.2

Solve Example 3.1 if the beam is post-tensioned and the prestressing operation is such that

(a) Two tendons are jacked at a time.
(b) One tendon is jacked at a time.
(c) All tendons are simultaneously tensioned.

Solution:

(a) From Example 3.1, $\Delta f_{pE} = 8{,}659.2$ psi. Clearly, the last tendon suffers no loss of prestress due to elastic shortening. So only the first four pairs have losses, with the first pair suffering the maximum loss of 8,659.2 psi. From Equation 3.5, the loss due to elastic shortening in the post-tensioned beam is

$$\Delta f_{pES} = \frac{4/4 + 3/4 + 2/4 + 1/4}{5}(8{,}659.2)$$

$$= \frac{10}{20} \times (8{,}659.2) = 4{,}330 \text{ psi (29.9 MPa)}$$

(b)

$$\Delta f_{pES} = \frac{9/9 + 8/9 + \cdots + 1/9}{10}(8{,}659.2)$$

$$= \frac{45}{90} \times (8{,}659.2) = 4{,}330 \text{ psi (29.9 MPa)}$$

In both cases the loss in prestressing in the post-tensioned beam is half that of the pretensioned beam.

(c) $\Delta f_{pES} = 0$

3.3 STEEL STRESS RELAXATION *(R)*

Stress-relieved tendons suffer loss in the prestressing force due to constant elongation with time, as discussed in Chapter 2. The magnitude of the decrease in the prestress depends not only on the duration of the sustained prestressing force, but also on the ratio

f_{pi}/f_{py} of the initial prestress to the yield strength of the reinforcement. Such a loss in stress is termed *stress relaxation.* The ACI 318-99 Code limits the tensile stress in the prestressing tendons to the following:

(a) For stresses due to the tendon jacking force, $f_{pJ} = 0.94\, f_{py}$, but not greater than the lesser of $0.80\, f_{pu}$ and the maximum value recommended by the manufacturer of the tendons and anchorages.
(b) Immediately after prestress transfer, $f_{pi} = 0.82\, f_{py}$, but not greater than $0.74\, f_{pu}$.
(c) In post-tensioned tendons, at the anchorages and couplers immediately after force transfer $= 0.70\, f_{pu}$.

The range of values of f_{py} is given by the following:

Prestressing bars: $f_{py} = 0.80\, f_{pu}$
Stress-relieved tendons: $f_{py} = 0.85\, f_{pu}$
Low-relaxation tendons: $f_{py} = 0.90\, f_{pu}$

If f_{pR} is the remaining prestressing stress in the steel after relaxation, the following expression defines f_{pR} for stress relieved steel:

$$\frac{f_{pR}}{f_{pi}} = 1 - \left(\frac{\log t_2 - \log t_1}{10}\right)\left(\frac{f_{pi}}{f_{py}} - 0.55\right) \tag{3.6}$$

In this expression, $\log t$ in hours is to the base 10, f_{pi}/f_{py} exceeds 0.55, and $t = t_2 - t_1$. Also, for low-relaxation steel, the denominator of the log term in the equation is divided by 45 instead of 10. A plot of Equation 3.6 is given in Figure 3.3.

An approximation of the term $(\log t_2 - \log t_1)$ can be made in Equation 3.6 so that $\log t = \log(t_2 - t_1)$ without significant loss in accuracy. In that case, the stress-relaxation loss becomes

$$\Delta f_{pR} = f'_{pi}\frac{\log t}{10}\left(\frac{f'_{pi}}{f_{py}} - 0.55\right) \tag{3.7}$$

where f'_{pi} is the initial stress in steel to which the concrete element is subjected.

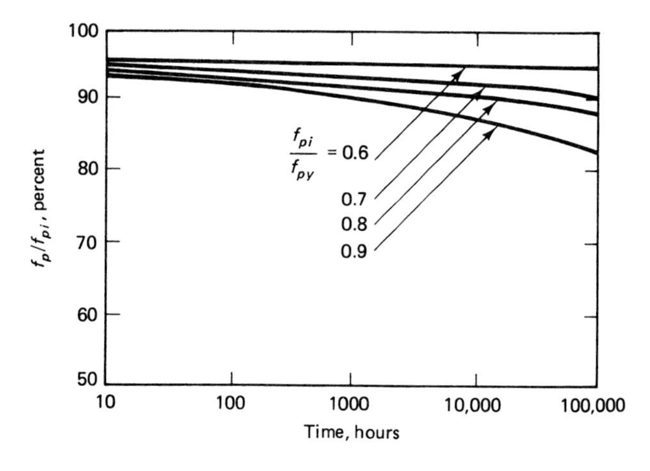

Figure 3.3 Stress-relaxation relationship in stress-relieved strands. (*Courtesy,* Post-Tensioning Institute.)

If a step-by-step loss analysis is necessary, the loss increment at any particular stage can be defined as

$$\Delta f_{pR} = f'_{pi}\left(\frac{\log t_2 - \log t_1}{10}\right)\left(\frac{f'_{pi}}{f_{py}} - 0.55\right) \tag{3.8}$$

where t_1 is the time at the beginning of the interval and t_2 is the time at the end of the interval from jacking to the time when the loss is being considered.

3.3.1 Relaxation Loss Computation

Example 3.3

Find the relaxation loss in prestress at the end of 5 years in Example 3.1, assuming that relaxation loss from jacking to transfer, from elastic shortening, and from long-term loss due to creep and shrinkage over this period is 20 percent of the initial prestress. Assume also that the yield strength $f_{py} = 230{,}000$ psi (1,571 MPa).

Solution: From Equation 3.1b for this stage

$$f_{pi} = f_{pJ} - \Delta f_{pR}(t_0, t_{tr})$$
$$= 0.75 \times 270{,}000 = 202{,}500 \text{ psi } (1{,}396 \text{ MPa})$$

The reduced stress for calculating relaxation loss is

$$f'_{pi} = (1 - 0.20) \times 202{,}500 = 162{,}000 \text{ psi } (1{,}170 \text{ MPa})$$

The duration of the stress-relaxation process is

$$5 \times 365 \times 24 \cong 44{,}000 \text{ hours}$$

From Equation 3.7,

$$\Delta f_{pR} = f'_{pi}\frac{\log t}{10}\left(\frac{f'_{pi}}{f_{py}} - 0.55\right)$$
$$= 162{,}000\frac{\log 44{,}000}{10}\left(\frac{162{,}000}{230{,}000} - 0.55\right)$$
$$= 162{,}000 \times 0.4643 \times 0.1543 = 11{,}606 \text{ psi } (80.0 \text{ MPa})$$

3.3.2 ACI-ASCE Method of Accounting for Relaxation Loss

The ACI-ASCE method uses the separate contributions of elastic shortening, creep, and shrinkage in the evaluation of the steel stress-relaxation loss by means of the equation

$$\Delta f_{pR} = [K_{re} - J(f_{pES} + f_{pCR} + f_{pSH})] \times C$$

The values of K_{re}, J, and C are given in Tables 3.4 and 3.5.

3.4 CREEP LOSS *(CR)*

Experimental work over the past half century indicates that flow in materials occurs with time when load or stress exists. This lateral flow or deformation due to the longitudinal stress is termed *creep*. A more detailed discussion is given in Ref. 3.9. It must be emphasized that creep stresses and stress losses result *only* from *sustained* loads during the loading history of the structural element.

Table 3.4 Values of C

f_{pi}/f_{pu}	Stress-relieved strand or wire	Stress-relieved bar or low-relaxation strand or wire
0.80		1.28
0.79		1.22
0.78		1.16
0.77		1.11
0.76		1.05
0.75	1.45	1.00
0.74	1.36	0.95
0.73	1.27	0.90
0.72	1.18	0.85
0.71	1.09	0.80
0.70	1.00	0.75
0.69	0.94	0.70
0.68	0.89	0.66
0.67	0.83	0.61
0.66	0.78	0.57
0.65	0.73	0.53
0.64	0.68	0.49
0.63	0.63	0.45
0.62	0.58	0.41
0.61	0.53	0.37
0.60	0.49	0.33

Source: Post-Tensioning Institute.

The deformation or strain resulting from this time-dependent behavior is a function of the magnitude of the applied load, its duration, the properties of the concrete including its mixture proportions, curing conditions, the age of the element at first loading, and environmental conditions. Since the stress-strain relationship due to creep is essentially linear, it is feasible to relate the creep strain ϵ_{CR} to the elastic strain ϵ_{EL} such that a creep coefficient C_u can be defined as

Table 3.5 Values of K_{RE} and J

Type of tendon[a]	K_{RE}	J
270 Grade stress-relieved strand or wire	20,000	0.15
250 Grade stress-relieved strand or wire	18,500	0.14
240 or 235 Grade stress-relieved wire	17,600	0.13
270 Grade low-relaxation strand	5,000	0.040
250 Grade low-relaxation wire	4,630	0.037
240 or 235 Grade low-relaxation wire	4,400	0.035
145 or 160 Grade stress-relieved bar	6,000	0.05

[a]In accordance with ASTM A416-74, ASTM A421-76, or ASTM A722-75.

Source: Prestressed Concrete Institute.

$$C_u = \frac{\epsilon_{CR}}{\epsilon_{EL}} \tag{3.9a}$$

Then the creep coefficient at any time t in days can be defined as

$$C_t = \frac{t^{0.60}}{10 + t^{0.60}} C_u \tag{3.9b}$$

As discussed in Chapter 2, the value of C_u ranges between 2 and 4, with an average of 2.35 for ultimate creep. The loss in prestressed members due to creep can be defined for bonded members as

$$\Delta f_{pCR} = C_t \frac{E_{ps}}{E_c} f_{cs} \tag{3.10}$$

where f_{cs} is the stress in the concrete at the level of the centroid of the prestressing tendon. In general, this loss is a function of the stress in the concrete at the section being analyzed. In post-tensioned, nonbonded members, the loss can be considered essentially uniform along the whole span. Hence, an average value of the concrete stress $\bar{f}_{cs}$ between the anchorage points can be used for calculating the creep in post-tensioned members.

The ACI-ASCE Committee expression for evaluating creep loss has essentially the same format as Equation 3.10, viz.,

$$\Delta f_{pCR} = K_{CR} \frac{E_{ps}}{E_c} (\bar{f}_{cs} - \bar{f}_{csd}) \tag{3.11a}$$

or

$$\Delta f_{pCR} = n K_{CR} (\bar{f}_{cs} - \bar{f}_{csd}) \tag{3.11b}$$

where K_{CR} = 2.0 for pretensioned members
= 1.60 for post-tensioned members (both for normal concrete)
$\bar{f}_{cs}$ = stress in concrete at level of steel cgs immediately after transfer
$\bar{f}_{csd}$ = stress in concrete at level of steel cgs due to all superimposed dead loads applied after prestressing is accomplished
n = modular ratio

Note that K_{CR} should be reduced by 20 percent for lightweight concrete.

3.4.1 Computation of Creep Loss

Example 3.4

Compute the loss in prestress due to creep in Example 3.1 given that the total superimposed load, excluding the beam's own weight after transfer, is 375 plf (5.5 kN/m).

Solution: At full concrete strength,

$$E_c = 57{,}000\sqrt{6{,}000} = 4.415 \times 10^6 \text{ psi } (30.4 \times 10^3 \text{ MPa})$$

$$n = \frac{E_s}{E_c} = \frac{27.0 \times 10^6}{4.415 \times 10^6} = 6.12$$

$$M_{SD} = \frac{375(50)^2}{8} \times 12 = 1{,}406{,}250 \text{ in.-lb } (158.9 \text{ kN-m})$$

$$\bar{f}_{csd} = \frac{M_{SD}e}{I_c} = \frac{1{,}406{,}250 \times 11}{33{,}750} = 458.3 \text{ psi } (3.2 \text{ MPa})$$

From Example 3.1,

$$\bar{f}_{cs} = 1{,}226.4 \text{ psi } (8.5 \text{ MPa})$$

Also, for normal concrete use, $K_{CR} = 2.0$ (pretensioned beam); so from Equation 3.11a,

$$\begin{aligned}\Delta f_{pCR} &= nK_{CR}(\bar{f}_{cs} - \bar{f}_{csd}) \\ &= 6.12 \times 2.0(1{,}226.4 - 458.3) \\ &= 9{,}401.5 \text{ psi } (64.8 \text{ MPa})\end{aligned}$$

3.5 SHRINKAGE LOSS *(SH)*

As with concrete creep, the magnitude of the shrinkage of concrete is affected by several factors. They include mixture proportions, type of aggregate, type of cement, curing time, time between the end of external curing and the application of prestressing, size of the member, and the environmental conditions. Size and shape of the member also affect shrinkage. Approximately 80 percent of shrinkage takes place in the first year of life of the structure. The average value of ultimate shrinkage strain in both moist-cured and steam-cured concrete is given as 780×10^{-6} in./in. in ACI 209 R-92 Report. This average value is affected by the length of initial moist curing, ambient relative humidity, volume-surface ratio, temperature, and concrete composition. To take such effects into account, the average value of shrinkage strain should be multiplied by a correction factor γ_{SH} as follows

$$\epsilon_{SH} = 780 \times 10^{-6}\,\gamma_{SH} \tag{3.12}$$

Components of γ_{SH} are factors for various environmental conditions and tabulated in Ref. 3.12, Sec. 2.

The Prestressed Concrete Institute stipulates for standard conditions an average value for nominal ultimate shrinkage strain $(\epsilon_{SH})_u = 820 \times 10^{-6}$ in./in. (mm/mm), (Ref. 3.4). If ϵ_{SH} is the shrinkage strain after adjusting for relative humidity at volume-to-surface ratio *V/S,* the loss in prestressing in pretensioned member is

$$\Delta f_{pSH} = \epsilon_{SH} \times E_{ps} \tag{3.13}$$

For post-tensioned members, the loss in prestressing due to shrinkage is somewhat less since some shrinkage has already taken place before post-tensioning. If the relative humidity is taken as a percent value and the *V/S* ratio effect is considered, the PCI general expression for loss in prestressing due to shrinkage becomes

$$\Delta f_{pSH} = 8.2 \times 10^{-6} K_{SH} E_{ps}\left(1 - 0.06\frac{V}{S}\right)(100 - RH) \tag{3.14}$$

Table 3.6 Values of K_{SH} for Post-Tensioned Members

Time from end of moist curing to application of prestress, days	**1**	**3**	**5**	**7**	**10**	**20**	**30**	**60**
k_{sh}	0.92	0.85	0.80	0.77	0.73	0.64	0.58	0.45

Source: Prestressed Concrete Institute.

Photo 3.1 101/280/680 interchange connectors, South San Jose, California.

where $K_{SH} = 1.0$ for pretensioned members. Table 3.6 gives the values of K_{SH} for post-tensioned members.

Adjustment of shrinkage losses for standard conditions as a function of time t in days after 7 days for moist curing and 3 days for steam curing can be obtained from the following expressions

(a) Moist curing, after 7 days

$$(\epsilon_{SH})_t = \frac{t}{35 + t}(\epsilon_{SH})_u \tag{3.15a}$$

where $(\epsilon_{SH})_u$ is the ultimate shrinkage strain, t = time in days after shrinkage is considered.

(b) Steam curing, after 1 to 3 days

$$(\epsilon_{SH})_t = \frac{t}{55 + t}(\epsilon_{SH})_u \tag{3.15b}$$

It should be noted that separating creep from shrinkage calculations as presented in this chapter is an accepted engineering practice. Also, significant variations occur in the creep and shrinkage values due to variations in the properties of the constituent materials from the various sources, even if the products are plant-produced such as pretensioned beams. Hence it is recommended that information from actual tests be obtained especially on manufactured products, large span-to-depth ratio cases and/or if loading is unusually heavy.

3.5.1 Computation of Shrinkage Loss

Example 3.5

Compute the loss in prestress due to shrinkage in Examples 3.1 and 3.2 at 7 days after moist curing using both the ultimate K_{SH} method of Equation 3.14 and the time-dependent method of Equation 3.15. Assume that the relative humidity RH is 70 percent and the volume-to-surface ratio is 2.0.

Solution A

K_{SH} method

(a) Pretensioned beam, $K_{SH} = 1.0$:
From Equation 3.14,

$$\Delta f_{pSH} = 8.2 \times 10^{-6} \times 1.0 \times 27 \times 10^{6}(1 - 0.06 \times 2.0)(100 - 70)$$
$$= 5{,}845.0 \text{ psi } (40.3 \text{ MPa})$$

(b) Post-tensioned beam, from Table 3.6, $K_{SH} = 0.77$:

$$\Delta f_{pSH} = 0.77 \times 5{,}845 = 4{,}500.7 \text{ psi } (31.0 \text{ MPa})$$

Solution B

Time-dependent method

From Equation 3.15a,

$$\epsilon_{SH,t} = \frac{t}{35 + t}\epsilon_{SH} = \frac{7}{35 + 7} \times 780 \times 10^{-6} = 130 \times 10^{-6} \text{ in/in}$$

$$\Delta f_{pSH} = \epsilon_{SH,t} E_s = 130 \times 10^{-6} \times 27 \times 10^{6} = 3{,}510 \text{ psi } (24.0 \text{ MPa})$$

3.6 LOSSES DUE TO FRICTION *(F)*

Loss of prestressing occurs in post-tensioning members due to friction between the tendons and the surrounding concrete ducts. The magnitude of this loss is a function of the tendon form or alignment, called the *curvature effect,* and the local deviations in the alignment, called the *wobble effect.* The values of the loss coefficients are often refined while preparations are made for shop drawings by varying the types of tendons and the duct alignment. Whereas the curvature effect is predetermined, the wobble effect is the result of accidental or unavoidable misalignment, since ducts or sheaths cannot be perfectly placed.

It should be noted that the maximum frictional stress loss would be at the far end of the beam if jacking is from one end. Hence frictional loss varies linearly along the beam span and can be interpolated for a particular location if such refinement in the computations is warranted.

3.6.1 Curvature Effect

As the tendon is pulled with a force F_1 at the jacking end, it will encounter friction with the surrounding duct or sheath such that the stress in the tendon will vary from the jacking plane to a distance L along the span as shown in Figure 3.4. If an infinitesimal length of the tendon is isolated in a free-body diagram as shown in Figure 3.5, then, assuming that μ denotes the coefficient of friction between the tendon and the duct due to the curvature effect, we have

$$dF_1 = -\mu F_1 \, d\alpha$$

or

$$\frac{dF_1}{F_1} = -\mu d\alpha \tag{3.16a}$$

Integrating both sides of this equation yields

$$\log_e F_1 = -\mu\alpha \tag{3.16b}$$

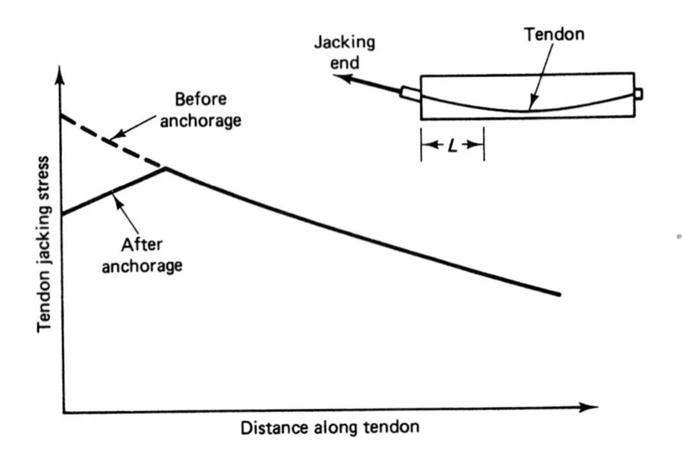

Figure 3.4 Frictional force stress distribution in tendon.

If $\alpha = L/R$, then

$$F_2 = F_1 e^{-\mu\alpha} = F_1 e^{-\mu(L/R)} \tag{3.17}$$

3.6.2 Wobble Effect

Suppose that K is the coefficient of friction between the tendon and the surrounding concrete due to wobble effect or length effect. Friction loss is caused by imperfection in alignment along the length of the tendon, regardless of whether it has a straight or draped alignment. Then by the same principles described in developing Equation 3.16,

$$\log_e F_1 = -KL \tag{3.18}$$

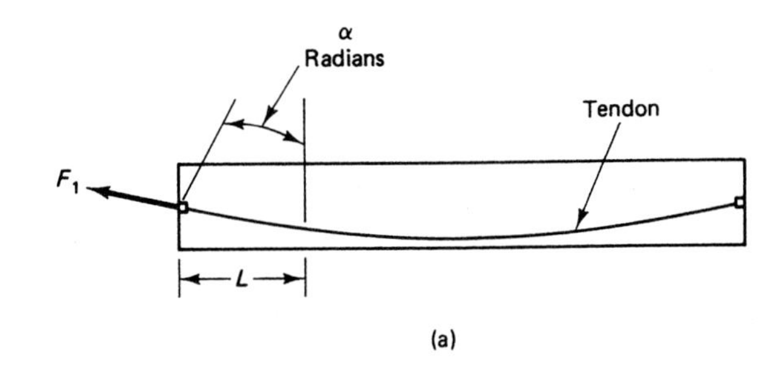

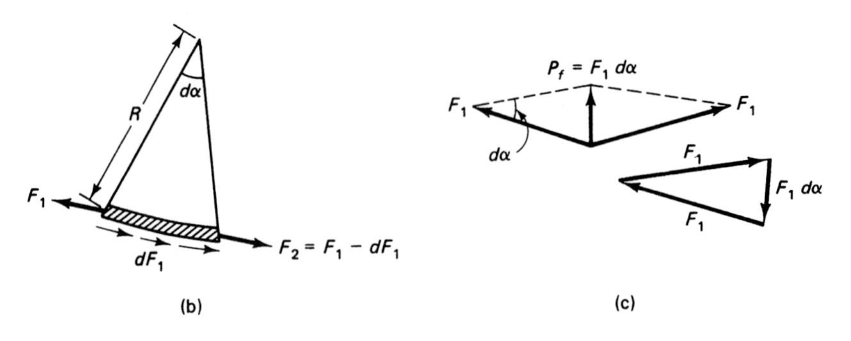

Figure 3.5 Curvature friction loss. (a) Tendon alignment. (b) Forces on infinitesimal length where F_1 is at the jacking end. (c) Polygon of forces assuming $F_1 = F_2$ over the infinitesimal length in (b).

or

$$F_2 = F_1 e^{-KL} \tag{3.19}$$

Superimposing the wobble effect on the curvature effect gives

$$F_2 = F_1 e^{-\mu\alpha - KL}$$

or, in terms of stresses,

$$f_2 = f_1 e^{-\mu\alpha - KL} \tag{3.20}$$

The frictional loss of stress Δf_{pF} is then given by

$$\Delta f_{pF} = f_1 - f_2 = f_1(1 - e^{-\mu\alpha - KL}) \tag{3.21}$$

Assuming that the prestress force between the start of the curved portion and its end is small ($\cong$ 15 percent), it is sufficiently accurate to use the initial tension for the entire curve in Equation 3.21. Equation 3.21 can thus be simplified to yield

$$\Delta f_{pF} = -f_1(\mu\alpha + KL) \tag{3.22}$$

where L is in feet.

Since the ratio of the depth of beam to its span is small, it is sufficiently accurate to use the projected length of the tendon for calculating α. Assuming the curvature of the tendon to be based on that of a circular arc, the central angle α along the curved segment in Figure 3.6 is twice the slope at either end of the segment. Hence,

$$\tan\frac{\alpha}{2} = \frac{m}{x/2} = \frac{2m}{x}$$

If

$$y \cong \frac{1}{2}m \quad \text{and} \quad \alpha/2 = 4y/x$$

then

$$\alpha = 8y/x \text{ radian} \tag{3.23}$$

Table 3.7 gives the design values of the curvature friction coefficient μ and the wobble or length friction coefficient K adopted from the ACI 318 Commentary.

3.6.3 Computation of Friction Loss

Example 3.6

Assume that the alignment characteristics of the tendons in the post-tensioned beam of Example 3.2 are as shown in Figure 3.7. If the tendon is made of 7-wire uncoated strands in flexible metal sheathing, compute the frictional loss of stress in the prestressing wires due to the curvature and wobble effects.

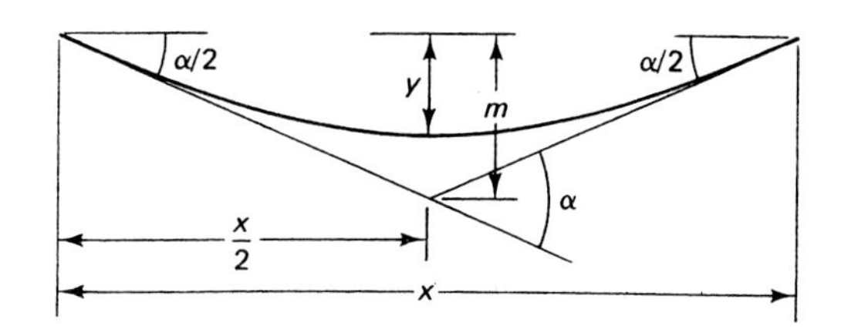

Figure 3.6 Approximate evaluation of the tendon's central angle.

Table 3.7 Wobble and Curvature Friction Coefficients

Type of tendon	Wobble coefficient, K per foot	Curvature coefficient, μ
Tendons in flexible metal sheathing		
Wire tendons	0.0010–0.0015	0.15–0.25
7-wire strand	0.0005–0.0020	0.15–0.25
High-strength bars	0.0001–0.0006	0.08–0.30
Tendons in rigid metal duct		
7-wire strand	0.0002	0.15–0.25
Mastic-coated tendons		
Wire tendons and 7-wire strand	0.0010–0.0020	0.05–0.15
Pregreased tendons		
Wire tendons and 7-wire strand	0.0003–0.0020	0.05–0.15

Source: Prestressed Concrete Institute.

Solution:

$$P_i = 309{,}825 \text{ lb}$$

$$f_1 = \frac{309{,}825}{1.53} = 202{,}500 \text{ psi}$$

From Equation 3.23,

$$\alpha = \frac{8y}{x} = \frac{8 \times 11/12''}{50} = 0.1467 \text{ radian}$$

From Table 3.7, use $K = 0.0020$ and $\mu = 0.20$. From Equation 3.22, the prestress loss due to friction is

$$\begin{aligned}\Delta f_{pF} &= f_{pi}(\mu\alpha + KL)\\ &= 202{,}500(0.20 \times 0.1467 + 0.0020 \times 50)\\ &= 202{,}500 \times 0.1293 = 26{,}191 \text{ psi (180.6 MPa)}\end{aligned}$$

This loss due to friction is 12.93 percent of the initial prestress.

3.7 ANCHORAGE-SEATING LOSSES *(A)*

Anchorage-seating losses occur in post-tensioned members due to the seating of wedges in the anchors when the jacking force is transferred to the anchorage. They can also occur in the prestressing casting beds of pretensioned members due to the adjustment

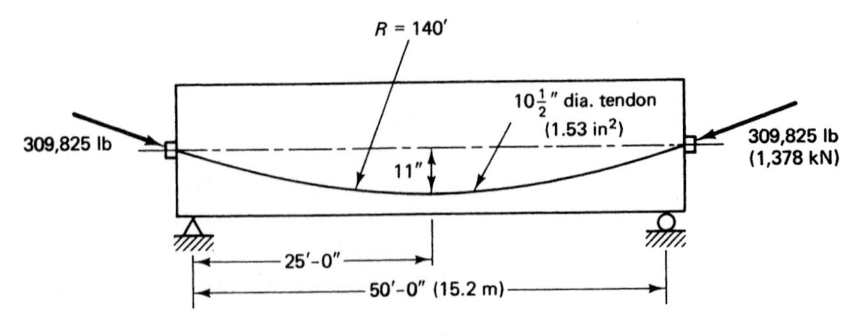

Figure 3.7 Prestressing tendon alignment.

expected when the prestressing force is transferred to these beds. A remedy for this loss can be easily effected during the stressing operations by overstressing. Generally, the magnitude of anchorage-seating loss ranges between $\frac{1}{4}$ in. and $\frac{3}{8}$ in. (6.35 mm and 9.53 mm) for the two-piece wedges. The magnitude of the overstressing that is necessary depends on the anchorage system used since each system has its particular adjustment needs, and the manufacturer is expected to supply the data on the slip expected due to anchorage adjustment. If Δ_A is the magnitude of the slip, L is the tendon length, and E_{ps} is the modulus of the prestressing wires, then the prestress loss due to anchorage slip becomes

$$\Delta f_{pA} = \frac{\Delta_A}{L} E_{ps} \tag{3.24}$$

3.7.1 Computation of Anchorage-Seating Loss

Example 3.7

Compute the anchorage-seating loss in the post-tensioned beam of Example 3.2 if the estimated slip is $\frac{1}{4}$ in. (6.35 m).

Solution:

$$E_{ps} = 27 \times 10^6 \text{ psi}$$

$$\Delta_A = 0.25 \text{ in.}$$

$$\Delta f_{pA} = \frac{\Delta_A}{L} E_{ps} = \frac{0.25}{50 \times 12} \times 27 \times 10^6 = 11{,}250 \text{ psi (77.6 MPa)}$$

Note that the percentage of loss due to anchorage slip becomes very high in short-beam elements and thus becomes of major significance in short-span beams. In such cases, it becomes difficult to post-tension such beams with high accuracy.

Photo 3.2 Terracentre, Denver, Colorado. (*Courtesy,* Post-Tensioning Institute.)

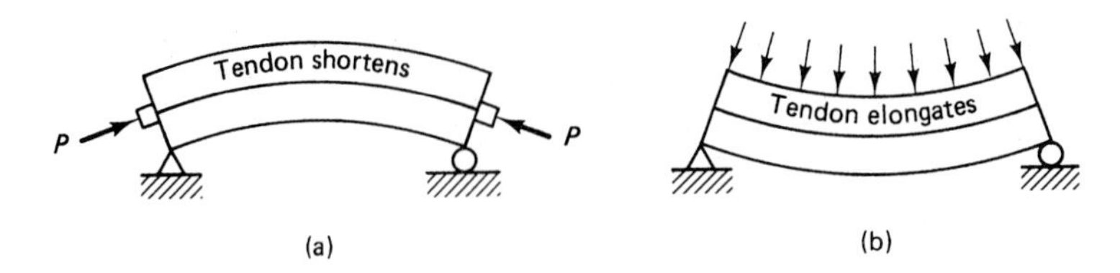

Figure 3.8 Change in beam longitudinal shape. (a) Due to prestressing. (b) Due to external load.

3.8 CHANGE OF PRESTRESS DUE TO BENDING OF A MEMBER (Δf_{pB})

As the beam bends due to prestress or external load, it becomes convex or concave depending on the nature of the load, as shown in Figure 3.8. If the unit compressive strain in the concrete along the level of the tendon is ϵ_c, then the corresponding change in prestress in the steel is

$$\Delta f_{pB} = \epsilon_c E_{ps}$$

where E_s is the modulus of the steel. Note that any loss due to bending need not be taken into consideration if the prestressing stress level is measured after the beam has already bent, as is usually the case.

Figure 3.9 presents a flowchart for step-by-step evaluation of time-dependent prestress losses without deflection.

3.9 STEP-BY-STEP COMPUTATION OF ALL TIME-DEPENDENT LOSSES IN A PRETENSIONED BEAM

Example 3.8

A simply supported pretensioned 70-ft-span lightweight steam-cured double T-beam as shown in Figure 3.10 is prestressed by twelve $\frac{1}{2}$-in. diameter (twelve 12.7 mm dia) 270-K grade stress-relieved strands. The tendons are harped, and the eccentricity at midspan is 18.73 in. (476 mm) and at the end 12.98 in. (330 mm). Compute the prestress loss at the critical section in the beam of 0.40 span due to dead load and superimposed dead load at

(a) stage I at transfer
(b) stage II after concrete topping is placed
(c) two years after concrete topping is placed

Suppose the topping is 2 in. (51 mm) normal-weight concrete cast at 30 days. Suppose also that prestress transfer occurred 18 h after tensioning the strands. Given

$$f'_c = 5{,}000 \text{ psi, lightweight (34.5 MPa)}$$

$$f'_{ci} = 3{,}500 \text{ psi (24.1 MPa)}$$

and the following noncomposite section properties.

$$A_c = 615 \text{ in.}^2 \text{ (3,968 cm}^2\text{)}$$

$$I_c = 59{,}720 \text{ in.}^4 \text{ (}2.49 \times 10^6 \text{ cm}^4\text{)}$$

$$c_b = 21.98 \text{ in. (55.8 cm)}$$

$$c^t = 10.02 \text{ in. (25.5 cm)}$$

$$S_b = 2{,}717 \text{ in.}^3 \text{ (44,520 cm}^3\text{)}$$

$$S^t - 5{,}960 \text{ in.}^3 \text{ (97,670 cm}^3\text{)}$$

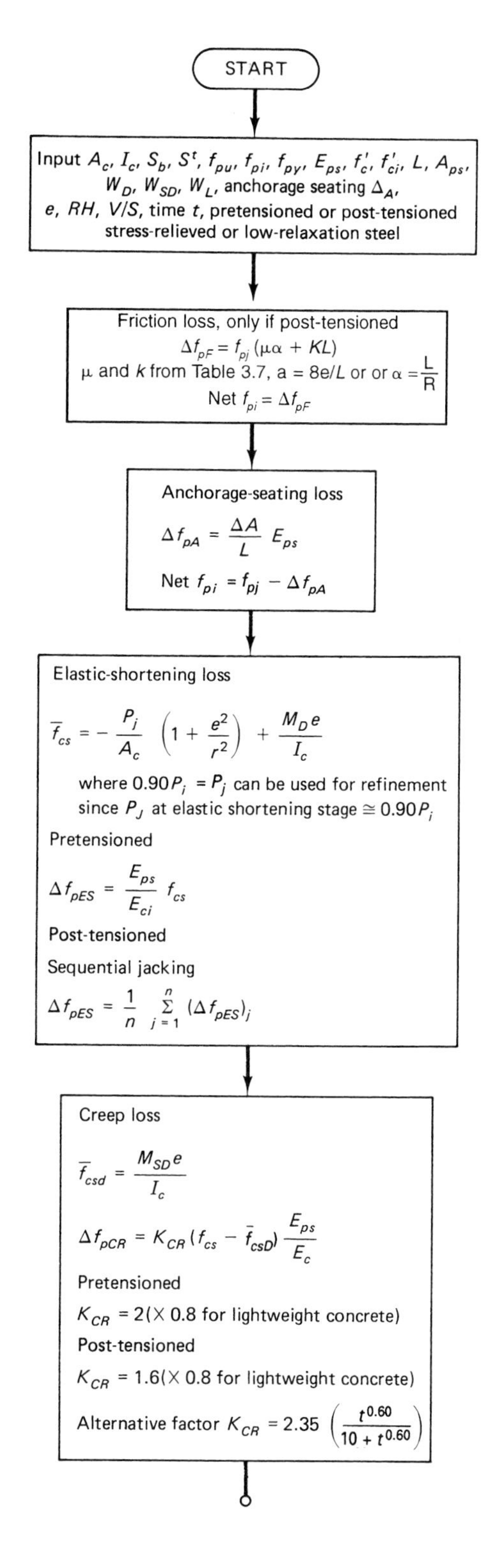

Figure 3.9 Flowchart for step-by-step evaluation of prestress losses.

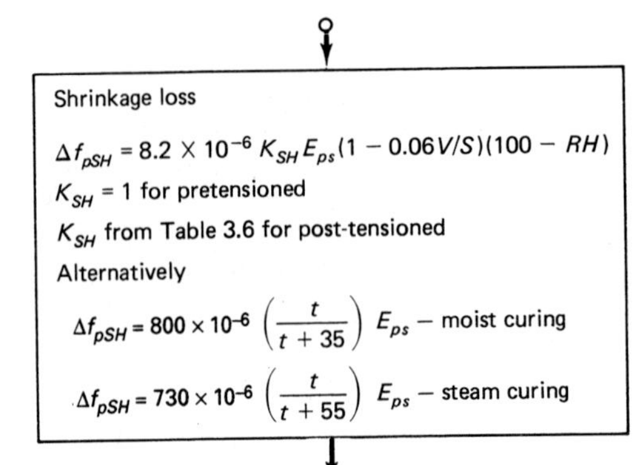

Relaxation of steel loss

(i) Stress-relieved strands

Pretensioned

$$f_{pi} = f_{pJ} - \Delta f_{pR}(t_0, t_{tr}) - \Delta f_{pES}$$

$$f_{pJ} - \Delta f_{pR}(t_0, t_{tr}) \cong 0.90 f_{pJ}$$

$$\Delta f_{pR} = f_{pi} \frac{(\log t_2 - \log t_1)}{10}\left(\frac{f_{pi}}{f_{py}} - 0.55\right)$$

where t_2 and t_1 are in hours

Post-tensioned

$$f_{pi} = f_{pJ} - \Delta f_{pF} - \Delta f_{pES}$$

where f_{pES} is for case of sequential jacking

$$\Delta f_{pR} = f_{pi} \frac{\log t}{10}\left(\frac{f_{pi}}{f_{py}} - 0.55\right)$$

where $\log t = \log (t_2 - t_1)$

(ii) Low-relaxation strands

Replace the denominator (10) in the $(\log t_2 - \log t_1)$ term for pretensioned and the $(\log t)$ term for post-tensioned by a denominator value of 45.

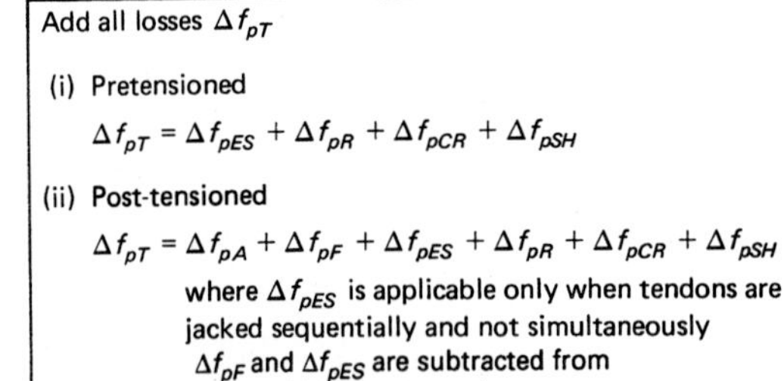

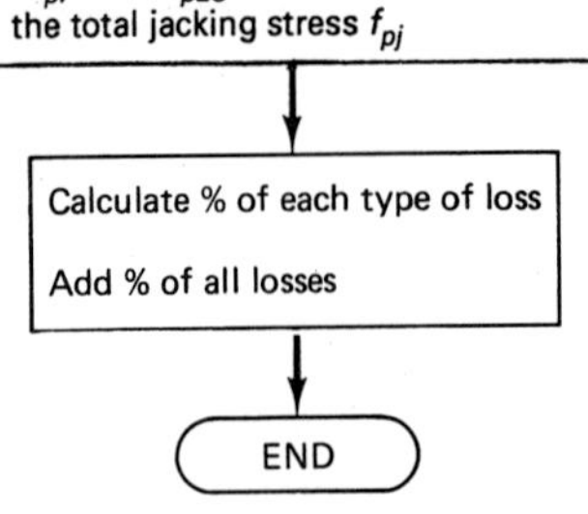

Figure 3.9 *Continued*

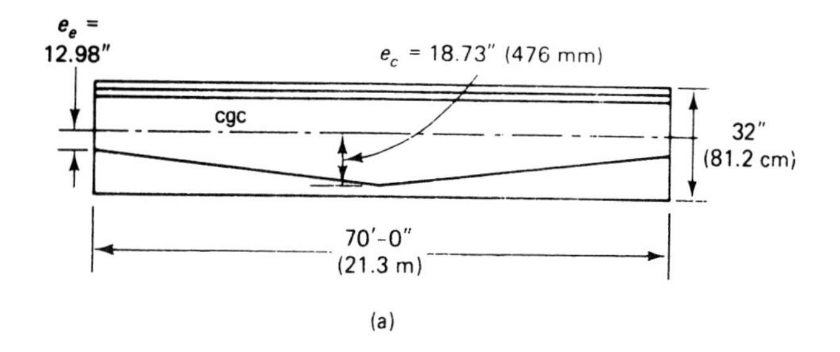

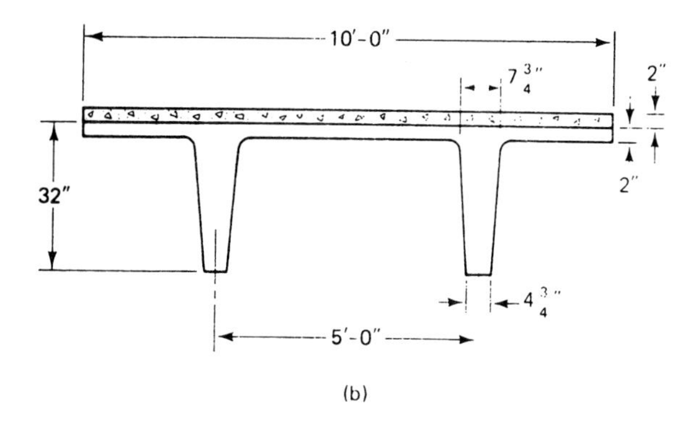

Figure 3.10 Double-T pretensioned beam. (a) Elevation. (b) Pretensioned section.

$$
\begin{aligned}
W_D \text{ (no topping)} &= 491 \text{ plf } (7.2 \text{ kN/m}) \\
W_{SD} \text{(2-in. topping)} &= 250 \text{ plf } (3.65 \text{ kN/m}) \\
W_L &= 40 \text{ psf } (1{,}915 \text{ Pa})\text{—Transient} \\
f_{pu} &= 270{,}000 \text{ psi } (1{,}862 \text{ MPa}) \\
f_{py} &= 0.85 f_{pu} \approx 230{,}000 \text{ psi } (1{,}589 \text{ MPa}) \\
f_{pi} &= 0.70 f_{pu} = 0.82 f_{py} = 0.82 \times 0.85 f_{pu} \cong 0.70 f_{pu} \\
&= 189{,}000 \text{ psi } (1{,}303 \text{ MPa}) \\
E_{ps} &= 28 \times 10^6 \text{ psi } (193.1 \times 10^3 \text{ MPa})
\end{aligned}
$$

Solution:

$$
\begin{aligned}
f'_{ci} &= 3{,}500 \text{ psi} \\
E_{ci} &= 115^{1.5}(33\sqrt{3{,}500}) = 2.41 \times 10^6 \text{ psi} \\
E_c &= 115^{1.5}(33\sqrt{5{,}000}) = 2.88 \times 10^6 \text{ psi}
\end{aligned}
$$

Stage 1: Stress Transfer

(a) *Elastic shortening.* Given critical section distance from support = $0.40 \times 70 = 28$ ft, e at critical section = $12.98 + 0.8(18.73 - 12.98) = 17.58$ in. Dead-load moment M_D at 0.40 of the span is

$$M_D = W_D\frac{X}{2}(L - X) = 491\left(\frac{28}{2}\right)(70 - 28)$$

$$= 288{,}708 \text{ ft-lb} = 3{,}464{,}496 \text{ in.-lb } (391 \text{ kN-m})$$

$$f_{pi} = 0.70f_{pu} = 0.70 \times 270{,}000 = 189{,}000 \text{ psi}$$

Assume elastic-shortening loss and steel-relaxation loss $\cong$ 18,000 psi; then net-steel stress f_{pi} = 189,000 – 18,000 = 171,000 psi, and

$$P_i = A_{ps}f_{pi} = 12 \times 0.153 \times 171{,}000 = 313{,}956 \text{ lb}$$

$$r^2 = \frac{I_c}{A_c} = \frac{59{,}720}{615} = 97.11 \text{ in}^2$$

$$\bar{f}_{cs} = -\frac{P_i}{A_c}\left(1 + \frac{e^2}{r^2}\right) + \frac{M_D e}{I_c}$$

$$= -\frac{313{,}956}{615}\left(1 + \frac{(17.58)^2}{97.11}\right) + \frac{3{,}464{,}496 \times 17.58}{59{,}720}$$

$$= -2{,}135.2 + 1{,}019.9 = -1{,}115.3 \text{ psi } (7.7 \text{ MPa})$$

$$n = \frac{E_{ps}}{E_{ci}} = \frac{28 \times 10^6}{2.41 \times 10^6} = 11.62$$

$$\Delta f_{pES} = n\bar{f}_{CS} = 11.62 \times 1{,}115.3 = 12{,}958 \text{ psi } (85.4 \text{ MPa})$$

If f_{pi} = 189,000 psi is used, then the net f_{pi} = 189,000 – 12,958 = 176,042 psi, and we have

$$\bar{f}_{cs} = -2{,}135.20 \times \frac{176{,}042}{171{,}000} + 1{,}019.90 = -1178.3 \text{ psi}$$

$$\Delta f_{pes} = nf_{CS} = 11.62 \times 1178.3 = 13{,}690 \text{ psi } (94 \text{ MPa})$$

vs. 12,985 psi in the refined solution, a small difference of ~6 percent. Thus, an assumption of 10-percent loss at the beginning in estimating $P_i \cong 0.9P_J$ would have been adequate.

(b) *Steel-Stress Relaxation.* Calculate the steel relaxation at transfer.

$$f_{py} = 230{,}000 \text{ psi}$$

$$f_{pi} = 189{,}000 \text{ psi (or net } f_{pi} = 171{,}000 \text{ psi could be used)}$$

$$t = 18 \text{ hours}$$

$$\Delta f_{pR} = f_{pi}\left(\frac{\log t_2 - \log t_1}{10}\right)\left(\frac{f_{pi}}{f_{py}} - 0.55\right)$$

$$= 189{,}000\left(\frac{\log 18}{10}\right)\left(\frac{189{,}000}{230{,}000} - 0.55\right)$$

$$= 6{,}446.9 \text{ psi} \cong 6{,}447 \text{ psi}$$

$$\Delta f_{pES} + \Delta f_{pR} = 12{,}958 + 6{,}447 = 19{,}405 \cong 18{,}000 \text{ psi, assumed OK.}$$

(c) *Creep Loss*

$$\Delta f_{pCR} = 0$$

(d) *Shrinkage Loss*

$$\Delta f_{pSH} = 0$$

The stage-I total losses are

$$\Delta f_{pT} = \Delta f_{pES} + \Delta f_{pR} + \Delta f_{pCR} + \Delta f_{pSH}$$

$$= 12{,}958 + 6{,}447 + 0 + 0 = 19{,}405 \text{ psi } (134 \text{ MPa})$$

The strand stress f_{pi} at the end of stage I = 189,000 – 19,405 = 169,595 psi (1,169 MPa), giving P_i = 311,376 psi.

Stage II: Transfer to Placement of Topping after 30 Days

(a) *Creep Loss*

$$E_c = 2.88 \times 10^6 \text{ psi}$$

$$E_{ps} = 28 \times 10^6 \text{ psi}$$

$$n = \frac{E_{ps}}{E_c} = \frac{28 \times 10^6}{2.88 \times 10^6} = 9.72$$

$$\bar{f}_{cs} = 1{,}115.3 \text{ psi}$$

Intensity of 2-in. normal-weight concrete topping:

$$W_{SD} = \frac{2}{12} \times 10 \times 150 = 250 \text{ plf}$$

The moment due to the 2-in. topping is

$$M_{SD} = W_{SD}\left(\frac{x}{2}\right)(L - x) = 250\left(\frac{28}{2}\right)(70 - 28) \times 12 = 1{,}764{,}000 \text{ in.-lb (199 kN-m)}$$

$$\bar{f}_{csd} = \frac{M_{SD}e}{I_c} = \frac{1{,}764{,}000 \times 17.58}{59{,}720} = 519.3 \text{ psi}$$

Although 30 days' duration is short for long-term effects, sufficient approximation can be justified in stage II using the creep factor K_{CR} of Equation 3.11 to account for stage III as well (see stage-III creep calculations).

For lightweight concrete, use $K_{CR} = 2.0 \times 80\% = 1.6$. Then, from Equation 3.10, the prestress loss due to long-term creep is

$$\Delta f_{pCR} = nK_{CR}(f_{cs} - f_{csd}) = 9.72 \times 1.6(1{,}115.3 - 519.3) = 9{,}269 \text{ psi (63.3 MPa)}$$

(b) *Shrinkage Loss.* Assume relative humidity $RH = 70\%$. Then, from Equation 3.14, the prestress loss due to long-term shrinkage is

$$\Delta f_{pSH} = 8.2 \times 10^{-6} K_{SH} E_{ps}\left(1 - 0.06\frac{V}{S}\right)(100 - RH)$$

$$\frac{V}{S} = \frac{615}{364} = 1.69$$

$K_{SH} = 1.0$ for pretensioned members; hence,

$$\Delta f_{pSH} = 8.2 \times 10^{-6} \times 1.0 \times 28 \times 10^6 (1 - 0.06 \times 1.69)(100 - 70)$$
$$= 6{,}190 \text{ psi (42.7 MPa)}$$

(c) *Steel Relaxation Loss at 30 Days*

$$t_1 = 18 \text{ hours}$$

$$t_2 = 30 \text{ days} = 30 \times 24 = 720 \text{ hours}$$

$$f_{ps} = 169{,}595 \text{ psi from stage I}$$

$$\Delta f_{pR} = 169{,}595\left(\frac{\log 720 - \log 18}{10}\right)\left(\frac{169{,}595}{230{,}000} - 0.55\right) = 5{,}091 \text{ psi (35.1 MPa)}$$

Stage-II total loss is

$$\Delta f_{pT} = \Delta f_{pCR} + \Delta f_{pSH} + \Delta f_{pR} = 9{,}269 + 6{,}190 + 5{,}091 = 20{,}550 \text{ psi (142 MPa)}$$

The increase in stress in the strands due to the addition of topping is

$$f_{SD} = n\bar{f}_{csd} = 9.72 \times 519.3 = 5{,}048 \text{ psi (34.8 MPa)}$$

Hence, the strand stress at the end of stage II is

$$f_{pe} = f_{ps} - \Delta f_{pT} + f_{SD} = 169{,}595 - 20{,}550 + 5{,}048 = 154{,}093 \text{ psi (1,062 MPa)}$$

Stage III: At End of Two Years

The values for long-term creep and long-term shrinkage evaluated for stage II are assumed not to have increased significantly, since the long-term values of K_{CR} for creep and K_{SH} for shrinkage were used in stage-II computations. Accordingly,

$$f_{pe} = 154{,}093 \text{ psi (1,066 MPa)}$$

$$t_1 = 30 \text{ days} = 720 \text{ hours}$$

$$t_2 = 2 \text{ years} \times 365 \times 24 = 17{,}520 \text{ hours}$$

The steel relaxation stress loss is

$$\Delta f_{pR} = 154{,}093\left(\frac{\log 17{,}520 - \log 720}{10}\right)\left(\frac{154{,}093}{230{,}000} - 0.55\right) = 2{,}563 \text{ psi (17.7 MPa)}$$

So the strand stress f_{pe} at the end of stage III $\cong$ 154,093 – 2,563 = 151,530 psi (1,033 MPa).

Summary of Stresses

Stress level at various stages	Steel stress, psi	Percent
After tensioning ($0.70\,f_{pu}$)	189,000	100.0
Elastic shortening loss	–12,958	–6.9
Creep loss	–9,269	–4.9
Shrinkage loss	–6,190	–3.3
Relaxation loss (6,447 + 5,091 + 2,563)	–14,101	–7.5
Increase due to topping	5,048	2.7
Final net stress f_{pe}	151,530 psi (1,045 MPa)	80.1

Percentages of total losses = 100 – 80.1 = 19.9%, say, 20% for this pretensioned beam.

3.10 STEP-BY-STEP COMPUTATION OF ALL TIME-DEPENDENT LOSSES IN A POST-TENSIONED BEAM

Example 3.9

Solve Example 3.8 assuming that the beam is post-tensioned. Assume also that the anchorage seating loss is $\frac{1}{4}$ in. and that all strands are simultaneously tensioned in a flexible duct. Also assume that the total jacking force prior to the friction and anchorage seating losses resulted in f_{pi} = 189,000 psi ($f_{pJ} = f_{pi}$ of Equation 3.1d in this case).

Solution:

(a) *Anchorage seating loss*

$$\Delta_A = \frac{1''}{4} = 0.25'' \qquad L = 70 \text{ ft}$$

From Equation 3.24, the anchorage slip stress loss is

$$\Delta f_{pA} = \frac{\Delta_A}{L} E_{Ps} = \frac{0.25}{70 \times 12} \times 28 \times 10^6 \cong 8{,}333 \text{ psi (40.2 MPa)}$$

(b) *Elastic shortening.* Since all jacks are simultaneously tensioned, the elastic shortening will precipitate during jacking. As a result, no elastic shortening stress loss takes place in the tendons. Hence, $\Delta f_{pES} = 0$.

(c) *Frictional loss.* Assume that the parabolic tendon approximates the shape of an arc of a circle. Then, from Equation 3.23,

$$\alpha = \frac{8y}{x} = \frac{8(18.73 - 12.98)}{70 \times 12} = 0.0548 \text{ radian}$$

From Table 3.7, use $K = 0.001$ and $\mu = 0.25$. Then, from Example 3.8,

$$f_{pi} = 189{,}000 \text{ psi } (1{,}303 \text{ MPa})$$

From Equation 3.22, the stress loss in prestress due to friction is

$$\begin{aligned} \Delta f_{pF} &= f_{pi}(\mu\alpha + KL) \\ &= 189{,}000(0.25 \times 0.0548 + 0.001 \times 70) \\ &= 15{,}819 \text{ psi } (109 \text{ MPa}) \end{aligned}$$

The stress remaining in the prestressing steel after all initial instantaneous losses is

$$f_{pi} = 189{,}000 - 8{,}333 - 0 - 15{,}819 = 164{,}848 \text{ psi } (1{,}136 \text{ MPa})$$

Hence, the net prestressing force is

$$P_i = 164{,}848 \times 12 \times 0.153 = 296{,}726 \text{ lb}$$

compared to $P_i = 311{,}376$ lb in the pretensioned case of Example 3.8.

Photo 3.3 Linn Cove Viaduct, Grandfather Mountain, North Carolina. A 90° cantilever and a 10 percent superelevation in one direction to a full 10 percent in the opposite direction within 180 ft. Designed by Figg and Muller Engineers, Inc., Tallahassee, Florida. (*Courtesy,* Figg and Muller Engineers, Inc.)

Stage I: Stress at Transfer

(a) *Anchorage Seating Loss*

$$\text{Loss} = 8{,}333 \text{ psi}$$

$$\text{Net stress} = 164{,}848 \text{ psi}$$

(b) *Relaxation Loss*

$$\Delta f_{pR} = 164{,}848\left(\frac{\log 18}{10}\right)\left(\frac{164{,}848}{230{,}000} - 0.55\right)$$

$$\cong 3{,}450 \text{ psi } (23.8 \text{ MPa})$$

(c) *Creep Loss*

$$\Delta f_{pCR} = 0$$

(d) *Shrinkage Loss*

$$\Delta f_{pSH} = 0$$

So the tendon stress f_{pi} at the end of stage I is

$$164{,}848 - 3{,}450 = 161{,}398 \text{ psi } (1{,}113 \text{ MPa})$$

Stage II: Transfer to Placement of Topping after 30 Days

(a) *Creep Loss*

$$P_i = 161{,}398 \times 12 \times 0.153 = 296{,}327 \text{ lb}$$

$$\bar{f}_{cs} = -\frac{P_i}{A_c}\left(1 + \frac{e^2}{r^2}\right) + \frac{M_D e}{I_c}$$

$$= -\frac{296{,}327}{615}\left(1 + \frac{(17.58)^2}{97.11}\right) + \frac{3{,}464{,}496 \times 17.58}{59{,}720}$$

$$= -2{,}016.20 + 1{,}020.00 = 996.2 \text{ psi } (6.94 \text{ MPa})$$

Hence, the creep loss is

$$\Delta f_{pCR} = nK_{CR}(\bar{f}_{cs} - \bar{f}_{csd})$$

$$= 9.72 \times 1.6(996.2 - 519.3) \cong 7{,}417 \text{ psi } (51.2 \text{ MPa})$$

(b) *Shrinkage Loss.* From Example 3.8, for $K_{SH} = 0.58$ at 30 days, Table 3.6,

$$\Delta f_{pSH} = 6{,}190 \times 0.58 = 3{,}590 \text{ psi } (24.8 \text{ MPa})$$

(c) *Steel Relaxation Loss at 30 Days*

$$f_{ps} = 161{,}398 \text{ psi}$$

The relaxation loss in stress becomes

$$\Delta f_{pR} = 161{,}398\left(\frac{\log 720 - \log 18}{10}\right)\left(\frac{161{,}398}{230{,}000} - 0.55\right)$$

$$\cong 3{,}923 \text{ psi } (27.0 \text{ MPa})$$

Stage II: Total Losses

$$\Delta f_{pT} = \Delta f_{pCR} + \Delta f_{pSH} + \Delta f_{pR}$$

$$= 7{,}417 + 3{,}590 + 3{,}923 = 14{,}930 \text{ psi } (103.0 \text{ MPa})$$

From Example 3.8, the increase in stress in the strands due to the addition of topping, is $f_{SD} = 5{,}048$ psi (34.8 MPa); hence, the strand stress at the end of stage II is

$$f_{pe} = f_{ps} - \Delta f_{pT} + \Delta f_{SD} = 161{,}398 - 14{,}930 + 5{,}048 = 151{,}516 \text{ psi } (1{,}045 \text{ MPa})$$

Stage III: At End of 2 Years

$$f_{pe} = 151{,}516 \text{ psi}$$
$$t_1 = 720 \text{ hours}$$
$$t_2 = 17{,}520 \text{ hours}$$

The steel relaxation stress loss is

$$\Delta f_{pR} = 151{,}516\left(\frac{\log 17{,}520 - \log 720}{10}\right)\left(\frac{151{,}516}{230{,}000} - 0.55\right)$$
$$\cong 2{,}284 \text{ psi } (15.8 \text{ MPa})$$

Using the same assumptions for stage III creep and shrinkage as in Example 3.8, the strand stress f_{pe} at the end of stage III is approximately

$$151{,}516 - 2{,}284 = 149{,}232 \text{ psi } (1{,}029 \text{ MPa})$$

Summary of Stresses

Stress level at various stages	Steel stress psi	Percent
After tensioning ($0.70f_{pu}$)	189,000	100.0
Elastic shortening loss	0	0.0
Anchorage loss*	−8,333	−4.4
Frictional loss*	−15,819	−8.4
Creep loss	−7,417	−3.9
Shrinkage loss	−3,590	−1.9
Relaxation loss (3,450 + 3,923 + 2,284)	−9,657	−5.1
Increase due to topping	+5,048	+2.7
Final net stress f_{pe}	149,232	79.0
Percentage of total losses = 100 − 79.0 = 21.0% beam.		

*Frictional and anchorage seating losses are included in this table since the total jacking stress is given as 189,000 psi; otherwise the tendons would have to be jacked an additional stress of such a magnitude as to neutralize the frictional and anchorage seating losses.

3.11 LUMP-SUM COMPUTATION OF TIME-DEPENDENT LOSSES IN PRESTRESS

Example 3.10

Solve Examples 3.8 and 3.9 by the approximate lump-sum method, and compare the results.

Solution for Example 3.8. From Table 3.1, the total loss $\Delta f_{pT} = 45{,}000$ psi (228 MPa). So the net final strand stress by this method is

$$f_{pe} = 189{,}000 - 45{,}000 = 144{,}000 \text{ psi } (993 \text{ MPa})$$
$$\text{Step-by-step } f_{pe} \text{ value} = 151{,}530$$
$$\text{Percent difference} = \frac{151{,}530 - 144{,}000}{189{,}000} = 3.9\%$$

Solution for Example 3.9. From Table 3.2, the total loss $\Delta_{pT} = 35{,}000$ psi (241 MPa). So the net final strand stress by the lump-sum method is

$$f_{pe} = 189{,}000 - 35{,}000 = 154{,}000 \text{ psi } (1{,}062 \text{ MPa})$$

$$\text{Step-by-step } f_{pe} \text{ value} = 149{,}232$$

$$\text{Percent difference} = \frac{154{,}000 - 149{,}232}{189{,}000} = 2.5\%$$

In both cases, the difference between the step-by-step "exact" method and the approximate lump-sum method is quite small, indicating that in normal, standard cases both methods are equally reliable.

3.12 SI PRESTRESS LOSS EXPRESSIONS

$$\Delta f_{pR} = f_{pi}\left(\frac{\log t_2 - \log t_1}{10}\right)\left(\frac{f'_{pi}}{f_{py}} - 0.55\right) \tag{3.8}$$

for stress-relieved tendons where t is in hours. The denominator 10 becomes 45 for low-relaxation tendons.

$$\Delta f_{pCR} = nK_{CR}(\bar{f}_{cs} - \bar{f}_{csd}) \tag{3.11b}$$

where for normal concrete

$$K_{CR} = 2.0 \text{ for pretensioned}$$
$$= 1.6 \text{ for post-tensioned}$$

reduced by 20% for lightweight concrete.

$$n = \text{modular ratio} = \frac{E_{ps}}{E_c}$$

$$\Delta f_{pSH} = 8.2 \times 10^{-6} K_{SH} E_{ps}\left(1 - 0.06\frac{V}{S}\right)(100 - RH) \tag{3.14}$$

$$K_{SH} = 1.0, \text{ pretensioned}$$
$$= \text{range of } 0.92 \text{ (1 day) to } 0.45 \text{ (60 days)}$$

Equation 3.15a, moist curing for 7 days

$$\epsilon_{SH,t} = \left[\frac{t}{t + 35}\right]\epsilon_{SH,u}$$

where $\epsilon_{SH,t} = 800 \times 10^{-6}$ mm/mm

Equation 3.15b, steam curing 1 to 3 days max

$$\epsilon_{SH,t} = \left[\frac{t}{t + 55}\right]\epsilon_{SH,u}$$

where $\epsilon_{SH,t} = 730 \times 10^{-6}$ mm/mm

$$\Delta f_{pF} = -f_1(\mu\alpha + 3.28KL) \tag{3.22}$$

where L, meter.

$$\Delta f_{pA} = \left(\frac{\Delta_A}{L}\right)E_{ps} \tag{3.24}$$

$$E_c = w^{1.5}0.043\sqrt{f'_c} \qquad w \text{ (lightweight)} \simeq 1830 \text{ Kg/m}^3.$$
$$E_{ci} = w^{1.5}0.043\sqrt{f'_{ci}}$$

$$MPa = 10^6 \text{ N/m}^2 = \text{N/mm}^2$$

$$\text{(psi) } 0.006895 = \text{MPa}$$

$$\text{(lb/ft) } 14.593 = \text{N/m}$$

$$\text{(in.-lb)} = 0.113 = \text{N-m}$$

3.12.1 SI Prestress Loss Example

Example 3.11

Solve Example 3.9 using SI units for losses in prestress, considering self-weight and superimposed dead load only.

Data

$f'_c = 34.5$ MPa

$f'_{ci} = 24.1$ MPa

$A_c = 3{,}968$ cm^2 $\qquad S^t = 97{,}670$ cm^3

$I_c = 2.49 \times 10^6$ cm^4 $\qquad S_b = 44{,}520$ cm^3

$r^2 = I_c/A_c = 626$

$c_b = 55.8$ cm $\qquad c^t = 25.5$ cm

$e_c = 47.6$ cm $\qquad e_e = 33.0$ cm

$f_{pu} = 1{,}860$ MPa

$f_{py} = 0.85 f_{pu} = 1{,}580$ MPa

$f_{pi} = 0.82 f_{py} = (0.82 \times 0.85) f_{pu} = 0.7 f_{pu} = 1{,}300$ MPa

$E_{ps} = 193{,}000$ MPa

Span $l = 21.3$ m

$A_{ps} =$ twelve tendons, 12.7-mm diameter (99 mm^2)

$= 12 \times 99 = 1{,}188$ mm^2

$M_D = 391$ kN-m $\qquad M_{SD} = 199$ kN-m

$\Delta_A = 0.64$ cm

$V/S = 1.69$ $\qquad RH = 70\%$

Solution:

(a) *Anchorage seating loss*

$$\Delta A = 0.64 \text{ cm} \qquad l = 21.3 \text{ m}$$

$$\Delta f_{pA} = \frac{\Delta_A}{l}(E_{ps}) = \frac{0.64}{21.3 \times 100} \times 193{,}000 = 58.0 \text{ MPa}$$

(b) *Elastic Shortening*

Since all jacks are simultaneously tensioned, the elastic shortening will simultaneously precipitate during jacking. As a result, no elastic shortening loss takes place in the tendons.

Hence

$$\Delta f_{pES} = 0.$$

(c) *Frictional Loss*

Assume that the parabolic tendon approximates the shape of an arc of a circle. Then, from Equation 3.23,

$$\alpha = \frac{8y}{x} = \frac{8(e_c - e_e)}{l} = \frac{8(47.6 - 33.0)}{21.3 \times 100}$$

$$= 0.055 \text{ radians}$$

From Table 3.7, $K = 0.001$, $\mu = 0.25$, $f_{pi} = 1{,}300$ MPa
From Equation 3.22,

$$\Delta f_{pF} = f_{pi}(\mu\alpha + 3.28KL)$$
$$= 1{,}300(0.25 \times 0.055 + 0.001 \times 3.28 \times 21.3)$$
$$= 109 \text{ MPa}$$

The stress remaining in the prestressing steel after all instantaneous stresses

$$f_{pi} = 1{,}300 - (58 + 0 + 109) = 1{,}133 \text{ MPa}$$

Hence, the net prestressing force is

$$P_i = f_{pi}A_{ps} = 1{,}133 \times 1{,}188 = 1.35 \times 10^6 \text{ N}$$

Stage I: Stress at Transfer

(a) *Anchorage Seating Loss* $\Delta_{fA} = 58$ MPa

(b) *Relaxation Loss*
From Equation 3.8,

$$\Delta f_{pR} = f_{pi}\left(\frac{\log t_2 - \log t_1}{10}\right)\left(\frac{f'_{pi}}{f_{py}} - 0.55\right)$$

$$= 1{,}133\left(\frac{\log 18 - \log 0}{10}\right)\left(\frac{1{,}133}{1{,}580} - 0.55\right)$$

$$= 24.4 \text{ MPa}$$

(c) *Creep Loss* $\Delta f_{cR} = 0$
(d) *Shrinkage Loss* $\Delta f_{SH} = 0$
Tendon stress at the end of stage I

$$f_{ps} = 1{,}133 - 24.4 \cong 1{,}108 \text{ MPa}$$

Stage II: Transfer to Placement of Topping After 30 Days

(a) *Creep Loss*

$P_i = 1{,}108 \times 1{,}188 = 1.32 \times 10^6$ N

$$\bar{f}_{cs} = -\frac{P_i}{A_c}\left(1 + \frac{e^2}{r^2}\right) + \frac{M_D e_b}{I_c} \quad \text{at tendons centroid}$$

e at 0.4 of span = 17.58 in. = 44.7 cm

$$\bar{f}_{cs} = -\frac{1.32 \times 10^6}{3{,}968 \times 10^2}\left[1 + \frac{(44.6)^2}{626}\right] + \frac{3.91 \times 10^7 \text{ N-cm} \times 44.6 \text{ N/mm}^2}{2.49 \times 10^6} \times \frac{1}{100}$$

$$= -13.90 + 7.00 \text{ N/mm}^2 = 6.90 \text{ MPa at cgs}$$

w (lightweight) = 1,800 Kg/m^3

$$E_c\,(\text{lightweight}) = w^{1.5}\,0.043\sqrt{34.5}$$

$$= 1{,}830^{1.5} \times 0.043\sqrt{34.5} = 19{,}770 \text{ MPa}$$

$$n = \frac{E_{ps}}{E_c} = \frac{193{,}000}{19{,}770} = 9.76$$

$\bar{f}_{csd}$ = stress in concrete at cgs due to all superimposed dead loads after prestressing is accomplished.

$$\bar{f}_{csd} = \frac{M_{SD}e}{I_c} = \frac{1.99 \times 10^7 \text{ N-cm} \times 44.7}{2.49 \times 10^6} \times \frac{1}{100}\text{N/mm}^2$$

$$= 3.57 \text{ MPa}$$

$$K_{CR} = 1.6 \text{ for post-tensioned beam}$$

From Equation 3.11b,

$$\Delta f_{pCR} = nK_{CR}(\bar{f}_{cs} - \bar{f}_{csd})$$

$$= 9.76 \times 1.6(6.90 - 3.57) = 52.0 \text{ MPa}$$

(b) *Shrinkage Loss at 30 Days*
From Equation 3.14,

$$\Delta f_{pSH} = 8.2 \times 10^{-6}\,K_{SH}\,E_{ps}\left(1 = 0.06\frac{V}{S}\right)(100 - RH)$$

K_{SH} at 30 days = 0.58 (Table 3.6)

$$\Delta f_{pSH} = 8.2 \times 10^{-6} \times 0.58 \times 193{,}000(1 - 0.06 \times 1.69)(100 - 70)$$

$$= 24.7 \text{ MPa}$$

(c) *Relaxation Loss at 30 Days (720 Hrs)*

$$f_{ps} = 1{,}108 \text{ MPa}$$

$$\Delta f_{pR} = 1{,}108\left(\frac{\log 720 - \log 18}{10}\right)\left(\frac{1{,}108}{1{,}580} - 0.55\right)$$

$$= 110.8(2.85 - 1.25)0.151 = 26.8 \text{ MPa}$$

Stage II: Total Losses

$$\Delta f_{pT} = \Delta f_{pCR} + \Delta f_{pSH} + \Delta f_R$$

$$= 52.0 + 24.7 + 26.8 = 104 \text{ MPa}$$

Increase of tensile stress at bottom cgs fibers due to addition of topping is from before,

$$\Delta f_{SD} = nf_{CSD} = 9.76 \times 3.57 = 34.8 \text{ MPa}$$

$$f_{pe} = f_{ps} - \Delta f_{pT} + \Delta f_{SD}$$

$$= 1{,}108 - 103.5 + 34.5 = 1{,}039 \text{ MPa}$$

Stage III: At End of Two Years

$f_{pe} = 1{,}039$ MPa

$t_1 = 720$ hrs. $\qquad t_2 = 17{,}520$ hrs.

$$\Delta f_{pR} = 1{,}039\left(\frac{\log 17{,}520 - \log 720}{10}\right)\left(\frac{1{,}039}{1{,}580} - 0.55\right)$$

$$= 103.9(4.244 - 2.857)0.108 = 15.6 \text{ MPa}$$

On the assumption that Δf_{pCR} and Δf_{pSH} were stable in this case, the stress in the tendons at end of stage III can approximately be $f_{ps} = 1{,}039 - 15.6 \cong 1{,}020$ MPa.

REFERENCES

3.1 PCI Committee on Prestress Loss. "Recommendations for Estimating Prestress Loss." *Journal of the Prestressed Concrete Institute 20* (1975): 43–75.

3.2 ACI-ASCE Joint Committee 423. "Tentative Recommendations for Prestressed Concrete." *Journal of the American Concrete Institute 54* (1957): 548–578.

3.3 AASHTO Subcommittee on Bridges and Structures. *Interim Specifications Bridges.* American Association of State Highway and Transportation Officials, Washington, D.C., 1975.

3.4 Prestressed Concrete Institute. *PCI Design Handbook.* 5th Ed. PCI, Chicago: 1999.

3.5 Post-Tensioning Institute. *Post-Tensioning Manual.* 5th ed. Post-Tensioning Institute, Phoenix, Ariz., 2000.

3.6 Lin, T. Y. "Cable Friction in Post-Tensioning." *Journal of Structural Division.* ASCE, New York, November 1956, pp. 1107–1 to 1107–13.

3.7 Tadros, M. K., Ghali, A., and Dilger, W. H. "Time Dependent Loss and Deflection in Prestressed Concrete Members." *Journal of the Prestressed Concrete Institute 20,* 1975, 86–98.

3.8 Branson, D. E. "The Deformation of Noncomposite and Composite Prestressed Concrete Members." In *Deflection of Structures.* American Concrete Institute, Farmington Hills, 1974, pp. 83–128.

3.9 Nawy, E. G. *Reinforced Concrete—A Fundamental Approach.* 4th ed. Prentice-Hall, Upper Saddle River, N.J.: 2000, 776 pp.

3.10 Cohn, M. Z. *Partial Prestressing, From Theory to Practice.* NATO-ASI Applied Science Series, vols 1 and 2. Dordrecht, The Netherlands: Martinus Nijhoff, in Cooperation with NATO Scientific Affairs Division, 1986.

3.11 ACI Committee 318, *Building Code Requirements for Structural Concrete (ACI 318-99)* and *Commentary (ACI 318 R-99),* American Concrete Institute, Farmington Hills, MI, 2000, pp. 392.

3.12 ACI Committee 435, "Control of Deflection in Concrete Structures," ACI Committee Report R435-95, E. G. Nawy, Chairman, American Concrete Institute, Farmington Hills, MI, 1995, pp. 77.

PROBLEMS

3.1 A simply supported pretensioned beam has a span of 75 ft (22.9 m) and the cross section shown in Figure P3.1. It is subjected to a uniform gravitational live-load intensity $W_L = 1{,}200$ plf (17.5 kN/m) in addition to its self-weight and is prestressed with 20 stress-relieved $\frac{1}{2}$in. dia (12.7 mm dia) 7-wire strands. Compute the total prestress losses by the step-by-step method, and compare them with the values obtained by the lump-sum method. Take the following values as given:

$f'_c = 6{,}000$ psi (41.4 MPa), normal-weight concrete

$f'_{ci} = 4{,}500$ psi (31 MPa)

$f_{pu} = 270{,}000$ psi (1,862 MPa)

$f_{pi} = 0.70 f_{pu}$

Relaxation time $t = 5$ years

$e_c = 19$ in. (483 mm)

Relative humidity $RH = 75\%$

$V/S = 3.0$ in. (7.62 cm)

Assume SD load = 30% LL.

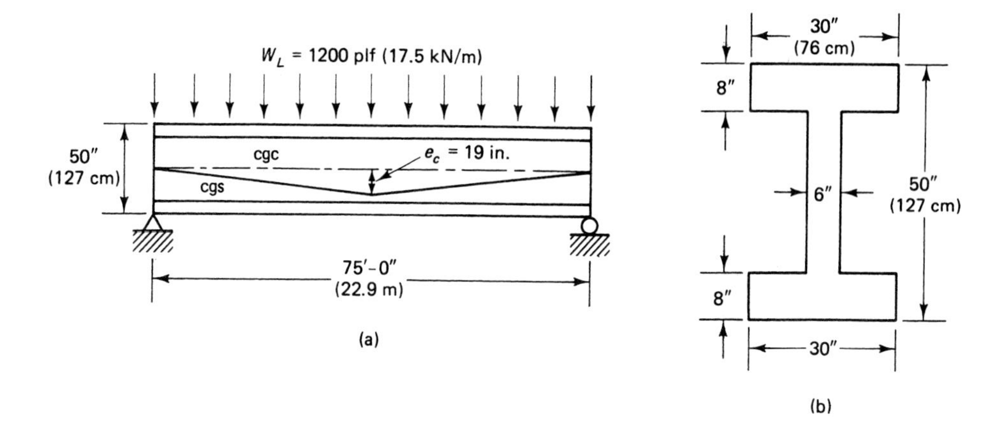

Figure P3.1 (a) Elevation. (b) Section.

3.2 Compute, by the detailed step-by-step method, the total losses in prestress of the 10-ft (3.28-m)-wide flange double T-beam in Example 1.1 which has a span of 64 ft (19.5 m) for a steel relaxation period of 7 years. Use $RH = 70\%$ and $V/S = 3.5$ in. (8.9 cm), and solve for both pretensioned and post-tensioned prestressing conditions. Assume SD load = 30% LL. In the post-tensioned case, assume that the total jacking stress prior to the friction and anchorage seating losses is 189,000 psi.

3.3 Compute, by the detailed step-by-step method, the total losses of prestress in the AASHTO 36-in. (91.4 cm)-deep beam used in Problem 1.1 and which has a span of 34 ft (10.4 m) for both the pretensioned and the post-tensioned case. Use all the data of Problem 1.1 in your solution, and assume that the relative humidity $RH = 70\%$ and the volume-to-surface ratio $V/S = 3.2$. Determine the steel relaxation losses at the end of the first year after erection and at the end of 4 years.

3.4 Compute, by the detailed step-by-step method, the total prestress losses of the simply supported double T-beam of Example 3.9 if it was post-tensioned using flexible ducts for the tendon. Assume that the tendon profile is essentially parabolic. Assume also that all strands are tensioned simultaneously and that the anchorage slip $\Delta_A = \frac{3}{8}$ in. (9.5 mm). All the data are identical to those of Example 3.8; the critical section is determined to be at a distance 0.4 times the span from the face of the support.

FLEXURAL DESIGN OF PRESTRESSED CONCRETE ELEMENTS

4.1 INTRODUCTION

Flexural stresses are the result of external, or imposed, bending moments. In most cases, they control the selection of the geometrical dimensions of the prestressed concrete section regardless of whether it is pretensioned or post-tensioned. The design process starts with the choice of a preliminary geometry, and by trial and adjustment it converges to a final section with geometrical details of the concrete cross section and the sizes and alignments of the prestressing strands. The section satisfies the flexural (bending) requirements of concrete stress and steel stress limitations. Thereafter, other factors such as shear and torsion capacity, deflection, and cracking are analyzed and satisfied.

While the input data for the analysis of sections differ from the data needed for design, every design is essentially an analysis. One assumes the geometrical properties of the section to be prestressed and then proceeds to determine whether the section can safely carry the prestressing forces and the required external loads. Hence, a good understanding of the fundamental principles of analysis and the alternatives presented thereby significantly simplifies the task of designing the section. As seen from the discussion in Chapter 1, the basic mechanics of materials, principles of equilibrium of internal couples, and elastic principles of superposition have to be adhered to in all stages of loading.

Maryland Concert Center parking garage, Baltimore. (*Courtesy,* Prestressed Concrete Institute.)

It suffices in the flexural design of reinforced concrete members to apply only the limit states of stress at failure for the choice of the section, provided that other requirements such as serviceability, shear capacity, and bond are met. In the design of prestressed members, however, additional checks are needed at the load transfer and limit state at service load, as well as the limit state at failure, with the failure load indicating the reserve strength for overload conditions. All these checks are necessary to ensure that at service load cracking is negligible and the long-term effects on deflection or camber are well controlled.

In view of the preceding, this chapter covers the major aspects of both the service-load flexural design and the ultimate-load flexural design check. The principles and methods presented in Chapter 1 for service load computations are extended into step-by-step procedures for the design of prestressed concrete linear elements, taking into consideration the impact of the magnitude of prestress losses discussed in Chapter 3. Note that a logical sequence in the design process entails *first* the service-load design of the section required in flexure, and then the analysis of the available moment strength M_n of the section for the limit state at failure. Throughout the book, a negative sign (−) is used to denote compressive stress and a positive sign (+) is used to denote tensile stress in the concrete section. A convex or hogging shape indicates negative bending moment; a concave or sagging shape denotes positive bending moment, as shown in Figure 4.1.

Unlike the case of reinforced concrete members, the external dead load and partial live load are applied to the prestressed concrete member at varying concrete strengths at various loading stages. These loading stages can be summarized as follows:

- Initial prestress force P_i is applied; then, at transfer, the force is transmitted from the prestressing strands to the concrete.
- The full self-weight W_D acts on the member together with the initial prestressing force, provided that the member is simply supported, i.e., there is no intermediate support.
- The full superimposed dead load W_{SD} including topping for composite action, is applied to the member.
- Most short-term losses in the prestressing force occur, leading to a reduced prestressing force P_{eo}.
- The member is subjected to the full service load, with long-term losses due to creep, shrinkage, and strand relaxation taking place and leading to a net prestressing force P_e.
- Overloading of the member occurs under certain conditions up to the limit state at failure.

A typical loading history and corresponding stress distribution across the depth of the critical section are shown in Figure 4.2, while a schematic plot of load versus defor-

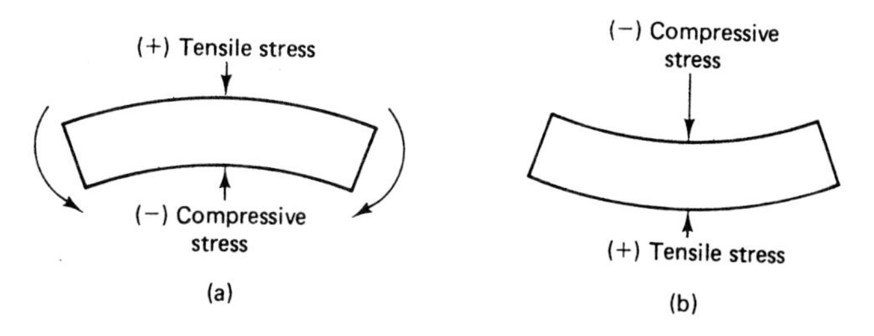

Figure 4.1 Sign convention for flexure stress and bending moment. (a) Negative bending moment. (b) Positive bending moment.

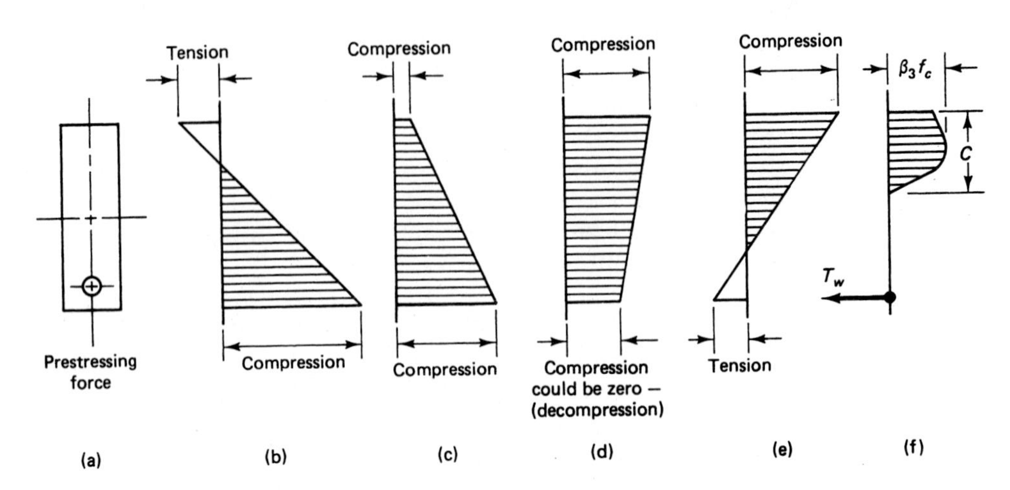

Figure 4.2 Flexural stress distribution throughout loading history. (a) Beam section. (b) Initial prestressing stage. (c) Self-weight and effective prestress. (d) Full dead load plus effective prestress. (e) Full service load plus effective prestress. (f) Limit state of stress at ultimate load for underreinforced beam.

mation (camber or deflection) is shown in Figure 4.3 for the various loading stages from the self-weight effect up to rupture.

4.2 SELECTION OF GEOMETRICAL PROPERTIES OF SECTION COMPONENTS

4.2.1 General Guidelines

Under service-load conditions, the beam is assumed to be homogeneous and elastic. Since it is also assumed (because expected) that the prestress compressive force transmitted to the concrete closes the crack that might develop at the tensile fibers of the beam, beam sections are considered uncracked. Stress analysis of prestressed beams under these conditions is no different from stress analysis of a steel beam, or, more accurately, a beam column. The axial force due to prestressing is always present regardless of whether bending moments do or do not exist due to other external or self-loads.

As seen from Chapter 1, it is advantageous to have the alignment of the prestressing tendon eccentric at the critical sections, such as the midspan section in a simple beam and the support section in a continuous beam. As compared to a rectangular solid section, a nonsymmetrical flanged section has the advantage of efficiently using the concrete material and of concentrating the concrete in the compressive zone of the section where it is most needed.

Equations 4.1, 4.2, and 4.3 to be subsequently presented are stress equations that are convenient in the analysis of stresses in the section once the section is chosen. For design, it is necessary to transpose the three equations into geometrical equations so that the student and the designer can readily choose the concrete section. A logical transposition is to define the minimum section modulus that can withstand all the loads after losses.

4.2.2 Minimum Section Modulus

To design or choose the section, a determination of the required minimum section modulus, S_b and S^t, has to be made first. If

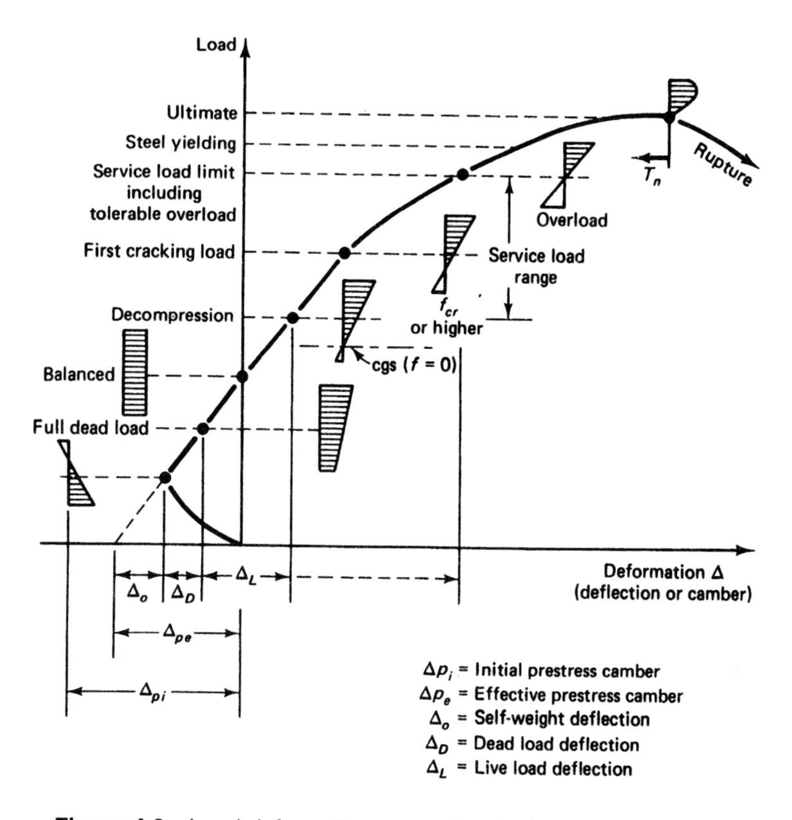

Figure 4.3 Load-deformation curve of typical prestressed beam.

f_{ci} = maximum allowable compressive stress in concrete immediately after transfer and prior to losses
$= 0.60 f'_{ci}$

f_{ti} = maximum allowable tensile stress in concrete immediately after transfer and prior to losses
$= 3\sqrt{f'_{ci}}$ (the value can be increased to $6\sqrt{f'_{ci}}$ at the supports for simply supported members)

f_c = maximum allowable compressive stress in concrete after losses at service-load level
$= 0.45 f'_c$ or $0.60 f'_c$ when allowed by the code

f_t = maximum allowable tensile stress in concrete after losses at service load level
$= 6\sqrt{f'_c}$ (the value can be increased in one-way systems to $12\sqrt{f'_c}$ if long-term deflection requirements are met)

then the *actual* extreme fiber stresses in the concrete cannot exceed the values listed.

Using the uncracked unsymmetrical section, a summary of the equations of stress from Section 1.3 for the various loading stages is as follows.

Stress at Transfer

$$f^t = -\frac{P_i}{A_c}\left(1 - \frac{ec_t}{r^2}\right) - \frac{M_D}{S^t} \leq f_{ti} \tag{4.1a}$$

Photo 4.1 Ninian Central oil drilling platform, Cheiron, England, and C. G. Doris, Scotland. (*Courtesy,* Ben C. Gerwick.)

$$f_b = -\frac{P_i}{A_c}\left(1 + \frac{ec_b}{r^2}\right) + \frac{M_D}{S_b} \leq f_{ci} \tag{4.1b}$$

where P_i is the initial prestressing force. While a more accurate value to use would be the horizontal component of P_i, it is reasonable for all practical purposes to disregard such refinement.

Effective Stresses after Losses

$$f^t = -\frac{P_e}{A_c}\left(1 - \frac{ec_t}{r^2}\right) - \frac{M_D}{S^t} \leq f_t \tag{4.2a}$$

$$f_b = -\frac{P_e}{A_c}\left(1 + \frac{ec_b}{r^2}\right) + \frac{M_D}{S_b} \leq f_c \tag{4.2b}$$

Service-load Final Stresses

$$f^t = -\frac{P_e}{A_c}\left(1 - \frac{ec_t}{r^2}\right) - \frac{M_T}{S^t} \leq f_c \tag{4.3a}$$

$$f_b = -\frac{P_e}{A_c}\left(1 + \frac{ec_b}{r^2}\right) + \frac{M_T}{S_b} \leq f_t \tag{4.3b}$$

where $M_T = M_D + M_{SD} + M_L$
P_i = initial prestress
P_e = effective prestress after losses
t denotes the top, and b denotes the bottom fibers
e = eccentricity of tendons from the concrete section center of gravity, cgc
r^2 = square of radius of gyration
S^t/S_b = top/bottom section modulus value of concrete section

The *decompression stage* denotes the increase in steel strain due to the increase in load from the stage when the effective prestress P_e acts *alone* to the stage when the additional load causes the compressive stress in the concrete at the cgs level to reduce to zero (see Figure 4.3). At this stage, the *change* in concrete stress due to decompression is

$$f_{\text{decomp}} = \frac{P_e}{A_c}\left(1 + \frac{e^2}{r^2}\right) \tag{4.3c}$$

This relationship is based on the assumption that the strain between the concrete and the prestressing steel bonded to the surrounding concrete is such that the gain in the steel stress is the same as the decrease in the concrete stress.

4.2.2.1 Beams With Variable Tendon Eccentricity.

Beams are prestressed with either draped or harped tendons. The maximum eccentricity is usually at the midspan controlling section for the simply supported case. Assuming that the effective prestressing force is

$$P_e = \gamma P_i$$

where γ is the residual prestress ratio, the loss of prestress is

$$P_i - P_e = (1 - \gamma)P_i \tag{a}$$

If the actual concrete extreme fiber stress is equivalent to the maximum allowable stress, the change in this stress after losses, from Equations 4.1a and b, is given by

$$\Delta f^t = (1 - \gamma)\left(f_{ti} + \frac{M_D}{S^t}\right) \tag{b}$$

$$\Delta f_b = (1 - \gamma)\left(-f_{ci} + \frac{M_D}{S_b}\right) \tag{c}$$

From Figure 4.4(a), as the superimposed dead-load moment M_{SD} and live-load moment M_L act on the beam, the net stress at top fibers is

$$f_n^t = f_{ti} - \Delta f^t - f_c$$

or

$$f_n^t = \gamma f_{ti} - (1 - \gamma)\frac{M_D}{S^t} - f_c \tag{d}$$

The net stress at the bottom fibers is

$$f_{bn} = f_t - f_{ci} - \Delta f_b$$

or

$$f_{bn} = f_t - \gamma f_{ci} - (1 - \gamma)\frac{M_D}{S_b} \tag{e}$$

From Equations (d) and (e), the chosen section should have section moduli values

$$S^t \geq \frac{(1 - \gamma)M_D + M_{SD} + M_L}{\gamma f_{ti} - f_c} \tag{4.4a}$$

and

$$S_b \geq \frac{(1 - \gamma)M_D + M_{SD} + M_L}{f_t - \gamma f_{ci}} \tag{4.4b}$$

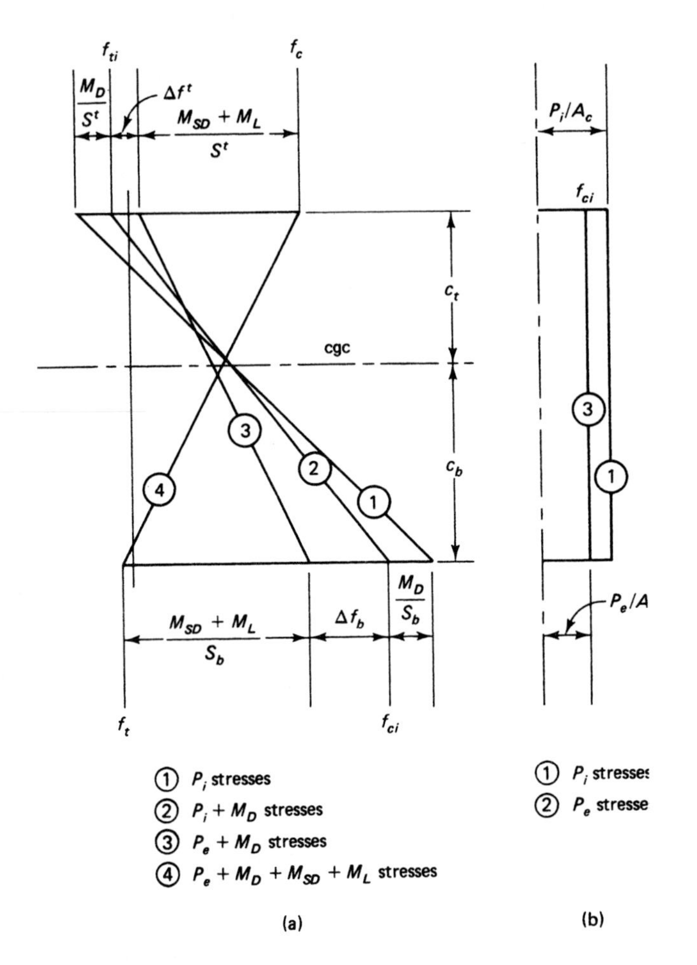

Figure 4.4(a) Maximum fiber stresses in beams with draped or harped tendons. (a) Critical section such as midspan. (b) Support section of simply-supported beam ($e_e = 0$ as tendon moves to cgc).

The required eccentricity of the prestressing tendon at the critical section, such as the midspan section, is

$$e_c = (f_{ti} - \bar{f}_{ci})\frac{S^t}{P_i} + \frac{M_D}{P_i} \tag{4.4c}$$

where $\bar{f}_{ci}$ is the concrete stress at transfer at the level of the centroid cgc of the concrete section and

$$P_i = \bar{f}_{ci} A_c$$

Thus,

$$\bar{f}_{ci} = f_{ti} - \frac{c_t}{h}(f_{ti} - f_{ci}) \tag{4.4d}$$

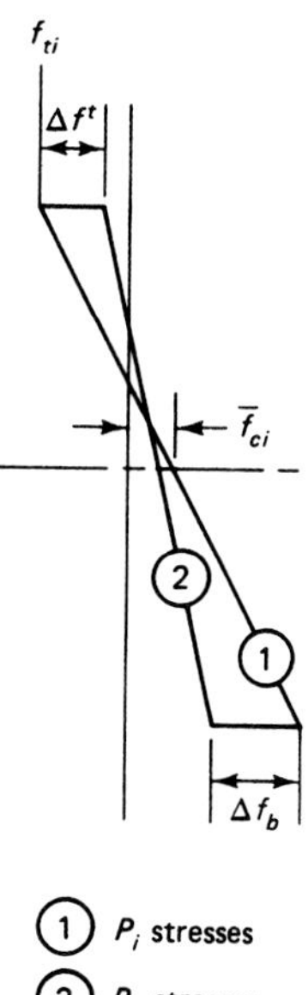

Figure 4.4(b) Maximum fiber stresses at support section of beams with straight tendons (stress distribution at midspan section similar to that of Figure 4.4a).

4.2.2.2 Beams with Constant Tendon Eccentricity. Beams with constant tendon eccentricity are beams with straight tendons, as is normally the case in precast moderate-span simply supported beams. Because the tendon has a large eccentricity at the support, creating large tensile stresses at the top fibers without any reduction due to superimposed $M_D + M_{SD} + M_L$, in such beams smaller eccentricity of the tendon at midspan has to be used as compared to a similar beam with a draped tendon. In other words, the controlling section is the support section, for which the stress distribution at the support is shown in Figure 4.4(b). Hence,

$$\Delta f^t = (1 - \gamma)(f_{ti}) \qquad \text{(a}'\text{)}$$

and

$$\Delta f_b = (1 - \gamma)(-f_{ci}) \qquad \text{(b}'\text{)}$$

The net stress at the service-load condition after losses at the top fibers is

$$f_n^t = f_{ti} - \Delta f^t - f_c$$

or

$$f_n^t = \gamma f_{ti} - f_{cs} \qquad \text{(c}'\text{)}$$

where f_{cs} is the actual service-load stress in concrete. The net stress at service load after losses at the bottom fibers is

$$f_{bn} = f_t - f_{ci} - \Delta f_b$$

or

$$f_{bn} = f_t - \gamma f_{ci} \qquad \text{(d}'\text{)}$$

From Equations (c) and (d), the chosen section should have section moduli values

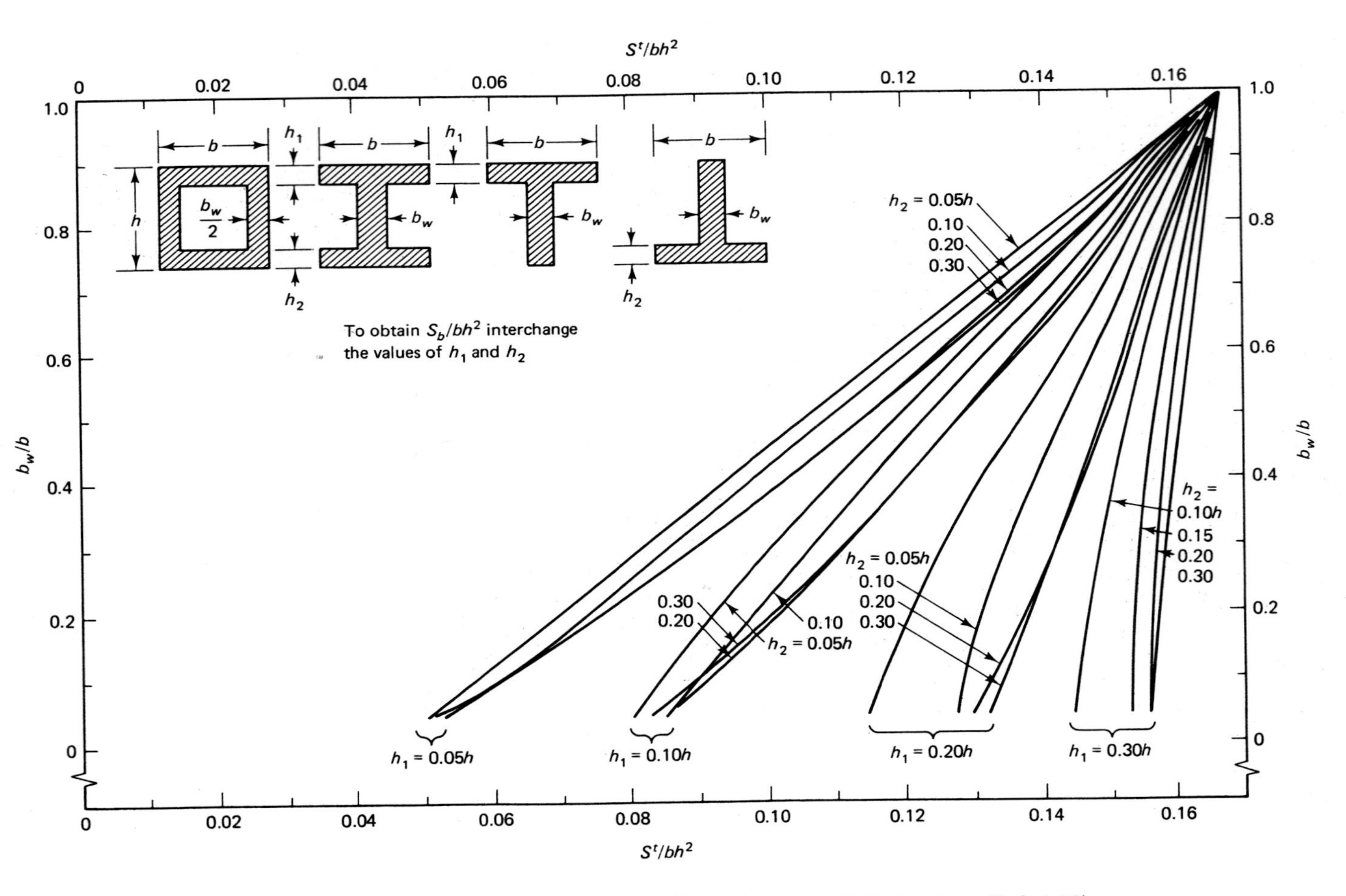

Figure 4.5 Section moduli of flanged and boxed sections (adapted from Ref. 4.16).

$$S^t \geq \frac{M_D + M_{SD} + M_L}{\gamma f_{ti} - f_c} \tag{4.5a}$$

and

$$S_b \geq \frac{M_D + M_{SD} + M_L}{f_t - \gamma f_{ci}} \tag{4.5b}$$

The required eccentricity value at the critical section, such as the support for an ideal beam section having properties close to those required by Equations 4.5a and b, is

$$e_e = (f_{ti} - \bar{f}_{ci})\frac{S^t}{P_i} \tag{4.5c}$$

A graphical representation of section moduli of nominal sections is shown in Figure 4.5. It may be used as a speedy tool for the choice of initial trial sections in the design process.

Table 4.1 gives the section moduli of standard PCI rectangular sections. Tables 4.2 and 4.3 give the geometrical outer dimensions of standard PCI T-sections and AASHTO I-sections, respectively, as well as the top-section moduli of those sections needed in the preliminary choice of the section in the service-load analysis. Table 4.4(a) gives dimensional details of the *actual* "as-built" geometry of the standard PCI and AASHTO sections, and Table 4.4(b) gives girder properties of optimized sections used in different states. Properties of bulb sections are given in Appendix C. Additional details are presented in Refs. 4.9 and 4.12.

4.3 SERVICE-LOAD DESIGN EXAMPLES

4.3.1 Variable Tendon Eccentricity

Example 4.1

Design a simply supported pretensioned double-T-beam for a parking garage with harped tendon and with a span of 60 ft (18.3 m) using the ACI 318 Building Code allowable stresses. The beam has to carry a superimposed service load of 1,100 plf (16.1 kN/m) and superimposed dead load of 100 plf (1.5 kN/m), and has no concrete topping. Assume the beam is made of normal-weight concrete with f'_c = 5,000 psi (34.5 MPa) and that the concrete strength f'_{ci} at transfer is 75 percent of the cylinder strength. Assume also that the time-dependent losses of the initial prestress are 18 percent of the initial prestress, and that f_{pu} = 270,000 psi (1,862 MPa) for stress-relieved tendons, $f'_t = 12\sqrt{f'_c}$.

Table 4.1 Section Properties and Moduli of Standard PCI Rectangular Sections

Designation	12RB16	12RB20	12RB24	12RB28	12RB32	12RB36	16RB32	16RB36	16RB40
Section modulus S, (in.^{3})	512	800	1,152	1,568	2,048	2,592	2,731	3,456	4,267
Width, b (in.)	12	12	12	12	12	12	16	16	16
Depth, h (in.)	16	20	24	28	32	36	32	36	40

Table 4.2 Geometrical Outer Dimensions and Section Moduli of Standard PCI Double T-Sections

Designation	Top-/bottom-section modulus, in.^{3}	Flange width b_f, in.	Flange depth t_f, in.	Total depth h, in.	Web width $2\ b_w$, in.
8DT12	1,001/315	96	2	12	9.5
8DT14	1,307/429	96	2	14	9.5
8DT16	1,630/556	96	2	16	9.5
8DT20	2,320/860	96	2	20	9.5
8DT24	3,063/1,224	96	2	24	9.5
8DT32	5,140/2,615	96	2	32	9.5
10DT32	5,960/2,717	120	2	32	12.5
*12DT34	10,458/3,340	144	4	34	12.5
*15DT34	13,128/4,274	180	4	34	12.5

*Pretopped

Table 4.3 Geometrical Outer Dimensions and Section Moduli of Standard AASHTO Bridge Sections

	AASHTO sections					
Designation	Type 1	Type 2	Type 3	Type 4	Type 5	Type 6
Area A_c in.^{2}	276	369	560	789	1,013	1,085
Moment of inertia I_{ge}, in.^{4}	22,750	50,979	125,390	260,741	521,180	733,320
Top-/bottom-section modulus, in.^{3}	1,476 1,807	2,527 3,320	5,070 6,186	8,908 10,544	16,790 16,307	20,587 20,157
Top flange width, b_f (in.)	12	12	16	20	42	42
Top flange average thickness, t_f (in.)	6	8	9	11	7	7
Bottom flange width, b_2 (in.)	16	18	22	26	28	28
Bottom flange average thickness, t_2 (in.)	7	9	11	12	13	13
Total depth, h (in.)	28	36	45	54	63	72
Web width, b_w (in.)	6	6	7	8	8	8
c_t/c_b (in.)	15.41 12.59	20.17 15.83	24.73 20.27	29.27 24.73	31.04 31.96	35.62 36.38
r^2, in.^{2}	82	132	224	330	514	676
Self-weight w_D lb/ft	287	384	583	822	1055	1130

Table 4.4(a) Geometrical Details of As-Built PCI and AASHTO Sections

Designation	b_f (in.)	h_f (in.)	b_{w1} (in.)	$bw2$ (in.)	h (in.)	b (in.)
8DT12	96	2	5.75	3.75	12	48
8DT14	96	2	5.75	3.75	14	48
8DT16	96	2	5.75	3.75	16	48
8DT18	96	2	5.75	3.75	18	48
8DT20	96	2	5.75	3.75	20	48
8DT24	96	2	5.75	3.75	24	48
8DT32	96	2	7.75	4.75	32	48
10DT32	120	2	7.75	4.75	32	60
12DT34	144	4	7.75	4.75	34	60
15DT34	180	4	7.75	4.75	34	90

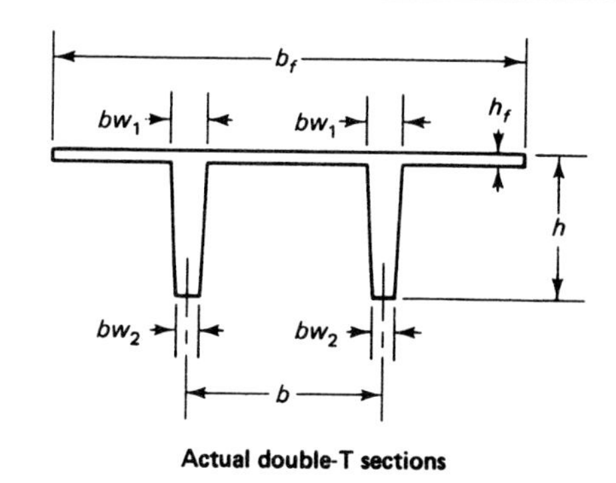

Actual double-T sections

Designation	b_f (in.)	x_1 (in.)	x_2 (in.)	b_2 (in.)	x_3 (in.)	x_4 (in.)	b_w (in.)	h (in.)
AASHTO 1	12	4	3	16	5	5	6	28
AASHTO 2	12	6	3	18	6	6	6	36
AASHTO 3	16	7	4.5	22	7.5	7	7	45
AASHTO 4	20	8	6	26	9	8	8	54
AASHTO 5	42	5	7	28	10	8	8	63
AASHTO 6	42	5	7	28	10	8	8	72

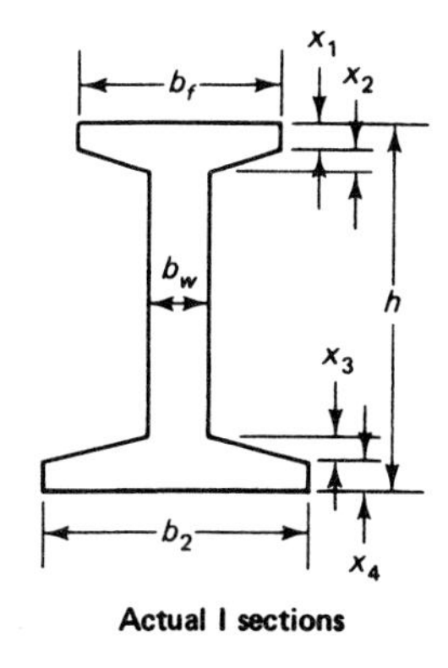

Actual I sections

Solution:

$$\gamma = 100 - 18 = 82\%$$

$$f'_{ci} = 0.75 \times 5{,}000 = -3{,}750 \text{ psi (25.9 MPa)}$$

Use $f'_t = 12\sqrt{5{,}000} = 849$ psi (5.9 MPa) as the maximum stress in tension, and assume a self-weight of approximately 1,000 plf (14.6 kN/m). Then the self-weight moment is given by

$$M_D = \frac{wl^2}{8} = \frac{1{,}000(60)^2}{8} \times 12 = 5{,}400{,}000 \text{ in.-lb (610 kN-m)}$$

and the superimposed load moment is

$$M_{SD} + M_L = \frac{(1{,}100 + 100)\,(60)^2}{8} \times 12 = 6{,}480{,}000 \text{ in.-lb (732 kN-m)}$$

Since the tendon is harped, the critical section is close to the midspan, where dead-load and superimposed dead-load moments reach their maximum. The critical section is in many cases taken at 0.40 L from the support, where L is the beam span. From Equations 4.4a and b,

Table 4.4(b) Girder Properties of Optimized Sections

Agency	Girder Type	Depth (in.)	Web Width (in.)	area (in.^{2})	Inertia (in.^{4})	y_t (in.)	y_b (in.)	S_t (in.^{3})	S_b (in.^{3})	ρ	α
CTL	BT-48	48	6	557	177,736	23.53	24.47	7,553	7,264	0.554	0.940
	BT-60	60	6	629	308,722	29.59	30.41	10,432	10,154	0.545	0.931
	BT-72	72	6	701	484,993	35.64	36.36	13,606	13,340	0.534	0.914
PCI	BT-54	54	6	659	268,077	26.37	27.63	10,166	9,702	0.558	0.943
	BT-63	63	6	713	392,638	30.82	32.12	12,715	12,224	0.556	0.942
	BT-72	72	6	767	545,894	35.40	36.60	15,421	14,915	0.549	0.934
AASHTO	Type VI	72	8	1,085	733,320	35.62	36.38	20,587	20,157	0.522	0.893
	Mod. Type VI	72	6	941	671,088	35.56	36.44	18,871	18,417	0.550	0.941
Washington	80/6	50	6	513	159,191	27.24	22.76	5,844	6,994	0.501	0.943
	100/6	58	6	591	256,560	30.01	27.99	8,549	9,166	0.517	0.925
	120/6	73.5	6	688	475,502	37.68	35.82	12,619	13,275	0.512	0.908
	14/6	73.5	6	736	534,037	35.30	38.20	15,122	13,985	0.538	0.894
Colorado	G54/6	54	6	631	242,592	27.33	26.67	8,877	9,095	0.527	0.924
	G68/6	68	6	701	426,575	33.99	34.01	12,548	12,544	0.526	0.911
Nebraska	1600	63	5.9	852	494,829	32.64	30.36	15,159	16,300	0.586	1.051
	1800	70.9	5.9	898	659,505	36.72	34.18	17,959	19,297	0.585	1.049
	2000	78.7	5.9	944	849,565	40.74	37.96	20,854	22,380	0.582	1.042
	2400	94.5	5.9	1,038	1,323,985	48.84	45.66	27,106	28,999	0.572	1.023
Florida	BT-54	54	6.5	785	311,765	28.11	25.89	11,091	12,042	0.546	0.983
	BT-63	63	6.5	843	458,521	32.88	30.12	13,945	15,223	0.549	0.992
	BT-72	72	6.5	901	638,672	37.64	34.36	16,968	18,588	0.548	0.991
Texas	U54A	54	10.2	1,022	379,857	30.12	23.90	12,612	15,895	0.516	0.996
	U54B	54	10.2	1,118	403,878	31.54	22.48	12,807	17,966	0.509	1.029

1 in. = 25.4 mm; 1 in.^{2} = 645 mm^2; 1 in.^{3} = 16,390 mm^3; 1 in.^{4} = 416,000 mm^4

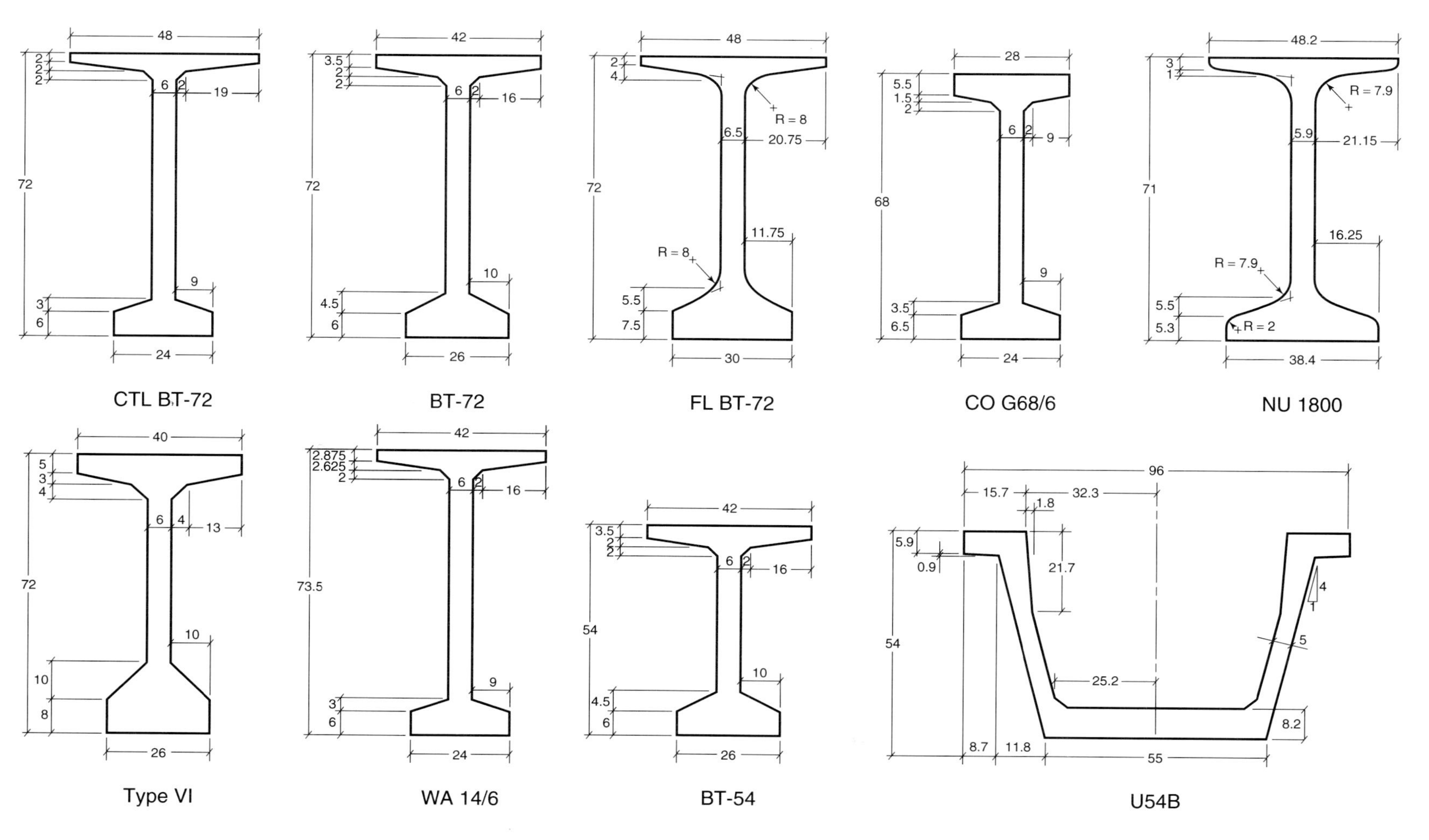

Figure 4.6 Cross Sections of Optimized Bridge Girder Sections [See Table 4.4(b)]

$$S^t \geq \frac{(1-\gamma)M_D + M_{SD} + M_L}{\gamma f_{ti} - f_c}$$

$$\geq \frac{(1-0.82)5{,}400{,}000 + 6{,}480{,}000}{0.82 \times 184 \times 2{,}250} = 3{,}104 \text{ in}^3 \text{ } (50{,}860 \text{ cm}^3)$$

$$S_b \geq \frac{(1-\gamma)M_D + M_{SD} + M_L}{f_t - \gamma f_{ci}}$$

$$\geq \frac{(1-0.82)\,5{,}400{,}000 + 6{,}480{,}000}{849 + (0.82 \times 2{,}250)} = 2{,}766 \text{ in.}^3 \text{ } (45{,}330 \text{ cm})^3$$

From the PCI design handbook, select a nontopped normal weight concrete double-T 12 DT 34 168-D1, since it has the bottom-section modulus value S_b closest to the required value.

The section properties of the concrete are as follows:

$$A_c = 978 \text{ in.}^2 \qquad c_t = 8.23 \text{ in.}$$

$$I_c = 86{,}072 \text{ in.}^4 \qquad c_b = 25.77 \text{ in.}$$

$$r^2 = \frac{I_c}{A_c} = 88.0 \text{ in.}^2 \qquad e_c = 22.02 \text{ in.}$$

$$S^t = 10{,}458 \text{ in.}^3 \qquad e_e = 12.77 \text{ in.}$$

$$S_b = 3{,}340 \text{ in.}^3 \qquad W_D = 1{,}019 \text{ plf}$$

$$\frac{V}{S} = 2.39 \text{ in.}$$

Design of Strands and Check of Stresses. From Figure 4.7, the assumed self-weight is close to the actual self-weight. Hence, use

$$M_D = \frac{1{,}019}{1{,}000} \times 5{,}400{,}000 = 5{,}502{,}600 \text{ in.-lb}$$

$$f_{pi} = 0.70 \times 270{,}000 = 189{,}000 \text{ psi}$$

$$f_{pe} = 0.82 f_{pi} = 0.82 \times 189{,}000 = 154{,}980 \text{ psi}$$

(a) *Analysis of Stresses at Transfer.* From Equation 4.1a,

$$f^t = -\frac{P_i}{A_c}\left(1 - \frac{ec_t}{r^2}\right) - \frac{M_D}{S^t} \leq f_{ti} = 184 \text{ psi}$$

Then

$$184 = -\frac{P_i}{978}\left(1 - \frac{22.02 \times 8.23}{88.0}\right) - \frac{5{,}502{,}600}{10{,}458}$$

$$P_i = (184 + 526.16)\frac{978}{1.06} = 655{,}223 \text{ lb.}$$

$$\text{Required number of tendons} = \frac{655{,}223}{189{,}000 \times 0.153} = 22.66\ \tfrac{1}{2}\text{-in. dia. tendons.}$$

Try sixteen $\frac{1}{2}$in. dia. strands for the standard section:

$$A_{ps} = 16 \times 0.153 = 2.448 \text{ in}^2 \text{ } (15.3 \text{ cm}^2)$$

$$P_i = 2.448 \times 189{,}000 = 462{,}672 \text{ lb } (2{,}058 \text{ kN})$$

$$P_e = 2.448 \times 154{,}980 = 379{,}391 \text{ lb } (1{,}688 \text{ kN})$$

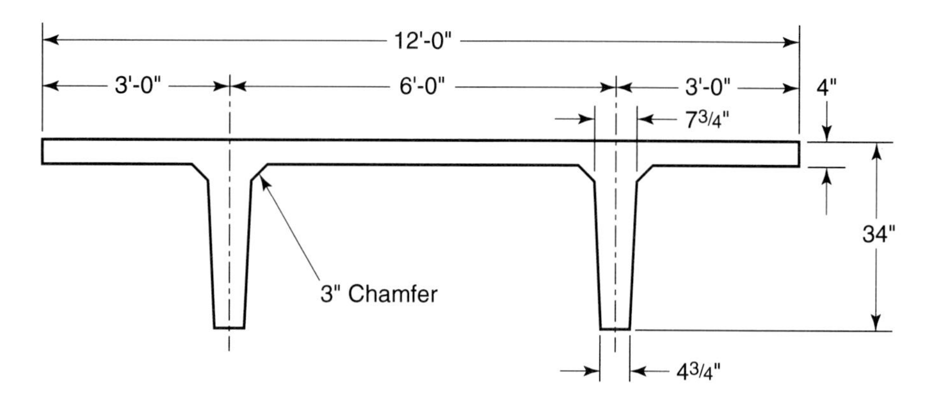

Figure 4.7 Double-T Section

(b) *Analysis of Stresses at Service Load at Midspan*

$$P_e = 379{,}391 \text{ lb}$$

$$M_{SD} = \frac{100(60)^2 12}{8} = 540{,}000 \text{ in.-lb (61 kN-m)}$$

$$M_L = \frac{1{,}100(60)^2 12}{8} = 5{,}940{,}000 \text{ in.-lb (788 kN-m)}$$

Total moment $M_T = M_D + M_{SD} + M_L = 5{,}502{,}600 + 6{,}480{,}000$
$= 11{,}982{,}600$ in.-lb (1,354 kN-m)

From Equation 4.3a,

$$f^t = -\frac{P_e}{A_c}\left(1 - \frac{ec_t}{r^2}\right) - \frac{M_T}{S^t}$$

$$= -\frac{379{,}391}{978}\left(1 - \frac{22.02 \times 8.23}{88.0}\right) - \frac{11{,}982{,}600}{10{,}458}$$

$$= +411 - 1146 = -735 < f_c = -2{,}250 \text{ psi, O.K.}$$

From Equation 4.3b,

$$f_b = -\frac{P_e}{A_c}\left(1 + \frac{ec_b}{r^2}\right) + \frac{M_T}{S_b}$$

$$= -\frac{379{,}391}{978}\left(1 + \frac{22.02 \times 25.77}{88.0}\right) + \frac{11{,}982{,}600}{3{,}340}$$

$$= -2{,}889 + 3{,}587 = +698 \text{ psi (T)} < f_t = +849 \text{ psi, O.K.}$$

(c) *Analysis of Stresses at Support Section*

$$e_e = 12.77 \text{ in. (324 mm)}$$

$$f_{ti} = 6\sqrt{f'_{ci}} = 6\sqrt{3{,}750} \cong 367 \text{ psi}$$

$$f_t = 12\sqrt{f'_c} = 12\sqrt{5{,}000} = 849 \text{ psi}$$

(i) *At Transfer*

$$f^t = -\frac{462{,}672}{978}\left(1 - \frac{12.77 \times 8.23}{88.0}\right) - 0 = +92 \text{ psi } (T)$$

$$f_b = -\frac{462{,}672}{978}\left(1 + \frac{12.77 \times 25.77}{88.0}\right) + 0 = -2{,}240 \text{ psi } (C)$$

$$< f_{ci} = -2{,}250 \text{ psi, O.K.}$$

If $f_b > f_{ci}$, the support eccentricity has to be changed.

(ii) *At Service Load*

$$f^t = -\frac{379{,}391}{978}\left(1 - \frac{12.77 \times 8.23}{88.0}\right) - 0 = +75 \text{ psi } (T) \; < f_t = 849 \text{ psi, O.K.}$$

$$f_b = \frac{379{,}391}{978}\left(1 + \frac{12.77 \times 25.77}{88.0}\right) + 0 = -1{,}840 \text{ psi } (C)$$

$$< f_c = -2{,}250 \text{ psi, O.K.}$$

Adopt the section for service-load conditions using sixteen $\frac{1}{2}$-in. (1.7 mm) strands with midspan eccentricity $e_c = 22.02$ in. (560 mm) and end eccentricity $e_e = 12.77$ in. (324 mm).

4.3.2 Variable Tendon Eccentricity with No Height Limitation

Example 4.2

Design an I-section for a beam having a 65-ft (19.8 m) span to satisfy the following section modulus values: Use the same allowable stresses as in Example 4.1.

$$\text{Required } S^t = 3570 \text{ in}^3 \text{ } (58{,}535 \text{ cm}^3)$$

$$\text{Required } S_b = 3{,}780 \text{ in}^3 \text{ } (61{,}940 \text{ cm}^3)$$

Photo 4.2 Crack development in prestressed T-beam (Nawy et al.).

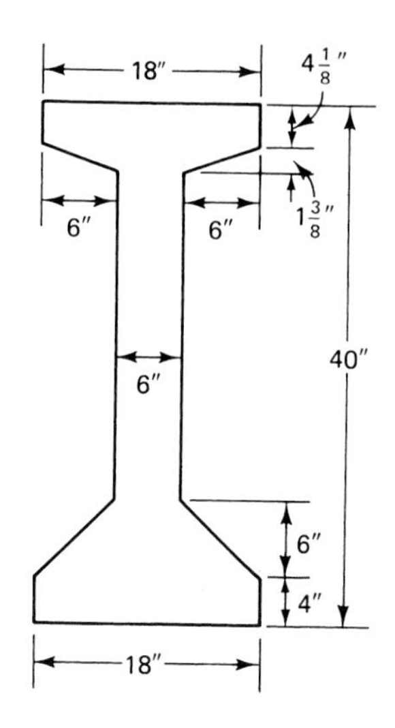

Figure 4.8 I-beam section in Example 4.2.

Since the section moduli at the top and bottom fibers are almost equal, a symmetrical section is adequate. Next, analyze the section in Figure 4.8 chosen by trial and adjustment.

Analysis of Stresses at Transfer. From Equation 4.4d,

$$\bar{f}_{ci} = f_{ti} - \frac{c_t}{h}(f_{ti} - f_{ci})$$

$$= +184 - \frac{21.16}{40}(+184 + 2{,}250) \cong -1{,}104 \text{ psi } (C)\ (7.6 \text{ MPa})$$

$$P_i = A_c\bar{f}_{ci} = 377 \times 1{,}104 = 416{,}208 \text{ lb } (1{,}851 \text{ kN})$$

$$M_D = \frac{393(65)^2}{8} \times 12 = 2{,}490{,}638 \text{ in.-lb } (281 \text{ kN-m})$$

From Equation 4.4c, the eccentricity required at the section of maximum moment at midspan is

$$e_c = (f_{ti} - \bar{f}_{ci})\frac{S^t}{P_i} + \frac{M_D}{P_i}$$

$$= (184 + 1{,}104)\frac{3{,}572}{416{,}208} + \frac{2{,}490{,}638}{416{,}208}$$

$$= 11.05 + 5.98 = 17.04 \text{ in. } (433 \text{ mm})$$

Since $c_b = 18.84$ in., and assuming a cover of 3.75 in., try $e_c = 18.84 - 3.75 \cong 15.0$ in. (381 mm).

$$\text{Required area of tendons } A_p = \frac{P_i}{f_{pi}} = \frac{416{,}208}{189{,}000} = 2.2 \text{ in}^2\ (14.2 \text{ cm}^2)$$

$$\text{Number of tendons} = \frac{2.2}{0.153} = 14.38$$

Try thirteen $\frac{1}{2}$-in. tendons, $A_p = 1.99$ in.^{2} (12.8 cm^2), and an actual $P_i = 189{,}000 \times 1.99 =$ 3.76,110 lb (1,673 kN), and check the concrete extreme fiber stresses. From Equation 4.1a

$$f^t = -\frac{P_i}{A_c}\left(1 - \frac{ec_t}{r^2}\right) - \frac{M_D}{S^t}$$

$$= -\frac{376{,}110}{377}\left(1 - \frac{15.0 \times 21.16}{187.5}\right) - \frac{2{,}490{,}638}{3{,}340}$$

$$= +691.2 - 745.7 = -55 \text{ psi } (C), \text{ no tension at transfer, O.K.}$$

From Equation 4.1b

$$f_b = -\frac{P_i}{A_c}\left(1 + \frac{ec_b}{r^2}\right) + \frac{M_D}{S_b}$$

$$= -\frac{376{,}110}{377}\left(1 + \frac{15 \times 18.84}{187.5}\right) + \frac{2{,}490{,}638}{3{,}750}$$

$$= -2{,}501.3 + 664.2 = -1{,}837 \text{ psi } (C) < f_{ci} = 2{,}250 \text{ psi, O.K.}$$

Analysis of Stresses at Service Load. From Equation 4.3a

$$f^t = -\frac{P_e}{A_c}\left(1 - \frac{ec_t}{r^2}\right) - \frac{M_T}{S^t}$$

$$P_e = 13 \times 0.153 \times 154{,}980 = 308{,}255 \text{ lb } (1{,}371 \text{ kN})$$

$$\text{Total moment } M_T = M_D + M_{SD} + M_L = 2{,}490{,}638 + 7{,}605{,}000$$

$$= 10{,}095{,}638 \text{ in.-lb } (1{,}141 \text{ kN-m})$$

$$f^t = -\frac{308{,}225}{377}\left(1 - \frac{15.0 \times 21.16}{187.5}\right) - \frac{10{,}095{,}638}{3{,}340}$$

$$= +566.5 - 3{,}022.6 = -2{,}456 \text{ psi } (C) > f_c = -2{,}250 \text{ psi}$$

Hence, either enlarge the depth of the section or use higher strength concrete. Using $f'_c = 6{,}000$ psi,

$$f_c = 0.45 \times 6{,}000 = -2{,}700 \text{ psi, O.K.}$$

$$f_b = -\frac{P_e}{A_c}\left(1 + \frac{ec_b}{r^2}\right) + \frac{M_T}{S_b} = -\frac{308{,}255}{377}\left(1 + \frac{15.0 \times 18.84}{187.5}\right) + \frac{10{,}095{,}638}{3{,}750}$$

$$= 2{,}050 + 2{,}692.2 = 642 \text{ psi } (T), \text{ O.K.}$$

Check Support Section Stresses

$$\text{Allowable } f'_{ci} = 0.75 \times 6{,}000 = 4{,}500 \text{ psi}$$

$$f_{ci} = 0.60 \times 4{,}500 = 2{,}700 \text{ psi}$$

$$f_{ti} = 3\sqrt{f'_{ci}} = 201 \text{ psi for midspan}$$

$$f_{ci} = 6\sqrt{f'_{ci}} = 402 \text{ psi for support}$$

$$f_c = 0.45 f'_c = 2{,}700 \text{ psi}$$

$$f_{t1} = 6\sqrt{f'_c} = 465 \text{ psi}$$

$$f_{t2} = 12\sqrt{f'_c} = 930 \text{ psi}$$

(a) *At Transfer.* Support section compressive fiber stress.

$$f_b = -\frac{P_i}{A_c}\left(1 + \frac{ec_b}{r^2}\right) + 0$$

$$P_i = 376{,}110 \text{ lb}$$

or

$$-2{,}700 = -\frac{376{,}110}{377}\left(1 + \frac{e \times 18.84}{187.5}\right)$$

so that

$$e_e = 16.98 \text{ in.}$$

Accordingly, try $e_e = 12.49$ in.:

$$f^t = -\frac{376{,}110}{377}\left(1 - \frac{12.49 \times 21.16}{187.5}\right) - 0$$

$$= 409 \text{ psi } (T) > f_{ti} = 402 \text{ psi}$$

$$f_b = 2250 \text{ psi}$$

Thus, use mild steel at the top fibers at the support section to take all tensile stresses in the concrete, or use a higher strength concrete for the section, or reduce the eccentricity.

(b) *At Service Load*

$$f^t = -\frac{308{,}255}{377}\left(1 - \frac{12.49 \times 21.16}{187.5}\right) - 0 = 335 \text{ psi } (T) < 930 \text{ psi, O.K.}$$

$$f_b = -\frac{308{,}255}{377}\left(1 + \frac{12.49 \times 18.84}{187.5}\right) + 0 = -1{,}844 \text{ psi } (C) < -2{,}700 \text{ psi, O.K.}$$

Hence, adopt the 40-in. (102-cm)-deep I-section prestressed beam of f'_c equal to 6,000 psi (41.4 MPa) normal-weight concrete with thirteen $\frac{1}{2}$-in. tendons having midspan eccentricity $e_c = 15.0$ in. (381 mm) and end section eccentricity $e_e = 12.5$ in. (318 m).

An alternative to this solution is to continue using $f'_c = 5{,}000$ psi, but change the number of strands and eccentricities.

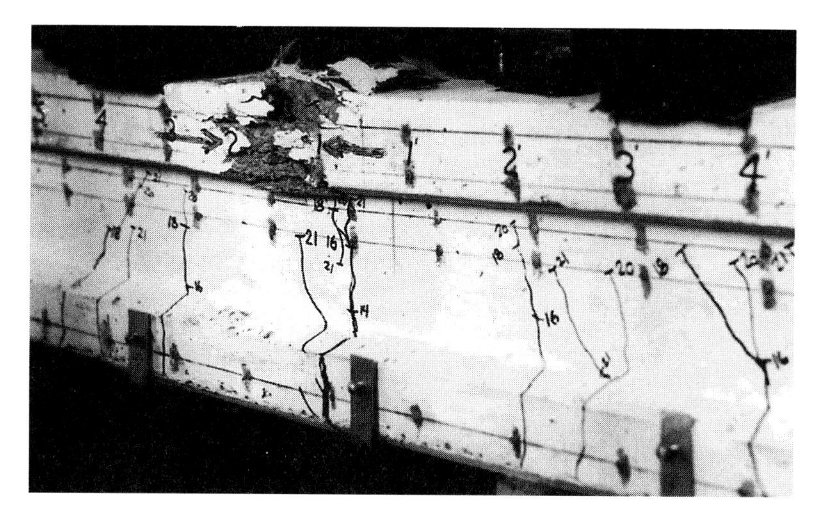

Photo 4.3 Prestressed beam at failure. Note crushing of concrete on top fibers (Nawy, Potyondy, et al.).

4.3.3 Constant Tendon Eccentricity

Example 4.3

Solve Example 4.2 assuming that the prestressing tendon has constant eccentricity. Use $f'_c = 5{,}000$ psi (34.5 MPa) normal-weight concrete, permitting a maximum concrete tensile stress $f_t = 12\sqrt{f'_c} = 849$ psi.

Solution: Since the tendon has constant eccentricity, the dead-load and superimposed dead- and live-load moments at the support section of the simply supported beam are zero. Hence, the support section controls the design. The required section modulus at the support, from Equation 4.5a, is

$$S^t \geq \frac{M_D + M_{SD} + M_L}{\gamma f_{ti} - f_c}$$

$$S_b \geq \frac{M_D + M_{SD} + M_L}{f_t - \gamma f_{ci}}$$

Assume $W_D = 425$ plf. Then

$$M_D = \frac{425 \times (65)^2}{8} \times 12 = 2{,}693{,}438 \text{ in.-lb } (304 \text{ kN-m})$$

$$M_{SD} + M_L = 7{,}605{,}000 \text{ in.-lb } (859 \text{ kN-m})$$

Thus, the total moment $M_T = 10{,}298{,}438$ in.-lb (1,164 kN-m), and we also have

$$\text{Allowable } f_{ci} = -2{,}250 \text{ psi}$$

$$f'_{ci} = -3{,}750 \text{ psi}$$

$$f_{ti} = 6\sqrt{f'_{ci}} \text{ for support section} = 367 \text{ psi}$$

$$f_c = -2{,}250 \text{ psi } (15.5 \text{ MPa})$$

$$f_t = +849 \text{ psi}$$

$$\gamma = 0.82$$

$$\text{Required } S^t = \frac{10{,}298{,}438}{0.82 \times 367 + 2{,}250} = 4{,}035.8 \text{ in}^3 \ (61{,}947 \text{ cm}^3)$$

$$\text{Required } S_b = \frac{M_D + M_{SD} + M_L}{f_t - \gamma f_{ci}} = \frac{10{,}298{,}438}{849 + 0.82 \times 2{,}250}$$

$$= 3{,}823.0 \text{ in}^3 \ (62{,}713 \text{ cm}^3)$$

First Trial. Since the required $S^t = 4{,}035.8$, which is greater than the available S^t in Example 4.2, choose the next larger I-section with $h = 44$ in. as shown in Figure 4.9. The section properties are:

$$I_c = 92{,}700 \text{ in}^4$$

$$r^2 = 228.9 \text{ in}^2$$

$$A_c = 405 \text{ in}^2$$

$$c_t = 23.03 \text{ in.}$$

$$S^t = 4{,}030 \text{ in}^3$$

$$c_b = 20.97 \text{ in}$$

$$S_b = 4{,}420 \text{ in}^3$$

$$W_D = 422 \text{ plf}$$

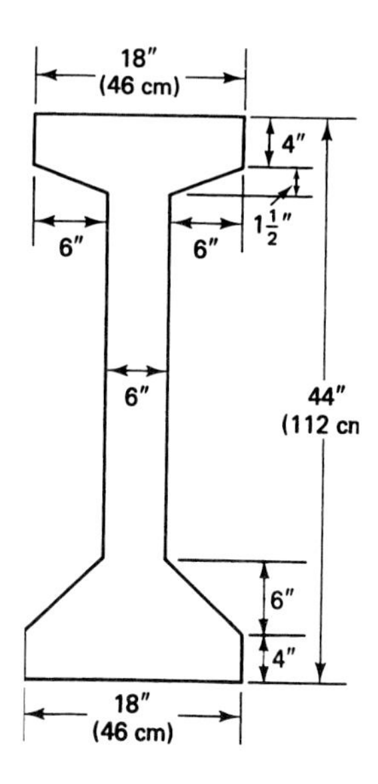

Figure 4.9 I-beam section in Example 4.3.

From Equation 4.5c, the required eccentricity at the critical section at the support is

$$e_e = (f_{ti} - \bar{f}_{ci})\frac{S^t}{P_i}$$

where

$$\bar{f}_{ci} = f_{ti} - \frac{c_t}{h}(f_{ti} - f_{ci})$$

$$= 367 - \frac{23.03}{44}(367 + 2{,}250) = -1{,}002 \text{ psi (6.9 MPa)}$$

and

$$P_i = A_c\bar{f}_{ci} = 405 \times 1{,}002 = 405{,}810 \text{ lb (1,805 kN)}$$

Hence,

$$e = (367 + 1{,}002)\frac{4{,}030}{405{,}810} = 13.60 \text{ in. (346 mm)}$$

The required prestressed steel area is

$$A_p = \frac{P_i}{f_{pi}} = \frac{405{,}810}{189{,}000} = 2.15 \text{ in}^2 \text{ (14.4 cm}^2\text{)}$$

So we try $\frac{1}{2}$in. tendons. The required number of tendons is 2.15/0.153 = 14.05. Accordingly, use fourteen $\frac{1}{2}$in. (12.7 mm) tendons. As a result,

$$P_i = 14 \times 0.153 \times 189{,}000 = 404{,}838 \text{ lb (1,801 kN)}$$

(a) *Analysis of Stresses at Transfer at End Section.* From Equation 4.1a,

$$f^t = -\frac{P_i}{A_c}\left(1 - \frac{ec_t}{r^2}\right) - \frac{M_D}{S^t} = -\frac{404{,}838}{405}\left(1 - \frac{13.60 \times 23.03}{228.9}\right) - 0$$

$$= +368.2 \text{ psi } (T) \cong f_{ti} = 367, \text{ O.K.}$$

From Equation 4.2b,

$$f_b = -\frac{P_i}{A_c}\left(1 + \frac{ec_b}{r^2}\right) + \frac{M_D}{S_b} = -\frac{404{,}838}{405}\left(1 + \frac{13.6 \times 20.97}{228.9}\right) + 0$$

$$= -2{,}245 \text{ psi } (C) \cong f_{ci} = -2{,}250, \text{ O.K.}$$

(b) *Analysis of Final Service-Load Stresses at Support*

$$P_e = 14 \times 0.153 \times 154{,}980 = 331{,}967 \text{ lb } (1{,}477 \text{ kN})$$

$$\text{Total moment } M_T = M_D + M_{SD} + M_L = 0$$

From Equation 4.3a,

$$f^t = -\frac{P_e}{A_c}\left(1 - \frac{ec_t}{r^2}\right) - \frac{M_T}{S^t}$$

$$= -\frac{331{,}967}{405}\left(1 - \frac{13.60 \times 23.03}{228.9}\right) - 0 = 302 \text{ psi } (T) < f_t = 849 \text{ psi, O.K.}$$

This is also applicable to midspan since eccentricity e is constant. From Equation 4.3b

$$f_b = -\frac{P_e}{A_c}\left(1 + \frac{ec_b}{r^2}\right) + \frac{M_T}{S_b}$$

$$= -\frac{331{,}967}{405}\left(1 + \frac{13.60 \times 20.97}{228.9}\right) + 0$$

$$= -1{,}841 \text{ psi } (12.2 \text{ MPa}) (C) < f_c = -2{,}250 \text{ psi, O.K.}$$

(c) *Analysis of Final Service-Load Stresses at Midspan.* From before, the total moment $M_T = M_D + M_{SD} + M_L = 10{,}298{,}438$ in.-lb. So the extreme concrete fiber stress due to M_T is

$$f_1^t = \frac{M_T}{S^t} = -\frac{10{,}298{,}438}{4{,}030} = -2{,}555 \text{ psi } (C) (17.6 \text{ MPa})$$

$$f_{1b} = \frac{M_T}{S_b} = \frac{10{,}298{,}438}{4{,}420} = +2{,}330 \text{ psi } (T) (16.1 \text{ MPa})$$

Hence, the final midspan fiber stresses are

$$f^t = +302 - 2{,}555 = -2{,}253 \text{ psi } (C) \cong f_c = -2{,}250 \text{ psi, accept}$$

$$f_b = -1{,}841 + 2{,}330 = +489 \text{ psi } (T) < f_t = 849 \text{ psi, O.K.}$$

Consequently, accept the trial section with a constant eccentricity $e = 13.60$ in. (345 mm) for the fourteen $\frac{1}{2}''$ (12.7 mm dia.) tendons.

4.4 PROPER SELECTION OF BEAM SECTIONS AND PROPERTIES

4.4.1 General Guidelines

Unlike steel-rolled sections, prestressed sections are not yet fully standardized. In most cases, the design engineer has to select the type of section to be used in the particular project. In the majority of simply supported beam designs, the distance between the cgc and cgs lines, viz., the eccentricity e, is proportional to the required prestressing force.

Since the midspan moment usually controls the design, the larger the eccentricity at midspan the smaller is the needed prestressing force, and consequently the more economical is the design. For a large eccentricity, a large concrete area at the top fibers is needed. Hence, a T-section or a wide-flange I-section becomes suitable. The end section is usually solid in order to avoid large eccentricities at planes of zero moment, and also in order to increase the shear capacity of the support section, and prevent anchorage zone failures.

Another popular section in wide use is the double-T-section. This section adds the advantages of the single-T-section to its own ease of handling and erection inherent in its stability. Figure 4.10 shows typical sections in general usage. Other sections such as hollow-core slabs and nonsymmetrical sections are also commonly used. Note that flanged sections can replace rectangular solid sections of the same depth without any loss of flexural strength. Rectangular sections, however, are used as short-span supporting girders or ledger beams.

I-sections are used as typical floor beams with composite slab topping action in long-span parking structures. T-sections with heavy bottom flanges, such as that in Figure 4.10(d), are generally used in bridge structures. Double-T-sections are widely used in floor systems in buildings and also in parking structures, particularly because of the composite action advantage of the top wide flange, which is 10 to 15 feet wide in many cases.

Hollow-core cast and extruded sections are shallow one-way beam strips that serve as easily erectable floor slabs. Large, hollow box girders are used as bridge girders for very large spans in what are known as *segmental bridge deck systems.* These segmental girders have large torsional resistance, and their flexural strength-to-weight ratio is relatively higher than in other types of prestressing systems.

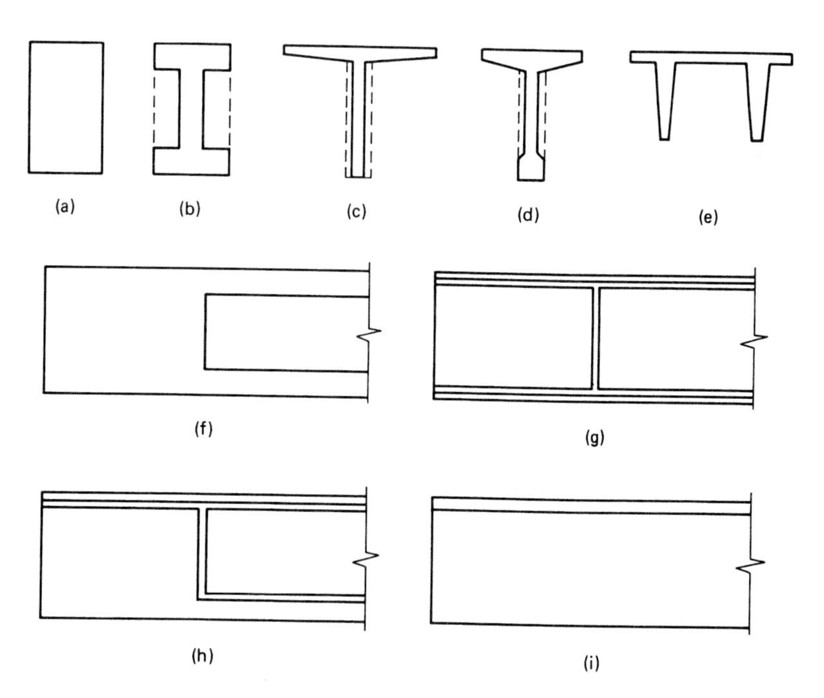

Figure 4.10 Typical prestressed concrete sections. (a) Rectangular beam section. (b) I-beam section. (c) T-beam section. (d) T-section with heavy bottom flange. (e) Double-T-section. (f) End part of beam in (b). (g) End part of beam in (c). (h) End part of beam in (d). (i) End part of beam in (e).

4.4.2 Gross Area, the Transformed Section, and the Presence of Ducts

In general, the gross cross-sectional area of the concrete section is adequate for use in the service-load design of prestressed sections. While some designers prefer refining their designs through the use of the transformed section in their solutions, the accuracy gained in accounting for the contribution of the area of the reinforcement to the stiffness of the concrete section is normally not warranted. In post-tensioned beams, where ducts are grouted, the gross cross section is still adequate for all practical design considerations. It is only in cases of large-span bridge and industrial prestressed beams, where the area of the prestressing reinforcement is large, that the transformed section or the net concrete area excluding the duct openings has to be used.

4.4.3 Envelopes for Tendon Placement

The tensile stress in the extreme concrete fiber under service-load conditions cannot exceed the maximum allowable by codes such as the ACI, PCI, AASHTO, or CEB-FIP. It is therefore important to establish the limiting zone in the concrete section, i.e., an envelope within which the prestressing force can be applied without causing tension in the extreme concrete fibers. From Equation 4.1a, we have

$$f_t = 0 = -\frac{P_i}{A_c}\left(1 - \frac{ec_t}{r^2}\right)$$

for the prestressing force part only, giving $e = r^2/c_t$. Hence, the *lower kern point*

$$k_b = \frac{r^2}{c_t} \tag{4.6a}$$

Similarly, from Equation 4.1b, if $f_b = 0$, $-e = r^2/c_b$, where the negative sign represents measurements upwards from the neutral axis, since positive eccentricity is positive downwards. Hence, the *upper kern point*

$$k_t = \frac{r^2}{c_b} \tag{4.6b}$$

Photo 4.4 Super CIDS offshore platform under tow to Arctic, Global Marine Development. (*Courtesy,* Ben C. Gerwick.)

From the determination of the upper and lower kern points, it is clear that

(a) If the prestressing force acts below the lower kern point, tensile stresses result at the extreme upper concrete fibers of the section.
(b) If the prestressing force acts above the upper kern point, tensile stresses result at the extreme lower concrete fibers of the section.

In a similar manner, kern points can be established for the right and left of the vertical line of symmetry of a section so that a central kern or core area for load application can be established, as Figure 4.11 shows for a rectangular section.

4.4.4 Advantages of Curved or Harped Tendons

Although straight tendons are widely used in precast beams of moderate span, the use of curved tendons is more common in in-situ-cast post-tensioned elements. Nonstraight tendons are of two types:

(a) Draped: gradually curved alignment such as parabolic forms, used in beams subjected primarily to uniformly distributed external loading.
(b) Harped: inclined tendons with a discontinuity in alignment at planes of concentrated load applications, used in beams subjected primarily to concentrated transverse loading.

Figures 4.12, 4.13, and 4.14 describe the alignment bending moment and stress distribution for beams that are prestressed with straight, draped, and harped tendons, respectively. These diagrams are intended to illustrate the economic advantages of the draped and harped tendons over the straight tendons. In Figure 4.12, at section 1-1, undesirable tensile stress in the concrete is shown at the top fibers. Section 1-1 in Figures 4.13 and 4.14 shows the uniform compression if the tendon acts at the cgc of the section at the support. Another advantage of draped and harped tendons is that they allow the prestressed beams to carry heavy loads because of the balancing effect of the vertical component of the prestressing nonstraight tendon. In other words, the required prestressing forces P_p for the parabolic tendon in Figure 4.13 and P_h for the harped tendon in Figure 4.14 are smaller at the midspan than the force required in the straight tendon of Figure 4.12. Hence, for the same stress level, a smaller number of strands are needed in the case of draped or harped tendons, and sometimes smaller concrete sections can be used with the resulting efficiency in the design. (Compare Examples 4.2 and 4.3 again.)

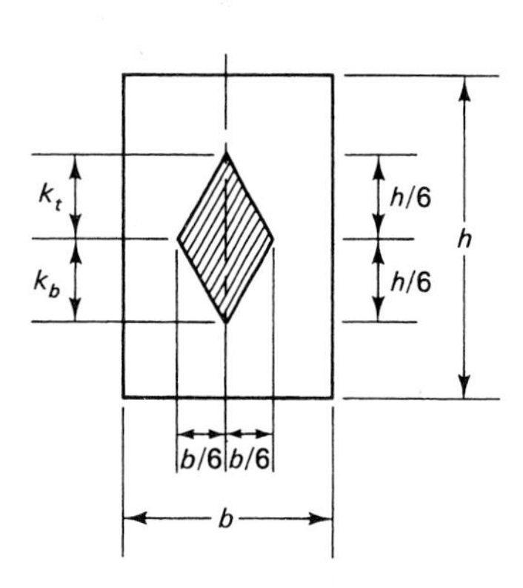

Figure 4.11 Central kern area for a rectangular section.

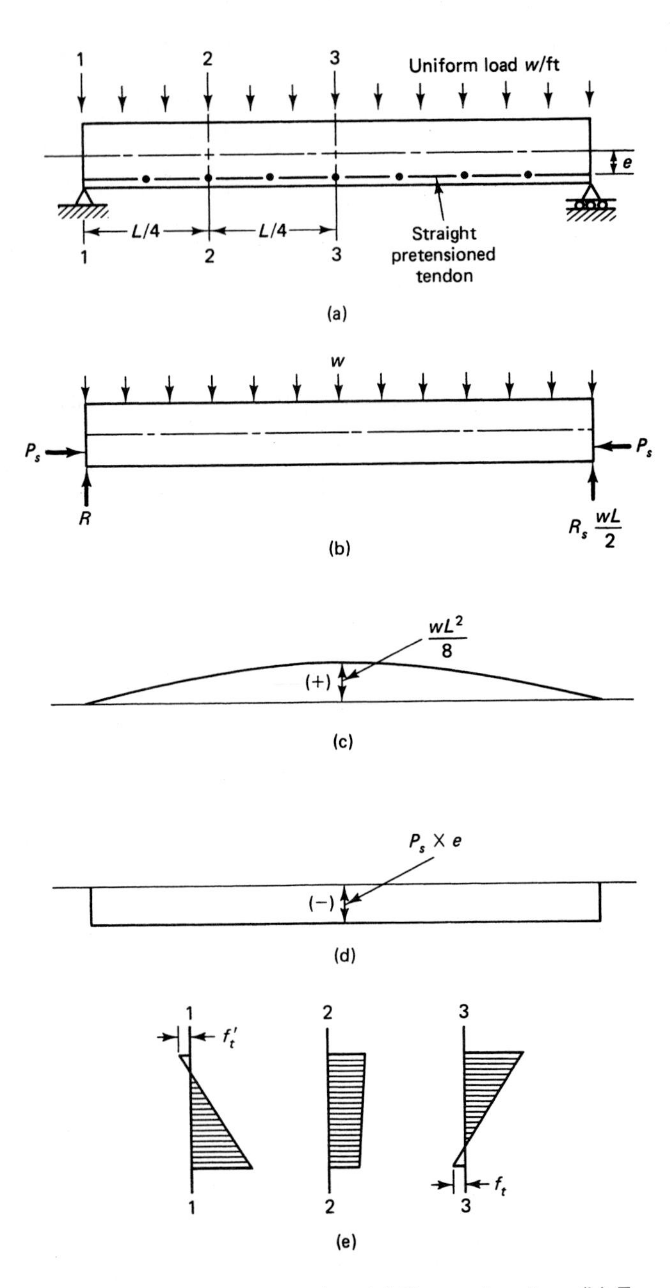

Figure 4.12 Beam with straight tendon. (a) Beam elevation. (b) Free-body diagram. (c) External load balancing moment diagram. (d) Prestressing force bending moment diagram. (e) Typical stress distribution in sections 1, 2, and 3 (Equation 4.3).

4.4.5 Limiting-Eccentricity Envelopes

It is desirable that the designed eccentricities of the tendon along the span be such that limited or no tension develops at the extreme fibers of the beam controlling sections. If it is desired to have no tension along the span of the beam in Figure 4.15 with the draped tendon, the controlling eccentricities have to be determined at the sections that follow along the span. If M_D is the self-weight dead-load moment and M_T is the total moment

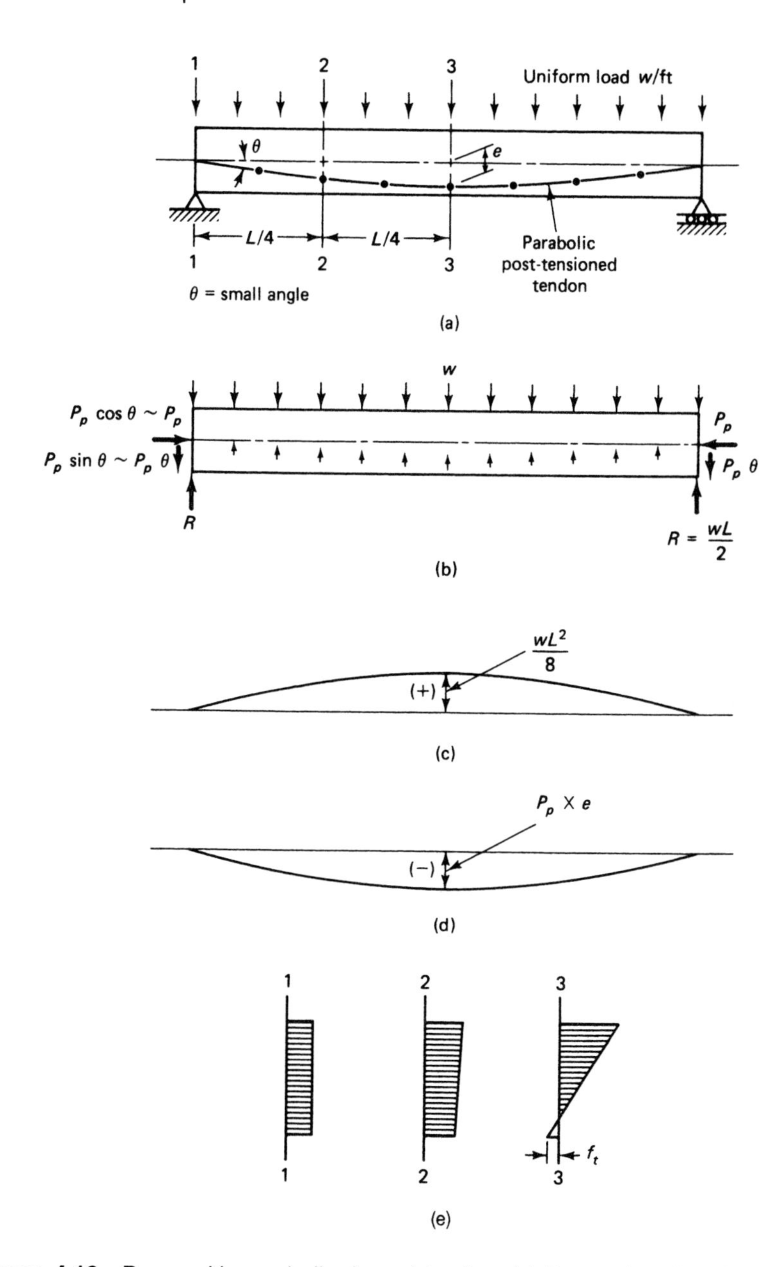

Figure 4.13 Beam with parabolic draped tendon. (a) Beam elevation. (b) Free-body diagram. (c) External load bending moment diagram. (d) Prestressing force bending moment diagram. (e) Typical stress distribution on sections 1, 2, and 3 (Equation 4.3).

due to all transverse loads, then the arms of the couple composed of the center-of-pressure line (C-line) and the center of the prestressing tendon line (cgs line) due to M_D and M_T are a_{min} and a_{max}, respectively, as shown in Figure 4.15.

Lower cgs Envelope. The minimum arm of the tendon couple is

$$a_{min} = \frac{M_D}{P_i} \tag{4.7a}$$

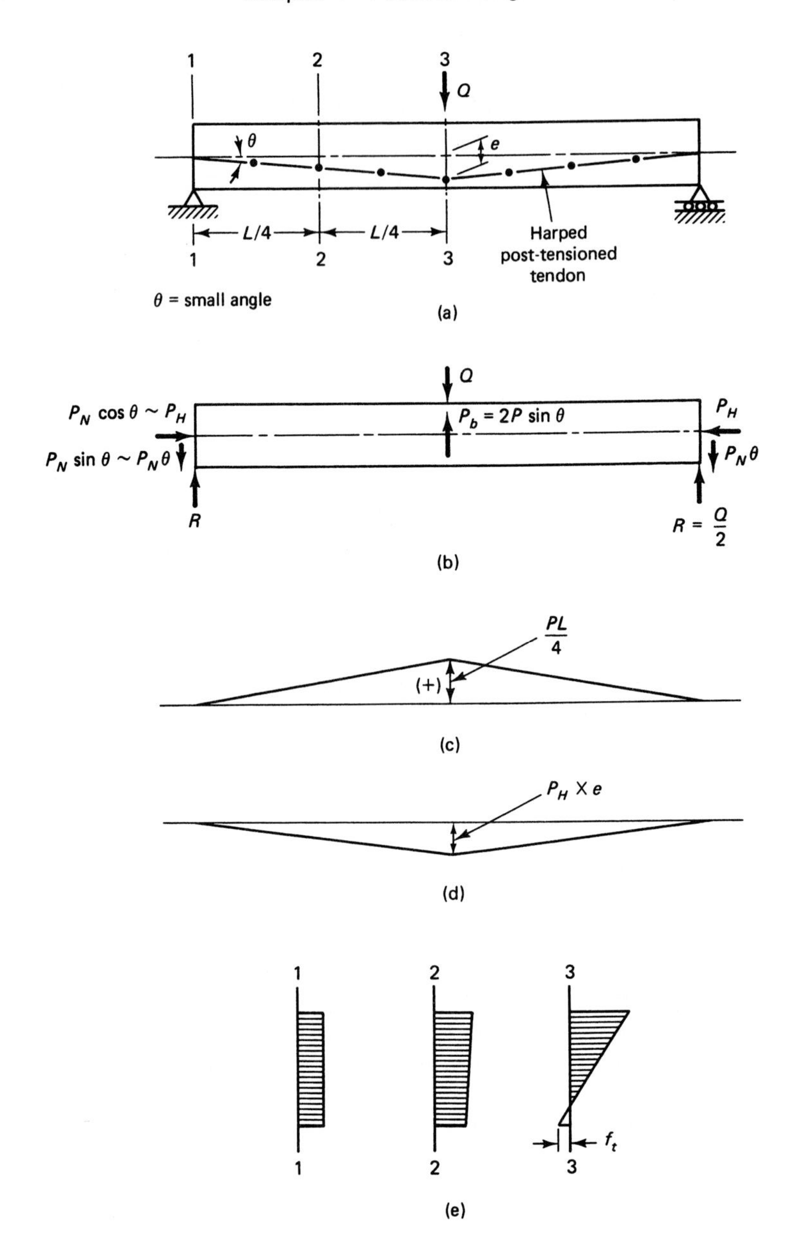

Figure 4.14 Beam with harped tendon. (a) Beam elevation. (b) Free-body diagram. (c) External load bending moment diagram. (d) Prestressing force bending moment diagram. (e) Typical stress distribution in sections 1, 2, and 3 (Equation 4.3).

This defines the maximum distance below the *bottom* kern where the cgs line is to be located so that the C-line does not fall *below* the *bottom* kern line, thereby preventing tensile stresses at the *top* extreme fibers. Hence, the limiting bottom eccentricity is

$$e_b = (a_{\min} + k_b) \tag{4.7b}$$

Upper cgs Envelope. The maximum arm of the tendon couple is

$$a_{\max} = \frac{M_T}{P_e} \tag{4.7c}$$

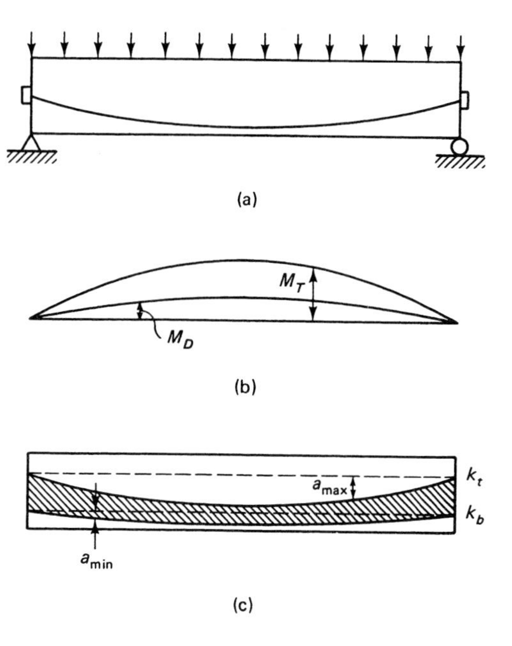

Figure 4.15 Cgs envelope determination. (a) One tendon location in beam. (b) Bending moment diagram. (c) Limiting cgs envelope.

This defines the minimum distance below the *top* kern where the cgs line is to be located so that the C-line does not fall *above* the *top* kern, thereby preventing tensile stresses at the *bottom* extreme fibers. Hence, the limiting top eccentricity is

$$e_t = (a_{\max} - k_t) \tag{4.7d}$$

Limited tensile stress is allowed in some codes both at transfer and at service-load levels. In such cases, it is possible to allow the cgs line to fall slightly outside the two limiting cgs envelopes described in Equations 4.7a and c.

If an additional eccentricity e'_b, e'_t is superimposed on the cgs-line envelope that results in limited tensile stress at both the top and bottom extreme concrete fibers, the additional top stress $f^{(t)}$ and bottom stress $f_{(b)}$ would be

$$f^{(t)} = \frac{P_i e'_b c_t}{I_c} \tag{4.8a}$$

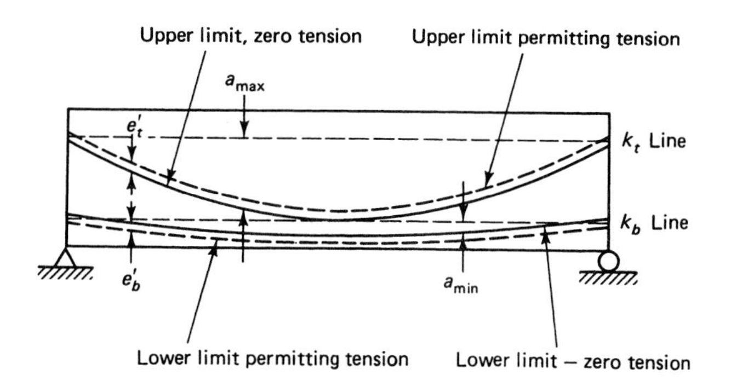

Figure 4.16 Envelope permitting tension in concrete extreme fibers.

and

$$f_{(b)} = \frac{P_e e_t' c_b}{I_c} \tag{4.8b}$$

where t and b denote the top and bottom fibers, respectively. From Equation 4.6, the additional eccentricities to be added to Equations 4.7b and d would be

$$e_b' = \frac{f^{(t)} A_c k_b}{P_i} \tag{4.9a}$$

and

$$e_t' = \frac{f_{(b)} A_c k_t}{P_e} \tag{4.9b}$$

The envelope allowing limited tension is shown in Figure 4.16. It should be noted that the upper envelope is outside the section, but the stresses are within the allowable limits, indicates a non-economical section. A change in eccentricity or prestressing force improves the design.

4.4.6 Prestressing Tendon Envelopes

Example 4.4

Suppose that the beam in example 4.2 is a post-tensioned bonded beam and that the prestressing tendon is draped in a parabolic shape. Determine the limiting envelope for tendon location such that the limiting concrete fiber stresses are at no time exceeded. Consider the midspan, quarter-span, and beam ends as the controlling sections. Assume that the magnitude of prestress losses is the same as in Example 4.2 but that $P_i = 549{,}423$ lb, $P_e = 450{,}526$ lb, $f_c' = 6000$ psi, $e_c = 13$ in. and $e_e = 6$ in.

Solution: The design moments of the I-beam in Example 4.2 can be summarized together with the section properties needed here:

$$P_i = 549{,}423 \text{ lb } (2{,}431 \text{ kN})$$

$$P_e = 450{,}526 \text{ lb } (2{,}004 \text{ kN})$$

$$M_D = 2{,}490{,}638 \text{ in.-lb } (281 \text{ kN-m})$$

$$M_{SD} + M_L = 7{,}605{,}000 \text{ in.-lb } (859 \text{ kN-m})$$

$$M_T = M_D + M_{SD} + M_L = 10{,}095{,}638 \text{ in.-lb } (1{,}141 \text{ kN-m})$$

$$A_c = 377 \text{ in}^2 \ (2{,}536 \text{ cm}^2)$$

$$f_c' = 6{,}000 \text{ psi}$$

$$r^2 = 187.5 \text{ in}^2 \ (1{,}210 \text{ cm}^2)$$

$$c_t = 21.16 \text{ in. } (537 \text{ mm})$$

$$c_b = 18.84 \text{ in. } (479 \text{ mm})$$

Since bending moments in this example are due to a uniformly distributed load, the shape of the bending moment diagram is parabolic, with the moment value being zero at the simply supported ends. Hence, quarter-span moments are

$$M_D = 0.75 \times 2{,}490{,}638 = 1{,}867{,}979 \text{ in.-lb } (211 \text{ kN-m})$$

$$M_T = 0.75 \times 10{,}095{,}638 = 7{,}571{,}729 \text{ in.-lb } (856 \text{ kN-m})$$

From Equations 4.6a and b, the kern point limits are

$$k_t = \frac{r^2}{c_b} = \frac{187.5}{18.84} = 9.95 \text{ in. (253 mm)}$$

$$k_b = \frac{r^2}{c_t} = \frac{187.5}{21.16} = 8.86 \text{ in. (225 mm)}$$

Lower Envelope

From Equation 4.7a, the maximum distance that the cgs line is to be placed below the *bottom* kern to prevent tensile stress at the top fibers is determined as follows:

(i) *Midspan*

$$a_{\min} = \frac{M_D}{P_i} = \frac{2{,}490{,}638}{549{,}423} = 4.53 \text{ in. (115 mm)}$$

giving

$$e_1 = k_b + a_{\min} = 8.86 + 4.53 = 13.39 \text{ in. (340 mm)}$$

(ii) *Quarter span*

$$a_{\min} = \frac{1{,}867{,}979}{549{,}423} = 3.40 \text{ in. (86 mm)}$$

giving

$$e_2 = 8.86 + 3.40 = 12.26 \text{ in. (311 mm)}$$

(iii) *Support*

$$a_{\min} = 0$$

giving

$$e_3 = 8.86 + 0 = 8.86 \text{ in. (225 mm)}$$

Upper Envelope

From Equation 4.7b, the maximum distance that the cgs line is to be placed below the *top* kern to prevent tensile stress at the bottom extreme fibers is determined as follows:

(i) *Midspan*

$$a_{\max} = \frac{M_T}{P_e} = \frac{10{,}095{,}638}{450{,}526} = 22.41 \text{ in. (569 mm)}$$

$$e_1 = a_{\max} - k_t = 22.41 - 9.95 = 12.46 \text{ in. (316 mm)}$$

Clear minimum cover = 3.0 in.
Note that e_1 cannot exceed c_b otherwise tendon is outside the section.

(ii) *Quarter span*

$$a_{\max} = \frac{7{,}571{,}729}{450{,}526} = 16.80 \text{ in. (427 mm)}$$

$$e_2 = 16.80 - 9.95 = 6.85 \text{ in. (174 mm)}$$

(iii) *Support*

$$a_{\max} = 0$$

$$e_3 = 0 - 9.95 = -9.95 \text{ in. } (-253 \text{ mm}) \quad \text{(9.95 in. above cgc line)}$$

Now, assume for practical purposes that the maximum fiber tensile stresses under working-load conditions for the purpose of constructing the cgs envelopes does not exceed $f_t = 6\sqrt{f'_c} = 465$ psi for both top and bottom fibers, since $f'_c = 6{,}000$ psi from Example 4.2. From Equation 4.9a, this additional eccentricity to add to the *lower* cgs envelope in order to allow limited tension at the *top* fibers is

$$e'_b = \frac{f^{(t)} A_c k_b}{P_i} = \frac{465 \times 377 \times 8.86}{549{,}423} = 2.83 \text{ in. (72 mm)}$$

Similarly, from Equation 4.9b, the additional eccentricity to add to the *upper* cgs envelope in order to allow limited tension at the *bottom* fibers is

$$e'_t = \frac{f_{(b)} A_c k_t}{P_e} = \frac{465 \times 377 \times 9.95}{450{,}526} = 3.87 \text{ in. (98 mm)}$$

We thus have the following summary of cgs envelope eccentricities:

	Zero tension, in.	Increment	Allowable tension, in.
Midspan			
Lower envelope	13.39	+2.83	16.22
Upper envelope	12.46	−3.87	8.59
Quarter span			
Lower envelope	12.26	+2.83	15.09
Upper envelope	6.86	−3.87	2.99
Support			
Lower envelope	8.86	+2.83	11.69
Upper envelope	−9.95	−3.87	−13.82

Actual midspan eccentricity $e_c = 13$ in. < 16.22 in.
Hence, tendon is inside envelope at midspan.
Actual support eccentricity $e_e = 6$ in. < 11.69 in.
Hence, tendon is also inside envelope at support.

Figure 4.17 illustrates the band of the cgs envelopes for both zero and limited tension in the concrete.

4.4.7 Reduction of Prestress Force Near Supports

As seen from Example 4.3 and Sections 4.4.3 and 4.4.5, straight tendons in pretensioned members can cause high-tensile stresses in the concrete extreme fibers at the support sections because of the absence of bending moment stresses due to self-weight and superimposed loads and the dominance of the moment due to the prestressing force alone. Two common and practical methods of reducing the stresses at the support section due to the prestressing force are:

1. Changing the eccentricity of some of the cables by raising them towards the support zone, as shown in Figure 4.18(a). This reduces the moment values.
2. Sheathing some of the cables by plastic tubing towards the support zone, as shown in Figure 4.18(b). This eliminates the prestress transfer of part of the cables at some distance from the support section of the simply supported prestressed beam.

Note that raised cables are also used in long-span post-tensioned prestressed beams, theoretically discontinuing part of the tendons where they are no longer needed

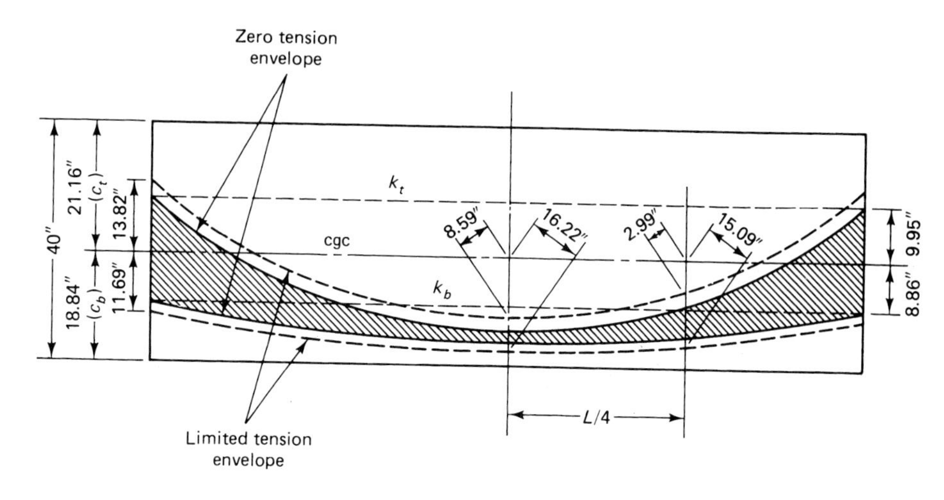

Figure 4.17 Cgs-line envelopes for the prestressing tendon (1 in. = 25.4 mm).

by raising them upwards. Additional frictional losses due to these additional curvatures have to be accounted for in the design (analysis) of the section.

4.5 END BLOCKS AT SUPPORT ANCHORAGE ZONES

4.5.1 Stress Distribution

A large concentration of compressive stress in the longitudinal direction occurs at the support section on a small segment of the face of the beam end, both in pretensioned and

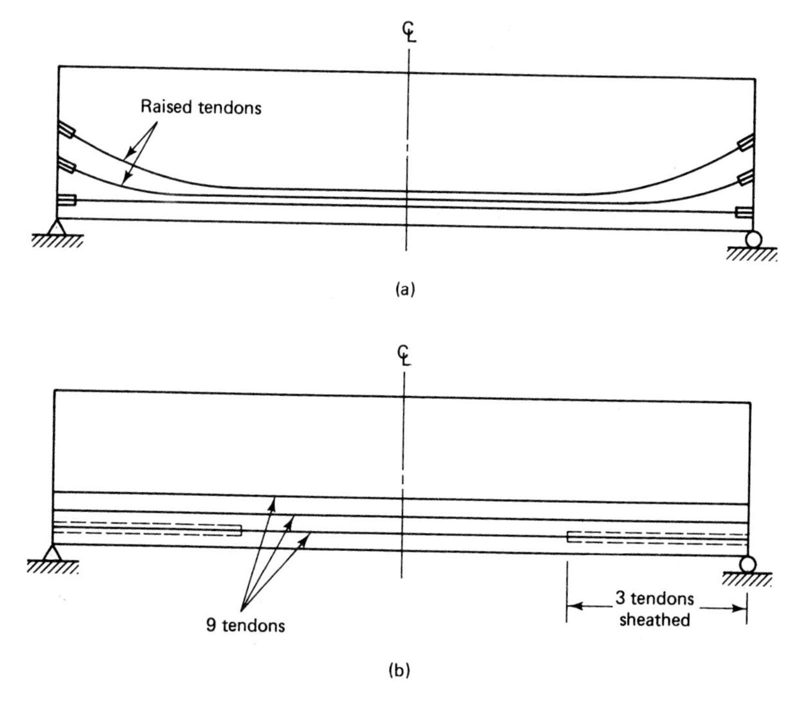

Figure 4.18 Reduction of prestressing force near supports. (a) Raising part of the tendons. (b) Sheathing part of the tendons.

post-tensioned beams, due to the large tendon prestressing forces. In the *pretensioned beams* the concentrated load transfer of the prestressing force to the surrounding concrete gradually occurs over a length l_t from the face of the support section until it becomes essentially uniform.

In *post-tensioned beams,* this manner of gradual load distribution and transfer is not possible since the force acts directly on the face of the end of the beam through bearing plates and anchors. Also, some or all of the tendons in the post-tensioned beams are raised or draped towards the top fibers through the web part of the concrete section.

As the nongradual transition of the longitudinal compressive stress from concentrated to linearly distributed produces high transverse *tensile* stresses in the vertical (transverse) direction, longitudinal bursting cracks develop at the anchorage zone. When the stresses exceed the modulus of rupture of the concrete, the end block has to split (crack) longitudinally unless appropriate vertical reinforcement is provided. The location of the concrete-bursting stresses and the resulting bursting cracks as well as the surface-spalling cracks would thus have to depend on the location and distribution of the horizontal concentrated forces applied by the prestressing tendons to the end bearing plates.

It is sometimes necessary to increase the area of the section towards the support by a gradual transmission of the web to a width at the support equal to the flange width, in order to accommodate the raised tendons [see Figure 4.19(a)]. Such an increase in the cross-sectional area does not contribute, however, to preventing bursting or spalling cracks, and has no effect on reducing the transverse tension in the concrete. In fact, both test results and the theoretical analysis of this three-dimensional stress problem demonstrate that the tensile stresses could increase.

Consequently, it is essential to provide the necessary anchorage reinforcement in the load transfer zone in the form of *closed* ties, stirrups, or anchorage devices enclosing all the main prestressing and mild nonprestressed longitudinal reinforcement. However, it is at the same time advisable to insert reinforcing vertical mats and confining hoops close to the end face behind the bearing places in the case of post-tensioned beams. If the design has to follow AASHTO requirements for bridges, properly reinforced end blocks are required. Typical stress contours of equal vertical stress based on three-dimensional

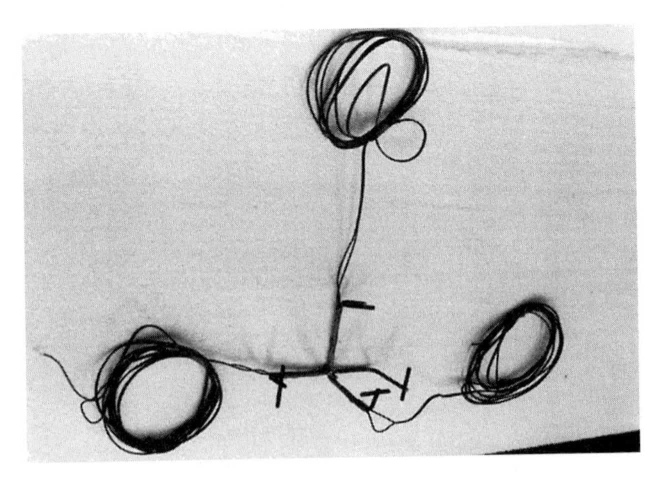

Photo 4.5 Three-dimensional instrumentation for end-block stress determination (Nawy et al.).

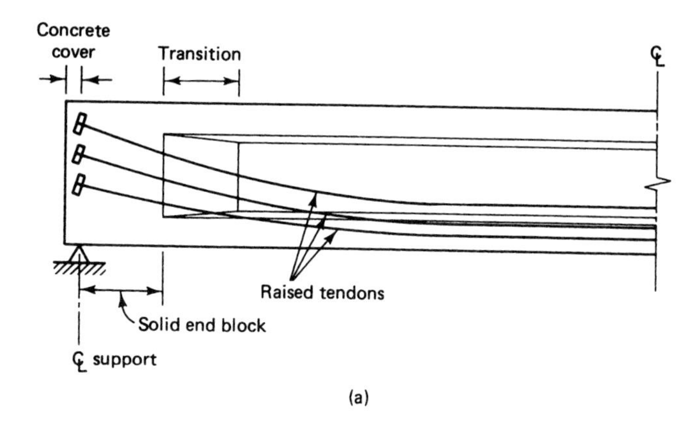

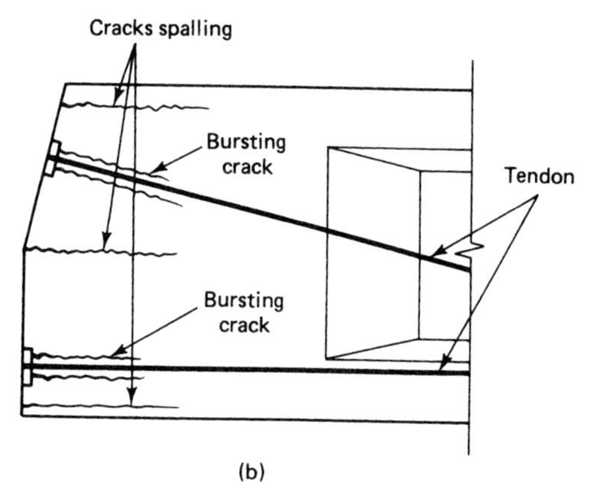

Figure 4.19 End anchorage zones for bonded tendons. (a) Transition to solid section at support. (b) End-zone bursting and spalling cracks.

analysis and test results from Refs. 4.5, 4.7, 4.28. An idealization of the tensile and compressive stress paths is shown in Figure 4.20.

4.5.2 Development and Transfer Length in Pretensioned Members and Design of Their Anchorage Reinforcement

As the jacking force is released in pretensioned members, the prestressing force is dynamically transferred through the bond interface to the surrounding concrete. The interlock or adhesion between the prestressing tendon circumference and the concrete over a finite length of the tendon gradually transfers the concentrated prestressing force to the entire concrete section at planes away from the end block and towards the midspan. The length of embedment determines the magnitude of prestress that can be developed along the span: the larger the embedment length, the higher is the prestress developed.

As an example for $\frac{1}{2}$-in. 7-wire strand, an embedment of 40 in. (102 cm) develops a stress of 180,000 psi (1,241 MPa), whereas an embedment of 70 in. (178 cm) develops a stress of 206,000 psi (1,420 MPa). From Figure 4.21, it is plain that the embedment length l_d that gives the full development of stress is a combination of the transfer length l_t and the flexural bond length l_f. These are given respectively by

$$l_t = \frac{1}{1000}\left(\frac{f_{pe}}{3}\right)d_b \qquad (4.10a)$$

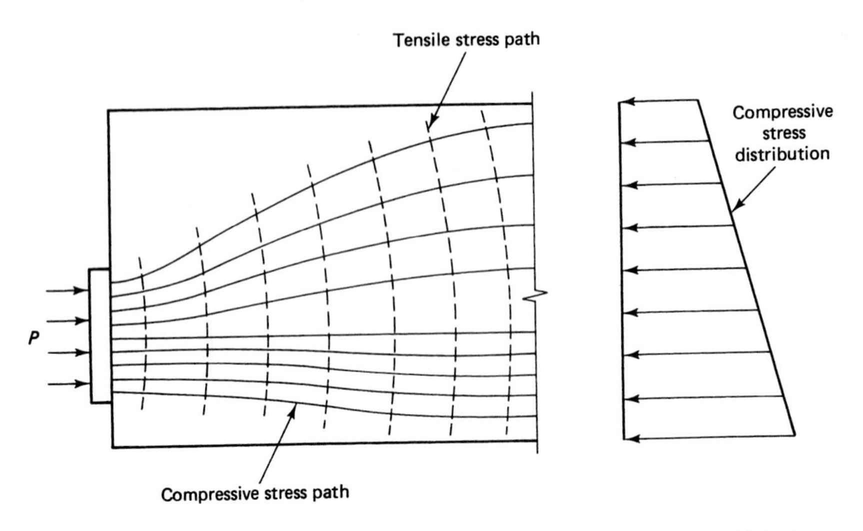

Figure 4.20 Idealized tensile and compressive stress paths at end blocks.

or

$$l_t = \frac{f_{pe}}{3000} d_b \tag{4.10b}$$

and

$$l_f = \frac{1}{1,000} (f_{ps} - f_{pe}) d_b \tag{4.10c}$$

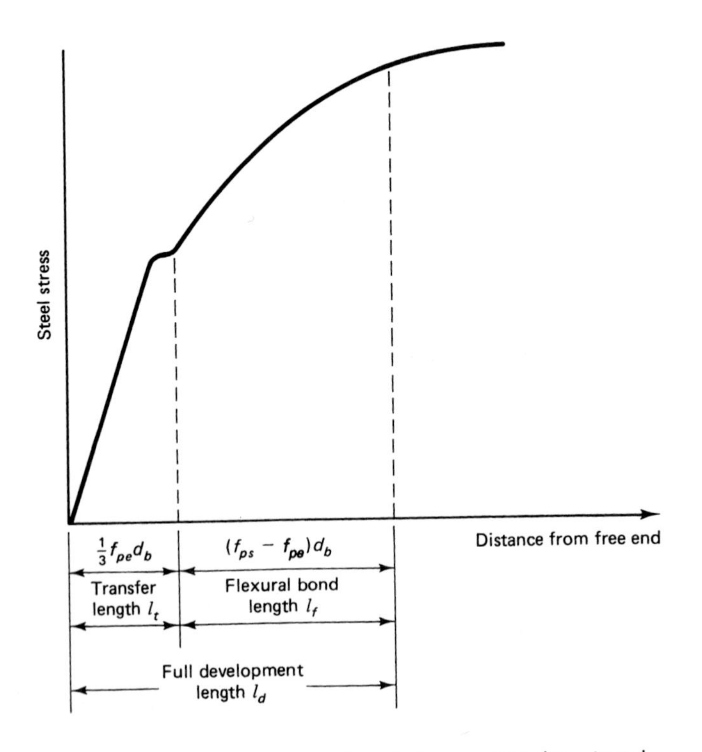

Figure 4.21 Development length for prestressing strand.

where f_{ps} = stress in prestressed reinforcement at nominal strength (psi)
f_{pe} = effective prestress after losses (psi)
d_b = nominal diameter of prestressing tendon (in.).

Combining Equations 4.10b and 4.10c gives

$$\text{Min } l_d = \frac{1}{1{,}000}\left(f_{ps} - \frac{2}{3}f_{pe}\right)d_b \tag{4.10d}$$

Equation 4.10d gives the minimum required development length for prestressing strands. If part of the tendon is sheathed towards the beam end to reduce the concentration of bond stresses near the end, the stress transfer in that zone is eliminated and an increased adjusted development length l_d is needed.

4.5.2.1 Design of Transfer Zone Reinforcement in Pretensioned Beams.

Based on laboratory tests, empirical expressions developed by Mattock et al. give the total stirrup force F as

$$F = 0.0106\frac{P_i h}{l_t} \tag{4.11}$$

where h is the pretensioned beam depth and l_t is the transfer length. If the average stress in a stirrup is taken as *half* the maximum permissible steel f_s, then $F = \frac{1}{2}A_t f_s$. Substituting this for F in Equation 4.11 gives

$$A_t = 0.021\frac{P_i h}{f_s l_t} \tag{4.12}$$

where A_t is the total area of the stirrups and $f_s \leq 20{,}000$ psi (138 MPa) for crack-control purposes.

4.5.2.2 Reinforcement Selection in Pretensioned Beams

Example 4.5

Design the anchorage reinforcement needed to prevent bursting or spalling cracks from developing in the beam of Example 4.2, pretensioned.

Solution:

$$P_i = 376{,}110 \text{ lb } (1{,}673 \text{ kN})$$

From Equation 4.12,

$$A_t = 0.021\frac{P_i h}{f_s l_t}$$

From Equation 4.10b, the transfer length is $l_t = (f_{pe}/3{,}000)d_b$. So since f_{pe} = 154,980 psi and $d_b = \frac{1}{2}$ in., we have

$$l_t = \frac{154{,}980}{3{,}000} \times 0.5 = 25.83 \text{ in. } (66 \text{ cm})$$

Now,

$$A_t = 0.021\frac{P_i h}{f_s l_t}$$

So since $f_s \leq 20{,}000$ psi, we get

$$A_t = 0.021 \frac{376{,}110 \times 40}{20{,}000 \times 25.83}$$

$$= 0.61 \text{ in.}^2 \; (3.9 \text{ cm}^2)$$

Trying #3 closed ties,

$$2 \times 0.11 = 0.22 \text{ in.}^2 \; (9.5 \text{ mm dia) ties}$$

$$\text{Min no. of stirrups} = \frac{0.61}{0.22} = 2.78$$

Use three #3 ties to provide the envelope for all the main longitudinal reinforcement. Wrap the tendons with helical steel wire through the development length, l_t, in order to effect good transfer.

4.5.3 Post-Tensioned Anchorage Zones: Linear Elastic and Strut-and-Tie Theories

The anchorage zone can be defined as the volume of concrete through which the concentrated prestressing force at the anchorage device spreads transversely to a linear distribution across the entire cross-section depth along the span (Ref. 4.5, 4.7, 4.12, 4.30). The length of this zone follows St. Venant's principle, namely, that the stress becomes uniform at an approximate distance ahead of the anchorage device equal to the depth, h, of the section. The entire prism which would have a transfer length, h, is the total anchorage zone.

This zone is thus composed of two parts:

1. *General Zone:* The general extent of the zone is identical to the total anchorage zone. Its length extent along the span is therefore equal to the section depth, h, in standard cases.
2. *Local Zone:* This zone is the insert prism of concrete surrounding and immediately ahead of the anchorage device and the confining reinforcement it contains. See the shaded area in Figure 4.22(c) and its magnification in Fig. 4.22(a). Also shown are the distribution of tensile and compressive stresses in the local zone and their stress

Photo 4.6 End block of post-tensioned I-beam at ultimate load (Nawy et al.).

Photo 4.7 Anchorage block instrumentation (Nawy et al.).

contours obtained from the finite element analysis of the Rutgers tests (Ref. 4.7). The length of the local zone has to be considered as the larger of either its maximum width or the length of the anchorage device confining reinforcement.

The confining reinforcement throughout the entire anchorage zone has to be so chosen as to prevent bursting and splitting which are the result of the high concentrated compressive forces transmitted through the anchorage devices. In addition, checks have to be made of the bearing stresses on the concrete in the local zone due to these high compressive forces to ensure that the allowable compressive bearing capacity of the concrete is never exceeded.

4.5.3.1 Design Methods for the General Zone

Essentially, three methods are applicable to the design of the anchorage zone.

1. *Linear Elastic Stress Analysis Approach Including Use of Finite Elements:* This involves computing the detailed state of stresses as linearly elastic. The application of the finite element method is somewhat limited by the difficulty of developing adequate models that can correctly model the cracking in the concrete (Ref. 4.30). Nevertheless, appropriate assumptions can always be made to get reasonable results.
2. *Equilibrium-Based Plasticity Approach such as the Strut-and-Tie Models:* The strut-and-tie method provides for idealizing the path of the prestressing forces as a truss structure with its forces following the usual equilibrium principles. The ultimate load predicted by this method is controlled by failure of any one of the component struts or ties. The method usually gives conservative results for this application.
3. *Approximate Methods:* These apply to rectangular cross sections without discontinuities.

4.5.3.2 Linear Elastic Analysis Method for Confining Reinforcement Determination

The anchorage zone is subjected to three levels of stress as seen in Figure 4.22(a) and the stress contour zones:

(a) High bearing stresses ahead of the anchorage devices. Proper confinement of the concrete is necessary in order to prevent the compressive failure of the compressive segment shown in the darkly shaded area of Figures 4.22(a) and 4.22(b).

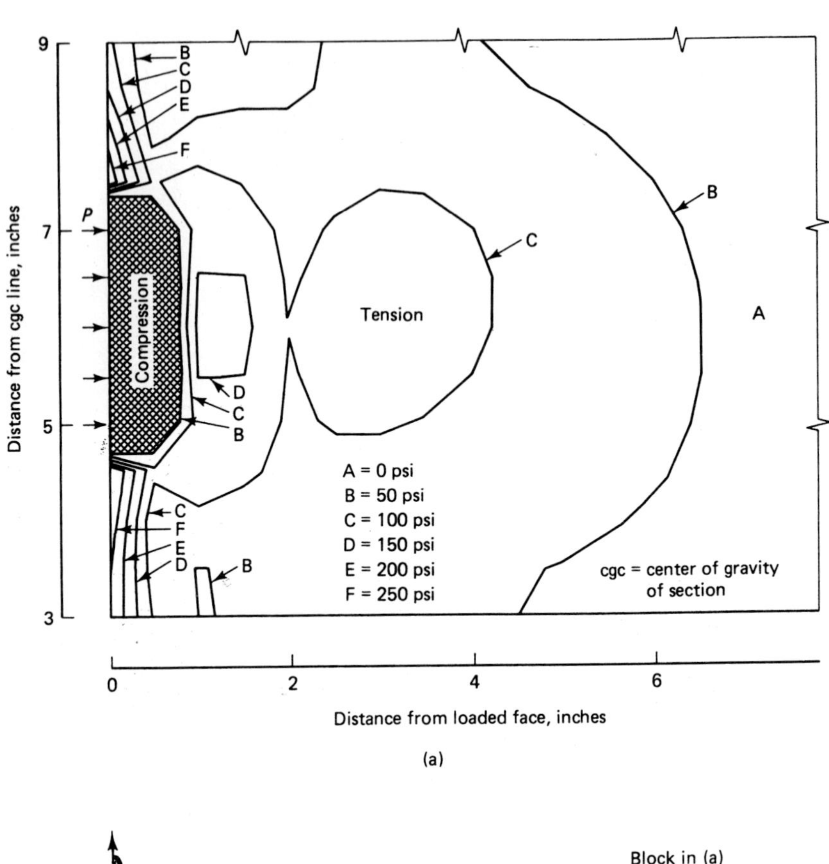

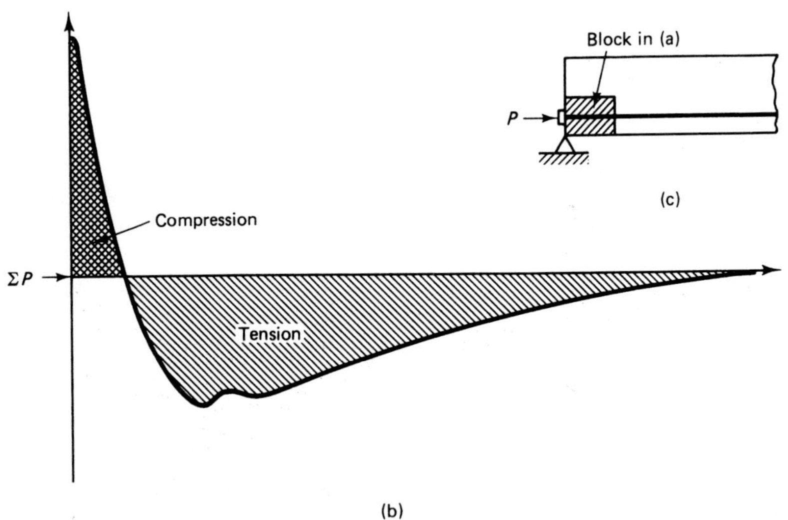

Figure 4.22 Principal tensile stress contours of equal vertical stress at anchorage zone (eccentricity $e_e = 6$ in.). (a) Contours of stress. (b) Stress distribution at 4.5 in. above base. (c) Segment of beam elevation. (Nawy et al., Refs. 4.7, 4.28)

(b) Extensive tensile-bursting stresses in the tension contour areas, normal to the tendon axis as shown in Figures 4.22(a) and (b) and in Figure 4.23(b).

(c) High compressive in the stress field areas marked D and E in Figure 4.22(a).

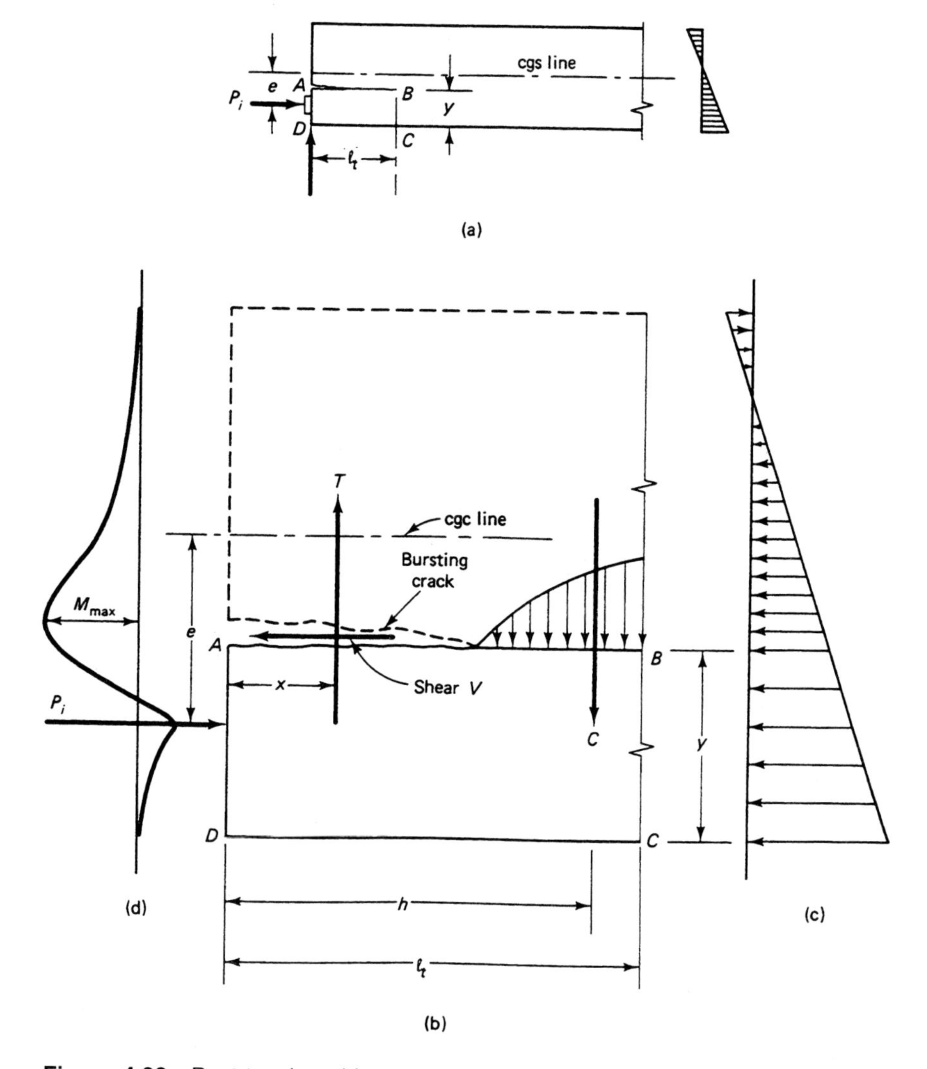

Figure 4.23 Post-tensioned beam end-block forces. (a) Beam elevation, showing transfer length l_t. (b) Free-body diagram ABCD. (c) Fiber stress distribution across beam block depth. (d) Moment values on crack surface AB for all possible locations y across beam block depth.

The following discussion illustrates that a linear elastic stress analysis can predict the cracking locations and give a reasonably reliable approximate estimate of the flow of stresses after cracking. The area of the tensile reinforcement is computed to carry the total tensile force obtained through integrating the tensile stresses in the concrete. In compressive stress regions, if the compressive force is very high, the provision of additional compressive reinforcement would become necessary.

On a parallel approach, a linearly elastic finite element analysis as shown in Figure 4.22 results in more accurate determination of the state of stresses in the anchorage zone. However, the process of computation is time-consuming and costly. The results can be limited because of the difficulty of developing adequate models that can correctly model the cracking in the concrete. A nonlinear finite element analysis to predict the post-cracking response could resolve this discrepancy. Yet, the design engineer expects less rigor and faster answers in the routine day-to-day office applications.

Figure 4.23 schematically illustrates the linearly elastic end block forces. It shows the end-block forces and the fiber stresses due to the prestressing force P_i, as well as the bending moment value for each possible crack height y above the beam bottom CD. The maximum moment value M_{max} determines the potential position of the horizontal bursting crack. This moment is resisted by the couple provided by the tensile force T of the vertical anchorage zone reinforcement and the compressive force C provided by the end-block concrete, while the horizontal shear force V at the crack split surface is resisted by the aggregate interlock forces. From practical observations, the vertical anchorage zone stirrups that provide the force T should be distributed over a zone width $h/2$ from the end-face of the beam such that X in Figure 4.23 can vary between $h/4$ and $h/5$.

From equilibrium of moments,

$$T = \frac{M_{\max}}{h - x} \tag{4.13}$$

and the total required area of vertical steel reinforcement becomes

$$A_t = \frac{T}{f_s} \tag{4.14}$$

where the steel stress f_s used in the calculation should not exceed 20,000 psi (138.5 MPa) for crack-width control purposes.

In summary and in lieu of a linear elastic finite element analysis, the procedure outlined can reasonably though less precisely give a detailed anchorage design as given in Example 4.6 part (a).

4.5.3.3 Strut-and-Tie Method for Confining End-Block Reinforcement

The strut-and-tie concept is based on a plasticity approach approximating the flow of forces in the anchorage zone by a series of straight compression struts and straight tension ties connected at discrete points that are called nodes to form truss units. The compressive forces are carried by the plastic compression struts and the tensile forces are carried either by non-prestressed reinforcement such as mild steel bars as confining ties or by prestressing steel reinforcement. The yield strength of the anchorage confining reinforcement is used to determine the total area of reinforcement needed in the anchorage block. Figure 4.24 (adapted from Ref. 4.18) illustrates the flow of the concentric and eccentric prestressing forces P ahead of the point of application of these forces through the anchorage device towards the end of general zone where the stresses become uniform by St. Venant's principle.

After significant cracking is developed, compressive stress trajectories in the concrete tend to congregate into straight lines that can be idealized as straight compressive struts in uniaxial compression. These struts would become part of truss units where the principal tensile stresses are idealized as tension ties in the truss unit with the nodal locations determined by the direction of the idealized compression struts. Figure 4.25(a) shows the development of a strut and Figure 4.25(b) sketches the resulting strut-and-tie trusses for multiple anchorage in a flanged T-section (Ref. 4.30). Figure 4.26 summarizes the concept of the idealized struts and ties in the anchorage zone. Figure 4.27 sketches standard strut-and-tie idealized trusses for concentric and eccentric cases both for solid and flanged sections as given in ACI 318-99 Code.

The tension tie in the ensuing truss analogy can be reasonably assumed to be at a distance $h/2$ from the anchorage device. This assumption is essentially consistent with the approximated location of the tensile force T in Figure 4.23 of the elastic stress-analysis approach. It is clear from all these diagrams that the designer has to make an engineering

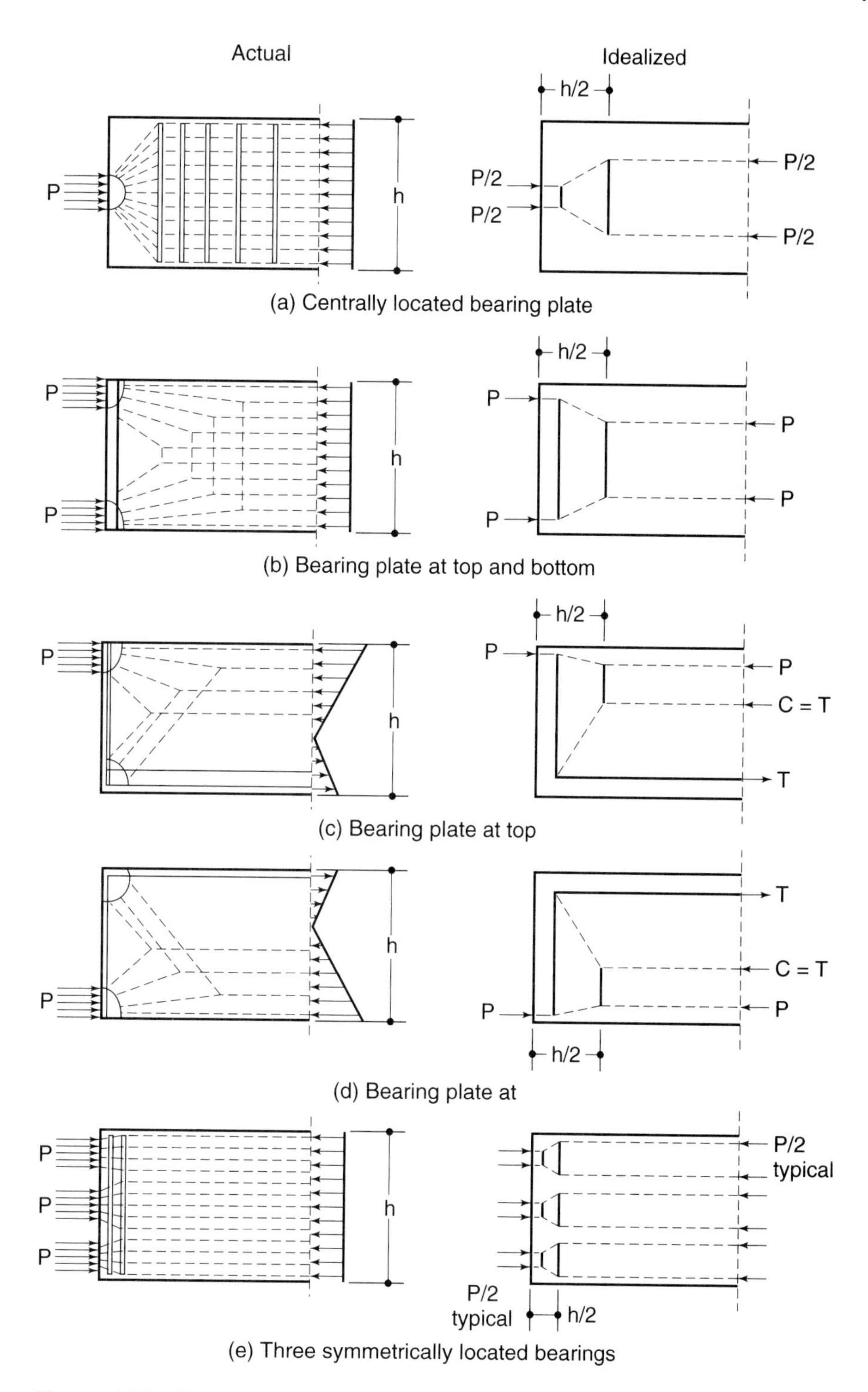

Figure 4.24 Schematic of Compression Strut-and-Tie Force Paths (Adapted from Ref. 4.18).

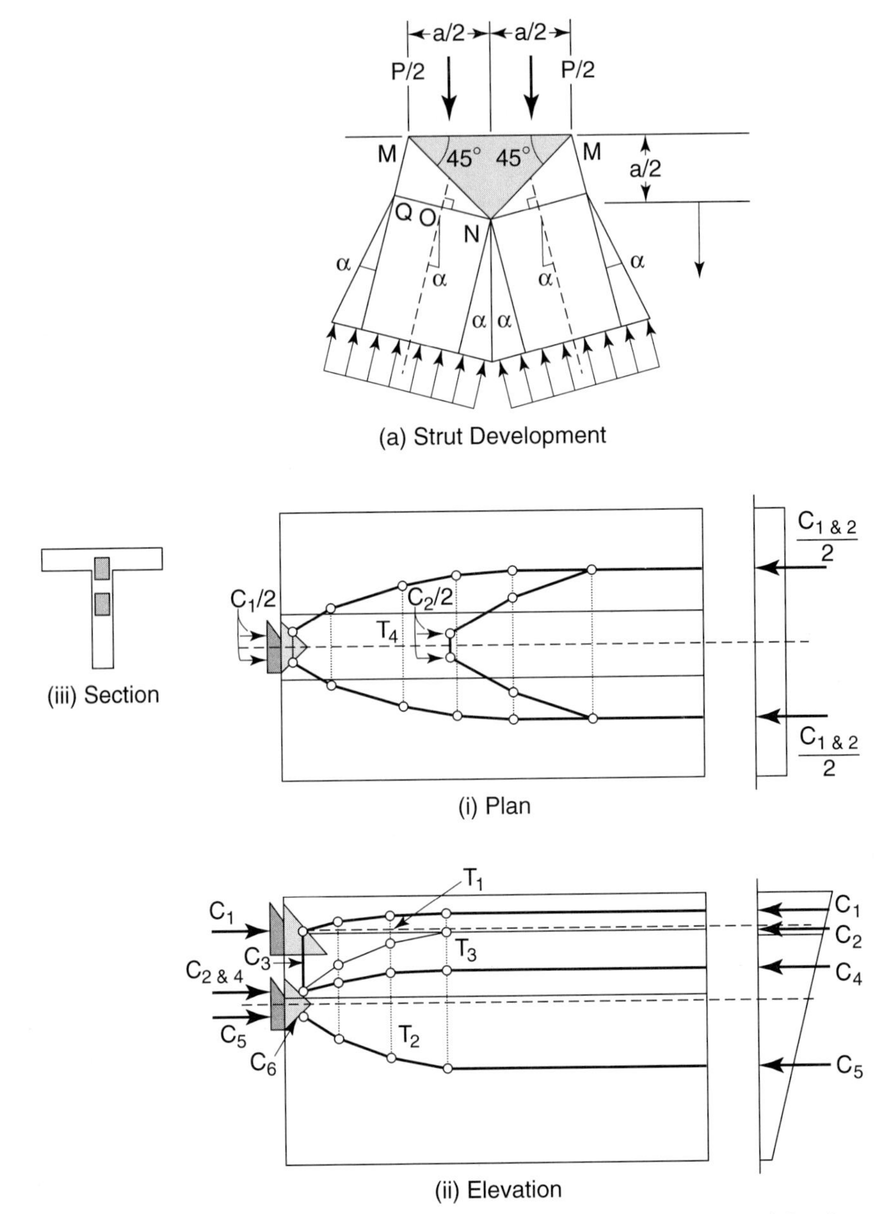

Figure 4.25 Strut-and-Tie Development.

judgment on the number of paths of struts, resulting ties, and ensuing nodes, particularly in the usual case of multiple anchorage devices. Part (b) of Example 4.6 illustrates the assumed idealized paths for the anchorage zone in the I-beam under consideration.

$$T_{\text{burst}} = 0.25\ \Sigma P_{su}\left(1 - \frac{a}{h}\right) \tag{4.15(a)}$$

$$d_{\text{burst}} = 0.5\,(h - 2e) \tag{4.15(b)}$$

where $\Sigma\, P_{su}$ = sum of the total *factored* tendon loads
a = depth of the anchorage device or single group of closely-spaced devices

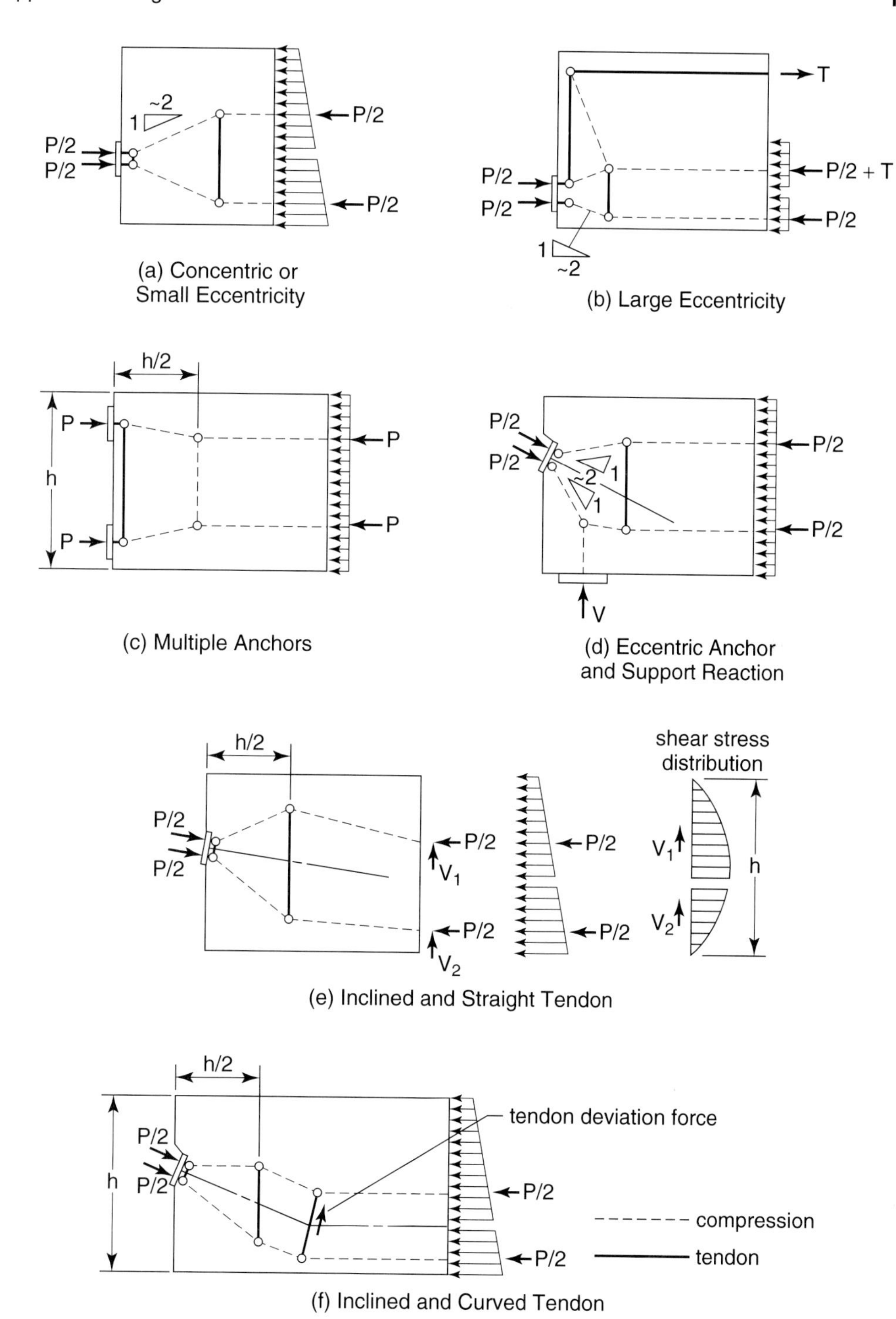

Figure 4.26 Typical Strut-and-Tie Models For End-Block Anchorage Zones.

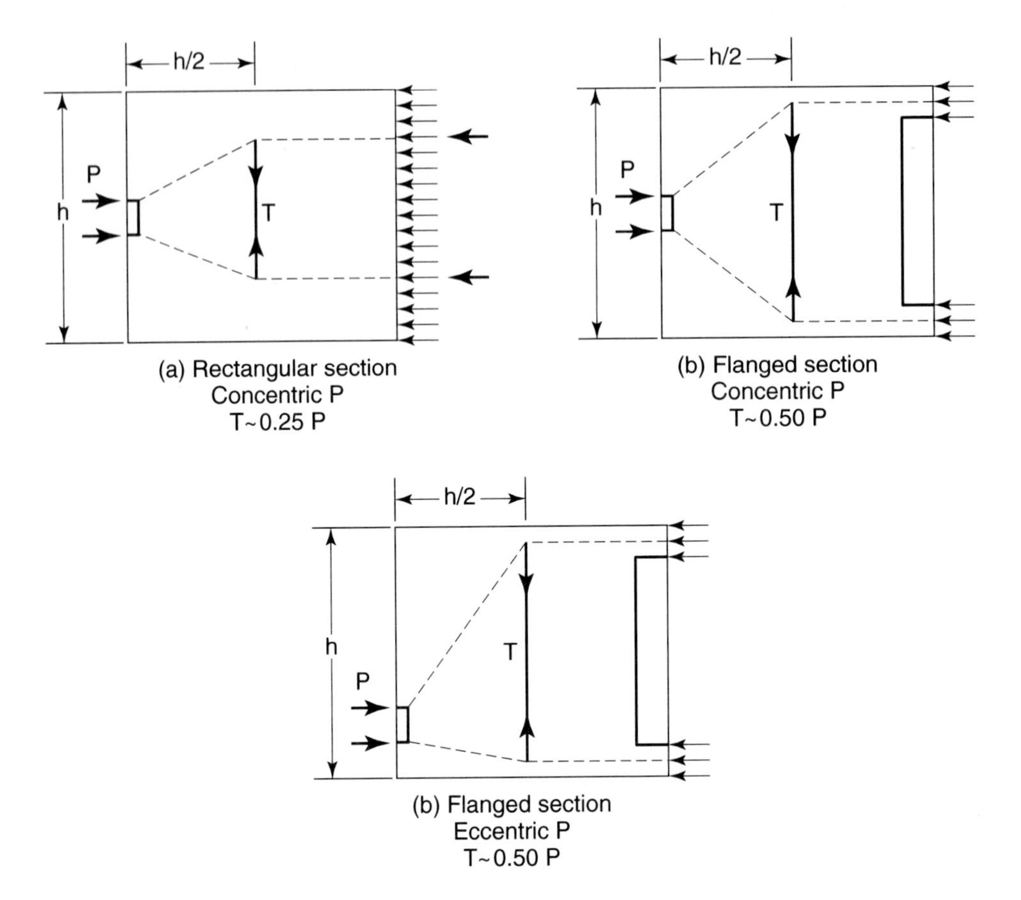

Figure 4.27 Strut-and-Tie Idealized Trusses in Standard Concentric and Eccentric Cases, ACI 318-99.

e = eccentricity of the anchorage device or group of closely-spaced devices from the centroid of the beam cross section
h = depth of the cross section.

Anchorage devices are treated as closely spaced if their center-to-center spacing does not exceed 1.5 times the width of the anchorage device

4.5.3.5 Allowable Bearing Stresses

The maximum allowable bearing stress at the anchorage device seating should not exceed the smaller of the two values obtained from Equations 4.16(a) and 4.16(b) as follows:

$$f_b \leq 0.7\,\phi\, f'_{ci}\sqrt{A/A_g} \qquad 4.16\text{ (a)}$$

$$f_b \leq 2.25\,\phi\, f'_{ci} \qquad 4.16\text{ (b)}$$

where f_b = maximum factored tendon load, P_u, divided by the effective bearing area A_b
f'_{ci} = concrete compressive strength at stressing
A = maximum area of portion of the supporting surface that is geometrically similar to the loaded area and concentric with it
A_g = gross area of the bearing plate
A_h = effective net area of the bearing plate calculated as the area A_g minus the area of openings in the bearing plate.

Equations 4.16(a) and 4.16(b) are valid only if general zone reinforcement is provided and if the extent of concrete along the tendon axis ahead of the anchorage device is at least twice the length of the local zone.

4.5.4 Design of End Anchorage Reinforcement for Post-tensioned Beams

Example 4.6

Design an end anchorage reinforcement for the post-tensioned beam in Example 4.2, giving the size, type, and distribution of reinforcement. Use $f'_c = 5{,}000$ psi (34.5 MPa) normal-weight concrete.

Assume that the beam ends are rectangular blocks extending 40 in. (104 cm.) into the span beyond the anchorage devices then transitionally reduce to the 6-in. thick web. Solve the problem using (a) the linear elastic stress analysis method, (b) the plastic strut-and-tie method. Sketch the truss model you determine.

(a) *Solution by the Linear Elastic Stress Method:*

1. *Establish the configuration of the tendons to give eccentricity* $e_e = 12.49$ in. (317 mm)
   From Example 4.2,
   $c_b = 18.84$ in., hence distance from the beam bottom fibers $= c_b - e_e = 6.35$ in. (161 mm)

   For a centroidal distance of the $13 - \frac{1}{2}$ in. size tendons = 6.35 in. from the beam bottom fibers, try the following row arrangement:

   1st row : 5 tendons at 2.5 in.
   2nd row : 5 tendons at 7.0 in.
   3rd row : 3 tendons at 11.5 in.

   $$\text{distance of the centroid of tendons} = \frac{5 \times 2.5 + 5 \times 7.0 + 3 \times 11.5}{13} \cong 6.35 \text{ in., O.K.}$$

2. *Ultimate forces in tendon rows and bearing capacity of the concrete*
   1st-row force $P_{u1} = 5 \times 0.153 \times 270{,}000 = 206{,}550$ lb. (919 kN)
   2nd-row force $P_{u2} = 5 \times 0.153 \times 270{,}000 = 206{,}550$ lb. (919 kN)
   3rd-row force $P_{u3} = 3 \times 0.153 \times 270{,}000 = 123{,}930$ lb. (551 kN)

3. *Elastic analysis of forces*
   Divide the beam depth into 4-in. increments of height as shown in Figure 4.28, and assume that the concrete stress at the center of each increment is uniform across the depth of the increment. Then calculate the incremental moments due to these internal stresses and due to the external prestressing force P_i about each horizontal plane in order to determine the *net* moment on the section. The net maximum moment will determine the position of the potential horizontal bursting crack and the reinforcement that has to be provided to prevent the crack from developing. Using a plus (+) sign for clockwise moment, the initial prestressing force before losses, from Example 4.2, is $P_i = 376{,}110$ lb (1,673 kN). From Figure 4.28, the concrete internal moment at the plane 4 in. from the bottom fibers is

   $$M_{c4} = 2{,}117 \times 4 \times 18 \times (2 \text{ in.}) = 304{,}848 \text{ in.-lb}$$
   $$= 0.3 \times 10^6 \text{ in.-lb (34.4 kN-m)}$$

   and that at the plane 8 in. from the bottom fibers is

   $$M_{c8} = 2{,}117 \times 4 \times 18 \times (6 \text{ in.}) + 1{,}851 \times 4 \times \frac{18 + 10}{2} \times (2 \text{ in.})$$
   $$= 1{,}121{,}856 \text{ in.-lb} = 1.12 \times 10^6 \text{ in.-lb (127 kN-m)}$$

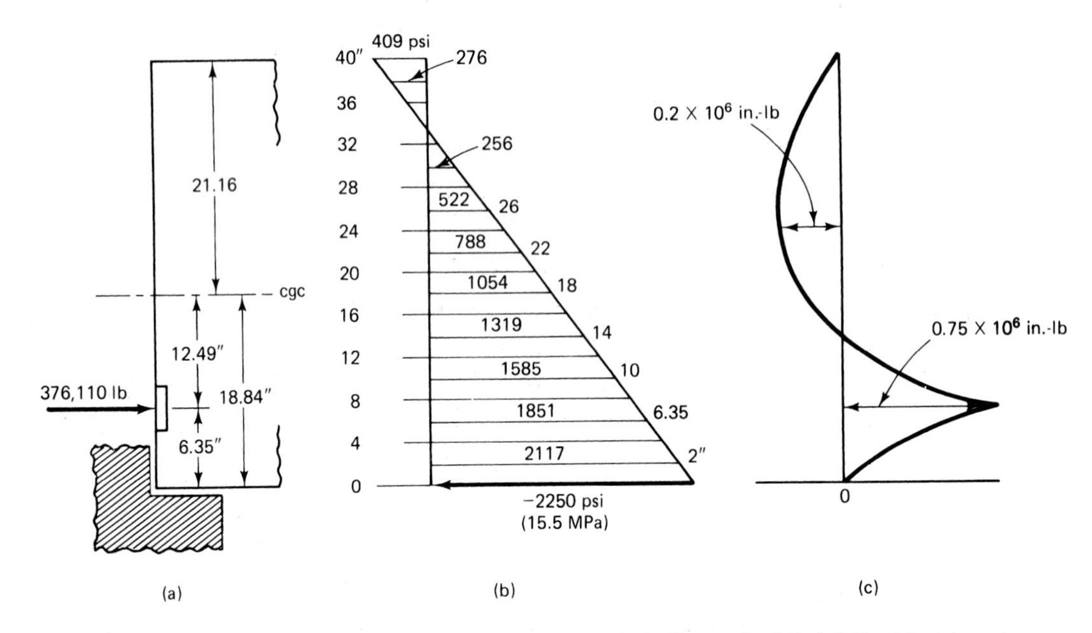

Figure 4.28 Anchorage zone stresses and moments in Example 4.6. (a) Post-tensioned end-block zone. (b) Transfer concrete stress distribution across depth. (c) Net anchorage zone moments on potential horizontal cracking planes along the beam depth.

The prestressing force moment at the plane 8 in. from the bottom fibers is

$$M_{c8} = 376{,}110 \times (8 - 6.35) = -620{,}582 \text{ in.-lb}$$

$$= -0.62 \times 10^6 \text{ in.-lb (70.1 kN-m)}$$

The net moment is then $1.12 \times 10^6 - 0.62 \times 10^6 = 0.50 \times 10^6$ in.-lb (56.6 kN-m).

In a similar manner, we can find the net moment for all the other incremental planes at 4-in. increments to get the values tabulated in Table 4.5. From the table, the maximum net moments are $+M_{max} = 0.75 \times 10^6$ in.-lb (84.6 kN-m) at the horizontal plane 6.35 in. above the beam bottom fibers (bursting potential crack effect) and $-M_{max} = -0.20 \times 10^6$ in.-lb at the horizontal plane 24 in. (61 cm) above the beam bottom fibers (spalling potential crack effect).

4. *Design of anchorage reinforcement*

From Equation 4.11, assuming that the center of the tensile vertical force T is at a distance $x \approx 15$ in. , we obtain

$$T = \frac{M_{max}}{h - x} = \frac{+0.75 \times 10^6}{40 - 15} = 30{,}000 \text{ lb (133 kN)}$$

Allowing a maximum steel stress, $f_s = 20{,}000$ psi, (the Code allows $0.60 f_y = 36{,}000$ psi)

The bursting zone reinforcement is

$$A_t = \frac{T_b}{f_s} = \frac{30{,}000}{20{,}000} = 1.50 \text{ in.}^2 \text{ (968 mm}^2\text{)}$$

So

Try No. 3 closed ties ($A_s = 2 \times 0.11 = 0.22$ in^2)

$$\text{Required no. of stirrups} = \frac{1.50}{0.22} = 6.82$$

Use six No. 3 ties in addition to what is required for shear.

The spalling zone force

Table 4.5 Anchorage Zone Moments for Example 4.6

Moment plane dist. d from bottom in.	Section width in.	Stress at plane $(d - 2.0)^*$ psi	Concrete resistance force at $(d - 2.0)^*$ lb	Moment M_p for a P_i about horiz. plane in col. (1) in.-lb × 10^6	Moment M_c of concrete in col. (4) about horiz. plane in col. (1) in.-lb × 10^6	Net moment $(M_c - M_p)$ col. (6) – col. (5) in.-lb × 10^6
(1)	(2)	(3)	(4)	(5)	(6)	(7)
0	18	−2,250	162,000	0	0	0
4	18	−2,117	152,424	0	+0.30	+0.30
6.35	13.3	−1,851	103,656	0	+0.75	+0.75
8	10	−1,585	63,400	−0.62	+1.12	+0.50
12	6	−1,319	31,656	−2.13	+2.25	+0.12
16	6	−1,054	25,296	−3.63	+3.54	−0.09
20	6	−788	18,912	−5.13	+4.94	−0.19
24	6	−522	12,528	−6.64	+6.44	−0.20
28	6	−256	6,144	−8.14	+7.99	−0.15
32	6	~0	~0	−9.65	+9.61	−0.04
36	18	+276	−19,872	−11.15	+11.13	−0.02
40	18	+409	−29,448	−12.66	+12.65	~0

*d = distance from plane about which moment is taken less half the depth of one slice (in this example slice depth = 4 in.).

$$T_s = \frac{-0.2 \times 10^6}{40 - 15} = 8{,}000 \text{ lb}$$

So

$$A_s = \frac{T_s}{f_s} = \frac{8{,}000}{20{,}000} = 0.40 \text{ in}^2 \ (250 \text{ mm}^2)$$

Hence, we have

$$\text{Req no. of No. 3 stirrups} = \frac{0.40}{0.22} = 1.82$$

So use two No. 3 additional stirrups. Then

Total number of stirrups = 6.82 + 1.82 + 4 ≈ 12.64

Use 12 No. 3 closed ties. Extend the stirrups into the compression zone in Figure 4.23.

Space the No. 3 closed ties at 3 in. center to center with the first stirrup starting 3 in. from the beam end. Also, provide four No. 3 bars 10 in. long at 3 in. center to center each way 2 in. from the end face at the anchor location, since cracking can occur vertically and horizontally. Add spiral reinforcement under the anchors if the manufacturer specifies that such a reinforcement is useful.

Next, check the bearing plate stresses.

(b) Solution by the Plastic Strut-and-Tie Method:

1. *Establish the configuration of the tendons to give eccentricity* e_e = 12.49 in. (317 mm)
From Example 4.2,
c_b = 18.84 in., hence distance from the beam fibers = $c_b - e_e$ = 6.35 in. (161 mm)

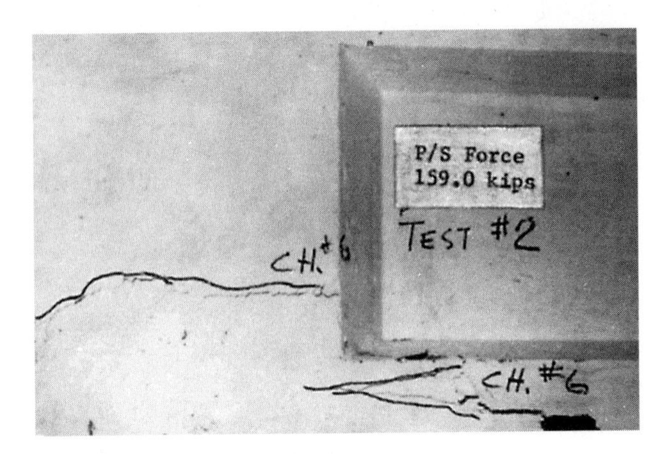

Photo 4.8 Typical bursting crack of the anchorage zone (Nawy et al.).

For a centroidal distance of the 13-$\frac{1}{2}$-in.-size strands = 6.35 in. from the beam bottom fibers, try the following row arrangement of tendons with the indicated distances from the bottom fibers:

1st row : 5 tendons at 2.5 in.
2nd row : 5 tendons at 7.0 in.
3rd row : 3 tendons at 11.5 in.

$$\text{distance of the centroid of tendons} = \frac{5 \times 2.5 + 5 \times 7.0 + 3 \times 11.5}{13} \cong 6.35 \text{ in., O.K.}$$

2. *Ultimate forces in tendon rows and bearing capacity of the concrete*
1st row force $P_u1 = 5 \times 0.153 \times 270{,}000 = 206{,}550$ lb. (919 kN)
2nd row force $P_u2 = 5 \times 0.153 \times 270{,}000 = 206{,}550$ lb. (919 kN)
3rd row force $P_u3 = 3 \times 0.153 \times 270{,}000 = 123{,}930$ lb. (551 kN)
Total ultimate compressive force = 206,550 + 206,550 + 123,930 = 537,830 lb. (2389 kN)
Total area of rigid bearing plates supporting the Supreme 13-chucks anchorage

$$\text{devices} = 14 \times 11 + 6 \times 4 = 178 \text{ in.}^2 \ (113 \text{ cm}^2)$$

$$\text{Actual bearing stress } f_b = \frac{537{,}380}{178} = 3020 \text{ psi (20.8 MPa)}$$

From Equations 4.16(a) and (b), the maximum allowable bearing pressure on the concrete is

$$f_b \leq 0.7\,\phi f'_{ci}\sqrt{A/A_g}$$

$$f_b \leq 2.25\,\phi f'_{ci}$$

Assume that the initial concrete strength at stressing is $f'_{ci} = 0.75 f'_c$

$$= 0.75 \times 5{,}000 = 3750 \text{ psi}$$

concentric area, A, of concrete with the bearing plates $\cong 18 \times 14 + 10 \times 7 = 322$ in.^{2}

$$\text{Allowable bearing stress, } f_b = 0.7 \times 0.90 \times 3750\sqrt{\frac{322}{178}} = 3{,}178 \text{ psi.} > 3020 \text{ psi, O.K.}$$

The bearing stress from Eq. 4.14 (b) does not control.

3. *Draw the strut-and-tie model*
Total length of distance a, as in Figure 4.25 between forces $P_{u1} - P_{u3} = 11.5 - 2.5 = 9.0$ in.

Hence depth $a/2$ ahead of the anchorages = 9.0/2 = 4.5 in.

Construct the strut-and-tie model assuming it to be as shown in Figure 4.29.

The geometrical dimensions for finding the horizontal force components from the ties 1–2 and 3–2 have cotangent values of 26.5/15.5 and 13.0/15.5 respectively. From statics, truss analysis in Figure 4.29 gives the member forces as follows:

$$\text{tension tie 1–2} = 123{,}990 \times \frac{26.5}{15.5} = 211{,}982 \text{ lb (942 kN)}$$

$$\text{tension tie 3–2} = 206{,}550 \times \frac{13}{15.5} = 173{,}235 \text{ lb (728 kN)}$$

Use the larger of the two values for choice of the closed tension tie stirrups. Try No. 3 closed ties, giving a tensile strength per tie = $\phi f_y A_v$

$$= 0.90 \times 60{,}000 \times 2(0.11) = 11{,}880 \text{ lb}$$

$$\text{required number of stirrup ties} = \frac{211{,}982}{11{,}880} = 17.8$$

For the tension tie a-b-c in Figure 4.29, use the force P_u = 173,235 lb to concentrate additional No. 4 vertical ties ahead of the anchorage devices. Start the first tie at a distance of $1\frac{1}{2}$ in. from the end rigid steel plate transferring the load from the anchorage devices to the concrete.

$$\text{Number of ties} = \frac{173{,}235}{0.90 \times 60{,}000(2 \times 0.20)} = 8.0$$

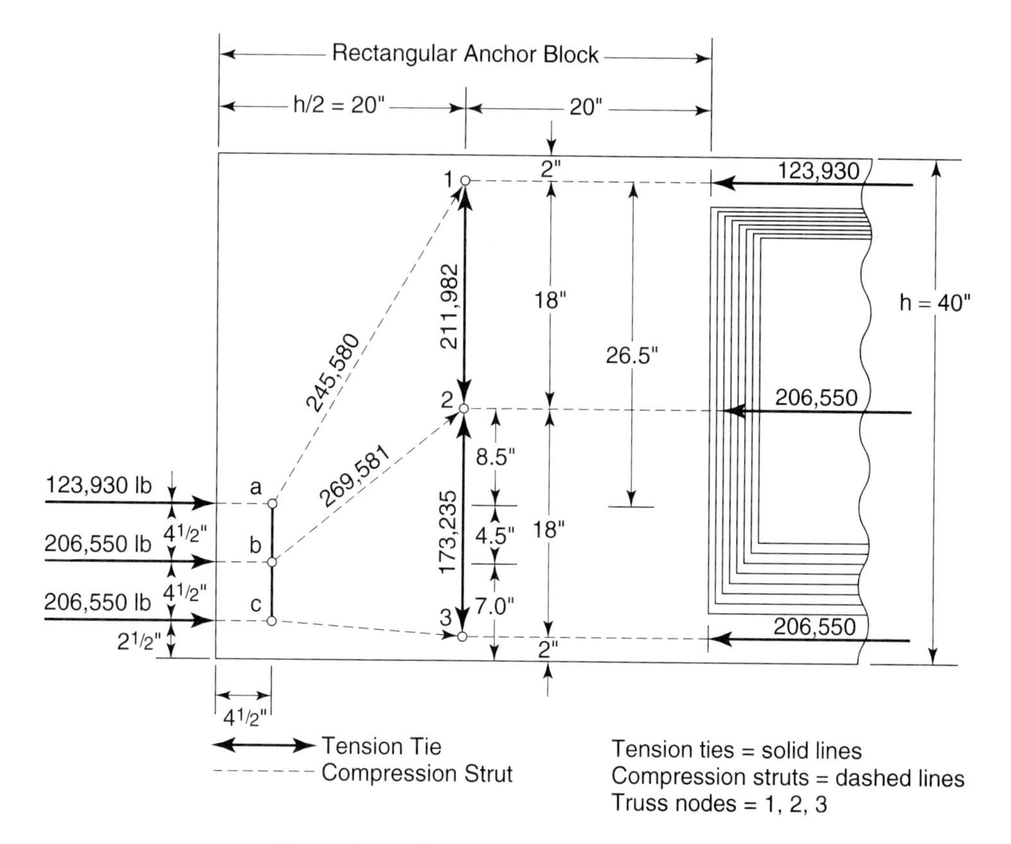

Figure 4.29 Struts-and-Ties in Example 4.6

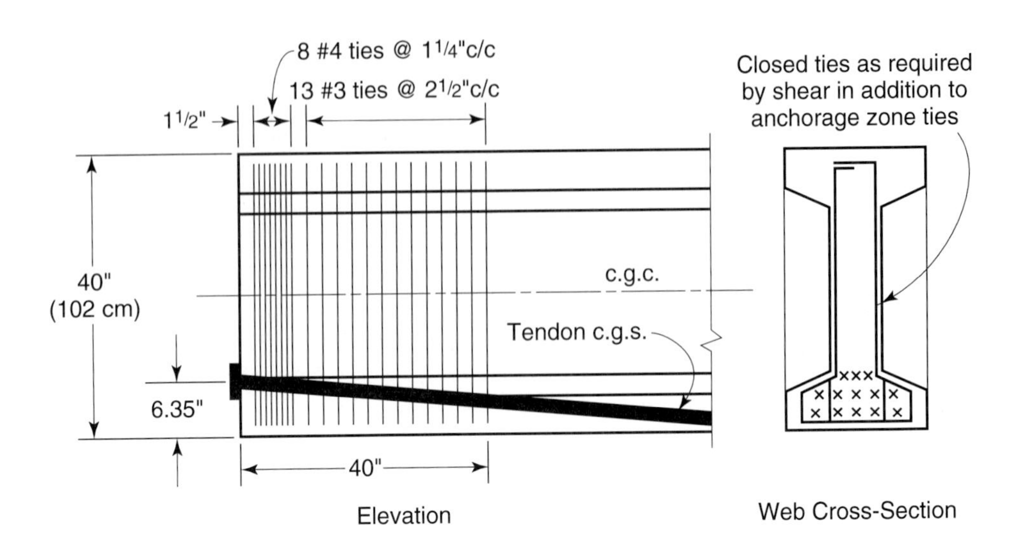

Figure 4.30 End anchorage reinforcement in Example 4.6. (a) Anchorage zone, (b) Beam cross section.

Use eight No. 4 closed ties @ $1\frac{1}{4}$in. (12.7 mm @ 32 mm) center to center with the first tie to start at $1\frac{1}{2}$ in. *ahead* of the anchorage devices.

Only thirteen ties in lieu of the 17.8 calculated are needed since part of the zone is covered by the No. 4 ties. Use 13 No. 3 closed ties @ $2\frac{1}{2}$ in. (9.5 mm @ 57 mm) center to center beyond the last No. 4 tie so that a total distance of 40 in. (104 cm) width of the rectangular anchor block is confined by the reinforcement closed ties.

Note that this solution requires a larger number of confining ties than the elastic solution in part (a). Adopt this design of the anchorage zone. Figure 4.30 shows a schematic of the anchorage zone confining reinforcement details resulting from the strut-and-tie analysis.

It should also be noted that the idealized paths of the compression struts for cases where there are several layers of prestressing strands should be such that at each layer level a stress path is assumed in the design. This can be seen in Example 4.7 and Figure 4.39(b). If layers are combined, a more conservative solution with more confining reinforcement area results, which is not necessarily justified.

4.6 FLEXURAL DESIGN OF COMPOSITE BEAMS

Composite sections are normally precast, prestressed supporting elements over which sit situ-cast top slabs that act integrally with them. Composites have the advantage of the precast part becoming in many cases the falsework for supporting the situ-cast top slab and topping in bridges and industrial buildings, as shown in Figure 4.31. Sometimes the precast, prestressed element is shored during the placement and curing of the situ-cast top slab. In such a case, the slab weight acts only on the composite section, which has a substantially larger section modulus than the precast section. Hence, the concrete stress calculations have to take this situation into account in the design. The concrete stress distribution due to composite action can be seen in Figure 4.32, in part (e) of which the load taken by the cured composite section is the sum of $W_{SD} + W_L$.

4.6.1 Unshored Slab Case

From Equations 4.2a and b, the extreme concrete fiber stress equations before casting the top slab are

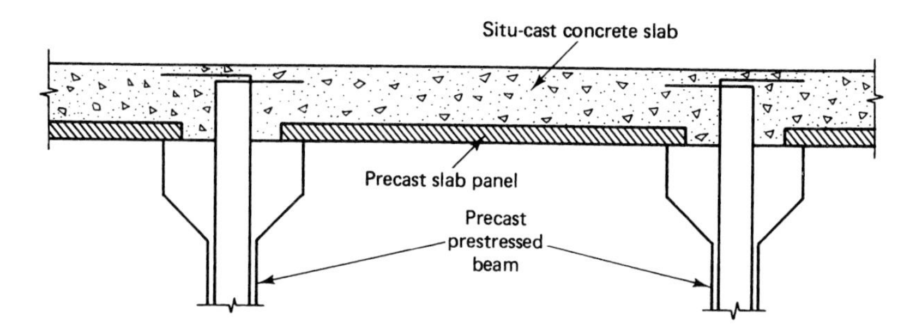

Figure 4.31 Composite prestressed concrete construction.

$$f^t = -\frac{P_e}{A_c}\left(1 - \frac{ec_t}{r^2}\right) - \frac{M_D + M_{SD}}{S^t} \tag{4.17}$$

and

$$f_b = -\frac{P_e}{A_c}\left(1 + \frac{ec_b}{r^2}\right) + \frac{M_D + M_{SD}}{S_b} \tag{4.18}$$

where S^t and S_b are the section moduli of the precast section only, and M_{SD} is any additional superimposed moment such as the web slab concrete.

After the situ-cast slab hardens and composite action takes place, new higher moduli S_c^t and S_{cb} are available, with the cgc line moving upwards towards the top fibers. The concrete fiber stress counterparts to Equations 4.18a and b for the extreme top and bottom fibers of the *precast* part of the composite section (level AA in Figure 4.32(e) are

$$f^t = -\frac{P_e}{A_c}\left(1 - \frac{ec_t}{r^2}\right) - \frac{M_D + M_{SD}}{S^t} - \frac{M_{CSD} + M_L}{S_c^t} \tag{4.19a}$$

and

$$f_b = -\frac{P_e}{A_c}\left(1 + \frac{ec_b}{r^2}\right) + \frac{M_D + M_{SD}}{S_b} + \frac{M_{CSD} + M_L}{S_{cb}} \tag{4.19b}$$

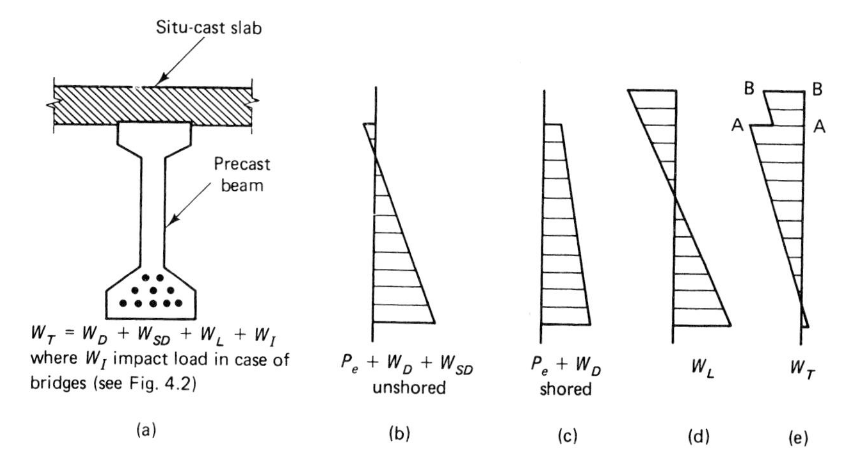

Figure 4.32 Flexural stress distribution in composite beams. (a) Composite beam. (b) Concrete stress distribution. (c) Concrete stress distribution with precast beam shored. (d) Live-load stress for shored case, or live load plus superimposed dead load for unshored case. (e) Final service-load stress due to all loads.

where M_{CSD} is the additional composite superimposed dead load after erection, i.e., at service, and $S_c^{\,t}$ and S_{cb} are the section moduli of the composite section at the level of the top and bottom fibers, respectively, of the precast section.

The fiber stresses at the level of the top and bottom fibers of the situ-cast slab (levels BB and AA of Figure 4.32(e)) are

$$f^{ts} = -\frac{M_{CSD} + M_L}{S_{cb}^t} \tag{4.20a}$$

and

$$f_{bs} = -\frac{M_{CSD} + M_L}{S_{bcb}} \tag{4.20b}$$

where $M_{CSD} + M_L$ are the incremental moments added after composite action has developed, and S_{cb}^t and S_{bcb} are the section moduli of the composite section for the top and bottom fibers AA and BB, respectively, of the slab in Figure 4.32(e).

4.6.2 Fully Shored Slab Case

In cases where the situ-cast slab is fully shored until composite action develops, the concrete fiber stresses before shoring and top slab casting become, from Equations 4.18 and 4.19,

$$f^t = -\frac{P_e}{A_c}\left(1 - \frac{ec_t}{r^2}\right) - \frac{M_D}{S^t} \tag{4.21a}$$

and

$$f_b = -\frac{P_e}{A_c}\left(1 + \frac{ec_b}{r^2}\right) + \frac{M_D}{S_b} \tag{4.21b}$$

After the top slab is situ cast and full composite action is developed when the concrete hardens, Equations 4.19a and b become, for the beam shored after erection,

$$f^t = -\frac{P_e}{A_c}\left(1 - \frac{ec_t}{r^2}\right) - \frac{M_D}{S^t} - \frac{M_{SD} + M_{CSD} + M_L}{S_c^t} \tag{4.22a}$$

Photo 4.9 Jacking tendons of post-tensioned beam (Nawy et al.).

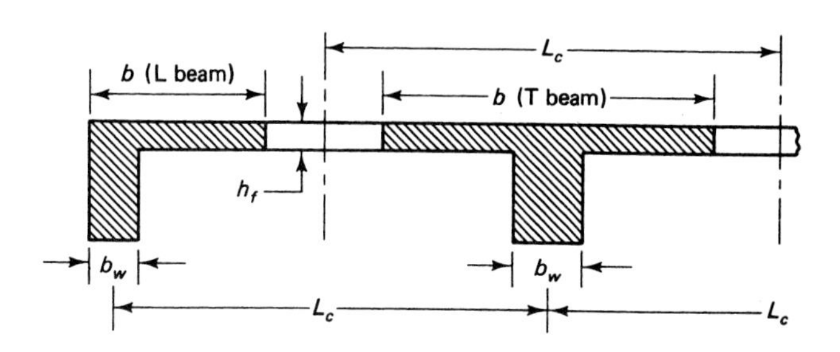

Figure 4.33 Effective flange width of composite section.

and

$$f_b = -\frac{P_e}{A_c}\left(1 + \frac{ec_b}{r^2}\right) + \frac{M_D}{S_b} + \frac{M_{SD} + M_{CSD} + M_L}{S_{cb}} \tag{4.22b}$$

Note that adequate check has to be made for the horizontal interface shear stresses between the situ-cast and the precast beams, as will be discussed in Chapter 5.

4.6.3 Effective Flange Width

In order to determine the theoretical composite action that resists the flexural stresses, a determination has to be made of the slab width that can effectively contribute to the stiffness increase resulting from composite action.

Figure 4.33 and Table 4.6 give the ACI and AASHTO requirements for determining the effective top flange width of the composite section. If the topping concrete is of different strength than that of the precast section, the width b has to be modified to account for the difference in the moduli of the two concretes in order to ensure that the strains in both materials at the interface are compatible. The modified width of the composite topping for calculating the composite I_{cc} is

$$b_m = \frac{E_{ct}}{E_c}(b) = n_c b \tag{4.23}$$

Table 4.6 Values of Effective Flange Width

	Width b as the least of the tabulated values (Modify to $b_M = n_c b$, where $n_c = E_{ct}/E_c$ when flange concrete is of different strength from that of the precast web)	
	End beam	**Intermediate beam**
ACI	$b_w + 6H_f$	$b_w + 16h_f$
	$\frac{1}{2}(b_w + L_c)$	L_c
	$b_w + \frac{L}{12}$	$\frac{L}{4}$
AASHTO	$b_w + 6h_f$	$b_w + 12h_f$
	$\frac{1}{2}(b_w + L_c)$	L_c
	$b_w + \frac{L}{12}$	$\frac{L}{4}$

L = span of end or intermediate beam

where E_{ct} = modulus of the topping concrete
E_c = modulus of the precast concrete

Once the modified width b_m is defined, the entire composite section is considered to be of the higher strength concrete.

4.7 SUMMARY OF STEP-BY-STEP TRIAL-AND-ADJUSTMENT PROCEDURE FOR THE SERVICE-LOAD DESIGN OF PRESTRESSED MEMBERS

1. Given the superimposed dead-load intensity W_{SD}, the live-load intensity W_L, the span and the height limitation, the material strengths f_{pu}, f'_c, the type of concrete, and whether the prestress type is pretensioning or post-tensioning,
2. Assume the intensity of self-weight W_D, and calculate moments M_D, M_{SD}, and M_L.
3. Calculate f_{pi}, f'_{ci}, f_{ti}, f_t and f_c, where f_{pi} = 0.70 f_{pu}, f_{ci} = 0.60 f'_c, f_{ti} = , and $f_{ti} = 6\sqrt{f'_{ci}}$, for the support section, $f_c = 0.45\, f'_c$ or $0.60\, f'_c$ as allowed and $f_t = 6\sqrt{f'_c}$ to $12\sqrt{f'_c}$.
4. Calculate the prestress losses $\Delta f_{pT} = \Delta f_{pES} + \Delta f_{pR} + \Delta f_{pSH} + \Delta f_{pCR} + \Delta f_{pE} + \Delta f_{pA} + \Delta f_{pB}$ for the type of prestressing used. Determine net stresses $f_{pe} = f_{ps} - \Delta f_{pT}$.
5. Find the minimum required section modulus of the minimum efficient section for evaluating the concrete fiber stresses at the top and bottom fibers.
   (a) For harped or draped tendons, use the midspan controlling section:

$$S^t \geq \frac{(1-\gamma)M_D + M_{SD} + M_L}{\gamma f_{ti} - f_c}$$

$$S_b \geq \frac{(1-\gamma)\,M_D + M_{SD} + M_L}{\gamma f_t - \gamma f_{ci}}$$

   where

$$\gamma = \frac{P_e}{P_i}$$

   (b) For straight tendons, use the end-support controlling section:

$$S^t \geq \frac{M_D + M_{SD} + M_L}{\gamma f_{ti} - f_c}$$

$$S_b \geq \frac{M_D + M_{SD} + M_L}{f_t - \gamma f_{ci}}$$

6. Select a trial section with section modulus properties close to those required in step 5 to be checked later for composite section fiber stress requirements.
7. For (a) the controlling section in the span (usually the midspan or at 0.4 of span), (b) the controlling section at the support, and (c) any other section along the span if both straight and draped tendons are used, analyze the concrete fiber stresses expected at stress transfer immediately before such transfer:

$$f^t = -\frac{P_i}{A_c}\left(1 - \frac{ec_t}{r^2}\right) - \frac{M_D}{S^t}$$

$$f_b = -\frac{P_i}{A_c}\left(1 + \frac{ec_b}{r^2}\right) + \frac{M_D}{S_b}$$

If the stresses exceed the allowable values, enlarge the section, or change the eccentricity e_c or e_e, or both.

8. Analyze the concrete fiber stresses for the service-load conditions, as in step 7:

$$f^t = -\frac{P_e}{A_c}\left(1 - \frac{ec_t}{r^2}\right) - \frac{M_T}{S^t}$$

$$f_b = -\frac{P_e}{A_c}\left(1 + \frac{ec_b}{r^2}\right) + \frac{M_T}{S_b}$$

where $M_T = M_D + M_{SD} + M_L$. If the stresses exceed the allowable values, enlarge the section or change the eccentricity e_c or e_e, or both.

9. For cases where many strands have to be used, establish the envelopes of limiting eccentricities for zero tension $e_b = (k_b + a_{\min})$ and $e_t = (a_{\max} - k_t)$, where $a_{\min} = M_D/P_i$ and $a_{\max} = M_T/P_e$. If the tension in the concrete is used in the design, add

$$e'_b = \frac{f_{(t)}A_c k_b}{P_i}$$

and

$$e'_t = \frac{f_{(b)}A_c k_t}{P_e}$$

to the bottom and top envelopes, respectively, where $f_{(t)}$ and $f_{(b)}$ are the extreme fiber stresses calculated to be in the concrete.

10. Investigate the end-block anchorage zone stresses, and design the necessary reinforcement to prevent bursting or spalling cracks. Determine the minimum development length

$$l_d = \frac{1}{1{,}000}\left(f_{ps} - \frac{2}{3}f_{pe}\right)d_b$$

of which the transfer length $l_t = \dfrac{f_{pe}}{3{,}000}\, d_b$, where f_{pe} is in psi units.

Post-tensioned Anchorage

Design the anchorage block reinforcement. Use the strut-and-tie plastic truss units to compute the ultimate tension force in the tie for confining reinforcement selection.

Pretensional Prestress Transfer Zone

$$A_t = 0.021\,\frac{P_i h}{f_s l_t}$$

11. Determine the composite action stresses, and revise the section if these stresses exceed the maximum allowable concrete fiber stresses both in the precast section and the situ-cast top slab. Use the modified effective width $b_m = (E_{ct}/E_c)b$ for the top composite flange when calculating the section modulus of the composite section.

(a) *Unshored Slab Case.* Before the top slab is situ cast:

$$f^t = -\frac{P_e}{A_c}\left(1 - \frac{ec_t}{r^2}\right) - \frac{M_D + M_{SD}}{S^t}$$

$$f_b = -\frac{P_e}{A_c}\left(1 + \frac{ec_b}{r^2}\right) + \frac{M_D + M_{SD}}{S_b}$$

After the top slab is cast and cured to develop full composite action, the stresses at top and bottom fibers of the *precast* part of the composite section will be

$$f^t = -\frac{P_e}{A_c}\left(1 - \frac{ec_t}{r^2}\right) - \frac{M_D + M_{SD}}{S^t} - \frac{M_{CSD} + M_L}{S_c^t}$$

$$f_b = -\frac{P_e}{A_c}\left(1 + \frac{ec_b}{r^2}\right) + \frac{M_D + M_{SD}}{S_b} + \frac{M_{CSD} + M_L}{S_{cb}}$$

where S_c^t and S_{cb} are the section moduli of the composite section at the level of the top and bottom fibers of the precast section. Also, M_L includes M_I if impact stresses exist.

The fibers stresses at the level of the top and bottom fibers of the situ-cast hardened slab are

$$f^{ts} = -\frac{M_{CSD} + M_L}{S_{cb}^t}$$

$$f_{bs} = -\frac{M_{CSD} + M_L}{S_{bcb}}$$

where S_{ch}^t and S_{bcb} are the section moduli of the composite section at the level of the top and bottom of the situ-cast slab.

(b) ***Fully Shored Slab Case.*** Before shoring, and before the topping is situ-cast,

$$f^t = -\frac{P_e}{A_c}\left(1 - \frac{ec_t}{r^2}\right) - \frac{M_D}{S^t}$$

$$f_b = -\frac{P_e}{A_c}\left(1 + \frac{ec_b}{r^2}\right) + \frac{M_D}{S_b}$$

After the situ-cast slab is cured and *full composite action develops,*

$$f^t = -\frac{P_e}{A_c}\left(1 - \frac{ec_t}{r^2}\right) - \frac{M_D}{S^t} - \frac{M_{SD} + M_{CSD} + M_L}{S_c^t}$$

$$f_b = -\frac{P_e}{A_c}\left(1 + \frac{ec_b}{r^2}\right) + \frac{M_D}{S_c} + \frac{M_{SD} + M_{CSD} + M_L}{S_{cb}}$$

The effective width of the top flange of the composite section is determined in accordance with the applicable ACI or AASHTO specifications.

M_D = moment due to self-weight of the precast element, M_{SD} = moment due to situ-cast slab and any other construction load, and M_{CSD} = moment due to additional composite superimposed load.

12. Proceed to determine the strength of the section for the limit state at failure and for shear and torsional strength.

Figure 4.34 shows a flowchart for the service-load flexural design of prestressed beams.

4.8 DESIGN OF COMPOSITE POST-TENSIONED PRESTRESSED SIMPLY SUPPORTED SECTION

Example 4.7

Supported Section. A two-lane simply supported bridge has a 64 ft (19.5 m) span, center to center, of bearings. The width of the bridge is such that the exterior beams are 28 ft (8.54 m) center to center. The spacing of the interior beams is at 7 ft center to center. Design the supporting interior post-tensioned beams, with deck slab *unshored* during construction, to carry AASHTO HS20-44 loading; establish the tendon eccentricities and tendon envelopes; and design the anchorage block and reinforcement, given the following information:

Concrete

$$\text{Precast beam } f'_c = 5{,}000 \text{ psi normal-weight concrete}$$

$$5\tfrac{3}{4} \text{ in. deck } f'_c = 3{,}000 \text{ psi normal-weight concrete}$$

$$f'_{ci} = 4{,}000 \text{ psi (27.6 MPa)}$$

$$f_{ci} = 0.55 f'_{ci} = -2{,}200 \text{ psi (15.2 MPa)}$$

$$f_c = 0.40 f'_c = -2{,}000 \text{ psi (13.8 MPa)}$$

$$f_{ti} = 212 \text{ psi } = 3\sqrt{f'_c}$$

$$f_t = 6\sqrt{f'_c} = 424 \text{ psi}$$

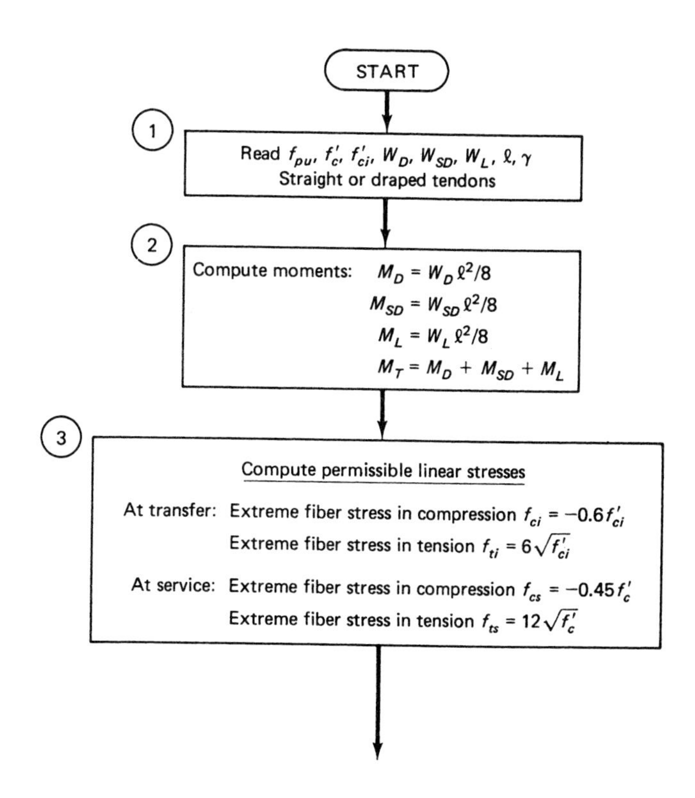

Figure 4.34 Flowchart for service-load flexural design of prestressed beams.

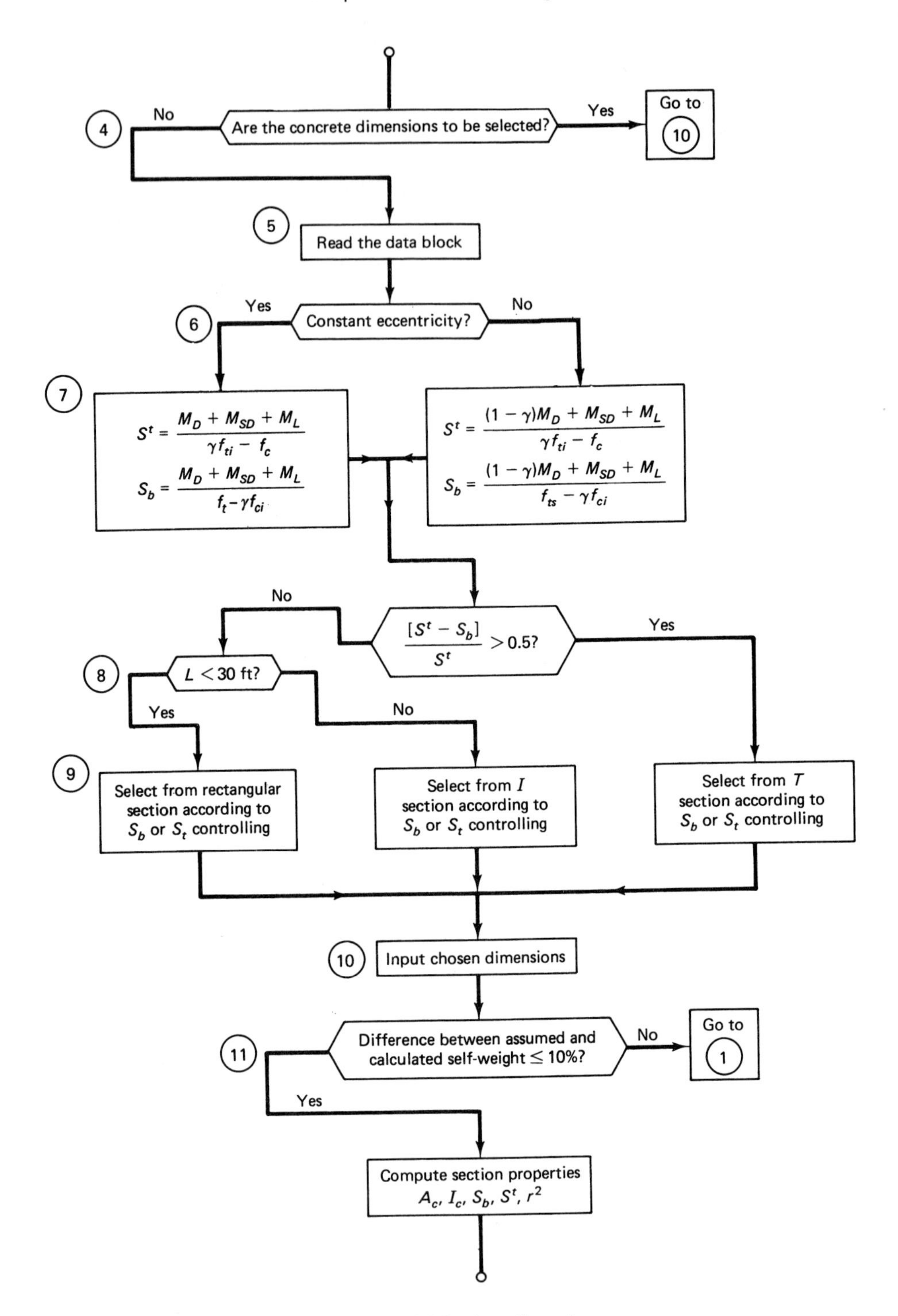

Figure 4.34 *(continued)*

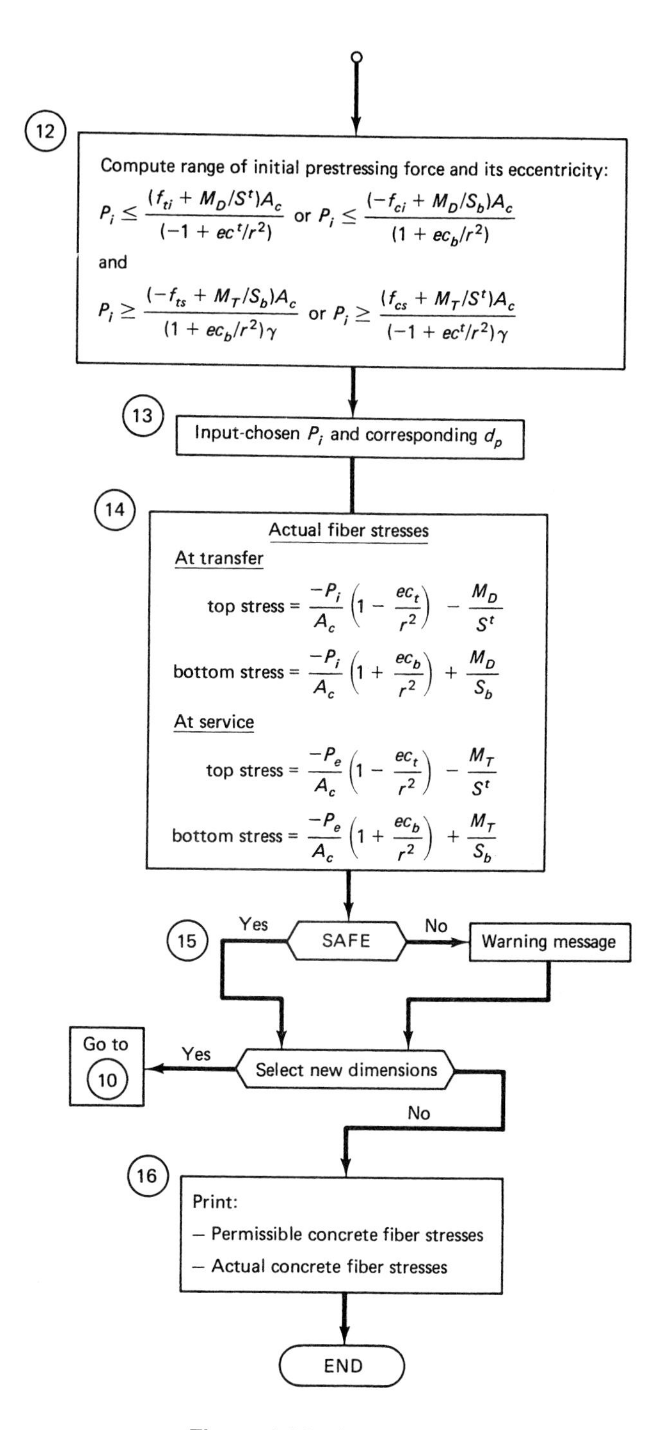

Figure 4.34 *(continued)*

Photo 4.10 Photo layout of bridge deck prestressing tendon.

Prestressing Steel

$$f_{pu} = 270{,}000 \text{ psi } (1{,}862 \text{ MPa})$$

$$f_{py} = 0.90f_{pu} = 243{,}000 \text{ psi } (1{,}675 \text{ MPa})$$

$$f_{pi} = 0.70f_{pu} = 189{,}000 \text{ psi } (1{,}303 \text{ MPa})$$

$$f_{pe} \text{ after losses} = 0.80f_{pi} = 151{,}200 \text{ psi } (1{,}043 \text{ MPa})$$

Locate and draw the distribution of tendons in both the midspan and end sections. Use a live-load moment value including impact due to AASHTO HS20-44 loading for one interior beam of the 64 ft span bridge = 9,300,000 in.-lb (1,051 kN-m).

Solution:

Bending Moments and New Allowable Stresses (Steps 1–4). Since the beam spacing and span length are known, the moments due to situ-cast slab and diaphragms can be initially determined. The clear distance between the webs of the beams is 7 ft, 0 in. – 6 in. = 6 ft, 6 in. Assume a deck slab $5\frac{3}{4}$ in. (14.6 cm) thick made of $1\frac{3}{4}$ in. precast panels and 4 in. situ-cast topping and a diaphragm 8 in. (20 cm) thick at midspan and 45 in. (122 cm) deep, cast integrally with the deck slab. We have:

$$\text{Diaphragm weight} = \frac{8}{12} \times \frac{45}{12} \times 6.5 \times 150 = 2{,}500 \text{ lb } (11.6 \text{ kN})$$

$$1\tfrac{3}{4} \text{ in. precast formwork panel weight} = \frac{1.75}{12} \times 7 \times 150 = 153 \text{ plf } (2.2 \text{ kN/m})$$

$$4 \text{ in. situ-cast topping weight} = \frac{4}{12} \times 7 \times 150 = 350 \text{ plf } (5.1 \text{ kN/m})$$

$$M_{SD1} = \frac{153(64)^2}{8} \times 12 = 940{,}032 \text{ in.-lb}$$

$$M_{CSD} = \text{composite superimposed dead load} = 0$$

$$M_{SD2} = \frac{PL}{4} + \frac{W_{SD}L^2}{8}$$

$$= \frac{2{,}500 \times 64 \times 12}{4} + \frac{350(64)^2\, 12}{8}$$

$$= 499{,}200 + 2{,}150{,}400 = 2{,}649{,}600 \text{ in.-lb}$$

$$\text{Total } M_{SD} = 940{,}032 + 2{,}649{,}000 = 3{,}589{,}632 \text{ in.-lb (406 kN-m)}$$

$$M_{CSD} = 0 \text{ in this case}$$

Minimum Section Moduli and Choice of Trial Section (Steps 5–6)

$$\gamma = \frac{151{,}200}{189{,}000} = 0.80$$

$$S^t \geq \frac{(1 - \gamma)M_D + M_{SD} + M_L}{\gamma f_{ti} - f_c}$$

$$S_b \geq \frac{(1 - \gamma)M_D + M_{SD} + M_L}{f_t - \gamma f_{ci}}$$

Assume that the self-weight of the precast beam element is approximately 583 plf (8.5 kN/m). Then

$$M_D = \frac{583(64)^2 \times 12}{8} = 3{,}581{,}952 \text{ in.-lb (405 kN-m)}$$

$$\text{Min } S^t = \frac{(1 - 0.80)3{,}581{,}952 + 3{,}589{,}632 + 9{,}300{,}000}{0.80(212) - (-2{,}000)} = 6.271 \text{ in.}^3 \text{ (103,222 cm}^{3)}$$

$$\text{Min } S_b = \frac{(1 - 0.80)3{,}581{,}952 + 3{,}589{,}632 + 9{,}300{,}000}{424 - 0.80(-2{,}200)} = 6{,}229 \text{ in.}^3 \text{ (102,075 cm}^3\text{)}$$

The expected actual section modulus for the top fibers is usually considerably larger than the section modulus for the bottom fibers of the composite section. So choose the precast element based on $S_b = 6{,}299$ in.^{3} (100,813 cm^3).

AASHTO type III is chosen as the closest trial section for $S_b = 6{,}229$ in.^{3} since AASHTO type IV has a much larger section modulus. We have as in Figure 4.35.

$$S^t = 5{,}070 \text{ in.}^3 \text{ (83,082 cm}^3\text{)}$$

$$S_b = 6{,}186 \text{ in.}^3 \text{ (101,370 cm}^3\text{)}$$

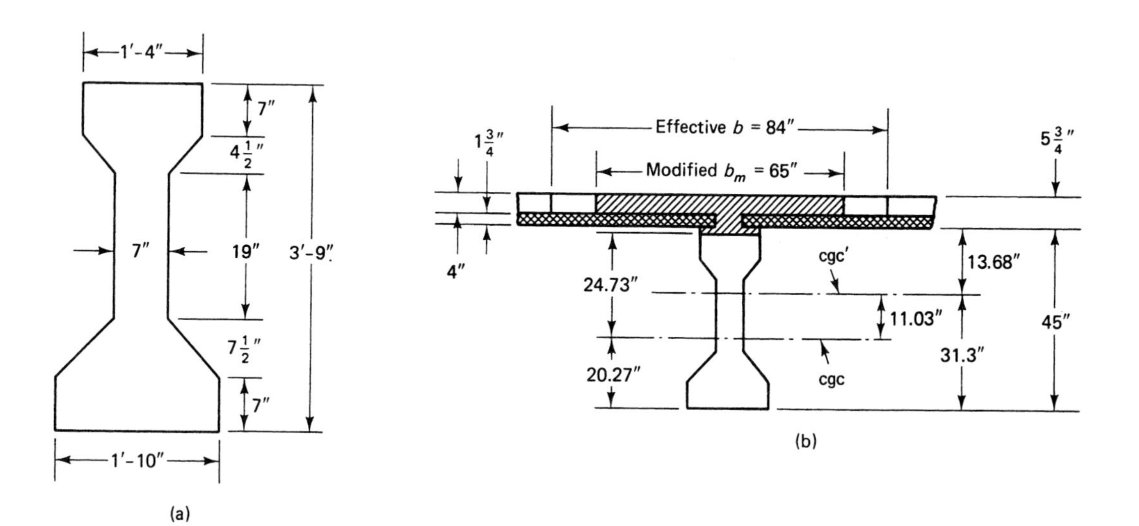

Figure 4.35 Example 4.7. (a) Section (AASHTO-III). (b) Composite section properties.

$$I_c = 125{,}390 \text{ in.}^4 \ (5.2 \times 10^6 \text{ cm}^4)$$

$$A_c = 560 \text{ in.}^2 \ (3{,}613 \text{ cm}^2)$$

$$r^2 = 223.9 \text{ in.}^2 \ (1{,}445 \text{ cm}^2)$$

$$c_t = 24.73 \text{ in. } (62.8 \text{ cm})$$

$$c_b = 20.27 \text{ in. } (51.5 \text{ cm})$$

$$W_D = 583 \text{ plf } (8.2 \text{ kN/m})$$

Try A_{ps} = twenty-two $\frac{1}{2}$-in. dia (12.7 mm) 7-wire stress-relieved tendons.

Stresses at Transfer

$$M_T = 3{,}581{,}952 + 3{,}589{,}632 + 9{,}300{,}000 = 16{,}471{,}584 \text{ in.-lb } (1{,}861 \text{ kN-m})$$

$$A_{ps} = 22 \times 0.153 = 3.366 \text{ in.}^2 = (21.7 \text{ cm}^2)$$

$$P_i = A_{ps} f_{pi} = 3.366 \times 0.70 \times 270{,}000 = 636{,}174 \text{ lb } (2{,}830 \text{ kN})$$

$$P_e = 0.80 P_i = 0.80 \times 636{,}174 = 508{,}940 \text{ lb } (2{,}264 \text{ kN})$$

$$k_t = \frac{r^2}{c_b} = \frac{223.91}{20.27} = 11.05 \text{ in. } (28.1 \text{ cm})$$

$$k_b = \frac{r^2}{c_t} = \frac{223.91}{24.73} = 9.05 \text{ in. } (23.0 \text{ cm})$$

$$a_{\min} = \frac{M_D}{P_i} = \frac{3{,}581{,}952}{636{,}174} = 5.63 \text{ in.}$$

$$a_{\max} = \frac{M_T}{P_e} = \frac{16{,}471{,}584}{508{,}940} = 32.36 \text{ in.}$$ (Upper envelope, outside section, hence section can be improved)

$$e_b = a_{\min} + k_b = 5.63 + 9.05 = 14.68 \text{ in.}$$

$$e_t = a_{\max} - k_t = 32.36 - 11.05 = 21.31 \text{ in.}$$

Hence, use reduced tendon eccentricities

$$e_c = 16.27 \text{ in. } (413 \text{ mm}) \qquad e_e = 10 \text{ in. } (254 \text{ mm})$$

(a) *Midspan Section*

$$f^t = -\frac{P_i}{A_c}\left(1 - \frac{e_c c_t}{r^2}\right) - \frac{M_D}{S^t}$$

$$= -\frac{636{,}174}{560}\left(1 - \frac{16.27 \times 24.73}{223.91}\right) - \frac{3{,}581{,}952}{5{,}070}$$

$$= 905.4 - 706.5 = 198.9 \text{ psi } (T) < 212 \text{ psi, O.K.}$$

$$f_b = -\frac{P_i}{A_c}\left(1 + \frac{e_c c_b}{r^2}\right) + \frac{M_D}{S_b}$$

$$= -\frac{636{,}174}{560}\left(1 + \frac{16.27 \times 20.27}{223.91}\right) + \frac{3{,}581{,}952}{6{,}186}$$

$$= -2{,}809.3 + 579.0 = -2{,}230.3 \ (C) \ (15.4 \text{ MPa}) \cong f_{ci} = -2{,}200 \text{ psi, O.K.}$$

(b) *Support Section*

$$f^t = -\frac{P_i}{A_c}\left(1 - \frac{e_c c_t}{r^2}\right) = -\frac{636{,}174}{560}\left(1 - \frac{10 \times 24.73}{223.91}\right)$$

$$= +118.7 \text{ psi } (T) < 212 \text{ psi, O.K.}$$

$$f_b = -\frac{P_i}{A_c}\left(1 + \frac{e_c c_b}{r^2}\right) = -\frac{636{,}174}{560}\left(1 + \frac{10 \times 20.27}{223.9}\right)$$

$$= -2{,}164.4 \text{ psi } (C) < 2{,}200 \text{ psi, O.K.}$$

Composite Section Properties

$$\frac{E_c\text{ (topping)}}{E_c\text{ (precast)}} = \frac{57{,}000\sqrt{3{,}000}}{57{,}000\sqrt{5{,}000}} = 0.77$$

Effective flange width = 7 ft = 84 in. (213 cm)

Modified effective flange width = 0.77 × 84 = 65 in. (165 cm)

$$c_b' = \frac{(5.75 \times 65)(47.875) + (560 \times 20.27)}{(5.75)(65) + 560} = 31.32 \text{ in.}$$

$$I_c' = 125{,}390 + 560(31.32 - 20.27)^2 + \frac{65(5.75)^3}{12} + 65 \times 5.75(16.56)^2$$

$$= 297{,}044 \text{ in.}^4$$

$$r^2 = 318.12 \text{ in.}^2$$

$$S_{cb} = 9{,}490 \text{ in.}^3$$

$$S_c^t = 21{,}714 \text{ in.}^3 \text{ at top of precast section}$$

$$\text{Precast } c^t = 45 - 31.32 = 13.68 \text{ in.}$$

$$S_c^t = \frac{297{,}044}{13.68} = 21{,}714 \text{ in.}^3$$

$$\text{slab top } c^t = 13.68 + 5.75 = 19.43 \text{ in.}$$

$$S_{cs}^t = \frac{297{,}044}{19.43} = 15{,}288 \text{ in.}^3 \text{ at top of slab}$$

$$S_{bcs} = \frac{297{,}044}{(19.43 - 4)} = 19{,}251 \text{ in.}^3 \text{ at bottom of slab}$$

Stresses After $1\frac{3}{4}$ In. Precast Panel Is Erected As Formwork, W_{SD} (Step 11a) Before Diaphragm and Slab Are Cast,

(a) *Midspan Section*

$$f^t = -\frac{P_e}{A_c}\left(1 - \frac{e_c c_t}{r^2}\right) - \frac{M_D + M_{SD}}{S^t}$$

$$M_D + M_{SD} = 3{,}581{,}952 + 940{,}032 = 4{,}521{,}984 \text{ in.-lb}$$

$$f^t = \frac{508{,}904}{560}\left(1 - \frac{16.27 \times 24.73}{223.9}\right) - \frac{4{,}521{,}984}{5{,}070}$$

$$= +724.3 - 891.9 = -167.6 \text{ psi } (C)\text{, no tension, O.K.}$$

$$f_b = -\frac{P_e}{A_c}\left(1 + \frac{e_c c_b}{r^2}\right) + \frac{M_D + M_{SD}}{S_b}$$

$$= -\frac{508{,}940}{560}\left(1 + \frac{16.27 \times 20.27}{223.9}\right) + \frac{4{,}521{,}984}{6{,}186}$$

$$= -2{,}247.4 + 731.0 = -1{,}516.4 \text{ psi } (C) < 2{,}000 \text{ psi, O.K.}$$

(b) *Support Section*

$$f^t = -\frac{508{,}940}{560}\left(1 - \frac{10 \times 24.73}{223.9}\right) = 94.9 \text{ psi } (T) < f_t = 424 \text{ psi, O.K.}$$

$$f_b = -\frac{508{,}940}{560}\left(1 + \frac{10 \times 20.27}{223.9}\right) = -1{,}731.6 \text{ psi (11.9 MPa) } (C)$$

$$< -2{,}000 \text{ psi, O.K.}$$

Stresses Immediately After Casting Concrete Slab Topping: Midspan Section (Slab Concrete Not Hardened)

$$f^t = -\frac{P_e}{A_c}\left(1 - \frac{e_c c_t}{r^2}\right) - \frac{M_D + M_{SD}}{S^t}$$

$$M_D + M_{SD} = 4{,}521{,}984 + 2{,}649{,}600 = 7{,}171{,}584 \text{ in.-lb}$$

$$f^t = \frac{508{,}940}{560}\left(1 - \frac{16.27 \times 24.73}{223.9}\right) - \frac{7{,}171{,}584}{5{,}070}$$

$$= 724.3 - 1{,}414.5 = -690.2 \text{ psi } (C) < -2{,}000 \text{ psi, O.K.}$$

$$f_b = -\frac{P_e}{A_c}\left(1 + \frac{e_c c_b}{r^2}\right) + \frac{M_D + M_{SD}}{S_b}$$

$$= -\frac{508{,}940}{560}\left(1 + \frac{16.27 \times 20.27}{223.9}\right) + \frac{7{,}171{,}584}{6{,}186}$$

$$= -2{,}247.4 + 1{,}159.3 = -1{,}088.1 \text{ psi } (C)$$

$$< = -2{,}000 \text{ psi (7.5 MPa} < 13.8 \text{ MPa), O.K.}$$

Stresses at Service Load (Step 11) Add the Effect of M_{SD2} Due to Unshored Slab

(a) *Midspan Section*

$$f^t = -\frac{P_e}{A_c}\left(1 - \frac{e_c c_t}{r^2}\right) - \frac{M_D + M_{SD}}{S^t} - \frac{M_{CSD} + M_L}{S_c^t}$$

$$M_D + M_{SD} = 7{,}171{,}584 \text{ in.-lb}$$

$$M_{CSD} + M_L = 0 + 9{,}300{,}000 = 9{,}300{,}000$$

From the previous stage, $f^t = -690.2$ psi (C), $f_b = -1088.1$ psi (C). Stress at top fibers of precast section in composite action is

$$f_c^t = -690.2 - \frac{9{,}300{,}000}{21{,}714} = -690.2 - 428.3 = -1{,}118.5 \text{ psi } (C) < -2000 \text{ psi, O.K.}$$

$$f_{bc} = -1088.1 + \frac{9{,}300{,}000}{9{,}492} = -1088.1 + 979.8 = -108.3 \text{ psi } (C)$$

Modular ratio $n = 0.77$ from before. Stress at top fibers of slab after concrete hardened is

$$f_{cs}^t = -\frac{9{,}300{,}000}{15{,}288} \times 0.77 = -468 \text{ psi } (C)$$

stress at bottom fibers of the slab is

$$f_{bcs} = -\frac{9{,}300{,}000}{19{,}251} \times 0.77 = -372 \text{ psi } (C)$$

(*b*) *Support Section.* Same kind of calculations as in previous step. Result is $f^t = +94.9$ psi (T) and $f_b = -1{,}731.6$ psi (C).

Tendon Envelope

(*a*) *Midspan Section*

$$a_{\max} = 32.36 \text{ in.}$$

$$a_{\min} = 5.63 \text{ in.}$$

(*b*) *Quarter Section*

$$M_D = 2{,}686{,}464$$

$$a_{\min} = \frac{M_D}{P_i} = 4.22 \text{ in.}$$

$$M_T = 12{,}355{,}248$$

$$a_{\max} = \frac{M_T}{P_e} = 24.28 \text{ in.}$$

See Figure 4.36 for the tendon envelope and Figure 4.37 for the stress distribution. Figure 4.38 gives the anchorage zone stresses and the net moments along the depth of the beam.

Design of End-Block Anchorage

(*a*) *Solution by Linear Elastic Method:*

$$P_i = 636{,}174 \text{ lb}$$

$$e_e = 10 \text{ in.}$$

$$f^t = +119 \text{ psi } (T)$$

$$f_b = -2{,}164 \text{ psi } (C)$$

(*i*) *Bursting Crack Reinforcement*

$$h = 45 \text{ in.}$$

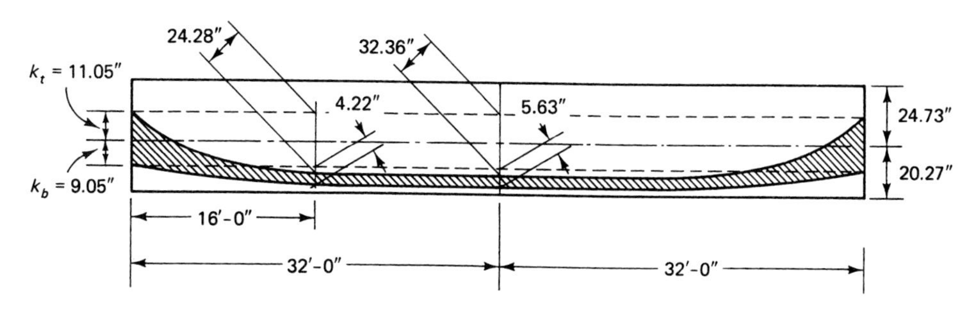

Figure 4.36 Prestressing tendon envelope.

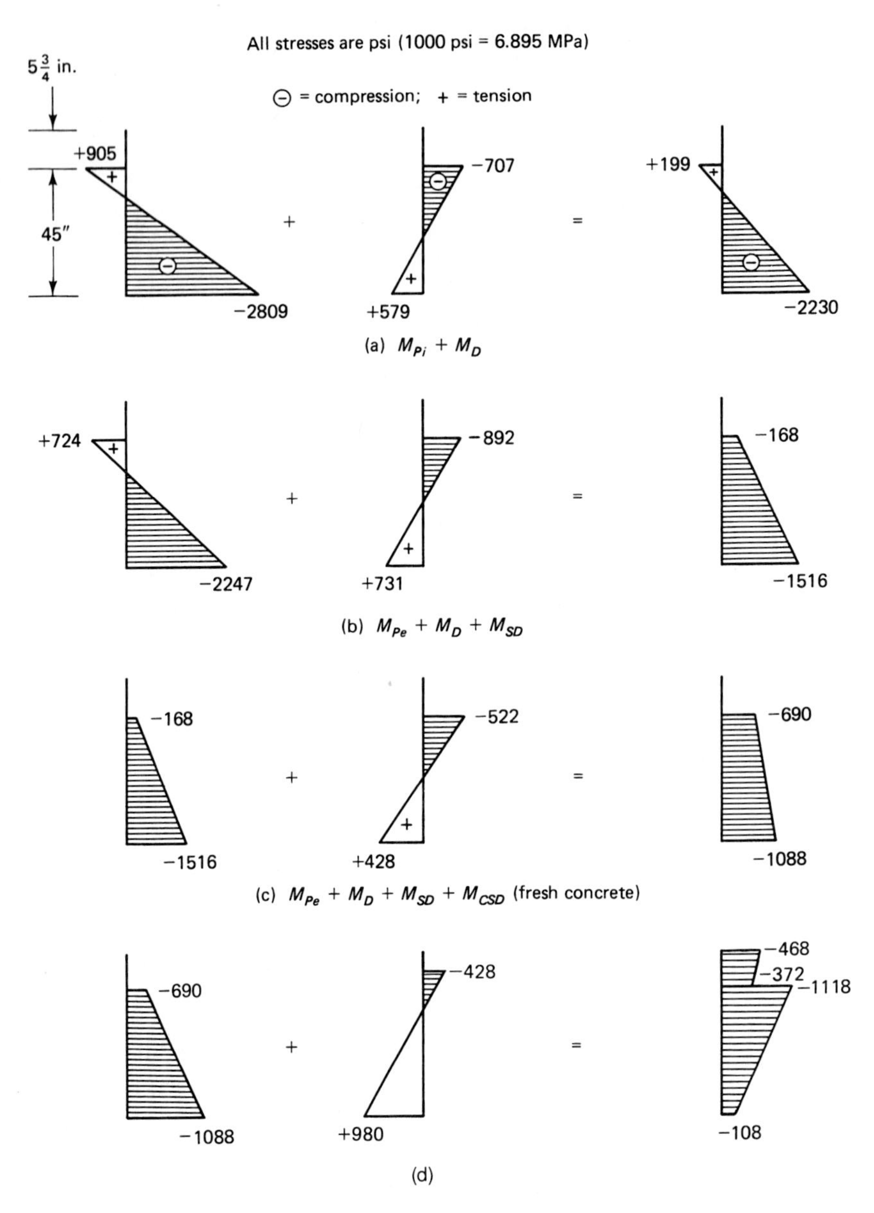

Figure 4.37 Midspan concrete fiber stresses in Example 4.7. (a) Transfer. (b) Erection. (c) Topping cast. (d) Service-load stage.

Use

$$x = h/3 = 15 \text{ in.}$$

$$T_b = \frac{M_{max}}{h - x} = \frac{2.15 \times 10^6}{45 - 15} = 71{,}670 \text{ lb (316 kN)}$$

$$A_s = \frac{T_b}{f_s} = \frac{71{,}670}{20{,}000} = 3.58 \text{ in}^2 \text{ (23.1 cm}^2\text{)}$$

(ii) Spalling Crack Reinforcement

$$T_{sp} = \frac{M_{min}}{h - x} = \frac{+0.04 \times 10^6}{45 - 15} = 1{,}330$$

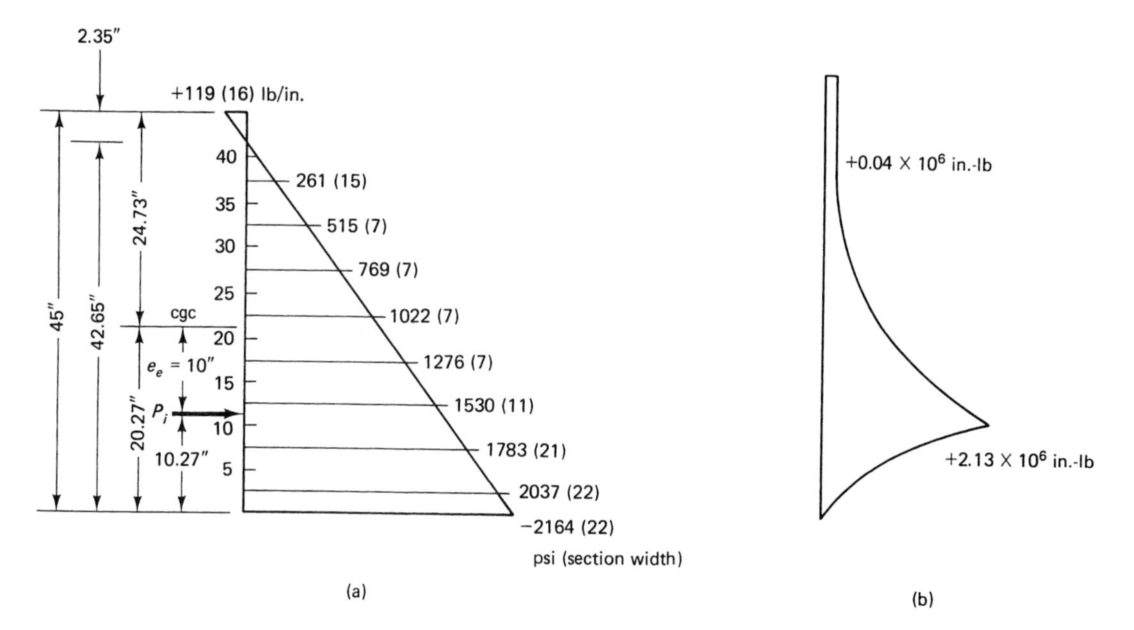

Figure 4.38 Anchorage zone elastic stresses and net moments in Example 4.7. (a) Anchorage zone stresses. Bracketed values are section widths (in.) along which anchorage stresses are acting. (b) Net bursting or splitting moments ($M_p - M_c$) at 5-in. depth intervals.

$$A_s = \frac{T_{sp}}{20{,}000} = \frac{1{,}330}{20{,}000} = 0.07 \text{ in}^2$$

$$\text{Total reinforcement} = 3.58 + 0.07 = 3.65 \text{ in}^2 \text{ (23.5 cm}^2)$$

Try #4 vertical reinforcement (12.7 mm dia.):

$$\text{Number required} = \frac{3.65}{0.20 \times 2} = 9.13$$

Arrangement of Strands. Use a 2 in. × 2 in. (25 mm × 25 mm) grid for arranging the distribution of strands. Eccentricities are:

$$e_c = 16.27 \text{ in. (41.3 cm)}$$

$$e_e = 10.0 \text{ in. (25.4 cm)}$$

The arrangement of the strands to develop the required tendon eccentricities are shown in Fig. 4.39(a).

The anchorage zone moments at the various planes at 5-in. intervals along the height of the section are given in Table 4.7.

(b) Solution by the Plastic Strut-and-Tie Method:

(1) *compute distance of the centroid of tendons:*

$$= \frac{4(2) + 6(4) + 4(6) + 3(5.17) + 2(20.27) + 3(27.77)}{22} = 10.27 \text{ in.}$$

(2) *determine concrete bearing capacity at the anchorage devices plane:*

$$\text{row forces: } P_{u1},\ P_{u3} = 4(0.153)(270{,}000) = 165{,}240 \text{ lb}$$

$$P_{u2} = 6(0.153)(270{,}000) = 247{,}860 \text{ lb}$$

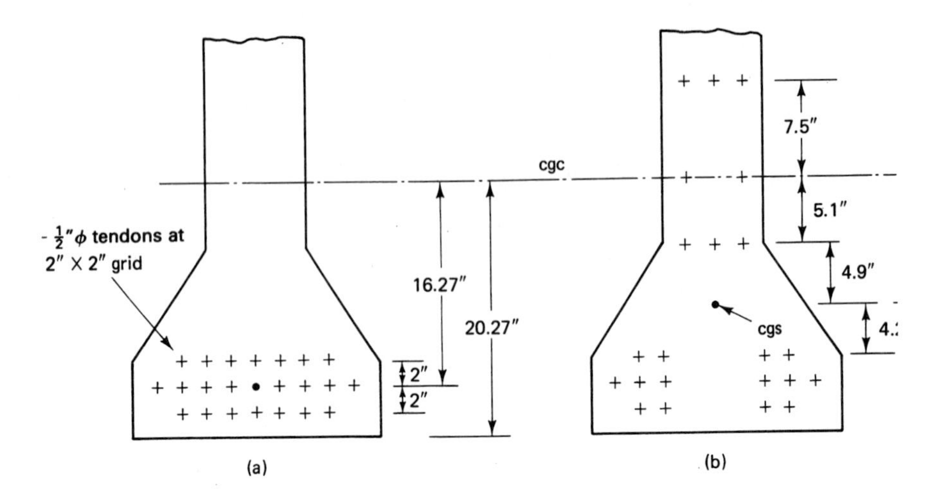

Figure 4.39(a) Strand arrangement in Example 4.7. (a) Midspan ($e_c = 16.27$ in.). (b) End section ($e_e = 10.0$ in.).

$$P_{u4}, P_{u6} = 8(0.153)(270{,}000) = 123{,}930 \text{ lb}$$

$$P_{u5} = 2(0.153)(270{,}000) = 82{,}620 \text{ lb}$$

Total ultimate compressive force = 165,250 (2) + 247,860 + 123,930 (2) + 82,620 = 908,820 lb

Total area of rigid bearing plates supporting the anchorage chucks:

$$A_b = 16 \times 12 + 8 \times 16 = 320 \text{ in.}^2$$

$$f_b = \frac{908{,}820}{320} = 2840 \text{ psi}$$

Table 4.7 Anchorage Zone Moments for Example 4.7

Moment plane dist. d from bottom (in.)	Section width (in.)	Stress at plane ($d - 2.5$) (psi)	Concrete resist. force at ($d - 2.5$) (lb)	Moment M_p of force P_i about horiz. plane in col. (1) (in.-lb × 10^6)	Moment M_c of concrete in col. (4) about horiz. plane in col. (1) (in.-lb × 10^6)	Net moment ($M_c - M_p$) col. (6) – col. (5) (in.-lb × 10^6)
(1)	(2)	(3)	(4)	(5)	(6)	(7)
0	22	−2,164	0	0	0	0
5	22	−2,037	237,040	0	+0.56	+0.56
10.27	15	−1,783	133,725	0	+2.15	+2.15
15	7	−1,530	53,550	−3.01	+4.42	+1.41
20	7	−1,276	44,660	−6.19	+7.00	+0.81
25	7	−1,022	35,770	−9.37	+9.80	+0.43
30	7	−769	26,915	−12.55	+12.76	+0.19
35	12	−515	25,750	−15.73	+15.80	+0.07
40	16	−216	20,880	−18.91	+18.95	+0.04
45	16	+119	+9,520	−22.09	+22.15	+0.06

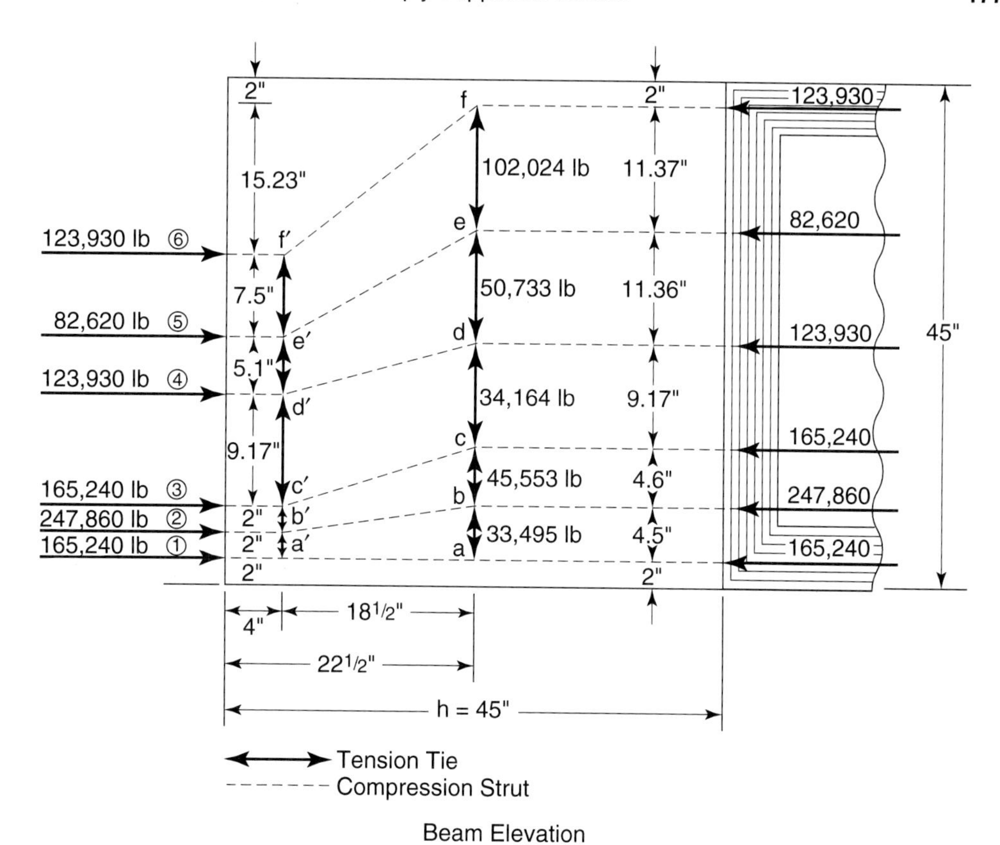

Figure 4.39(b) Strut-and-Tie Forces in Example 4.7.

From Equations 4.16(a) and (b), the maximum allowable bearing pressure on the concrete is

$$f_b \leq 0.7\,\phi f'_{ci}\sqrt{A/A_g}$$

$$f_b \leq 2.25\,\phi f'_{ci}$$

Assume that the initial concrete strength at stressing is $f'_{ci} = 0.75\,f'_c$

$$= 0.75 \times 5{,}000 = 3750 \text{ psi}$$

concentric area, A, of concrete with the bearing plates $\cong 18 \times 16 + 10 \times 18 = 468$ in.^{2}

$$\text{Allowable bearing stress } f_b = 0.7 \times 0.90 \times 3750\sqrt{\frac{468}{320}} = 2860 \text{ psi.} > 2840 \text{ psi, O.K.}$$

The bearing stress from Eq. 4.14 (b) does not control.

(3) *Draw the strut-and-tie model and select the anchorage reinforcement* choose distance a, as in Figure 4.39(b) between two forces $P_{u5} - P_{u6} = 7.5$ in.

$$\text{hence, depth } a/2 \text{ ahead of the anchorages} = \frac{7.5}{2} = 3.75 \text{ in., say 4 in.}$$

Construct the strut-and-tie model assuming it to be as shown in Figure 4.39 (b). The tension tie forces range between 33,495 and 102,024 lb. Choosing the larger value of 102,204 lb and using No. 3 closed U-stirrups confining reinforcement within the anchorage zone area:

$$\text{tensile strength per tie} = \phi f_y A_v$$

$$= 0.90 \times 60{,}000 \times 2(0.11) = 11{,}880 \text{ lb}$$

$$\text{required number of stirrup ties} = \frac{102{,}024}{11{,}880} = 8.59, \text{ use nine No. 3 closed U-stirrups}$$

Trying No. 4 closed U-Stirrups in the compression zone adjacent to the anchorage devices plane, the applicable force can also be assumed in this case to be approximately 102,024 lb.

$$\text{tensile strength of one No. 4 confining tie} = 0.90 \times 60{,}000 \times 20(0.20) = 21{,}600 \text{ lb.}$$

$$\text{number of ties} = \frac{102{,}024}{21{,}600} = 4.7, \text{ used five No. 4 closed U-stirrups.}$$

Comparing solutions (a) and (b), adopt the following confining reinforcement in the anchorage zone over a distance $h = 45$ in. from the beam end:

Use five No. 4 closed U-stirrups starting at $1\frac{1}{2}$in. from the anchorage devices plane and spaced at $1\frac{1}{2}$in. on centers, then continue with the nine stirrups at 5 in. on centers over a distance of 40 in. It should be noted that if a smaller number of path lines are assumed in the idealization of the compression strut paths, the tension tie forces would have been larger resulting in more confining reinforcement.

4.9 ULTIMATE-STRENGTH FLEXURAL DESIGN

4.9.1 Cracking-Load Moment

As mentioned in Chapter 1, one of the fundamentals differences between prestressed and reinforced concrete is the continuous shift in the prestressed beams of the compressive C-line away from the tensile cgs line as the load increases. In other words, the moment arm of the internal couple continues to increase with the load without any appreciable change in the stress f_{pe} in the prestressing steel. As the flexural moment continues to increase when the full superimposed dead load and live load act, a loading stage is reached where the concrete compressive stress at the bottom-fibers reinforcement level of a simply supported beam becomes zero. This stage of stress is called the limit state of *decompression:* Any additional external load or overload results in cracking at the bottom face, where the modulus of rupture of concrete f_r is reached due to the cracking moment M_{cr} caused by the first cracking load. At this stage, a sudden increase in the steel stress takes place and the tension is dynamically transferred from the concrete to the steel.

Figure 4.40 relates load to steel stress at the various loading stages. It shows not only the load-deformation curve, including its abrupt change of slope at the first cracking load, but also the dynamic dislocation in the load-stress diagram at the first cracking load after decompression in a bonded prestressed beam. Beyond that dislocation point the beam can no longer be considered to behave elastically, and the rise in the compressive C-line stabilizes and stops so that the section starts to behave like a reinforced concrete section with *constant* moment resistance arm.

It is important to evaluate the first cracking load, since the section stiffness is reduced and hence an increase in deflection has to be considered. Also, the crack width has to be controlled in order to prevent reinforcement corrosion or leakage in liquid containers.

The concrete fiber stress at the tension face is

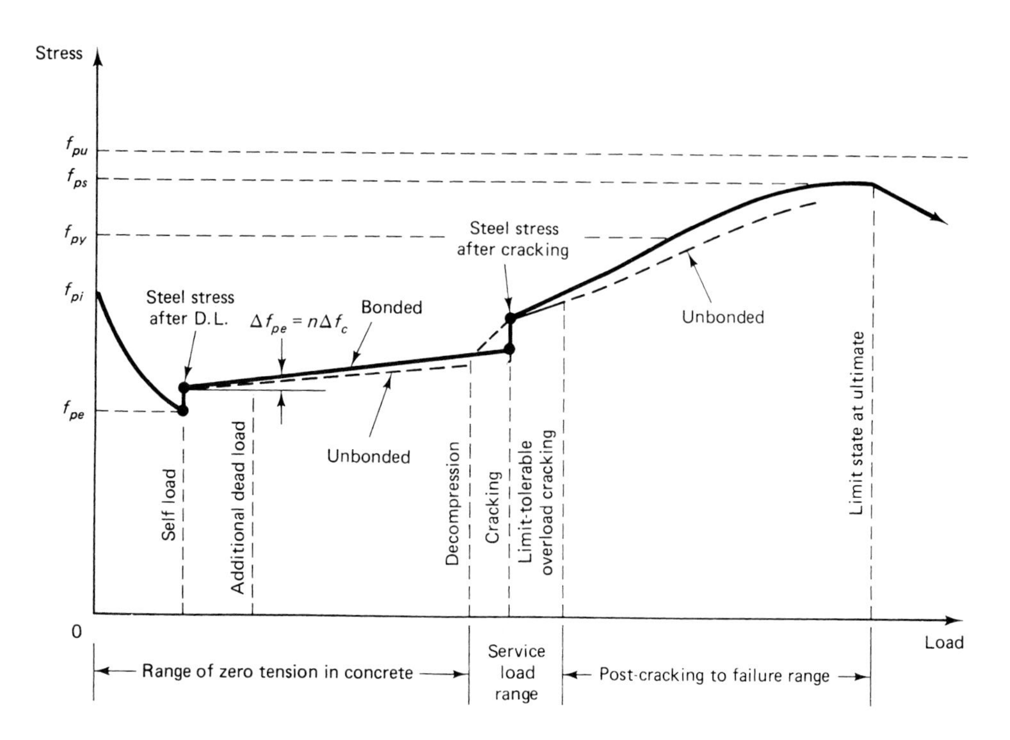

Figure 4.40 Prestressing steel stress at various load levels.

$$f_b = -\frac{P_e}{A_c}\left(1 + \frac{ec_b}{r^2}\right) + \frac{M_{cr}}{S_b} = f_r \tag{4.24}$$

where the modulus of rupture $f_r = 7.5\sqrt{f'_c}$ and the cracking moment M_{cr} is the moment due to all loads at that load level $(M_D + M_{SD} + M_L)$. From Eq. 4.24,

$$M_{cr} = f_r S_b + P_e\left(e + \frac{r^2}{c_b}\right) \tag{4.25}$$

Note that the term r^2/c_b is the upper kern value k_t so that $P_e r^2/C_b$ denotes the elastic moment required to raise the C-line from the prestressing steel level to the upper kern point giving zero tension at the bottom fibers. Consequently, the term $f_r S_b$ is that additional moment required to cause the development of the first crack at the extreme tension fibers due to overload, such as at the bottom fibers at midspan of a simply supported beam.

4.9.2 Partial Prestressing

"Partial prestressing" is a controversial term, since it is *not* intended to denote that a beam is prestressed partially, as might seem to be the case. Rather, partial prestressing describes prestressed beams wherein limited cracking is permitted through the use of additional mild nonprestressed reinforcement to control the extent and width of the cracks and to assume part of the ultimate flexural moment strength. Two major advantages of partial prestressing are the efficient use of all constituent materials and the control of excessive camber due to the long-term creep of concrete under compression.

Reinforced concrete beams always have to be designed as underreinforced to ensure ductile failure by yielding of the reinforcement. Prestressed beams can be either *un-*

derreinforced using a relatively small percentage of mild nonprestressed steel, leading to rupture of the tensile steel at failure, or essentially *overreinforced,* using a large percentage of steel, resulting in crushing of the concrete at the compression top fibers in a somewhat less ductile failure.

Another type of, say, premature failure occurs at the first cracking load level, where M_{cr} approximates the nominal moment strength M_n of the section. This type of failure can occur in members that are prestressed and reinforced with very small amounts of steel, or in members that are concentrically prestressed with small amounts of steel, or in hollow members.

It is generally advisable to evaluate the magnitude of the cracking moment M_{cr} in order to determine the reserve strength and overload limits that the designed section has.

4.9.3 Cracking Moment Evaluation

Example 4.8

Calculate the cracking moment M_{cr} in the I-beam of Example 4.2, and evaluate the magnitude of the overload moment that the beam can tolerate at the modulus of rupture of concrete. Also, determine what safety factor the beam has against cracking due to overload. Given is $f_r = 7.5\sqrt{f'_c} = 7.5\sqrt{5{,}000} = 530$ psi (3.7 MPa).

Solution: From Example 4.2,

$$P_e = 308{,}225 \text{ lb } (1{,}371 \text{ kN})$$

$$r^2/c_b = 187.5/18.84 = 9.95 \text{ in. } (25.3 \text{ cm})$$

$$S_b = 3{,}750 \text{ in.}^3 \ (61{,}451 \text{ cm}^3)$$

$$e_c = 14 \text{ in. } (35.6 \text{ cm})$$

$$M_D + M_{SD} = 2{,}490{,}638 \text{ in.-lb } (281 \text{ kN-m})$$

$$M_L = 7{,}605{,}00 \text{ in.-lb } (859 \text{ kN-m})$$

From Equation 4.25,

$$M_{cr} = f_r S_b + P_e\left(e + \frac{r^2}{c_b}\right)$$

$$= 530 \times 3750 + 308{,}255(14 + 9.95)$$

$$= 9{,}370{,}207 \text{ in.-lb } (1{,}509 \text{ kN-m})$$

$$M_T = M_D + M_{SD} + M_L = 2{,}490{,}638 + 7{,}605{,}000$$

$$= 10{,}095{,}638 \text{ in.-lb } (1{,}141 \text{ kN-m})$$

$$\text{Overload moment} = M_T - M_{cr} = 10{,}095{,}638 - 9{,}370{,}207$$

$$= 725{,}431 \text{ in.-lb } (83 \text{ kN-m})$$

Since $M_T > M_{cr}$, the beam had tensile cracks at service load, as the design in Example 4.2 presupposed. The safety factor against cracking is given by

$$\frac{M_{cr}}{M_T} = \frac{9{,}370{,}207}{10{,}095{,}638} = 0.93$$

If the service load M_T is less than M_{cr}, the safety factor against cracking will be greater than 1, as in nonpartially prestressed members.

Note that where nonprestressed reinforcement is used to develop a partially prestressed section, the total factored moment $\phi M_n \geq 1.2M_{cr}$, as required by the ACI code.

4.10 LOAD AND STRENGTH FACTORS

4.10.1 Reliability and Structural Safety of Concrete Components

Three developments in recent decades have majorly influenced present and future design procedures: the vast increase in the experimental and analytical evaluation of concrete elements, the probabilistic approach to the interpretation of behavior, and the digital computational tools available for rapid analysis of safety and reliability of systems. Until recently, most safety factors in design have had an empirical background based on local experience over an extended period of time. As additional experience is accumulated and more knowledge is gained from failures as well as familiarity with the properties of concrete, factors of safety are adjusted and in most cases lowered by the codifying bodies.

In 1956, Baker (Ref. 4.24) proposed a simplified method of safety factor determination, as shown in Table 4.8, based on probabilistic evaluation. This method expects the design engineer to make critical choices regarding the magnitudes of safety margins in a design. The method takes into consideration that different weights should be assigned to the various factors affecting a design. The weighted failure effects W_t for the various factors of workmanship, loading conditions, results of failure, and resistance capacity are tabulated in the table.

The safety factor against failure is

$$\text{S.F.} = 1.0 + \frac{\Sigma W_t}{10} \tag{4.26}$$

where the maximum total weighted value of all parameters affecting performance equals 10.0. In other words, for the worst combination of conditions affecting structural performance, the safety factor S.F. = 2.0.

This method assumes adequate prior performance data similar to a design in progress. Such data in many instances are not readily available for determining safe weighted values W_t in Eq. 4.26. Additionally, if the weighted factors are numerous, a probabilistic determination of them is more difficult to codify. Hence, an undue value-

Table 4.8 Baker's Weighted Safety Factor

Weighted Failure Effect		Maximum W_t
1. Results of failure: 1.0 to 4.0		
Serious, either human or economic		4.0
Less serious, only the exposure of nondamageable material	1.0	
2. Workmanship: 0.5 to 2.0		
Cast in place		2.0
Precast "factory manufactured"	0.5	
3. Load conditions: 1.0 to 2.0 (high for simple spans and overload possibilities; low for load combinations such as live loads and wind)		2.0
4. Importance of member in structure (beams may use lower value than columns)		0.5
5. Warning of failure		1.0
6. Depreciation of strength		0.5
		Total = ΣW_t = 10.0

$$\text{S.F.} = 1.0 + \frac{\Sigma W_t}{10}$$

judgment burden is probably placed on the design engineer if the full economic benefit of the approach is to be achieved.

Another method with a smaller number of probabilistic parameters deals primarily with loads and resistances. Its approaches for steel and concrete structures are generally similar: both the load-and-resistance-factor-design method (LRFD) and first-order second-moment method (FOSM) propose general reliability procedures for evaluating probability-based factored load design criteria (see Refs. 4.25 and 4.26). They are intended for use in proportioning structural members on the basis of load types such that the resisting strength levels are *greater* than the factored load or moment distributions. As these approaches are basically load oriented, they reduce the number of individual variables that have to be considered, such as those listed in Table 4.8.

Assume that ϕ_i represents the resistance factors of a concrete element and that γ_i represents the load factors for the various types of load. If R_n is the nominal resistance of the concrete element and W_i represents the load effect for various types of superimposed load,

$$\phi_i R_n \geq \gamma_i W_i \tag{4.27}$$

where i represents the load in question, such as dead, live, wind, earthquake, or time-dependent effects.

Figures 4.41(a) and (b) show a plot of the separate frequency distributions of the actual load W and the resistance R with means values $\overline{R}$ and $\overline{W}$. Figure 4.41(c) gives the two distributions superimposed and intersecting at point C.

It is recognized that safety and reliable integrity of the structure can be expected to exist if the load effect W falls at a point to the left of intersection C on the W curve, and to the right of intersection C on the resistance curve R. Failure, on the other hand, would be expected to occur if the load effect or the resistance fall within the shaded area in Fig. 4.41(c). If β is a safety index, then

$$\beta = \frac{\overline{R} - \overline{W}}{\sqrt{\sigma_R^2 + \sigma_W^2}} \tag{4.28}$$

where σ_R and σ_W are the standard deviations of the resistance and the load, respectively.

A plot of the safety index β for a hypothetical structural system against the probability of failure of the system is shown in Fig. 4.42. One can observe that such a probability is reduced as the difference between the mean resistance $\overline{R}$ and load effect $\overline{W}$ is

Photo 4.11 Post-tensioning bridge deck conduits.

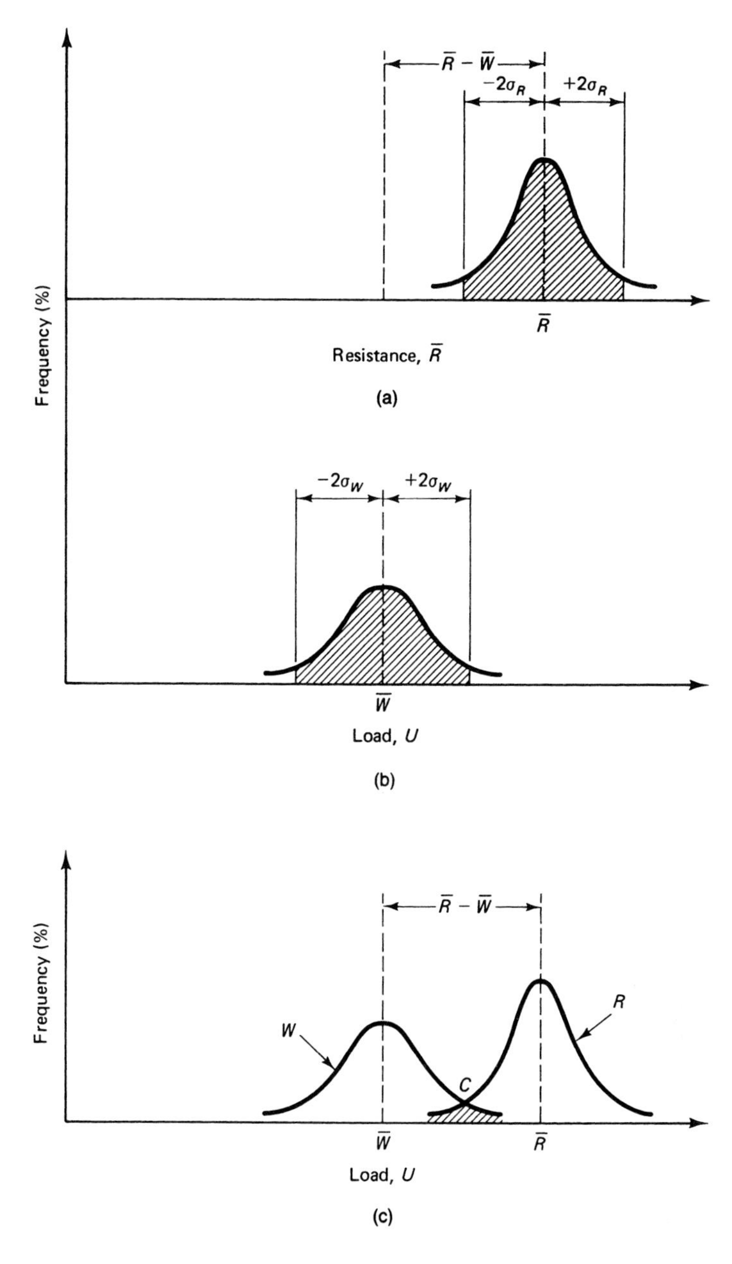

Figure 4.41 Frequency distribution of loads vs. resistance.

increased, or the variability of resistance as measured by their standard deviations σ_R and σ_W is decreased, thereby reducing the shaded area under intersection C in Fig. 4.41(c).

The extent of increasing the difference $(\bar{R} - \bar{W})$ or decreasing the degree of scatter of σ_R or σ_W is naturally dictated by economic considerations. It is economically unreasonable to design a structure for zero failure, particularly since types of risk other than load are an accepted matter, such as the risks of severe earthquake, hurricane, volcanic eruption, and fire. Safety factors and corresponding load factors would thus have to disregard those types or levels of load, stress, and overstress whose probability of occurrence is very low. In spite of this, it is still possible to achieve reliable safety conditions by choosing such a safety index value β through a proper choice of R_n and W_i using the appropri-

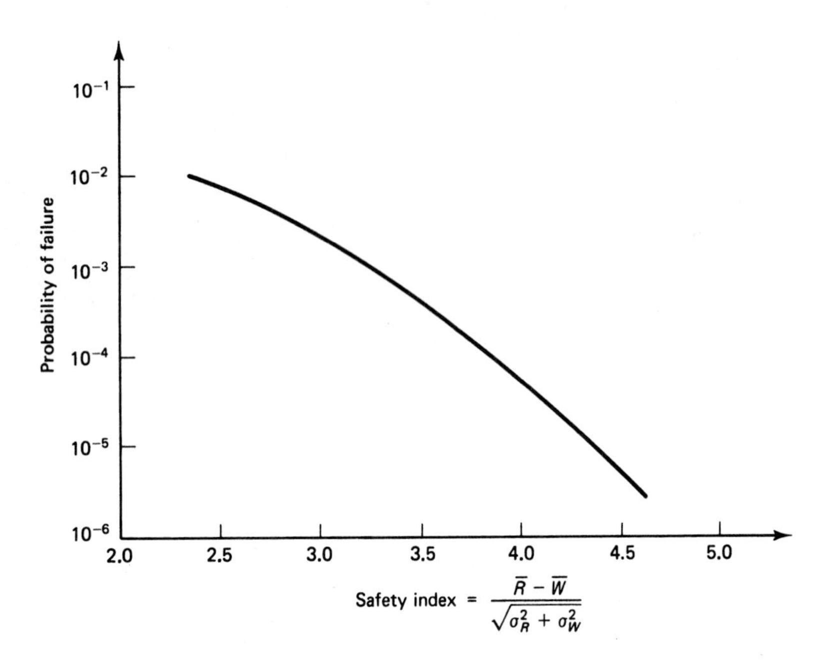

Figure 4.42 Probability of failure vs. safety index.

ate resistance factors ϕ_i and load factors γ_i in Equation 4.27. A safety index β having the value 1.75 to 3.2 for concrete structures is suggested where the lower value accounts for load contributions from wind and earthquake.

If the factored external load is expressed as U_i, then $\Sigma\,\gamma_i W = U_i$ for the different loading combinations. The following are values of U_i recommended in Ref. 4.25 to choose a maximum U for use in Eq. 4.27, that is, $\phi_i R_n \geq \gamma_i\, W_i \geq U_{i(\max)}$:

$$\gamma_i U_w = \max \begin{cases} 1.4D_n \\ 1.2D_n + 1.6L_n \\ 1.2D_n + 1.6S_n + (0.5L_n \text{ or } 0.8W_n) \\ 1.2D_n + 1.3W_n + 0.5L_n \\ 1.2D_n + 1.5E_n + (0.5L_n \text{ or } 0.2S_n) \\ 0.9D_n - (1.3W_n \text{ or } 1.5E_n) \end{cases} \tag{4.29}$$

where the subscript n stands for the nominal value of the variable working load, i.e.,

D_n = dead load L_n = live load S_n = snow load
W_n = wind load E_n = earthquake load

ϕ_i and γ_i are considered to have optimal values; a resistance factor ϕ of 0.7 to 0.85 is recommended. In this probabilistic approach, the present ACI code would require that

$$\phi_i\, R_n = \max \begin{cases} 1.4D_n + 1.7L_n \\ 0.9D_n - 1.3W_n \end{cases} \tag{4.30}$$

As more substantive records of performance are compiled, the details of the foregoing approach to reliability, safety, and reserve strength evaluation of structural components will be more universally accepted and extended beyond the treatment of the component elements to the treatment of the total structural system.

Photo 4.12 Tendon stressing with Freyssinet jack.

4.10.2 ACI Load Factors and Safety Margins

The general concepts of safety and reliability of performance presented in the preceding section are inherent in a more simplified but less accurate fashion in the ACI code. The load factors γ and the strength reduction factors ϕ give an overall safety factor based on load types such that

$$S.F. = \frac{\gamma_1 D + \gamma_2 L}{D + L} \times \frac{1}{\phi} \tag{4.31}$$

where ϕ is the strength reduction factor and γ_1 and γ_2 are the respective load factors for the dead load D and the live load L. Basically, a single common factor is used for dead load and another for live load. Variation in resistance capacity is accounted for in ϕ. Hence, the method is a simplified empirical approach to safety and the reliability of structural performance that is not economically efficient for every case and not fully adequate in other instances, such as combinations of dead and wind loads.

The ACI factors are termed *load factors,* as they restrict the estimation of reserve strength to the loads only as compared to the other parameters listed in Table 4.8. The estimated service or working loads are magnified by the coefficients, such as a coefficient of 1.4 for dead loads and 1.7 for live load. The types of normally occurring loads can be identified as dead load, D; live load, L; wind load, W; loads due to lateral pressure such as from soil in a retaining wall, H; lateral fluid pressure loads, F; loads due to earthquake, E; and loads due to time-dependent effects, such as creep or shrinkage.

The basic combination of vertical gravity loads is dead load plus live load. The *dead load,* which constitutes the weight of the structure and other relatively permanent features, can be estimated more accurately than the live load. The *live load* is estimated using the weight of nonpermanent loads, such as people and furniture. The transient nature of live loads makes them difficult to estimate more accurately. Therefore, a higher load factor is normally used for live loads then for dead loads. If the combination of loads consists only of live and dead loads, the ultimate load can be taken as

$$U = 1.4D + 1.7L \tag{4.32a}$$

Structures are seldom subjected to dead and live loads alone; wind load is often present. For structures in which wind load should be considered, the recommended combination is

$$U = 0.75(1.4D + 1.7L + 1.7W) \tag{4.32b}$$

Maximum dead, live, and wind loads rarely, if ever, occur simultaneously. Hence, the total factored load has to be reduced using a reduction factor of 0.75. Since wind load is applied laterally, it is possible that the absence of vertical live load while wind load is present can produce maximum stress. The following load combination should also be used to arrive at the maximum value of the factored load U:

$$U = 0.9D + 1.3W \tag{4.32c}$$

Structures that have to resist lateral pressure due to earth fill or fluid pressure should be designed for the worst or the following combination of factored loads:

$$U = 1.4D + 1.7L + 1.7H \tag{4.33a}$$

$$U = 0.9D + 1.7H \tag{4.33b}$$

$$U = 1.4D + 1.7L \tag{4.33c}$$

$$U = 1.4D + 1.7L + 1.4F \tag{4.33d}$$

$$U = 0.9D + 1.4F \tag{4.33e}$$

$$U = 1.4D + 1.7L \tag{4.33f}$$

The largest of the following combinations must be considered for earthquake loading:

$$U = 0.75(1.4D + 1.7L + 1.87E) \tag{4.34a}$$

$$U = 0.9D + 1.43E \tag{4.34b}$$

$$U \geq 1.4D + 1.7L \tag{4.34c}$$

The philosophy used for combining the various load components for earthquake loading is essentially the same as that used for wind loading.

4.10.3 Design Strength vs. Nominal Strength: Strength-Reduction Factor ϕ

The strength of a particular structural unit calculated using the current established procedures is termed *nominal strength*. For example, in the case of a beam, the resisting moment capacity of the section calculated using the equations of equilibrium and the properties of concrete and steel is called the *nominal strength moment* M_n of the section. This nominal strength is reduced using a strength reduction factor ϕ to account for inaccuracies in construction, such as in the dimensions or position of reinforcement or variations in properties. The reduced strength of the member is defined as the design strength of the member.

For a beam, the design moment strength ϕM_n should be at least equal to, or better, slightly greater than, the external factored moment M_u for the worst condition of factored load U. The factor ϕ varies for the different types of behavior and for the different types of structural elements. For beams in flexure, for instance, ϕ is 0.9.

For tied columns that carry dominant compressive loads, the factor ϕ equals 0.7. The smaller strength-reduction factor used for columns is due to the structural importance of the columns in supporting the total structure compared to other members, and to guard against progressive collapse and brittle failure with no advance warning of collapse. Beams, on the other hand, are designed to undergo excessive deflections before failure. Hence, the inherent capability of the beam for advanced warning of failure permits the use of a higher strength reduction factor or resistance factor.

Table 4.9 summarizes the resistance factors ϕ for various structural elements as given in the ACI code. A comparison of these values to those given in Ref. 4.24 indicates

Table 4.9 Resistance or Strength Reduction Factor ϕ

Structural Element	Factor ϕ
Beam or slab: bending or flexure	0.9
Columns with ties	0.7
Columns with spirals	0.75
Columns carrying very small axial loads (refer to Chapter 9 for more details)	0.7–0.9, or 0.75–0.9
Beam: shear and torsion	0.85

that the ϕ values in this table, as well as the load factors of Equation 4.33, are in some cases more conservative than they should be. In cases of earthquakes, wind, and shear forces, the probability of load magnitude and reliability of performance are subject to higher randomness, and hence a higher coefficient of variation, than the other types of loading.

4.10.4 AASHTO Strength-Reduction Factors

Flexure

For factory-produced precast prestressed concrete members, $\phi = 1.0$
For post-tensioned cast-in-place concrete members, $\phi = 0.95$

Shear and Torsion

Reduction factor for prestressed members, $\phi = 0.90$
See LRFD and Standard AASHTO other factors in Chapter 12.

4.10.5 ANSI Alternate Load and Strength-Reduction Factors

The ANSI/ASCE Standard, "Minimum Design Loads for Buildings and Other Structures," presents a set of load factors and load combinations for use in the strength design of concrete, steel, or masonry buildings. These factors can be summarized as follows:

$$U = 1.4D + 1.0L \tag{4.35a}$$

$$U = 1.2D + 1.6L \tag{4.35b}$$

$$U = 1.2D + 1.3W + 0.5L \tag{4.35c}$$

$$U = 1.2D + 1.5E + 0.5L \tag{4.35d}$$

$$U = 0.9D + (1.3W \quad \text{or} \quad 1.5E) \tag{4.35e}$$

The load factors for live load L in Equation 4.35c and d are to be taken as 1.0 for areas occupied in places of public assembly and all areas where specified live loads exceed 100 psf.

Where deformation-induced forces T are significant in design, structural members and connections are to be designed for the following load combinations, which are modified versions of Equations 4.35a and b:

$$U = 1.4D + 1.0L + 1.2T \tag{4.36a}$$

$$U = 1.2(D + T) + 1.6L \tag{4.36b}$$

Estimates of differential settlement, creep, shrinkage, and temperature changes are to be based on realistic assessment of such effects occurring in service.

The corresponding strength reduction factors to be used in conjunction with Equations 4.35 and 4.36 are as follows:

Flexure and or axial tensions: 0.85
Axial compression members spirally reinforced: 0.70
Other axial compression members: 0.65
Shear and torsion: 0.75
Bearing on concrete: 0.60

There is little significant difference between using the ACI Code factors and the alternative ANSI Code factors in the design of members. As an example comparing the flexural strength by both methods, we have

$$\text{ACI Code: } N = \frac{U}{\phi} = \frac{1.4D + 1.7L}{0.90} = 1.56D + 1.89L$$

$$\text{ANSI Code: } N = \frac{U}{\phi} = \frac{1.2D + 1.6L}{0.85} = 1.41D + 1.88L$$

The comparison shows that the reserve strength for live load in both methods is almost the same (a factor of 1.89 in the ACI versus 1.88 in the ANSI). For the dead load, which normally does not exceed 20 to 30 percent of the live load, the difference is less than 10 percent, with the cumulative effect on the total for all types of load being less than 2 to 3 percent. Hence, design by either approach would be almost identical to that by the other for all practical purposes. The ACI Code factors are the mandatory ones to use.

4.11 LIMIT STATE IN FLEXURE AT ULTIMATE LOAD IN BONDED MEMBERS: DECOMPRESSION TO ULTIMATE LOAD

4.11.1 Introduction

As discussed in Section 4.9.1, the prestressed concrete beam starts to behave like a reinforced concrete beam when the value of the flexural moment is well beyond the cracking moment M_{cr} and the total service load moment M_T. The ultimate theory in flexure and the principles and concepts underlying it are thus equally applicable to prestressed concrete. A detailed fundamental treatment of this subject is given in chapter 5 of Ref. 4.2. The same fundamental format of equations will be given here, modified to reflect the characteristics of the different reinforcing materials and the geometry peculiar to prestressed concrete.

Cracking develops when the tensile stress in the concrete at the extreme fibers of the critical section exceeds the maximum stress level $f_r \cong 7.5\sqrt{f'_c}$. Prior to attaining this level, the overload causes the compressive stress in the concrete *at the level of the prestressing steel* to continually decrease until it becomes zero at a load level termed the *decompression load.* The stress level in the tendons is correspondingly termed the *decompression stress* (see Figures 4.3 and 4.43).

Some investigators define the decompression load as the load at which the first crack appears at the extreme fibers of the critical section, such as the bottom midspan fibers of a simply supported beam. A minor difference results in the analysis using this definition of the decompression load.

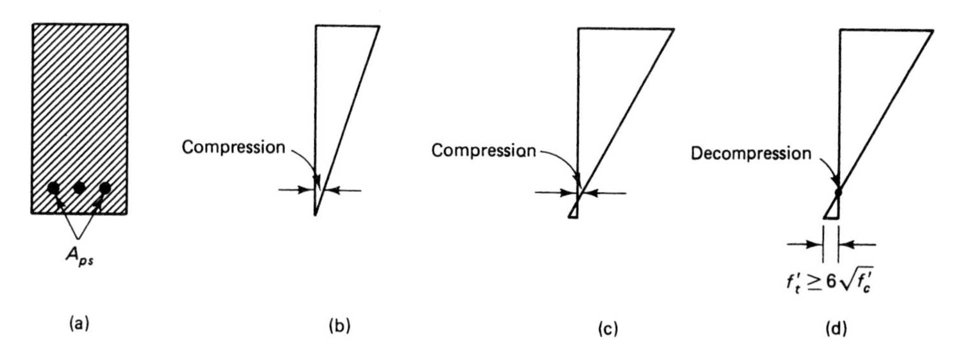

Figure 4.43 Strain in the concrete up to the decompression stage at the tension reinforcement level. (a) Cross section. (b) Entire section in compression. (c) Tension at the lower fibers below the modulus-of-rupture level. (d) Decompression stage with zero stress in the concrete at the level of the prestressing reinforcement.

To follow the loading step-by-step states, suppose that the effective prestress f_{pe} at service load due to all loads results in a strain ϵ_1 such that

$$\epsilon_1 = \epsilon_{pe} = \frac{f_{pe}}{E_{ps}} \tag{4.37a}$$

At decompression, i.e., when the compressive stress in the surrounding concrete at the level of the prestressing tendon is neutralized by the tensile stress due to overload, a decompression strain $\epsilon_{\text{decomp}} = \epsilon_2$ results such that

$$\epsilon_2 = \epsilon_{\text{decomp}} = \frac{P_e}{A_c E_c}\left(1 + \frac{e^2}{r^2}\right) \tag{4.37b}$$

Figures 4.43 and 4.44 illustrate the stress distribution at and after the decompression stage where the behavior of the prestressed beam starts to resemble that of a reinforced concrete beam.

As the load approaches the limit state at ultimate, the additional strain ϵ_3 in the steel reinforcement follows the linear triangular distribution shown in Fig. 4.44(b), where the maximum compressive strain at the extreme compression fibers is $\epsilon_c = 0.003$ in./in. In such a case, the steel strain increment due to overload above the decompression load is

$$\epsilon_3 = \epsilon_c\left(\frac{d - c}{c}\right) \tag{4.37c}$$

where c is the depth of the neutral axis. Consequently, the total strain in the prestressing steel at this stage becomes

$$\epsilon_s = \epsilon_1 + \epsilon_2 + \epsilon_3 \tag{4.37d}$$

The corresponding stress f_{ps} at nominal strength can be easily obtained from the stress-strain diagram of the steel supplied by the producer.

4.11.2 The Equivalent Rectangular Block and Nominal Moment Strength

It is important to be able to evaluate the reserve strength in the prestressed beam up to failure, as discussed in Chapter 1. Hence, the total design would have to incorporate the moment strength of the prestressed section in addition to the service-load level checks

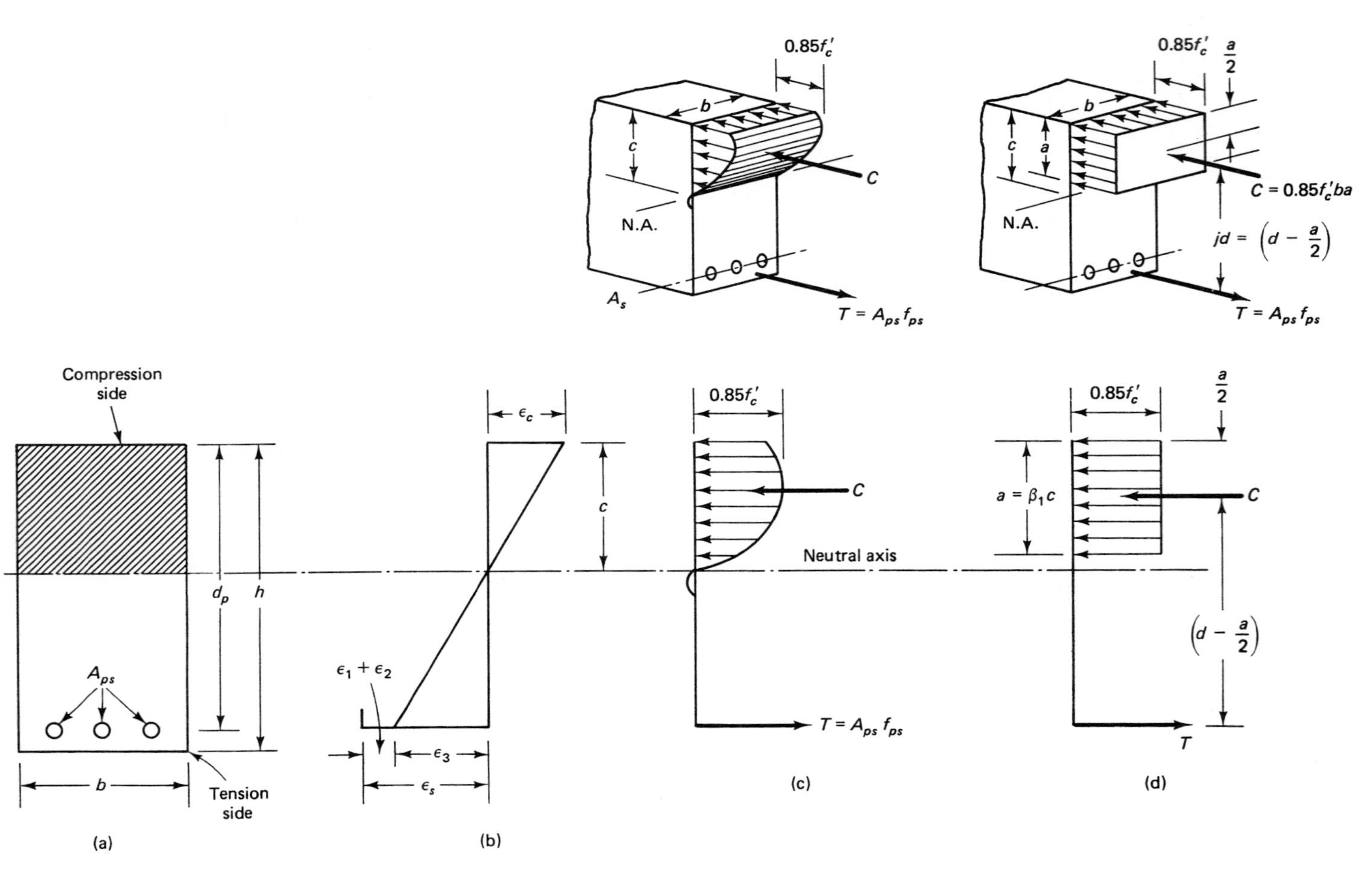

Figure 4.44 Stress and strain distribution across beam depth. (a) Beam cross section. (b) Strains. (c) Actual stress block. (d) Assumed equivalent stress block.

described in detail in Sections 4.1 through 4.7. The following assumptions are made in defining the behavior of the section at ultimate load:

1. The strain distribution is assumed to be linear. This assumption is based on Bernoulli's hypothesis that plane sections remain plane before bending and perpendicular to the neutral axis after bending.
2. The strain in the steel and the surrounding concrete is the same prior to cracking of the concrete or yielding of the steel as after such cracking or yielding.
3. Concrete is weak in tension. It cracks at an early stage of loading at about 10 percent of its compressive strength limit. Consequently, concrete in the tension zone of the section is neglected in the flexural analysis and design computations, and the tension reinforcement is assumed to take the total tensile force.

To satisfy the equilibrium of the horizontal forces, the compressive force C in the concrete and the tensile force T in the steel should balance each other—that is,

$$C = T \tag{4.38}$$

The terms in Figure 4.44 are defined as follows:

b = width of the beam at the compression side

d = depth of the beam measured from the extreme compression fiber to the centroid of steel area

h = total depth of the beam

4.11.2.1 Nominal Moment Strength of Rectangular Sections. The actual distribution of the compressive stress in a section at failure has the form of a rising parabola, as shown in Figure 4.44(c). It is time-consuming to evaluate the volume of the compressive stress block if it has a parabolic shape. An equivalent rectangular stress block due to Whitney can be used with ease and without loss of accuracy to calculate the compressive force and hence the flexural moment strength of the section. This equivalent stress block has a depth a and an average compressive strength $0.85f'_c$. As seen from Figure 4.44(d), the value of $a = \beta_1 c$ is determined by using a coefficient β_1 such that the area of the equivalent rectangular block is approximately the same as that of the parabolic compressive block, resulting in a compressive force C of essentially the same value in both cases.

The value $0.85f'_c$ for the average stress of the equivalent compressive block is based on the core test results of concrete in the structure at a minimum age of 28 days. Based on exhaustive experimental tests, a maximum allowable strain of 0.003 in./in. was adopted by the ACI as a safe limiting value. Even though several forms of stress blocks, including the trapezoidal, have been proposed to date, the simplified equivalent rectangular block is accepted as the standard in the analysis and design of reinforced concrete. The behavior of the steel is assumed to be elastoplastic.

Using all the preceding assumptions, the stress distribution diagram shown in Figure 4.44(c) can be redrawn as shown in Figure 4.44(d). One can easily deduce that the compression force C can be written $0.85f'_c\, ba$—that is, the *volume* of the compressive block at or near the ultimate when the tension steel has yielded ($\epsilon_s > \epsilon_y$). The tensile force T can be written as $A_{ps}f_{ps}$; thus, the equilibrium Equation 4.38 can be rewritten as

$$A_{ps}f_{ps} = 0.85f'_c\, ba \tag{4.39}$$

A little algebra yields

$$a = \beta_1 c = \frac{A_{ps}\, f_{ps}}{0.85f'_c\, b}$$

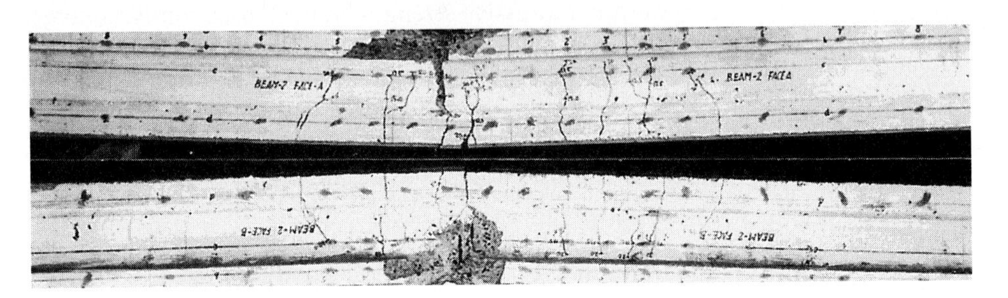

Photo 4.13 Flexural cracks at failure of prestressed T-beams (Nawy, Potyondy).

The nominal moment strength is obtained by multiplying C or T by the moment arm $(d_p - a/2)$, yielding

$$M_n = A_{ps} f_{ps}\left(d_p - \frac{a}{2}\right) \tag{4.40a}$$

where d_p is the distance from the compression fibers to the center of the prestressed reinforcement. The steel percentage $\rho_p = A_{ps}/bd_p$ gives nominal strength of the prestressing steel only as follows

$$M_n = \rho_p f_{ps} bd_p^2\left(1 - 0.59\rho_p \frac{f_{ps}}{f'_c}\right) \tag{4.40b}$$

If ω_p is the reinforcement index $= \rho_p(f_{ps}/f'_c)$, Equation 4.40b becomes

$$M_n = \rho_p f_{ps} bd_p^2(1 - 0.5\omega_p) \tag{4.40c}$$

The contribution of the mild steel tension reinforcement should be similarly treated, so that the depth a of the compressive block is

$$a = \frac{A_{ps} f_{ps} + A_s f_y}{0.85 f'_c b} \tag{4.41a}$$

If $c = a/\beta_1$, the strain at the level of the mild steel is (Fig. 4.44)

$$\epsilon_3 = \epsilon_c\left(\frac{d - c}{c}\right) \tag{4.41b}$$

Equation 4.45(b), for rectangular sections but with mild tension steel and no compression steel accounted for, becomes

$$M_n = \rho_p f_{ps} bd_p^2\left(1 - 0.59\rho_p \frac{f_{ps}}{f'_c}\right) + \rho f_y bd^2\left(1 - 0.59\frac{f_y}{f'_c}\right) \tag{4.42a}$$

or can be rewritten as either

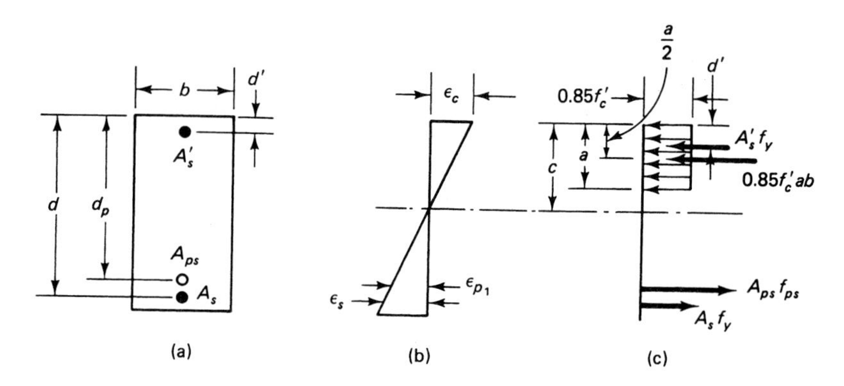

Figure 4.45 Strain, stress, and forces across beam depth of rectangular section. (a) Beam section. (b) Strain. (c) Stresses and forces.

$$M_n = A_{ps}f_{ps}\left\{1 - 0.59\left(\omega_p + \frac{d}{d_p}\omega\right)\right\} + A_s f_y\left\{1 - 0.59\left(\frac{d_p}{d}\omega_p + \omega\right)\right\} \quad (4.42b)$$

where $\omega = \rho(f_y/f'_c)$, or

$$M_n = A_{ps}f_{ps}\left(d_p - \frac{a}{2}\right) + A_s f_y\left(d - \frac{a}{2}\right) \quad (4.42c)$$

The contribution from compression reinforcement can be taken into account provided it has been found to have yielded,

$$a = \frac{A_{ps}f_{ps} + A_s f_y - A'_s f_y}{0.85f'_c b} \quad (4.43)$$

where b is the section width of the compression face of the beam.

Taking moments about the center of gravity of the compressive block in Figure 4.45, the nominal moment strength in Equation 4.42b becomes

$$M_n = A_{ps}f_{ps}\left(d_p - \frac{a}{2}\right) + A_s f_y\left(d - \frac{a}{2}\right) + A'_s f_y\left(\frac{a}{2} - d'\right) \quad (4.44)$$

4.11.2.2 Nominal Moment Strength of Flanged Sections. When the compression flange thickness h_f is less than the neutral axis depth c and equivalent rectangular block depth a, the section can be treated as a flanged section as in Figure 4.46. From the figure,

$$T_p + T_s = T_{pw} + T_{pf} \quad (4.45)$$

where T_p = total prestressing force = $A_{ps}f_{ps}$
T_s = ultimate force in the nonprestressed steel = $A_s f_y$
T_{pw} = part of the total force in the tension reinforcement required to develop the web = $A_{pw}f_{ps}$
A_{pw} = total reinforcement area corresponding to the force T_{pw}
T_{pf} = part of the total force in the tension reinforcement required to develop the flange = $C_f = 0.85\,f'_c(b - b_w)h_f$
$C_w = 0.85f'_c\, b_w a$

Substituting in Equation 4.45, we obtain

$$T_{pw} = A_{ps}f_{ps} + A_s f_y - 0.85f'_c(b - b_w)h_f \quad (4.46)$$

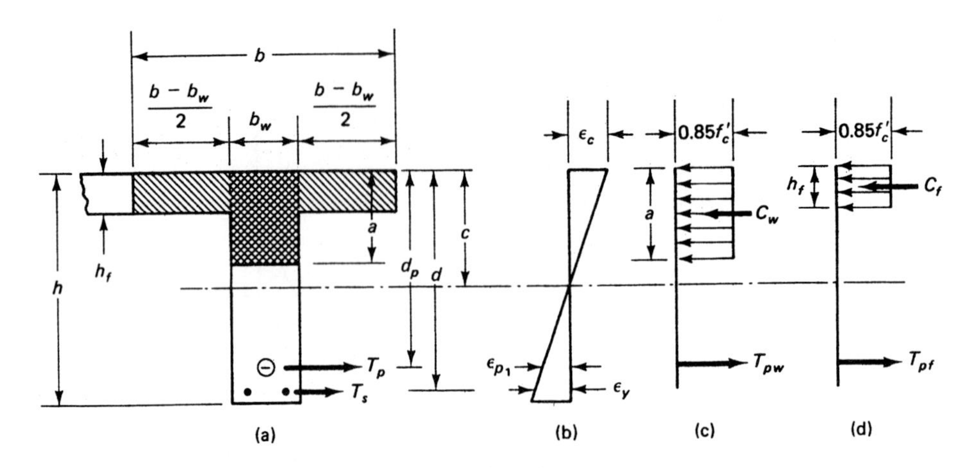

Figure 4.46 Strain, stress, and forces in flanged sections. (a) Beam section. (b) Strain. (c) Web stress and forces. (d) Flange stress and force.

Summing up all forces in Figures 4.46(c) and (d), we have

$$T_{pw} + T_{pf} = C_w + C_f$$

giving

$$a = \frac{A_{pw} f_{ps}}{0.85 f'_c b_w} \tag{4.47a}$$

or

$$a = \frac{A_{ps} f_{ps} + A_s f_y - 0.85 f'_c (b - b_w) h_f}{0.85 f'_c b_w} \tag{4.47b}$$

Eq. 4.44 for a beam with compression reinforcement can be rewritten to give the nominal moment strength for a flanged section where the neutral axis falls outside the flange and $a > h_f$ as follows, taking moments about the center of the prestressing steel:

$$M_n = A_{pw} f_{ps}\left(d_p - \frac{a}{2}\right) + A_s f_y (d - d_p) + 0.85 f'_c (b - b_w) h_f \left(d_p - \frac{h_f}{2}\right) \tag{4.48}$$

The design moment in all cases would be

$$M_u = \phi M_n \tag{4.49}$$

where $\phi = 0.90$ for flexure.

In order to determine whether the neutral axis falls outside the flange, requiring a flanged section analysis, one has to determine, as discussed in Ref. 4.2, whether the total compressive force C_n is larger or smaller than the total tensile force T_n. If $T_p + T_s$ in Figure 4.46 is larger than C_f, the neutral axis falls outside the flange and the section has to be treated as a flanged section. Otherwise, it should be treated as a rectangular section of the width b of the compression flange.

Another method of determining whether the section can be considered flanged is to calculate the value of the equivalent rectangular block depth a from Eq. 4.47b, thereby determining the neutral axis depth $c = a/\beta_1$.

4.11.2.3 Determination of Prestressing Steel Nominal Failure Stress f_{ps}. The value of the stress f_{ps} of the prestressing steel at failure is not readily available. However, it can be determined by *strain compatibility* through the various loading stages up to the limit state at failure. Such a procedure is required if

$$f_{pe} = \frac{P_e}{A_{ps}} < 0.50 f_{pu} \tag{4.50a}$$

Approximate determination is allowed by the ACI 318 building code provided that

$$f_{pe} = \frac{P_e}{A_{ps}} \geq 0.50 f_{pu} \tag{4.50b}$$

with separate equations for f_{ps} given for bonded and nonbonded members.

Bonded Tendons. The empirical expression for bonded members is

$$f_{ps} = f_{pu}\left(1 - \frac{\gamma_p}{\beta_1}\left[\rho_p \frac{f_{pu}}{f'_c} + \frac{d}{d_p}(\omega - \omega')\right]\right) \tag{4.51}$$

where the reinforcement index for the compression nonprestressed reinforcement is $\omega' = \rho'(f_y/f'_c)$. If the compression reinforcement is taken into account when calculating f_{ps} by Eq. 4.51, the term $[\rho_p(f_{pu}/f'_c) + (d/d_p)(\omega - \omega')]$ should not be less than 0.17 and d' should not be greater than $0.15d_p$. Also,

$$\begin{aligned} \gamma_p &= 0.55 \text{ for } f_{py}/f_{pu} \text{ not less than } 0.80 \\ &= 0.40 \text{ for } f_{py}/f_{pu} \text{ not less than } 0.85 \\ &= 0.28 \text{ for } f_{py}/f_{pu} \text{ not less than } 0.90 \end{aligned}$$

The value of the factor γ_p is based on the criterion that $f_{py} = 0.80 f_{pu}$ for high-strength prestressing bars, 0.85 for stress-relieved strands, and 0.90 for low-relaxation strands.

Unbonded Tendons. For a span-to-depth ratio of 35 or less,

$$f_{ps} = f_{pe} + 10{,}000 + \frac{f'_c}{100\rho_p} \tag{4.52a}$$

where f_{ps} shall not be greater than f_{py} or $(f_{pe} + 60{,}000)$.

For a span-to-depth ratio greater than 35,

$$f_{ps} = f_{pe} + 10{,}000 + \frac{f'_c}{300\rho_p} \tag{4.52b}$$

where f_{ps} shall not be greater than f_{py} or $(f_{pe} + 30{,}000)$. Figure 4.47, from Ref. 4.9, shows seating losses for typical unbonded tendons.

4.11.2.4 Limiting Values of the Reinforcement Index. The reinforcement index ω_p, a measure of the percentage of reinforcement in the section, is given by

$$\omega_p = \frac{A_{ps} f_{ps}}{b d_p f'_c} = \rho_p \frac{f_{ps}}{f'_c} \tag{4.53}$$

Minimum Reinforcement. If the percentage of reinforcement is too small, the concrete section will be too weak to resist the tensile stress level after cracking and the section will behave almost as a plain section, with premature abrupt failure through rupture of the reinforcement. Hence, a minimum percentage $\rho_{\min}$ with a minimum $\omega_{p\,\min}$ has to be observed in the design in order to prevent such a failure. The total amount of pre-

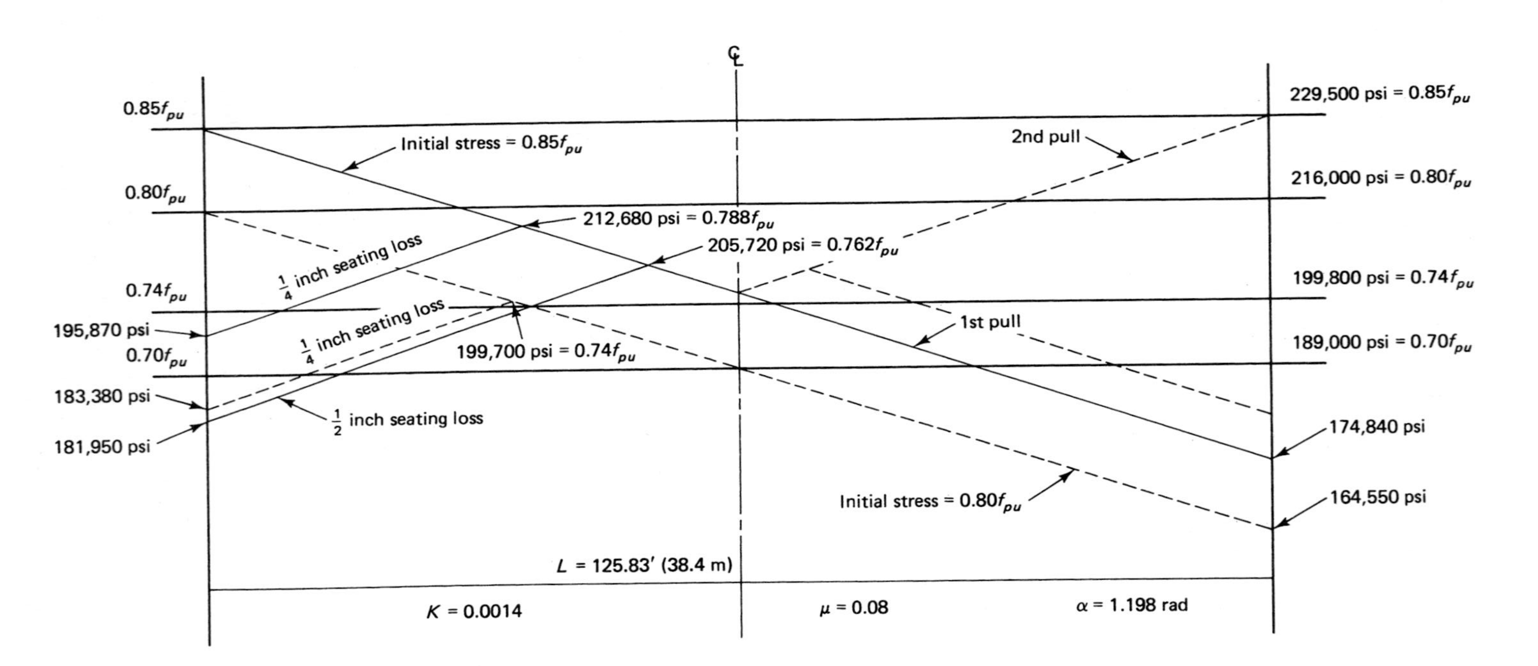

Figure 4.47 Stress diagram for unbonded tendons with various values of initial stress and seating loss (100,000 psi = 689.5 MPa).

stressed and nonprestressed reinforcement required by the ACI should not be less than that required to develop a factored moment $M_u = \phi M_n$ such that

$$M_u \geq 1.2M_{cr} \tag{4.54a}$$

where M_{cr} is based on a modulus of rupture $f_r = 7.5\sqrt{f'_c}$. An exception can be made where the flexural member has shear and flexural strengths at least twice those of the factored loads in Equations 4.33. Also, the minimum area of bonded nonprestressed reinforcement in beams in accordance with the ACI code has to be computed as

$$\text{Min } A_s = 0.004A \tag{4.54b}$$

where A is that part of the cross section between the flexural tension face and the center of gravity cgc of the gross section (in.^{2}). This reinforcement has to be uniformly distributed over the precompressed tensile zone as close as possible to the extreme tension fibers.

In two-way flat plates, where the tension stress in the concrete at service load exceeds $2\sqrt{f'_c}$, bonded nonprestressed steel is required such that

$$A_s = \frac{N_c}{0.5f_y} \tag{4.55a}$$

where $f_y \leq 60{,}000$ psi
N_c = tensile force in the concrete due to unfactored dead plus live loads ($D + L$).

In the negative-moment areas of slabs at column support, the minimum area of nonprestressed steel in each direction should be

$$A_s = 0.00075hl \tag{4.55b}$$

where h = total slab thickness
l = span length in the direction parallel to that of the reinforcement being determined.

The reinforcement A_s has to be distributed with a slab width between lines that are $1.5h$ outside the faces of the column support, with at least four bars or wires to be provided in each direction and a spacing not to exceed 12 in.

Maximum Reinforcement. If the percentage of reinforcement is too large, the concrete section behaves as if it were overreinforced. As a result, a nonductile failure would occur by initial crushing of the concrete at the compression fibers since the reinforcement at the tension side cannot yield first. Reinforced concrete beams always have to be designed as underreinforced, with a maximum reinforcement percentage ρ not exceeding 75 percent of the balanced percentage ρ_b denoting the conditions where the tensile steel yields simultaneously with crushing of the concrete at the compression fibers.

In prestressed beams, however, it is not always possible to impose an underreinforced condition. The prestressing forces P_i and P_e at transfer and service load also control the value of the area of the tensile reinforcement needed, including the area of the nonprestressed reinforcement. Additionally, the yield strength, and in turn the yield strain value of the prestressing steel is not well defined. Consequently, the prestressed beam designed to satisfy all the service-load requirements could behave as either underreinforced or overreinforced at the limit state of ultimate-load design, particularly if it is a partially prestressed beam.

In order to ensure ductility of behavior, the ACI code limits the percentage of reinforcement in such manner that the reinforcement index does not exceed $0.36\beta_1$ as a measure of underreinforced behavior, where the concrete strength effect

$$\beta_1 = 0.85 - \frac{0.05(f'_c - 4{,}000)}{1{,}000} \geq 0.65 \tag{4.56}$$

For a rectangular section with prestressing steel only,

$$\omega_p = \rho_p \frac{f_{ps}}{f'_c} \leq 0.36\beta_1 \tag{4.57a}$$

For rectangular sections with tensile and compressive mild steel,

$$\left[\omega_p + \frac{d}{d_p}(\omega - \omega')\right] \leq 0.36\beta_1 \tag{4.57b}$$

where

$$\omega = \frac{A_s f_y}{bdf'_c}$$

and

$$\omega' = \frac{A'_s f_y}{bdf'_c}$$

Finally, for flanged sections,

$$\left[\omega_{pw} + \frac{d}{d_p}(\omega_w - \omega'_w)\right] \leq 0.36\beta_1 \tag{4.57c}$$

where ω_{pw}, ωw, and ω'_w are computed in the same manner as in Equation 4.57a, b, except that the web width b_w is used in the denominators of these equations. Note that the terms ω_p, $(\omega_p + (d/d_p)(\omega - w'))$, and $(\omega_{pw} + (d/d_p)(\omega_w - \omega'_w)$ are each equal to $0.85a/d_p$, where a is the depth of the equivalent rectangular concrete compressive block as follows:

(a) In rectangular sections and in flanged sections in which $a \leq h_f$,

$$\left[\omega_p + \frac{d}{d_p}(\omega - \omega')\right] = \left[\frac{A_{ps}}{bd_p}\cdot\frac{f_{ps}}{f'_c} + \frac{d}{d_p}\left(\frac{A_s}{bd}\cdot\frac{f_y}{f'_c} - \frac{A'_s}{bd}\cdot\frac{f_y}{f'_c}\right)\right]$$

$$= \frac{A_{ps} f_{ps} + A_s f_y - A'_s f_y}{bd_p \cdot f'_c} = \frac{0.85 f'_c ab}{bd_p f'_c} = \frac{0.85a}{d_p}$$

(b) In flanged sections in which $a > h_f$, let C_F be the resultant concrete compression force in outstanding flanges. Then,

$$\left[\omega_{pw} + \frac{d}{d_p}(\omega_w - \omega'_w)\right] = \left[\frac{(A_{ps} f_{ps} - C_F)}{b_w d_p f'_c} + \frac{d}{d_p}\left(\frac{A_s}{b_w d}\cdot\frac{f_y}{f'_c} - \frac{A'_s}{b_w d}\cdot\frac{f_y}{f'_c}\right)\right]$$

$$= \frac{A_{ps} f_{ps} + A_s f_y - A'_s f_y - C_F}{b_w d_p f'_c} \tag{4.57d}$$

$$= \frac{\text{compression force in web}}{b_w d_p f'_c}$$

$$= \frac{0.85 f'_c b_w a}{b_w d_p f'_c} = \frac{0.85a}{d_p}$$

An exception can be made to the ACI requirements in Equations 4.57a, b, and c, provided that the design moment strength does not exceed the moment strength based on the compression portion of the moment couple. In other words, unless a strain compatibility analysis is performed, the overreinforced prestressed beam moment strength should be determined from the empirical expression

$$M_n = 0.25 f'_c bd^2 \quad (4.58a)$$

for rectangular sections, and

$$M_n = 0.25 f'_c b_w d^2 + 0.85 f'_c (b - b_w) h_f (d - 0.5 h_f) \quad (4.58b)$$

for flanged sections. These equations can be modified as follows:

(a) For the overreinforced rectangular section,

$$M_n = f'_c bd_p^2 (0.36\beta_1 - 0.08\beta_1^2) \quad (4.59a)$$

(b) For the overreinforced flanged section,

$$M_n = f'_c b_w d_p^2 (0.36\beta_1 - 0.08\beta_1^2) + 0.85 f'_c (b - b_w) h_f (d_p - 0.5 h_f) \quad (4.59b)$$

4.11.2.5 Limit State in Flexure at Ultimate Load in Nonbonded Tendons. The discussion presented in Sections 4.11.1 and 4.11.2 defines the design and analysis process for pretensioned beams, where the concrete is cast around the prestressed tendons, thereby achieving full bond, as well as for post-tensioned beams, where the tendons are fully grouted under pressure after the tendons are prestressed.

Post-tensioned tendons that are not grouted or that are asphalt coated (many in the United States) are nonbonded tendons. Consequently, as the superimposed load acts on the beam, slip results between the tendons and the surrounding concrete, permitting a uniform deformation along the entire length of the prestressing tendon. As cracks develop at the critical high-moment zones, the increase in the steel tensile stress is not concentrated at the cracks, but is uniformly distributed along the freely sliding tendon. As a result, the net increase in strain and stress is less in the nonbounded case than in the case of bonded tendons as the load continues to increase to the ultimate. Hence a lesser number of cracks, but of larger width, develops in nonbonded prestressing (Ref 4.4). The final stress in the prestressed tendons at ultimate load would be only slightly higher than the effective prestress f_{pe}.

In order to ensure a structure with good serviceability performance, a reasonable percentage of nonprestressed steel has to be used, within the limitations mentioned in Section 4.11.2.4. The nonprestressed reinforcement controls the flexural crack development and width, and contributes to substantially increasing the moment strength capacity M_n of the section. It undergoes a strain larger than its yield strain, since its deformation at the postelastic range has to be compatible with the deformation of the adjacent prestressing strands. Hence, the stress level in the nonprestressed steel will always be higher than its yield strength at ultimate load. Figure 4.48 shows a typical stress-strain diagram for a 270-K 7-wire $\frac{1}{2}$in. prestressing strand, while Fig. 4.49 schematically illustrates the relative stresses of the prestressed and the nonprestressed steel and seating losses.

From this discussion, it can be concluded that the expressions presented for the M_n calculations of the nominal moment strength for bonded beams can be equally used for nonbonded elements. Note that while it is always advisable to grout the post-tensioned tendons, it is sometimes not easy to do so, as, for example, in two-way slab systems or shallow-box elements, where the concrete thickness is small. Also, consideration has to be given to the cost of pressure grouting in cases where there is a congestion of tendons.

4.12 PRELIMINARY ULTIMATE-LOAD DESIGN

If the preliminary design starts at the ultimate-load level, the required design moment $M_u = \phi M_n$ has to be at least equal to the factored moment M_u. The first trial depth has to be based on a reasonable span-to-depth ratio, with the top flange width determined by whether the beam is for residential floors or parking garages, where a double-T-section

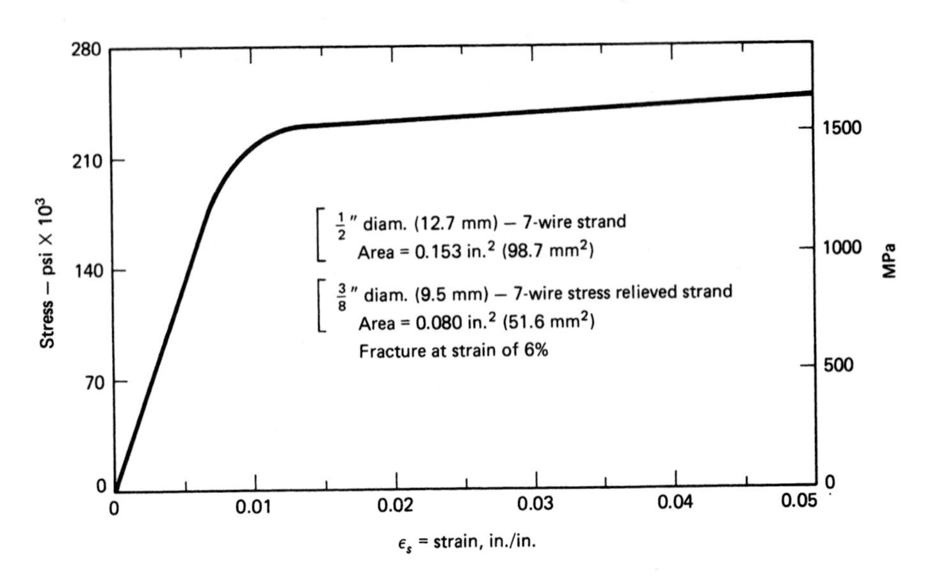

Figure 4.48 Typical stress-strain relationship of 7-wire 270-K prestressing strand.

or a hollow-box shallow section is preferable, or whether the beam is intended to support a bridge deck with spacing decided by load and the number of lanes, where an I-section might be preferable.

As a rule of thumb, the average depth of a prestressed beam is about 75 percent of the depth of a comparably loaded reinforced concrete beam. Another guideline for an initial trial is to use 0.6 in. of depth per foot of span. Once a first-trial depth is chosen, a determination is made of the other geometrical properties of the section.

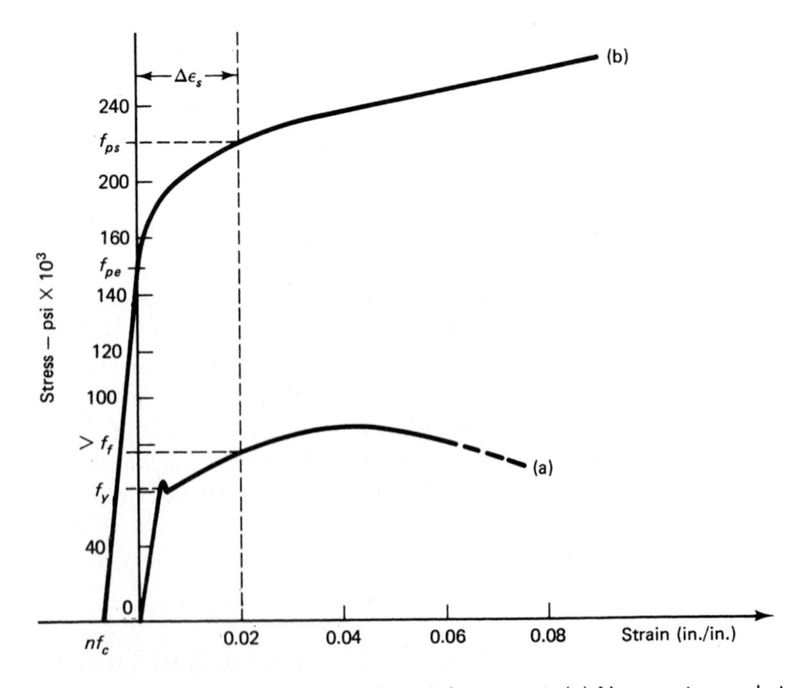

Figure 4.49 Stress-strain diagrams for reinforcement. (a) Nonprestressed steel. (b) Prestressed steel (100,000 psi = 689.5 MPa).

Assume that the center of gravity of the prestressing steel is approximately $0.85/h$ from the middepth of the flange. Then the lever arm of the moment couple $jd \cong 0.80h$. Assume also that the nominal strength of the prestressed steel is $f_{ps} \cong 0.90\, f_{pu}$. Then the area A_{ps} of the prestressing tendons is

$$A_{ps} = \frac{M_u/\phi}{0.9 f_{pu}(0.80h)} \tag{4.60a}$$

or

$$A_{ps} = \frac{M_n}{0.72 f_{pu} h} \tag{4.60b}$$

If the compressive block depth a equals the flange thickness h_f, the volume of the compressive block of Figure 4.44(d) in terms of the area $ba = A'_c$ is

$$C = 0.85 f'_c A'_c$$

$$T = 0.9 f_{pu} A_{ps} = \frac{M_n}{0.8h}$$

From the equilibrium of forces, $C = T$. Hence, the area of the compression flange is

$$A'_c = \frac{M_n}{0.85 f'_c(0.8h)} = \frac{M_n}{0.68 f'_c h} \tag{4.61}$$

Once the width of the flange is chosen for the first trial and the beam depth is known, the web thickness can be chosen based on the shear requirements to be discussed in Chapter 5. Thereafter, by trial and adjustment, one can select the ideal section for the particular design requirement conditions and proceed to analyze the stresses for the service-load conditions.

4.13 SUMMARY STEP-BY-STEP PROCEDURE FOR LIMIT-STATE-AT-FAILURE DESIGN OF THE PRESTRESSED MEMBERS

1. Determine whether or not partial prestressing is to be chosen, using an effective percentage of nonprestressed steel. Choose a trial depth h based on either 0.6 in. per ft of span or 75 percent of the depth needed for reinforced concrete sections after calculating the required nominal strength $M_n = M_u/\phi$.

Photo 4.14 Bridge deck prestressing reinforcement set up prior to concreting.

Photo 4.15 Full-scale bridge test (Nawy, Goodkind).

2. Select a trial flange thickness such that the total concrete area of the flange $A'_c \cong M_n/0.68f'_c h$, based on choosing a flange width dictated by planning requirements and spacing of beams. Choose a preliminary area of prestressing steel $A_{ps} = M_n/0.72f_{pu}h$.
3. Use a reasonable value for the steel stress f_{ps} at failure for a first trial. If $f_{pe} < 0.5f_{pu}$, strain compatibility analysis would thereafter be needed. Determine whether the tendons are bonded or nonbonded. Use the value of the effective prestress f_{pe} from the service-load analysis if that design was already made. If $f_{pe} > 0.5f_{pu}$, use the approximate values from the following applicable cases:

   (a) *Bonded tendons*

$$f_{ps} = f_{pu}\left(1 - \frac{\gamma_p}{\beta_1}\left\{\rho_p \frac{f_{pu}}{f'_c} + \frac{d}{d_p}(\omega - \omega')\right\}\right)$$

   (b) *Nonbonded tendons, span/depth ratio ≤ 35*

$$f_{ps} = f_{pe} + 10{,}000 + \frac{f'_c}{100\rho_p}$$

   (c) *Nonbonded tendons, span/depth ratio > 35*

$$f_{ps} = f_{pe} + 10{,}000 + \frac{f'_c}{300\rho_p}$$

4. Determine whether the trial section chosen should be considered rectangular or flanged by determining the position of the neutral axis, $c = a/\beta_1$. If rectangular,

$$a = \frac{A_{ps} f_{ps} + A_s f_y - A'_s f_y}{0.85 f'_c b}$$

   If flanged,

$$a = \frac{A_{pw} f_{ps}}{0.85 f'_c b_w}$$

   where $A_{pw} f_{ps} = A_{ps} f_{ps} + A_s f_y - 0.85 f'_c (b - b_w) h_f$.
5. If h_f is larger than c and a, analyze the element as a rectangular section singly or doubly reinforced.
6. Find the reinforcement indices ω_p, ω, and ω' for the case $a < h_f$ (neutral axis within the flange; hence, use for a rectangular section).

   (a) Rectangular sections with prestressing steel only:

$$\omega_T = \omega_p = \rho_p \frac{f_{ps}}{f'_c} = \frac{A_{ps}}{bd_p} \frac{f_{ps}}{f'_c}$$

(b) Rectangular sections with compression steel in addition to nonprestressed tension steel:

$$\omega_T = \omega_p + \frac{d}{d_p}(\omega - \omega')$$

If the total index in (a) or (b) is less than or equal to $0.36\beta_1$, then the moment strength is

$$M_n = A_{ps} f_{ps}\left(d_p - \frac{a}{2}\right) + A_s f_y\left(d - \frac{a}{2}\right) + A'_s f_y\left(\frac{a}{2} - d'\right)$$

7. Find the reinforcement indices ω_{pw}, ω_w and ω'_w for the case $a > h_f$ (neutral axis outside the flange), with the total index

$$\omega_T = \omega_{pw} + \frac{d}{d_p}(\omega_w - \omega'_w)$$

The indices are calculated on the basis of the web width b_w. If the total index $\omega_T < 0.36\beta_1$, then

$$M_n = A_{pw} f_{ps}\left(d_p - \frac{a}{2}\right) + A_s f_y (d - d_p) + 0.85 f'_c (b - b_w) h_f \left(d_p - \frac{h_f}{2}\right)$$

where

$$a = \frac{A_{pw} f_{ps}}{0.85 f'_c b_w}$$

and

$$A_{pw} f_{ps} = A_{ps} f_{ps} + A_s f_y - 0.85 f'_c (b - b_w) h_f$$

If the total index $\omega_T > 0.36\beta_1$, the section is overreinforced and the nominal strength is

$$M_n = f'_c b_w d_p^2 (0.36\beta_1 - 0.08\beta_1^2) + 0.85 f'_c (b - b_w) h_f (d_p - 0.5h_f)$$

8. Check for the minimum required reinforcement $A_s > 0.004A$. Also, check whether $M_u \geq 1.2M_{cr}$ to ensure the use of adequate nonprestressed tension steel, particularly in nonbonded tendons.

9. Select the size and spacing of the nonprestressed tension reinforcement, and compression reinforcement where applicable.

10. Verify that the design moment $M_u = \phi M_n$ is equal to or larger than the factored moment M_u. If not, adjust the design.

A flowchart for programming the step-by-step trial-and-adjustment procedure in analyzing the nominal flexural strength of rectangular and flanged prestressed sections taking d_p as the single-layer cgs depth of tendon is shown in Fig. 4.50. Similarly, a flowchart for programming the nominal flexural strength of prestressed beams using strain-compatibility analysis of *multilayered* strand depths d_{p_1} to d_{p_n} is given in Fig. 4.51. Both charts are applicable to fully prestressed beams that use no mild steel and that allow no tension in the concrete, as well as to "partially prestressed" beams where limited tensile stress is permitted in the concrete through the use of nonprestressed reinforcement. A

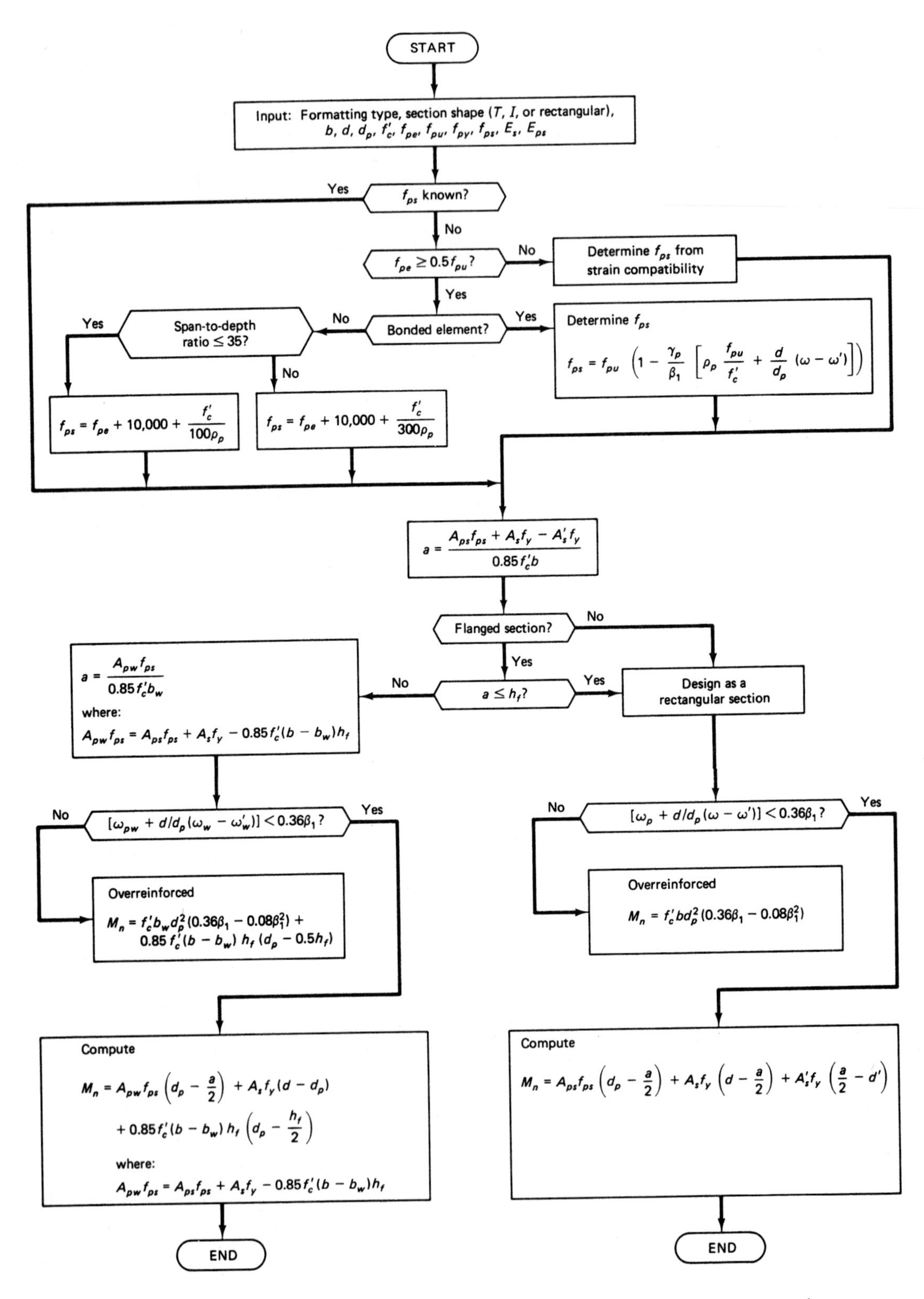

Figure 4.50 Flowchart for flexural analysis of rectangular and flanged prestressed sections based on cgs profile depth.

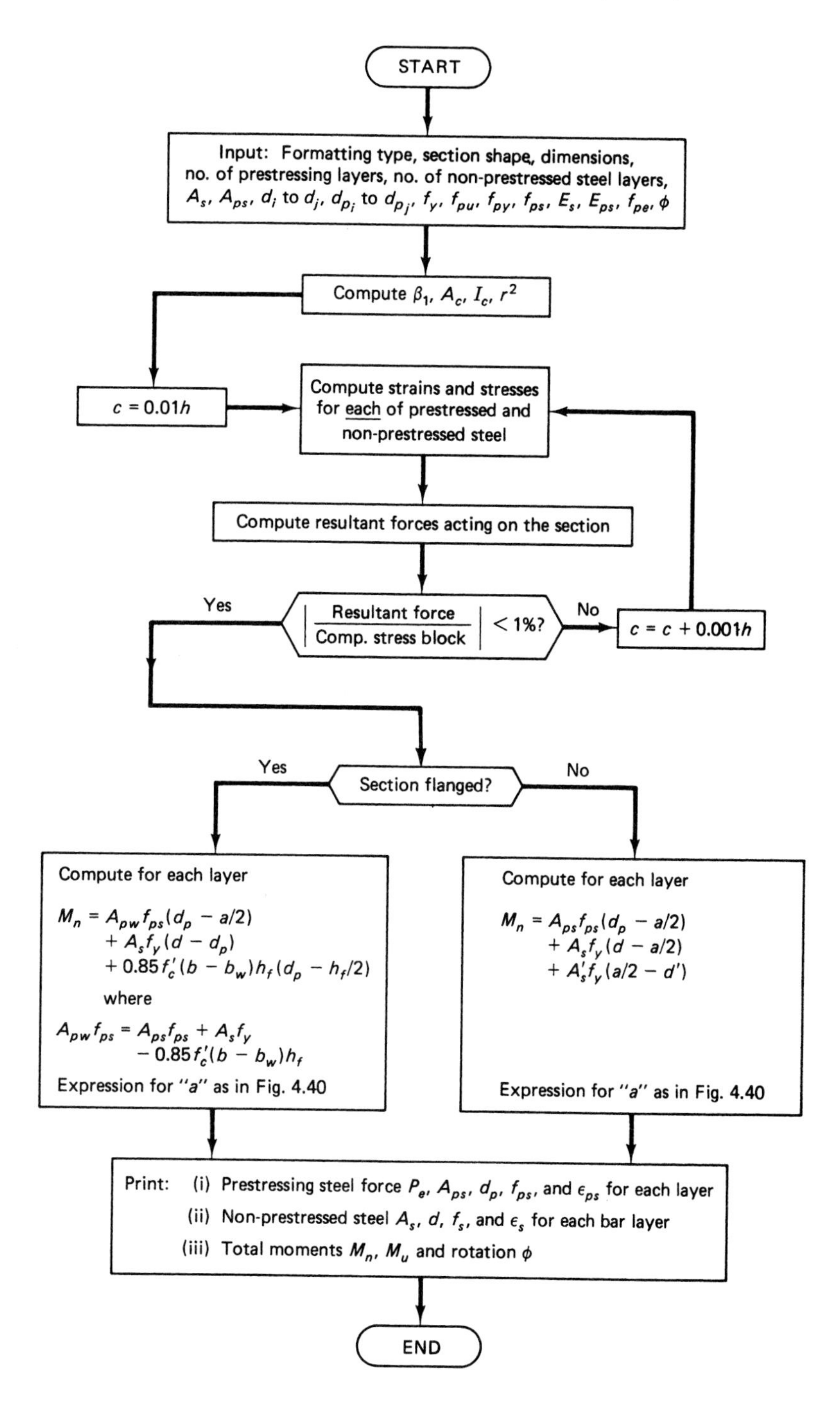

Figure 4.51 Flowchart for flexural analysis of rectangular and flanged prestressed sections using compatibility analysis for individual layers of strands and bars.

computer program based on the flowcharts in the two figures can be equally used for a single effective depth d_p of the cgs tendon profile.

4.14 ULTIMATE-STRENGTH DESIGN OF PRESTRESSED SIMPLY SUPPORTED BEAM BY STRAIN COMPATIBILITY

Example 4.9

Design the bonded beam in Example 4.2 by the ultimate-load theory using nonprestressed reinforcement to *partially* carry part of the factored loads. Use strain compatibility to evaluate f_{ps}, given the modified section in Fig. 4.52 with a composite 3 in. top slab and

$$f_{pu} = 270{,}000 \text{ psi } (1{,}862 \text{ MPa})$$

$$f_{py} = 0.85 f_{pu} \text{ for stress-relieved strands}$$

$$f_y = 60{,}000 \text{ psi } (414 \text{ MPa})$$

$$f'_c = 5{,}000 \text{ psi normal-weight concrete } (34.5 \text{ MPa})$$

Use 7-wire $\frac{1}{2}$-in. dia tendons. The nonprestressed partial mild steel is to be placed with a $1\frac{1}{2}$-in. clear cover, and no compression steel is to be accounted for. No wind or earthquake is taken into consideration.

Solution: From Example 4.2,

$$\text{Service } W_L = 1{,}100 \text{ plf } (16.1 \text{ kN/m})$$

$$\text{Service } W_{SD} = 100 \text{ plf } (1.46 \text{ kN/m})$$

$$\text{Assumed } W_D = 393 \text{ plf } (5.74 \text{ kN/m})$$

$$\text{Beam span} = 65 \text{ ft } (19.8 \text{ m})$$

1. **Factored moment (step 1)**

$$W_u = 1.4(W_D + W_{SD}) + 1.7W_L$$

$$= 1.4(100 + 393) + 1.7(1{,}100) = 2{,}560 \text{ plf } (37.4 \text{ kN/m})$$

The factored moment is given by

$$M_u = \frac{w_u l^2}{8} = \frac{2{,}560(65)^2 12}{8} = 16{,}224{,}000 \text{ in.-lb } (1{,}833 \text{ kN-m})$$

and the required nominal moment strength is

$$M_n = \frac{M_u}{\phi} = \frac{16{,}224{,}000}{0.9} = 18{,}026{,}667 \text{ in.-lb } (2{,}037 \text{ kN-m})$$

2. **Choice of preliminary section (step 2)**

Assuming a depth of 0.6 in./ft of span, we can have a trial section depth $h = 0.6 \times 65 \cong 40$ in. (102 cm). Then assume a mild partial steel 4 #6 = $4 \times 0.44 = 1.76$ in.^{2} (11.4 cm^2). From Equation 4.61,

$$A'_c = \frac{M_n}{0.68 f'_c h} = \frac{18{,}026{,}667}{0.68 \times 5{,}000 \times 40} = 132.5 \text{ in.}^2 \ (855 \text{ cm}^2)$$

Assume a flange width of 18 in. Then the average flange thickness = 132.5/18 ≅ 7.5 in. (191 mm). So suppose the web $b_w = 6$ in. (152 mm), to be subsequently verified for shear requirements. Then from Equation 4.60b,

$$A_{ps} = \frac{M_n}{0.72 f_{pu} h} = \frac{18{,}026{,}667}{0.72 \times 270{,}000 \times 40} - 2.32 \text{ in.}^2 \ (15 \text{ cm}^2)$$

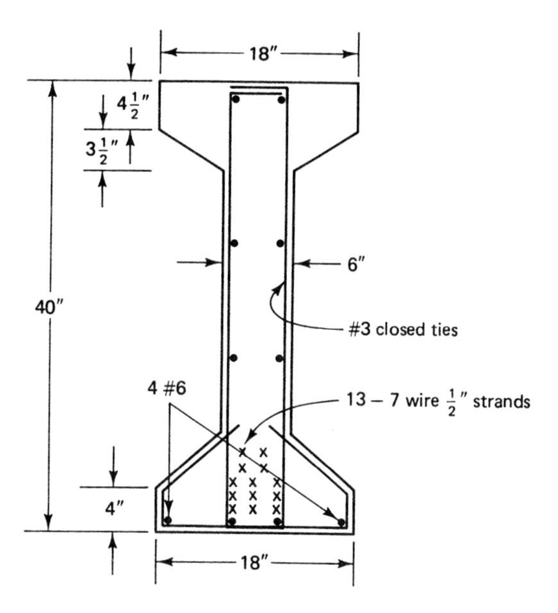

Figure 4.52 Midspan section of the beam in Example 4.9.

and the number of $\frac{1}{2}$-in. stress-relieved wire strands = 2.32/0.153 = 15.16. So try thirteen $\frac{1}{2}$-in. tendons.

$$A_{ps} = 13 \times 0.153 = 1.99 \text{ in.}^2 \text{ (12.8 cm}^2\text{)}$$

3. Calculate the stress f_{ps} in the prestressing tendon at nominal strength using the strain-compatibility approach (step 3)

The geometrical properties of the trial section are very close to the assumed dimensions for the depth h and the top flange width b. Hence, use the following data for the purpose of the example:

$$A_c = 377 \text{ in.}^2$$

$$c_t = 21.16 \text{ in.}$$

$$d_p = 15 + c_t = 15 + 21.16 = 36.16 \text{ in.}$$

$$r^2 = 187.5 \text{ in.}^2$$

$$e = 15 \text{ in. at midspan}$$

$$e^2 = 225 \text{ in.}^2$$

$$e^2/r^2 = 225/187.5 = 1.20$$

$$E_c = 57{,}000\sqrt{5{,}000} = 4.03 \times 10^6 \text{ psi (27.8} \times 10^3 \text{ MPa)}$$

$$E_{ps} = 28 \times 10^6 \text{ psi (193} \times 10^3 \text{ MPa)}$$

The maximum allowable compressive strain ϵ_c at failure = 0.003 in./in. Assume that the effective prestress at service load is $f_{pe} \cong 155{,}000$ psi (1,069 MPa).

(a)

$$\epsilon_1 = \epsilon_{pe} = \frac{f_{pe}}{E_{ps}} = \frac{155{,}000}{28 \times 10^6} = 0.0055 \text{ in./in.}$$

$$P_e = 13 \times 0.153 \times 155{,}000 = 308{,}295 \text{ lb}$$

The increase in prestressing steel strain as the concrete is decompressed by the increased external load (see Figure 4.3 and Equation 4.3c) is given as

$$\epsilon_2 = \epsilon_{\text{decomp}} = \frac{P_e}{A_c E_c}\left(1 + \frac{e^2}{r^2}\right)$$

$$= \frac{308{,}295}{377 \times 4.03 \times 10^6}(1 + 1.20) = 0.0004 \text{ in./in.}$$

(b) Assume that the stress $f_{ps} \cong 205{,}000$ psi as a first trial. Suppose the neutral axis inside the flange is verified on the basis of $h_f = 3 + 4\frac{1}{2} + 3\frac{1}{2}/2 = 9.25$ in. Then, from Equation 4.41a

$$a = \frac{A_{ps} f_{ps} + A_s f_y}{0.85 f'_c b} = \frac{1.99 \times 205{,}000 + 1.76 \times 60{,}000}{0.85 \times 5{,}000 \times 18}$$

$$= 6.71 \text{ in. (17 cm)} < h_f = 9.25 \text{ in.}$$

Hence, the equivalent compressive block is inside the flange and the section has to be treated as rectangular.

Accordingly, for 5,000 psi concrete,

$$\beta_1 = 0.85 - 0.05 = 0.8$$

$$c = \frac{a}{\beta_1} = \frac{6.71}{0.80} = 8.39 \text{ in. (22.7 cm)}$$

$$d = 40 - (1.5 + \tfrac{1}{2} \text{ in. for stirrups } + \tfrac{5}{16} \text{ in. for bar}) \cong 37.6 \text{ in.}$$

The increment of strain due to overload to the ultimate, from Equation 4.37(c) is

$$\epsilon_3 = \epsilon_c\left(\frac{d - c}{c}\right) = 0.003\left(\frac{37.6 - 8.39}{8.39}\right) = 0.0104 \text{ in./in.}$$

and the total strain is

$$\epsilon_{ps} = \epsilon_1 + \epsilon_2 + \epsilon_3$$

$$= 0.0055 + 0.0004 + 0.0104 = 0.0163 \text{ in./in.}$$

From the stress-strain diagram in Figure 4.48 the f_{ps} corresponding to $\epsilon_{ps} = 0.0163$ is 230,000 psi.

Second trial for f_{ps} value

Assume

$$f_{ps} = 229{,}000 \text{ psi}$$

$$a = \frac{1.99 \times 229{,}000 + 1.76 \times 60{,}000}{0.85 \times 5{,}000 \times 18} = 7.34 \text{ in., consider section as a rectangular beam.}$$

$$c = \frac{7.34}{0.80} = 9.17 \text{ in.}$$

$$\epsilon_3 = 0.003\left(\frac{37.6 - 9.17}{9.17}\right) = 0.0093$$

Then the total strain is $\epsilon_{ps} = 0.0055 + 0.0004 + 0.0093 = 0.0152$ in./in. From Figure 4.48, $f_{ps} = 229{,}000$ psi (1.579 MPa); use

$$A_s = 4 \#6 = 1.76 \text{ in.}^2$$

4. Available moment strength (steps 6 through 10)

From Equation 4.42c, if the neutral axis were to fall within the flange,

$$M_n = 1.99 \times 229{,}000\left(36.16 - \frac{7.34}{2}\right) + 1.76 \times 60{,}000\left(37.6 - \frac{7.34}{2}\right)$$

$$= 14{,}806{,}017 + 3{,}583{,}008 = 18{,}389{,}025 \text{ in.-lb (2,078 kN-m)}$$

$> \text{required } M_n = 18{,}026{,}667$ in.-lb, O.K.

The percentage of moment resisted by the nonprestressed steel is

$$\frac{3{,}583{,}008}{18{,}026{,}667} \cong 20\%$$

5. Check for minimum and maximum reinforcement (steps 6 and 9)

(a) Min $A_s = 0.004A$

where A is the area of the part of the section between the tenison face and the cgc. From the cross section of Figure 4.8,

$$A = 377 - 18\left(4.125 + \frac{1.375}{2}\right) - 6(21.16 - 5.5) \cong 201 \text{ in}^2$$

$$\text{Min } A_s = 0.004 \times 201 = 0.80 \text{ in}^2 < 1.76 \text{ used, O.K.}$$

(b) The maximum steel index, from Equation 4.57b, is

$$\omega_p + \frac{d}{d_p}(\omega - \omega') \leq 0.36\beta_1 < 0.29 \text{ for } \beta_1 = 0.80$$

and the actual total reinforcement index is

$$\omega_T = \frac{1.99 \times 229{,}000}{18 \times 36.16 \times 5{,}000} + \frac{37.6}{36.16}\left(\frac{1.76 \times 60{,}000}{18 \times 37.6 \times 5{,}000}\right)$$

$$= 0.14 + 0.03 = 0.17 < 0.29, \text{ O.K.}$$

6. Choice of section for ultimate load (step 11)

From steps 1–5 of the design, the section in Example 4.2 with the modifications shown in Fig. 4.52 has the normal moment strength M_n that can carry the factored load, provided that four #6 nonprestressed bars are used at the tension side as a partially prestressed section.

So one can adopt the section for flexure, as it also satisfies the service-load flexural stress requirements both at midspan and at the support. Note that the section could only develop the required nominal strength $M_n = 18{,}026{,}667$ in.-lb by the addition of the nonprestressed bars at the tension face to resist 20 percent of the total *required* moment strength. Note also that this section is adequate with a concrete $f'_c = 5{,}000$ psi, while the section in Example 4.2 has to have $f'_c = 6{,}000$ psi strength in order not to exceed the allowable service-load concrete stresses. Hence, ultimate-load computations are necessary in prestressed concrete design to ensure that the constructed elements can carry all the factored load and are thus an integral part of the total design.

4.15 STRENGTH DESIGN OF BONDED PRESTRESSED BEAM USING APPROXIMATE PROCEDURES

Example 4.10

Design the beam in Example 4.9 as a partially prestressed beam using the ACI approximate procedures if permissible. Use the exact standard section used in Example 4.2 with (a) bonded prestressing steel, and (b) nonbonded prestressing steel. Neglect the contribution of the compressive nonprestressed steel.

Solution:

1. Section properties (steps 1 and 2)

The width of the top flange in Example 4.2 is $b = 18$ in., and its average thickness from Figure 4.8 is

$$h_f = 4\tfrac{1}{2} + \frac{3\tfrac{1}{2}}{2} = 6.25 \text{ in}$$

Try four #6 (four 12.7 mm dia) nonprestressed tension steel bars in this cycle in addition to the prestressing reinforcement.

2. **Stress f_{ps} in prestressing steel at nominal strength (step 3)**
From Example 4.9,

$$f_{pe} \cong 155{,}000 \text{ psi}$$

$$0.5f_{pu} = 0.50 \times 270{,}000 = 135{,}000 \text{ psi}$$

$$f_{pe} > 0.5f_{pu}$$

Hence, one can use the ACI approximate procedure for determining f_{ps}.

(A) *BONDED CASE*

If the position of the neutral axis is not known, analyze as a rectangular section as follows: From Equation 4.51,

$$f_{ps} = f_{pu}\left(1 - \frac{\gamma_p}{\beta_1}\left[\rho_p \frac{f_{pu}}{f'_c} + \frac{d}{d_p}(\omega - \omega')\right]\right)$$

$$\frac{f_{py}}{f_{pu}} = \frac{229{,}500}{270{,}000} = 0.85, \text{ use } \gamma_p = 0.40$$

$$A_{ps} = 13 \times 0.153 = 1.99 \text{ in}^2$$

$$A_s = 4 \times 0.44 = 1.76 \text{ in}^2$$

$$\rho_p = \frac{A_{ps}}{bd_p} = \frac{1.99}{18 \times 36.16} = 0.0032$$

$$\omega = \frac{A_s}{bd} \times \frac{f_y}{f'_c} = \frac{1.76}{18 \times 37.6} \times \frac{60{,}000}{5{,}000} - 0.00132$$

For $\omega' = 0$,

$$f_{ps} = 270{,}000\left(1 - \frac{0.40}{0.80}\left[0.0032 \times \frac{270{,}000}{5{,}000} + \frac{37.6}{36.16}(0.0132)\right]\right)$$

$$= 270{,}000 \times 0.897 = 242{,}190 \text{ psi } (1{,}670 \text{ MPa})$$

$$a = \frac{1.99 \times 242{,}190 + 1.76 \times 60{,}000}{0.85 \times 5{,}000 \times 18} = 7.68 \text{ in } > h_f = 6.25 \text{ in}$$

Hence, the neutral axis is outside the flange, and analysis has to be based on a T-section. Using in such a case the web width b_w.

$$\rho_p = \frac{A_{ps}}{b_w d_p} = \frac{1.99}{6 \times 36.16} = 0.0092$$

$$\omega_w = \frac{A_s}{b_w d} \times \frac{f_y}{f'_c} = \frac{1.76}{6 \times 37.6} \times \frac{60{,}000}{5{,}000} = 0.0936$$

$$f_{ps} = 270{,}000\left(1 - \frac{0.40}{0.80}\left[0.0092 \times \frac{270{,}000}{5{,}000} + \frac{37.6}{36.16}(0.0936 - 0)\right]\right)$$

$$= 189{,}793 \text{ psi } (1{,}309 \text{ MPa})$$

$$A_{pw} f_{ps} = A_{ps} f_{ps} + A_s f_y - 0.85 f'_c (b - b_w) h_f$$

$$= 1.99 \times 189{,}793 + 1.76 \times 60{,}000 - 0.85 \times 5{,}000(18 - 6) \times 6.25$$

$$= 377{,}688 + 105{,}600 - 318{,}750 = 164{,}538 \text{ lb}$$

$$a = \frac{164{,}538}{0.85 \times 5{,}000 \times 6} = 6.45 \text{ in. } (16.4 \text{ cm})$$

3. Available nominal moment strength (steps 4–8)

$$M_n = A_{pw} f_{ps}\left(d_p - \frac{a}{2}\right) + A_s f_y(d - d_p) + 0.85f_c'(b - b_w)h_f\left(d_p - \frac{h_f}{2}\right)$$

$$= 164{,}538\left(36.16 - \frac{6.45}{2}\right) + 1.76(60{,}000)(37.6 - 36.16)$$

$$+ 0.85(5{,}000)(18 - 6) \times 6.25\left(36.16 - \frac{6.25}{2}\right) = 16{,}071{,}226 \text{ in.-lb}$$

(1,816 kN-m) < required M_n = 18,026,667 in.-lb (2,037 kN-m), hence the section is inadequate.

Proceed to another trial and adjustment cycle using more nonprestressed reinforcement. Try five #8s (five 25.4 mm dia), $A_s = 3.95$ in.^{2} (25.5 cm^2). We have

$$\omega_w = \frac{3.95}{6 \times 37.6} \times \frac{60{,}000}{5{,}000} = 0.21$$

giving f_{ps} = 173,458 psi and $A_{pw}f_{ps}$ = 263,421 lb (1,172 kN). So

$$a = \frac{263{,}421}{0.85 \times 5{,}000 \times 6} = 10.3 \text{ in. (26.2 cm)}$$

$$M_n = 263{,}421\left(36.16 - \frac{10.3}{2}\right) + 3.95(60{,}000)(37.6 - 36.16)$$

$$+ 0.85(5{,}000)(18 - 6) \times 6.25\left(36.16 - \frac{6.25}{2}\right)$$

$$= 19{,}039{,}871 \text{ in.-lb (2,152 kN-m)} > \text{Required } M_n = 18{,}026{,}667 \text{ in.-lb, O.K.}$$

Hence, use five #8 nonprestressed bars at the bottom fibers, and adopt the design for the bonded case.

(B) *NONBONDED CASE*

$$\text{Span-to-Depth ratio} = \frac{65 \times 12}{40} = 19.5 < 35$$

Hence, from Equation 4.52a,

$$f_{ps} = f_{pe} + 10{,}000 + \frac{f_c'}{100\rho_p} = 155{,}000 + 10{,}000 + \frac{5{,}000}{100 \times 1.99/(6 \times 36.16)}$$

$$= 170{,}451 \text{ psi (1,175 MPa)}$$

Notice that $b_w = 6$ in. is used here for ρ_p, since it is now known that the section behaves like a T-beam, as the neutral axis is below the flange. Thus,

$$f_{ps} = 170{,}451 \text{ psi (1,175 MPa)}$$

1. Selection of nonprestressed steel

Try five #8 nonprestressed tension reinforcements to resist part of the factored moment:

$$A_s = 5 \times 0.79 = 3.95 \text{ in}^2 \text{ (25.4 cm}^2\text{)}$$

$$A_{pw} f_{ps} = 1.99 \times 170{,}421 + 3.95 \times 60{,}000 - 0.85 \times 5{,}000(18 - 6)6.25$$

$$= 257{,}388 \text{ lb}$$

$$a = \frac{A_{pw} f_{ps}}{0.85f_c'b_w} = \frac{257{,}388}{0.85 \times 5{,}000 \times 6} = 10.1 \text{ in. (25.6 cm)}$$

2. Available moment strength (steps 4–8)

From Equation 4.48,

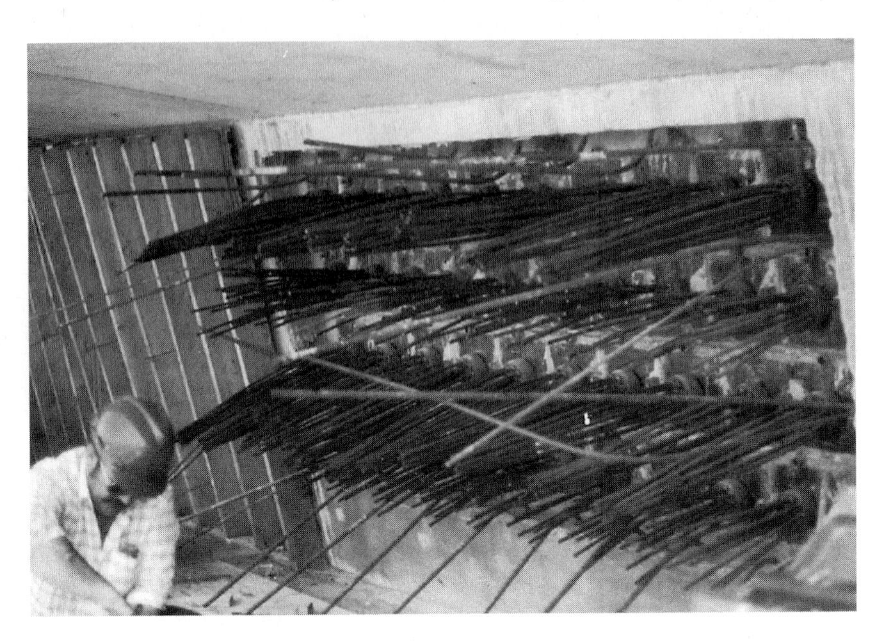

Photo 4.16 Diaphragm anchorage.

$$\text{Available } M_n = 257{,}388\left(36.16 - \frac{10.1}{2}\right) + 3.95 \times 60{,}000(37.6 - 36.16)$$

$$+ 0.85 \times 5{,}000(18 - 6) \times 6.25 \times \left(36.16 - \frac{6.25}{2}\right)$$

$$= 18{,}878{,}527 \text{ in.-lb } (2{,}133 \text{ kN-m})$$

$$> \text{Req. } M_n = 18{,}026{,}667 \text{ in.-lb, O.K.}$$

(C) *CHECK FOR REINFORCEMENT LIMITS*

1. **Minimum reinforcement**

From Equation 4.25, the cracking moment, M_{cr} is given by

$$M_{cr} = f_r S_b + P_e\left(e + \frac{r^2}{c_b}\right)$$

From Example 4.2, $f_r = 7.5\sqrt{5{,}000} = 530.3$ psi (3.7 MPa). So since $S_b = 3{,}750$ in.^{3}, $e = 15$ in., $r^2/c_b = 187.5/18.84 = 9.95$ in., and $P_e = 308{,}255$ lb (1,371 kN), we get

$$M_{cr} = 530.3 \times 3{,}750 + 308{,}295(15 + 9.95)$$

$$= 9{,}680{,}585 \text{ in.-lb } (1{,}090 \text{ kN-m})$$

$$1.2M_{cr} = 1.2 \times 9{,}680{,}585 = 11{,}616{,}702 \text{ in.-lb } (1{,}313 \text{ kN-m})$$

$$M_u = \phi M_n = 0.90 \times 18{,}026{,}667$$

$$= 16{,}224{,}000 \text{ in.-lb } (1{,}833 \text{ kN-m})$$

Finally, from Equation 4.54a,

$$M_u > 1.2M_{cr}$$

Hence, the requirement for minimum reinforcement is satisfied for both the non-bonded and the bonded case.

2. **Maximum allowable reinforcement index**

$$\text{Max. allow. } \omega_p = 0.36\beta_1 = 0.36 \times 0.80 = 0.288$$

$$\text{Actual } \omega_p = \rho_p(f_{ps}/f'_c) = 0.0032 \times \frac{170{,}451}{5{,}000} = 0.109$$

$$\text{Actual } \omega = \rho(f_y/f'_c) = \frac{3.96}{18 \times 37.6} \times \frac{60{,}000}{5{,}000} = 0.0702$$

Hence,

$$\omega_p + \omega = 0.109 + 0.0702 = 0.1862 < 0.288, \text{ O.K.}$$

Accordingly, adopt the design that uses the concrete section in Example 4.2 and include five #8 nonprestressed steel bars at the tension side. Note that the moment strength capacity of the non-bonded section for the same area of nonprestressed steel is less than the moment strength capacity of the bonded section, which is expected (19,039,871 in.-lb vs. 18,875,527 in.-lb).

If $f'_c = 6{,}000$ psi would have been used in the strength design in this example, as it was in the service-load design of this section in Example 4.2, less mild steel reinforcement would have been needed.

4.16 USE OF THE ANSI LOAD AND STRENGTH REDUCTION FACTORS IN EXAMPLE 4.10

Using ANSI load factors for an alternate solution,

$$U = 1.2D + 1.6L$$

$$\phi = 0.85 \text{ for flexure}$$

$$w_u = 1.2(100 + 397) + 1.6(1{,}100) = 2{,}352 \text{ plf}$$

$$M_u = \frac{2{,}352(65)^2 \times 12}{8} = 14{,}905{,}800 \text{ in.-lb}$$

$$M_n = \frac{M_u}{\phi} = \frac{14{,}905{,}800}{0.85} = 17{,}536{,}236 \text{ in.-lb}$$

$$M_n \text{ by the ACI Code load factors} = 18{,}026{,}667 \text{ in.-lb}$$

$$\text{Percentage difference} = \frac{18{,}026{,}667 - 17{,}536{,}236}{18{,}026{,}667} = 2.7\%$$

Such a small percentage difference has no significant effect on the design.

4.17 SI FLEXURAL DESIGN EXPRESSION

$$f'_{ci} = 0.8f'_c$$

$$f_{ci} = 0.60f'_{ci}$$

$$f_{ti} = \tfrac{1}{4}\sqrt{f'_c} \text{ (midspan)}$$

$$= \tfrac{1}{2}\sqrt{f'_{ci}} \text{ (support)}$$

$$f_c = 0.45f'_c \text{ due to prestress + sustained load}$$

$$f_c = 0.6f'_c \text{ due to prestress + total load if it includes transient load}$$

Stress at transfer

$$f^t = -\frac{P_i}{A_c}\left(1 - \frac{ec_t}{r^2}\right) - \frac{M_d}{S^t} \leq f_{ti} \tag{4.1a}$$

$$f_b = -\frac{P_i}{A_c}\left(1 + \frac{ec_b}{r^2}\right) + \frac{M_D}{S_b} \leq f_{ci} \tag{4.16}$$

Effective stress after losses

$$f^t = -\frac{P_e}{A_c}\left(1 - \frac{ec_t}{r^2}\right) - \frac{M_D}{S^t} \leq f_t \tag{4.2a}$$

$$f_b = -\frac{P_e}{A_c}\left(1 + \frac{ec_b}{r^2}\right) + \frac{M_D}{S_b} \leq f_c \tag{4.2b}$$

Service load final stress

$$f^t = -\frac{P_e}{A_c}\left(1 - \frac{ec_t}{r^2}\right) - \frac{M_T}{S^t} \leq f_c \tag{4.3a}$$

$$f_b = -\frac{P_e}{A_c}\left(1 + \frac{ec_b}{r^2}\right) + \frac{M_T}{S_b} \leq f_t \tag{4.36}$$

$$f_{\text{decomp}} = \frac{P_e}{A_c}\left(1 + \frac{e^2}{r^2}\right) \tag{4.3c}$$

$$S^t \geq \frac{(1-\gamma)M_D + M_{SD} + M_L}{\gamma f_{ti} - f_c} \tag{4.4a}$$

$$S_b \geq \frac{(1-\gamma)M_D + M_{SD} + M_L}{f_t - \gamma f_{ci}} \tag{4.4b}$$

$$e_e = (f_{ti} - \bar{f}_{ci})\frac{S^t}{P_i} \tag{4.5c}$$

$$k_b = \frac{r^2}{c_t}, \qquad k_t = \frac{r^2}{c_b} \tag{4.6}$$

$$b_m = \frac{E_{ct}}{E_c}(b) = n_c b \tag{4.23}$$

Unshored case

$$f^t = -\frac{P_e}{A_c}\left(1 - \frac{ec_t}{r^2}\right) - \frac{M_D + M_{SD}}{S_t} - \frac{M_{CSD} + M_L}{S_c^t} \tag{4.19a}$$

$$f_b = -\frac{P_e}{A_c}\left(1 + \frac{ec_b}{r^2}\right) + \frac{M_D + M_{SD}}{S_b} + \frac{M_{CSD} + M_L}{S_{cb}} \tag{4.19b}$$

Shored case

$$f^t = -\frac{P_e}{A_c}\left(1 - \frac{ec_t}{r^2}\right) - \frac{M_D}{S^t} - \frac{M_{SD} + M_{CSD} + M_L}{S_c^t} \tag{4.22a}$$

$$f_b = -\frac{P_e}{A_c}\left(1 + \frac{ec_b}{r^2}\right) + \frac{M_d}{S_b} + \frac{M_{SD} + M_{CSD} + M_L}{S_{cb}} \quad (4.22b)$$

Equation 4.51 for bonded tendons

$$f_{ps} = f_{pu}\left(1 - \frac{\gamma_p}{\beta_1}\left[\rho_p \frac{f_{pu}}{f'_c} + \frac{d}{d_p}(\omega - \omega')\right]\right) \text{MPa}$$

where $\gamma_p = 0.55$ for $f_{py}/f_{pu} \geq 0.80$
$= 0.40$ for $f_{py}/f_{pu} \geq 0.85$
$= 0.28$ for $f_{py}/f_{pu} \geq 0.90$

Equation 4.52 for nonbonded tendons

$$f_{ps} = f_{pe} + 70 + \frac{f'_c}{100\rho_p} \text{ MPa for } \frac{\text{span}}{\text{depth}} \leq 35$$

$$f_{ps} = f_{pe} + 70 + \frac{f'_c}{300\rho_p} \text{ MPa for } \frac{\text{span}}{\text{depth}} > 35$$

$$\text{MPa} = \text{N/mm}^2 = 106 \text{ N/m}^2$$

$$\text{(lb) } 4.448 = \text{N}$$

$$\text{(psi) } 0.006895 = \text{MPa}$$

$$\text{(lb/ft) } 14.593 = \text{N/m}$$

$$\text{(in.-lb) } 0.113 = \text{N-m}$$

$$1 \text{ Kg force} = 9.806 \text{ N}$$

4.17.1 SI Flexural Design of Prestressed Beams

Example 4.11

Solve Example 4.10 using SI units. Tendons are bonded.
Data:

$A_c = 5045 \text{ cm}^2$ $b = 45.7$ cm $b_w = 15.2$ cm

$I_c = 7.04 \times 10^6 \text{ cm}^4$

$r^2 = 1{,}394 \text{ cm}^2$

$c_b = 89.4$ cm $c^t = 32.5$ cm

$e_e = 84.2$ cm $e_c = 60.4$ cm

$S_b = 78{,}707 \text{ cm}^3$ $S_t = 216{,}210 \text{ cm}^3$

$w_D = 11.9 \times 10^3$ kN/m $w_{SD} = 1{,}459$ N/m $w_L = 16.1$ kN/m

$l = 19.8$ m

$f'_c = 34.5$ MPa $f_{pi} = 1{,}300$ MPa

$f_{pu} = 1{,}860$ MPa $f_{py} = 1{,}580$ MPa

Prestress loss $\gamma = 18\%$ $f_y = 414$ MPa

$A_{ps} = 13$ tendons, diameter 12.7 mm ($A_{ps} = 99$ min^2)

$= 13 \times 99 = 1{,}287 \text{ mm}^2$

Required $M_n = 18.03 \times 10^6$ in.-lb $= 2{,}037$ kN-m

Solution:

1. *Section properties (Steps 1 and 2)*
Flange width $b = 18$ in. = 45.7 cm
Average thickness $h_f = 4.5 + \frac{1}{2}(3.5) \cong 6.25$ in. = 15.7 cm
Try 4 No. 20 M mild steel bars for partial prestressing (diameter = 19.5 mm, $A_s = 300$ mm^2).

$$A_s = 4 \times 300 = 1{,}200 \text{ mm}^2$$

2. *Stress f_{ps} in prestressing steel at nominal strength and neutral axis position (Step 3)*

$$f_{pe} = \gamma f_{pi} = 0.82 \times 1{,}300 \cong 1{,}066 \text{ MPa}$$

Verify Neutral Axis Position
If outside flange, its depth has to be greater than $a = A_{pw}f_{ps}/0.85f'_c b_w$
$0.5f_{pu} = 0.50 \times 1{,}860 = 930$ MPa < 1,066, hence, one can use ACI approximate procedure for determining f_{ps}. From equation 4.51,

$$f_{ps} = f_{pu}\left(1 - \frac{\gamma_p}{\beta_1}\left[\rho_p \frac{f_{pu}}{f'_c} + \frac{d}{d_p}(\omega - \omega')\right]\right)$$

$$d_p = 36.16 \text{ in.} = 91.8 \text{ cm}, \; d = 37.6 \text{ in.} = 95.5 \text{ cm}$$

$$\frac{f_{py}}{f_{pu}} = \frac{1{,}580}{1{,}860} = 0.85, \text{ use } \gamma_p = 0.40$$

$$\rho_p = \frac{A_{ps}}{bd_p} = \frac{1{,}287}{457 \times 918} = 0.00306$$

$$\rho = \frac{A_s}{bd} = \frac{1{,}200}{457 \times 955} = 0.00275$$

$$\omega_p = \frac{A_{ps}}{bd_p} \times \frac{f_{ps}}{f'_c} = 0.00306 \times \frac{1{,}674}{34.5} = 0.14$$

$$\omega = \frac{A_s}{bd} \times \frac{f_y}{f'_c} = 0.00275 \times \frac{414}{34.5} = 0.033$$

$$\omega' = 0$$

For $f'_c = 34.5$ MPa, $\beta_1 = 0.80$

$$f_{ps} = 1{,}860\left(1 - \frac{0.40}{0.80}\left[0.00306 \times \frac{1{,}860}{34.5} + \frac{955}{918} \times 0.033\right]\right)$$

$$= 1{,}860(1 - 0.1) = 1{,}674 \text{ MPa}$$

From Equation 4.47a,

$$a = \frac{A_{pw}f_{ps}}{0.85f'_c b_w}$$

where $A_{pw}f_{ps} = A_{ps}f_{ps} + A_s f_y - 0.85f'_c(b - b_w)h_f$

$$A_{pw}f_{ps} = 1{,}287 \times 1{,}674 + 1{,}200 \times 414$$
$$- 0.85 \times 34.5 \times (45.7 - 15.2)15.7 \times 10^2$$
$$= 10^6(2.15 + 0.5 - 1.14) \text{ N} = 1{,}240 \text{ kN}$$

$$a = \frac{1{,}240 \times 10}{0.85 \times 34.7 \times 15.2} = 24.7 \text{ cm} > h_f = 15.7 \text{ cm}$$

Hence neutral axis is outside the flange and analysis has to be based on a T-section.

3. *Available nominal moment strength (Step 4–8)*

$$\omega_T = \omega_p + \omega = 0.14 + 0.033 = 0.173 < 0.36\beta_1, \text{ hence, O.K.}$$

hence, maximum reinforcement index ω is satisfied.

$$M_n = A_{pw} f_{ps}\left(d - \frac{a}{2}\right) + A_s f_y(d - d_p) + 0.85f'_c(b - b_w)h_f\left(d_p - \frac{h_f}{2}\right)$$

$$\text{Available } M_n = 1.24 \times 10^6\left(91.8 + \frac{27.7}{2}\right) + 1{,}200 \times 414(95.5 - 91.8)$$

$$+0.85 \times 34.5(45.7 - 15.2)15.7\left(91.8 - \frac{15.2}{2}\right) \times 10^2$$

$$= 10^6(96.6 + 1.83 + 118.2) \text{ N-cm} = 2{,}166 \text{ kN-m}$$

$$> \text{Required } M_n = 2{,}037 \text{ kN-m}$$

hence, section is O.K.

REFERENCES

4.1 ACI Committee 318. *Building Code Requirements for Structural Concrete* (ACI 318-99) and Commentary, ACI 318R-95. American Concrete Institute, Farmington Hills, MI, 1996, pp. 392.

4.2 Nawy, E. G., *Reinforced Concrete—A Fundamental Approach,* 4th Ed., Prentice Hall, Upper Saddle River, NJ.: 2000, pp. 776.

4.3 Nawy, E. G., and Chiang, J. Y., "Serviceability Behavior of Post-Tensioned Beams." *Journal of the Prestressed Concrete Institute* 25, Chicago Jan.–Feb. 1980: 74–85.

4.4 Nawy, E. G., and Potyondy, G. J., "Moment Rotation, Cracking, and Deflection of Spirally Bound Pretensioned Prestressed Concrete Beams." *Engineering Research Bulletin No. 51.* New Brunswick, N.J.: Bureau of Engineering Research, Rutgers University, 1970, pp. 1–97.

4.5 Post-Tensioning Institute, *Post-Tensioning Manual,* 6th Ed., Phoenix, AZ, 2000.

4.6 Nawy, E. G., and Goodkind, H. "Longitudinal Crack in Prestressed Box Beam." *Journal of the Prestressed Concrete Institute* Chicago, (1969): 38–42.

4.7 Yong, Y. K., Gadebeku, C. and Nawy, E. G., "Anchorage Zone Stresses of Post-tensioned Prestressed Beams Subjected to Shear Forces," *ASCE Structural Division Journal,* Vol. 113, No. 8, August 1987, pp. 1789–1805.

4.8 Nawy, E. G., and Huang, P. T. "Crack and Deflection Control of Pretensioned Prestressed Beams." *Journal of the Prestressed Concrete Institute* 22 (1977): 30–47.

4.9 Prestressed Concrete Institute. *PCI Design Handbook, Precast and Prestressed Concrete.* 5th ed. Prestressed Concrete Institute, Chicago, 1999.

4.10 Gerwick B. C., *Construction of Prestressed Concrete Structures.* John Wiley, New York, 1993.

4.11 Nawy, E. G. "Flexural Cracking Behavior of Pretensioned and Post-Tensioned Beams—The State of the Art." *Journal of the American Concrete Institute,* December 1985, pp. 890–900.

4.12 American Association of State Highway and Transportation Officials. *AASHTO Standard Specifications for Highway Bridges.* Washington, D.C.: AASHTO, 16th Ed., 1996 and Supplements, 1997, 1998.

4.13 Nilson, A. H. "Flexural Design Equations for Prestressed Concrete Members." *Journal of the Prestressed Concrete Institute* 14 (1969): 62–71.

4.14 Lin, T. Y., and Burns, N. H. *Design of Prestressed Concrete Structures.* John Wiley & Sons, New York, 1981.

4.15 Nilson, A. H. *Design of Prestressed Concrete.* John Wiley & Sons, New York, 1987.

4.16 Naaman, A. E. *Prestressed Concrete Analysis and Design.* McGraw Hill, New York, 1982.

4.17 Federal Highway Administration, "Optimized Sections for High Strength Concrete Bridge Girders," FHWA Publication No. RD-95-180, Washington, D.C., August 1997, pp. 156.

4.18 Collins, M. P. and Mitchell, D., *Prestressed Concrete Structures,* Prentice Hall, Upper Saddle River, NJ, 1991, 1–766 p.

4.19 Abeles, P.W., and Bardhan-Roy, B. K. *Prestressed Concrete Designer's Handbook.* 3d Ed. Viewpoint Publications, London, 1981.

4.20 Guyon, Y. *Limit State Design of Prestressed Concrete; Vol. 1, Design of Section.* John Wiley & Sons, New York, 1972.

4.21 Marshall, W. T., and Mattock, A. H. "Control of Horizontal Cracking in the Ends of Pretensioned Prestressed Concrete Girders." *Journal of the Prestressed Concrete Institute* 7 (1962): 56–74.

4.22 Ezeldin, A., and Balaguru, P. N. "Analysis of Partially Prestressed Beams for Strength and Serviceability Using Microcomputers." Paper presented at the First Canadian Conference on Computer Applications in Civil Engineering, MacMaster University, Hamilton, Ontario, Canada, May 1986.

4.23 Gergely, P., and Sozen, M. A., "Design of Anchorage Zone Reinforcement in Prestressed Concrete Beams." *Journal of the Prestressed Concrete Institute* 12 (1967): 63–75.

4.24 Baker, A. L. L. *The Ultimate Load Theory Applied to the Design of Reinforcement and Prestressed Concrete Frames.* London: Concrete Publications, 1956.

4.25 Ellingwood, B., McGregor, J. G., Galambos, T. V., and Cornell, C. A. "Probability Based Load Criteria: Load Factors and Load Combinations." *Journal of the Structural Division, ASCE* 108 (1982): 978–997.

4.26 American National Standards Institute. *Minimum Design Loads for Buildings and Other Structures,* ANSI–ASCE 7-95, 1995, pp. 214.

4.27 Stone, W. C., and Breen, J. E. "Design of Post-Tensioned Girder Anchorage Zones." *Journal of the Prestressed Concrete Institute* 29 (1984): 64–109.

4.28 Nawy, E. G., Yong, Y. K., and Gadebeku, B. K. "Anchorage Zone Stresses of Post-Tensioned Prestressed Beams." *Proceedings of the International Conference on Structural Mechanics of Reinforced and Prestressed Concrete,* Sept. 1996, Chinese Academy of Sciences, Nanjing Institute of Technology, 1986.

4.29 ACI Committee 435, "Control of Deflection in Concrete Structures," ACI Committee Report, E. G. Nawy, Chairman, American Concrete Institute, Farmington Hills, 1995, pp. 77.

4.30 Breen J. E., Burdet, O., Roberts, C., Sanders, D., and Wollman, G. "Anchorage Zone Reinforcement for Post-tensioned Concrete Girders," NCHRP Report 356, Transportation Research Board, Washington, D.C., 1994, 204 p.

4.31 Nawy E. G., "Cracking of Concrete—ACI and CEB Approaches," Proceedings, *International Conference on Advances in Concrete Technology,* 2nd Edition, V. M. Malhotra, ed., CANMET/ACI, Farmington Hills, MI, 1992, pp 203–242.

4.32 Nawy, E. G., *Fundamentals of High Performance Concrete,* 2nd ed. John Wiley and Sons, New York, 2000.

4.32 Nawy, E. G., editor-in-chief, *Concrete Construction Engineering Handbook,* CRC Press, Boca Raton, FL, 1998, pp. 1250.

PROBLEMS

4.1 Design, for service-load and ultimate-load conditions, a pretensioned symmetrical I-section beam to carry a superimposed dead load of 750 plf (10.95 kN/m) and a service live load of 1,500 plf (21.90 kN/m) on a 50 ft (15.2 m) simply supported span. Assume that the sectional properties are $b = 0.5h$, $h_f = 0.2h$, and $b_w = 0.40b$, using the following data:

$$f_{pu} = 270{,}000 \text{ psi } (1{,}862 \text{ MPa})$$

$$E_{ps} = 28.5 \times 10^6 \text{ psi } (196 \times 10^3 \text{ MPa})$$

$$f'_c = 5{,}000 \text{ psi (34.5 MPa) normal-weight concrete}$$

$$f'_{ci} = 3{,}500 \text{ psi (24.1 MPa)}$$

$$f_t = 12\sqrt{f'_c} \text{ assuming deflection is not critical}$$

Sketch the design details, including the anchorage zone reinforcement and arrangement of strands for (a) straight-tendon case, and (b) a harped tendon at the third span points with end eccentricity zero. Assume total prestress losses of 22 percent.

4.2 Solve Problem 4.1 if the beam is post-tensioned bonded and the tendon is draped. Use strain compatibility to determine the value of the tendon stress f_{ps} at nominal strength. Design the anchorage zone reinforcement by the strut-and-tie method.

4.3 A double-T pretensioned roof beam is shown in Figure P4.3. It has a simple span of 74 ft (22.6 m) and carries superimposed service live and dead loads of 60 psf (2,873 Pa; W_{SD} part of load = 25 psf). It also carries a 2-in. (5.1 cm) concrete topping. Design the prestressing reinforcement and the appropriate eccentricities using 270-grade prestressing strands (f_{pu} = 1,862 MPa) with a total prestress loss of 20 percent. Use the appropriate percentage of nonprestressed mild steel for partial prestressing behavior at the limit state at failure. Assume the strands to be harped at midspan, and sketch the reinforcing details including the anchorage zone strands. Also, draw the distribution of stresses for the various loading stages in your solution. The following data are given:

$$f_{pu} = 270{,}000 \text{ psi, stress-relieved strands (1,862 MPa)}$$

$$E_{ps} = 28 \times 10^6 \text{ psi } (193 \times 10^3 \text{ MPa})$$

$$f'_c = 5{,}000 \text{ psi (34.5 MPa) normal-weight concrete}$$

$$f'_c \text{ for topping} = 3{,}000 \text{ psi (20.7 MPa) normal-weight concrete}$$

$$f'_{ci} = 4{,}000 \text{ psi}$$

$$V/S = 1.79 \text{ in.}$$

$$e_c = 17.71 \text{ in.}$$

	Untopped	Topped
A_c	567 in.^{2} (3,658 cm^2)	—
T_c	55,464 in.^{4}	71,886 in.^{4}
C_b	21.21 in.	23.66 in.
C_t	10.79 in.	10.34 in.
S_b	2,615 in.^{3}	3,038 in.^{3}
S^t	5,140 in.^{3}	6,952 in.^{3}
W_D	591 plf	791 plf

4.4 A bridge girder has a simple span of 55 ft (16.8 m). It is subjected to a total superimposed service load of 4,600 plf (67.2 kN/m). Design the section as a post-tensioned bonded beam using an AASHTO standard section with parabolically draped tendons. Assume a 7 in. situ-cast slab over the precast section and the following data:
Beams spaced at 7′ –6″ c. to c.

$$f_{pu} = 270{,}000 \text{ psi stress relieved (1,862 MPa)}$$

$$E_{ps} = 28 \times 10^6 \text{ psi } (193 \times 10^3 \text{ MPa})$$

$$f'_c = 5{,}000 \text{ psi (34.5 MPa) normal-weight concrete}$$

$$\text{slab } f'_c = 3{,}000 \text{ psi (20.7 MPa) normal-weight concrete}$$

$$f'_{ci} = 4{,}000 \text{ psi (27.6 MPa)}$$

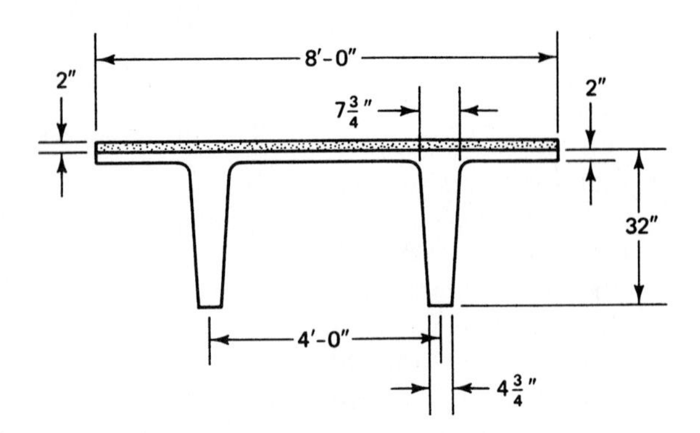

Figure P4.3 Double-T cross section.

Design the bridge section for service-load and ultimate-load conditions, and detail all the reinforcement including the anchorage zone steel and the nonprestressed tensile steel. Assume the section to be constant throughout the span, and, for the limit state at failure analysis, find the stress f_{ps} at nominal strength by (a) the ACI approximate procedure, and (b) strain compatibility. Assume a total prestress loss of 20 percent. Design the anchorage zone reinforcement by the strut-and-tie method and compare the results with those obtained using the linear elastic analysis approach.

4.5 Solve problem 4.4 if the draped tendons are nonbonded, and compare the two solutions. Use the tie-and-strut method to design the anchorage zone reinforcement. Assume the tendons are harped at midspan.

5.1 INTRODUCTION

This chapter presents procedures for the design of prestressed concrete sections to resist shear and torsional forces resulting from externally applied loads. Since the strength of concrete in tension is considerably lower than its strength in compression, design for shear and torsion becomes of major importance in all types of concrete structures.

The behavior of prestressed concrete beams at failure in shear or combined shear and torsion is distinctly different from their behavior in flexure: They fail abruptly without sufficient advance warning, and the diagonal cracks that develop are considerably wider than the flexural cracks. Both shear and torsional forces result in shear stress. Such a stress can result in principal tensile stresses at the critical section which can exceed the tensile strength of the concrete.

Photos in this chapter show typical beam shear failure and torsion failure. Notice the curvelinear plane of twist depicting torsional failure caused by the imposed torsional moments. As will be discussed in subsequent sections, the shearing stresses in regular beams are caused, not by direct shear or pure torsion, but by a combination of external

Empire State Performing Arts Center, Albany, New York, Ammann & Whitney design, prestressed concrete shell ring. (*Courtesy,* New York Office of General Services.)

loads and moments. This leads to *diagonal tension,* or flexural shear stresses, in the member. Only in special applications in certain structural systems are direct shear or pure torsion applied. Examples of such cases are corbels or brackets involving direct shear, or a cantilever balcony involving essentially a direct twist on the supporting beam.

5.2 BEHAVIOR OF HOMOGENEOUS BEAMS IN SHEAR

Consider the two infinitesimal elements, A_1 and A_2 of a rectangular beam in Figure 5.1(a) made of homogeneous, isotropic, and linearly elastic material. Figure 5.1(b) shows the bending stress and shear stress distributions across the depth of the section. The tensile normal stress f_t and the shear stress v are the values in element A_1 across plane a_1–a_1 at a distance y from the neutral axis. From the principles of classical mechanics, the normal stress f and the shear stress v for element A_1 can be written as

$$f = \frac{My}{I} \tag{5.1}$$

and

$$v = \frac{VA\bar{y}}{Ib} = \frac{VQ}{Ib} \tag{5.2}$$

where M and V = bending moment and shear force at section a_1–a_1
A = cross-sectional area of the section at the plane passing through the centroid of element A_1

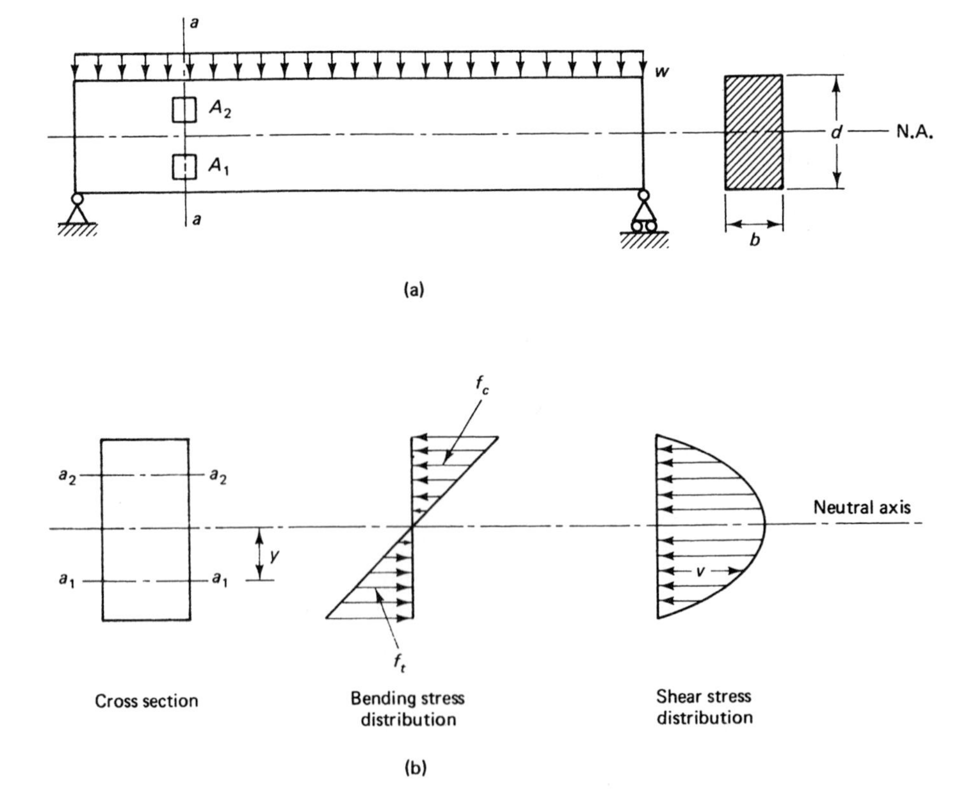

Figure 5.1 Stress distribution for a typical homogeneous rectangular beam.

Photo 5.1 Typical diagonal tension (flexure shear) failure at rupture load level. (Test by Nawy et al.)

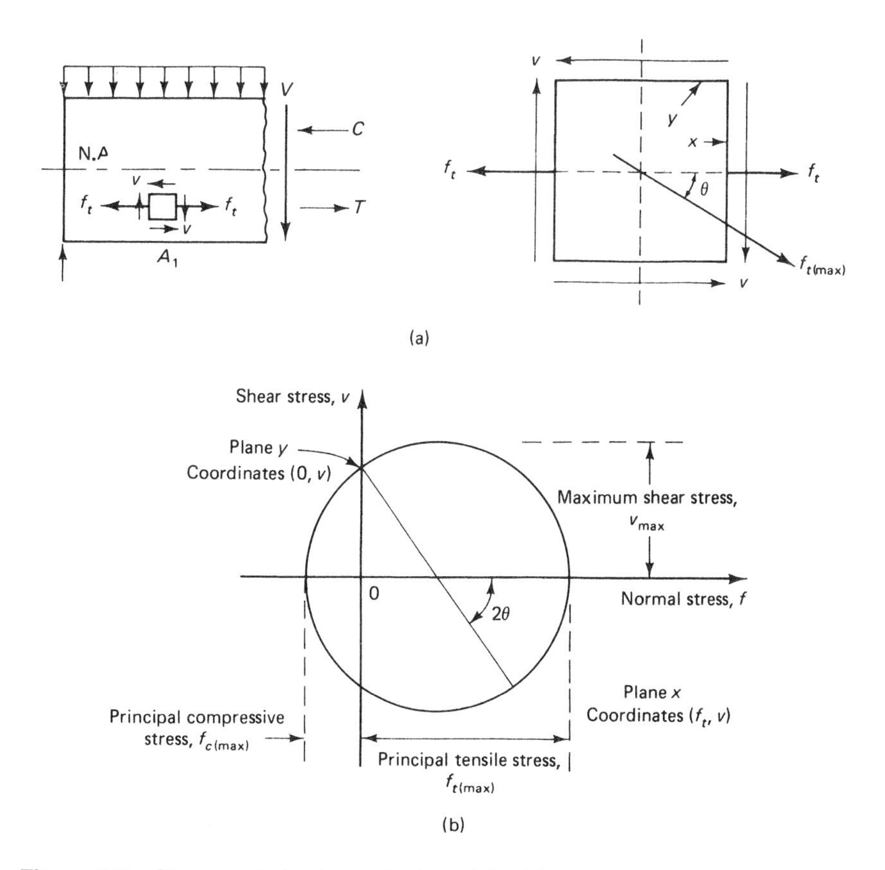

Figure 5.2 Stress state in elements A_1 and A_2. (a) Stress state in element A_1. (b) Mohr's circle representation, element A_1. (c) Stress state in element A_2. (d) Mohr's circle representation, element A_2.

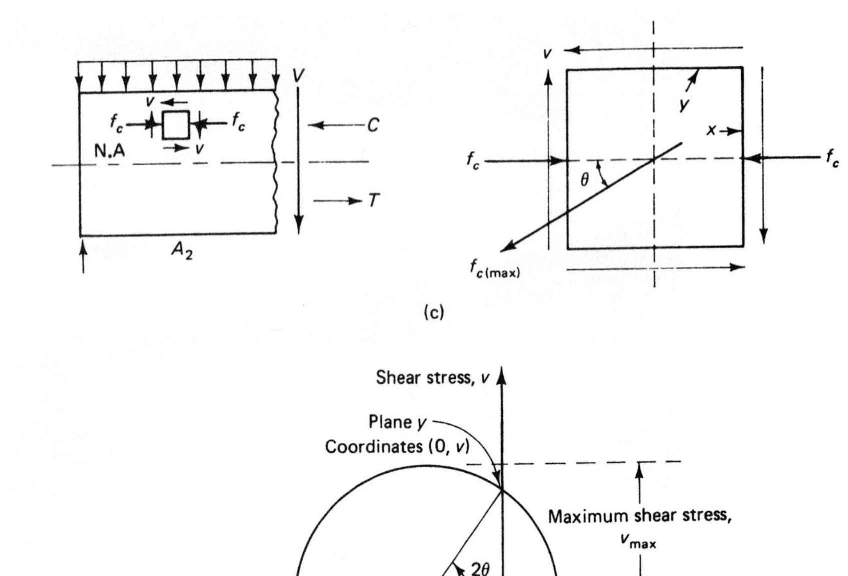

Figure 5.2 *Continued*

y = distance from the element to the neutral axis
$\bar{y}$ = distance from the centroid of A to the neutral axis
I = moment of inertia of the cross section
Q = statical moment of the cross-sectional area above or below that level about the neutral axis
b = width of the beam

Figure 5.2 shows the internal stresses acting on the infinitesimal elements A_1 and A_2. Using Mohr's circle in Figure 5.2(b), the principal stresses for element A_1 in the tensile zone below the neutral axis become

$$f_{t(\max)} = \frac{f_t}{2} + \sqrt{\left(\frac{f_t}{2}\right)^2 + v^2} \quad \text{principal tension} \tag{5.3a}$$

$$f_{c(\max)} = \frac{f_t}{2} - \sqrt{\left(\frac{f_t}{2}\right)^2 + v^2} \quad \text{principal compression} \tag{5.3b}$$

and

$$\tan 2\theta_{\max} = \frac{v}{f_t/2} \tag{5.3c}$$

Photo 5.2 Simply supported beam prior to developing diagonal tension crack in flexure shear (load stage 11). (Test by Nawy et al.)

5.3 BEHAVIOR OF CONCRETE BEAMS AS NONHOMOGENEOUS SECTIONS

The behavior of reinforced and prestressed concrete beams differs from that of steel beams in that the tensile strength of concrete is about one-tenth of its strength in compression. The compression stress f_c in element A_2 of Figure 5.2(b) above the neutral axis prevents cracking, as the maximum principal stress in the element is in compression. For element A_1 below the neutral axis, the maximum principal stress is in tension; hence

Photo 5.3 Principal diagonal tension crack at failure of beam in the preceding photograph (load stage 12).

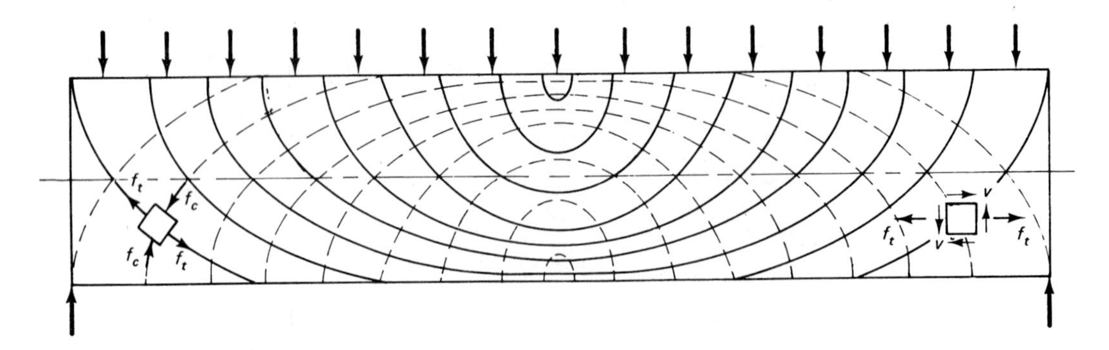

Figure 5.3 Trajectories of principal stresses in a homogeneous isotropic beam. Solid lines are tensile trajectories, dashed lines are compressive trajectories.

cracking ensues. As one moves toward the support, the bending moment and hence f_t decreases, accompanied by a corresponding increase in the shear stress. The principal stress $f_{t(\max)}$ in tension acts at an approximately 45° plane to the normal at sections close to the support, as seen in Figure 5.3. Because of the low tensile strength of concrete, diagonal cracking develops along planes perpendicular to the planes of principal tensile stress—hence the term *diagonal tension cracks.* To prevent such cracks from openings, special diagonal tension reinforcement has to be provided.

If f_t close to the support of Figure 5.3, is assumed equal to zero, the element becomes nearly in a state of pure shear, and the principal tensile stress, using Equation 5.3b, would be equal to the shear stress v on a 45° plane. It is this diagonal tension stress that causes the inclined cracks.

Definitive understanding of the correct shear mechanism in reinforced concrete is still incomplete. However, the approach of ACI-ASCE Joint Committee 426 gives a systematic empirical correlation of the basic concepts developed from extensive test results.

5.4 CONCRETE BEAMS WITHOUT DIAGONAL TENSION REINFORCEMENT

In regions of large bending moments, cracks develop almost perpendicular to the axis of the beam. These cracks are called *flexural cracks.* In regions of high shear due to the diagonal tension, the inclined cracks develop as an extension of the flexural crack and are termed *flexure shear cracks.* Figure 5.4 portrays the types of cracks expected in a reinforced concrete beam with or without adequate diagonal tension reinforcement.

In prestressed beams, the section is mostly in compression at service load. From Figures 5.2(c) and (d), the principal stresses for element A_2 would be

$$f_{t(\max)} = -\frac{f_c}{2} + \sqrt{(f_c/2)^2 + v^2} \qquad \text{principal tension} \tag{5.4a}$$

$$f_{c(\max)} = -\frac{f_c}{2} - \sqrt{(f_c/2)^2 + v^2} \qquad \text{principal compression} \tag{5.4b}$$

and

$$\tan 2\theta_{\max} = \frac{v}{f_c/2} \tag{5.4c}$$

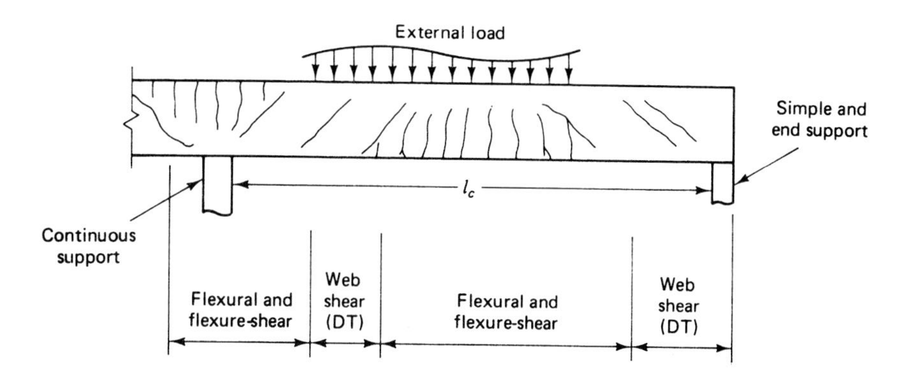

Figure 5.4 Crack categories.

5.4.1 Modes of Failure of Beams Without Diagonal Tension Reinforcement

The slenderness of the beam, that is, its shear span-to-depth ratio, determines the failure mode of the beam. Figure 5.5 demonstrates schematically the failure patterns for the different slenderness ratio limits. The shear span a for concentrated load is the distance between the point of application of the load and the face of support. For distributed loads, the shear span l_c is the clear beam span. Fundamentally, three modes of failure or their combinations occur: flexural failure, diagonal tension failure, and shear compression failure (web shear). The more slender the beam, the stronger the tendency toward flexural behavior, as seen from the following discussion.

5.4.2 Flexural Failure [*F*]

In the region of flexural failure, cracks are mainly vertical in the middle third of the beam span and perpendicular to the lines of principal stress. These cracks result from a very small shear stress v and a dominant flexural stress f which results in an almost horizontal principal stress $f_{t(\max)}$. In such a failure mode, a few very fine vertical cracks start to develop in the midspan area at about 50 percent of the failure load in flexure. As the external load increases, additional cracks develop in the central region of the span and the initial cracks widen and extend deeper toward the neutral axis and beyond, with a marked increase in the deflection of the beam. If the beam is underreinforced, failure occurs in a ductile manner by initial yielding of the main longitudinal flexural reinforcement. This type of behavior gives ample warning of the imminence of collapse of the beam. The shear span-to-depth ratio for this behavior exceeds a value of 5.5 in the case of concentrated loading, and in excess of 16 for distributed loading.

5.4.3 Diagonal Tension Failure [Flexure Shear, *FS*]

Diagonal tension failure precipitates if the strength of the beam in diagonal tension is lower than its strength in flexure. The shear span-to-depth ratio is of *intermediate* magnitude, varying between 2.5 and 5.5 for the case of concentrated loading. Such beams can be considered of intermediate slenderness. Cracking starts with the development of a few fine vertical flexural cracks at midspan, followed by the destruction of the bond between the reinforcing steel and the surrounding concrete at the support. Thereafter, without ample warning of impending failure, two or three diagonal cracks develop at about $1\frac{1}{2}d$ to $2d$ distance from the face of the support in the case of reinforced concrete beams, and usually at about a quarter of the span in the case of prestressed concrete beams. As they stabilize, one of the diagonal cracks widens into a principal diagonal tension crack and extends to the top compression fibers of the beam, as seen in Figure 5.5(b) and 5.5(c).

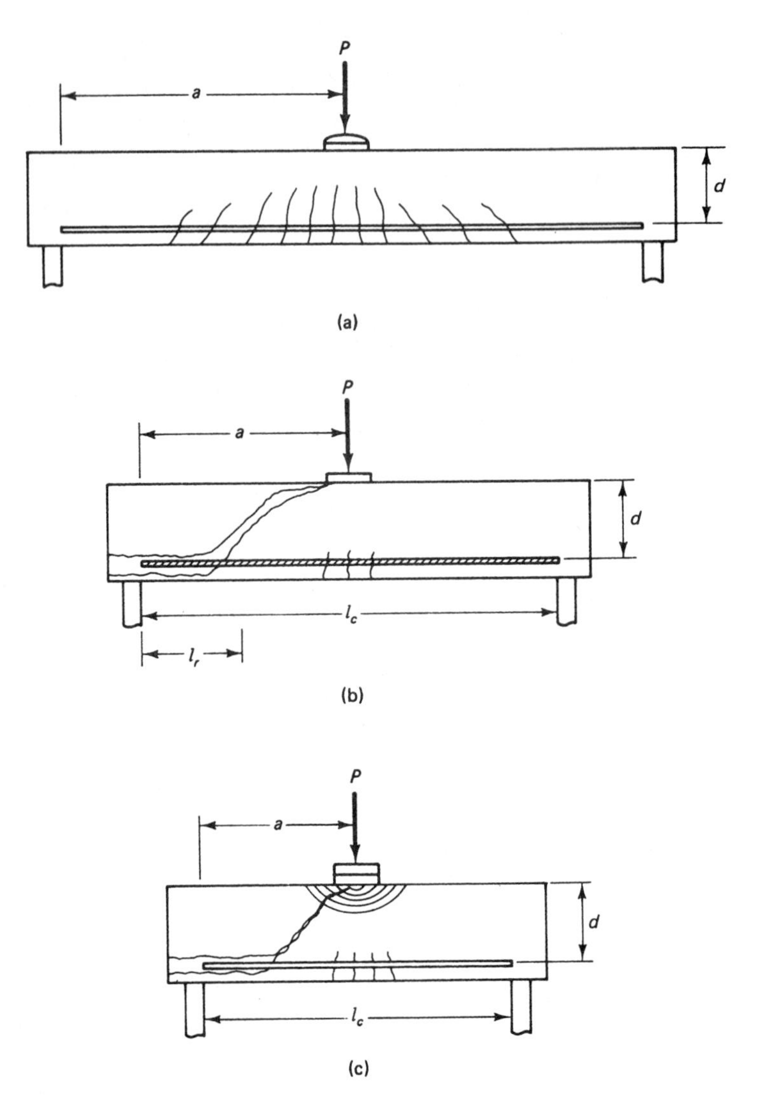

Figure 5.5 Failure patterns as a function of beam slenderness. (a) Flexural failure. (b) Diagonal tension failure (flexure shear). (c) Shear compression failure (web shear).

Notice that the flexural cracks do not propagate to the neutral axis in this essentially brittle failure mode, which has relatively small deflection at failure.

Although the maximum external shear is at the support, the critical location of the maximum principal stress in tension is not. It is considerably reduced at that section because of the high compression force of the prestressing tendon, in addition to the vertical compression force of the beam reaction at the supports. This is the reason why the stabilized diagonal crack is located further into the span, depending on the magnitude of the prestressing force and the variation in its eccentricity, with an average value of about one-quarter of the span in flanged prestressed beams. In sum, the diagonal tension failure is the result of the combination of the flexural and shear stresses, taking into account the balancing contribution of the vertical component of the prestressing force, and noted by a combination of flexural and diagonal cracks. It is best termed *flexure shear* in the case of prestressed beams, and is more common to account for than *web shear,* discussed next.

Photo 5.4 Shear failure in prestressed I-beam (Nawy et al.).

5.4.4 Shear Compression Failure [Web Shear, *WS*]

Beams that are most subject to shear compression failure have a small span-to-depth ratio of magnitude 2.5 for the case of concentrated loading and less than 5.0 for distributed loading. As in the diagonal tension case, a few fine flexural cracks start to develop at midspan and stop propagating as destruction of the bond occurs between the longitudinal bars and the surrounding concrete at the support region. Thereafter, an inclined crack steeper than in the diagonal tension case suddenly develops and proceeds to propagate toward the neutral axis. The rate of its progress is reduced with the crushing of the con-

Photo 5.5 Installation of precast prestressed double-T floor beams in a multifloor office structure.

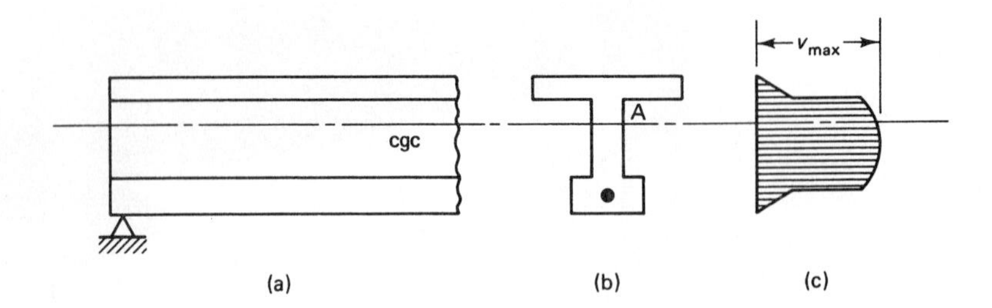

Figure 5.6 Maximum horizontal shear stress distribution across depth. (a) Beam elevation. (b) Beam cross section. (c) Shear stress.

crete in the top compression fibers and a redistribution of stresses within the top region. Sudden failure takes place as the principal inclined crack dynamically joins the crushed concrete zone, as illustrated in Figure 5.5(c). This type of failure can be considered relatively less brittle than the diagonal tension failure due to the stress redistribution. Yet it is, in fact, a brittle type of failure with limited warning, and such a design should be avoided completely.

A concrete beam or element is not homogeneous, and the strength of the concrete throughout the span is subject to a normally distributed variation. Hence, one cannot expect that a stabilized failure diagonal crack occurs at both ends of the beam. Also, because of these properties, overlapping combinations of flexure–diagonal tension failure and diagonal tension–shear compression failure can occur at overlapping shear span-to-depth ratios. If the appropriate amount of shear reinforcement is provided, brittle failure of horizontal members can be eliminated with little additional cost.

It should be emphasized that most failures tend to occur by diagonal tension, which is a combination of flexure and shear effects. The shear-compression type of failure, with the resulting crushing of the top compressive area of the concrete and failure to resist the flexural forces, leads to separation of the tension flange from the web in the flanged section as the inclined crack extends towards the support. Crushing of the web of the section causes the beam to resemble a tied arch. This type of failure in prestressed beams can be better described as web-shear failure. It is important to evaluate *both* the flexure-shear capacity and the web-shear capacity of each critical section in order to determine which type predominates in determining the shear strength of the concrete section.

The distribution of the maximum horizontal shearing stress in an uncracked flanged section is shown in Figure 5.6. Because of the abrupt change of section width at the corner A, a check of the capacity of the section at critical locations along the span becomes necessary, particularly for web-shear failure.

5.5 SHEAR AND PRINCIPAL STRESSES IN PRESTRESSED BEAMS

As mentioned in Section 5.4, flexure shear in prestressed concrete beams includes the effect of the externally applied compressive prestressing force that the reinforced concrete beam does *not* have. The vertical component of the prestressing tendon force reduces the vertical shear caused by the external transverse load, and the *net* transverse load to which a beam is subjected is markedly less in prestressed than in reinforced concrete beams.

Additionally, the compressive force of the prestressing tendon, even in cases of straight tendons, considerably reduces the effect of the tensile flexural stresses, so that the extent and magnitude of flexural cracking in prestressed members are reduced. As a

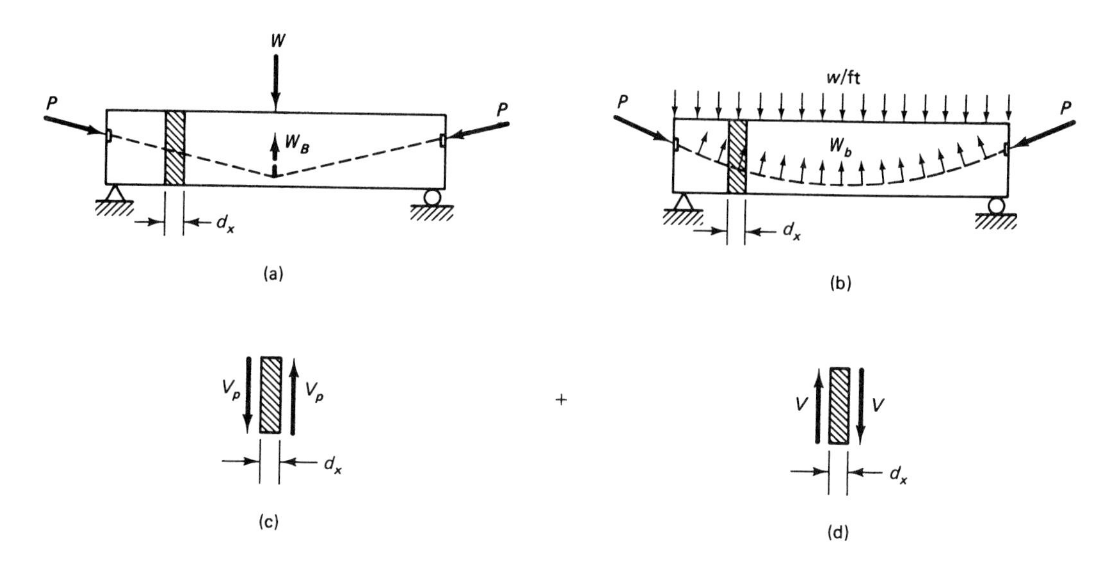

Figure 5.7 Balancing load to counteract vertical shear. (a)Beam with harped tendon. (b) Beam with draped tendon. (c) Internal shear vector V_p due to prestressing force *P* on infinitesimal element *dx*. (d) Internal shear vector *V* due to external load *W* on infinitesimal element *dx*.

result, the shear forces and the resulting principal stresses in a prestressed beam are considerably lower than those same forces and stresses in reinforced concrete beams, all else being equal. Consequently, the *basic* equations developed for prestressed concrete in shear are identical to those developed for reinforced concrete and described in detail in Refs. 5.3 and 5.4. Figure 5.7 illustrates the contributions of the vertical component of the tendon force in counterbalancing part or most of the vertical shear V caused by the external transverse load. The net shearing force V_c carried by the concrete is

$$V_c = V - V_p \tag{5.5}$$

From Equation 5.2, the net unit shearing stress v at any depth of the cross section is

$$v_c = \frac{V_c Q}{Ib} \tag{5.6}$$

The compressive fiber-stress distribution f_c due to the external bending moment is

$$f_c = -\frac{P_e}{A_c} \pm \frac{P_e ec}{I_c} \mp \frac{M_T c}{I_c} \tag{5.7}$$

and the principal tensile stress, from Equation 5.4a, is

$$f_t' = \sqrt{(f_c/2)^2 + v_c^2} - \frac{f_c}{2} \tag{5.8}$$

5.5.1 Flexure-Shear Strength [V_{ci}]

To design for shear, it is necessary to determine whether flexure shear or web shear controls the choice of concrete shear strength V_c. The inclined stabilized crack at a distance $d/2$ from a flexural crack that develops at the first cracking load in flexure shear is shown in Figure 5.8. If the effective depth is d_p, the depth from the compression fibers to the

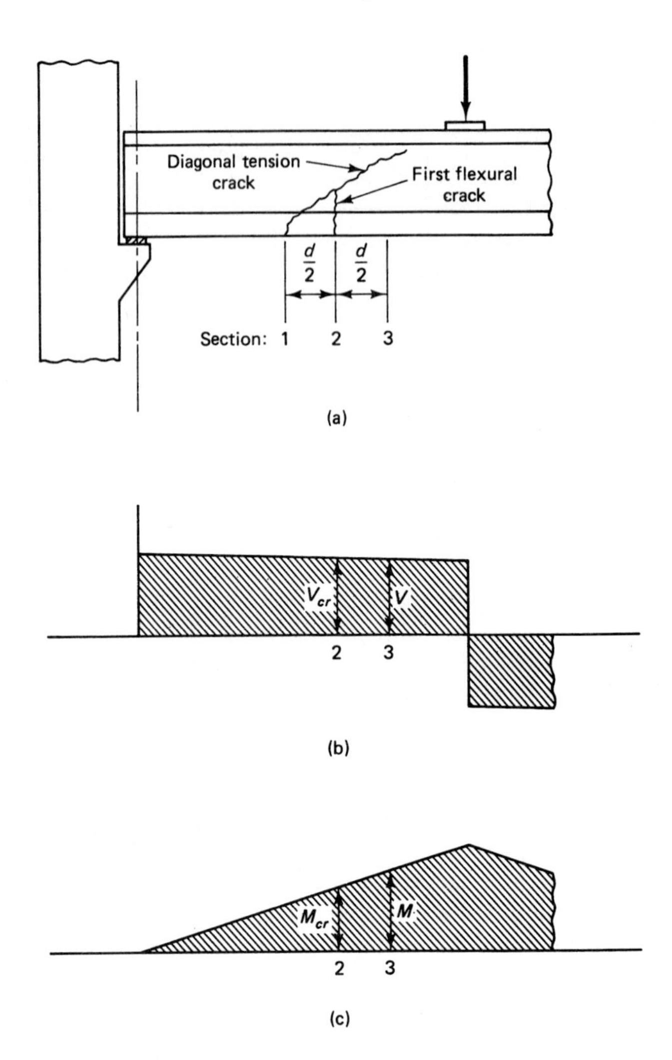

Figure 5.8 Flexure-shear crack development. (a) Crack pattern and types. (b) Shear diagram due to external load with frictional shear force V_{cr} ordinate at section 2. (c) Moment diagram with first cracking moment M_{cr} ordinate at section 2.

centroid of the longitudinal prestressed reinforcement, the change in moment between sections 2 and 3 is

$$M - M_{cr} \cong \frac{Vd_p}{2} \tag{5.9a}$$

or

$$V = \frac{M_{cr}}{M/V - d_p/2} \tag{5.9b}$$

where V is the shear at the section under consideration. Extensive experimental tests indicate that an additional vertical shear force of magnitude $0.6b_w d_p \sqrt{f'_c}$ is needed to fully develop the inclined crack in Figure 5.8 (Ref. 5.5). Hence, the total vertical shear acting at plane 2 of Figure 5.8 is

$$V_{ci} = \frac{M_{cr}}{M/V - d_p/2} + 0.6 b_w d_p \sqrt{f'_c} + V_d \tag{5.10}$$

where V_d is the vertical shear due to self-weight. The vertical component V_p of the prestressing force is disregarded in Equation 5.10, since it is small along the span sections where the prestressing tendon is not too steep.

The value of V in Equation 5.10 is the factored shear force V_i at the section under consideration due to externally applied loads occurring simultaneously with the maximum moment M_{max} occurring at that section, i.e.,

$$V_{ci} = 0.6\lambda \sqrt{f'_c}\, b_w d_p + V_d + \frac{V_i}{M_{max}}(M_{cr}) \geq 1.7\lambda \sqrt{f'_c}\, b_w d_p \tag{5.11}$$

$$\leq 5.0\lambda \sqrt{f'_c}\, b_w d_p$$

where λ = 1.0 for normal-weight concrete
= 0.85 for sand-lightweight concrete
= 0.75 for all-lightweight concrete
V_d = shear force at section due to unfactored dead load
V_{ci} = nominal shear strength provided by the concrete when diagonal tension cracking results from combined vertical shear and moment
V_i = factored shear force at section due to externally applied load occurring simultaneously with M_{max}.

For lightweight concrete, $\lambda = f_{ct}/6.7 \sqrt{f'_c}$ if the value of the tensile splitting strength f_{ct} is known. Note that the value $\sqrt{f'_c}$ should not exceed 100.

The equation for M_{cr}, the moment causing flexural cracking due to external load, is given by

$$M_{cr} = \frac{I_c}{y_t}(6\sqrt{f'_c} + f_{ce} - f_d) \tag{5.12}$$

In the ACI code, f_{ce} is termed as f_{pe}

where f_{ce} = concrete compressive stress due to effective prestress after losses at *extreme* fibers of section where tensile stress is caused by external load, psi. At the centroid, $f_{ce} = \bar{f}_c$
f_d = stress due to unfactored dead load at extreme fiber of section resulting from self-weight only where tensile stress is caused by externally applied load, psi
y_t = distance from centroidal axis to extreme fibers in tension.

and M_{cr} = that portion of the applied *live load* moment that causes cracking. For simplicity, S_b may be substituted for I_c/y_t.

A plot of Equation 5.10 is given in Figure 5.9 with experimental data from Ref. 5.6. Compare this plot with an analogous one in Figure 6.6 of Ref. 5.3 for reinforced concrete where an asymptotic horizontal value of shear is achieved along the span.

Note that in shear design of composite sections, the same design stipulations used for precast sections apply. This is because design for shear is based on the limit state at failure due to factored loads. Although the entire composite section resists the factored shear as a monolithic section, calculation of the shear strength V_c should be based on the properties of the *precast* section since most of the shear strength is provided by the web of the precast section. Consequently, f_{ce} and f_d in Equation 5.12 are calculated using the precast section geometry.

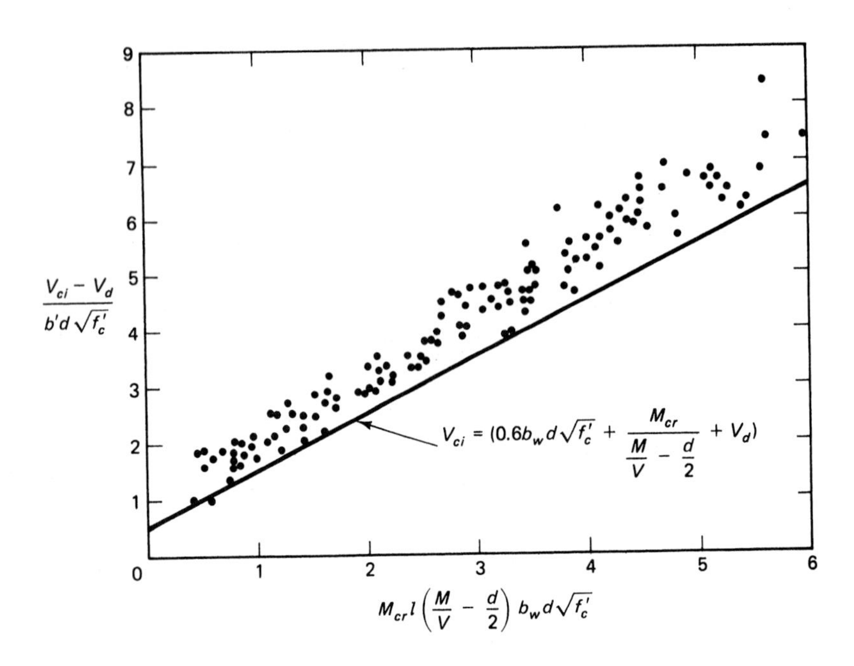

Figure 5.9 Moment-shear relationship in flexure-shear cracking.

5.5.2 Web-Shear Strength [V_{cw}]

The web-shear crack in the prestressed beam is caused by an indeterminate stress that can best be evaluated by calculating the principal tensile stress at the critical plane from Equation 5.8. The shear stress v_c can be defined as the web shear stress v_{cw} and is maximum near the centroid cgc of the section where the actual diagonal crack develops, as extensive tests to failure have indicated. If v_{cw} is substituted for v_c and $\bar{f}_c$, which denotes the concrete stress f_c due to effective prestress *at the cgc level,* is substituted for f_c in the equation, the expression equating the principal tensile stress in the concrete to the direct tensile strength becomes

$$f'_t = \sqrt{(\bar{f}_c/2) + v_{cw}^2} - \frac{\bar{f}_c}{2} \tag{5.13}$$

where $v_{cw} = V_{cw}/(b_w d_p)$ is the shear stress in the concrete due to all loads causing a nominal strength vertical shear force V_{cw} in the web. Solving for v_{cw} in Equation 5.13 gives

$$v_{cw} = f'_t \sqrt{1 + \bar{f}_c/f'_t} \tag{5.14a}$$

Using $f'_t = 3.5\sqrt{f'_c}$ as a reasonable value of the tensile stress on the basis of extensive tests, Equation 5.14(a) becomes

$$v_{cw} = 3.5\sqrt{f'_c}\,(\sqrt{1 + \bar{f}_c/3.5\sqrt{f'_c}}) \tag{5.14b}$$

which can be further simplified to

$$v_{cw} = 3.5\sqrt{f'_c} + 0.3\bar{f}_c \tag{5.14c}$$

In the ACI code, $\bar{f}_c$ is termed f_{pc}. The notation used herein is intended to emphasize that this is the stress in the concrete, and not the prestressing steel. The nominal shear strength V_{cw} provided by the concrete when diagonal cracking results from *excessive principal tensile stress* in the web becomes

$$V_{cw} = (3.5\lambda\sqrt{f'_c} + 0.3\bar{f}_c)b_w d_p + V_p \tag{5.15}$$

where V_p = the vertical component of the effective prestress at the particular section contributing to added nominal strength

λ = 1.0 for normal-weight concrete, and less for lightweight concrete

d_p = distance from the extreme compression fiber to the centroid of prestressed steel, or $0.8h$, whichever is greater.

The ACI code stipulates the value of $\bar{f}_c$ to be the resultant concrete compressive stress at either the centroid of the section or the junction of the web and the flange when the centroid lies within the flange. In case of composite sections, $\bar{f}_c$ is calculated on the basis of stresses caused by prestress and moments resisted by the precast member acting *alone.* A plot relating the nominal web shear stress v_{cw} to the centroidal compressive stress in the concrete is given in Figure 5.10. Note the similarity between the plots of Equations 5.14b and c, showing that the approximation used in the latter linearized equation is justified. The code also allows using a value of 1.0 instead of 0.3 in the second term inside the bracket in Equation 5.15.

5.5.3 Controlling Values of V_{ci} and V_{cw} for the Determination of Web Concrete Strength V_c

The ACI code has the following additional stipulations for calculating V_{ci} and V_{cw} in order to choose the required value of V_c in the design:

(a) In pretensioned members where the section at a distance $h/2$ from the face of the support is closer to the end of the member than the transfer length of the prestressing tendon, a reduced prestressed value has to be considered when computing V_{cw}. This value of V_{cw} has to be taken as the maximum limit of V_c in the expression

$$V_{cw} = \left(0.6\lambda\sqrt{f'_c} + 700\frac{V_u d_p}{M_u}\right)b_w d_p \geq 2\lambda\sqrt{f'_c}\,b_w d_p \leq 5\lambda\sqrt{f'_c}\,b_w d_p \tag{5.16}$$

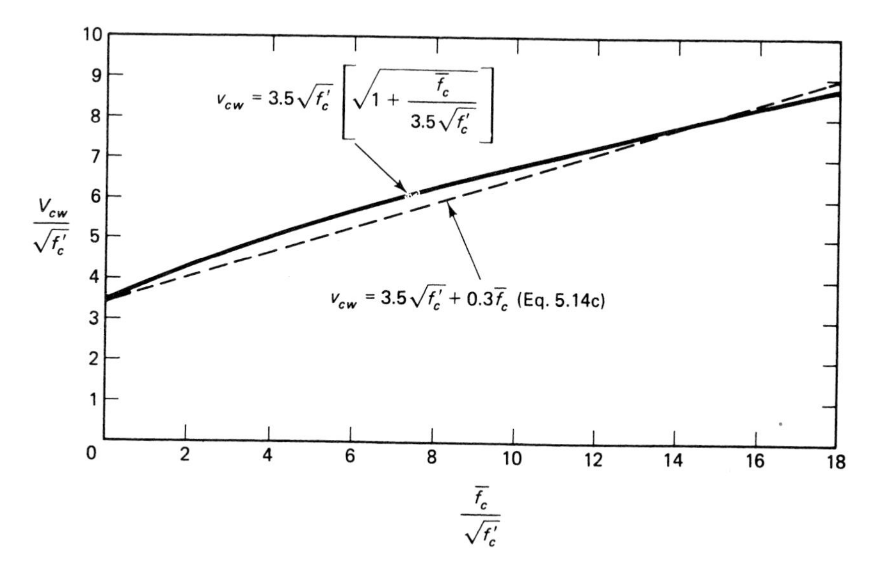

Figure 5.10 Centroidal compressive stress vs. nominal shear stress in web-shear cracking.

the value $V_u d_p / M_u$ cannot exceed 1.0.

(b) In pretensioned members where bonding of some tendons does not extend to the end of the member, a reduced prestress has to be considered when computing V_c in accordance with Equation 5.16 or with the *lesser* of the two values of V_c obtained from Equations 5.11 and 5.15. Also, the value of V_{cw} calculated using the reduced prestress must consequently be taken to be the maximum limit of Equation 5.16.

(c) Equation 5.16 can be used in determining V_c for members where the effective prestress force is not less than 40 percent of the tensile strength of the flexural reinforcement, *unless* a more detailed analysis is performed using Equations 5.11 for V_{ci} and 5.15 for V_{cw} and choosing the lesser of these two as the limiting V_c value to be used as the capacity of the web in designing the web reinforcement.

(d) The first plane for the total required nominal shear strength $V_n = V_u/\phi$ to be used for web steel calculation is also at a distance $h/2$ from the face of the support.

5.6 WEB-SHEAR REINFORCEMENT

5.6.1 Web Steel Planar Truss Analogy

In order to prevent diagonal cracks from developing in prestressed members, *whether due to flexure-shear or web-shear action,* steel reinforcement has to be provided, ideally in the form of the solid lines depicting tensile stress trajectories in Figure 5.3. However, practical considerations preclude such a solution, and other forms of reinforcement are improvised to neutralize the tensile stresses at the critical shear failure planes. The mode of failure in shear reduces the beam to a simulated arched section in compression at the top and tied at the bottom by the longitudinal beam tension bars, as seen in Figure 5.11(a). If one isolates the main concrete compression element shown in Figure 5.11(b), it can be considered as the compression member of a triangular truss, as shown in Figure 5.11(c), with the polygon of forces C_c, T_b, and T_s representing the forces acting on the truss members—hence the expression *truss analogy.* Force C_c is the compression in the simulated concrete strut, force T_b is the tensile force increment of the main longitudinal tension bar, and T_s is the force in the bent bar. Figure 5.12(a) shows the analogy truss for the case of using vertical stirrups instead of inclined bars, with the forces polygon having a vertical tensile force T_s instead of the inclined one in Figure 5.11(c).

As can be seen from the previous discussion, the shear reinforcement basically performs four main functions:

1. It carries a portion of the external factored shear force V_u.
2. It restricts the growth of the diagonal cracks.
3. It holds the longitudinal main reinforcing bars in place so that they can provide the dowel capacity needed to carry the flexural load.
4. It provides some confinement to the concrete in the compression zone if the stirrups are in the form of closed ties.

5.6.2 Web Steel Resistance

If V_c, the nominal shear resistance of the plain web concrete, is less than the nominal total vertical shearing force $V_u/\phi = V_n$, web reinforcement has to be provided to carry the difference in the two values; hence,

$$V_s = V_n - V_c \tag{5.17}$$

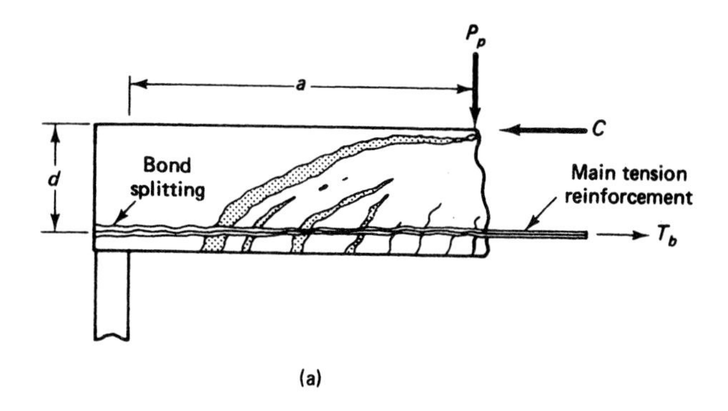

(a)

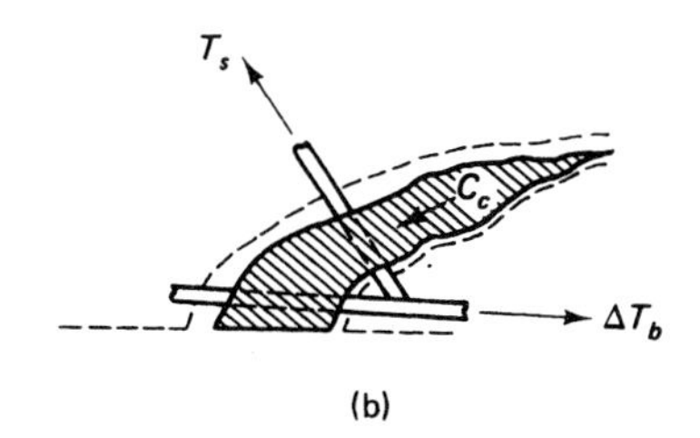

(b)

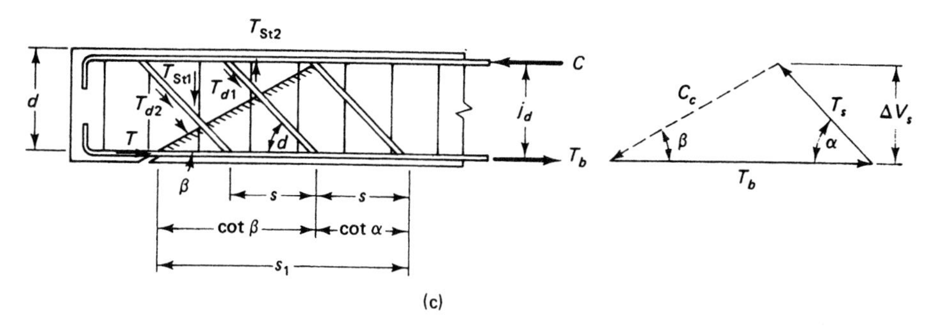

(c)

Figure 5.11 Diagonal tension failure mechanism. (a) Failure pattern. (b) Concrete simulated strut. (c) Planar truss analogy.

Here, V_c is the *lesser* of V_{ci} and V_{cw}. V_c can be calculated from Equation 5.11 or 5.15, and V_s can be determined from equilibrium analysis of the bar forces in the analogous triangular truss cell. From Figure 5.11(c),

$$V_s = T_s \sin \alpha = C_c \sin \beta \tag{5.18a}$$

where T_s is the force resultant of all web stirrups across the diagonal crack plane and n is the number of spacings s. If $s_1 = ns$ in the bottom tension chord of the analogous truss cell, then

$$s_1 = jd(\cot \alpha + \cot \beta) \tag{5.18b}$$

Assuming that moment arm $jd \simeq d$, the stirrup force per unit length from Equations 5.18a and b, where $s_1 = ns$, becomes

$$\frac{T_s}{s_1} = \frac{T_s}{ns} = \frac{V_s}{\sin \alpha} \frac{1}{d(\cot \beta + \cot \alpha)} \tag{5.18c}$$

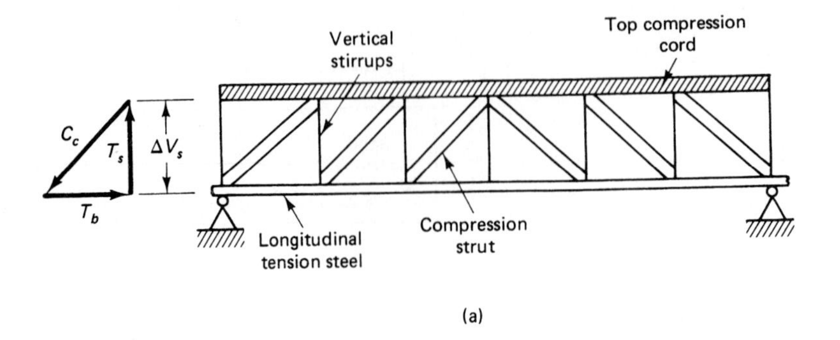

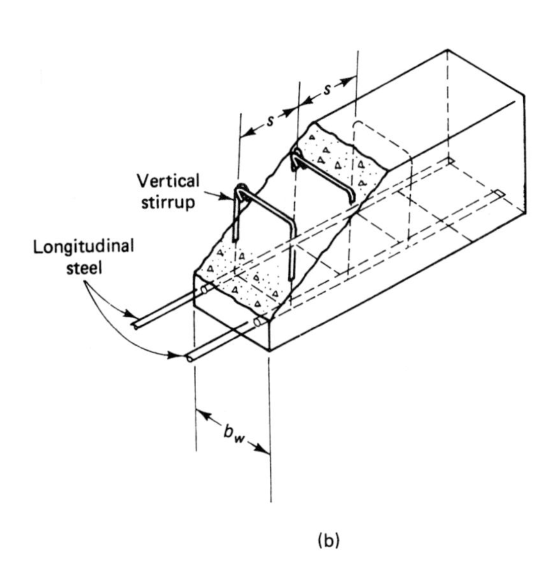

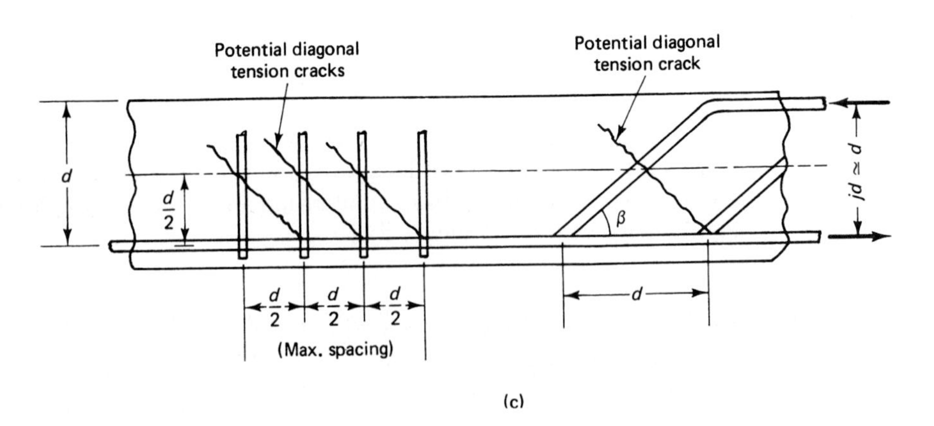

Figure 5.12 Web steel arrangement. (a) Truss analogy for vertical stirrups. (b) Three-dimensional view of vertical stirrups. (c) Spacing of web steel.

If there are n inclined stirrups within the length s_1 of the analogous truss chord, and if A_v is the area of one inclined stirrup, then

$$T_s = nA_v f_y \tag{5.19a}$$

Hence,

$$nA_v = \frac{V_s ns}{d \sin \alpha(\cot \beta + \cot \alpha) f_y} \tag{5.19b}$$

But it can be assumed that in the case of diagonal tension failure the compression diagonal makes an angle $\beta = 45°$ with the horizontal; so Equation 5.19b becomes

$$V_s = \frac{A_v f_y d}{s}[\sin \alpha(1 + \cot \alpha)]$$

or

$$V_s = \frac{A_v f_y d}{s}(\sin \alpha + \cos \alpha) \tag{5.20a}$$

or, solving for s and using the fact that $V_s = V_n - V_c$,

$$s = \frac{A_v f_y d}{V_n - V_c}(\sin \alpha + \cos \alpha) \tag{5.20b}$$

If the inclined web steel consists of a single bar or a single group of bars all bent at the same distance from the face of the support, then

$$V_s = A_v f_y \sin \alpha \leq 3.0\sqrt{f_c'}\, b_w d$$

If vertical stirrups are used, angle α becomes 90°, giving

$$V_s = \frac{A_v f_y d}{s} \tag{5.21a}$$

or

$$s = \frac{A_v f_y d}{(V_u/\phi) - V_c} = \frac{A_v \phi f_y d}{V_u - \phi V_c} \tag{5.21b}$$

In Equations 5.21a and b, d_p is the distance from the extreme compression fibers to the centroid of the prestressing reinforcement, and d is the corresponding distance to the centroid of the nonprestressed reinforcement. The value of d_p need not be less than $0.80h$.

5.6.3 Limitation on Size and Spacing of Stirrups

Equations 5.20 and 5.21 give an inverse relationship between the spacing of the stirrups and the shear force or shear stress they resist, with the spacing s decreasing with the increase in $(V_n - V_c)$. In order for every *potential* diagonal crack to be resisted by a vertical stirrup, as shown in Figure 5.11(c), maximum spacing limitations are to be applied for the vertical stirrups as follows:

(a) $s_{max} \leq \frac{3}{4}h \leq 24$ in., where h is the total depth of the section.
(b) If $V_s > 4\lambda\sqrt{f_c'}\, b_w d_p$, the maximum spacing in (a) shall be reduced by half.
(c) If $V_s > 8\lambda\sqrt{f_c'}\, b_w d_p$, enlarge the section.

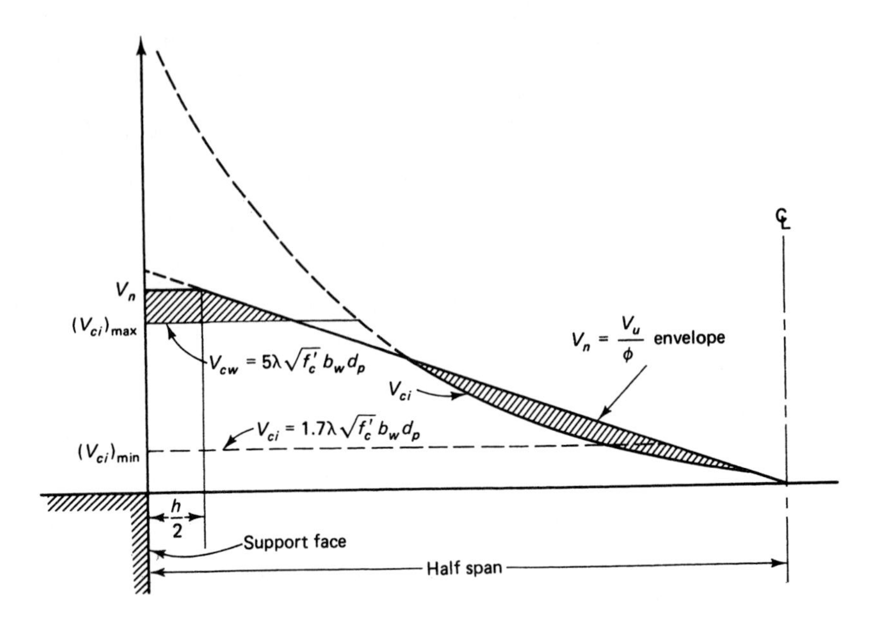

Figure 5.13 Web steel envelope for uniformly loaded prestressed beam.

(d) If $V_u = \phi V_n > \frac{1}{2}\phi V_c$, a minimum area of shear reinforcement has to be provided. This area may be computed by the equation

$$A_v = \frac{50 b_w s}{f_y} \tag{5.22a}$$

If the effective prestress force P_e is equal to or greater than 40 percent of the tensile strength of the flexural reinforcement, the equation

$$A_v = \frac{A_{ps} f_{pu}\, s}{80 f_y d}\sqrt{\frac{d_p}{b_w}} \tag{5.22b}$$

which gives a lesser required minimum A_v, may be used instead.

(e) The web reinforcement must develop the full required development length in order to be effective. This means that the stirrups or mesh should extend to the compression and tension surfaces of the section, less the clear concrete cover requirement and a 90° or 135° hook used at the compression side.

A typical qualitative diagram showing the zone along the span of a uniformly loaded prestressed beam for which web reinforcement has to be provided is given in Figure 5.13. The shaded area is the envelope of excess shear V_s requiring web steel.

5.7 HORIZONTAL SHEAR STRENGTH IN COMPOSITE CONSTRUCTION

Full transfer of horizontal shear forces has to be assumed at the contact surfaces of the interconnected elements.

5.7.1 Service-Load Level

The maximum horizontal shear stress v_h can be evaluated from the basic principles of mechanics and the equation

$$v_h = \frac{VQ}{I_c b_v} \tag{5.23}$$

where V = unfactored design vertical shear acting on the composite section
Q = moment of area about cgc of the segment above or below cgc
I_c = moment of inertia of entire composite section
b_v = contact width of precast section web, or width of section at which horizontal shear is being calculated.

Equation 5.23 can be simplified to

$$v_h = \frac{V}{b_v\, d_{pc}} \tag{5.24}$$

where d_{pc} is the effective depth from the extreme compression fibers of the composite section to the centroid cgs of the prestressing reinforcement.

5.7.2 Ultimate-Load Level

Direct Method. In the limit state at failure, Equation 5.24 can be modified such that the factored load V_u can be substituted for V to give

$$v_{uh} = \frac{V_u}{b_v d_{pc}} \tag{5.25a}$$

or, in terms of the nominal vertical shear strength V_n,

$$v_{nh} = \frac{V_u/\phi}{b_v\, d_{pc}} = \frac{V_n}{b_v\, d_{pc}} \tag{5.25b}$$

where $\phi = 0.85$. If V_{nh} is the nominal horizontal shear strength, then $V_u \le V_{nh}$ and the total nominal shear strength is

$$V_{nh} = v_{nh}\, b_v\, d_{pc} \tag{5.25c}$$

The ACI code limits v_{nh} to 80 psi if no dowels or vertical ties are provided and the contact surface is roughened, or if minimum vertical ties are provided but there is no roughening of the surface of contact. v_{nh} can go up to 500 psi; otherwise the friction theory, with the following assumptions, has to be used:

(a) When no vertical ties are provided, but the contact surface of the precast element is intentionally roughened, use

$$V_{nh} \le 80A_c \le 80b_v d_{pc} \tag{5.26a}$$

where A_c is the area of concrete resisting shear $= b_v d_{pc}$.

(b) When minimum vertical ties are provided, where $A_v = 50(b_w s)/f_y$, but the contact surface of precast elements is *not* roughened, use

$$V_{nh} \le 80b_v d_{pc}$$

(c) If the contact surface of the precast element is roughened to a full amplitude of $\frac{1}{4}$ in., and minimum vertical steel in (b) is provided, use

$$V_{nh} \le 500\, b_v d_{pc} \tag{5.26b}$$

(d) If the factored shear $V_u > \phi(500\, b_v d_p)$, the shear friction theory can be used to design the dowel reinforcement. In this case, all horizontal shear has to be taken by ties in the pependicular plane such that

$$V_{nh} = \mu A_{vf} f_y \tag{5.27}$$

where A_{vf} = area of shear-friction reinforcement, in.^{2}
f_y = design yield strength, not to exceed 60,000 psi
μ = coefficient of friction
= 1.0λ for concrete placed against intentionally roughened concrete surface
= 0.60λ for concrete placed against unroughened concrete surface
λ = factor for type of concrete.

In all cases, the nominal shear strength $V_n \le 0.20 f'_c A_{cc} \le 800\, A_{cc}$, where A_{cc} is the concrete contact area resisting shear transfer. Note that in most cases, the shear stress v_{nh} resulting from the factored shear force does not exceed 500 psi. Hence, the shear friction theory is not normally necessary in designing the dowel reinforcement for composite action.

The maximum allowable spacing of the dowels or ties for horizontal shear is the smaller of four times the least dimension of the supported section and 24 inches.

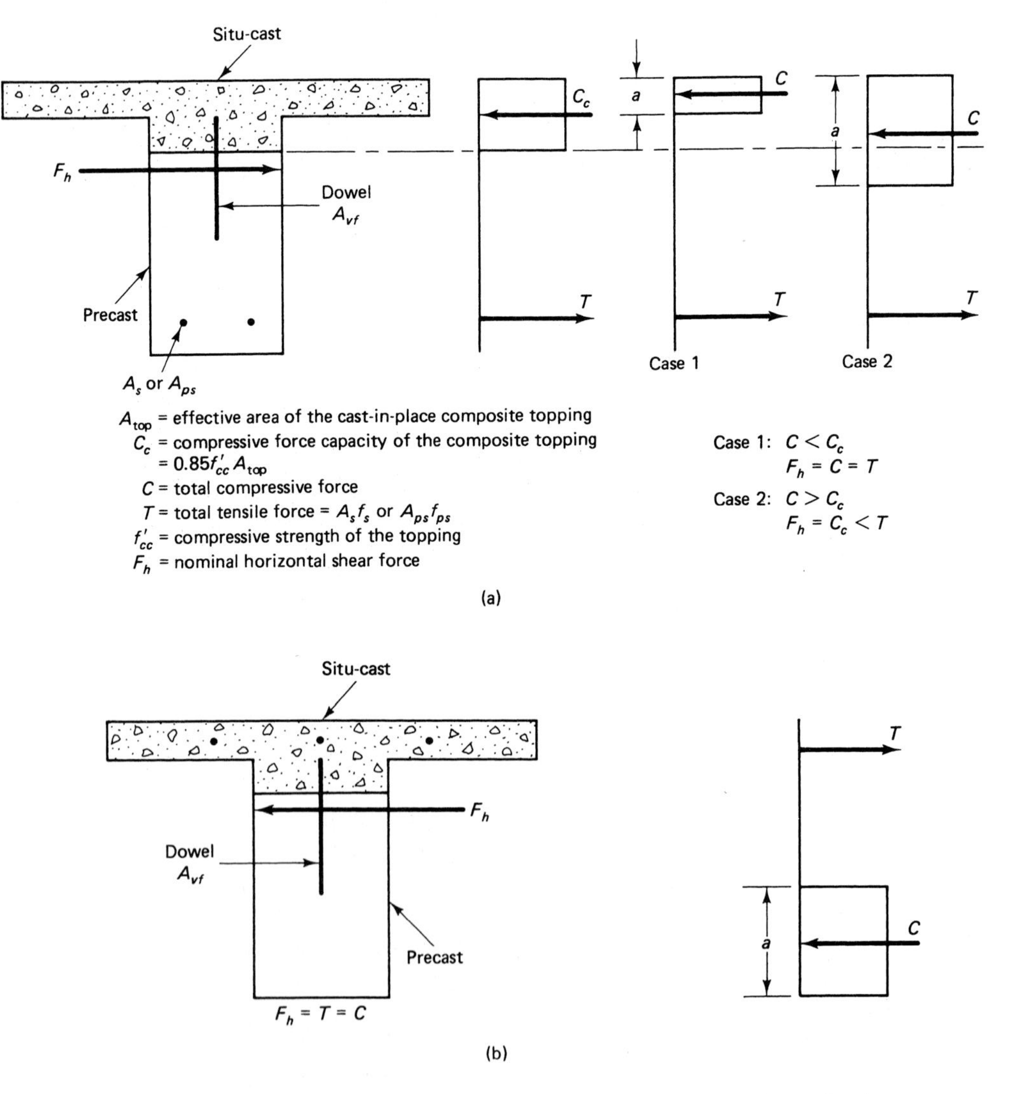

Figure 5.14 Composite action forces (F_h acts longitudinally along the beam span). (a) Positive-moment section. (b) Negative-moment section.

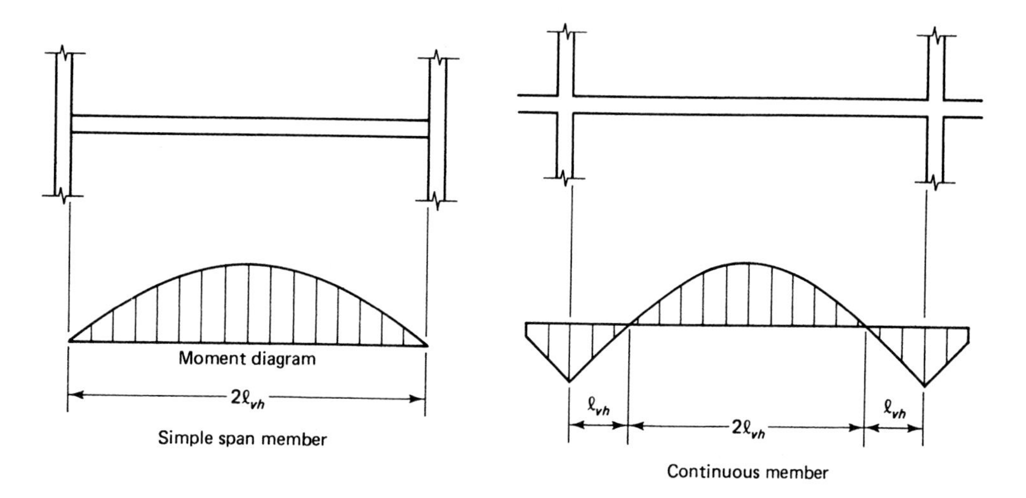

Figure 5.15 Horizontal shear length in composite action.

Basic Method. The ACI code allows an alternative method wherein horizontal shear is investigated by computing the actual change in compressive or tensile force in any plane and transferring that force as *horizontal* shear to the supporting elements. The area-of-contact surface A_{cc} is substituted for $b_v d_{pc}$ in Equations 5.25b and c to give

$$V_{nh} = v_{nh} A_{cc} \tag{5.28}$$

where $V_{nh} \geq F_h$, the horizontal shear force, and is at least equal to the compressive force C or tensile force T in Figure 5.14. (See Equation 5.30 for the value of F_h.)

The value of the contact area A_{cc} can be defined as

$$A_{cc} = b_v l_{vh} \tag{5.29}$$

where l_{vh} is the horizontal shear length defined in Figures 5.15(a) and (b) for simple span and continuous span members, respectively.

5.7.3 Design of Composite-Action Dowel Reinforcement

Ties for horizontal shear may consist of single bars or wires, multiple leg stirrups, or vertical legs of welded wire fabric. The spacing cannot exceed four times the least dimension of the support element or 24 in., whichever is less. If μ is the coefficient of friction, then the nominal horizontal shear force F_h in Figure 5.14 can be defined as

$$F_h = \mu A_{vf} f_y \leq V_{nh} \tag{5.30}$$

The ACI values of μ are based on a limit shear-friction strength of 800 psi, a quite conservative value as demonstrated by extensive testing (Ref. 5.7). The Prestressed Concrete Institute (Ref. 5.8) recommends, for concrete placed against an intentionally roughened concrete surface, a maximum $\mu_e = 2.9$ instead of $\mu = 1.0\lambda$, and a maximum design shear force

$$V_u \leq 0.25\lambda^2 f'_c A_c \leq 1{,}000\lambda^2 A_{cc} \tag{5.31a}$$

with a required area of shear-friction steel of

$$A_{vf} = \frac{V_{uh}}{\phi f_y \mu_e} \tag{5.31b}$$

or

$$A_{vh} = \frac{V_{nh}}{\mu_e f_y} = \frac{F_h}{\mu_e f_y} \tag{5.31c}$$

Using the PCI less conservative values, Equation 5.31c becomes

$$F_h \leq \mu_e A_{vf} f_y \leq V_{nh} \tag{5.32}$$

with

$$\mu_e = \frac{1{,}000\lambda^2 \, b_v I_{vh}}{F_h} \leq 2.9$$

where $b_v l_{vh} = A_{cc}$. The minimum reinforcement is

$$A_v = \frac{50 b_v s}{f_y} = \frac{50 b_v l_{vh}}{f_y} \tag{5.33}$$

5.8 WEB REINFORCEMENT DESIGN PROCEDURE FOR SHEAR

The following is a summary of a recommended sequence of design steps:

1. Determine the required nominal shear strength value $V_n = V_u/\phi$ at a distance $h/2$ from the face of the support.
2. Calculate the nominal shear strength V_c that the web has by one of the following two methods.

   (a) *ACI conservative method if* $f_{pe} > 0.40 f_{pu}$

$$V_c = \left(0.60\lambda\sqrt{f'_c} + \frac{700 V_u d_p}{M_u}\right) b_w d_p$$

where $2\lambda\sqrt{f'_c} b_w d_p \leq V_c \leq 5\lambda\sqrt{f'_c} b_w d_p$ and where $V_u d_p / M_u \leq 1.0$ and V_u is calculated at the same section for which M_u is calculated.

If the average tensile splitting strength f_{ct} is specified for lightweight concrete, then $\lambda = f_{ct}/6.7\sqrt{f'_c}$ with $\sqrt{f'_c}$ not to exceed a value of 100.

(b) *Detailed analysis where* V_c *is the lesser of* V_{ci} *and* V_{cw}

$$V_{ci} = 0.60\lambda\sqrt{f'_c} \, b_w d_p + V_d + \frac{V_i}{M_{\max}}(M_{cr}) \geq 1.7\lambda\sqrt{f'_c} \, b_w d_p$$

$$V_{cw} = (3.5\lambda\sqrt{f'_c} + 0.3\bar{f}_c) b_w d_p + V_p$$

using d_p or $0.8h$, whichever is larger, and

where $M_{cr} = (I_c/y_t)(6\lambda\sqrt{f'_c} + f_{ce} - f_d)$

or $M_{cr} = S_b(6\lambda\sqrt{f'_c} + f_{ce} - f_d)$

V_i = factored shear force at section due to externally applied loads occurring simultaneously with $M_{\max}$

f_{ce} = compressive stress in concrete after occurrence of all losses at *extreme fibers* of section where external load causes tension. f_{ce} becomes $\bar{f}_c$ for the stress at the centroid of the section.

3. If $V_u/\phi \le \frac{1}{2}V_c$, no web steel is needed. If $V_u/\phi > \frac{1}{2}V_c < V_c$, provide minimum reinforcement. If $V_u/\phi > V_c$ and $V_s = V_u/\phi - V_c \le 8\lambda\sqrt{f'_c}b_w d_p$, design the web steel. If $V_s = V_u/\phi - V_c > 8\lambda\sqrt{f'_c}f_w d_p$, or if $V_u > \phi(V_c + 8\lambda\sqrt{f'_c}b_w d_p)$, enlarge the section.

4. Calculate the required minimum web reinforcement. The spacing is $s \le 0.75h$ or 24 in., whichever is smaller.

$$\text{Min } A_v = \frac{50 b_w s}{f_y} \qquad \text{(conservative)}$$

If $f_{pe} \ge 0.40 f_{pu}$, a less conservative Min A_v is the smaller of

$$A_v = \frac{A_{ps} f_{pu} s}{80 f_y d_p}\sqrt{\frac{d_p}{b_w}}$$

where $d_p \ge 0.80h$, and

$$A_v = 50 b_w s / f_y$$

5. Calculate the required web reinforcement size and spacing. If $V_s = (V_u/\phi - V_c) \le 4\lambda\sqrt{f'_c}\, b_w d_p$, then the stirrup spacing s is as required by the design expressions in step 6, to follow. If $V_s = (V_u/\phi - V_c) > 4\lambda\sqrt{f'_c}\, b_w d_p$, then the stirrup spacing s is half the spacing required by the design expressions in step 6.

6.

$$s = \frac{A_v f_y d_p}{(V_u \phi) - V_c} = \frac{A_v \phi f_y d_p}{V_u - \phi V_c} \le 0.75h \le 24 \text{ in.} \ge \text{minimum } s \text{ from step 4}$$

7. Draw the shear envelope over the beam span, and mark the band requiring web steel.

8. Sketch the size and distribution of web stirrups along the span using #3- or #4-size stirrups as preferable, but no larger size than #6 stirrups.

9. Design the vertical dowel reinforcement in cases of composite sections.

(a) $V_{nh} \le 80 b_v d_{pc}$ for both roughened contact and no vertical ties or dowels, and nonroughened but with minimum vertical ties, use

$$A_v = \frac{50 b_w s}{f_y} = \frac{50 b_v I_{vh}}{f_y}$$

(b) V_{nh} $500 b_v d_{pc}$ for a roughened contact surface with full amplitude $\frac{1}{4}$ in.

(c) For cases where $V_{nh} > 500 b_v d_{pc}$, design vertical ties for $V_{nh} = A_{vf} f_y\, \mu$,

where A_{vf} = area of frictional steel dowels
μ = coefficient of friction = 1.0λ for intentionally roughened surface, where $\lambda = 1.0$ for normal-weight concrete. In all cases, $V_n \le V_{nh} \le 0.2 f'_c A_{cc} \le 800\, A_{cc}$, where $A_{cc} = b_v l_{vh}$.

An alternative method of determining the dowel reinforcement area A_{vf} is by computing the horizontal force F_h at the concrete contact surface such that

$$F_h \le \mu_e A_{vf} f_y \le V_{nh}$$

where

$$\mu_e = \frac{1{,}000\lambda^2 b_v l_{vh}}{F_h} \le 2.9$$

Figure 5.16 outlines the foregoing steps in flowchart form.

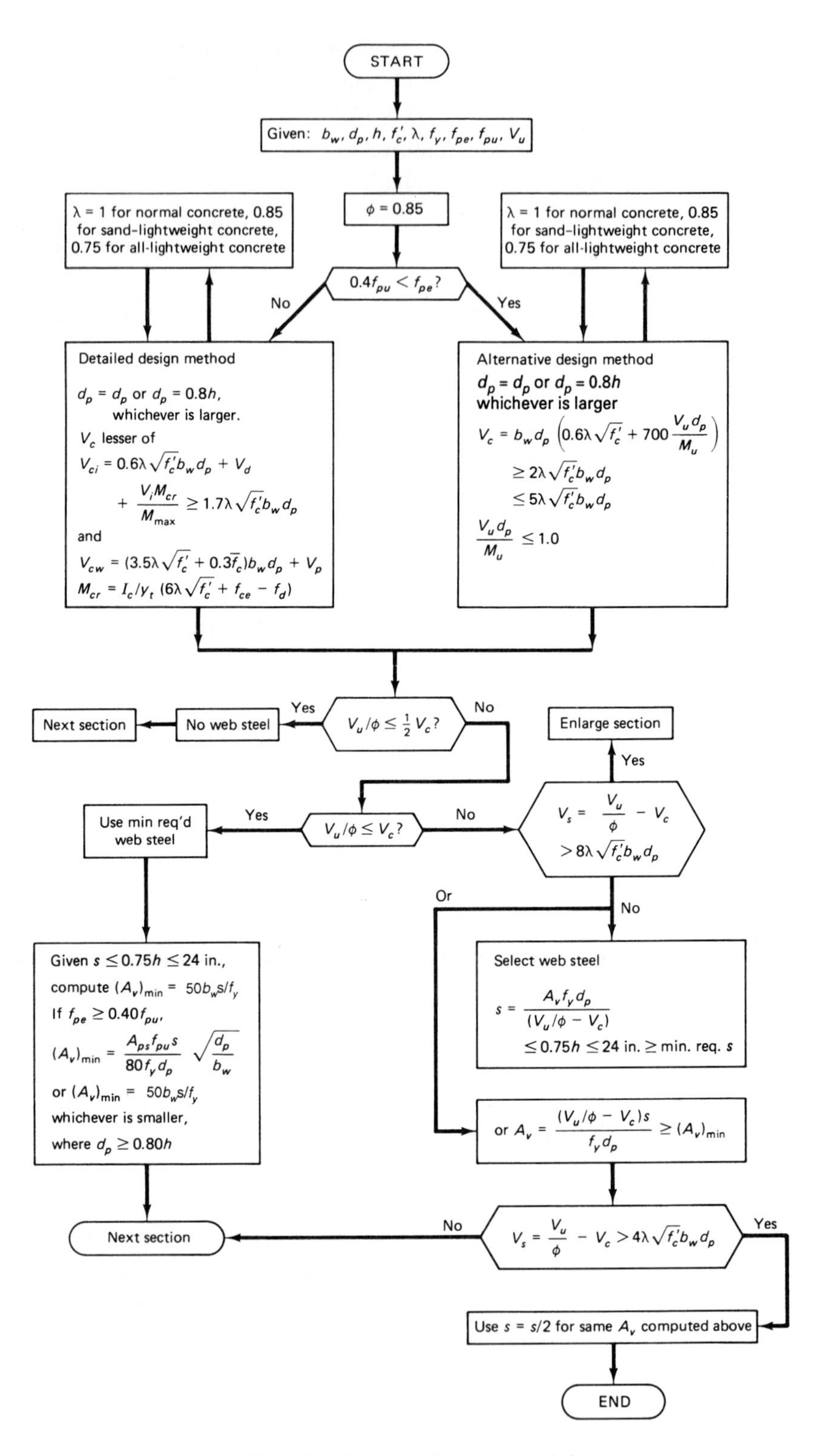

Figure 5.16 Flowchart for shear-web reinforcement.

5.9 PRINCIPAL TENSILE STRESSES IN FLANGED SECTIONS AND DESIGN OF DOWEL-ACTION VERTICAL STEEL IN COMPOSITE SECTIONS

Example 5.1

A prestressed concrete T-beam section has the distribution of compressive service load shown in Figure 5.17. The unfactored design external vertical shear $V = 120{,}000$ lb (554 kN), and the factored vertical shear $V_u = 190{,}000$ lb (845 kN).

(a) Compute the principal tensile stress at the centroidal cgc axis and at the reentrant corner A of the web-flange junction, and calculate the maximum horizontal shearing stresses at service load for these locations.

(b) Compute the required nominal horizontal shear strength at the interaction surface A–A between the precast web and the situ-cast flange, and design the necessary vertical ties or dowels to prevent fracture slip at A–A, thereby ensuring complete composite action. Use the ACI direct method, and assume that the contact surface is intentionally roughened. Given data are as follows:

f'_c for web = 6,000 psi normal-weight concrete

f'_c for flange = 3,000 psi normal-weight concrete

Effective width of flange b_m = 60 in. (152.4 cm)

where b_m is the modified width to account for the difference in moduli of the concrete of the precast and topping parts.

Solution:

Service-Load Horizontal Shear Stresses. The maximum horizontal shear stress is

$$v_h = \frac{VQ}{Ib_v}$$

Now,

$$Q_A = 60 \times 12(19.32 - 6) = 9{,}590 \text{ in.}^3 \ (157{,}172 \text{ cm}^3)$$

$$Q_{cgc} = 60 \times 12(19.32 - 6) + \frac{12 \times (7.32)^2}{2} = 9{,}912 \text{ in.}^3 \ (162{,}429 \text{ cm}^3)$$

So the horizontal shear stresses at service load are

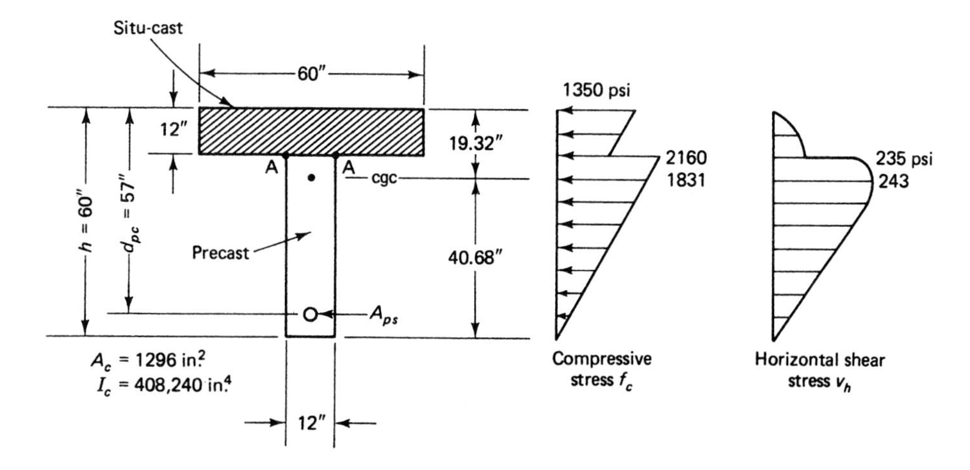

Figure 5.17 Beam cross section in Example 5.1.

$$v_h \text{ at A} = \frac{120{,}000 \times 9590}{408{,}240 \times 12} = 235 \text{ psi (1.6 MPa)}$$

and

$$v_h \text{ at cgc} = \frac{120{,}000 \times 9{,}912}{408{,}240 \times 12} = 243 \text{ psi (1.7 MPa)}$$

From Equation 5.13, the corresponding principal tensile stresses are, at A,

$$f_t' = \sqrt{\left(\frac{f_{CA}}{2}\right)^2 + v_h^2} - \frac{f_{CA}}{2}$$

$$= \sqrt{\left(\frac{2{,}160}{2}\right)^2 + (235)^2} - \frac{2{,}160}{2} = 25 \text{ psi (111 Pa)}$$

and at cgc,

$$f_t' = \sqrt{\left(\frac{1{,}831}{2}\right)^2 + (243)^2} - \frac{1{,}831}{2} = 32 \text{ psi (221 Pa)}$$

Thus, the principal tensile stresses are low and should not cause any cracking at service load.

The horizontal shear stress $v_h = 235$ psi at contact surface A–A has to be checked to verify whether it is within acceptable limits. In accordance with AASHTO, the maximum allowed is 160 psi < 235 psi, and hence special provision for added vertical ties or dowels has to be made if AASHTO requirements are applicable.

Dowel Reinforcement Design

$$V_u = 190{,}000 \text{ lb}$$

$$\text{Req. } V_{nh} = \frac{V_u}{\phi} = \frac{190{,}000}{0.85} = 223{,}530 \text{ lb (994 kN)}$$

$$b_v = 12 \text{ in. (30.5 cm} \qquad d_{pc} = 57 \text{ in. (145 cm)}$$

From Equation 5.26b,

$$\text{Available } V_{nh} = 500 b_v d_{pc} = 500 \times 12 \times 57 = 342{,}000 \text{ lb (1520 kN)} > 223{,}530 \text{ lb}$$

Hence, specify roughening the contact surface of the precast web fully to $\frac{1}{4}$in. amplitude.

Now,

$$\text{min unit} \frac{A_v}{l_{vh}} = \frac{50 b_v}{f_y} = \frac{50 \times 12}{60{,}000} = 0.01 \text{ in.}^2\text{/in. along span}$$

So using #3 vertical stirrups, $A_v = 2 \times 0.11 = 0.22$ in.^{2}, and $s = 0.22/0.01 = 22$ in. (56 cm) center to center < 24 in., O.K. Vertical-web steel reinforcement for shear in the web would most probably require smaller spacing. Hence, extend all web stirrups into the situ-cast top slab.

5.10 DOWEL STEEL DESIGN FOR COMPOSITE ACTION

Example 5.2

Using (a) the ACI friction coefficient and (b) the PCI friction coefficient, design the dowel reinforcement of Example 5.1 for full composite action by the alternative method, assuming a simply supported beam of effective 65 ft (19.8 m) span.

Solution: From Figures 5.14 and 5.17,

$$A_{top} = 60 \times 12 = 720 \text{ in}^2 \text{ (4,645 cm}^2\text{)}$$

$$C_c = 0.85 f'_{cc} A_{top} = 0.85 \times 3{,}000 \times 70 = 1{,}836{,}000 \text{ lb (8,167 kN)}$$

Assume $A_{ps}f_{ps} > C_c$, since the prestressing force is not given. Then

$$F_h = 1{,}836{,}000 \text{ lb}$$

$$l_{vh} = \frac{65 \times 12}{2} = 390 \text{ in.}$$

$$b_v = 12 \text{ in.}$$

$$80b_v l_{vh} = 80 \times 12 \times 390 = 374{,}400 \text{ lb } (1{,}665 \text{ kN}) < 1{,}836{,}000 \text{ lb}$$

Hence, vertical ties are needed.

For roughened surface to full $\frac{1}{4}$in. amplitude and minimum reinforcement,

$$V_{nh} = 500\, b_v d = 500 \times 12 \times 57 = 342{,}000$$

$$\text{Req. } \frac{V_u}{\phi} = \frac{190{,}000}{0.85} = 223{,}529 \text{ lb} < V_{nh} = 342{,}000 < \text{available } F_h = 1{,}836{,}000 \text{ lb}$$

Use $F_h = 223{,}529$ lb for determining the required composite action reinforcement.

Using ACI μ Value. From Equation 5.27 and $\mu = 1.0$, with $l_{vh} = 390$ in.,

$$\text{Total } A_{vf} = \frac{223{,}529}{1.0 \times 60{,}000} = 3.73 \text{ in}^2 \ (24.1 \text{ cm}^2)$$

$$\text{Min } A_{vf} = \frac{50 b_v\, l_{vh}}{f_y} = \frac{50 \times 12 \times 390}{60{,}000} = 3.90 \text{ in}^2 \ (25.1 \text{ cm}^2)$$

From Example 5.1, min $A_{vf} = 0.01$ in.^{2}/in. $= 0.12$ in.^{2}/12 in., controls. So, trying #3 stirrups, we obtain $A_v = 2 \times 0.11 = 0.22$ in.^{2}, and

$$s = \frac{l_{vh}\, A_v}{A_{vf}} = \frac{390 \times 0.22}{3.90} = 22 \text{ in. center-to-center } < 24 \text{ in. } < \text{max. allow. } 4 \times 12 = 48 \text{ in.}$$

So use #3 U ties at 22 in. center to center.

Using PCI μ_e Value

$$\lambda = 1.0$$

$$\mu_e = \frac{1{,}000\lambda^2\, b_v\, l_{vh}}{F_h} \le 2.9$$

$$= \frac{1{,}000 \times 1 \times 12 \times 390}{1{,}836{,}000} = 2.55 < 2.9$$

So use $\mu_e = 2.55$; then, from Equation 5.32,

$$\text{Req. } A_{vf} = \frac{223{,}529}{2.55 \times 60{,}000} = 1.46 \text{ in.}^2 < \text{min. } A_{vf} = 3.90 \text{ in.}^2\text{, controls}$$

So use #3 U ties at 22 in. center-to-center (9.5 mm dia at 56 cm).

5.11 DOWEL REINFORCEMENT DESIGN FOR COMPOSITE ACTION IN AN INVERTED T-BEAM

Example 5.3

A simply supported inverted T-beam has an effective 24 ft (7.23 m) span length. The beam, shown in cross section in Figure 5.18, has a 2 in. (5.1 cm) situ-cast topping on a nonroughened surface. Design the required dowel action stirrups to develop full composite behavior, assuming that the factored shear V_u to which the beam is subjected at the critical section is 160,000 lb (712 kN). Given data are as follows:

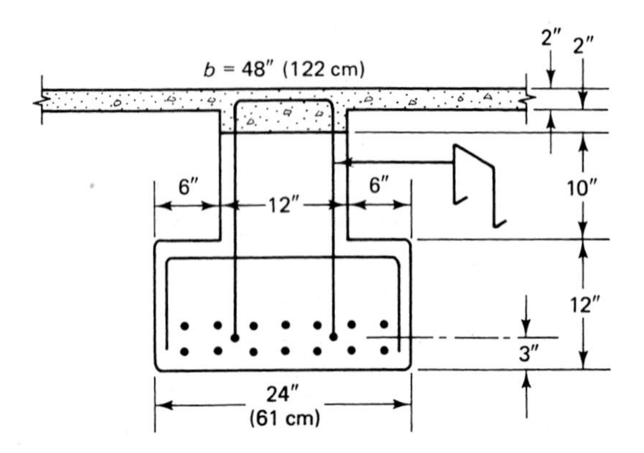

Figure 5.18 Beam cross section in Example 5.3.

f'_c (precast) = 6,000 psi (41.4 MPa), normal-weight concrete

f'_{cc} (topping) = 3,000 psi (20.7 MPa), normal-weight concrete

Prestressing steel:

Twelve $\frac{1}{2}$ in. dia 270 k tendons

f_{pu} = 270,000 psi (1,862 MPa)

f_{ps} = 242,000 psi (1,669 MPa)

Tie steel f_y = 60,000 psi (414 MPa)

Use both the ACI direct method and the alternative method with effective μ_e to carry out your design.

Solution:

$$d_p = 2 + 2 + 10 + 12 - 3 = 23 \text{ in.}$$

$$A_{ps} = 12 \times 0.153 = 1.836 \text{ in.}^2$$

$$T_n = A_{ps}f_{ps} = 1.836 \times 242{,}000 = 444{,}312 \text{ lb } (1{,}976 \text{ kN})$$

$$b_v = 12 \text{ in.}$$

$$l_{vh} = \frac{24 \times 12}{2} = 144 \text{ in.}$$

$$A_{\text{top}} = 2 \times 48 + 2 \times 12 = 120 \text{ in.}^2$$

$$C_c = 0.85f'_{cc}A_{\text{top}} = 0.85 \times 3{,}000 \times 120 = 306{,}000 \text{ lb } (1{,}361 \text{ kN}) < T_n = 444{,}312 \text{ lb}$$

Accordingly, use F_h = 306,000 lb (1,361 kN). Then, for a nonroughened surface,

$$\text{Available } V_{nh} = 80b_v l_{vh} = 80 \times 12 \times 144 = 138{,}240 \text{ lb } (615 \text{ kN}) < C_c = 306{,}000 \text{ lb}$$

Hence, ties are required for developing full composite action using $\lambda = 1.0$.

ACI Direct Method

$$\text{Req. } V_{nh} = \frac{V_u}{\phi} = \frac{160{,}000}{0.85} = 188{,}235 \text{ lb } (837 \text{ kN})$$

Use $\mu = 1.0$. Then, from Equation 5.26a, with dowel reinforcement, we have

$$\text{Available } V_{nh} = 80b_v d_{pc} = 80 \times 12 \times 23 = 22{,}000 \text{ lb} \ll \text{required } V_{nh}$$

From Equation 5.27, for an unroughened surface $\mu = 0.6\lambda = 0.60$. Then

$$\text{Req. total } A_{vf} = \frac{V_n}{\mu f_y} = \frac{188{,}235}{0.60 \times 60{,}000} = 5.23 \text{ in}^2$$

$$\text{Req. min } A_{vf} = \frac{50 b_v l_{vh}}{f_y} = \frac{50 \times 12 \times 144}{60{,}000} = 1.44 \text{ in}^2 < 5.23 \text{ in}^2$$

So use $A_{vf} = 5.23$ in.^{2} (33.7 cm^2), and try #3 inverted U ties. Then $A_{vf} = 2 \times 0.11 = 0.22$ in.^{2} (1.4 cm^2) and the spacing is

$$s = \frac{l_{vh} A_v}{A_{vf}} = \frac{144 \times 0.22}{5.23} = 6.05 \text{ in. (15.4 cm)}$$

The maximum allowable spacing is $s = 4(2 + 2) = 16$ in., or $0.75h = 0.75 \times 26 = 19.5$ in. < 24 in. Thus, use #3 inverted U ties 6 in. (15 cm) center-to-center over the entire simply supported span.

Alternative Method Using μ_e

$$F_h = 306{,}000 \text{ lb}$$

$$\mu_e = \frac{1{,}000\lambda^2 b_v l_{vh}}{F_h} = \frac{1{,}000 \times 1.0 \times 12 \times 144}{306{,}000} = 5.65 > 2.9$$

So use $\mu_e = 2.9$; then, from Equation 5.31c, we get

$$\text{Req. } A_{vf} = \frac{F_h}{\mu_e f_y} = \frac{306{,}000}{2.9 \times 60{,}000} = 1.76 \text{ in}^2$$

$$\text{Req. min. } A_{vf} \text{ from (a)} = 1.44 \text{ in}^2 < 1.76 \text{ in}^2$$

So use $A_{vf} = 1.76$ in.^{2} Then the spacing is

$$s = \frac{l_{vh} A_v}{A_{vf}} = \frac{144 \times 0.22}{1.76} = 18 \text{ in. center-to-center}$$

and the maximum allowable spacing is

$$s = 4(2 + 2) = 16 \text{ in.} < 24 \text{ in.}$$

Hence, use #3 inverted U ties at 16 in. center to center over the entire simply supported span.

5.12 SHEAR STRENGTH AND WEB-SHEAR STEEL DESIGN IN A PRESTRESSED BEAM

Example 5.4

Design the bonded beam of Example 4.2 to be safe against shear failure, and proportion the required web reinforcement.

Solution:

Data and Nominal Shear Strength Determination

$$f_{pu} = 270{,}000 \text{ psi (1,862 MPa)}$$

$$f_y = 60{,}000 \text{ psi (414 MPa)}$$

$$f_{pe} = 155{,}000 \text{ psi (1,069 MPa)}$$

$$f'_c = 5{,}000 \text{ psi normal-weight concrete}$$

$$A_{ps} = 13 \text{ 7-wire } \tfrac{1}{2}\text{-in. tendons} = 1.99 \text{ in.}^2 \text{ (12.8 cm}^2\text{)}$$

$A_s = 4\ \#6\ \text{bars} = 1.76\ \text{in}^2\ (11.4\ \text{cm}^2)$

$\text{Span} = 65\ \text{ft}\ (19.8\ \text{cm})$

$\text{Service}\ W_L = 1{,}100\ \text{plf}\ (16.1\ \text{kN/m})$

$\text{Service}\ W_{SD} = 100\ \text{plf}\ (1.46\ \text{kN/m})$

$\text{Service}\ W_D = 393\ \text{plf}\ (5.7\ \text{kN/m})$

$h = 40\ \text{in.}\ (101.6\ \text{cm})$

$d_p = 36.16\ \text{in.}\ (91.8\ \text{cm})$

$d = 37.6\ \text{in.}\ (95.5\ \text{cm})$

$b_w = 6\ \text{in.}\ (15\ \text{cm})$

$e_c = 15\ \text{in.}\ (38\ \text{cm})$

$e_e = 12.5\ \text{in.}\ (32\ \text{cm})$

$I_c = 70{,}700\ \text{in.}^4\ (18.09 \times 10^6\ \text{cm}^4)$

$A_c = 377\ \text{in.}^2\ (2{,}432\ \text{cm}^2)$

$r^2 = 187.5\ \text{in.}^2\ (1{,}210\ \text{cm}^2)$

$c_b = 18.84\ \text{in.}\ (48\ \text{cm})$

$c_t = 21.16\ \text{in.}\ (54\ \text{cm})$

$P_e = 308{,}255\ \text{lb}\ (1{,}371\ \text{kN})$

$$\text{Factored load } W_u = 1.4D + 1.7L$$
$$= 1.4(100 + 393) + 1.7 \times 1{,}100 = 2{,}560\ \text{plf}$$

$$\text{Factored shear force at face of support} = V_u = W_U L/2$$
$$= (2{,}560 \times 65)/2 = 83{,}200\ \text{lb}$$

$$\text{Req. } V_n = V_u/\phi = 83{,}200/0.85 = 97{,}882\ \text{lb at support}$$

Plane at $\frac{1}{2}d_p$ from Face of Support

1. *Nominal shear strength V_c of web (steps 2, 3)*

$$\frac{1}{2}d_p = \frac{36.16}{2 \times 12} \cong 1.5\ \text{ft}$$

$$V_n = 97{,}882 \times \frac{[(65/2) - 1.5]}{65/2} = 93{,}364\ \text{lb}$$

$$V_u \text{ at } \tfrac{1}{2}d_p = 0.85 \times 93{,}364 = 79{,}359\ \text{lb}$$

$$f_{pe} = 155{,}000\ \text{psi}$$

$$0.40 f_{pu} = 0.40 \times 270{,}000 = 108{,}000\ \text{psi}\ (745\ \text{MPa})$$
$$< f_{pe} = 155{,}000\ \text{psi}\ (1{,}069\ \text{MPa})$$

Use ACI alternate method
Since $d_p > 0.8h$, use $d_p = 36.16$ in., assuming that part of the prestressing strands continue straight to the support. From Equation 5.16,

$$V_c = \left(0.60\lambda\sqrt{f'_c} + 700\frac{V_u d_p}{M_u}\right)b_w d_p \geq 2\lambda\sqrt{f'_c}b_w d_p \leq 5\lambda\sqrt{f'_c}b_w d_p$$

$\lambda = 1.0$ for normal-weight concrete

$$M_u \text{ at } d/2 \text{ from face} = \text{reaction} \times 1.5 - \frac{W_u(1.5)^2}{2}$$

$$= 83{,}200 \times 1.5 - \frac{2{,}560(1.5)^2}{2} = 121{,}920 \text{ ft-lb} = 1{,}463{,}040 \text{ in.-lb}$$

$$\frac{V_u d_p}{M_u} = \frac{79{,}359 \times 36.16}{1{,}463{,}040} = 1.96 > 1.0$$

So use $V_u d_p / M_u = 1.0$. Then

$$\text{Min. } V_c = 2\lambda\sqrt{f_c'}\,b_w d_p = 2 \times 1.0\sqrt{5{,}000} \times 6 \times 36.16 = 30{,}683 \text{ lb}$$

$$\text{Max. } V_c = 5\lambda\sqrt{f_c'}\,b_w d_p = 76{,}707 \text{ lb}$$

$$V_c = (0.60 \times 1.0\sqrt{5{,}000} + 700 \times 1.0)6 \times 36.16$$

$$= 161{,}077 \text{ lb} > \max V_c = 76{,}707 \text{ lb}$$

Then $V_c = 76{,}707$ lb and controls (341 kN). Also, $V_u/\phi > \frac{1}{2}V_c$; hence, web steel is needed. Accordingly,

$$V_s = \frac{V_u}{\phi} - V_c = 93{,}363 - 76{,}707 = 16{,}656 \text{ lb}$$

$$8\lambda\sqrt{f_c'}\,b_w d_p = 8 \times 1.0\sqrt{5{,}000} \times 6 \times 36.16 = 122{,}713 \text{ lb (546 kN)}$$

$$> V_s = 16{,}656 \text{ lb}$$

So the section depth is adequate.

2. *Minimum web steel (step 4)*
From Equation 5.22b,

$$\text{Min. } \frac{A_v}{s} = \frac{A_{ps} f_{pu}}{80 f_y d_p}\sqrt{\frac{d_p}{b_w}}$$

$$= \frac{1.99 \times 270{,}000}{80 \times 60{,}000 \times 36.16}\sqrt{\frac{36.16}{6}} = 0.0076 \text{ in.}^2/\text{in.}$$

3. *Required web steel (steps 5, 6)*
From Equation 5.21b,

$$s = \frac{A_v f_y d_p}{V_u/\phi - V_c} \le 0.75h \le 24 \text{ in.}$$

or

$$\frac{A_v}{s} = \frac{V_s}{f_y d_p} = \frac{16{,}656}{60{,}000 \times 36.16} = 0.0077 \text{ in}^2/\text{in.}$$

(prestressing force is > 0.4 × tensile strength)
Then the minimum required web-shear steel $A_v/s = 0.0077$ in.²/in. So trying #3 U stirrups, $A_v = 2 \times 0.11 = 0.22$ in.², and we get $0.0077 = 0.22/s$. so that the maximum spacing is

$$s = \frac{0.22}{0.0077} = 28.9 \text{ in. (73 cm)}$$

and

$$4\lambda\sqrt{f_c'}\,b_w d_p = 4 \times 1.0\sqrt{5{,}000} \times 6 \times 36.16 = 61{,}366 \text{ lb} > V_s$$

Hence, we do not need to use $\frac{1}{2}s$. Now,

$$0.75h = 0.75 \times 40.0 = 30.0 \text{ in.}$$

Thus, use #3 & web-shear reinforcement at 24 in. center to center (9.5 mm dia at 62 cm center to center).

Plane at Which No Web Steel Is Needed. Assume such a plane is at distance x from support. By similar triangles,

$$\frac{1}{2}V_c = \frac{76{,}707}{2} = 93{,}364 \times \frac{65/2 - x}{65/2}$$

or

$$\frac{65}{2} - x = \frac{76{,}707}{93{,}364} \times \frac{65}{4}$$

giving

$$x = 19.15 \text{ ft } (5.84 \text{ m}) \approx 230 \text{ in.}$$

Therefore, adopt the design in question, using #3 U at 24 in. center-to-center over a stretch length of approximately 230 in., with the first stirrup to start at 18 in. from the face of support. Extend the stirrups to the midspan if composite action doweling is needed.

5.13 WEB-SHEAR STEEL DESIGN BY DETAILED PROCEDURES

Example 5.5

Solve Example 5.4 by detailed procedures, determining the value of V_c as the smaller of the flexure shear V_{ci} and the web shear V_{cw}. Assume that the tendons are harped at midspan and not draped. Also assume $f'_c = 6000$ psi.

Solution: The profile of the prestressing strands is shown in Figure 5.19.

Plane at d/2 from Face of Support. From Example 5.4, $V_n = 93{,}364$ lb.

1. *Flexure-shear cracking, V_{ci} (step 2)*
   From Equation 5.11,

$$V_{ci} = 0.60\lambda\sqrt{f'_c}\,b_w d_p + V_d + \frac{V_i}{M_{\max}}(M_{cr}) \geq 1.7\lambda\sqrt{f'_c}\,b_w d_p$$

From Equation 5.12, the cracking moment is

$$M_{cr} = \frac{I_c}{y_t}(6\lambda\sqrt{f'_c} + f_{ce} - f_d)$$

where $I_c/y_t = S_b$ since y_t is the distance from the centroid to the extreme tension fibers. Now,

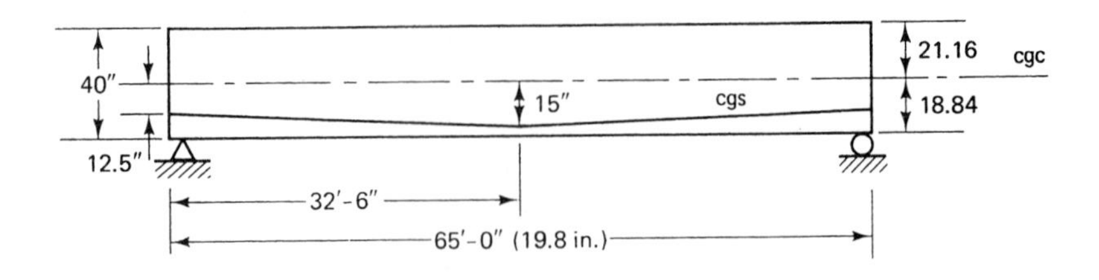

Figure 5.19 Tendon profile in Example 5.5.

$$I_c = 70{,}700 \text{ in.}^4$$
$$c_b = 18.84 \text{ in.}$$
$$P_e = 308{,}255 \text{ lb}$$
$$S_b = 3{,}753 \text{ in.}^3$$
$$r^2 = 187.5 \text{ in.}^2$$

So from Equation 4.3b, the concrete stress at the extreme bottom fibers *due to prestress only* is

$$f_{ce} = -\frac{P_e}{A_c}\left(1 + \frac{ec_b}{r^2}\right)$$

and the tendon eccentricity at $d_p/2 \cong 1.5$ ft from the face of the support is

$$e = 12.5 + (15 - 12.5)\frac{1.5}{65/2} = 12.62 \text{ in.}$$

Thus,

$$f_{ce} = -\frac{308{,}255}{377}\left(1 + \frac{12.62 \times 18.84}{187.5}\right) \cong -1{,}855 \text{ psi (12.8 MPa)}$$

From Example 4.2, the unfactored dead load due to self-weight $W_D = 393$ plf (5.7 kN/m) is

$$M_{d/2} = \frac{W_D x(l - x)}{2} = \frac{393 \times 1.5(65 - 1.5) \times 12}{2} = 224{,}600 \text{ in.-lb (25.4 kN-m)}$$

and the stress due to the unfactored dead load at the extreme concrete fibers where tension is created by the external load is

$$f_d = \frac{M_{d/2}c_b}{I_c} = \frac{224{,}600 \times 18.84}{70{,}700} = 60 \text{ psi}$$

Also,

$$M_{cr} = 3{,}753(6 \times 1.0 \times \sqrt{6{,}000} + 1{,}855 - 60)$$
$$= 8{,}480{,}872 \text{ in.-lb (958 kN-m)}$$
$$V_d = W_D\left(\frac{l}{2} - x\right) = 393\left(\frac{65}{2} - 1.5\right) = 12{,}183 \text{ lb (54.2 kN)}$$
$$W_{SD} = 100 \text{ plf}$$
$$W_L = 1{,}100 \text{ plf}$$
$$W_U = 1.4 \times 100 + 1.7 \times 1100 = 2{,}010 \text{ plf}$$

The factored shear force at the section due to externally applied loads occurring simultaneously with M_{max} is

$$V_i = W_U\left(\frac{l}{2} - x\right) = 2{,}010\left(\frac{65}{2} - 1.5\right) = 62{,}310 \text{ lb (277 kN)}$$

and

$$M_{max} = \frac{W_U x(l - x)}{2} = \frac{2{,}010 \times 1.5(65 - 1.5)}{2} \times 12$$
$$= 1{,}148{,}715 \text{ in.-lb (130 kN-m)}$$

Hence,

$$V_{ci} = 0.6 \times 1.0\sqrt{6{,}000} \times 6 \times 36.16 + 12{,}183 + \frac{62{,}310}{1{,}148{,}715}(8{,}480{,}872)$$

$$= 482{,}296 \text{ lb } (54.5 \text{ kN-m})$$

$$1.7\lambda\sqrt{f'_c}b_w d_p = 1.7 \times 1.0\sqrt{6{,}000} \times 6 \times 36.16 = 28{,}569 \text{ lb } (127 \text{ kN})$$

$$< V_{ci} = 482{,}296 \text{ lb}$$

Hence, $V_{ci} = 482{,}296$ lb (214.5 kN).

2. *Web-shear cracking, V_{cw} (step 2)*
From Equation 5.15,

$$V_{cw} = (3.5\sqrt{f'_c} + 0.3\bar{f}_c)b_w d_p + V_p$$

$$\bar{f}_c = \text{compressive stress in concrete at the cgc}$$

$$= \frac{P_e}{A_c} = \frac{308{,}255}{377} \cong 818 \text{ psi } (5.6 \text{ MPa})$$

$$V_p = \text{vertical component of effective prestress at section}$$

$$= P_e \tan\theta$$

where θ is the angle between the inclined tendon and the horizontal. So

$$V_p = 308{,}255\,\frac{(15 - 12.5)}{65/2 \times 12} = 1{,}976 \text{ lb } (8.8 \text{ kN})$$

Hence, $V_{cw} = (3.5\sqrt{6{,}000} + 0.3 \times 818) \times 6 \times 36.16 + 1{,}976 = 114{,}038$ lb (507 kN). In this case, web-shear cracking controls (i.e., $V_c = V_{cw} = 114{,}038$ lb (507 kN) is used for the design of web reinforcement). Compare this value with $V_c = 76{,}707$ lb (341 kN) obtained in Example 5.4 by the more conservative alternative method.

Now, from Example 5.4,

$$V_s = \frac{V_u}{\phi} - V_c = (93{,}364 - 114{,}038) \text{ lb}$$

So no web steel is needed unless $V_u/\phi > \frac{1}{2}V_c$. Accordingly, we evaluate the latter:

$$\frac{1}{2}V_c = \frac{114{,}038}{2} = 57{,}019 \text{ lb } (254 \text{ kN}) < 93{,}364 \text{ lb } (415 \text{ kN})$$

Since $V_u/\phi > \frac{1}{2}V_c$ but $< V_c$, use minimum web steel in this case.

3. *Minimum web steel (step 4)*
From Example 5.4,

$$\text{Req. } \frac{A_v}{s} = 0.0077 \text{ in.}^2/\text{in.}$$

So, trying #3 U stirrups, we get $A_v = 2 \times 0.11 = 0.22$ in.^{2}, and it follows that

$$s = \frac{A_v}{\text{Req. } A_v/s} = \frac{0.22}{0.0077} = 28.94 \text{ in. } (73 \text{ cm})$$

We then check for the minimum A_v as the lesser of the two values given by

$$A_v = 50b_w s/f_y$$

or

$$A_v = \frac{A_{ps}}{80}\frac{f_{pu}s}{f_y d_p}\sqrt{\frac{d_p}{b_w}}$$

So the maximum allowable spacing $\leq 0.75h \leq 24$ in. Then use #3 U stirrups at 24 in. center-to-center over a stretch length of 84 in. from the face of the support, as in Example 5.4.

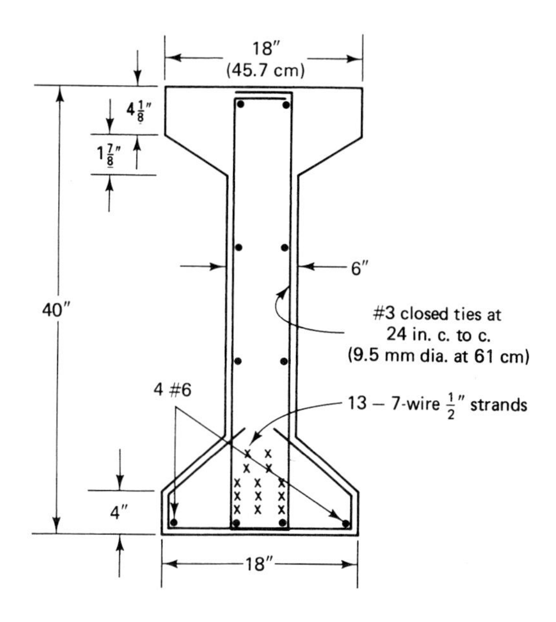

Figure 5.20 Web reinforcement details in Example 5.5.

Details of the section reinforcement (step 8) are shown in Figure 5.20.

5.14 DESIGN OF WEB REINFORCEMENT FOR A PCI DOUBLE T-BEAM

Example 5.6

A simply supported PCI 12 DT 34 pretopped double-T-beam has a span of 70 ft (21.3 m). It is subjected to a superimposed service dead load of 200 plf (2.9 kN/m), including a 2-in. additional topping placed sometime after service, and a service live load $W_L = 720$ plf. Design the web reinforcement needed to prevent shear cracking at the quarter-span section 17 ft 6 in. (5.3 m) from the support, calculating the nominal web-shear strength V_c by the detailed design method. Also, design any dowel reinforcement if necessary, assuming that the top surface of the precast T-beam is intentionally unroughened. The section properties are shown in Figure 5.21 and are as follows:

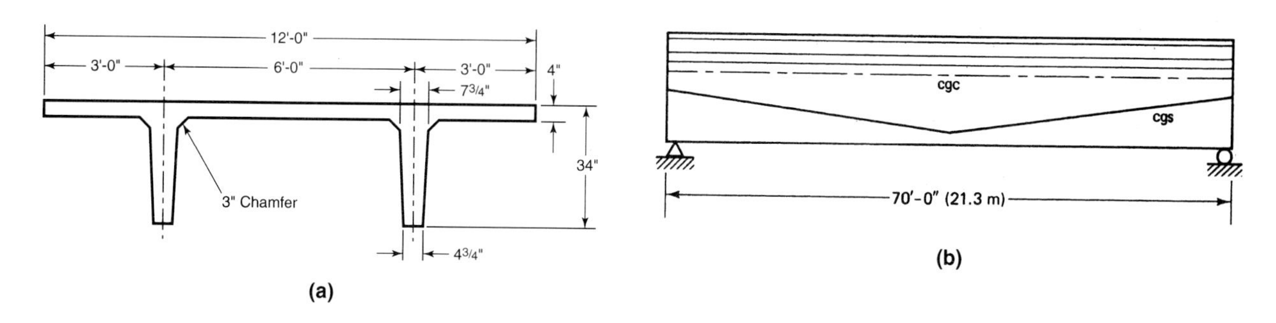

Figure 5.21 Beam geometry in Example 5.5. (a) Section. (b) Elevation.

Section property	Pretopped
A_c	978 in.^{2}
I_c	86,072 in.^{4}
r^2	88.0 in.^{2}
c_b	25.77 in.
c_t	8.23 in.
S_b	3,340 in.^{4}
S^t	10,458 in.^{3}
W_D	1,019 plf
$2b_w$	12.50 in.

Other data are:

f'_c (precast) = 5,000 psi (34.5 MPa), normal-weight concrete
f'_{cc} (topping) = 3,000 psi (20.7 MPa), normal-weight concrete, for future topping if used
f'_{ci} = 4,000 psi (27.6 MPa)
f_{pu} = 270,000 psi (1,862 MPa), low-relaxation steel
f_{ps} = 240,000 psi (1,655 MPa)
f_{pe} = 148,000 psi (1,020 MPa)
e_e = 11.38 in. (28.3 cm)
e_c = 21.77 in. (57.2 cm)
A_{ps} = 18 $\frac{1}{2}$-in. (12.7 mm) dia strands
f_{yv} for stirrups = 60,000 psi (414 MPa)

Use the same value for the effective depth d_p for the midspan as well as other sections. Note that b_w for *both* webs = 2 (4.75 + 7.75)/2 = 12.50 in. (32 cm).

Solution:

$$W_u = 1.4\,(200 + 1{,}019) + 1.7 \times 720 \cong 2{,}930 \text{ plf } (42.8\text{kN/m})$$

$$V_u \text{ at face of support} = \frac{2{,}930 \times 70}{2} = 102{,}550 \text{ lb}$$

$$V_n \text{ at 17 ft 6 in. from the face of the support} = \frac{1}{0.85}\left(102{,}550 \times \frac{(35 - 17.5)}{35}\right)$$

$$= 60{,}324 \text{ lb } (268 \text{ kN})$$

1. *Flexure-shear cracking, V_{ci} (step 2)*

$$d_p = 34 - 25.77 + 21.77 = 30.0 \text{ in. } (76 \text{ cm})$$

$$P_e = 18 \times 0.153 \times 148{,}000 = 407{,}592 \text{ lb } (18{,}176 \text{ kN})$$

$$e \text{ at 17 ft, 6 in. from support} = 11.38 + (21.77 - 11.38)\frac{17.5}{35} = 16.58 \text{ in.}$$

Use the precast section properties for computing f_{ce} and f_d as discussed in Section 5.5:

$$f_{ce} = -\frac{P_e}{A_c}\left(1 + \frac{ec_b}{r^2}\right) = -\frac{407{,}592}{978}\left(1 + \frac{16.58 \times 25.77}{88.0}\right)$$

$$= 2{,}440 \text{ psi } (16.8 \text{ MPa})$$

Use allowable extreme compressive stresses as follows:

(a) prestress + sustained load: $f_c = 0.45f'_c$

(b) prestress + total load (allowing 33% increase due to transient load: $f_c = 0.60\,f'_c$)
Note that although $f_{ce} = 0.45\,f'_c$, this should not affect the shear strength since f_{ce} is due to prestress only, and the inclusion of self-weight reduces it to less than $0.45\,f'_c$.
We thus have

Self-Weight $W_D = 1{,}019$ plf

$$M_{17.5} = \frac{W_D x(l - x)}{2} = \frac{1{,}019 \times 17.5(70 - 17.5)}{2} \times 12$$

$$= 5{,}617{,}238 \text{ in.-lb } (634 \text{ kN-m})$$

$$f_d = \frac{Mc_b}{I_c} = \frac{5{,}617{,}238 \times 25.77}{86{,}072} = 1{,}682 \text{ psi } (11.6 \text{ MPa})$$

$$M_{cr} = S_b(6.0\lambda\sqrt{f'_c} + f_{ce} - f_d)$$

$$= 3{,}340\,(6.0 \times 1.0\sqrt{5{,}000} + 2{,}440 - 1{,}682)$$

$$= 3{,}948{,}762 \text{ in.-lb } (445 \text{ kN-m})$$

It should be noted that the factor 6.0 in the cracking moment expression is low, since the modulus of rupture is taken 7.5. If 7.5 is used in the expression the cracking moment value would have become 4,303,022 in.-lb, thereby reducing the number of stirrups needed in this design.

Unfactored shear due to self-weight dead load is:

$$V_d = W_D\left(\frac{l}{2} - x\right) = 1{,}019\left(\frac{70}{2} - 17.5\right) = 17{,}833 \text{ lb}$$

$$W_{SD} = 200 \text{ plf}$$

$$W_L = 720 \text{ plf}$$

Factored external load intensity is:

$$W_U = 1.4 \times 200 + 1.7 \times 720 = 1{,}504 \text{ plf } (22.0 \text{ kN/m})$$

$$V_i = W_U\left(\frac{l}{2} - x\right) = 1{,}504\left(\frac{70}{2} - 17.5\right) = 26{,}320 \text{ lb } (117 \text{ kN})$$

$$M_{max} = W_U x\left(\frac{l - x}{2}\right) = \frac{1{,}504 \times 17.5(70 - 17.5)}{2} \times 12$$

$$= 8{,}290{,}800 \text{ in.-lb } (937 \text{ kN-m})$$

$$V_{ci} = 0.6\lambda\sqrt{f'_c}\,b_w d_p + V_d + \frac{V_i}{M_{max}} \times (M_{cr}) \geq 1.7\lambda\sqrt{f'_c}\,b_w d_p$$

$$= 0.6 \times 1.0 \times \sqrt{5{,}000} \times 12.5 \times 30.0 + 17{,}833$$

$$+ \frac{26{,}320}{8{,}290{,}800}(3{,}948{,}762)$$

$$= 47{,}403 \text{ lb } (206 \text{ kN})$$

$$1.7\lambda\sqrt{f'_c}\,b_w d_p = 1.7 \times 1.0\sqrt{5{,}000} \times 12.5 \times 30.0 = 45{,}078 \text{ lb} < 47{,}403 \text{ lb}$$

Hence, $V_{ci} = 47{,}403$ controls.

2. *Web-shear cracking, V_{cw} (step 2)*

$$\bar{f}_c = \frac{P_e}{A_c} = \frac{407{,}592}{978} = 417 \text{ psi } (2.9 \text{ MPa})$$

For the vertical component of the prestress force,

$$V_p = P_e \tan\theta = 407{,}592 \frac{(21.77 - 11.38)}{70/2 \times 12} = 10{,}083 \text{ lb } (44.0 \text{ kN})$$

$$V_{cw} = (3.5\lambda\sqrt{f'_c} + 0.3\bar{f}_c)b_w d_p + V_p$$

$$= (3.5\sqrt{5{,}000} + 0.3 \times 417) \times 12.5 \times 30.0 + 10{,}083$$

$$= 149{,}803 \text{ lb vs. } V_{ci} = 47{,}403 \text{ lb}$$

Now, V_c is the smaller of V_{ci} and V_{cw}; hence,

$$V_c = V_{ci} = 47{,}403$$

3. *Design of web reinforcement (steps 3–8)*
From above,

$$V_c = 47{,}403 \text{ lb}$$

So

$$\frac{1}{2}V_c = 23{,}702 \text{ lb}$$

Now, V_u / ϕ at a section 17.5 ft from support = 60,324 lb > $V_c > \frac{1}{2}V_c$; hence design of stirrups is necessary. If $V_u / \phi < V_c > \frac{1}{2}V_c$, only minimum-web steel is needed.

$$\text{Req.} \frac{A_v}{s} = \frac{V_n - V_c}{f_y d_p} = \frac{(V_u/\phi) - V_c}{f_y d_p} = \frac{60{,}324 - 47{,}403}{60{,}000 \times 30.0}$$

$$= 0.0078 \text{ in.}^2\text{/in. spacing}$$

Using $d \cong d_p = 30.0$ in. and $b_w = 12.5$ in.:

$$\text{Min.}\left(\frac{A_v}{s}\right) = \frac{A_{ps}}{80}\frac{f_{pu}}{f_y d_p}\sqrt{\frac{d_p}{b_w}} = \frac{18 \times 0.153}{80} \times \frac{270{,}000}{60{,}000 \times 30.0}\sqrt{\frac{30.0}{12.5}} = 0.0080$$

Or

$$\text{Min.}\left(\frac{A_v}{s}\right) = \frac{50 b_w}{f_y} = \frac{50 \times 12.5}{60{,}000} = 0.0104 \text{ in.}^2\text{/in.}$$

The *lesser* of the two minimum value applies, consequently, min. $A_v = 0.0080$ in.^{2}/ in. applies as the lesser of the two values.

$$\text{Hence controlling } \frac{A_v}{s} = 0.0080 \text{ in.}^2\text{/in.} = 0.010 \text{ in.}^2\text{/ft for both webs or}$$

$$0.005 \text{ in.}^2\text{/ft per web}$$

Try one row of D5 deformed welded wire fabric at 10 in. center-to-center weld spacing.

The maximum allowable spacing is, then $0.75h \le 24$ in. So we have

$$0.75h = 0.75 \times 34 = 25.5 \text{ in.}$$

Accordingly, adopt one row D5 WWF web reinforcement in one layer at 10 in. center-to-center weld spacing per web at the quarter-span section.

Note, in comparing the solution for V_{ci} and V_{cw} in Example 5.6, that V_{ci} has its highest value close to the support and *rapidly* decreases toward the midspan, while V_{cw} has a lesser variation in its value, as can be seen from Figure 5.13. It is important to calculate the flexure shear V_{ci} and web shear V_{cw} at several sections along the span in order to determine the most efficient distribution of the web steel. A computer program facilitates finding these values at constant intervals of, say, $\frac{1}{10}$th of the span, and a plot can be made similar to the one in Figure 5.13 showing the variation of the shear strengths of the web along the span.

4. *Design of dowel steel for full composite section of the additional 2-in. topping (step 9), if such topping is added later to the pretopped section.*

Section at $\frac{1}{2}d_p$ from face of support

$$\text{Used } d_p = 30.0 + 2.0 = 32 \text{ in.}$$

V_u at support = 102,550 (457 kN)

$$\frac{1}{2}d_p = \frac{32.0}{2 \times 12} = 1.33 \text{ ft (40 cm)} \qquad h/2 = 17 \text{ in.} = 1.33 \text{ ft}$$

$$V_u = 102{,}550 \times \left(\frac{35 - 1.33}{35}\right) = 98{,}653 \text{ lb (440 kN)}$$

$$\text{Req } V_{nh} = \frac{V_u}{\varphi} = \frac{98{,}653}{0.85} = 116{,}068 \text{ (516 kN)}$$

$$b_v = 12 \text{ ft.} \qquad \text{topping } h = 2 \text{ in.}$$

From Figure 5.14

$$C_c = 0.85 f'_{cc} A_{top} = 0.85 \times 3000 \times 12 \times 12 \times 2 = 734{,}400 \text{ lb (3,267 kN)}$$

$$T_s = A_{ps} f_{ps} = 18 \times 0.153 \times 240{,}000 = 660{,}960 \text{ (2,940 kN)}$$

$$< C_c = 734{,}400 \text{ lb}$$

Hence,

$$F_h = 660{,}960 \text{ lb (2,178 kN)}$$

$$l_{vh} = \frac{70 \times 12}{2} = 420 \text{ in. (1,067 cm)}$$

$$b_v = 144 \text{ in. (366 cm)}$$

$$80 b_v l_{vh} = 80 \times 144 \times 420 = 4{,}838{,}400 \text{ lb (21,520 kN)} >> 660{,}960 \text{ lb}$$

No dowel reinforcement is needed to extend to future additional 2-in. topping for full composite action to be developed. The section is adopted when it satisfies the flexural, deflection, and cracking requirements.

5.15 BRACKETS AND CORBELS

Brackets and corbels are short-haunched cantilevers that project from the inner face of columns or concrete walls to support heavy concentrated loads or beam reactions. They are very important structural elements for supporting precast beams, gantry girders, and any other forms of precast structural systems. Precast and prestressed concrete is becoming increasingly dominant, and larger spans are being built, resulting in heavier shear loads at supports. Hence, the design of brackets and corbels has become increasingly important. The safety of the total structure could depend on the sound design and construction of the supporting element, in this case the corbel, necessitating a detailed discussion of this subject.

In brackets and corbels, the ratio of the shear arm or span to the corbel depth is often less than 1.0. Such a small ratio changes the state of stress of a member into a two-dimensional one. Shear deformations would hence affect the nonlinear stress behavior of the bracket or corbel in the elastic state and beyond, and the shear strength becomes a major factor. Corbels also differ from deep beams in the existence of potentially large horizontal forces transmitted from the supported beam to the corbel or bracket. These

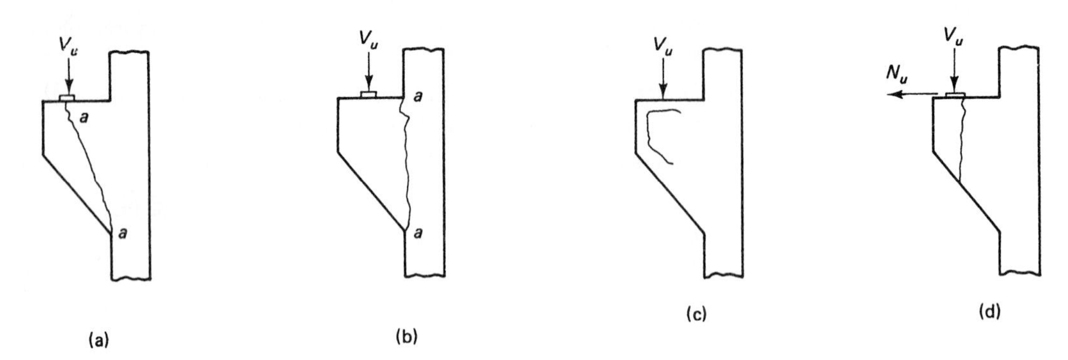

Figure 5.22 Failure patterns. (a) Diagonal shear. (b) Shear friction. (c) Anchorage splitting. (d) Vertical splitting.

horizontal forces result from long-term shrinkage and creep deformation of the supported beam, which in many cases is anchored to the bracket.

The cracks are usually mostly vertical or steeply inclined pure shear cracks. They often start from the point of application of the concentrated load and propagate toward the bottom reentrant corner junction of the bracket to the column face, as in Figure 5.22(a). Or they start at the upper reentrant corner of the bracket or corbel and proceed almost vertically through the corbel toward its lower fibers, as shown in Figure 5.22(b). Other failure patterns in such elements are shown in Figure 5.22(c) and (d). They can also develop through a combination of the ones illustrated. Bearing failure can also occur by crushing of the concrete under the concentrated load-bearing plate, if the bearing area is not adequately proportioned.

As will be noticed in the subsequent discussion, detailing of the corbel or bracket reinforcement is of major importance. Failure of the element can be attributed in many cases to incorrect detailing that does not realize full anchorage development of the reinforcing bars.

5.15.1 Shear Friction Hypothesis for Shear Transfer in Corbels

Corbels cast at different times than the main supporting columns can have a potential shear crack at the interface between the two concretes through which shear transfer has to develop. The smaller the ratio *a/d,* the larger the tendency for pure shear to occur through essentially vertical planes. This behavior is accentuated in the case of corbels with a potential interface crack between two dissimilar concretes.

The shear friction approach in this case is recommended by the ACI, as shown in Figure 5.22(b). An assumption is made of an already cracked vertical plane (*a–a* in Figure 5.23) along which the corbel is considered to slide as it reaches its limit state of failure. A coefficient of friction μ is used to transform the horizontal resisting forces of the well-anchored closed ties into a vertical nominal resisting force larger than the external factored shear load. Hence, the nominal vertical resisting shear force

$$V_n = A_{vf} f_y \mu \tag{5.34a}$$

to give

$$A_{vf} = \frac{V_n}{f_y \mu} \tag{5.34b}$$

where A_{vf} is the total area of the horizontal anchored closed shear ties.

The external factored vertical shear has to be $V_u \le \phi V_n$, where for normal concrete,

$$V_n \le 0.2 f'_c b_w d \tag{5.35a}$$

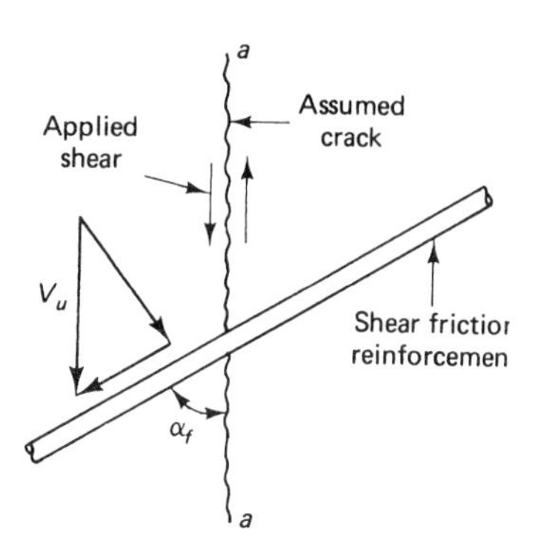

Figure 5.23 Shear-friction reinforcement at crack.

or

$$V_n \leq 800 b_w d \tag{5.35b}$$

whichever is smaller. The required effective depth d of the corbel can be determined from Equation 5.35a, or b, whichever gives a larger value.

For all-lightweight or sand-lightweight concretes, the shear strength V_n should not be taken greater than $(0.2 - 0.07a/d)f'_c b_w d$, or $(800 - 280a/d)b_w d$ in pounds.

If the shear friction reinforcement is inclined to the shear plane such that the shear force produces some tension in the shear friction steel,

$$V_n = A_{vf} f_y(\mu \sin \alpha_f + \cos \alpha_f) \tag{5.35c}$$

where α_f is the angle between the shear friction reinforcement and the shear plane. The reinforcement area becomes

$$A_{vf} = \frac{V_n}{f_y(\mu \sin \alpha_f + \cos \alpha_f)} \tag{5.35d}$$

Photo 5.6 High-strength concrete corbel at failure (Nawy et al.).

The assumption is made that all the shear resistance is due to the resistance at the crack interface between the corbel and the column. The ACI coefficient of friction μ has the following values:

Concrete cast monolithically	1.4λ
Concrete placed against hardened roughened concrete	1.0λ
Concrete placed against unroughened hardened concrete	0.6λ
Concrete anchored to structural steel	0.7λ

$\lambda = 1.0$ for normal-weight concrete, 0.85 for sand-lightweight concrete, and 0.75 for all-lightweight concrete. The PCI values are less conservative than the ACI values based on comprehensive tests.

If considerably higher strength concretes, such as polymer-modified concretes, are used in the corbels to interface with the normal concrete of the supporting columns, higher values of μ could logically be used for such cases as those listed above. Work in the field (Ref. 5.7) substantiates the use of higher values.

Part of the horizontal steel A_{vf} is incorporated in the top tension tie, and the remainder of A_{vf} is distributed along the depth of the corbel as in Figure 5.24. Evaluation of the top horizontal primary reinforcement layer A_s will be discussed in the next section.

5.15.2 Horizontal External Force Effect

When the corbel or bracket is cast monolithically with the supporting column or wall and is subjected to a large horizontal tensile force N_{uc} produced by the beam supported by the corbel, a modified approach is used, often termed the *strut theory approach.* In all cases, the horizontal factored force N_{uc} cannot exceed the vertical factored shear V_u. As shown in Figure 5.25, reinforcing steel A_n has to be provided to resist the force N_{uc}.

where

$$A_n = \frac{N_{uc}}{\phi f_y} \tag{5.36}$$

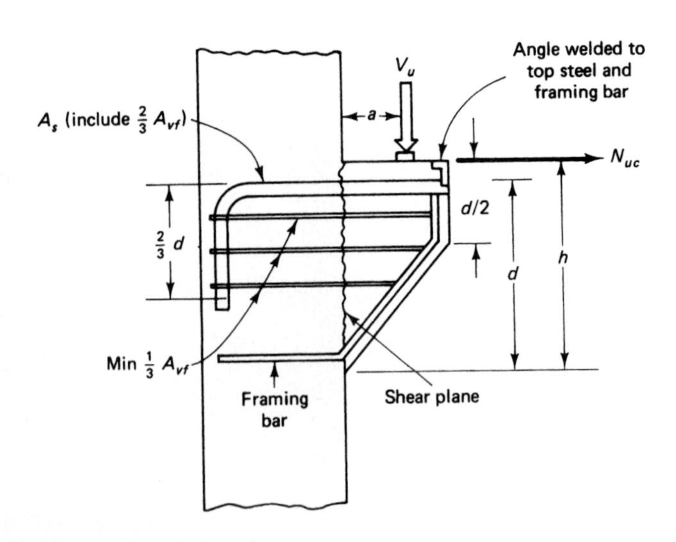

Figure 5.24 Reinforcement schematic for corbel design by the shear friction hypothesis.

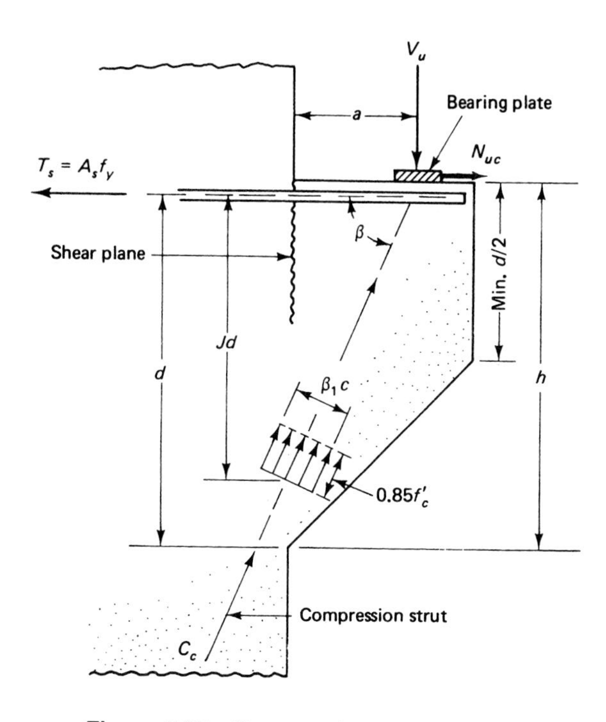

Figure 5.25 Compression strut in corbel.

and

$$A_f = \frac{V_u a + N_{uc}(h - d)}{\phi f_y\, jd} \tag{5.37}$$

Reinforcement A_f also has to be provided to resist the bending moments caused by V_u and N_{uc}.

The value of N_{uc} considered in the design should not be less than 0.20 V_u. The flexural steel area A_f can be obtained approximately by the usual expression for the limit state at failure of beams, that is,

$$A_f = \frac{M_u}{\phi f_y\, jd} \tag{5.38}$$

where $M_u = V_u a + N_{uc}(h - d)$. The axis of such an assumed section lies along a compression strut inclined at an angle β to the tension tie A_s, as shown in the figure. The volume of the compressive block is

$$C_c = 0.85 f'_c \beta_1 c b = \frac{T_s}{\cos \beta} = \frac{A_s f_y}{\cos \beta} = \frac{V_u}{\sin \beta} \tag{5.39a}$$

for which the depth $\beta_1 c$ of the block is obtained perpendicular to the direction of the compressive strut, i.e.,

$$\beta_1 c = \frac{A_s f_y}{0.85 f'_c\, b \cos \beta} \tag{5.39b}$$

The effective depth d minus $\beta_1 c/2 \cos \beta$ in the vertical direction gives the lever arm jd between the force T_s and the horizontal component of C_c in Figure 5.25. Therefore,

$$jd = d - \frac{\beta_1 c}{2 \cos \beta} \tag{5.39c}$$

If the right-hand side is substituted for jd in Equation 5.38, then

$$A_f = \frac{M_u}{\phi f_y (d - \beta_1 c/2 \cos \beta)} \tag{5.40}$$

To eliminate several trials and adjustments, the lever arm jd from Equation (5.39c) can be approximated for all practical purposes in most cases as

$$jd \cong 0.85d \tag{5.41a}$$

so that

$$A_f = \frac{M_u}{0.85 \phi f_y d} \tag{5.41b}$$

The area A_s of the primary tension reinforcement (tension tie) can now be calculated and placed as shown in Figure 5.26:

$$A_s \geq \tfrac{2}{3} A_{vf} + A_n \tag{5.42}$$

or

$$A_s \geq A_f + A_n \tag{5.43}$$

whichever is larger. Then

$$\rho = \frac{A_s}{bd} \geq 0.04 \frac{f'_c}{f_y}$$

If A_h is assumed to be the total area of the closed stirrups or ties parallel to A_s, then

$$A_h \geq 0.5(A_s - A_n) \tag{5.44}$$

The bearing area under the external load V_u on the bracket should not project beyond the straight portion of the primary tension bars, A_s, nor should it project beyond the interior face of the transverse welded anchor bar shown in Figure 5.26.

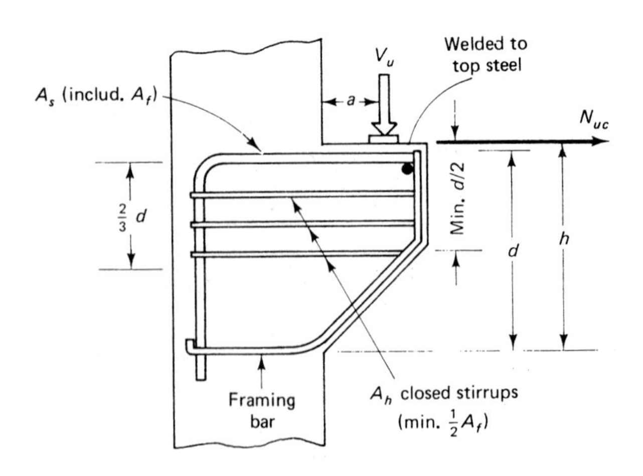

Figure 5.26 Reinforcement schematic for corbel design by strut theory.

5.15.3 Sequence of Corbel Design Steps

As discussed in the preceding section, a horizontal factored force N_{uc}, a vertical factored force V_u, and a bending moment $[V_u a + N_{uc}(h - d)]$ basically act on the corbel. To prevent failure, the corbel has to be designed to resist these three parameters simultaneously by one of the following two methods, depending on the type of corbel construction sequence, that is, whether the corbel is cast monolithically with the column or not:

(a) For a monolithically cast corbel with the supporting column, by evaluating the steel area A_h of the closed stirrups which are placed below the primary steel ties A_s. Part of A_h is due to the steel area A_n from Equation 5.36 resisting the horizontal force N_{uc}.

(b) By calculating the steel area A_{vf} by the shear friction hypothesis if the corbel and the column are *not* cast simultaneously, using part of A_{vf} along the depth of the corbel stem and incorporating the balance in the area A_s of the primary top steel reinforcing layer.

The primary tension steel area A_s is the major component of both methods. Calculations of A_s depend on whether Equation 5.42 or 5.43 governs. If Equation 5.42 controls, $A_s = \frac{2}{3} A_{vf} + A_n$ is used and the remaining $\frac{1}{3} A_{vf}$ is distributed over a depth $\frac{2}{3} d$ adjacent to A_s. If Equation 5.43 controls, $A_s = A_f + A_n$, with the addition of $\frac{1}{2} A_f$ provided as closed stirrups parallel to A_s and distributed within $\frac{2}{3} d$ vertical distance adjacent to A_s.

In both cases, the primary tension reinforcement plus the closed stirrups automatically yield the total amount of reinforcement needed for either type of corbel. Since the mechanism of failure is highly indeterminate and randomness can be expected in the propagation action of the shear crack, it is sometimes advisable to choose the larger calculated value of the primary top steel area A_s in the corbel regardless of whether the corbel element is cast simultaneously with the supporting column.

As seen from the foregoing discussions, the horizontal closed stirrups are also a major element in reinforcing the corbel. Occasionally, additional inclined closed stirrups are also used.

The following sequence of steps is proposed for the design of the corbel:

1. Calculate the factored vertical force V_u and the nominal resisting force V_n of the section such that $V_n \geq V_u/\phi$, where $\phi = 0.85$ for all calculations. V_u/ϕ should be $\leq 0.20 f'_c b_w d$, or $\leq 800 b_w d$ for normal-weight concrete. If not, the concrete section at the support should be enlarged.
2. Calculate $A_{vf} = V_n / f_y \mu$ for resisting the shear friction force, and use in the subsequent calculation of the primary tension top steel A_s.
3. Calculate the flexural steel area A_f and the direct tension steel area A_n, where

$$A_f = \frac{v_u a + N_{uc}(h - d)}{\phi f_y Jd}$$

and

$$A_n = \frac{N_{uc}}{\phi f_y}$$

4. Calculate the primary steel area from (a) $A_s = \frac{2}{3} A_{vf} + A_n$ and (b) $A_s = A_f + A_n$, whichever is larger. If case (a) controls, the remaining $\frac{1}{3} A_{vf}$ has to be provided as closed stirrups parallel to A_s and distributed with a $\frac{2}{3} d$ distance adjacent to A_s, as in Figure 5.24.

If case (b) controls, use in addition $\frac{1}{2}A_f$ as closed stirrups distributed within a distance $\frac{2}{3}d$ adjacent to A_s, as in Figure 5.26. Then

$$A_h \geq 0.5(A_s - A_n)$$

and

$$\rho = \frac{A_s}{bd} \geq 0.04\frac{f'_c}{f_y}$$

or

$$\text{Min. } A_s = 0.04\frac{f'_c}{f_y}bd$$

5. Select the size and spacing of the corbel reinforcement with special attention to the detailing arrangements, as many corbel failures are due to incorrect detailing.

Figure 5.27 shows a flowchart for proportioning corbels.

5.15.4 Design of a Bracket or Corbel

Example 5.7

Design a corbel to support a factored vertical load V_u = 90,000 lb (180 kN) acting at a distance a = 5 in. (127 mm) from the face of the column. The corbel has a width b = 10 in. (254 mm), a total thickness h = 18 in. (457 mm), and an effective depth d = 14 in. (356 mm). The following data are given:

f'_c = 5,000 psi (34.5 MPa), normal-weight concrete

f_y = 60,000 psi (414 MPa)

Assume the corbel to be either cast after the supporting column was constructed, or cast simultaneously with the column. Neglect the weight of the corbel.

Solution:

Step 1

$$V_n \geq \frac{V_u}{\phi} = \frac{90{,}000}{0.85} = 105{,}882 \text{ lb}$$

$$0.2f'_c b_w d = 0.2 \times 5{,}000 \times 10 \times 14 = 140{,}000 \text{ lb} > V_n$$

$$800 b_w d = 800 \times 10 \times 14 = 112{,}000 \text{ lb} > V_n, \text{ O.K.}$$

Step 2

(a) Monolithic construction, normal-weight concrete $\mu = 1.4\lambda$:

$$A_{vf} = \frac{V_u}{\phi f_y \mu} = \frac{105{,}882}{60{,}000 \times 1.4} = 1.261 \text{ in.}^2 \text{ (813 mm}^2\text{)}$$

(b) Nonmonolithic construction, $\mu = 1.0\lambda$:

$$A_{vf} = \frac{105{,}882}{60{,}000 \times 1.0} = 1.765 \text{ in.}^2 \text{ (1138 mm}^2\text{)}$$

Choose the larger A_{vf} = 1.765 in.^{2} as controlling.

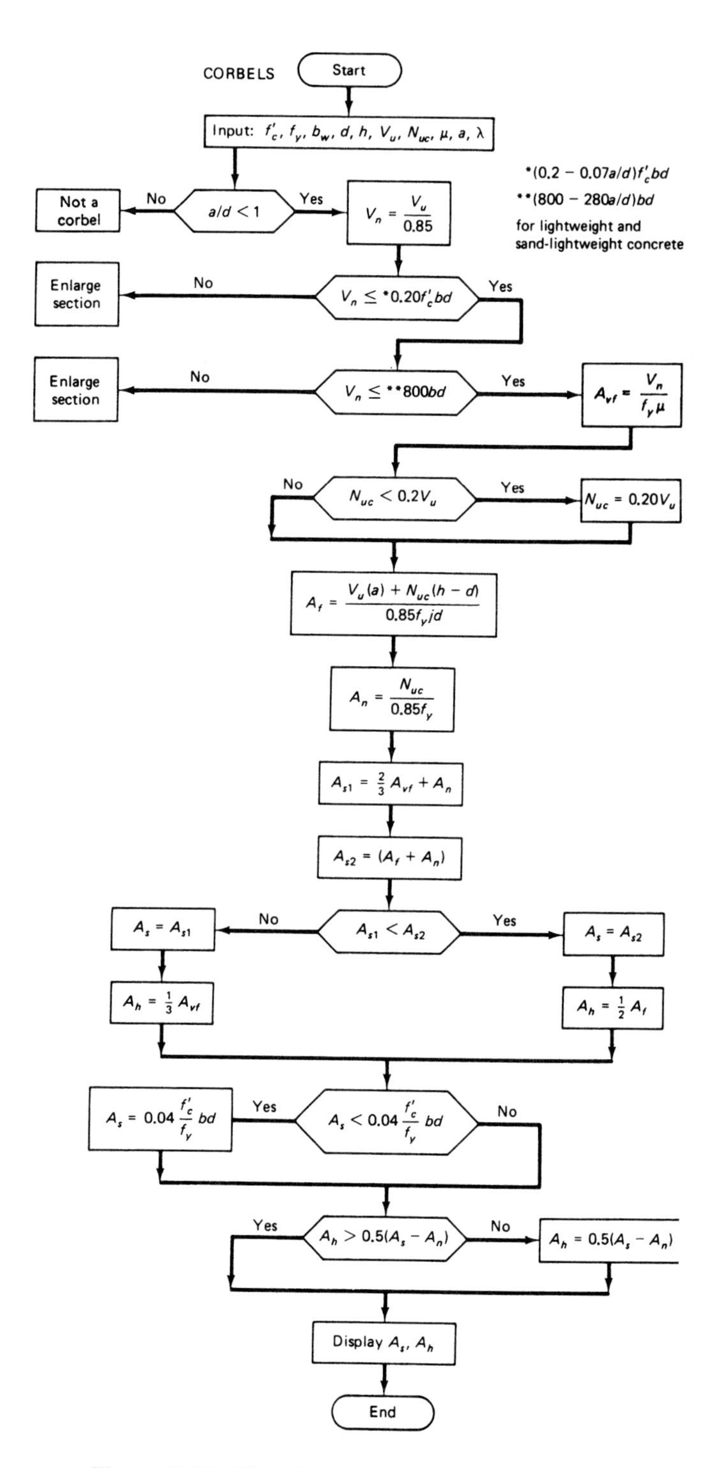

Figure 5.27 Flowchart for proportioning corbels.

Step 3. Since no value of the horizontal external force N_{uc} transmitted from the superimposed beam is given, use

$$\text{Min } N_{uc} = 0.20V_u = 0.2 \times 90{,}000 = 18{,}000 \text{ lb}$$

$$A_f = \frac{M_u}{\phi f_y jd} = \frac{V_u a + N_{uc}(h - d}{\phi f_y jd} \qquad (\text{where } jd \cong 0.85d)$$

$$= \frac{90{,}000 \times 518{,}000(18 - 14)}{0.90 \times 60{,}000(0.85 \times 14)} = 0.812 \text{ in}^2 \text{ (524 mm}^2\text{)}$$

$$A_n = \frac{N_{uc}}{\phi f_y} = \frac{18{,}000}{0.85 \times 60{,}000} = 0.353 \text{ in}^2 \text{ (278 mm}^2\text{)}$$

Step 4. Check the controlling area of primary steel A_s:

(a) $A_s = (\frac{2}{3} A_{vf} + A_n) = \frac{2}{3} \times 1.765 + 0.353 = 1.529$ in.^{2}
(b) $A_s = A_f + A_n = 0.812 + 0.353 = 1.165$ in.^{2}

$$\text{Min } A_s = 0.04\frac{f'_c}{f_y}bd = 0.04 \times \frac{5{,}000}{60{,}000} \times 10 \times 14 = 0.47 \text{ in.}^2$$

$$< 1.529, \text{ O.K.}$$

Provide $A_s = 1.529$ in.^{2} (986.2 mm^2).

Horizontal closed stirrups:
Since case (a) controls,

$$\tfrac{1}{3}A_{vf} = \tfrac{1}{3} \times 1.765 = 0.588 \text{ in.}^2$$

$$A_h = 0.5(A_s - A_n) = 0.5(1.529 - 0.353) = 0.588 \text{ in.}^2$$

Use the larger of the two values $\frac{1}{3}A_{vf}$ and A_h.

Step 5. Select bar sizes:

(a) Required $A_s = 1.529$ in.^{2}; use three No. 7 bars = 1.80 in.^{2} (three bars of diameter 22 mm gives 1,161 mm^2) A_s.
(b) Required $A_h = 0.588$ in^2; use three No. 3 closed stirrups = $2 \times 3 \times 0.11 = 0.66$ in.^{2} spread over $\frac{2}{3}d = 9.33$ in. vertical distance. Hence, use three No.-3 closed stirrups at 3 in. center to center. Also use three framing size No.-3 bars and one welded No.-7 anchor to bar.

Details of the bracket reinforcement are shown in Figure 5.28. The bearing area under the load has to be checked, and the bearing pad designed such that the bearing stress at the factored load V_u should not exceed $\phi(0.85f'_c A_1)$, where A_1 is the pad area. We have

$$V_u = 90{,}000 \text{ lb} = 0.70(0.85 \times 5{,}000)\, A_1$$

$$A_1 = \frac{90{,}000}{0.70 \times 0.85 \times 5000} = 30.25 \text{ in}^2 \text{ (19,516 mm}^2\text{)}$$

Use a plate $5\frac{1}{2}$in. × $5\frac{1}{2}$in. Its thickness has to be designed based on the manner in which V_u is applied as an undeformable plate.

5.15.5 S1 Expressions for Shear in Prestressed Concrete Beams

$$V_{ci} = \left[\frac{\lambda\sqrt{f'_c}}{20}b_w d + V_d + V_i\left(\frac{M_{cr}}{M_{\max}}\right)\right] \geq \left(\frac{\sqrt{f'_c}}{7}\right)b_w d \tag{5.11}$$

$$M_{cr} = S_b(0.5\lambda\sqrt{f'_c} + f_{ce} - f_d) \tag{5.12}$$

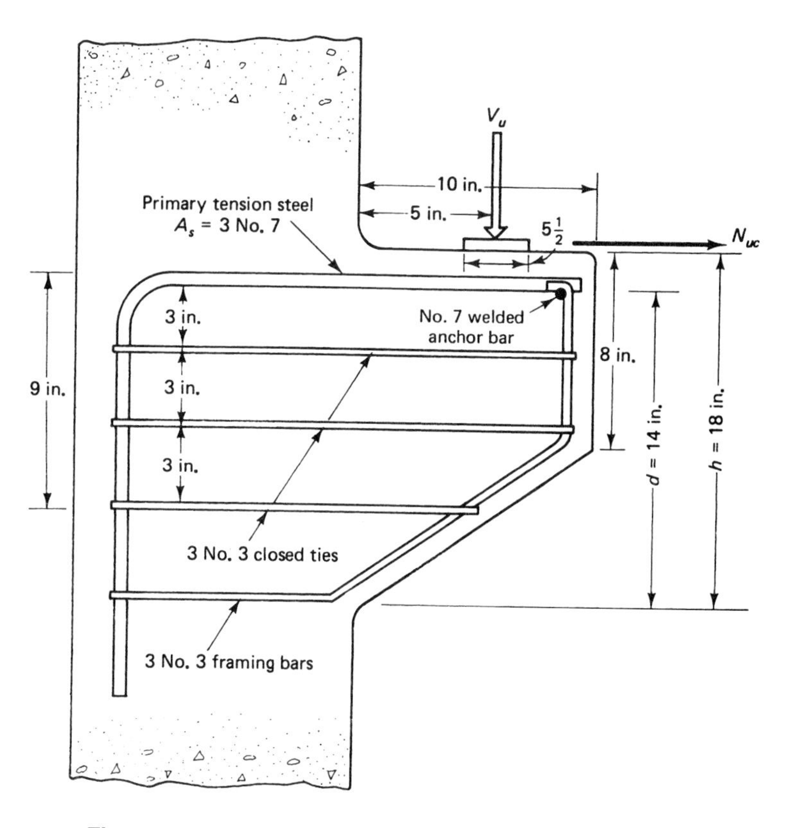

Figure 5.28 Corbel reinforcement details (Example 5.7).

$$V_{cw} = 0.3(\lambda\sqrt{f_c'} + 0.3\bar{f_c})b_w d + V_p \tag{5.15}$$

(See Sec. 5.5.1 for explanation on f_c vs. f_{ce} of the ACI code). M_{cr} for shear = moment causing flexural cracking at section due to externally applied load.

$$V_c = \left(\frac{\lambda\sqrt{f_c'}}{20} + 5\frac{V_u d}{M_u}\right)b_w d; \qquad \frac{V_u d}{M_u} \leq 1.0$$

$$\geq \left[\lambda\frac{\sqrt{f_c'}}{5}\right] b_w d$$

$$\leq [0.4\lambda\sqrt{f_c'}\, b_w d]$$

$$s = \frac{A_v f_y d}{(V_u/\phi) - V_c} = \frac{A_v f_y d}{V_s} \tag{5.21b}$$

Max $s = \frac{3}{4}h \leq 600$ mm

when $V_s > (\lambda\sqrt{f_c'}/3)b_w d$, max $s \leq 3/16h \leq 300$ mm

$V_s > (2\lambda\sqrt{f_c'}/3)b_w d$, enlarge section.

Min. A_v: the smaller of

$$A_v \geq \frac{0.35 b_w s}{f_y} \quad \text{or} \quad \frac{A_{ps} f_{pu} s}{80 f_{yv} d}\sqrt{\frac{d}{b_w}}$$

where b_w, s and d are in millimeters and f_y is in MPa.

Max. allowable shear friction force without dowels, $F_h = 0.55 b_v l_{vh}$

$$1 \text{ lb} \times 4.448 = \text{N}$$
$$\text{psi} \times 0.006895 = \text{MPa}$$
$$(\text{in.-lb}) \times 0.1130 = \text{N-m}$$
$$1 \text{ Pa} = \text{N/m}^2$$
$$1 \text{ MPa} = \text{N} \times 10^6/\text{m}^2$$
$$(\text{lb/ft}) \times 14.593 = \text{N/m}$$

5.15.6 SI Shear Design of Prestressed Beams

Example 5.8

Solve Example 5.6 using the SI units system. The sectional geometric properties of the beam are as follows:

Section property	Pretopped
A_c	6,310 cm^2
I_c	3.58×10^6 cm^4
r^2	568 cm^2
c_b	65.5 cm
c_t	20.9 cm.
S_b	54,733 cm^3
S^t	171,376 cm^3
W_D	14,870 N/m
$W_D + W_{SD}$	17,789 N/m
W_L	10,507 N/m
$2b_w$	32 cm

Other data are:

f'_c (precast) = 34.5 MPa, normal-weight concrete, section pretopped
f'_{cc} (topping) = 20.7 MPa, normal-weight concrete, at a later stage if used
$f'_{ci} = 27.6$ MPa
$f_{pu} = 1{,}862$ MPa, low-relaxation steel
$f_{ps} = 1{,}655$ MPa
$f_{pe} = 1{,}020$ MPa
$e_e = 28.3$ cm
$e_c = 57.2$ cm
$A_{ps} = 18 - 12.7$ mm diameter tendons = 99 mm^2
f_{yv} for stirrups = 414 MPa

Use the same value for the effective depth d_p for the midspan as well as other sections. Note that b_w for *both* webs = 32 cm.

Solution:

$$W_u = 1.4 \times 17{,}789 + 1.7 \times 10{,}507 \cong 42.8\text{kN/m}$$

$$V_u \text{ at face of support} = \frac{42.8 \times 21.3}{2} = 456 \text{ kN}$$

$$\tfrac{1}{4} \text{ of the span} = 21.3/4 = 5.33 \text{ m.}$$

$$V_n \text{ at 5.33m. from the face of the support} = \frac{1}{0.85}\left(456 \times \frac{(10.67 - 5.33)}{10.67}\right)$$

$$= 268 \text{ kN}$$

(1) *Flexure-shear cracking, V_{ci} (step 2)*

$$d_p = 76.2 \text{ cm}$$

$$P_e = 18 \times 99 \times 1{,}020 \text{ MPa} = 18{,}176 \text{ kN}$$

$$e \text{ at 5.33 m from support} = 42.1 \text{ cm}$$

Use the precast section properties for computing f_{ce} and f_d as discussed in Section 5.5:

$$f_{ce} = -\frac{P_e}{A_c}\left(1 + \frac{ec_b}{r^2}\right) = -\frac{18{,}176}{6{,}310}\left(1 + \frac{42.1 \times 65.46}{568}\right)$$

$$= 16.8 \text{ MPa}$$

Use allowable extreme compressive stresses as follows:

(a) prestress + sustained load: $f_c = 0.45\,f'_c$

(b) prestress + total load (allowing 33% increase due to transient load: $f_c = 0.60\,f'_c$)
Note that although $f_{ce} = 0.45\,f'_c$, this should not affect the shear strength since f_{ce} is due to prestress only, and the inclusion of self-weight reduces it to less than $0.45\,f'_c$. We thus have:

$$\text{Self-weight } W_D = 14{,}870 \text{ N/m}$$

$$M_{5.33} = \frac{W_D x(l - x)}{2} = \frac{14{,}870 \times 5.33(21.3 - 5.33)}{2}$$

$$= 634 \text{ kN-m}$$

$$f_d = \frac{Mc_b}{I_c} = \frac{6{,}340{,}000 \times 65.46}{3.58 \times 10^6} = 11.6 \text{ MPa}$$

$$M_{cr} = S_b(0.50\lambda\sqrt{f'_c} + f_{ce} - f_d)$$

$$= 54{,}733\,(0.50 \times 1.0\sqrt{34.5} + 16.8 - 11.6)$$

$$= 445 \text{ kN-m}$$

Unfactored shear due to self-weight is

$$V_d = W_D\left(\frac{1}{2} - x\right) = 14{,}870\left(\frac{21.3}{2} - 5.33\right) = 79.1\text{kN}$$

$$W_{SD} = 2{,}919 \text{ N/m}$$

$$W_L = 10{,}507 \text{ N/m}$$

Factored external load intensity is

$$W_U = 1.4 \times 2{,}919 + 1.7 \times 10{,}507 = 22.0 \text{ kN/m}$$

$$V_i = W_U\left(\frac{1}{2} - x\right) = 22\left(\frac{21.3}{2} - 5.33\right) = 117 \text{ kN}$$

$$M_{\max} = W_U x\left(\frac{l - x}{2}\right) = \frac{22 \times 5.33(21.3 - 5.33)}{2}$$

$$= 937 \text{ kN-m}$$

$$V_{ci} = 0.6\lambda\sqrt{f'_c}\,b_w d_p + V_d + \frac{V_i}{M_{\max}}(M_{cr}) \geq 0.33\lambda\sqrt{f'_c}\,b_w d_p$$

$$= 0.6 \times 1.0 \times \sqrt{34.5} \times 3.18 \times 7.62 + 79.1$$

$$+ \frac{117}{937}(445)$$

$$= 220 \text{ kN}$$

$$0.33\lambda\sqrt{f_c'}\,b_w d_p = 0.33 \times 1.0\sqrt{34.5} \times 31.8x76.2 = 202 \text{ kN} < 220 \text{ kN}$$

Hence, $V_{ci} = 220kN$ controls.

(2) *Web-shear cracking, V_{cw} (step 2)*

$$\bar{f_c} = \frac{P_e}{A_c} = \frac{18{,}176}{6{,}310} = 2.9 \text{ MPa}$$

For the vertical component of the prestress force,

$$V_p = P_e \tan\theta = 18{,}176\frac{(55.3 - 28.9)}{21.3/2} = 44.0 \text{ kN}$$

$$V_{cw} = [0.3(\lambda\sqrt{f_c'} + \bar{f_c})]b_w d_p + V_p$$

$$= [0.3(1.0x\sqrt{34.5} + 2.9)] \times 318 \times 762 + 44$$

$$= 681 \text{ kN vs.} V_{ci} = 211 \text{ kN}$$

Now, V_c is the smaller of V_{ci} and V_{cw}; hence,

$$V_c = V_{ci} = 211 \text{ kN}$$

(3) *Design of web reinforcement (steps 3–8)*
From above,

$$V_c = 211 \text{ kN},$$

So

$$\frac{1}{2}V_c = 106 \text{ kN}$$

Now, V_u / ϕ at a section 5.33 m. from support = 268 > $V_c > \frac{1}{2}V_c$; hence design of stirrups is necessary. If $V_u / \phi < V_c > \frac{1}{2}V_c$, only minimum-web steel is need.

$$\text{Req.}\frac{A_v}{s} = \frac{V_n - V_c}{f_y d_p} = \frac{(V_c/\phi) - V_c}{f_y d_p} = \frac{(268 - 211)}{414x762}$$

$$= 0.21 \text{ mm}^2/\text{mm spacing}$$

Using $d \cong d_p = 762$ mm and $b_w = 318$ mm:

$$\text{Min.}\left(\frac{A_v}{s}\right) = \frac{A_{ps}\ f_{pu}}{80\ f_y d_p}\sqrt{\frac{d_p}{b_w}} = \frac{18 \times 99}{80} \times \frac{1{,}860}{414x762}\sqrt{\frac{762}{318}} = 0.20 \text{ mm}^2/\text{mm}$$

Or

$$\text{Min.}\left(\frac{A_v}{s}\right) = \frac{b_w}{3\,f_y} = \frac{318}{3x414} = 0.25 \text{ mm}^2/\text{mm}$$

The *lesser* of the two minimum value applies; consequently, min. $A_v = 0.20$ mm²/mm applies as the lesser of the two values.
Controlling web steel is hence

$$\frac{A_v}{s} = 0.21 \text{ mm}^2/\text{mm for both webs or}$$

0.105 mm² / mm per web.

Try one row of D5 deformed welded wire fabric at 254 mm center-to-center weld spacing.

The maximum allowable spacing is, then $0.75h$ where h = 864 mm but not to exceed 610 mm. So we have

$$0.75 \times 864 = 648$$

Accordingly, adopt using one row D5 WWF in one layer at 254 mm center-to-center of welds per web at the quarter-span section.

Note, in comparing the solution for V_{ci} and V_{cw} that V_{ci} has its highest value close to the support and *rapidly* decreases toward the midspan, while V_{cw} has a lesser variation in its value, as can be seen from Figure 5.13. It is important to calculate the flexure shear V_{ci} and web shear V_{cw} at several sections along the span in order to determine the most efficient distribution of the web steel. A computer program facilitates finding these values at constant intervals of, say, 1/10th of the span, and a plot can be made similar to the one in Figure 5.13 showing the variation of the shear strengths of the web along the span.

(4) *Design of dowel steel for full composite section of the additional 2-in. topping (step 9), if such topping is added later to the pretopped section.*

Section at $\frac{1}{2}d_p$ from face of support

$$\text{Use } d_p = 76.2 + 5.1 = 81.3 \text{ cm}$$

V_u at support = 457 kN

$$\frac{1}{2}d_p = \frac{81.3}{2} = 41 \text{ cm} \qquad h/2 = 0.4 \text{ m}$$

$$V_u = 457 \times \left(\frac{10.7 - 0.4}{10.7}\right) = 440 \text{ kN}$$

$$\text{Req } V_{nh} = \frac{V_u}{\varphi} = \frac{440}{0.85} = 516 \text{ kN}$$

$$b_v = 366 \text{ cm} \qquad \text{topping } h = 5.08 \text{ cm}$$

From Figure 5.14

$$C_c = 0.85f'_{cc}A_{top} = 0.85 \times 20.7 \times 366 \times 5.08 \times 10^{-3} = 3{,}267 \text{ kN}$$

$$T_s = A_{ps}f_{ps} = 18 \times 99 \times 1655x10^{-3} = 2{,}940 \text{ kN}$$

$$< C_c = 3{,}267 \text{ kN}$$

Then

$$F_h = 2{,}940 \text{ kN}$$

$$l_{vh} = \frac{21.3}{2} = 10.67\text{m} = 1{,}067 \text{ cm}$$

$$b_v = 366 \text{ cm}$$

$$0.55b_v l_{vh} = 0.55 \times 366 \times 1067 = 21{,}520 \text{ kN} >> 2940 \text{ kN}$$

No dowel reinforcement is needed to extend to the additional 5 cm topping for full composite action to be developed. The section is adopted when it satisfies the flexural, deflection, and cracking requirements.

5.16 TORSIONAL BEHAVIOR AND STRENGTH

5.16.1 Introduction

Torsion occurs in monolithic concrete construction primarily where the load acts at a distance from the longitudinal axis of the structural member. An end beam in a floor panel, a spandrel beam receiving load from one side, a canopy or a bus-stand roof projecting from a monolithic beam on columns, and peripheral beams surrounding a floor opening are all examples of structural elements subjected to twisting moments. These moments occasionally cause excessive shearing stresses. As a result, severe cracking can develop well beyond the allowable serviceability limits unless special torsional reinforcement is provided. Photos in this section illustrate the extent of cracking at failure of a beam in torsion. They show the curvilinear plane of twist caused by the imposed torsional moments. In actual spandrel beams of a structural system, the extent of damage due to torsion is usually not as severe. This is due to the redistribution of stresses in the structure. However, loss of integrity due to torsional distress should always be avoided by proper design of the necessary torsional reinforcement.

An introduction to the subject of torsional stress distribution has to start with the basic elastic behavior of simple sections, such as circular or rectangular sections. Most concrete beams subjected to twist are components of rectangles, for example, flanged sections such as T-beams and L-beams. Although circular sections are rarely a consideration in normal concrete construction, a brief discussion of torsion in circular sections serves as a good introduction to the torsional behavior of other types of sections.

Shear stress is equal to shear strain times the shear modulus at the elastic level in circular sections. As in the case of flexure, the stress is proportional to its distance from the neutral axis (i.e., the axis through the center of the circular section) and is maximum at the extreme fibers. If r is the radius of the element, $J = \pi r^4/2$, its polar moment of inertia, and v_{te} the elastic shearing stress due to an elastic twisting moment T_e, then

$$v_{te} = \frac{T_e r}{J} \tag{a}$$

When deformation takes place in the circular shaft, the axis of the circular cylinder is assumed to remain straight. All radii in a cross section also remain straight (i.e., there is no warping) and rotate through the same angle about the axis. As the circular element starts to behave plastically, the stress in the plastic outer ring becomes constant while the stress in the inner core remains elastic, as shown in Figure 5.29. As the whole cross section becomes plastic, $b = 0$ and the shear stress

$$v_{tf} = \frac{3}{4}\frac{T_p r}{J} \tag{b}$$

where v_{tf} is the nonlinear shear stress due to an ultimate twisting moment T_p. (The subscript f denotes failure.)

In rectangular sections, the torsional problem is considerably more complicated. The originally plane cross sections undergo warping due to the applied torsional moment. This moment produces axial as well as circumferential shear stresses with zero values at the corners of the section and the centroid of the rectangle, and maximum values on the periphery at the middle of the sides, as shown in Figure 5.30. The maximum torsional shearing stress would occur at midpoints A and B of the larger dimension of the cross section. These complications plus the fact that the reinforced and prestressed concrete sections are neither homogeneous nor isotropic make it difficult to develop exact

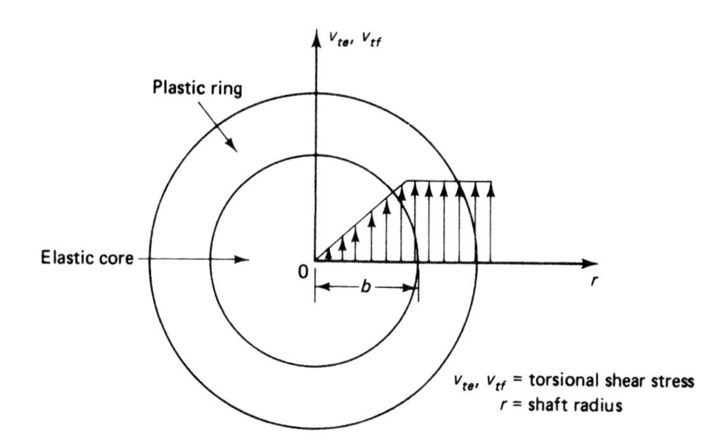

Figure 5.29 Torsional stress distribution through circular section.

mathematical formulations based on physical models such as Equations (a) and (b) for circular sections.

For over seventy years, the torsional analysis of concrete members has been based on either (1) the classical theory of elasticity developed through mathematical formulations coupled with membrane analogy verifications (St.-Venant's), or (2) the theory of plasticity represented by the sand-heap analogy (Nadai's). Both theories were applied essentially to the state of pure torsion. But it was found experimentally that the elastic theory is not entirely satisfactory for the accurate prediction of the state of stress in concrete in pure torsion. The behavior of concrete was found to be better represented by the plastic approach. Consequently almost all developments in torsion as applied to prestressed concrete and to reinforced concrete have been in the latter direction.

5.16.2 Pure Torsion in Plain Concrete Elements

5.16.2.1 Torsion in elastic materials. In 1853, St.-Venant presented his solution to the elastic torsional problem with warping due to pure torsion which develops in noncircular sections. In 1903, Prandtl demonstrated the physical significance of the mathematical formulations by his membrane analogy model. The model establishes particular

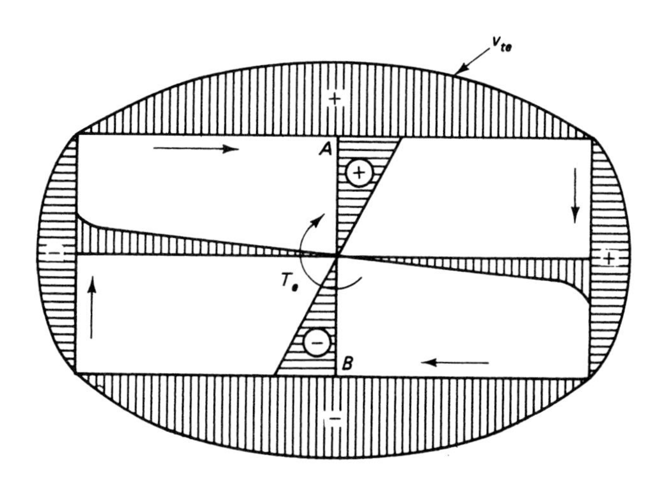

Figure 5.30 Pure torsional stress distribution in a rectangular section.

relationships between the deflected surface of the loaded membrane and the distribution of torsional stresses in a bar subjected to twisting moments. Figure 5.31 shows the membrane analogy behavior for rectangular as well as L-shaped forms.

For small deformations, it can be proved that the differential equation of the deflected membrane surface has the same form as the equation that determines the stress distribution over the cross section of the bar subjected to twisting moments. Similarly, it can be demonstrated that (1) the tangent to a contour line at any point of a deflected membrane gives the direction of the shearing stress at the corresponding cross section of the actual membrane subjected to twist; (2) the maximum slope of the membrane at any point is proportional to the magnitude of shear stress τ at the corresponding point in the actual member; and (3) the twisting moment to which the actual member is subjected is proportional to *twice* the volume under the deflected membrane.

It can be seen from Figure 5.30 and 5.31(b) that the torsional shearing stress is inversely proportional to the distance between the contour lines. The closer the lines, the higher the stress, leading to the previously stated conclusion that the maximum torsional shearing stress occurs at the middle of the longer side of the rectangle. From the membrane analogy, this maximum stress has to be proportional to the steepest slope of the tangents at points A and B.

If δ is the maximum displacement of the membrane from the tangent at point A, then from basic principles of mechanics and St.-Venant's theory,

$$\delta = b^2 G\theta \tag{5.45a}$$

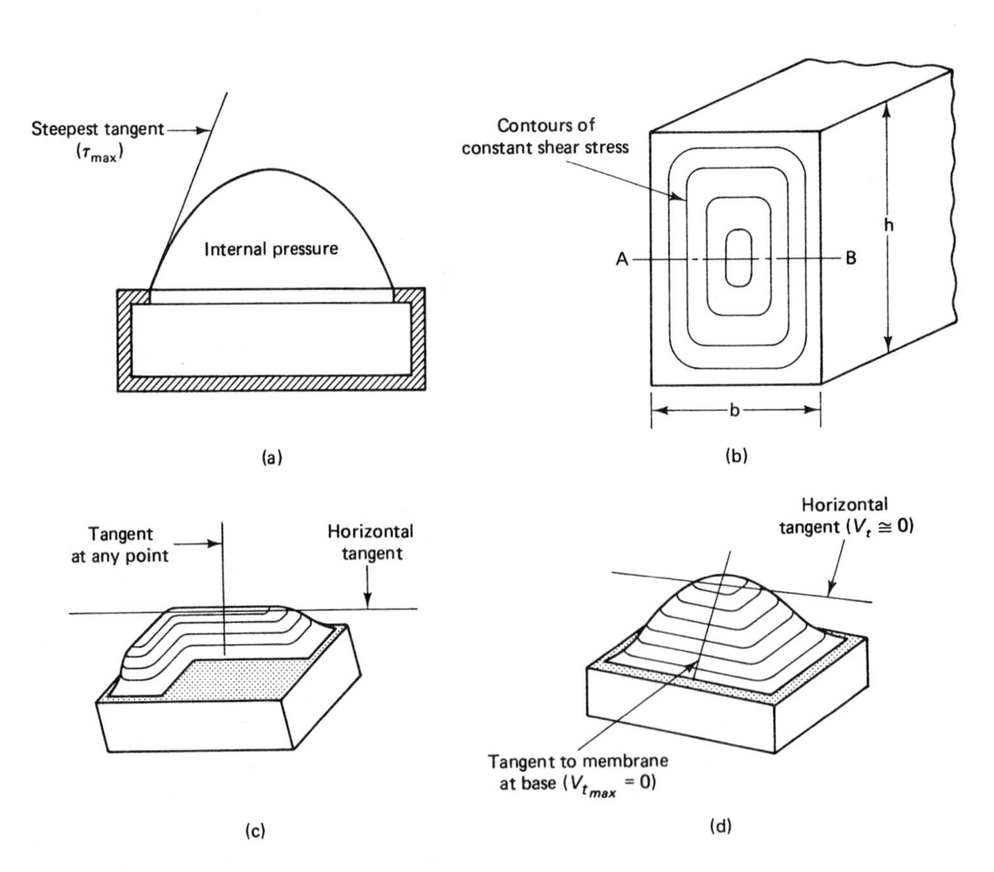

Figure 5.31 Membrane analogy in elastic pure torsion. (a) Membrane under pressure. (b) Contours in a real beam or in a membrane. (c) L-section. (d) Rectangular section.

where G is the shear modulus and θ is the angle of twist. But $v_{t(\max)}$ is proportional to the slope of the tangent; hence,

$$v_{t(\max)} = k_1 bG\theta \tag{5.45b}$$

where k_1 is a constant. The corresponding torsional moment T_e is proportional to *twice* the volume under the membrane, or

$$T_e \propto 2(\tfrac{2}{3}\delta bh) = k_2 \delta bh$$

where, again, k_2 is a constant. Or yet again,

$$T_e = k_3 b^3 hG\theta \tag{5.45c}$$

with k_3 constant. From Equations 5.45a and b,

$$v_{t(\max)} = \frac{T_e b}{kb^3 h} \simeq \frac{T_e b}{J_1} \tag{5.45d}$$

The denominator kb^3h in 5.45d represents the polar moment of inertia J_1 of the section. Comparing this equation to Equation (a) for the circular section shows the similarity of the two expressions, except that the factor k in the equation for the rectangular section takes into account the shear strains due to warping. Equation 5.45d can be further simplified to give

$$v_{t(\max)} = \frac{T_e}{kb^2 h} \tag{5.46}$$

It can also be written to give the stress at planes inside the section, such as an inner concentric rectangle of dimensions x and y, where x is the shorter side, so that

$$v_{t(\max)} = \frac{T_e}{kx^2 y} \tag{5.47}$$

It is important to note in using the membrane analogy approach that the torsional shear stress changes from one point to another along the same axis as AB in Figure 5.31, because of the changing slope of the analogous membrane, rendering the torsional shear stress calculations lengthy.

5.16.2.2 Torsion in plastic materials. As indicated earlier, the plastic sand-heap analogy provides a better representation of the behavior of brittle elements such as concrete beams subjected to pure torsion than does the elastic analogy. The torsional moment is also proportional to *twice* the volume under the heap, and the maximum torsional shearing stress is proportional to the slope of the sand heap. Figure 5.32 is a two- and three-dimensional illustration of the sand heap. The torsional moment T_p in part (d) of the figure is proportional to twice the volume of the rectangular heap shown in parts (b) and (c). It can also be recognized that the slope of the sand-heap sides as a measure of the torsional shearing stress is *constant* in the sand-heap analogy approach, whereas it is continuously variable in the membrane analogy approach. This characteristic of the sand heap considerably simplifies the solutions.

5.16.2.3 Sand-heap analogy applied to L-beams. Most concrete elements subjected to torsion are flanged sections, most commonly L-beams comprising the external wall beams of a structural floor. The L-beam in Figure 5.33 is chosen in applying the plas-

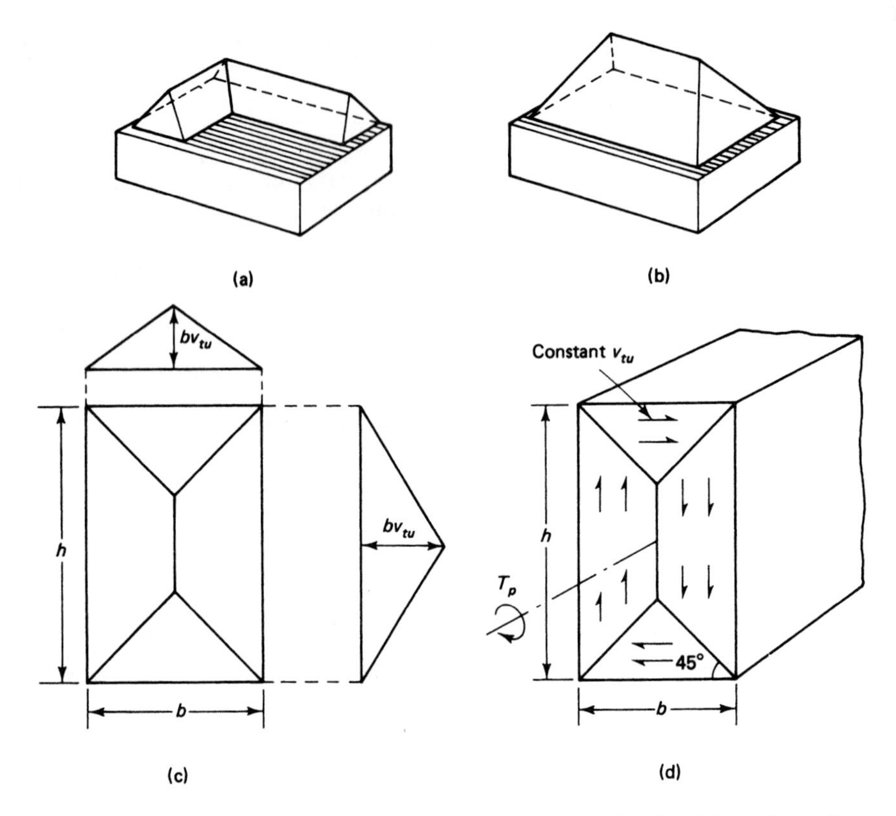

Figure 5.32 Sand-heap analogy in plastic pure torsion. (a) Sand-heap L-section. (b) Sand-heap rectangular section. (c) Plan of rectangular section. (d) Torsional shear stress.

tic sand-heap approach to evaluate its torsional moment capacity and shear stress to which it is subjected.

The sand heap is broken into three volumes:

V_1 = pyramid representing a square cross-sectional shape $= y_1 b_w^2/3$
V_2 = tent portion of the web, representing a rectangular cross-sectional shape $= y_1 b_w (h - b_w)/2$
V_3 = tent representing the flange of the beam, transferring part *PDI* to *NQM* $= y_2 h_f (b - b_w)/2$

The torsional moment is proportional to twice the volume of the sand heaps; hence,

$$T_p \cong \left[\frac{y_1 b_w^2}{3} + \frac{y_1 b_w (h - b_w)}{2} + \frac{y^2 h_f (b - b_w)}{2}\right] 2 \tag{5.48}$$

Also, the torsional shear stress is proportional to the slope of the sand heaps; hence,

$$y_1 = \frac{v_t b_w}{2} \tag{5.49}$$

$$y_2 = \frac{v_t h_f}{2} \tag{5.50}$$

Substituting y_1 and y_2 from Equations 5.49 and 5.50 into Equation 5.48 give us

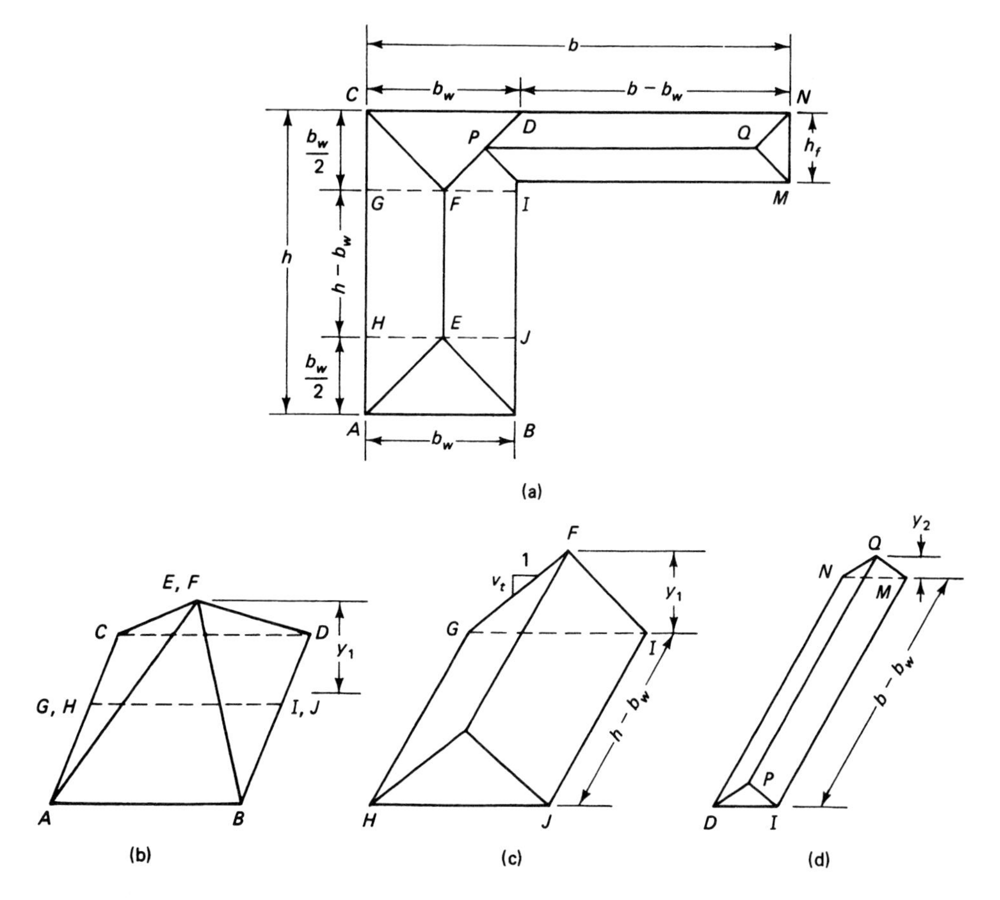

Figure 5.33 Sand-heap analogy of flanged section. (a) Sand heap on L-shaped cross section. (b) Composite pyramid from web (V_1). (c) Tent segment from web (V_2). (d) Transformed tent of beam flange (V_3).

$$v_{t(\max)} = \frac{T_p}{(b_w^2/6)(3h - b_w) + (h_f^2/2)(b - b_w)} \tag{5.51}$$

If both the numerator and denominator of Equation 5.51 are divided by $(b_w h)^2$ and the terms rearranged, we have

$$v_{t(\max)} = \frac{T_p h/(b_w h)^2}{[\frac{1}{6}(3 - b_w/h)] + \frac{1}{2}(h_f/b_w)2(b/h - b_w/h)} \tag{5.52a}$$

If one assumes that C_t is the denominator in this equation and that $J_E = C_t/(b_w h)^2$, the equation becomes

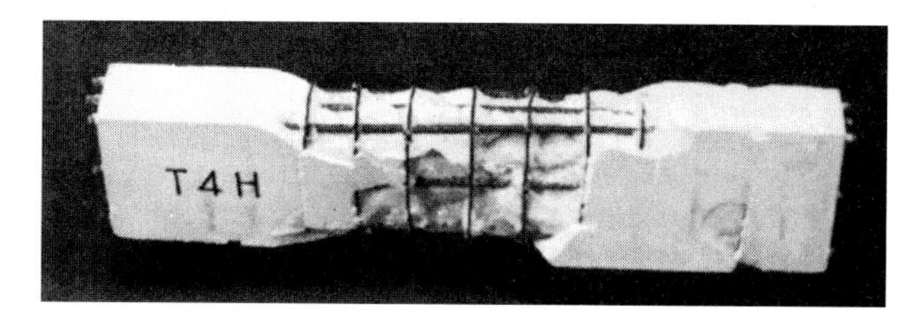

Photo 5.7 Reinforced plaster beam at failure in pure torsion. (Rutgers tests: Law, Nawy, et al.)

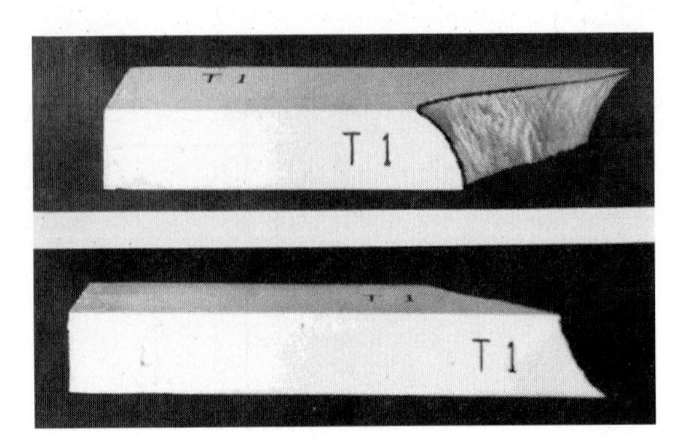

Photo 5.8 Plain mortar beam in pure torsion. (a) Top view. (b) Bottom view. (Rutgers tests: Law, Nawy et al.)

$$v_{t(\max)} = \frac{T_p h}{J_E} \tag{5.52b}$$

where J_E is the equivalent polar moment of inertia, a function of the shape of the beam cross section. Note that Equation 5.52b is similar in format to Equation 5.45d from the membrane analogy, except for the different values of the denominators J and J_E. Equation 5.52a can be readily applied to rectangular sections by setting $h_f = 0$.

It must also be recognized that concrete is not a perfectly plastic material; hence, the actual torsional strength of the plain concrete section has a value lying between the membrane analogy and the sand-heap analogy values.

Equation 5.52b can be rewritten designating $T_p = T_c$ as the nominal torsional resistance of the plain concrete and $v_{t(\max)} = v_{tc}$ using ACI terminology, so that

$$T_c = k_2 b^2 h v_{tc} \tag{5.53a}$$

or

$$T_c = k_2 x^2 y v_{tc} \tag{5.53b}$$

where x is the smaller dimension of the rectangular section.

Extensive work on reinforced concrete beams by Hsu and confirmed by others has established that k_2 can be taken as $\frac{1}{3}$. This value originated from research in the skew-bending theory of plain concrete. It was also established that $6\sqrt{f'_c}$ can be considered as a limiting value of the pure torsional strength of a member without torsional reinforcement. Using a reduction factor of 2.5 for the first cracking torsional load $v_{tc} = 2.4\sqrt{f'_c}$, and using $k_2 = \frac{1}{3}$ in Equation 5.53, results in

$$T_c = 0.8\sqrt{f'_c}\,x^2 y \tag{5.54a}$$

where x is the shorter side of the rectangular section. The high reduction factor of 2.5 is used to offset any effect of bending moments that might be present.

If the cross section is a T- or L-section, the area can be broken into component rectangles as in Figure 5.34, such that

$$T_c = 0.8\sqrt{f'_c}\,\Sigma x^2 y \tag{5.54b}$$

5.17 TORSION IN REINFORCED AND PRESTRESSED CONCRETE ELEMENTS

Torsion rarely occurs in concrete structures without being accompanied by bending and shear. The foregoing should give a sufficient background on the contribution of the plain concrete in the section toward resisting *part* of the combined stresses resulting from tor-

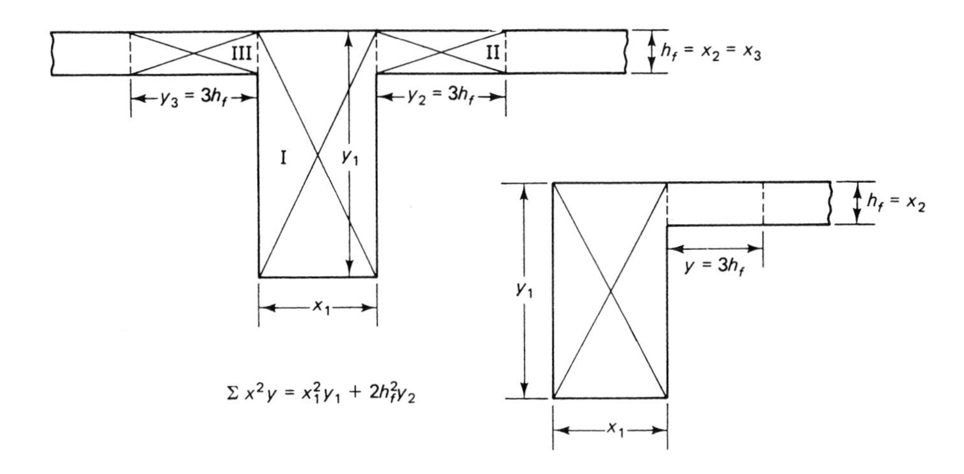

Figure 5.34 Component rectangles for calculation of T_c.

sional, axial, shear, or flexural forces. The capacity of the plain concrete to resist torsion when in combination with other loads could, in many cases, be lower than when it resists the same factored external twisting moments alone. Consequently, torsional reinforcement has to be provided to resist the excess torque.

Inclusion of longitudinal and transverse reinforcement to resist part of the torsional moments introduces a new element in the set of forces and moments in the section. If

T_n = required total nominal torsional resistance of the section, including the reinforcement

T_c = nominal torsional resistance of the plain concrete

and

T_s = torsional resistance of the reinforcement

then

$$T_n = T_c + T_s \tag{5.55}$$

Several theories have been proposed over the past half century. A general discussion presented here concentrates on (a) the skew bending theory, (b) the space truss analogy theory, (c) the compression field theory, and (d) the plasticity equilibrium truss theory. Except for the skew bending theory, the other models consider the shear flow in hollow box sections as the principal element in evaluating the torsional capacity of solid and hollow sections.

5.17.1 Skew-Bending Theory

Skew-bending theory considers in detail the internal deformational behavior of the series of transverse warped surfaces along the beam. Initially proposed by Lessig, it had subsequent contributions from Collins, Hsu, Zia, Gesund, Mattock, and Elfgren among the several researchers in this field. Hsu made a major contribution experimentally to the development of the skew-bending theory as it presently stands. In his book (Ref. 5.9), Hsu details the development of the theory of torsion as applied to concrete structures and how the skew-bending theory formed the basis of the initial ACI code provisions on torsion. The complexity of the torsional problem permits here only the brief discussion that follows.

The failure surface of the normal beam cross section subjected to bending moment M_u remains plane after bending, as shown in Figure 5.35(a). If a twisting moment T_u is also applied exceeding the capacity of the section, cracks develop on three sides of the beam cross section, and compressive stresses appear on portions of the fourth side along the beam. As torsional loading proceeds to the limit state at failure, a skewed failure surface results due to the combined torsional moment T_u and bending moment M_u. The neutral axis of the skewed surface and the shaded area in Figure 5.35(b) denoting the compression zone would no longer be straight, but subtend a varying angle θ with the original plane cross sections.

Prior to cracking, neither the longitudinal bars nor the closed stirrups have any appreciable contribution to the torsional stiffness of the section. At the postcracking stage of loading, the stiffness of the section is reduced, but its torsional resistance is considerably increased, depending on the amount and distribution of *both* the longitudinal bars and the transverse *closed* ties. It has to be emphasized that little additional torsional strength can be achieved beyond the capacity of the plain concrete in the beam unless both longitudinal torsion bars and transverse ties are used.

The skew-bending theory idealizes the compression zone by considering it to be of uniform depth. It assumes the cracks on the remaining three faces of the cross section to be uniformly spread, with the steel ties (stirrups) at those faces carrying the tensile forces at the cracks and the longitudinal bars resisting shear through dowel action with the concrete. Figure 5.36(a) shows the forces acting on the skewly bent plane. The polygon in Figure 5.36(b) gives the shear resistance F_c of the concrete, the force T_t of the active longitudinal steel bars in the compression zone, and the normal compressive block force C_c.

The torsional moment T_c of the resisting shearing force F_c generated by the shaded compressive block area in Figure 5.36(a) is thus

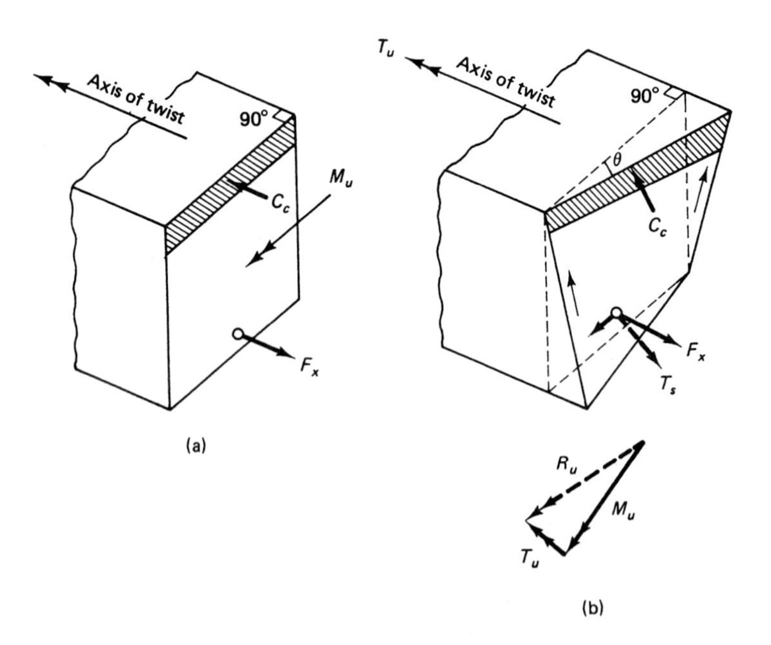

Figure 5.35 Skew bending due to torsion. (a) Bending before twist. (b) Bending and torsion.

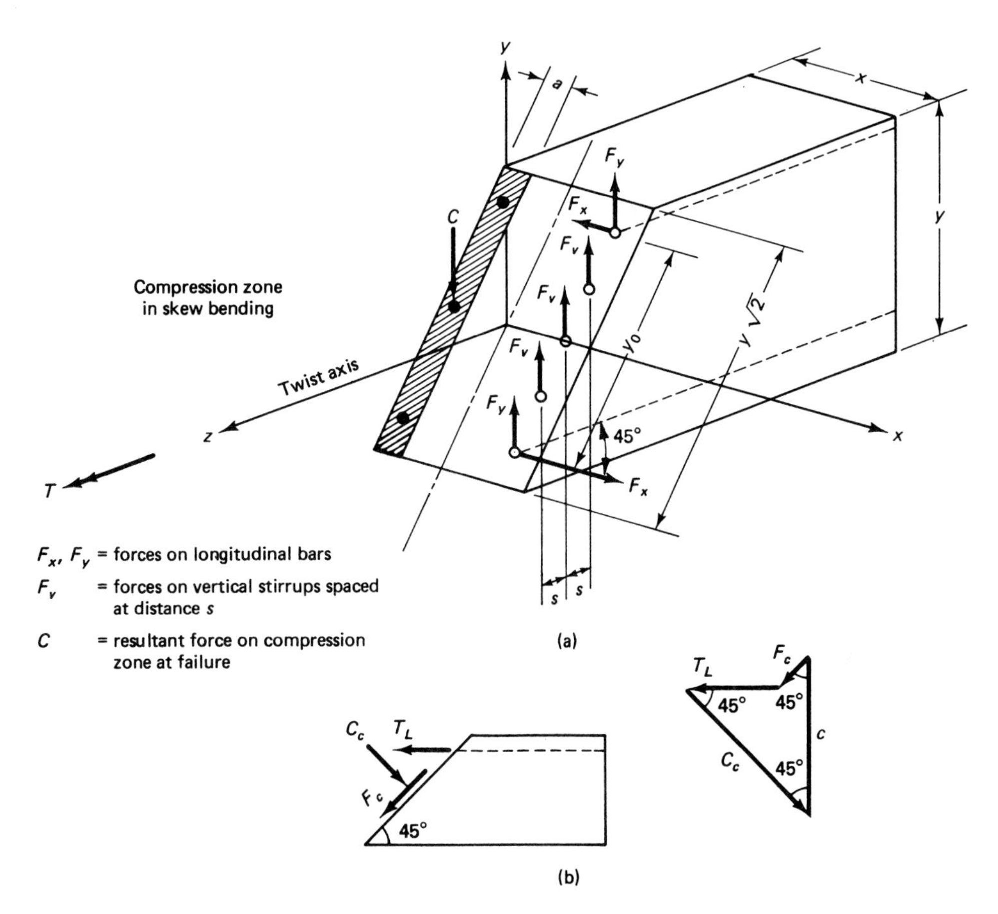

Figure 5.36 Forces on the skew-bent planes. (a) All forces acting on skew plane at failure. (b) Vector forces on compression zone.

$$T_c = \frac{F_c}{\cos 45^\circ} \times \text{its arm about forces } F_v \text{ in the figure}$$

or

$$T_c = \sqrt{2}F_c(0.8x) \tag{5.56a}$$

where x is the shorter side of the beam. Extensive tests (Refs. 5.9 and 5.10) to evaluate F_c in terms of the internal stress in concrete $k_1\sqrt{f'_c}$ and the geometrical torsional constants of the section k_2x^2y led to the expression

$$T_c = \frac{2.4}{\sqrt{x}}x^2y\sqrt{f'_c} \tag{5.56b}$$

5.17.2 Space Truss Analogy Theory

Space truss analogy theory was originally developed by Rausch and later extended by Lampert and Collins, with additional work by Hsu, Thurliman, Elfgren, and others. Further refinement was introduced by Rabbat and Collins (Ref. 5.13) on the variable angle space truss and Collins and Mitchell (Ref. 5.11).

Hsu (Refs. 5.18, and 5.19) proposed combining the equilibrium, compatibility and the softened constitutive laws of concrete in a unified theory that can predict with reasonable accuracy the shear and torsional behavior of beams (the softened truss model). The shear flow concept was utilized in deriving the relevant expressions for shear equilibrium. The space truss analogy is an extension of the model used in the design of the shear-resisting stirrups, in which the diagonal tension cracks, once they start to develop, are resisted by the stirrups. Because of the nonplanar shape of the cross sections due to the twisting moment, a space truss composed of the stirrups is used as the diagonal tension members, and the idealized concrete strips at a variable angle θ between the cracks are used as the compression members (struts), as shown in Figure 5.37.

It is assumed in this theory that the concrete beam behaves in torsion similarly to a thin-walled box with a constant shear flow in the wall cross section, producing a constant torsional moment. The use of hollow-walled sections rather than solid sections proved to give essentially the same ultimate torsional moment, provided that the walls were not too thin. Such a conclusion is borne out of tests which have shown that the torsional strength of the solid sections is composed of the resistance of the closed stirrup cage, consisting of the longitudinal bars and transverse stirrups, and the idealized concrete inclined compression struts in the plane of the cage wall. The compression struts are the inclined concrete strips between the cracks in Figure 5.37.

The CEB-FIP code is based on the space truss model. In this code, the effective wall thickness of the hollow beam is taken as $\frac{1}{6}D_0$, where D_0 is the diameter of the circle inscribed in the rectangle connecting the corner longitudinal bars, namely, $D_0 = x_0$ in Figure 5.37. In summary, the absence of the core does not affect the strength of such members in torsion—hence, the acceptability of the space truss analogy approach based on hollow sections.

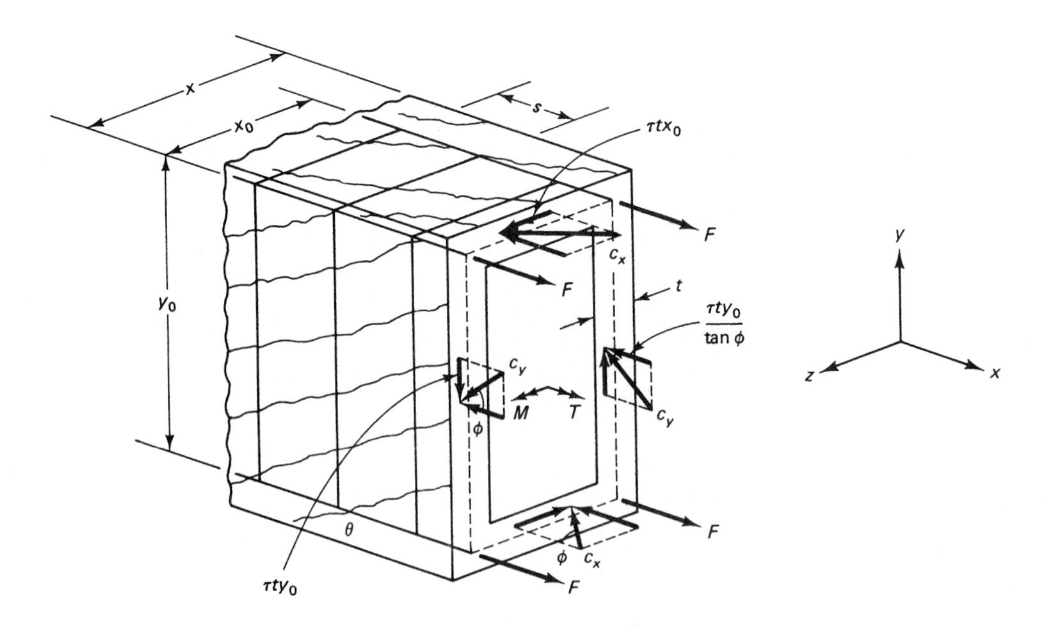

F = tensile force in each longitudinal bar

c_x = inclined compressive force on horizontal side

c_y = inclined compressive force on vertical side

τt = shear flow force per unit length of wall

Figure 5.37 Forces on hollow-box concrete surface by truss analogy.

5.17.3 Compression Field Theory

The compression field theory can be considered a special case of the general truss model theory. Elfgren proposed the compression fields to describe components of the plasticity truss model currently used in the European Code, and Collins and Mitchell modified the approach, proposing the terms to be subsequently discussed. The angle of inclination θ in Fig. 5.37 of the diagonal cracks or the compression struts between the diagonal cracks is not idealized to 45°, but rather uses limits based on the areas of the longitudinal tension steel and the transverse torsional web steel (inclined or vertical closed stirrups or ties). Figure 5.38 points up the fact that the torsional force is resisted by the tangential components of the diagonal compression struts, which produce a shear flow q around the perimeter (Ref. 5.11).

Assuming that concrete carries no tension after cracking, and that torsional shear is carried by the field of diagonal compression struts, the angle of inclination θ of these struts can be defined as

$$\tan^2 \theta = \frac{\epsilon_l + \epsilon_d}{\epsilon_t + \epsilon_d} \tag{5.57}$$

where ϵ_l = longitudinal tensile strain in the main bars A
ϵ_t = transverse tensile strain in bars B
ϵ_d = diagonal compression strain

The area A_o in the figure enclosed by the shear flow q can be obtained as

$$A_o = A_{oh} - \frac{a_0}{2} p_h \tag{5.58}$$

where A_{oh} = area enclosed by the centerline of the hoop
p_h = hoop centerline perimeter

Photo 5.9 Reinforced concrete beams in torsion, testing setup. (*Courtesy,* Thomas T. C. Hsu.)

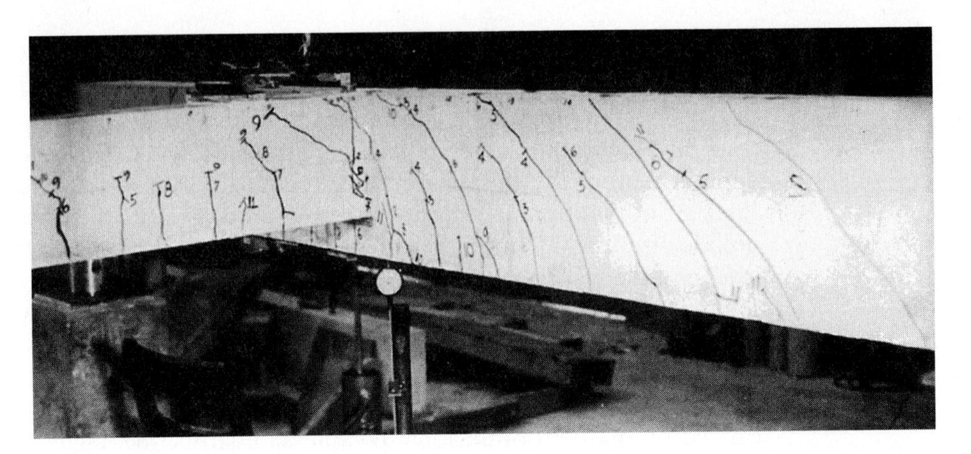

Photo 5.10 Closeup of torsional cracking of beams in the preceding photograph. (*Courtesy,* Thomas T. C. Hsu.)

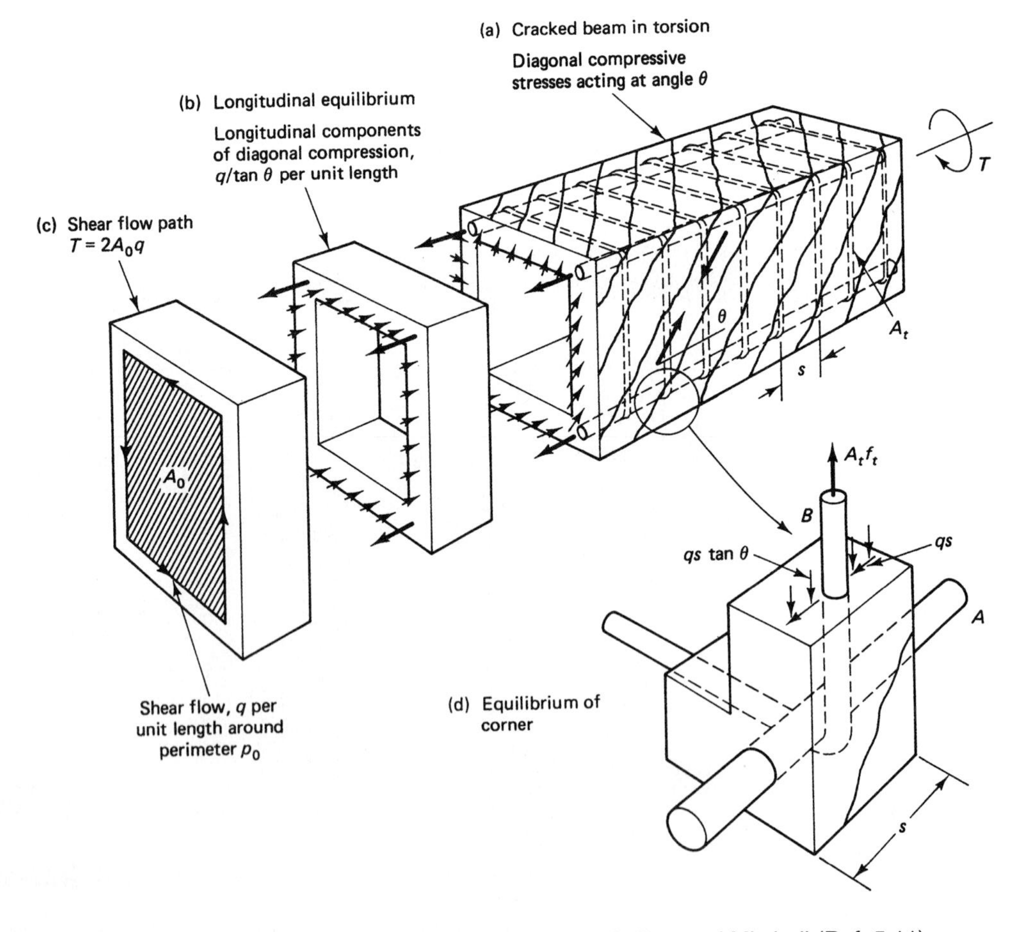

Figure 5.38 Compression field truss model by Collins and Mitchell (Ref. 5.11).

a_o = compression block depth (identical to the depth a of the equivalent rectangular block in flexure)

The equivalent wall thickness t_d in the analysis of the twisted beam is shown in Figure 5.39, and the depth a_0 of the compressive block is defined in Equation 5.61.

The diagonal torsional cracks, as well as the exposed transverse ties after spalling of the concrete cover at torsion failure, are demonstrated in Figure 5.40.

The transverse and longitudinal strains in the steel at the nominal torsional moment T_n can be respectively defined as

$$\epsilon_t = \left(\frac{0.85\beta_1 f_c' A_o}{\tau_n A_{oh} \tan\theta} - 1\right) 0.003 \tag{5.59a}$$

$$\epsilon_l = \left(\frac{0.85\beta_1 f'_c A_o}{\tau_n A_{oh}} \tan\theta - 1\right) 0.003 \tag{5.59b}$$

where the nominal torsional shear stress is

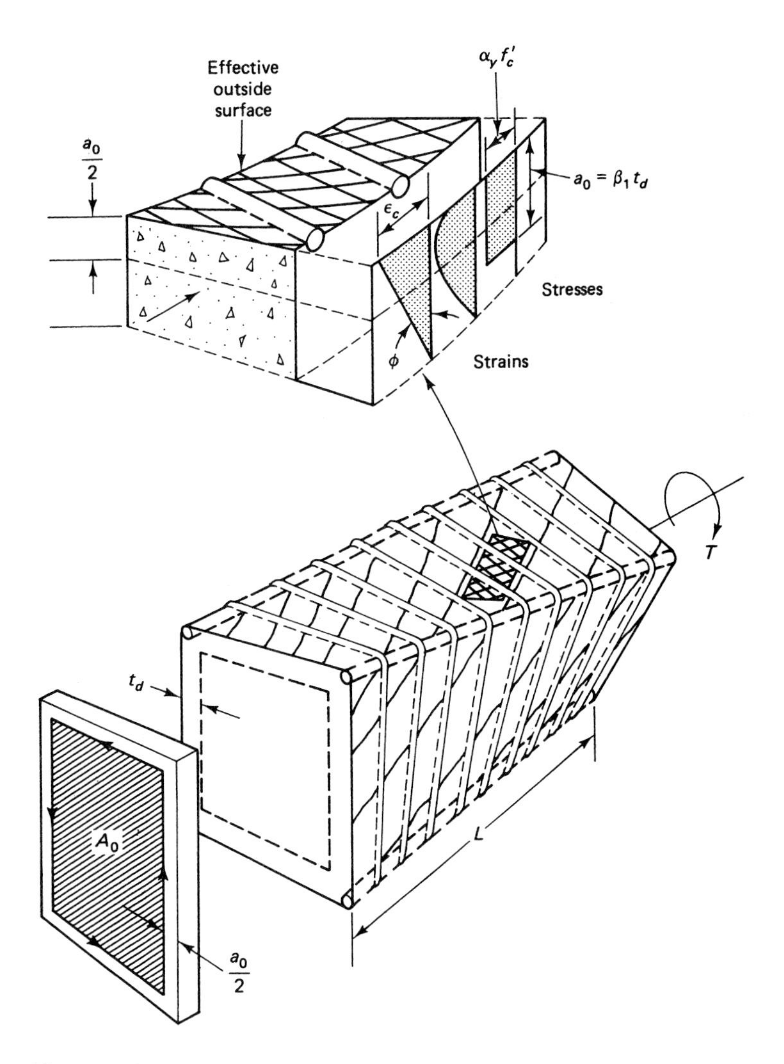

Figure 5.39 Effective thickness t_d and compression block depth a_o.

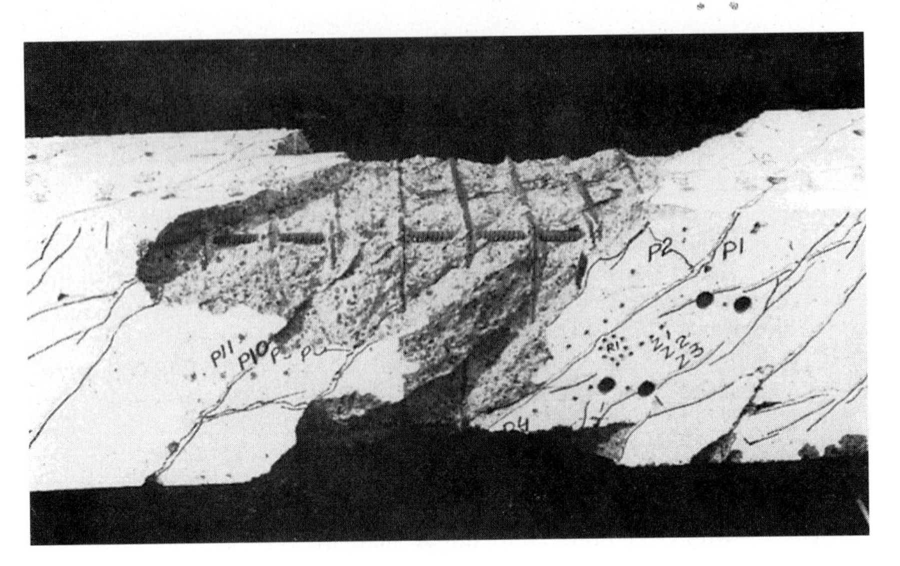

Figure 5.40 Torsional failure of web-reinforced beam after spalling of cover (Collins and Mitchell, Ref. 5.11).

$$\tau_n = \frac{T_n p_h}{A_{oh}^2} \tag{5.60}$$

The area A_0 enclosed by the shear flow can be obtained from Equation 5.60 and the following expression for the compression block depth a_0 in torsion according to Collins and Mitchell (Ref. 5.11):

$$a_0 = \frac{A_{oh}}{p_h}\left[1 - \sqrt{1 - \frac{T_n p_h}{0.85 f_c' A_{oh}^2}\left(\tan\theta + \frac{1}{\tan\theta}\right)}\right] \tag{5.61}$$

where T_n is the nominal torsional moment strength at the limit state at failure. It should be noted that Equations 5.58, 5.59, 5.60, 5.61, and 5.63 are based on the assumptions: (1) spalling of concrete cover, and (2) using non-softened stress-strain curve of the concrete. These assumptions do not seem to be correct, considering the actual torsional behavior of the concrete element.

For combined torsion and shear, the shearing stress at nominal strengths T_n and V_n in Equation 5.61 becomes

$$\tau_n = \frac{T_n p_h}{A_{oh}^2} + \frac{V_n - V_p}{b_v d_v} \tag{5.62}$$

where V_p = vertical component on the prestressing force.
b_v = minimum effective *web* width within shear depth d_v after spalling of cover. Subtract the diameters of ducts from the web width if ungrouted, or half the diameter of ducts for grouted tendons.
d_v = effective shear depth. This can be taken as the flexural lever arm, but *not less* than the vertical distance between the centers of bars or prestressing tendons at the corners of the stirrups.

The predicted values of the compressive strut inclination θ in Figure 5.38 range between 24° for pure torsion and 90° for pure flexure. Hence, the lower the value of θ selected for a given torque, the less is the transverse hoop steel needed and the more is the required

Photo 5.11 Dauphin Island Bridge, Alabama, assembly of segmental bridge units. (*Courtesy*, Prestressed Concrete Institute.)

area of longitudinal steel. Since transverse closed stirrups or ties are more expensive than longitudinal bars, a choice of lower values of θ is more economical in design.

The compression field theory assumes that the geometrical properties of the designed section are chosen on the basis of yielding of the transverse web reinforcement and longitudinal steel *prior* to diagonal crushing of the concrete. Consequently, the transverse strain ϵ_t in Equations 5.57 and 5.58a should be taken as the yield strain ϵ_{ty}.

The range of the compression strut angle θ in degrees can be evaluated from

$$10 + \frac{35(\tau_n/f_c')}{0.42 - 50\epsilon_l} < \theta < 80 - \frac{35(\tau_n/f_c')}{0.42 - 65\epsilon_{ty}} \tag{5.63}$$

5.17.4 Plasticity Equilibrium Truss Theory

Hsu (Ref. 5.17, 5.18) proposed combining the equilibrium, compatibility and the softened constitutive laws of concrete in a unified theory that can predict with reasonable accuracy the shear and torsional behavior of beams (The softened truss model). The shear flow concept is utilized in deriving the relevant expressions for shear equilibrium.

5.17.4.1 Equilibrium in element shear. A unit square membrane element of thickness t is subjected to shear flow q due to pure shear in Figure 5.41 (Hsu, Ref. 5.18). Reinforcement in both the longitudinal (E-W) direction l and transverse (N-S) direction t is subjected to a unit stress f_l/s_l and f_v/s respectively such that the shear flow q can be defined by the equilibrium equations

$$q = (F_l) \tan \theta \tag{5.64a}$$

where unit $F_l = A_l f_l / s_l$ and

$$q = (F_l) \cot \theta \tag{5.64b}$$

where unit $F_t = A_t f_v / s$

A_l and A_t are the cross-sectional areas of the reinforcement, and s_l and s are the spacings in the l and t directions, respectively.

From the geometry of the triangles in Figure 5.41 the shear flow can also be defined as

$$q = (f_D t) \sin \theta \cos \theta \tag{5.65}$$

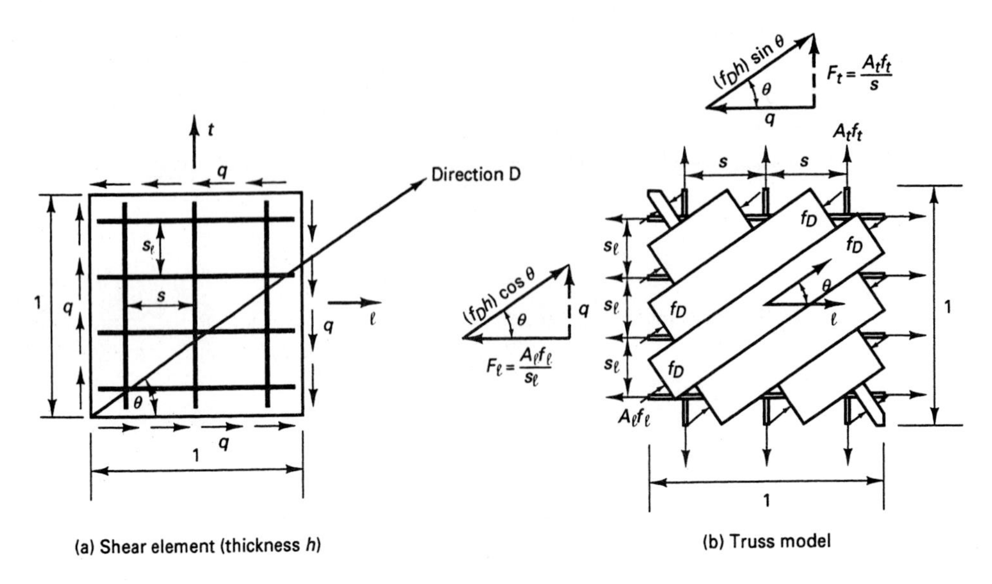

Figure 5.41 Equilibrium forces in element shear (Ref. 5.18).

If the reinforcement in both directions is assumed to have yielded, Equations 5.64a, b and 5.65 give

$$\tan\theta = \sqrt{\frac{F_{ty}}{F_{ly}}} \tag{5.66a}$$

and

$$q_y = \sqrt{F_{ly}F_{ty}} \tag{5.66b}$$

where the subscript y denotes the yielding of the reinforcement.

5.17.4.2 Equilibrium in element torsion. The case of a hollow tube of any shape and variable thickness is considered (Figure 5.42). It is subjected to pure torsion. St.-Venant's theory stipulates that the cross-sectional shape remains unchanged in elastic small defor-

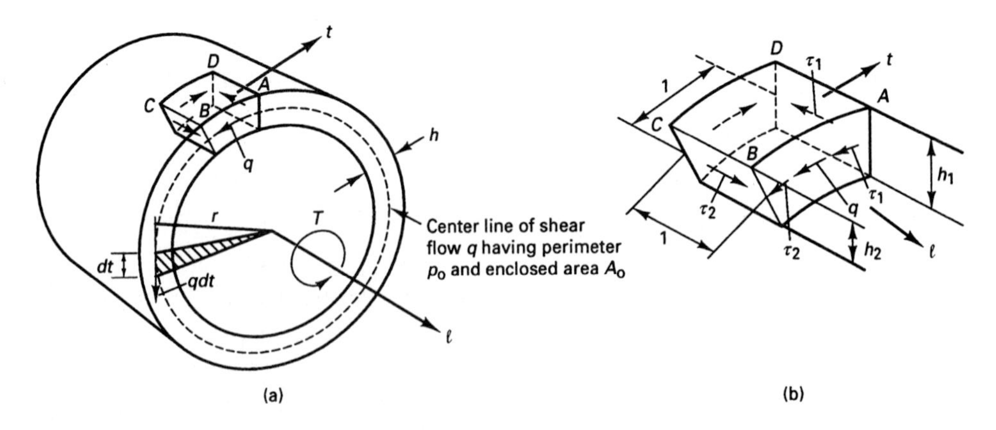

Figure 5.42 Hollow tube equilibrium torsion forces. (a) Section of tube subjected to torsion T. (b) Unit shear element from tube wall of varying thickness h. Note: l and t denote the longitudinal and transverse directions.

mations and the warping deformation perpendicular to the cross-section would be the same along the member's axis. Hence, it can be assumed that only shear stresses develop in the tube wall in the form of shear flow q in Figure 5.42a and that the in-plane normal stresses in the wall vanish. If an infinitesimal wall element ABCD is isolated as in Figure 5.42b, the shear flow in the l direction has to be equal to the shear flow in the t direction or

$$\tau_l t_1 = \tau_t t_2 \tag{5.67}$$

On this basis, the shear flow q is considered constant throughout the cross-section (Ref. 5.18). The torsional force over an infinitesimal distance dt along the shear flow path is qdt so that the torsional resistance to the external torsional moment T in Figure 5.42a becomes

$$T = q\oint r\,dt \tag{5.68}$$

It can be seen from Figure 5.42a that rdt in the integral is equal to *twice* the area of the shaded triangle formed by r and dt. A summation of the total area around the cross section gives

$$\oint r\,dt = 2A_0 \tag{5.69}$$

where A_0 = cross-sectional area bounded by the shear flow center line. Substituting $2A_0$ into Equation 5.68 gives

$$q = \frac{T}{2A_0} \tag{5.70}$$

By neglecting warping, the shear element subjected to pure torsion in the tube wall of Figure 5.42a becomes identical to the membrane shear element in Figure 5.41a. Hence, substituting for the shear flow q from Equation 5.70 into Equations 5.64a, b, and 5.65, the following three equations of equilibrium for torsion result in

$$T = \frac{\overline{F}_l}{p_0}(2A_0)\tan\theta \tag{5.71a}$$

where $\overline{F}_l = F_l\,p_0$ and p_0 = perimeter of the shear flow path. $\overline{F}_l$ is the *total* longitudinal force due to torsion.

$$T = F_t(2A_0)\cot\theta \tag{5.71b}$$

$$T = (f_D t)(2A_0)\sin\theta\cos\theta \tag{5.71c}$$

Equation 5.71b can be written at yield as

$$T_n = \frac{2A_0\,A_t\,f_{yv}}{s}\cot\theta \tag{5.72}$$

where T_n is the maximum torsional moment strength.

The required torsional reinforcement in the transverse and longitudinal directions become

$$A_t = \frac{T_n s}{2A_0\,f_{yv}\cot\theta} \tag{5.73}$$

$$A_{l_1} = \frac{A_t}{s}\left(\frac{f_{yv}}{f_{yl}}\right)(s_l\cot^2\theta) \tag{5.74a}$$

where A_{l_1} is the area of one longitudinal bar. If s_l as the longitudinal reinforcement spacing represents the perimeter p_h of the center-line of the outermost closed transverse torsional reinforcement, then

$$A_l = \frac{A_t}{s} p_h \left(\frac{f_{yv}}{f_{yl}} \right) \cot^2 \theta \tag{5.74b}$$

where A_l = *total* area of all longitudinal torsional steel in the section.

5.17.4.3 Shear-torsion-bending interaction. Consider the rectangular box in Figures 5.37 and 5.43. The shear flow q will not be the same on the four walls of the box when subjected to combined shear and torsion as shown in Figure 5.43. Failure can precipitate in two distinct modes:

(a) yielding of the longitudinal bottom tension steel and the transverse stirrups,
(b) yielding of the longitudinal top compression steel and the transverse stirrups,

(a) Bottom Tension Steel Yielding

If the failure mode is caused by yielding of the longitudinal bottom stringer (tensile steel) and the transverse stirrups due to combined shear and torsion, the following expression can be derived from equilibrium (Ref. 5.18)

$$\frac{M}{F_B y_0} + \left(\frac{V}{2y_0} \right)^2 \frac{y_0}{F_B} \frac{s}{A_t f_v} + \left(\frac{T}{2A_0} \right)^2 \frac{(y_0 + x_0)}{F_B} \frac{s}{A_t f_v} = 1 \tag{5.75}$$

if M_0, V_0, and T_0 are the moments and forces acting *alone,* they can be defined as follows

$$M_0 = F_B y_0 \tag{5.76a}$$

$$V_0 = 2y_0 \sqrt{\left(\frac{F_T}{y_0} \right) \frac{A_t f_v}{s}} \quad \text{for a two-web box} \tag{5.76b}$$

$$T_0 = 2A_0 \sqrt{\left(\frac{2F_T}{p_0} \right) \frac{A_t f_v}{s}} \tag{5.76c}$$

where $p_0 = 2(y_0 + x_0)$

$$R = \frac{F_T}{F_B} \tag{5.76d}$$

A nondimensional interaction surface relationship can be obtained by introducing Equation 5.76 into Equation 5.75 such that

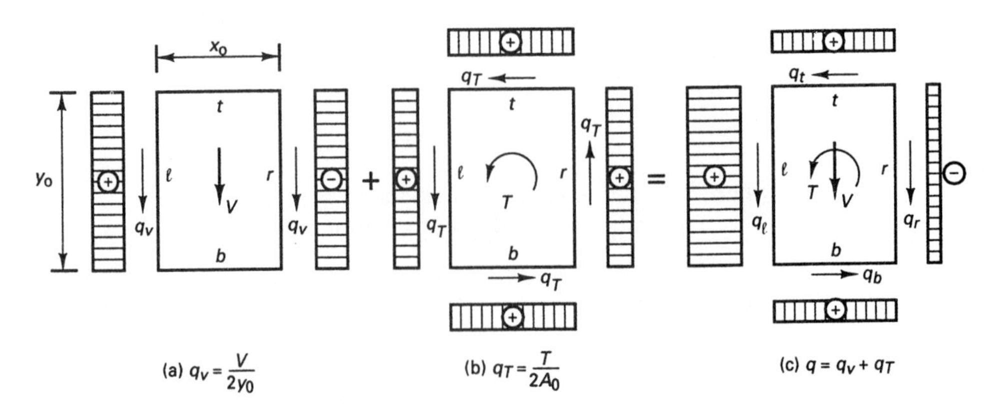

Figure 5.43 Hollow section shear flow q due to combined shear and torsion.

Photo 5.12 The Woodley Park Zoo station, Washington, D.C. (*Courtesy,* H. Wilden & Assoc.).

$$\left(\frac{M}{M_0}\right) + \left(\frac{V}{V_0}\right)^2 R + \left(\frac{T}{T_0}\right)^2 R = 1 \tag{5.77a}$$

(b) Top Compression Steel Yielding

If the failure mode is caused by yielding of the longitudinal top chord (compression steel) and the transverse stirrups, Equation 5.77a becomes

$$-\left(\frac{M}{M_0}\right)\frac{1}{R} + \left(\frac{V}{V_0}\right)^2 + \left(\frac{T}{T_0}\right)^2 = 1 \tag{5.77b}$$

From both Equations 5.77a and 5.77b the interaction of V and T is *circular* for a constant bending moment M for both failure surfaces. The intersection of the two failure surfaces for these two failure modes forms a peak interaction curve between V and T that Equations 5.77a and b give

$$\left(\frac{V}{V_0}\right)^2 + \left(\frac{T}{T_0}\right)^2 = \frac{1 + R}{2R} \tag{5.78a}$$

Equation 5.78 for $R = 0.25$, 0.5, and 1.0 on the peak planes gives the circular plots shown in Figure 5.44.

A third mode of failure is caused by yielding in the top bar, in the bottom bar, and in the transverse reinforcement, all on the side where shear flows due to shear and torsion are additive i.e. left wall (Ref. 5.15). A modified form of Equation 5.78 results as follows

$$\left(\frac{V}{V_0}\right)^2 + \left(\frac{T}{T_0}\right)^2 + \sqrt{2}\left(\frac{VT}{V_0 T_0}\right) = \frac{1 + R}{2R} \tag{5.78b}$$

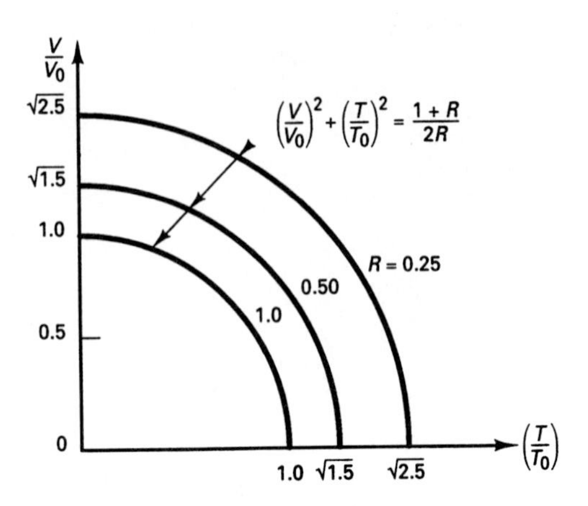

Figure 5.44 Shear-torsion interaction diagram.

The factored torsional moment strength, ϕT_n, must equal or exceed the external torsion, T_u, due to the factored loads. In the calculation of T_n(ACI 318–99, Ref. 5.1), all the torque is assumed to be resisted by the closed stirrups and longitudinal steel with the torsional moment T_c resisted by the concrete compression struts assumed as zero. At the same time, the shear resisted by concrete, V_c is assumed to be unchanged by the presence of torsion. This simplification eliminates the need for the rigor of the lengthy interaction expressions for V, T, and M used in the previous codes. In summary, the web reinforcement for shear is determined by the value of $V_s = V_n - V_c$ while the web reinforcement for torsion by the T_n value alone, where $T_n = T_u/\phi$ with $\phi = 0.85$.

5.17.5 Design of Prestressed Concrete Beams Subjected to Combined Torsion, Shear, and Bending in Accordance with the ACI 318–99 Code

Adjusting in the equilibrium truss model of Section 5.17.4, the following are the ACI 318 Code provisions for designing the longitudinal and transverse reinforcement in prestressed elements.

5.17.5.1 Compatibility torsion. In statically indeterminate systems, stiffness assumptions, compatibility of strains at the joints and redistribution of stresses may affect the stress resultants, leading to a reduction of the resulting torsional shearing stresses. A reduction is permitted in the value of the factored moment used in the design of the member if part of this moment can be redistributed to the intersecting members. The ACI Code permits a maximum factored torsional moment at the critical section $h/2$ from the face of the supports for prestressed concrete members as follows:

where A_{cp} = area enclosed by outside perimeter of concrete cross section $= x_0 y_0$
p_{cp} = outside perimeter of concrete cross section A_{cp}, in. $= 2(x_0 + y_0)$

$$T_u = \phi 4\sqrt{f'_c}\left(\frac{A_{cp}^2}{p_{pc}}\right)\sqrt{1 + \frac{\bar{f}_c}{4\sqrt{f'_c}}} \tag{5.79}$$

where $\bar{f}_c$ = average compressive stress in the concrete at the centroidal axis due to effective prestress only after allowing for all losses. $\bar{f}_c$ is denoted in the ACI Code as f_{pc}.

Neglect of the full effect of the total value of external torsional moment in this case does not, in effect, lead to failure of the structure but may result in excessive cracking if $\phi 4\sqrt{f'_c}(A^2_{cp}/p_{cp})$ is considerably smaller in value than the actual factored torque.

If the actual factored torque is less than that given in Equation 5.79, the beam has to be designed for the lesser torsional value. Torsional moments are neglected however if for prestressed concrete

$$T_u < \phi\sqrt{f'_c}\left(\frac{A^2_{cp}}{p_{cp}}\right)\sqrt{1+\frac{\bar{f}_c}{4\sqrt{f'_c}}} \tag{5.80}$$

5.17.5.2 Torsional moment strength. The size of the cross section is chosen on the basis of reducing unsightly cracking and preventing the crushing of the surface concrete caused by the inclined compressive stresses due to shear and torsion defined by the left hand side of the expressions in Equation 5.81. The geometrical dimensions for torsional moment strength in both reinforced and prestressed members are limited by the following expressions.

(a) Solid Sections

$$\sqrt{\left(\frac{V_u}{b_w d}\right)^2+\left(\frac{T_u p_h}{1.7A^2_{0h}}\right)} \le \phi\left(\frac{V_c}{b_w d}+8\sqrt{f'_c}\right) \tag{5.81}$$

(b) Hollow Sections

$$\left(\frac{V_u}{b_w d}\right)+\left(\frac{T_u p_h}{1.7A^2_{0h}}\right) \le \phi\left(\frac{V_c}{b_w d}+8\sqrt{f'_c}\right) \tag{5.82}$$

where A_{0h} = area enclosed by the centerline of the outermost closed transverse torsional reinforcement, sq. in.

p_h = perimeter of centerline of outermost closed transverse torsional reinforcement, in.

The area A_{0h} for different sections are given in Figure 5.45. Figures 5.46 and 5.47 give guidance to the determination of the area A_{0h} and the shear flow area $A_0 \cong 0.85A_{0h}$ in Equation 5.84(a).

The sum of the stresses at the left hand side of Equation 5.82 should not exceed the stresses causing shear cracking plus $8\sqrt{f'_c}$. This is similar to the limiting strength $V_s \le 8\sqrt{f'_c}$ for shear without torsion.

5.17.5.3 Hollow sections wall thickness. The shear stresses due to shear and to torsion both develop in the walls of the hollow section as seen in Figure 5.48a. Note that in a solid section the shear stresses due to torsion still concentrate in the outer zones of the section as in Figure 5.48b and as discussed in Section 5.17.3.1.

If the wall thickness in the hollow section varies around its perimeter, the section geometry has to be evaluated at such a location where the left-hand side of Equation 5.82 has a maximum value. Also, if the wall thickness $t < A_{0h}/p_h$ the left-hand side of Equation 5.82 should be taken as

$$\left(\frac{V_u}{b_w d}\right)+\left(\frac{T_u}{1.7A_{0h}t}\right)$$

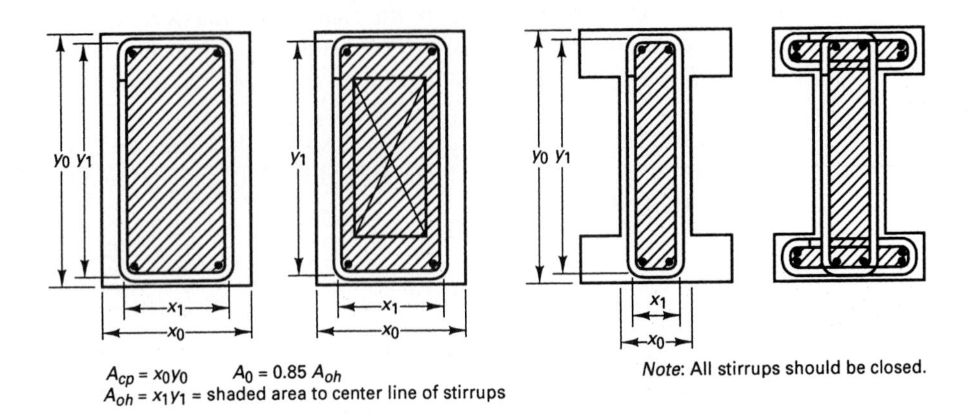

Figure 5.45 Torsional geometric parameters.

The wall thickness t is the thickness where stresses are being checked.

$$V_c = \left(0.6\lambda\sqrt{f'_c} + 700\frac{V_u d}{M_u}\right)b_w d; \qquad \frac{V_u d}{M_u} \le 1.0$$

$$\le 1.7\lambda\sqrt{f'_c}\,b_w d \tag{5.83}$$

$$\ge 5.0\lambda\sqrt{f'_c}\,b_w d$$

where $f_{pe} > 0.4f_{pu}$.

5.17.5.4 Torsional web reinforcement. As indicated in Section 5.17.3, meaningful additional torsional strength due to the addition of torsional reinforcement can be achieved only by using both stirrups and longitudinal bars. Ideally, *equal* volumes of steel in both the closed stirrups and the longitudinal bars should be used so that both participate equally in resisting the twisting moments. This principle is the basis of the ACI expressions for proportioning the torsional web steel. If s is the spacing of the stirrups, A_l is the total cross-sectional area of the longitudinal bars, and A_t is the cross section of one stirrup leg, the transverse reinforcement for torsion has to be based on the full external torsional moment strength value T_n, namely, (T_u/ϕ) where

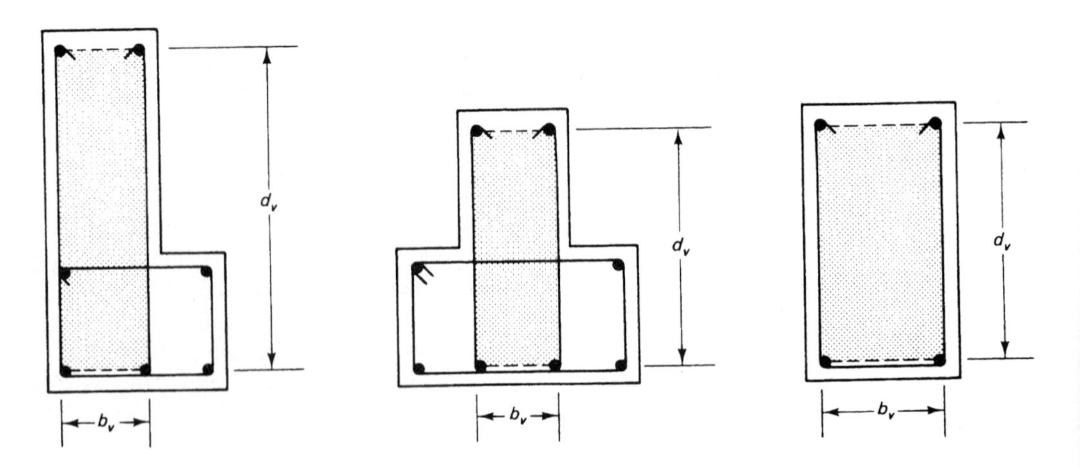

Figure 5.46 Shear-flow geometry and effective shear area.

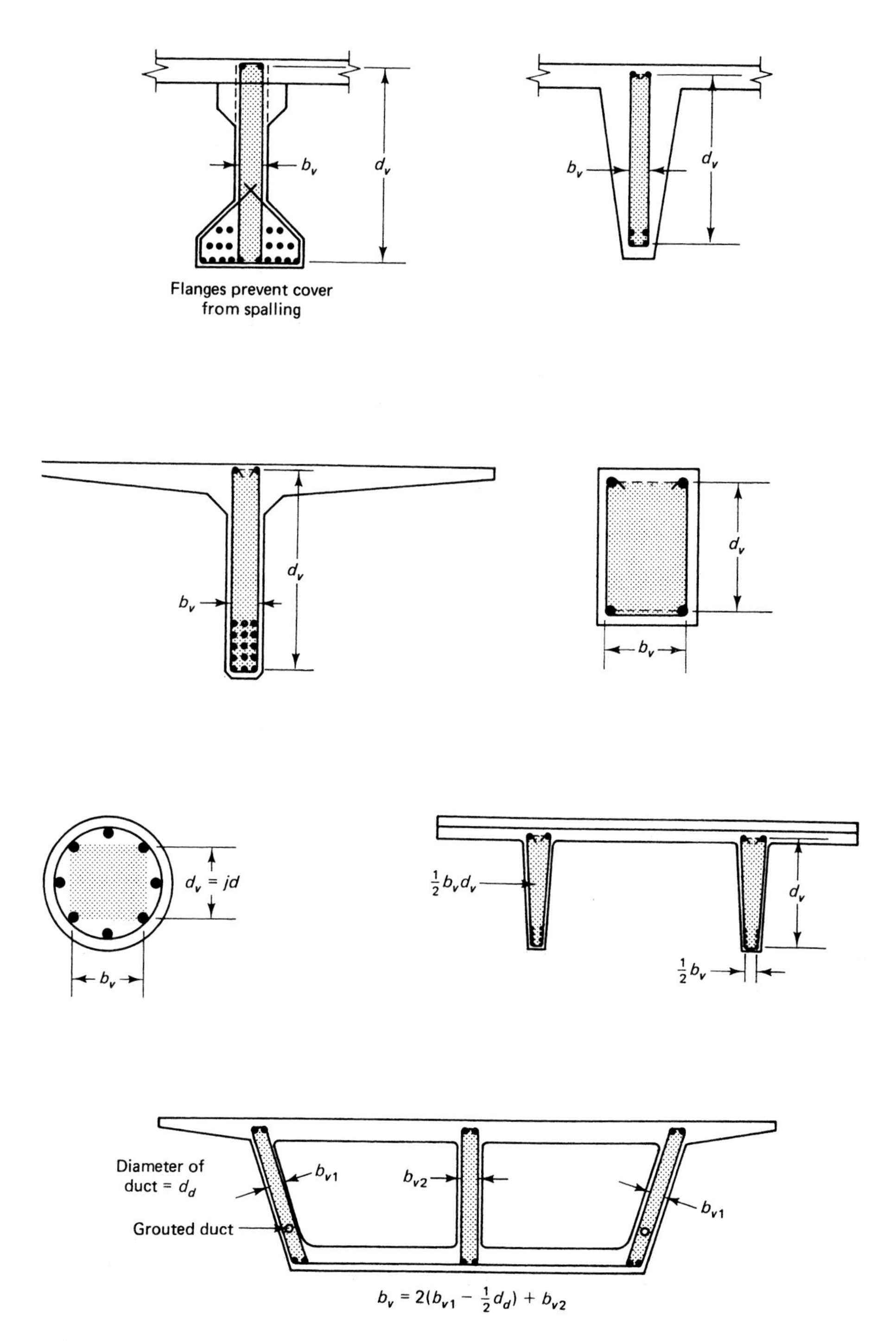

Figure 5.47 Effective shear width and depth of typical prestressed concrete sections.

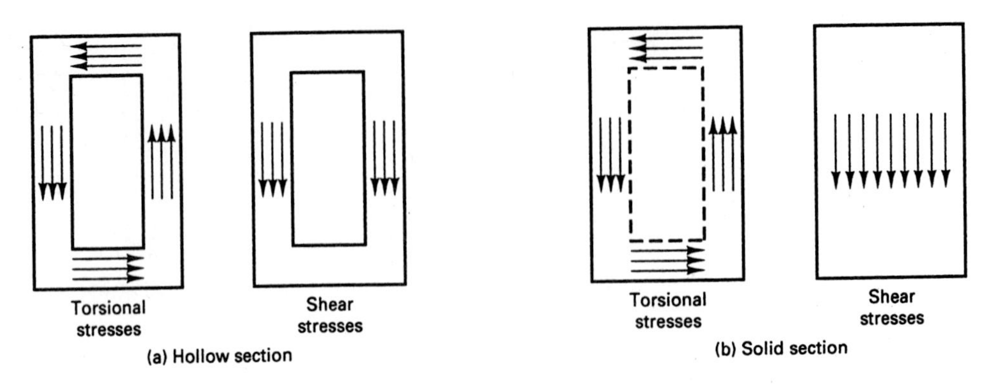

Figure 5.48 Superposition of torsional and shear stresses. (a) Directly additive occurring in the left wall of the box (Equation 7.30b). (b) Torsion acts on "tubular" outer-wall section while shear stress acts on the full width of solid section: stresses combined using square root of sum of squares (Equation 7.30b).

$$T_n = \frac{2A_0 A_t f_{yv}}{s} \cot\theta \tag{5.84a}$$

(See the derivation of Equation 5.72)

A_0 = gross area enclosed by the shear flow path, sq. in.
A_t = cross-sectional area of one leg of the transverse closed stirrups, sq. in.
f_{yv} = yield strength of closed transverse torsional reinforcement not to exceed 60,000 psi.
θ = angle of the compression diagonals (struts) in the space truss analogy for torsion (See Figure 5.39).

Transposing terms in Equation 5.84b, the transverse reinforcement area becomes

$$\frac{A_t}{s} = \frac{T_n}{2A_0 f_{yv} \cot\theta} \tag{5.84b}$$

The area A_0 has to be determined by analysis (Ref. 7.14 and 7.15) except that the ACI 318 Code permits taking $A_0 = 0.85A_{0h}$ in lieu of the analysis.

As discussed in Section 5.17.3, the factored torsional resistance ϕT_n must equal or exceed the factored external torsional moment T_u. All the torsional moment is assumed in the ACI 318–99 code to be resisted by the closed stirrups and the longitudinal steel with the torsional resistance, T_c, of the concrete disregarded, namely $T_c = 0$. The shear V_c resisted by the concrete is assumed to be unchanged by the presence of torsion (see Section 5.17.3.2).

The angle θ subtended by the concrete compression diagonals (struts) should not be taken smaller than 30° nor larger than 60°. It can be obtained by analysis as detailed in Ref. 5.17 and 5.18 by Hsu. The additional longitudinal reinforcement for torsion should not be less than

$$A_l = \frac{A_t}{s} p_h \left(\frac{f_{yv}}{f_{yl}}\right) \cot^2\theta \tag{5.85}$$

where f_{yl} = yield strength of the longitudinal torsional reinforcement, not to exceed 60,000 psi.

The same angle θ should be used in both Equations 5.84 and 5.85. It should be noted that as θ gets smaller, the amount of stirrups required by Equation 5.84 decreases. At the same time the amount of longitudinal steel required by Equation 5.85 increases.

In lieu of determining the angle θ by analysis, the ACI Code allows a value of θ equal to

(i) 45° for non-prestressed members or members with less prestress than in (ii),

(ii) 37.5° for prestressed members with an effective prestressing force larger than 40 percent of the tensile strength of the longitudinal reinforcement.

The PCI (Ref. 5.12) recommends computing the value of θ from the expression:

$$\cot\theta = \frac{T_{u}/\phi}{1.7A_{0h}(A_t/s)f_{yv}} \tag{5.86}$$

5.17.5.5 Minimum torsional reinforcement. It is necessary to provide a minimum area of torsional reinforcement in all regions where the factored torsional moment T_u exceeds the value given by Equation 5.80. In such a case, the minimum area of the required transverse closed stirrups is

$$A_v + 2A_t \geq \frac{50b_w s}{f_{yv}} \tag{5.87}$$

The maximum spacing should not exceed the smaller of $p_h/8$ or 12 in.

The maximum total area of the additional longitudinal torsional reinforcement should be determined by

$$A_{l,\min} = \frac{5\sqrt{f'_c}A_{cp}}{f_{yl}} - \left(\frac{A_t}{s}\right)p_h\frac{f_{yv}}{f_{yl}} \tag{5.88}$$

where A_t/s should not be taken less than $25b_w/f_{yv}$. The additional longitudinal reinforcement required for torsion should be distributed around the perimeter of the closed stirrups with a maximum spacing of 12 in. The longitudinal bars or tendons should be placed inside the closed stirrups and at least one longitudinal bar or tendon in each corner of the stirrup. The bar diameter should be at least $\frac{1}{16}$ of the stirrup spacing but not less than a No. 3 bar. Also, the torsional reinforcement should extend for a minimum distance of $(b_t + d)$ beyond the point theoretically required for torsion because torsional diagonal cracks develop in a helical form extending beyond the cracks caused by shear and flexure. b_t is the width of that part of cross section containing the stirrups resisting torsion. The critical section in beams is at a distance d from the face of the support for reinforced concrete elements and at $h/2$ for prestressed concrete elements, d being the effective depth and h the total depth of the section.

5.17.6 SI–Metric Expressions for Torsion Equations

In order to design for combined torsion and shear using the SI (System International) method, the following equations replace the corresponding expressions in the PI (Pound-Inch) method

$$T_u \leq \frac{\phi\sqrt{f'_c}}{3}\left(\frac{A_{cp}^2}{p_{cp}}\right)\sqrt{1 + \frac{3f_c}{\sqrt{f'_c}}} \tag{5.79}$$

$$T_u \leq \frac{\phi\sqrt{f'_c}}{12}\left(\frac{A_{cp}^2}{p_{cp}}\right)\sqrt{1+\frac{3f_c}{\sqrt{f'_c}}} \tag{5.80}$$

$$\sqrt{\left(\frac{V_u}{b_w d}\right)^2 + \left(\frac{T_u p_h}{1.7A_{0h}^2}\right)^2} \leq \phi\left(\frac{V_c}{b_w d} + \frac{8\sqrt{f'_c}}{12}\right) \tag{5.81}$$

$$\left(\frac{V_u}{b_w d}\right) + \left(\frac{T_u p_n}{1.7A_{0h}^2}\right) \leq \phi\left(\frac{V_c}{b_w d} + \frac{8\sqrt{f'_c}}{12}\right) \tag{5.82}$$

$$V_c = \left(\lambda\sqrt{f'_c}/20 + \frac{5V_u d}{M_u}\right)b_w d \tag{5.83}$$

$$\geq (0.17\lambda\sqrt{f'_c})b_w d$$

$$\leq (0.4\lambda\sqrt{f'_c})b_w d$$

and

$$V_u d/M_u \leq 1.0$$

$$T_n = \frac{2A_0 A_t f_{yv}}{s}\cot\theta \tag{5.84a}$$

where f_{yv} is in MPa, s in millimeter, A_0, A_t in mm^2 and T_n in kN-m.

$$A_t = \frac{T_n}{2A_0 f_{yv}\cot\theta} \tag{5.84b}$$

$$A_l = \frac{A_t}{s}p_h\left(\frac{f_{yv}}{f_{yl}}\right)\cot^2\theta \tag{5.85}$$

where f_{yv} and f_{yl} are in MPa, p_h and s in millimeters and A_l, A_t in mm^2.

$$A_v = \frac{0.35b_w s}{f_y} \tag{5.86}$$

$$A_{l,\min} = \frac{5\sqrt{f'_c}A_{cp}}{12f_{yl}} - \left(\frac{A_t}{s}\right)p_h\left(\frac{f_{yv}}{f_{yl}}\right) \tag{5.87}$$

where A_t/s should not be taken less than 0.175 b_w/f_{yv}. Maximum allowable spacing of transverse stirrups is the smaller of $\frac{1}{8}P_h$ or 300 mm, and bars should have a diameter of at least $\frac{1}{16}$ of the stirrups spacing but not less than No. 10 M bar size. Max. f_{yv} or f_{yl} should not exceed 400 MPa. Min. A_{vt} the smaller of

$$A_{vt} \geq \frac{0.35b_w}{f_y} \text{ where } b_w, d \text{ and } s \text{ are in millimeters}$$

$$\geq \frac{A_{ps}f_{pu}}{80f_y d}\sqrt{\frac{d}{b_w}}$$

5.18 DESIGN PROCEDURE FOR COMBINED TORSION AND SHEAR

The following is a summary of the recommended sequence of design steps. A flowchart describing the sequence of operations in graphical form is shown in Figure 5.49.

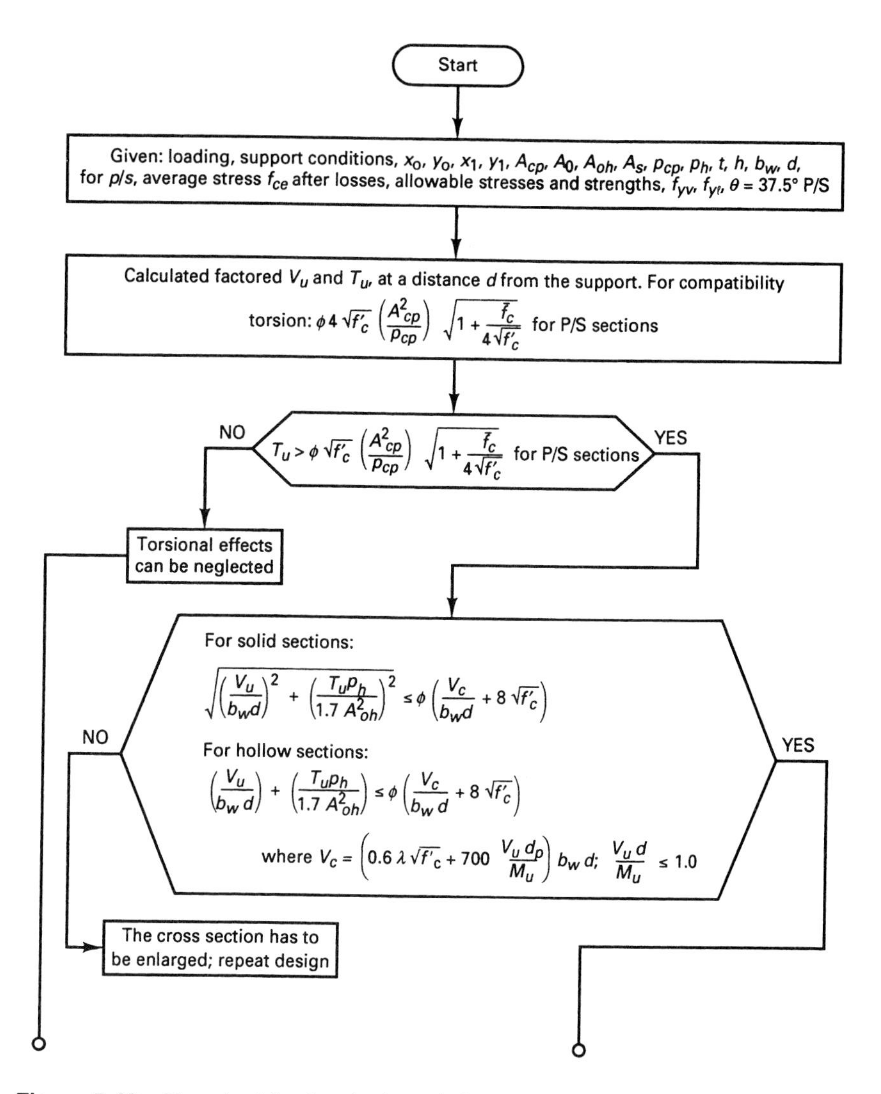

Figure 5.49 Flowchart for the design reinforcement for combined shear and torsion: (a) torsional web steel, (b) shear web steel.

1. Classify whether the applied torsion is equilibrium or compatibility torsion. Determine the critical section and compute the factored torsional moment T_u. The critical section is taken at $h/2$ from the face of the support in prestressed concrete beams. If T_u is less than $\phi\sqrt{f'_c}(A^2_{cp}/p_{cp})\sqrt{1+\bar{f}_c/4\sqrt{f'_c}}$ for prestressed members, torsional effects are neglected. $\bar{f}_c$ is the compressive stress in the concrete after prestress losses at the centroid of the section resisting externally applied loads (termed as f_{pc} in the ACI Code).
2. Check whether the factored torsional moment T_u causes equilibrium or compatibility torsion. For compatibility torsion, limit the design torsional moment to the lesser of the actual moment T_u or $T_u = \phi 4\sqrt{f'_c}(A^2_{cp}/p_{cp})\sqrt{1+\bar{f}_c/4\sqrt{f'_c}}$ for prestressed concrete members. The value of the design nominal strength T_n has to be at least equivalent to the factored T_u/ϕ, proportioning the section such that

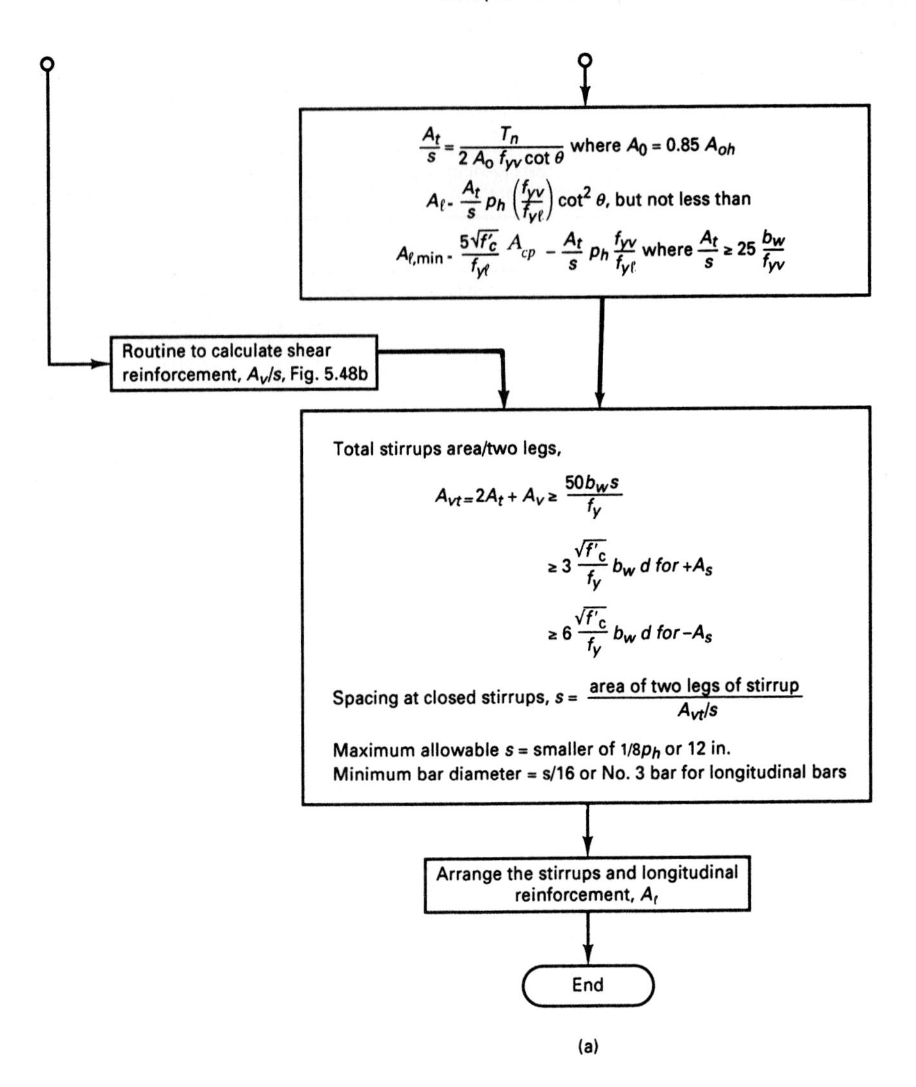

Figure 5.49 *Continued*

(a) for solid sections:

$$\sqrt{\left(\frac{V_u}{b_w d}\right)^2 + \left(\frac{T_u p_h}{1.7A_{0h}^2}\right)^2} \leq \phi\left(\frac{V_c}{b_w d} + 8\sqrt{f'_c}\right)$$

(b) for hollow sections:

$$\left(\frac{V_u}{b_w d}\right) + \left(\frac{T_u p_h}{1.7A_{0h}^2}\right) \leq \phi\left(\frac{V_c}{b_w d} + 8\sqrt{f'_c}\right)$$

If the wall thickness is less than A_{0h}/p_h, the second term should be taken as $T_u/1.7A_{0h}t$.

$$V_c = \left(0.6\lambda\sqrt{f'_c} + 700\frac{V_u d}{M_u}\right) b_w d; \qquad \frac{V_u d}{M_u} \leq 1.0$$

$$\geq 1.7\lambda\sqrt{f'_c}\, b_w d$$

$$\leq 5.0\lambda\sqrt{f'_c}\, b_w d \quad \text{and} \quad f_{pe} \geq 0.4 f_{pu}$$

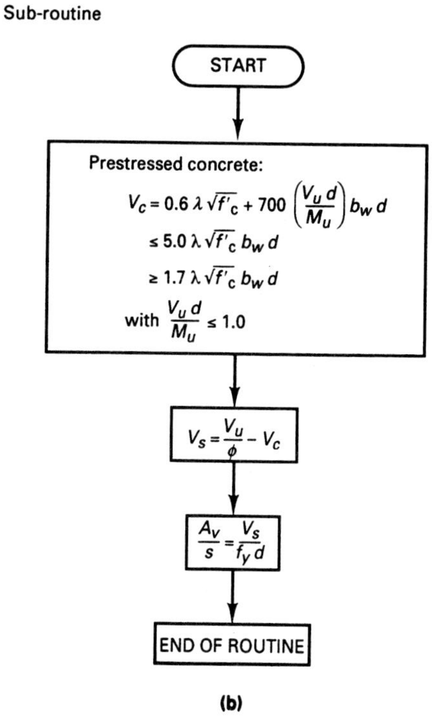

Sub-routine

START

Prestressed concrete:

$$V_c = 0.6\,\lambda\sqrt{f'_c} + 700\left(\frac{V_u d}{M_u}\right) b_w d$$

$$\leq 5.0\,\lambda\sqrt{f'_c}\, b_w d$$

$$\geq 1.7\,\lambda\sqrt{f'_c}\, b_w d$$

with $\frac{V_u d}{M_u} \leq 1.0$

$$V_s = \frac{V_u}{\phi} - V_c$$

$$\frac{A_v}{s} = \frac{V_s}{f_y d}$$

END OF ROUTINE

(b)

Figure 5.49 *Continued*

3. Select the required *torsional* closed stirrups to be used as transverse reinforcement, using a maximum yield strength of 60,000 psi, such that

$$\frac{A_t}{2} = \frac{T_n}{2A_0 f_{yv} \cot\theta}$$

Unless using A_0 and θ values obtained from analysis (Ref. 5.18) or from Equation 5.86, use $A_0 = 0.85A_{0h}$ and $\theta = 45°$ for non-prestressed members and 37.5° for prestressed members with an effective prestress not less than the tensile strength of the longitudinal reinforcement. The additional longitudinal reinforcement should be

$$A_l = \left(\frac{A_t}{s}\right) p_h \left(\frac{f_{yv}}{f_{yl}}\right) \cot^2\theta$$

but not less than

$$A_{l,\min} = \frac{5\sqrt{f'_c}\,A_{cp}}{f_{yl}} - \left(\frac{A_t}{s}\right) p_h \frac{f_{yv}}{f_{yl}}$$

where A_t/s shall not be less than $25b_w/f_{yv}$.

Maximum allowable spacing of transverse stirrups in the smaller of $\frac{1}{8}p_h$ or 12 in., and bars should have a diameter of at least $\frac{1}{16}$ of the stirrup spacing but not less than a No. 3 bar size.

4. Calculate the required *shear* reinforcement A_v per unit spacing in a transverse section. V_u is the factored external shear force at the critical section, V_c is the nominal shear resistance of the concrete in the web, and V_s is the shearing force to be resisted by the stirrups:

$$\frac{A_v}{s} = \frac{V_s}{f_y d}$$

where $V_s = V_n - V_c$ and

$$V_c = \left(0.6\lambda\sqrt{f_c'} + \frac{700V_u d}{M_u}\right) b_w d$$

$$V_c \leq 1.7\lambda\sqrt{f_c'}\, b_w d \geq 5.0\lambda\sqrt{f_c'}\, b_w d; \qquad \frac{V_u d}{M_u} \leq 1.0$$

$$\lambda = 1.0 \text{ for normal-weight concrete}$$
$$= 0.85 \text{ for sand lightweight concrete}$$
$$= 0.75 \text{ for all lightweight concrete}$$

The value of V_n has to be at least equal to the factored V_u/ϕ.

5. Obtain the total A_{vt}, the area of closed stirrups for torsion and shear, and design the stirrups such that

$$A_{vt} = 2A_t + A_v$$
$$\geq \text{the lesser of } \frac{50b_{ws}}{f_{yv}} \quad \text{or} \quad \frac{A_{ps} f_{pu}}{80 f_y\, d}\sqrt{\frac{d}{b_w}}$$

Extend the stirrups a distance $(b_t + d)$ beyond the point theoretically no longer required, where b_t = width of the cross-section containing the closed stirrup resisting torsion.

5.19 DESIGN OF WEB REINFORCEMENT FOR COMBINED TORSION AND SHEAR IN PRESTRESSED BEAMS

Example 5.9

A parking garage floor for medium-size cars has the prestressed concrete flooring system shown in Figure 5.50. The floor panels are 36 ft × 54 ft (11m × 16.5 m) on centers, and 54 ft (16.5m) span precast double-T's are supported by typical precast prestressed concrete spandrel L-beams spanning 36 ft (11m) on centers (Figures 5.50(a) and (b)). The spandrel beams are torsionally restrained by their connections to the supporting columns. The floor is subjected to a service superimposed dead load due to the double-T's of W_{SD} = 77 psf (3,687 Pa) and a service live load of W_L = 50 psf (2,394 Pa). The depth of the L-beam is chosen as 6′–3″ so as to provide a parapet wall for the roof on top of the double-Tee beams.

Design the spandrel beam web reinforcement to resist the combined torsion and shear to which it is subjected. Given data are the following:

Beam Properties

A_c = 696 in.^{2} (4,491 cm^2)
I_c = 364,520 in.^{4} (93.3 × 10^6 cm^4)
c_b = 33.2 in. (84.3 cm)
c_t = 41.8 in. (106 cm)

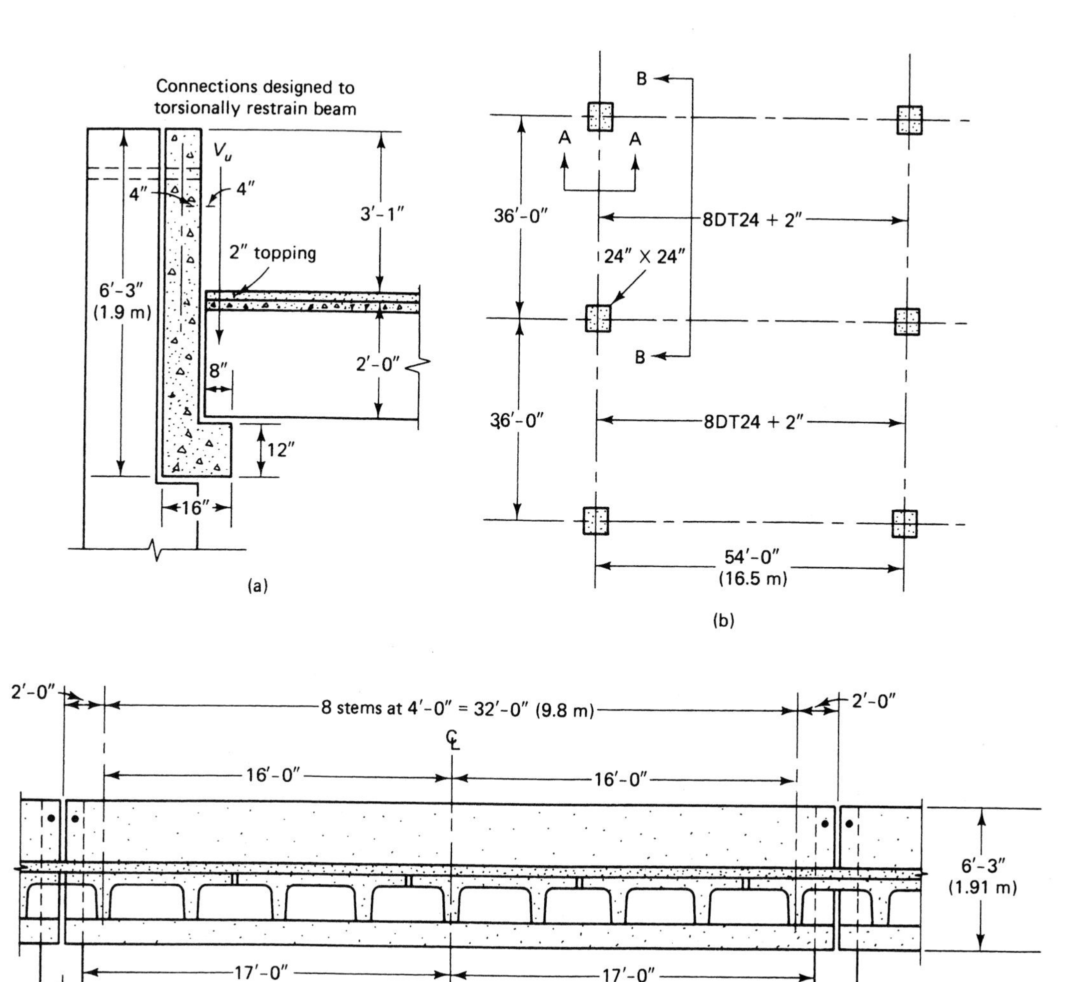

Figure 5.50 Geometrical details of structure in Example 5.9. (a) Section A-A. (b) Partial plan. (c) Section B-B.

$S^t = 8{,}720$ in.^{3} (142,895 cm^3)
$S_b = 10{,}990$ in.^{3} (180,094 cm^3)
$W_D = 725$ plf (10.6 kN.m)
$f'_c = 5{,}000$ psi (34.5 MPa), normal-weight concrete
$f_y = 60{,}000$ psi (418 MPa) for stirrups

Prestressing

A_{ps} = six $\frac{1}{2}$in. dia, 270 K stress-relieved tendons
$f_{pu} = 270{,}000$ psi (1,862 MPa)
$f_{ps} = 255{,}000$ psi (1,758 MPa)
$f_{pe} = 155{,}000$ psi (1,069 MPa)

$E_{ps} = 28 \times 10^6$ psi (193×10^3 MPa)
$d_p = 71.5$ in. (190 cm)
$e = 71.5 - 41.8 = 29.7$ in. (75 cm), straight tendon

Disregard the effects of winds or earthquake.

Solution:

1. *Calculate T_u, V_u, M_u, T_{SL}, V_{SL} acting on L-beam (step 1)*
   (a) Service load

$$W_D = 725 \text{ plf } (10.6 \text{ kN/m})$$

$$W_{SD} = \frac{77 \times 54}{2} \times 4 \text{ ft} = 8{,}317 \text{ lb/stem } (37.0 \text{ kN})$$

$$W_L = \frac{50 \times 54}{2} \times 4 \text{ ft} = 5{,}400 \text{ lb/stem } (24.0 \text{ kN})$$

$$\text{Total } P_{SL} \text{ per stem} = 8{,}316 + 5{,}400 = 13{,}716 \text{ lb } (61.0 \text{ kN})$$

   (b) Factored loads

$$W_{Du} = 1.4 \times 725 = 1{,}015 \text{ plf } (14.8 \text{ kN/m})$$

$$W_{SDu} = 1.4 \times 8{,}316 = 11{,}642 \text{ lb/stem } (51.8 \text{ kN/m})$$

$$W_{Lu} = 1.7 \times 5{,}400 = 9{,}180 \text{ lb/stem } (40.8 \text{ kN})$$

$$\text{Total } P_u \text{ per stem} = 11{,}642 + 9{,}180 = 20{,}822 \text{ lb } (92.6 \text{ kN})$$

$$T_u \text{ at face of support} = \tfrac{1}{2} P_u \times \text{arm} \times \text{no. of stems}$$

$$= \frac{20{,}822}{2} \times \frac{8}{12} \times 9 = 62{,}466 \text{ ft-lb } (84.7 \text{ kN-m})$$

$$T_{SL} \text{ at face of support} = \frac{13{,}716}{20{,}822} \times 62{,}466 = 41{,}148 \text{ ft-lb } (55.8 \text{ kN-m})$$

$$V_u \text{ at face of support} = \tfrac{1}{2}(P_u \times \text{no. of stems} + \text{factored } W_D \times \text{span})$$

$$= \tfrac{1}{2}(20{,}822 \times 9 + 1{,}015 \times 34) = 110{,}954 \text{ lb } (494 \text{ kN})$$

$$V_{SL} \text{ at face of support} = \tfrac{1}{2}(13{,}716 \times 9 + 725 \times 34) = 74{,}047 \text{ lb } (329 \text{ kN})$$

$$M_u \text{ at face of support} = 0$$

Similarly, calculate the values of T_u, V_u, and M_u, and the corresponding service-load values at each transverse stem contact point along the span of the L-beam, and construct the torsion, shear, and moment diagrams as shown in Figure 5.51.

$$A_{ps} = 6 \times 0.153 = 0.918 \text{ in.}^2$$

$$P_e = A_{ps} f_{pe} = 0.918 \times 155{,}000 = 142{,}290 \text{ lb } (633 \text{ kN})$$

2. *L-Beam torsional geometrical details (Step 1)*

A_{cp} = area enclosed by outside perimeter of concrete cross section $= 8 \times 75 = 600$ in.^{2} (3871 cm^2)
p_{cp} = outside perimeter of concrete cross section $= 2(8 + 75) = 166$ in. (422 cm)
x_1 = smaller dimension to center of tie
$= 8 - 2(1.5 + 0.25) = 4.5$ in. (11.4 cm)
$y_1 = 75 - 2(1.5 + 0.25) = 71.5$ in. (181.6 cm)

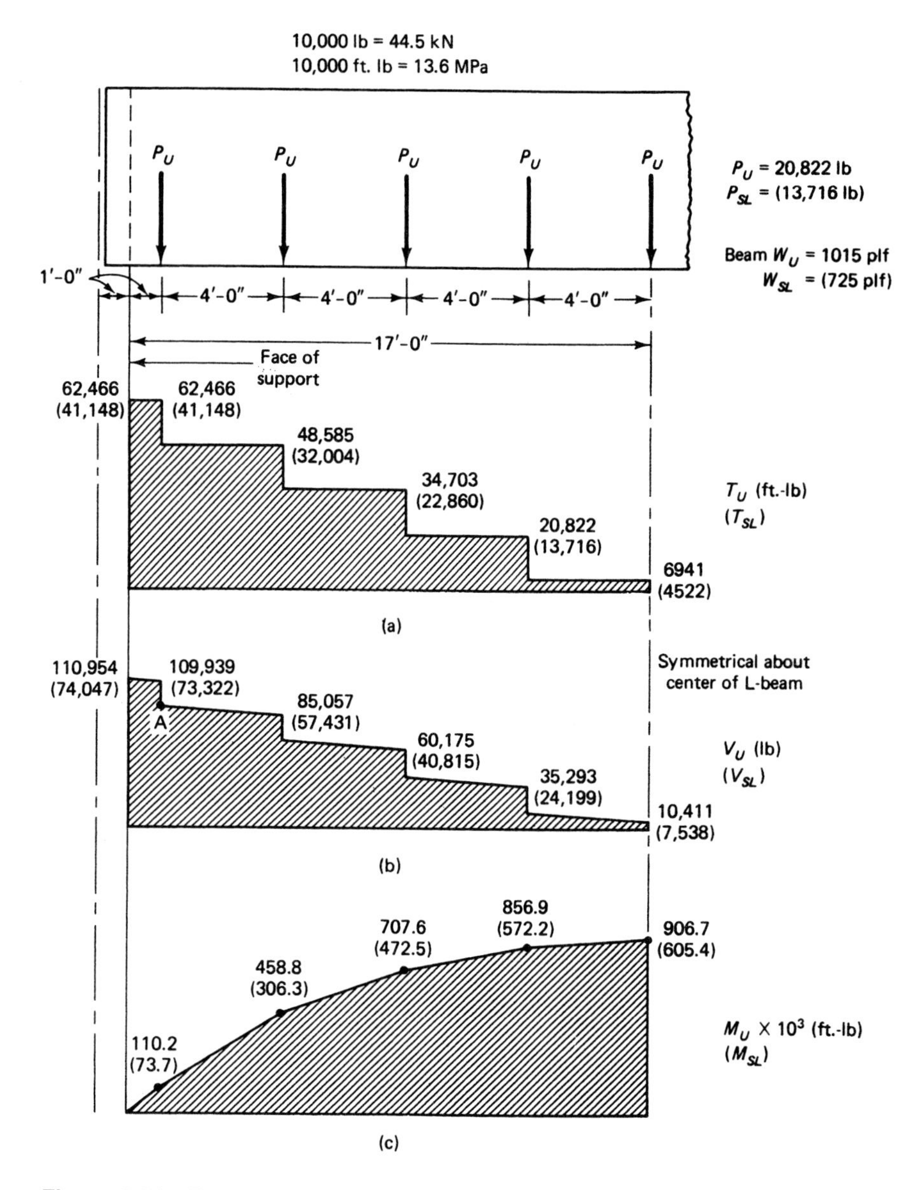

Figure 5.51 Force and moment diagrams for beams in Example 5.9. (a) Torsional moment. (b) Shear. (c) Flexural moment. Bracketed values are for service-load level.

h = total depth = 75 in. (191 cm)

b_w = web width = 8 in. (20.3 cm)

p_h = perimeter of center line of outermost closed transverse torsional reinforcement
$= 2(x_1 + y_1) = 2(4.5 + 71.5) = 152$ in.

d_p = effective depth $= 75 - (1.5 + 0.5 + 0.5 + 1.0) = 71.5$ in. (182 cm)

A_{0h} = area enclosed by centerline of the outermost closed transverse torsional ties
$= (x_1)(y_1) = 4.5 \times 71.5 = 322$ in.^{2} (2077 cm^2)

A_0 = gross area enclosed by shear flow path
$= 0.85A_{0h} = 0.85 \times 322 = 274$ in.^{2} (1766 cm^2)

θ = angle of compression diagonals in truss analogy for torsion ≅ 37.5° for prestressed beams.

$\cot \theta = 1.3$

3. *Cracking moment capacity (Step 1)*

$$f_d = \text{unfactored dead load stress}$$

From Figure 5.51 and $w_D = 725$ lb/ft

$$f_d = \frac{M_D}{S_b} = \frac{725(36)^2}{8} \times 12 \times \frac{1}{10{,}990} = 128 \text{ psi (0.9 MPa)}$$

At the extreme fibers of the section,

$$f_{ce} = -\left(\frac{P_e}{A_c} + \frac{P_e e}{S_b}\right) = -\left(\frac{142{,}290}{696} + \frac{142{,}290 \times 29.7}{10{,}990}\right)$$

$$= -(204.4 + 384.5) = 588.9 \text{ psi say } 589 \text{ psi } (C) \text{ (4.1 MPa)}$$

At the centroid of the section, $\bar{f}_c = -204.4$ psi

$$M_{cr} = S_b(6\lambda\sqrt{f'_c} + f_{ce} - f_d)$$

$$= 10{,}990(6 \times 1.0\sqrt{5000} + 589 - 128)$$

$$= 9.73 \times 10^6 \text{ in.-lb (1,100 kN-m)}$$

$$1.2M_{cr} = 1.2 \times 9.37 \times 10^6 = 11.2 \times 10^6 \text{ in.-lb}$$

$$a = \frac{A_{ps} f_{ps}}{0.85 f'_c b} = \frac{0.918 \times 255{,}000}{0.85 \times 5000 \times 8} = 6.9 \text{ in. (175 mm)}$$

$$M_n = A_{ps} f_{ps}\left(d_p - \frac{a}{2}\right) = 0.918 \times 255{,}000 \times \left(7.15 - \frac{6.9}{2}\right)$$

$$= 15{,}930{,}000 = 15.9 \times 10^6 \text{ in.-lb (1800 kN-m)}$$

$$> 1.2M_{cr} = 11.2 \times 10^6 \text{ in.-lb}$$

hence, minimum flexural reinforcement is satisfied for flexure.

4. *Verify whether torsional reinforcement is needed (Step 2)*
From Equation 5.80,

$$\text{Min. for disregarding torsion } T_u \le \phi\sqrt{f'_c}\left(\frac{A_{cp}^2}{p_{cp}}\right)\sqrt{1 + \frac{\bar{f}_c}{4\sqrt{f'_c}}}$$

$$= 0.85\sqrt{5000}\left(\frac{600^2}{166}\right)\sqrt{1 + \frac{204.4}{4\sqrt{5000}}}$$

$$= 171{,}080 \text{ in.-lb (19.3 kN-m)}$$

Considering section at $h/2$ from support face in Figure 5.51, namely at 3 ft from the face of the support,

$$\text{Rqd. } T_u = \tfrac{1}{2}(62{,}466 + 48{,}585) \times 12$$

$$= 666{,}310 \text{ in.-lb (75 kN-m)} > 171{,}080 \text{ in.-lb}$$

The average value was used instead of 48,585 in.-lb as a conservative value of the torsional moment.

Hence, torsion has to be considered and appropriate torsional reinforcement provided. The garage elements are all precast. Thus, assume equilibrium torsion condition and no redistribution of moment, using the total applied factored $T_u = 666{,}450$ in.-lb (Equation 5.79 is therefore inapplicable).

5. *Check adequacy of section for torsion*

(a) *Determine V_c as the smaller value obtained for V_{ci} from Equation 5.11 and V_{cw} from Equation 5.15.*

From Fig. 5.51 for section at $h/2 = 3$ ft from face of support,

$$\text{Rqd. } V_u = \tfrac{1}{2}(109{,}939 + 85{,}057) = 97{,}500 \text{ lb (434 kN)}$$

$$\text{Rqd. } M_u = \tfrac{1}{2}(110{,}200 + 458{,}000)12$$

$$= 3{,}409{,}200 \text{ in.-lb (385 kN-m)}$$

$$V_{ci} = \left[0.6\lambda\sqrt{f_c'}\,b_w d + V_d + V_i\frac{M_{cr}}{M_{\max}}\right]$$

$$\geq 1.7\sqrt{f_c'}\,b_w d$$

$$\leq 5.0\sqrt{f_c'}\,b_w d$$

$$V_d = \text{shear at section due to unfactored dead load}$$

$$V_d \text{ at face of support} = \tfrac{1}{2}(8316 \times 9 + 725 \times 34) = 49{,}749 \text{ lb}$$

$$V_d \text{ at point A in Figure 5.51} = 49{,}749 - 725 \times 1 \text{ ft} - 8316 = 40{,}708 \text{ lb}$$

$$V_d \text{ at 5 ft. from support} = 40{,}708 - 725 \times 5 - 8316 = 28{,}767 \text{ lb}$$

At the required section at $h/2$ from face of support,

$$V_d = \tfrac{1}{2}(40{,}708 + 28{,}767 = 34{,}738 \text{ lb (154.5 kN)}$$

V_i = factored shear force at section due to externally applied loads occurring simultaneously with $M_{\max}$

$$= \frac{20{,}822 \text{ per stem}}{20{,}822 + 1{,}015 \times 4 \text{ per stem}} \times 97{,}500$$

$$= 81{,}891 \text{ lb}$$

$M_{\max}$ = maximum factored moment at section due to externally applied load, namely due to live load and superimposed dead load

Factored M_u at point A in Figure 5.51 due to the live load and

$$SDL = \tfrac{1}{2}(20{,}822 \times 9) = 93{,}699 \text{ ft-lb}$$

Factored M_u at 5 ft from face

$$= \tfrac{1}{2}(20{,}822 \times 9)5 - 20{,}822 \times 4 = 385{,}207 \text{ ft-lb}$$

Hence

$$M_{\max} \text{ at } h/2 = 3 \text{ ft from face of support}$$

$$= \tfrac{1}{2}(93{,}699 + 385{,}307)12 = 2.87 \times 10^6 \text{ in.-lb}$$

$$M_{cr} = 7.96 \times 10^6 \text{ in.-lb from before}$$

hence,

$$V_{ci} = \left[0.6 \times 1.0\sqrt{5000} \times 8 \times 71.5 + 34{,}738 + 81{,}591\,\frac{7.96}{2.87}\right]$$

$$= 24{,}269 + 34{,}738 + 226{,}294 = 285{,}301 \text{ lb}$$

$$v_{ci} = \frac{285{,}301}{8 \times 71.5} = 499 \text{ psi (3.46 MPa)}$$

$$1.7\lambda\sqrt{f'_c} = 1.7\sqrt{5000} = 120 \text{ psi} < 499 \text{ psi}$$

$$5\lambda\sqrt{f'_c} = 5.0\sqrt{5000} = 354 \text{ psi} < 499 \text{ psi}$$

Use $v_{ci} = 354$ psi (2.4 MPa)

From Equation 5.15,

$$V_{cw} = (3.5\sqrt{f'_c} + 0.3\bar{f}_c)b_w d + V_p$$

$V_p = 0$ since tendons are straight, hence

$$V_{cw} = 3.5\sqrt{5000} + 0.3\left(\frac{142{,}290}{696}\right) = 310 \text{ psi} < v_{ci} = 354 \text{ psi}$$

Use $v_c = 310$ psi in this solution

$$V_c = v_c b_w d = 310 \times 8 \times 71.5 = 176{,}750 \text{ lb (786 kN)}$$

(b) *Alternate method for evaluating V_c*
If $f_{pe} > 0.4 f_{pu}$, the ACI allows using a more conservative expression as in Equation 5.16

$$V_c = \left(0.6\lambda\sqrt{f'_c} + 700\frac{V_u d_p}{M_u}\right)b_w d_p$$

$$\geq 2\lambda\sqrt{f'_c}\, b_w d$$

$$\leq 5\lambda\sqrt{f'_c}\, b_w d$$

$$\frac{V_u d_p}{M_u} = \frac{97{,}500 \times 71.5}{3{,}409{,}200} = 2.03 > 1.0,$$

$$\text{use } \frac{V_u d}{M_u} = 1.0$$

$$v_c = \frac{V_c}{b_w d} = (0.6 \times 1.0\sqrt{5000} + 700 \times 1.0) = 742 \text{ psi (5.1 MPa)}$$

$$> 5\sqrt{f'_c} = 354 \text{ psi}$$

$$v_c = 354 \text{ psi (2.4 MPa)}$$

(c) *Check section adequacy*
v_c in solution (a) will be used in this check = 310 psi (2.1 MPa). From Equation 5.81 for solid sections,

$$\sqrt{\left(\frac{V_u}{b_w d}\right)^2 + \left(\frac{T_u p_h}{1.7A_{0h}^2}\right)^2} = \sqrt{\left(\frac{97{,}500}{8 \times 71.5}\right)^2 + \left(\frac{666{,}310 \times 152}{1.77 \times (322)^2}\right)^2}$$

$$\sqrt{29{,}054 + 304{,}558} = 580 \text{ psi (4.0 MPa)}$$

$$\phi\left(\frac{V_c}{b_w d} + 8\sqrt{f'_c}\right) = 0.85(310 + 8\sqrt{5000})$$

$$= 744 \text{ psi (5.1 MPa) available}$$

$$> 580 \text{ psi (3.9 MPa) actual,}$$

hence section is adequate.

6. *Torsional reinforcement (Step 3)*

$$T_n = T_u/\phi = 666{,}310/0.85 = 783{,}390 \text{ in.-lb (86.6 kN-m)}$$

From Equation 5.38b,

$$\frac{A_t}{s} = \frac{T_n}{2A_0 f_{yv} \cot\theta} = \frac{783{,}390}{2 \times 274 \times 60{,}000 \times 1.3}$$

$$= 0.0183 \text{ in.}^2/\text{in./one leg } (0.046 \text{ cm}^2/\text{cm/one leg})$$

Using the PCI method in whict $\cot\theta$ is computed,

$$\cot\theta = \frac{T_u/\phi}{1.7A_{0h}(A_t/s)f_{yv}} = \frac{783{,}390}{1.7(322)(0.0183)} = 1.303$$

by assuming a value of s, finding (A_t/s) for the tie size chosen and entering the value of (A_t/s) into the expression.

7. *Shear reinforcement (Step 4)*

$$V_c = 310 \times 8 \times 71.5 = 177{,}320 \text{ lb (788 kN)}$$

$$V_n = \frac{V_u}{\phi} = \frac{97{,}500}{0.85} = 114{,}705 \text{ lb (510 kN)}$$

$$V_s = (V_n - V_c); \qquad \text{but } V_n < V_c$$

Use minimum shear web reinforcement.

$$\frac{A_v}{s} = \frac{50b_w}{f_y} = \frac{50 \times 8}{60{,}000}$$

$$= 0.0067 \text{ in.}^2/\text{in./two legs } (0.017 \text{ cm}^2/\text{cm/two legs})$$

$$\frac{A_{vi}}{s} = 2\left(\frac{A_T}{s}\right) + \frac{A_v}{s} = 2 \times 0.0183 + 0.0067$$

$$= 0.0433 \text{ in.}^2/\text{in./two legs } (0.110 \text{ cm}^2/\text{cm/two legs})$$

Assuming No. 4 closed ties (12.7 mm diameter), $A_v = 2 \times 0.20 = 0.40$ in.^{2}

$$s = \frac{\text{cross-sectional tie area}}{A_{vt}/s} = \frac{0.40}{0.0433} = 9.2 \text{ in.}$$

Maximum allowable spacing $s_{max} = p_h/8$ or 12 in. = 152/8 = 19 in. or 12 in.

$$\text{Min. } \frac{A_v}{s} = \text{lesser of } \frac{50b_w}{f_y} \quad \text{or} \quad \frac{A_{ps} f_{pu}}{80 f_y d}\sqrt{\frac{d}{b_w}}$$

$$\frac{A_v}{s} = \frac{50b_w}{f_y} = \frac{50 \times 8}{60{,}000}$$

$$= 0.0067 \text{ in.}^2/\text{in./two legs } (0.017 \text{ cm}^2/\text{cm/two legs})$$

$$\frac{A_v}{s} = \frac{A_{ps} f_{pu}}{80 f_y d}\sqrt{\frac{d}{b_w}} = \frac{0.918 \times 270{,}000}{80 \times 60{,}000 \times 71.5}\sqrt{\frac{71.5}{8}}$$

$$= 0.0022 \text{ in.}^2/\text{in./two legs } (0.006 \text{ cm}^2/\text{cm/two legs})$$

Available $\dfrac{A_v}{s} = 0.0433 > 0.0022$, O.K.

Minimum bar diameter = $s/16$ or No. 3 bar = 9.2/16 = 0.58 in. > 0.5 in. for No. 4 bar, use No. 5 closed stirrups. $A_v = 0.31 \times 2 = 0.62$ in.^{2}

$$s = \frac{0.62}{0.0433} = 14.3 \text{ in.} > 12 \text{ in.}$$

Another alternative for the transverse reinforcement is to use No. 4 bars but reduce the computed spacing of 9.2 in. to $9.2 \times (0.50/0.58) \simeq 8.0$ in. This gives No. 4 closed stirrups spaced at 8 in. center to center instead of No. 5 at 12 in. center to center. Using No. 4 closed stirrups at 8 in. center to center is more preferable as it is easier to bend them than the No. 5 bars. Therefore, for this design, use closed No. 4 stirrups (12.7 mm diameter) at 8 in. center to center for transverse shear + torsion web reinforcement.

8. *Longitudinal reinforcement (Steps 5–6)*
From Equation 5.85,

$$A_l = \frac{A_t}{s} p_h \left(\frac{f_{yv}}{f_{yl}}\right) \cot^2 \theta$$

$$= 0.018 \times 152 \left(\frac{60{,}000}{60{,}00}\right)(1.3)^2 = 4.62 \text{ in.}^2 \text{ (30 cm}^2\text{)}$$

From Equation 5.86,

$$A_{l,\min} = \frac{5\sqrt{f'_c}\, A_{cp}}{f_{yl}} - \left(\frac{A_t}{s}\right) p_h \frac{f_{yv}}{f_{yl}}$$

$$= \frac{5\sqrt{5000} \times 600}{60{,}000} - 0.018 \times 152 \times \frac{60{,}000}{60{,}000}$$

$$= 0.80 \text{ in}^2 \text{ (5.2 cm}^2\text{)}; \qquad A_l = 4.62 \text{ in.}^2 \text{ controls}$$

Using No. 4 longitudinal bars = 0.20 in.^{2}

$$\text{No. of bars} = \frac{4.62}{0.20} = 23.1 \text{ bars}$$

Use 12 No. 4 bars on each face equally spaced (12 bars 12.7 mm diameter/face) and add another 4 bars for reinforcing the ledge to give a total of 28 No. 4 bars. Note that maximum allowable spacing = 12 in. In this case $s \simeq 6.5$ in. c. to c., O.K. Adopt the design. Details of reinforcement and cross-sectional geometry of the L-beam are given in Fig. 5.52.

For the reinforcing details to be complete, a design of the ledge and hanger reinforcement would be required, as well as details of the anchorage of the longitudinal rein-

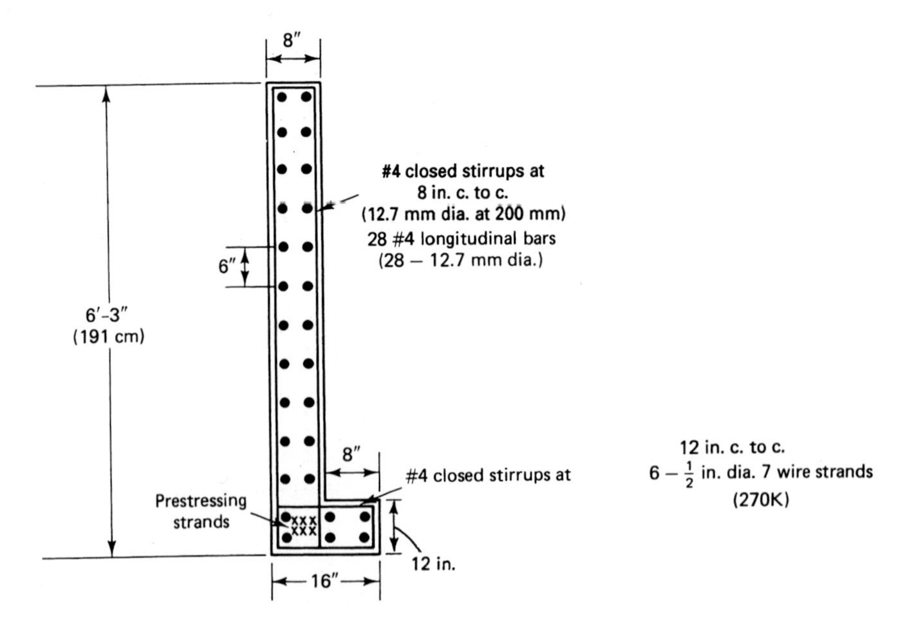

Figure 5.52 Reinforcement details of beam in Example 5.9.

forcement at the supports. Chapter 10 on the design of connections provides th
tails.

5.20 SI COMBINED TORSION AND SHEAR DESIGN OF PRESTRESSED BEAM

Example 5.10

Solve Example 5.9 using the SI procedure and ACI Shear value obtained from Equation 5.16 in Example 5.9.

Solution:

1. *See calculation step 1 in Example 5.9.*
2. *See calculation step 2 in Example 5.9.*
3. *L-Beam torsional geometrical details*

$$A_{cp} = 3870 \text{ cm}^2 \qquad p_{cp} = 422 \text{ cm}$$

$$x_1 = 11.7 \text{ cm} \qquad y_1 = 181.6 \text{ cm} \qquad p_h = 2(x_1 + y_1) = 386 \text{ cm}$$

$$h = 191 \text{ cm} \qquad b_w = 20.3 \text{ cm} \qquad d = 182 \text{ cm}$$

$$A_{0h} = 2077 \text{ cm}^2 \qquad A_0 = 0.85A_{0h} = 1766 \text{ cm}^2$$

$$\theta = 37.5° \qquad \cot\theta = 1.3 \qquad \cot^2\theta = 1.69$$

From Figure 5.51

$$\text{factored } T_u = 75 \text{ kN-m}$$

$$\text{factored } V_u = 434 \text{ kN}$$

$$\text{factored } M_u = 385 \text{ kN-m}$$

$$\frac{V_u d}{M_u} = \frac{434 \times 182 \text{ cm}}{38{,}500 \text{ kN-m}} = 2.05 > 1.0, \text{ use } \frac{V_u d}{M_u} = 1.0$$

$$P_e = 142{,}290 \text{ lb} = 633 \text{ kN} \qquad e = 29.7 \text{ in.} = 75.4 \text{ cm}$$

$$A_c = 696 \text{ in}^2 = 4490 \text{ cm}^2$$

$$c = 26.2 \text{ in.} = 66.5 \text{ cm}$$

$$S_b = 10{,}990 \text{ in}^3 = 180{,}094 \text{ cm}^3$$

$$f'_c = 34.5 \text{ MPa} \qquad f_{pu} = 1860 \text{ MPa}$$

$$E_s = 200{,}000 \text{ MPa} \qquad f_y = 414 \text{ MPa} \qquad f_{ps} = 1760 \text{ MPa}$$

$$A_{ps} = 6 \text{ tendons } 12.7 \text{ mm diameter} = 5.92 \text{ cm}^2$$

$$\phi \text{ for shear and torsion} = 0.85$$

$$1 \text{ Pa} = \text{N/m}^2$$

4. *Cracking moment capacity*

$$S_b = 180{,}094 \text{ cm}^2$$

$$\bar{f}_c = \frac{P_e}{A_c} = \frac{633{,}000}{4490 \times 10^2} = 1.41 \text{ MPa}$$

$$f_{ce} = \frac{P_e}{A_c} \pm \frac{P_e e}{S_b} = 1.41 + \frac{633{,}000 \times 75.4}{180{,}094 \times 10^2}$$

$$= 1.41 + 2.65 = 4.06 \text{ say } 4.1 \text{ MPa}$$

From Example 5.9, $\bar{f}_c = 1.14$ MPa

$$f_d = \text{unfactored dead load stress} = 289 \text{ psi} = 2.0 \text{ MPa}$$

$$M_{cr} = S_b(\tfrac{1}{2}\sqrt{f'_c} + f_{ce} - f_d)$$

$$= 180{,}094\left(\frac{\sqrt{34.5}}{2} + 4.06 - 2.0\right) \times 10^{-3} = 900 \text{ kN-m}$$

$$1.2M_{cr} = 1.2 \times 900 = 1080 \text{ kN-m}$$

$$a = \frac{A_{ps} f_{ps}}{0.85 f'_c} = \frac{5.9 \text{ cm}^2 \times 1760}{0.85 \times 34.5 \times 20.3} = 17.5 \text{ cm}$$

Nominal moment strength,

$$M_n = A_{ps} f_{ps}\left(d_p - \frac{a}{2}\right)$$

$$= 5.92 \times 1760\left(182 - \frac{17.5}{2}\right) \text{N-m} = 1800 \text{ kN-m}$$

$$> 1.2M_{cr} = 1080 \text{ kN-m}$$

hence flexural reinforcement is satisfied for flexure.

5. *Verify whether torsional reinforcement is needed*
From Equation 5.80,

$$T_u \leq \frac{\phi\sqrt{f'_c}}{12}\left(\frac{A_{cp}^2}{p_{cp}}\right)\sqrt{1 + \frac{3\bar{f}_c}{\sqrt{f'_c}}}$$

$$T_u = \frac{0.85\sqrt{34.5}}{12}\left(\frac{(3870)^2}{422}\right)\sqrt{1 + \frac{3 \times 1.14}{\sqrt{34.5}}} \times 10^{-3} \text{ kN-m}$$

$$= 18.6 \text{ kN-m}$$

From Figure 5.51 and the acting torsional moment value from Example 5.9, $T_u = 75$ kN-m > 18.6 kN-m. Hence, torsional reinforcement is required. The garage elements are all precast; assume equilibrium torsion condition with no redistribution of torsional moment using the total applied factored $T_u = 75$ kN-m.

6. *Check adequacy of section for torsion*
From Figure 5.51 at $h/2$ from face of support and values computed in Example 5.9, $V_u = 434$ kN, $M_u = 385$ kN-m.
From Equation 5.80 for $f_{pe} > 0.4 f_{pu}$

$$v_c = \frac{V_c}{b_w d} = \left(\frac{\sqrt{f'_c}}{20} + 5\frac{V_u d}{M_u}\right), \qquad \frac{V_u d}{M_u} \leq 1.0$$

where f'_c is in MPa

$$\frac{V_u d}{M_u} = \frac{434 \times 1.82}{385} = 2.05 > 1.0, \text{ use } 1.0$$

$$v_c = \frac{V_c}{b_w d} = \left(\frac{\sqrt{34.5}}{20} + 5 \times 1.0\right) = 5.3 \text{ MPa}$$

$$\text{Max. allow. } v_c = 0.4\sqrt{f'_c} = 0.4\sqrt{34.5} = 2.4 \text{ MPa controls.}$$

From Equation 5.81 for solid section,

$$\sqrt{\left(\frac{V_u}{b_w d}\right)^2 + \left(\frac{T_u p_h}{1.7A_{0h}^2}\right)^2} \leq \phi\left(\frac{V_c}{b_w d} + \frac{8\sqrt{f_c'}}{12}\right)$$

$$\sqrt{\left(\frac{V_u}{b_w d}\right)^2 + \left(\frac{T_u p_h}{1.7A_{0h}^2}\right)^2} = \sqrt{\left(\frac{434 \times 10^3}{0.2 \times 1.82}\right)^2 + \left(\frac{75 \times 10^3 \times 3.86}{1.7 \times 0.208^2}\right)^2}$$

$$= \sqrt{1422 \times 10^6 + 15.5 \times 10^{12}}\ \text{N/m}^2 = 4.0\ \text{MPa}$$

$$\phi\left(\frac{V_c}{b_w d} + \frac{8\sqrt{f_c'}}{12}\right) = 0.85\left(2.4 + \frac{8\sqrt{34.5}}{12}\right) = 5.4\ \text{MPa}$$

available > 4.0 MPa actual, hence section is adequate.

7. *Torsional reinforcement*

$$T_n = T_u/\phi = 75/0.85 = 88\ \text{kN-m}$$

From Equation 5.83b,

$$\frac{A_t}{s} = \frac{T_n}{2A_0 f_{yv} \cot\theta} = \frac{88 \times 10^3}{2 \times 1766 \times 414 \times 1.3}$$

$$= 0.046\ \text{cm}^2/\text{cm/one leg}$$

8. *Shear reinforcement*
From Example 5.9,

$V_c = 1796$ kN $> V_n = 510$ kN, hence provide only minimum reinforcement for shear.

$$\frac{A_v}{s} = \frac{0.35 b_w}{f_y} = \frac{0.35 \times 20.3}{414} = 0.017\ \text{cm}^2/\text{cm/two legs}$$

$$\frac{A_{vt}}{s} = 2\left(\frac{A_t}{s}\right) + \frac{A_v}{s} = 2 \times 0.046 + 0.017 = 0.110\ \text{cm}^2/\text{cm/two legs}$$

Assuming No. 10 M closed stirrups are used

$$A_v = 2 \times 100 = 200\ \text{mm}^2 = 2.0\ \text{cm}^2$$

$$s = \frac{\text{cross-sectional area}}{A_{vt}/s} = \frac{2.0}{0.11} = 18\ \text{cm (7.2 in. c. to c.)}$$

Maximum allowable $s = p_h/8$ or 30 cm $= p_h/8 = 386/8 = 48$ cm

$$\text{Min. } \frac{A_v}{s} = \text{lesser of } \frac{0.35 b_w}{f_y} \text{ or } \frac{A_{ps} f_{pu}}{80 f_y d}\sqrt{\frac{d}{d_p}}$$

where b_w, d, and s are in millimeters

$$\frac{0.35 b_w}{f_y} = \frac{0.35 \times 203}{414} = 0.17\ \text{mm}^2/\text{mm/two legs}$$

$$= 0.017\ \text{cm}^2/\text{cm/two legs}$$

$$\frac{A_{ps} f_{pc}}{80 f_y d}\sqrt{\frac{d}{b_w}} = \frac{592 \times 1860}{80 \times 414 \times 1820}\sqrt{\frac{1820}{203}}$$

$$= 0.06\ \text{mm}^2/\text{mm/two legs}$$

$$= 0.006\ \text{cm}^2/\text{cm/two legs controls}$$

Available $A_v = 0.17 > 0.06$, O.K.
Miminum bar diameter = $s/16$ or No. 10 M bar = (18/16) × 10 mm = 11.3 mm = available No. 10 M bars (11.3 mm), O.K.
Use No. 10 bars at 18 cm c. to c.

9. *Longitudinal reinforcement*
From Equation 5.85,

$$A_l = \frac{A_t}{s} p_h \left(\frac{f_{yv}}{f_{yl}}\right) \cot^2 \theta$$

$$= 0.046 \times 386 \left(\frac{414}{414}\right)(1.3)^2 = 30\text{cm}^2$$

From Equation 5.86,

$$A_{l,\min} = \frac{5\sqrt{f'_c} A_{cp}}{12 f_{yl}} - \frac{A_t}{s} p_h \left(\frac{f_{yv}}{f_{yl}}\right)$$

$$= \frac{5\sqrt{34.5} \times 3870}{12 \times 414} - 0.046 \times 386 \left(\frac{414}{414}\right)$$

$$= 22.9 - 17.7 = 5.2\text{cm}^2$$

$$A_l = 30 \text{ cm}^2 \text{ controls}$$

Using No. 10 M bars, $A_s = 1.0$ cm²
No. of bars = 30/1.0 = 30 bars
Use 15 No. 10 M bars on each face of the L-beam equally spaced.
Note that maximum allowable spacing $s = 30$ cm
In this case $s = \dfrac{191 - 2(1.5)}{15} \simeq 12$ cm O.K.

Adopt the design.

For the reinforcing details to be complete, a design of the ledge and hanger reinforcement would be required, as well as details of the anchorage of the longitudinal reinforcement at the supports. Chapter 10 on the design of connections provides these details.

REFERENCES

5.1 ACI Committee 318. *Building Code Requirements for Structural Concrete,* (ACI 318–99), and *Commentary to the Building Code Requirements for Reinforced Concrete* (ACI 318R–99) American Concrete Institute, Farmington Hills, MI: 2000, 392 pp.

5.2 Nawy, E. G. *Reinforced Concrete—A Fundamental Approach,* 4th Ed., Upper Saddle River, N.J.: Prentice Hall, 2000, pp. 786.

5.3 Mattock, A. H., Chen, K. C., and Soogswang, K. "The Behavior of Reinforced Concrete Corbels." *Journal of the Prestressed Concrete Institute* 21 (1976): 52–77.

5.4 Nawy, E. G. *Simplified Reinforced Concrete.* Upper Saddle River, N.J.: Prentice Hall, 1986.

5.5 Sozen, M. A., Zwoyer, E. M., and Siess, C. P. Strength in Shear of Beams without Web Reinforcement. Urbana, Illinois: Bulletin No. 452, Engineering Experiment Station, University of Illinois, April 1959.

5.6 ACI Committee 318. *Building Code Requirements for Reinforced Concrete,* 318–63, and *Commentary on the Building Code Requirements for Reinforced Concrete,* 318–63. Farmington Hills, MI: American Concrete Institute, 1963.

5.7 Nawy, E. G., and Ukadike, M. M. "Shear Transfer in Concrete and Polymer Modified Concrete Members Subjected to Shear Load." *Journal of the American Society for Testing and Materials,* March 1983, pp. 83–97.

5.8 Prestressed Concrete Institute. *Manual of Design and Detailing of Precast and Prestressed Connections.* Chicago: Prestressed Concrete Institute, 1987.

5.9 Hsu, T. T. C. *Torsion in Reinforced Concrete.* Van Nostrand Reinhold, New York, 1983.

5.10 Hsu, T. T. C. "Torsion in Structural Concrete—Uniformly Prestressed Members without Web Reinforcement." *Journal of the Prestressed Concrete Institute* 13 (1968): 34–44.

5.11 Collins, M. P., and Mitchell, D. "Shear and Torsion Design of Prestressed and Non-Prestressed Concrete Beams." *Journal of the Prestressed Concrete Institute* 25 (1980): 32–100.

5.12 Prestressed Concrete Institute. *PCI Design Handbook* 5th Ed. Prestressed Concrete Institute, Chicago, 1999.

5.13 Rabbat, B. G. and Collins, M. P. "A Variable Angle Space Truss Model For Structural Concrete Members Subjected to Complex Loading," SP 55–22, American Concrete Institute, Farmington Hills, 1978, pp 547–587.

5.14 McGee, W. D., and Zia, P. "Prestressed Concrete Members under Torsion, Shear and Bending." *Journal of the American Concrete Institute* 73 (1976): 26–32.

5.15 Zia, P., and Hsu, T. T. C. *Design for Torsion and Shear in Prestressed Concrete.* ASCE Annual Convention, Reprint No. 3423, 1979.

5.16 Abeles, P. W., and Bardhan-Roy, B. K. *Prestressed Concrete Designer's Handbook.* 3d Ed. Viewpoint Publications, London, 1981.

5.17 Hsu, T. T. C. "Shear Flow Zone in Torsion of Reinforced Concrete," Vol. 116 No. 11, Journ. of Structural Division, ASCE, New York, Nov. 1990, pp 3206–3225.

5.18 Hsu, T. T. C. *Unified Theory of Reinforced Concrete,* CRC Press, Boca Raton, 1993, pp. 313.

PROBLEMS

5.1 A post-tensioned bonded prestressed beam has the cross section shown in Figure P5.1. It has a span of 75 ft (22.9 m) and is subjected to a service superimposed dead load $W_{SD} = 450$ plf (6.6 kN/m) and a superimposed service live load $W_L = 2{,}300$ plf (33.6 kN/m). Design the web reinforcement necessary to prevent shear cracking (a) by the detailed design method and (b) by the alternative method at a section 15 ft (4.6 m) from the face of the support. The profile of the prestressing tendon is parabolic. Use #3 stirrups in your design, and detail the section. The following data are given:

$$A_c = 876 \text{ in.}^2 \ (5{,}652 \text{ cm}^2)$$

$$I_c = 433{,}350 \text{ in.}^4 \ (18.03 \times 10^6 \text{ cm}^4)$$

$$r^2 = 495 \text{ in.}^2 \ (3{,}194 \text{ cm}^2)$$

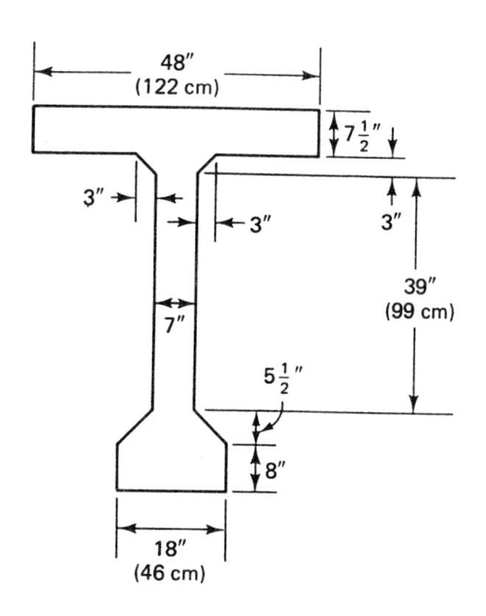

Figure P5.1.

$c_t = 25$ in. (63.5 cm)

$S^t = 17{,}300$ in.^{3} (2.83×10^5 cm^3)

$c_b = 38$ in. (96.5 cm)

$S_b = 11{,}400$ in.^{3} (1.86×10^5 cm^3)

$W_d = 910$ plf (13.3 kN/m)

$e_c = 32$ in. (81.3 cm)

$e_e = 2$ in. (5 cm)

$f'_c = 5{,}000$ psi (44.5 MPa), normal-weight concrete

$f'_{ci} = 3{,}500$ psi (24.1 MPa)

f_y for stirrups $= 60{,}000$ psi (41.8 MPa)

$f_{pu} = 270{,}000$ psi (1,862 MPa) low-relaxation strands

$f_{ps} = 243{,}000$ psi (1,675 MPa)

$f_{pe} = 157{,}500$ psi (1,086 MPa)

$A_{ps} =$ twenty-four $\frac{1}{2}$-in. dia (12.7 mm dia) 7-wire tendons

5.2 Find the shear strengths V_c, V_{ci}, and V_{cw} for the beam in Problem 5.1 at 1/10 span intervals along the entire span, and plot the variations in their values along the span in a manner similar to the plot in Figure 5.13.

5.3 Assume that a 4 in. (10 cm) topping of width $b = 8$ ft 6 in. (2.6 m) is situ cast on the precast section of Problem 5.1. If the top surface of the precast section is unroughened, design the necessary dowel reinforcement to ensure full composite action. Use the ACI coefficient of friction for determining the area and spacing of the shear-friction reinforcement, and use $f'_c = 3{,}000$ psi (20.7 MPa) for the topping. Compare the results with those obtained using the PCI coefficient of friction.

5.4 A 14 in. (35.6 cm) standard PCI double-T simply supported beam is shown in Figure P5.4. It has a span of 40 ft (12.2 m) and is subjected to a service dead load $W_{SD} = 25$ psf (1,197 Pa) plus self-weight $W_D = 31$ psf (1,484 Pa) and a service live load $W_L = 45$ psf (2,155 Pa). Design the web-shear reinforcement at $\frac{1}{2}d_p$ from the support and at quarter span by (a) the detailed method and (b) the alternative method, and then compare the two designs. The tendon is harped at midspan. Given data are as follows:

f'_c (precast) $= 5{,}000$ psi, lightweight concrete

$f'_{ci} = 3{,}500$ psi

f'_c (topping) $= 3{,}000$ psi, normal weight

$f_{pu} = 270{,}000$ psi, low-relaxation strand

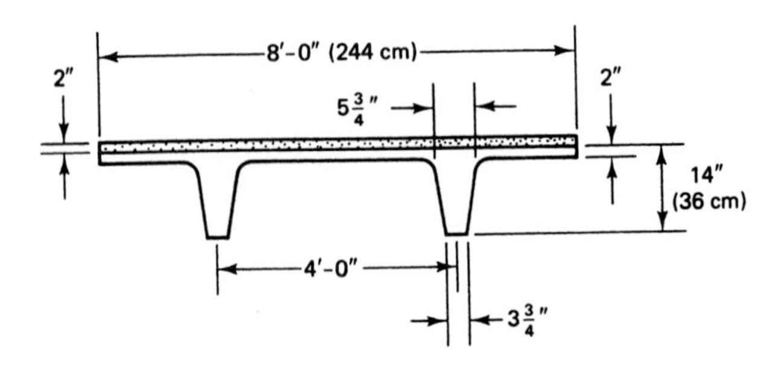

Figure P5.4.

$$f_{ps} = 189{,}000 \text{ psi } (1{,}303 \text{ MPa})$$
$$f_{pe} = 156{,}000 \text{ psi } (1{,}076 \text{ MPa})$$
$$\text{Stirrups } f_y = 60{,}000 \text{ psi } (41.8 \text{ MPa})$$
$$A_{ps} = \text{six } \tfrac{1}{2}\text{-in. } (12.7 \text{ mm}) \text{ dia 7-wire strands}$$
$$e_c = 8.01 \text{ in. } (20.3 \text{ cm})$$
$$e_e = 4.51 \text{ in. } (11.5 \text{ cm})$$

Use $d_p = 10$ in. in the solution. The values of the section properties are as follows:

Section properties	Untopped	Topped
A_c	306 in.^{2}	
I_c	4,508 in.^{4}	7,173 in.^{4}
c_b	10.51 in.	12.40 in.
c_t	3.49 in.	3.60 in.
S_b	429 in.^{3}	578 in.^{3}
S^t	1,292 in.^{3}	1,992 in.^{3}
W_D	31 psf	56 psf

5.5 Design a bracket to support a concentrated factored load $V_u = 125{,}000$ lb (556 kN) acting at a lever arm $a = 4$ in. (101.6 mm) from the column face. The horizontal factored force $N_{uc} = 40{,}000$ lb (177.9 kN). Given data are:

$$b = 14 \text{ in. } (355.6 \text{ mm})$$
$$f'_c = 5{,}000 \text{ psi } (34.47 \text{ MPa}), \text{ normal-weight concrete}$$
$$f_y = 60{,}000 \text{ psi } (413.7 \text{ MPa})$$

Assume that the bracket was cast after the supporting column cured, and that the column surface at the bracket location was not roughened before casting the bracket. Detail the reinforcing arrangements for the bracket.

5.6 Solve Problem 5.5 if the structural system was made from monolithic sand-lightweight concrete in which the corbel or bracket was cast simultaneously with the supporting column.

5.7 Design the transverse and longitudinal reinforcement in Example 5.9 for combined torsion and shear assuming that the L-beam concrete is made of sand-lightweight concrete.

5.8 Design the web reinforcement for the beam in Example 5.9 for combined shear and torsion assuming that the centerline dimensions of the interior floor panels are 30 ft x 56 ft (9.1 m × 17.1 m). The floor is subjected to a service superimposed dead load due to the double T's of $W_{SD} = 77$ psf (3,687 Ma) and a service live load of 60 psf (2,873 MPa).

INDETERMINATE PRESTRESSED CONCRETE STRUCTURES

6.1 INTRODUCTION

As in reinforced concrete and other structural materials, continuity can be achieved at intermediate supports and knees of portal frames. The reduction of moments and stresses at midspans through the design of continuous systems results in shallower members that are stiffer than simply supported members of equal span and of comparable loading and are of lesser deflection.

Consequently, lighter structures with lighter foundations reduce the cost of materials and construction. In addition, the structural stability and resistance to longitudinal and lateral loads are usually improved. As a result, the span-to-depth ratio is also improved, depending on the type of continuous system being considered. For continuous flat plates, a ratio of 40 to 45 is reasonable, while in box girders this ratio can be 25 to 30.

An additional advantage of continuity is the elimination of anchorages at intermediate supports through continuous post-tensioning over several spans, thereby reducing further the cost of materials and labor.

Continuous prestressed concrete is widely applied in the United States in the construction of flat plates for floors and roofs with continuity in one or both directions and with prestressing in one or both directions. Also, continuity is widely used in long-span

Lincoln Executive Plaza, Arlington Heights, Illinois. (*Courtesy,* Prestressed Concrete Institute.)

prestressed concrete bridges, particularly situ-cast post-tensioned spans. Cantilevered box girder bridges, widely used in Europe as segmental bridges, are increasingly being used in the United States for very large spans, and cable-stayed bridges with prestressed decks are increasingly built as well.

The success of prestressed concrete construction is largely due to the economy of using precast elements, with the associated high-quality control during fabrication. This desirable feature has been widely achieved by imposing continuity on the precast elements through placement of situ-cast reinforced concrete at the intermediate supports. The situ-cast concrete tends to resist the superimposed dead load and the live load that act on the spans after the concrete hardens. Note that forming, shoring, and reshoring can also be avoided in this type of construction, thereby reducing the costs further as compared with the costs of reinforced concrete.

6.2 DISADVANTAGES OF CONTINUITY IN PRESTRESSING

There are several disadvantages to having continuously prestressed elements:

1. Higher frictional losses due to the larger number of bends and longer tendons.
2. Concurrence of moment and shear at the support sections, which reduces the moment strength of those sections.
3. Excessive lateral forces and moments in the supporting columns, particularly if they are rigidly connected to the beams. These forces are caused by the elastic shortening of the long-span beams under prestress.
4. Effects of higher secondary stresses due to shrinkage, creep temperature variations, and settlement of the supports.
5. Secondary moments due to induced reactions at the supporting columns caused by the prestressing force (to be subsequently discussed).
6. Possible serious reversal of moments due to alternate loading of spans.
7. Moment values at the interior supports that require additional reinforcement at these supports, which might otherwise not be needed in simply supported beams.

All these factors can be accounted for through appropriate design and construction of the final system, including special provisions for bearings at the supporting columns.

6.3 TENDON LAYOUT FOR CONTINUOUS BEAMS

The construction system used, the length of the adjacent spans, and the engineering judgment and ingenuity of the design engineer determine the type of layout and method of framing to be used for achieving continuity. Basically, there are two categories of continuity in beams:

1. Monolithic continuity, where all the tendons are generally continuous throughout all or most of the spans and all tendons are prestressed at the site. Such prestressing is accomplished by post-tensioning.
2. Nonmonolithic continuity, where precast elements are used as simple beams on which continuity is imposed at the support sections through situ-cast reinforced concrete which provides the desired level of continuity to resist the superimposed dead load and live load after the concrete hardens.

Figure 6.1 schematically demonstrates the various systems and combinations of systems to achieve monolithic continuity. Figure 6.2 illustrates how continuity is achieved in nonmonolithic construction. Figure 6.1(a) presents a simple continuity system in which all the spans are situ cast and post-tensioning is accomplished after the concrete hardens. Problems are encountered, however, in the accurate evaluation of frictional losses due to the large number of bends. The system shown in Figure 6.1(b), using variable-depth beams, namely, nonprismatic sections, can add to the cost of formwork.

Frictional losses in the post-tensioned straight cables are easier to evaluate accurately. Additional costs are incurred due to the necessity of several anchorages. The system shown in Figure 6.1(a) has the advantage on that of Figure 6.1(b) in that the cost of formwork will in general be less because the continuous beam has a constant depth, although architectural considerations sometimes require nonprismatic continuous sections.

Continuity achieved through the use of precast pretensioned beams with situ-cast concrete connecting joints can in many cases be easier to erect, and considerable savings may accrue since formwork and shoring at the site are generally not needed. Figures 6.2(a) and (c) are essentially comparable in the degree of their accuracy in estimating frictional and other losses.

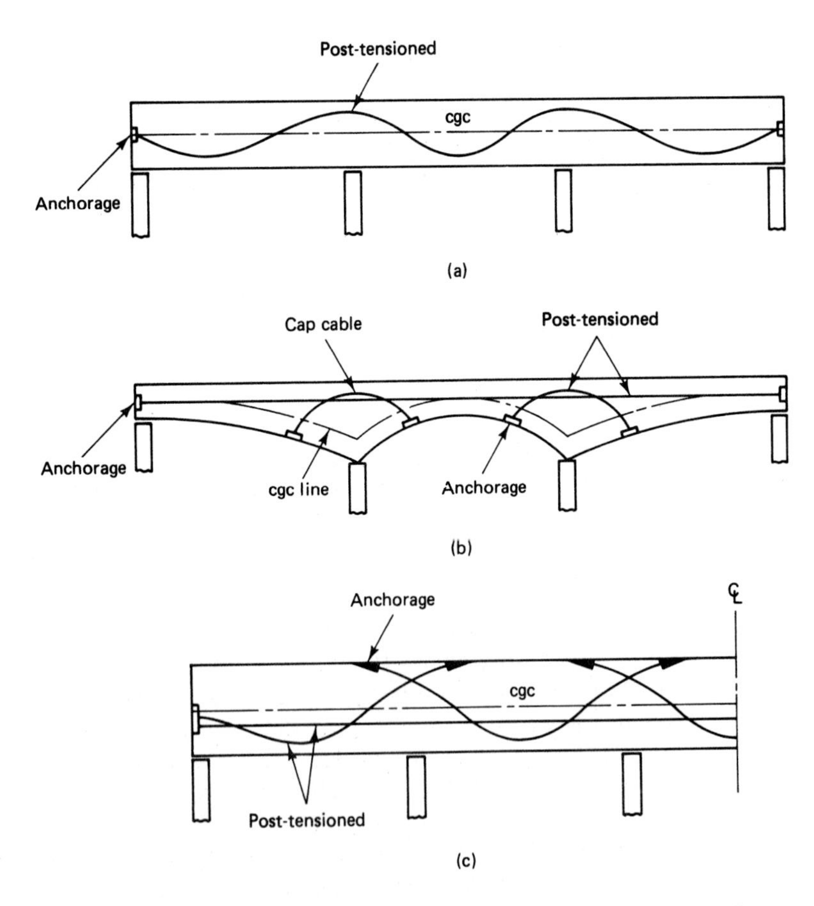

Figure 6.1 Tendon geometry in beams with monolithic continuity. (a) Beam with constant depth. (b) Nonprismtic beam with overlapping tendons. (c) Prismatic beam with overlapping tendons.

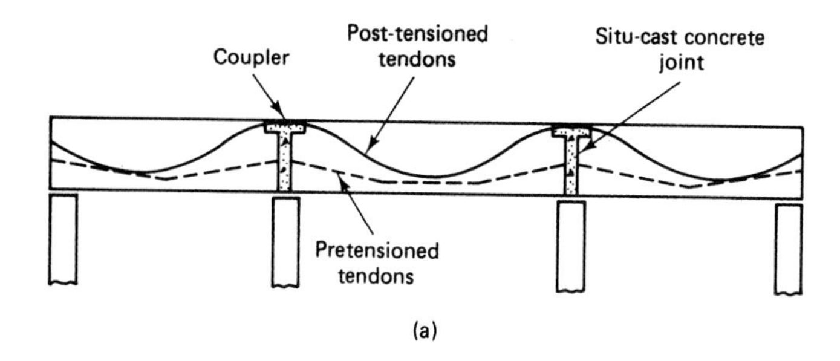

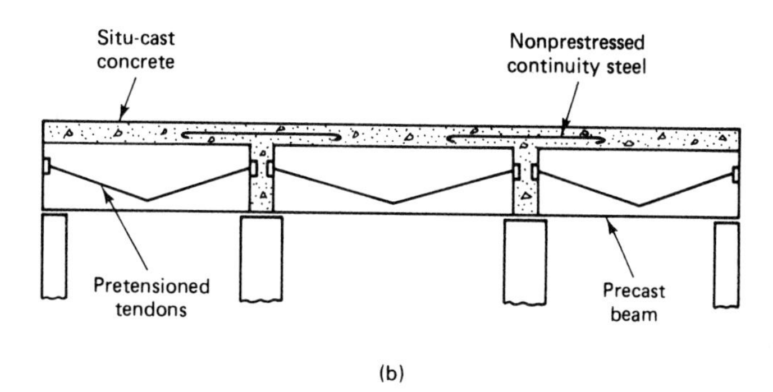

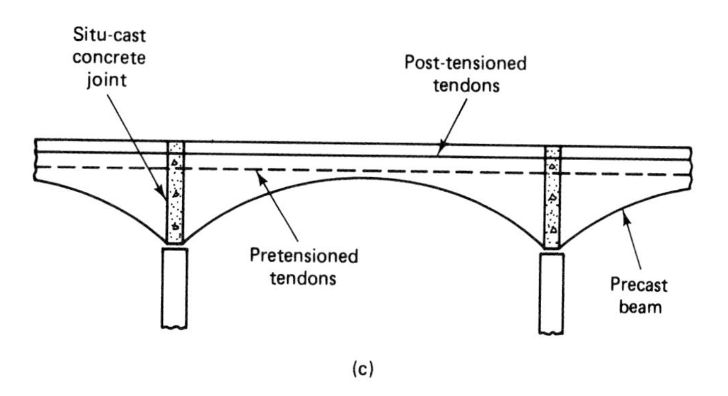

Figure 6.2 Continuity using precast pretensioned beams. (a) Post-tensioned continuity using couplers. (b) Continuity using nonprestressed steel. (c) Continuity in post-tensioning for nonprismatic beams.

The system illustrated in Figure 6.2(b) is probably the simplest for achieving continuity in prestressed concrete composite construction. The precast pretensioned elements are designed to carry the prestressing and self-weight moments, while the nonprestressed steel at the negative moment region at the support is designed to resist the additional superimposed dead load and the applied live load moments. If design for continuity due to the total dead load is to be achieved, the precast beams have to be shored before placing the composite concrete topping.

In general, precast elements, including those shown in Figures 6.2(a) and (c), are designed to resist their own weight as well as handling and transportation stresses by the strength provided in pretensioning. Post-tensioning or the use of nonprestressed steel at the supports provides the strength required to resist the live load and superimposed load stresses, and no shoring is used in the construction process.

6.4 ELASTIC ANALYSIS FOR PRESTRESS CONTINUITY

6.4.1 Introduction

Reinforced concrete structures are usually statically indeterminate due to the continuity provided by monolithic construction. Advantageously, the bending moments are always smaller than those of comparable statically determinate beams, leading to shallower, more economical sections. Deformations due to axial loads are usually ignored except in very stiff members, and the settlement of the supports is also rarely considered since creep and shrinkage do not cause major stresses.

In prestressed concrete, continuity also leads to reduced bending moments. However, the bending moments due to the eccentric prestressing forces cause *secondary reactions* and secondary bending moments. These secondary forces and moments increase or decrease the primary effect of the eccentric prestressing forces. Also, the effects of elastic shortening, shrinkage, and creep become considerable as compared to those in reinforced concrete continuous structures.

Because prestressed elements, including those that are partially prestressed, have very limited flexural cracking as compared with reinforced concrete elements, the elastic theory for indeterminate structures can be applied with sufficient accuracy at the limit state of service load. In other words, the prestressed elements can be essentially considered homogenous elastic material because of the limited cracking level, whereas in reinforced concrete it would not be rational to make such an assumption since flexural cracks start to generate at almost 5 to 10 percent of the failure load.

6.4.2 Support Displacement Method

Figure 6.3(a) shows a two-span continuous prestressed concrete beam. In part (b), the central support is assumed to have been removed. Because of the induced *secondary* force or reaction R at the internal support caused by the eccentric prestress, the original moments due to prestressing, namely, $M_1 = P_e e_1$, will be called *primary moments,* and the moments M_2 caused by the induced reactions will be called *secondary moments.* The effect of the secondary moment is to shift the location of the line of thrust, the C-line, at the intermediate supports of the continuous structure, and to return the beam section at the support to its original position before prestressing [see Figure 6.3(c)]. The line of thrust is the center line of compressive force acting along the beam span. The secondary reaction R causes the camber Δ to be neutralized and the beam to be held down at the intermediate support by an equal but opposite reaction R, provided that the C-line at the intermediate support is above the cgc line. If the two lines coincide, the reaction R will be zero, as explained in Section 6.6

The primary structure bending moment diagram M_1 due to the prestressing force is shown in Figure 6.4(a). If it is superimposed on the secondary moment diagram M_2 in Figure 6.4(b), a resulting moment diagram $M_3 = (M_1 + M_2)$ [Figure 6.4(c)] is generated due to the prestressing force for the condition where the beam lower fibers just touch the intermediate support, with the thrust line (C-line) moving a distance y from the tendon cgs profile, i.e., the T-line [Figure 6.4(d)]. As a sign convention, the bending moments diagrams are drawn on the *tension* side of the columns. Such a convention can help eliminate errors in superposition in the analysis of portal frames and other systems whose vertical members are subjected to moments.

The deviation of the C-line from the cgs line is

$$y = \frac{M_2}{P_e} \tag{6.1}$$

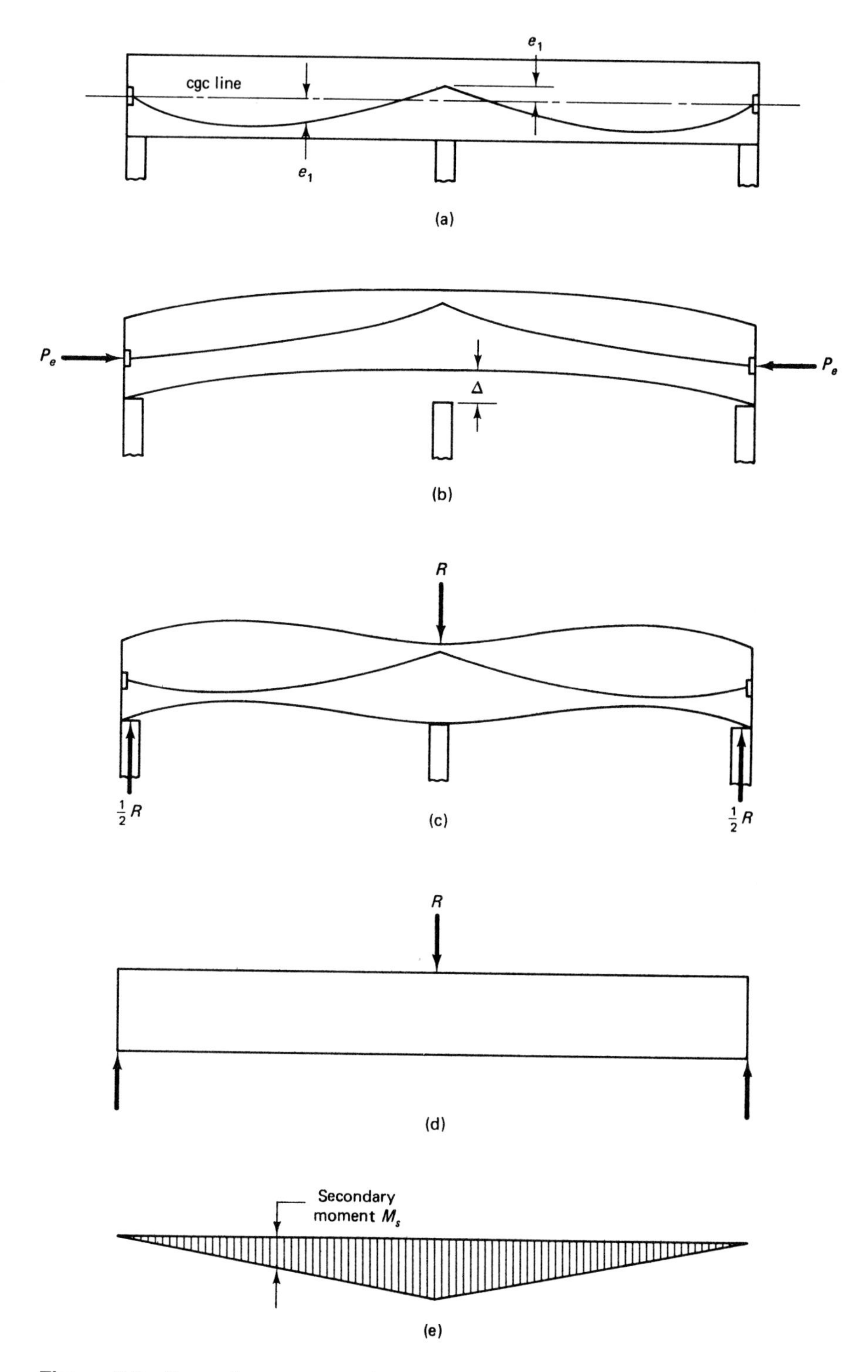

Figure 6.3 Secondary moments in continuous prestressed beams. (a) Tendon profile prior to prestressing. (b) Profile after prestressing if beam is not restrained by central support. (c) Secondary reaction to eliminate uplift or camber. (d) Reaction R on theoretically simply supported beam. (e) Secondary moment diagram due to R.

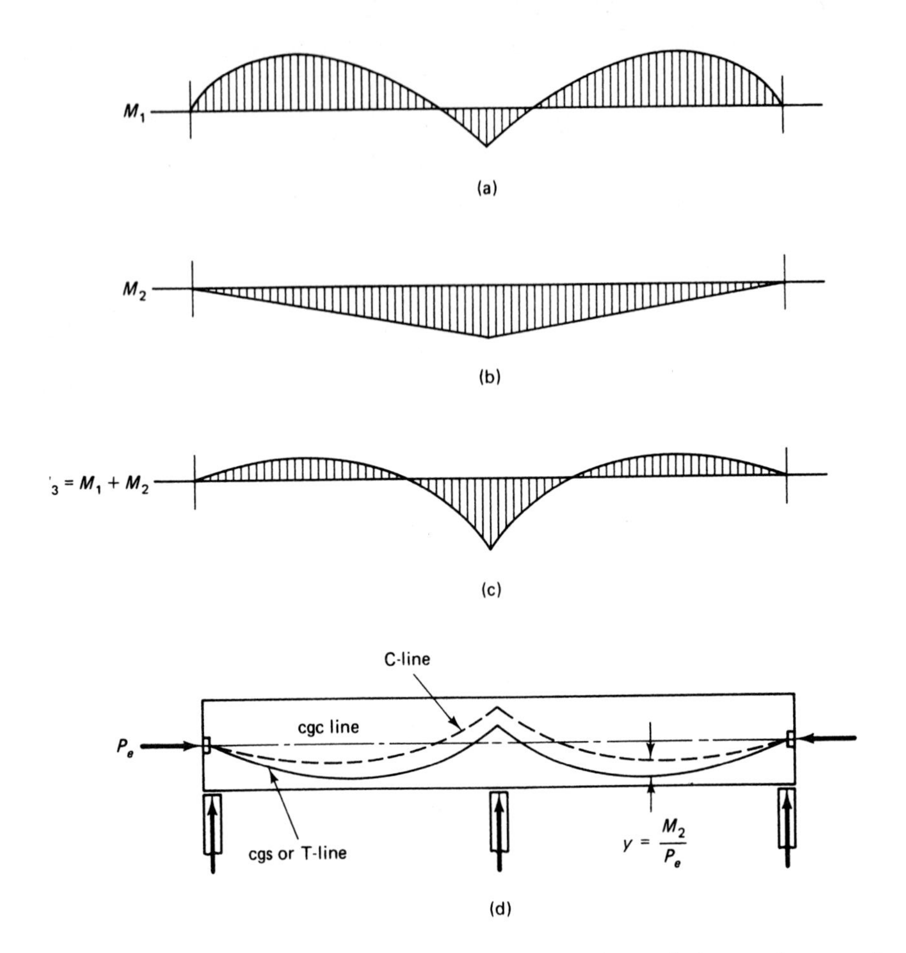

Figure 6.4 Superposition of secondary moments due only to prestress and transformation of the thrust C-line. (a) Primary moments M_1. (b) Secondary moments M_2. (c) Superposition of (b) on (a) to give resulting moment M_3. (d) Transformation of the C-line from the T-line.

and the new location of the tendon profile cgs is determined from the net moment $M_3 = M_1 + M_2$ using the appropriate moment sign, positive (+) above and negative (−) below the base line. The resulting limit eccentricity of the C-line is

$$e' = e_3 = \frac{M_3}{P_e} \tag{6.2}$$

where P_e is the effective prestressing force after losses. Note that e' is negative when the thrust line is *above* the neutral axis, as is the case for the intermediate support section. The concrete fiber stresses due to prestress only at an intermediate support become, from Equations 1.4a and b,

$$f^t = -\frac{P_e}{A_c}\left(1 + \frac{e'_e c_t}{r^2}\right) \tag{6.3a}$$

and

$$f_b = -\frac{P_e}{A_c}\left(1 - \frac{e'_c c_b}{r^2}\right) \tag{6.3b}$$

The concrete fiber stresses at the support due to prestressing and the self-weight support moment are

$$f^t = -\frac{P_e}{A_c}\left(1 + \frac{e'_e c_t}{r^2}\right) + \frac{M_D}{S^t} \tag{6.4a}$$

and

$$f_b = -\frac{P_e}{A_c}\left(1 - \frac{e'_e c_t}{r^2}\right) - \frac{M_D}{S_b} \tag{6.4b}$$

Alternatively, using the M_3 moment values in Equations 6.4a and b, the net moment at the section is $M_4 = M_3 - M_D$, and the concrete fiber stresses at the support where the tendons are above the neutral axis are evaluated from

$$f^t = -\frac{P_e}{A_c} - \frac{M_4}{S^t} \tag{6.5a}$$

and

$$f_b = -\frac{P_e}{A_c} + \frac{M_4}{S_b} \tag{6.5b}$$

Both Equations 6.4 and 6.5 should give the same results whether applied to support, midspan sections, or any other sections along the span provided that the appropriate sign convention is maintained.

6.4.3 Equivalent Load Method

The equivalent load method is based on theoretically replacing the effects of the prestressing force by equivalent loads produced by the prestressing moments profile along the span due to the primary moment M_1 in Figure 6.5(b). If the shear diagram causing moments M_1 is constructed as in Figure 6.5(c), and the load producing this shear is evaluated as in Figure 6.5(d), the reaction R is the same as the displacement reaction R in the method described in Section 6.4.1. Calculation of the moment distribution due to the loading on the continuous beam in Figure 6.5(d) produces the moment diagram of moment M_3 in part (e) of the figure. This moment is the same as the net moment M_3 in Section 6.4.1, so that the resulting limit eccentricity of the cgs line will be $e_3 = M_3/P_e$. The prestress interior support reaction R is obtained from Figure 6.5(d) in order to determine the secondary moment M_2 caused by a load R acting at point c of a simple span AB. The deviation of the C-line from the cgs line is then $y = M_2/P_e$, as in the previous method.

6.5 EXAMPLES INVOLVING CONTINUITY

6.5.1 Effect of Continuity on Transformation of C-Line for Draped Tendons

Example 6.1

A bonded post-tensioned prestressed prismatic beam is continuous on three supports. It has two equal spans of 90 ft (27.4 m), and the tendon profile is shown in Figure 6.6. The effective prestressing force P_e after losses is 300,000 lb (1,334 kN). The beam overall dimensions are b = 12 in. (30 cm) and h = 34 in. (86 cm). Compute the primary and secondary moments due to prestressing, and find the concrete fiber stresses at the intermediate support C due to the prestressing force. Use both the support displacement method and the equivalent load method; assume that the variation in tension force along the beam can be neglected.

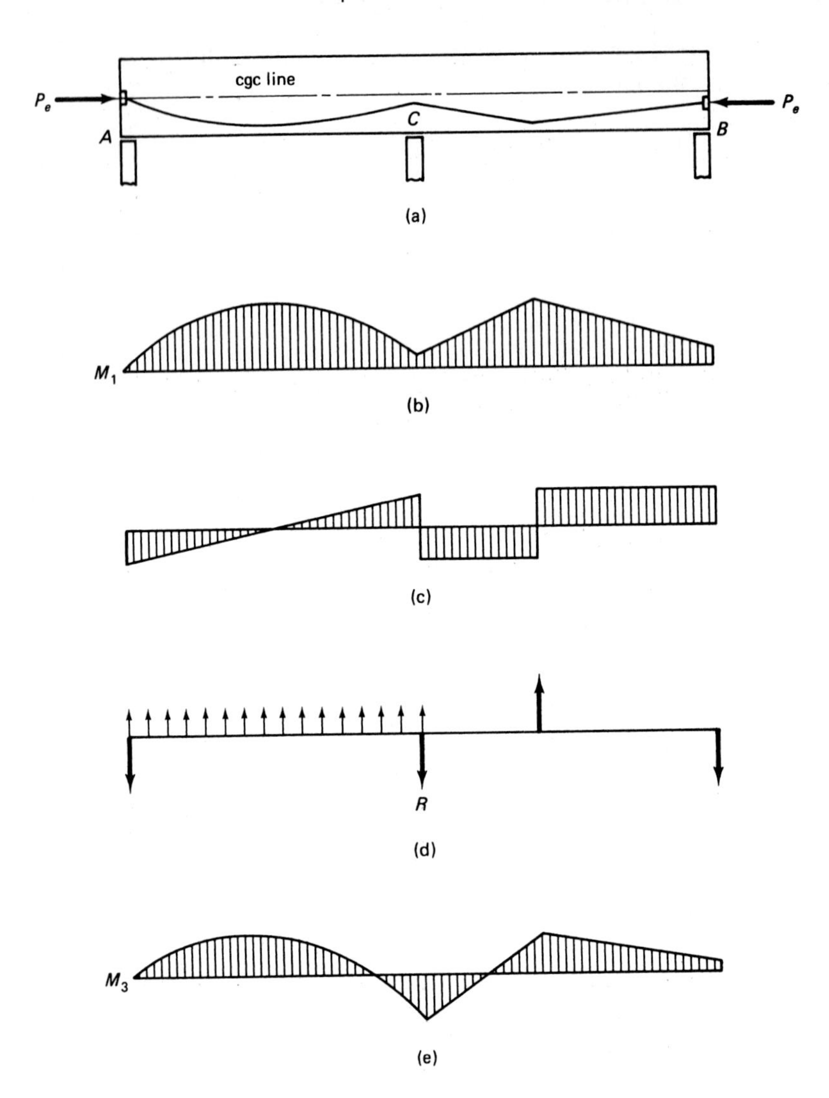

Figure 6.5 Equivalent load method of C-line transformation. (a) Primary structure after prestressing. (b) Primary moment M_1 due to prestressing. (c) Shear diagram for moments M_1. (d) Load-causing moment in (b) and shear in (c). (e) Moment diagram for loads in (d) after moment distributions.

Solution (a):

Support Displacement Method. The primary moment M_1 due to prestressing causes upward *camber* or deflection at the intermediate support C. This camber, Δ_c, can be readily obtained from basic mechanics by the moment area method, taking the moments of areas AEC and ADC about point A in Figure 6.7(a) to get the tangential deviation of the elastic curve at A from the horizontal at C as the displacement at C. From the figure,

$$EI\Delta_c = \left[(3.0 \times 10^6 + 1.05 \times 10^6)\frac{90 \times 2}{3}\right]\frac{90}{2} \times 144$$

$$- \left(\frac{2.1 \times 10^6 \times 90}{2}\right)\frac{90 \times 2}{3} \times 144$$

$$= 7.58 \times 10^{11} \text{ in}^3\text{-lb}$$

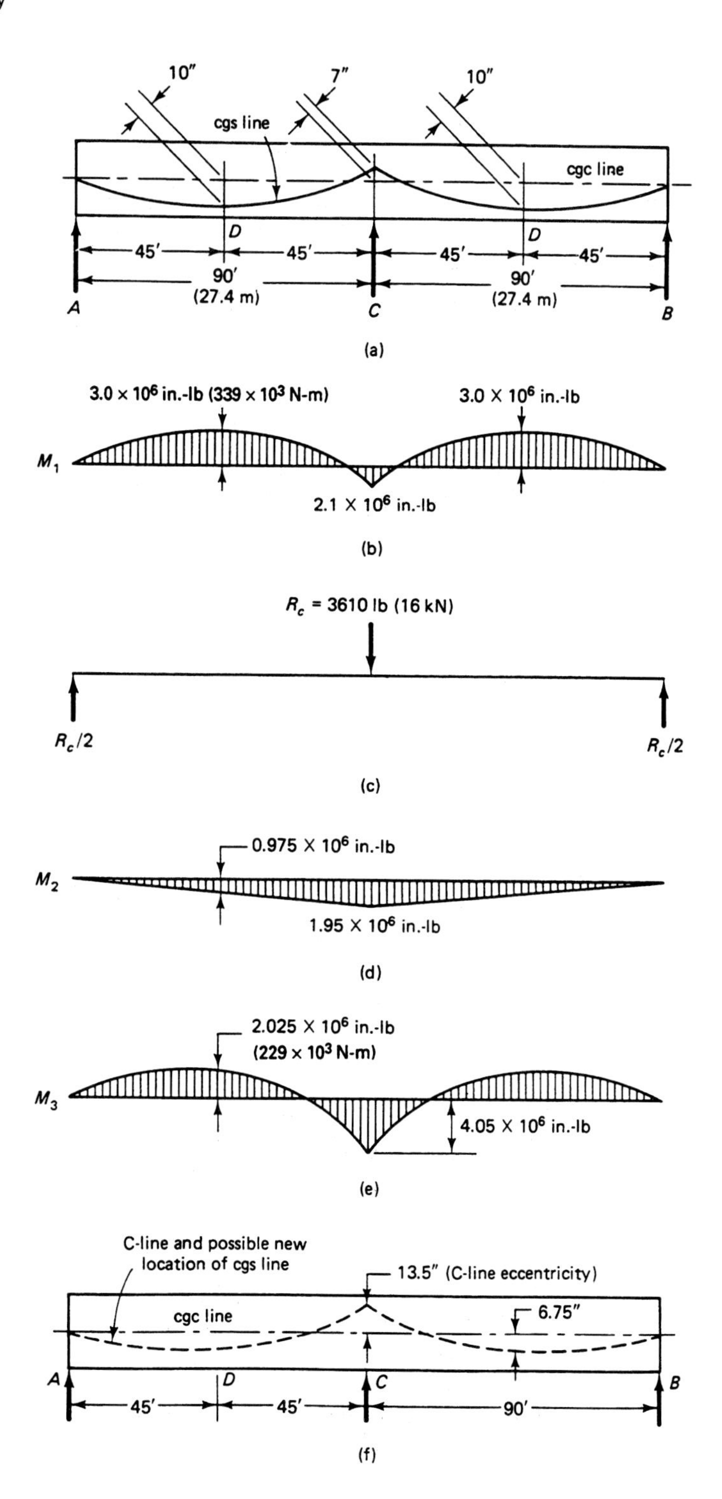

Figure 6.6 Transformation of thrust line in Example 6.1 due to continuity. (a) Tendon geometry: one possible location. (b) Primary moment M_1 due to prestress P_e. (c) Reaction R on theoretically simple beam. (d) Secondary moment M_2 due to R. (e) Final moment $M_3 = M_1 + M_2$. (f) New location of C-line and possible cgs line.

Similarly, from Figure 6.7(c),

$$EI\Delta_c = \frac{45R \times 12 \times 90}{2} \times \frac{90 \times 2}{3} \times 144 = 2.1 \times 10^8\ R \text{ in.}^3\text{-lb}$$

Equating the right sides of these equations to each other yields

$$7.58 \times 10^{11} = 2.1 \times 10^8\ R$$

Then

$$R_c = \frac{7.58 \times 10^{11}}{2.1 \times 10^8} = 3{,}610 \text{ lb} \downarrow (16 \text{ kN})$$

$$R_A = R_B = 1{,}805 \text{ lb} \uparrow (8 \text{ kN})$$

The secondary moment M_2 due to concentrated load R_c varies linearly from interior support C to end supports A and B in Figures 6.6(d) and 6.7(c) . From Figure 6.6(c),

$$M_2 = \frac{R_c}{2} \times 90 \times 12 = \frac{3{,}610}{2} \times 90 \times 12 = 1.95 \times 10^6 \text{ in.-lb}$$

The total moment M_3 at C due to prestress continuity is

$$M_1 + M_2 = 2.1 \times 10^6 + 1.95 \times 10^6 = 4.05 \times 10^6 \text{ in.-lb}$$

From Equation 6.1, the distance through which the C-line has to be transformed upwards at support C is

$$y_c = \frac{M_2}{P_e} = \frac{1.95 \times 10^6}{300{,}000} = 6.5 \text{ in. } (16.5 \text{ cm})$$

From Equation 6.2, the distance of the C-line above the cgc line, i.e., the eccentricity of the C-line above the cgc line at interior support C, is

$$e_c = \frac{M_3}{P_e} = \frac{4.05 \times 10^6}{300{,}000} = 13.5 \text{ in. } (34.3 \text{ cm})$$

The midspan total moment is

$$M_3 = 3.0 \times 10^6 - \frac{1}{2} \times 1.95 \times 10^6$$

$$= 2.025 \times 10^6 \text{ in.-lb}$$

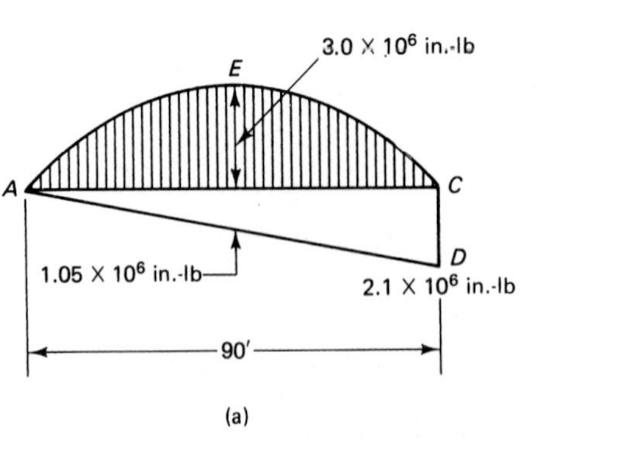

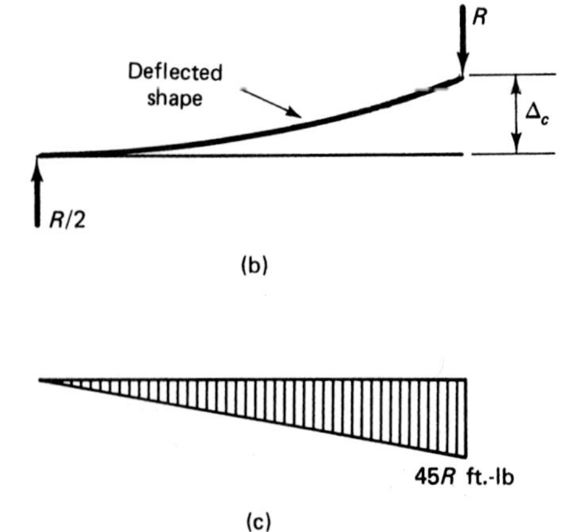

Figure 6.7 Camber Δ_c due to M_1. (a) Primary moment M_1. (b) Deflected shape due to R. (c) Secondary moment M_2 due to R.

and the C-line eccentricity is

$$e_D = \frac{2.025 \times 10^6}{300{,}000} = 6.75 \text{ in. (17.1 cm)}$$

Concrete Fiber Stresses at Interior Support C due to Prestress Only

$$c_t = c_b = \frac{34}{2} = 17 \text{ in.}$$

$$e_C = 13.5 \text{ in.}$$

$$A_c = bh = 12 \times 34 = 408 \text{ in.}^2$$

$$I_c = \frac{bh^3}{12} = \frac{12(34)^3}{12} = 39{,}304 \text{ in.}^4$$

$$r^2 = \frac{I_c}{A_c} = \frac{39{,}304}{408} = 96.33 \text{ in.}^2$$

From Equations 6.3a and b, the top concrete fiber stress is

$$f^t = -\frac{P_e}{A_c}\left(1 + \frac{ec_t}{r^2}\right)$$

$$= -\frac{300{,}000}{408}\left(1 + \frac{13.5 \times 17}{96.33}\right)$$

$$= -2{,}487 \text{ psi } (C) \text{ (17.1 MPa)}$$

and the bottom concrete fiber stress is

$$f_b = -\frac{P_e}{A_c}\left(1 - \frac{ec_b}{r^2}\right)$$

$$= -\frac{300{,}000}{408}\left(1 - \frac{13.5 \times 17}{96.33}\right)$$

$$= +1{,}016 \text{ psi } (T) \text{ (7.0 MPa)}$$

Although the bottom fiber stress in tension is higher and well beyond the maximum allowable, it is only due to prestress. Once self-weight is considered, it diminishes considerably.

Solution (b):

Equivalent Load Method. From Equation 1.16 for load balancing,

$$W_b = \frac{8Pa}{l^2}$$

where a is the eccentricity of the tendon from the cgc line. So

$$W_b = \frac{8 \times 300{,}000 \times 13.5}{(90)^2 \times 12} = 333.3 \text{ lb/ft}$$

$$\text{FEM} = \frac{W_b l^2}{12} = \frac{333.3(90)^2}{12} = 224{,}978 \text{ ft-lb}$$

From the moment distribution operation in Figure 6.8, the final moment at the interior support C is $M_3 = M_1 + M_2 = 337{,}467$ ft-lb $= 4.05 \times 10^6$ in.-lb, which is the same value as solution (a).

Since the M_1 diagram is the primary moment diagram as in Figure 6.6(b), the diagram for the secondary moment M_2 can be constructed from $M_3 - M_1$ as shown in Figure 6.8(b), which is identical to Figure 6.6(d). Thereafter, all other steps for calculation of fiber stresses and location of both fiber stresses and the C-line are identical and give the same results as solution (a).

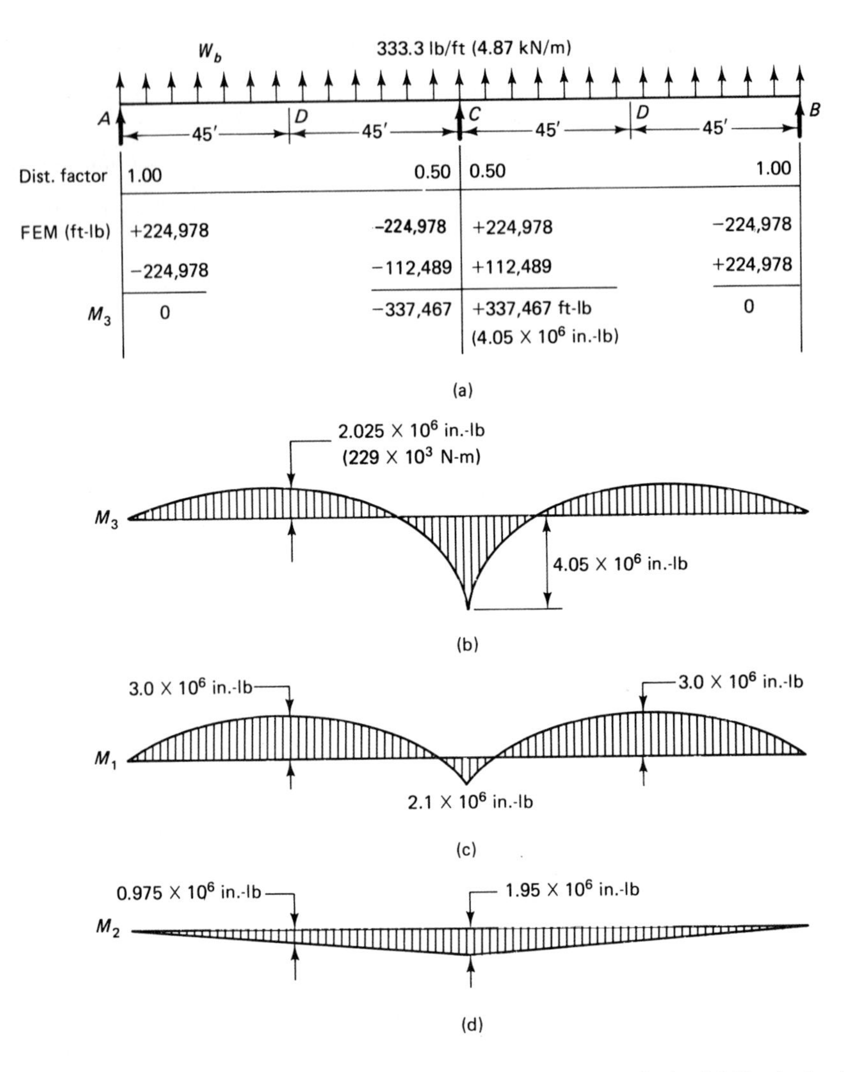

Figure 6.8 Equivalent load method for continuous beam analysis. (a) Equivalent load and moment distribution. (b) Total moment M_3. (c) Primary moment M_1. (d) Secondary moment M_2.

6.5.2 Effect of Continuity on Transformation of C-line for Harped Tendons

Example 6.2

Solve Example 6.1 assuming that the prestressing tendon is harped at the midspan of both adjacent spans. Use the support displacement method in your solution.

Solution: Construct the primary and secondary moment diagram shown in Figure 6.9. Then

$$EI\Delta_c = (3.0 \times 10^6 + 1.05 \times 10^6)90 \times \frac{1}{2} \times \frac{90}{2} \times 144 - \frac{2.1 \times 10^6 \times 90}{2}$$

$$\times \frac{90 \times 2}{3} \times 144 = 3.65 \times 10^{11} \text{ in.}^3\text{-lb}$$

From Figure 6.7 (c),

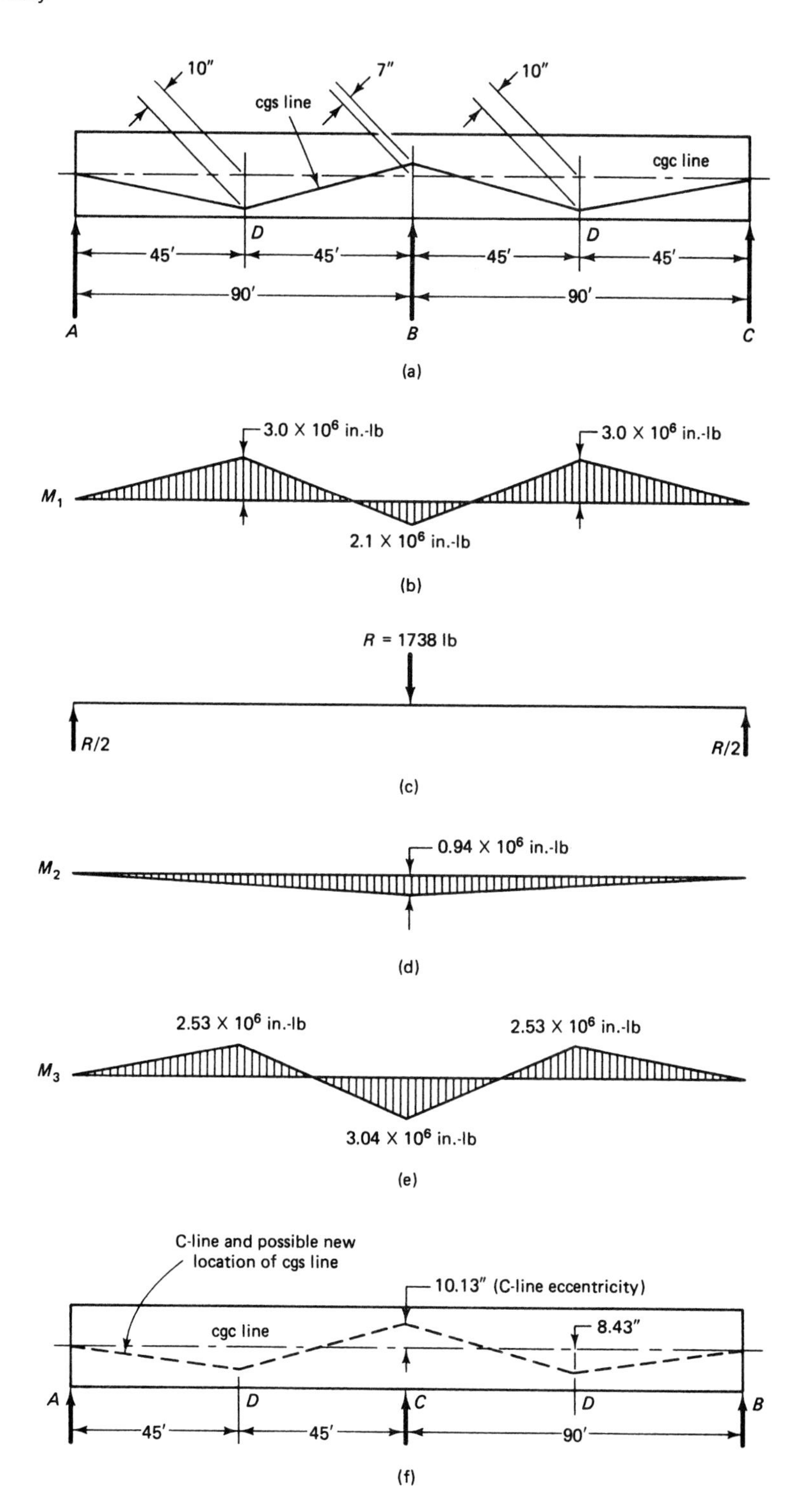

Figure 6.9 Transformation of thrust line in Example 6.2 due to continuity. (a) Tendon geometry: one possible location. (b) Primary moment M_1 due to prestress P_e. (c) Reaction R on theoretically simple beam. (d) Secondary moment M_2 due to R. (e) Final moment $M_3 = M_1 + M_2$. (f) New location of C-line and possible cgs line.

$$EI\Delta_c = 2.1 \times 10^8 R_c$$

$$R_c = \frac{3.65 \times 10^{11}}{2.1 \times 10^8} = 1{,}738 \text{ lb} \downarrow$$

The secondary moment ordinate at interior support C is

$$M_2 = \frac{R}{2} \times 90 \times 12 = \frac{1{,}738}{2} \times 90 \times 12 = 0.94 \times 10^6 \text{ in.-lb}$$

Thus, the total moment $M_3 = 2.1 \times 10^6 + 0.94 \times 10^6 = 3.04 \times 10^6$ in.-lb (344×10^3 N-m).

From Equation 6.1, the distance through which the C-line has to be transformed upwards at support C is

$$y_c = \frac{M_2}{P_e} = \frac{0.94 \times 10^6}{300{,}000} = 3.13 \text{ in. (8 cm)}$$

From Equation 6.2, the distance of the C-line above the cgc line, i.e., the eccentricity of the C-line above the cgc line at interior support C, is

$$e_c = \frac{M_3}{P_e} = \frac{3.04 \times 10^6}{300{,}000} = 10.13 \text{ in. (25.7 cm)}$$

So the midspan total moment is

$$M_3 = 3.0 \times 10^6 - \frac{0.94 \times 10^6}{2} = 2.53 \times 10^6 \text{ in.-lb}$$

and the C-line eccentricity is

$$e_D = \frac{2.53 \times 10^6}{300{,}000} = 8.43 \text{ in. (21.4 cm)}$$

Concrete Fiber Stresses at Interior Support C Due to Prestress Only. $e_c = 10.13$ in. for the C-line or thrust line. So the top concrete fiber stress is

$$f^t = -\frac{P_e}{A_c}\left(1 + \frac{ec_t}{r^2}\right)$$

$$= -\frac{300{,}000}{408}\left(1 + \frac{10.13 \times 17}{96.33}\right)$$

$$= -2{,}049 \text{ psi } (C)\ (14.1 \text{ MPa})$$

and the bottom concrete fiber stress is

$$f_b = -\frac{P_e}{A_c}\left(1 - \frac{ec_b}{r^2}\right)$$

$$= -\frac{300{,}000}{408}\left(1 - \frac{10.13 \times 17}{96.33}\right)$$

$$= +579 \text{ psi } (T)\ (4.0 \text{ MPa})$$

Comparing the results of the harped tendon case of this example to the draped parabolic tendon of Example 6.1 reveals that smaller total continuity moments M_3 resulted at the intermediate support since the triangular area of the moment diagram is one-half the product of the span and the moment ordinate while the parabolic area is two-thirds of that same product. Consequently, the concrete fiber stresses are lower.

6.6 LINEAR TRANSFORMATION AND CONCORDANCE OF TENDONS

It can be recognized from the discussion in Section 6.4 that the profile of the line of thrust (the C-line) follows the profile of the prestressing tendon (the cgs line). This is to be expected since the C-line ordinates are the moment ordinates resulting from the product of the prestressing force P_e and the tendon eccentricity from the cgc line varying along the span. The deflection behavior of any beam is a function of the variation in moment along the span, and the shape of the moment diagram is a function of the type of load, namely, concentrated or distributed. Consequently, tendon profiles are often draped for distributed loads, while they are harped for concentrated loads, as illustrated in Chapters 1 and 4.

Examples 6.1 and 6.2 show that in a continuous beam, the deviation of the C-line from the cgc line is directly proportional to the secondary moment M_2. Since the M_2 diagram varies *linearly* with the distance from the support, it is possible to *linearly* transform the C-line by raising or lowering its position at the *interior* support while preferably maintaining its original position at the exterior simple supports. The profile of the C-line remains the same because of the linearity of the transformation. Consequently, it is possible to linearly transform the cgs line, i.e., the prestressing tendon profile along the beam span, without changing the profile positions of the C-line. This flexibility has major practical significance in the design of continuous prestressed concrete beams.

Example 6.3 demonstrates such a flexibility. Compare the tendon profile it presents with that of Example 6.1, and note that the C-line in Figure 6.6(f) is the same as the C-line in Figures 6.10(f) and 6.11(f), although the cgs line locations along the span are not the same as those in Figures 6.6(a), 6.10(a), and 6.11(a). Also, note that in solution b, where the tendon profile coincides with the C-line profile, the reaction $R = 0$ and the secondary moment $M_2 = 0$. This means that when the cgs line *coincides* with C-line, the beam just touches the intermediate support and behaves like a simply supported beam. Such a beam is called a *concordant beam,* and the prestressing tendon is called a *concordant tendon.*

6.6.1 Verification of Tendon Linear Transformation Theorem

Example 6.3

The continuous beam of Example 6.1 has a new tendon profile as shown in

(a) Figure 6.10(a) with eccentricities $e_c = 0$ and an eccentricity $e_D = 13.5$ in. (24.3 cm) at midspan similar to the intermediate support eccentricity in Example 6.1, and
(b) Figure 6.11(a) with eccentricities $e_c = 13.5$ in. (24.3 cm) and $e_D = 6.75$ in. (17.1 cm) that are the same C-line eccentricities as in Example 6.1.

Verify that the profile and alignment of the C-line in (a) and (b) are the same as the C-line geometry in Example 6.1 and Figure 6.6(f). The total moment at the intermediate support C in both cases is $M_3 = 4.05 \times 10^6$ in.-lb, and at midspan D the moment is $M_3 = 2.025 \times 10^6$ in.-lb, as in Example 6.1.

Solution (a): The secondary moment is

$$M_2 = 4.05 \times 10^6 \text{ in.-lb } (458 \times 10^3 \text{ N-m})$$

Also,

$$\frac{R_c}{2} \times 90 \times 12 = 4.05 \times 10^6$$

$$R_c = \frac{4.05 \times 10^6 \times 2}{90 \times 12} = 7{,}500 \text{ lb } (33.4 \text{ kN}) \downarrow$$

$$R_A = R_B = 3{,}750 \text{ lb } (16.7 \text{ kN}) \uparrow$$

Solution (b): Both $M_2 = 0$ and $R = 0$.

It can be seen from both solution (a) in Figure 6.10 and solution (b) in Figure 6.11 that the C-line coordinates are the same and close to the values obtained in Example 6.1. The

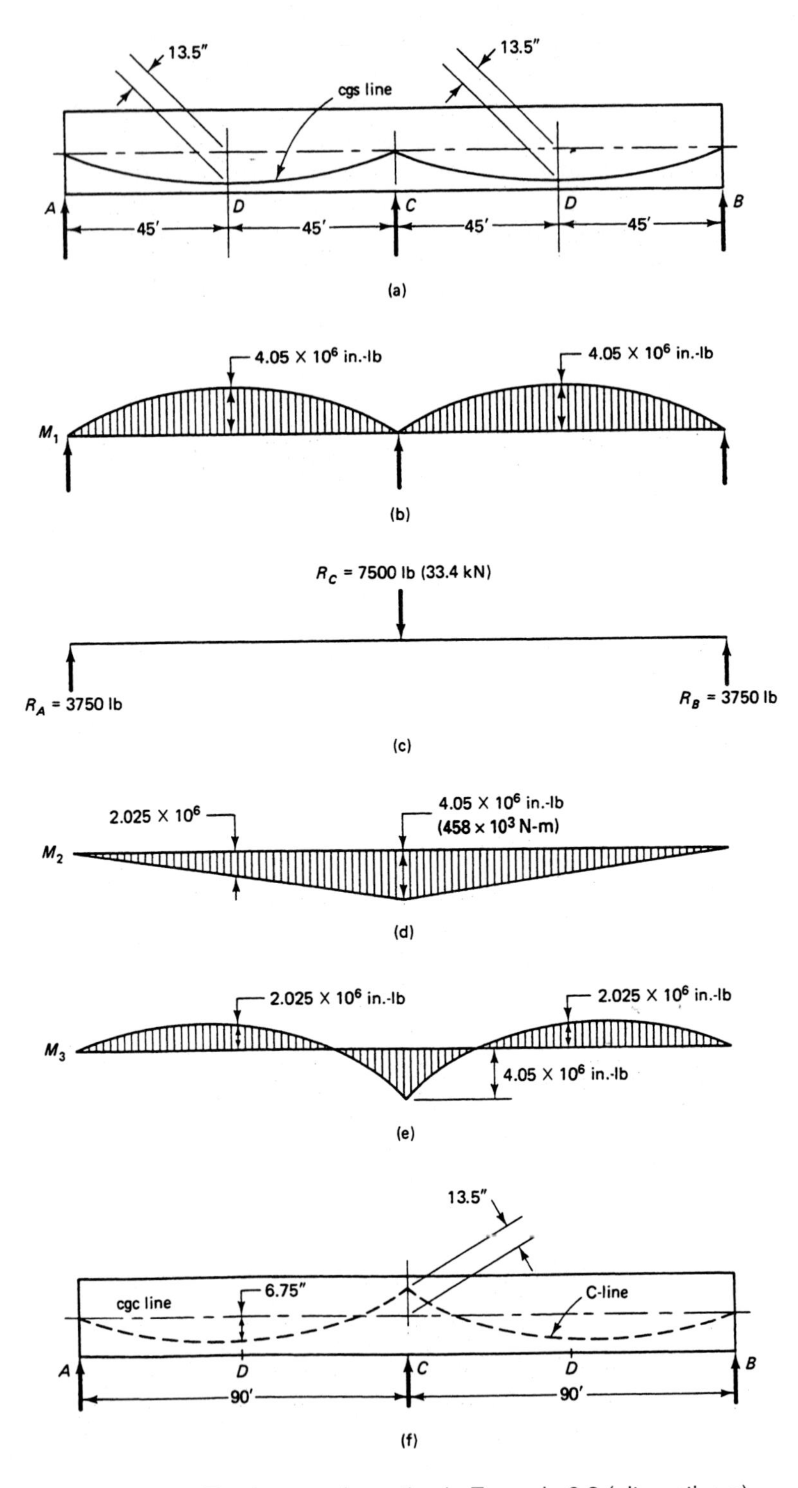

Figure 6.10 Tendon transformation in Example 6.3 (alternative a).

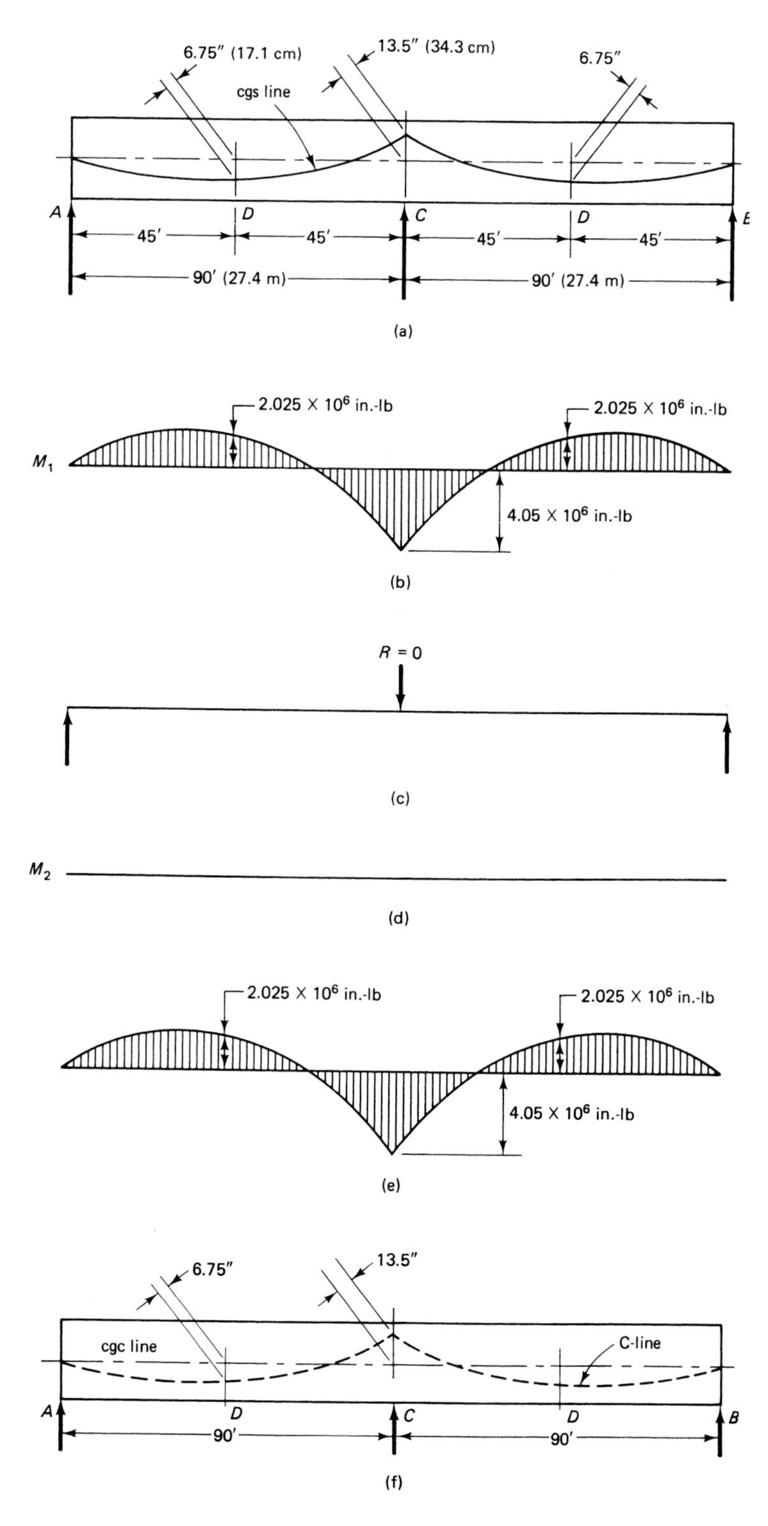

Figure 6.11 Tendon transformation in Example 6.3 (alternative b).

fiber stresses are also the same, viz., $f^t = -2{,}487$ psi (C) (17.1 MPa) and $f_b = +1{,}016$ psi (T) (7.0 MPa).

6.6.2 Concordance Hypotheses

The following list summarizes the hypotheses defining the transformation and concordance of tendons in continuous prestressed beams:

1. Any tendon profile can be linearly transformed without affecting the C-line position.
2. A beam with a concordant tendon is a continuous beam whose C-line coincides with its cgs line.
3. A concordant tendon induces no reactions on intermediate supports.
4. The eccentricity of any concordant tendon measured from the cgc line produces a moment diagram representing a profile similar in form to the moment profile due to the superimposed load.
5. Any line of thrust (C-line) is a profile for a concordant tendon.
6. Superposition of several concordant tendons produces a concordant tendon, but superposition of concordant and nonconcordant tendons produces a nonconcordant tendon.
7. A change in eccentricity at one or both *end* supports results in a shift of the C-line, but a change in eccentricity at *intermediate* supports does not affect the position of the C-line.
8. The choice of concordance or nonconcordance is determined by concrete cover and efficiency in beam depth selection. Bending moments and shear diagrams for superposition of transverse loads on continuous beams is shown in Figure 6.12.

It is advisable to start a design assuming concordance in order to eliminate the need for calculating the secondary moment M_2. By trial and adjustment, one can arrive at the final beam section depth that fulfills the design requirements with the cgs location either concordant or nonconcordant, as the final design dictates.

6.7 ULTIMATE STRENGTH AND LIMIT STATE AT FAILURE OF CONTINUOUS BEAMS

The service-load design of continuous prestressed beams assumes *elastic* behavior of the material up to the limit of allowable tensile stress in the concrete due to all loads. This limit in tensile stress is based on allowing some limited cracking beyond the first cracking load as determined by the modulus of rupture of concrete. As cracking becomes more effective during overload conditions, internal plastic deformation at the critical regions of maximum or peak moments and plastic redistribution of elastic moments from the negative to the positive moment regions are generated. At this stage of overload and beyond, up to the limit state at failure, plastic hinging develops at the most highly stressed regions in the continuous beam.

Total redistribution and full development of plastic hinges at the continuous supports of a fully bonded prestressed beam render the beam statically determinate, as if concordance of tendons were present with zero moments at the supports. Theoretically, in such cases the secondary moments M_2 can be disregarded beyond the first cracking load. However, such an assumption can result in an unsafe design unless a concordant tendon is assumed from the beginning and is executed in the final design with no overload conditions permitted. Otherwise, it is important to consider the secondary moment

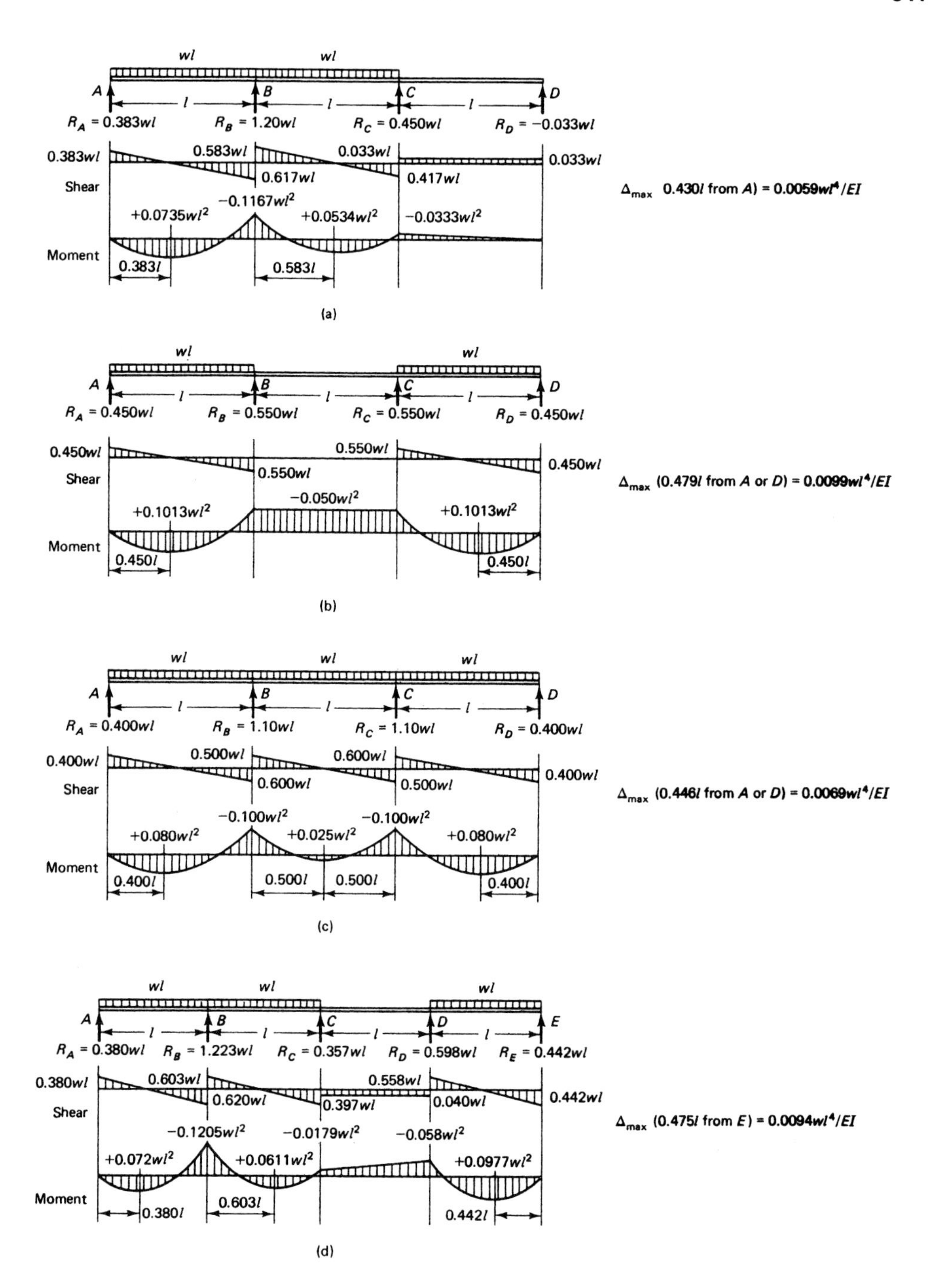

Figure 6.12 Bending moments and shear diagrams for continuous beams. (a) Continuous beam, three equal spans, one end span unloaded. (b) Continuous beam, three equal spans, end spans loaded. (c) Continuous beam, three equal spans, all spans loaded. (d) Continuous beam, four equal spans, third span unloaded. (e) Continuous beam, four equal spans, first and third spans loaded. (f) Continuous beam, four equal spans, all spans loaded. (g) Continuous beam, two equal spans, concentrated load at center of one span. (h) Continuous beam, two equal spans, concentrated load at any point.

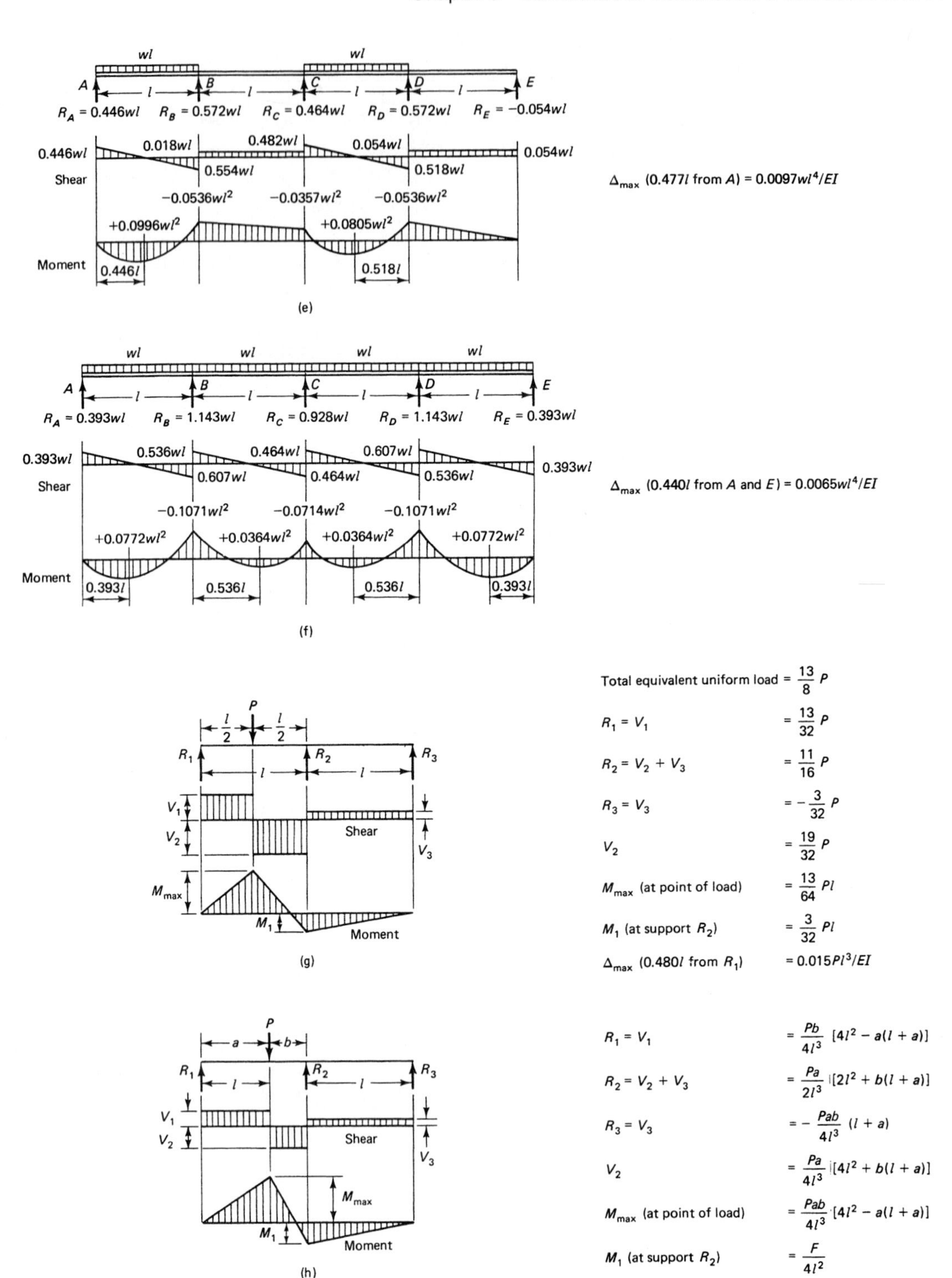

Figure 6.12 *Continued*

Phote 6.1 Seven Mile Bridge, Florida Keys. (*Courtesy,* Post-Tensioning Institute.)

M_2 due to prestressing up to the limit state of the failure load, but with a load factor of 1.0. Considering the secondary moments is mandated by the fact that the elastic deformation caused by the nonconcordant tendons changes the amount of *inelastic* rotation required to obtain a given amount of redistribution. Conversely, for a beam with a given elastic rotational capacity, the amount by which the moment at the support may be varied is changed by an amount equal to the secondary moment at the support due to prestressing (Ref. 6.2).

In order to determine the moments to be used in the design, the following sequence of steps is recommended:

1. Determine the moments due to the dead and live loads at factored load level.
2. Modify by algebraic addition of secondary moments M_2 due to prestressing.
3. Redistribute as permitted. A positive secondary moment at the support caused by transforming a tendon downwards from a concordant profile will therefore *reduce* the negative moments near the supports and *increase* the positive moment in the midspan region. A tendon that is transformed upwards will have a reverse effect.

The ACI code permits an increase or decrease in negative moments calculated by the elastic theory for any assumed loading arrangements up to a limit defined by the percentage

$$p_d = 20\left[1 - \frac{\omega_p + \dfrac{d}{d_p}(\omega - \omega')}{0.36\beta_1}\right] \tag{6.6}$$

where $\omega_p = \rho_p f_{ps}/f'_c$
$\omega' = \rho' f_y/f'_c$
$\omega = \rho f_y/f'_c$
for $\rho_p = A_p/bd_p$
and $\rho' = A'_s/bd$.

The modified negative moments are to be used for calculating moments at sections within spans for the same loading arrangements.

The ACI Code also requires that redistribution of moments be made only when the section at which the moment is reduced is so designed that whichever reinforcement index ω applies is $\leq 0.24\beta_1$, i.e., either

$$\omega_p \leq 0.24\beta_1 \tag{6.7a}$$

or

$$\left[\omega_p + \frac{d}{d_p}(\omega - \omega')\right] \leq 0.24\beta_1 \tag{6.7b}$$

or

$$\left[\omega_{pw} + \frac{d}{d_p}(\omega_w - \omega'_w)\right] \leq 0.24\beta_1 \qquad 6.7c)$$

where ω_{pw} is for flanged sections using the web width.

6.8 TENDON PROFILE ENVELOPE AND MODIFICATIONS

The envelope for limiting tendon eccentricities for continuous beams can be constructed in the same manner as discussed in Section 4.4.3 for simply supported beams. A determination has to be made as to whether tension is to be allowed in the design in order to establish the limiting maximum and minimum ordinates of the upper and lower envelopes relative to the top and bottom kerns.

The tendon profile at the intermediate supports cannot have the same peak configuration as that of the negative bending moment diagram, due to both design and practical considerations. These considerations include the magnitude of frictional losses that increase with the decrease in the radius of curvature, the high level of compressive stress concentration in cases of abrupt changes in the tendon, and the additional difficulties that can be encountered in post-tensioning. Consequently, it is advisable to modify the tendon profile at the support so as to have a curvilinear transition at the support zone. Such modification has to be accounted for by modifying the primary moment M_1 diagram and the total moment M_3 diagram.

Typical tendon alternative profiles with equal upper and lower eccentricities at the peak moment sections are shown in Figure 6.13, where the chain-dotted line gives the average bending moments used in the service-load design. Selection of the tendon profile should be based on the following considerations:

1. The eccentricity should be as large as possible at the point where the largest bending moment develops at the limit state at failure.
2. Where possible, the total prestressing moment at any section should be sufficient to counteract the average service-load bending moment at that section. The test for this condition should be based on the resulting stress values rather than moment values, since the prestressing force causes axial load stress as well as bending moment stress.
3. A tendon profile alternative that produces the *least* frictional losses should be chosen in the design.
4. It is essential to consider the ultimate-load requirements when selecting the tendon profile; a tendon chosen only on the basis of linear transformation and concordance is not necessarily satisfactory, as it might neither totally fulfill the service-load stress

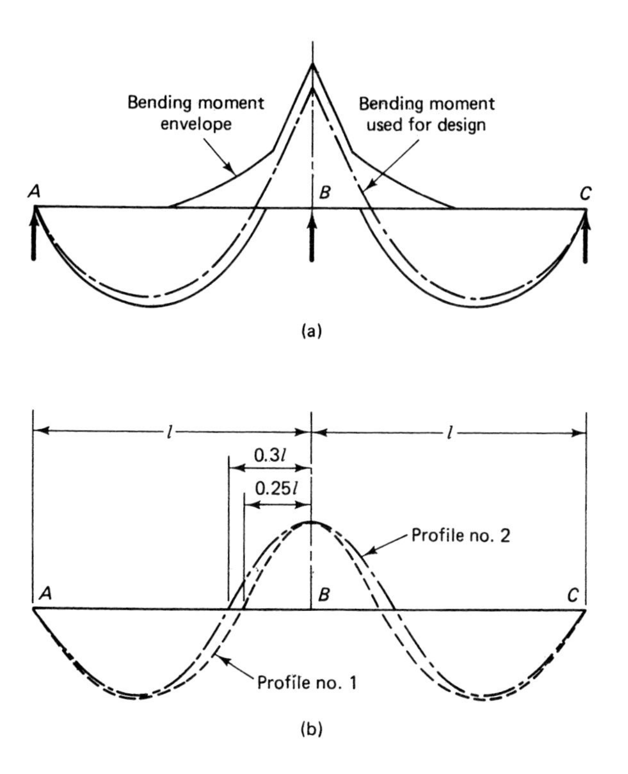

Figure 6.13 Tendon profile modification. (a) Bending moment diagram for continuous beam. (b) Tendon profile alternatives.

requirements nor fully satisfy the ultimate-load requirements at both the midspan and interior supports.

5. A decrease in eccentricities at the intermediate supports through additional tendon transformation decreases the service-load compressive stresses and can result in allowing additional live-load moments, and hence more live load, *provided* that the midspan profile eccentricity and moments produce satisfactory concrete stresses.

6.9 TENDON AND C-LINE LOCATION IN CONTINUOUS BEAMS

Example 6.4

Design the tendon profile in the beam of Example 4.7, assuming the unshored beam to be continuous over three spans 64 ft (19.5 m) each, with only uniformly distributed live load $w_L = 1{,}514$ plf (22.1 kN/m). Consider the post-tensioned prestressing tendon to be continuous throughout the structure and fully grouted. Disregard tension force variation due to frictional losses in the bends, and assume that a maximum allowable concrete compressive fiber stress $f_c = 0.45f'_c = 2{,}250$ psi (15.5 MPa) is reached at the extreme top fibers of the composite section when the live load acts on the section. Use a modified effective width $b_m = 65$ in. (165 cm) for the compression flange to account for the modular ratio of the topping and precast concrete.

Solution:

1. *Input data from Example 4.7*
   (a) *Stress data*

Precast $f'_c = 5{,}000$ psi (34.5 MPa), normal-weight concrete

Topping $f'_c = 3{,}000$ psi (20.7 MPa), normal-weight concrete

$f'_{ci} = 4{,}000$ psi (27.6 MPa)

$f_{pu} = 270{,}000$ psi (1,862 MPa)

$f_{py} = 243{,}000$ psi (1,655 MPa)

$f_{pi} = 189{,}000$ psi (1,303 MPa)

$f_{pe} = 0.8(0.7f_{pu}) = 151{,}200$ psi (1,043 MPa)

$\gamma = 0.8$

Midspan $f_t = 12\sqrt{f'_c} = 12\sqrt{5{,}000} = 849$ psi (5.85 MPa)

Support $f_t = 6\sqrt{f'_c} = 425$ psi (2.93 MPa)

(b) *Load data*

$W_D = 583$ plf (8.5 kN/m)

$W_{SD} = (1\frac{3}{4}/12) \times 7 \text{ ft} \times 150 = 153$ plf (2.2 kN/m) for $1\frac{3}{4}$ in. precast slab formwork

$W_{CSD} = (4/12) \times 7 \text{ ft} \times 150 = 350$ plf (5.1 kN/m) for 4 in. situ-cast topping

$W_L = 1{,}514$ plf (22.1 kN/m) (obtained from $M_L = 9{,}300{,}000$ in.-lb)

Span = 64 ft (19.5 cm)

(c) *Section properties*

AASHTO Type III

Property	**Precast**	**Composite**
A_c, in.^{2}	560	934
I_c, in.^{4}	125,390	297,045
r^2, in.^{2}	223.9	318.1
c_b, in.	20.27	31.32
c_t, in.	24.73	13.86
S_b, in.^{3}	6,186	9,490
S^t, in.^{3}	5,070	21,174
S_{cbs}, in.^{3} (bottom of slab)		19,251
S^t_{cs}, in.^{3} (top of slab)		15,288

$n = E_c\,(\text{topping})/E_c\,(\text{precast}) = 0.77$

Precast $h = 45$ in. (114.3 cm)

Transformed $b = 65$ in. (165 cm)

Situ-cast $h_f = 5.75$ in. (14.6 cm)

(d) *Prestressing steel*

Twenty-two $\frac{1}{2}$-in. dia (12.7 mm dia) 7-wire low-relaxation strands

$A_{ps} = 22 \times 0.153 = 3.366 \text{ in.}^2$ (21.7 cm^2)

2. *Assume a trial tendon profile location as shown in Figure 6.14*

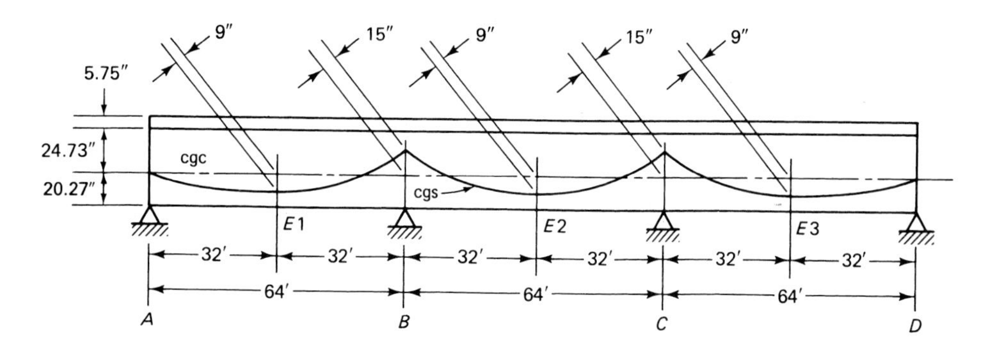

Figure 6.14 Trial profile of the prestressing tendon.

The prestressing force after losses is

$$P_e = 3.366 \times 151{,}200 = 508{,}940 \text{ lb } (2{,}264 \text{ kN})$$

while the primary moment at support B is

$$M_B = P_e \times e_B = 508{,}940 \times 15 = 7.63 \times 10^6 \text{ in.-lb } (862 \times 10^3 \text{ N-m})$$

The primary moment at midspan E_1 is

$$M_{E1} = 508{,}940 \times 9 = 4.58 \times 10^6 \text{ in.-lb}$$

For simplification, the tangential deviation used for computing Δ_B is based on assuming the tangent to the elastic curve at B as horizontal.

Taking moments of areas about A in Figure 6.15 yields

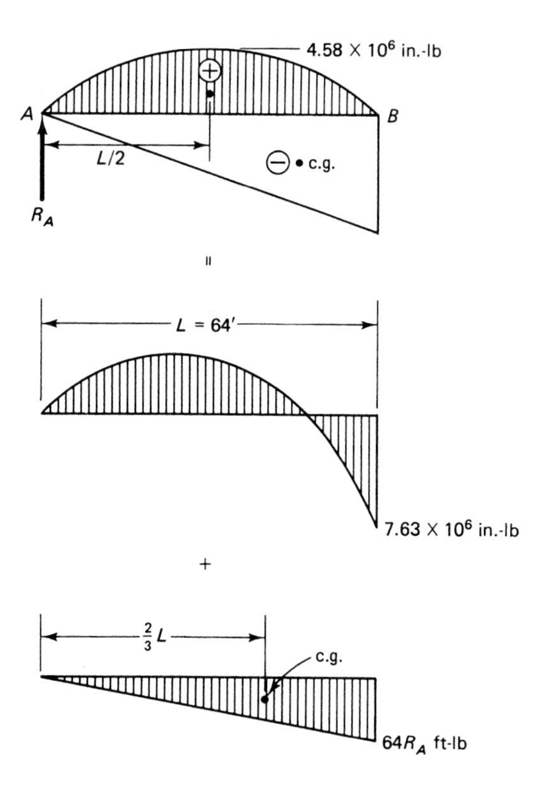

Figure 6.15 Computation of displacement by taking moments of area about support A.

$$E_cI_c\Delta_B = E_cI_c\Delta_c = \left[\left(\frac{7.63}{2} + 4.58\right)10^6 \times \frac{64 \times 2}{3}\right] \times \frac{64}{2} \times 144$$

$$- \left[\frac{7.63 \times 10^6 \times 64}{2}\right] \times \frac{64 \times 2}{3} \times 144$$

$$= 15.0 \times 10^{10} \text{ in.-lb}$$

Also,

$$E_cI_c\Delta_B = (R_A \times 64 \times 12)\left(\frac{64 \times 12}{2}\right)\left(\frac{64 \times 12 \times 2}{3}\right)$$

$$= 151 \times 10^6 R_B \text{ in-lb}$$

$$R_A = R_D = \frac{15.0 \times 10^{10}}{151 \times 10^6} = 994 \text{ lb} \uparrow \text{(By exact calculation} = 798 \text{ lb)}$$

The secondary moment M_2 at support B is

$$R_A \times 64 \times 12 = 994 \times 64 \times 12 = 0.76 \times 10^6 \text{ in.-lb}$$

The total moment at support B due to prestressing only is

$$\text{Support } M_3 = (7.63 + 0.76) \times 10^6 = 8.39 \times 10^6 \text{ in.-lb (see Figure 6.16)}$$

The C-line eccentricity is

$$e'_B = -\frac{8.39 \times 10^6}{508{,}940} = -16.49 \text{ in. (41.9 cm)}$$

(The eccentricity e' of the C-line when it is above the neutral axis is considered negative in order to conform to Equations 6.3a and b.) Finally, the total moment at midspans, E_1, E_2, and E_3 due to prestressing only is

$$\text{Midspan } M_3 = (4.58 - 0.38)10^6 = 4.20 \times 10^6 \text{ in.-lb}$$

and the C-line eccentricity is

$$e'_{E1} = +\frac{4.20 \times 10^6}{508{,}940} = +8.25 \text{ in. (21.0 cm)}$$

3. *Concrete fiber stresses due to prestress and self-weight (583 plf)*
Using the moment factors and reaction factors from Figure 6.12, we have

$$M_D \text{ at } B = M_{B1} = 0.10 \times 583(64)^2 \times 12 = 2.87 \times 10^6 \text{ in.-lb}$$
(see Figure 6.17)

$$P_e = A_{ps}f_{pe} = 3.366 \times 151{,}200 = 508{,}940 \text{ lb}$$

M_D at E_1 = 2.12×10^6 in.-lb (Max. + M is not at midspan; hence, calculate M_{E1} from the area of the shear diagram)

$$\text{Net } R_B = 1.1W\uparrow - 994 = 1.1 \times 583 \times 64 - 994 = 40{,}048 \text{ lb}$$

The total support B moment due to prestressing and self-weight is, then,

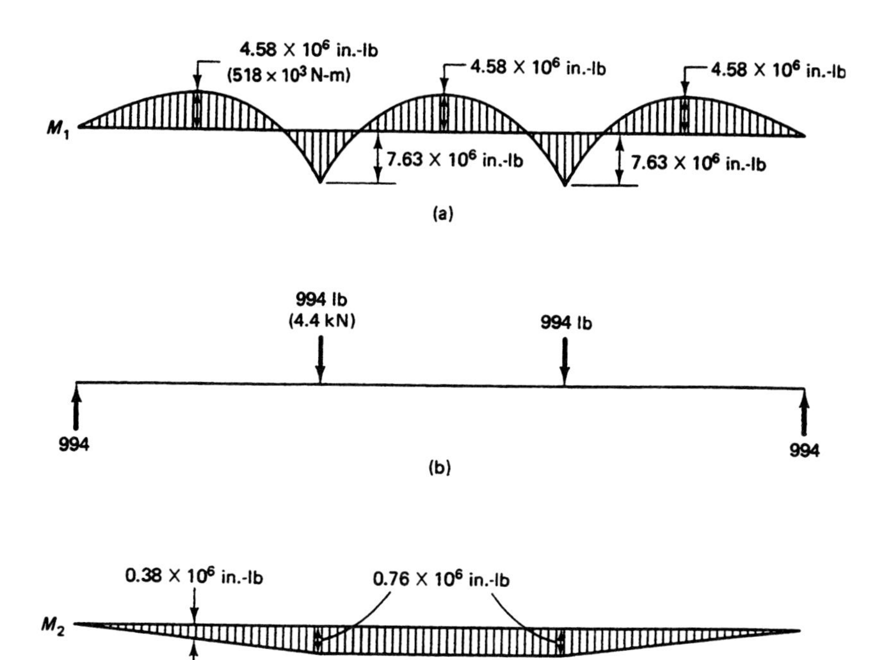

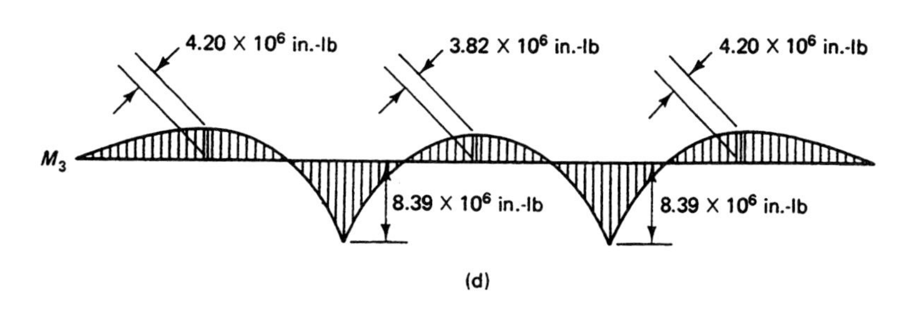

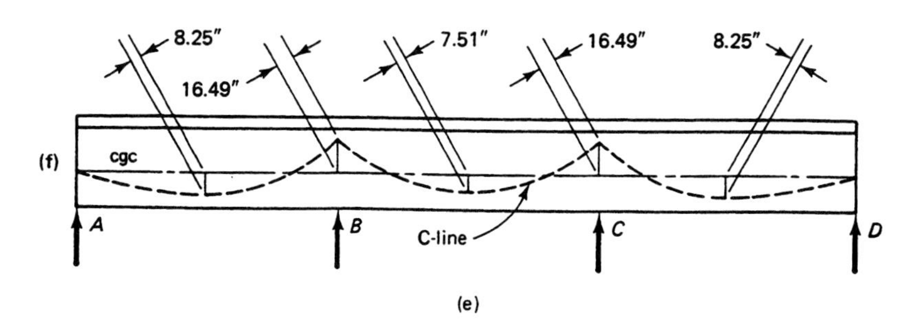

Figure 6.16 Tendon C-line profile in continuous beam of Example 6.4.

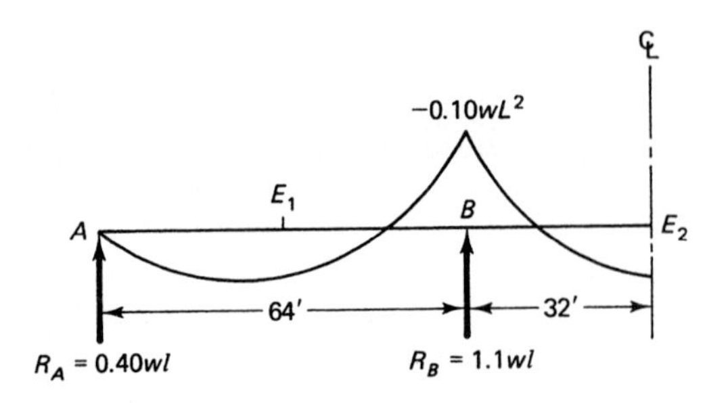

Figure 6.17 Moment diagram due to external load.

$$M_4 = M_3 - 2.87 \times 10^6 = (8.39 - 2.87)10^6$$
$$= 5.52 \times 10^6 \text{ in.-lb } (624 \times 10^3 \text{ N-m})$$

(i) *Support section B or C*
The construction process in this stage involves mounting the precast I-beams and prestressing them. The fiber stresses due to prestressing and self-weight, from Equations 6.5a and b, are as follows:

$$f_1^t = -\frac{P_e}{A_c} - \frac{M_4}{S^t} = -\frac{508{,}940}{560} - \frac{5.52 \times 10^6}{5{,}070} = -908.8 - 1{,}088.8$$
$$= -1997.6 \text{ psi } (C)\ (13.8 \text{ MPa}) < 2{,}250 \text{ psi, O.K.}$$
$$f_{1b} = -\frac{P_e}{A_c} + \frac{M_4}{S_b} = -\frac{508{,}940}{560} + \frac{5.52 \times 10^6}{6{,}186} = -908.8 + 892.3$$
$$= -16.5 \text{ psi } (C), \text{ no tension, O.K.}$$

Alternatively, the concrete fiber stresses at the support can be computed from Equations 4.1a and b or Equations 6.4a and b using the C-line eccentricities $e'_e = -16.49$ in. We obtain

$$f_1^t = -\frac{P_e}{A_c}\left(1 + \frac{e'_e c_t}{r^2}\right) + \frac{M_D}{S^t}$$
$$= -\frac{508{,}940}{560}\left(1 + \frac{16.49 \times 24.73}{223.9}\right) + \frac{2.87 \times 10^6}{5{,}070}$$
$$= -2{,}564.1 + 566.1 = -1{,}998 \text{ psi } (C)\ (13.8 \text{ MPa})$$
$$= -\frac{P_e}{A_c}\left(1 - \frac{e'_e c_b}{r^2}\right) - \frac{M_D}{S_b}$$
$$= -\frac{508{,}940}{560}\left(1 - \frac{16.49 \times 20.27}{223.91}\right) - \frac{2.87 \times 10^6}{6{,}186}$$
$$= +447.9 - 463.9 = -16.0 \text{ psi } (C)$$

(ii) *Outer span midspan E_1*
The eccentricity at the midspan is $e'_{E1} = +8.25$ in. Hence,

$$f_1^t = -\frac{P_e}{A_c} + \frac{M_4}{S^t}$$
$$f_{1b} = -\frac{P_e}{A_c} - \frac{M_4}{S_b}$$

Total moment M_4 at $E_1 = (4.20 - 2.12)10^6 = 2.08 \times 10^6$ in.-lb

$$f_1^t = -\frac{P_e}{A_c} + \frac{M_4}{S^t} = -\frac{508{,}940}{560} + \frac{2.08 \times 10^6}{5{,}070}$$

$$= -908.8 + 410.3 = -498.5 \text{ psi } (C)\ (3.4 \text{ MPa}), \text{ O.K.}$$

$$f_{1b} = -908.8 - \frac{2.08 \times 10^6}{6{,}186} = -1{,}245 \text{ psi } (C)\ (8.6 \text{ MPa}), \text{ O.K.}$$

Figure 6.16 gives the moments, eccentricities, and C-line geometry of the continuous beam in this example.

4. *Effect of adding the superimposed dead loads on support sections B and C* At this stage of the construction process the $1\frac{3}{4}$ in. thick precast slabs (W_{SD}) are erected, followed by placing the 4 in. thick layer of wet concrete (W_{SD}). Thus, $W_{SD} = 153 + 350 = 503$ plf (7.3 kN/m). The support moment due to superimposed load is

$$M_{B2} = 0.1wl^2 = 0.1 \times 503(64)^2 \times 12 = -2.47 \times 10^6 \text{ in.-lb}$$

Also,

$$f_2^t = -\frac{P_e}{A_c} - \frac{M_5}{S^t}$$

$$f_{2b} = -\frac{P_e}{A_c} + \frac{M_5}{S_b}$$

$$M_5 = (M_4 - 2.47 \times 10^6) \text{ in.-lb} = 3.05 \times 10^6 \text{ in.-lb}$$

Hence,

$$f_2^t = -908.8 - \frac{3.05 \times 10^6}{5{,}070} = -1{,}510.4 \text{ psi } (C$$

$$f_{2b} = -908.8 + \frac{3.05 \times 10^6}{6{,}186} = -415.8 \text{ psi } (C)$$

5. *Effects of adding the superimposed live load on support sections B and C* After the concrete cures, resulting in full composite action for supporting the total service live load, the section moduli for the precast beam become

$$S_c^t = 21{,}714 \text{ in.}^3$$

and

$$S_{cb} = 9{,}490 \text{ in.}^3$$

The loading combination which causes the highest stress condition is when the load acts on only two *adjacent* spans AB and BC. We have

$$W_L = 1{,}514 \text{ plf } (22.1 \text{ kN/m})$$

including W_{CSD}. From Figure 6.12, the support moments due to live load on two adjacent spans is

$$M_{B3} = 0.1167W_L l^2 = 0.1167 \times 1{,}514(64)^2 \times 12 = -8.68 \times 10^6 \text{ in.-lb}$$

The live-load moment causes tension at the top, and compression at the bottom fibers of the support section. The resulting total fiber stresses in the precast section at the support due to all loads become

$$f_T^t = f_2^t + \frac{M_{B3}}{S_c^t}$$

and

$$f_{bT} = f_{2b} - \frac{M_{B3}}{S_{cb}}$$

Hence,

$$f_T^t = -1{,}510.4 + \frac{8.68 \times 10^6}{21{,}714}$$

$$\cong -1{,}111 \text{ psi } (C)\ (7.7 \text{ MPa}) < 2{,}250 \text{ psi, O.K.}$$

$$f_{bT} = -414.3 - \frac{8.68 \times 10^6}{9{,}490} = -1{,}329 \text{ psi (9.2 MPa)} < 2{,}250 \text{ psi, O.K.}$$

Alternative one-step solution using Equations 4.19a and b

$$f_T^t = -\frac{P_e}{A_c}\left(1 - \frac{ec_t}{r^2}\right) - \frac{M_D + M_{SD}}{S^t} - \frac{M_{CSD} + M_L}{S_c^t}, \text{where } S_c^t \text{ is at the top of the precast section}$$

$$f_{bT} = -\frac{P_e}{A_c}\left(1 + \frac{ec_b}{r^2}\right) + \frac{M_D + M_{SD}}{S_b} + \frac{M_{CSD} + M_L}{S_{cb}}$$

where $e = e'_B = M_3/P_e$ and M_{CSD} is the additional superimposed dead load at service after erection, assumed zero here.

From before, the C-line eccentricity is $e'_B = M_3/P_e = -8.39 \times 10^6/508{,}940 = -16.49$ in. (41.9 cm). Also, the support moments, using the appropriate signs, are

$$M_D = -2.87 \times 10^6 \text{ in.-lb}$$

$$M_{SD} = -2.47 \times 10^6 \text{ in.-lb}$$

and

$$M_L = -8.68 \times 10^6 \text{ in.-lb (including } M_{CSD})$$

Hence, at top of the precast section,

$$f_T^t = -\frac{508{,}940}{560}\left[1 - \frac{(-16.49) \times 24.73}{223.91}\right] + \frac{(2.87 + 2.47) \times 10^6}{5{,}070} + \frac{8.68 \times 10^6}{21{,}714}$$

$$= -2{,}560 + 1{,}053.2 + 399.7 = -1{,}111 \text{ psi } (C)$$

$$f_{bT} = -\frac{508{,}940}{560}\left[1 + \frac{(-16.49) \times 20.27}{223.91}\right] - \frac{(2.87 + 2.47) \times 10^6}{6{,}186} - \frac{8.68 \times 10^6}{9{,}490}$$

$$= +447.9 - 863.2 - 914.6$$

$$= -1{,}330 \text{ psi}(C) \text{ (9.2 MPa)} < 2{,}250 \text{ psi, O.K.}$$

Stresses at top and bottom fibers of the situ-cast 4 in. slab

From Equations 4.20a and b, the maximum fiber stresses at the top and bottom fibers on the composite slab at the support section are evaluated using section moduli S^t_{CS} and S_{bCS}, where

$$f_{CS}^t = +\frac{M_L}{S_{CS}^t} \times \text{modular ratio } n = 0.77$$

$$f_{bCS} = +\frac{M_L}{S_{bCS}} \times n$$

Hence,

$$f_{CS}^t = \frac{8.68 \times 10^6}{15{,}288} \times 0.77 \cong +437 \text{ psi } (T) \text{ (3 MPa)} < 849 \text{ psi, O.K.}$$

$$f_{bCS} = +\frac{8.68 \times 10^6}{19{,}251} \times 0.77 \cong +347 \text{ psi } (T) \text{ (2.4 MPa)}$$

Superposition of the tensile stress at the bottom fibers of the slab and the compressive stress of –1,111 psi at the extreme top fibers of the precast section can result in a net compressive stress at the bottom of the slab = –1,111 + 347 = –764 psi.

The fiber stress at the extreme lower fibers of the precast section at support B or C is

$$f_{bT} = -414.3 - \frac{8.68 \times 10^6}{9{,}490} \cong -1{,}329 \text{ psi (9.2 MPa)}$$

The final distribution of stress is shown in Figure 6.18.

Note that the concrete fiber stresses are considerably below the maximum allowable stresses at service load for the same live load and spans as the simply supported beam of Example 4.7. Consequently, the selected continuous tendon profile is not the most efficient in this example, since the superimposed live load is the same in both cases and the section is the same for the span lengths used. Example 6.5 demonstrates preferable modifications.

6. *Limit state at failure*

(a) *Degree of ductility for moment redistribution*

From Equations 6.6 and 6.7, the maximum redistribution percentage is

$$p_d = 20\left[1 - \frac{\omega_p + \dfrac{d}{d_p}(\omega - \omega')}{0.36\beta_1}\right]$$

where

$$\omega_p + \frac{d}{d_p}(\omega - \omega') \leq 0.24\beta_1$$

Also,

$$h = 45 + 5.75 = 50.75 \text{ in.}$$

$$d = 50.75 - (1.5 + 0.5 \text{ for stirrup} + 0.25) \cong 48.5 \text{ in.}$$

$$\text{Support } d_p = c_b + e_B = 20.27 + 15 \text{ from bottom fibers}$$
$$= 35.27 \text{ in. (89.6 cm)}$$

Since the width of the compression flange at support $b = 22$ in., try two #4 bars as compression steel and four #4 bars at tension steel. We obtain

$$\omega' = \frac{A_s' f_y}{bd f_c'} = \frac{2 \times 0.20}{22 \times 48.5} \times \frac{60{,}000}{5{,}000} = 0.0045$$

$$\omega = \frac{4 \times 0.2}{22 \times 48.5} \times \frac{60{,}000}{5{,}000} = 0.0090$$

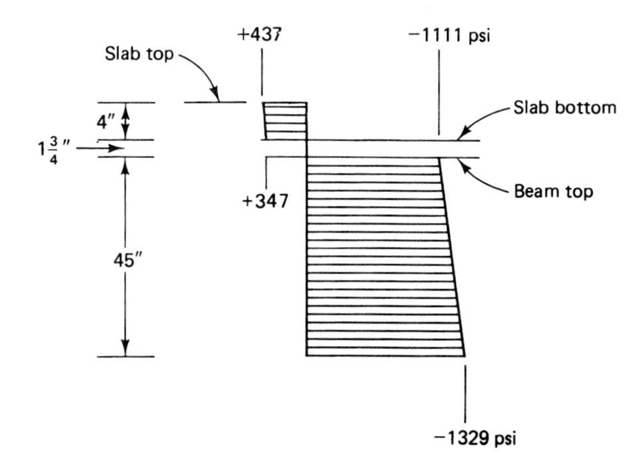

Figure 6.18 Stress distribution in the concrete at service load.

$$\omega_p = \frac{A_{ps} f_{ps}}{bd_p f'_c} = \frac{3.366 \times 240{,}000}{22 \times 35.27 \times 5{,}000} = 0.2082$$

$$\beta_1 = 0.80 \text{ for } 5{,}000 \text{ psi concrete}$$

$$\frac{d}{d_p}(\omega - \omega') = \frac{48.5}{35.27}(0.0090 - 0.0045) = 0.0062$$

$$\omega_p + \frac{d}{d_p}(\omega - \omega') = 0.2082 + 0.0062 = 0.2144$$

$$0.24\beta_1 = 0.24 \times 0.80 = 0.19 < 0.2144 \text{ at the support}$$

Hence, no redistribution of moment from support to midspan is permitted, i.e., the redistribution factor ρ_D in Equation 6.6 is zero. For maximum allowable ω_T, $0.36\beta_1 = 0.36 \times 0.80 = 0.45 > 0.2144$, O.K.

(b) *Flexural moments modifications*

It is advisable to apply the redistribution modifications separately to the dead and live loads since alternate span loading for live load has to be considered for worst loading conditions while dead load acts simultaneously on all spans. Since no redistribution occurs here, the elastic moments at support are

$$M_D + M_{SD} = (2.87 + 2.47)10^6 = 5.34 \times 10^6 \text{ in.-lb}$$

and

$$M_L = 8.68 \times 10^6 \text{ in.-lb}$$

(c) *Nominal moment strength*

The support section factored M_u is

$$1.4 \times 5.34 \times 10^6 + 1.7 \times 8.68 \times 10^6 = 22.23 \times 10^6 \text{ in.-lb}$$

From step 2, the factored elastic secondary moment induced by reactions due to prestress, using a load factor of 1.0 as stipulated in the ACI code, is $M_2 = 0.76 \times 10^6$ in.-lb, and the total factored moment $M_u = (22.23 - 0.76) \times 10^6$ in.-lb $= 21.47 \times 10^6$ in.-lb (2.42 kN-m). Also, the required nominal moment strength $M_n = M_u/\phi = 21.47 \times 10^6/0.90. = 23.86 \times 10^6$ in.-lb (2.70 kN-m). If A'_s yielded, $f_y = f'_s = 60{,}000$ psi, width b at bottom = 22 in. So

$$a = \frac{A_{ps}f_{ps} + A_s f_y - A'_s f_y}{0.85 f'_c b}$$

$$= \frac{3.366 \times 240{,}000 + 0.8 \times 60{,}000 - 0.4 \times 60{,}000}{0.85 \times 5{,}000 \times 22}$$

$$= 8.90 \text{ in. } (22.6 \text{ cm})$$

The depth of the flange up to the 7-in.-wide web section at support is

$$7 + 7.5/2 = 10.8 \text{ in. } > 8.90 \text{ in.}$$

The neutral axis is inside the flange, and the section behaves like a rectangular section.

From Equation 4.44,

$$\text{Available } M_n = A_{ps} f_{ps}\left(d_p - \frac{a}{2}\right) + A_s f_y\left(d - \frac{a}{2}\right) + A'_s f_y\left(\frac{a}{2} - d'\right)$$

$$= 3.366 \times 240{,}000\left(35.27 - \frac{8.90}{2}\right)$$

$$+ 0.80 \times 60{,}000\left(48.5 - \frac{8.9}{2}\right) + 0.40 \times 60{,}000\left(\frac{8.9}{2} - 3\right)$$

$$= 27.05 \times 10^6 \text{ in.-lb } (3.1 \times 10^3 \text{ kN-m}) > 23.86 \times 10^6 \text{ in.-lb}$$

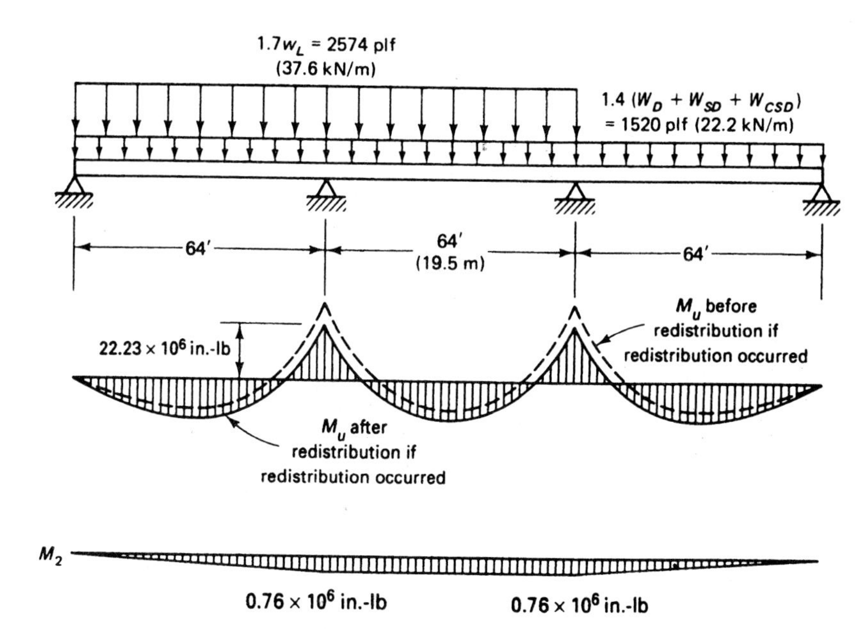

Figure 6.19 Loading and factored moment in Example 6.4.

Hence, the section is safe but not efficient. Figure 6.19 shows the load and moment distributions.

In sum, the section is efficient neither at service load nor at ultimate load. It could be reduced in size by either changing the tendon eccentricities or enlarging the span or allowing a higher live load with new tendon eccentricities.

6.10 TENDON TRANSFORMATION TO UTILIZE ADVANTAGES OF CONTINUITY

Example 6.5

Linearly transform the prestressing tendon in the continuous prestressed beam in Example 6.4 such that the superimposed live load W_L on the 64 ft (19.8 m) spans can be increased by at least 50 percent.

Solution:

1. *Transformation of tendon*
The tendon eccentricity at support B in Example 6.4 produces compressive fiber stress at the support precast section due to all loads of 1,111 psi at the top, and 1,329 psi at the bottom fibers. These are lower than the maximum allowable $f_c = -2{,}250$ psi. Therefore, the beam can sustain more load if the concrete compressive stress capacity is to be utilized. In order to allow the beam to carry more live load, more compressive stress at the midspan bottom fibers due to prestress needs to be developed through an increase in the tendon eccentricity.

Accordingly, assume that the tendon is linearly transformed throughout all the spans such that $e_B = e_C = 11$ in. (27.9 cm), as shown in Figure 6.20. Then the transformation vertical distance $= 15 - 11 = 4$ in. (10 cm), the midspan eccentricity $e_{E1} = 9 + 4 = 13$ in. (33 cm), and the primary support B moment $M_1 = 508{,}940 \times 11 = 5.60 \times 10^6$ in.-lb $(0.63 \times 10^3$ kN-m). Also, the primary midspan E_1 moment $M_1 = 508{,}940 \times 13 = 6.62 \times 10^6$ in.-lb $(0.75 \times 10^3$ kN-m), and we have

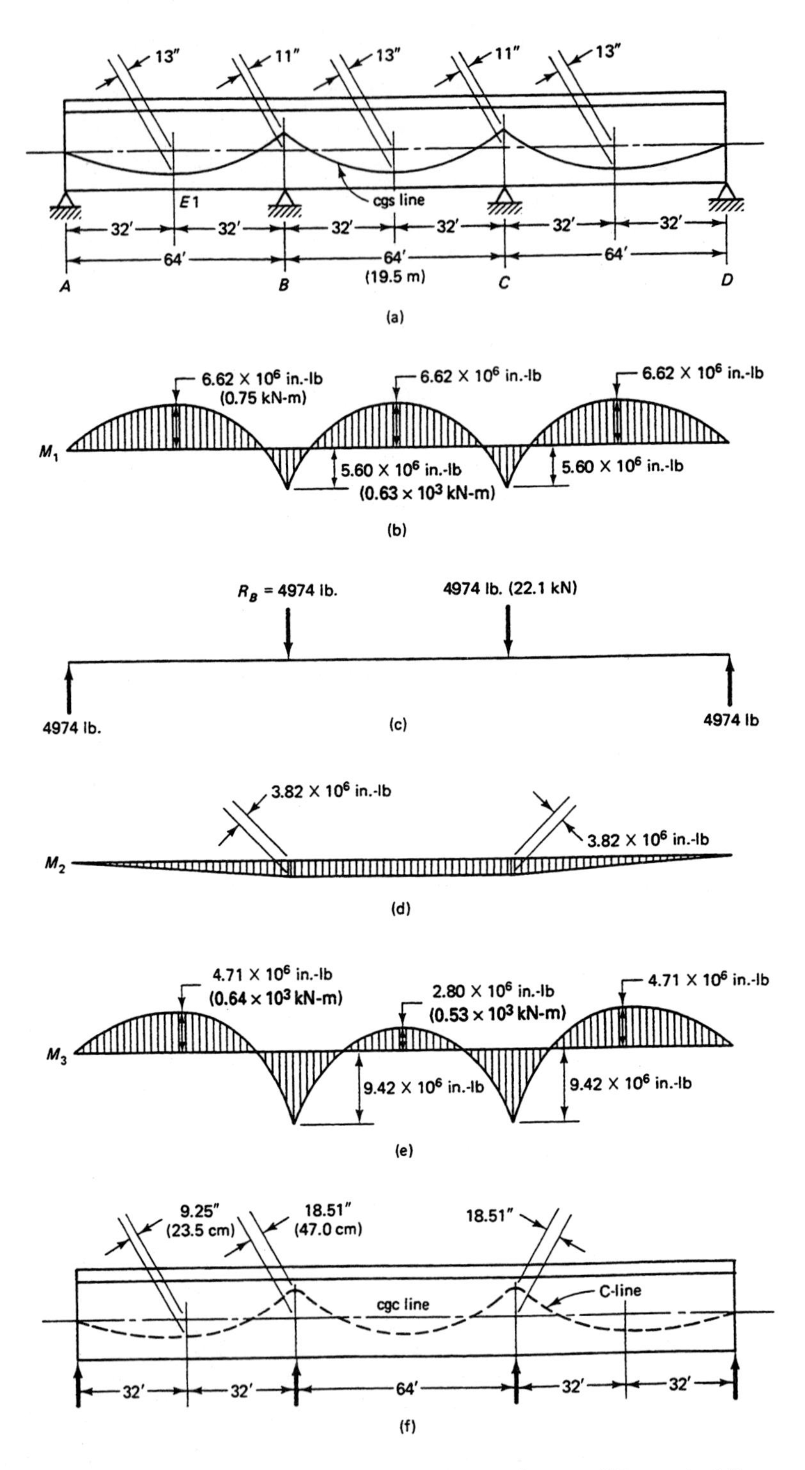

Figure 6.20 Tendon transformation in continuous beam of Example 6.5.

$$E_c I_c \Delta_B = \left[\left(\frac{5.60}{2} + 6.62\right)10^6 \times \frac{64 \times 2}{3}\right] \times \frac{64}{2} \times 144$$

$$- \left[\frac{5.60 \times 10^6 \times 64}{2}\right] \times \frac{64 \times 2}{3} \times 144 = 75.1 \times 10^{10}$$

Also,

$$E_c I_c \Delta_c = \left[64R_A \times 12 \times \frac{64 \times 12}{2}\right]\frac{64 \times 2}{3} \times 12$$

$$= 15.1 \times 10^7 R_A \text{ in.-lb}$$

$$R_A = R_B = \frac{75.1 \times 10^{10}}{15.1 \times 10^7} = 4{,}974 \text{ lb}$$

$$M_2 = R_A \times 64 \times 12 = 4{,}974 \times 64 \times 12$$

$$= 3.82 \times 10^6 \text{ in.-lb } (0.44 \times 10^3 \text{ kN-m})$$

$$\text{Support } M_3 = (5.60 + 3.82) \times 10^6 = 9.42 \times 10^6 \text{ in.-lb (see Figure 6.20)}$$

$$\text{Midspan } M_3 = \left(6.62 - \frac{1}{2} \times 3.82\right)10^6 = 4.71 \times 10^6 \text{ in.-lb}$$

From Example 6.4, $P_e = 508{,}940$ lb. So calculate the final C-line eccentricities = M_3/P_e as shown in Figure 6.20(f).

2. *Concrete fiber stresses due to prestress and self-weight (583 plf)*
From step 3 of the solution to Example 6.4, the moment due to self-weight is 2.87×10^6 in.-lb so the total moment $M_4 = M_3 - 2.87 \times 10^6 = (9.42 - 2.87)\,10^6 = 6.55 \times 10^6$ in.-lb.

(i) *Support section B or C*

$$f^t = -\frac{P_e}{A_c} - \frac{M_4}{S^t} = -\frac{508{,}940}{560} - \frac{6.55 \times 10^6}{5{,}070}$$

$$= -908.8 - 1{,}291.9$$

$$= -2{,}200.7 \text{ psi } (C)\ (15.2 \text{ MPa}) < 2{,}250 \text{ psi, O.K.}$$

$$f_b = -908.8 + \frac{6.55 \times 10^6}{6{,}186} = -908.8 + 1{,}058.8$$

$$= +150.0 \text{ psi } (T), \text{ O.K.}$$

(ii) *Outer span midspan section E_1*
From Example 6.4, $M_D = 2.12 \times 10^6$ in.-lb. So the net moment $M_4 = M_3 - 2.12 \times 10^6 = (4.71 - 2.12)10^6 = 2.59 \times 10^6$ in.-lb (0.4×10^3 kN-m), and

$$f^t = -908.8 + \frac{2.59 \times 10^6}{5{,}070} = -398.0 \text{ psi } (C)\ (2.7 \text{ MPa}), \text{ O.K.}$$

$$f_b = -908.8 - \frac{2.59 \times 10^6}{6{,}186}$$

$$= -1{,}327.5 \text{ psi } (C)\ (9.2 \text{ MPa}), \text{ no tension, O.K.}$$

3. *Determination of live-load intensity for new tendon profile for unshored construction*

$$f_T^t = -\frac{P_e}{A_c}\left(1 - \frac{e'_B c_t}{r^2}\right) - \frac{M_D + M_{SD}}{S^t} - \frac{M_{CSD} + M_L}{S_c^t}$$

$$f_{bT} = -\frac{P_e}{A_c}\left(1 + \frac{e'_B c_b}{r^2}\right) + \frac{M_D + M_{SD}}{S_b} + \frac{M_{CSD} + M_L}{S_{cb}}$$

(i) *Support section (at B or C)*
From Figure 6.20(e), $M_3 = -9.42 \times 10^6$ in.-lb Hence,

$$e'_B = -\frac{9.42 \times 10^6}{508{,}940} = -18.5 \text{ in. } (47.0 \text{ cm})$$

From Example 6.4, $M_D + M_{SD} = -(2.87 + 2.47)(10^6 = -5.34 \times 10^6$ in.-lb. Also, from Figure 6.12, the live-load moment for a three-span beam with one span unloaded is

$$M_L = -0.1167 W_L l^2 = -0.1167 W_L (64)^2 (12)$$
$$= -5{,}736 W_L \text{ in.-lb, including } M_{CSD}$$

The maximum allowable tensile stress $f_t = +849$ psi at midspan and $f_t = +425$ psi at support.

$$f_T^t = +425 = -\frac{508{,}940}{560}\left(1 - \frac{(-18.5) \times 24.73}{223.91}\right)$$
$$+ \frac{5.34 \times 10^6}{5{,}070} + \frac{5{,}736 W_L}{21{,}714}$$

giving $W_L = 8{,}092$ plf.

(ii) *Midspan section (at E_1)*
Since $W_D = 583$ plf, by proportioning we obtain

$$M_D + M_{SD} = +(2.12 + 1.83)10^6 = +3.95 \times 10^6 \text{ in.-lb}$$
$$M_L = +2.12 \times 10^6 \times \frac{W_L}{583} = +3{,}636 W_l \text{ in.-lb}$$
$$f_T^t = -2{,}250 = -\frac{508{,}940}{560}\left(1 - \frac{9.25 \times 24.73}{223.91}\right)$$
$$-\frac{3.95 \times 10^6}{5{,}070} - \frac{3{,}636 W_L}{21{,}714}$$
$$\frac{3{,}636 W_L}{21{,}714} = 20 - 779 + 2{,}250$$
$$W_L = 8{,}904 \text{ plf}$$
$$f_{bT} = 849 = -\frac{508{,}940}{560}\left(1 + \frac{9.25 \times 20.27}{223.91}\right)$$
$$+ \frac{3.95 \times 10^6}{6{,}186} + \frac{3{,}636 W_L}{9{,}490}$$
$$\frac{3{,}636 W_L}{9{,}490} = 849 + 1{,}669.8 - 638.5$$
$$W_L = 4{,}908 \text{ plf}$$

Hence, $W_L = 4{,}908$ plf controls for service load levels, to be verified by checking the ultimate moment strength available.

4. *Available nominal moment strength*

$$A_{ps} = 3.366 \text{ in.}^2$$
$$\text{Support } d_p = 11 + 20.27 = 31.27 \text{ in.}$$

Assume that A_s at supports B and C is increased to four #8 bars in order to facilitate an increased live load. Then

$$A_s = 4 \times 0.79 = 3.16 \text{ in.}^2$$

$$\omega = \frac{3.16}{22 \times 48.5} \times \frac{60{,}000}{5{,}000} = 0.0355$$

$$\omega_p + \frac{d}{d_p}(\omega - \omega') = 0.2064 + \frac{48.5}{31.27}(0.0355 - 0.0045)$$

$$= 0.2545 > 0.24\beta_1 = 0.19$$

Hence, there is no redistribution of moment from support to midspan.

Now, from before, the elastic $M_2 = 3.82 \times 10^6$ in.-lb and we have

$$M_u = 1.4(M_D + M_{SD}) + 1.7M_L - M_2$$

or

$$M_u = 1.4 \times 5.34 \times 10^6 + 1.7M_L = -3.82 \times 10^6$$

$$= 3.66 \times 10^6 + 1.7M_L$$

Next,

$$\text{Required } M_n = \frac{M_u}{\phi} = \frac{3.66 \times 10^6}{0.9} = 4.07 \times 10^6 + 1.89M_L$$

$$a = \frac{A_{ps} f_{ps} + A_s f_y - A_s' f_y}{0.85 f_c' b}$$

If A_s' yielded $f_y = f_s' = 60{,}000$ psi, then

$$a = \frac{3.366 \times 240{,}000 + (3.16 - 0.40)60{,}000}{0.85 \times 5{,}000 \times 22} = 10.41 \text{ in. (26.4 cm)}$$

which is less than the flange depth up to the 7-in.-wide web section. Hence, the neutral axis is inside the flange, and the section behaves like a rectangular section. Accordingly, we have

$$\text{Available } M_n = A_{ps} f_{ps}\left(d_p - \frac{a}{2}\right) + A_s f_y\left(d - \frac{a}{2}\right) + A_s' f_y\left(\frac{a}{2} - d'\right)$$

$$= 3.366 \times 240{,}000\left(31.27 - \frac{10.41}{2}\right) + 3.16$$

$$\times 60{,}000\left(48.5 - \frac{10.41}{2}\right) + 0.40 \times 60{,}000\left(\frac{10.41}{2} - 3\right)$$

$$= 29.32 \times 10^6 \text{ in.-lb} = 4.07 \times 10^6 + 1.89M_L$$

Hence,

$$M_L = \frac{(29.32 - 4.07) \times 10^6}{1.89} = 13.36 \times 10^6 \text{ in.-lb}$$

If $M_L = 0.1167 W_L l^2$, then

$$13.36 \times 10^6 = 0.1167 W_L (64)^2 \times 12$$

So $W_L = 2{,}329$ plf $< W_L = 4{,}908$ plf from the service-load analysis. Hence, $W_L = 2{,}329$ plf controls, and the percent increase in live load is

$$\frac{(2{,}329 - 1{,}514)}{1{,}514} \times 100 = 53.8\% \cong 50\%, \text{ O.K.}$$

Thus, we can adopt the new profile of the tendon with four #8 bars at the support top fibers in the situ-cast slab and two #4 bars at the bottom precast section fibers.

6.11 DESIGN FOR CONTINUITY USING NONPRESTRESSED STEEL AT SUPPORT

Example 6.6

Design the beam in Example 6.5 such that the section and tendon profile of the AASHTO type-III bridge beam used in Example 4.7 is made continuous through the use of nonprestressed mild steel reinforcement to carry the superimposed service dead load $W_{SD} = 503$ plf and service live load $W_L = 2{,}290$ plf (33.4 kN/m). Assume that the tendon profile in the precast simply supported section is the same as the one in Example 4.7, namely, with midspan eccentricity $e_c = 16.27$ in. (41.3 cm) and end eccentricity $e_e = 10.0$ in. (25.4 cm). Sketch the prestressing tendon and other reinforcement details if the maximum allowable concrete compressive fiber stress at service load is $f_c = 2{,}250$ psi (15.5 MPa).

Solution:

1. *Data for strength design at support*
Because continuity is obtained in this case through the use of reinforced concrete at the supports, it is suggested that the topping concrete also be of $f'_c = 5{,}000$ psi compressive strength. Thus, we have

$$f'_c = 5{,}000 \text{ psi (34.5 MPa), normal weight}$$

$$f_y = 60{,}000 \text{ psi (413.7 MPa)}$$

Design the continuity to resist the superimposed dead and live loads only, and *not* the self-weight:

$$d = 50.75 - (1.5 \text{ in. cover} + 0.5 \text{ in. for stirrup} + 0.5 \text{ in. for half-bar dia})$$

$$= 48.25 \text{ in. (123 cm)}$$

$$b_b = 22 \text{ in. at bottom}$$

$$b_t = 84 \text{ in. at top (modified } b_m = 65 \text{ in.)}$$

2. *Nominal moment strength*
From Example 6.5, the required moment strength due to $(W_{SD} + W_L)$ at support B is $M_n = 1.4 \times 2.47 \times 10^6 + 1.7 \times 13.36 \times 10^6 = 26.17 \times 10^6$ in.-lb. The bonded prestressed steel does not extend through the supports; hence, consider $\omega_p = 0$ for calculating the moment redistribution factor. Assume

$$A_s = 9.0 \text{ in.}^2$$

$$\omega = \frac{9.0}{22 \times 48.25} \times \frac{60{,}000}{5{,}000} = 0.1012$$

$$A'_s = \text{two \#4} = 0.40 \text{ in.}^2$$

and

$$\omega' = 0.0961 \times \frac{0.4}{8.5} = 0.045$$

Then from Example 6.4, $d_p = 35.87$ in., $d = 48.25$ in., $\omega_p + [d/d_p]\,(\omega - \omega') = 0 + [48.25/35.87](0.1012 - 0.0045) = 0.1300$, and $0.24\beta_1 = 0.192 > 0.1300$. Hence, redistribution is permitted and the moment redistribution factor is

$$\rho_d = 20\left[1 - \frac{\omega_p + [d/d_p]\,(\omega - \omega')}{0.36\beta_1}\right]$$

$$= 20\left[1 - \frac{0.1300}{0.36 \times 0.80}\right] = 10.97\%$$

So use a distribution factor of 0.10:

$$\text{Rqd. } M_n = (1 - 0.10) \times 26.17 \times 10^6 = 23.55 \times 10^6 \text{ in.-lb } (2.2 \times 10^6 \text{ kN-m})$$

$$M_n = A_s f_y \left(d = \frac{a}{2} \right)$$

Assume for the first trial that $d - a/2 \cong 0.9d$. Then

$$23.55 \times 10^6 = A_s \times 60{,}000 \times 0.9(48.25)$$

$$A_s = \frac{23.55 \times 10^6}{60{,}000 \times 0.9 \times 48.25} = 9.04 \text{ in.}^2 \ (48.1 \text{ cm}^2)$$

$$a = \frac{A_s f_y}{0.85 f'_c b} = \frac{9.04 \times 60{,}000}{0.85 \times 5{,}000 \times 22} = 5.80 \text{ in.}$$

The depth of the flange to the web is 5.75 + 7 + 4.5/2 for the AASHTO type-3 section = 15 in. > 5.80/0.9 = 6.44 in. The neutral axis is inside the flange, and the section behaves like a rectangular section. Therefore,

$$A_s = \frac{M_n}{f_y (d - a/2)} = \frac{23.55 \times 10^6}{60{,}000(48.25 - 5{,}80/2)} = 8.65 \text{ in.}^2 \ (70.2 \text{ cm}^2)$$

Since the assumed $A_s = 9.0$ in.^{2} is very close to 8.65 in.^{2}, the moment distribution factor is satisfactory.

The area A_s is to be distributed over the total actual flange width of 84 in. A_s per ft width = (8.65/84) × 12 = 1.24 in.^{2}/12 in. Using #6 bars, A_s per bar = 0.44 in.^{2} (2.82 cm^2), and the spacing is

$$s = \frac{\text{Bar } A_s}{\text{Rqd. } A_s/12 \text{ in.}} = \frac{0.44}{1.24/12} = 4.25 \text{ in.}$$

Thus, use #6 bars at $4\frac{1}{4}$in. center to center over the 84-in. width (19.1 mm dia. bars at 10.8 cm). The total number of bars over the 84-in.-width flange is

$$\frac{84 - (2 \times 15\text{-in. cover})}{4.25} + 1 \cong 20$$

so that we have

$$\text{Total } A_s = 20 \times 0.44 = 8.80 \text{ in.}^2, \text{ O.K.}$$

Accordingly, adopt the design for flexure. Note that the complete design would involve dowel design for composite action, stirrup design for web shear, end block design, and design for serviceability requirements in deflection and crack control as detailed in earlier examples. Note also that continuity on three spans in this example using mild steel only at the supports allowed a 50% increase in the live-load intensity from 1,514 plf to 2,329 plf.

3. *Beam geometry schematic details*
Figure 6.21 gives the reinforcing and tendon profile details of the continuous beam of this example. Note how the normal-weight concrete and mild steel provide continuity at the supports for the superimposed dead and live loads.

6.12 INDETERMINATE FRAMES AND PORTALS

6.12.1 General Properties

Concrete frames are indeterminate structures consisting of horizontal, vertical, or inclined members joined in such a manner that the connection can withstand the stresses and bending moments that act on it. The degree of indeterminacy depends upon the number of spans, the number of vertical members, and the type of end reactions. Typical frame configurations are shown in Figure 6.22. If n is the number of joints, b the number

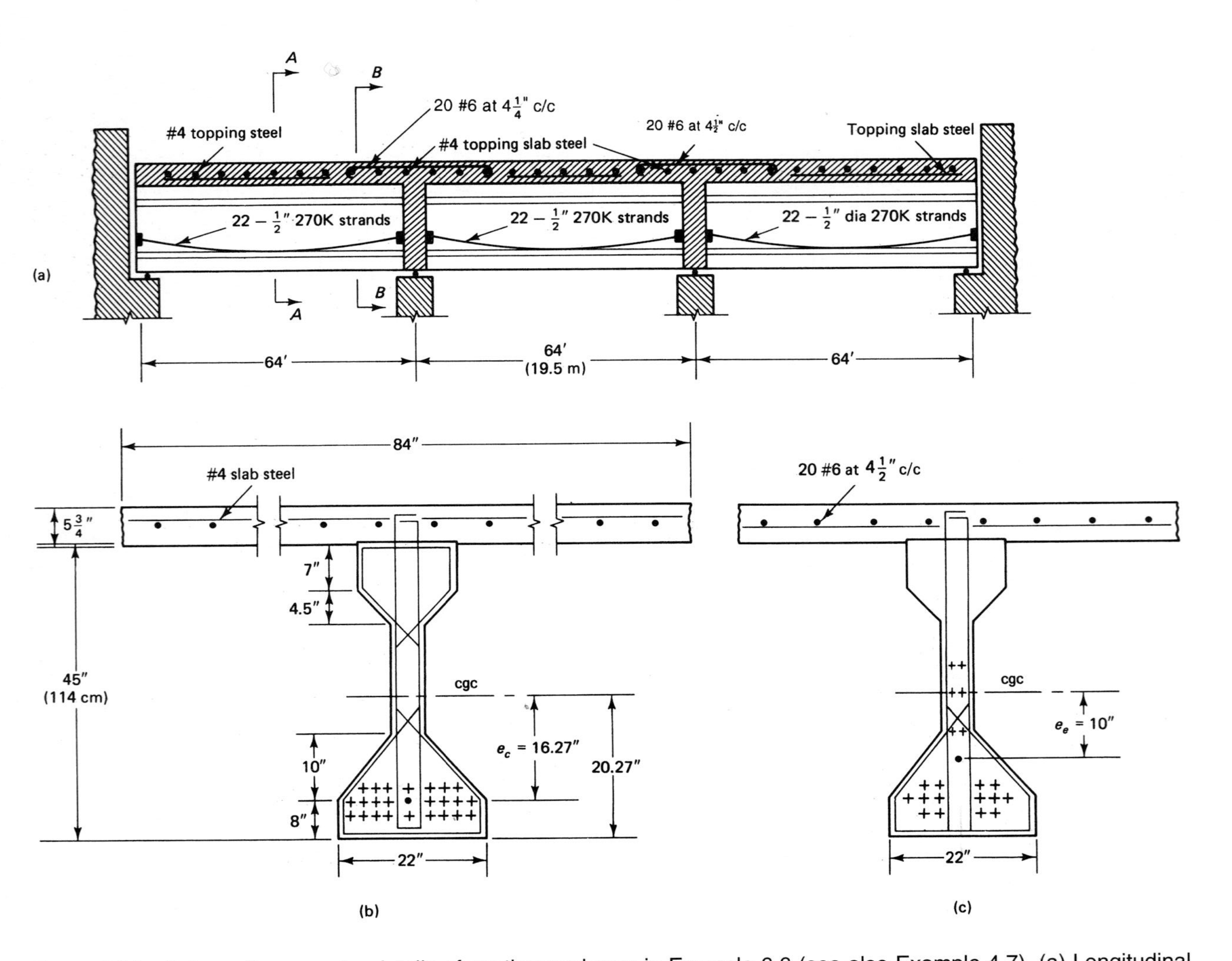

Figure 6.21 Schematic geometry details of continuous beam in Example 6.6 (see also Example 4.7). (a) Longitudinal section of bridge beam (not to scale). (b) Midspan section A–A. (c) Support section B–B.

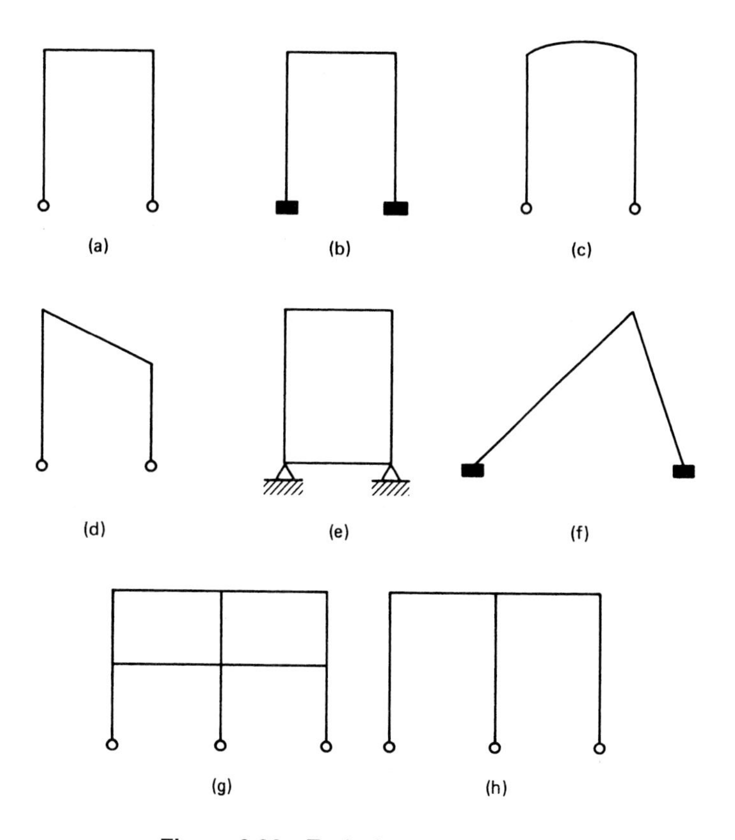

Figure 6.22 Typical structure frames.

of members, r the number of reactions, and s the number of indeterminacies, the degree of indeterminancy is determined from the following inequalities:

$$3n + s > 3b + r \qquad \text{(unstable)} \tag{6.8a}$$

$$3n + s = 3b + r \qquad \text{(statically determinate)} \tag{6.8b}$$

$$3n + s < 3b + r \qquad \text{(statically indeterminate)} \tag{6.8c}$$

The degree of indeterminacy is

$$s = 3b + r - 3n \tag{6.8d}$$

where $3n$ equations of static equilibrium are always available and the total number of unknowns is $3b + r$. As an example, the degree of indeterminancy of the frame in Figure 6.22(a) is

$$s = 3 \times 3 + 2 \times 2 - 3 \times 4 = 1$$

and for the frame in part (g) of the same figure it is

$$s = 3 \times 10 + 2 \times 3 - 3 \times 9 = 9$$

Note that in order for a frame to perform satisfactorily, the following conditions have to be satisfied:

1. The design must be based on the most unfavorable moment and shear combinations. If moment reversal is possible due to reversal of live-load direction, the highest values of positive and negative bending moments have to be considered in the design.

2. Proper foundation support for horizontal thrust has to be provided. If the frame is designed as hinged, an expensive construction procedure, an actual hinge system has to be provided.

6.12.2 Forces and Moments in Portal Frames

The behavior of concrete frames before cracking can be considered reasonably elastic, as was done in the case of a continuous beam at service-load and slight-overload conditions. Consequently, well before the development of plastic hinges, the bending moment diagrams shown in Figures 6.23 and 6.24 will be used in the design of indeterminate prestressed concrete frames. The usual methods of analyses of indeterminate structures including frames, such as virtual work, stiffness matrix, and flexibility matrix procedures, as well as the clapeyron three- or four-moment equations, are assumed familiar in this text, so that only the minimum guidelines and simplifications are presented.

6.12.2.1 Uniform Gravity Loading on Single-Bay Portal. Suppose that the moments of inertia I_c of the vertical columns and I_b of the horizontal beam of the portal in Figure 6.25(a) are not equal. The following values of the moments and thrusts can be inferred:

End Shear in Beam

$$V_B = V_C = \frac{1}{2} Wl \tag{6.9a}$$

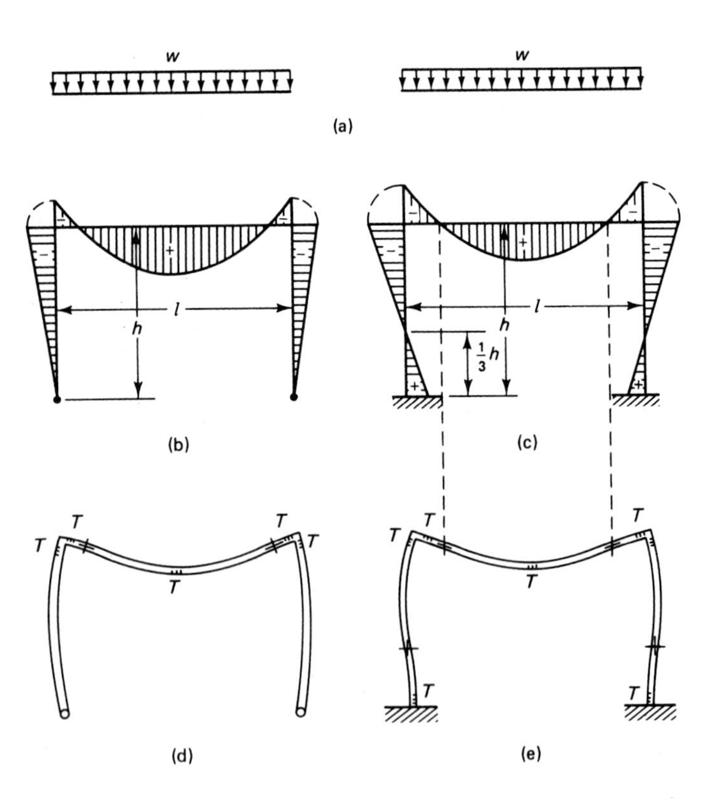

Figure 6.23 Right-angled portal frame loaded with gravity load intensity w (T indicates tension fibers). (a) Load intensity. (b) Bending moment (hinged-base frame). (c) Bending moment (fixed-base frame). (d) Deformation of frame in (b). (c) Deformation of frame in (c).

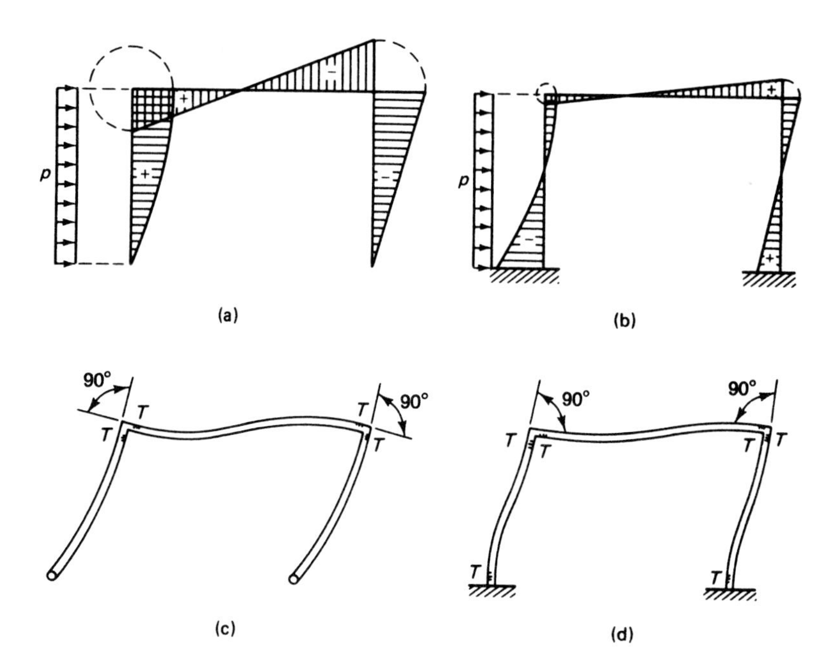

Figure 6.24 Right-angled portal frame loaded with wind load intensity p (T indicates tension fibers). (a) Bending moment (hinged-base frame). (b) Bending moment (fixed-base frame). (c) Deformation of frame in (a). (d) Deformation of frame in (b).

Horizontal Thrust

$$H = \frac{1}{h} C_1 \, wl^2 \tag{6.9b}$$

where

$$C_1 = \frac{1}{12\left(\dfrac{2}{3}\dfrac{I_b}{I_c}\dfrac{h}{l} + 1\right)} \tag{6.9c}$$

Maximum Negative Moment at Corner

$$M_B = M_c = -Hh = -C_1 \, wl^2 \tag{6.9d}$$

Maximum Positive Moment at Midspan

$$M_{\max} = \frac{1}{8} wl^2 - Hh = \left(\frac{1}{8} - C_1\right) wl^2 \tag{6.9e}$$

Bending Moments at Any Point x

$$M_x = \frac{1}{2} x(l - x)w - C_1 \, wl^2 \tag{6.9f}$$

where the points of contraflexure from either corner of the portal are

$$x_1 = \frac{1}{2}\left(1 - \sqrt{1 - 8C_1}\right) l = C_2 l \tag{6.9g}$$

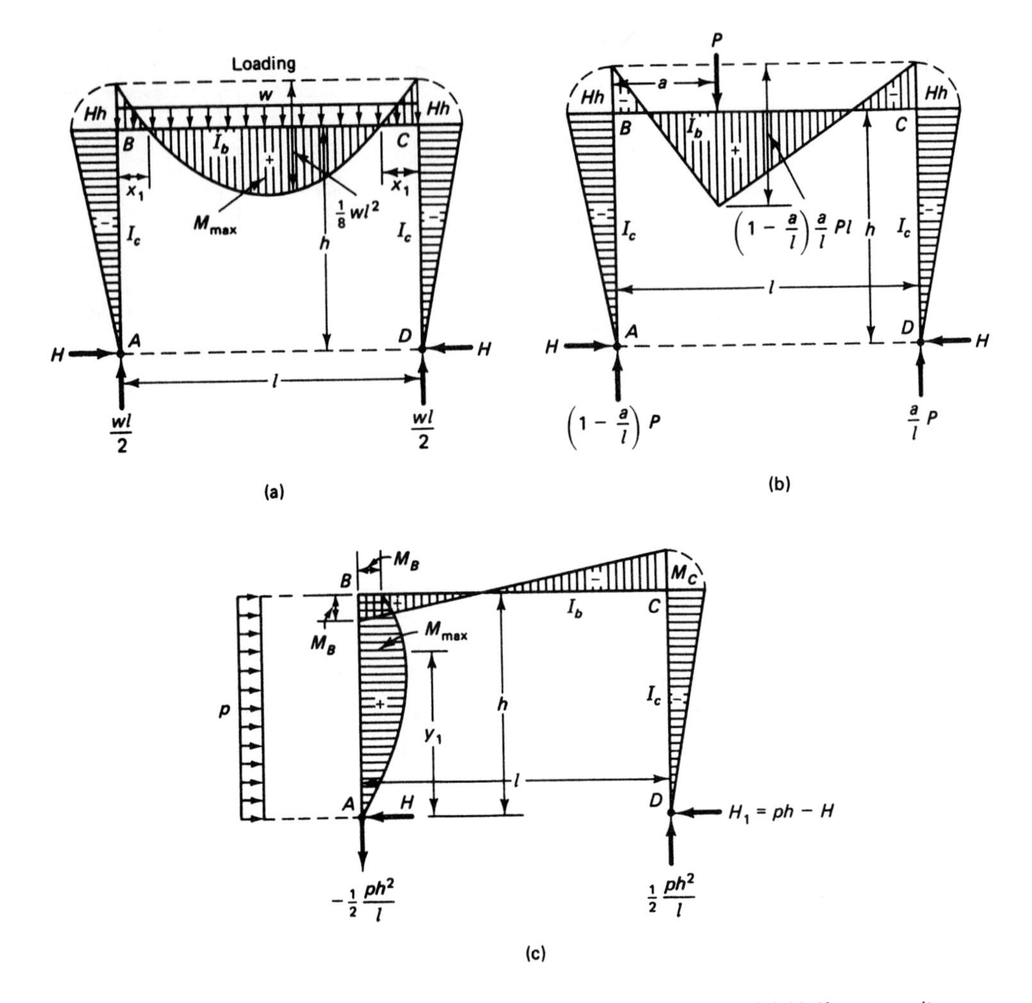

Figure 6.25 Bending moment ordinates in single-bay frame. (a) Uniform gravity loading. (b) Concentrated gravity loading. (c) Uniform horizontal pressure.

and

$$C_2 = \frac{1}{2}(1 - \sqrt{1 - 8C_1}) \tag{6.9h}$$

6.12.2.2 Concentrated Gravity Loading on Single-Bay Portal. Since the concentrated load P does not have to act at midspan, nonsymmetry of shears results. From Figure 6.25(b), the end shear are

$$V_B = \left(1 - \frac{a}{l}\right)P$$

and

$$V_c = \frac{a}{l}P \tag{6.10a}$$

Horizontal Thrust

$$H = C_3\frac{a}{l}\left(1 - \frac{a}{l}\right)P\frac{l}{h} \tag{6.10b}$$

where

$$C_3 = \frac{1}{2\left(\frac{2}{3}\frac{I_b}{I_c}\frac{h}{l} + 1\right)} \tag{6.10c}$$

Bending Moments at Corners

$$M_B = M_c = -Hh = -C_3\frac{a}{l}\left(1 - \frac{a}{l}\right)Pl \tag{6.10d}$$

Bending Moments at Any Point along BC. For $x < a$,

$$M_x = \left(1 - \frac{a}{l}\right)\left(\frac{x}{l} - \frac{a}{l}C_3\right)Pl \tag{6.10e}$$

For $x > a$,

$$M_x = \frac{a}{l}\left[1 - \frac{x}{l} - \left(1 - \frac{a}{l}\right)\right]C_3\,Pl \tag{6.10f}$$

Maximum Positive Moment at x = a

$$M_{\max} = \frac{a}{l}\left(1 - \frac{a}{l}\right)Pl - Hh = (1 - C_3)\frac{a}{l}\left(1 - \frac{a}{l}\right)Pl \tag{6.10g}$$

Horizontal Thrust for Several Concentrated Gravity Loads

$$H = \frac{1}{h}C_3\left[P_1\frac{a_1}{l}\left(1 - \frac{a_1}{l}\right) + P_2\frac{a_2}{l}\left(1 - \frac{a_2}{l}\right) + \ldots\right] \tag{6.10h}$$

or

$$H = \frac{1}{h}C_3\Sigma P\frac{a}{l}\left(1 - \frac{a}{l}\right) \tag{6.10i}$$

6.12.2.3 Uniform Horizontal Pressure on Single-Bay Portal. From Figure 6.25(c), we have the following:

Vertical Reactions at Supports

$$R_A = -\frac{1}{2}ph\frac{h}{l}$$

and

$$R_D = +\frac{1}{2}ph\frac{h}{l} \tag{6.11a}$$

Horizontal Reactions. For windward hinge A,

$$H_A = \frac{1}{8}\frac{11\frac{I_b}{I_c}\frac{h}{l} + 18}{2\frac{I_b}{I_c}\frac{h}{l} + 3}ph = C_4\,ph \tag{6.11b}$$

where

$$C_4 = \frac{1}{8}\frac{11\frac{I_b}{I_c}\frac{h}{l} + 18}{2\frac{I_b}{I_c}\frac{h}{l} + 3} \tag{6.11c}$$

For leeward hinge D,

$$H_D = ph - H_A = (1 - C_4)ph \tag{6.11d}$$

The bending moments at any point y along the column height due to horizontal pressure, with y being measured from the *bottom*, are

$$M_y = H_A y - \frac{1}{2}py^2 \tag{6.11e}$$

Maximum Moment at Windward Column

$$M_{\max} = \frac{1}{2}\left(\frac{1}{8}\frac{11\frac{I_b}{I_c}\frac{h}{l} + 18}{2\frac{I_b}{I_c}\frac{h}{l} + 3}\right)ph^2 = \frac{1}{2}C_4 ph^2 \tag{6.11f}$$

Point of Maximum Bending Moment above Support A

$$y_1 = \frac{1}{8}\left(\frac{11\frac{I_b}{I_c}\frac{h}{l} + 18}{2\frac{I_b}{I_c}\frac{h}{l} + 3}\right)h = C_4 h \tag{6.11g}$$

Bending Moments in Corners of Portal

$$M_B = H_A h - \frac{1}{2}ph^2 = \frac{3}{8}\frac{\frac{I_b}{I_c}\frac{h}{l} + 2}{2\frac{I_b}{I_c}\frac{h}{l} + 3}ph^2$$

$$= (C_4 - 0.5)ph^2 \tag{6.11h}$$

$$M_c = -H_D h = -(1 - C_4)ph^2 \tag{6.11i}$$

The constants C_1, C_2, C_3, and C_4 in Equations 6.9, 6.10, and 6.11 can be graphically represented as shown in Figure 6.26. Canned computer programs for the analysis of indeterminate beams and frames render the use of charts such as this unnecessary except for a quick check of numerical values.

6.12.3 Application to Prestressed Concrete Frames

As with continuous beams, a tendon profile has to be assumed at the start in order to determine the secondary bending moments M_2 for the portal frame horizontal beam and vertical legs. A concordant tendon is assumed for the horizontal beam for symmetrical

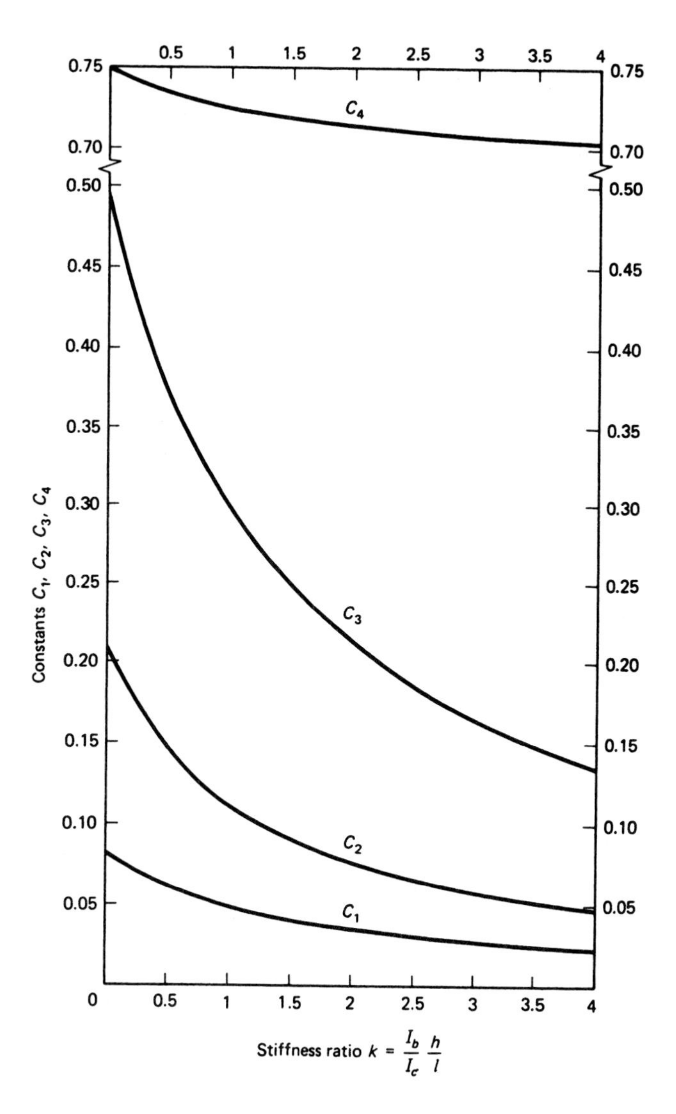

Figure 6.26 Constants C_1 through C_4 in Equations 6.9, 6.10, and 6.11.

gravity loading, and the vertical columns or legs are proportioned to resist the horizontal pressure and the extra moment caused by the shortening of the beam.

Longitudinal shortening of the horizontal beam caused by the prestressing force results in tensile stresses at the outside face of the frame columns. The prestressing vertical tendon should be designed to resist these stresses as well as others. The longitudinal shortening also results in horizontal reactions at the column's supports. Consequently, in order to obtain a prestressing force P in the longitudinal member, a force $P + \Delta P$ has to be applied to the frame. The incremental force ΔP can be evaluated by means of the following expressions.

Frame with Two Hinges at Supports

$$\Delta P = \frac{M_B}{h} = \frac{3}{2k + 3} \frac{E_c I_c}{h^2} \epsilon_{BC} \tag{6.12}$$

where $k = (I_b/I_c)(h/l)$ and ϵ_{BC} is the total strain due to elastic shortening and movement due to shrinkage and creep. The subscripts B and C denote the member extremities of the frame in Figures 6.25 and 6.27.

Frame with Fixed Supports

$$\Delta P = \frac{M_B - M_A}{h} = \frac{E_c I_c}{h}\left(\frac{3}{k+2} + \frac{k+3}{k(k+2)}\right)$$

$$= \frac{3(2k+1)}{k(k+2)}\frac{EI_c}{h^2} \tag{6.13}$$

Figure 6.27 shows the axial deformation due to the strain $\epsilon_{BC} = \Delta l/l$ caused by shortening, creep, and shrinkage.

The *tributary* moments M_A and M_B due to the longitudinal shortening of member BC in Figure 6.27 are as follows.

Frame with Two Hinges at Supports

$$M_B = \frac{6}{2k+3}\frac{E_c I_b \theta}{l} = \frac{3}{2k+3}\frac{E_c I_c}{h}\epsilon_{BC} \tag{6.14}$$

as $k \to 0$, $M_B \to \dfrac{EI_c}{h}\epsilon_{BC}$.

Frame with Fixed Supports

$$M_B = \frac{6}{k+2}\frac{E_c I_c \theta}{h} = \frac{3}{k+2}\frac{E_c I_c}{h}\epsilon_{BC} \tag{6.15}$$

as $k \to 0$, $M_B \to \dfrac{1.5E_c I_c}{h}\epsilon_{BC}$ and $M_A \to \infty$.

The reason for the drastic change in moment values M_B and M_A is that as k approaches zero, ΔP approaches infinity. In such a case, the stiffness of the vertical members relative to the horizontal member approaches infinity, and the horizontal member becomes very flexible, as shown in Figure 6.28. The effects of the horizontal reactions on

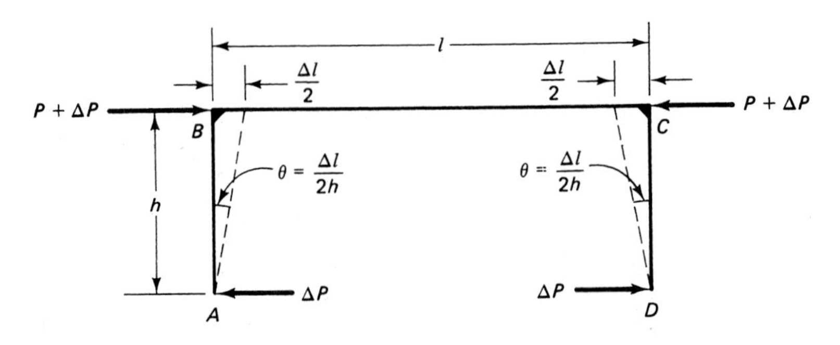

Figure 6.27 Longitudinal deformation of beam BC due to elastic shortening, creep, and shrinkage.

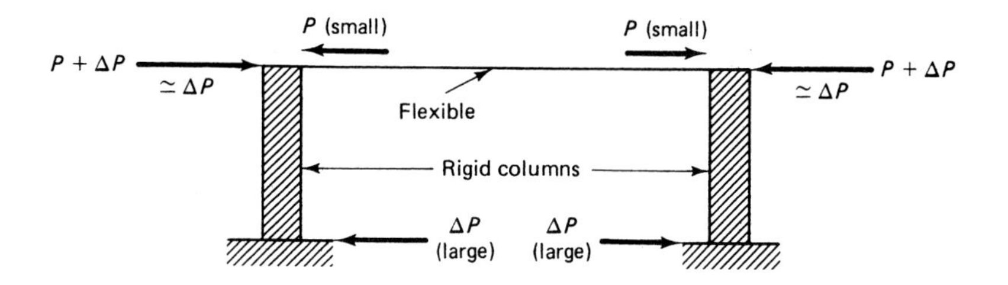

Figure 6.28 Effect of tributary moments due to elastic shortening.

the prestressing force are schematically shown in Figure 6.29 for both hinged-base and fixed-base frames.

Note that the discussion here and in the previous section applies equally to continuously cast and precast prestressed composite frames. Continuity at the corner of the frames has to be accomplished in the construction process. A typical prestressing tendon profile for a frame is shown in Figure 6.30. The prestressing force P_1 is assumed to be less than P_2 in order to allow for the frictional losses in prestress.

6.12.4 Design of Prestressed Concrete Bonded Frame

Example 6.7

A warehouse structure is made of a prestressed single-bay hinged-base portal frame made of standard double-T-sections for both the horizontal beam and the two vertical columns. The units are 8-ft. wide (2.44 m). The frame has a clear span of 80 ft (24.4 m) and is subjected to a uniform gravity live-load intensity W_L = 240 plf (3.5 kN/m) and a uniform horizontal wind pressure of intensity p_w = 65 plf (0.95 kN/m) at the windward side and a suction of intensity p_L = 40 plf (0.58 kN/m) at the leeward side, as shown in Figure 6.31. Design the frame, the profile, and the location of the prestressing tendons for service-load and ultimate load conditions given the following data:

$$f_{pu} = 270{,}000 \text{ psi } (1{,}862 \text{ MPa}) \text{ for low-relaxation tendons}$$

$$f_{ps} = 235{,}000 \text{ psi } (1{,}620 \text{ MPa})$$

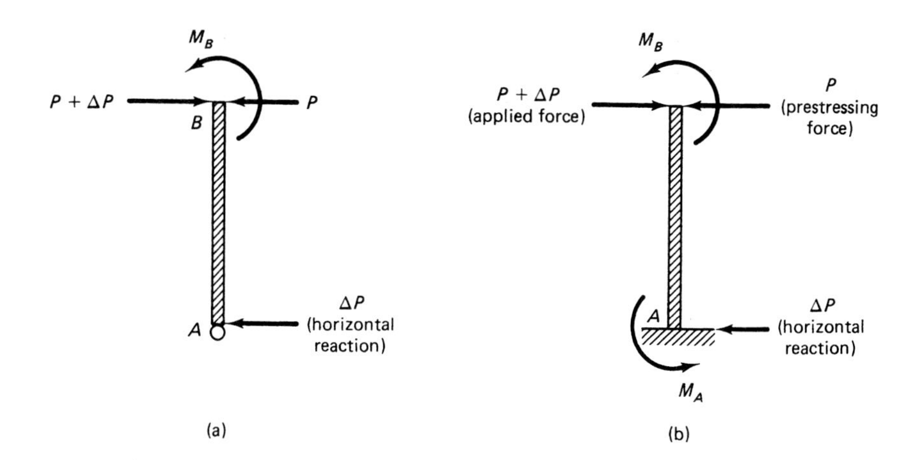

Figure 6.29 Horizontal reaction effect on prestressing force. (a) Hinged-base frame. (b) Fixed-base frame.

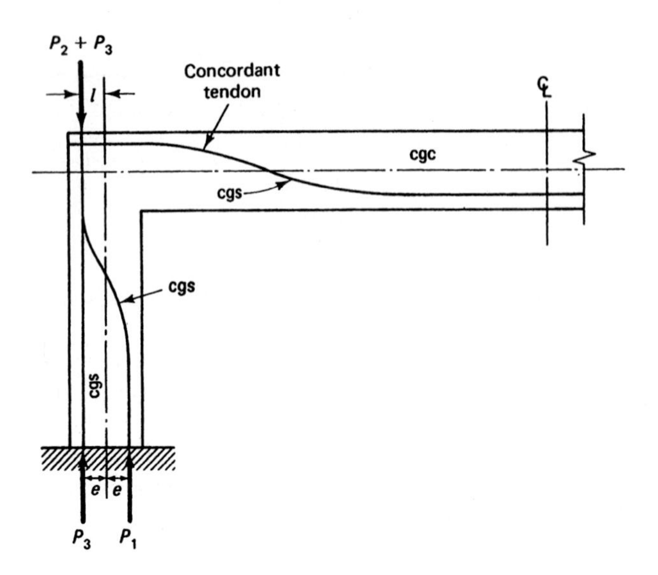

Figure 6.30 Tendon profile in a prestressed frame.

f_{pi} = 189,000 psi (1,303 MPa)

Total losses = 21%, losses after one month of prestressing = 17%

f_{pe} = (one month) = (1 − 0.17)189,000 = 156,870 psi (1,082 MPa)

f_{pe} (final) = (1 − 0.21)189,000 = 149,310 psi (1,029 MPa)

f'_c = 5,000 psi (34.5 MPa)

$f_c = 0.45f'_c$ = 2,250 psi (15.5 MPa)

$f'_{ci} \cong 0.70f'_c$ = 3,500 psi (24.1 MPa)

$f_{ci} = 0.6f'_{ci}$ = 2,100 psi (14.5 MPa)

f_{ti} (midspan) $= 3\sqrt{f'_{ci}}$ = 177 psi

f_{ti} (support) $= 6\sqrt{f'_{ci}}$ = 355 psi

$f_t = 6\sqrt{f'_c}$ to $12\sqrt{f'_c}$ = 849 psi (5.85 MPa)

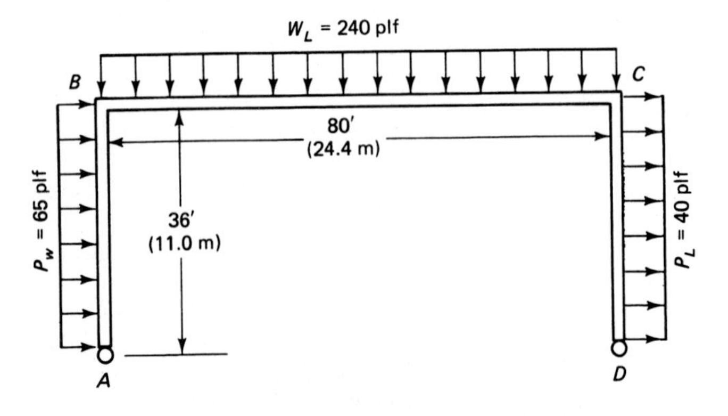

Figure 6.31 Portal frame in Example 6.7.

Solution:

Frame Horizontal Beam BC

Preliminary Analysis. Assume that self-weight $W_D = 600$ plf (8.8 kN/m). Then from Equation 6.9e, the stiffness coefficient is

$$C_1 = \frac{1}{12\left(\frac{2}{3}k + 1\right)}$$

where

$$k = \frac{I_b}{I_c}\frac{h}{l}$$

Assume at this stage that $I_b = I_c$, where I_b is the moment of inertia of the beam BC and I_c is the moment of inertia of the column AB or DC. Then

$$k = \frac{h}{l} \times 1.0 = \frac{36}{80} = 0.45$$

From the chart for C_1 in Figure 6.26, $C_1 = 0.064$. So given total losses = 21%, it follows that $\gamma = 0.79$.

Now, assume 2 in. of concrete topping ($f'_c = 3{,}000$ psi lightweight), 5 psf insulation, and a waterproofing width of the segment = 8 ft. Then

$$W_{SD} = \left(\frac{2}{12} \times 110 + 5\right)8 \text{ ft} = 187 \text{ plf}$$

Beam BC is to be designed as a concordant cable, i.e., it is to behave as a simply supported beam *for self-weight* W_D. But it would be considered continuous for the superimposed dead load W_{SD} and live load W_L as part of the rigid portal frame. The C-line would then coincide with the cgs line due to concordance. Accordingly, if $W_D \cong 600$ plf, then from Equation 6.9e, the *midspan* moment is

$$M = \left(\frac{1}{8} - C_1\right)wl^2$$

so that

$$M_D = \frac{wl^2}{8} = \frac{600(80)^2}{8} \times 12 = 5.760 \times 10^6 \text{ in.-lb}$$

$$M_{SD} = \left(\frac{1}{8} - 0.064\right)187(80)^2 \times 12 = 0.876 \times 10^6 \text{ in.-lb}$$

$$M_L = \left(\frac{1}{8} - 0.064\right)240(80)^2 \times 12 = 1.124 \times 10^6 \text{ in.-lb}$$

Assume $f_t = 0$. Then from Equation 4.4b, the minimum section modulus at the bottom fibers for an efficient section is given by

$$S_b \geq \frac{(1-\gamma)M_D + M_{SD} + M_L}{f_t - \gamma f_{ci}}$$

or

$$S_b = \frac{(1-0.79)5.760 \times 10^6 + 0.876 \times 10^6 + 1.124 \times 10^6}{0 - 0.79(-2{,}100)} = 1{,}935 \text{ in.}$$

The closest section is PCI 8DT32 + 2 Double-T type 168-DI with 2 in. concrete topping (Ref. 6.15).

Properties of Preliminary Section

Property	Untopped	Topped
A_c (in.^{2})	567	759
I_c (in.^{4})	55,464	71,886
$r^2 = I_c/A_c$ (in.^{2})	97.8	94.7
c_b (in.)	21.21	23.66
c_t (in.)	10.79	10.34
S^t (in.^{3})	5,140	6,952
S_b (in.^{3})	2,615	3,038
W_D (plf)	591	738

$e_e = 8.21$ in. (20.9 cm)

$e_c = 17.46$ in. (44.3 cm)

sixteen $\frac{1}{2}$-in. dia (12.7-mm dia) 270-K, stress-relieved strands

$A_{ps} = 16 \times 0.153 = 2.45$ in.^{2} (15.8 cm^2)

Analysis of Section at Transfer

(a) *Midspan Section* ($e_c = 17.46$ in.)

$$M_D = 5.760 \times 10^6 \times \frac{591}{600} = 5.673 \times 10^6 \text{ in.-lb}$$

where

$$591 = \frac{567}{12 \times 12} \times 150 \text{ plf}$$

From Equation 4.1a,

$$f^t = -\frac{P_i}{A_c}\left(1 - \frac{ec_t}{r^2}\right) - \frac{M_D}{S^t} \leq f_{ti}$$

$$P_i = A_{ps} f_{pi} = 2.45 \times 189{,}000 = 463{,}050 \text{ lb}$$

$$f^t = -\frac{463{,}050}{567}\left(1 - \frac{17.46 \times 10.79}{97.8}\right) - \frac{5.673 \times 10^6}{5{,}140}$$

$$= -347 \text{ psi } (C)\text{, no tension, O.K.}$$

Provide nonprestressed steel at the top fibers at midspan to account for any possible tensile stresses. Then, from Equation 4.1b,

$$f_b = -\frac{P_i}{A_c}\left(1 + \frac{ec_b}{r^2}\right) + \frac{M_D}{S_b}$$

$$= -\frac{463{,}050}{567}\left(1 + \frac{17.46 \times 21.21}{97.8}\right) + \frac{5.673 \times 10^6}{2{,}615}$$

$$= -1{,}740 \text{ psi } (C) < f_{ci} = 2{,}100 \text{ psi, O.K.}$$

(b) *Support Section* ($e_e = 8.21$ in.)

$$f^t = \frac{463{,}050}{567}\left(1 - \frac{8.21 \times 10.79}{97.8}\right) - 0$$

$$= -76.9 \text{ psi } (C)\text{, no tension, O.K.}$$

$$f_b = \frac{463{,}050}{567}\left(1 + \frac{8.21 \times 21.21}{97.8}\right) + 0$$

$$= -2{,}271 \text{ psi } (C)\ (15.7 \text{ MPa}) > f_{ci} = 2{,}100 \text{ psi, unsatisfactory}$$

Hence, lower the magnitude of the prestressing force by *debonding* some strands over a length of 15% of span from the support face, or change the eccentricity of the tendon. If the former technique is employed, debond four strands over a length = 0.15 × 80 ft = 12 ft (366 m) from the support, releasing the anchorage of the four grouted strands. We obtain

$$A_{ps} = (16 - 4)0.153 = 1.836 \text{ in.}^2$$

$$P_i = 1.836 \times 189{,}000 = 347{,}004 \text{ lb } (1{,}543 \text{ kN})$$

$$f_b = -\frac{347{,}004}{567}\left(1 + \frac{8.21 \times 21.21}{97.8}\right) + 0$$

$$= -1{,}701.6 \text{ psi } < 2{,}100 \text{ psi, O.K.}$$

Frame Vertical Column Analysis. Choose a double-T as walls for the frame and suppose that e_b and S_b refer to the outer face and that e_t and S_t refer to the inner face of the vertical T-section. Since it was assumed, in calculating the stiffness coefficient k in the previous section, that $I_b = I_c$, choose also 8DT32, hinged at the base. This vertical member will act as a compression member subject to large axial load and bending. The bending moments are caused by wind load and moments from the frame horizontal beam *BC*. In such a case it is preferable to spread the tendon across the section, as shown in Figure 6.32 comparing the beam section and the column section.

Assume that the center of gravity of the prestressing strands coincides with the cgc line, and design the distribution of the strands according to

$$e_c = e_e = 0$$

Try using twenty $\frac{1}{2}$-in. dia 270-K low-relaxation strands:

$$A_{ps} = 20 \times 0.153 = 3.06 \text{ in.}^2$$

$$P_i = A_{ps} f_{pi} = 3.06 \times 189{,}000 = 587{,}340 \text{ lb}$$

$$f^t = f_b = -\frac{P_i}{A_c} \pm 0 = -\frac{578{,}340}{567} = -1{,}020 \text{ psi } (C) < f_{ci} = 2{,}100 \text{ psi, O.K.}$$

Frame Moments and Reactions at Service-Load Level

Horizontal Portal Beam BC

Free Support W_D Stage. Assume that the length of precast beams is 80 − 1.3 = 78.7 ft. Then the midspan moment $M_E = wl^2/8 = [591(78.7)^2/8] \times 12 = 5.491 \times 10^6$ in.-lb and the reaction at the column-wall bracket support is $R_D = 591 \times 78.7/2 = 23{,}256$ lb.

Composite Topping W_{SD} Stage. From before, the midspan moment is $M_{SD} = 0.876 \times 10^6$ in.-lb. The support moment is then

$$M_B = M_c = \frac{wl^2}{8} - 0.876 \times 10^6$$

$$= \frac{187(80)^2 \times 12}{8} - 0.876 \times 10^6 = 0.919 \times 10^6 \text{ in.-lb}$$

Redistribution of Moments. From Equations 6.6 and 6.7, the maximum distribution percentage is

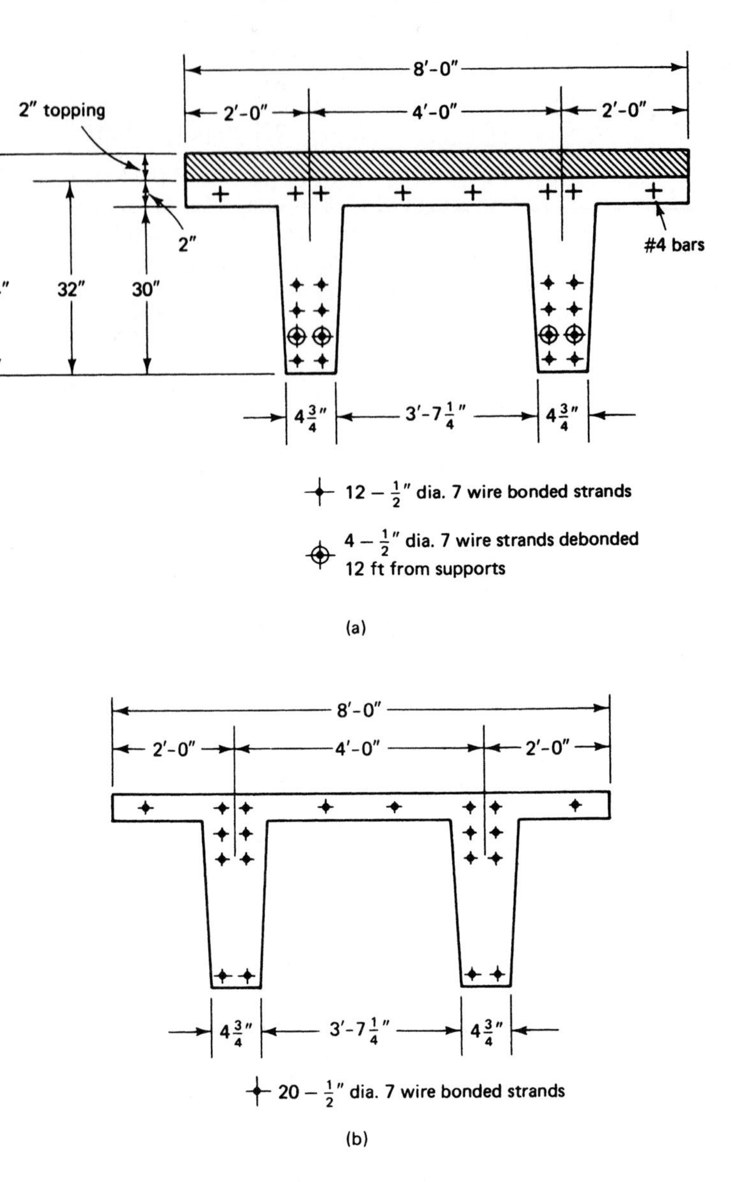

Figure 6.32 Details of beam and wall double-T's in Example 6.7. (a) Horizontal beam standard PCI section 8DT32 + 2 (168-D1). (b) Vertical wall sePection 8DT32 with twenty $\frac{1}{2}$-in. dia. strands with $e_c = e_e = 0$.

$$p_d = 20\left[1 - \frac{\omega_p + \dfrac{d}{d_p}(\omega - \omega')}{0.36\beta_1}\right]$$

$$d = 32 + 2 - 2.5 \cong 31.5 \text{ in.}$$

$$\text{compression side } b = 2 \times 4.75 = 9.50 \text{ in.}$$

$d_p = c_b + e_e = 21.21 + 8.21 = 29.42$ in. or $d_p = 0.8h = 0.8 \times 32 = 25.6$ in. whichever is larger. Use $d_p = 29.42$ in., and assume two #5 bars per rib at the compression side and two #7 bars per rib at the tension side of both the horizontal roof beams and the vertical wall beams. We obtain

$$A'_s = 4 \times 0.305 = 1.22 \text{ in.}^2$$

$$\omega' = \frac{A'_s}{bd}\frac{f_y}{f'_c} = \frac{1.22}{9.5 \times 31.5} \times \frac{60{,}000}{5{,}000} = 0.0489$$

$$A_s = 4 \times 0.60 = 2.40 \text{ in.}^2$$

$$\omega = \frac{A_s}{bd}\frac{f_y}{f'_c} = \frac{2.40}{9.5 \times 31.5} \times \frac{60{,}000}{5{,}000} = 0.0962$$

Use $\omega_p = (A_{ps}/bd_p)(f_{ps}/f'_c) = 0$ since the prestressing steel is not continuous over corners of the portal frame. Then $\omega_p + (d/d_p)(\omega - \omega') = 0 + (31.5/29.42) \times (0.0962 - 0.0489) = 0.0506$. Also, $0.24\beta_1 = 0.24 \times 0.80 = 0.192 > 0.0506$. Hence, moment redistribution is permissible and we have

$$\text{Maximum distribution factor } \rho_D = 20\left[1 - \frac{\omega_p + (d/d_p)(\omega - \omega')}{0.36\beta_1}\right]$$

$$= 20\left(1 - \frac{0 + 0.0506}{0.36 \times 0.80}\right) = 16.49\%$$

Accordingly, use a moment distribution factor of 0.12 for transferring 12% of the moment from the frame corners B and C to midspan BC. Also, rigid connecting steel plates should be used at the portal upper joints and be so designed to provide a moment connection capable of transferring at least 12% of the support moment to the midspan. Then the adjusted $M_B = M_c = (1 - 0.12)0.919 \times 10^6 = 0.809 \times 10^6$ in.-lb, the adjusted midspan moment $M_E = 0.876 \times 10^6 \times 1.12 = 0.981 \times 10^6$ in.-lb, and the superimposed dead-load reaction R_{SD} at the support $= (187 \times 80)/2 = 7{,}480$ lb.

Live-load W_L Stage. From before, the midspan is $M_L = 1.124 \times 10^6$ in.-lb. So the support moment is

$$M_B = M_c = \frac{240(80)^2 \times 12}{8} - 1.124 \times 10^6$$

$$= 1.180 \times 10^6 \text{ in.-lb}$$

The adjusted $M_n = M_c = (1 - 0.12) \times 1.180 \times 10^6 = 1.038 \times 10^6$ in.-lb, and the adjusted midspan $M_L = 1.124 \times 10^6 \times 1.12 = 1.259 \times 10^6$ in.-lb. The live-load reaction at the vertical support is

$$R_L = \frac{240 \times 80}{2} = 9{,}600 \text{ lb}$$

Wind Pressure Stage. From Equations 6.11h and i,

$$M_B = (C_4 - 0.5)ph^2$$

$$M_c = -(1 - C_4)ph^2$$

From before,

$$k = \frac{I_b}{I_c}\frac{h}{l} = 0.45 \text{ for } I_b = I_c$$

From the chart for C_4 in Figure 6.26, $C_4 = 0.73$.

Windward side moment M_B

$$M_{B1} = (0.73 - 0.5)65(36)^2 \times 12 = 232{,}502 \text{ in.-lb}$$

$$M_{B2} = (1 - 0.73)40(36)^2 \times 12 = 167{,}961 \text{ in.-lb}$$

$$\text{Total } M_B = 232{,}502 + 167{,}961 = 400{,}463 \text{ in.-lb}$$

Leeward side moment M_c

$$M_{c1} = -(1 - 0.73)65(36)^2 \times 12 = -272{,}938 \text{ in.-lb}$$

$$M_{c2} = -(0.73 - 0.5)40(36)^2 \times 12 = -143{,}078 \text{ in.-lb}$$

$$\text{Total } M_c = -272{,}938 - 143{,}078 = -416{,}016 \text{ in.-lb}$$

The controlling wind moment $M_W = 416{,}016$ in.-lb, since wind can blow from either the left or the right.

From Equation 6.11a, the vertical reactions at A and D due to wind are

$$R_{WA} = -\tfrac{1}{2}\, ph\frac{h}{l} = -\frac{(65 + 40)(36)^2}{2 \times 80} = -851 \text{ lb}$$

$$R_{WD} = +\tfrac{1}{2}\, ph\frac{h}{l} = +851 \text{ lb}$$

Loads and Moments Due to Long-Term Effects

Moments to Restrain End Rotations at B and C Due to Long-Term Prestress Losses. Figure 6.33 shows the moment distributions on horizontal member BC. One month after prestressing we have:

$$f_{pe1} = 156{,}870 \text{ psi}$$

$$\text{Midspan } P_e = 16 \times 0.153 \times 156{,}870 = 384{,}018 \text{ lb}$$

$$\text{Support } P_e = (16 - 4) \times 0.153 \times 156{,}870 = 288{,}013 \text{ lb}$$

$$\text{Midspan monent } M_E = 384{,}018 \times 17.46 = 6.705 \times 10^6 \text{ in.-lb}$$

$$\text{Support moment } M_B = 288{,}013 \times 8.21 = 2.364 \times 10^6 \text{ in.-lb}$$

The eccentricity at section F, where four strands were debonded, is

$$e_F = (17.46 - 8.21)\frac{12}{40} + 8.21 = 10.99 \text{ in.}$$

and the moment at section F is

$$M_F = 384{,}018 \times 10.99 = 4.220 \times 10^6 \text{ in.-lb}$$

The reduced M_F due to debonding is $288{,}013 \times 10.99 = 3.165 \times 10^6$ in.-lb.

The service load after all losses have occurred is as follows:

$$f_{pe} = 149{,}310 \text{ psi}$$

$$f_{pe}/f_{pe1} = \frac{149{,}310}{156{,}870} = 0.952$$

$$M_E = 6.705 \times 10^6 \times 0.952 = 6.383 \times 10^6 \text{ in.-lb}$$

$$M_F = 4.220 \times 10^6 \times 0.952 = 4.017 \times 10^6 \text{ in.-lb}$$

$$M_{F1} = 3.165 \times 10^6 \times 0.952 = 3.013 \times 10^6 \text{ in.-lb}$$

$$M_B = M_C = 2.364 \times 10^6 \times 0.952 = 2.251 \times 10^6 \text{ in.-lb}$$

Slopes at B and C at Beam Erection One Month After Prestressing

$$\text{Slope } \theta = \frac{1}{E_c I_b}[MI]$$

To find the areas of the moment diagrams for half the span due to symmetry, (i) add half of Figure 6.33(b) to half of Figure 6.33(d), and (ii) add half of Figure 6.33(c) to half of Figure 6.33(d). Then subtract (ii) from (i) to get the rotation of the beam at B or

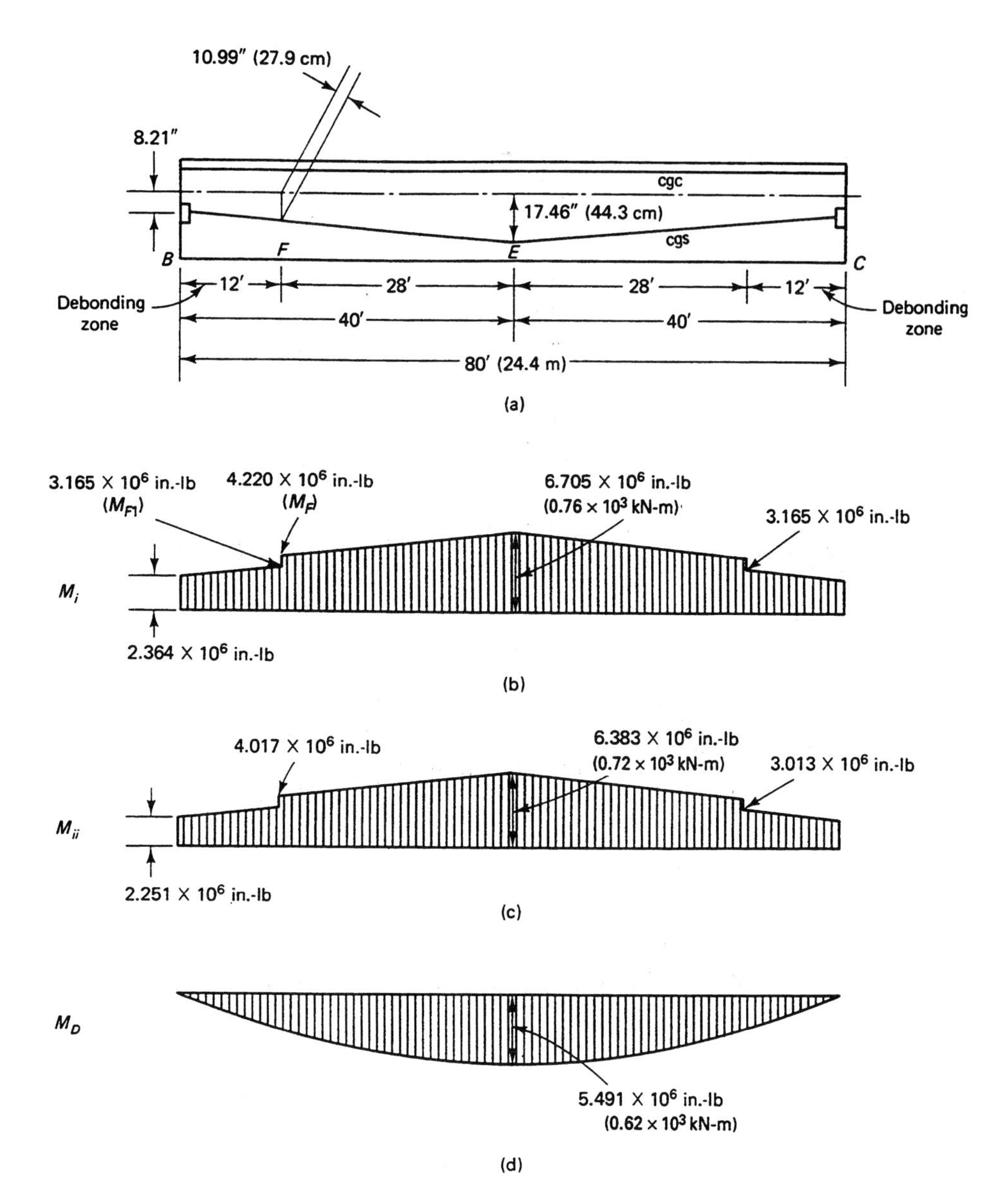

Figure 6.33 Bending moment diagrams for primary and self-weight moments for beam BC. (a) Tendon profile. (b) Prestressing moments one month after initial prestress. (c) Effective prestressing moment after all losses. (d) Beam BC self-weight moments.

C that would have to be restrained by a welded connection to develop continuity at the portal frame corners *B* and *C*. We have:

(i)

$$\theta E_c I_b \times 10^{-6} = M_{(i)} l \text{ at beam erection}$$

$$= 2.364 \times 12 \times 12 + (3.165 - 2.364) \times 12 \times 12 \times \frac{1}{2}$$

$$+ 4.220 \times 28 \times 12 + (6.705 - 4.220) \times 28 \times 12 \times \frac{1}{2}$$

$$- 5.491 \times 40 \times 12 \times \frac{2}{3} = 2{,}233.49 - 1{,}757.12 = 476.47$$

(ii)

$$\theta E_c I_b \times 10^{-6} = M_{(ii)} l \text{ at service load}$$

$$= 2.251 \times 12 \times 12 + (3.013 - 2.251) \times 12 \times 12 \times \frac{1}{2}$$

$$+4.017 \times 28 \times 12 + (6.383 - 4.017) \times 28 \times 12 \times \frac{1}{2}$$

$$-5.491 \times 40 \times 12 \times \frac{2}{3} = 2{,}126.21 - 1{,}757.12 = 369.09$$

The rotational angle θ at B or C caused by the reduction in the prestressing force due to long-term losses is

$$\frac{1}{E_c I_b}(476.37 - 369.09)10^6 = \frac{107.3 \times 10^6}{E_c I_b}$$

If M_r is the resisting moment at the connection weld to restrain the member against this rotation,

$$\text{Slope } \theta = \frac{M_r l/2}{E_c I_b} = \frac{M_r \times 480}{E_c I_b}$$

Equating the right sides of the preceding equations yields

$$\frac{107.3 \times 10^6}{E_c I_b} = \frac{480 M_r}{E_c I_b}$$

$$M_r = \frac{107.3 \times 10^6}{480} = 0.224 \times 10^6 \text{ in.-lb}$$

Moments Resulting from Creep and Shrinkage Long-Term Losses

(a) *Creep*

$$P_i = 463{,}050 \text{ lb}$$

$$P_e \text{ at erection} = 384{,}018 \text{ lb}$$

It is reasonable to take the creep force as the average of P_i and P_e. Thus,

$$\epsilon_{CR} = \frac{1}{A_c E_c}\left[\left(\frac{P_i + P_e}{2}\right) C_u\right]$$

Use the creep coefficient $C_u = 2.25$:

$$E_c = 57{,}000\sqrt{f'_c} = 57{,}000\sqrt{5{,}000} = 4.03 \times 10^6 \text{ psi}$$

$$\epsilon_{CR} = \frac{1}{567 \times 4.03 \times 10^6}\left(\frac{463{,}050 + 384{,}018}{2} \times 2.25\right)$$

$$= 417 \times 10^{-6} \text{ in./in.}$$

(b) *Shrinkage*
From Equation 3.14, the shrinkage strain from the time of erection (30 days after prestressing) to one year later is

$$\epsilon_{SH} = 8.2 \times 10^{-6} K_{SH}\left(1 - 0.06\frac{V}{S}\right)(100 - RH)$$

Now, $V/S = 1.79$, and if we assume that $RH = 75\%$, then, from Table 3.6, which states that after 30 days to within one year $K_{SH} = 0.45$, we have

$$\epsilon_{SH} = 8.2 \times 10^{-6} \times 0.45(1 - 0.06 \times 1.79)(100 - 75)$$
$$= 82.3 \times 10^{-6} \text{ in./in.}$$

So the total deformation strain due to creep and shrinkage is $(417 + 82.3)10^{-6} = 499.3 \times 10^{-6}$ in./in.

From Equation 6.12,

$$M_B = M_c = \frac{3}{(2k + 3)} \frac{E_c I_c}{h} \epsilon_{BC}$$

From before, the stiffness coefficient $k = 0.45$, and $E_c = 4.03 \times 10^6$ psi. Also, precast $I_c = 55{,}464$ in.[4] Consequently,

$$M_B = M_C = \frac{3}{(2 \times 0.45 + 3)} \times \frac{4.03 \times 10^6 \times 55{,}464}{36 \times 12} \times 499.3 \times 10^{-6}$$
$$= 198{,}724 \text{ in.-lb} = 0.199 \times 10^6 \text{ in.-lb}$$

These moments due to long-term effects will produce tensile stresses at the inside face of the vertical member and bottom face of the horizontal member. Elastic shortening should also be considered often for accuracy.

Final Moments and Stresses in the Horizontal Beam BC

Midspan Section (e_c = 17.46 in.)

$$M_D = 5.491 \times 10^6 \text{ in.-lb } (0.62 \times 10^3 \text{ kN-m})$$
$$M_{SD} = 0.981 \times 10^6 \text{ in.-lb}$$
$$M_L = 1.259 \times 10^6 \text{ in.-lb}$$
$$M_r = 0.224 \times 10^6 \text{ in.-lb}$$
$$M_{CR+SH} = 0.199 \times 10^6 \text{ in.-lb}$$
$$P_e \text{ after all losses} = 2.45 \times 149{,}310 = 365{,}810 \text{ lb}$$

The total superimposed moments are

$$M_T = M_{SD} + M_L + M_r + M_{CR+SH}$$
$$= (0.981 + 1.259 + 0.244 + 0.199) \times 10^6$$
$$= 2.663 \times 10^6 \text{ in.-lb}$$

$$f_b = -\frac{P_e}{A_c}\left(1 + \frac{e_c c_b}{r^2}\right) + \frac{M_D}{S_b} + \frac{M_T}{S_{bc}}$$
$$= -\frac{365{,}810}{567}\left(1 + \frac{17.46 \times 21.21}{97.8}\right) + \frac{5.491 \times 10^6}{2{,}615} + \frac{2{,}663 \times 10^6}{3{,}038}$$
$$= -112 \text{ psi } (C), \text{ no tension, O.K.}$$

$$f^t = -\frac{P_e}{A_c}\left(1 - \frac{e_c c_t}{r^2}\right) - \frac{M_D}{S^t} - \frac{M_T}{S_c^t}$$
$$= -\frac{365{,}810}{567}\left(1 - \frac{17.46 \times 10.79}{97.8}\right) - \frac{5.491 \times 10^6}{5{,}140} - \frac{2.663 \times 10^6}{6{,}952}$$
$$= -854 \text{ psi } (C) < f_c = 2{,}250 \text{ psi, O.K.}$$

Support Section (e_e = 8.21 in.)

$$M_D = 0$$
$$M_{SD} = 0.809 \times 10^6 \text{ in.-lb}$$

$$M_L = 1.038 \times 10^6 \text{ in.-lb}$$

$$M_W = 0.416 \times 10^6 \text{ in.-lb}$$

Not including the relief moments due to rotation, creep, and shrinkage, which cause compressive stresses, the total negative moments at supports B or C are

$$-M_T = (0.809 + 1.038 + 0.416)10^6 = 2.26 \times 10^6 \text{ in.-lb}$$

The sections at supports B and C are virtually reinforced concrete. P_e for 12 strands at either support after all losses = 274,133 lb, and

$$f^t = -\frac{247{,}133}{567}\left(1 - \frac{8.21 \times 10.79}{97.8}\right) + 0 + \frac{2.26 \times 10^6}{6{,}952}$$

$$= +280 \text{ psi } (T) < f_t = 12\sqrt{f_c'} = 849 \text{ psi, O.K.}$$

Provide nonprestressed steel to accommodate all the tensile stress. Also,

$$f_b = -\frac{274{,}133}{567}\left(1 + \frac{8.21 \times 21.21}{97.8}\right) - 0 - \frac{2.26 \times 10^6}{3{,}038}$$

$$= -2{,}088 \text{ psi } (C) < f_c = 2{,}250 \text{ psi, O.K.}$$

$$M_u = 1.4 \times 0.809 \times 10^6 + 1.7(1.038 \times 10^6 + 0.416 \times 10^6)$$

$$= 3.61 \times 10^6 \text{ in.-lb}$$

$$\text{Rqd. } M_n = \frac{M_u}{\phi} = \frac{3.61 \times 10^6}{0.90} = 4.01 \times 10^6 \text{ in.-lb}$$

$$M_n = A_s f_y\left(d - \frac{a}{2}\right)$$

Assume a moment arm $d - a/2 \cong 0.9d = 0.9 \times 31.5 = 28.35$ in. Then

$$4.01 \times 10^6 = A_s \times 60{,}000 \times 28.35$$

$$A_s = \frac{4.01 \times 10^6}{60{,}000 \times 28.35} = 2.36 \text{ in}^2 \text{ } (15.2 \text{ cm}^2)$$

$$a = \frac{A_s f_y}{0.85 f_c' b} = \frac{2.36 \times 60{,}000}{0.85 \times 5{,}000 \times 96}$$

$$= 0.35 \text{ in. } (0.89 \text{ cm}) < h_f = 4 \text{ in.}$$

Hence, treat as a rectangular section:

$$d - \frac{a}{2} = 31.5 - \frac{0.35}{2} = 31.3 \text{ in. } (79.5 \text{ cm})$$

$$A_s = \frac{4.01 \times 10^6}{60{,}000 \times 31.3} = 2.14 \text{ in.}^2 \text{ } (13.8 \text{ cm}^2)$$

Use two #7 bars (22-mm dia) in each rib. Then

$$A_s = 4 \times 0.60 = 2.40 \text{ in.}^2 > 2.14 \text{ in.}^2, \text{ O.K.}$$

Final Moments and Stresses in the Vertical Column Walls AB and DC. The direct load on the column is

$$R_D + R_{SD} + R_L + R_W = 23{,}256 + 7{,}480 + 9{,}600 + 851 = 41{,}187 \text{ lb (183 kN)}$$

Assuming 15 in eccentricity, the moment M_D becomes

$$M_D = 41{,}187 \times 15 = 0.617 \times 10^6 \text{ in.-lb}$$

$$M_{SD} = 0.809 \times 10^6$$
$$M_L = 1.038 \times 10^6$$
$$M_W = 0.416 \times 10^6$$

The total moment is

$$M_T = (0.617 + 0.809 + 1.038 + 0.416)10^6$$
$$= 2.880 \times 10^6 \text{ in.-lb } (0.33 \times 10^3 \text{ kN-m})$$

For 20 strands in the wall units,

$$P_e = A_{ps} f_{pe} = 3.06 \times 149{,}310 = 456{,}887 \text{ lb } (2{,}032 \text{ kN})$$

$$f_b \text{ (outer face)} = -\frac{P_e}{A_c} - \frac{P}{A_c} + \frac{M_T}{S_b}$$

$$= -\frac{456{,}887}{567} - \frac{41{,}187}{567} + \frac{2{,}880 \times 10^6}{2{,}615}$$

$$= +223 \text{ psi } (T) \leq 849 \text{ psi, O.K.}$$

$$f_t \text{ (innerface)} = -\frac{P_e}{A_c} - \frac{P}{A_c} - \frac{M_T}{S^t}$$

$$= -\frac{456{,}887}{567} - \frac{41.187}{567} - \frac{2{,}880 \times 10^6}{5{,}140}$$

$$= -1{,}439 \text{ psi } (C) \ (9.9 \text{ MPa}) < f_c = 2{,}250 \text{ psi } (15.5 \text{ MPa})$$

Consequently, adopt the double-T section 8DT32 for the walls with twenty $\frac{1}{2}$-in. dia 7-wire 270-K low-relaxation strands arranged as shown in Figure 6.32. Also, adopt the double-T section 8DT32 + 2(168 – D1) for the horizontal top beam BC with sixteen $\frac{1}{2}$-in. dia 7-wire 270-K low-relaxation strands with four strands debonded 12 ft (3.66 m) from the face of the supports.

Figure 6.34 gives a schematic of the configuration details of the prestressed concrete portal frame. The total design would involve designing the vertical wall brackets, shear strength, flexural strength, and serviceability checks as well as detailing the welded connections between the horizontal beam and the supporting wall columns.

6.13 LIMIT DESIGN (ANALYSIS) OF INDETERMINATE BEAMS AND FRAMES

The discussions presented so far deal with proportioning the controlling sections in the design process, such as the midspan and support sections, with redistribution factors p_D for continuity empirically provided by the code. The continuity factors assume that adequate longitudinal reinforcement is provided at the critical continuity zones to properly control the cracking levels of those zones.

Such a procedure does not necessarily give the most efficient solution to a statically indeterminate continuous beam or frame, since full redistribution at ultimate load is not considered. As the applied load is gradually increased until the structure as a whole reaches its limit capacity, the critical sections, such as the supports or corners of frames, develop severe cracking, and the rotation becomes so large that, for all practical purposes, rotating *plastic hinges* have developed. If the number of plastic hinges that develop equals the number of indeterminacies, the structure becomes determinate, as *full redistribution* of moments would have taken place throughout it. With the development of an additional hinge, the structure becomes a mechanism tending toward collapse.

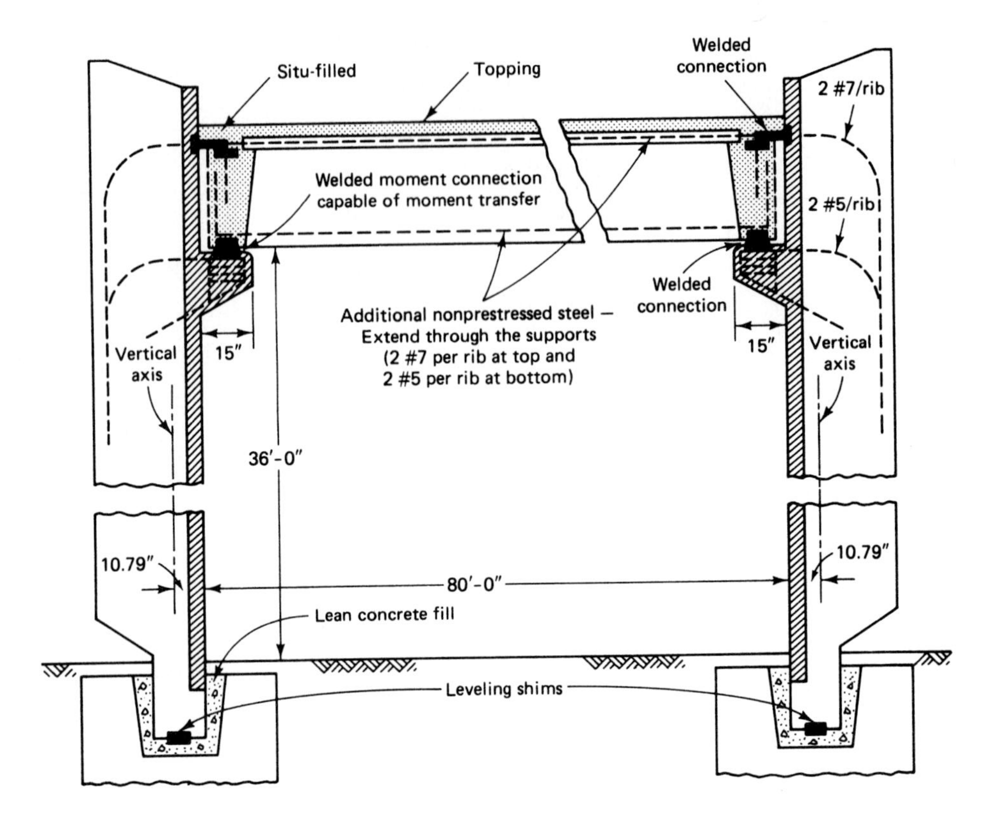

Figure 6.34 Sectional elevation and connection details of frame in Example 6.7.

Analysis of the structure at *full* moment redistribution is termed as *plastic* or *limit* analysis. Since concrete cracks severely at high overloads, it is possible for the designer to *impose* the desirable locations of the plastic hinges by making the concrete member fail or making it adequately strong at any section by decreasing or increasing the reinforcement percentage without appreciably altering the stiffness of the member. This flexibility in proportioning is not available in the plastic design of steel structures, where the resulting locations of the plastic hinges are obtained from mechanisms determined by upper and lower bound solutions. Details of Baker's *theory of imposed rotations* are presented in Refs. 6.5, 6.6, and 6.7.

6.13.1 Method of Imposed Rotations

The imposed locations of the plastic hinges coincide with the locations of the maximum elastic moments for combined gravity loads and horizontal wind loads. These locations occur at the intermediate supports of continuous beams and beam-column corners of frames, as seen in the portal frame of Figure 6.35. By superposing part (a) on part (b), one plainly sees that the maximum elastic moment occurs at corner C. Since plastic moments are a magnification of the elastic moments, the natural location for the development of a plastic hinge is at that corner.

Because the structure is indeterminate to the first degree, only one hinge develops, resulting in a *basic* frame ABC, which is the fundamental frame for the imposed hinges seen in Figure 6.35(e), numbered in the order in which they are expected to form.

The structure in Figure 6.35(e) has nine indeterminacies; hence, nine plastic hinges are formed. A tenth hinge reduces the structure to a mechanism resulting in collapse. Note that no plastic hinges are permitted to form at midspan of the horizontal members.

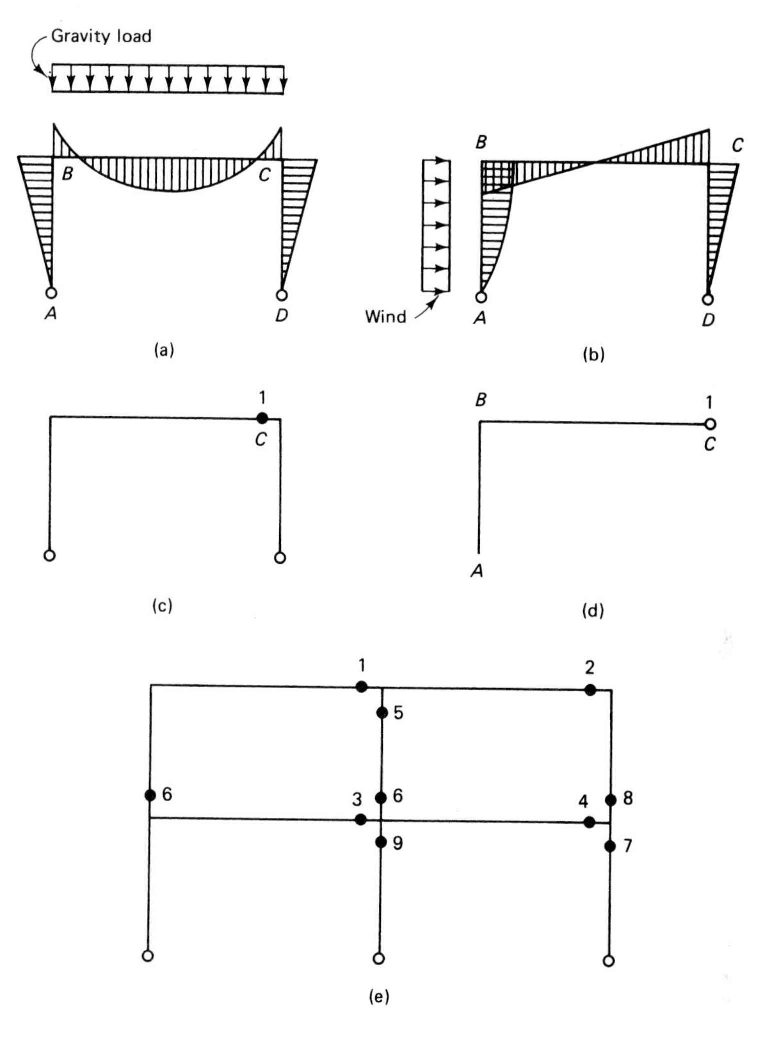

Figure 6.35 Imposed plastic hinges in concrete frames. (a) Gravity-load elastic moment. (b) Wind-load intensity moments. (c) Hinge 1 at C reducing frame to statically determinate. (d) Basic plastic frame. (e) Succession of plastic hinges in two-span, two-level frame.

The plastic moments resulting in hinges 1,2,3, . . . ,n are denoted $\overline{X}_1, \overline{X}_2, \overline{X}_3, \ldots, \overline{X}_n$ and are assumed to remain constant throughout the progressive deformation of the structure. Hence, the derivative of the total strain energy U with respect to the assumed plastic moments $\overline{X}_i$ at any hinge i is set equal to the plastic rotation at the hinge, i.e.,

$$\frac{\delta U}{\delta \overline{X}_i} = -\theta_i \tag{6.16}$$

If δ_{ik} is assumed to represent the relative rotation of the ith hinge due to a unit moment at the kth hinge, $\delta_{ik} = \delta_{ki}$ from Maxwell's reciprocal theorem. The coefficients δ_{ik} are called *influence coefficients,* because they represent the displacement or rotation at a particular section due to a unit moment at *another* section, i.e., $\delta_{ik} = -\theta_i$.

From the principle of virtual work,

$$\delta_{ik} = \Sigma \int_0^l \frac{M_i M_k}{E_c I}\, ds \tag{6.17}$$

Consequently,

$$\Sigma \int_0^l \frac{M_i M_k}{E_c I} ds = -\theta_i \tag{6.18}$$

The left-hand side of Equation 6.18 represents the integration of the products of the areas of the M_i diagrams and the ordinates of M_k diagrams at their centroids along the horizontal distance s along the span. Substituting δ_{i0} and δ_{ik} for M_k, we obtain

$$\delta_{i0} + \sum_{k=1}^{k=n} \delta_{ik} \overline{X}_k = -\theta_i \tag{6.19}$$

This is a structure having n plastic hinges to reduce it to statically determinate:

$$\begin{aligned} \delta_{10} + \delta_{11}\overline{X}_1 + \delta_{12}\overline{X}_2 + \cdots + \delta_{1n}\overline{X}_n &= -\theta_1 \\ \delta_{20} + \delta_{21}\overline{X}_1 + \delta_{22}\overline{X}_2 + \cdots + \delta_{2n}\overline{X}_n &= -\theta_2 \\ \delta_{n0} + \delta_{n1}\overline{X}_1 + \delta_{n2}\overline{X}_2 + \cdots + \delta_{nn}\overline{X}_n &= -\theta_n \end{aligned} \tag{6.20}$$

The number of equations is equal to the number of redundancies or indeterminacies. By trial and adjustment of the redundant plastic moments $\overline{X}_1, \ldots, \overline{X}_n$ in the solution of Equations 6.20 for controlled maximum allowable rotation of the largest rotating hinge θ_1, the plastic moments at the beam supports and column ends are obtained for the plastic design of the concrete structure. The arbitrary plastic moment values $\overline{X}_1, \overline{X}_2, \ldots, \overline{X}_n$ are chosen to result in plastic rotations $\theta_1, \theta_2, \ldots, \theta_n$ that give *full redistribution* of moments throughout the structure.

It can be proven that the influence coefficient δ_{ik} in Equations 6.20 is

$$\delta_{ik} = \frac{A_i}{E_i}\eta \tag{6.21}$$

where A_i is the area under the primary M_i bending moment diagram and η is the ordinate of the M_k moment diagram under the centroid of the M_i diagram (Ref. 6.5). As an example, in Figure 6.36 the influence coefficient δ_{01} is obtained by superposing the moment diagram M_0 of the primary structure on the diagram $\overline{X}_1$ of the redundant structure created by the development of hinge 1. We have

$$A_i = \frac{2}{3} la$$

and η under the centroid of the M_0 diagram $= c/2$, resulting in

$$\delta_{01} = -\frac{1}{EI}\left(\frac{2}{3} la\right)\left(\frac{c}{2}\right) = \frac{1}{3EI} lac$$

δ_{11} is obtained by superposing the redundant structure $\overline{X}_1$ on itself:

$$A_i = \frac{1}{2} la$$

$$\eta = \frac{2}{3} c$$

$$\delta_{11} = -\frac{1}{EI}\left(\frac{1}{2} la \times \frac{2}{3} c\right) = -\frac{1}{3EI} lac$$

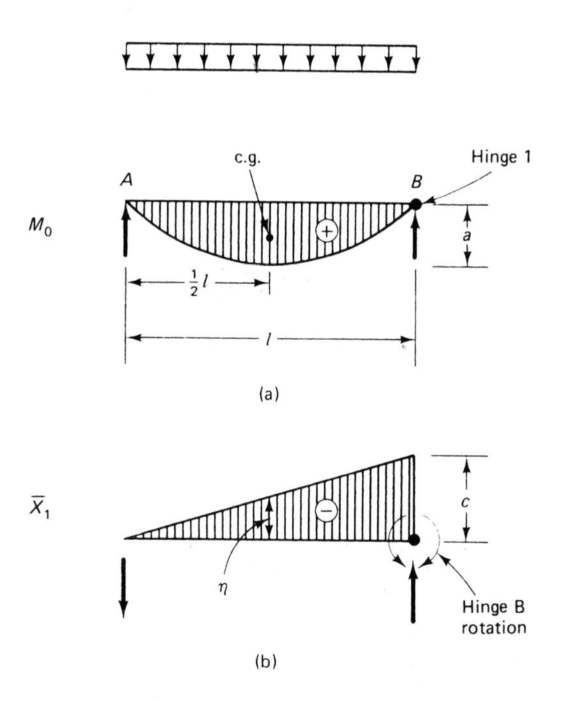

Figure 6.36 Influence coefficient determination from superposing M_0 and $\overline{X}_1$. (a) Primary structure moment. (b) Redundant structure moment.

Table 6.1 gives the values $\int M_i\, M_k\, ds$ for evaluating the influence coefficient values δ_{ik} for various combinations of primary and redundant moment diagrams. It can aid the designer in easily forming and solving sets of Equations 6.20 for any indeterminate structural system.

6.13.2 Determination of Plastic Hinge Rotations in Continuous Beams

Example 6.8

Determine the required plastic hinge rotation in the four-span beam of Figure 6.37. The beam is subjected to simple-span plastic moment M_0 so that the midspan moment is equal to the support moment $= \frac{1}{2} M_0$ before full rotation of the hinges and full moment redistribution take place.

Solution: The structure is statically indeterminate to the third degree, so that three hinges will develop at the plastic limit. Assume the maximum ordinate c of the redundant moment at hinge location to be unity. Then, from Table 6.1 and Figure 6.38,

$$EI\delta_{10} = -\frac{2}{3} M_0\, l$$

$$\delta_{11} = \frac{2}{3} l$$

$$\delta_{12} = \frac{1}{6} l$$

$$\delta_{13} = 0$$

From Equation 6.19,

$$-\delta_1 = \delta_{10} + \delta_{11}\overline{X}_1 + \delta_{12}\overline{X}_2 + \delta_{13}\overline{X}_3$$

Table 6.1 Product Integral Values $\int M_i M_k \, ds$ for Various Moment Combinations $EI\,\delta_{ik}$

M_k \ M_i	l, a	a	a	Parabolic a	a	a, b
l, c	lac	$\frac{1}{2}lac$	$\frac{1}{2}lac$	$\frac{2}{3}lac$	$\frac{1}{2}lac$	$\frac{1}{2}l(a+b)c$
c	$\frac{1}{2}lac$	$\frac{1}{3}lac$	$\frac{1}{6}lac$	$\frac{1}{3}lac$	$\frac{1}{4}lac$	$\frac{1}{6}l(2a+b)c$
c	$\frac{1}{2}lac$	$\frac{1}{6}lac$	$\frac{1}{3}lac$	$\frac{1}{3}lac$	$\frac{1}{4}lac$	$\frac{1}{6}l(a+2b)c$
Parabolic c	$\frac{2}{3}lac$	$\frac{1}{3}lac$	$\frac{1}{3}lac$	$\frac{8}{15}lac$	$\frac{5}{12}lac$	$\frac{1}{3}l(a+b)c$
c	$\frac{1}{2}lac$	$\frac{1}{4}lac$	$\frac{1}{4}lac$	$\frac{5}{12}lac$	$\frac{1}{3}lac$	$\frac{1}{4}l(a+b)c$
c, d	$\frac{1}{2}la(c+d)$	$\frac{1}{6}la(2c+d)$	$\frac{1}{6}la(c+2d)$	$\frac{1}{3}la(c+d)$	$\frac{1}{4}la(c+d)$	$\frac{1}{6}l[a(2c+d)+b(2d+c)]$

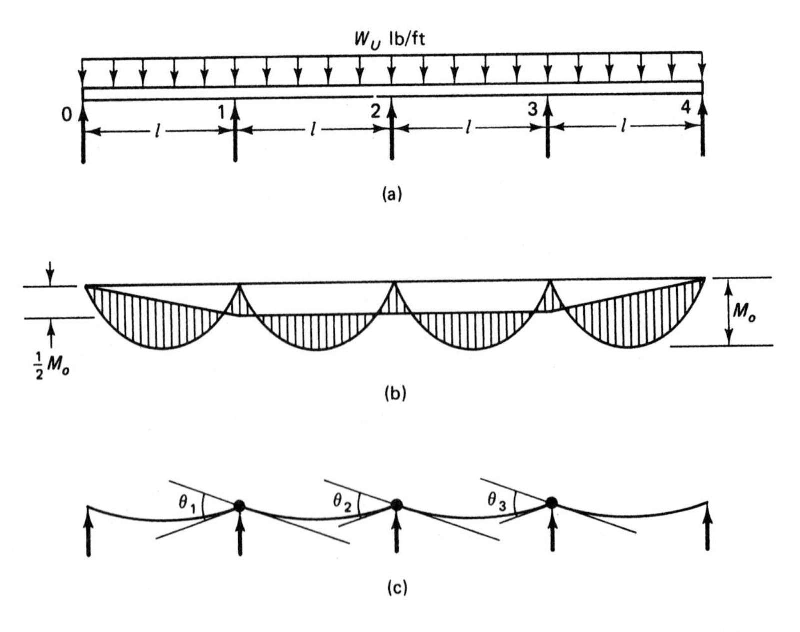

Figure 6.37 Primary moments and plastic hinge rotations in Example 6.8.

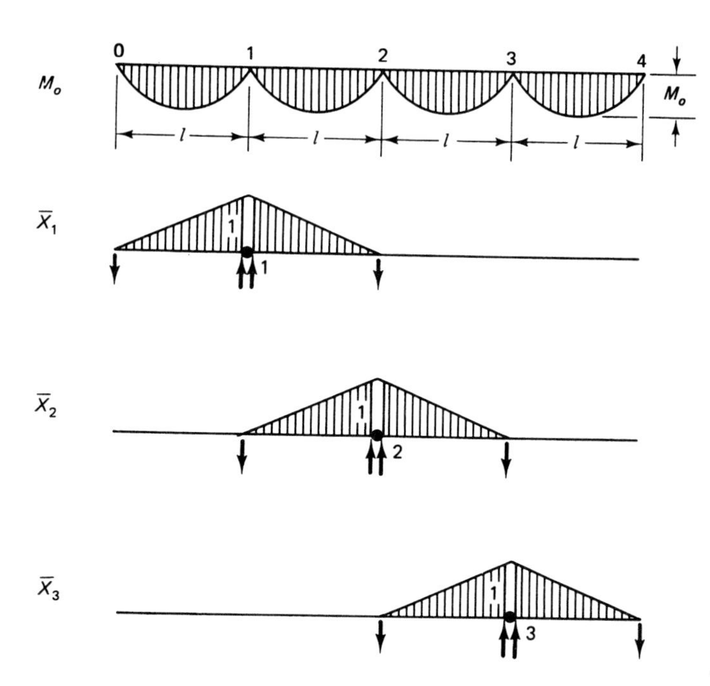

Figure 6.38 Primary and redundant moments in Example 6.8.

$$-EI\theta_1 = -\frac{2}{3}M_0 l + 0.5M_0\left(\frac{2l}{3}\right) + 0.5M_0\left(\frac{l}{6}\right) + 0 = -\frac{M_0 l}{4}$$

Also, again from Table 6.1 and Figure 6.38,

$$EI\delta_{20} = \frac{2}{3}M_0 l\left(-\frac{1}{2}\right) + \frac{2}{3}M_0 l\left(-\frac{1}{2}\right) = -\frac{2}{3}M_0 l$$

$$EI\delta_{21} = \left(-\frac{l}{2}\right)\left(-\frac{1}{3}\right) = +\frac{l}{6}$$

$$EI\delta_{22} = 2\left(-\frac{l}{2}\right)\left(-\frac{2}{3}\right) = +\frac{2l}{3}$$

$$EI\delta_{23} = \left(-\frac{l}{2}\right)\left(-\frac{1}{3}\right) = +\frac{l}{6}$$

From Equation 6.19,

$$-\theta_2 = \delta_{20} + \delta_{21}\overline{X}_1 + \delta_{22}\overline{X}_2 + \delta_{23}\overline{X}_3$$

$$-EI\theta_2 = -\frac{2}{3}M_0 l + 0.5M_0\left(+\frac{l}{6}\right) + 0.5M_0\left(+\frac{2}{3}l\right) + 0.5M_0\left(+\frac{l}{6}\right) = -\frac{M_0 l}{6}$$

From symmetry, $\theta_3 = \theta_1$. Therefore, the required plastic hinge rotations at the support are

$$\theta_1 = \frac{M_0 l}{4EI} = \theta_3$$

and

$$\theta_2 = \frac{M_0 l}{6EI}$$

Since $\theta_2 < \theta_1$, the first hinge to develop, and the controlling one in the design, is $\theta_1 = M_0 l/4EI$.

Note that the procedure used in Example 6.8 can be used in the limit design of any continuous beam or multistory frame. Also, it is important to maintain the correct sign convention by drawing all moments at the *tension* side of the member, as noted earlier.

The preceding discussion gives the basic *imposed rotations approach* embodied in Baker's theory. Other modified approaches have been proposed by Cohn (Ref. 6.17), Sawyer (Ref. 6.18), and Furlong (Ref. 6.19). Cohn's method is based on the requirements of limit equilibrium and serviceability, with a subsequent check of rotational compatibility. Sawyer's method is based on the simultaneous requirements of limit equilibrium and rotational compatibility, with a subsequent check of serviceability.

Furlong's method is based on assigning ultimate moments for various loading patterns on the continuous spans that would satisfy serviceability and limit equilibrium for the worst case. The sections are reinforced in such a manner that the ultimate moment strengths for each span are equal to or greater than the *product* of the maximum ultimate moment M_o in the span when the ends are free to rotate and a moment coefficient k_1 for various boundary conditions as listed in Table 6.2.

6.13.3 Rotational Capacity of Plastic Hinges

Rotation is the *total* change in slope along the short plasticity length concentrated at the hinge zone. It can also be described as the angle of discontinuity between the plastic parts of the member on either side of the plastic hinge. As Figure 6.39 shows, there are two types of hinges—tensile and compressive. In order that the first hinge that develops in the structure, usually the critical hinge, can rotate without rupture until the *n*th hinge develops, the concrete section at the first hinge has to be made ductile enough through section core confinement to be able to sustain the necessary rotation. This is equally applicable to both tension and compression hinges, where confinement of the concrete core is obtained through concentration of closed stirrups at the supports and column ends. A typical plot showing increase in rotation through increase in confining reinforcement is shown in Figure 6.40 (Ref. 6.14).

The plasticity length l_p determines the extent of severe cracking and the magnitude of rotation of the hinge. Therefore, it is important to limit the magnitude of l_p through the use of *closely spaced* ties or closed stirrups. In this manner, the strain capacity of the concrete at the confined section can be significantly increased, as experimentally demonstrated by several investigators, including Nawy (Refs. 6.12, 6.13, and 6.14). Several empirical expressions have been developed; see, for example, Baker (Ref. 6.5), Corley (Ref. 6.11), Nawy (Ref. 6.14), Sawyer (Ref. 6.18), and Mattock (Ref. 6.20). Two of them, for the plasticity length l_p and the concrete strain ϵ_c (Ref. 6.20), are

$$l_p = 0.5d + 0.5Z \tag{6.22}$$

Table 6.2 Beam Moment Coefficients for Assigned Moments

Boundary condition	Moment type	Beam loaded by one concentrated load at midspan	All other beams
Span with ends restrained	Negative	0.37	0.50
	Positive	0.42	0.33
Span with one end restrained	Negative	0.56	0.75
	Positive	0.50	0.46

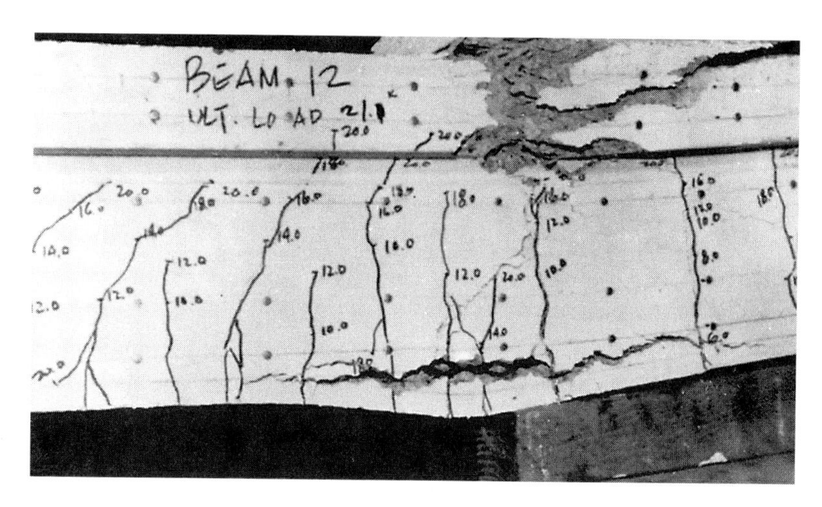

Photo 6.2 Pretensioned T-beam with rectangular confining reinforcement at failure (Nawy, Potyondy).

and

$$\epsilon_c = 0.003 + 0.02\frac{b}{Z} + 0.2\rho_s \tag{6.23}$$

where d = effective depth of the beam (in.)
Z = distance from the critical section to the point of contraflexure
ρ_s = ratio of volume of confining binder steel (including the compression steel) to the volume of the concrete core
I_p = *half* the plasticity length on each side of the centerline of plastic hinge.

Equation 6.22 can be more conservative for high values of ρ_s.

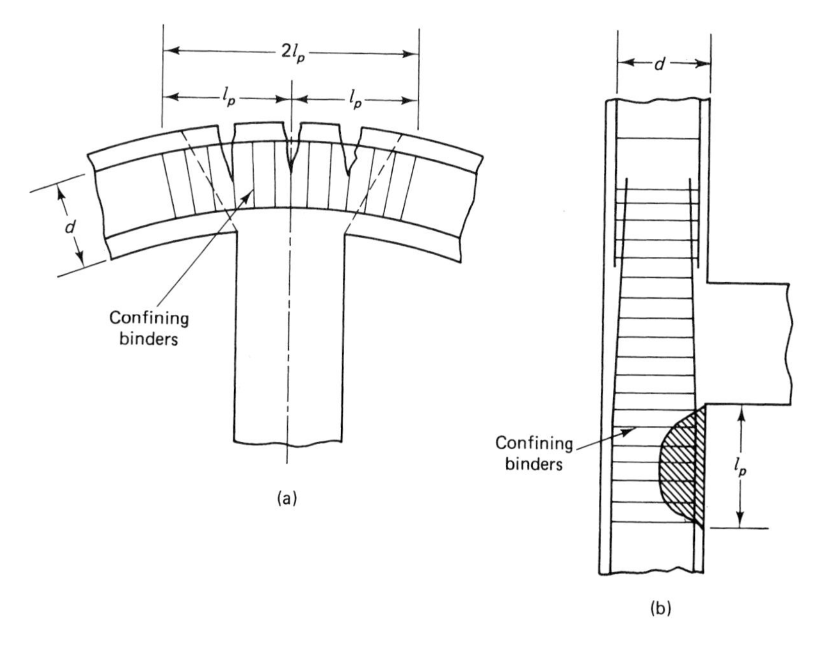

Figure 6.39 Plasticity zones l_p in plastic hinges. (a) Tensile hinge. (b) Compressive hinge.

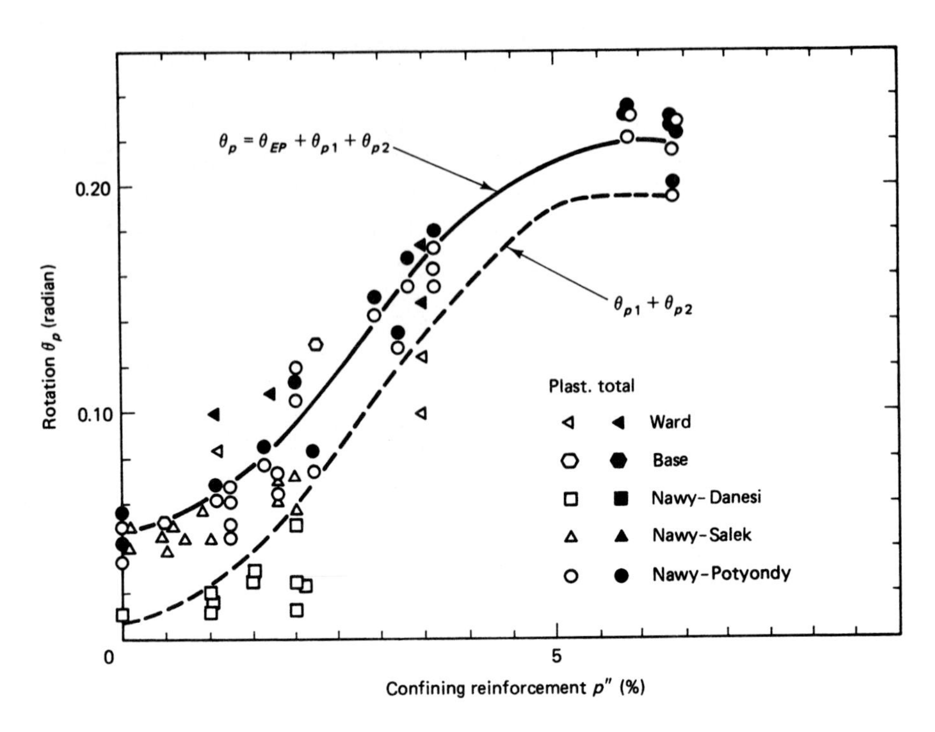

Figure 6.40 Comparison of plastic rotation with results of other authors.

Once the concrete strain ϵ_c is determined, the angle of rotation of the plastic hinge is readily determined from the expression

$$\theta_p = \left(\frac{\epsilon_c}{c} - \frac{\epsilon_{ce}}{kd}\right) l_p \tag{6.24}$$

where c = neutral axis depth at the limit state at failure
ϵ_{ce} = strain in the concrete at the extreme compression fibers when the yield curvature is reached
kd = neutral axis depth corresponding to ϵ_{ce}
ϵ_c = concrete compressive strain at the end of the inelastic range or at the limit state at failure.

The strain ϵ_{ce} can usually be taken at the load level when the strain in the tension reinforcement reaches the yield strain $\epsilon_y = f_y/E_s$. It can be taken to be approximately 0.001 in./in. or higher, depending on whether the tension steel yields before the concrete crushes at the extreme compression fibers in cases of overreinforced beams, as is sometimes the case in prestressed beams. If concrete crushes first, the value of ϵ_{ce} will have to be higher than 0.001 in./in. A limit of allowable $\epsilon_c = 1.0\%$ for confined concrete is recommended in determining the maximum allowable plastic rotation θ_p, although strains of confined concrete as high as 13% could be obtained, as shown in Ref. 6.14.

The discussion in this entire section (6.13) is equally applicable to reinforced and prestressed concrete indeterminate structures at the plastic loading range where full redistribution of moments has taken place. As the prestressed concrete section is cracked and decompression in the prestressing steel has taken place, the structural system gradually starts to behave similarly to a reinforced concrete system. As the load reaches the limit state at failure, the flexural behavior of the prestressed concrete elements is expected to closely resemble that of reinforced concrete elements.

6.13.4 Calculation of Available Rotational Capacity

Example 6.9

Determine the required and available rotational capacities of the critical plastic hinges in the continuous prestressed concrete beam in Example 6.8 for both confined and unconfined concrete. Given data are as follows:

$$M_u = \tfrac{1}{2} M_0 = 400bd^2$$

$$c = 0.28d$$

$$kd = 0.375d$$

$\epsilon_{ce} = 0.001$ in./in. at end of the elastic range

$\epsilon_c = 0.004$ in./in. at end of the inelastic range for unconfined sections

Max. allowable $\epsilon_c = 0.01$ in./in. for confined sections

$E_cI_c = 150{,}000bd^3$ in.^{3}-lb

$$\frac{Z}{d} = 5.5$$

$f'_c = 5{,}000$ psi

$f_y = 60{,}000$ psi for the mild steel

Also, calculate the maximum allowable span-to-depth ratio l/d for the beam if full redistribution of moments is to occur at the limit state at failure.

Solution:

$$M_0 = 2 \times 400bd^2 = 800bd^2$$

From Example 6.8,

$$\text{Required } \theta_1 = \theta_3 = \frac{M_0 l}{4E_cI_c} = \frac{800bd^2\, l}{4 \times 150{,}000bd^3} = \frac{1}{750}\frac{l}{d} \text{ radian}$$

$$\text{Required } \theta_2 = \frac{M_0 l}{6EI} = \frac{800bd^2}{6 \times 150{,}000bd^3} = \frac{1}{1{,}125}\frac{l}{d} \text{ radian}$$

From Equation 6.22,

$$l_p = 0.5d + 0.05Z = 0.5d + 0.05 \times 5.5d = 0.775d$$

The total plasticity length on both sides of the hinge centerline is $2 \times 0.775d = 1.55d$.

Unconfined Section. From Equation 6.24,

$$\text{Available } \theta_p = \left(\frac{\epsilon_c}{c} - \frac{\epsilon_{ce}}{kd}\right) l_p = \left(\frac{0.004}{0.28d} - \frac{0.001}{0.375d}\right) 1.55d = 0.018 \text{ radian}$$

For full moment redistribution,

$$\frac{1}{750}\frac{l}{d} \leq 0.018 \qquad \text{and} \qquad \frac{1}{1{,}125}\frac{1}{\mathrm{d}} \leq 0.018$$

or

$$\frac{l}{d} \leq 13.5 \qquad \text{and} \qquad \frac{l}{d} \leq 20.3$$

Confined Sections

Max. allow $\epsilon_c = 0.01$ in./in.

$$\text{Available } \theta_p = \left(\frac{0.01}{0.28d} - \frac{0.001}{0.375d}\right) 1.55d = 0.51 \text{ radian}$$

For full moment redistribution,

$$\frac{1}{750}\frac{l}{d} \leq 0.051 \qquad \text{and} \qquad \frac{1}{1{,}125}\frac{l}{d} \leq 0.051$$

or

$$\frac{l}{d} \leq 38.3 \qquad \text{and} \qquad \frac{l}{d} = 57.4$$

Comparing the results of the unconfined sections in the first case to the confined sections in the second case, one sees that confinement of the concrete at the plastic hinging zone permits more slender sections for full plasticity and, hence, a more economical indeterminate structural system.

6.13.5 Check for Plastic Rotation Serviceability

Example 6.10

If closed-stirrup binders are used in Example 6.9 with binder ratio $\rho_s = 0.025$ and $l/d = 35$ with c at failure $= 0.25d$, verify whether the continuous beam satisfies the rotation serviceability criteria given that $b = \frac{1}{2}d$.

Solution:

$$\frac{Z}{d} = 5.5$$

Hence,

$$\frac{b}{Z} = \frac{1}{11}$$

Also,

$$\text{Available } \epsilon_c = 0.003 + 0.02\frac{b}{Z} + 0.2\rho_s$$

$$= 0.003 + 0.02 \times \frac{1}{11} + 0.2 \times 0.025 = 0.0098, \text{ say } 0.01 \text{ in./in.}$$

The maximum allowable to be utilized is $\epsilon_c = 0.01$ in./in. So use, for $\epsilon_c = 0.01$, the corresponding available plastic rotation:

$$\theta_p = \left(\frac{0.01}{0.25d} - \frac{0.001}{0.375d}\right)1.55d = 0.058 \text{ radian}$$

$$\text{Rqd. } \theta_1 = \frac{1}{750}\frac{l}{d} = \frac{35}{750} = 0.046 \text{ radian}$$

$$\text{Rqd. } \theta_2 = \frac{1}{1{,}125}\frac{l}{d} = \frac{35}{1{,}125} = 0.031 \text{ radian}$$

Available $\theta_p = 0.058$ radian > required $\theta = 0.046$ radian. Thus, the beam satisfies the serviceability criteria for plastic rotation.

The foregoing discussion for the limit design of reinforced and prestressed concrete indeterminate beams and frames permits the design engineer to provide ductile connections at beam-column supports and generate full moment redistribution throughout the structure, resulting in full utilization of the strength of the prestressed system. Also, continuity in both pretensioned and post-tensioned systems to withstand seismic loading can be effectively utilized through the appropriate confinement of the connecting zones by means of the procedures presented in this section.

6.13.6 Transverse Confining Reinforcement for Seismic Design

Transverse reinforcement in the form of closely spaced hoops (ties) or spirals has to be adequately provided for concrete frame structural elements in seismic regions. The aim is to produce *adequate rotational capacity* within the plastic hinges that may develop as a result of the seismic forces. The Uniform Building Code, the International Building Code (IBC2000), and the ACI Code on seismic design require design and detailing of closed ties at the beam-column connection zones and in shear walls to be governed by the following (See Chapter 15 of Ref. 6.8 by the author and Chapter 13 to follow):

1. For column spirals, the minimum volumetric ratio of the spiral hoops needed for the concrete core confinement is

$$\rho_s \geq \frac{0.12f'_c}{f_{yh}} \tag{6.25}$$

or

$$\rho_s \geq 0.45\left(\frac{A_g}{A_{ch}} - 1\right)\frac{f'_c}{f_{yh}} \tag{6.26}$$

whichever is greater, where

ρ_s = ratio of volume of spiral reinforcement to the core volume measured out to out.
A_g = gross area of the column section.
A_{ch} = core area of section measured to the outside of the transverse reinforcement (sq. in.).
f_{yh} = specified yield of transverse reinforcement, psi.

2. For column rectangular hoops, the total cross-sectional area within spacing s is

$$A_{sh} \geq 0.09sh_c\frac{f'_c}{f_{yh}} \tag{6.27}$$

or

$$A_{sh} \geq 0.3sh_c\left(\frac{A_g}{A_{ch}} - 1\right)\frac{f'_c}{f_{yh}} \tag{6.28}$$

whichever is greater, where

A_{sh} = total cross-sectional area of transverse reinforcement (including cross ties) within spacing s and perpendicular to dimension h_c.
h_c = cross-sectional dimension of column core measured center to center of confining reinforcement, in.
A_{ch} = cross-sectional area of structural member, measured out-to-out of transverse reinforcement.
s = spacing of transverse reinforcement measured along the longitudinal axis of the member, in.
s_{max} = $\frac{1}{4}$ of the smallest cross-sectional dimension of the member or 4 in., whichever is smaller (IBC requires 4 in.).

3. The confining transverse reinforcement in *columns* should be placed on *both* sides of a potential hinge over a distance l_0. The largest of the following three conditions govern:
   (a) depth of member at joint face
   (b) $\frac{1}{6}$ of the clear span
   (c) 18 in.
4. For beam confinement, the confining transverse reinforcement at *beam* ends should be placed over a length equal to *twice* the member depth h from the face of the joint on either side or of any other location where plastic hinges can develop. The maximum hoop spacing should be the smallest of the following four conditions:
   (a) $\frac{1}{4}$ effective depth d.
   (b) $8 \times$ diameter of longitudinal bars.
   (c) $24 \times$ diameter of the hoop.
   (d) 12 in. (300 mm).

   Figure 6.41 from Reference 6.8 gives a typical detailing example of confining reinforcement at a joint to resist seismic forces.
5. Reduction in confinement at joints: A 50% reduction in confinement and an increase in the minimum tie spacing to 6 in. are allowed by the ACI Code if a joint is confined on all *four* faces by adjoining beams with each beam wide enough to cover three quarters of the adjoining face.

6.13.7 Selection of Confining Reinforcement

Example 6.11

Design the confining reinforcement in the column at the beam-column joint of Figure 6.41. Given:

column size = 15×24 in. (380×610 mm)

f'_c = 4,000 psi (27.6 MPa), normal weight

f_{yh} = 60,000 psi (414 MPa)

clear cover = $1\frac{1}{2}$ in. (38 mm)

Solution: From Equations 6.27 and 6.28, whichever is greater,

$$A_{sh} \geq 0.09 s h_c \frac{f'_c}{f_{yh}}$$

or

$$A_{sh} \geq 0.3 s h_c \left(\frac{A_g}{A_{ch}} - 1\right) \frac{f'_c}{F_{yh}}$$

h_c = column core dimension = $24 - 2(1.5 + 0.5) = 20$ in.

$$A_{sh} = 0.09 \times 3.5 \times 20 \left(\frac{4{,}000}{60{,}000}\right) = 0.42 \text{ in.}^2$$

$$A_{sh} = 0.3 \times 3.5 \times 20 \left(\frac{15 \times 24}{11 \times 20} - 1\right)\left(\frac{4{,}000}{60{,}000}\right) = 0.89 \text{ in}^2 \text{ controls}$$

Trying $s = 3\frac{1}{2}$ in., maximum allowance $s = \frac{1}{4}$ smallest column dimension or 4 in., $b/4 = 0.25 \times 15 = 3.75$ in.

Use No. 4 hoops plus two No. 4 crossties at $3\frac{1}{2}$ in. center to center. Place the confining hoops in the column on both sides of potential hinge over a distance l_0 being the largest of

(a) depth of member = 24 in. (610 mm)
(b) $\frac{1}{6}$ × clear span = $(24 \times 12)/6 = 48$ in. (1220 mm)
(c) 18 in. (450 mm)

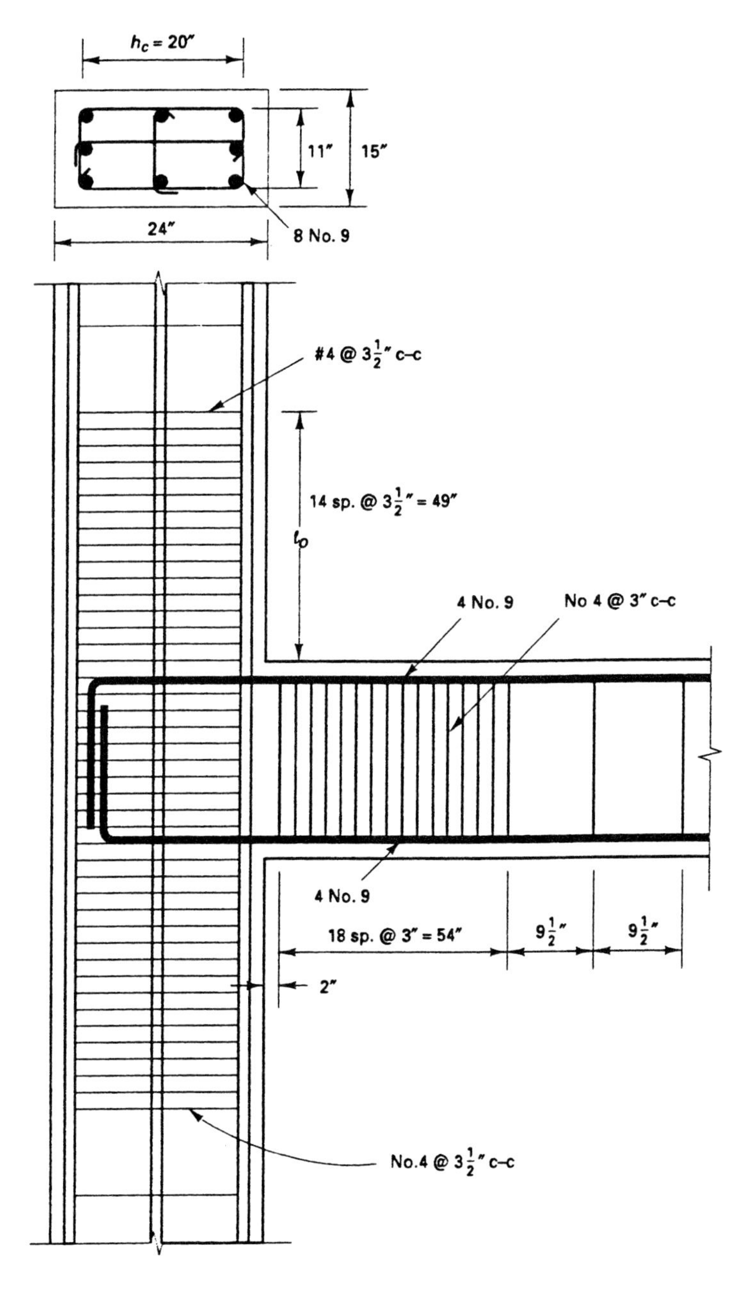

Figure 6.41 Confining reinforcement for seismic resistance (Example 6.11).

Use $l_0 = 48$ in. (1220 mm), spacing the No. 4 hoops and crossties at 3.5 in. center to center over this distance (12.7-mm dia. bars at 89 mm center to center) as shown in Figure 6.41.

REFERENCES

6.1 ACI Committee 318. *Building Code Requirements for Structural Concrete (ACI 318-99) and Commentary (ACI 318 R-99).* Farmington Hills, MI: American Concrete Institute, 2000, pp. 392.

6.2 Gerwick, B. C., *Construction of Prestressed Concrete Structures,* 2nd ed. New York: John Wiley & Sons, 1993.

6.3 Taylor, F. W., Thompson, S. E., and Smulski, E. *Concrete Plain and Reinforced,* vol. 2. John Wiley & Sons, New York, 1947.

6.4 Abeles, P. W., and Bardhan-Roy, B. K. *Prestressed Concrete Designer's Handbook,* 3d ed. Viewpoint Publications, London, 1981.

6.5 Baker, A. L. L. *The Ultimate Load Theory Applied to the Design of Reinforced and Prestressed Concrete Frames.* Concrete Publications Ltd., London, 1956.

6.6 Baker, A. L. L. *Limit State Design of Reinforced Concrete.* London: Cement and Concrete Association, 1970.

6.7 Ramakrishnan, V., and Arthur, P. D. *Ultimate Strength Design of Structural Concrete.* London: Wheeler, 1977.

6.8 Nawy, E. G. *Reinforced Concrete—A Fundamental Approach,* 4th ed. Prentice Hall, Upper Saddle River, N.J., 2000, pp. 786.

6.9 Lin, T. Y., and Thornton, K. "Secondary Moments and Moment Redistribution in Continuous Prestressed Concrete Beams." *Journal of the Prestressed Concrete Institute,* January–February 1972, pp. 8–20.

6.10 Nilson, A. H. *Design of Prestressed Concrete.* New York: John Wiley & Sons, 1987.

6.11 Corley, W. G. "Rotational Capacity of Reinforced Concrete Beams." *Journal of the Structural Division, ASCE* 92 (1966): 121–146.

6.12 Nawy, E. G., and Salek, F. "Moment-Rotation Relationships of Non-Bonded Prestressed Flanged Sections Confined with Rectangular Spirals." *Journal of the Prestressed Concrete Institute,* August 1968, pp. 40–55.

6.13 Nawy, E. G., Danesi, R., and Grosco, J. "Rectangular Spiral Binders Effect on the Rotation Capacity of Plastic Hinges in Reinforced Concrete Beams." *Journal of the American Concrete Institute,* Farmington Hills, MI, December 1968, pp. 1001–1010.

6.14 Nawy, E. G., and Potyondy, J. G. "Moment Rotation, Cracking and Deflection of Spirally Bound Pretensioned Prestressed Concrete Beams." *Engineering Research Bulletin No. 51,* N.J.: Bureau of Engineering Research, Rutgers University, New Brunswick, 1970, pp. 1–97.

6.15 Prestressed Concrete Institute. *PCI Design Handbook.* Chicago: Prestressed Concrete Institute, 5th ed., 1999.

6.16 Park, R., and Paulay, J. *Reinforced Concrete Structures.* John Wiley & Sons, New York, 1975.

6.17 Cohn, M. Z., "Rotational Compatibility in the Limit Design of Reinforced Concrete Beams." *Proceedings of the International Symposium on the Flexural Mechanics of Reinforced Concrete,* ASCE-ACI. Miami, Nov. 1964, pp. 359–382.

6.18 Sawyer, H. A. "Design of Concrete Frames for Two Failure Stages." *Proceedings of the International Symposium on the Flexural Mechanics of Reinforced Concrete,* ASCE-ACI. Miami, Nov. 1964, pp. 405–431.

6.19 Furlong, R. W. "Design of Concrete Frames by Assigned Limit Moments." *Journal of the American Concrete Institute* 67, Farmington Hills, MI, 1970, 341–353.

6.20 Mattock, A. H. "Discussion of Rotational Capacity of Reinforced Concrete Beams by W. G. Corley." *Journal of the Structural Division, ASCE* 93, 1967, 519–522.

6.21 International Conference of Building Officials, *Uniform Building Code (UBC),* Vol. 2, ICBO, Whittier, California, 1997.

6.22 International Code Council, International Building Code 2000 (IBC), Joint UBC, BOCA, SBCCI, Whittier, California, 2000.

PROBLEMS

6.1 A two-span continuous beam has a parabolic tendon profile shown in Figure P6.1. The prestressing force P_e after all losses is 450,000 lb (2,002 kN). The beam has a rectangular section 15-in. (38.1 cm) wide.

(a) Find the final profile of the thrust C-line and the beam reactions at all supports.

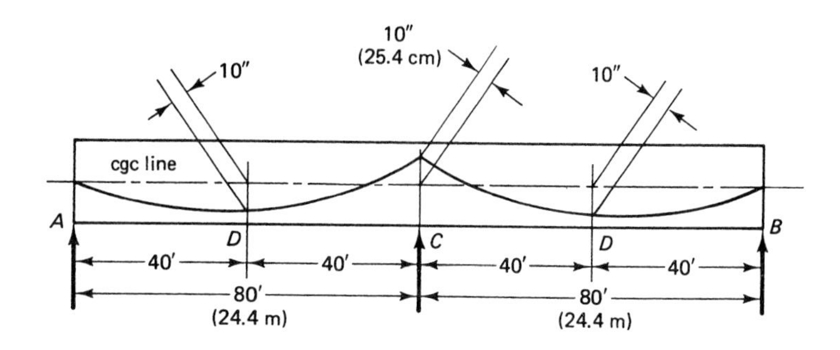

Figure P6.1.

(b) Design the beam depth such that the concrete fiber stresses due only to prestressing do not exceed the maximum allowable for normal-weight concrete having cylinder strength $f'_c = 6{,}000$ psi (41.4 MPa).

(c) Determine the shape of the concordant tendon, and draw a beam elevation of the tendon profile.

6.2 Solve Problem 6.1 for a tendon profile harped at midspan points D, but having the same eccentricities. Compare the results with those of Problem 6.1.

6.3 Solve Problem 6.1 for a tendon profile which has eccentricities $e_A = e_B = 3$ in. (7.6 cm) at the exterior supports above the cgc line.

6.4 Develop the tendon profile for the continuous beam in Example 6.4 if the beam is continuous over two equal spans of 64 ft (19.4 m).

6.5 Solve Problem 6.4 if the beam is continuous over four equal spans of 64 ft (19.4 m).

6.6 Design, for service loading, the frame in Example 6.7 using the same loading conditions if the span of the horizontal beam is 90 ft (27.4 m) and the height of the portal is 25 ft (7.6 m).

6.7 Design the portal frame of an aircraft hangar having the dimensions and the loading shown in Figure P6.7. Detail the connections and the configuration of the prestressing tendons of the horizontal member. Use the same allowable stresses as in Example 6.7.

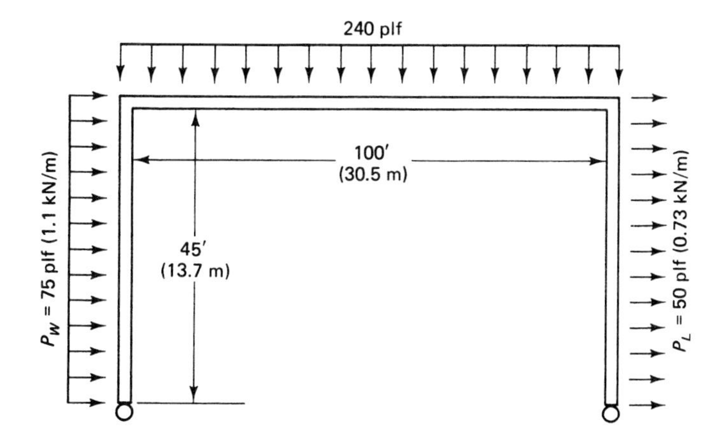

Figure P6.7.

7 CAMBER, DEFLECTION, AND CRACK CONTROL

7.1 INTRODUCTION

Serviceability of prestressed concrete members in their deflection and cracking behavior is at least as important a criterion in design as serviceability of reinforced concrete elements. The fact that prestressed concrete elements are more slender than their counterparts in reinforced concrete, and their behavior more affected by flexural cracking, makes it more critical to control their deflection and cracking. The primary design involves proportioning the structural member for the limit state of flexural stresses at service load and for limit states of failure in flexure, shear, and torsion, including anchorage development strength. Such a design can only become complete if the magnitudes of long-term deflection, camber (reverse deflection), and crack width are determined to be within allowable serviceability values.

Prestressed concrete members are continuously subjected to sustained eccentric compression due to the prestressing force, which seriously affects their long-term creep deformation performance. Failure to predict and control such deformations can lead to high reverse deflection, i.e., camber, which can produce convex surfaces detrimental to proper drainage of roofs of buildings, to uncomfortable ride characteristics in bridges and aqueducts, and to cracking of partitions in apartment buildings, including misalignment of windows and doors.

Transamerica Pyramid, San Francisco, California.

The difficulty of predicting very accurately the total long-term prestress losses makes it more difficult to give a precise estimate of the magnitude of expected camber. Accuracy is even more difficult in partially prestressed concrete systems, where limited cracking is allowed through the use of additional nonprestressed reinforcement. Creep strain in the concrete increases camber, as it causes a negative increase in curvature which is usually more dominant than the decrease produced by the decrease in prestress losses due to creep, shrinkage, and stress relaxation. A best estimate of camber increase should be based on accumulated experience, span-to-death ratio code limitations, and a correct choice of the modulus E_c of the concrete. Calculation of the moment-curvature relationships at the major incremental stages of loading up to the limit state at failure would also assist in giving a more accurate evaluation of the stress-related load deflection of the structural element.

The cracking aspect of serviceability behavior in prestressed concrete is also critical. Allowance for limited cracking in "partial prestressing" through the additional use of nonprestressed steel is prevalent. Because of the high stress levels in the prestressing steel, corrosion due to cracking can become detrimental to the service life of the structure. Therefore, limitations on the magnitudes of crack widths and their spacing have to be placed, and proper crack width evaluation procedures used. The presented discussion of the state of the art emphasizes the extensive work of the author on cracking in pretensioned and post-tensioned prestressed beams.

7.2 BASIC ASSUMPTIONS IN DEFLECTION CALCULATIONS

Deflection calculations can be made either from the moment diagrams of the prestressing force and the external transverse loading, or from the moment-curvature relationships. In either case, the following basic assumptions have to be made:

1. The concrete gross cross-sectional area is accurate enough to compute the moment of inertia except when refined computations are necessary.
2. The modulus of concrete $E_c = 33w^{1.5}\sqrt{f'_c}$, where the value of f'_c corresponds to the cylinder compressive strength of concrete at the age at which E_c is to be evaluated.
3. The principle of superposition applies in calculating deflections due to transverse load and camber due to prestressing.
4. All computations of deflection can be based on the center of gravity of the prestressing strands (cgs), where the strands are treated as a single tendon.
5. Deflections due to shear deformations are disregarded.
6. Sections can be treated as *totally elastic* up to the decompression load. Thereafter, the cracked moment of inertia I_{cr} can give a more accurate determination of deflection and camber.

7.3 SHORT-TERM (INSTANTANEOUS) DEFLECTION OF UNCRACKED AND CRACKED MEMBERS

7.3.1 Load-Deflection Relationship

Short-term deflections in prestressed concrete members are calculated on the assumption that the sections are homogeneous, isotropic, and elastic. Such an assumption is an approximation of actual behavior, particularly that the modulus E_c of concrete varies with the age of the concrete and the moment of inertia varies with the stage of loading, i.e., whether the section is uncracked or cracked.

Photo 7.1 Supporting base of the Transamerica Pyramid, San Francisco, California.

Ideally, the load-deflection relationship is trilinear, as shown in Figure 7.1. The three regions prior to rupture are:

Region I. Precracking stage, where a structural member is crack free.

Region II. Postcracking stage, where the structural member develops acceptable controlled cracking in both distribution and width.

Region III. Postserviceability cracking stage, where the stress in the tensile reinforcement reaches the limit state of yielding.

7.3.1.1 Precracking stage: region I. The precracking segment of the load-deflection curve is essentially a straight line defining full elastic behavior, as in Figure 7.1. The maximum tensile stress in the beam in this region is less than its tensile strength in flexure, i.e., it is less than the modulus of rupture f_r of concrete. The flexural stiffness EI of the beam can be estimated using Young's modulus E_c of concrete and the moment of inertia

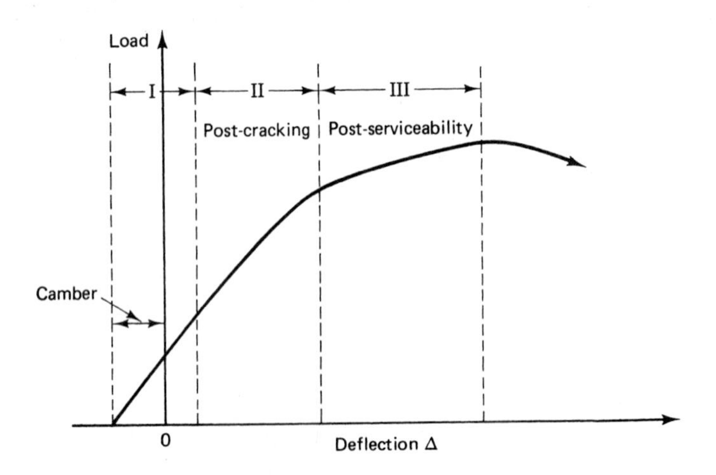

Figure 7.1 Beam load-deflection relationship. Region I, precracking stage; region II, postcracking stage; region III, postserviceability stage.

of the uncracked concrete cross section. The load-deflection behavior significantly depends on the stress-strain relationship of the concrete. A typical stress-strain diagram of concrete is shown in Figure 7.2.

The value of E_c can be estimated using the ACI empirical expression given in Chapter 2, viz.,

$$E_c = 33w^{1.5}\sqrt{f'_c} \tag{7.1a}$$

or

$$E_c = 57{,}000\sqrt{f'_c} \quad \text{for normal-weight concrete}$$

The precracking region stops at the initiation of the first flexural crack, when the concrete stress reaches its modulus of rupture strength f_r. Similarly to the direct tensile splitting strength, the modulus of rupture of concrete is proportional to the square root of its compressive strength. For design purposes, the value of the modulus of rupture for concrete may be taken as

$$f_r = 7.5\lambda\sqrt{f'_c} \tag{7.1b}$$

where $\lambda = 1.0$ for normal-weight concrete. If all-lightweight concrete is used, then $\lambda = 0.75$, and if sand-lightweight concrete is used, $\lambda = 0.85$.

If one equates the modulus of rupture f_r to the stress produced by the cracking moment M_{cr} (decompression moment), then

$$f_b = f_r = -\frac{P_e}{A_c}\left(1 + \frac{ec_b}{r^2}\right) + \frac{M_{cr}}{S_b} \tag{7.2a}$$

where subscript b stands for the bottom fibers at midspan of a simply supported beam. If the distance of the extreme *tension* fibers of concrete from the center of gravity of the concrete section is y_t, then the cracking moment is given by

$$M_{cr} = \frac{I_g}{y_t}\left[\frac{P_e}{A_c}\left(1 + \frac{ec_b}{r^2}\right) + 7.5\lambda\sqrt{f'_c}\right] \tag{7.2b}$$

or

$$M_{cr} = S_b\left[7.5\lambda\sqrt{f'_c} + \frac{P_e}{A_c}\left(1 + \frac{ec_b}{r^2}\right)\right] \tag{7.2c}$$

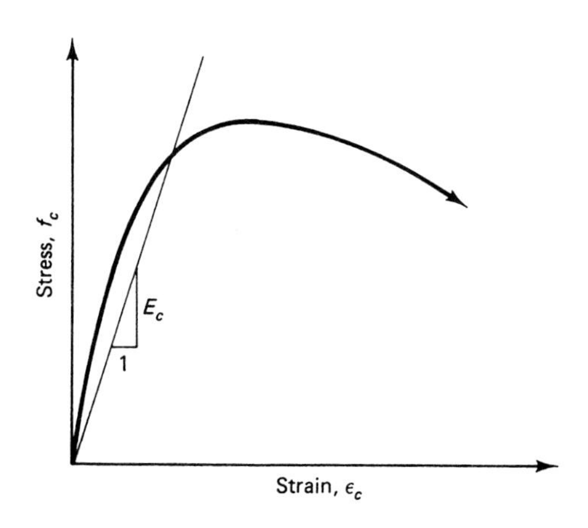

Figure 7.2 Stress-strain diagram of concrete.

where S_b = section modulus at the bottom fibers. More conservatively, from Equation 5.12, the cracking moment due to that portion of the applied *live load* that causes cracking is

$$M_{cr} = S_b[6.0\lambda\sqrt{f'_c} + f_{ce} - f_d] \tag{7.3a}$$

where f_{ce} = compressive stress at the center of gravity of concrete section due to effective prestress *only* after losses when tensile stress is caused by applied external load

f_d = concrete stress at extreme tensile fibers due to unfactored *dead* load when tensile stresses and cracking are caused by the external load.

A factor 7.5 can also be used instead of 6.0 for deflection purposes. Equation 7.3 can be transformed to the PCI format (Ref. 7.7) giving identical results:

$$\frac{M_{cr}}{M_a} = 1 - \left(\frac{f_{tl} - f_r}{f_L}\right) \tag{7.3b}$$

where M_a = maximum service unfactored live load moment

f_{tl} = final calculated total service load concrete stress in the member

f_r = modulus of rupture

f_L = service live load of concrete stress in the member.

7.3.1.2 Calculation of cracking moment M_{cr}

Example 7.1

Compute the cracking moment M_{cr} for a prestressed rectangular beam section having a width b = 12 in. (305 mm) and a total depth h = 24 in. (610 mm), given that f'_c = 4,000 psi (27.6 MPa). The concrete stress f_b due to eccentric prestressing is 1,850 psi (12.8 MPa) in compression. Use a modulus of rupture value of 7.5 $\sqrt{f'_c}$.

Solution: The modulus of rupture $f_r = 7.5\sqrt{f'_c} = 7.5\sqrt{4{,}000}$ = 474 psi (3.27 MPa). Also, $I_g = bh^3/12 = 12(24)^3/12$ = 13,824 in^4 (575,400 cm^4); y_t = 24/2 = 12 in. (305 mm) to the tension fibers; and $S_b = I_g/y_t$ = 13,824/12 = 1,152 in^3 (18,878 cm^3).

$$M_{cr} = S_b\left[7.5\lambda\sqrt{f'_c} + \frac{P_e}{A_c}\left(1 + \frac{ec_b}{r^2}\right)\right] = 1{,}152[474 + 1850]$$

$$= 2.68 \times 10^6 \text{ in.-lb } (302.9 \text{ kN-m})$$

If the beam were not prestressed, the moment would be $M_{cr} = f_r I_g/y_t = 474 \times 13{,}824/12 = 0.546 \times 10^6$ in.-lb (61.7 kN-m).

7.3.1.3 Postcracking service-load stage: region II. The precracking region ends at the initiation of the first crack and moves into region II of the load-deflection diagram of Figure 7.1. Most beams lie in this region at service loads. A beam undergoes varying degrees of cracking along the span corresponding to the stress and deflection levels at each section. Hence, cracks are wider and deeper at midspan, whereas only narrow, minor cracks develop near the supports in a simple beam.

When flexural cracking develops, the contribution of the concrete in the tension area diminishes substantially. Hence, the flexural rigidity of the section is reduced, making the load-deflection curve less steep in this region than in the precracking stage segment. As the magnitude of cracking increases, stiffness continues to decrease, reaching a lower bound value corresponding to the reduced moment of inertia of the cracked section. The moment of inertia I_{cr} of the cracked section can be calculated from the basic principles of mechanics.

7.3.1.4 Postserviceability cracking stage and limit state of deflection behavior at failure: region III. The load-deflection diagram of Figure 7.1 is considerably flatter in region III than in the preceding regions. This is due to substantial loss in stiffness of the section because of extensive cracking and considerable widening of the stabilized cracks throughout the span. As the load continues to increase, the strain ϵ_s in the steel at the tension side continues to increase beyond the yield strain ϵ_y with no additional stress. The beam is considered at this stage to have structurally failed by initial yielding of the tension steel. It continues to deflect without additional loading, the cracks continue to open, and the neutral axis continues to rise toward the outer compression fibers. Finally, a secondary compression failure develops, leading to total crushing of the concrete in the maximum moment region followed by rupture.

7.3.2 Uncracked Sections

7.3.2.1 Deflection calculations. Deflection calculations for uncracked prestressed sections tend to be more accurate than those for cracked sections since the assumptions of elastic behavior are more applicable. The use of the moment of inertia of the gross section rather than the transformed section does not appreciably affect the accuracy sought in the calculations.

Suppose a beam is prestressed with a constant eccentricity tendon as shown in Figure 7.3. Use the sign convention of plotting the primary moment diagram on the tension side of the beam, and employ the elastic weight method by converting the moment diagram ordinates to elastic weights $M_1/(E_cI_c)$ on a beam span l. Then the moment of the weight intensity $(Pe)/E_cI_c$ of the half-span AC in Figure 7.3(c) about the midspan point C gives

$$\delta_c = \frac{Pel}{2E_cI_c}\left(\frac{l}{2}\right) - \frac{Pe}{E_cI_c}\left(\frac{l}{2} \times \frac{l}{4}\right) = \frac{Pel^2}{8E_cI_c} \tag{7.4}$$

Notice that the deflection diagram in Figure 7.3(d) is drawn *above* the base line, as the beam cambers upwards due to prestressing.

Similar computations can be performed for any tendon profile and any type of transverse loading regardless of whether the tendon geometry or loading is symmetrical or not. The final camber or deflection is the superposition of the deflections due to prestressing on the deflections due to external loads.

7.3.2.2 Strain and curvature evaluation. The distribution of strain across the depth of the section at the controlling stages of loading is linear, as is shown in Figure 7.4, with the angle of curvature dependent on the top and bottom concrete extreme fiber strains ϵ_{ct} and ϵ_{cb}. From the strain distributions, the curvature at the various stages of loading can be expressed as follows:

(1) Initial prestress:

$$\phi_i = \frac{\epsilon_{cbi} - \epsilon_{cti}}{h} \tag{7.5a}$$

(2) Effective prestress after losses:

$$\phi_e = \frac{\epsilon_{cbe} - \epsilon_{cte}}{h} \tag{7.5b}$$

(3) Service load:

$$\phi = \frac{\epsilon_{ct} - \epsilon_{cb}}{h} \tag{7.5c}$$

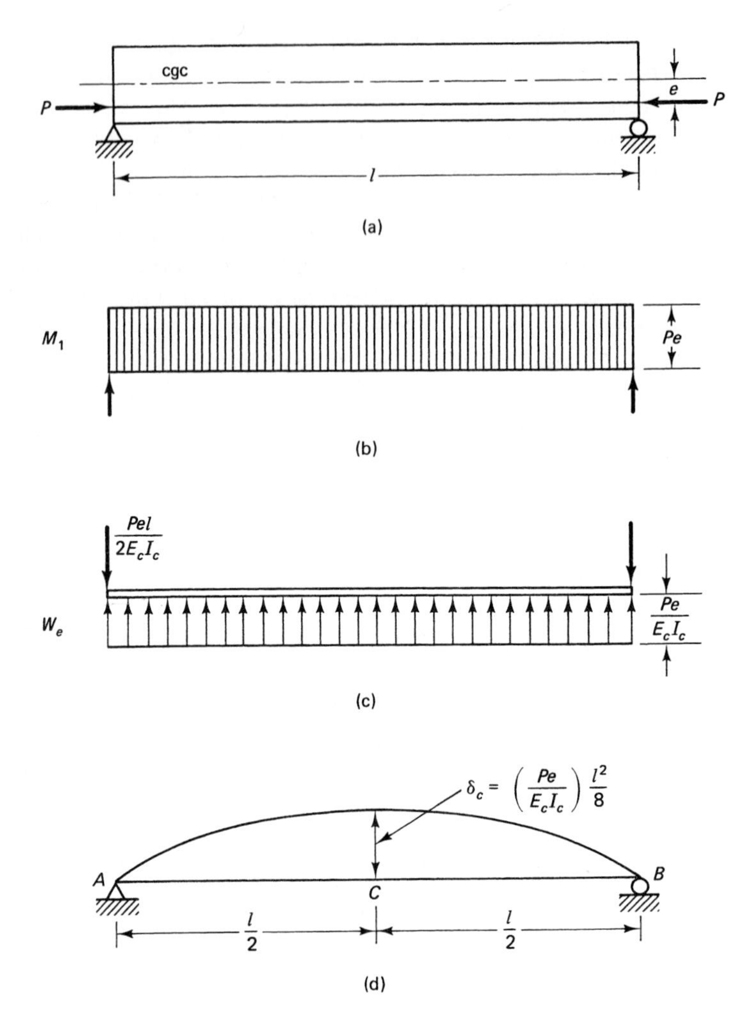

Figure 7.3 Calculation of deflection by elastic weight or moment-area method. (a) Prestressing force. (b) Primary moment M_1. (c) Elastic weight $W_e = M/E_cI_c$. (d) Deflection.

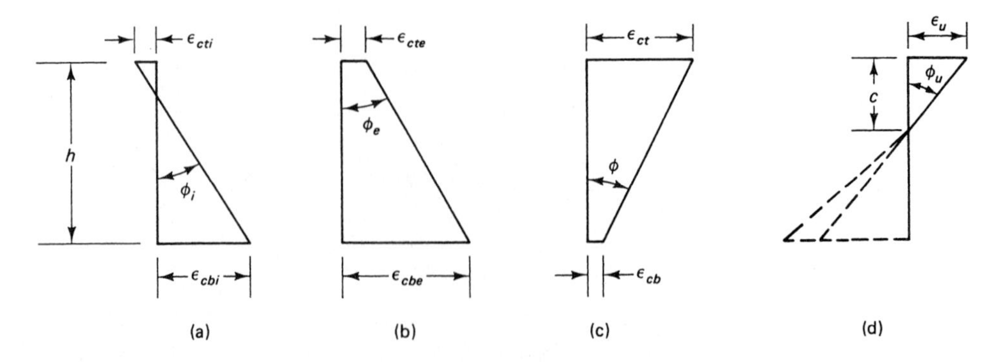

Figure 7.4 Strain distribution and curvature at controlling stages. (a) Initial prestress, $\phi_i = (\epsilon_{cbi} - \epsilon_{cti})/h$. (b) Effective prestress after losses, $\phi_e = (\epsilon_{cbe} - \epsilon_{cte})/h$. (c) Service load, $\phi = (\epsilon_{ct} - \epsilon_{cb}/h)$. (d) Failure, $\phi_u = \epsilon_u/C$.

(4) Failure:

$$\phi_u = \frac{\epsilon_u}{c} \tag{7.5d}$$

Use a *plus* sign for tensile strain and a *minus* sign for compressive strain. Figure 7.4c denotes the stress distribution for uncracked section. It has to be modified to show tensile stress at the bottom fibers if the section is cracked.

The effective curvature ϕ_e in Figure 7.4(b) after losses is the sum, using the appropriate sign, of the initial curvature ϕ_i, the change in curvature $d\phi_1$ due to loss of prestress from creep, relaxation, and shrinkage, and the change in curvature $d\phi_2$ due to creep of concrete under sustained prestressing force, i.e.,

$$\phi_e = \phi_i + d\phi_1 + d\phi_2 \tag{7.6}$$

where, from the basic mechanics of materials,

$$\phi = \frac{M}{E_c I_c} \tag{7.7a}$$

For the primary moment, $M_1 = P_e e$, so that

$$\phi = \frac{P_e e}{E_c I_c} \tag{7.7b}$$

Substituting into Equation 7.4 for simply supported beams with constant-eccentricity tendons yields

$$\delta_c = \frac{\phi l^2}{8} \tag{7.8}$$

The general expression for deflection in terms of curvature as proposed by Tadros in Ref. 7.3 gives

$$\delta = \phi_c \frac{l^2}{8} - (\phi_e - \phi_c)\frac{a^2}{6} \tag{7.9}$$

Photo 7.2 Priest Point Park Bridge in Olympia, Washington, a cast-in-place prestressed concrete structure. (*Courtesy,* Arvid Grant and Associates, Inc.)

where ϕ_c = curvature at midspan
ϕ_e = curvature at the support
a = length parameter as a function of the tendon profile.

7.3.2.3 Immediate deflection of simply supported beam prestressed with parabolic tendon

Example 7.2

Find the immediate midspan deflection of the beam shown in Figure 7.5 prestressed by a parabolic tendon with maximum eccentricity e at midspan and effective prestressing force P_e. Use both the elastic weight method and the equivalent weight method. The span of the beam is l ft, and its stiffness is E_cI_c.

Solution:

Elastic Weight Method. From Figure 7.5(b),

$$R'_e = \frac{1}{2}\left(\frac{P_e el}{E_cI_c} \times \frac{2}{3}\right) = \frac{P_e el}{3E_cI_c}$$

The moment due to the elastic weight W_e about the midspan point C is

$$M_C = \delta_c = R'_e\left(\frac{1}{2}\right) - \left[\frac{P_e el}{E_cI_c} \times \frac{2}{6}\left(\frac{3}{8} \times \frac{l}{2}\right)\right]$$

$$= \frac{1}{E_cI_c}\left(\frac{P_e el^2}{6} - \frac{3P_e el^2}{48}\right) = \frac{5P_e el^2}{48E_cI_c}$$

Then

$$\delta_c = \frac{5}{48}\frac{P_e el^2}{E_cI_c} \tag{a}$$

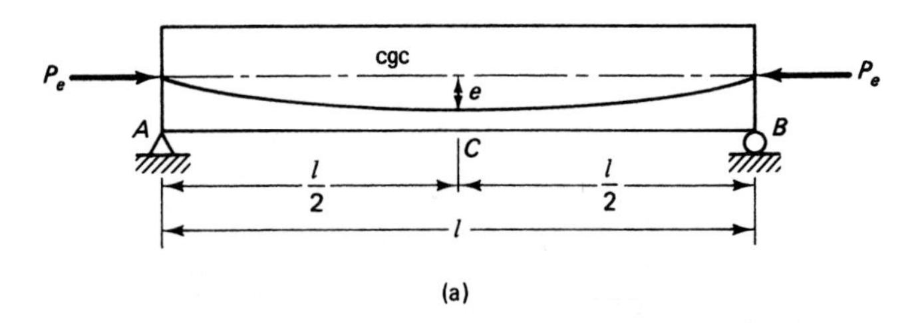

(a)

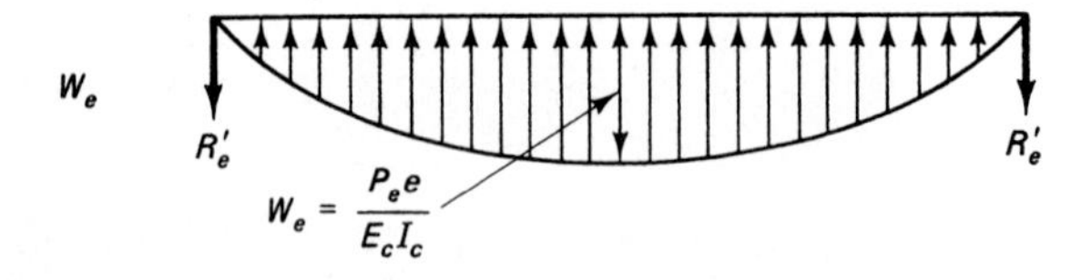

(b)

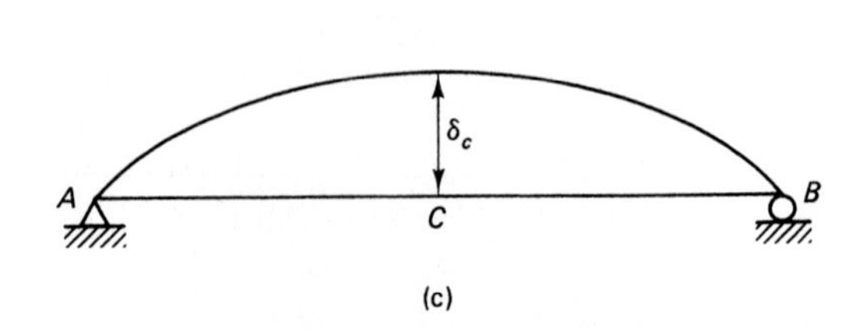

(c)

Figure 7.5 Deflection of beam in Example 7.2. (a) Tendon profile. (b) Elastic weight M/E_cI_c. (c) Deflection.

Equivalent Weight Method. From Chapter 1, the equivalent balancing load intensity W resulting from the pressure of the parabolic tendon on the concrete is

$$W = \frac{8P_e e}{l^2}$$

Also, from the basic mechanics of materials, the midspan deflection of a uniformly loaded simply supported beam is

$$\delta_c = \frac{5}{384}\frac{wl^4}{E_c I_c} \tag{b}$$

Substituting for the load intensity W from the previous equation into this one yields

$$\delta_c = \frac{5}{48}\frac{P_e e l^2}{E_c I_c} \tag{c}$$

As expected, Equation (c) is identical to Equation (a) for the midspan deflection of the beam.

Figure 7.6 shows typical midspan deflection expressions for simply supported beams, complementing the shear and moment expressions for continuous beams given earlier in Figure 6.12.

7.3.3 Cracked Sections

7.3.3.1 Effective-moment-of-inertia computation method. As the prestressed element is overloaded, or in the case of partial prestressing where limited controlled cracking is allowed, the use of the gross moment of inertia I_g underestimates the camber or deflection of the prestressed beam. Theoretically, the cracked moment of inertia I_{cr} should be used for the section across which the cracks develop while the gross moment of inertia I_g should be used for the beam sections between the cracks. However, such refinement in the numerical summation of the deflection increases along the beam span is sometimes unwarranted because of the accuracy difficulty of deflection evaluation. Consequently, an effective moment of inertia I_e can be used as an average value along the span of a simply supported bonded tendon beam, a method developed by Branson in Refs. 7.4 and 7.5. According to this method,

$$I_e = I_{cr} + \left(\frac{M_{cr}}{M_a}\right)^3 (I_g - I_{cr}) \le I_g \tag{7.10a}$$

Equation 7.10a can also be written in the form

$$I_e = \left(\frac{M_{cr}}{M_a}\right)^3 I_g + \left[1 - \left(\frac{M_{cr}}{M_a}\right)^3\right] I_{cr} \le I_g \tag{7.10b}$$

The ratio (M_{cr}/M_a) from Equation 7.3b can be substituted into Equations 7.10a and b to get the effective moment of inertia

$$\frac{M_{cr}}{M_a} = 1 - \left(\frac{f_{tl} - f_r}{f_L}\right) \tag{7.11}$$

where I_{cr} = moment of inertia of the cracked section, from Equation 7.13 to follow
I_g = gross moment of inertia

Note that both M_{cr} and M_a are the unfactored moments due to *live load only* such that M_{cr} is taken as that portion of the live load moment which causes cracking. The effective moment of inertia I_e in Equations 7.10a and b thus depends on the maximum moment M_a along the span in relation to the cracking moment capacity M_{cr} of the section.

In the case of uncracked continuous beams with both ends continuous,

$$\text{Avg. } I_e = 0.70 I_m + 0.15(I_{e1} + I_{e2}) \tag{7.12a}$$

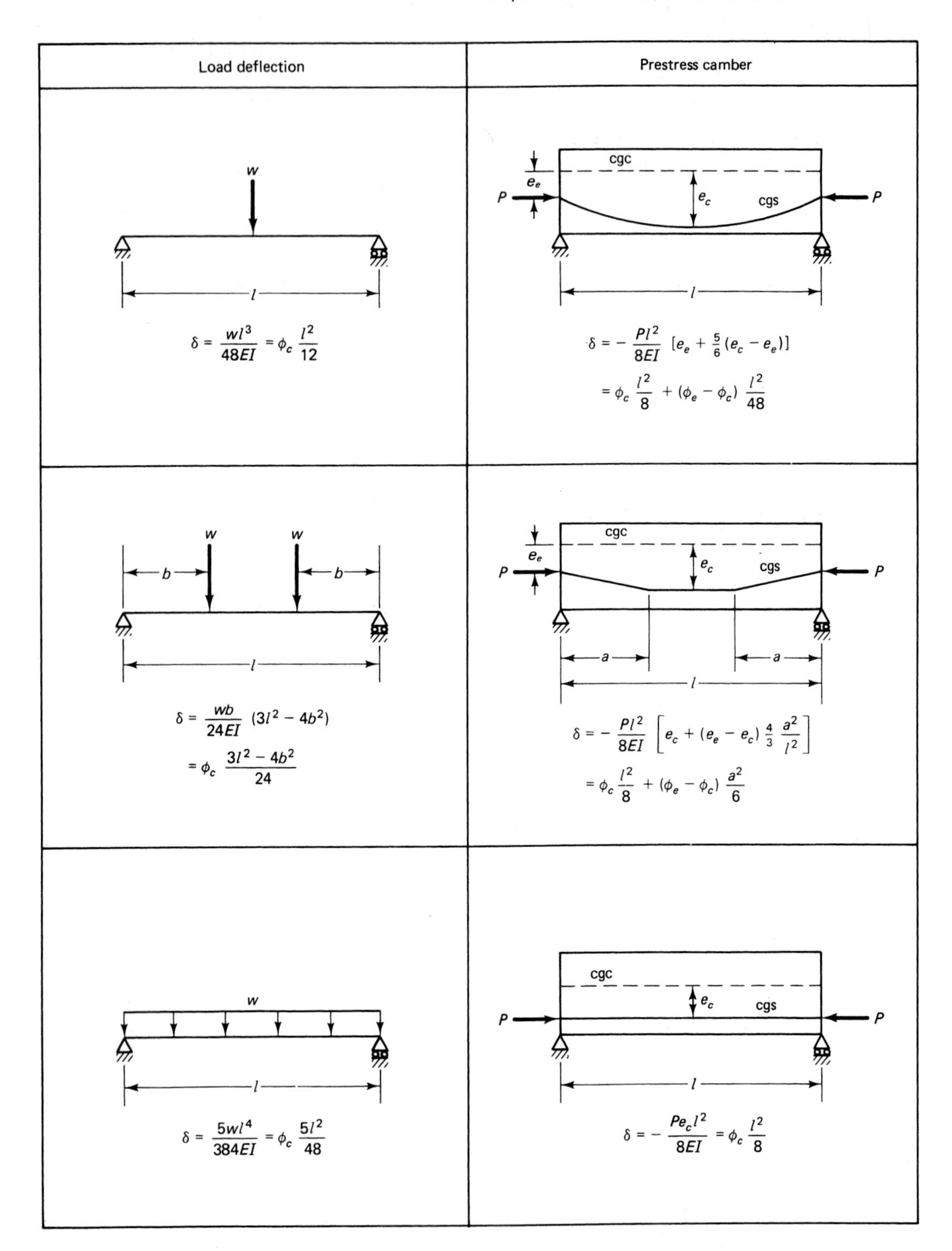

Figure 7.6 Short-term deflection in prestressed beams. Subscript c indicates midspan; subscript e indicates support.

and for continuous uncracked beams with one end continuous,

$$\text{Avg. } I_e = 0.85 I_m + 0.15(I_{\text{cont. end}}) \tag{7.12b}$$

where I_m is the midspan section moment of inertia and I_{e1} *and* I_2 are the end-section moments of inertia.

7.3.3.2 Billnear computation method. In graphical form, the bilinear moment-deflection relationship follows stages I and II described in Section 7.3.1 in accordance with ACI code. The idealized diagram for the I_g and I_{cr} zones is shown in Figure 7.7. Branson's effective I_e gives the average total *immediate* deflection $\delta_{\text{tot}} = \delta_e + \delta_{cr}$ described in the previous section.

The ACI code requires that computation of deflection in the cracked zone in the bonded tendon beams be based on the transformed section whenever the tensile stress f_t in the concrete exceeds $6\sqrt{f'_c}$. Hence, δ_{cr} in Figure 7.7 is evaluated using the transformed I_{cr} utilizing the contribution of the reinforcement in the bilinear method of deflection computation. The cracking moment of inertia can be calculated by the PCI approach (Ref. 7.7) for fully prestressed members by means of the equation

$$I_{cr} = n_p A_{ps} d_p^2 (1 - 1.6\sqrt{n_p \rho_p}) \tag{7.13a}$$

where $n_p = E_{ps}/E_c$. If nonprestressed reinforcement is used to carry tensile stresses, namely, in "partial prestressing," Equation 7.13 can be modified to give

$$I_{cr} = (n_p A_{ps} d_p^2 + n_s A_s d^2)(1 - 1.6\sqrt{n_p \rho_p + n_s \rho}) \tag{7.13b}$$

where $n_s = E_s/E_c$ for the nonprestressed steel, d = effective depth to center of mild steel or nonprestressed strand steel.

7.3.3.3 Incremental moment-curvature method. The cracked moment of inertia can be calculated more accurately from the moment-curvature relationship along the beam span and from the stress and, consequently, strain distribution across the depth of the critical sections. As shown in Figure 7.4(d) for strain ϵ_{cr} at first cracking,

$$\phi_{cr} = \frac{\epsilon_{cr}}{c} = \frac{M}{E_c I_{cr}} \tag{7.14}$$

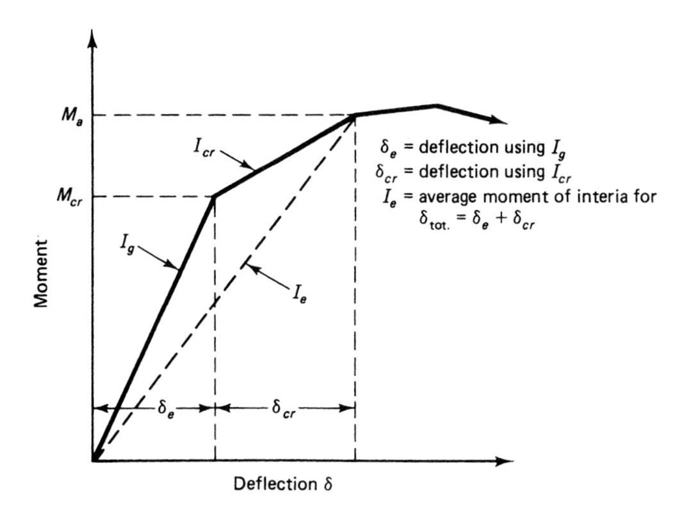

Figure 7.7 Moment-deflection relationship.

where ϵ_{cr} is the strain at the extreme concrete compression fibers and M is the total moment, including the prestressing primary moment M_1, about the centroid cgc of the section under consideration. Equation 7.14 can be rewritten to give

$$I_{cr} = \frac{Mc}{E_c \epsilon_{cr}} = \frac{Mc}{f} \tag{7.15}$$

where f is the concrete stress at the extreme compressive fibers of the section.

A flowchart for instantaneous deflection calculation and construction of the moment-curvature diagram in step-by-step increments is given in Figure 7.8.

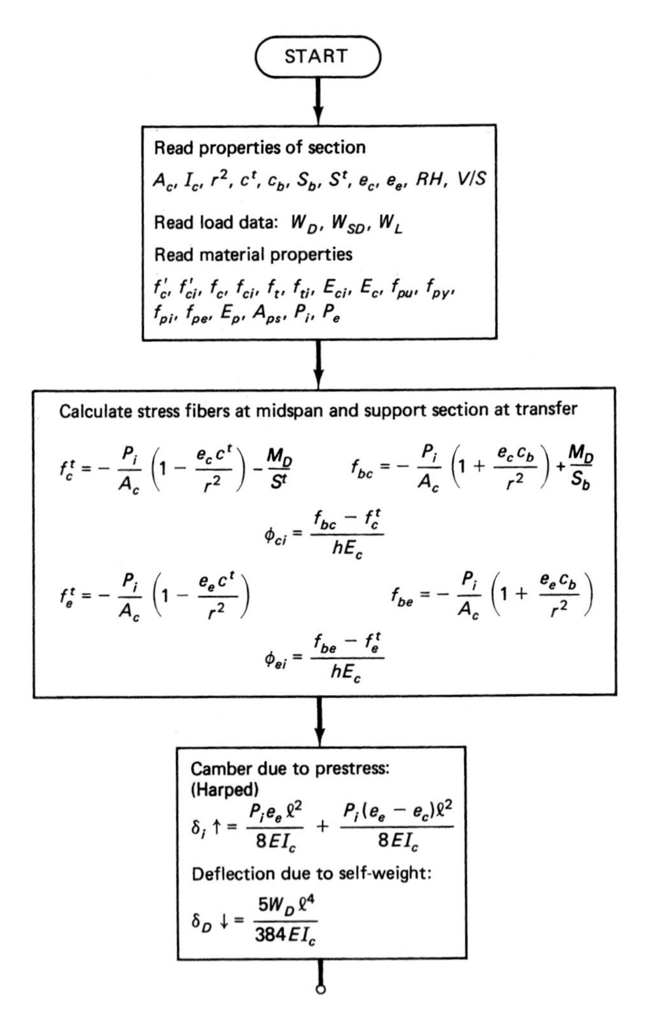

Figure 7.8 Flowchart for immediate moment-curvature camber and deflection.

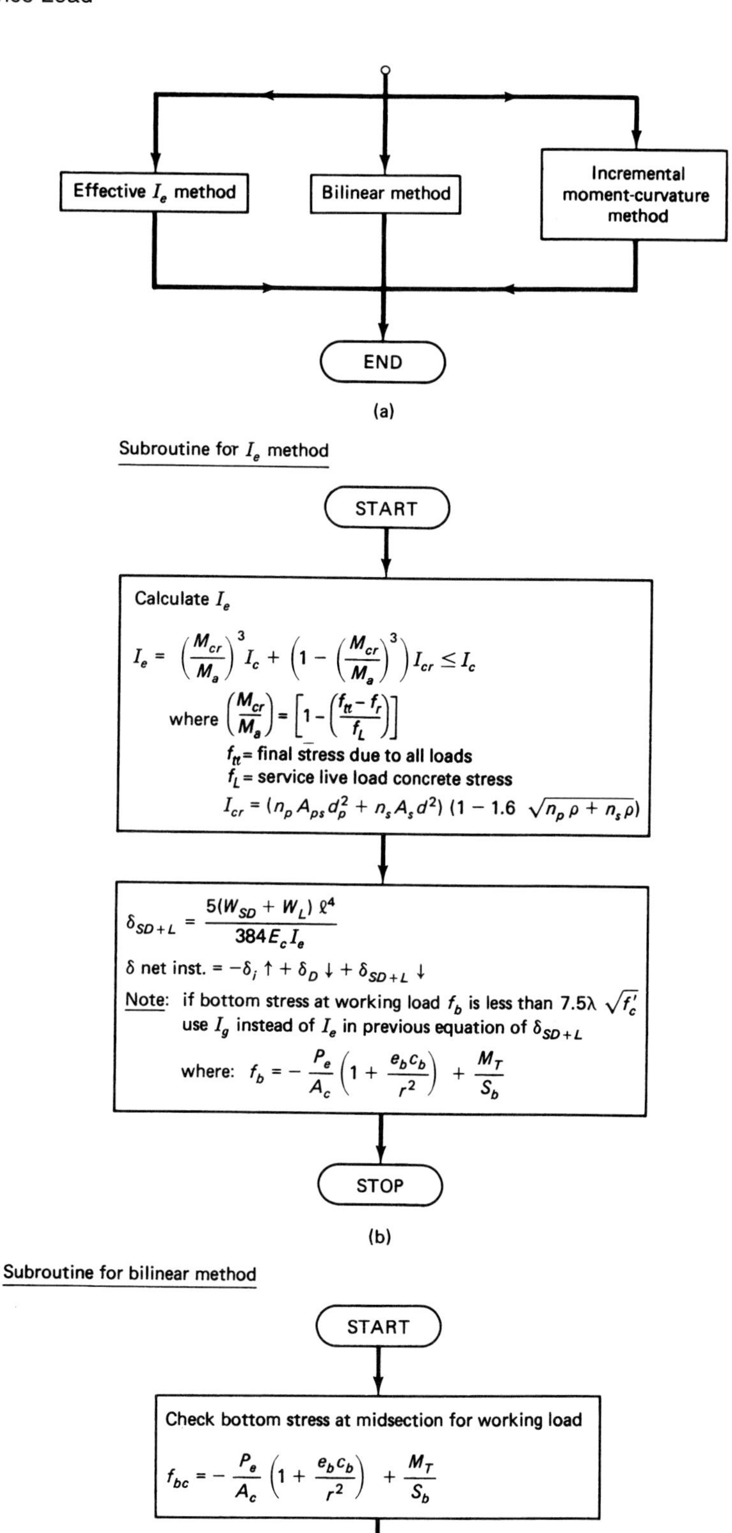

Figure 7.8 *Continued*

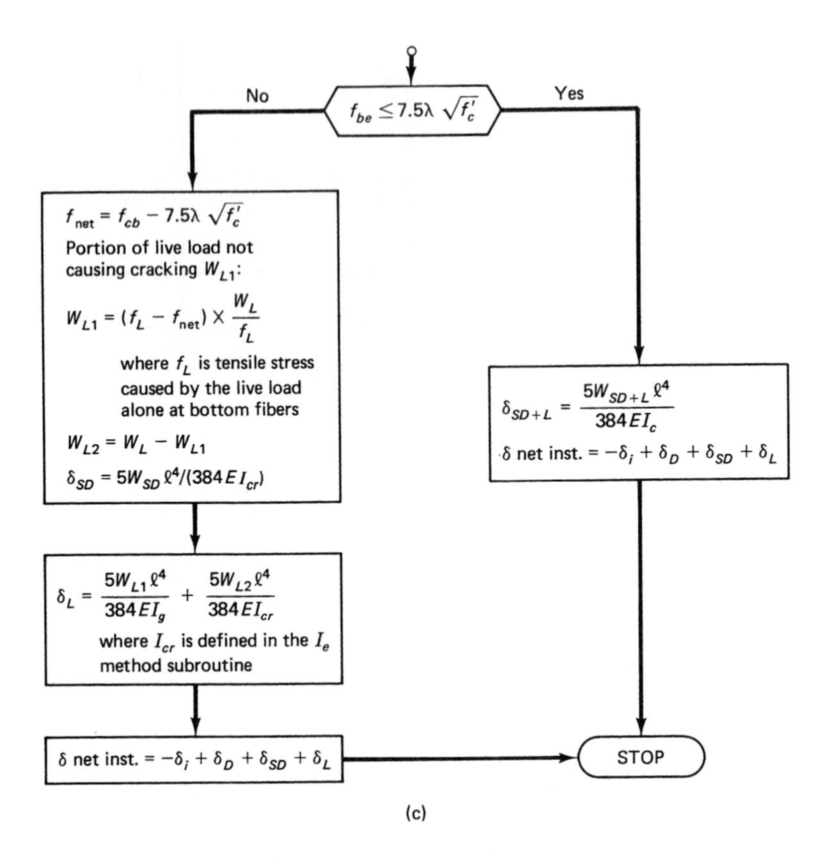

(c)

Subroutine for incremental moment-curvature method

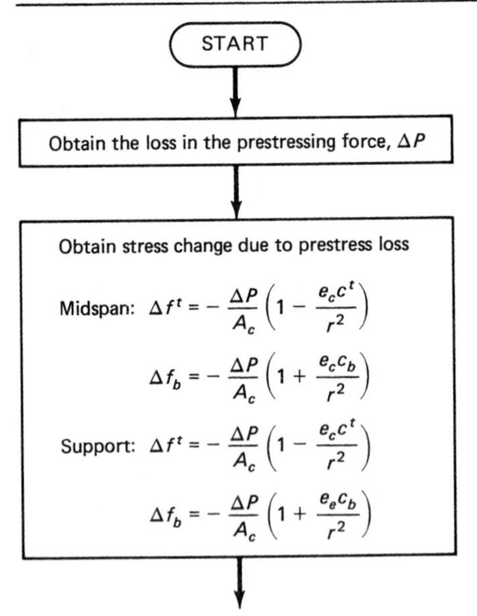

Figure 7.8 *Continued*

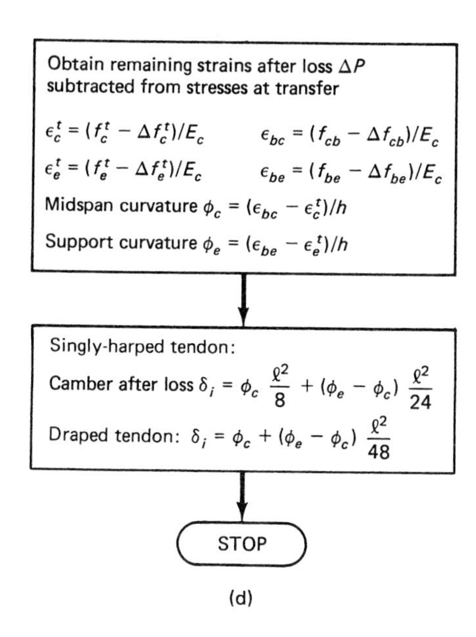

Figure 7.8 *Continued*

7.4 SHORT-TERM DEFLECTION AT SERVICE LOAD

Example 7.3 Non-Composite Uncracked Double T-Beam Deflection

Evaluate the total short-term (immediate) elastic deflection of the beam in Example 4.1 using (a) applicable moment of inertia I_g or I_e method, (b) incremental moment-curvature method. The beam carries a superimposed service live load of 1,100 plf (16.1 kN/m) and superimposed dead load of 100 plf (1.5 kN/m). It is bonded pretensioned, with A_{ps} = sixteen $\frac{1}{2}$-in. diameter 7-wire 270-ksi (f_{pu} = 270 ksi = 1,862 MPa) stress-relieved strands = 2.448 in.^{2} Disregard the contribution of the nonprestressed steel in calculating the moment of inertia in this example. Assume that strands are jacked to $0.70f_{pu}$ resulting in the initial prestress P_i = 462,272 lb. The effective prestress P_e = 379,391 lb occurs at the first load application 30 days after erection and does not include all the time-dependent losses.

Data

(a) Geometrical Properties (Fig. 7.9)

$A_c = 978$ in.^{2} (6,310 cm^2)
$I_c = 86{,}072$ in.^{4} (3.59×10^6 cm^4)
$S_b = 3{,}340$ in.^{3} (5.47×10^4 cm^3)
$S^t = 10{,}458$ in.^{3}
$W_D = 1{,}019$ plf, self-weight
$W_{SD} = 100$ plf (1.46 kN/m)
$W_L = 1{,}100$ plf (16.05 kN/m)
$e_c = 22.02$ in.
$e_e = 12.77$ in.
$c_b = 25.77$ in.
$c_t = 8.23$ in.
$A_{ps} = 16 \times 0.153 = 2.448$ in.^{2} (15.3 cm^2)
$P_i = 462{,}672$ (2,058 kN) at transfer
$P_e = 379{,}391$ 1 lb (1,688 kN)

(b) Material Properties

$V/S = 2.39$ in
$RH = 70\%$

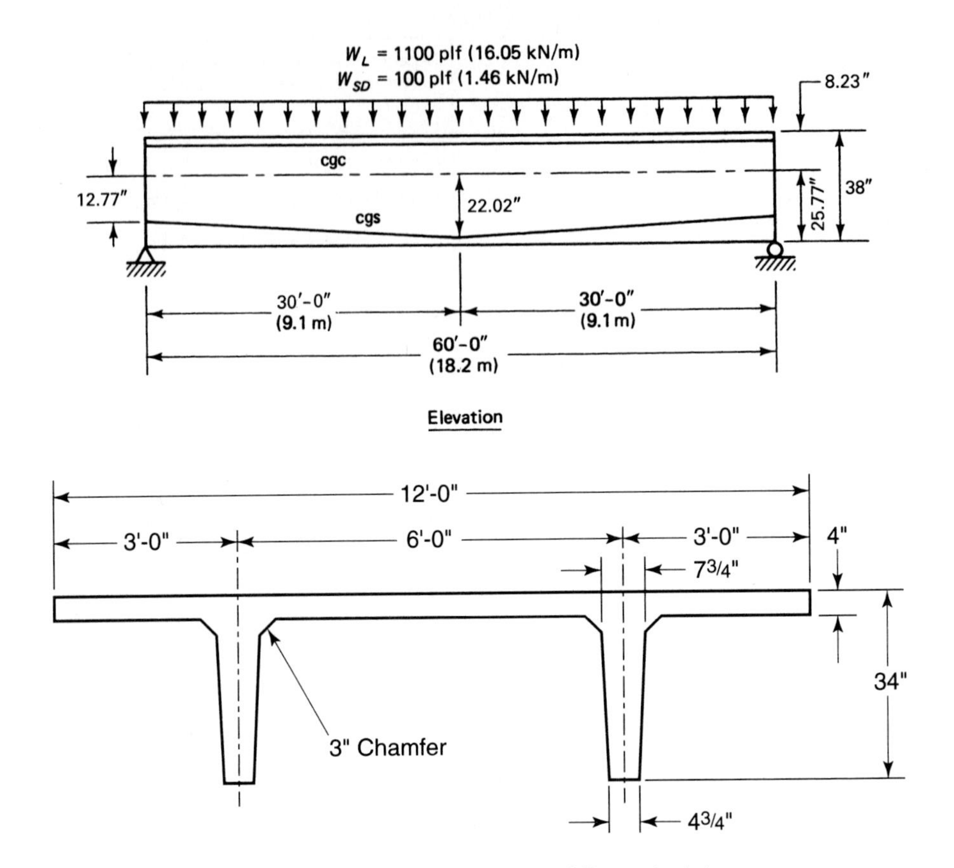

Figure 7.9 Beam geometry of Example 4.1.

$f'_c = 5{,}000$ psi
$f'_{ci} = 3{,}750$ psi
$f_{pu} = 270{,}000$ psi (1,862 MPa)
$f_{pi} = 189{,}000$ psi (1,303 MPa)
$f_{pe} = 154{,}980$ psi (1,067 MPa)
$f_{py} = 230{,}000$ psi
$E_{ps} = 28.5 \times 10^6$ psi (196 GPa)

(c) Allowable Stresses
$f_{ti} = 2{,}250$ psi
$f_c = 2{,}250$ psi
$f_{ti} = 184$ psi (midspan)
$f_t = 849$ psi (midspan)

Solution: (a)

1. *Midspan Section Stresses*

$$e_c = 22.02 \text{ in. (559 mm)}$$

Maximum self-weight moment

$$M_D = \frac{1{,}019(60)^2}{8} \times 12 = 5{,}502{,}600 \text{ in.-lb}$$

(a) At transfer, calculated fiber stresses are
From Equation 4.1a,

$$f^t = -\frac{P_i}{A_c}\left(1 - \frac{e_c c_t}{r^2}\right) - \frac{M_D}{S^t}$$

$$= -\frac{462{,}672}{978}\left(1 - \frac{22.02 \times 8.73}{88.0}\right) - \frac{5{,}502{,}600}{10{,}458}$$

$$= +501 - 526 = -25 \text{ psi } (C) > f_t = +184 \text{ psi}(T), \text{ O.K.}$$

$$f_b = -\frac{P_i}{A_c}\left(1 + \frac{e_c c_b}{r^2}\right) + \frac{M_D}{S_b}$$

$$= -\frac{462{,}672}{978}\left(1 + \frac{22.02 \times 25.77}{88.0}\right) + \frac{5{,}502{,}600}{3{,}340}$$

$$= -3{,}524 + 1{,}647 = -1{,}877 \text{ psi } (C) < -2{,}250 \text{ psi, O.K.}$$

(b) At service load

$$M_{SD} = \frac{100(60)^2 12}{8} = 540{,}000 \text{ in.-lb (61 kN-m)}$$

$$M_L = \frac{1{,}100(60)^2 12}{8} = 5{,}940{,}000 \text{ in.-lb (672 kN-m)}$$

$$\text{Live-load } f^t = \frac{5{,}940{,}000}{10{,}458} = -568 \text{ psi } (C)$$

$$\text{Live-load } f_b = \frac{5{,}940{,}000}{3{,}340} = 1{,}778 \text{ psi } (T)$$

Total Moment $M_T = M_D + M_{SD} + M_L = 5{,}502{,}600 + 6{,}480{,}000 = 11{,}982{,}600$ in-lb (1,354 kN-m).
From Equation 4.3a,

$$f^t = -\frac{P_e}{A_c}\left(1 - \frac{ec_t}{r^2}\right) - \frac{M_T}{S^t}$$

$$= -\frac{379{,}391}{978}\left(1 - \frac{22.02 \times 8.23}{88.0}\right) - \frac{11{,}982{,}600}{10{,}458}$$

$$f^t = +411 - 1146 = -735 \text{ psi} < f_c = -2{,}250 \text{ psi, O.K.}$$

From Equation 4.3b,

$$f_b = -\frac{P_e}{A_c}\left(1 + \frac{ec_b}{r^2}\right) + \frac{M_T}{S_b}$$

$$= -\frac{379{,}391}{978}\left(1 + \frac{22.02 \times 25.77}{88.0}\right) + \frac{11{,}982{,}600}{3{,}340}$$

$$= -2{,}689 + 3{,}587 = +698 \text{ psi } (T) < 849 \text{ psi, O.K.}$$

Hence, the section is uncracked and the gross moment of inertia I_g would have to be used for deflection calculations. In such a case, the effective moment of inertia I_e is taken as I_g.

2. *Support Section Stresses*
From Example 4.1,

$$f_{ti} = 6\sqrt{f'_{cr}} = 6\sqrt{3{,}750} = 367 \text{ psi}$$

$$f_t = 12\sqrt{f'_c} = 12\sqrt{5{,}000} = 849 \text{ psi}$$

$$e_e = 12.77 \text{ in.}$$

Follow the same steps as in the midspan section, with the moment $M = 0$ in the above steps. A check of support section stresses at transfer gave stresses below the allowable, hence O.K.

Summary of Fiber Stresses (psi)

	Midspan		Support	
	f^t	f_b	f^t	f_b
Prestress P_i only	+501	−3,524	+92	−2,242
At transfer and W_d	−25	−1,877	+92	−2,242
Live load W_L only	−568	+1,778	0	0
At service load	−735	+698	+75	−1,839

(1 psi = 6.895 kPa)

3. *Deflection and Camber Calculation at Transfer*
From basic mechanics or from Figure 7.6, for $a = l/2$, the camber at midspan due to a single harp or depression of the prestressing tendon is

$$\delta\uparrow = \frac{Pe_c l^2}{8EI} + \frac{P(e_e - e_c)l^2}{24EI}$$

So

$$E_{ci} = 57{,}000\sqrt{f'_c} = 57{,}000\sqrt{3{,}750} = 3.49 \times 10^6 \text{ psi (24.1 MPa)}$$

$$E_c = 57{,}000\sqrt{f'_c} = 57{,}000\sqrt{5{,}000} = 4.03 \times 10^6 \text{ psi (27.8 MPa)}$$

$$\delta_{pi} = \uparrow = \frac{462{,}672 \times 22.02 \times (60 \times 12)^2}{8 \times 3.49 \times 10^6 \times 86.702}$$

$$+ \frac{462{,}672 \times (12.77 - 22.02)(60 \times 12)^2}{24 \times 3.49 \times 10^6 \times 86{,}072}$$

$$= -2.20 + 0.31 = -1.89 \text{ in. (48 mm)}\uparrow$$

This upward deflection (camber) is due to prestress only. The self-weight per inch is 1,019/12 = 84.9 lb/in., and the deflection caused by self-weight is $\delta_D\downarrow = 5wl^4/384EI$,

$$\delta_D = 5 \times 84.9\,(60 \times 12)^4/384 \times 3.49 \times 10^6 \times 86{,}072 = 0.99 \text{ in.}\downarrow$$

Thus, the net camber at transfer is −1.89 ↑ +0.99 ↓ = −0.90 in. ↑ (25 mm).

4. *Total Immediate Deflection at Service Load of Uncracked Beam*

(a) Superimposed dead load deflection, using $E_c = 4.03 \times 10^6$ psi

$$\delta_{SD} = 0.99\frac{E_{ci}}{E_c}\left(\frac{100}{1{,}019}\right) = 0.99\left(\frac{3.49}{4.03}\right)\left(\frac{100}{1{,}019}\right) = 0.08 \text{ in. (2.0 mm)}\downarrow$$

(b) Live load deflection

$$\delta_L = \frac{5wl^4}{384E_cI_c} = \frac{5(1100)(60 \times 12)^4}{384 \times 4.03 \times 10^6 \times 86{,}072} \times \frac{1}{12} = 0.93 \text{ in.}\downarrow$$

A summary of the short-term cambers and deflections at service load is as follows:

Camber due to initial prestress = 1.89 in. (48 mm) ↑
Deflections due to self-weight = 0.99 in. (25 mm) ↓
Deflection due to superimposed dead load = 0.08 in. (2 mm) ↓
Deflection due to live load = 0.93 in. (23 mm) ↓
Net deflection at transfer = −1.89 + 0.99 = −0.90 in. ↑

If deflection due to prestress loss from the transfer stage to erection at 30 days is considered, reduced camber is

$$= 1.89\left(\frac{462{,}672 - 379{,}391}{462{,}672}\right)$$

$$= 1.89\left(\frac{83{,}281}{462{,}672}\right) = 0.34 \text{ in.} \downarrow$$

Solution: (b)

Alternate Solution by Incremental Moment Curvature Method

P_e at 30 days after transfer is 331,967 lb. So 30 days' prestress loss

$$\Delta P = P_i - P_e = 462{,}672 - 379{,}391 = 83{,}281 \text{ lb } (370 \text{ kN})$$

Strains at Transfer Due to Prestressing

$$E_{ci} \text{ at 7 days} = 3.49 \times 10^6 \text{ psi}$$

(i) Due to prestressing force (P_i)
Midspan :

$$f^t = +501 \text{ psi}$$

$$f_b = -3{,}524 \text{ psi}$$

$$\epsilon_c^t = \frac{+501}{3.49 \times 10^6} = +144 \times 10^{-6} \text{ in./in.}$$

$$\epsilon_{cb} = -1{,}010 \times 10^{-6} \text{ in./in.}$$

Support:

$$f^t = +92 \text{ psi}$$

$$f_b = -2{,}242 \text{ psi}$$

$$\epsilon_e^t = 26 \times 10^{-6} \text{ in./in.}$$

$$\epsilon_{eb} = -642x\ 10^{-6} \text{ in./in.}$$

$$(1 \text{ psi} = 6.895 \text{ KPa})$$

(ii) Due to prestressing force and self-weight ($P_i + W_D$)
Midspan:

$$f^t = -25 \text{ psi} \qquad \epsilon_c^t = -7.2 \times 10^{-6} \text{ in./in.}$$

$$f_b = -1{,}877 \text{ psi} \qquad \epsilon_{cb} = -537.8 \times 10^{-6} \text{ in./in.}$$

Support: same as in (i)
Strain change due to prestress loss

$$-\Delta P = 82{,}281 \text{ lb.}$$

$$E_{ci} = 3.49 \times 10^{-6} \text{ psi}$$

Midspan Section

$$\Delta f^t = -\frac{(-\Delta P)}{A_c}\left(1 - \frac{ec_t}{r^2}\right) = +\frac{83{,}281}{978}\left(1 - \frac{22.02 \times 8.23}{88.0}\right)$$

$$= -90 \text{ psi } (C)$$

$$\Delta\epsilon_c^t = \frac{-90}{3.49 \times 10^6} = -26 \times 10^{-6} \text{ in./in.}$$

$$\Delta f_b = -\frac{(-\Delta P)}{A_c}\left(1 + \frac{ec_b}{r^2}\right) = +\frac{83{,}281}{978}\left(1 + \frac{22.02 \times 25.77}{88.0}\right) = +634 \text{ psi } (T)$$

$$\Delta\epsilon_{cb} = \frac{634}{3.49 \times 10^6} = +182 \times 10^{-6} \text{ in./in.}$$

Support Section

$$\Delta f^t = -\frac{(-\Delta P)}{A_c}\left(1 - \frac{ec_t}{r^2}\right) = +\frac{83{,}281}{978}\left(1 - \frac{12.77 \times 8.23}{88.0}\right)$$

$$= -16.5 \text{ psi } (C)$$

$$\Delta\epsilon_e^t = \frac{-16.5}{3.49 \times 10^6} = -5 \times 10^{-6} \text{ in./in.}$$

$$\Delta f_b = -\frac{(-\Delta P)}{A_c}\left(1 + \frac{ec_b}{r^2}\right) = +\frac{83{,}281}{978}\left(1 + \frac{12.77 \times 25.77}{88.0}\right)$$

$$= 404 \text{ psi } (T)$$

$$\Delta\epsilon_{be} = \frac{+404}{3.49 \times 10^6} = +116 \times 10^{-6} \text{ in./in.}$$

Superimposing the strain at transfer on the strain due to prestress loss gives the strain distributions at service load after prestress *due to prestress only,* as shown in Figure 7.10.

From Figure 7.10

Midspan curvature

$$\phi_c = \frac{-828 - 118}{34} \times 10^{-6} = -27.82 \times 10^{-6} \text{ rad/in.}$$

Support curvature

$$\phi_e = \frac{-526 - 21}{34} \times 10^{-6} = -16.09 \times 10^{-6} \text{ rad/in.}$$

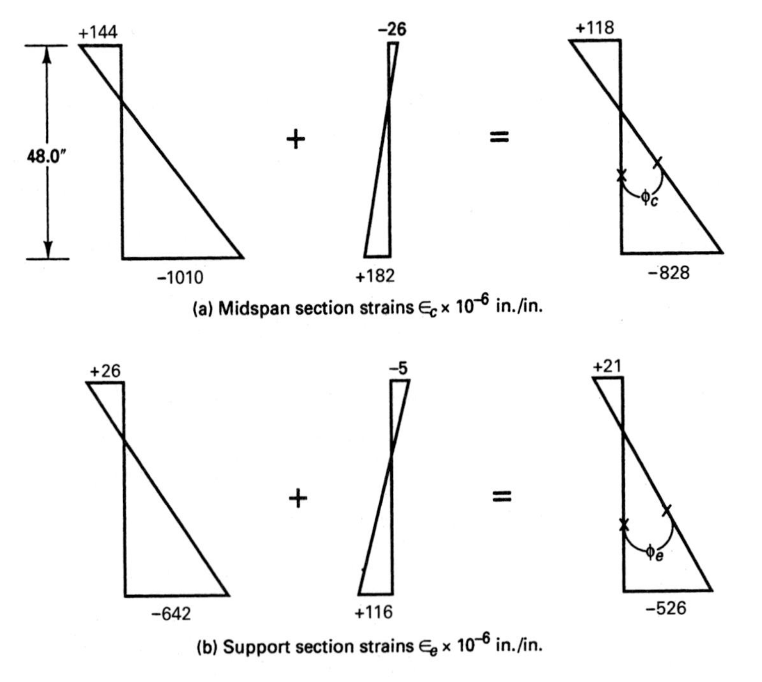

Figure 7.10 Strain distribution across section depth at prestress transfer in Example 7.4.

From Figure 7.6 for $a = l/2$, the beam camber after losses due only to P_e is

$$\delta_e \uparrow = \phi_c\left(\frac{l^2}{8}\right) + (\phi_e - \phi_c)\frac{l^2}{24}$$

$$= -27.82 \times 10^{-6}\frac{(60 \times 12)^2}{8} + (-16.09 + 27.82)$$

$$\times 10^{-6}\frac{(60 \times 12)^2}{24} = -1.80 + 0.25$$

$$= -1.55 \text{ in.} \uparrow (39 \text{ mm}) \text{ (camber)}$$

which is identical to $(-1.89 + 0.34) = -1.55$ in. ↑ after losses in the previous solution. The deflections due to self-weight W_D, superimposed dead load W_{SD} and live load W_L are the same as in the previous solution.

Note that the computed deflection values can differ by 20 to 40 percent from the actual values because of the several parameters which affect the modulus of concrete. Hence, all computational values in the various steps of the solution can be rounded to three significant figures without appreciably affecting the final results.

7.5 SHORT-TERM DEFLECTION OF CRACKED PRESTRESSED BEAMS

7.5.1 Short-Term Deflection of the Beam in Example 4.3 If Cracked

Example 7.4

Solve Example 7.3 by (a) the bilinear method, (b) the effective moment of inertia method for a condition of tensile stress of $f_b = 750$ psi at midspan bottom fibers at service load instead of $f_b = -56$ psi (*C*) in the previous example, i.e., the tensile stress exceeding the modulus of rupture $f_r = 7.5\sqrt{f'_c} = 530$ psi. Assume that the net beam camber due to prestress and self-weight is $\delta = 0.95$ in.

Solution: The net tensile stress beyond the first cracking load at the modulus of rupture is $f_{net} = f_b - f_r = 750 - 530 = +220$ psi (*T*). From Example 7.3, the tensile stress caused by the live load alone at the bottom fibers is +1,778 psi. Now, since W_L = 1,100 plf, the portion of the load that would not result in tensile stress at the bottom fibers is

$$w_1 = \frac{(1{,}778 - 220)}{(1{,}778)} \times 1{,}100 = 964 \text{ plf}$$

$$= \frac{964}{12} = 80 \text{ lb/in.}$$

The deflection determined by the uncracked I_g is

$$\delta_g = \frac{5w_1l^4}{384E_cI_g} = \frac{5 \times 80(60 \times 12)^4}{384 \times 4.03 \times 10^6 \times 169{,}020} = 0.8 \text{ in.} \downarrow (20 \text{ mm})$$

(a) *Bilinear Method*

$$I_{cr} = n_pA_{ps}d_p^2(1 - 1.6\sqrt{n_p\rho_p})$$

$$n_p = \frac{E_{ps}}{E_c} = \frac{28.5 \times 10^6}{4.03 \times 10^6} = 7.07$$

$$d_p = e_c + c_t = 22.02 + 8.23 = 30.25 \text{ in.} > 0.8h = 27.2 \text{ in.}$$

Used $d_p = 30.25$ in. and $A_{ps} = 2.448$ in.^{2} Then

$$\rho_p = \frac{A_{ps}}{bd_p} = \frac{2.448}{144 \times 30.25} = 0.0006$$

$$I_{cr} = 7.07 \times 2.448\,(30.25)^2(1 - 1.6\sqrt{7.07 \times 0.0006}$$
$$= 14{,}806 \text{ in.}^4\ (5.9 \times 10^5 \text{ cm}^4)$$

Balance of the total load that results in cracking of the section is

$$w_2 = \frac{1{,}100 - 964}{12} = 11.3 \text{ lb/in.}$$

$$\delta_{cr} = \frac{5w_2 l^4}{384E_c I_{cr}} = \frac{5 \times 11.3(60 \times 12)^4}{384 \times 4.03 \times 10^6 \times 14{,}806}$$
$$= 0.66 \text{ in.} \downarrow (17 \text{ mm})$$

Thus, the total deflection due to live load

$$\delta_L = 0.80 + 0.66 = +1.46\text{in.} \downarrow (37 \text{ mm})$$

(b) *Effective Moment of Inertia Method I_e Method:*
From Equation 7.10b,

$$I_e = \left(\frac{M_{cr}}{M_a}\right)^3 I_g + \left[1 - \left(\frac{M_{cr}}{M_a}\right)^3\right] I_{cr} \leq I_g$$

From Equation 7.11,

$$\left(\frac{M_{cr}}{M_a}\right) = 1 - \left(\frac{f_{ti} - f_r}{f_L}\right)$$

$$f_{tl} = \text{final total stress} = +750 \text{ psi } (T)$$

$$f_r = \text{modulus of rupture} = 530 \text{ psi from before}$$

$$f_L = \text{live load stress} = 1{,}451 \text{ psi}$$

$$\left(\frac{M_{cr}}{M_a}\right) = 1 - \left(\frac{750 - 530}{1{,}778}\right) = 1 - 0.124 = 0.876$$

$$\left(\frac{M_{cr}}{M_a}\right)^3 = 0.67$$

$$I_e = 0.67 \times 86{,}072 + (1 - 0.67)14{,}806$$
$$= 62{,}554 \text{ in.}^4$$

Total live-load intensity = 1,100/12 = 92 lb/in.

Deflection due to live load

$$\delta_L = \frac{5 \times 92(60 \times 12)^4}{384 \times 4.03 \times 10^6 \times 62{,}554} = 1.28 \text{ in.} \downarrow (33 \text{ mm})$$

as compared to 1.46 in. in Solution (a). Choose $\delta_L = +1.46$ in. ↓. Use this value for the final net long-term deflection after losses as tabulated in Example 7.6.

7.6 CONSTRUCTION OF MOMENT-CURVATURE DIAGRAM

Example 7.5

Construct the moment-curvature diagram for the midspan section of the bonded double-T beam in Example 7.3 for the following incremental strain steps:

Photo 7.3 Deflection of continuous beam (Nawy et al.).

1. Strain at transfer $f_{pi} = 189{,}000$ psi due to P_i only.
2. Strain at $f_{pe} = 154{,}980$ psi prior to gravity loads.
3. Decompression at tendon cgs level.
4. Modulus of rupture level.
5. Cracked section, strain ϵ_{c1} at top = 0.001 in./in.
6. Cracked section, strain ϵ_{c1} at top = 0.003 in./in.

Note the strain distribution in Figure 7.11 due to the prestressing force P_e. Use the stress-strain diagram of Figure 7.12 for the prestressing steel and that of Figure 7.13 for the concrete to determine the actual stresses through strain compatibility.

Solution:

1. *Prestress Transfer Stage*
   From the data for Example 7.3, the midspan stresses due only to prestress Pi are as follows:

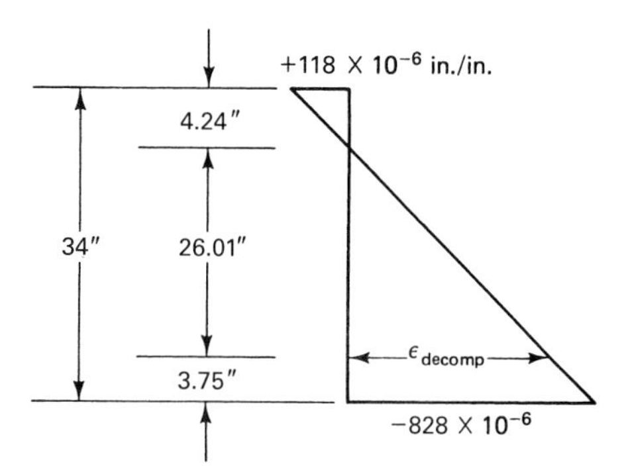

Figure 7.11 Strain distribution due only to prestress P_e.

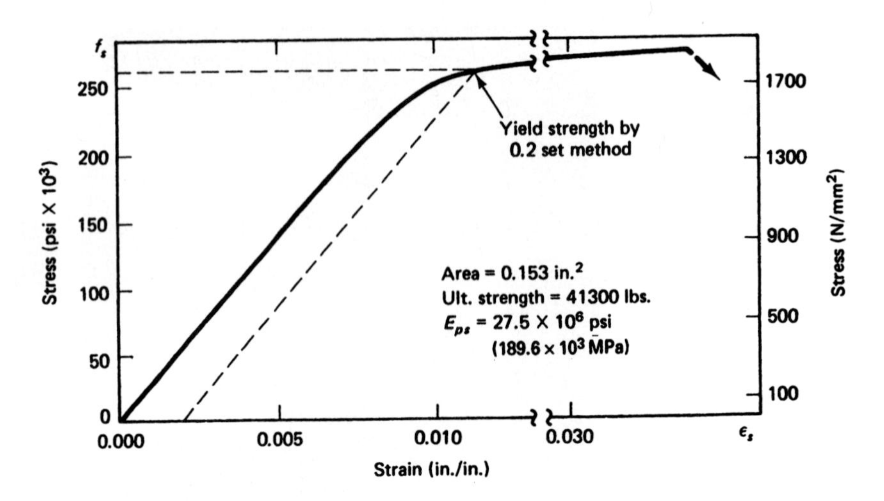

Figure 7.12 Stress-strain diagram for $\frac{1}{2}$ in. (12.7 mm) dia prestressing tendons.

$$f^t = +501 \text{ psi}$$

$$f_b = -3{,}524 \text{ psi}$$

$$\epsilon_c^t = \frac{+501}{3.49 \times 10^6} = +144 \times 10^{-6} \text{ in./in.}$$

$$\epsilon_{cb} = \frac{-3524}{3.49 \times 10^6} = 1{,}010 \times 10^{-6} \text{ in./in.}$$

$$\phi_i = \frac{(\epsilon_{cb} - \epsilon_c^t)}{h} = \frac{(-1010 - 144)}{34} \times 10^6 = -33.94 \times 10^{-6} \text{ rad/in.}$$

From Example 7.3, the corresponding moments due to $P_i + M_D$ are $M_i = -462{,}672 \times 22.02 + 5{,}502{,}600 = -4.96 \times 10^6$ in.-lb.

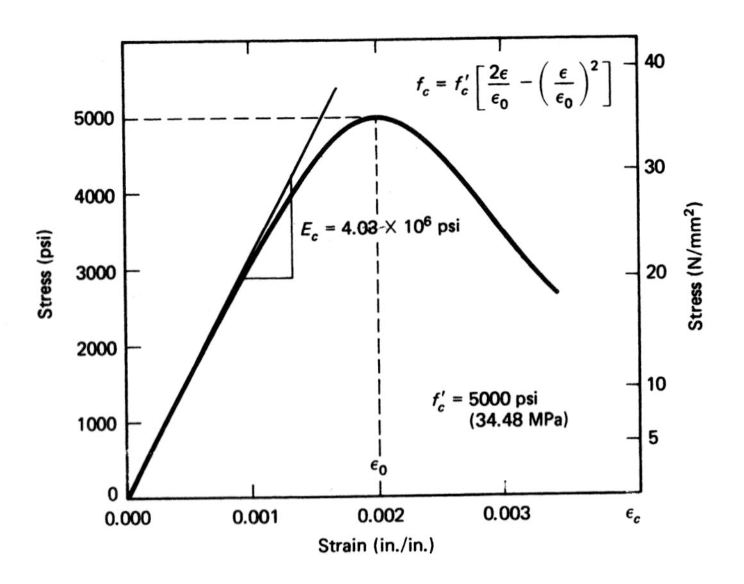

Figure 7.13 Stress-strain diagram for $f'_c = 5{,}000$ psi concrete.

2. *Prestress Stage after Losses*
In the subsequent decompression stage a moment value M_g due to gravity loads has to be found which would reduce the stress in the prestressing steel to zero. From Example 4.1, $P_e = 379{,}391$ lb. Hence,

$$\frac{P_e}{P_i} = \frac{379{,}391}{462{,}672} = 0.82$$

The stresses and strains at midspan at transfer prestress P_i are

$$f_c^t = +501\text{psi}$$
$$f_b = -3{,}524 \text{ psi}$$
$$\epsilon_c^t = +144 \times 10^{-6} \text{ in./in.}$$
$$\epsilon_{cb} = -1{,}010 \times 10^{-6} \text{ in./in.}$$

Reduce the strains up to the P_e stage as follows:

$$\epsilon_c^t = 0.82(+144 \times 10^{-6}) = +118 \times 10^{-6} \text{ in./in.}$$
$$\epsilon_{cb} = 0.82(-1010 \times 10^{-6}) = -828 \times 10^{-6} \text{ in./in.}$$

The strain distribution becomes, as shown in Figure 7.11,

$$\phi_2 = \frac{(\epsilon_{cb} - \epsilon_c^t)}{h} = \frac{(-828 - 118)10^{-6}}{34} = -27.82 \times 10^{-6} \text{ rad/in.}$$

The corresponding gravity-load moment $M_g = 0$.

3. *Decompression Stage with Zero Concrete Stress at Tendon cgs*
From Figure 7.11, the decompression strain at the cgs level is

$$\epsilon_{\text{decomp.}} = -828 \times 10^{-6} \times \frac{26.01}{26.01 + 3.75} = 723 \times 10^{-6} \text{ in./in.}$$

and

$$\epsilon_{pe} = \frac{f_{pe}}{E_{ps}} = \frac{154{,}980}{27.5 \times 10^6} = 5{,}636 \times 10^{-6} \text{ in./in.}$$

Compatibility of strain requires that the prestressing tendons in the *bonded* beam undergo the same change in strain as the surrounding concrete, *increasing* the tensile strain in the tendon in order to reduce the compressive stress in the concrete at the cgs level to zero. Thus,

$$\text{Total } \epsilon_{pe} = 5{,}636 \times 10^{-6} + 723 \times 10^{-6} = 6{,}359 \times 10^{-6} \text{ in./in.}$$

From the stress-strain diagram in Figure 7.12 the corresponding stress $f_{pe} = 177{,}000$ psi. Consequently, we have

$$\text{Adjusted } P_e = 177{,}000 \times 0.153 \times 16 = 433{,}296$$

$$\text{Adjusted } f^t = -\frac{433{,}296}{978}\left(1 - \frac{22.02 \times 8.23}{88.0}\right) \cong +469 \text{ psi } (T)$$

$$\epsilon_c^t = -\frac{+469}{4.03 \times 10^6} = 116 \times 10^{-6} \text{ in./in.}$$

$$\text{Adjusted } f_b = -\frac{433{,}296}{978}\left(1 + \frac{22.02 \times 25.77}{88.0}\right) \cong -3{,}300 \text{ psi } (C)$$

$$\epsilon_{cb} = \frac{-3{,}300}{4.03 \times 10^6} = -819 \times 10^{-6} \text{ in./in.}$$

$$f_{\text{decomp.}} = \frac{M_{\text{decomp.}} \times y}{I_c} = \frac{M_{\text{decomp.}} \times 22.02}{86{,}072} = 2{,}884 \text{ psi}$$

$$M_{\text{decomp.}} = \frac{2{,}884 \times 86{,}072}{22.02} = 11.27 \times 10^6 \text{ in.-lb } (1.27 \times 10^6 \text{ N-m})$$

$$f_t = \frac{M_{\text{decomp.}}}{S^t} = \frac{11.27 \times 10^6}{10{,}458} = -1{,}078 \text{ psi } (C)$$

$$\text{Net stress } f^t = -1{,}078 + 469 = -609 \text{ psi } (C)\ (4.16 \text{ MPa})$$

$$\epsilon_c^t = \frac{-609}{4.03 \times 10^6} = -151.1 \times 10^{-6} \text{ in./in.}$$

$$f_b = \frac{11.27 \times 10^6}{S_b} = \frac{11.27 \times 10^6}{3{,}340} = +3{,}374 \text{ psi } (T)$$

$$\text{Net stress } f_b = +3{,}374 - 3{,}300 = +74 \text{ psi } (T)$$

$$\epsilon_{cb} = \frac{+74}{4.03 \times 10^6} = +18.4 \times 10^{-6} \text{ in./in.}$$

$$\phi_{\text{decomp.}} = \frac{(\epsilon_{cb} - \epsilon_c^t)}{h} = \frac{(18.4 + 151.1)}{34} = 10^{-6} = +4.99 \times 10^{-6} \text{ rad/in.}$$

Corresponding $M = 11.27 \times 10^{-6}$ in.-lb
Figure 7.14 gives the stress and strain distributions in this beam at the decompression state.

4. *Modulus of Rupture Stage*

$$f_r = 7.5\lambda\sqrt{f_c'} = 7.5\sqrt{5{,}000} = 530 \text{ psi}$$

$$M_{cr} = S_b\left[7.5\lambda\sqrt{f_c'} + \frac{P_e}{A_c}\left(1 + \frac{ec_b}{r^2}\right)\right]$$

From before, the second part of the above expression for moment gives a stress of 3,300 psi.

$$\text{Therefore } M_{cr} = 3{,}340(530 + 3{,}300) = 12.8 \times 10^6 \text{ in.-lb}$$

Net bottom concrete stress = modulus of rupture f_r for this stage = +530 psi (T)

$$\epsilon_{cb} = \frac{+530}{4.03 \times 10^6} = +132 \times 10^{-6} \text{ in./in.}$$

$$f^t = \frac{12.8 \times 10^6}{10{,}458} = -1{,}224 \text{ psi } (C)$$

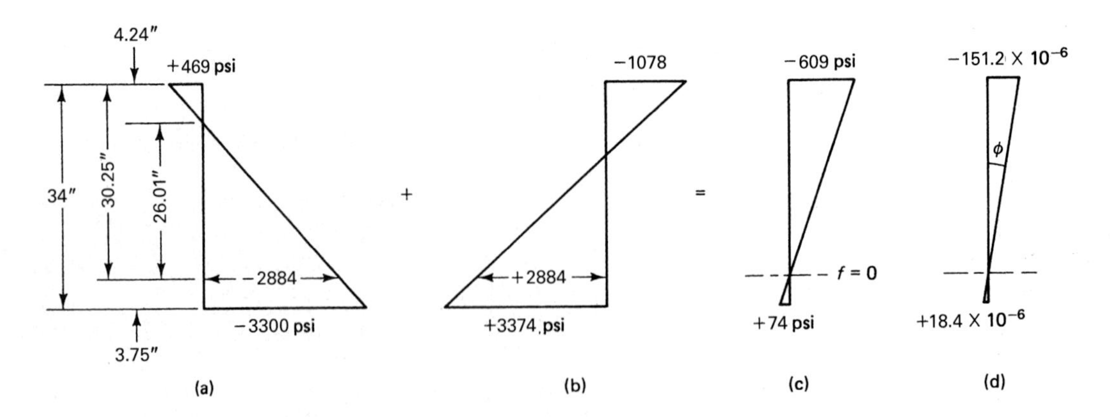

Figure 7.14 Stress distribution at decompression in Example 7.6. (a) Loading stress. (b) Decompression stress. (c) Final stress. (d) Unit strain.

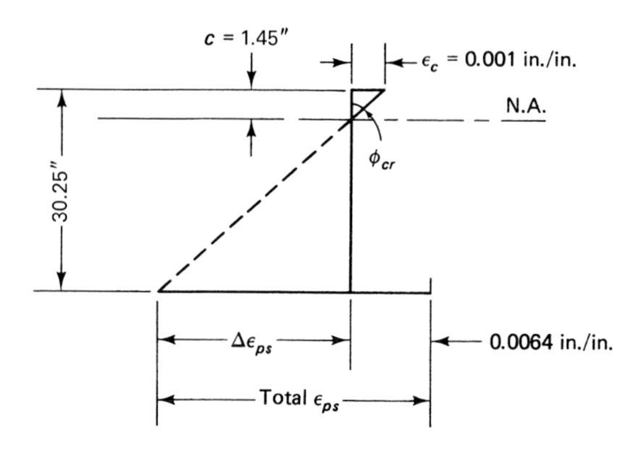

Figure 7.15 Strain distribution at $\epsilon_c = 0.001$ in./in. in Example 7.5.

$$\text{Net stress } f^t = -1{,}224 + 469 = -755 \text{ psi } (C)$$

$$\epsilon_c^t = \frac{-755}{4.03 \times 10^6} = -187 \times 10^{-6} \text{ in./in.}$$

$$\phi_5 = \frac{(\epsilon_{cb} - \epsilon_c^t)}{h} = \frac{(132 + 187)}{34} \times 10^{-6}$$

$$= +9.38 \times 10^{-6} \text{ rad./in.}$$

5. *Cracked Section Stage,* $\epsilon_c = 0.001$ in./in.
From before, $\epsilon_{pe} = 6{,}359 \times 10^{-6} = 0.0064$ in./in. By trial and adjustment, assume a neutral axis depth $c = 1.5$ in. below the top fibers of the flange. Then $\Delta\epsilon_{ps}$ is the additional strain in bonded prestressing strands due to $\epsilon_c = 0.001$ in./in. at the top fibers, and from similar triangles in Figure 7.15,

$$\Delta\epsilon_{ps} = \frac{(30.25 - 1.5)}{1.5} \times 0.001 = 0.0192 \text{ in./in.}$$

So the total $\epsilon_{ps} = 0.0192 + 0.0064 = 0.0256$ in./in.

From the stress-diagram of the prestressing steel in Figure 7.12, the corresponding stress is

$$f_{ps} \cong 260{,}000 \text{ psi}$$

and

$$A_{ps} = 16 \times 0.153 = 2.448 \text{ in.}^2$$

Thus, the tensile force $T_p = 260{,}000 \times 2.448 = 636{,}480$ lb.

From Figure 7.13, $f_c = 3{,}000$ psi corresponds to $\epsilon_c = 0.001$ in./in. The compressive force is then $C_c = (12 \times 12 \times 1.5)3000 = 648{,}000 > T = 636{,}480$ lb.
Hence, the neutral axis depth should be reduced.

Second Trial

Assume $c_c = 1.45$ in. Then

$$\Delta\epsilon_{ps} = \frac{(30.25 - 1.45)}{1.45} \times 0.001 = 0.0199 \text{ in./in.}$$

and

$$\text{Total } \epsilon_{ps} = 0.0199 + 0.0064 = 0.0263 \text{ in./in.}$$

From Figure 7.12, $f_{ps} \cong 255{,}000$ psi, $T_p = 255{,}000 \times 2.448 = 624{,}240$ lb., and $C_c = (12 \times 12 \times 1.45)3000 = 624{,}400 \text{ lb} \cong T_p$. Hence, assumed $c = 1.45$ in., O.K.

$$\text{Moment } M = 636{,}480\left(30.25 - \frac{1.45}{3}\right) = 19.0 \times 10^6 \text{ in.-lb}$$

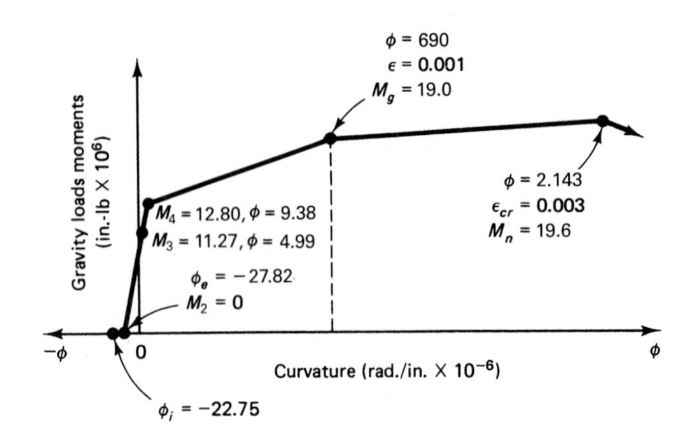

Figure 7.16 Moment-curvature diagram of Example 7.6.

and from Equation 7.5 d,

$$\phi_u = \frac{\epsilon_u}{c} = \frac{0.001}{1.45} = 690 \times 10^{-6} \text{ rad/in.}$$

6. *Fully Cracked Section Stage, $\epsilon_c = 0.003$ in./in.* (Ultimate Load)
$\epsilon_c = 0.003$ in./in. is the maximum unit strain allowed by the ACI Code at ultimate load. Assume $f_{ps} = 263{,}000$ psi. Then

$$a = \frac{A_{ps}f_{ps}}{0.85f'_c b} = \frac{2.448 \times 263{,}000}{0.85 \times 5{,}000 \times 144} = 1.1 \text{ in.}$$

$$c = \frac{a}{\beta_1} = \frac{1.1}{0.80} = 1.38 \text{ in.}$$

From Figure 7.15,

$$\epsilon_{ps} = \frac{30.25 - 1.38}{1.38} \times 0.003 = 0.0628 \text{ in./in.}$$

$$\text{Total } \epsilon_{ps} = 0.0628 + 0.0064 = 0.0692 \text{ in./in.}$$

From the stress diagram in Figure 7.12, $f_{ps} \cong f_{pu} = 270{,}000$ psi. So use $a \cong 1.1$ in., giving

$$M_n = A_{ps}f_{ps}\left(d_p - \frac{a}{2}\right) = 2.448 \times 270{,}000\left(30.25 - \frac{1.1}{2}\right)$$
$$= 19.6 \times 10^6 \text{ in.-lb}$$

Use $c \cong 1.4$ in.

$$\phi_u = \frac{\epsilon_u}{c} = \frac{0.003}{1.4} = 2{,}143 \times 10^{-6} \text{ rad/in.}$$

A schematic plot of the moment-curvature diagram is shown in Figure 7.16. The load-deflection diagram has the same form and can be inferred from the moment-curvature diagram.

7.7 LONG-TERM EFFECTS ON DEFLECTION AND CAMBER

7.7.1 PCI Multipliers Method

The ACI Code provides the following equation for estimating the time-dependent factor for deflection of nonprestressed concrete members:

$$\lambda = \frac{\xi}{1 + 50\rho'} \tag{7.16}$$

where ξ = time-dependent factor for sustained load
ρ' = compressive reinforcement ratio
λ = multiplier for *additional* long-term deflection

In a similar manner, the PCI multipliers method provides a multiplier C_1 which takes account of long-term effects in prestressed concrete members, C_1 differs from λ in Equation 7.16, because the determination of long-term cambers and deflections in prestressed members is more complex due to the following factors:

1. The long-term effect of the prestressing force and the prestress losses.
2. The increase in strength of the concrete after release of prestress due to losses.
3. The camber and deflection effect during erection.

Because of these factors, Equation 7.16 cannot be readily used.

Table 7.1, based on Refs. 7.7 and 7.8, can provide reasonable multipliers of immediate deflection and camber provided that the upward and downward components of the initial calculated camber are separated in order to take into account the effects of loss of prestress, *which only apply to the upward component.*

Shaikh and Branson, in Ref. 7.6, propose that substantial reduction can be achieved in long-term camber by the addition of nonprestressed steel. In that case, a reduced multiplier C_2 can be used given by

$$C_2 = \frac{C_1 + A_s/A_{ps}}{1 + A_s/A_{ps}} \tag{7.17}$$

Table 7.1 C_1 Multipliers for Long-Term Camber and Deflection

	Without composite topping	**With composite topping**
At erection:		
(1) Deflection (downward) component—apply to the elastic deflection due to the member weight at release of prestress	1.85	1.85
(2) Camber (upward) component—apply to the elastic camber due to prestress at the time of release of prestress	1.80	1.80
Final:		
(3) Deflection (downward) component—apply to the elastic deflection due to the member weight at release of prestress	2.70	2.40
(4) Camber (upward) component—apply to the elastic camber due to prestress at the time of release of prestress	2.45	2.20
(5) Deflection (downward)—apply to the elastic deflection due to the superimposed dead load only	3.00	3.00
(6) Deflection (downward)—apply to the elastic deflection caused by the composite topping	—	2.30

where C_1 = multiplier from Table 7.1
A_s = area of nonprestressed reinforcement
A_{ps} = area of prestressed strands.

7.7.2 Incremental Time-Steps Method

The incremental time-steps method is based on combining the computations of deflections with those of prestress losses due to time-dependent creep, shrinkage, and relaxation. The design life of the structure is divided into several time intervals selected on the basis of specific concrete strain limits, such as unit strain levels $\epsilon_{c1} = 0.001$ and 0.002 in./in., and ultimate allowable strain $\epsilon_c = 0.003$ in./in. The strain distributions, curvatures, and prestressing forces are calculated for each interval together with the incremental shrinkage, creep, and relaxation strain losses during the particular time interval. The procedure is repeated for all subsequent incremental intervals, and an integration or summation of the incremental steps is made to give the total time-dependent deflection at the particular section along the span. These calculations should be made for a sufficient number of points along the span, such as midspan and quarter-span points, to be able to determine with sufficient accuracy the form of the moment-curvature diagram.

The general expression for the total rotation at the end of a time interval can be expressed as

$$\phi_t = -\frac{P_i e_x}{E_c I_c} + \sum_0^t (P_{n-1} - P_n)\frac{e_x}{E_c I_c} - \sum_0^t (C_n - C_{n-1})P_{n-1}\frac{e_x}{E_c I_c} \qquad (7.18a)$$

where P_i = initial prestress before losses
e_x = eccentricity of tendon at any section along the span
Subscript $n-1$ = beginning of a particular time step
Subscript n = end of the aforementioned time step
C_{n-1}, C_n = creep coefficients at beginning and end, respectively, of a particular time step
$P_n - P_{n-1}$ = prestress loss at a particular time interval from all causes.

Obviously, this elaborate procedure is justified only in the evaluation of deflection and camber of very large-span bridge systems such as segmental bridges, where the erec-

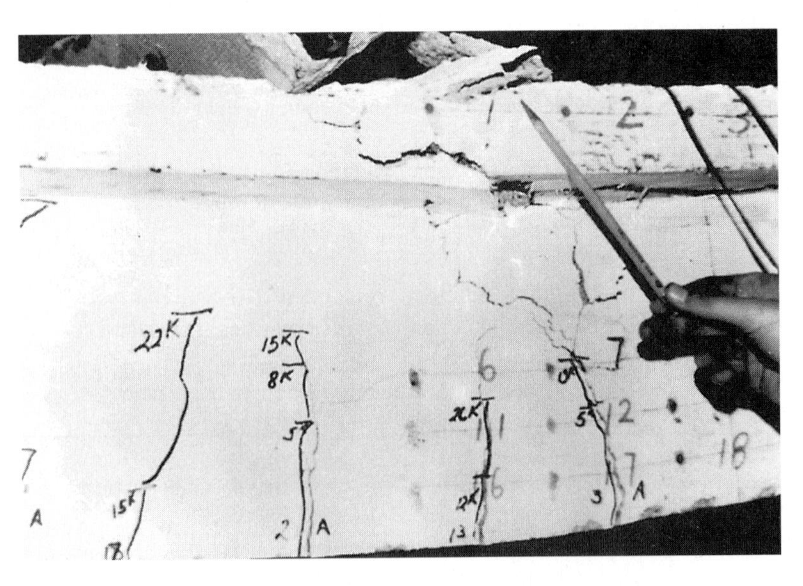

Photo 7.4 Prestressed I-beam at the limit state of failure load (Nawy et al.).

tion and assembly of the segments require a relatively accurate estimate of deflections. From Equation 7.18a, the total deflection at a particular section is

$$\delta_x = \phi_t k l^2 \tag{7.18b}$$

Now, suppose that the following strains from subsequent Example 7.7 are used to illustrate calculation of incremental and total rotations:

ϵ^t_{n-1} = gross strain due only to prestress at the top fibers, e.g., $\epsilon^t_c = +144 \times 10^{-6}$ in./in. (Figure 7.19)

$\epsilon_{b,n-1}$ = gross strain due only to prestress at the bottom fibers, e.g., $\epsilon_{cb} = -1010 \times 10^{-6}$ in./in. (Figure 7.19)

$\Delta\epsilon^t_{CR,n}$ = gross creep incremental strain at the top fibers, e.g., $\Delta\epsilon^t_{CRc} = +127 \times 10^{-6}$ in./in. (Figure 7.20)

$\Delta\epsilon_{CRb,n}$ = gross creep incremental strain at the bottom fibers, e.g., $\Delta\epsilon_{CRcb} = -895 \times 10^{-6}$ in./in. (Figure 7.20)

$\Delta\epsilon_{ps,n}$ = strain reduction due to prestress loss caused by creep force ΔP, n (such as 169×10^{-6} in./in., as seen from Figure 7.20)

Then the net incremental creep strain that will result in incremental rotation ϕ_n is

$$\Delta\epsilon^t_{CR,\text{net}} = (\Delta\epsilon^t_{CR,n} - \Delta\epsilon^t_{ps,n}) \tag{7.19a}$$

for the top fibers and

$$\Delta\epsilon_{CRb,\text{net}} = (\Delta\epsilon_{CRb,n} - \Delta\epsilon_{psb,n}) \tag{7.19b}$$

for the bottom fibers.

The incremental rotation is, then,

$$\Delta\phi_n = \frac{\Delta\epsilon^t_{CR,\text{net}} - \Delta\epsilon_{CRb,\text{net}}}{h} \tag{7.19c}$$

and the total rotation becomes

$$\phi_T = \phi_{n-1} + \Delta\phi_n \tag{7.20}$$

A schematic of the changes in strains and rotations from time step $n-1$ to time step n is shown in Figure 7.17.

The selection of the time intervals depends on the refinement level desired in the computation of cambers. For each time step, the incremental creep and shrinkage strains and relaxation loss in prestress are computed as shown in Example 7.7 to give a curvature increment $\Delta\phi$. Thereafter, new values of stress, strain, and curvature are obtained at the end of the time interval, adding the curvature increment $\Delta\phi_n$ to the total curvature ϕ_{n-1} at the beginning of the desired intervals, as given in Equation 7.18. Clearly, the incremental time-step procedure is lengthy and, hence, justified only in evaluation and assembly of segments requiring relatively accurate estimates of deformations.

The total camber (↑) or deflection (↓) due to the prestressing force can be obtained from Equation 7.20 as

$$\delta_T = \phi_T k l^2 \tag{7.21}$$

where k is a function of the span and geometry of the section and the prestressing tendon.

Several investigators have proposed different formats for estimating the additional time-dependent deflection $\Delta\delta$ from the moment-curvature relationship ϕ modified for creep. Both Tadros and Dilger recommend integrating the modified curvature along the beam span, while Naaman expresses the long-term deflection in terms of midspan and support curvatures at a time interval t (Refs. 7.10, 7.11, 7.12). As an example, Naaman's expression gives, for a parabolic tendon,

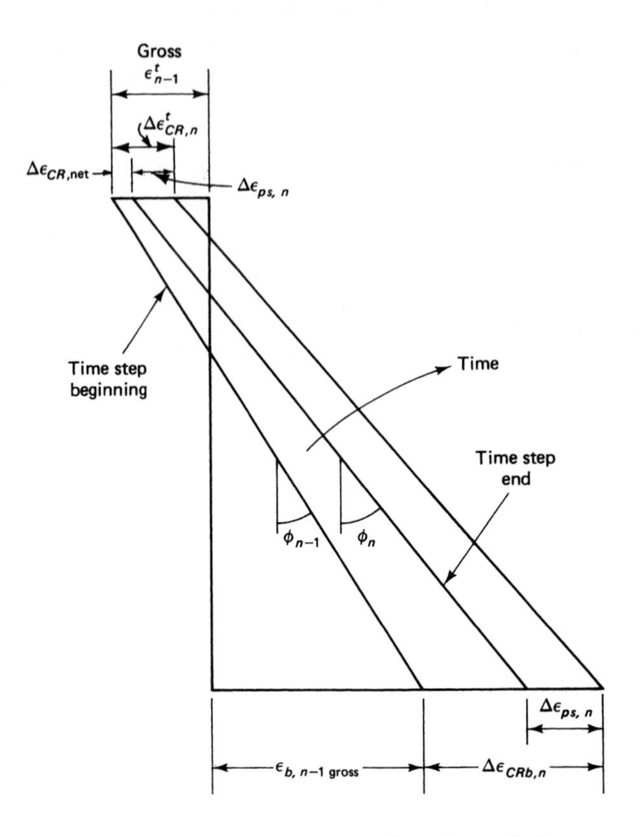

Figure 7.17 Strain changes and rotations at step n.

$$\Delta\delta(t) = \phi_1(t)\frac{l^2}{8} + [\phi_2(t) - \phi_1(t)]\frac{l^2}{48}$$

where $\phi_1(t)$ = midspan curvature at time t
$\phi_2(t)$ = support curvature at time t

in which

$$\phi(t) = \frac{M}{E_{ce}(t)I_c}$$

where $E_{ce}(t)$ = time adjusted modulus

$$= \frac{E_c(t_1)}{1 + KC_c(t)}$$

in which $E_c(t_1)$ = modulus of concrete at start of interval
$C_c(t)$ = creep coefficient at end of time interval.

7.7.3 Approximate Time-Steps Method

The approximate time-steps method is based on a simplified form of summation of constituent deflections due to the various time-dependent factors. If C_u is the long-term creep coefficient, the curvature at effective prestress P_e can be defined as

$$\phi_e = \frac{P_i e_x}{E_c I_c} + (P_i - P_e)\frac{e_x}{E_c I_c} - \left(\frac{P_i + P_e}{2}\right)\frac{e_x}{E_c I_c}C_u \tag{7.22}$$

The final deflection under P_e is

$$\delta_{et} = -\delta_i + (\delta_i - \delta_e) - \left(\frac{\delta_i + \delta_e}{2}\right)C_u \qquad (7.23a)$$

or

$$\delta_{et} = -\delta_e - \left(\frac{\delta_i + \delta_e}{2}\right)C_u \qquad (7.23b)$$

Adding the deflection due to self-weight δ_D and superimposed dead load δ_{SD}, which are affected by creep gives the final time-dependent *increase* in deflection due to prestressing and sustained loads as

$$\Delta\delta = -\delta_e - \left(\frac{\delta_i + \delta_e}{2}\right)C_u + (\delta_D + \delta_{SD})(1 + C_c) \qquad (7.24a)$$

and the final total *net* deflection including live-load deflection is

$$\delta_T = -\delta_e - \left(\frac{\delta_i + \delta_e}{2}\right)C_u + (\delta_D + \delta_{SD})(1 + C_u) + \delta_L \qquad (7.24b)$$

Intermediate deflections are found by substituting C_t for C_u in Equations 7.24a and b, where

$$C_t = \frac{t^{0.60}}{10 + t^{0.60}}C_u \qquad (7.25)$$

in which $t^{0.60}/(10 + t^{0.60})$ is the creep ratio α.

Branson et al., in Refs. 7.4 to 7.6, proposed the following expression for predicting the time-dependent increase in deflection $\Delta\delta$ of Equation 7.24a:

$$\Delta\delta = -\left[\eta + \frac{(1 + \eta)}{2}k_r C_t\right]\delta_{i(P_i)} + k_r C_t\,\delta_{i(D)} + K_a k_r C_t \delta_{i(SD)} \qquad (7.26)$$

where $\eta = P_e/P_i$
C_t = creep coefficient at time t
K_a = factor corresponding to age of concrete at superimposed load application
= $1.25t^{-0.118}$ for moist-cured concrete
= $1.13t^{-0.095}$ for steam-cured concrete
t = age, in days, at loading
$k_r = 1/(1 + A_s/A_{ps})$ when $A_s/A_{ps} << 1.0$
$\cong 1$ for all practical purposes.

For the final deflection increment, C_u is used in place of C_t in Equation 7.26.

For noncomposite beams, the total deflection $\delta_{T,t}$ becomes (Ref. 7.9)

$$\delta_{T,t} = -\delta_{pi}\left[1 - \frac{\Delta P}{P_0} + \lambda(k_r C_t)\right] + \delta_D[1 + k_r C_t] + \delta_{SD}[1 + K_a k_r C_t] + \delta_L \qquad (7.27)$$

where δ_p = deflection due to prestressing
ΔP = total loss of prestress excluding initial elastic loss
$\lambda = 1 - \Delta P/2P_0$
in which P_0 = prestress force at transfer after elastic loss
= P_i less elastic loss.

For composite beams, the total deflection is

$$\delta_T = -\delta_{pi}\left[1 - \frac{\Delta P}{P_0} + K_a k_r C_u \lambda\right] + \delta_D[1 + K_a k_r C_u]$$
$$+ \delta_{pi}\frac{I_e}{I_{\text{comp.}}}\left[1 - \frac{\Delta P - \Delta P_C}{P_0} + k_r C_u(\lambda - \alpha\lambda')\right]$$
$$+ (1 - \alpha)k_r C_u \delta_D \frac{I_c}{I_{\text{comp}}} + \delta_D\left[1 + \alpha k_r C_u \frac{I_c}{I_{\text{comp.}}}\right] \tag{7.28}$$
$$+ \delta_{df} + \delta_L$$

where $\lambda' = 1 - (\Delta P_c/2P_0)$

ΔP_c = loss of prestress at time composite topping slab is cast, excluding initial elastic loss

$I_{\text{comp.}}$ = moment of inertia of composite section

δ_{df} = deflection due to differential shrinkage and creep between precast section and composite topping slab
$= Fy_{cs}l^2/8E_{cc}I_{\text{comp.}}$ for simply supported beams (for continuous beams, use the appropriate factor in the denominator)

y_{cs} = distance from centroid of composite section to centroid of slab topping

F = force resulting from differential shrinkage and creep

E_{cc} = modulus of composite section

α = creep strain at time t divided by ultimate creep strain
$= t^{0.60}/(10 + t^{0.60})$.

In sum, comparing the relative rigor involved in applying the three methods of Sections 7.7.1, 7.7.2, and 7.7.3, it is important to recognize that the degree of spread can be very large. Engineering judgment has to be exercised in determining a reasonable accurate concrete modulus E_c value at the various loading stages and in achieving values of creep coefficients that are neither under- nor overestimated.

7.7.4 Computer Methods for Deflection Evaluation

Several computer approaches and canned programs are available for deflection calculations. They lend themselves handily for such more refined methods as the time-step method in Section 7.7.2. Keep in mind, however, that the deflection under short and long-term loading is governed by a variety of possible conditions too numerous to be covered by a single set of rules for calculating deflections. These conditions are related to all properties of the concrete constituent materials which affect deflection, particularly long-term deflection. Hence, except in cases of very large span bridges such as cable-stayed bridges, deflection calculation procedures and methods should be viewed within a ±40 percent variability, if not more. The material properties input to any computer program should be carefully scrutinized based on laboratory tests if large span structures are involved.

7.7.5 Deflection of Composite Beams

Computing deflections for composite prestressed beams is similar to that for noncomposite sections. The process becomes more rigorous if the incremental time-steps method is used. The additional steps at the various construction stages of the precast element and the situ-cast top slab require consideration of the changes in the moments of inertia from the precast to the composite values at the appropriate stages. Moreover, the difference in the shrinkage characteristics and time-step increments due to the difference in shrinkage values of the precast section and the added concrete topping increase the rigor of the computational process. Fortunately, the use of computer programs facilitates speedy evaluation of camber and deflection in composite elements.

7.8 PERMISSIBLE LIMITS OF CALCULATED DEFLECTION

The ACI Code requires that the calculated deflection has to satisfy the serviceability requirement of maximum permissible deflection for the various structural conditions listed in Table 7.2. Note that long-term effects cause measurable increases in deflection and camber with time and result in excessive overstress in the concrete and the reinforcement, *requiring* computation of deflection and camber.

AASHTO permissible deflection requirements, shown in Table 7.3, are more rigorous because of the dynamic impact of moving loads on bridge spans.

Following is a step-by-step procedure for computing deflection:

1. Determine the properties of the concrete, including the concrete modulus E_c, concrete creep, and the shrinkage and prestress relationship at the various loading stages.
2. Choose the time increments to be used in the deflection calculations.
3. Compute the concrete fiber stresses at the top and bottom extreme fibers due to all loads.
4. Compute the initial strains ϵ_{ci} at the top and bottom fibers and the corresponding rotations, as well as subsequent strains and rotations. Use the equations

Table 7.2 ACI Minimum Permissible Ratios of Span (l) to Deflection (δ) (l = Longer Span)

Type of member	Deflection δ to be considered	$(l/\delta)_{min}$
Flat roofs not supporting and not attached to nonstructural elements likely to be damaged by large deflections	Immediate deflection due to live load L	180[a]
Floors not supporting and not attached to nonstructural elements likely to be damaged by large deflections	Immediate deflection due to live load L	360
Roof or floor construction supporting or attached to nonstructural elements likely to be damaged by large deflections	That part of total deflection occurring after attachment of nonstructural elements; sum of long-term deflection due to all sustained loads (dead load plus any sustained portion of live load) and immediate deflection due to any additional live load[b]	240[c]
Roof or floor construction supporting or attached to nonstructural elements not likely to be damaged by large deflections		

[a]Limit not intended to safeguard against ponding. Ponding should be checked by suitably calculating deflection, including added deflections due to ponded water, and considering long-term effects of all sustained loads, camber, construction tolerances, and reliability of provisions for drainage.

[b]Long-term deflection has to be determined, but may be reduced by the amount of deflection calculated to occur before attachment of nonstructural elements. This reduction is made on the basis of accepted engineering data relating to time-deflection characteristics of members similar to those being considered.

[c]Ratio limit may be lower if adequate measures are taken to prevent damage to supported or attached elements, but should not be lower than tolerance of nonstructural elements.

Table 7.3 AASHTO Maximum Permissible Deflection (l = Longer Span)

Type of member	Deflection considered	Maximum permissible deflection: Vehicular traffic only	Vehicular and pedestrian traffic
Simple or continuous spans	Instantaneous due to service live load plus impact	$\frac{l}{800}$	$\frac{l}{1000}$
Cantilever arms		$\frac{l}{300}$	$\frac{l}{375}$

$$\phi_i = \frac{\epsilon_{cbi} - \epsilon_{ci}^t}{h}$$

$$\phi_e = \frac{\epsilon_{cbe} - \epsilon_{cte}}{h}$$

$$\phi = \frac{\epsilon_c^t - \epsilon_{cb}}{h}$$

$$\phi_u = \frac{\epsilon_u}{c}$$

Also, compute the strains at the cgs line and compute the relaxation of the strands during the first time interval.

5. Compute the total change of stress in the prestressing steel due to creep, shrinkage, and relaxation acting as a force F at the cgs. Then compute the concrete fiber stresses at the cgs level due to F.
6. Add the result of step 5 to the result of step 3.
7. Repeat the same procedure for all time intervals, and add the effect of superimposed dead loads.
8. Add the deflections due to live load to get the total deflection δ_T.
9. Verify whether the computed δ_T is within the permissible limits. If not, change the section.

Figure 7.18 presents a flowchart for computation of deflection by the approximate time-step method.

7.9 LONG-TERM CAMBER AND DEFLECTION CALCULATION BY THE PCI MULTIPLIERS METHOD

Example 7.6

Given f_{pi} = 189,000 psi, evaluate the long-term camber and deflection of the bonded T-beam in Example 7.3 by the PCI multipliers method, and verify whether the deflection values satisfy the ACI permissible limits. If the beam were to be post-tensioned, assume that f_{pi} would be equal to 189,000 psi after anchorage losses and after eliminating frictional losses by jacking from both beam ends and then rejacking so as to maintain the net f_{pi} = 189,000 psi prior to erection. Also, assume that the nonstructural elements attached to the structure will not be damaged by deflections and that live load is transient. Use $E_c = 4.03 \times 10^6$ psi for all loads in the solution.

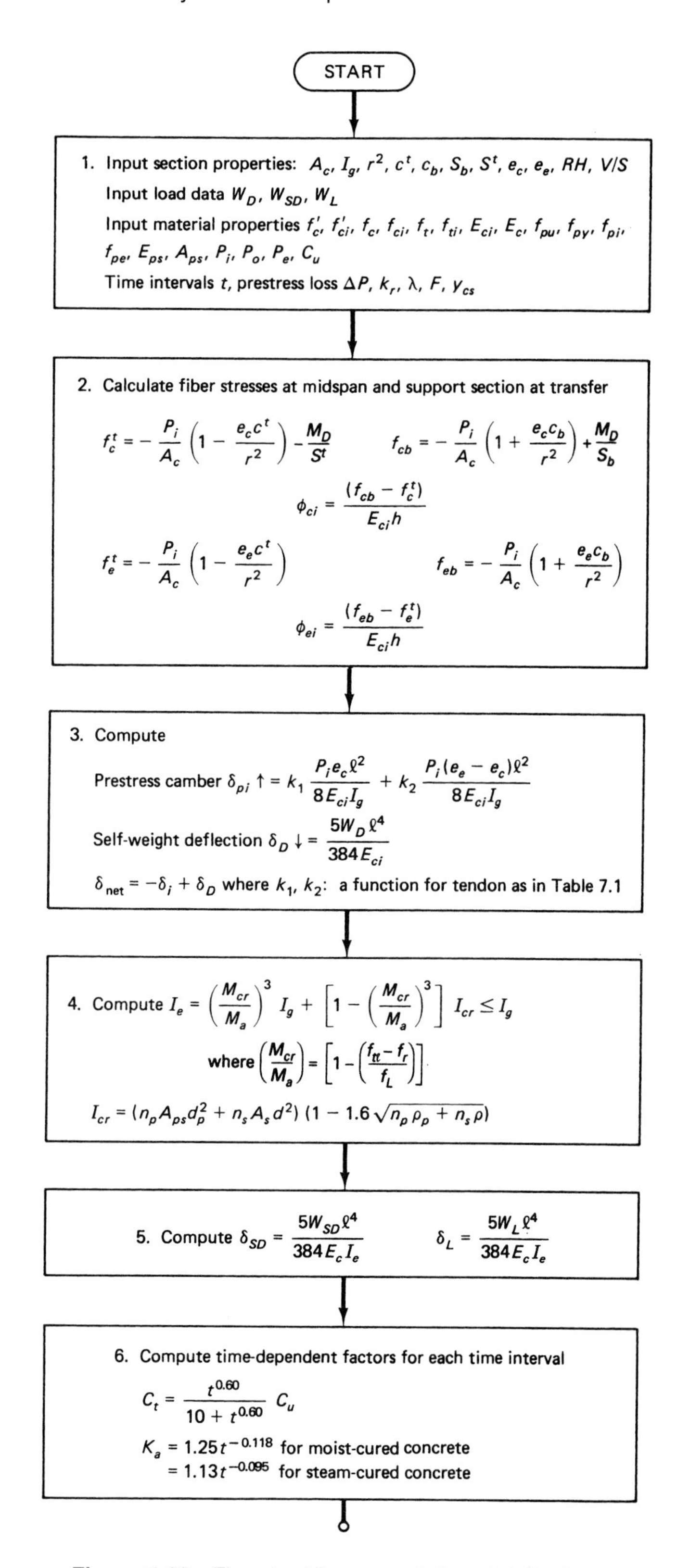

Figure 7.18 Flowchart for computation of deflection.

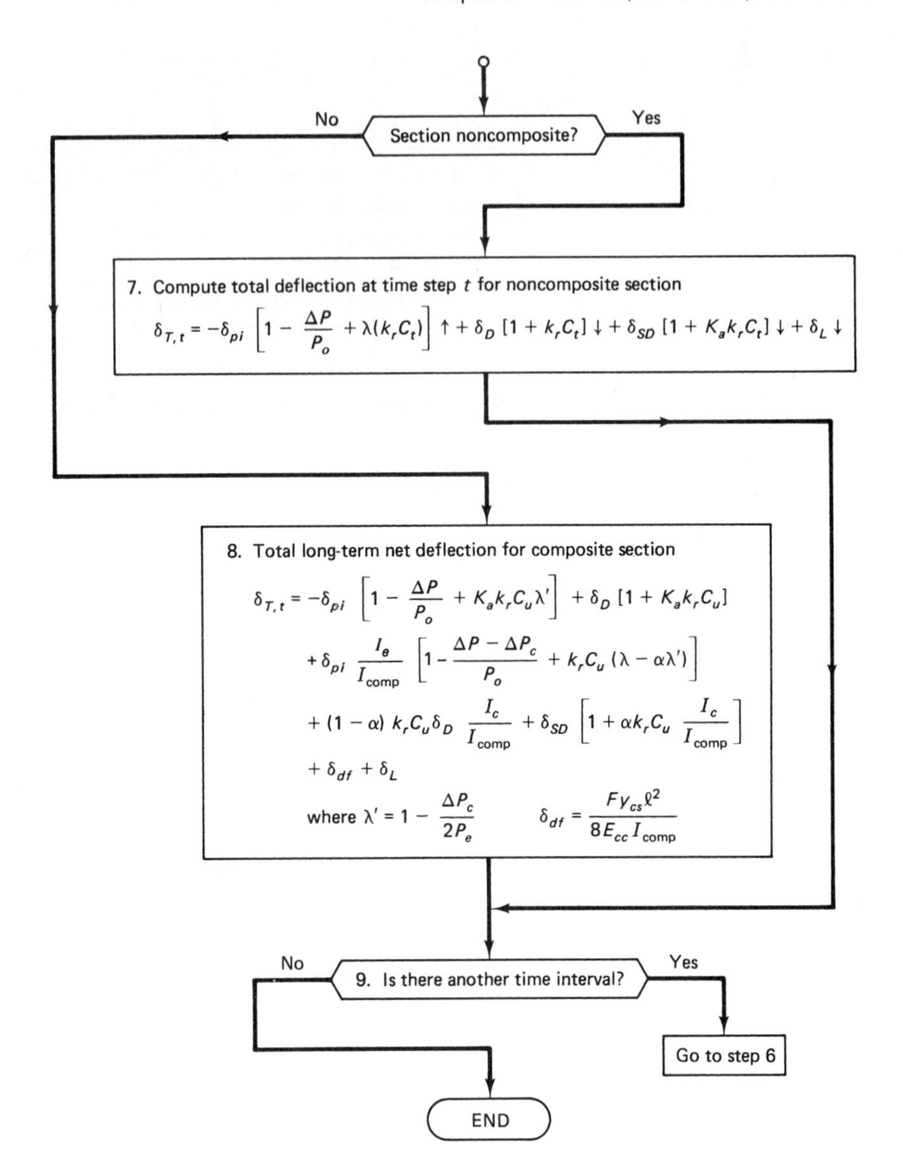

Figure 7.18 *Continued*

Solution:

$$I_g = 86{,}072 \text{ in.}^4$$

$$W_D = 1{,}019 \text{ plf} = 84.9 \text{ lb/in.}$$

$$\delta_D = \frac{5Wl^4}{384E_{ci}I_g} = \frac{5 \times 84.9\,(60 \times 12)^4}{384 \times 3.49 \times 10^6 \times 86{,}072} = 0.99 \text{ in.} \downarrow (14 \text{ mm})$$

$$W_{SD} = 100 \text{ plf} = 8.3 \text{ lb/in.}$$

$$\delta_{SD} = \frac{5 \times 8.3(60 \times 12)^4}{384 \times 4.03 \times 10^6 \times 86{,}072} = 0.08 \text{ in.} \downarrow (2.0 \text{ mm})$$

$$W_L = 1{,}100 \text{ plf} = 91.7 \text{ lb/in.}$$

the section did not crack (See Example 7.3)

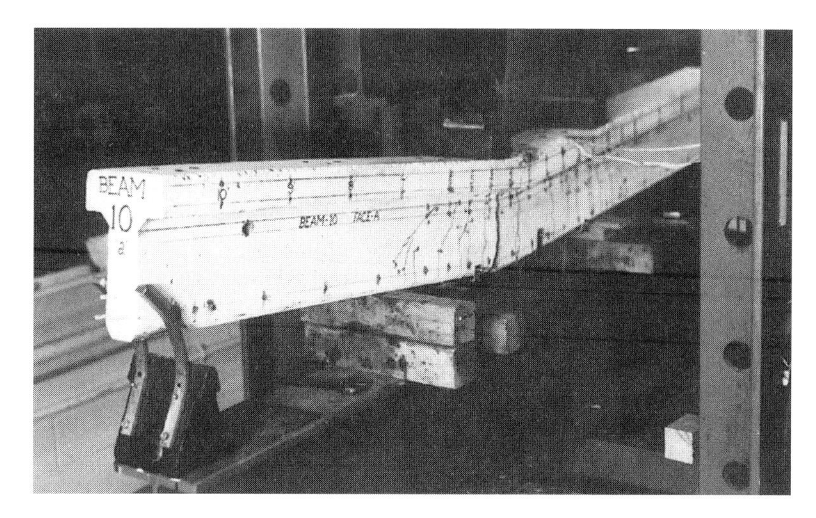

Photo 7.5 Typical deflection prior to limit state at failure (Nawy et al.).

$$I_e = I_g = 86{,}072 \text{ in.}^4 \text{ (max. } f_t < f_r = 530 \text{ psi}$$

$$\delta_L = \frac{5 \times 91.7\,(60 \times 12)^4}{384 \times 4.03 \times 10^6 \times 86{,}072} = 0.93 \text{ in.} \downarrow (24 \text{ mm})$$

If the section were cracked, the effective I_e would have had to be used instead of I_g. Using the PCI multipliers method for calculating deflection at construction erection time (30 days) and at the final service-load deflection (5 years), the following are the tabulated values of long-term deflection and camber obtained using the applicable PCI multipliers from Table 7.1. If the section is composite after erection, I_{comp} has to be used in calculating δ_L and δ_{SD} if the beam is shored during placement of concrete topping. If mild steel reinforcement A_s was also used in this prestressed beam, the reduced multiplier would be used. The C_1 multiplier is reduced by the factor C_2 where

$$C_2 = \frac{C_1 + A_s/A_{ps}}{1 + A_s/A_{ps}}$$

Table 7.4 Long-Term Camber and Deflection Values in Example 7.6 by PCI Multipliers Method

Load	Transfer δ_p	Erection δ_{er} (in.)			Final δ_{net} (in.)
		Multiplier		Multiplier (noncomposite)	
	(1)		(2)		(3)
Prestress	1.89in. ↑	1.80	3.40 in. ↑	2.45	4.63 in. ↑
W_D	0.99 in. ↓	1.85	1.82 in. ↓	2.70	2.67 in. ↓
Net δ	0.90 in. ↑		1.57 in. ↑		1.96 in. ↑
W_{SD}			0.08 in. ↑	3.00	0.24 in. ↓
Net δ			1.49 in. ↑		1.72 in. ↑
W_L					0.93 in. ↓
Final δ	0.90 in. ↑		1.49 in. ↑		0.79 in. ↑

due to the mild steel reinforcement controlling propagation or widening of the flexural cracks at long-term loading, hence enhancing stiffness. As an example, assume three No. 5 bars were also used in the prestressed beam,

$$\frac{A_s}{A_{ps}} = \frac{3 \times 0.31}{2.142} = 0.43 \qquad \text{giving } C_2 = 2.01$$

As an example of the adjustment of values previously tabulated, the value of the original camber becomes 3.80 in. ↑ instead of 4.63 in. ↑ shown in the table as a multiplier value of 2.01 is used instead of the previous 2.45 multiplier. Similar adjustment for all deflection components can be made applying the relevant correction factor.

From Table 7.4 the camber after erection and installation of the superimposed dead load at 30 days = 1.49 in. ↑ (38 mm). Also, the final net camber after 5 years = 0.79 in. ↑ (20 mm), the live-load deflection = 0.93 in. ↓ (24 mm), and the allowable deflection = $l/240$ = $(65 \times 12)/240 = 3.25$ in. (82.5 mm) > 0.79 in. camber if the live load is assumed all transient in this case, which is satisfactory.

7.10 LONG-TERM CAMBER AND DEFLECTION CALCULATION BY THE INCREMENTAL TIME-STEPS METHOD

Example 7.7

Solve Example 7.6 by the incremental time-steps method assuming that $f_{pi} = 189{,}000$ psi and that prestress losses are incrementally evaluated at prestressing (7 days after casting), 30 days after transfer (completion of erection and application of the superimposed dead load), 90 days, and 5 years. Assume that the ultimate creep coefficient $C_u = 2.35$ for the concrete and $f_{py} = 230{,}000$ psi for the prestressing steel used in the beam. Plot the camber-time and deflection-time relationships for the beam, using $E_c = 4.03 \times 10^6$ for all incremental steps in this solution, except at transfer, where $f'_{ci} = 3{,}750$ psi. Assume the beam to be post-tensioned.

Solution:

Instantaneous Stresses, Strains, and Deflections

$$E_{ci} = 57{,}000\sqrt{3{,}750} = 3.49 \times 10^6 \text{ psi}$$

From Example 7.3 and Figure 7.9, the initial fiber stresses (psi) and strains (in./in.) for the beam at transfer due to prestress force P_i and $P_i + W_D$ are as follows:

Prestress Force P_i

Midspan:

$$f^t = +501 \text{ psi (3.1 MPa)}$$

$$f_b = -3{,}524 \text{ psi (24.3 MPa)}$$

$$\epsilon_c^t = \frac{+501}{3.49 \times 10^6} = +144 \times 10^{-6} \text{ in./in.}$$

$$\epsilon_{cb} = -1{,}010 \times 10^{-6} \text{ in./in.}$$

Support:

$$f^t = +92 \text{ psi (0.7 MPa)}$$

$$f_b = -2{,}242 \text{ psi (15.5 MPa)}$$

$$\epsilon_e^t = +26 \times 10^{-6} \text{ in./in.}$$

$$\epsilon_{eb} = -642 \text{ in./in.}$$

Note that unless otherwise stated, the modulus E_c of concrete should be calculated for the time change at each incremental time stop.

Continuing, we have

$$\text{Midspan } \phi_{ci} = \frac{-1010 - 144}{34} \times 10^{-6} = -33.94 \times 10^{-6} \text{ rad/in.}$$

$$\text{Support } \phi_{ei} = \frac{-642 - 26}{34} \times 10^{-6} = -19.65 \times 10^{-6} \text{ rad/in.}$$

From Figure 7.6,

$$\delta_i \uparrow = \phi_c\left(\frac{l^2}{8}\right) + (\phi_e - \phi_c)\frac{l^2}{24}$$

$$\delta_i \uparrow = -33.94 \times 10^{-6}\frac{(60 \times 12)^2}{8} + (-19.65 + 33.94) \times 10^{-6} \times \frac{(60 \times 12)^2}{24}$$

$$= \frac{(60 \times 12)^2}{24} \times 10^{-6}(-33.94 \times 2 - 19.65)$$

$$= -1.89 \text{ in.} \uparrow (48 \text{ mm})$$

Notice that this value is the same as that obtained by the moment expression in Example 7.3.

Finally,

$$\text{Self-wt. } \delta_D = +\frac{5wl^4}{384E_cI_g} = \frac{5 \times \left(\frac{1019}{12}\right)(60 \times 12)^4}{384 \times 3.49 \times 10^6 \times 86{,}072} = +0.99 \text{ in.} \downarrow (25 \text{ mm})$$

$$\text{Net camber at transfer} = -1.89 \uparrow + 0.99 \downarrow = -0.90 \text{ in.} \uparrow (23 \text{ mm})$$

Time Dependent Factors

(a) *Creep.* From Equation 3.10,

$$\epsilon_{CR} = \frac{C_t}{E_c}(f_{cs}) = C_1\epsilon_{cs}$$

where f_{cs} = concrete stress at cgs level
ϵ_{cs} = concrete strain at cgs level
ϵ_{CR} = unit creep strain per unit stress at ultimate creep = C_u/E_c
= $2.35 / 4.03 \times 10^6 = 0.583 \times 10^{-6}$ in./in. per unit stress.

Note that creep strain has to be calculated at the centroid of the reinforcement in order to calculate the creep loss in prestress.
From Equation 3.9b, the creep coefficient at any time, in days, is

$$C_t = \frac{t^{0.60}}{10 + t^{0.60}} C_u$$

As an example, at 30 days after transfer

$$\epsilon'_{CR,t} = \epsilon'_{CR}\left(\frac{t^{0.60}}{10 + t^{0.60}}\right) = 0.583 \times 10^{-6}\left(\frac{30^{0.60}}{10 + 30^{0.60}}\right)$$

$$= 0.254 \times 10^{-6} \text{ in./in. per unit stress}$$

Creep strains at other time intervals are similarly computed.

(b) *Shrinkage of Concrete.* From Equation 3.15a for moist-cured concrete,

$$\epsilon_{SH,t} = \frac{t}{t + 35}\epsilon_{SH}$$

where $\epsilon_{SH} = 800 \times 10^{-6}$ in./in. for moist-cured concrete

Thirty days after transfer, the shrinkage time $t = 30$ days if the beam is posttensioned and $t = 30 + 7 = 37$ days if it is pretensioned. Hence,

$$\epsilon_{SH,30} = \frac{30}{30 + 35} \times 800 \times 10^{-6} = 369 \times 10^{-6} \text{ in./in.}$$

In a similar manner, ϵ_{SH} may be calculated for all other steps tabulated in Table 7.5.

(c) *Relaxation of Strands.* From Equation 3.6,

$$\frac{f_{pR}}{f_{pi}} = 1 - \left(\frac{\log t_2 - \log t_1}{10}\right)\left(\frac{f_{pi}}{f_{py}} - 0.55\right)$$

where log t, in hours, is to the base 10, f_{pi}/f_{py} exceeds 0.55, and f_{pR} is the remaining stress in the steel after relaxation for 30 days = 720 hr after prestressing. If the relaxation loss ratio is

$$R = 1 - \frac{f_{pR}}{f_{pi}} = \left(\frac{\log 720 - 0}{10}\right)\left(\frac{189{,}000}{230{,}000} - 0.55\right) = 0.078$$

we must find the R values for all time-steps using $t_I = 0$ as a base.

Table 7.5 gives the incremental time-dependent parameters for prestress loss factors in this example for time steps 7, 30, 90, and 365 days, and 5 years after prestressing.

Table 7.5 Time-Dependent Incremental Prestress Loss Factors in Example 7.7

Time	Creep $\times 10^{-6}$		Shrinkage $\times 10^{-6}$		Relaxation	
Days	$\epsilon'_{CR,t}$	$\Delta\epsilon'_{CR}$	$\epsilon_{SH,t}$	$\Delta\epsilon_{SH}$	R	ΔR
(1)	(2)	(3)	(4)	(5)	(6)	(7)
−7	—	—	0	—	—	—
P/S	0	—	133	133	0	—
30	0.254	0.254	369	236	0.0776	0.0776
90	0.349	0.095	576	207	0.0906	0.0130
365	0.452	0.103	730	154	0.1071	0.0165
5 yrs	0.525	0.073	785	55	0.1261	0.0190

Δ = Incremental increase. Columns 2, 4, and 6 give cumulative values.

Transfer to Erection (Step end = 30 days)

(a) *Concrete Fiber Stresses at the cgs Level for Calculation of Creep*

The tendon eccentricities are $e_c = 22.02$ in. and $e_e = 12.77$ in. Figure 7.19 gives the instantaneous stresses and the corresponding gross strains before losses due to prestress. From the figure,

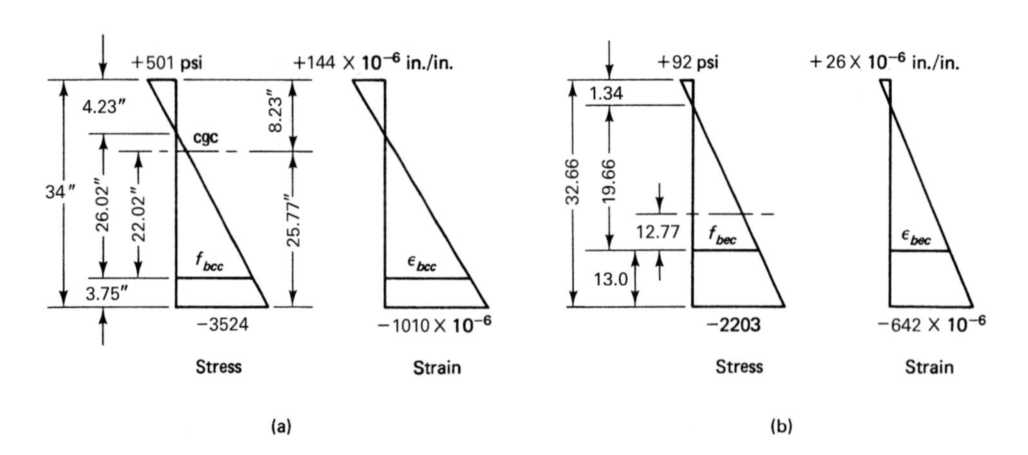

Figure 7.19 Stress and strain at transfer due only to prestress before losses in Example 7.7. (a) Midspan section. (b) Support section.

$$f_{bcc} = -3{,}524 \times \frac{26.02}{26.02 + 3.75} = -3{,}080 \text{ psi } (C)\ (21.2 \text{ MPa})$$

$$\epsilon_{bcc} = \frac{-3{,}080}{4.03 \times 10^6} = -764 \times 10^{-6} \text{ in./in.}$$

$$f_{bec} = -2{,}242 \times \frac{19.66}{19.66 + 13.00} = -1{,}350 \text{ psi } (C)\ (9.3 \text{ MPa})$$

$$\epsilon_{bec} = \frac{-1{,}350}{4.03 \times 10^6} = -335 \times 10^{-6} \text{ in./in.}$$

Creep Incremental Strain

$$\Delta\epsilon_{CR} = \Delta\epsilon'_{CR} \times \text{stress } f \text{ at cgs}$$

From Table 7.5 $\Delta\epsilon'_{CR} = 0.254 \times 10^{-6}$ in./in. per unit stress. So

$$\text{Midspan } \Delta\epsilon_{CR} = \Delta\epsilon'_{CR} \times f_{bcc} = 0.254 \times 10^{-6}(-3{,}080) = -782 \times 10^{-6} \text{ in./in.}$$

$$\text{Support } \Delta\epsilon_{CR} = \Delta\epsilon'_{CR} \times f_{bec} = 0.254 \times 10^{-6}(-1{,}350) = -343 \times 10^{-6} \text{ in./in.}$$

Shrinkage Incremental Strain

$$\Delta\epsilon_{SH} = -236 \times 10^{-6} \text{ in./in.}$$

Relaxation Stress Loss

$$\Delta f_{R30} = 0.0776 \times 189{,}000 = 14{,}666 \text{ psi } (101.0 \text{ MPa})$$

Total Steel Stress Loss

$$\Delta f_T = (\Delta\epsilon_{CR} + \Delta\epsilon_{SH})E_{ps} + \Delta f_R$$

$$\text{Midspan } \Delta f_{T30} = (782 + 236) \times 10^{-6} \times 27.5 \times 10^6 + 14{,}666 = 42{,}661 \text{ psi}$$

$$\text{Support } \Delta f_{T30} = (343 + 236) \times 10^{-6} \times 27.5 \times 10^6 + 14{,}666 = 30{,}589 \text{ psi}$$

Hence, use an average $\Delta f_{T30} = \frac{1}{2}(42{,}661 + 30{,}589) = 36{,}625$ psi (253 MPa).

(b) *Corresponding Change in Concrete Fiber Stresses and Strains* Prestress force loss $\Delta P_{30} = \Delta f_{T30} A_{ps} = 36{,}625 \times 2.448 = 89{,}658$ lb. (399 kN)

(i) *Midspan Section* (1 psi = 6.895×10^{-3} MPa)

$$\Delta f^t = -\frac{\Delta P_{30}}{A_c}\left(1 - \frac{ec_t}{r^2}\right) = +\frac{89{,}658}{978}\left(1 - \frac{22.02 \times 8.23}{88.0}\right)$$

$$= +97 \text{ psi } (T)$$

$$\Delta\epsilon_c^t = \frac{+97}{4.03 \times 10^6} = +24 \times 10^{-6} \text{ in./in.}$$

$$\Delta f_b = -\frac{\Delta P_{30}}{A_c}\left(1 + \frac{ec_b}{r^2}\right) = -\frac{89{,}658}{978}\left(1 + \frac{22.02 \times 25.77}{88.0}\right) = -683 \text{ psi } (C)$$

$$\Delta\epsilon_{cb} = \frac{-683}{4.03 \times 10^6} = -169 \times 10^{-6} \text{ in./in.}$$

(ii) Support Section

$$\Delta f^t = -\frac{89{,}658}{978}\left(1 - \frac{12.77 \times 8.23}{88.0}\right) = +18 \text{ psi } (T)$$

$$\Delta\epsilon_e^t = \frac{+18}{4.03 \times 10^6} = +4 \times 10^{-6} \text{ in./in.}$$

$$\Delta f_b = -\frac{89{,}658}{978}\left(1 + \frac{12.77 \times 25.77}{88.0}\right) = -435 \text{ psi } (C)$$

$$\Delta\epsilon_{eb} = \frac{-435}{4.03 \times 10^6} = -108 \times 10^{-6} \text{ in./in.}$$

(c) *Net Strains, Resulting Curvatures, and Camber*
Net creep strain (in./in.)

(i) *Fiber gross strain*
Midspan

$$\Delta\epsilon^t_{CRc} = f^t_7 \times \Delta\epsilon'_{CR} = +501 \times 0.254 \times 10^{-6} = +127 \times 10^{-6} \text{ in./in.}$$

$$\Delta\epsilon_{CRcb} = f_{7b} \times \Delta\epsilon'_{CR} = -3{,}524 \times 0.254 \times 10^{-6} = -895 \times 10^{-6} \text{ in./in.}$$

Support

$$\Delta\epsilon^t_{C\,\text{Re}} = +92 \times 0.254 \times 10^{-6} = +23 \times 10^{-6} \text{ in./in.}$$

$$\Delta\epsilon_{C\,\text{Re}b} = -2{,}242 \times 0.254 \times 10^{-6} = -569 \times 10^{-6} \text{ in./in.}$$

Net strains (in./in.)

$$\Delta_{\text{net}}\,\epsilon_{CR} = \Delta\epsilon_{CR} - \Delta\epsilon_{ps}$$

where $\Delta\epsilon_{ps}$ is the strain loss due to prestress loss Δf in part (b) of the solution. From Figure 7.20, we have the following:
Midspan

$$\Delta\epsilon^t_{CRc,\text{net}} = \Delta\epsilon^t_{CRc} - \Delta\epsilon^t_{psc} = (+127 - 24)10^{-6} \text{ in./in.} = +103 \times 10^{-6}$$

$$\Delta\epsilon^t_{C\,\text{Re}\,b,\text{net}} = \Delta\epsilon_{CRcb} - \Delta\epsilon_{pscb} = (-895 + 169)10^{-6} \text{ in./in.} = -726 \times 10^{-6}$$

Support

$$\Delta\epsilon^t_{C\,\text{Re}\,b,\text{net}} = \Delta\epsilon^t_{C\,\text{Re}} - \Delta\epsilon^t_{pse} = (+23 - 4)10^{-6} \text{ in./in.} = +19 \times 10^{-6}$$

$$\Delta\epsilon_{C\,\text{Re}\,b,\text{net}} = \Delta\epsilon_{C\,\text{Re}b} - \Delta\epsilon_{pseb} = (-569 + 108)10^{-6} \text{ in./in.} = -461 \times 10^{-6}$$

(ii) *Curvatures (rad/in.)*
$\Delta\phi_{30}$ is the added curvature due to losses at the end of 30 days after transfer based on the adjusted net strains, in other words the curvature increment for this step.
Midspan

$$\Delta\phi_{c30} = \frac{\Delta\epsilon_{cRcb,\text{net}} - \Delta\epsilon^t_{cRc,\text{net}}}{h} = \frac{(-726 - 103)10^{-6}}{34}$$

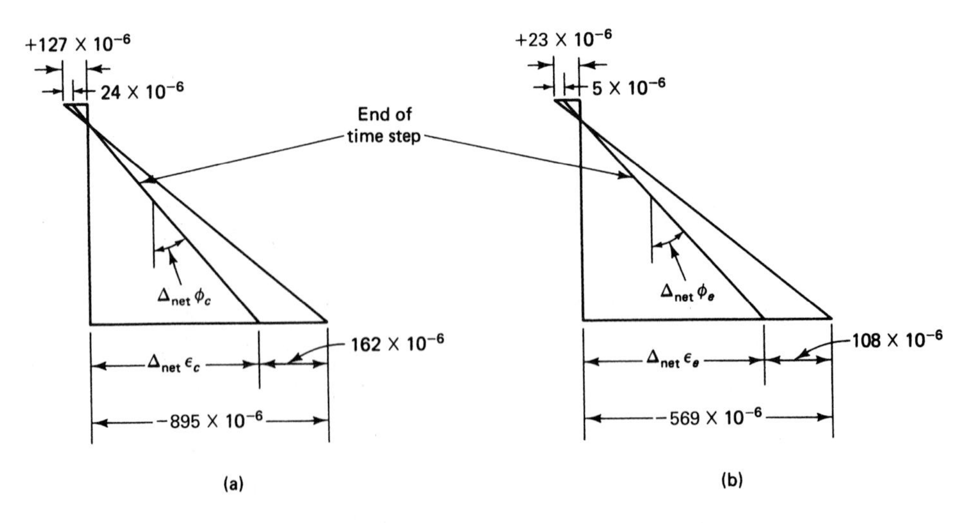

Figure 7.20 Creep incremental strains at 30 days in Example 7.7.

$$= -24.38 \times 10^{-6}\ \text{rad/in.}\ (0.96 \times 10^{-6}\ \text{rad/mm})$$

Support

$$\Delta\phi_{e30} = \frac{\Delta\epsilon_{cRe\,b,\text{net}} - \Delta\epsilon^{t}_{cRe,\text{net}}}{h} = \frac{(-461 - 19)10^{-6}}{34}$$

$$= -14.12 \times 10^{-6}\ \text{rad/in.}\ (0.54 \times 10^{-6}\ \text{rad/mm})$$

(iii) *Total Curvature (rad/in.) and camber*
From before, $\phi_{ci} = -33.94 \times 10^{-6}$ rad/in. and $\phi_{ei} = -19.65 \times 10^{-6}$ rad/in. So the total curvature at 30 days after transfer is

$$\phi_T = \phi_i + \Delta\phi_{30}$$

Midspan

$$\phi_{Tc30} = (-33.94 - 24.38) \times 10^{-6}\ \text{rad/in.} = -58.32 \times 10^{-6}(2.30 \times 10^{-6}\ \text{rad/mm})$$

Support

$$\phi_{Te30} = (-19.65 - 14.12) \times 10^{-6} = -33.77 \times 10^{-6}(1.33 \times 10^{-6}\ \text{rad/mm})$$

From Figure 7.6, the camber due to prestress at the end of 30 days for singly harped tendon beam is

$$\delta_{P30}\uparrow = \phi_{Tc30}\left(\frac{l^2}{8}\right) + (\phi_{Te30} - \phi_{Tc30})\frac{l^2}{24}$$

$$= -58.32 \times 10^{-6}\frac{(60 \times 12)^2}{8} + (-33.77 + 58.32)10^{-6} \times \frac{(60 \times 12)^2}{24}$$

namely,

$$\delta_{30(\text{camb.})}\uparrow = \frac{(60 \times 12)^2}{24}(-58.32 \times 2 + 33.77) \times 10^{-6}$$

$$= -3.24\ \text{in.}\uparrow (82\ \text{mm})$$

(d) *Long-term Deflections Due to Gravity Loads at 30 Days after Transfer.* Assume $E_c = 4.03 \times 10^{-6}$ psi as a reasonable value for the modulus of concrete for the rest of the example. We have $W_D = 1019$ plf $= 84.9$ lb/in. Also,

$$\text{Self-weight deflection } \delta_D = \frac{5Wl^4}{384E_cI_g} = +0.99\ \text{in.}\downarrow (25\ \text{mm})\ \text{from before}$$

$$W_{SD} = 100\ \text{plf}$$

$$\delta_{SD} = \frac{5 \times \frac{100}{12}(60 \times 12)^4}{384 \times 4.03 \times 10^6 \times 169{,}020}$$

$$= +0.08\ \text{in.}\downarrow (1.5\ \text{mm})$$

$$C_t = \frac{t^{0.60}}{10 + t^{0.60}}C_u = \frac{30^{0.60}}{10 + 30^{0.60}} \times 2.35 = 1.02$$

(C_t for W_{SD} at ~ 15 days = 0.80)

$$\delta_{D30} = 0.99(1 + 1.02) = +2.00\ \text{in.}\downarrow (51\ \text{mm})$$

$$\delta_{SD30} = 0.08(1 + 0.80) = +0.14\ \text{in.}\downarrow (4\ \text{mm})$$

$$\delta_L = 0\ \text{(building occupied at 90 days)}$$

$$\text{Total gravity load deflections} = 2.00 + 0.14 + 0 = +2.14\ \text{in.}\downarrow (31\ \text{mm})$$

$$\delta_{\text{net}30} = -3.24\uparrow + 2.14\downarrow = -1.10\ \text{in.}\uparrow (28\ \text{mm})\ \text{(camber)}$$

Service-Load Step—90 Days after Transfer

(a) *New Reduced Concrete Fiber Stresses and Strains Due to Prestress Losses in the Previous Stage*

Prestressing force change = ΔP_e

(i) *Midspan*

$$f^t = +501\text{-}97 = +404 \text{ psi}$$

$$f_b = -3{,}524 + 683 = -2{,}841 \text{ psi}$$

$$\epsilon_c^t = \frac{+404}{4.03 \times 10^6} = +100 \times 10^{-6} \text{ in./in.}$$

$$\epsilon_{cb} = -705 \times 10^{-6} \text{ in./in.}$$

(ii) *Support*

$$f^t = +92 - 18 = +74 \text{ psi}$$

$$f_b = -2{,}242 + 435 = -1{,}807 \text{ psi}$$

$$\epsilon_e^t = +18 \times 10^{-6} \text{ in./in.}$$

$$\epsilon_{eb} = -448 \times 10^{-6} \text{ in./in.}$$

From Figure 7.21,

$$f_{bcc} = -2{,}841 \times \frac{26.02}{26.02 + 3.75} = -2{,}483 \text{ psi}$$

$$\epsilon_{bcc} = -616 \times 10^{-6} \text{ in./in.}$$

$$f_{bec} = -1{,}807 \times \frac{19.66}{19.66 + 13.0} = -1{,}088 \text{ psi}$$

$$\epsilon_{bec} = -270 \times 10^{-6} \text{ in./in.}$$

Creep Incremental Strain

$$\Delta\epsilon'_{CR} = 0.095 \times 10^{-6} \text{ in./in. per unit stress (Table 7.5)}$$

$$\text{Midspan } \Delta\epsilon_{CR} = \Delta\epsilon'_{CR} f_{bcc} = 0.095 \times 10^{-6} (-2{,}483) = -236 \times 10^{-6} \text{ in./in.}$$

$$\text{Support } \Delta\epsilon_{CR} = \Delta\epsilon'_{CR} f_{bec} = 0.095 \times 10^{-6} (-1{,}088) = -103 \times 10^{-6} \text{ in./in.}$$

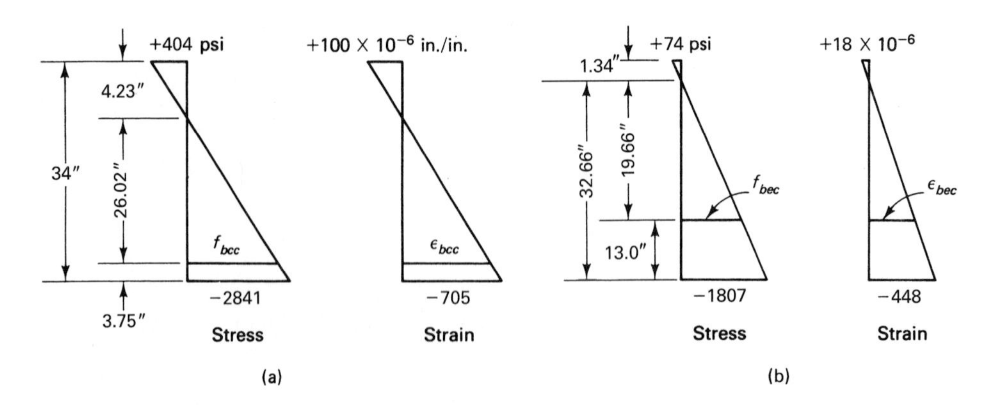

Figure 7.21 Adjusted stress and strain at 30 days due to prestress only in Example 7.7. (a) Midspan section. (b) Support section.

Shrinkage Incremental Strain

$$\Delta\epsilon_{SH} = -207 \times 10^{-6} \text{ in./in.}$$

Relaxation Stress Loss

$$\Delta f_R = 0.0130(189{,}000 - 36{,}625) = 1{,}981 \text{ psi}$$

Total Steel Stress Loss

$$\Delta f_T = (\Delta\epsilon_{CR} + \Delta E_{SH})E_{ps} + \Delta f_R$$

$$\text{Midspan } \Delta f_{T90} = (236 + 207) \times 10^{-6} \times 27.5 \times 10^6 + 1{,}981 = 14{,}164 \text{ psi}$$

$$\text{Support } \Delta f_{T90} = (103 + 207) \times 10^{-6} \times 27.5 \times 10^6 + 1{,}981 = 10{,}506 \text{ psi}$$

Hence, use an average $\Delta f_{T90} = \frac{1}{2}(14{,}164 + 10{,}506) = 12{,}335$ psi

(b) *Correspnding Change in Concrete Fiber Stresses and Strains.* Prestress force loss $\Delta P_{90} = \Delta f_{T90} A_{ps} = 12{,}335 \times 2.448 = 30{,}196$ lb (134 kN)

(i) *Midspan Section* (1 psi = 6.895×10^{-3} MPa)

$$\Delta f^t = +97 \times \frac{30{,}196}{89{,}658} = +33 \text{ psi}$$

$$\Delta f_b = -230 \text{ psi}$$

$$\Delta\epsilon_c^t = +8 \times 10^{-6} \text{ in./in.}$$

$$\Delta\epsilon_{cb} = -57 \times 10^{-6} \text{ in./in.}$$

(ii) Support Section

$$\Delta f^t = +6 \text{ psi}$$

$$\Delta f_b = -147 \text{ psi}$$

$$\Delta\epsilon_e^t = +2 \times 10^{-6} \text{ in./in.}$$

$$\Delta\epsilon_{eb} = -36 \times 10^{-6} \text{ in./in.}$$

(c) *Net Strains, Resulting Curvatures, and Camber*

(i) *Net creep strain (in./in.)*
Fiber gross strain
Midspan

$$\Delta\epsilon_{CRc}^t = f_{30}^t \times \Delta\epsilon_{CR}' = +404 \times 0.095 \times 10^{-6} = +38 \times 10^{-6} \text{ in./in.}$$

$$\Delta\epsilon_{CRcb} = f_{30b} \times \Delta\epsilon_{CR}' = -2{,}841 \times 0.095 \times 10^{-6} = -270 \times 10^{-6} \text{ in./in.}$$

Support

$$\Delta\epsilon_{C\,\mathrm{Re}}^t = +74 \times 0.095 \times 10^{-6} = +7 \times 10^{-6} \text{ in./in.}$$

$$\Delta\epsilon_{C\,\mathrm{Re}b} = -1{,}807 \times 0.095 \times 10^{-6} = -172 \times 10^{-6} \text{ in./in.}$$

Net strains (in./in.)
Midspan

$$\Delta\epsilon_{CRc,\mathrm{net}}^t = \Delta\epsilon_{CRc}^t - \Delta\epsilon_{psc}^t = (+38 - 8) \times 10^{-6} = +30 \times 10^{-6}$$

$$\Delta\epsilon_{CRcb,\mathrm{net}} = \Delta\epsilon_{CRcb} - \Delta\epsilon_{pscb} = (-270 + 57) \times 10^{-6} = -213 \times 10^{-6}$$

Support

$$\Delta\epsilon_{C\,\mathrm{Re,net}}^t = \Delta\epsilon_{C\,\mathrm{Re}}^t - \Delta\epsilon_{pse}^t = (+7 - 2) \times 10^{-6} = +5 \times 10^{-6}$$

$$\Delta\epsilon_{C\,\mathrm{Re}\,b,\,\mathrm{net}} = \Delta\epsilon_{C\,\mathrm{Re}\,b} = \Delta\epsilon_{pseb} = (-172 + 36) \times 10^{-6} = -136 \times 10^{-6}$$

(ii) *Curvature (rad/in.)*
Midspan

$$\Delta\phi_{c90} = \frac{\Delta\epsilon_{Crc,\text{net}} - \Delta\epsilon^t_{cRcb,\text{net}}}{h} = \frac{(-213 - 30) \times 10^{-6}}{34}$$

$$= -7.15 \times 10^{-6} \text{ rad/in.}$$

Support

$$\Delta\phi_{e90} = \frac{\Delta\epsilon_{C\,\text{Reb,net}} - \Delta\epsilon^t_{C\,\text{Re,net}}}{h} = \frac{(-136 - 5)10^{-6}}{34}$$

$$= -4.15 \times 10^{-6} \text{ rad/in.}$$

(iii) *Total Curvature (rad/in.) and camber*
From before, $\phi_{c30} = -58.32 \times 10^{-6}$ rad/in. and $\phi_{e30} = -33.77 \times 10^{-6}$. So the total curvature is

$$\phi_T = \Delta\phi_{30} + \Delta\phi_{90}$$

and we also have the following:

$$\textit{Midspan } \phi_{Tc90} = (-58.32 - 7.15) \times 10^{-6} = -65.47 \times 10^{-6} \text{ (1.72 rad/mm)}$$

$$\textit{Support } \phi_{Te90} = (-33.77 - 4.15) \times 10^{-6} = -37.92 \times 10^{-6} \text{ (1.04 rad/mm)}$$

$$\delta_{P90} \text{ (camber)} \uparrow = \phi_{Tc90}\left(\frac{l^2}{8}\right) + (\phi_{Te90} - \phi_{Tc90})\frac{l^2}{24}$$

$$= -65.47 \times 10^{-6}\frac{(60 \times 12)^2}{8} + (-37.92 + 65.47) \times 10^{-6} \times \frac{(60 \times 12)^2}{24}$$

$$= \frac{(60 \times 12)^2}{24}(-65.47 \times 2 + 37.92) \times 10^{-6}$$

$$= -3.65 \text{ in.} \uparrow \text{(93 mm)}$$

(d) *Long-term Deflections Due to Gravity Loads at 90 Days after Transfer*

$$t = 90 \text{ days}$$

$$C_t = \frac{t^{0.60}}{10 + t^{0.60}} C_u = \frac{90^{0.60}}{10 + 90^{0.60}} \times 2.35 = 1.41$$

$$(C_t = 1.27 \text{ for } t = 60 \text{ days})$$

$$\delta_{D90} = \delta_i(1 + C_t) = 0.99(1 + 1.41) = +2.39 \text{ in.} \downarrow \text{(51 mm)}$$

$$\delta_{SD90} = 0.08(1 + 1.27) = +0.17 \text{ in.} \downarrow$$

δ_L (from Example 7.6) = +0.93 in. ↓
So the total deflection at 90 days due to gravity loads is

$$\delta_g = +2.39 + 0.18 + 0.93 = +3.50 \text{ in.} \downarrow \text{(89 mm)}$$

and the net deflection is

$$\text{Net } \delta_{\text{net},90} = -3.65 \uparrow +3.5 \downarrow = -0.51 \text{ in.} \uparrow \text{(4 mm) (camber)}$$

Service-Load Deflection at 5 Years
The same steps as the previous give the results tabulated in Tables 7.6 and 7.7, while a plot of the cambers and deflections as a function of time is shown in Figure 7.22. The deflection-time relationship becomes almost asymptotic.

From Table 7.7, the final net deflection (camber) at five years is 0.87 in., which is much less than the maximum allowable deflection or camber. From Table 7.2, $\delta_T = l/240 = 60 \times 12 /$

Table 7.6 Long-Term Stress and Strain Changes in Example 7.7

Time at step end	Gross fiber stress in concrete at start of time increment		Creep strain increment $\times 10^{-6}$ in./in. per unit stress	Shrinkage strain increment $\times 10^{-6}$ in./in.	Steel relaxation increment psi	Total steel stress loss increment psi	Concrete stress change due to losses psi		Concrete strain change due to losses $\times 10^{-6}$ in./in.	
	Midspan	End					Midspan	End	Midspan	End
Days	f^t / f_b	f^t / f_b	$\Delta\epsilon_{CRc}$ / $\Delta\epsilon_{CRb}$	$\Delta\epsilon_{SH}$	Δf_R	$\Delta f_{TR,t}$	Δf^t / Δf_b	Δf^t / Δf_b	$\Delta\epsilon^t_c$ / $\Delta\epsilon_{cb}$	$\Delta\epsilon^t_e$ / $\Delta\epsilon_{cb}$
(1)	(2)	(3)	(4)	(5)	(6)	(7)	(8)	(9)	(10)	(11)
P/S*	0	0	0	133	—	—	—	—	—	—
30	+501 −3524	+92 −2242	−782 −343	236	14,666	36,625	+97 −683	+18 −435	+24 −169	+4 −108
90	+404 −2841	+74 −1807	−236 −103	207	1,981	12,335	+33 −230	+6 −147	+8 −57	+2 −36
365	+371 −2611	+68 −1660	−235 −103	154	2,311	11,194	+30 −209	+6 −133	+7 −52	+2 −33
5 yr	+341 −2402	+62 −1527	−153 −67	55	2,448	6,986	+19 −130	+4 −83	+5 −32	+1 −21

*P/S = step at transfer of prestress at 7 days after concrete is cast; 1,000 psi = 6,895 MPa.

Table 7.7 Long-Term Curvatures and Deflections in Example 7.7

Time at step end	Net creep strain increment × 10^{-6} in./in.		Curvature increment × 10^{-6} rad/in.		Total cumulative curvature × 10^{-6} rad/in.		Deflection		
	Midspan	End	Midspan	End	Midspan	End	Prestress camber force P in.	Gravity loads $W_D + W_{SD} + W_L$ in.	Net in.
Days	$\Delta\epsilon^t_{CR,net}$ / $\Delta\epsilon_{CRb,net}$	$\Delta\epsilon^t_{CR,net}$ / $\Delta\epsilon_{CRb,net}$	$\Delta\phi_c$	$\Delta\phi_e$	$\Sigma\phi_c$	$\Sigma\phi_e$	$\delta_{p,t}$	δ_g	$\delta_{net,t}$
(1)	(2)	(3)	(4)	(5)	(6)	(7)	(8)	(9)	(10)
P/S*	0	0	−33.94	−19.65	−33.94	−19.65	−1.89	+0.99	−0.90↑
30	+103 −726	+19 −461	−24.38	−14.12	−58.92	−34.07	−3.28	+2.14	−1.14↑
90	+38 −270	+7 −172	−7.15	−4.15	−66.07	−38.22	−3.68	+3.50	−0.16↑
365	+31 −217	+5 −138	−7.29	−4.21	−73.36	−42.43	−4.09	+3.93	−0.16↑
5 yr	+20 −143	+4 −90	−4.79	−2.76	−78.15	45.19	−4.35	+4.26	−0.09↑

*P/S = step at transfer of prestress at 7 days after concrete is cast; 1 in. = 25.4 mm.

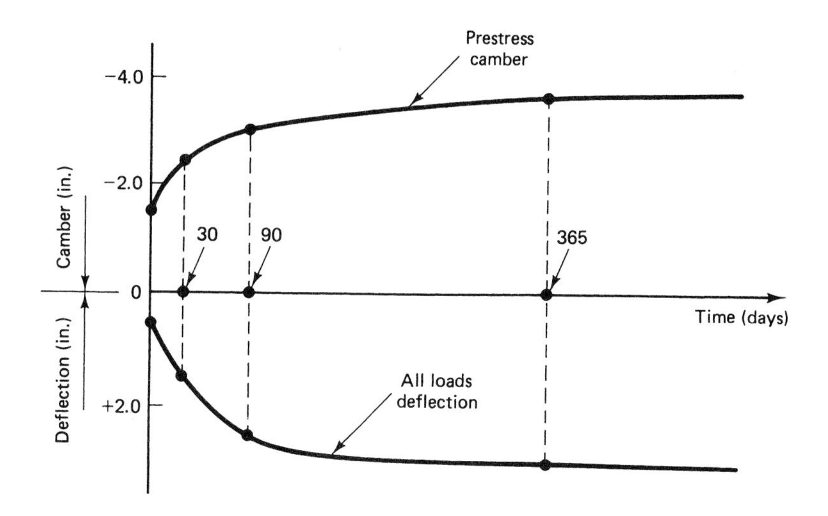

Figure 7.22 Prestressed camber and load deflections vs. time in Example 7.7.

240 = 3.09 in. >> 0.09 in. Hence, the beam satisfies the serviceability requirements for time-dependent deflection. Note that long-term creep losses can be considerably reduced by the addition of nonprestressed reinforcement to the section at the compression side.

7.11 LONG-TERM CAMBER AND DEFLECTION CALCULATION BY THE APPROXIMATE TIME-STEPS METHOD

Example 7.8

Solve Example 7.6 by the approximate time-steps method using the same allowable steel and concrete stresses. Compare this solution with those of Examples 7.6 and 7.7.

Photo 7.6 Deflection at failure of prestressed T-beam with confining reinforcement (Nawy, Potyondy).

Solution:

Data for This Alternative Solution. From Example 7.7,

$$P_i = 462{,}672 \text{ lb}$$

$$P_e = 379{,}391 \text{ lb}$$

$$A_{ps} = 2.448 \text{ in.}^2$$

$$C_u = 2.35$$

$$C_{t30} = 1.02$$

$$C_{t90} = 1.41$$

$$C_{t365} = 1.82$$

$$C_{t5yr} = 2.12$$

Use the same C_t value for δ_{SD} as for δ_D, and consider it accurate enough. Then

$$K_a = 1.25t^{-0.118} \text{ for moist-cured concrete}$$

$$t = \text{age at loading, in days} = 30, 90, 356, 5 \text{ yr}$$

$$k_r = \frac{1}{1 + A_s/A_{ps}}$$

where $A_s/A_{ps} << 1.0$ under normal conditions

Use $k_r \cong 1$ as accurate enough for practical purposes, since usually $A_s/A_{ps} << 1$.

Instantaneous Camber and Deflections. From Example 7.3,

$$\delta_{i(P_i)} = \delta_{pi} = \text{instanteous intital prestress camber } = -1.89 \text{ in.} \uparrow$$

$$\delta_{i(D)} = \text{instanteous dead-load deflection} = +0.99 \text{ in.} \downarrow$$

$$\delta_{i(SD)} = \text{instantaneous superimposed dead-load deflection} = +0.08 \text{ in.} \downarrow$$

$$\delta_L = \text{live-load deflection} = +0.93 \text{ in.} \downarrow$$

From Equation 7.26, the incremental time-step *net* deflection is

$$\Delta\delta_T = -\left[\eta + \left(\frac{1+\eta}{2}\right)k_r C_u\right]\delta_{i(P_i)} \uparrow + k_r C_t \delta_{i(D)} \downarrow + K_a k_r C_t \delta_{i(SD)} \downarrow$$

δ_p is the deflection (camber) due to prestressing = $\delta_{i(P_i)}$ and $\eta = P_e/P_i$. From Equation 7.27, the total *net* deflection due to loads is

$$\delta_T = -\delta_p\left[1 - \frac{\Delta P}{P_o} + \lambda(k_r C_t)\right] \uparrow + \delta_D[1 + k_r C_t] \downarrow + \delta_{SD}[1 + K_a k_r C_t] \downarrow + \delta_L \downarrow$$

where $\Delta P = (P_o - P_e)$ is the total loss of prestress excluding any initial elastic loss, $\lambda = 1 - \Delta P/2P_o$, and δ_p and η are as before.

Transfer to Erection (30 days). Assume $P_o \cong P_i$. Then $\Delta P = P_i - P_e = 462{,}672 - 379{,}391 = 83{,}281$ lb, and $\Delta P/P_i = 83{,}281/462{,}672 = 0.18$. So

$$\lambda = 1 - \frac{83{,}281}{2 \times 462{,}672} = 0.91$$

$$k_r = 1$$

$$K_a = 1.25(30)^{-0.118} = 0.84$$

$$C_t = 1.02$$

$$\delta_{T30} = -1.89(1 - 0.81 + 0.91 \times 1.02)\uparrow +0.99(1 + 1.02)\downarrow$$
$$+ 0.08(1 + 0.84 \times 1.02)\downarrow +0$$
$$= -3.30\uparrow + 2.15\downarrow = -1.15 \text{ in.}\uparrow (29 \text{ mm})$$

Erection to Service-Load Deflection (90 days). The total interval from transfer is $t = 90$ days. So we have

$$K_a = 1.25(90)^{-0.118} = 0.74$$
$$C_t = 1.41$$
$$\delta_L = +0.93 \text{ in.}\downarrow$$
$$\delta_{T90} = -1.89(1 - 0.18 + 0.91 \times 1.41)\uparrow +0.99(1 + 1.41)\downarrow$$
$$+0.08(1 + 0.74 \times 1.41)\downarrow +0.93\downarrow$$
$$= -3.97\uparrow + 3.48\downarrow = -0.49 \text{ in.}\uparrow (12 \text{ mm})$$

Service-Load Deflection at 365 days

$$K_a = 1.25(365)^{-0.118} = 0.62$$
$$C_t = 1.82$$
$$\delta_{T365} = -1.89(1 - 0.18 + 0.91 \times 1.82)\uparrow +0.99(1 + 1.82)\downarrow$$
$$+ 0.08(1 + 0.62 \times 1.82)\downarrow +0.93\downarrow$$
$$= -4.68\uparrow + 3.89\downarrow = -0.79 \text{ in.}\uparrow (20 \text{ mm})$$

Service-Load Deflections at 5 Years

$$K_a = 1.25(1{,}825)^{-0.118} = 0.52$$
$$C_t = 2.12$$
$$\delta_{T5yr} = -1.89(1 - 0.18 + 0.91 \times 2.12)\uparrow +0.99(1 + 2.12)\downarrow$$
$$+ 0.08(1 + 0.52 \times 2.12)\downarrow + 0.93\downarrow$$
$$= -5.20\uparrow + 4.19\downarrow = -1.01 \text{ in.}\uparrow (26 \text{ mm})$$

Comparison of Deflection Calculations by the Three Methods. Table 7.8 gives the calculated values of camber and deflection using the three methods in Examples 7.6, 7.7, and 7.8. The minus sign (–) indicates upward camber ↑, and the plus sign (+) indicates downward

Table 7.8 Camber and Deflection Comparisons (δ Inch)

Time at step end, days	Methods								
	PCI multipliers			Incremental time-step			Approximate time-step		
	Camber	δ_g	δ_{net}	Camber	δ_g	δ_{net}	Camber	δ_g	δ_{net}
(1)	(2)	(3)	(4)	(5)	(6)	(7)	(8)	(9)	(10)
7	–1.89	+0.99	–0.90	–1.89	+0.99	–0.90	–1.89	+0.99	–0.90
30	–3.40	+1.91	–1.49	–3.24	+2.14	–1.10	–3.30	+2.15	–1.15
90				–3.65	+3.50	–0.15	–3.97	+3.48	–0.49
365				–4.09	+3.93	–0.16	–4.68	+3.89	–0.79
5 yrs	–3.63	+3.84	–0.79	–4.35	+4.26	–0.09	–5.20	+4.19	–1.01

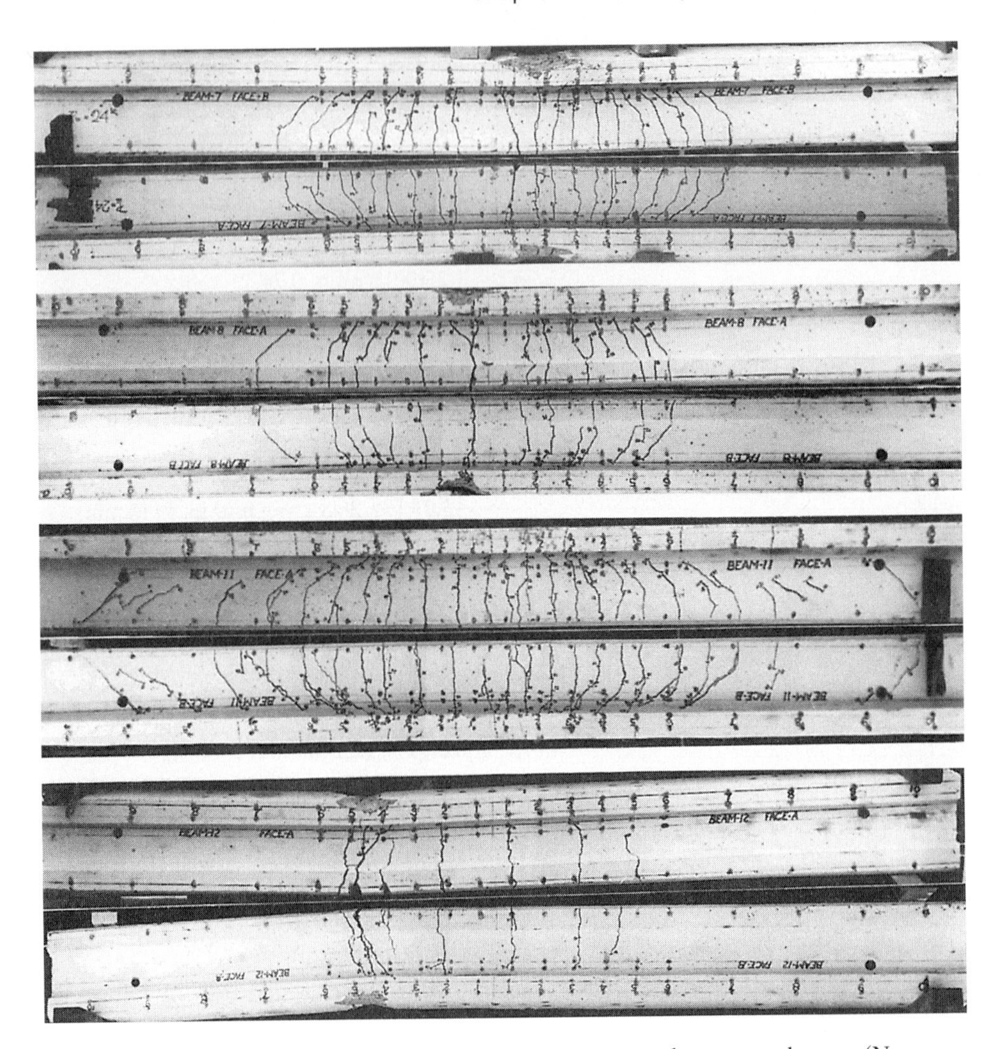

Photo 7.7 Typical cracking propagation in prestressed concrete beams (Nawy et al.).

deflection ↓. The camber is the upward deflection due to the prestress force *less* the reduction in deflection due to self-weight.

Comparison of the net deflections shows that the multipliers method and the approximate time-steps method give essentially comparable results, while the incremental time-steps method gives slightly lower camber values (approximately $\frac{3}{4}$-in. difference). This variation is expected because incremental prestress losses are determined at each step rather than as a single lump-sum loss taken at the final stage. The incremental time step method is time-consuming, and use of computers is necessary to justify its use. A large number of incremental time steps need to be investigated in large-span major structures such as segmental or cable-stayed bridges where accuracy of deflection and computations of camber are of a major concern.

7.12 LONG-TERM DEFLECTION OF COMPOSITE DOUBLE-T CRACKED BEAM

Example 7.9

A 72-ft (21.9 m) span simply supported roof normal weight concrete double-T-beam (Figure 7.23) is subjected to a superimposed topping load W_{SD} = 250 plf (3.65 kN/m) and a service live load W_L = 280 plf (4.08 kN/m). Calculate the short-term (immediate) camber and deflection of this beam by (a) the I_e method, (b) the bilinear method as well as the time-dependent

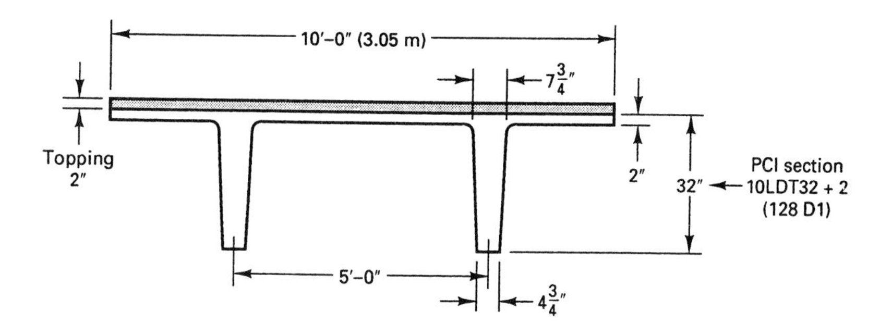

Figure 7.23 Double-T composite beam in Example 7.9.

deflections after 2-in. topping is cast (30 days) and the final deflection (5 years), using the PCI multipliers method. Given prestress losses 18%.

	Noncomposite	Composite
A_c, in.^{2}	615 (3,968 cm^2)	855 (5,516 cm^2)
I_c, in.^{4}	59,720 (24.9×10^5 cm^4)	77,118 (32.1×10^5 cm^4)
r^2, in.^{2}	97 (625 cm^2)	90 (580 cm^2)
c_b, in.	21.98 (558 mm)	24.54 (623 mm)
c_t, in.	10.02 (255 mm)	9.46 (240 mm)
S_b, in.^{3}	2,717 (4.5×10^4 cm^3)	3,142 (5.1×10^4 cm^3)
S^t, in.^{3}	5,960 (9.8×10^4 cm^3)	8,152 (13.4×10^4 cm^3)
W_d, plf	641 (9.34 kN/m)	891 (13.0 kN/m)

$V/S = 615/364 = 1.69$ in. (43 mm)
$RH = 75\%$
$e_c = 18.73$ in. (476 mm)
$e_e = 12.81$ in. (325 mm)
$f'_c = 5{,}000$ psi (34.5 MPa)
$f'_{ci} = 3{,}750$ psi (25.9 MPa)
topping $f'_c = 3{,}000$ psi (20.7 MPa)
f_t at bottom fibers $= 12\sqrt{f'_c} = 849$ psi (5.9 MPa)
A_{ps} = twelve $\frac{1}{2}$-in. dia low-relaxation prestressing steel depressed at midspan only

$f_{pu} = 270{,}000$ psi (1,862 MPa), low relaxation
$f_{pi} = 189{,}000$ psi (1,303 MPa)
$f_{PJ} = 200{,}000$ psi (1,380 MPa)
$f_{py} = 260{,}000$ psi (1,793 MPa)
$E_{ps} = 28.5 \times 10^6$ psi (19.65×10^4 MPa)

Solution by the I_e Method

1. *Midspan Section Stresses*

$$f_{PJ} = 200{,}000 \text{ psi at Jacking}$$

$$f_{pi} \text{ assumed} = 0.945 f_{PJ} = 189{,}000 \text{ psi at transfer}$$

$$e_c = 18.73 \text{ in. (475 mm)}$$

$$P_i = 12 \times 0.153 \times 189{,}000 = 347{,}004 \text{ lbs (1,540 kN)}$$

Self-Weight Moment

$$M_D = \frac{641(72)^2}{8} \times 12 = 4{,}984{,}416 \text{ in.-lb}$$

(a) *At Transfer*

From Equation 4.1a,

$$f^t = -\frac{P_i}{A_c}\left(1 - \frac{e_c c_t}{r^2}\right) - \frac{M_D}{S^t}$$

$$= -\frac{347{,}004}{615}\left(1 - \frac{18.73 \times 10.02}{97}\right) - \frac{4{,}984{,}416}{5{,}960}$$

$$= +527.44 - 836.31$$

$$= -308.87 \text{ psi } (C), \text{ say } 310 \text{ psi } (C) \ (2.1 \text{ MPa}) < 0.60f'_{ci} = 0.60(3{,}750)$$

$$= 2{,}250 \text{ psi, O.K.}$$

From Equation 4.1b,

$$f_b = -\frac{P_i}{A_c}\left(1 + \frac{e_c c_b}{r^2}\right) + \frac{M_D}{S_b}$$

$$= -\frac{347{,}004}{615}\left(1 + \frac{18.73 \times 21.98}{97}\right) + \frac{4{,}984{,}416}{2{,}717}$$

$$= -2{,}958.95 + 1{,}834.53$$

$$= -1{,}124.42 \text{ psi } (C), \text{ say } 1{,}125 \text{ psi } (C) < -2{,}250 \text{ psi, O.K.}$$

(b) *After Slab Is Cast*

At this load level assume 18 percent prestress loss

$$f_{pe} = 0.82 f_{pi} = 0.82 \times 189{,}000 = 154{,}980 \text{ psi}$$

$$P_e = 12 \times 0.153 \times 154{,}980 = 284{,}543 \text{ lb}$$

For the 2-in. slab,

$$W_{SD} = \frac{2}{12} \times 10 \text{ ft} \times 150 = 250 \text{ plf } (3.6 \text{ kN/m})$$

$$M_{SD} = \frac{250(72)^2}{8} \times 12 = 1{,}944{,}000 \text{ in.-lb}$$

$$M_D + M_{SD} = 4{,}984{,}416 + 1{,}944{,}000 = 6{,}928{,}416 \text{ in.-lb } (783 \text{ kN-m})$$

From Equation 4.18a,

$$f^t = -\frac{P_e}{A_c}\left(1 - \frac{e_c c_t}{r^2}\right) - \frac{M_D + M_{SD}}{S^t}$$

$$= -\frac{284{,}543}{615}\left(1 - \frac{18.73 \times 10.02}{97}\right) - \frac{6{,}928{,}416}{5{,}960}$$

$$= +432.5 - 1{,}162.5 = -730 \text{ psi } (5.0 \text{ MPa}) < 0.45f'_c = -2{,}250 \text{ psi, O.K.}$$

From Equation 4.18b,

$$f_b = -\frac{P_e}{A_c}\left(1 + \frac{e_c c_b}{r^2}\right) + \frac{M_D + M_{SD}}{S_b}$$

$$= -\frac{284{,}543}{615}\left(1 + \frac{18.73 \times 21.98}{97}\right) + \frac{6{,}928{,}416}{2{,}717}$$

$$= -2{,}426.33 + 2{,}550.02 = +123.7 \text{ (0.85 MPa), say 124 psi } (T)\text{, O.K.}$$

This is a very low tensile stress when the unshored slab is cast and before the service load is applied, $<<12\sqrt{f'_c} = 849$ psi.

(c) *At Service Load for the Precast Section*
Section modulus for composite section at the top of the precast section is

$$S_c^t = \frac{77{,}118}{9.46 - 2} = 10{,}337 \text{ in}^3$$

$$M_L = \frac{280(72)^2}{8} \times 12 = 2{,}177{,}288 \text{ in.-lb (246 kN-m)}$$

from Equation 4.19a,

$$f^t = -\frac{P_e}{A_c}\left(1 - \frac{e_c c_t}{r^2}\right) - \frac{M_D + M_{SD}}{S^t} - \frac{M_{CSD} + M_L}{S_c^t}$$

$$M_{CSD} = \text{superimposed dead load} = 0 \text{ in this case}$$

$$f^t = -730 - \frac{2{,}177{,}288}{10{,}337}$$

$$= -730 - 210 = -940 \text{ psi (6.5 MPa) } (C)\text{, O.K.}$$

from Equation 4.19b,

$$f_b = +123.7 + \frac{2{,}177{,}288}{3{,}142} = +123.7 + 693.0$$

$$= +816.7, \text{ say 817 psi } (T) \text{ (5.4 MPa)} < f_t = 849 \text{ psi, O.K.}$$

(d) *Composite Slab Stresses*
Precast double-T concrete modulus is

$$E_c = 57{,}000\sqrt{f'_c} - 57{,}000\sqrt{5{,}000} = 4.03 \times 10^6 \text{ psi } (2.8 \times 10^4 \text{ MPa})$$

Situ-cast slab concrete modulus is

$$E_c = 57{,}000\sqrt{3{,}000} = 3.12 \times 10^6 \text{ psi } (2.2 \times 10^4 \text{ MPa})$$

Modular ratio

$$n_p = \frac{3.12 \times 10^6}{4.03 \times 10^6} = 0.77$$

S_c^t for 2-in. slab top fibers = 8,152 in.^{3} from data.
S_{cb} for 2-in. slab bottom fibers = 10,337 in.^{3} from before for top of precast section.

$$\text{Stress } f_{cs}^t \text{ at top slab fibers} = n\frac{M_L}{S_C^t}$$

$$= -0.77 \times \frac{2{,}177{,}288}{8{,}152} = -207 \text{ psi (1.4 MPa) } (C)$$

Stress f_{csb} at bottom slab fibers

$$= -0.77 \times \frac{2{,}177{,}288}{10{,}337} = -162 \text{ psi (1.1 MPa) } (C)$$

2. *Support Section Stresses*
Check is made at the support face (a slightly less conservative check can be made at $50d_b$ from end).

$$e_c = 12.81 \text{ in.}$$

(a) *At Transfer*

$$f^t = -\frac{347{,}004}{615}\left(1 - \frac{12.81 \times 10.02}{97}\right) - 0$$

$$= +182.\ \text{psi}\ (T)\ (1.26\ \text{MPa}) << -\ 2{,}250\ \text{psi, O.K.}$$

$$f_b = -\frac{347{,}004}{615}\left(1 + \frac{12.81 \times 21.98}{97}\right) + 0$$

$$= -2{,}202\ \text{psi}\ (C)\ (15.2\ \text{MPa}) < 0.60f'_{ci} = -2{,}250\ \text{psi, O.K.}$$

(b) After slab is cast and at service load, the support section stresses both at top and bottom extreme fibers were found to be below the allowable, hence, O.K.

Summary of Midspan Stresses (psi)

	f^t	f_b
Transfer P_e only	+433	–2,426
W_D at transfer	–1,163	+2,550
Net at transfer	–730	+124
External load (W_L)	–210	+693
Net total at service	–940	+817

3. *Camber and Deflection Calculation*

At Transfer

Initial

$$E_{ci} = 57{,}000\sqrt{3{,}570} = 3.49 \times 10^6\ \text{psi}\ (2.2 \times 10^4\ \text{MPa})$$

From before, 28 days

$$E_c = 4.03 \times 10^6\ \text{psi}\ (2.8 \times 10^4\ \text{MPa})$$

Due to initial prestress only, from Figure 7.6

$$\delta_i = \frac{P_i e_c l^2}{8E_{ci}I_g} + \frac{P_i(e_e - e_c)l^2}{24E_{ci}I_g}$$

$$= \frac{(-347{,}004)(18.73)(72 \times 12)^2}{8(3.49 \times 10^6)59{,}720}$$

$$+ \frac{(-347{,}004)(12.81 - 18.73)(72 \times 12)^2}{24(3.49 \times 10^6)59{,}720}$$

$$= -2.90 + 0.30 = -2.6\ \text{in.}\ (66\ \text{mm}) \uparrow$$

Self-weight intensity $w = 641/12 = 53.42$ lb/in.

$$\text{Self-weight}\ \delta_D = \frac{5wl^4}{384E_{ci}I_g}\ \text{for uncracked section}$$

$$= \frac{5 \times 53.42(72 \times 12)^4}{384(3.49 \times 10^6)59{,}720} = 1.86\ \text{in.}\ \times 47\ \text{mm}) \downarrow$$

Thus the net camber at transfer

$$= -2.6 + 1.86 = -0.74\ \text{in.}\ (19\ \text{mm}) \uparrow$$

4. *Immediate Service Load Deflection*

(a) *Effective I_e Method*

Modulus of Rupture

$$f_r = 7.5\sqrt{f'_c} = 7.5\sqrt{5{,}000} = 530 \text{ psi}$$

f_b at service load = 817 psi (5.4 MPa) in tension (from before). Hence, the section is cracked and the effective I_e from Eqs. 3.19(a) or (b) should be used.

$$d_p = 18.73 + 10.02 + 2\,(\text{topping}) = 30.75 \text{ in. (780 mm)}$$

$$\rho_p = \frac{A_{ps}}{bd_p} = \frac{12(0.153)}{120 \times 30.75} = 4.98 \times 10^{-4}$$

From Equation 7.13,

$$I_{cr} = n_p A_{ps} d_p^2 (1 - 1.6\sqrt{n_p \rho_p})$$

$$n_p = 28.5 \times 10^6 / 4.03 \times 10^6 = 7 \text{ to be used in Equation 7.13}$$

Equation 7.13 gives $I_{cr} = 11{,}110$ in.^{4} (4.63×10^5 cm^4), use.

From Equation 7.3a and the stress f_{pe} and f_d values already calculated for the bottom fibers at midspan with $f_r = 7.5\sqrt{f'_c} = 530$ psi.

Moment M_{cr} due to that portion of live load that causes cracking is

$$M_{cr} = S_b(7.5\sqrt{f'_c} + f_{ce} - f_d)$$

$$= 3{,}142(530 + 2{,}426 - 2{,}550)$$

$$= 1{,}275{,}652 \text{ in.-lb}$$

M_a, unfactored maximum live load moment = 2,177,288 in.-lb

$$\frac{M_{cr}}{M_a} = \frac{1{,}275{,}652}{2{,}177{,}288} = 0.586$$

where M_{cr} is the moment due to that portion of the *live load* that causes cracking and M_a is the maximum service *unfactored live load.*

Using the preferable PCI expression of *(M_{cr}/M_a)* from Equation 7.11, and the stress values previously tabulated,

$$\frac{M_{cr}}{M_a} = 1 - \frac{f_{tl} - f_r}{f_L} = 1 - \left(\frac{817 - 530}{693}\right) = 0.586$$

$$\left(\frac{M_{cr}}{M_a}\right)^3 = (0.586)^3 = 0.20$$

Hence, from Equation 7.10b,

$$I_e = \left(\frac{M_{cr}}{M_a}\right)^3 I_g + \left[1 - \left(\frac{M_{cr}}{M_a}\right)^3\right] I_{cr} \leq I_g$$

$$I_e = 0.2(77{,}118) + (1 - 0.2)11{,}110$$

$$= 15{,}424 + 8{,}888 = 24{,}312 \text{ in}^4$$

$$w_{SD} = \frac{1}{12}(891 - 641) = 20.83 \text{ lb/in.}$$

$$w_L = \frac{1}{12} \times 280 = 23.33 \text{ lb/in.}$$

$$\delta_L = \frac{5wl^4}{384E_cI_e} = \frac{5 \times 23.33(72 \times 12)^4}{384(4.03 \times 10^6)24{,}312}$$

$$= +1.73 \text{ in. } (45 \text{ mm}) \downarrow \text{ (as an average value)}$$

When the concrete 2-in. topping is placed on the precast section, the resulting topping deflection with $I_g = 59{,}720$ in.^{4}

$$\delta_{SD} = \frac{5 \times 20.83(72 \times 12)^4}{384(4.03 \times 10^6)59{,}720} = +0.63 \text{ in.} \downarrow$$

Solution by Bilinear Method

$$f_{\text{net}} = f_{cb} - 7.5\lambda\sqrt{f'_c} = 817 - 530$$

$$= +287 \text{ psi } (T) \text{ causing cracking}$$

$$f_L = \text{tensile stress caused by live load alone}$$

$$= +693 \text{ psi } (T)$$

$$w_{L1} = \text{Portion of live load not causing cracking}$$

$$= (f_L - f_{\text{net}})\frac{w_L}{f_L} = \frac{693 - 287}{693} \times 280 \text{ plf}$$

$$= 0.586 \times 280 = 164.1 \text{ plf} = 13.68 \text{ lb/in.}$$

δ_{L1} due to uncracked I_g

$$\delta_{L1} = \frac{5w_{L1}l^4}{384E_cI_g} = \frac{5 \times 13.68(72 \times 12)^4}{384(4.03 \times 10^6)77{,}118}$$

$$= 0.32 \text{ in.} \downarrow$$

$$w_{L2} = w_L - w_{L1} = \frac{1}{12}(280 - 164.1) = 9.66 \text{ lb/in.}$$

δ_{L2} due to cracked I_{cr}

$$\delta_{cr} = \frac{5w_{L2}l^4}{384E_cI_{cr}} = \frac{5 \times 9.66(72 \times 12)^4}{384(4.03 \times 10^6)11{,}110}$$

$$= 1.57 \text{ in.} \downarrow$$

Total live-load deflection prior to prestress losses

$$= \delta_{L1} + \delta_{cr} = 0.32 + 1.57 = 1.89 \text{ in.} \downarrow$$

versus 1.73 in. $\downarrow$ obtained by the I_e method.

From before, $\delta_i = -0.74 \uparrow$

Net short-term deflection prior to prestress loss is

$$\delta_{\text{Total}} = -0.74 + 1.89 = 1.15 \text{ in.} \downarrow$$

5. *Long-term Deflection (Camber) by PCI Multipliers*
When the 2-in. concrete topping is placed on the precast section, the resulting topping deflection with $I_g = 59{,}720$ in.^{4} is

$$\delta_{SD} = \frac{5 \times 20.83(72 \times 12)^4}{384(4.03 \times 10^6)59{,}720} = +0.63 \text{ in. } (16 \text{ mm})$$

Using PCI multipliers at slab topping completion stage (30 days) and at the final service load (5 years), the following are the tabulated deflection values:

Load	Transfer δ_p in. (1)	PCI Multipliers	δ_{30} in. (2)	PCI Multiplier (Composite)	δ_{Final} in. (3)
Prestress	−2.60	1.80	−4.68	2.20	−5.71 ↑
w_D	+1.86	1.85	+3.44	2.40	+4.46 ↓
	−0.74 ↑		−1.24 ↑		−1.25 ↑
w_{SD}			+0.63 ↓	2.30	+1.45 ↓
w_L			+1.89 ↓		+1.89 ↓
Final δ	−0.74 ↑		+1.28 ↓		+2.09 ↓

Hence, final deflection ≈ 2.1 in. (53 mm) ↓

$$\text{Allowable deflection} = \text{span}/180 = \frac{72 \times 12}{180} = 4.8 \text{ in.} > 2.1 \text{ in., O.K.}$$

7.13 CRACKING BEHAVIOR AND CRACK CONTROL IN PRESTRESSED BEAMS

7.13.1 Introduction

The increased use of partial prestressing, allowing limited tensile stresses in the concrete under service-load and overload conditions while allowing nonprestressed steel to carry the tensile stresses, is becoming prevalent due to practicality and economy. Consequently, an evaluation of the flexural crack widths and spacing and control of their development become essential. Work in this area is relatively limited because of the various factors affecting crack width development in prestressed concrete. However, experimental investigations support the hypothesis that the major controlling parameter is the reinforcement stress change beyond the decompression stage. Nawy, et al., have undertaken extensive research since the 1960s on the cracking behavior of prestressed pretensioned and post-tensioned beams and slabs because of the great vulnerability of the highly stressed prestressing steel to corrosion and other environmental effects and the resulting premature loss of prestress (Refs. 7.13–7.17). Serviceability behavior under service and overload conditions can be controlled by the design engineer through the application of the criteria presented in this section.

7.13.2 Mathematical Model Formulation for Serviceability Evaluation

Crack Spacing. Primary cracks form in the region of maximum bending moment when the external load reaches the cracking load. As loading is increased, additional cracks will form and the number of cracks will be stabilized when the stress in the concrete no longer exceeds its tensile strength at further locations regardless of load increase. This condition essentially produces the absolute minimum crack spacing that can occur at high steel stresses, here termed *stabilized minimum crack spacing.* The maximum possible crack spacing under this stabilized condition is twice the minimum and is termed the *stabilized maximum crack spacing.* Hence, the stabilized mean crack spacing a_{cs} is the mean value of the two extremes.

The total tensile force T transferred from the steel to the concrete over the stabilized mean crack spacing can be defined as

$$T = \gamma a_{cs} \, \mu \Sigma o \tag{7.29}$$

where γ is a factor reflecting the distribution of bond stress, μ is the maximum bond stress which is a function of $\sqrt{f'_c}$, and Σo is the sum of the reinforcing elements' circumferences. The resistance R of the concrete area in tension, A_t can be defined (see Figure 7.24) as

$$R = A_t f'_t \tag{7.30}$$

By equating Equations 7.29 and 7.30, the following expression for a_{cs} is obtained, where c is a constant to be developed from the tests:

$$a_{cs} = c\frac{A_t f'_t}{\Sigma o \sqrt{f'_c}} \tag{7.31}$$

From extensive tests (see Refs. 7.13, 7.14, and 7.15), $cf'_t/\sqrt{f'_c}$ is found to have an average value of 1.2 for pretensioned, and 1.54 for post-tensioned prestressed beams.

Crack Width. If Δf_s is the net stress in the prestressed tendon or the magnitude of the tensile stress in the normal steel at any crack width load level in which the decompression load (decompression here means $f_c = 0$ at the level of the reinforcing steel) is taken as the reference point, then for the prestressed tendon

$$\Delta f_s = f_{nt} - f_d \text{ ksi } (=1{,}000 \text{ psi}) \tag{7.32}$$

where f_{nt} is the stress in the prestressing steel at any load level beyond the decompression load and f_d is the stress in the prestressing steel corresponding to the decompression load.

The unit strain $\epsilon_s = \Delta f_s/E_s$. Because it is logical to disregard as insignificant the unit strains in the concrete due to the effects of temperature, shrinkage, and elastic shortening, the maximum crack width can be defined as

$$w_{\max} = k a_{cs} \epsilon_s^{\alpha} \tag{7.33}$$

or

$$w_{\max} = k' a_{cs} (\Delta f_s)^{\alpha} \tag{7.34}$$

where k and α are constants to be established by tests.

7.13.3 Expressions for Pretensioned Beams

Equation 7.34 is rewritten in terms of Δf_s so that the following expression at the reinforcement level is obtained based on large numbers of tests:

$$w_{\max} = 1.4 \times 10^{-5} a_{cs} (\Delta f_s)^{1.31} \tag{7.35}$$

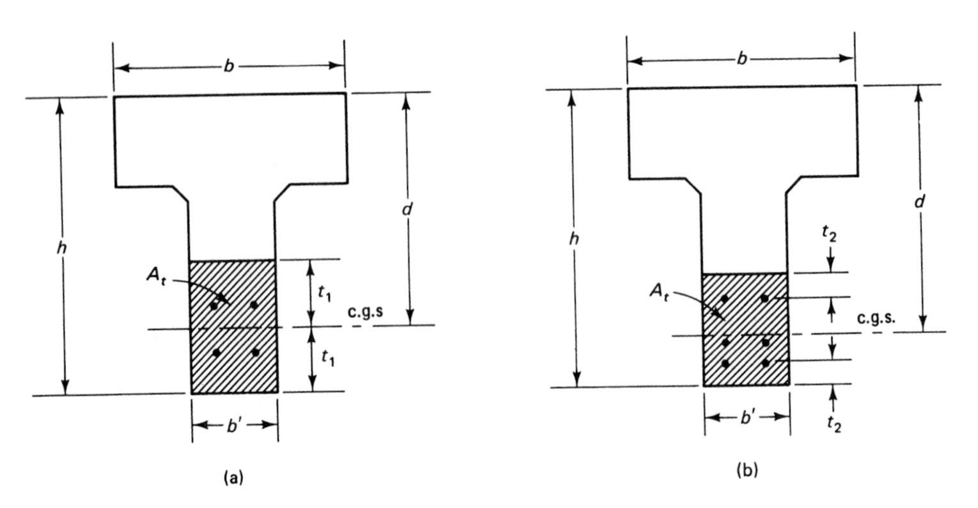

Figure 7.24 Effective concrete area in tension. (a) For even distribution of reinforcement in concrete. (b) For noneven distribution of reinforcement in concrete.

A 40-percent band of scatter envelops all the data for the expression in Equation 7.35 for $\Delta f_s = 20$ to 80 ksi.

Linearizing Equation 7.35 for easier use by the design engineer leads to the simplified expression

$$w_{max} = 5.85 \times 10^{-5} \frac{A_t}{\Sigma o}(\Delta f_s) \tag{7.36a}$$

of maximum crack width at the reinforcing steel level, and a maximum crack width (in.) at the tensile face of the concrete of

$$w'_{max} = 5.85 \times 10^{-5} R_i \frac{A_t}{\Sigma o}(\Delta f_s) \tag{7.36b}$$

where R_i is the ratio of distance from neutral axis to tension face to the distance from neutral axis to centroid of reinforcement.

A plot of the data and the best-fit expression for Equation 7.36a is given in Figure 7.25 with a 40-percent spread, which is reasonable in view of the randomness of crack development and the linearization of the original expression (Equation 7.35).

7.13.4 Expressions for Post-Tensioned Beams

The expression developed for the crack width in post-tensioned bonded beams which contain mild steel reinforcement is

$$w_{max} = 6.51 \times 10^{-5} \frac{A_t}{\Sigma o}(\Delta f_s) \tag{7.37a}$$

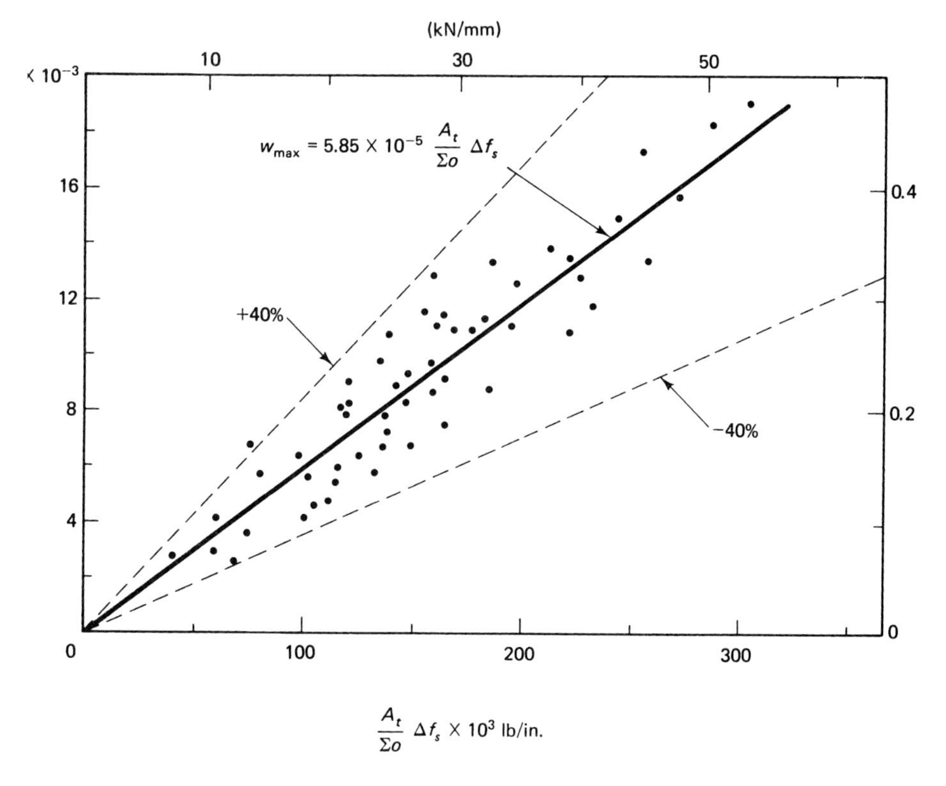

Figure 7.25 Linearized maximum crack width versus $(A_t/\Sigma_o)\Delta f_s$, pretensioned beams.

Photo 7.8 Flexural cracking propagation in pretensioned prestressed T-beam (Nawy et al.).

for the width at the reinforcement level closest to the tensile face, and

$$w_{\max} = 6.51 \times 10^{-5} R_i \frac{A_t}{\Sigma o} (\Delta f_s) \tag{7.37b}$$

at the tensile face of the concrete lower fibers.

For nonbonded beams, the factor 6.51 in Equations 7.37a and 7.37b becomes 6.83. Figure 7.26 gives a regression plot of Equation 7.37a that shows a scatter band of ±40 percent, which is not unexpected in flexural cracking behavior. The crack spacing stabilizes itself beyond an incremental stress Δf_s of 30,000 psi to 35,000 psi, as shown in Figure 7.27, depending on the *total* reinforcement percent ρ_T of both the prestressed and the nonprestressed steel.

Recent work by Nawy et al., on the cracking performance of high strength prestressed concrete beams, both pretensioned and post-tensioned has shown that the factor 5.85 in Equation 7.36a is considerably reduced. For concrete strengths in the range of 9,000 to 14,000 psi (60 to 100 MPa), this factor reduces to 2.75, so that the expression for the maximum crack width at the reinforcement level (inch) becomes

$$w_{\max} = 2.75 \times 10^{-5} \frac{A_t}{\Sigma o} (\Delta f_s) \tag{7.38a}$$

In SI units, the expression is

$$w_{\max} = 4.0 \times 10^{-5} \frac{A_t}{\Sigma o} (\Delta f_s) \tag{7.38b}$$

where A_t, cm^2; Σo, cm; Δf_s, MPa.

7.13.5 Long-Term Effects on Crack-Width Development

Limited studies on crack-width development and increase with time show that both sustained and cyclic loadings increase the amount of microcracking in the concrete. Also, microcracks formed at service-load levels in partially prestressed beams do not seem to have a recognizable effect on the strength or serviceability of the concrete element. Macroscopic cracks, however, do have a detrimental effect, particularly in terms of corrosion of the reinforcement and appearance. Hence, an increase of crack width due to sustained loading significantly affects the durability of the prestressed member regardless of

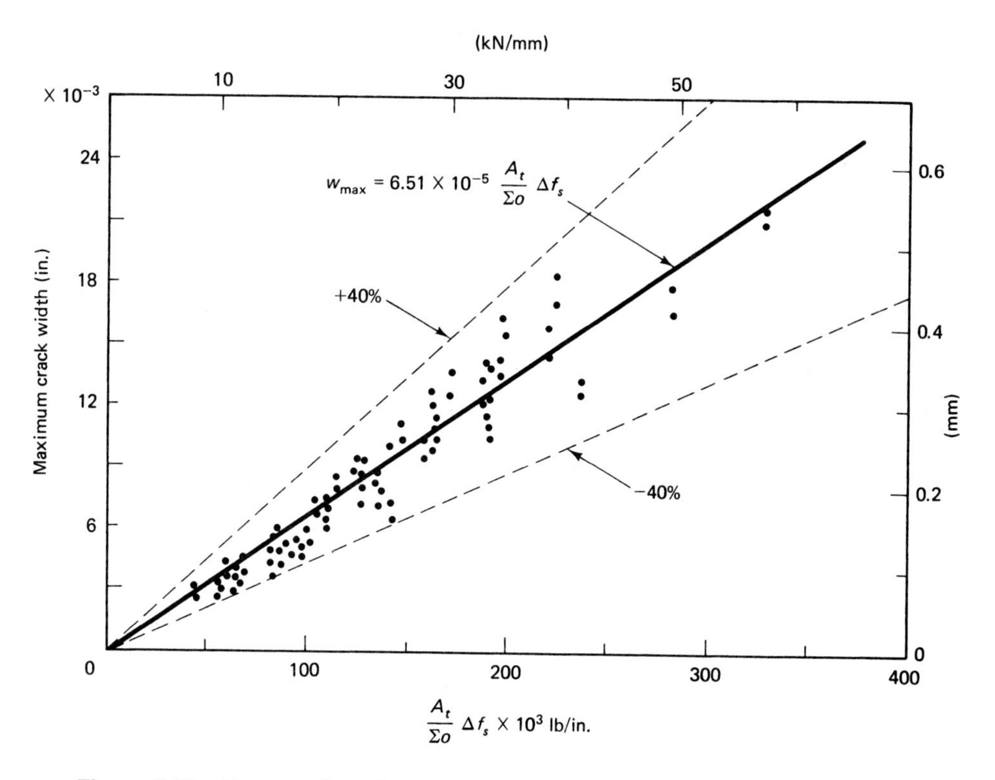

Figure 7.26 Linearized maximum crack width versus $(A_t/\Sigma_o)\Delta f_s$, post-tensioned beams.

whether prestressing is circular, such as in tanks, or linear, such as in beams. Information obtained from sustained load tests of up to two years and fatigue tests of up to one million cycles indicates that a *doubling* of crack width with time can be expected. Therefore, engineering judgment has to be exercised as regards the extent of tolerable crack width under long-term loading conditions.

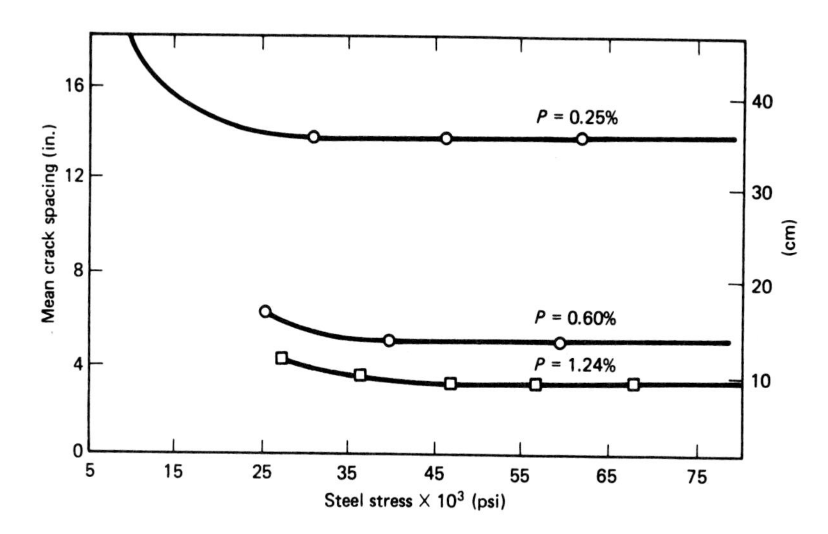

Figure 7.27 Reinforcement percentage effect on the relationship between crack spacing and incremental reinforcement stress.

Table 7.9 Maximum Tolerable Flexural Crack Widths

Exposure condition	Crack width in.	Crack width mm
Dry air or protective membrane	0.016	0.41
Humidity, moist air, soil	0.012	0.30
De-icing chemicals	0.007	0.18
Seawater and seawater spray; wetting and drying	0.006	0.15
Water-retaining structures (excluding nonpressure pipes)	0.004	0.10

7.13.6 Tolerable Crack Widths

The maximum crack width that a structural element should tolerate depends on the particular function of the element and the environmental conditions to which the structure is liable to be subjected. Table 7.9 from the ACI Committee 224 report on cracking serves as a reasonable guide on the acceptable crack widths in concrete structures under the various environmental conditions encountered.

7.14 CRACK WIDTH AND SPACING EVALUATION IN PRETENSIONED T-BEAM WITHOUT MILD STEEL

Example 7.10

A pretensioned prestressed concrete beam has a T-section as shown in Figure 7.28. It is prestressed with fifteen $\frac{7}{16}$-in. dia 7-wire strand 270-K grade. The locations of the neutral axis and center of gravity of steel are shown in the figure. $f'_c = 5{,}000$ psi, $E_c = 57{,}000\sqrt{f'_c}$, and $E_s = 28 \times 10^6$ psi. Find the mean stabilized crack spacing and the crack widths at the steel level as well as at the tensile face of the beam at $\Delta f_s = 30 \times 10^3$ psi. Assume that no failure in shear or bond takes place.

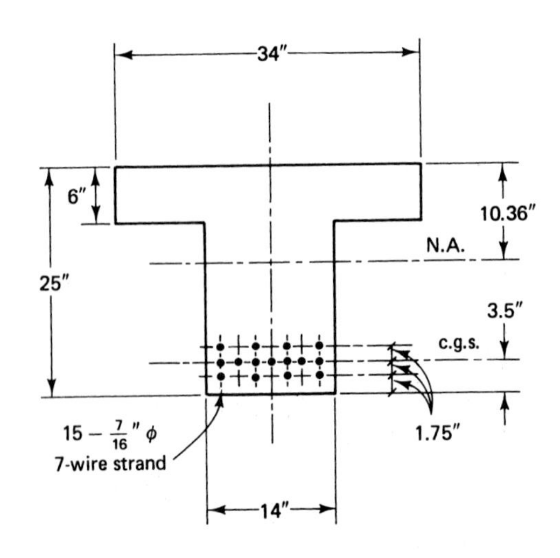

Figure 7.28 Beam cross section in Example 7.10.

Solution:

$$\Delta f_s = 30{,}000 \text{ psi} = 30 \text{ ksi}$$

Mean Stabilized Crack Spacing

$$A_t = 7 \times 14 = 98 \text{ sq in.}$$

$$\Sigma o = 15\pi D = 15\pi\left(\frac{7}{16}\right) = 20.62 \text{ in.}$$

$$a_{cs} = 1.2\left(\frac{A_t}{\Sigma o}\right) = 1.2\left(\frac{98}{20.62}\right) = 5.7 \text{ in. (145 mm)}$$

Maximum Crack Width at Steel Level

$$w_{\max} = 5.85 \times 10^{-5}\left(\frac{A_t}{\Sigma o}\right)\Delta f_s = 5.85 \times 10^{-5}\left(\frac{98}{20.62}\right)30$$

$$= 834.1 \times 10^{-5} \text{ in.} \cong 0.0083 \text{ in. (0.21 mm)}$$

Maximum Crack Width at Tensile Face of Beam

$$R_i = \frac{25 - 10.36}{25 - 10.36 - 3.5} = 1.31$$

$$w'_{\max} = w_{\max} R_i = 0.0083 \times 1.31 = 0.011 \text{ in. (0.28 mm)}$$

7.15 CRACK WIDTH AND SPACING EVALUATION IN PRETENSIONED T-BEAM CONTAINING NONPRESTRESSED STEEL

Example 7.11

The beam in Example 7.10 also contains three #6 nonprestressed mild steel bars as shown in Figure 7.29. Find the crack spacing and width for an incremental steel stress $\Delta f_s = 30{,}000$ psi = 30 ksi (207 MPa)

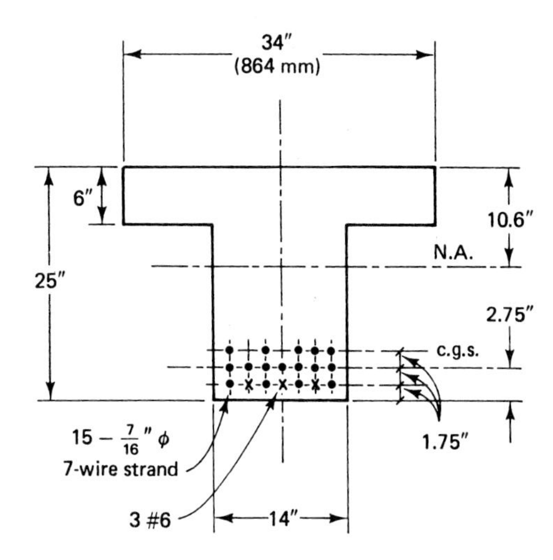

Figure 7.29 Beam cross section in Example 7.11.

Solution:

Mean Stabilized Crack Spacing

$$A_t = 14(3 \times 1.75 + \tfrac{1}{2} \times \tfrac{7}{16} + 1\tfrac{3}{8}) = 14 \times 6.84 = 95.8 \text{ in.}^2$$

$$\Sigma o = 20.62 + 3 \times 2.36 = 27.70 \text{ in.}$$

$$a_{cs} = 1.2\left(\frac{A_t}{\Sigma o}\right) = 1.2\left(\frac{95.8}{27.7}\right) = 4.15 \text{ in. (105 mm)}$$

Maximum Crack Width at Steel Level

$$w_{\max} = 5.85 \times 10^{-5}\left(\frac{A_t}{\Sigma o}\right)\Delta f_s = 5.85 \times 10^{-5}\left(\frac{95.8}{27.7}\right)30$$

$$= 606.9 \times 10^{-5} \cong 0.0061 \text{ in. (0.15 mm)}$$

Maximum Crack Width at Tensile Face of Beam

$$R_i = \frac{25 - 10.6}{25 - 10.6 - 2.75} = 1.24$$

$$w'_{\max} = w_{\max}R_i = 0.0061 \times 1.24 = 0.007 \text{ in. (0.18 mm)}$$

7.16 CRACK WIDTH AND SPACING EVALUATION IN PRETENSIONED I-BEAM CONTAINING NONPRESTRESSED MILD STEEL

Example 7.12

A pretensioned prestressed concrete I-beam has the geometry shown in Figure 7.30. It is prestressed with twenty $\frac{7}{16}$-in. dia 7-wire 270-K grade low-relaxation strands and four #7 mild steel bars having yield strength $f_y = 60{,}000$ psi. Find the mean stabilized crack spacing and the crack widths at the steel level as well as at the tensile face of the beam at incremental steel stress $\Delta f_s = 20{,}000$ psi (138 MPa). Assume that no failure in shear or bond takes place, and check whether the crack widths that develop satisfy the crack control criteria for deicing chemicals.

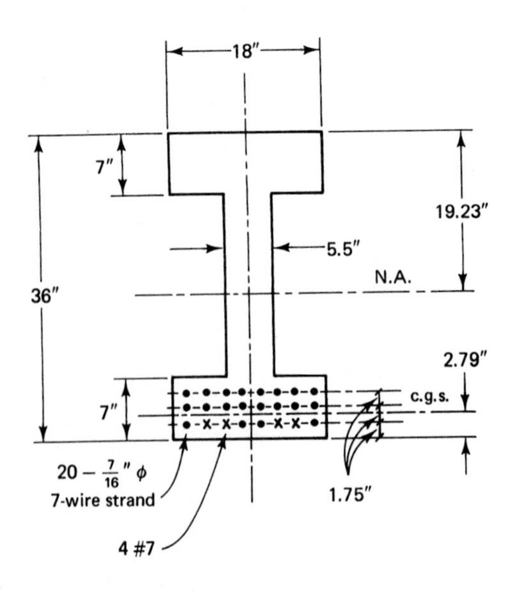

Figure 7.30 Beam cross section in Example 7.12.

Solution: $\Delta f_s = 20{,}000$ psi = 20 ksi (138 MPa)

Mean Stabilized Crack Spacing

$$A_t = 18 \times (3 \times 1.75 + \tfrac{1}{2} \times \tfrac{7}{16} + 1\tfrac{5}{16}) = 122.06 \text{ in.}^2$$

$$\Sigma o = 20\pi D + 4 \times 2.75 = 20\pi \times \tfrac{7}{16} + 4 \times 2.75 = 38.49 \text{ in.}$$

$$a_{cs} = 1.2\left(\frac{A_t}{\Sigma o}\right) = 1.2\left(\frac{122.06}{38.49}\right) = 3.8 \text{ in. (97 mm)}$$

Maximum Crack Width at Steel Level

$$w_{\max} = 5.85 \times 10^{-5}\left(\frac{A_t}{\Sigma o}\right)\Delta f_s = 5.85 \times 10^{-5}\left(\frac{122.06}{38.49}\right)20$$

$$= 371.0 \times 10^{-5} \cong 0.0037 \text{ in. (0.1 mm)}$$

Maximum Crack Width at Tensile Face of Beam

$$R_i = \frac{36 - 19.23}{36 - 19.23 - 2.79} = 1.2$$

$$w'_{\max} = w_{\max} R_i = 0.0037 \times 1.2 = 0.004 \text{ in. (0.1 mm)}$$

Maximum Allowable Crack Width for Deicing. From Table 7.9, the maximum tolerable crack width for deicing is $W_{\max} = 0.007 > 0.004$ in. (0.1 mm). Hence, serviceability requirement is satisfied.

7.17 CRACK WIDTH AND SPACING EVALUATION FOR POST-TENSIONED T-BEAM CONTAINING NONPRESTRESSED STEEL

Example 7.13

A post-tensioned prestressed concrete beam has a T-section as shown in Figure 7.31. It is prestressed with twelve $\frac{7}{16}$-in. dia 7-wire strands of 270-K grade and additionally reinforced with four #6 nonprestressed steel bars. The locations of the neutral axis and center of gravity

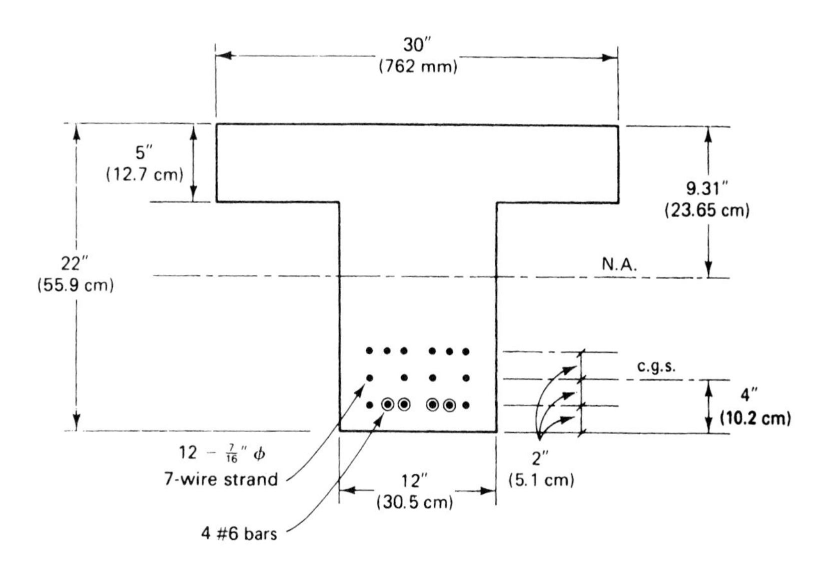

Figure 7.31 Beam cross section in Example 7.13.

of steel are shown in the figure. Assume that $f'_c = 5{,}000$ psi, $E_c = 57{,}000\sqrt{f'_c}$ psi, and $E_s = 28{,}000$ ksi. Find the mean stabilized crack spacing and the crack widths at the steel level as well as at the tensile face of the beam at $\Delta f_s = 30{,}000$ psi, assuming there is no failure in shear or bond. Then determine whether the beam satisfies the serviceability criteria for crack control for humidity and moist air.

Solution:

$$\Delta f_s = 30{,}000 \text{ psi} = 30 \text{ ksi}$$

Mean Stabilized Crack Spacing

$$A_t = 8 \times 12 = 96 \text{ in.}^2$$

$$\Sigma o = 12 \times \pi \times \tfrac{7}{16} + 4 \times 2.36 = 25.93 \text{ in.}$$

$$a_{cs} = 1.54\frac{A_t}{\Sigma o} = 1.54 \times \frac{96}{25.93} = 5.70 \text{ in. (145 mm)}$$

Maximum Crack Width at Steel Level

$$w_{\max} = 6.51 \times 10^{-5}\frac{A_t}{\Sigma o}(\Delta f_s) = 6.51 \times 10^{-5} \times \frac{96}{25.93} \times 30$$

$$= 0.0072 \text{ in. (0.18 mm)}$$

Maximum Crack Width at Tensile Face of Beam

$$R_i = \frac{22 - 9.31}{22 - 9.31 - 4} = 1.46$$

$$w'_{\max} = w_{\text{mas}}R_i = 0.007 \times 1.46 = 0.0102 \text{ in. (0.26 mm)}$$

Maximum Tolerable Crack Width for Humidity. From Table 7.9, the maximum tolerable crack width for the stated humidity conditions is 0.012 in. (0.3 mm) > 0.0102 in. (0.26 mm), which is satisfactory.

7.18 SI DEFLECTION AND CRACKING EXPRESSIONS

$$E_c = w_c^{1.5}\, 0.043\sqrt{f'_c} \text{ MPa} \tag{7.1a}$$

where f'_c is in MPa units and w_c is in Kg/m^3 ranging between 1,500 to 2,500 Kg/m^3.

For $f'_c > 35$ MPa, <80 MPa

$$E_c = 3.32\sqrt{f'_c} + 6{,}895\left(\frac{w_c}{2{,}320}\right)^{1.5} \text{ MPa}$$

For normal-weight concrete, $E_c = 3.32\sqrt{f'_c} + 6{,}895$ MPa

$$f_r = 0.62\sqrt{f'_c} \tag{7.1b}$$

$$I_e = \left(\frac{M_{cr}}{M_a}\right)^3 I_g + \left[1 - \left(\frac{M_{cr}}{M_a}\right)^3\right] I_{cr} \tag{7.10b}$$

$$\left(\frac{M_{cr}}{M_a}\right) = \left[1 - \left(\frac{f_{tl} - f_r}{f_L}\right)\right] \tag{7.11}$$

$$I_{cr} = n_p A_{ps} d_p^2 (1 - 1.6\sqrt{n_p \rho_p}) \tag{7.13}$$

$$I_{cr} = (n_p A_{ps} d_p^2 + n_s A_s d^2)(1 - 1.6\sqrt{n_p \rho_p + n_s \rho} \tag{7.14}$$

$$\lambda = \frac{\xi}{1 + 50\rho'} \tag{7.16}$$

Equations 7.36 and 7.37 on crack control

$$w_{max} = \alpha_w \times 10^{-5} \frac{A_t}{\Sigma o} (\Delta f_s), \text{ millimeters}$$

where A_t, cm², Σo, cm; Δf_s, MPa

$$\begin{aligned} \alpha_w &= 8.48 \times 10^{-5} \text{ for pretensioned} \\ &= 9.44 \times 10^{-5} \text{ for post-tensioned} \\ &= 4.0 \times 10^{-5} \text{ for concretes with } f'_c > 70 \text{ MPa} \\ \text{MPa} &= \text{N/mm}^2 \\ \text{(psi) } 0.006895 &= \text{MPa} \\ \text{(lb/ft) } 14.593 &= \text{N/m} \\ \text{(in.-lb) } 0.113 &= \text{N-m} \end{aligned}$$

7.19 SI DEFLECTION CONTROL

Example 7.14

Solve Example 7.9 for short-term deflection using the SI procedure.

Data

(a) *Section Geometry*	*Noncomposite*	*Composite*
A_c, cm²	3,968	5,516
I_c, cm⁴	24.9×10^5	32.2×10^5
r^2, cm²	626	581
c_b, cm	55.8	62.3
c_t, cm	25.5	24.0
e_c, cm	47.5	
e_e, cm	32.5	
S_b, cm³	4.5×10^4	5.2×10^4
S^t, cm³	9.8×10^4	13.4×10^4
w_D, kN/m	9.34	13.0
$l = 21.95$ m	topping $t = 5$ cm	flange width $b = 3.05$ m
$w_L = 4.09$ kN/m		

(b) *Material properties*

$V/S = 615/364 = 0.43$ cm

$RH = 75\%$

$f'_c = 34.5$ MPa, normal weight (2,370 kg/m³)

$f_c = 0.45 f'_c = 15.5$ MPa

$f'_{ci} = 25.9$ MPa

$f_{ci} = 0.6 \times 25.9 = 15.5$ MPa

topping $f'_c = 20.7$ MPa

f_t at bottom fibers $= \sqrt{f'_c} = 5.9$ MPa at service load

f_t allowable before unshored slab cast $= 1.4$ MPa $(\frac{1}{4}\sqrt{f'_c})$

$A_{ps} =$ twelve tendons, 12.7-mm diameter, low-relaxation (0.99 mm^2/tendon)

$f_{pu} = 1{,}860$ MPa, low-relaxation

$f_{pi} = 1{,}300$ MPa

$f_{PJ} = 1{,}380$ MPa

$f_{py} = 1{,}790$ MPa

$E_{ps} = 19.7 \times 10^4$ MPa

Solution:

1. *Midspan section stresses*

$$f_{pJ} = 1{,}380 \text{ MPa}$$

$$f_{pi} \text{ assumed } = 0.945 f_{pJ} = 1{,}300 \text{ MPa at transfer}$$

$$e_c = 47.5 \text{ cm}$$

$$P_i = 12 \times 99 \times 1{,}300 \text{ MPa} = 1.54 \times 10^6 \text{ N} = 1{,}540 \text{ kN}$$

$$\text{self-weight moment } M_D = \frac{9{,}340(21.95)^2}{8} = 562{,}500 \text{ N-m}$$

From Equation 4.1a,

$$f^t = -\frac{P_i}{A_c}\left(1 - \frac{e_c c_t}{r^2}\right) - \frac{M_d}{S^t}$$

$$= -\frac{1{,}540{,}000}{3{,}968 \times 100}\left(1 - \frac{47.5 \times 25.5}{626}\right) - \frac{562{,}500}{9.8 \times 10^4}$$

$$= 3.6 - 5.7 = -2.1 \text{ MPa } (C)$$

From Equation 4.1b,

$$f_b = -\frac{P_i}{A_c}\left(1 + \frac{e_c c_b}{r^2}\right) + \frac{M_D}{S_b}$$

$$= -\frac{1{,}540{,}000}{3{,}968 \times 100}\left(1 + \frac{47.5 \times 55.8}{626}\right) + \frac{562{,}500}{4.5 \times 210^4}$$

$$= -20.3 + 12.5 = -7.8 \text{ MPa } (C) < \text{allow. } 15.5 \text{ MPa, O.K.}$$

2. *After unshored slab is cast*
At this load level, assume 18% prestress losses.

$$f_{pe} = 0.82 f_{pi} = 0.82 \times 1{,}300 = 1{,}066 \text{ MPa}$$

$$P_e = 12 \times 99 \times 1{,}066 = 1{,}266 \text{ kN}$$

For the 5 cm topping,

$$\text{concrete weight} = 2{,}370 \text{ kg/m} = 2{,}370 \times 9.81 \text{ N/m}^3$$

$$= 23.3 \text{ kN/m}^3$$

For 5 cm slab, $w_{SD} = 0.05 \times 3.05 \times 23.3 = 3.6$ kN/m

$$M_{SD} = \frac{3{,}600(21.95)^2}{8} = 216{,}800 \text{ N-m}$$

$$M_D + M_{SD} = 562{,}500 + 216{,}800 = 780{,}000 \text{ N-m}$$

(In Example 7.9, $M_D + M_{SD} = 782$ kN-m since 2 in. topping is slightly more than 5 cm.) For unshored case, from Equation 4.18a,

$$f^t = -\frac{P_e}{A_c}\left(1 - \frac{e_c c_t}{r^2}\right) - \frac{M_D + M_{SD}}{S^t}$$

$$= -\frac{1{,}266{,}000}{3{,}968 \times 100}\left(1 - \frac{47.5 \times 25.5}{626}\right) - \frac{780{,}000}{9.8 \times 10^4}$$

$$= +2.96 - 7.96 = -5.0 \text{ MPa } (C) < \text{allow. } f_c = 15.5 \text{ MPa, O.K.}$$

From Equation 4.18b,

$$f_b = -\frac{P_e}{A_c}\left(1 + \frac{e_c c_b}{r^2}\right) + \frac{M_D + M_{SD}}{S_b}$$

$$= -\frac{1{,}266{,}000}{3{,}968 \times 100}\left(1 + \frac{47.5 \times 55.8}{626}\right) + \frac{780{,}000}{4.5 \times 10^4}$$

$$= -16.61 + 17.34 = 0.64 \text{ MPa } (T) < \text{allow. } f_t = 1.4 \text{ MPa, O.K.}$$

3. *At service load for precast section*
Section modulus at top of precast section is

$$S_c^t = \frac{32.2 \times 10^5}{25.5 - 5.0} = 15.7 \times 10^4 \text{ cm}^3$$

$$M_L = \frac{4{,}090(21.95)^2}{8} = 246{,}320 \text{ N-m}$$

From Equation 4.19a,

$$f^t = -\frac{P_e}{A_c}\left(1 - \frac{e_c c^t}{r^2}\right) - \frac{M_D + M_{SD}}{S^t} - \frac{M_{CSD} + M_L}{S_c^t}$$

M_{SD} = superimposed dead load = 0 in this case

$$f^t = -5.0 - \frac{246{,}320}{15.7 \times 10^4} = -5.0 - 1.57 = -6.57 \text{ MPa } (C) \text{, O.K.}$$

From Equation 4.19b,

$$f_b = +0.64 + \frac{246{,}320}{5.2 \times 10^4} = +0.74 + 4.74$$

$$= +5.38 \text{ MPa } (T) \simeq \text{allow. } f_t = +5.9 \text{ MPa, O.K.}$$

4. *Composite slab stresses*
Precast double-T concrete modulus

$$E_c = w_c^{1.5}\, 0.043\sqrt{f'_c}, \qquad w_c = 2{,}370 \text{ Kg/m}^3$$

$$E_c = 2{,}370^{1.5} \times 0.043\sqrt{34.5} = 2.91 \times 10^4 \text{ MPa}$$

Situ-cast slab concrete modulus

$$E_c = 2{,}370^{1.5} \times 0.043\sqrt{20.7} = 2.25 \times 10^4 \text{ MPa}$$

Modular ratio

$$n_p = \frac{2.25 \times 10^4}{2.91 \times 10^4} = 0.77$$

S_c^t for 5 cm slab top fiber = 13.4×10^4 cm^3

S_{cb} for 5 cm slab bottom fiber = 15.7×10^4 cm^3 from before for top of precast section.

$$\text{Stress } f_{cs}^t \text{ at top slab fiber} = n\frac{M_L}{S_c^t} = -0.77 \times \frac{246{,}320}{13.4 \times 10^4} = 1.4 \text{ MPa } (C)$$

$$\text{Stress } f_{csb} \text{ at bottom slab fibers} = -0.77 \times \frac{246{,}320}{15.7 \times 10^4} - 1.2 \text{ MPa } (C)$$

5. *Support section stresses*
Check is made at the support face (a slightly less conservative check can be made at $50d_b$ from end).

$$e_e = 32.5 \text{ cm}$$

At transfer

$$f^t = \frac{1{,}540{,}000}{3{,}968 \times 100}\left(1 - \frac{32.5 \times 25.5}{626}\right) - 0$$

$$= -1.26 \text{ MPa } (C) < \text{allow. } f_c = 15.5 \text{ MPa, O.K.}$$

$$f_b = -\frac{1{,}540{,}000}{3{,}968 \times 100}\left(1 + \frac{32.5 \times 55.8}{626}\right) + 0$$

$$= -15.2 \text{ MPa } (C) < \text{allow. } f_c = 15.5 \text{ MPa, O.K.}$$

After the unshored slab was cast and at service load, the support section stresses both at top and bottom extreme fibers were found to be below the allowable, hence O.K.

Summary of Midspan Stresses (MPa)

Load Stage	f^t	f_b
Transfer P_e only	+2.96	–16.60
w_D at transfer	–7.96	+17.34
Net at transfer	–5.00	+0.74
External load w_L	–1.57	+4.74
Net total at service	–6.57	+5.48

6. *Short-term (immediate) deflection*
(a) *Deflection at transfer*

$$\text{Initial } E_{ci} = w_c^{1.5}\, 0.043\sqrt{f_c'}$$

$$= (2{,}370)^{1.5} \times 0.043\sqrt{25.9} = 2.52 \times 10^4 \text{ MPa}$$

from before, 28 days $E_c = 2.80 \times 10^4$ MPa

From Figure 7.6, deflection due to initial prestress only.

$$\delta_i = \frac{P_i e_c l^2}{8E_{ci}I_g} + \frac{P_i(e_e - e_c)l^2}{24E_{ci}I_g}$$

$$= 10^3\left[\frac{-1{,}540{,}00 \times 47.5(21.95)^2}{8(2.52 \times 10^4)(24.9 \times 10^5)}\right]$$

$$+ 10^3\left[\frac{-1{,}540{,}000(32.5 - 47.5)(21.95)^2}{24(2.52 \times 10^4)(24.9 \times 10^5)}\right]$$

$$= -70.2 + 7.3 = -62.9 \text{ say } 65 \text{ mm} \uparrow$$

$$\text{Self-weight } \delta_D = \frac{5w_D l^4}{384 E_{ci} I_g}$$

$$= \frac{5 \times 9{,}340(21.95)^4 \times 10^5}{384(2.52 \times 10^4)(24.9 \times 10^5)} = 45 \text{ mm} \downarrow$$

Thus, net camber at transfer = –(65 – 45) = –20 mm ↑.

(b) *Immediate service load deflection*
From Equation 7.13,

$$I_{cr} = n_p A_{ps} d_p^2 (1 - 1.6\sqrt{n_p \rho_p}$$

$$d_p = e_c + c_t + 5 \text{ cm (topping)}$$

$$= 47.5 + 25.5 + 5 = 78 \text{ cm}$$

$$A_p = 12 \times 99 = 1{,}188 \text{ mm}^2$$

$$\rho_p = \frac{A_p}{bd_p} = \frac{11.88 \text{ cm}^2}{305 \times 78} = 0.0047$$

$$n_p = \frac{E_{ps}}{E_c} = \frac{19.7 \times 10^4}{2.91 \times 10^4} = 6.8$$

$$I_{cr} = 6.8 \times 11.88(78)^2(1 - 1.6\sqrt{6.8 \times 4.94 \times 10^{-4}})$$

$$= 4.03 \times 10^5 \text{ cm}^4$$

From Equation 7.11,

$$\frac{M_{cr}}{M_a} = 1 - \left(\frac{f_{tl} - f_r}{f_L}\right)$$

$$f_r = 0.62\sqrt{f_c'} = 0.62 \times \sqrt{3.45} = 3.64 \text{ MPa}$$

$$f_{tl} = +5.48 \text{ MPa } (T) \qquad f_L = +4.74\ (T)$$

$$\left(\frac{M_{cr}}{M_a}\right) = 1 - \frac{(5.48 - 3.64)}{4.74} = 0.6$$

$$\left(\frac{M_{cr}}{M_a}\right)^3 = 0.216$$

from Equation 7.10b,

$$I_e = \left(\frac{M_{cr}}{M_a}\right) I_g + \left[1 - \left(\frac{M_{cr}}{M_a}\right)^3\right] I_{cr} \le I_g$$

$$= 0.216(24.9 \times 10^5) + (1 - 0.216)(4.03 \times 10^5)$$

$$= 5.378 \times 10^5 + 3.160 \times 10^5 = 8.54 \times 10^5 \text{ cm}^4$$

$$w_L = 4{,}090 \text{ N/m}$$

$$\delta_L = \frac{5w_L l^4}{384 E_c I_e} = \frac{5(4{,}090)(21.95)^4 \times 10^5}{384(2.91 \times 10^4)(8.54 \times 10^5)} = +50 \text{ mm} \downarrow$$

When the concrete 5 cm topping is placed on the precast section, the resulting topping deflection with $I_g = 24.9 \times 10^5$, $w_{SD} = 3{,}600$ N/m.

$$\delta_{SD} = \frac{5 \times 3{,}600(21.95)^4 \times 10^5}{384(2.91 \times 10^4)(24.9 \times 10^5)} = +15 \text{ mm} \downarrow$$

(c) *Summary of short-term deflections*

$$\text{Prestress Camber } \delta_i = 65 \text{ mm} \uparrow$$

$$\text{Dead Load } \delta_D = 45 \text{ mm} \downarrow$$

$$\text{Live Load } \delta_L = 50 \text{ mm} \downarrow$$

$$\text{5 cm topping Load } \delta_{SD} = 15 \text{ mm} \downarrow$$

7.20 SI CRACK CONTROL

Example 7.15

Solve Example 7.11 using SI procedure

Data

$$\Delta f_s = 207 \text{ MPa}$$

$$A_t = 618 \text{ cm}^2$$

$$\Sigma o = 70.4 \text{ cm}$$

(a) *Steel level*

$$w_{max} = 8.48 \times 10^{-5}\left(\frac{A_t}{\Sigma o}\right)\Delta f_s$$

$$= 8.48 \times 10^{-5}\left(\frac{618}{70.4}\right)207 = 0.15 \text{ mm}$$

(b) *Tensile beam face*

$$R_i = 1.24 \text{ from Example 7.11}$$

$$w_{max} = 1.24 \times 0.15 = 0.19 \text{ say } 0.2 \text{ mm}$$

REFERENCES

7.1 ACI Committee 318. *Building Code Requirements for Structured Concrete (ACI 318–99* and *Commentary 318R–99)*, American Concrete Institute, Farmington Hills, MI, 2000, pp. 392.

7.2 ACI Committee 435. *Control of Deflection in Concrete Structures,* Committee Report ACI 435R–95, Chairman, E. G. Nawy. American Concrete Institute, Farmington Hills, MI, 2000, 1995, p. 77.

7.3 Tadros, M. K. "Designing for Deflection." Paper presented at CIP Seminar on Advanced Design Concepts in Precast Prestressed Concrete, Prestressed Concrete Institute Convention, Dallas, October 1979.

7.4 Branson, D. E. *Deformation of Concrete Structures.* McGraw Hill, New York, 1977.

7.5 Branson, D. E. "The Deformation of Non-Composite and Composite Prestressed Concrete Members." In *Deflection of Concrete Structures,* ACI Special Publication SP-43. American Concrete Institute, Farmington Hills, MI, 1974, pp. 83–127.

7.6 Shaikh, A. F., and Branson, D. E. "Non-Tensioned Steel in Prestressed Concrete Beams. *Journal of the Prestressed Concrete Institute* 15, 1970, 14–36.

7.7 Prestressed Concrete Institute. *PCI Design Handbook.* 5th ed. Prestressed Concrete Institute, Chicago, 1999.

7.8 Martin, L. D. "A Rational Method of Estimating Camber and Deflections of Precast, Prestressed Concrete Members." *Journal of the Prestressed Concrete Institute* 22 (1977): 100–108.

7.9 Nilson, A. H. *Design of Prestressed Concrete.* John Wiley & Sons, New York, 1987.

7.10 Tadros, M. K., Ghali, A., and Dilger, W. H. "Effect of Non-Prestressed Steel on Prestress Loss and Deflection." *Journal of the Prestressed Concrete Institute* 22, 1977, 50–63.

7.11 Tadros, M. K. "Expedient Service Load Analysis of Cracked Prestressed Concrete Sections. *Journal of the Prestressed Concrete Institute* 27, 86–111, 1982, also Discussions and Author's Closure, 28 (1983): 137–158.

7.12 Naaman, A. E. "Partially Prestressed Concrete: Review and Recommendations. *Journal of the Prestressed Concrete Institute* 30, 1985, 30–71.

7.13 Nawy, E. G., and Potyondy, J. G. "Flexural Cracking Behavior of Pretensioned Prestressed Concrete I- and T-Beams." *Journal of the American Concrete Institute* 65, Farmington Hills, MI, 1971, 335–360.

7.14 Nawy, E. G., and Huang, P. T. "Crack and Deflection Control of Pretensioned Prestressed Beams." *Journal of the Prestressed Concrete Institute* 22, 1977, 30–47.

7.15 Nawy, E. G., and Chiang, J. Y. "Serviceability Behavior of Post-Tensioned Beams." *Journal of the Prestressed Concrete Institute* 25, 1980, 74–95.

7.16 Cohn, M. Z. "Partial Prestressing From Theory to Practice." *NATO-ASI Applied Science Series,* Vols. I and II, Publ. Martinus Nijhoff Publishers, in Cooperation with NATO Scientific Affairs Division, Dordrecht, 1986, Vol. 1, p. 405; Vol. II, p. 425.

7.17 Nawy, E. G. "Flexural Cracking Behavior of Pretensioned and Post-Tensioned Beams: The State of the Art." *Journal of the American Concrete Institute,* Farmington Hills, MI, December 1985, pp. 890–900.

7.18 Nawy, E. G. "Flexural Cracking Behavior and Crack Control of Pretensioned and Post-Tensioned Prestressed Beams." *Proceedings of the NATO-NSF Advanced Research Workshop,* vol. 2. Dordrecht-Boston: Martinus Nijhoff, 1986, pp. 137–156.

7.19 Neville, A. M. *Properties of Concrete.* 4th ed., Addison Wesley Longman, 1996.

7.20 Bazant, Z. P. "Prediction of Creep Effects Using Age Adjusted Effective Modulus Method." *Journal of the American Concrete Institute,* Farmington Hills, MI, April 1972, pp. 212–217.

7.21 Libby, J. R. *Modern Prestressed Concrete.* Van Nostrand Reinhold, New York, 1984.

7.22 Nawy, E. G., *High Performance Concrete,* 2nd Ed., John Wiley and Sons, New York, 2000.

7.23 Nawy, E. G., editor-in-chief, *Concrete Construction Engineering Handbook,* CRC Press, Boca Raton, FL, 1998, 1250 pp.

7.24 PCI, *Prestressed Concrete Bridge Design Handbook,* Precast/Prestressed Concrete Institute, Chicago, 1998.

7.25 AASHTO, *Standard Specifications for Highway Bridges,* 16th ed. and 97–98 Supplements, American Association of State Highway and Transportation Officials, Washington, D.C., 1996–1998.

PROBLEMS

7.1 Calculate the instantaneous and long-term cambers and deflections of the AASHTO beam of Example 4.2 for 7, 30, 180, and 365 days, and 5 years by (a) the PCI multipliers method, (b) the incremental time-steps method, and (c) the approximate time-steps method. Then tabulate and compare the results. Are the deflections within the AASHTO permissible limits on deflection? Given $f'_c = 6{,}000$ psi.

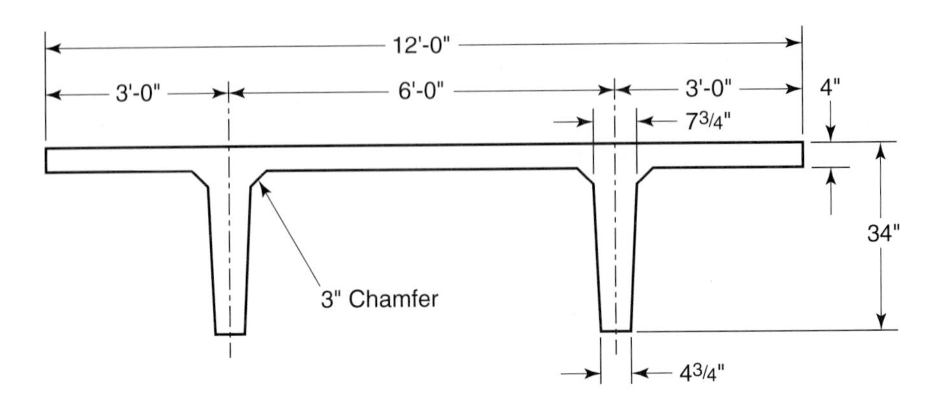

Figure P7.2

7.2 A 68-ft (20.7-m) span simply supported lightweight concrete double-T-beam is subjected to a superimposed topping load $W_{SD} = 250$ plf (3.65 kN/m) and a service live load $W_L = 300$ plf (4.38 kN/m). Calculate the immediate camber and deflection of this beam by the bilinear method and the time-dependent deflections at intervals of 7, 30, 90, and 365 days using the PCI multipliers method and verify whether they are within the permissible ACI limits on deflection for the conditions where nonstructural elements are not likely to be damaged by large deflections. Use Figure P7.2 and the following data.

	Noncomposite	Composite
A_c, in.^{2}	615 (3,968 cm^2)	855
I_c, in.^{4}	59,720 (24.9 × 10^5 cm^4)	82,723
r^2, in.^{2}	97	97
c_b, in.	21.98 (558 mm)	25.37
c_t, in.	10.02 (255 mm)	8.63
S_b, in.^{3}	2,717 (4.5 × 10^4)	3,261
S^t, in.^{3}	5,960 (9.8 × 10^4)	9,587
W_d, plf	491 (7.15 kN/m)	741

$V/S = 615/346 = 1.69$ in. (4.3 cm)

$RH = 75\%$

$e_c = 18.73$ in. (476 mm)

$e_e = 12.81$ in. (325 mm)

$f'_c = 5{,}000$ psi (34.5 MPa) (lightweight)

$f'_{ci} = 3{,}750$ psi (25.7 MPa)

f_t at bottom fibers $= 12\sqrt{f'_c} = 849$ psi (5.6 MPa)

A_{ps} = twelve $\frac{1}{2}$-in. dia low-relaxation steel depressed at midspan only

$f_{pu} = 270{,}000$ psi (1,862 MPa)

$f_{pi} = 189{,}000$ psi (1,303 MPa)

$f_{py} = 260{,}000$ psi (1,793 MPa)

$E_{ps} = 28.5 \times 10^6$ psi (196 GPa)

7.3 Determine the crack width and stabilized mean crack spacing in the double-T beam of Problem 7.2 for an incremental stress of 15,000 psi (103 MPa) beyond the decompression state. Also, determine whether the maximum crack width obtained satisfies the serviceability requirement for crack control for a humid and moist environment.

7.4 A simply supported bonded double-T beam has a 50-ft span and is subjected to a uniform live load of 1,250 plf and a superimposed dead load of 200 plf. Its geometrical properties and maximum allowable stresses are as follows:

$$A_c = 615 \text{ in.}^2$$
$$I_c = 59{,}720 \text{ in.}^4 \ (77{,}118 \text{ in}^4)$$
$$S_b = 2{,}717 \text{ in.}^3 \ (3{,}142 \text{ in}^3)$$
$$S^t = 5{,}960 \text{ in.}^3 \ (8{,}150 \text{ in}^3)$$
$$W_D = 641 \text{ plf} \ (491 \text{ plf})$$
$$V/S = 1.69 \text{ in.}$$
$$f_c' = 5{,}000 \text{ psi (normal weight)}$$
$$f_c = 2{,}250 \text{ psi}$$
$$f_{ci}' = 3{,}750 \text{ psi}$$
$$f_{ti} = 184 \text{ psi}$$
$$f_{pu} = 270{,}000 \text{ psi}$$
$$f_{py} = 235{,}000 \text{ psi}$$
$$f_{pi} = 195{,}000 \text{ psi}$$
$$E_{ps} = 28 \times 10^6 \text{ psi}$$
$$c_b = 21.98 \text{ in.} \ (24.54 \text{ in.})$$
$$c_t = 10.02 \text{ in.} \ (9.46 \text{ in.})$$
$$RH = 70\%$$
$$f_{ci} = 2{,}250 \text{ psi}$$
$$f_t = 849 \text{ psi}$$
$$f_{pe} = 150{,}000 \text{ psi}$$
$$f_{py} = 60{,}000 \text{ psi}$$
$$A_{ps} = \text{sixteen } \tfrac{1}{2}\text{-in. dia 7-wire low-relaxation strands}$$

Bracketed values are for the composite section due to 2-in. topping.

(a) Find the eccentricities e_c and e_e that would result in a tensile stress $f_t = 750$ psi at the lower fiber at midspan *at service load*, and tensile stresses within the allowable limits at the support section both at initial prestress and at service load. Use *nonprestressed* reinforcement where necessary.

(b) Find the long-term camber and deflection of the beam by the *approximate time-step procedure* for $t = 7$ days and $t = 180$ days, assuming that the ultimate creep coefficient $C_u = 2.35$. Use the moment-curvature approach to determine the initial camber at transfer. Are the values within the allowable ACI limits?

(c) Calculate the flexural crack width at service load for a stress increment $\Delta f_s = 15{,}000$ psi beyond the decompression stress.

7.5 Find the long-term camber and deflection in Example 7.9 by the incremental time-steps method, assuming that twelve $\tfrac{1}{2}$-in. dia 7-wire 270 K stress-relieved strands are used for prestressing the beam section. Calculate the flexural crack width at service load for a stress increment $\Delta f_s = 15{,}000$ psi beyond the decompression stress.

8

PRESTRESSED COMPRESSION AND TENSION MEMBERS

8.1 INTRODUCTION

Although prestressing is predominantly used in flexural members such as beams and slabs, it is also used in axially loaded members such as long columns (compression members) and ties for arches and truss elements (tension members). Yet another use is in pretensioned and post-tensioned prestressed piles and masts.

The theory, analysis, and design of prestressed compression members are similar to those of reinforced concrete members. The internal axial prestressing force in the *bonded* tendon produces no column action; hence, no buckling can result as long as the prestressing steel and the surrounding concrete are in direct contact along the total length of the element. As such, the bending tendency of the concrete at midlength is neutralized by the stretching effect of the axially embedded prestressing strands.

Columns are normally subjected to bending in addition to axial load, since external loads are rarely concentric. As a result, the concrete section is subjected to tension at the side farthest from the line of action of the longitudinal load. Cracking develops, but it can be prevented through the use of prestress in the columns. If the applied load is concen-

The George Moscone Convention Center, San Francisco, California, design by T.Y. Lin International. (*Courtesy,* Post-Tensioning Institute.)

tric, prestressing is inconsequential if not altogether disadvantageous, as the compressive stress on the concrete section is needlessly increased.

A compression member can be considered fully prestressed throughout its length if no loss in development of prestressing occurs at its ends. If partial loss occurs, the reinforcement segment in the development zone is considered nonpresstressed and the section at the end zone is treated as a reinforced concrete eccentrically loaded section.

Tension members are normally subjected to direct tension only. These elements are mostly linear, as for example, railroad ties, restraining ties for arch bridges, tension members in trusses, and foundation anchorages for earthwork retaining structures. Tension members can also be circular or parabolic in shape, as witness prestressed circular containers or catenary-shaped bridge elements. The fundamental function of the pure tension members is to prevent their cracking at service load and to enable them to sustain all the necessary deformation needed to develop full resistance to the external service loads and overloads. Being crack-free would fully protect the reinforcement from corrosion and other environmental conditions.

8.2 PRESTRESSED COMPRESSION MEMBERS: LOAD–MOMENT INTERACTION IN COLUMNS AND PILES

In order to evaluate the nominal strength of a column at various eccentric load levels, it is necessary to evaluate all possible combinations of ultimate nominal loads P_n and ultimate nominal moments M_n given by

$$M_n = P_n e_i \tag{8.1}$$

where e_i is the eccentricity of the load at the various load–moment combinations. A plot of the relationship between P_n and M_n is shown in the interaction diagram of Figure 8.1 for both nonslender columns (material failure) and slender columns (stability failure). In the nonslender column, failure occurs as the load reaches the value A along path OA, and the concrete arches at the compression side. In the slender column, the maximum load is reached at B along the path OBC, which intersects the interaction diagram at C. Instability occurs once the critical load is reached. A quantitative definition of slender and nonslender columns is given later.

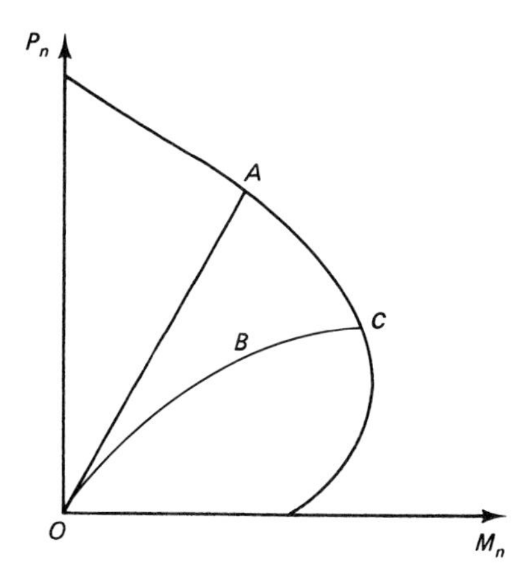

Figure 8.1 Basic interaction diagram for columns. (a) Path OA for material failure in nonslender column. (b) Path OBC for buckling failure of slender column.

The basic assumptions made in regard to prestressed concrete, similar to those made in regard to reinforced concrete columns, are as follows:

1. The strain distribution in the concrete varies linearly with depth.
2. The stress distribution in the compression zone is parabolic and is replaced in analysis and design by an equivalent rectangular block.

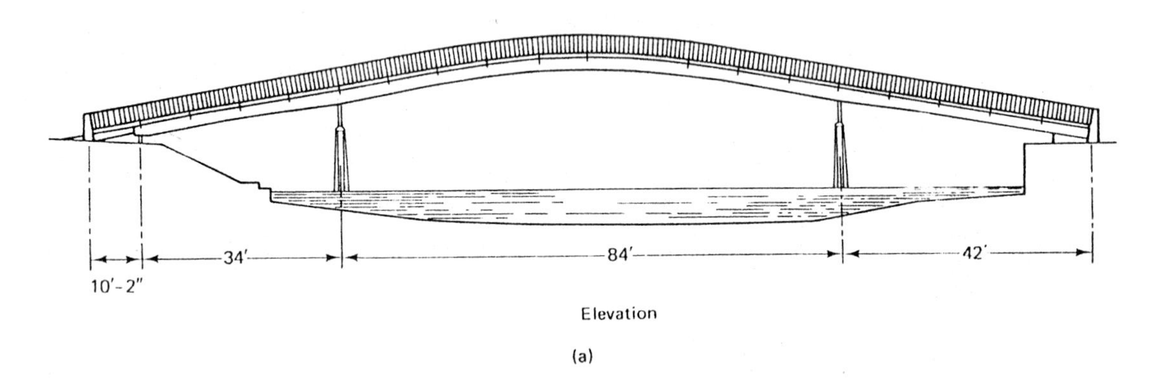

(a)

b

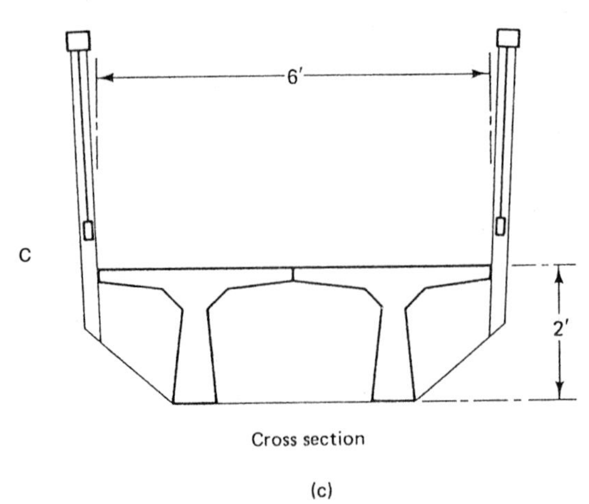

c

(c)

Photo 8.1 Eel Pie Island, Twickenham, England, prestressed concrete footbridge.

3. The stress-strain diagrams of the concrete and the prestressing steel are known.
4. The crushing strain of the concrete in combined bending and axial load at the extreme fibers is $\epsilon_c = 0.003$ in./in., and the average crushing strain at mid-depth of a concrete section subjected mainly to axial load is $\epsilon_0 = 0.002$ in./in.
5. The section is considered to have failed when the strain in the concrete at the extreme compression fibers reaches $\epsilon_c = 0.003$ in./in. or $\epsilon_0 = 0.002$ in./in. at mid-depth. Note that $\epsilon_c = 0.003$ is the value used in the ACI Code, whereas other codes use a higher value of 0.0035 or 0.0038.
6. Compatibility of strain is postulated between the concrete and the prestressing steel.

The modes of failure are also similar to those of reinforced concrete columns:

1. *Initial compression failure, small eccentricity.* This failure mode develops when the strain in the concrete at the loaded side reaches $\epsilon_{cu} = 0.003$ in./in. while the strain in the prestressing steel at the far side is below the yield strain. The eccentricity e of the axial load is smaller than the balanced eccentricity e_b.
2. *Initial tension failure, large eccentricity.* This failure mode is the reverse of the preceding one. The steel at the far side yields prior to the crushing of the concrete at the loaded side. The eccentricity e of the axial load is larger than the balanced eccentricity e_b.
3. *Balanced condition, balanced eccentricity.* This mode defines the condition of maximum moment value M_{nb} on the interaction curve corresponding to a maximum tensile strain in the tension layers equal to a strain increment $\Delta\epsilon_{ps} = 0.0012$ to 0.0020 in./in. beyond the service load level. The eccentricity of the axial load is defined as the balanced eccentricity e_b.

The three major controlling points on the interaction diagram are:

1. $M_u = 0$, corresponding to $\epsilon_0 = 0.002$ in./in. at failure due to the concentric load P_u. The neutral axis position is at infinity.
2. No tension at the extreme concrete tensile fibers and $\epsilon_{cu} = 0.003$ in./in. at the extreme concrete compression fibers. The neutral axis position is at the extreme tension fibers.
3. $P_u = 0$ and $\epsilon_{cu} = 0.003$ in./in. at the extreme compression fibers. The neutral axis is inside the section and is determined by trial and adjustment, assuming a depth c and then testing the assumption.

Figure 8.2 shows the strain and stress distribution for these three cases.

The remaining points on the interaction diagram are for cases that lie between stages (a), (b), and (c) of Figure 8.2, namely, from concentric loading to pure bending. In the case of columns, pure bending defines the state where the ratio of the factored axial load P_u to the factored flexural moment M_u is negligible. The parabolic distribution of stress for cases (b) and (c) is replaced by the equivalent rectangular block, where the block depth $a = \beta_1 c$, as is done in the case of flexural beams.

The typical case of a compression member lies between stages (b) and (c) of Figure 8.2. The strains, stresses, and forces for such a case are shown in Figure 8.3 for the critical section at the limit state of ultimate load by material failure. Cutting the free-body diagram at the column midheight above section 1–1, the cross-section of the member is shown in part (b) of the figure, and the strain and stress at failure in parts (c) and (d), re-

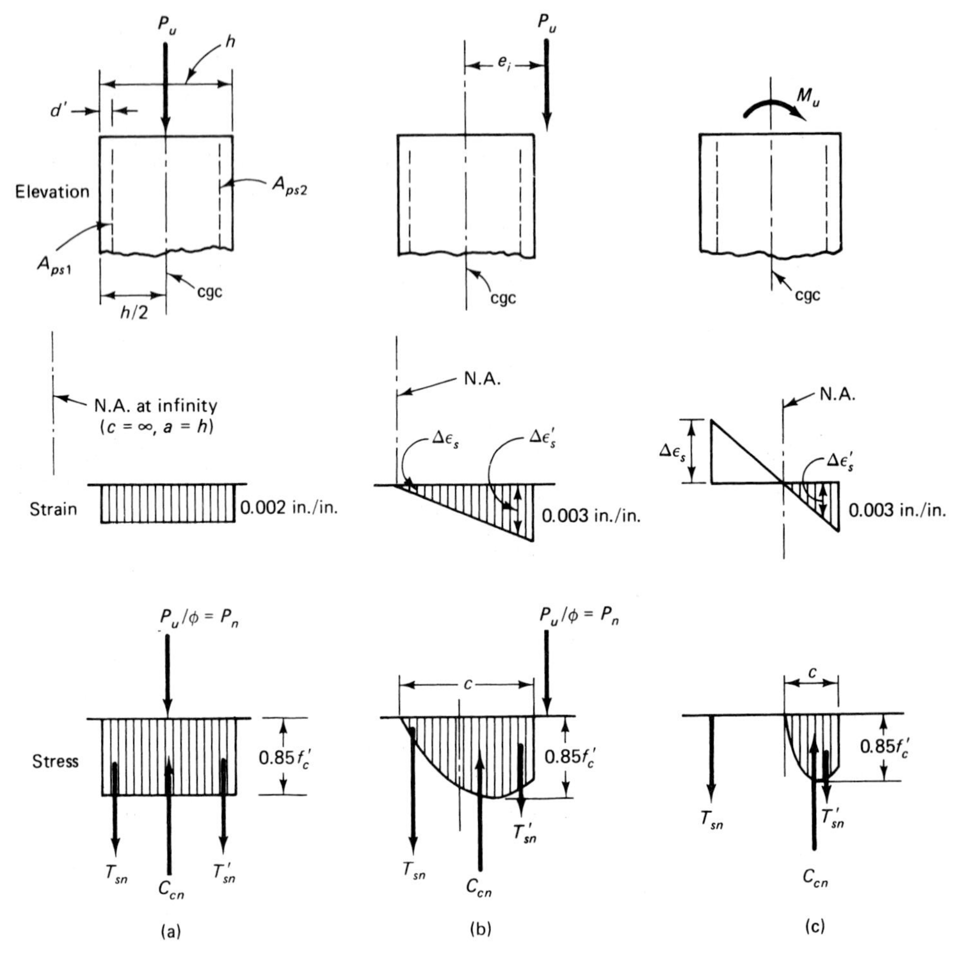

Figure 8.2 Strain and stress distribution alternatives across concrete section depth. (a) $M_u = 0$. (b) $\epsilon_c = 0.003$ in./in., and zero tension at the extreme tensile fibers. (c) $P_u = 0$ and $\epsilon_c = 0.003$ in./in. at the extreme compressive fibers.

spectively. The strain ϵ_{ce} is the *uniform* strain in the concrete under effective prestress after creep, shrinkage, and relaxation losses given respectively by

$$C_{cn} = 0.85 f'_c\, ba \tag{8.2a}$$

$$T'_{sn} = A'_{ps} f'_{ps} \tag{8.2b}$$

and

$$T_{sn} = f_{ps} A_{ps1} \tag{8.2c}$$

Equilibrium of forces then gives

$$P_n = C_{cn} - T'_{sn} - T_{sn} \tag{8.3}$$

If the effective prestressing force after all losses is P_e, the corresponding strain in the tendons prior to the application of the external load is

$$\epsilon_{pe} = \frac{f_{pe}}{E_{ps}} = \frac{P_e}{(A_{ps} + A'_{ps})E_{ps}} \tag{8.4a}$$

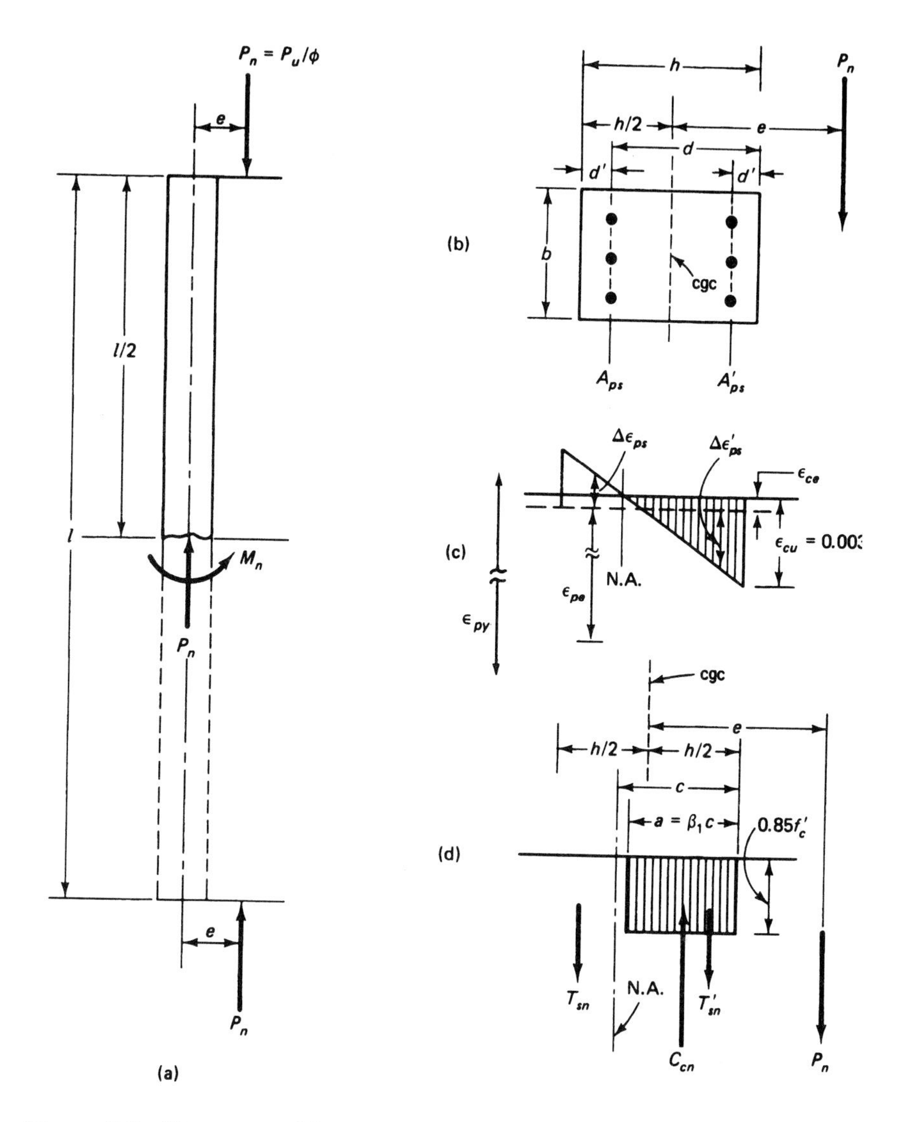

Figure 8.3 Stresses and forces in typical eccentrically loaded nonslender column. (a) Elevation. (b) Cross section. (c) Strain distribution. (d) Stresses and forces.

The change in strain in the prestressing steel area A'_{ps} as the compression member passes from the effective prestressing stage to the ultimate load can be defined as

$$\Delta\epsilon'_{ps} = \epsilon_{cu}\left(\frac{c - d'}{c}\right) - \epsilon_{ce} \tag{8.4b}$$

$$\Delta\epsilon_{ps} = \epsilon_{cu}\left(\frac{d - c}{c}\right) + \epsilon_{ce} \tag{8.4c}$$

$$\Delta\epsilon_p = \Delta_{ps} - \epsilon_{ce} \tag{8.4d}$$

$$T'_{sn} = A'_{ps}\, f'_{ps} = A'_{ps}\, E_{ps}(\epsilon_{pe} - \Delta\epsilon'_{ps})$$

Photo 8.2 Cable-stayed Sunshine Skyway Bridge, Tampa, Florida. (*Courtesy,* Figg Engineering Group Tallahassee, FL.)

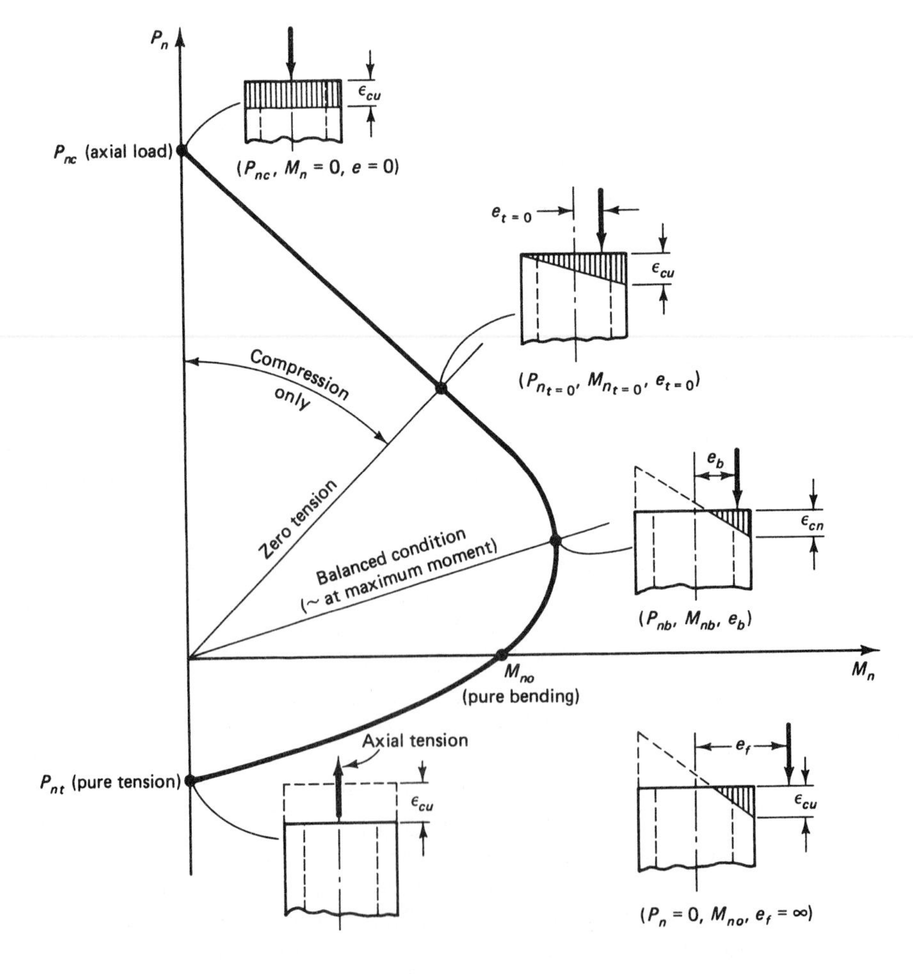

Figure 8.4 Load-moment interaction diagram controlling eccentricities.

or

$$T'_{sn} = A'_{ps}E_{ps}\left[\epsilon_{pe} - \epsilon_{cu}\left(\frac{c - d'}{c}\right) + \epsilon_{ce}\right] \tag{8.5}$$

Similarly,

$$T_{sn} = A_{ps}f_{ps} = A_{ps}E_{ps}(\epsilon_{pe} + \Delta\epsilon_{ps})$$

or

$$T_{sn} = A_{ps}E_{ps}\left[\epsilon_{pe} + \epsilon_{cu}\left(\frac{d - c}{c}\right) + \epsilon_{ce}\right] \tag{8.6}$$

Taking moments about the geometric centroid cgc of the section gives

$$M_n = P_n e = C_{cn}\left(\frac{h}{2} - \frac{a}{2}\right) - T'_{sn}\left(\frac{h}{2} - d'\right) + T_{sn}\left(d - \frac{h}{2}\right) \tag{8.7}$$

From Equations 8.2a, 8.5, 8.6, and 8.7, the nominal strengths P_n and M_n for several eccentricities e_i can be evaluated in order to construct the P–M interaction diagram for any section or develop nondimensional series of P–M interaction diagrams for various concrete strength levels. The design strengths are evaluated from the nominal strength values as

$$\text{Design } P_u = \phi P_n$$

and

$$\text{Design } M_u = \phi M_n = \phi P_n e$$

where ϕ is the strength reduction factor for compression members. Note that the *design* P_u and M_u have to have a value close to, but not less than, the *factored* values P_u and M_u. The load-moment interaction diagram for the controlling eccentricities is shown in Figure 8.4.

8.3 STRENGTH REDUCTION FACTOR ϕ

For members subject to flexure and relatively small axial loads, failure is initiated by yielding of the tension reinforcement and takes place in an increasingly ductile manner. Hence, for small axial loads it is reasonable to permit an increase in the ϕ factor from that required for pure compression members. When the axial load vanishes, the member is subjected to pure flexure, and the strength reduction factor ϕ becomes 0.90. Figure 8.5 shows the zone in which the value of ϕ can be increased from 0.7 to 0.9 for tied columns and 0.75 to 0.9 for spiral columns. In part (a), as the factored design compression load ϕP_n decreases beyond $0.1 A_g f'_c$, the ϕ factor increases from 0.7 to 0.9 for tied columns and 0.75 to 0.9 for spiral columns. For those cases where the value of P_{nb} is less than $0.1A_g f'_c$, ϕ values increase when the load $P_u < P_{ub}$, or $\phi P_n < \phi P_{nb}$, as in part (b) of the figure.

The value of $0.10f'_c A_g$ is chosen by the ACI Code as the design axial load value ϕP_n below which the ϕ factor can safely be increased for most compression members. In summary, if initial failure is in compression, the strength reduction factor ϕ is always 0.70 for tied columns and 0.75 for spirally reinforced columns.

The following expressions give variations in the value of ϕ for symmetrically reinforced compression members. The columns should have an effective depth not less than 70 percent of the total depth. For tied columns,

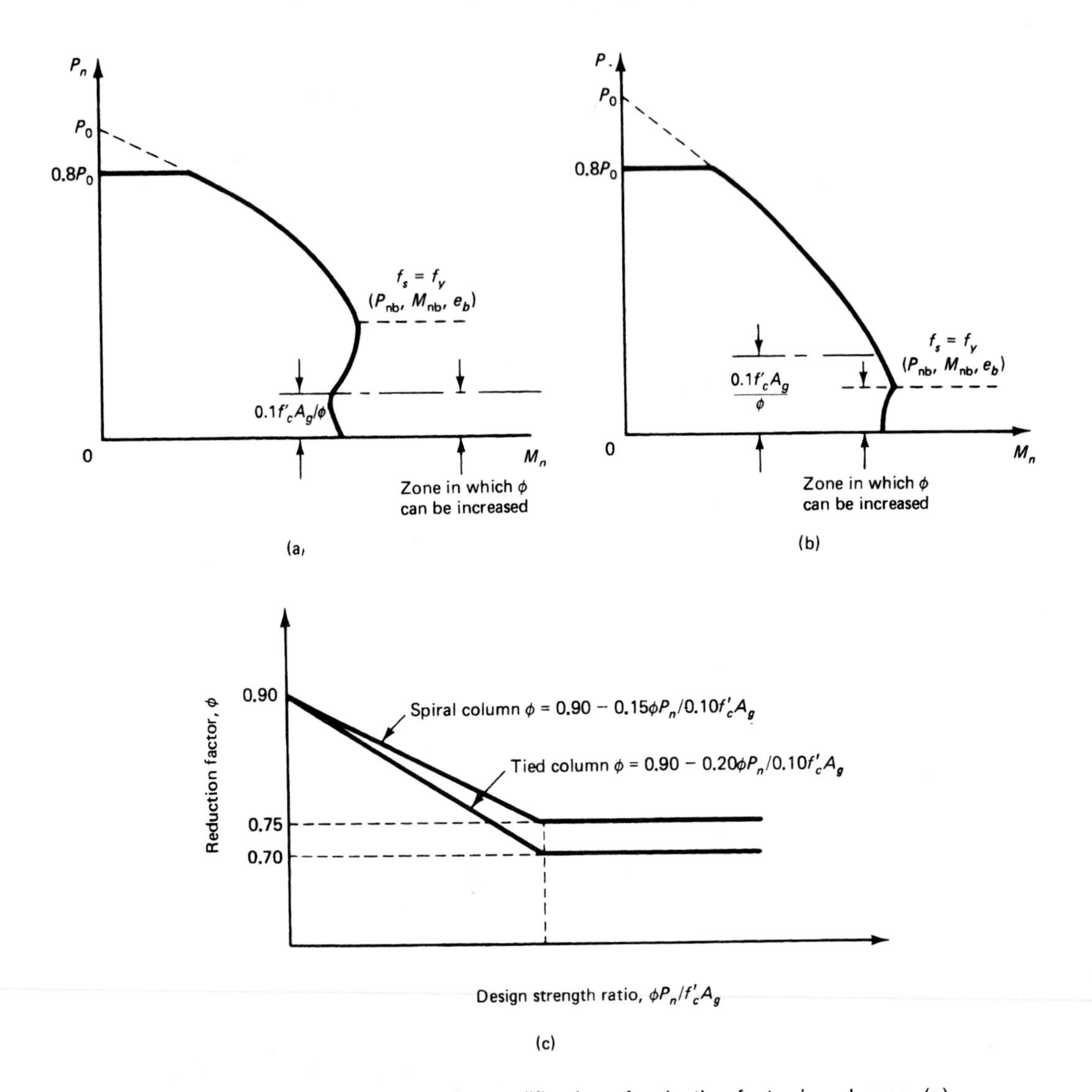

Figure 8.5 Controlling zones for modification of reduction factor in columns. (a) $0.1f'_cA_g < P_{nb}$. (b) $0.1f'_cA_g > P_{nb}$. (c) Variation of ϕ for symmetrically reinforced compression members (P_n is the nominal axial strength at given eccentricity).

$$\phi = 0.90 - \frac{0.20\phi P_n}{0.1f'_cA_g} \geq 0.70 \tag{8.8}$$

For spirally reinforced columns,

$$\phi = 0.90 - \frac{0.15\phi P_n}{0.1f'_cA_g} \geq 0.75 \tag{8.9}$$

In both of these equations, if ϕP_{nb} is less than $0.1f'_cA_g$, then ϕP_{nb} should be substituted for $0.1f'_cA_g$ in the denominator, using 0.7 P_{nb} for tied, and 0.75 P_{nb} for spirally reinforced, columns. Figure 8.5(c) presents the graphical representation of Equations 8.8 and 8.9.

It should be noted that the balanced condition in *prestressed* compression members is highly indeterminate. A reasonable approximation can be made by computing by trial

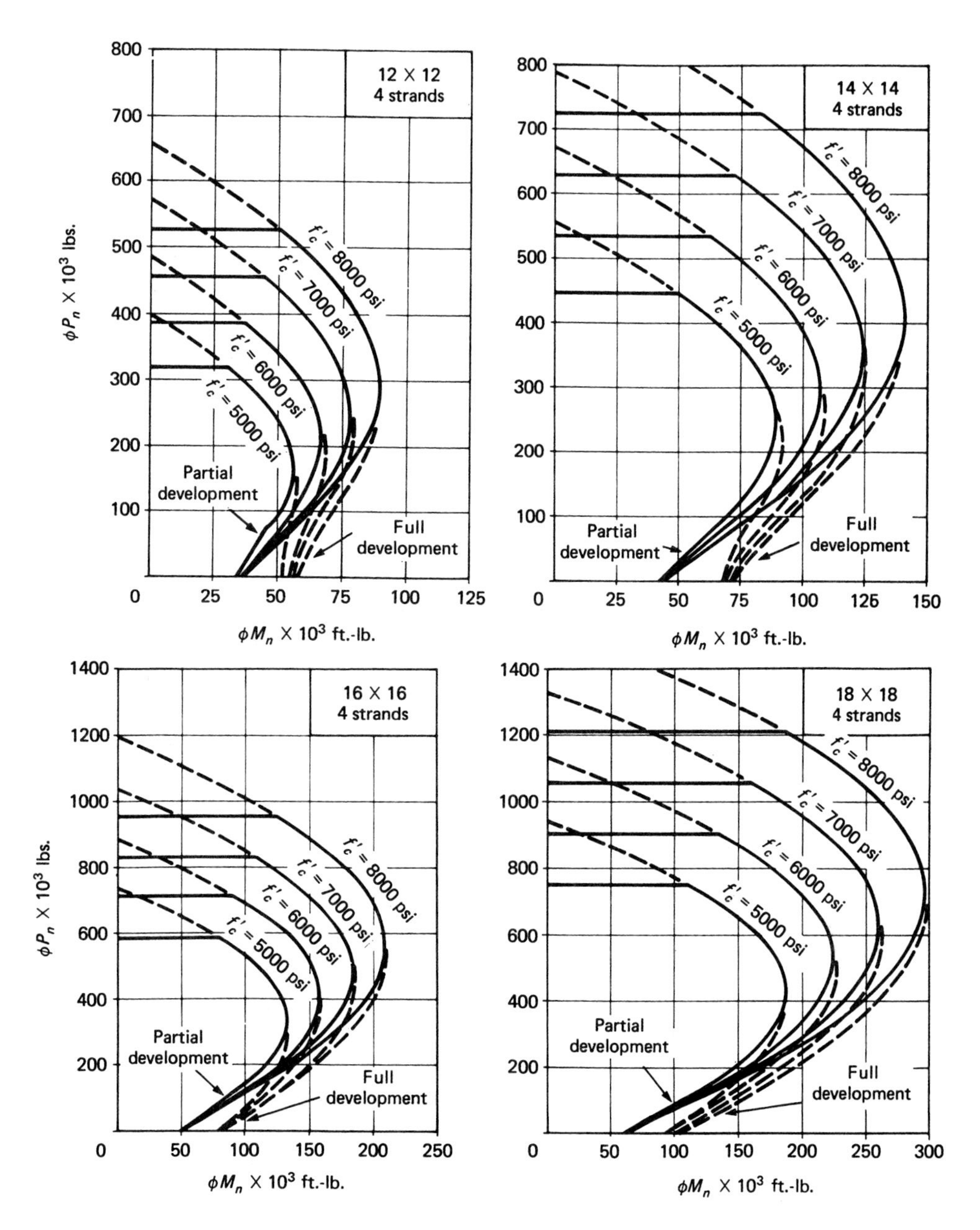

Figure 8.6 Typical design load–moment interaction plots for columns 12 in. × 12 in., 14 in. × 14 in., 16 in. × 16 in., and 24 in. × 24 in. (Ref. 8.3).

and adjustment the maximum moment value coordinate on the interaction diagram as the balanced moment M_{nb}, or by assuming that the tensile strain in the tension layers equals a strain increment $\Delta\epsilon_{ps} = 0.0012$ to 0.0020 in./in. beyond the service-load level and computing the compressive stress block depth a of the concrete accordingly. This assumption has to be verified and the M_{nb} value adjusted after the interaction diagram is plotted in order to determine the maximum moment ordinate in the diagram for which the neutral axis depth c_b will be used in the computations.

Typical load–moment interaction diagrams are given in Figure 8.6 (see Ref. 8.3).

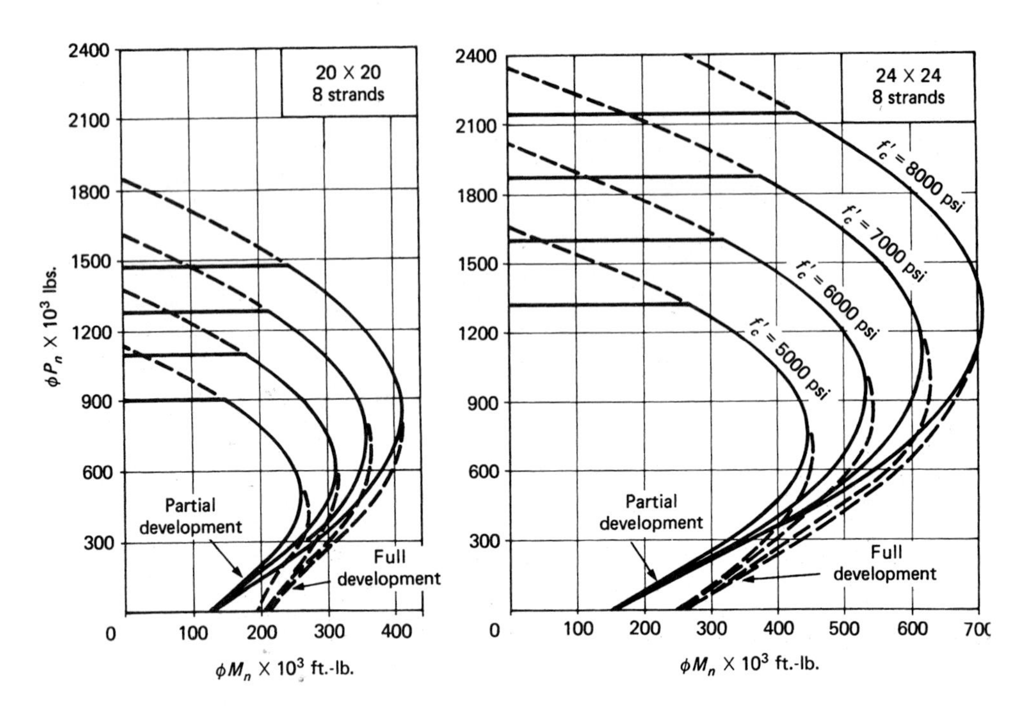

Figure 8.6 Continued

8.4 OPERATIONAL PROCEDURE FOR THE DESIGN OF NONSLENDER PRESTRESSED COMPRESSION MEMBERS

The following steps can be carried out for the design of nonslender (short) columns where the behavior is controlled by material failure:

1. Evaluate the factored external axial load P_u and the factored moment M_u. Compute the applied eccentricity $e = M_u/P_u$.
2. Assume a cross section and type of lateral reinforcement to be used, namely, tied or spiral. Avoid fractional quantities in selecting sectional dimensions.
3. Assume the number and size of strands.
4. Compute M_{nb} for the assumed section as the balanced moment and the corresponding P_{nb} and $e_b = M_{nb}/P_{nb}$ through evaluation of the maximum moment coordinate in the interaction diagram by trial and adjustment. Or, alternatively, assume that the strain in the extreme tensile fibers is equal to an assumed strain ϵ_{ps} of the prestressing steel, and then proceed to compute the balanced axial load P_{nb} and the balanced moment M_{nb}. This step also enables one to verify the value of the strength reduction factor. The balanced moment M_{nb} results from an ϵ_{ps} strain value to give a maximum moment in the interaction diagram.
5. Assume a neutral axis depth c, and find the corresponding P_n and M_n. Then check for the adequacy of the assumed section, i.e., whether $\phi P_n >$ the factored P_u and $\phi M_n >$ the factored M_u. If the section cannot support the factored load or is oversized and hence uneconomical, revise the cross section and the reinforcement through trial and adjustment by repeating steps 4 and 5 as necessary, including the construction of an interaction diagram.
6. Design the lateral reinforcement.

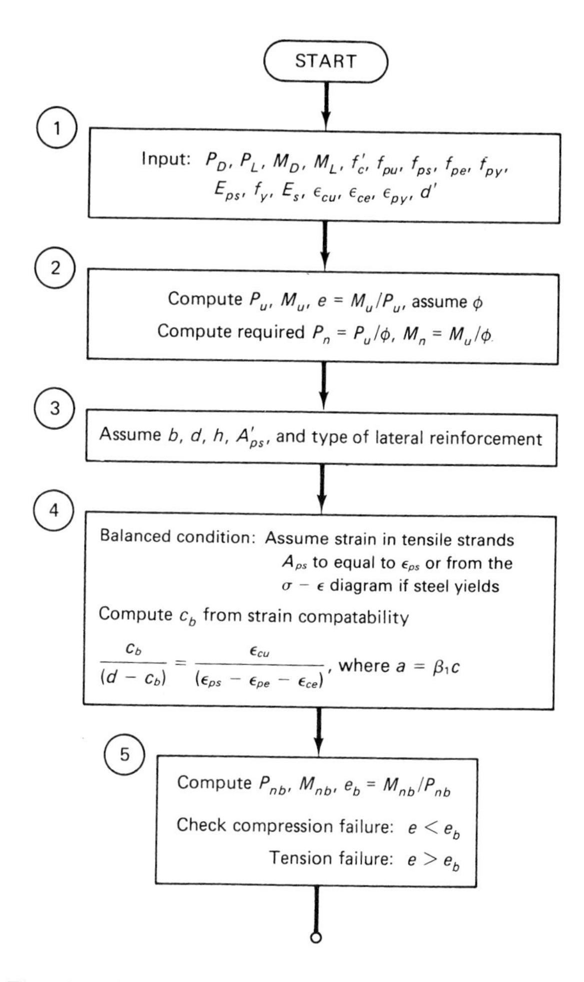

Figure 8.7 Flowchart for design or analysis of prestressed concrete nonslender compression members.

Figure 8.7 presents a flowchart of the trial-and-adjustment sequences of the design or analysis procedure.

8.5 CONSTRUCTION OF NOMINAL LOAD–MOMENT ($P_n - M_n$) AND DESIGN ($P_u - M_u$) INTERACTION DIAGRAMS

Example 8.1

Construct the nominal load–moment interaction diagram for a prestressed concrete compression member 14-in. (356 mm) wide and 14-in. deep. The member is reinforced with eight $\frac{1}{2}$-in. (12.7 mm) dia 7-wire stress-relieved 270-K strands, half on each side of the two faces parallel to the neutral axis as shown in Figure 8.8. The stress-strain diagram for the strands is shown in Figure 8.9. The effective prestress after all losses is f_{pe} = 150,000 psi (1,034 MPa). Additionally, draw the design interaction diagram using the appropriate strength reduction factor values. Consider the strands fully developed throughout the length of the member. Given data are as follows:

$$f'_c = 6{,}000 \text{ psi (47.5 MPa), normal-weight concrete}$$

$$E_{ps} = 29 \times 10^6 \text{ psi } (200 \times 10^3 \text{ MPa})$$

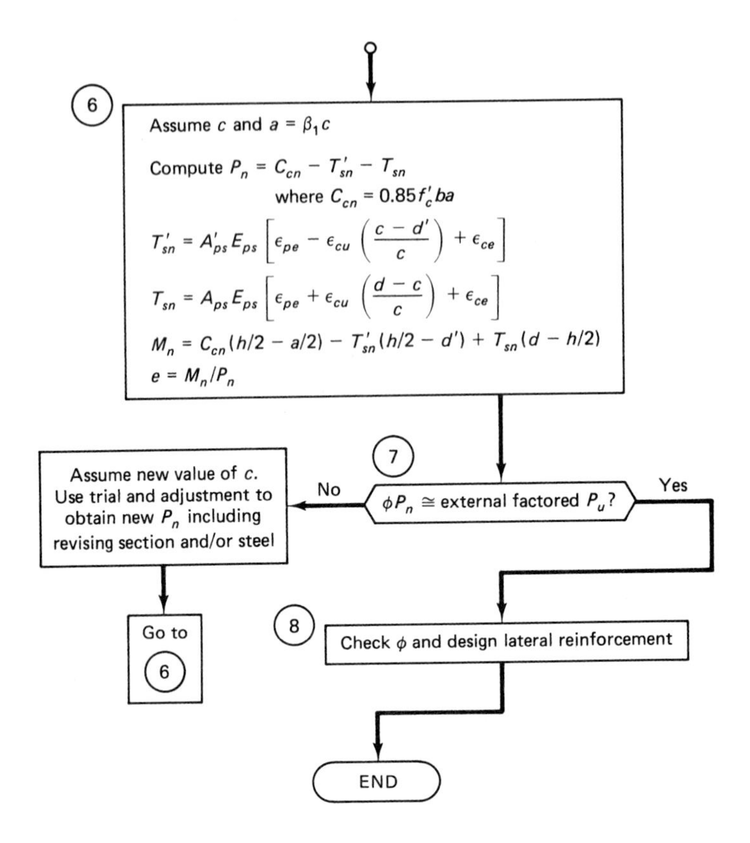

Figure 8.7 *Continued*

Photo 8.3 Interior of the George Moscone Convention Center, San Francisco, California, design by T. Y. Lin International. (*Courtesy,* Post-Tensioning Institute.)

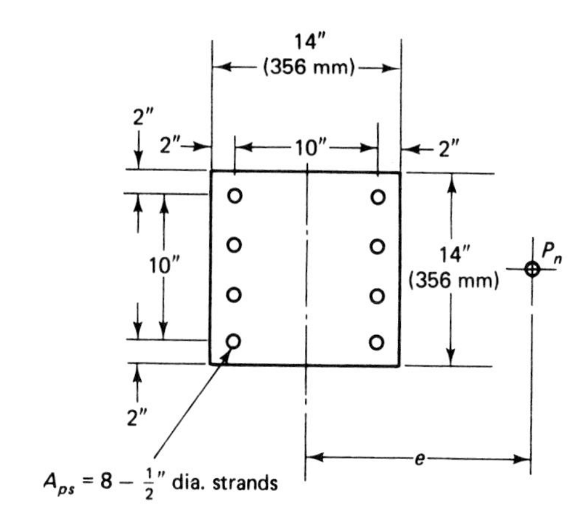

Figure 8.8 Section geometry.

f_{ps} = 240,000 psi (1,655 MPa)

ϵ_{cu} = 0.003 in./in. at failure

ϵ_{ce} = 0.0005 in./in. when P_e acts on the section $= \dfrac{P_e}{A_c E_c}\left(1 + \dfrac{e^2}{r^2}\right)$

ϵ_{py} = strand yield strain ≅ 0.012 in./in. from Figure 8.9

Assume a reasonable value of ϵ_p and adjust as necessary.

Solution:

Nominal Strength P_n–M_n Diagram

1. *Axial Compression:* $M_u = 0$, $c = \infty$ (Use $\epsilon_{cu} = 0.003$ since perfect axial compression is impossible.)

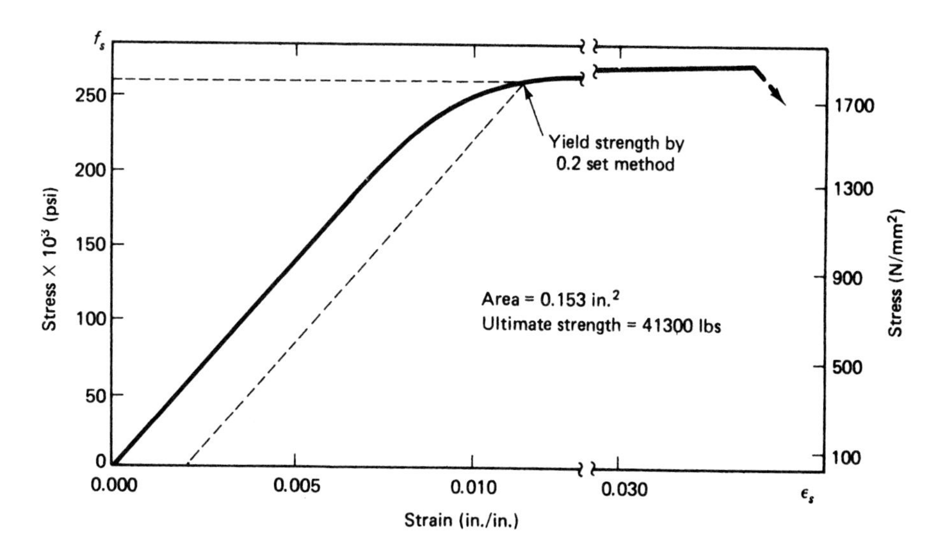

Figure 8.9 Stress-strain diagram for ½-in. (12.7 mm) dia 270-K prestressing tendon.

The compressive block depth a = 14 in. (356 mm), and the effective depth d = 14 − 2 = 12 in. (305 mm). So we have $C_{cn} = 0.85f'_c\,ba = 0.85 \times 6000 \times 14 \times 14 = 999{,}600$ lb (4,446 kN).

From Equation 8.5,

$$T'_{sn} = A'_{ps}E_{ps}\left[\epsilon_{pe} - \epsilon_{cu}\left(\frac{c - d'}{c}\right) + \epsilon_{ce}\right]$$

$$A'_{ps} = 4 \times 0.153 = 0.612 \text{ in.}^2\ (3.95 \text{ cm}^2)$$

From Figure 8.9, for $E_{ps} = 29 \times 10^6$ psi (200×10^3 MPa), $\epsilon_{pe} = 0.0052$ in./in. and ϵ_{cu} = 0.003 in./in. Thus,

$$T'_{sn} = 0.612 \times 29 \times 10^6\left[0.0052 - 0.003\left(\frac{\infty - 2}{\infty}\right) + 0.0005\right]$$

$$= 0.612 \times 29 \times 10^6(0.0052 - 0.003 + 0.0005)$$

$$= 47{,}920 \text{ lb } (213 \text{ kN})$$

From Equation 8.6,

$$T_{sn} = A_{ps}E_{ps}\left[\epsilon_{pe} + \epsilon_{cu}\left(\frac{d - c}{c}\right) + \epsilon_{ce}\right]$$

$$= 0.612 \times 29 \times 10^6\left[0.0052 + 0.003\left(\frac{12 - \infty}{\infty}\right) + 0.0005\right]$$

$$= 47{,}920 \text{ lb } (213 \text{ kN})$$

From Equation 8.2,

$$P_n = C_{cn} - T'_{sn} - T_{sn} = 999{,}600 - 47{,}920 = 47{,}920$$

$$= 903{,}760 \text{ lb } (4{,}020 \text{ kN})$$

From Equation 8.7,

$$M_n^{\curvearrowright} = C_{cn}\left(\frac{h}{2} - \frac{a}{2}\right)^{\curvearrowleft} - T'_{sn}\left(\frac{h}{2} - d'\right)^{\curvearrowright} + T_{sn}\left(d - \frac{h}{2}\right)^{\curvearrowleft}$$

$$= 999{,}600\left(\frac{14}{2} - \frac{14}{2}\right) - 47{,}920\left(\frac{14}{2} - 2\right) + 47{,}920\left(12 - \frac{14}{2}\right)$$

$$= 0$$

$$e_1 = \frac{M_n}{P_n} = 0$$

2. *Zero Tension at Tension Face, c = 14 in.*

$$\beta_1 = 0.85 - \frac{0.05(f'_c - 4{,}000)}{1{,}000} = 0.75$$

$$a = \beta_1 c = 0.75 \times 14 = 10.5 \text{ in. } (267 \text{ mm})$$

$$C_{cn} = 0.85 \times 6{,}000 \times 14 \times 10.5 = 749{,}700 \text{ lb } (3{,}335 \text{ kN})$$

$$T'_{sn} = 0.612 \times 29 \times 10^6\left[0.0052 - 0.003\left(\frac{14 - 2}{14}\right) + 0.0005\right]$$

$$= 55{,}526 \text{ lb } (247 \text{ kN})$$

$$T_{sn} = 0.612 \times 29 \times 10^6\left[0.0052 + 0.003\left(\frac{12 - 14}{14}\right) + 0.0005\right]$$

$$= 93{,}557 \text{ lb } (416 \text{ kN})$$

$$P_n = C_{cn} - T'_{sn} - T_{sn} = 749{,}700 - 55{,}526 - 93{,}557$$

$$= 600{,}617 \text{ lb } (2{,}672 \text{ kN})$$

$$M_n = 749{,}700\left(\frac{14}{2} - \frac{10.5}{2}\right)\curvearrowleft - 55{,}526\left(\frac{14}{2} - 2\right)\curvearrowright + 93{,}557\left(12 - \frac{14}{2}\right)\curvearrowleft$$

$$= 1{,}502{,}130 \text{ in.-lb } (170 \text{ kN-m})$$

$$e_2 = \frac{1{,}502{,}130}{600{,}617} = 2.50 \text{ in. } (63.5 \text{ mm})$$

3. *Pure Bending:* $P_u = 0$

Neglecting the effect of the compression steel A'_{ps}, we have

$$a = \frac{A_{ps} f_{ps}}{0.85 f'_c b} = \frac{0.612 \times 240{,}000}{0.85 \times 6{,}000 \times 14} = 2.06 \text{ in. } (52.3 \text{ mm})$$

$$c = \frac{2.06}{0.75} = 2.75 \text{ in. } (69.9 \text{ mm})$$

$$M_n = A_{ps} f_{ps}\left(d - \frac{a}{2}\right) = 0.612 \times 240{,}000\left(12 - \frac{2.06}{2}\right)$$

$$= 1{,}611{,}274 \text{ in.-lb}$$

$$e_3 = \frac{1{,}611{,}274}{0} = \infty$$

4. *Balanced Condition:* P_{nb}, M_{nb}, e_b

Assume the strain in the tensile strands A_{ps} to be equal to the incremental strain $\Delta\epsilon_p$ beyond the service-load level P_e. Taking a value $\Delta\epsilon_p \cong 0.0014$ to be modified by trial and adjustment, an from Figure 8.10, similar triangles give

$$\frac{c}{(d - c)} = \frac{\epsilon_{cu}}{\Delta_p} = \frac{0.003}{0.0014}$$

Hence $c = 8.15$ in. (207 mm). So

$$a_b = \beta_1 c = 0.75 \times 8.15 = 6.11 \text{ in. } (155 \text{ mm})$$

$$C_{cn} = 0.85 \times 6{,}000 \times 6.10 \times 14 = 435{,}540 \text{ lb } (1{,}937 \text{ kN})$$

$$T'_{sn} = 0.612 \times 29 \times 10^6 \left[0.0052 - 0.003\left(\frac{8.13 - 2}{8.13}\right) + 0.0005\right]$$

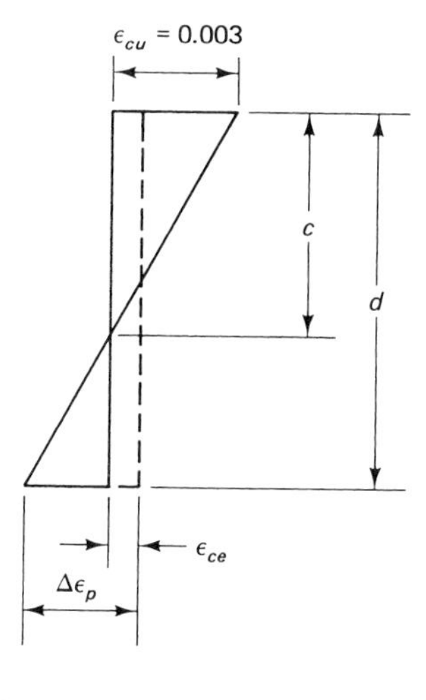

Figure 8.10 Strain distribution.

$$= 61{,}018 \text{ lb } (271 \text{ kN})$$

$$T_{sn} = 0.612 \times 29 \times 10^6 \left[0.0052 + 0.003\left(\frac{12 - 8.13}{8.13}\right) + 0.0005\right]$$

$$= 126{,}509 \text{ lb } (563 \text{ kN})$$

$$P_{nb} = 435{,}540 - 61{,}018 - 126{,}509 = 248{,}013 \text{ lb } (1{,}103 \text{ kN})$$

$$M_{nb} = 435{,}540\left(\frac{14}{2} - \frac{6.10}{2}\right)^{\curvearrowleft} - 61{,}018\left(\frac{14}{2} - 2\right)^{\curvearrowright} + 126{,}509\left(12 - \frac{14}{2}\right)^{\curvearrowleft}$$

$$= 2{,}047{,}838 \text{ in.-lb } (272 \text{ kN-m})$$

$$e_4 = e_b = \frac{2{,}047{,}838}{248{,}013} = 8.26 \text{ in. } (210 \text{ mm})$$

The coordinates for the preceding four cases are the controlling points on the P_n–M_n interaction diagram. Other points need to be computed as well, in order to develop an accurate diagram to cover the entire loading range. For example, additional points between the coordinates of the second and third cases have to be determined, assuming additional values of the neutral axis depth c and computing P_n, M_n, and e for the c-values assumed. Table 8.1 summarizes the values of the coordinates used for plotting the P_n–M_n interaction diagram as well as the P_u–M_u design diagram. From the diagram, it is seen that the maximum moment ordinate seems to have a value close to $M_n = 2{,}047{,}838$ in.-lb. Hence, an assumption of $c_b = 8.15$ in. is verified.

Design Load–Moment (P_u–M_u) Diagram. From Section 8.3, $0.1A_g f'_c = 0.1 \times 14 \times 14 \times 6{,}000 = 117{,}600$ lb (522 kN) and $P_{nb} = 248{,}013$ lb > 117,600 lb; hence, Figure 8.5a applies for the zone in which ϕ can be increased beyond the 0.70 value for tied columns. To increase ϕ for M_{n7}, we have, from Equation 8.8,

$$\phi = 0.90 - \frac{0.20\phi P_n}{0.1 f'_c A_g}$$

Assume $\phi = 0.77$. Then

$$\phi = 0.90 - \frac{0.20 \times 0.77 \times 101{,}224}{117{,}600} \cong 0.77$$

$$P_{u7} = \phi P_{n7} = 0.77 \times 101{,}224 = 77{,}638 \text{ lb } (345 \text{ kN})$$

$$M_{u7} = \phi M_{n7} = 0.77 \times 1{,}969{,}875 = 1{,}516{,}804 \text{ in.-lb } (171 \text{ kN-m})$$

$$M_{u3} \text{ for pure bending} = \phi M_{n3} = 0.90 \times 1{,}611{,}274$$

Table 8.1 Summary of P–M Interaction Diagram Coordinates in Example 8.1

Point	c in.	a in.	$P_n \times 10^3$ lb	$M_n \times 10^3$ in.-lb	ϕ	$P_u \times 10^3$ lb	$M_u \times 10^3$ in.-lb	e in.
1	∞	14	903.8	0	0.7	632.7*	0	0
5	18	13.5	826.6	388.9	0.7	578.7*	272.2	0.5
2	14	10.5	600.6	1,502.1	0.7	420.4	1,051.5	2.5
6	10	7.5	365.1	2,005.6	0.7	255.6	1,403.9	5.5
4	8.15	6.1	248	2,047.9	0.7	173.6	1,433.5	8.3
7	6	4.5	101.2	1,969.9	0.77	77.9	1,516.8	19.5
3	2.75	2.1	0	1,611.3	0.90	0	1,450.1	∞

*Max P_u allowed by the code for tied columns = $0.80\phi P_n$ = 498,906 lb (2,219 kN). Also,

1,000 lb = 4.448 kN

1,000 in.-lb = 1.1130 kN-m

1 in. = 25.4 mm.

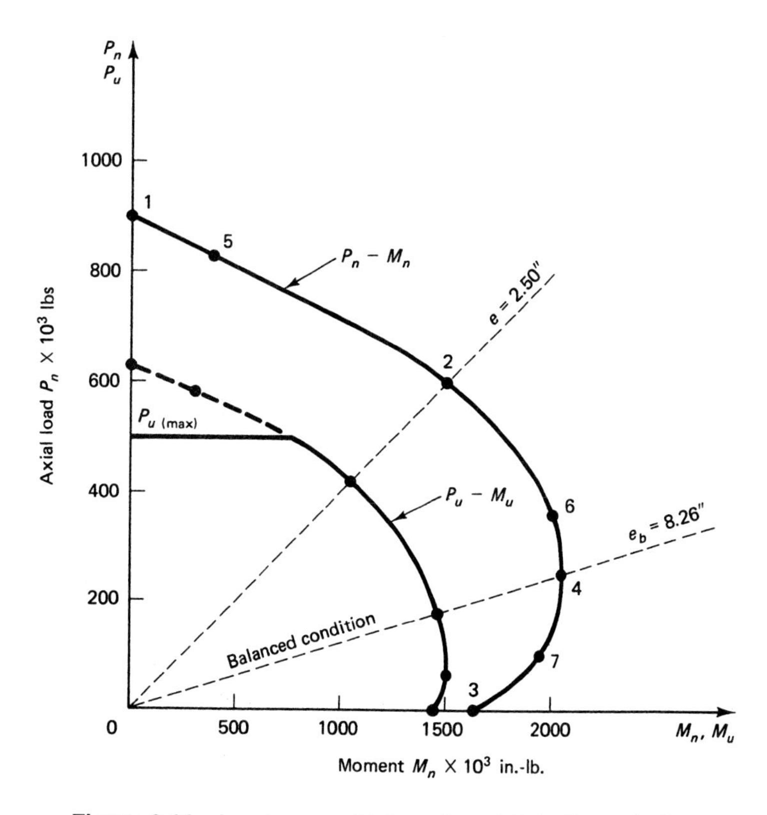

Figure 8.11 Load-moment interaction plots in Example 8.1.

$$= 1{,}450{,}147 \text{ in.-lb } (164 \text{ kN-m})$$

$$P_{u1} = \phi P_n = 0.70 \times 903{,}760 = 623{,}632 \text{ lb (not usable)}$$

The ACI Code requires that the maximum design axial load strength ϕP_n for tied prestressed columns should not exceed $0.80\phi P_n$, and for spirally reinforced prestressed columns should not exceed $0.85\phi P_n$. We thus have

$$\text{Max } P_u = 0.8\phi P_n = 0.8 \times 623{,}632$$

$$= 498{,}906 \text{ lb } (2{,}219 \text{ kN})$$

$$P_{u5} = 0.7 \times 826{,}648 = 578{,}654 \ (2{,}574 \text{ kN})$$

The remaining values of the coordinates are summarized in Table 8.1. Plots of the interaction diagrams for the nominal strength (P_n–M_n) and the design strength (P_u–M_u) are shown in Figure 8.11.

8.6 LIMIT STATE AT BUCKLING FAILURE OF SLENDER (LONG) PRESTRESSED COLUMNS

Considerable literature exists on the behavior of columns subjected to stability testing. If the column slenderness ratio exceeds the limits for short columns, the compression member will buckle prior to reaching its limit state of material failure. The strain in the compression face of the concrete at buckling load will then be less than the 0.003 in./in. shown in Figure 8.12. Such a column would be a slender member subjected to combined axial load and bending, deforming laterally and developing additional moment due to the $P\Delta$

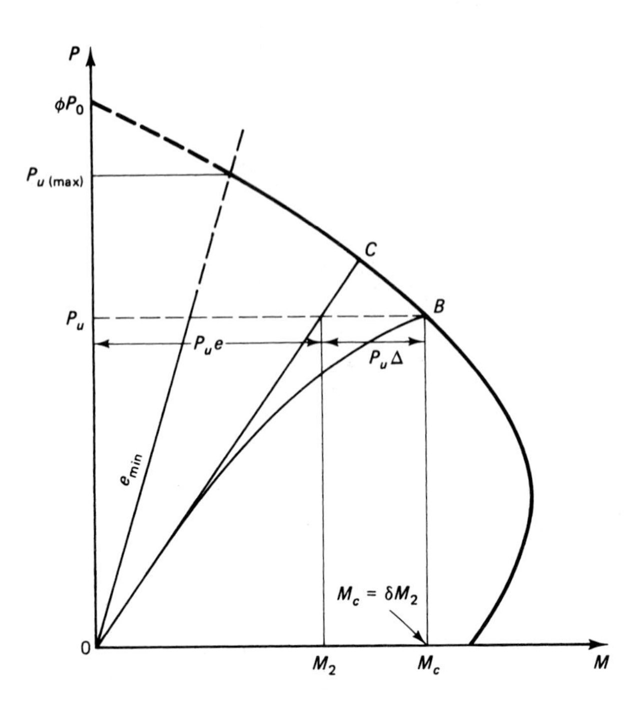

Figure 8.12 Loading moment (P–M) magnification interaction diagram.

effect, where P is the axial load and Δ is the deflection of the column's buckled shape at the section being considered.

Consider a slender column subjected to axial load P_u at an eccentricity e. The buckling effect produces an additional moment of $P_u\,\Delta$. This moment reduces the load capacity from point C to point B in the interaction diagram of Figure 8.12. The total moment

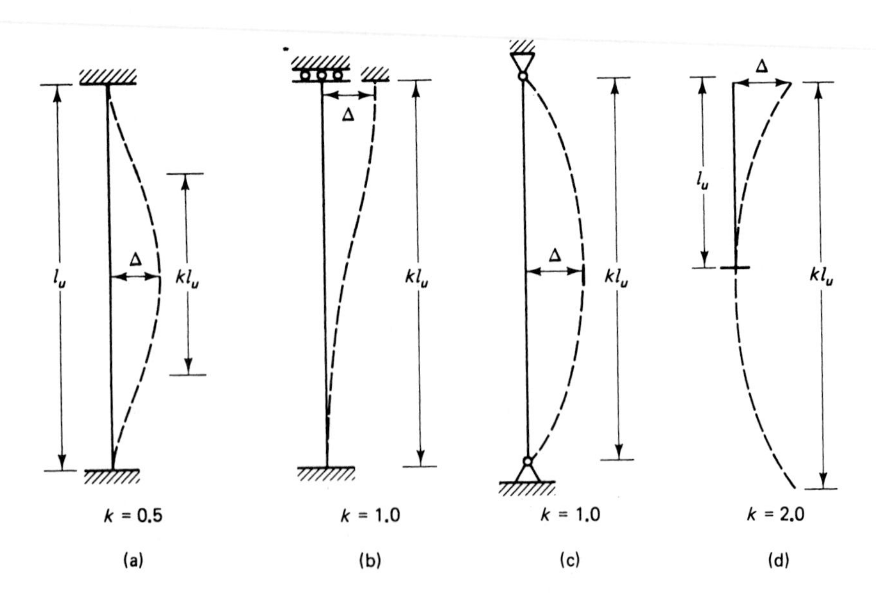

Figure 8.13 Values of column length factor k for typical end conditions. (a) Fixed-fixed. (b) Fixed-fixed with lateral motion. (c) Pinned. (d) Fixed-free.

$P_u e + P_u \Delta$ is represented by point B in the diagram, and the column can be designed for a larger or magnified moment M_c as a nonslender column.

The effective length kl_u shown in Figure 8.13 is used as the modified length of the column to account for end restraints other than being pinned. kl_u represents the length of an auxiliary pin-ended column which has an Euler buckling load equal to that of the column under consideration. Alternatively, it is the distance between the points of contraflexure of the member in its buckled form.

The value of the end restraint effective length factor k varies between 0.5 and 2.0 according to the nature of the restraint as follows:

Both column ends pinned, no lateral motion	$k = 1.0$
Both column ends fixed	$k = 0.5$
One end fixed, other end free	$k = 2.0$
Both column ends fixed, lateral motion exists	$k = 1.0$

Typical cases illustrating the buckled shape of the column for several end conditions and the corresponding length factors k are shown in Figure 8.13.

For members in a structural frame, the end restraint lies between the hinged and fixed conditions. The actual k value can be determined from the Jackson and Moreland alignment charts in Figure 8.14. In lieu of these charts, the following equations suggested in the ACI Code commentary can also be used for computing k:

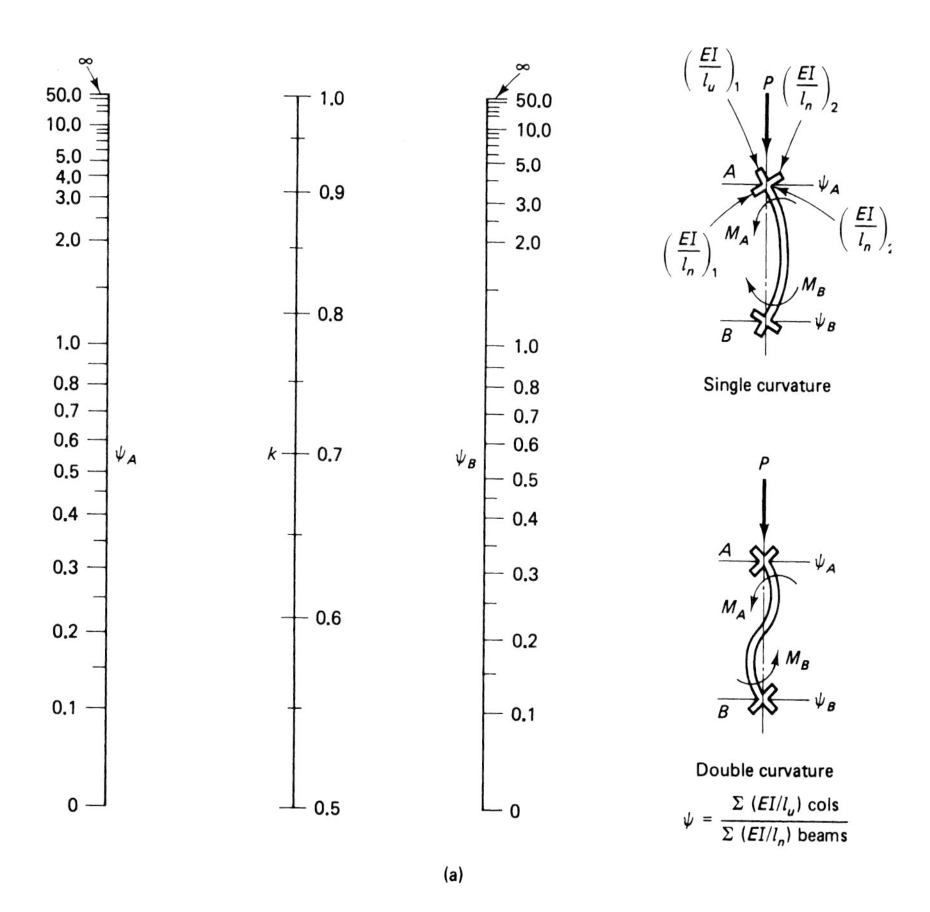

Figure 8.14 Effective length factor k for (a) braced and (b) unbraced frames.

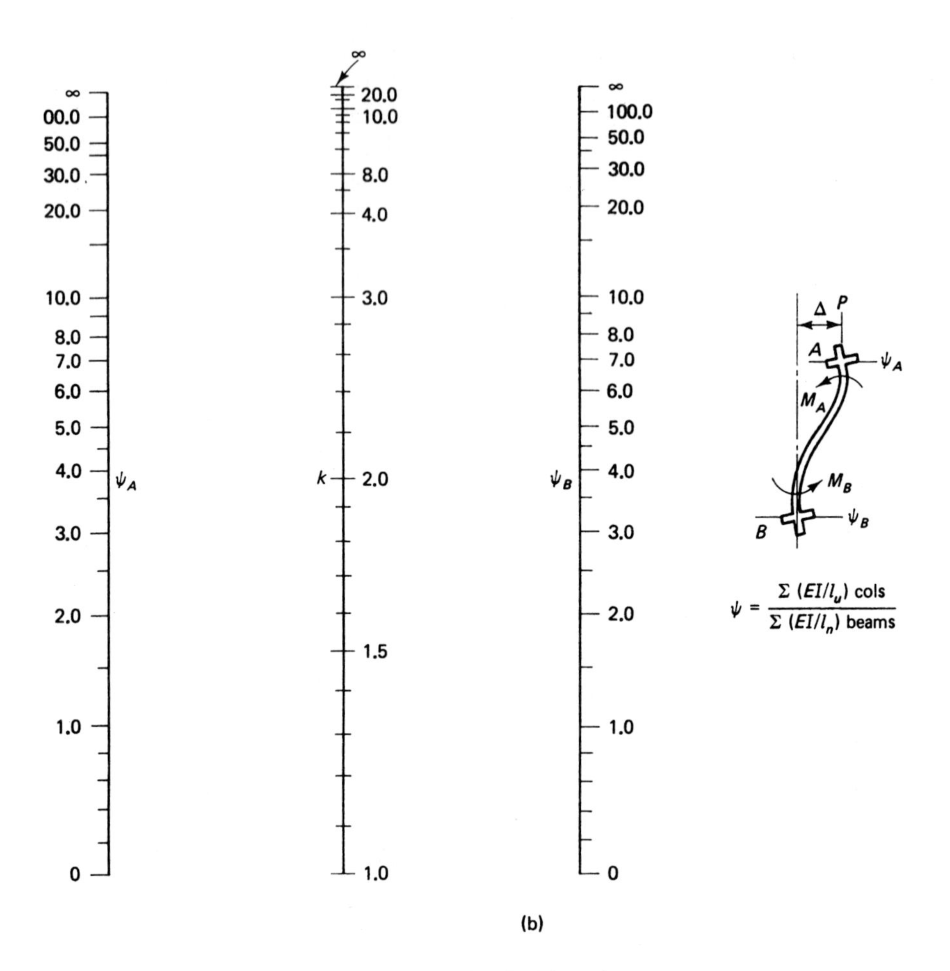

Figure 8.14 *Continued*

1. *Braced compression members.* An upper bound to the effective length factor may be taken as the smaller of the expressions

$$k = 0.7 + 0.05(\psi_A + \psi_B) \leq 1.0 \tag{8.10a}$$

and

$$k = 0.85 + 0.05\psi_{\min} \leq 1.0 \tag{8.10b}$$

where ψ_A and ψ_B are the values of ψ at the two ends of the column and $\psi_{\min}$ is the smaller of the two values. ψ is the ratio of the stiffness of all compression members to the stiffness of all flexural members in a plane at one end of the column. That is,

$$\psi = \frac{\Sigma\ EI/l_u \text{ columns}}{\Sigma\ EI/l_n \text{ beams}} \tag{8.11}$$

where l_u is the unsupported length of the column and l_n is the clear beam span.

2. *Unbraced compression members restrained at both ends.* The effective length may be taken as follows:

For $\psi_m < 2$,

$$k = \frac{20 - \psi_m}{20} \sqrt{1 + \psi_m} \tag{8.12a}$$

For $\psi_m \geq 2$,

$$k = 0.9\sqrt{1 + \psi_m} \tag{8.12b}$$

where ψ_m is the average of the ψ values at the two ends of the compression member.

3. *Unbraced compression members hinged at one end.* The effective length factor may be taken as

$$k = 2.0 + 0.3\psi \tag{8.13}$$

where ψ is the value at the restrained end.

The radius of gyration $r = \sqrt{I_g/A_g}$ can be taken as $r = 0.3h$ for rectangular sections, where h is the column dimension perpendicular to the axis of bending. For circular sections, r is taken as $0.25h$.

8.6.1 Buckling Considerations

Frames that do not have lateral bracing such as shear walls, diaphragms, or diagonal coupling beams are more flexible than those which are braced laterally. Lateral flexibility can cause the mass of a structure to sufficiently displace horizontally above the foundations that significant additional overturning moments can result leading to loss of stability of the structure. This behavior is particularly critical when nonslender columns support the floors.

The ACI 318 Code stipulates three methods for determining the forces on slender columns and members in frames that resist lateral forces in addition to the vertical gravity loads. However, for gravity loading *without* side-sway, a first-order analysis using moment magnification factors, δ_{ns}, is adequate. For combined gravity and side-sway forces causing the P-Δ effect, the three code methods are:

(a) Computer programs using a second-order analysis that determines iteratively the magnitudes of the additional overturning moments in a frame.

(b) Moment magnification factors δ_s are computed on the basis of first-order lateral displacements and the mass above each level.

(c) Moment magnification relationship similar in form to those required for computing the no-sway magnifier, δ_{ns}, for columns in braced frames using a stability index, Q. Horizontal displacement in this method need not be evaluated but moments that resist lateral forces have to be computed. This method is too cumbersome and least accurate. While the most accurate is the method in (a) using computer programs such as PCA's Frame Program, STAAD Pro, CSI Sap2000, and others.

Consider a slender column subjected to axial load P_u at an eccentricity e. The buckling effect produces an additional moment $P_u\,\Delta$ where Δ is the maximum lateral displacement of the compression member between its two ends from the vertical "plumb" position. This additional moment reduces the load capacity from point C to point B in the interaction diagram, Figure 8.12. The total moment $(P_u\,e + P_u\,\Delta)$ is represented by point B in the diagram and the column should be designed for a larger magnified moment M_c as a nonslender column by the usual *first-order analysis.*

In such an analysis, the moments and axial forces in a frame are obtained by the classical elastic procedures. These procedures do not consider the effects of the lateral displacement Δ on the axial force P_u and the bending moment M_c. Consequently the resulting load-deflection and load-moment relationships are linear. If the P-Δ effect is taken into account, a second-order analysis becomes necessary with a resulting nonlinear relationship of the load to the lateral displacement (deflection) and the moment. The ACI 381-99 Code permits using either a first- or a second-order analysis for columns of intermediate slenderness and requires a second-order analysis for long columns having a slenderness ratio of 100 or more. One ACI Code method where the P-Δ effect is ignored is termed the moment magnification method presented in Section 8.7.

8.7 MOMENT MAGNIFICATION METHOD: FIRST-ORDER ANALYSIS

The factored axial forces, P_u, the factored moments M_1 and M_2 at the column ends, and, where required, the relative story deflections, are computed in this method using an elastic first-order analysis with the section properties determined taking into account the influence of axial loads, the presence of cracked regions along the length of the member and the effects of duration of the load.

As discussed in Section 8.6 in conjunction with Figure 8.12, the moment M_2 is magnified by a magnification factor δ. The column is subjected to moments M_1 and M_2 at its ends where M_2 is considered larger than M_1. The factored axial force, P_u, and the factored moments, M_1 and M_2, are resisted by analytically chosen sectional properties taking into account the cracked regions along the compression member's length or height and the load duration. In lieu of these computations, the ACI 318-99 Code allows using the following average values for properties of members in a structure:

(a) Modulus of elasticity $E_c = 33w_c^{1.5}\sqrt{f'_c}$ and for concrete strength $f'_c > 5000$ psi $< 12{,}000$ psi $E_c = (40{,}000 + 1 \times 10^6)\,(w_c/145)^{1.5}$
(b) Moment of Inertia

Beams:	$0.35I_g$
Columns	$0.70I_g$
Walls—Uncracked	$0.70I_g$
—Cracked	$0.35I_g$
Flat plates and flat slabs:	$0.25I_g$

(c) Area: $1.0A_g$
(d) Radius of gyration $r = 0.30h$ for rectangular members where h is in the direction stability is being considered, or $r = 0.25D$ for circular members where D is the diameter of the compression member.

The moments of inertia should be divided by $(1 + \beta_d)$ when sustained lateral loads act, or for stability checks where β_d is a creep factor, thus

$$\beta_d = \frac{\text{maximum factored sustained axial load}}{\text{total factored axial load}}$$

The column load is assumed to act at an eccentricity $(e + \Delta)$ in Figure 8.12 to produce a moment M_c. The ratio M_c/M_2 is termed the magnification factor δ. The degree of magnification is dependent on the slenderness ratio kl_u/r where k is the effective length factor for compression members, a function of the relative stiffnesses at the joint of each end of the member.

The magnification factor is controlled by the type of the magnified moments δM_2 and δM_1 acting at the respective ends 2 and 1 of a column, namely, whether side-sway of the structural frame occurs or not. It should be noted that in the case of compression members subjected to bending about both principal axes, the moment about each axis should be *separately* considered based on the restraint condition corresponding to that axis.

8.7.1 Moment Magnification in Non-Sway Frames

In the case of compression members in non-sway frames, namely, braced frames, the effective length factor k can be taken as 1.0, unless analysis gives a lower value. In such a case, k values are computed on the basis of the EI values tabulated in Section 8.7 and the monograms in Figure 8.14.

The slenderness effects can be disregarded if

$$\frac{kl_u}{r} \leq 34 - 12\left(\frac{M_1}{M_2}\right) \tag{8.14}$$

kl_u = effective length between points of inflection and $[34 - 12\,(M_1/M_2)]$ cannot be taken greater than 40. The term (M_1/M_2) is positive if the member is bent in a single curvature so that the two terms subtract in Equation 8.14 and negative in double curvature so that the two terms add (see Figure 8.14a). If the non-sway magnification factor is δ_{ns} and the sway factor $\delta_s = 0$, the magnified moment becomes

$$M_c = \delta_{ns} M_2 \tag{8.15}$$

where

$$\delta_{ns} = \frac{C_m}{1 - \dfrac{P_u}{0.75P_c}} \geq 1.0 \tag{8.16a}$$

$$P_c = \frac{\pi^2 EI}{(kl_u)^2} \tag{8.16b}$$

where P_c is the Euler buckling load for pin-ended columns. Stiffness EI is to be taken as

$$EI = \frac{0.2E_cI_g + E_sI_{se}}{1 + \beta_d} \tag{8.16c}$$

or conservatively as

$$EI = \frac{0.4E_cI_g}{1 + \beta_d}$$

C_m = a factor relating the actual moment diagram to an equivalent uniform moment diagram. For members without transverse loads, namely, subjected to end loads only,

$$C_m = 0.6 + \frac{M_1}{M_2} \geq 0.4 \tag{8.17}$$

where $M_2 \leq M_1$ and $M_1/M_2 > 0$ if no inflection point exists between the column ends, Figure 8.14a (single curvature). For other conditions, such as members with transverse loads between supports, $C_m = 1.0$.

The minimum allowed value of M_2 is

$$M_{2,\min} = P_u(0.6 + 0.03h) \tag{8.18}$$

where h is in inches. In SI units $M_{2,\min} = P_u\,(15 + 0.03h)$ where h is in millimeters. In other words, the minimum eccentricity in the slender columns is $e_{\min} = 0.6 + 0.03h$. If $M_{2,\min}$ exceeds the applied moment M_2, the value of C_m in Equation 8.17 should either be taken as 1.0 or be based on the actual computed end moments M_1 and M_2.

Frames braced against side-sway or braced with shear walls, would normally have a lateral deflection less than total height $h_s/1500$. Once this ratio is exceeded, appropriate measures have to be taken to minimize the additional moments caused by side sway and hence reduce lateral drift of the frame and its constituent columns.

8.7.2 Moment Magnification in Sway Frames

For compression members not braced against side-sway, the effective length factor k can also be determined from the EI values presented in Section 8.7, but its value should not exceed 1.0. The slenderness effects can be disregarded if

$$\frac{kl_u}{r} < 22 \tag{8.19}$$

The end moments M_1 and M_2 should be magnified as follows:

$$\begin{aligned} M_1 &= M_{1ns} + \delta_s M_{1s} \\ M_2 &= M_{2ns} + \delta_s M_{2s} \end{aligned} \tag{8.20}$$

On the assumption that $M_2 > M_1$, the design moment should be

$$M_c = M_{2ns} + \delta_s M_{2s} \tag{8.21}$$

where M_{2ns} = factored end moment at the end of the compression member due to loads that cause no appreciable side-sway, computed using a first-order elastic frame analysis. M_{2s} = factored end moment at the end of the compression members due to loads that cause appreciable side-sway, computed using a first-order elastic frame analysis.

$$\delta_s M_s = \frac{M_s}{1 - \dfrac{\Sigma P_u}{0.75 \Sigma P_c}} \geq M_s \leq 2.5 \tag{8.22}$$

where ΣP_u is the summation for all the vertical loads in a story and ΣP_c is the summation of the Euler buckling loads, P_c, for pin-ended columns for all sway resisting columns in a story $[P_c = \pi^2 EI/(kl_u)^2]$ from Equation 8.16b) with the EI values obtained from Equations 8.16c or d.

In the case of an individual compression member having

$$\frac{l_u}{r} > \frac{35}{\sqrt{P_u/f'_c A_g}}$$

the member has to be designed for a factored axial load, P_u, and magnified moment $M_c = \delta_{ns} M_2$ where M_2 in this case is $M_2 = \delta_{ns} M_{2ns} + \delta_s M_{2s}$. This condition can develop in slender columns with high axial loads when the maximum moment may develop *between* the ends of the column so that the end moments might not necessarily be the maximum moments.

8.7.2.1 Moment magnification in sway frames using a stability index, *Q*. In this method (method c in Section 8.6.1), the code permits assuming a column in a braced structure as non-sway if the increase in column loads and moments due to second-order effects does not exceed 5 percent of the first-order end moments. A story within a structure can be considered non-sway if a stability index, Q, in the following expression does *not* exceed a value of 0.05:

$$Q = \frac{\Sigma P_u \Delta_o}{V_u l_c} \tag{8.23a}$$

where,

ΣP_u = total vertical load at a story
V_u = story shear
Δ_o = first-order relative deflection between the top and bottom of a particular story due to shear V_u
l_c = length of compression member in a frame measured from centers of joints.

The non-sway magnification factor in terms of Q is:

$$\delta_s = \frac{1}{1 - Q} \geq 1.0 \tag{8.23b}$$

When Q exceeds a value of 0.05, one has to proceed to a second-order analysis through computer program usage. Such a computer analysis would make it possible to efficiently compute the iterating values of moments and Δ_o sway values due to the P-Δ effect in a reasonably accurate and speedy manner.

It should be noted that the stability index Q method, while relatively adaptable to hand computations, is too cumbersome and least accurate for effective evaluation of the P-Δ effect on moments at the column joints in braced frames.

It is important to summarize that the moment magnification method, originally developed for prismatic columns, should work well for columns of slenderness ratio kl_u/r less than 100, particularly if the frame is braced. In the case of unbraced frames of comparable slenderness ratios, taking into account the P-Δ effect on the moments and deflections through a second-order analysis can give more accurate results. Such an analysis can be either

1. Execute several applications of the first-order analysis where the lateral load (from h_i in Figure 8.15 to follow) is incremented by $\Sigma P_u \Delta_i$ in each cycle, and consider the final result a second-order result, or
2. Use a real second-order analysis computer program in which the reduction in the relative side-sway resistance is used in a global stiffness matrix for the elements involved.

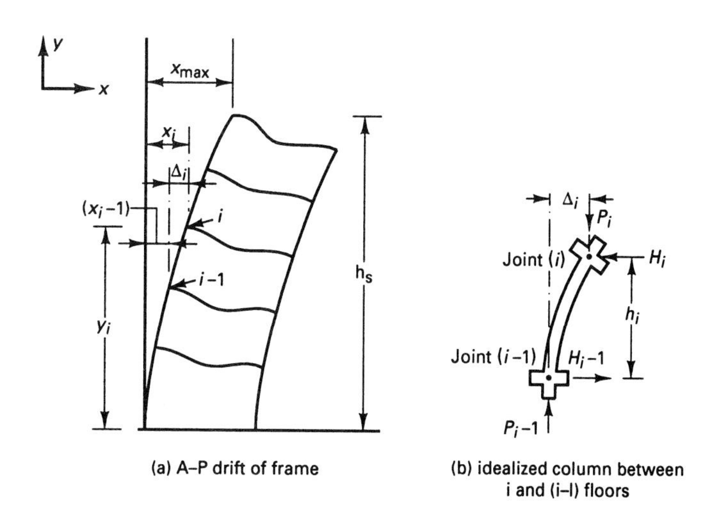

Figure 8.15 Second-order frame parameters.

8.8 SECOND-ORDER FRAMES ANALYSIS AND THE $P - \Delta$ EFFECTS

A second-order analysis is a frame analysis which includes the internal force effects resulting from lateral displacement (deflection) of a column. When such an analysis is performed in order to evaluate $\delta_s M_s$ in a non-braced frame, the deflections must be computed on the basis of fully cracked sections with reduced EI stiffness values. Approximations such as the use of several first-order analysis cycles and idealizations of nonprismatic sections can be made in the analysis. But the analysis should verify that the predicted strength of the compression members of a structural frame are in good agreement within a 15-percent range with results for columns in indeterminate reinforced concrete structures. The structure being analyzed should result in geometry of members similar to the geometry of the sections to be built. If the members in the final structure have cross-sectional dimensions differing by more than 10 percent from those assumed in the analysis, a new computation cycle has to be performed.

A second-order analysis is an iterative procedure of the $P - \Delta$ effects on the slender column, including shear deformations. Hence, it is reasonable to expect that canned computer programs have to be used rather than long-hand computations in the design of the slender columns of a frame structure. An attempt will be made here to illustrate the iteration procedure involved in the use of several cycles of lateral load increments to the $P - \Delta$ values. It must be stated, however, that the large majority of columns in concrete building frames do not necessitate such an analysis since the (kl_u/r) ratio is in most cases below 100.

Consider the column between the two floors $(i-1)$ and (i) in the frame shown in Figure 8.15. Assume that the maximum lateral displacement or drift at the upper end of the top column in the frame is x_{max} and that the total height of the building is h_s. A large drift, or lateral displacement of the building upper floors results in cracking of the masonry and interior finishes. Unless precautions are taken to permit movement of interior partitions without damage, the maximum lateral deflection limitation should be $h_s/500$. Hence, a good assumption is to choose x_{max} in the range of $h_s/350$ to $h_s/500$, considering that a *fully braced* frame has normally a ratio of maximum drift x_{max} to frame height h_s less than 1/1,500.

If x_i is the drift at floor level i, and y_i is the height of the column between floors $(i-1)$ and (i) in Figure 8.15a, it can be assumed that the proportional horizontal drift for a particular floor is directly proportional to the square of the ratio of the height h_i of the floor and the total height h_s of the entire frame. Hence,

$$x_i = x_{max}(h_i/h_s)^2 \tag{8.24}$$

The procedure can be summarized as follows:

1. Choose geometrical sections of the frame and its columns and their stiffness EI by approximate procedures.
2. Compute the drifts, namely, the lateral deflections Δ_i, and the corresponding ultimate loads $P_{u,i}$ at joints $i = 1, \ldots, n$ (Figure 8.15).
3. Find the equivalent horizontal forces H_i from $H_i = P_i\,\Delta_i/h_i$ (Figure 8.15b).
4. Add the values obtained in step 3 to the actual lateral loads acting on the frame.
5. Perform a frame analysis using the appropriate computer program.
6. The iterative computer program, using the stiffnesses, EI, chosen for the input data, gives Δ_i results that have to be compared with the x_i values allowed.
7. If all Δ_i values are $\leq$ all the x_i values, accept the solution and the design as a second order solution. If not, run additional computer cycles with modified stiffnesses until the desired results are achieved.

Any of several computer programs can be used to account for the P-Δ effects in frame side-sways. Strudel, PCA Frame, STAAD Pro, or CSI Sap 2000 are examples of such general-purpose programs.

8.9 OPERATIONAL PROCEDURE AND FLOWCHART FOR THE DESIGN OF SLENDER COLUMNS

1. Determine whether the frame has an appreciable side-sway. If it does, use the magnification factors δ_{ns} and δ_s. If the side-sway is negligible, assume that $\delta_s = 0$. Then assume a cross section, compute the eccentricity, using the greater of the end moments, and check whether it is more than the minimum allowable eccentricity, that is,

$$\frac{M_2}{P_u} \geq (0.6 + 0.03h) \text{ in.}$$

   If the given eccentricity is less than the specified minimum, use the minimum value.
2. Compute ψ_A and ψ_B using Equation 8.12 or 8.13, and then obtain k using Figure 8.14 or Equations 8.13. Compute kl_u/r, and determine whether the column is a short or long column. If the column is slender and kl_u/r is less than 100, compute the magnified moment M_c. Then, using the value obtained, compute the equivalent eccentricity to be used if the column is to be designed as a short column. If kl_u/r is greater than 100, perform a second-order analysis.
3. Design the equivalent nonslender column. The flowchart in Figure 8.16 presents the sequence of calculations. The necessary equations are provided in Section 8.2 and in the flowchart.

8.10 DESIGN OF SLENDER (LONG) PRESTRESSED COLUMN

Example 8.2

A square tied prestressed bonded column is part of a 5 × 3 bays frame building subjected to uniaxial bending. Its clear height is l_u = 15 ft (4.54 in.), and it is not braced against sidesway. The factored external load P_u = 300,000 lb (1,334 kN), and the factored end moments are M_1 = 425,000 in.-lb (48.0 kN-m) and M_2 = 750,000 in.-lb (84.8 kN-m). Design the column section and the reinforcement necessary for the following two conditions:

1. Consider gravity loads only, assuming negligible lateral sidesway due to wind.
2. Suppose sidesway wind effects cause a factored P_u = 24,000 lb (107 kN) and a factored M_u = 220,000 in.-lb (24.9 kN-m). The loads per floor of all columns at that level are $\Sigma P_u = 4.5 \times 10^6$ lb (20×10^3 kN) and $\Sigma P_c = 31.0 \times 10^6$ lb (138×10^3 kN).

Use $\frac{1}{2}$-in. dia 270-K stress-relieved prestressing strands. Given data are as follows:

$$\beta_d = 0.4$$

$$\psi_A = 1.0$$

$$\psi_B = 2.0$$

$$f'_c = 6{,}000 \text{ psi (41.4 MPa)}$$

$$f_{pu} = 270{,}000 \text{ psi (1,862 MPa)}$$

$$f_{ps} = 240{,}000 \text{ psi (1,655 MPa)}$$

$$f_{pe} = 150{,}000 \text{ psi (1,034 MPa)}$$

$$E_{ps} = 28 \times 10^6 \text{ psi } (200 \times 10^3 \text{ MPa})$$

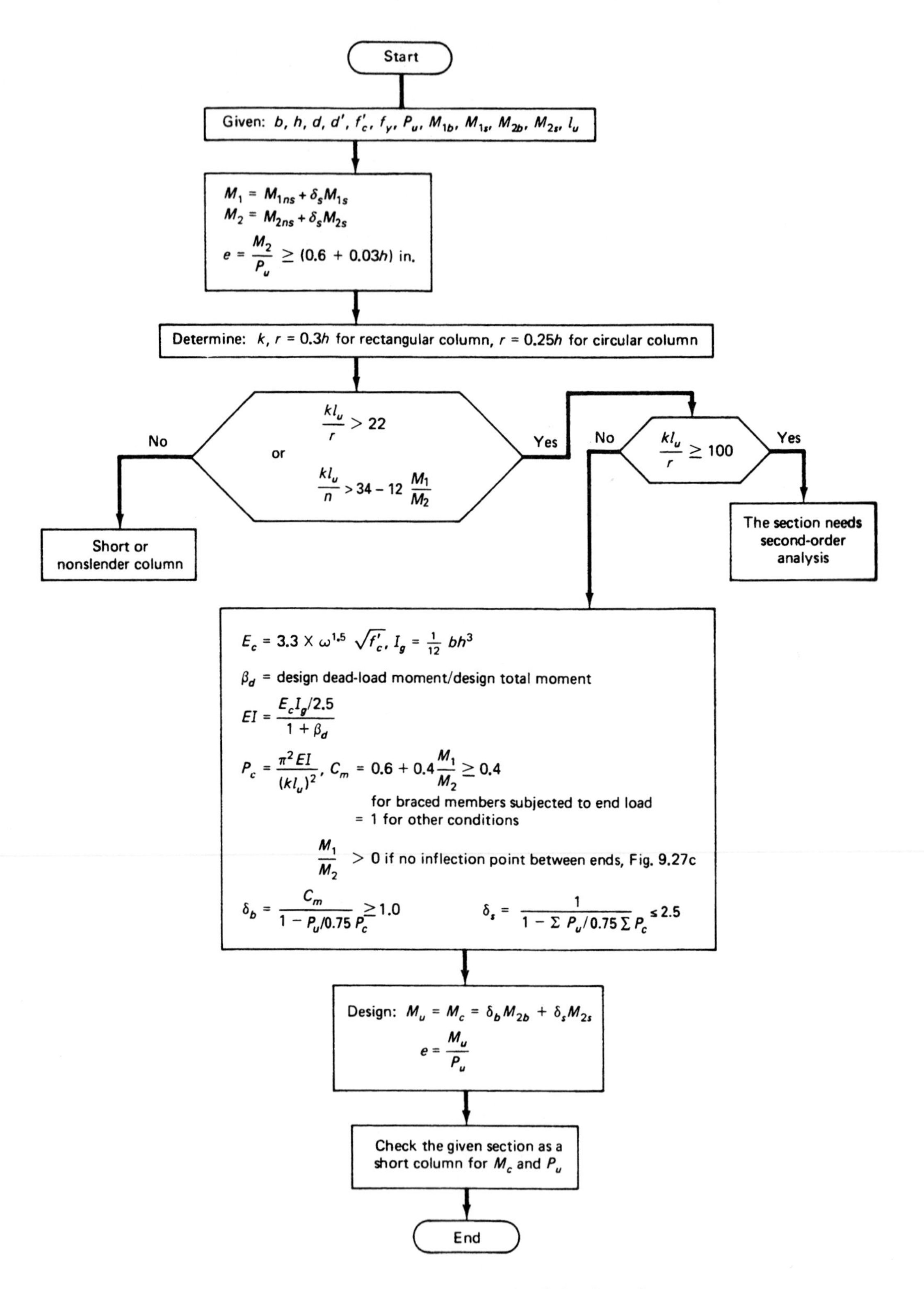

Figure 8.16 Flowchart for design of slender columns.

ϵ_{cu} at failure = 0.003 in./in.

ϵ_{ce} = 0.0005 in./in. when P_e acts on the section

d' = 2 in. (50.8 mm)

Ties f_y = 60,000 psi (414 MPa)

The stress-strain diagram of the prestressing steel is as in Figure 8.9.

Solution: ϵ_{pe} = 0.0052 in./in. from the stress-strain diagram of Figure 8.9, corresponding to f_{pe} = 150,000 psi. Similarly, $\epsilon_{py} \cong 0.012$ in./in. from the same figure, corresponding to f_{py} = 260,000 psi.

1. *Gravity Loads Only*

Check for No Sidesway and Minimum Eccentricity (Step 1). Since the frame has no appreciable sidesway, the entire M_2 is taken to be M_{2ns} and the magnification factor for sidesway, δ_s, is taken to be equal to zero in Equation 8.15. By trial and adjustment, a column section is assumed and analyzed. Accordingly, we try a section 15 in. × 15 in. (381 mm × 381 mm) as shown in Figure 8.17(a) and obtain

$$\text{Actual eccentricity} = \frac{M_{2ns}}{P_u} = \frac{750{,}000}{300{,}000} = 2.50 \text{ in. } (63.5 \text{ mm})$$

$$\text{Minimum allowable eccentricity} = 0.6 + 0.03h = 0.6 + 0.03 \times 15$$

$$= 1.05 \text{ in. } (2.67 \text{ mm}) < 2.50 \text{ in.}$$

Hence, use M_{2ns} = 750,000 in.-lb as the larger of the moments M_1 and M_2 on the column.

Compute the Eccentricity to Be Used for Equivalent Short Column (Step 2). From the chart in Figure 8.14(b), k = 1.45 and the slenderness ratio is

$$\frac{kl_u}{r} = \frac{1.45 \times 15 \times 12}{0.3 \times 15} = 58.0$$

Since 58.0 > 22 and < 100, use the moment magnification method. We obtain

$$E_c = 33w^{1.5}\sqrt{f_c'} = 33 \times 145^{1.5}\sqrt{6{,}000} = 4.46 \times 10^6 \text{ psi } (32 \times 10^3 \text{ MPa})$$

$$I_g = \frac{15(15)^3}{12} = 4{,}218.8 \text{ in.}^4$$

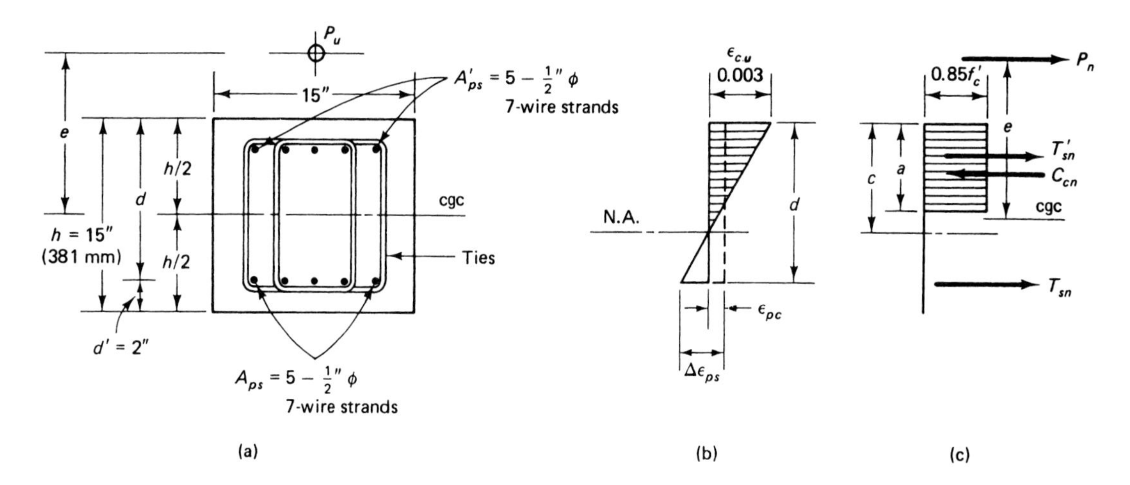

Figure 8.17 Proposed column section geometry in Example 8.2. (a) Cross-sectional details. (b) Strain distribution. (c) Stress block and forces.

$$EI = \frac{E_c I_g/2.5}{1 + \beta_d} = \frac{4.46 \times 10^6 \times 4{,}218.3}{2.5} \times \frac{1}{1 + 0.4}$$

$$= 5.34 \times 10^9 \text{ lb.-in.}^2$$

$$(kl_u)^2 = (1.45 \times 15 \times 12)^2 = 68.1 \times 10^3 \text{ in.}^2$$

Hence,

$$P_c = \text{Euler buckling load} = \frac{\pi^2 EI}{(kl_u)^2} = \frac{\pi^2 \times 5.34 \times 10^9}{68.1 \times 10^3}$$

$$= 773{,}132 \text{ lb} = 773.1 \text{ Kips } (3{,}439 \text{ kN})$$

Now, $C_m = 1.0$ for a nonbraced column. Assume $\phi = 0.7$. Then we have

$$\text{Moment magnifier } \delta_{ns} = \frac{C_m}{1 - P_u/0.75P_c} = \frac{1.0}{1 - \dfrac{300{,}000}{0.75 \times 773{,}132}} = 2.07$$

$$\text{Design moment } M_c = \delta_{ns} M_{2ns} = 2.07 \times 750{,}000$$

$$= 1{,}552{,}500 \text{ in.-lb } (184 \text{ kN-m})$$

$$\text{Required } P_n = \frac{P_u}{\phi} = \frac{300{,}000}{0.7} = 428{,}571 \text{ lb } (1{,}906 \text{ kN})$$

$$\text{Required } M_n = \frac{1{,}552{,}500}{0.7} = 2{,}217{,}860 \text{ in.-lb } (271 \text{ kN-m})$$

$$\text{Eccentricity } e = \frac{2{,}217{,}860}{428{,}571} = 5.18 \text{ in. } (131 \text{ mm})$$

Design of an Equivalent Nonslender Column (Step 3). The equivalent column has to carry a minimum nominal axial load $P_n = 428{,}571$ lb and a minimum nominal uniaxial moment $M_n = 2{,}217{,}860$ in.-lb.

To design the equivalent nonslender column, we analyze the assumed 15 in. × 15 in. column section assuming five $\frac{1}{2}$-in. dia. 7-wire stress-relieved strands on each of the two faces parallel to the neutral axis, as in Example 8.1. Then

$$A_{ps} = A'_{ps} = 5 \times 0.153 = 0.765 \text{ in.}^2 \ (4.94 \text{ cm}^2)$$

Balanced Failure Condition

$$d = h - 2 = 15 - 2 = 13 \text{ in. } (330 \text{ mm})$$

Comparing with Example 8.1 and using trial and adjustment, a reasonable assumption of the neutral axis depth for the balanced condition would be a value of $c_b = 8.3$ in. (211 mm). Then $a_b = \beta_1 \times c_b = 0.75 \times 8.3 = 6.23$ in. (158 mm).

Next, from Figure 8.3,

$$C_{cn} = 0.85 \times 6{,}000 \times 15 \times 6.23 = 476{,}595 \text{ lb } (2{,}119 \text{ kN})$$

From Equation 8.5,

$$T'_{sn} = 0.765 \times 28 \times 10^6 \left[0.0052 - 0.003\left(\frac{8.3 - 2}{8.3}\right) + 0.0005\right]$$

$$= 73{,}318 \text{ lb } (385 \text{ kN})$$

From Equation 8.6,

$$T_{sn} = 0.765 \times 28 \times 10^6 \left[0.0052 + 0.003\left(\frac{13 - 8.3}{8.3}\right) + 0.0005\right]$$

$$= 158{,}482 \text{ lb } (704 \text{ kN})$$

From Equation 8.2,

$$P_{nb} = C_{cn} - T'_{sn} - T_{sn}$$
$$= 476{,}595 - 73{,}318 - 158{,}482$$
$$= 229{,}310 \text{ lb } (1{,}020 \text{ kN})$$

From Equation 8.7,

$$M_{nb} = 476{,}595\left(\frac{15}{2} - \frac{6.23}{2}\right) = 73{,}318\left(\frac{15}{2} - 2\right) + 158{,}482\left(13 - \frac{15}{2}\right)$$
$$= 2{,}103{,}124 \text{ in.-lb } (237.7 \text{ kN-m})$$
$$e_b = \frac{M_{nb}}{P_{nb}} = \frac{2{,}103{,}124}{229{,}310} = 9.17 \text{ in. } (233 \text{ mm}) > \text{actual } e = 5.18 \text{ in.}$$

The prestressed column load has small eccentricity, and initial failure would be in compression. Also, $\phi = 0.7$, as assumed.

Assume Neutral Axis Depth c = 12 in.

$$a = \beta_1 c = 0.75 \times 12 = 9.0 \text{ in.}$$

From Equation 8.1a,

$$C_{cn} = 0.85 f'_c\, ba = 0.85 \times 6{,}000 \times 15 \times 9 = 688{,}500 \text{ lb}$$

From Equation 8.5,

$$T'_{sn} = A'_{ps} E_{ps}\left[\epsilon_{pe} - \epsilon_{cu}\left(\frac{c - d'}{c}\right) + \epsilon_{ce}\right]$$
$$= 0.765 \times 28 \times 10^6\left[0.0052 - 0.003\left(\frac{12 - 2}{12}\right) + 0.0005\right]$$
$$= 70{,}993 \text{ lb}$$

From Equation 8.6,

$$T_{sn} = A_{ps} E_{ps}\left[\epsilon_{pe} + \epsilon_{cu}\left(\frac{d - c}{c}\right) + \epsilon_{ce}\right]$$
$$= 0.765 \times 28 \times 10^6\left[0.0052 + 0.003\left(\frac{13 - 12}{12}\right) + 0.0005\right]$$
$$= 127{,}449 \text{ lb}$$

From Equation 8.2,

$$P_n = C_{cn} - T'_{sn} - T_{sn}$$
$$\text{Available } P_n = 688{,}500 - 70{,}993 - 127{,}449$$
$$= 490{,}058 \text{ lb} > \text{required } P_n = 428{,}571 \text{ lb}$$

Accordingly, we go on to a second trial-and-adjustment cycle.

Assume Neutral Axis Depth c = 11.2 in.

$$a = \beta_1 c = 0.75 \times 11.2 = 8.4 \text{ in.}$$
$$C_{cn} = 0.85 f'_c\, ba = 0.85 \times 6{,}000 \times 15 \times 8.4 = 642{,}600 \text{ lb}$$
$$T'_{sn} = 0.765 \times 28 \times 10^6\left[0.0052 - 0.003\left(\frac{11.2 - 2}{11.2}\right) + 0.0005\right]$$
$$= 69{,}309 \text{ lb}$$

$$T_{sn} = 0.765 \times 28 \times 10^6\left[0.0052 + 0.003\left(\frac{13 - 11.2}{11.2}\right) + 0.0005\right]$$

$$= 132{,}421 \text{ lb}$$

$$\text{Available } P_n = 642{,}600 - 6{,}309 - 132{,}421$$

$$= 440{,}870 \text{ lb} \cong \text{required } P_n = 428{,}571 \text{ lb, O.K.}$$

From Equation 8.7,

$$M_n = C_{cn}\left(\frac{h}{2} - \frac{a}{2}\right)^{\curvearrowleft} - T'_{sn}\left(\frac{h}{2} - d'\right)^{\curvearrowright} + T_{sn}\left(d - \frac{h}{2}\right)^{\curvearrowleft}$$

$$= 642{,}600\left(\frac{15}{2} - \frac{8.4}{2}\right) - 69{,}309\left(\frac{15}{2} - 2\right) + 132{,}421\left(13 - \frac{15}{2}\right)$$

$$= 2{,}467{,}696 \text{ in.-lb} > 2{,}217{,}860 \text{ in.-lb } (278.8 \text{ kN-m} > 250 \text{ kN-m}), \text{ O.K.}$$

$$e = \frac{2{,}467{,}696}{448{,}870} = 5.5 \simeq \text{actual } e = 5.18 \text{ in., accept.}$$

Consequently, adopt a section 15 in. × 15 in. with five $\frac{1}{2}$-in. dia. 7-wire stress-relieved 270-K strands at each of the two faces parallel to the neutral axis. Then design the necessary transverse ties.

2. *Gravity and Wind Loading (Sidesway).* From part 1, $P_c = 773{,}132$ lb and $U = 0.75$ $(1.4D + 1.7L + 1.7W)$. Also, $U = 0.9D + 1.3W$ (did not control), $P_u = 0.75\ (300{,}000 + 24{,}000) =$ 243,000 lb, $M_{2b} = 0.75 \times 750{,}000$ in.-lb $= 562{,}500$ in.-lb., and $M_{2s} = 0.75 \times 220{,}000 = 165{,}000$ in.-lb.

From Equation 8.16(a),

$$\delta_b = \frac{1.0}{1 - \dfrac{P_u}{0.75P_c}} = \frac{1.0}{1 - \dfrac{243{,}000}{0.75 \times 773{,}132}} = 1.72$$

Photo 8.4 One of the earliest prestressed concrete footbridges in England at the Festival of Britain, London, 1951.

From Equation 8.16(b),

$$\delta_s = \frac{1.0}{1 - \dfrac{\Sigma P_u}{\phi \Sigma P_c}} = \frac{1.0}{1 - \dfrac{4.5 \times 10^6}{0.75 \times 31.0 \times 10^6}} = 1.24$$

From Equation 8.15,

$$M_c = \delta_{ns} M_{2ns} + \delta_s M_{2s} = 1.72 \times 562{,}500 + 1.24 \times 165{,}000$$
$$= 1{,}172{,}100 \text{ in.-lb}$$

$$\text{Required } P_n = \frac{243{,}000}{0.7} = 347{,}143 \text{ lb}$$

$$\text{Required } M_n = \frac{1{,}172{,}100}{0.7} = 1{,}674{,}430 \text{ in.-lb}$$

$$\text{Eccentricity } e = \frac{1{,}674{,}430}{347{,}143} = 4.82 \text{ in.} < e_b = 9.17 \text{ in.}$$

Hence, initial compression failure occurs. We then have $P_n = 347{,}143 < P_n = 428{,}571$ lb in case 1.

The conditions for case 2 with sidesway do not control, since failure is still in compression and the required P_n is *less* than that for case 1. Accordingly, we adopt the same 15 in. × 15 in. section of case 1, with five ½-in. dia 270-K-stress-relieved prestressing strands on each of the two faces parallel to the neutral axis.

8.11 COMPRESSION MEMBERS IN BIAXIAL BENDING

8.11.1 Exact Method of Analysis

Columns in corners of buildings are compression members subjected to biaxial bending about both the x and the y axes as shown in Figure 8.18. Also, biaxial bending occurs due to imbalance of loads in adjacent spans and almost always in bridge piers. Such columns are subjected to moment M_{xx} about the x axis creating a load eccentricity e_y, and a moment M_{yy} about the y axis creating a load eccentricity e_x. Thus, the neutral axis is inclined at an angle θ to the horizontal.

The angle θ depends on the interaction of the bending moments about both axes and the magnitude of the total P_u. The compressive area in the column section can have one of the alternative shapes shown in Figure 8.18(c). Since such a column has to be designed from first principles, the trial-and-adjustment procedure has to be followed where compatibility of strain has to be maintained at all levels of the reinforcing bars. Additional computational effort is also needed, because of the position of the inclined neutral-axis plane and the four different possible forms of the concrete compression area.

Figure 8.19 shows the strain distribution and forces on a biaxially loaded rectangular column cross section. G_c is the center of gravity of the concrete compression area, having coordinates x_c and y_c from the neutral axis in the x and y directions, respectively. G_{sc} is the resultant position of the steel forces in the compression area having coordinates x_{sc} and y_{sc} from the neutral axis in the x and y directions, respectively. G_{st} is the resultant position of the steel forces in the tension area having coordinates x_{st} and y_{st} from the neutral axis in the x and y directions, respectively. From equilibrium of internal and external forces,

$$P_n = 0.85 f'_c A_c + F_{sc} - F_{st} \tag{8.25}$$

where A_c = area of the compression zone covered by the rectangular stress block
F_{sc} = resultant steel compressive forces ($\Sigma A'_s f_{sc}$)
F_{st} = resultant steel tensile force ($\Sigma A_s f_{st}$)

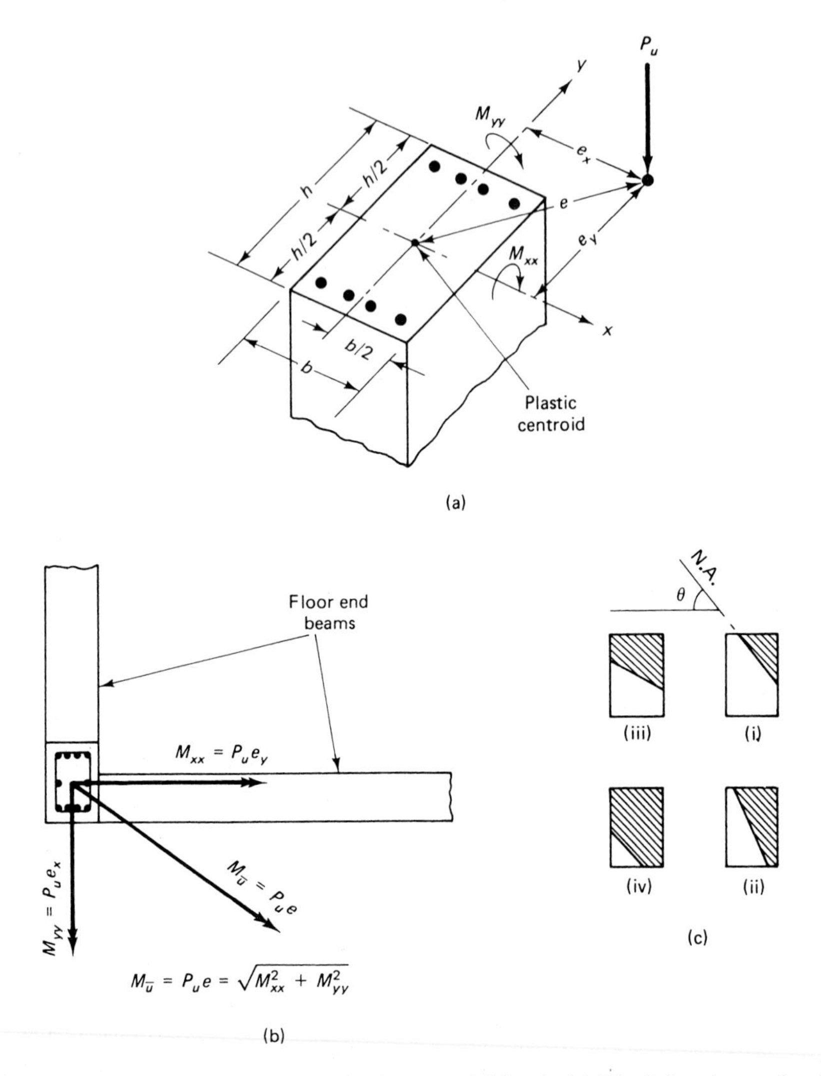

Figure 8.18 Corner column subjected to axial load. (a) Biaxially stressed column cross section. (b) Vector moments M_{xx} and M_{yy} in column plan. (c) Neutral axis direction.

Also, from equilibrium of internal and external moments,

$$P_n e_x = 0.85 f_c' A_c x_c + F_{sc} x_{sc} + F_{st} x_{st} \tag{8.26a}$$

$$P_n e_y = 0.85 f_c' A_c y_c + F_{sc} y_{sc} + F_{st} y_{st} \tag{8.26b}$$

The position of the neutral axis has to be assumed in each trial and the stress computed in *each* bar using

$$f_{si} = E_s \epsilon_{si} = E_c \epsilon_c \frac{s_i}{c} < f_y \tag{8.27}$$

8.11.2 Load Contour Method of Analysis

One method of arriving at a rapid solution is to design the column for the vector sum of M_{xx} and M_{yy} and use a circular reinforcing cage in a square section for the corner column. However, such a procedure cannot be economically justified in most cases. Another de

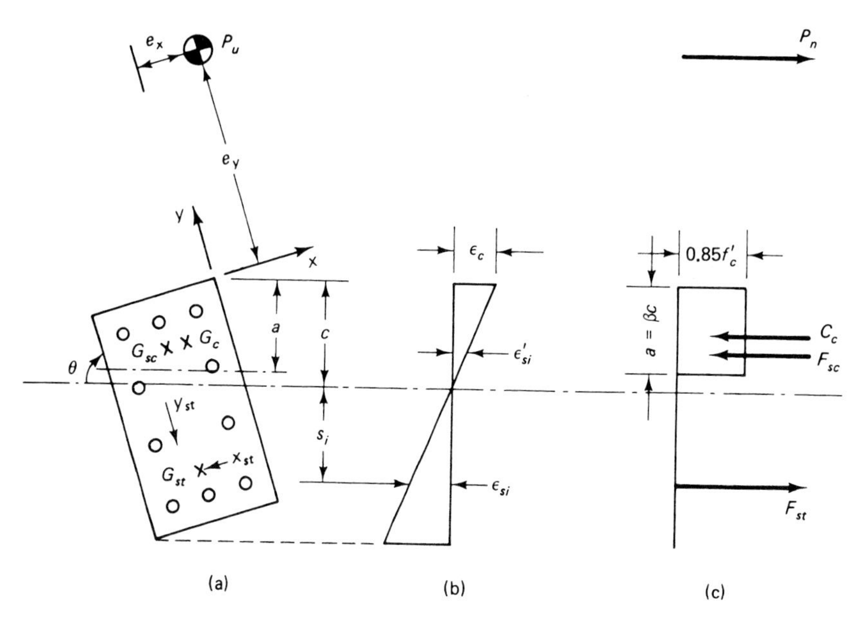

Figure 8.19 Strain compatibility and forces in biaxially loaded rectangular columns. (a) Cross section. (b) Strain. (c) Forces.

sign approach well proven by experimental verification is to transform the biaxial moments into an equivalent uniaxial moment and an equivalent uniaxial eccentricity. The section can then be designed for uniaxial bending, as previously discussed in this chapter, to resist the actual factored biaxial bending moments.

Such a method considers a failure surface instead of failure planes and is generally termed the *Bresler-Parme contour method.* The method involves cutting the three-dimensional failure surfaces in Figure 8.20 at a constant value P_n to give an interaction plane relating M_{nx} and M_{ny}. In other words, the contour surface S can be viewed as a curvilinear surface which includes a family of curves, termed the *load contours.*

The general nondimensional equation for the load contour at a constant load P_n may be expressed as

$$\left(\frac{M_{nx}}{M_{ox}}\right)^{\alpha_1} + \left(\frac{M_{ny}}{M_{oy}}\right)^{\alpha_2} = 1.0 \tag{8.28}$$

where $M_{nx} = P_n e_y$
$M_{ny} = P_n e_x$
$M_{ox} = M_{nx}$ at an axial load P_n such that M_{ny} or $e_x = 0$
$M_{oy} = M_{ny}$ at an axial load P_n such that M_{nx} or $e_y = 0$

The moments M_{ox} and M_{oy} are the *required* equivalent resisting moment strengths about the x and y axes, respectively, while α_1 and α_2 are exponents that depend on the cross-sectional geometry and the steel percentage and its location and material stress f'_c and f_y.

Equation 8.28 can be simplified using a common exponent and introducing a factor β for one particular axial load value P_n such that the ratio M_{nx}/M_{ny} would have the same value as the ratio M_{ox}/M_{oy} as detailed by Parme and associates. Such simplification leads to

$$\left(\frac{M_{nx}}{M_{ox}}\right)^{\alpha} + \left(\frac{M_{ny}}{M_{oy}}\right)^{\alpha} = 1.0 \tag{8.29}$$

where $M_{nx} = P_n e_y$
$M_{ny} = P_n e_x$

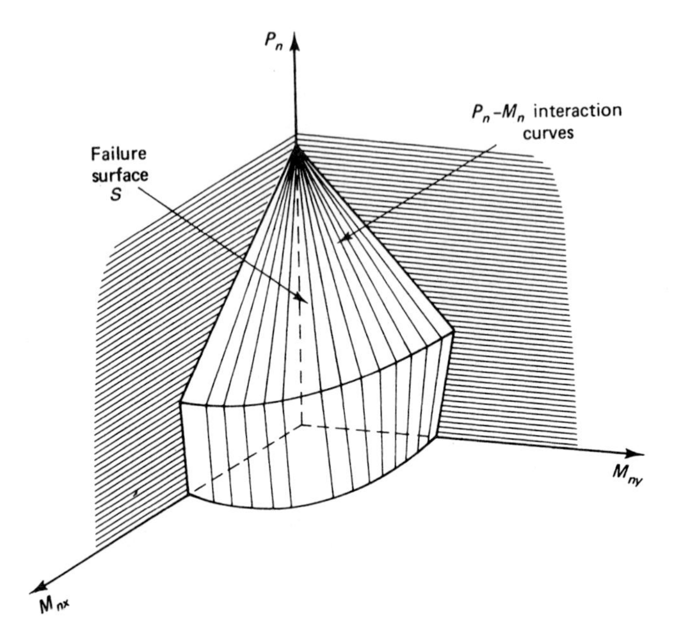

Figure 8.20 Failure interaction surface for biaxial column bending.

where $\alpha = \log 0.5/\log \beta$. Figure 8.21 gives a contour plot *ABC* from Equation 8.29.

For design purposes, the contour is approximated by two straight lines *BA* and *BC*, and Equation 8.29 can be simplified to two conditions:

1. For *AB* when $M_{ny}/M_{oy} < M_{nx}/M_{ox}$,

$$\frac{M_{nx}}{M_{ox}} + \frac{M_{ny}}{M_{oy}}\left[\frac{1-\beta}{\beta}\right] = 1.0 \tag{8.30a}$$

2. For *BC* when $M_{ny}/M_{oy} > M_{nx}/M_{ox}$,

$$\frac{M_{ny}}{M_{oy}} + \frac{M_{nx}}{M_{ox}}\left[\frac{1-\beta}{\beta}\right] = 1.0 \tag{8.30b}$$

In both of these equations, the *actual* controlling equivalent uniaxial moment strength M_{oxn} or M_{oyn} should be at least equivalent to the *required* controlling moment strength M_{ox} or M_{oy} of the chosen column section.

For rectangular sections where the reinforcement is evenly distributed along all the column faces, the ratio M_{oy}/M_{ox} can be taken to be approximately equal to *b/h*. In that case, Equations 8.30 can be modified as follows:

1. For $\dfrac{M_{ny}}{M_{nx}} > b/h$,

$$M_{ny} + M_{nx}\frac{b}{h}\frac{1-\beta}{\beta} \cong M_{oy} \tag{8.31a}$$

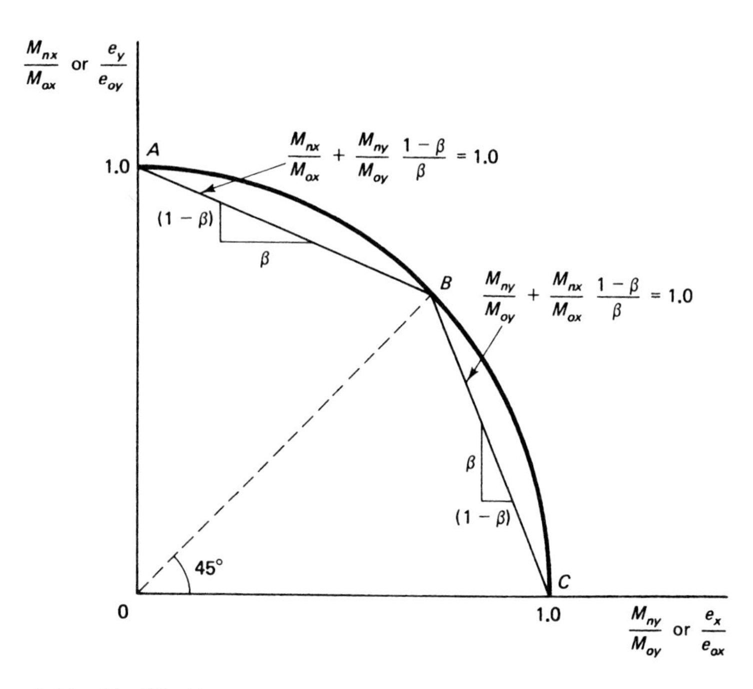

Figure 8.21 Modified interaction contour plot of constant P_n for biaxially loaded column.

2. For $\dfrac{M_{ny}}{M_{nx}} \leq b/h$,

$$M_{nx} = M_{ny}\frac{h}{b}\frac{1-\beta}{\beta} \cong M_{ox} \tag{8.31b}$$

The controlling required moment strength M_{ox} or M_{oy} for designing the section is the larger of the two values as determined from Equations 8.31.

Plots like those of Figure 8.22 are used in the selection of β in the analysis and design of the columns just described. In effect, the modified load-contour method can be summarized in Equation 8.31 as a method for finding equivalent required moment strengths M_{ox} and M_{oy} for designing the columns as if they were uniaxially loaded.

8.11.3 Step-by-Step Operational Procedure for the Design of Biaxially Loaded Columns

The following steps can be used as a guideline for the design of columns subjected to bending in both the *x* and *y* directions. The procedure assumes an equal area of reinforcement on all four faces.

1. Compute the uniaxial bending moments assuming an equal number of bars on each column face. Assume a value of an interaction contour β factor between 0.50 and 0.70 and a ratio of *h/b*. This ratio can be approximated to M_{nx}/M_{ny}. Using Equations 8.31 determine the equivalent required uniaxial moment M_{ox} or M_{oy}. If M_{nx} is larger than M_{ny}, use M_{ox} for the design and vice versa.
2. Assume a cross section for the column and a reinforcement ratio $\rho = \rho' \cong 0.01$ to 0.02 on each of the two faces parallel to the axis of bending of the larger equivalent

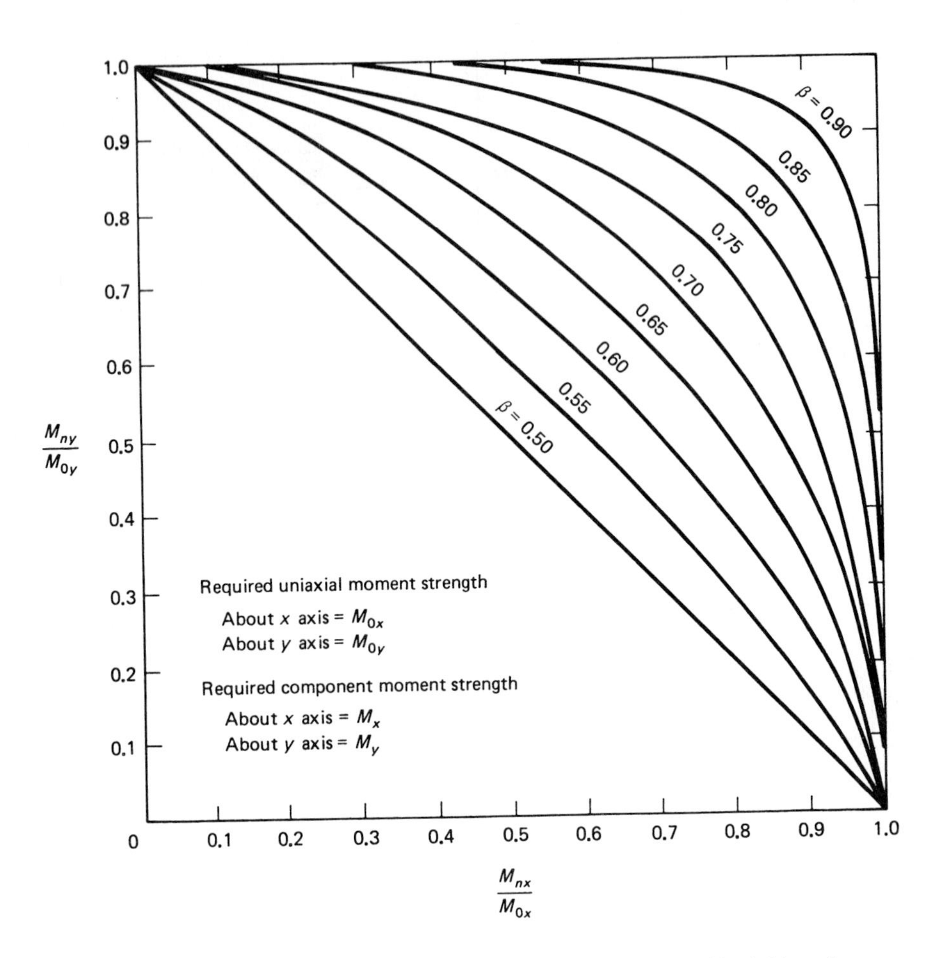

Figure 8.22 Contour β-factor chart for rectangular columns in biaxial bending.

moment. Then make a preliminary selection of the steel bars, and verify the capacity P_n of the assumed column cross section. In the completed design, the same amount of longitudinal steel should be used on all four faces.

3. Compute the *actual* nominal moment strength M_{oxn} for equivalent uniaxial bending about the x axis when $M_{ox} = 0$. Its value has to be at least equivalent to the *required* moment strength M_{ox}.
4. Compute the actual nominal moment strength M_{oyn} for the equivalent uniaxial bending moment about the y axis when $M_{oy} = 0$.
5. Find M_{ny} by entering M_{nx}/M_{oxn} and the trial β value into the β factor contour plots of Figure 8.22.
6. Make a second trial and adjustment, increasing the assumed β value if the M_{ny} value obtained from entering the chart is less than the required M_{ny}. Repeat this step until the two values of M_{ny} converge, either through changing β or changing the section.
7. Design the lateral ties and detail the section.

A flowchart for the primary steps in evaluating the controlling moment values in biaxially loaded columns is given in Figure 8.23. A detailed computational example, together with a discussion of biaxially loaded columns, is presented in Ref. 8.2.

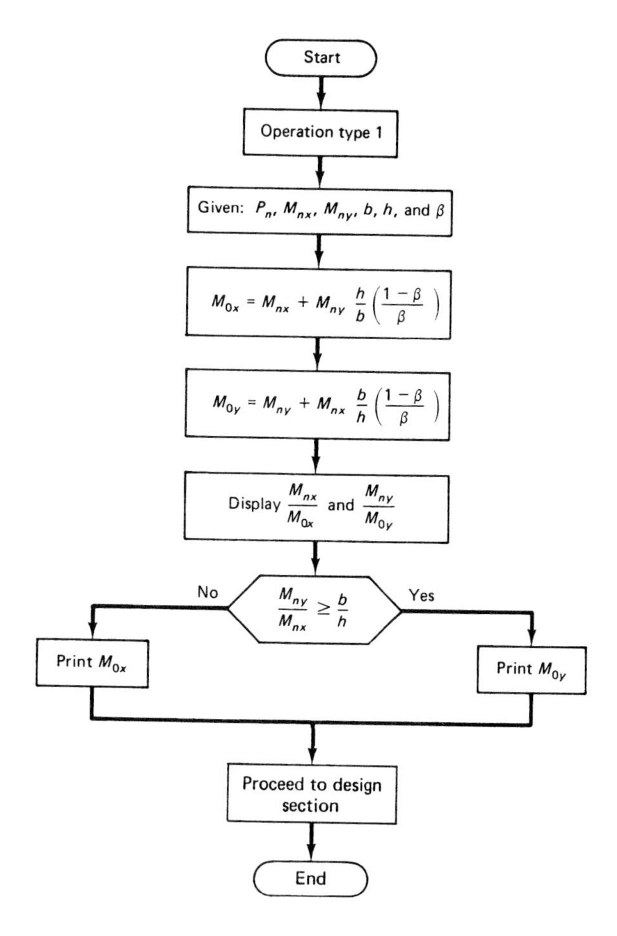

Figure 8.23 Flowchart for evaluation of controlling moment values in biaxially loaded columns.

8.12 PRACTICAL DESIGN CONSIDERATIONS

The following guidelines are presented for the design and arrangement of reinforcement to arrive at a practical design.

8.12.1 Longitudinal or Main Reinforcement

The average effective prestress in the concrete in prestressed compression members should not be less than 225 psi (1.55 MPa). This code requirement sets a minimum reinforcement ratio such that compressive members with lower prestress values will have a minimum nonprestressed reinforcement ratio of one percent.

8.12.2 Lateral Reinforcement for Columns

8.12.2.1 Lateral Ties. Lateral reinforcement is required to prevent spalling of the concrete cover or local buckling of the longitudinal bars. The reinforcement could be in the form of ties evenly distributed along the height of the column at specified intervals. Longitudinal bars spaced more than 6 in. apart should be supported by lateral ties, as shown in Figure 8.24.

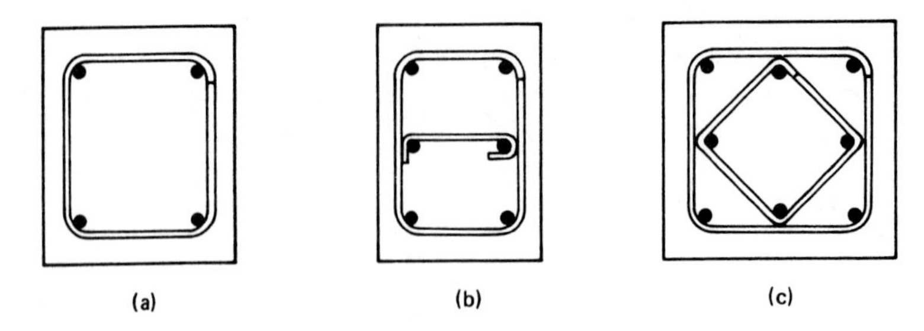

Figure 8.24 Typical arrangement of ties for four, six, and eight longitudinal bars in a column. (a) One tie. (b) Two ties. (c) Two ties.

The following guidelines are to be followed for the selection of the size and spacing of ties.

1. The size of the tie should not be less than a #3 (9.5 mm) bar.
2. The vertical spacing of the ties must not exceed
   (a) Forty-eight times the diameter of the tie
   (b) Sixteen times the diameter of the longitudinal bar
   (c) The least lateral dimension of the column

Figure 8.24 shows a typical arrangement of ties for four, six, and eight longitudinal bars in a column cross section.

8.12.2.2 Spirals. The other type of lateral reinforcement is spirals or helical lateral reinforcement, as shown in Figure 8.25. Spirals are particularly useful in increasing ductility or member toughness, and hence are mandatory in high-earthquake-risk regions. Normally, concrete outside the confined core of the spirally reinforced column can totally spall under unusual and sudden lateral forces such as earthquake-induced forces. The columns have to be able to sustain most of the load even after the spalling of the cover in order to prevent the collapse of the building. Hence, the spacing and size of spirals are designed to maintain most of the load-carrying capacity of the column, even under such severe load conditions.

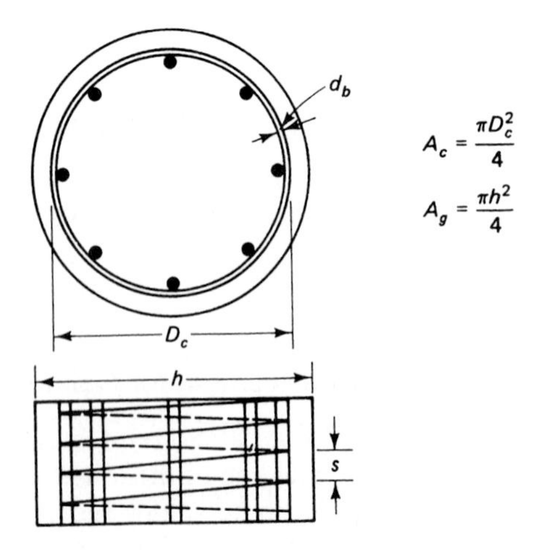

Figure 8.25 Helical or spiral reinforcement for columns.

Closely spaced spiral reinforcement increases the ultimate-load capacity of columns. The spacing or pitch of the spiral is so chosen that the load capacity due to the confining spiral action compensates for the loss due to spalling of the concrete cover.

Equating the increase in strength due to confinement to the loss of capacity in spalling, and incorporating a safety factor of 1.2, we obtain the minimum spiral reinforcement ratio

$$\rho_s = 0.45\left(\frac{A_g}{A_c} - 1\right)\frac{f'_c}{f_{sy}} \tag{8.32}$$

where $\rho_s = \dfrac{\text{volume of the spiral steel per one revolution}}{\text{volume of concrete core contained in one revolution}}$

$$A_c = \frac{\pi D_c^2}{4} \tag{8.33a}$$

$$A_g = \frac{\pi h^2}{4} \tag{8.33b}$$

h = diameter of the column
a_s = cross-sectional area of the spiral
d_b = nominal diameter of the spiral wire
D_c = diameter of the concrete core out-to-out of the spiral
and f_{sy} = yield strength of the spiral reinforcement

To determine the pitch s of the spiral, compute ρ_s using Equation 8.33, choose a bar diameter d_b for the spiral, compute a_s, and then obtain pitch b using Equation 8.35b below.

The spiral reinforcement ratio ρ_s can be written

$$\rho_s = \frac{a_s\pi(D_c - d_b)}{(\pi/4)D_c^2\, s} \tag{8.34}$$

Therefore, the pitch is given by

$$s = \frac{a_s\pi(D_c - d_b)}{(\pi/4)D_c^2\, \rho_s} \tag{8.35a}$$

or

$$s = \frac{4a_s(D_c - d_b)}{D_c^2\rho_s} \tag{8.35b}$$

The spacing or pitch of spirals is limited to a range of 1 to 3 in. (25.4 to 76.2 mm), and the diameter should be at least $\frac{3}{8}$ in. (9.53 mm). The spiral should be well anchored by providing at least $1\frac{1}{2}$ extra turns when splicing of spirals rather than welding is used.

8.12.2.3 Design of Spiral Lateral Reinforcement

Example 8.3

Design the lateral spiral reinforcement for a circular prestressed concrete column $h = 20$ in. (508 mm) and clear cover $d_c = 1.5$ in. (38 mm) given that $f_y = 60{,}000$ psi (414 MPa).

Solution: Using Equation 8.32,

$$\text{required } \rho_s = 0.45\left(\frac{A_g}{A_c} - 1\right)\frac{f'_c}{f_{sy}}$$

Using #3 spirals with a yield strength $f_y = 60{,}000$ psi, we obtain

$$\text{Clear concrete cover } d_c = 1.5 \text{ in. (38 mm)}$$

$$f_{sy} = 60{,}000 \text{ psi}$$

$$D_c = h - 2d_c = 20.0 - 2 \times 15 = 17.0 \text{ in. (432 mm)}$$

$$A_c = \frac{\pi(17.0)^2}{4} = 226.98 \text{ in.}^2$$

$$A_g = 314.0 \text{ in.}^2$$

$$\rho_s = 0.45\left(\frac{314.0}{226.98} - 1\right)\frac{4{,}000}{60{,}000} = 0.0115$$

For #3 spirals, $a_s = 0.11$ in.^{2} So using Equation 8.35b, we get

$$\text{pitch } s = \frac{4a_s(D_c - d_b)}{D_c^2\,\rho_s} = \frac{4 \times 0.11(17.0 - 0.375)}{(17.0)^2 \times 0.0115} = 2.20 \text{ in. (56 mm)}$$

Accordingly, provide #3 spirals at $2\frac{1}{4}$ in. pitch (9.53 mm dia spiral at 54.0 mm pitch).

8.13 RECIPROCAL LOAD METHOD FOR BIAXIAL BENDING

This method developed by Bressler relates the desired axial force P_u value to three other values on a reciprocal of the failure surface (Ref. 8.2). Assume S_1 denotes the coordinates on the failure surface in Figure 8.20 such that the values of the load and eccentricities as P_u, e_x and e_y. If S_2 is a point on the compatible reciprocal surface to that in Fig. 8.20, then S_2 would define the coordinates of that point as $1/P_u$, e_x and e_y, where $P_u = \phi P_n$, which is the factored (design) load.

If the desired axial load P_u under biaxial loading about the x and y axes is related to the P_u values devoted by P_{uy}, P_{ux}, and P_{uo}, then

$$\frac{1}{P_u} = \frac{1}{P_{ux}} + \frac{1}{P_{uy}} - \frac{1}{P_{uo}} \tag{8.36a}$$

or

$$\frac{1}{\phi P_n} = \frac{1}{\phi P_{nxo}} + \frac{1}{\phi P_{nyo}} - \frac{1}{\phi P_{no}} \tag{8.36b}$$

where, $P_{ux} = \phi P_{nxo}$ = design strength of the column having eccentricity e_x, provided $e_y = 0$

$P_{uy} = \phi P_{nyo}$ = design strength of the same column having eccentricity e_y, provided $e_x = 0$

$P_{uo} = \phi P_{no}$ = theoretical *axial* load design strength for the same column having eccentricity $e_y = e_x = 0$

M_{ux} = moment about the x-axis = $P_u\, e_y$

M_{uy} = moment about the y-axis = $P_u\, e_x$

e_x = eccentricity measures parallel to the y-axis as in Figure 8.26 a, namely $e_x = (M_{uy}/P_u) = (P_u e_x/P_u)$

e_y = eccentricity measured parallel to the y-axis = $(P_u e_x/P_u)$

x = column cross-section dimension parallel to the x-axis

y = column cross-section dimension parallel to the y-axis

The step-by-step operational procedure essentially follows the logic in the steps presented in Sec. 8.11.3.

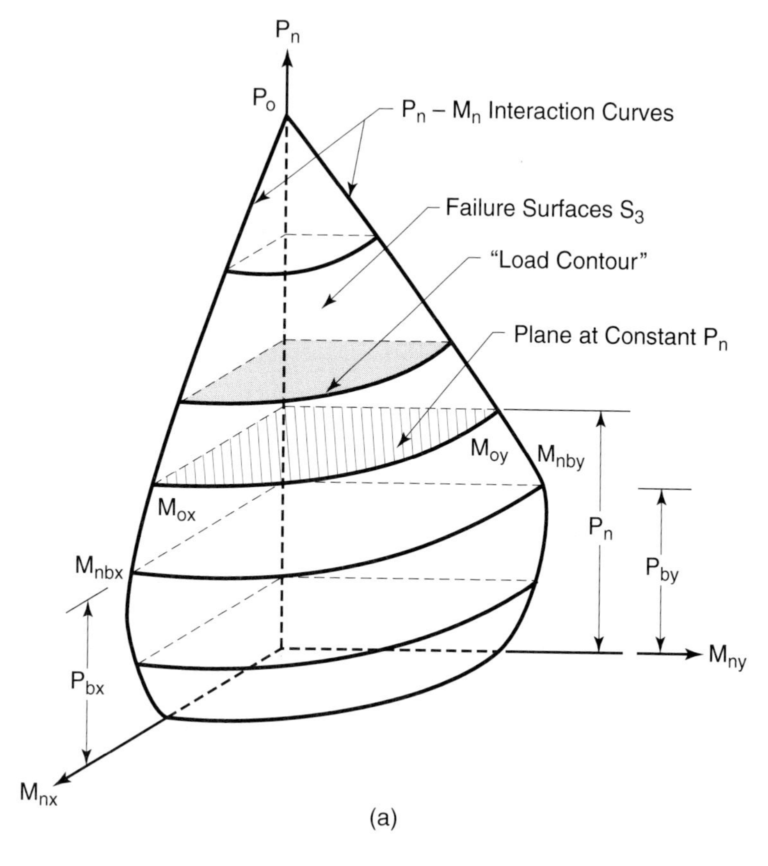

(a)

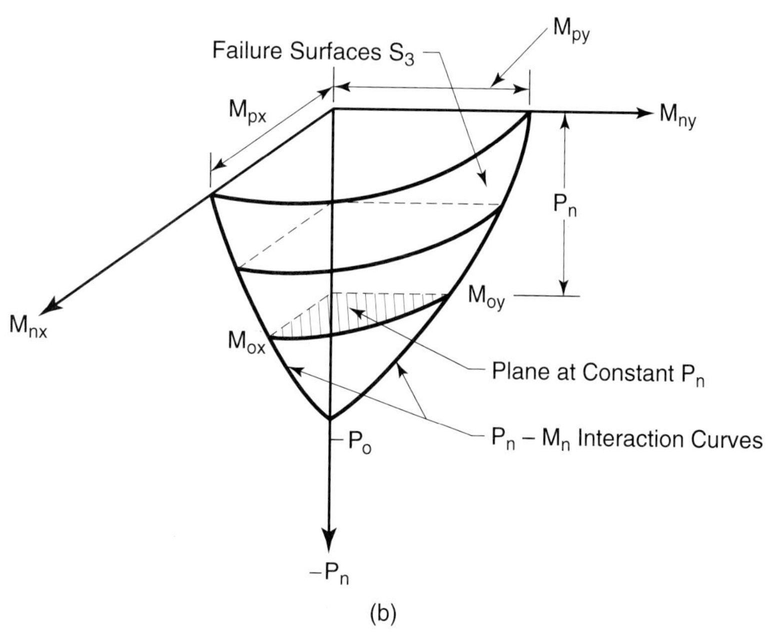

(b)

Figure 8.26 Failure Surface Interaction Diagram (Ref. 8.11) (a) Biaxial Bending and Compression, (b) Biaxial Bending and Tension.

8.14 MODIFIED LOAD CONTOUR METHOD FOR BIAXIAL BENDING

In lieu of Equation 8.32, Hsu in Ref. 8.11 proposed a modified expression which can represent both the strength interaction diagram and the failure surface of a reinforced concrete biaxially loaded columns as in Fig. 8.26 modifying the approach presented in Sec. 8.11.2. This method, as well as the reciprocal load method, seems to demand less computational rigor as can be seen from the two design examples to follow.

The interaction expression for the load and bending moments about the two axes is

$$\left(\frac{P_n - P_{nb}}{P_{no} - P_{nb}}\right) + \left(\frac{M_{nx}}{M_{nbx}}\right)^{1.5} + \left(\frac{M_{ny}}{M_{nby}}\right)^{1.5} = 1.0 \tag{8.37}$$

Where

P_n = nominal axial compression (positive), or tension (negative)

M_{nx}, M_{ny} = nominal bending moments about the x- and y-axis respectively

P_{no} = maximum nominal axial compression (positive) or axial tension (negative)
$= 0.85 f'_c(A_g - A_{st}) + f_y A_{st}$

P_{nb} = nominal axial compression at the balanced strain condition

M_{nbx}, M_{nby} = nominal bending moments about the x- and y-axis respectively, at the balanced strain condition

The value of P_{nb} and M_{nb} can be obtained from:

$$P_{nb} = 0.85 f'_c \beta_1 c_b b + A_{ps} f'_{ps} - A_{ps} f_{ps} \tag{8.38a}$$

and

$$M_{nb} = P_{nb} e_b = C_c\left(d - \frac{a}{2} - d''\right) + C_s(d - d' - d'') + T_s d'' \tag{8.38b}$$

where,

a_b = depth of the equivalent block $= \beta_1 c_b = (A_{ps}/f_{ps})/(0.85 f'_c b)$

$a = \beta_1 c$

f'_{ps} = stress in the compressive reinforcement closest to the load $= f_{py}$ if $f_{ps} \geq f_{py}$

T_s = force in the tensile side reinforcement

The step-by-step operational procedure for the design of biaxially loaded columns essentially follows the procedure of Sec. 8.11.3. This method seems to require less effort in the solution of biaxial bending problems.

8.14.1 Design of Biaxially Loaded Prestressed Concrete Column by the Modified Load Contour Method

Example 8.4

Assume the precast column section in Example 8.2 is a nonslender column subjected to biaxial bending without sidesway. Design the column for the following bending moments:

$$M_{ux} = M_{uy} = 825{,}000 \text{ in.-lb } (93.7 \text{ kN-m}) \text{ and } P_u = 300{,}000 \text{ lb } (1334 \text{ kN})$$

Given:

f'_c = 6,000 psi (41.4 MPa) normal weight concrete

f_{pu} = 270,000 psi (1863 MPa)

f_{ps} = 240,000 psi (1565 MPa)

The section is reinforced with five $\frac{1}{2}$-in. 7-wire (12-mm dia. 7 wire) tendons giving a total of sixteen tendons.

Solution:

$$P_u = 300{,}000 \text{ lbs}$$

$$M_{ux} = P_u e_y = 825{,}000 \text{ in.-lb about the } x\text{-axis}$$

$$M_{uy} = P_u e_x = 825{,}000 \text{ in.-lb about the } y\text{-axis}$$

$$f'_c = 6{,}000 \text{ psi}$$

$$f_{ps} = 240{,}000 \text{ psi}$$

Hence:

$$e_x = \frac{M_{ux}}{P_u} = \frac{825{,}000}{300{,}000} = 2.75 \text{ in.}$$

$$e_y = \frac{M_{uy}}{P_u} = \frac{825{,}000}{300{,}000} = 2.75 \text{ in.}$$

x = axis parallel to the shorter side b.

y = axis parallel to the longer side h.

The column section is 15 in. × 15 in.

$$b = 15 \text{ in.} \qquad h = 15 \text{ in.} \qquad d' = 2.5 \text{ in.}$$

On each face A_s = (five $\frac{1}{2}$-in. dia. 7-wire tendons) = 5 × 0.153 = 0.765 in.^{2}

Total reinforcement area $A_{st} = 16 \times 0.153 = 2.448$ in.^{2}

The small eccentricity of 2.75 in. suggests that it is compression failure. Try $\phi = 0.70$.

$$\text{Actual } P_n = \frac{300{,}000}{0.70} = 428{,}571 \text{ lb}$$

$$\text{Actual } M_n = \frac{825{,}000}{0.70} = 1{,}178{,}571 \text{ lb-in.}$$

From example 4.2,

$$P_{nb} = 229{,}310 \text{ lb.}$$

$$M_{nb} = P_{nb} e_b = 2{,}103{,}124 \text{ in.-lb (237 kN-m)}$$

$$e_b = \frac{M_{nb}}{P_{nb}} = \frac{2{,}103{,}124}{229{,}310} = 9.17 \text{ in.}$$

$e_b > e = 2.75$ in., hence compression failure and the strength reduction factor $\phi = 0.70$

$$P_{no} = 0.85 f'_c (A_g - A_{st}) + A_{st} f_{ps}$$

$$= 0.85 \times 6{,}000\,(225 - 2.448) + 1.53 \times 240{,}000$$

$$= 1{,}502{,}205 \text{ lb.}$$

Using the interaction surface expression for biaxial bending in equation 8.37,

$$\left(\frac{P_n - P_{nb}}{P_{no} - P_{nb}}\right) + \left(\frac{M_{nx}}{M_{nbx}}\right)^{1.5} + \left(\frac{M_{ny}}{M_{nby}}\right)^{1.5}$$

$$= \frac{428{,}571 - 229{,}310}{1{,}502{,}205 - 229{,}310} + \left(\frac{1{,}178{,}571}{2{,}103{,}124}\right)^{1.5} + \left(\frac{1{,}178{,}571}{2{,}103{,}124}\right)^{1.5}$$

$$= 0.156 + 0.420 + 0.420 = 0.996 \cong 1.0$$

(this section is very slightly overdesigned)

Accept the design, namely,

$b = 15$ in. $h = 15$ in. $d = 12.5$ in.
A_s = five $\frac{1}{2}$-in. dia. 7-wire strand tendons *on each face* as in Figure 8.17 giving a total of sixteen tendons.

8.15 PRESTRESSED TENSION MEMBERS

8.15.1 Service-Load Stresses

Tension elements and systems such as railroad ties, bridge truss tension members, foundation anchors for retaining walls, and ties in walls of liquid-retaining tanks combine the high strength of the prestressing strands with the stiffness of the concrete. As such, they provide tensile resistance and reduced deformations that could not be provided if the member were made of an all-steel section to carry the same load. The limited and controlled deformation of the prestressed tension member in spite of its slenderness makes it especially useful as a tie or as part of an overall structural system.

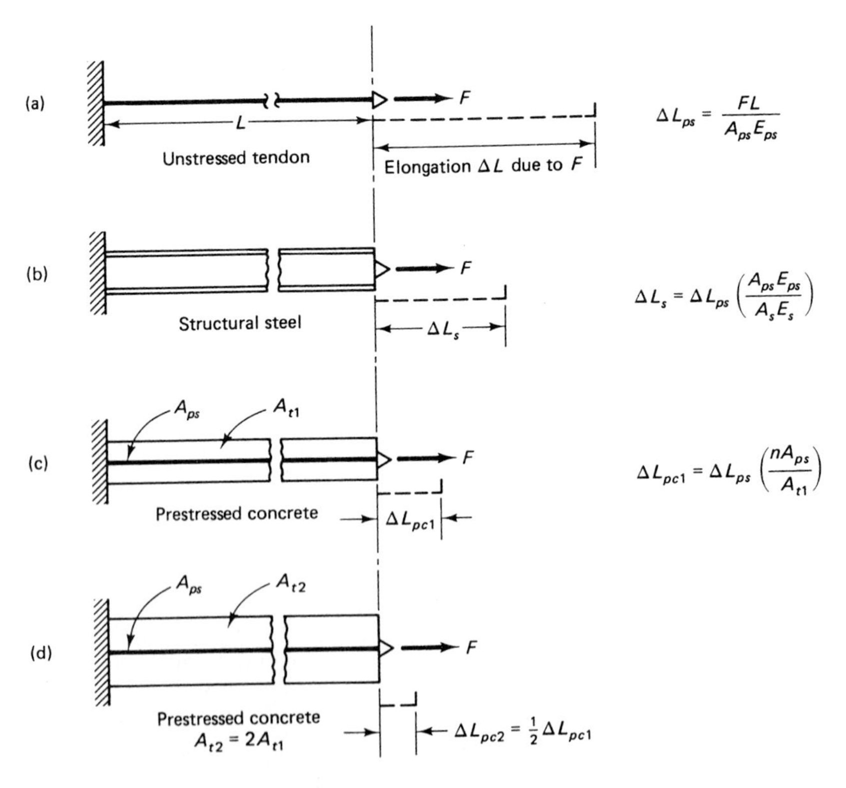

Figure 8.27 Relative elongation of tension members. (a) Unstressed tendon. (b) Structural steel. (c) Prestressed concrete, area $A_{t1} = A_g + (n - 1)A_{ps}$. (d) Prestressed concrete, area $A_{t2} = 2A_{t1}$.

Figure 8.27 compares the elongation of a prestressed concrete member in direct tension with a structural steel member of similar capacity. The elongation of the tension tie results from application of the external force F, while the elongation of the unstressed tendon in part (a) due to force F is, from basic mechanics,

$$\Delta L_{ps} = \frac{FL}{A_{ps}E_{ps}} \tag{8.39}$$

If the tendon is replaced by a rolled structural member, the change in the properties of the section results in a deformation

$$\Delta L_s = \Delta L_{ps}\left(\frac{A_{ps}E_{ps}}{A_sE_s}\right) \tag{8.40}$$

where A_s is considerably larger than A_{ps}. Hence, a considerably reduced elongation is seen as shown in Figure 8.27b. The transformed area of concrete in Figure 8.27c is

$$A_{t1} = A_g + (n_p - 1)A_{ps} \tag{8.41}$$

and if the change in stress in the concrete is

$$\Delta f_c = \frac{F}{A_{t1}} \tag{8.42}$$

the corresponding change in stress in the prestressing steel is

$$\Delta f_{ps} = \frac{n_pF}{A_{t1}} \tag{8.43}$$

where n_p is the modular ratio E_{ps}/E_c. Therefore,

$$\Delta L_{pc1} = \Delta L_{ps}\left(\frac{n_pA_{ps}}{A_{t1}}\right) \tag{8.44}$$

or

$$\Delta L = \frac{n_pFL}{A_tE_{ps}} = \frac{FL}{A_tE_c} \tag{8.45}$$

Equation 8.45 gives the elongation of the tension tie due to the external load F at service load. It is readily seen that if the area of concrete is doubled, the elongation is halved for the same tensile force F. In sum, through a judicious choice of the geometry of the section, it is possible to reduce the elongation of prestressed tension members considerably.

Shortening of the tension tie occurs due to both the prestressing force and the long-term effects. The stress in the concrete due to initial prestress after transfer is

$$f_{ci} = -\frac{P_i}{A_c} \tag{8.46}$$

where A_c is the net area of concrete in the section. The reduced stress in the concrete that results in an effective prestressing force P_e after all time-dependent losses have occurred is

$$f_{ce} = -\frac{P_e}{A_c} \tag{8.47a}$$

while the effective stress in the prestressing steel is

$$f_{pe} = -\frac{P_e}{A_{ps}} \tag{8.47b}$$

Superimposing Equation 8.42 on Equation 8.47a for the total effect of the external tensile force F and the prestressing force P_e, we have, for the total stress in the concrete,

$$f_c = -\frac{P_e}{A_c} + \frac{F}{A_t} \tag{8.48}$$

The corresponding stress in the tendon is

$$f_{ps} = f_{pe} + \frac{nF}{A_t} \tag{8.49}$$

If cracking is not allowed in the tension member, the change in length ΔL_{pc} (elongation) at service load, from Equation 8.49, is essentially equivalent to the reduction in length ΔP due to the effective prestressing force P_e (see next).

8.15.2 Deformation Behavior

Evaluation of the deformation of the tension member is critical to the overall design of the entire structure. Changes in the length of the member can induce severe stresses in the adjoining members, which could lead to structural failure. If the member is post-tensioned and fully bonded, then the reduction in length due to the initial prestress P_i alone is

$$\Delta_i = -\frac{P_i L}{A_c E_c} \tag{8.50a}$$

and the elastic reduction in length after losses is

$$\Delta P = -\frac{P_e L}{A_c E_c} \tag{8.50b}$$

Due to creep, the initial prestressing force P_i reduces to the effective force P_e. So using a creep coefficient C_u and assuming, as in Chapter 7, an average force $(P_i + P_e)/2$ as sufficiently accurate for evaluating creep loss, the change in length due to creep is

$$\Delta_{CR} = -\frac{L}{A_c E_c}\left[C_u\left(\frac{P_i + P_e}{2}\right)\right] \tag{8.51}$$

Photo 8.5 Laboratory prestressing bed. (*Courtesy,* Building Research Establishment, Garston, Watford, England.)

To account for shrinkage, the change in length is

$$\Delta_{SH} = \epsilon_{SH} L \tag{8.52}$$

so that the total effective reduction in length becomes

$$\Delta_e = -\left\{ \frac{L}{A_c E_c} \left[P_e + C_u \left(\frac{P_i + P_e}{2} \right) \right] + \epsilon_{SH} L \right\} \tag{8.53}$$

Conversely, the loss in tension due to creep and shrinkage alone, from Equations 8.51 and 8.52, is

$$\Delta P = \frac{\Delta_{CR} + \Delta_{SH}}{L} E_{ps} A_{ps} \tag{8.54}$$

8.15.3 Decompression and Cracking

In considering decompression and cracking, it has to be assumed that no cracking is allowed at service load. If the probability of cracking exists due to a possible overload, an additional prestressing force and the addition of nonprestressed reinforcement become necessary to control cracking.

Assuming that only prestressed steel is provided, the tensile stress in the concrete at the *first cracking load* F_{cr} should not exceed the direct tensile strength of the concrete, ranging between $f'_t = 3\lambda \sqrt{f'_c}$ and $f'_t = 5\lambda \sqrt{f'_c}$, where $\lambda = 1$, 0.85, and 0.75 for normal-weight, sand-lightweight and all-lightweight concrete, respectively. The cracking load F_{cr} can be evaluated from Equation 8.48 using the appropriate value of f'_t, viz.,

$$f'_t = -\frac{P_e}{A_c} + \frac{F_{cr}}{A_t} \tag{8.55}$$

Any overload beyond F_{cr} is expected to cause a dynamic increase in cracking such that all the applied load starts to be carried by the prestressing steel alone, with a consequent failure of the member. Therefore, provision of an adequate level of residual compressive stress in the concrete becomes necessary in major tension members.

A *decompression load,* namely, the load at $f'_t = 0$ in Equation 8.55, should be the maximum allowable service load to which the member can be subjected. Equation 8.55 then becomes

$$f'_t = 0 = -\frac{P_e}{A_c} + \frac{F_{\text{dec}}}{A_t} \tag{8.56}$$

8.15.4 Limit State at Failure and Safety Factors

After cracking, the entire tension in the member is assumed to be carried by the prestressing tendon. Consequently, the nominal strength of the linear tension member is

$$F_n = A_{ps} f_{pu} \tag{8.57}$$

and the design ultimate load is

$$F_u = \phi_n F_n = \phi A_{ps} f_{pu} \tag{8.58}$$

It is important to provide a minimum safety factor of 1.5 for the decompression load F_{dec}. The safety factor level is determined by the importance of the tension member in the structure, the importance of the structure itself, and the negative effects of long-term reductions in length of the tension member on the integrity of the overall structure. It is not unreasonable under certain conditions to use a safety factor of 2.0 or more in a particular design.

8.16 SUGGESTED STEP-BY-STEP PROCEDURE FOR THE DESIGN OF TENSION MEMBERS

1. Determine the factored load F_u and the corresponding required nominal strength $F_n = F_u/\phi$.
2. Choose a uniform concrete compressive stress value at service load due to P_e to range between $f_c = 0.20\,f'_c$ and $f_c = 0.30\,f'_c$. Select the area A_{ps} and the size of the prestressing strands, and then compute the net concrete area $A_c = A_g - A_{\text{duct}}$ and the transformed area $A_t = A_g + (n-1)A_{ps}$.
3. Compute the maximum allowable external force F based on decompression stress $f'_t = 0$, i.e., find F_{dec}. Then compute the first cracking load F_{cr}.
4. Find the factors of safety due to forces F, F_{dec}, and F_{cr} to verity that they exceed the value 1.5.
5. Compute the length-shortening deformations due to creep and shrinkage (Equation 8.53):

$$\Delta_e = -\left\{\frac{L}{A_c E_c}\left[P_e + C_u\left(\frac{P_i + P_e}{2}\right)\right] + \epsilon_{SH} L\right\}$$

   Then check whether the value obtained causes excessive stress in adjoining members. Next, compute the elongation

$$\Delta_L = \frac{FL}{A_t E_c}$$

   due to external load (Equation 8.45) and verify whether the value obtained is within the specified limits of the design.
6. Adopt the design if all requirements are satisfied; otherwise, proceed through another trial-and-adjustment cycle.

A flowchart for the step-by-step trial-and-adjustment procedure that can be used for the design and/or analysis of linear tension members is presented in Figure 8.28.

8.17 DESIGN OF LINEAR TENSION MEMBERS

Example 8.4

A linear post-tensioned fully grouted direct-tension tie for an underground shelter shown in Figure 8.29 has a length $L = 130$ ft (39.6 m). The tie is to be designed for a net horizontal roof and earthfill dead load of thrust $F_d = 115{,}000$ lb (512 kN) and live load of thrust $F_L = 55{,}000$ lb (245 kN). Design the tie with a safety factor not less than 1.5 against cracking and 1.2 against failure, using $\frac{1}{2}$-in. (12.7-mm) dia 270-K prestressing strands. The maximum allowable elongation due to external load is $\frac{1}{2}$-in., and given data are as follows:

$f'_c = 6{,}000$ psi, (41.4 MPa), normal weight

$f'_{ci} = 4{,}000$ psi (31.0 MPa)

$f'_t = 4\sqrt{f'_c} = 310$ psi (2.14 MPa)

$E_{ci} = 4.06 \times 10^6$ psi (28 MPa)

$E_c = 4.69 \times 10^6$ psi (32.2 MPa)

$f_{pu} = 270{,}000$ psi (1,862 MPa)

$f_{pi} = 190{,}000$ psi (1,310 MPa)

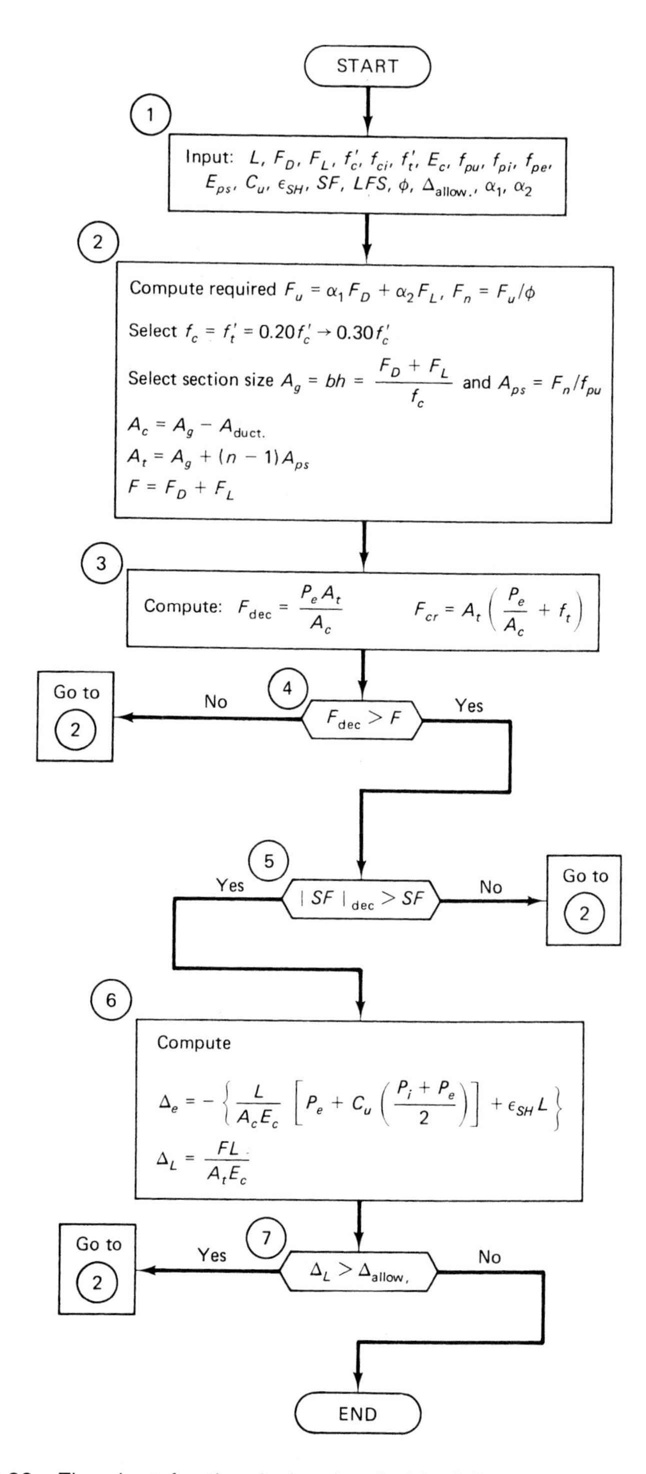

Figure 8.28 Flowchart for the design (analysis) of linear prestressed tension members.

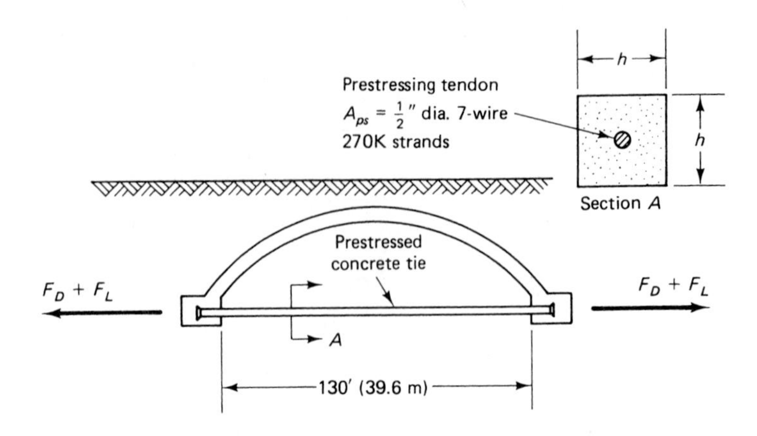

Figure 8.29 Prestressed concrete tie in Example 8.3.

$$f_{pe} = 150{,}000 \text{ psi } (1{,}034 \text{ MPa})$$

$$E_{ps} = 29 \times 10^6 \text{ psi } (200 \times 10^3 \text{ MPa})$$

$$n_p = E_{ps}/E_c = 29 \times 10^6/4.69 \times 10^6 = 6.18$$

$$C_u = 2.35$$

$$\epsilon_{SH} = 700 \times 10^{-6} \text{ in./in.}$$

The stress-strain diagram of the strands is given in Figure 8.9. Use a load factor of 1.4 for F_D and 1.7 for F_L.

Solution:

Factored Loads and Choice of Strands (Steps 1 and 2). The factored load is $F_u = 1.4F_D + 1.7F_L = 1.4 \times 115{,}000 + 1.7 \times 55{,}000 = 254{,}500$ lb (1,132 kN), and the required nominal strength is $F_n = F_u/\phi = 254{,}500/0.9 = 282{,}778$ lb (1,258 kN). So $A_{ps} = F_n/f_{pu} = 282{,}778/270{,}000 = 1.05$ in.^{2} (6.77 cm^2). Hence, trying ½-in. (12.7-mm) dia 7-wire strands, we find that the number of tendons = 1.05/0.153 = 6.86. So we use seven tendons and obtain $A_{ps} = 7 \times 0.153 = 1.07$ in.^{2} (6.9 cm^2). Then the available nominal strength is $F_n = 1.07 \times 270{,}000 = 288{,}900$ lb (1,285 kN).

Now, assume that the uniform compressive stress that is caused by the prestressing force P_e is $f_c = 0.25f'_c$. Then $P_e = 1.07 \times 150{,}000 = 160{,}500$ lb and $A_g = 160{,}500/(0.25 \times 6{,}000) = 107$ in.^{2} Accordingly, we try a tie rod section 11 in. × 11 in. (279 mm × 279 mm) and obtain $A_g = 121$ in.^{2} Then, assuming the tendon duct has a 2-in. diameter.

$$A_c = A_g - \frac{\pi(2)^2}{4} = 117.86 \text{ in.}^2\ (760 \text{ cm})^2$$

From Equation 8.41, the transformed concrete area is

$$A_t = A_g + (n-1)A_{ps} = 121 + (6.18 - 1)1.07 = 126.54 \text{ in.}^2\ (816.4 \text{ cm}^2)$$

Check of Forces F, F_{dec}, and F_{cr} (Steps 3 and 4). From Equation 8.56,

$$f'_t = 0 = -\frac{P_e}{A_c} + \frac{F_{dec}}{A_t}$$

or

$$0 = -\frac{160{,}500}{117.86} + \frac{F_{dec}}{126.54}$$

So the maximum allowable F is

$$\frac{160{,}500}{117.86} \times 126.54 = 172{,}320 \text{ lb (766 kN)}$$

and the actual $F = F_D + F_L = 115{,}000 + 55{,}000 = 170{,}000$ lb < 172,320 lb, which is satisfactory.

The first cracking load F_{cr} is obtained from Equation 8.55. Assume that the maximum tensile stress in the concrete at the first cracking load is $4\lambda\sqrt{f'_c} = 4 \times 1.0 \times \sqrt{6{,}000} = 310$ psi. Then

$$f'_t = -\frac{P_e}{A_c} + \frac{F_{cr}}{A_t}$$

or

$$310 = -\frac{160{,}500}{117.86} + \frac{F_{cr}}{126.54}$$

$$F_{cr} = 211{,}548 \text{ lb}$$

From before, the available nominal strength is $F_n = 288{,}900$ lb. So the design strength is $F_u = \phi F_n = 0.9 \times 288{,}900 = 260{,}010$ lb > required $F_u = 254{,}500$ lb, which is satisfactory. The cracking SF is $F_u/F_{cr} = 260{,}010/211{,}548 = 1.23$. This is close to 1.0, indicating that once the member is cracked it almost reaches its limit state at failure. Finally, the decompression SF is $F_u/F_{\text{dec}} = 260{,}010/172{,}320 = 1.51$ and the actual SF is $\phi F_n/F = 260{,}010/170{,}000 = 1.53$. Thus, the available SF is greater than the required SF = 1.5, which is satisfactory.

Deformation Check (Step 5). We know that $P_i = f_{pi}A_{ps} = 190{,}000 \times 1.07 = 203{,}300$ lb (904 kN). From Equation 8.50(a), the initial shortening at prestress transfer is

$$\Delta_i = -\frac{P_i L}{A_c E_c} = -\frac{203{,}300 \times 130 \times 12}{117.86 \times 4.69 \times 10^6} = 0.57 \text{ in. (15 mm)}$$

From Equation 8.53, the effective prestress shortening and long-term shortening due to creep and shrinkage is

$$\Delta_e = -\left\{\frac{L}{A_c E_c}\left[P_e + C_u\left(\frac{P_i + P_e}{2}\right)\right] + \epsilon_{SH}L\right\}$$

$$= -\left\{\frac{130 \times 12 \times 10^3}{117.86 \times 4.69 \times 10^6}\left[160.5 + 2.35\left(\frac{203.3 + 160.5}{2}\right)\right] + 700 \times 10^{-6}(130 \times 12)\right\}$$

$$= 2.75 \text{ in. (70 mm)}$$

From Equation 8.45, the elongation due to external load is

$$\Delta_L = \frac{FL}{A_t E_c} = \frac{170{,}000(130 \times 12)}{126.54 \times 4.69 \times 10^6} = 0.45 \text{ in. (12 mm)}$$

which is satisfactory since the allowable Δ is $\frac{1}{2}$ in. So the net shortening is $-2.75 + 0.45 = -2.30$ in. (58 mm). Thus, a deformation of this magnitude has to be imposed on the adjoining elements in order to determine whether the resulting stresses are within the allowable limits. Assuming they are, we can adopt the design of an 11 in. × 11 in. tie prestressed with seven $\frac{1}{2}$-in. 7-wire 270-K straight tendons.

REFERENCES

8.1 ACI Committee 318. *Building Code Requirements for Structural Concrete (ACI 318-99)* (ACI 318R-99), American Concrete Institute, Farmington Hills, MI, 2000, pp. 392.

8.2 Nawy, E. G., *Reinforced Concrete—A Fundamental Approach,* 4th ed. (1st ed., 1985). Upper Saddle River, N.J.: Prentice Hall, 1996, 876 pp.

8.3 Prestressed Concrete Institute. *PCI Design Handbook,* 5th ed. Prestressed Concrete Institute, Chicago, 1999.

8.4 Post-Tensioning Institute. *Post-Tensioning Manual.* 6th ed. Post-Tensioning Institute, Phoenix, 2000.

8.5 Lin, T. Y., and Lakhwara, T. R. "Ultimate Strength of Eccentrically Loaded Partially Prestressed Columns." *PCI Journal* 11 (1986): 37–49.

8.6 Gerwick, B. C., Jr. *Construction of Prestressed Concrete Structures.* Wiley-Interscience, New York, 1993.

8.7 Zia, P., and Moriadith, F. L. "Ultimate Load Capacity of Prestressed Concrete Columns." *Journal of the American Concrete Institute* 63 (1986): 767–788.

8.8 Gerwick, B. C., Jr. "Prestressed Concrete Developments in Japan." *Journal of the Prestressed Concrete Institute* 23 (1978): 66–76.

8.9 Wheen, R. J. "Prestressed Concrete Members in Direct Tension." *Journal of the Structural Division, American Society of Civil Engineers,* 105 (1979): 1471–1487.

8.10 Wilhelm, W. J., and Zia, P. "Effects of Creep and Shrinkage on Prestressed Concrete Columns." *Journal of the Structural Division, American Society of Civil Engineers,* Reston, Va, 96, 1970, 2103–2123.

8.11 Hsu, C. T. T., "Analysis and Design of Square and Rectangular Columns by Equation of Failure Surface," *ACI Struct. J.,* March–April 1988, American Concrete Institute, Farmington Hills, MI, March–April, 1988, pp. 167–189.

8.12 Nawy, E. G., editor-in-chief, *Concrete Construction Engineering Handbook,* CRC Press, Boca Raton, FL, 1998.

8.13 Nawy, E. G., *Fundamentals of Performance Concrete,* 2nd ed., John Wiley and Sons, New York, 2000.

PROBLEMS

8.1 Compute the nominal strengths P_n and M_n of the precast prestressed tied concrete nonslender column having the cross section shown in Figure P8.1 and an eccentricity $e = 9$ in. The column is prestressed with six $\frac{1}{2}$-in. dia 7-wire 270-K stress-relieved prestressing strands having the stress-strain properties shown in figure 8.9. Determine the type of initial failure of the column, and design the size and spacing of the necessary ties. Given data are as follows:

$$f'_c = 7{,}000 \text{ psi (48.6 MPa), normal-weight concrete}$$

$$f'_{ci} = 4{,}900 \text{ psi (33.8 MPa)}$$

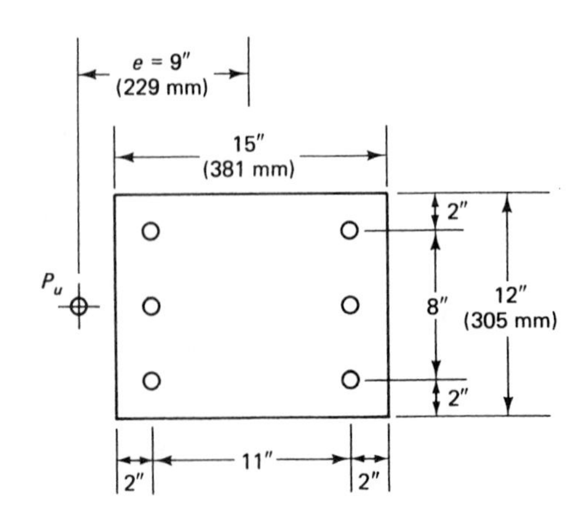

Figure P8.1 Column section.

$$f_{pu} = 270{,}000 \text{ psi } (1{,}862 \text{ MPa})$$

$$f_{py} = 255{,}000 \text{ psi } (1{,}758 \text{ MPa})$$

$$f_{pe} = 150{,}000 \text{ psi } (1{,}034 \text{ MPa})$$

$$\epsilon_{cu} = 0.0038 \text{ in./in.}$$

$$\epsilon_{ce} = 0.0008 \text{ in./in.}$$

$$E_{ps} = 29 \times 10^6 \text{ psi } (200 \times 10^3 \text{ MPa})$$

$$d' = 2 \text{ in. } (50.8 \text{ mm})$$

8.2 Construct the nominal load–moment $P_n - M_n$ and the design $P_u - M_u$ diagrams for the prestressed concrete columns in Example 8.1 if the overall sectional dimensions of the column are 20 in. (508 mm) × 20 in. (508 mm) and the prestressing reinforcement is twelve $\frac{1}{2}$-in. dia 7-wire 270-K, half of which are placed at each face parallel to the neutral axis.

8.3 Design a square tied prestressed bonded slender column having a clear height $l_u = 20$ ft (6.10 m) and which is not braced against sidesway. The column is loaded with the same magnitudes of loads and moments and is constructed of the same properties as the materials used in Example 8.2. The design should cover the following two loading conditions:

1. Gravity loading only, assuming that lateral sidesway due to wind is negligible.
2. Sidesway of the force and moment magnitudes of Example 8.2.

Compare the size needed in Problem 8.2 for $l_u = 20$ ft to the size of the column in Example 8.2 with $l_u = 15$ ft (4.57 m).

8.4 Design the column in Problem 8.3 as a biaxially loaded non-slender braced column subjected to factored moments $M_{ux} = M_{uy} = 1{,}100{,}000$ in.-lb. Use the Modified Load Contour Method in the solution.

8.5 Design the prestressed tension member in Example 8.4 if its length $L = 90$ ft (27.4 m), and compare the section obtained with the section of the 130-ft (39.6-m)-long tie in the example.

9.1 INTRODUCTION: REVIEW OF METHODS

Supported floor systems are usually constructed of reinforced concrete cast in place. Two-way slabs and plates are those panels in which the dimensional ratio of length to width is less than 2. The analysis and design of framed floor slab systems represented in Figure 9.1 encompasses more than one aspect of such systems. The present state of knowledge permits reasonable evaluation of (1) the moment capacity, (2) the slab–column shear capacity, and (3) serviceability behavior, as determined by deflection control and crack control. Note that flat plates are slabs supported directly on columns without beams, as shown in part (a) of the figure, compared to part (b) for slabs on beams, and part (c) for waffle slab floors.

Essentially the same principles are used in the analysis of continuous two-way prestressed concrete flat plate systems as in the analysis of reinforced concrete plate systems. The techniques of construction differ, however, and it is often unlikely that economic considerations alone can justify using prestressed two-way floor systems of the types shown in Figures 9.1(b) and (c). The prestressing is normally post-tensioned after the two-way plate is cast. Sometimes, on-site precast two-way slabs, called *lift slabs,* are used

University of Wyoming, addition. (*Courtesy,* Prestressed Concrete Institute.)

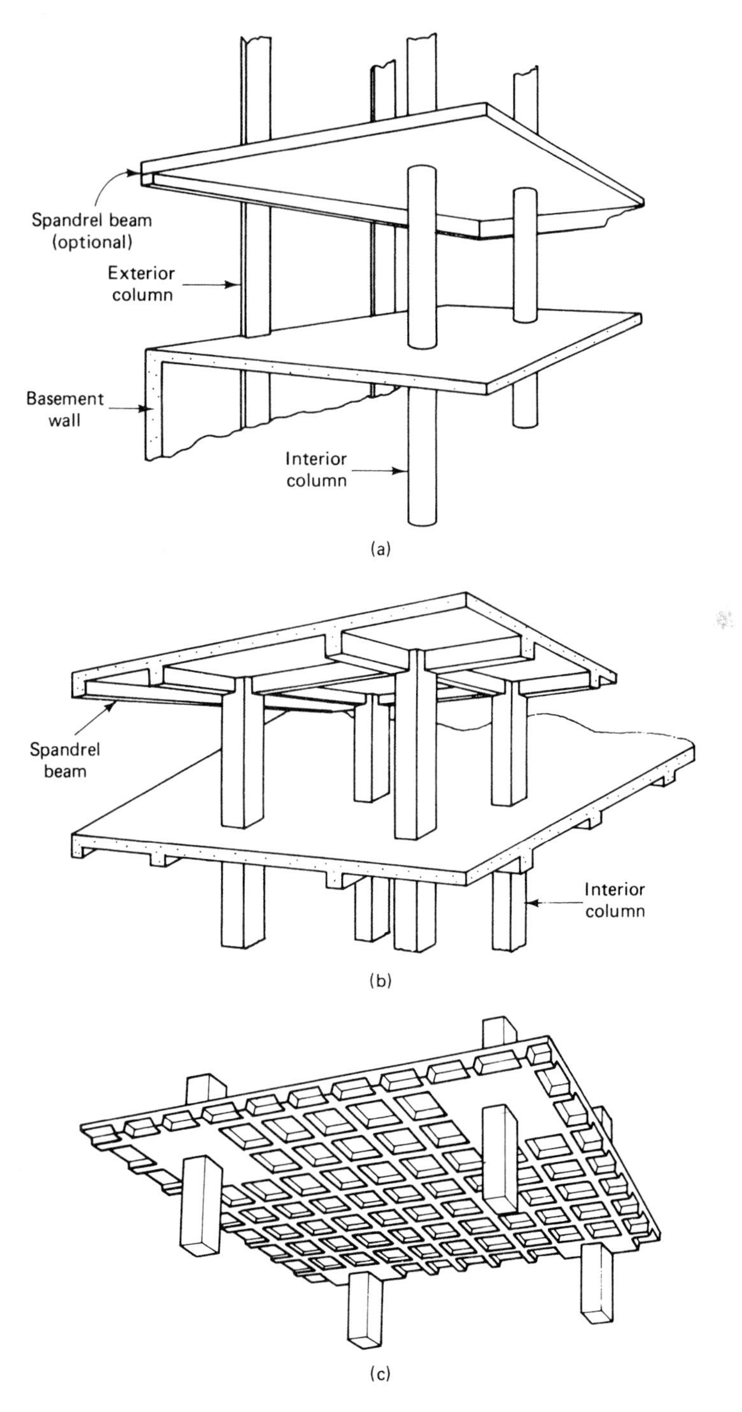

Figure 9.1 Two-way-action floor systems. (a) Two-way flat-plate floor. (b) Two-way slab floor on beams. (c) Waffle slab floor.

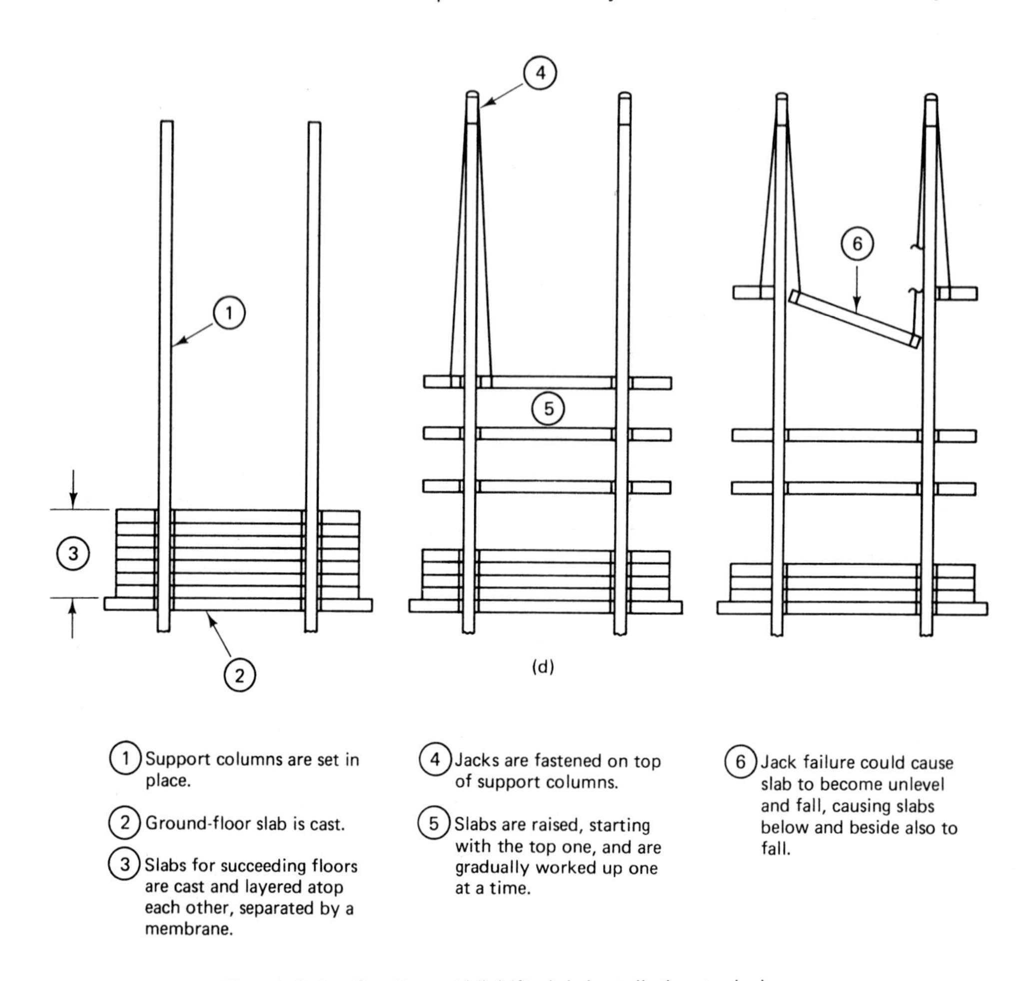

Figure 9.1 *Continued* (d) Lift-slab installation technique.

as a distinct structural system that is fast to construct and perhaps more economical than cast-in-place prestressed two-way slabs. However, the construction technique in lift slabs and the absence of the expertise required for such construction can create hazardous conditions which may result in loss of stability and structural collapse.

The technique for producing lift slabs involves casting a ground-level slab which can double as a casting bed over which all the other floor slabs are cast and stacked, separated by a membrane or sprayed parting agent. The columns, which can be steel or concrete, are built prior to the casting of the basic bottom slab extending to the building's height. All the other slabs are cast around the columns, with steel collars having sufficient clearance to permit *lifting* (jacking) of the slab or plate to the appropriate floor level, as shown in Figure 9.1(d). Lifting is accomplished through the use of jacks placed on top of the columns and connected to threaded rods extending down the faces of the columns to the lifting collars embedded in the slabs. Simultaneous activation of all the jacks is essential in order to maintain the slab in a totally horizontal state to avoid imbalance.

The analysis of slab behavior in flexure up to the 1940s and early 1950s followed the classical theory of elasticity, particularly in the United States. The small-deflections theory of plates, assuming the material to be homogeneous and isotropic, formed the basis of ACI Code recommendations with moment coefficient tables. The work, principally by Westergaard, which empirically allowed limited moment redistribution, guided the thinking of the code writers. Hence, the elastic solutions, complicated even for simple shapes and boundary conditions when no computers were available, made it mandatory to idealize, and sometimes render empirical, conditions beyond economic bounds.

In 1943, Johansen presented his yield-line theory for evaluating the collapse capacity of slabs. Since that time, extensive research into the ultimate behavior of reinforced concrete slabs has been undertaken. Studies by many investigators, such as those of Ockleston, Mansfield, Rzhanitsyn, Powell, Wood, Sawczuk, Gamble-Sozen-Siess, and Park, contributed immensely to a further understanding of the limit-state behavior of slabs and plates at failure as well as at serviceable load levels.

The various methods that are used for the analysis and design of two-way action slabs and plates are summarized in the following subsections.

9.1.1 The Semielastic ACI Code Approach

The ACI approach gives two alternatives for the analysis and design of a framed two-way action slab or plate system: the direct design method and the equivalent frame method. Both methods are discussed in more detail in Section 9.3. The equivalent frame method will be used in the design and analysis of prestressed slabs and plates.

9.1.2 The Yield-Line Theory

Whereas the semielastic code approach applies to standard cases and shapes and has an inherent excessively large safety factor with respect to capacity, the yield-line theory is a plastic theory easy to apply to irregular shapes and boundary conditions. Provided that serviceability constraints are applied, Johansen's yield-line theory represents the true behavior of concrete slabs and plates, permitting evaluation of the bending moments from an assumed collapse mechanism which is a function of the type of external load and the shape of the floor panel. This topic will be discussed in more detail in Section 9.14.

9.1.3 The Limit Theory of Plates

The interest in developing a limit solution became necessary due to the possibility of finding a variation in the collapse field which could give a lower failure load. Hence, an upper bound solution requiring a valid mechanism when applying the work equation was sought, as well as a lower bound solution requiring that the stress field satisfy everywhere the differential equation of equilibrium, that is,

$$\frac{\partial^2 M_x}{\partial x^2} - 2\frac{\partial^2 M_{xy}}{\partial x\,\partial y} + \frac{\partial^2 M_y}{\partial y^2} = -w \tag{9.1}$$

where M_x, M_y, and M_{xy} are the bending moments and w is the unit intensity of load. Variable reinforcement permits the lower bound solution still to be valid. Wood, Park, and other researchers have given more accurate semiexact predictions of the collapse load.

For limit-state solutions, the slab is assumed to be completely rigid until collapse. Further work at Rutgers by Nawy incorporated the deflection effect at high load levels as well as the compressive membrane force effects in predicting the collapse load.

9.1.4 The Strip Method

The strip method was proposed by Hillerborg, in attempting to fit the reinforcement to the strip fields. Since practical considerations require the reinforcement to be placed in orthogonal directions, Hillerborg set twisting moments equal to zero and transformed the slab into intersecting beam strips, hence the name "strip method."

Except for Johansen's yield-line theory, most of the other solutions are lower bound. Johansen's upper-bound solution can give the highest collapse load as long as a valid failure mechanism is used in predicting the collapse load.

9.1.5 Summary

The equivalent frame method will be the chief method discussed, because of the limitations of the use of the direct design method in its applicability to two-way prestressed floor systems and the need for more refined determination of the stiffnesses at the column-slab joints in the design process. The yield-line theory for the limit state at failure evaluation of slabs and plates will also be concisely presented.

9.2 FLEXURAL BEHAVIOR OF TWO-WAY SLABS AND PLATES

9.2.1 Two-Way Action

Consider a single rectangular panel supported on all four sides by unyielding supports such as shear walls or stiff beams. We seek to visualize the physical behavior of the panel under gravity load. The panel will deflect in a dishlike form under the external load, and its corners will lift if it is not monolithically cast with the supports. The contours shown in Figure 9.2(a) indicate that the curvatures and consequently the moments at the central area C are more severe in the shorter direction y with its steep contours than in the longer direction x.

Evaluation of the division of moments in the x and y directions is extremely complex, as the behavior is highly statically indeterminate. The simple case of the panel in part (a) of Figure 9.2 is expounded by taking strips AB and DE at midspan, as in part (b), such that the deflection of both strips at the central point C is the same.

The deflection of a simply supported uniformly loaded beam is $5wl^4/384EI$, i.e., $\Delta = kwl^4$, where k is a constant. If the thickness of the two strips is the same, the deflection of strip AB is $kw_{AB}L^4$ and the deflection of strip DE is $kw_{DE}S^4$, where w_{AB} and w_{DE} are the portions of the total load intensity w transferred to strips AB and DE, respectively, i.e., $w = w_{AB} + w_{DE}$. Equating the deflections of the two strips at the central point C, we get

$$w_{AB} = \frac{wS^4}{L^4 + S^4} \tag{9.2a}$$

and

$$w_{DE} = \frac{wL^4}{L^4 + S^4} \tag{9.2b}$$

It is seen from these equations that the shorter span S of strip DE carries the heavier portion of the load. Hence, the shorter span of such a slab panel on unyielding supports is subjected to the larger moment, supporting the foregoing discussion of the steepness of the curvature contours in Figure 9.2(a).

9.2.2 Relative Stiffness Effects

Alternatively, one has to consider a slab panel supported by flexible supports such as beams and columns, or flat plates supported by a grid of columns. In either case, the distribution of moments in the short and long directions is considerably more complex. This complexity arises from the fact that the degree of stiffness of the yielding supports determines the intensity of steepness of the curvature contours in Figure 9.2(a) in both the x and y directions and the redistribution of moments.

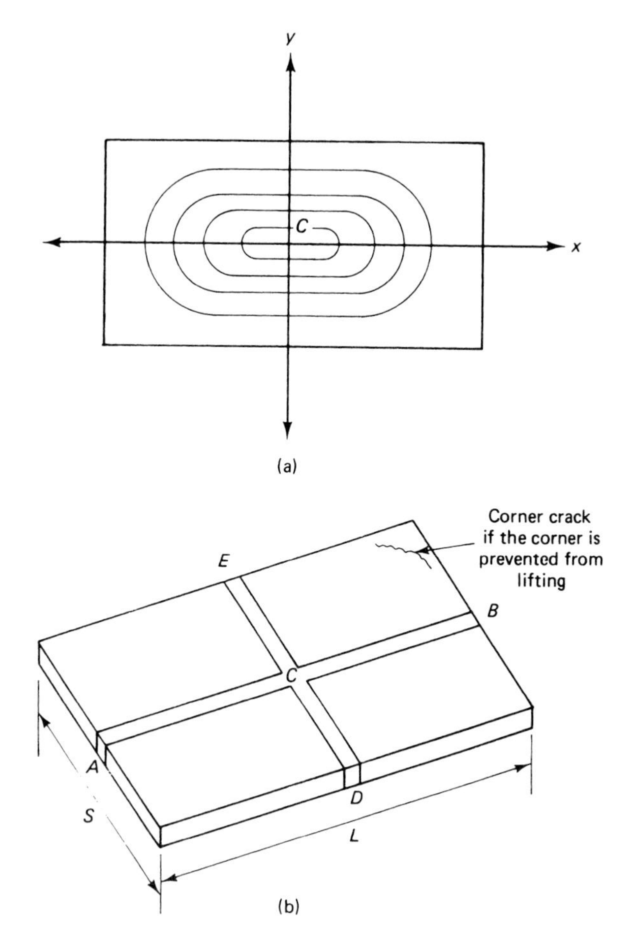

Figure 9.2 Deflection of panels and strips. (a) Curvature and deflection contours in a floor panel. (b) Central slips in a two-way slab panel.

The ratio of the stiffness of the beam supports to the slab stiffness can result in curvatures and moments in the long direction larger than those in the short direction, as the total floor behaves as an orthotropic *plate* supported on a grid of columns without beams. If the long span L in such floor systems of slab panels without beams is considerably larger than the short span S, the maximum moment at the center of a plate panel would approximate the moment at the middle of a uniformly loaded strip of span L that is clamped at both ends.

In sum, as slabs are flexible and highly underreinforced, redistribution of moments in both the long and short directions depends on the relative stiffnesses of the supports and the supported panels. Overstress in one region is reduced by such redistribution of moments to the lesser stressed regions.

9.3 THE EQUIVALENT FRAME METHOD

9.3.1 Introduction

The following discussion of the equivalent frame method of analysis for two-way systems summarizes the ACI Code approach to the evaluation and distribution of the total moments in a two-way slab panel. The Code assumes that vertical panels cut through an entire rectangularly planned multistory building along lines AB and CD in Figure 9.3 midway between columns. A rigid frame results in the x direction. Similarly, vertical planes EF and HG result in a rigid frame in the y direction. A solution of such an idealized frame consisting of horizontal beams or equivalent slabs and supporting columns enables the design of the slab as the beam part of the frame. The equivalent frame method thus treats the idealized frame in a manner similar to an actual frame, and hence is more exact and has fewer limitations than the direct design method. Basically, it involves a full moment distribution of many cycles as compared to the direct design method, which involves a one-cycle moment distribution approximation.

9.3.2 Limitations of the Direct Design Method

The following are the limitations of the direct design method:

1. There is a minimum of three continuous spans in each direction.
2. The ratio of the longer to the shorter span within a panel should not exceed 2.0.
3. Successive span lengths in each direction should not differ by more than one-third the length of the longer span.
4. Columns may be offset a maximum of 10 percent of the span in the direction of the offset from either axis between the center lines of successive columns.
5. All loads shall be due to gravity only and uniformly distributed over the entire panel. The live load shall not exceed three times the dead load.
6. If the panel is supported by beams on all sides, the relative stiffness of the beams in two perpendicular directions shall not be less than 0.2 or greater than 5.0.

Given these limitations, for prestressed concrete floor slabs, it is necessary to use the equivalent frame method.

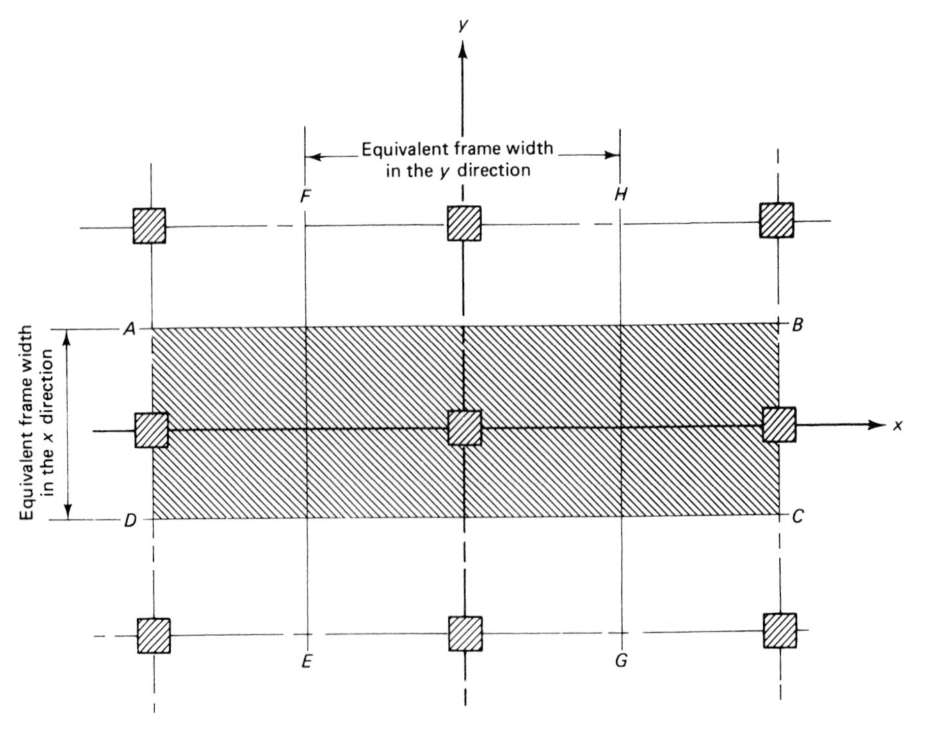

Figure 9.3 Floor plan with equivalent frame (shaded area in x direction).

9.3.3 Determination of the Statical Moment M_0

There are basically four major steps in the design of the floor panels:

1. Determine the total statical moment in each of the two perpendicular directions.
2. Distribute the total moment for the design of sections for negative and positive moment.
3. Distribute the negative and positive moments to the column and middle strips and to the panel beams, if any. A column strip is a width that is 25 percent of the equivalent frame width on each side of the column center line, and whose middle strip is the balance of the equivalent frame width.
4. Proportion the size and distribution of the reinforcement in the two perpendicular directions.

Given the above, correct determination of the values of the distributed moments becomes a principal objective. Consider typical interior panels having center line dimensions l_1 in the direction of the moments being considered and dimensions l_2 in the direction perpendicular to l_1, as shown in Figure 9.4. The clear span l_n extends from face to face of columns, capitals, or walls. Its value should not be less than $0.65l_1$, and circular supports shall be treated as square supports having the same cross-sectional area. The total statical moment of a uniformly loaded simply supported beam as a one-dimensional member is $M_0 = wl^2/8$. In a two-way slab panel as a two-dimensional member, the idealization of the structure through conversion to an equivalent frame makes it possible to compute M_0 once in the x direction and again in the orthogonal y direction. If one takes as a free-body diagram the typical interior panel shown in Figure 9.5(a), symmetry reduces the shears and twisting moments to zero along the edges of the cut segment. If no restraint existed at ends A and B, the panel would be considered simply supported in the span l_n direction. If one cuts at midspan, as in Figure 9.5(b), and considers half the panel as a free-body diagram, the moment M_0 at midspan would be

$$M_0 = \frac{wl_2\, l_{n1}}{2}\frac{l_{n1}}{2} - \frac{wl_2\, l_{n1}}{2}\frac{l_{n1}}{4}$$

or

$$M_0 = \frac{wl_2(l_{n1})^2}{8} \tag{9.3}$$

Due to the existence of restraint at the supports, M_0 in the x direction would be distributed to the supports and midspan such that

$$M_0 = M_C + \tfrac{1}{2}(M_A + M_B) \tag{9.4a}$$

The distribution would depend on the degree of stiffness of the support. In a similar manner, M_0 in the y direction would be the sum of the moments at midspan and the average of the moments at the supports in that direction.

In the orthogonal direction, Equation 9.4a becomes

$$M_0' = M_C' + \tfrac{1}{2}(M_A' + M_B') \tag{9.4b}$$

where M_0', M_A', M_B', and M_C' are at 90 degrees to M_0, M_A, M_B, and M_C, respectively. Also, in an analogous manner to Equation 9.3,

$$M'_0 = \frac{Wl_1(l_{n2})^2}{8} \tag{9.5}$$

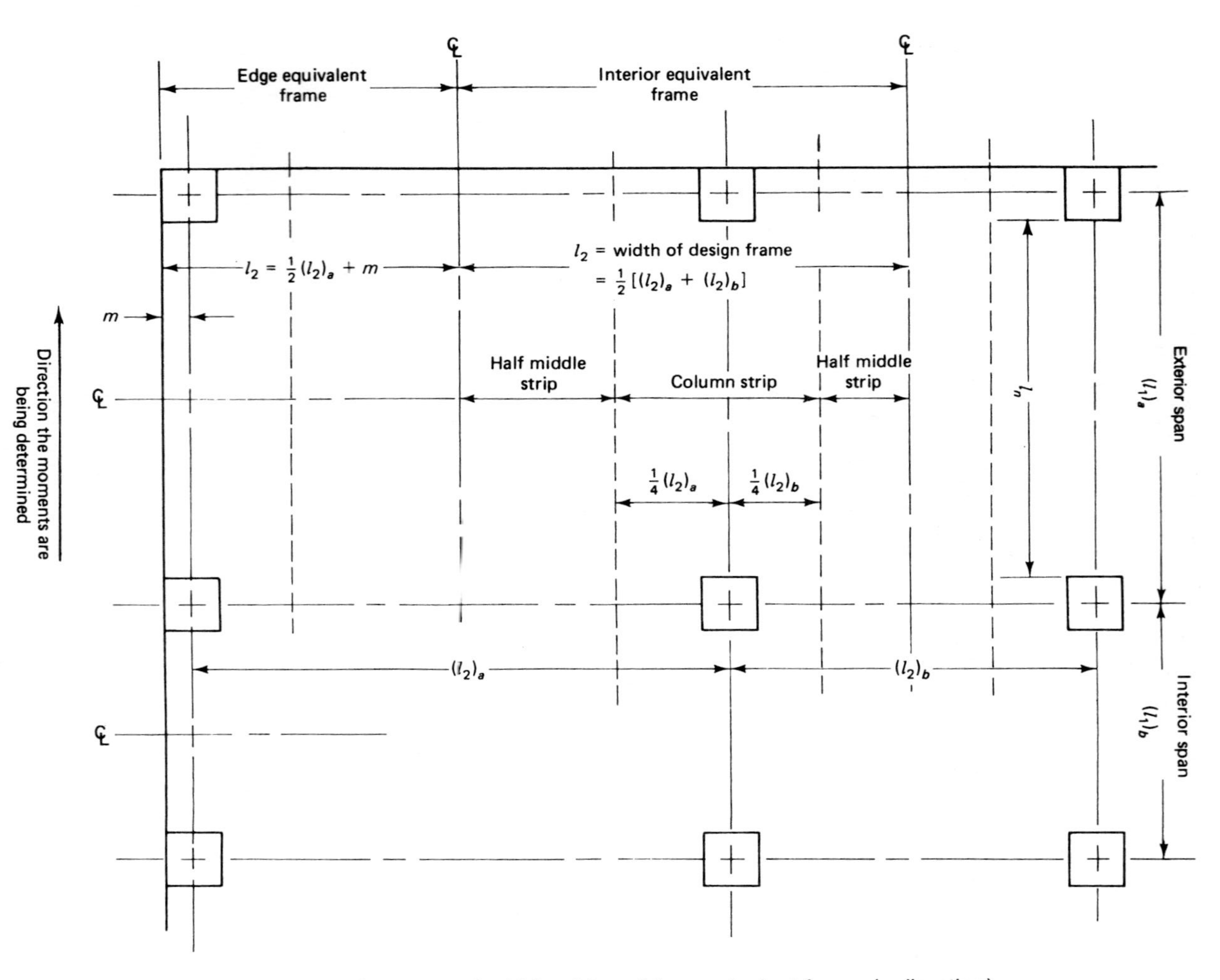

Figure 9.4 Column and middle strips of the equivalent frame (y direction).

Photo 9.1 Sydney Opera House during construction.

The load intensity W at service load in the prestressed concrete slab would be W_w per unit area.

9.3.4 Equivalent Frame Analysis

The structure, divided into continuous frames as shown in Figure 9.6 for frames in both orthogonal directions, would have the row of columns and a wide continuous beam (slab) ABCDE for gravity loading. Each floor is analyzed separately, whereby the columns are assumed fixed at the floors above and below. To satisfy statical and equilibrium considerations, each equivalent frame must carry the total applied load; alternate span loading has to be used for the worst live-load condition.

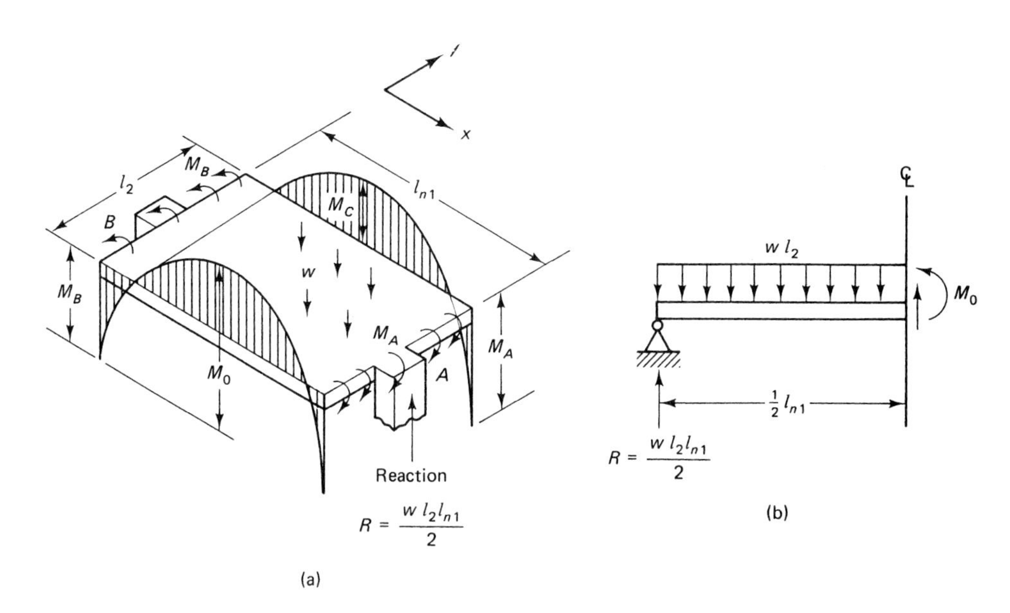

Figure 9.5 Simple moment M_0 acting on an interior two-way slab panel in x direction. (a) Moment on panel. (b) Free-body diagram.

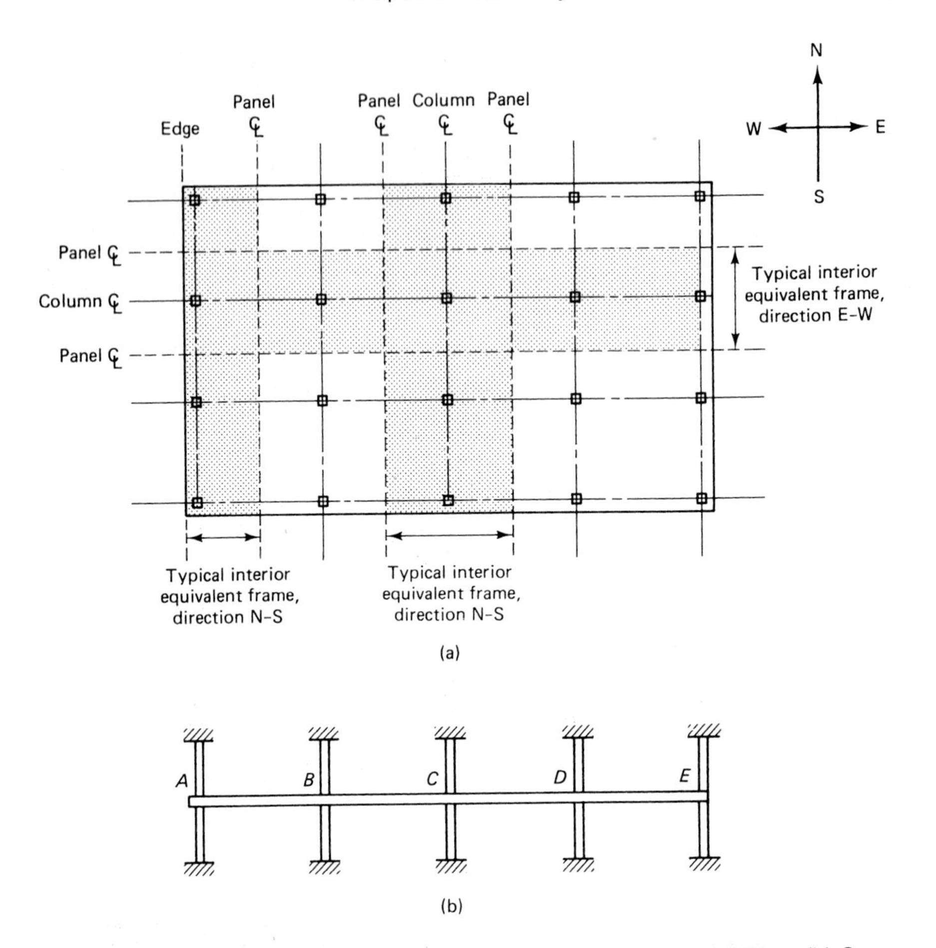

Figure 9.6 Idealized structure divided into equivalent frames. (a) Plan. (b) Section in E–W direction.

It is necessary to account for the rotational resistance of the column at the joint when running a moment relaxation or distribution, except when the columns are so slender as to have very small rigidity compared to the rigidity of the slab at the joint. In such cases, as, for example, in lift slab construction, only a continuous beam is necessary. A schematic illustration of the constituent elements of the equivalent frame is given in Figure 9.7. The slab strips are assumed to be supported by transverse slabs. The column provides a resisting torque M_T equivalent to the applied torsional moment intensity m_t. The exterior ends of the slab strip rotate more than the central section because of torsional deformation. In order to account for this rotation and deformation, the actual column and the transverse slab strip are conceptually replaced by an equivalent column such that the flexibility of the equivalent column is *equal* to the *sum* of the flexibilities of the actual column and the slab strip. This assumption is represented by the equation

$$\frac{1}{K_{ec}} = \frac{1}{\Sigma K_c} + \frac{1}{K_t} \tag{9.6}$$

where K_{ec} = flexural stiffness of the equivalent column, moment per unit rotation
ΣK_c = sum of flexural stiffnesses of the upper and lower columns at the joint, moment per unit rotation
K_t = torsional stiffness of the torsional beam, moment per unit rotation.

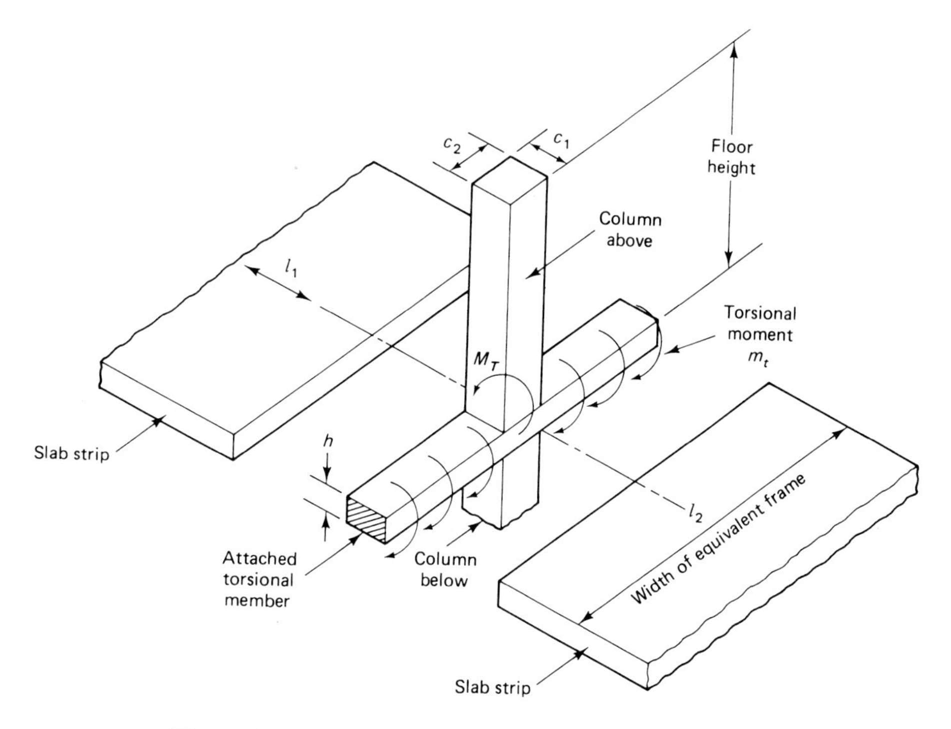

Figure 9.7 Constituent elements of the equivalent frame.

Alternatively, Equation 9.6 can be written as the stiffness equation

$$K_{ec} = \frac{\Sigma K_c}{1 + \dfrac{\Sigma K_c}{K_t}} \tag{9.7}$$

and the column stiffness for an equivalent frame (Ref. 9.9) can be defined as

$$K_c = \frac{EI}{l'}\left[1 + 3\left(\frac{L}{L'}\right)^2\right] \tag{9.8}$$

where I is the column moment of inertia, L is the centerline span, and L' is the clear span of the equivalent beam. The carryover factors are approximated by $-\frac{1}{2}(1 + 3h/L)$. An exact computation of the carryover factor can be made by the column-analogy method using the slab as an analogous column.

A simpler expression for K_c (Ref. 9.10) gives results within 5 percent of the more refined values from Equation 9.8, viz.,

$$K_c = \frac{4EI}{L_n - 2h} \tag{9.9}$$

where h is the slab thickness. The torsional stiffness of the slab in the column line is

$$K_t = \Sigma \frac{9E_{cs}C}{L_2\left(1 - \dfrac{c_2}{L_2}\right)^3} \tag{9.10a}$$

where L_2 = band width
L_n = span
c_2 = column dimensions in the direction parallel to the torsional beam and the torsional constant is

$$C = \Sigma(1 - 0.63x/y)x^3\, y/3 \tag{9.10b}$$

in which x = shorter dimension of the rectangular part of the cross section at the column junction (such as the slab depth)
y = longer dimension of the rectangular part of the cross section at the column junction (such as the column width).

The slab stiffness is given by the equation

$$K_s = \frac{4E_{cs}I_s}{L_n - c_1/2} \tag{9.11}$$

As the effective stiffness K_{ec} of the column and the slab stiffness K_s are established, the analysis of the equivalent frame can be performed by any applicable methods, such as relaxation or moment distribution.

The distribution factor for the fixed-end moment *FEM* is

$$DF = \frac{K_s}{\Sigma K} \tag{9.12}$$

where $\Sigma K = K_{ec} + K_{s(\text{left})} + K_{s(\text{right})}$. A carryover factor of $COF \cong \frac{1}{2}$ can be used without loss of accuracy since the nonprismatic section causes only very small effects on fixed-end moments and carryover factors. The fixed-end moment *FEM* for a uniformly distributed load is $wl_2(l_n)^2/12$ at the supports, such that after moment redistribution the sum of the negative distributed moment at the support and the midspan moment is always equal to the static moment $M_0 = wl_2(l_n)^2/8$.

9.3.5 Pattern Loading of Spans

Loading all spans simultaneously does not necessarily produce the maximum positive and negative flexural stresses. Consequently, it is advisable to analyze the multispan frame using alternative span loading patterns for the *live* load. For a three-span frame, the suggested patterns for the live load are shown in Figure 9.8.

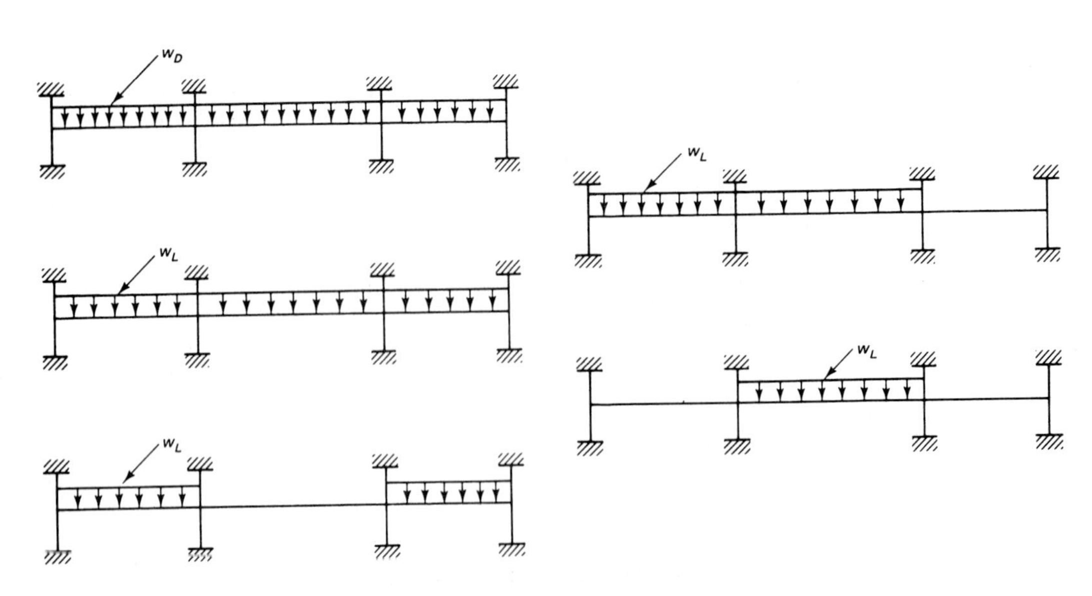

Figure 9.8 Loading patterns for live load.

9.4 TWO-DIRECTIONAL LOAD BALANCING

As mentioned in Chapter 1, load balancing represents the forces counteracting the external gravity load. These forces are produced by the transverse component of the longitudinal prestressing force in a parabolic or harped tendon. The load w in Equations 9.3 to 9.5 represents the *downward* external transverse load intensity, which can be either the working load intensity w_w or the factored load intensity w_u. The *upward* load intensity in the slab field due to the transverse component of the prestressing force, also mentioned in Chapter 1, would reduce the effect of w_w and can be chosen so as to *exactly balance* a particular downward load intensity. Under such a condition, the two-way slab would be subjected to neither bending nor twisting moments, and the analysis would be considerably simplified.

Two-directional load balancing in two-way slabs is different from one-directional load balancing in beams. The balancing load produced by the tendons in one direction either increases or decreases the balancing load produced by the tendons in the perpendicular direction. Hence, the prestressing forces and tendon profiles in the two orthogonal directions are totally *interrelated,* fully maintaining the basic principles of statics. The ultimate benefit of load balancing is in the creation of a design of a structural prestressed floor such that the upward component of the prestressing force results in a load intensity distribution in each direction that is equivalent to the downward external load intensity. Such a design is called a pure *balanced design.* Any deviation from the balanced condition should be analyzed as a load acting on the slab without being affected by the transverse upward component of prestress.

If a two-way slab on rigid supports such as walls is prestressed in both orthogonal directions having spans L_S and L_L in the short and long directions, respectively, as shown in Figure 9.9, the intensity of the upward balancing load required to produce balanced design loads is given, from Equation 1.15a, by the equations

$$W_{\text{bal}(S)} = \frac{8P_S e_S}{L_S^2} \tag{9.13a}$$

and

$$W_{\text{bal}(L)} = \frac{8P_L e_L}{L_L^2} \tag{9.13b}$$

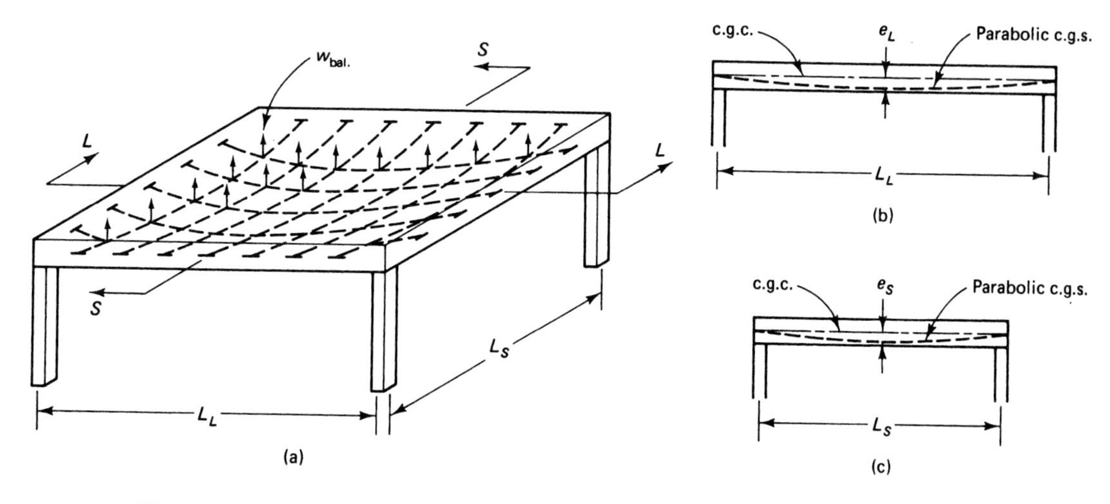

Figure 9.9 Balancing loads in two-way prestressed panel. (a) Three-dimensional view. (b) Section L–L in the long direction. (c) Section S–S in the short direction.

where P_S and P_L are the effective prestressing forces after losses in the short and long directions L_S and L_L, respectively, *per unit width of slab,* and e_S and e_L are the corresponding maximum eccentricities of the prestressing tendons. The total balancing load per unit width then becomes

$$W_{bal} = W_{bal(S)} + W_{bal(L)} = \frac{8P_S e_S}{L_S^2} + \frac{8P_L e_L}{L_L^2} \tag{9.14}$$

The designer would select the level of W_{bal} and determine the values of the prestressing forces P_S and P_L accordingly. Many combinations of P_S and P_L can satisfy the statics Equation 9.14. If the slab panels were supported on beams, or if simple panels were supported on a wall, the most economical design would be to carry the load W only in the short direction, or to carry $\frac{1}{2}$ W in each direction in the case of a square slab panel. The slab panel loaded by W_{bal} and stressed by prestressing forces P_S and P_L would be subjected to a uniform stress distribution P_S/h and P_L/h in the respective directions, h being the slab thickness. The slab panel would be totally level, with no deflection or cambers. Any deviation of the applied load from W_{bal} would require the use of the usual elastic theory for the analysis of two-way plates.

As prestressed post-tensioned two-way slabs are usually flat plates supported directly on columns, all of the load has to be carried in both directions using either uniformly distributed prestressing tendons or banded tendons, with a concentration of tendons at the columns strips of the two-way plate panels.

Uniform stress distribution and zero deflection/camber are not mandatory for proportioning the floor system. If they were, load balancing would not necessarily be the most economical way of determining the prestressing forces. Instead, the designer would often use a partial balancing load $W_{bal} < W_D + W_L$ for a multipanel floor system, as will be seen in Example 9.2. If the load intensity $W_w = W_D + W_L$ is larger than the balanced load W_{bal} from Equation 9.14, then unit moments M_S and M_L will result in the directions S and L, respectively.

The unit stress in the concrete in the short and long directions due to the unbalanced loading is obtained by superimposing the uniform compression due to balanced loading on the flexural stress in the concrete caused by the bending moments M_S and M_L that result from the unbalanced load $W_w - M_{bal}$. The resulting concrete stresses at the top and bottom fibers in each direction are given as follows:

Short direction

$$f^t = -\frac{P_S}{bh} - \frac{M_s c}{I_s} \tag{9.15a}$$

$$f_b = -\frac{P_S}{bh} + \frac{M_S c}{I_s} \tag{9.15b}$$

Long direction

$$f^t = -\frac{P_L}{bh} - \frac{M_L c}{I_L} \tag{9.16a}$$

$$f_b = -\frac{P_L}{bh} + \frac{M_L c}{I_L} \tag{9.16b}$$

In these equations, superscript t represents the top of the slab, subscript b represents the bottom of the slab, $c = \frac{1}{2}h$, width $b = 12$ in., and

$$P_s = \frac{\text{total } P_S}{L}$$

Photo 9.2 Kishwaukee River segmental bridge during construction. Winnebago County, Illinois. (*Courtesy,* H. Wilden & Associates, Macungie, Pennsylvania and the Prestressed Concrete Institute.)

and

$$P_L = \frac{\text{total } P_L}{S}$$

are the unit prestress forces. The service-load moment coefficient for evaluating M_S and M_L can be obtained from the chart in Figure 9.10 for any boundary condition (Ref. 9.13–CP 114). The bending moment coefficients there are for the maximum positive and negative bending moments, where βx_2 and $\beta x_2'$ apply to $+M$ and $-M$, respectively, on the short span L_x. Similarly, βy_2 and $\beta y_2'$ apply to the maximum positive and negative bending moments, respectively, on the long span L_y. In an analogous fashion, the charts in Figure 9.11 give a rapid method for the evaluation of the ultimate bending moment coefficients (Ref. 9.13–CP 110) in continuous two-way-action concrete plates.

9.5 FLEXURAL STRENGTH OF PRESTRESSED PLATES

9.5.1 Design Moments M_u

The design moments for statically indeterminate prestressed bonded members are determined by combining the frame distributed moments M_u due to the *factored* dead plus live loads with the secondary moments M_s induced into the frame by the tendons. The load balancing approach discussed in the previous section directly includes both the primary moments M_1 and the secondary moments M_s. Hence, for service-load intensity values, only the net loads M_{net} need to be considered in the calculation of the fixed-end factored moments, while W_{bal} needs to be considered for flexural strength analysis.

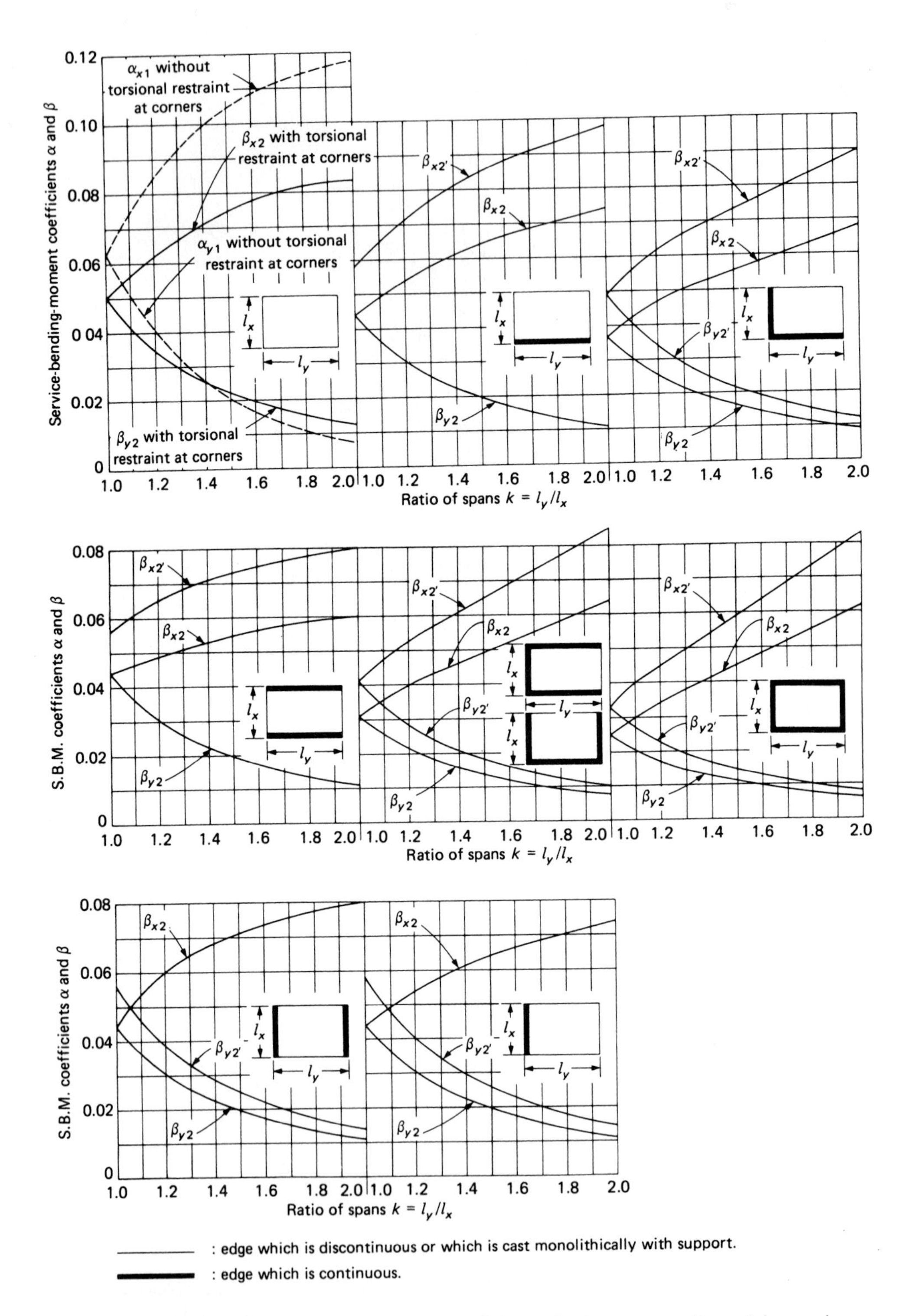

Figure 9.10 Service-load moment coefficients in two-way action slabs and plates (Ref. 9.13).

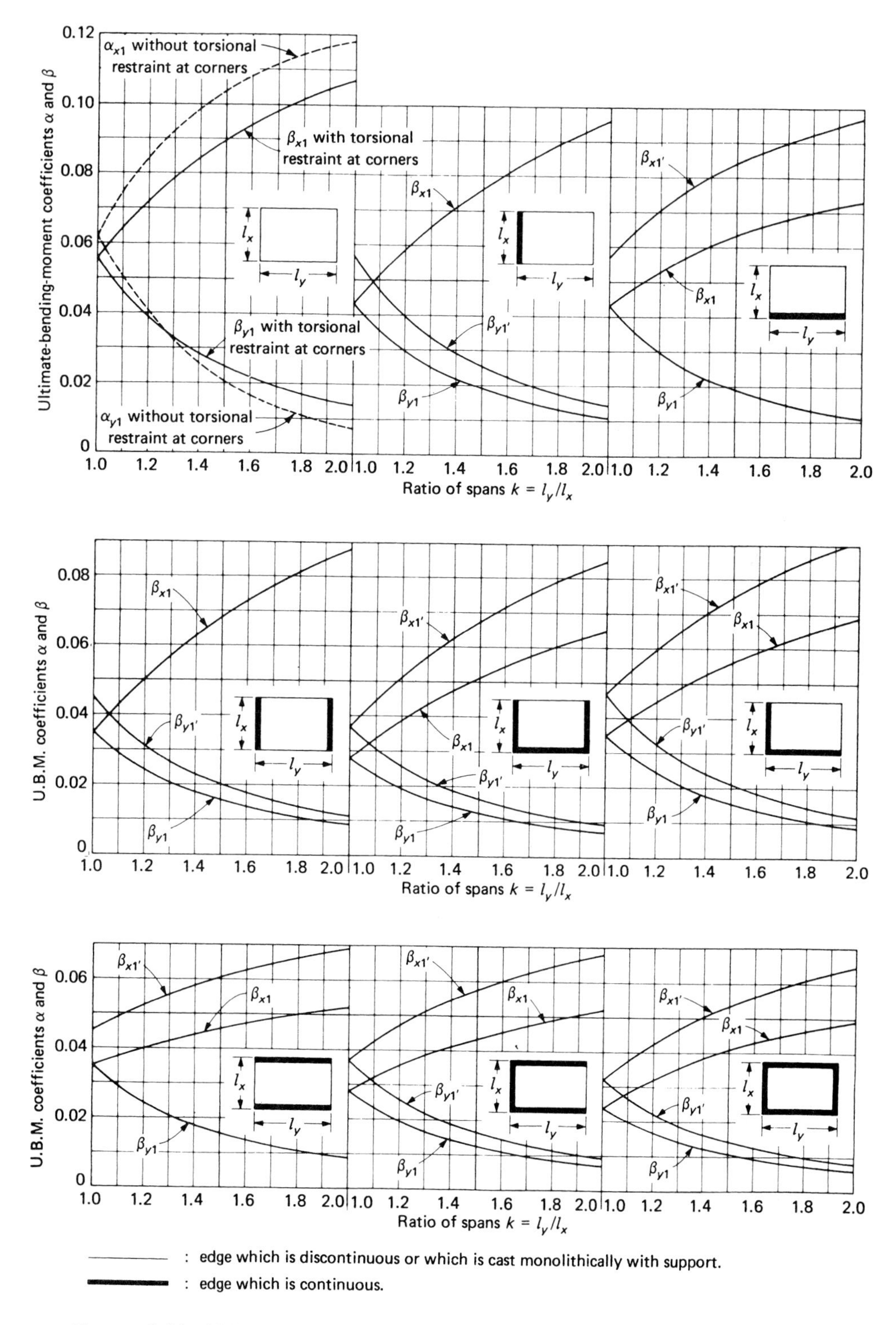

Figure 9.11 Ultimate-load moment coefficients in two-way action slabs and plates (Ref. 9.13).

Fixed-end M_u for moment distribution

If $M_1 = P_e\, e = Fe$ is the primary moment, M_{bal} is the balanced moment due to W_{bal}, M_S = distributed $M_{bal} - M_1$ is the secondary moment, and $\overline{M}_u$ is the fixed-end factored moment due to the factored load intensity W_u, then the design ultimate moment would be at least

$$\text{Design } M_u = \text{distributed } \overline{M}_u - M_S \tag{9.17a}$$

and the available moment strength would be

$$M_n = \frac{M_u}{\phi} \tag{9.17b}$$

Inelastic redistribution of moments due to continuity would be applied to the available moment strength M_n at the support towards the required M_n at midspan.

Where bonded reinforcement is provided at the supports with minimum nonprestressed steel provided in accordance with Equations 9.19 and 9.20, negative moments computed by the elastic theory for any assumed loading arrangement may be increased or decreased by not more than the percentage given by the inelastic moment redistribution factor

$$\rho_D = 20\left[1 - \frac{\omega_p + (d/d_p)(\omega - \omega')}{0.36\beta_1}\right] \text{percent} \tag{9.18}$$

The modified negative moment should be used for computing moments at sections within spans, viz., the positive moments, for the same loading arrangement. Inelastic moment redistribution of the negative moments can be made only when the section at which the moment is reduced is so designed that ω_p or $\omega_p + (d/d_p)(\omega - \omega')$ is not greater than $0.24\beta_1$.

Example 9.2 illustrates in detail the equivalent frame analysis procedure both for service-load and ultimate-load conditions, and the inelastic moment redistribution due to continuity that is utilized in the strength analysis.

9.6 BANDING OF PRESTRESSING TENDONS AND LIMITING CONCRETE STRESSES

9.6.1 Distribution of Prestressing Tendons

Each plate panel is assumed to be supported continuously along the transverse column center lines. The assumption is also made, as previously stated, that the slab panel behaves as an orthogonal panel of two wide beams whose width is equal to the panel width and which is supported along the column centerlines. Consequently, 100 percent of the load to be balanced is considered to be supported by the wide beam in *each* of the two perpendicular directions.

It is also known that the lateral distribution of moments is *not uniform* across the width of the panel, but tends to be more concentrated in the column strips. Consequently, it is not unreasonable to concentrate a large percentage of the tendons in the column strip, as defined in Figure 9.4, and to spread the remaining tendons in the middle strip. For continuous spans, 65 to 75 percent of the moment in each direction is carried by the respective column strip, while the total area and number of tendons required by the total applied moment are maintained.

The width of half the column strip on either side of the column is one-fourth of the *smaller* of the two panel dimensions. The middle strip is the slab band bound by the two column strips. Accordingly, the distribution or banding of the prestressing tendons follows the percentage distribution of the moment between the column and middle strips.

Photo 9.3 Bridge deck prestressing reinforcement.

Consequently, if 70 percent of the tendons are concentrated in the column strip, it is reasonable to expect that the column strip will carry *essentially* 70 percent, and the middle strip will carry the remaining 30 percent, of the total moment.

Since tendons exert downward loads at the high points which join adjacent parabolic profiles, the accompanying downward reactions should, as nearly as practicable, be resisted by columns, walls and/or upward tendon loads to achieve minimum deflections and maximum shear capacity. Consequently, a logical statical distribution of tendons can be one in which all tendons in one direction are placed through or immediately adjacent to the columns, and the tendons in the perpendicular direction are spaced uniformly across the bay width.

Figure 9.12 shows the resulting distribution of prestressing tendons in the two orthogonal directions. As a general guideline, the recommended spacing of the tendons in the column strip is about three to four times the slab thickness, while the maximum spacing of the tendons in the middle strip should not exceed six times the slab thickness. An average compressive stress in the concrete in each direction should be at least about 125 psi (0.90 MPa).

Investigations have verified (Ref. 9.4), through testing of the four-panel prestressed plates shown in Figure 9.13, that variation of distribution of prestressing strands does not alter the deflection behavior or magnitude and capacity for the *same* total prestressing steel percentage. Banding of the prestressing tendons as shown in part (b) of the figure, with about 65 to 75 percent of the tendons in the column strip, seems to be the most effective, particularly in enhancing the shear-moment transfer capacity at the column support section of the two-way slab.

9.6.2 Limiting Concrete Tensile Stresses at Service Load

9.6.2.1 Flexure. The ACI 318 Code limits the tensile stresses in the concrete for prestressed elements in order to control flexural cracking development. The following tabulated values give the maximum permissible tensile stress in the prestressed elements for the various moment regions.

1. Negative moment area with the addition of nonprestressed reinforcement $6\sqrt{f'_c}$

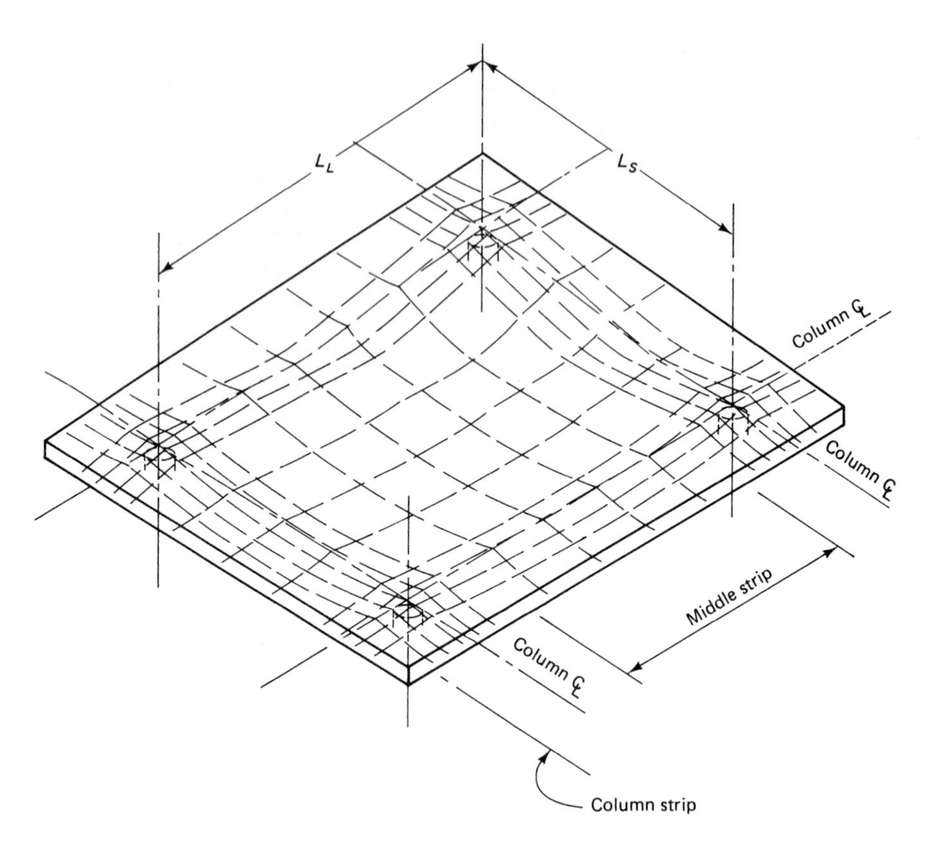

Figure 9.12 Bending of prestressing tendons in two-way-action panels.

2. Negative moment area without the addition of nonprestressed reinforcement — 0
3. Positive moment area with the addition of nonprestressed reinforcement — $2\sqrt{f'_c}$
4. Positive moment area without the addition of nonprestressed reinforcement — 0
5. Compressive stress in the concrete (Under certain conditions, $0.60f'_c$) — $f_c = 0.45f'_c$

9.6.2.2 Reinforcement. The minimum area of bonded reinforcement, except as required by Equation 9.20 below, is

$$A_s = 0.004\,A \tag{9.19a}$$

where A is the area in square inches of that part of the cross section between the flexural tension face and the center of gravity of the gross section. In positive-moment areas where the computed tensile stress in the concrete at service load exceeds $2\sqrt{f'_c}$, the minimum area of bonded reinforcement has to be computed from

$$A_s = \frac{N_c}{0.5f_y} \tag{9.19b}$$

where N_c is the tensile force in the concrete due to the unfactored dead plus live load, and $f_y = 60{,}000$ psi. In negative-moment areas at column supports, the minimum area of bonded reinforcement in each direction has to be determined from

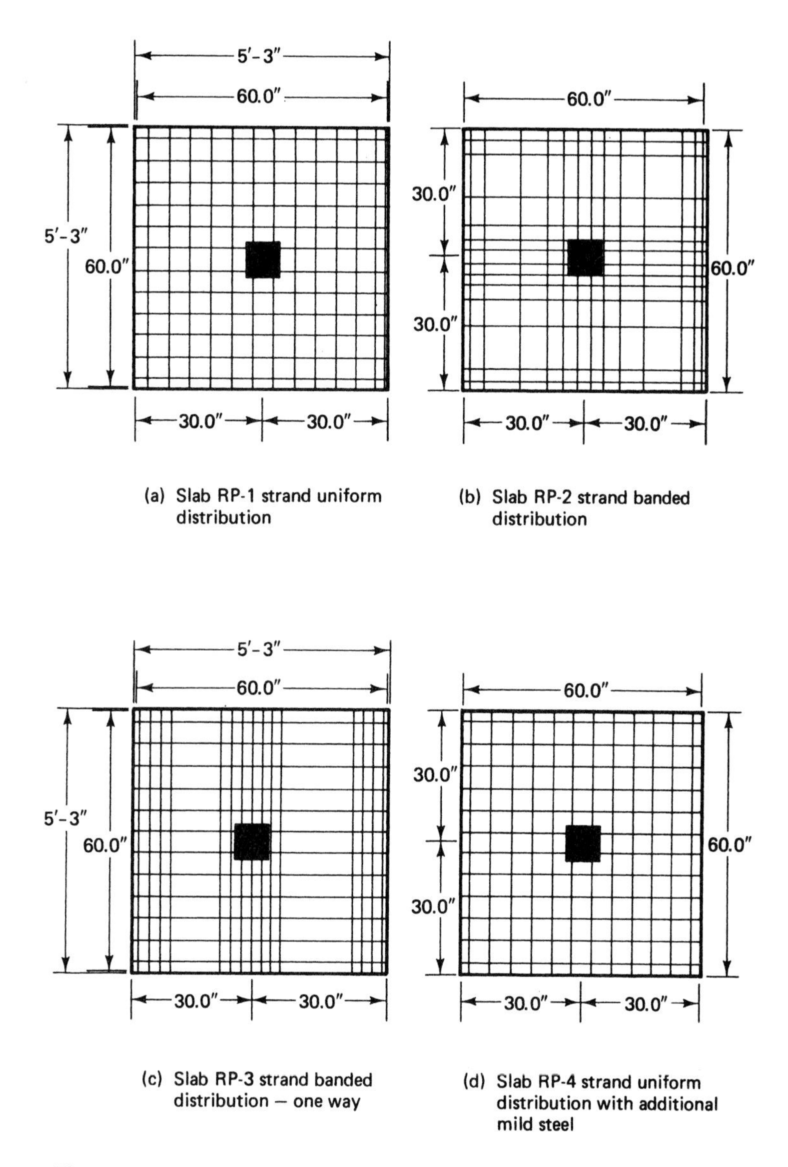

Figure 9.13 Various alternatives of tendon distribution (Ref. 9.4).

$$A_s = 0.00075hL \tag{9.20}$$

where L = span length in the direction parallel to the reinforcement being determined
h = slab thickness.

The reinforcement obtained from Equation 9.20 has to be distributed within a slab band width between the lines that are $1.5h$ outside opposite faces of the column support. At least four bars or wires should be provided in both directions.

The minimum length of bonded reinforcement in the *positive* area should be *one-third* the clear span, centered in the positive-moment area. The minimum length of bonded reinforcement in the *negative* area should extend *one-sixth* the clear span on each side of the support, placed at the *top* fibers. The stress f_{ps} in the prestressing reinforcement at nominal strength, as required by the ACI 318 Code, is governed by the following requirements.

Bonded Tendons. For bonded tendons,

$$f_{ps} = f_{pu}\left(1 - \frac{\gamma_p}{\beta_1}\left[\rho_p \frac{f_{pu}}{f'_c} + \frac{d}{d_p}(\omega - \omega')\right]\right) \tag{9.21}$$

where $\omega' = \rho' = f_y/f'_c$
and $\gamma_p = 0.40$ for $f_{py}/f_{pu} \geq 0.85$
$= 0.28$ for $f_{py}/f_{pu} \geq 0.90$.

If any compression reinforcement is considered, the term $[\rho_p f_{pu}/f'_c + (d/d_p)(\omega - \omega')]$ in Equation 9.21 shall be taken not less than 0.17, and d' shall be no greater than $0.15d_p$.

Nonbonded Tendons. For nonbonded tendons with a slab span-to-depth ratio ≤ 35,

$$f_{ps} = f_{pe} + 10{,}000 + \frac{f'_c}{100\rho_p} \tag{9.22}$$

where $f_{ps} \leq f_{py} \leq f_{pe} + 60{,}000$.

For nonbonded tendons with a slab span-to-depth ratio > 35,

$$f_{ps} = f_{pe} + 10{,}000 + \frac{f'_c}{300\rho_p} \tag{9.23}$$

where $f_{ps} \leq f_{py} \leq f_{pe} + 30{,}000$.

9.6.2.3 Shear

Column Support Sections in Flat Plates. The nominal shear strength provided by the concrete at the column junctions of two-way prestressed slabs is given by

$$V_c = (\beta_p \sqrt{f'_c} + 0.3\bar{f}_c)b_o d + V_p \tag{9.24a}$$

or the nominal unit shearing strength is

$$v_c = \beta_p \sqrt{f'_c} + 0.3\bar{f}_c + \frac{V_p}{b_o d} \tag{9.24b}$$

where b_o = perimeter of the critical shear section at distance $d/2$ from face of support

$\bar{f}_c$ = average value of the effective compressive stress in the concrete due to externally applied load for the two orthogonal directions computed at the section centroid after all prestress losses (termed f_{pc} by the ACI Code)

V_p = vertical component of all effective prestressing forces crossing the critical section

β_p = the smaller of the two values 3.5 or $(\alpha_s d/b_o + 1.5)$, where α_s is 40 for interior columns, 30 for edge columns, and 20 for corner columns.

In slabs with distributed tendons, the term V_p can be conservatively disregarded; otherwise it becomes necessary to use the actual reverse curvature tendon geometry in the calculations in order to assess the shear carried by the tendons crossing the critical section. According to the ACI 318 Code, no portion of the column cross section shall be closer to a discontinuous edge than four times the slab thickness, f'_c in Equations 9.24 shall not exceed 5,000 psi, and $\bar{f}_c$ in each direction shall be neither less than 125 psi nor greater than 500 psi.

If these requirements are not satisfied, V_c shall be computed by the smaller of the values obtained from the expressions

$$\text{(i)}\quad V_c = \left(2 + \frac{4}{\beta_c}\right)\sqrt{f'_c}\, b_o d \tag{9.25a}$$

$$\text{(ii)}\quad V_c = \left(\frac{\alpha_s d}{b_o} + 2\right)\sqrt{f'_c}\, b_o d \tag{9.25b}$$

$$\text{(iii)}\quad V_c = 4\sqrt{f'_c}\, b_o d \tag{9.25c}$$

where β_c = ratio of long to short side of the column or concentrated load area.

Equations 9.25(a) and (b) are the results of tests which indicate that as the ratio b_o/d increases, the available nominal shear strength V_c decreases so that in such situations Equation 9.25(c) would not control as it becomes unsafe.

Continuous Edge Support. For distributed load and continuous edge support such as supporting beams or wall support, if the effective prestress is not less than 40 percent of the tensile strength of the reinforcement, the maximum allowable shear stress is

$$V_c = \left[0.60\sqrt{f'_c} + 700\frac{V_u d}{M_u}\right] b_w d_p \geq 2\sqrt{f'_c}\, b_w d$$

$$< 5\sqrt{f'_c}\, b_w d \tag{9.26}$$

where b_w is taken as the strip width and $V_u d/M_u$ at a distance $d_p/2$ from the face of the support, $d_p \geq 0.80h$.

Values of $\sqrt{f'_c}$ in all the above equations are to be multiplied by a factor $\lambda = 1.0$ for normal-weight concrete, $\lambda = 0.85$ for sand-lightweight concrete, and $\lambda = 0.75$ for all-lightweight concrete.

Shear Force Coefficients. The maximum shear force at the edge of a two-way slab panel carrying uniformly distributed load and supported along the length of its perimeters can be approximated as follows

$$V = \tfrac{1}{3} w L_s \text{ (short edge)} \tag{9.27a}$$

$$V = kwL_s/(2k + 1) \text{ (long edge)} \tag{9.27b}$$

where k is the ratio of the long span L_L to the short span L_S. The same values are applicable to a panel that is fixed or continuous along all four edges. For other conditions, the distribution of shearing forces, the stresses due to which are rarely critical, have to be adjusted on the basis that the shearing force is slightly larger at a continuous edge than at a simply supported edge.

The ACI Code allows a 15 percent increase in the shear force at the first interior continuous support for one-way action.

9.7 LOAD-BALANCING DESIGN OF A SINGLE-PANEL TWO-WAY FLOOR SLAB

Example 9.1

A two-way single-panel prestressed warehouse lift slab 20 ft × 24 ft (6.10 m × 7.32 m) has the plan shown in Figure 9.14. It is supported on masonry walls on all sides, with negligible rotational restraint at these boundaries but that the corners are held down. The slab has to carry a superimposed service dead load of 15 psf (0.72 kPa) in addition to its self-weight and a service live load of 75 psf (3.59 kPa). No deflection is allowed under full dead load.

Design the slab as a post-tensioned nonbonded prestressed two-way floor using $\frac{1}{2}$-in. dia 7-wire 270-K tendons (12.7-mm dia strands). Given data are as follows:

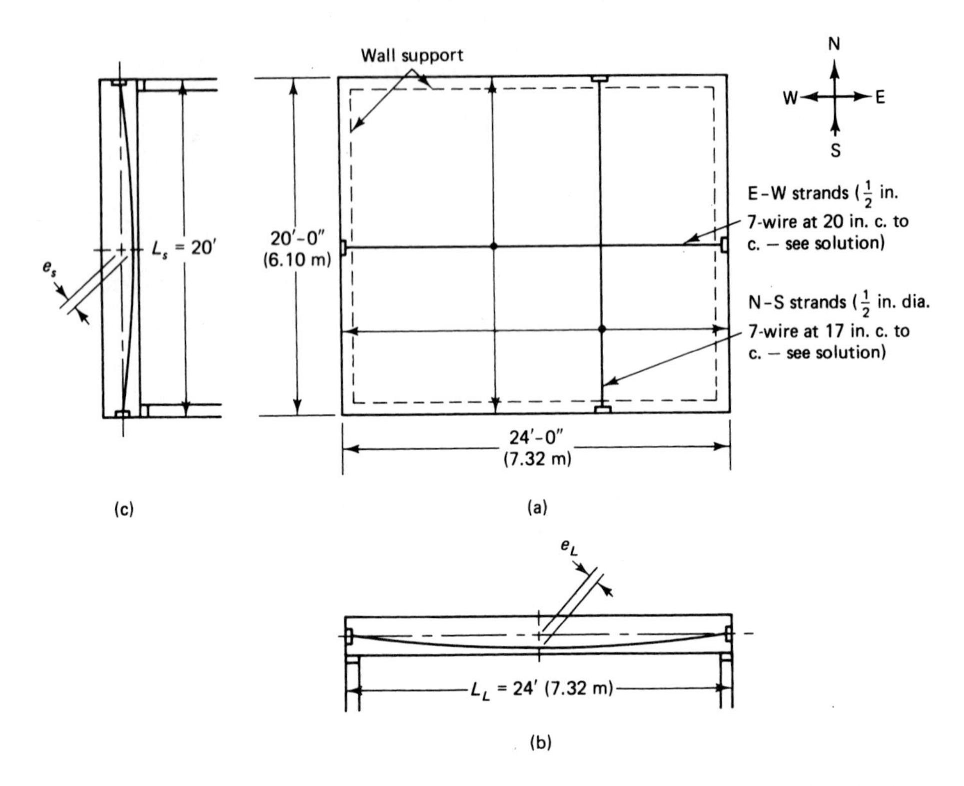

Figure 9.14 Two-way prestressed panel in Example 9.1. (a) Plan. (b) Section in longitudinal E–W direction. (c) Section in transverse N–S direction.

$f'_c = 5{,}000$ psi (34.5 MPa), normal weight

$f'_{ci} = 3{,}750$ psi (25.9 MPa)

E–W max. f_c due to net prestress after losses = 200 psi

N–S max. f_c due to net prestress after losses not to exceed 350 psi (ACI allows up to 500 psi)

Max. f_c due to combined stresses $= 0.45f'_c$

$E_c = 57{,}000\sqrt{f_c} = 4.03 \times 10^6$ psi (27.8×10^6 MPa)

$f_{ps} \leq 0.70 f_{pu} = 189{,}000$ psi (1,303 MPa), as required by the ACI Code

$f_{py} = 240{,}000$ psi (1,655 MPa)

$f_{pe} = 159{,}000$ psi (1,096 MPa)

$E_{ps} = 29 \times 10^6$ psi (200×10^3 MPa)

$f_y = 60{,}000$ psi (414 MPa)

$E_s = 29 \times 10^6$ psi (200×10^3 MPa)

Solution:

$$e_S = e_L = \frac{6}{2} - 1 = 2.0 \text{ in. (51 mm)}$$

Choose a trial slab thickness on the basis of a span-to-depth ratio ≅ 45:

$$h = \frac{(20 + 24) \times 12}{2} \times \frac{1}{45} = 5.87 \text{ in.}$$

So try a 6-in. (152-mm)-thick slab assuming a duct diameter ≅ 0.5 in. and an effective depth $d_p = 6.0 - (0.5/2 + \frac{3}{4}) = 5.0$ in. (127 mm).

Balancing Load

$$W_D = 15 \text{ psf} + \frac{6}{12} \times 150 = 90 \text{ psf (4.31 kPa)}$$

Since a balancing load is required for zero deflection or camber due to dead load, assume that $W_{\text{bal}} = W_D = 90$ psf (4.31 kPa). Also, since f_c due to prestressing = 200 psi (given), assume this to be the stress in the E–W direction. Then the effective prestressing force in the E–W direction is $P_L = 200 \times 6 \times 12 = 14{,}400$ lb per strip, and from Equation 9.13b,

$$W_{\text{bal}(L)} = \frac{8P_L e_L}{L_L^2} = \frac{8 \times 14{,}400 \times 2}{(24)^2 \times 12} \cong 33 \text{ psf (1.58 kPa)}$$

The uplift to be provided by the tendons in the short direction (preferably the load is to be carried by the short-direction span) becomes $W_{\text{bal}(S)} = W_D - W_{\text{bal}(L)} = 90 - 33 = 57$ psf (2.73 kPa). Then, from Equation 9.13a,

$$P_S = \frac{W_{\text{bal}(S)} L_S^2}{8e_S} = \frac{57 \times (20)^2 \times 12}{8 \times 2}$$

= 17,100 lb/ft (249.7 kN/m) after losses, and the compression in the concrete after losses in prestress in the N–S direction is

$$f_c = \frac{P_S}{bh} = \frac{17{,}100}{12 \times 6} = 238 \text{ psi} < 350 \text{ psi}$$

which is satisfactory. Hence, use $\frac{1}{2}$-in. dia 7-wire 270-K tendons with effective prestressing force $P_e = 159{,}000 \times 0.153 = 24{,}327$ lb (108.2 kN).

Required Spacing in the N–S Direction. The required spacing in the N–S direction is

$$s_S = \frac{24{,}327}{17{,}100} = 1.42 \text{ ft} = 17 \text{ in. (432 mm)}$$

Required Spacing in the E–W Direction. The required spacing in the E–W direction is

$$S_L = \frac{24{,}327}{14{,}400} = 1.69 \text{ ft} \cong 20 \text{ in. (508 mm)}$$

Note that both spacings correspond to the recommended spacing of 3 to 5 times the slab thickness. As an additional measure, to prevent splitting of the concrete in the anchorage zones at the wall, add two #4 nonprestressed mild steel bars (12.7-mm dia) along the anchorage line on the slab perimeter.

Service-load Stresses. The service live load $W_L = 75$ psf (3.59 kPa), and the aspect ratio $k = L_L/L_S = 24/20 = 1.20$. From Figure 9.10, the moment coefficients for the maximum midspan moments in the short and long directions are $\alpha_{\text{N–S}} = 0.062$ and $\alpha_{\text{E–W}} = 0.035$, respectively, assuming that the corners of the two-way slab are held down, i.e., torsionally restrained.

We assume an effective $L_S = 19.5$ ft and $L_L = 23.5$ ft.

Live-load Moments. The live-load moments are

$$M_S = 0.062 \times 75 \times (19.5)^2 \times 12 = 21{,}218 \text{ in.-lb/ft}$$

and

$$M_L = 0.035 \times 75 \times (23.5)^2 \times 12 = 17{,}396 \text{ in.-lb/ft}$$

The moment of inertia is

$$I_s = \frac{12(6)^3}{12} = 216 \text{ in}^4$$

Concrete Stresses Due to Live Load. In the short direction, we have

$$f = \frac{Mc}{I_s} = \frac{21{,}218 \times 3}{216} = 295 \text{ psi } (2.03 \text{ MPa})$$

while in the long direction,

$$f = \frac{17{,}396 \times 3}{216} = 242 \text{ psi } (1.67 \text{ MPa})$$

The combined axial stresses due to the balanced load and the flexural stresses due to the live load, from Equations 9.15 and 9.16, become, in the short direction (N–S),

$$f^t = -\frac{P_S}{bh} - \frac{M_S c}{I_s} = -238 - 295 = -533 \text{ psi } (C) \ (3.68 \text{ MPa})$$

and

$$f_b = -238 + 295 = +57 \text{ psi } (T) \text{ (very small and, hence, negligible)}$$

and in the long direction (E–W),

$$f^t = -200 - 242 = -442 \text{ psi } (C) \ (3.05 \text{ MPa})$$

and

$$f_b = -200 + 242 = +42 \text{ psi } (T) \text{ (again negligible)}$$

The ACI allowable compressive stress is $f_c = 0.45 \times 5{,}000 = 2{,}250$ psi, which is well above the actual stress and, hence, satisfactory. These low stress levels can justify making the slab thinner than 6 in., provided that the deflection due to live load is acceptable. Note that the slab develops zero deflection or camber under dead load in this example, due to the load balancing.

Deflection Check. Only the deflection due to live load is checked. From principles of mechanics, we have

$$\Delta = \frac{5}{48} \frac{ML^2}{E_c I_s}$$

$$I_s = 216 \text{ in.}^4$$

$$E_c = 4.03 \times 10^6 \text{ psi}$$

$$\Delta_{\text{E-W}} = \frac{5}{48} \frac{17{,}396(24 \times 12)^2}{4.03 \times 10^6 \times 216} = 0.17 \text{ in.}$$

$$\Delta_{\text{N-S}} = \frac{5}{48} \frac{21{,}218(20 \times 12)^2}{4.03 \times 10^6 \times 216} = 0.15 \text{ in.}$$

$$\text{Avg. midspan deflection } \Delta = \frac{0.17 + 0.15}{2} = 0.16 \text{ in. } (4.1 \text{ mm})$$

$$\text{Acceptable deflection} = \frac{L_s}{360} = \frac{20 \times 12}{360} = 0.67 \text{ in. } (17 \text{ mm}) \gg 0.16 \text{ in.}$$

Consequently, a second cycle reducing the slab thickness to $5\frac{1}{2}$ in. could also be used, provided that the resulting slab nominal moment strengths are adequate to carry the load. In this case, $h = 5\frac{1}{2}$ in. is not adequate for nominal moment strength, as the next part of the solution indicates.

Nominal Moment Strength

$$W_u = 1.4 \times 90 + 1.7 \times 75 = 254 \text{ psf } (12.2 \text{ kPa})$$

Also,

$$\text{Effective } L_S = 19.5 \text{ ft}$$

$$L_L = 23.5 \text{ ft}$$

From Figure 9.11, the moment coefficients for maximum factored moment are

$$\alpha_{\text{N-S}} = 0.072$$

and

$$\alpha_{\text{E-W}} = 0.038$$

N–S Direction. In the N–S direction, we have

$$\text{Factored } M_u = 0.072 \times 254(19.5)^2 \times 12 = 83{,}448 \text{ in.-lb/ft}$$

$$\text{Required } M_n = \frac{M_u}{\phi} = \frac{83{,}448}{0.9} = 92{,}720 \text{ in.-lb/ft}$$

Note that prestressing forces in this structure do not produce any secondary moments M_s since there is no continuity at the slab boundaries. We have $A_{ps} = 0.153$ in.^{2} on 1.42 ft center to center (from before), and A_{ps}/ft = 0.153/1.42 = 0.11 in.^{2}/ft. Also, the effective $f_{pe} = 159{,}000$ psi. So in cases where the A_{ps} used exceeds P_e/initial A_{ps}, reduce f_{pe} accordingly.

We have, further,

$$\rho_{\text{N-S}} = \frac{0.11}{12 \times 5} = 0.0018$$

$$\text{Slab span-to-depth ratio} = \frac{20 \times 12}{6} = 40$$

From Equation 9.23b,

$$f_{ps} = f_{pe} + 10{,}000 + \frac{f'_c}{300\rho_p} \leq f_{py} \leq f_{pe} + 30{,}000$$

$$f_{ps} = 159{,}000 + 10{,}000 + \frac{5{,}000}{300 \times 0.0018} = 178{,}259 \text{ psi}$$

$$< f_{py} = 240{,}000 \text{ psi } < f_{pe} + 30{,}000 = 189{,}000 \text{ psi}$$

$$< \text{limit } f_{ps} = 189{,}000 \text{ psi, O.K.}$$

$$a = \frac{A_{ps} f_{ps}}{0.85 f'_c b} = \frac{0.11 \times 178{,}259}{0.85 \times 5{,}000 \times 12} = 0.38 \text{ in.}$$

$$\text{Available } M_n = A_{ps} f_{ps}\left(d = \frac{a}{2}\right) = 0.11 \times 178{,}259\left(5 - \frac{0.38}{2}\right)$$

$$= 94{,}316 \text{ in.-lb/ft} > \text{req } M_n = 92{,}720 \text{ in.-lb, O.K.}$$

E–W Direction. In the E–W direction, we have

$$\text{Factored } M_u = 0.038 \times 254(23.5)^2 \times 12 = 63{,}964 \text{ in.-lb/ft}$$

$$\text{Required } M_n = \frac{M_u}{\phi} = \frac{63{,}964}{0.90} = 71{,}071 \text{ in.-lb/ft}$$

$$A_{ps} = 0.153 \text{ in.}^2 \text{ per } 1.69 \text{ ft center to center (from before)}$$

$$A_{ps}/\text{ft} = \frac{0.153}{1.69} = 0.09 \text{ in}^2/\text{ft}$$

$$\rho_{E\text{-}W} = \frac{0.09}{12 \times 5} = 0.0015$$

$$f_{ps} = 159{,}000 + 10{,}000 + \frac{5{,}000}{300 \times 0.0015} = 180{,}111 \text{ psi, O.K.}$$

$$a = \frac{0.09 \times 180{,}111}{0.85 \times 5{,}000 \times 12} = 0.32 \text{ in.}$$

$$\text{Available } M_n = 0.09 \times 180{,}111\left(5 - \frac{0.32}{2}\right)$$

$$= 78{,}456 \text{ in.-lb/ft} > \text{Req. } M_n \text{ of } 71{,}071 \text{ in.-lb/ft, O.K.}$$

$$(29.1 \text{ kN-m/m} > 26.3 \text{ kNm/m})$$

Shear Strength. From before, the aspect ratio $k = 1.2$, and from Equations 9.27,

$$V_u = \tfrac{1}{3} w_u L_S = \tfrac{1}{3} \times 254 \times 19.5 = 1{,}651 \text{ lb/ft} \qquad \text{(N–S)}$$

$$V_u = \frac{k w_u L_S}{2k + 1} \qquad \text{(E–W)}$$

$$= 1.2 \times 254 \times \frac{19.5}{2 \times 1.2 + 1} = 1{,}748 \text{ lb/ft (25.5 kN/m)}$$

From Equation 9.26,

$$2\sqrt{f'_c}\, b_w d_p \le V_c = \left(0.6\sqrt{f'_c} + 700\frac{V_u d}{M_u}\right) b_w d_p \le 5\sqrt{f'_c}\, b_w d_p$$

Consider $700\,(V_u d)/(M_{\bar{u}}) \simeq 0$ at the boundaries of the single-panel wall-supported slab in this example. In such a case,

$$V_c = 0.6\sqrt{5{,}000} \times 12 \times 5 = 2{,}546 \text{ lb/ft (37.2 kN/m)} \gg 1{,}748 \text{ lb/ft}$$

which is satisfactory. Hence, adopt the following design:

$$h = 6 \text{ in. (152 mm)}$$

$$d_p = 5 \text{ in. (127 mm)}$$

Use $\frac{1}{2}$-in. dia 7-wire 270-K tendons spaced 17 in. (432 mm) center-to-center in the N–S direction, and 20 in. (508 mm) center-to-center in the E–W direction. Also, use two #4 bars (12.7-mm dia) along the anchorage zone line around all the slab perimeter.

9.8 ONE-WAY SLAB SYSTEMS

One-way prestressed slabs behave similarly to beams regardless of whether they are simply supported or continuous over several supports. A one-way slab is therefore designed as a 12-in.-wide beam. The main prestressing strands are placed in the direction of the slab length, namely, spanning the continuous spans. The aspect ratio, the ratio of the span to the width of the slab band, has to have a value of 2 or greater in order for the slab to be considered one-way.

The same design and analysis procedures and examples as in Chapter 6 are to be used for the analysis and design of continuous one-way-action prestressed floor systems.

9.9 SHEAR-MOMENT TRANSFER TO COLUMNS SUPPORTING FLAT PLATES

9.9.1 Shear Strength

The shear behavior of two-way slabs and plates is a three-dimensional stress problem. The critical shear failure plane follows the perimeter of the loaded area and is located at a distance that gives a minimum shear perimeter b_0. Based on extensive analytical and experimental verification, the shear plane should not be closer than a distance $d/2$ from the concentrated load or reaction area.

If no special shear reinforcement is provided, the nominal shear strength V_c, as required by the ACI, is defined in Equations 9.24, 9.25, and 9.26. The coefficients for evaluating the factored external shear force V_u in a continuous two-way slab supported all along the panel perimeters can be evaluated approximately from Equations 9.27.

9.9.2 Shear-Moment Transfer

The unbalanced moment at the column face support of a slab without beams is one of the more critical design considerations in proportioning a flat plate or a flat slab. To ensure adequate shear strength requires moment transfer to the column by flexure across the perimeter of the column and by eccentric shearing stress such that approximately 60 percent is transferred by flexure and 40 percent by shear.

The fraction γ_v of the moment transferred by eccentricity of the shear stress decreases as the width of the face of the critical section resisting the moment increases in such a manner that

$$\gamma_v = 1 - \frac{1}{1 + \frac{2}{3}\sqrt{\frac{b_1}{b_2}}} \tag{9.28}$$

where $b_2 = c_2 + d$ is the width of the face of the critical section resisting the moment and $b_1 = c_1 + d$ is the width of the face at right angles to b_2.

The remaining portion γ_f of the unbalanced moment transferred by flexure is given by and acting on an effective slab width between lines that are $1\frac{1}{2}$ times the total slab thickness h on both sides of the column support.

$$\gamma_f = \frac{1}{1 + \frac{2}{3}\sqrt{\frac{b_1}{b_2}}} = 1 - \gamma_v \tag{9.29}$$

For exterior columns, $b_1 = c_1 + \frac{1}{2}d$. The value of γ_f can be increased to 1.0 provided that V_u is less than $0.75\phi V_c$. At interior supports, γ_f can be increased by 25 percent provided that $V_u \le 0.4\phi V_c$ and $\rho \le 0.375\rho_b$.

The distribution of shear stresses around the column edges is as shown in Figure 9.15. It is considered to vary linearly about the centroid of the critical section. The factored shear force V_u and the unbalanced factored moment M_u, both assumed to be acting at the column face, have to be transferred to the centroidal axis c–c of the critical section. Thus, the axis position has to be located, thereby obtaining the shear force arm g (the distance from the column face to the centroidal axis plane) of the critical section c–c for the shear moment transfer.

For computing the maximum shear stress sustained by the plate in the edge column region, the ACI Code requires using the full nominal moment strength M_n provided by the column strip in Equations 9.30 to follow as the unbalanced moment, multiplied by the transfer fraction factor γ_v. This unbalanced moment $M_n \ge M_{ue}/\phi$ is composed of two parts: the negative end panel moment $M_{ne} = M_e/\phi$ at the face of the column, and the mo-

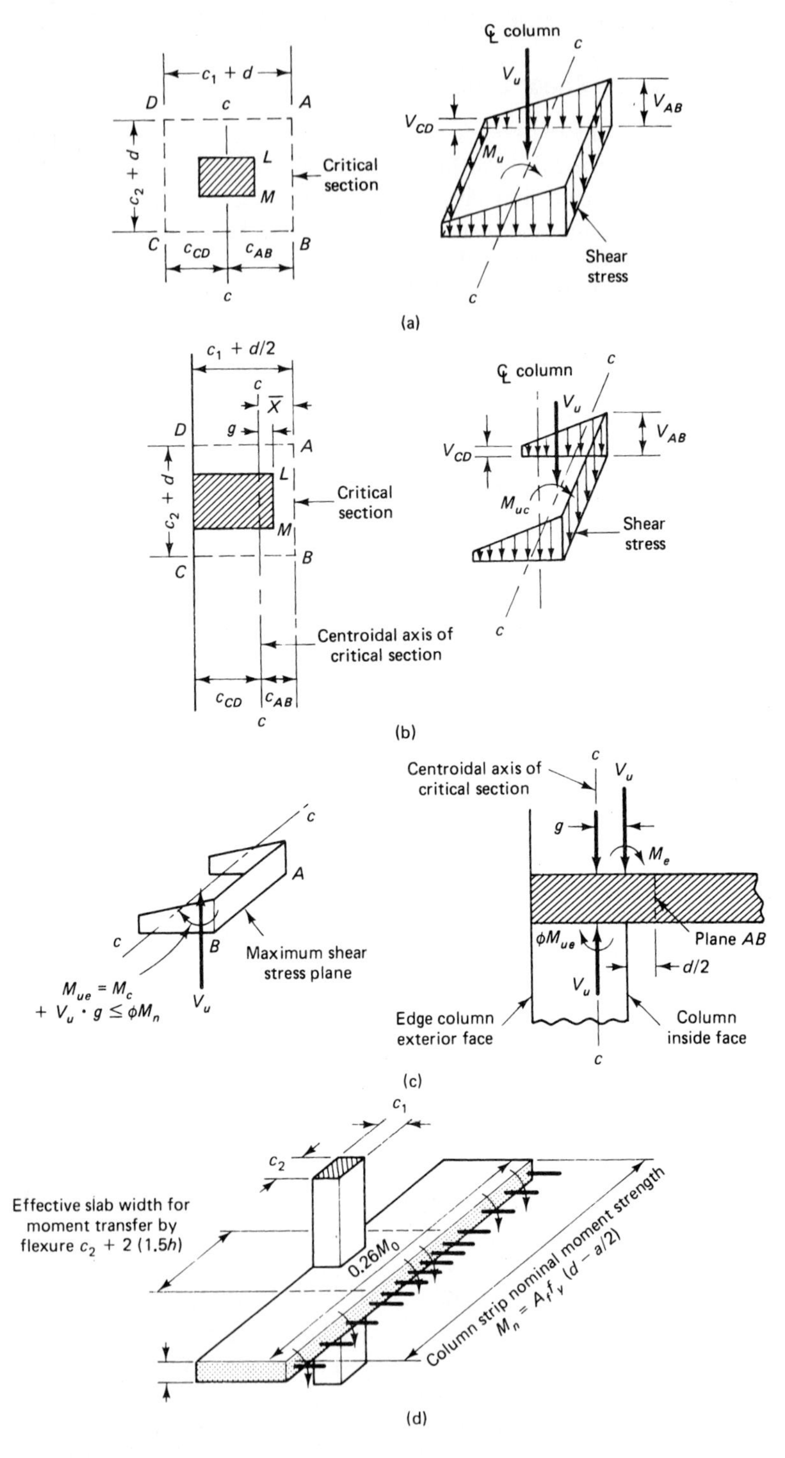

Figure 9.15 Shear stress distribution around column edge. (a) Interior column. (b) End column. (c) Critical surface. (d) Transfer nominal moment strength M_n.

ment $(V_u/\phi)g$ due to the eccentric factored perimetric shear force V_u. The limiting value of the shear stress intensity is expressed as

$$\frac{v_{u(AB)}}{\phi} = \frac{V_u}{\phi A_c} + \frac{\gamma_v M_{ue} c_{AB}}{\phi J_c} \tag{9.30a}$$

$$\frac{v_{u(CD)}}{\phi} = \frac{V_u}{\phi A_c} - \frac{\gamma_v M_{ue} c_{CD}}{\phi J_c} \tag{9.30b}$$

where the nominal shear strength intensity is

$$v_n = \frac{v_u}{\phi} \tag{9.30c}$$

and where A_c = area of concrete of assumed critical section
$= 2d(c_1 + c_2 + 2d)$ for an interior column
and J_c = property of assumed critical section analogous to polar moment of inertia.

The value of J_c for an interior column is

$$J_c = \frac{d(c_1 + d)^3}{6} + \frac{d^3(c_1 + d)}{6} + \frac{d(c_2 + d)(c_1 + d)^2}{2}$$

and the value of J_c for an edge column with bending parallel to the edge is

$$J_c = \frac{(c_1 + d/2)(d)^3}{6} + \frac{2(d)}{3}(c_{AB}^3 + c_{CD}^3) + (c_2 + d)(d)(c_{AB})^2$$

From basic principles of the mechanics of materials, the shearing stress is

$$v_u = \frac{V_u}{A_c} + \gamma_v \frac{Mc}{J}$$

where the second term on the right-hand side is the shearing stress resulting from the torsional moment at the column face.

If the nominal moment strength M_n of the shear moment transfer zone after the design of the reinforcement results in a larger value than M_{ue}/ϕ, this value of M_n should be used in Equations 9.30a and b in lieu of M_{ue}/ϕ. In such a case, where the moment strength value $M_n = M_{ne} + (V_u/\phi)g$ is increased because of the use of flexural reinforcement in excess of what is needed to resist M_{ue}/ϕ, the slab stiffness is raised, thereby increasing the transferred shear stress v_u computed from Equations 9.30a and b for development of full moment transfer. Consequently, it is advisable to maintain a design moment M_{ue} with a value close to the factored moment value M_{ue} if an increase in the shear stress due to additional moment transfer needs to be avoided and a resulting need for additional increase in the plate design thickness prevented.

Example 9.2 illustrates the procedure for computing the limit perimeter shear stress in the plate at the edge column region.

A higher perimetric shearing stress v_u can occur than that evaluated by Equation 9.30a or b when adjoining spans are unequal or unequally loaded in the case of an interior column. The ACI Code stipulates, in the slab section pertaining to factored moments in columns and walls, that the supporting element such as a column or a wall has to resist an unbalanced moment

$$M' = 0.07[(w_{nd} + 0.5w_{nl})l_2 l_{n2} - w'_{nd} l'_2 (l'_n)^2] \tag{9.31}$$

where w'_{nd}, l'_2, and l'_n refer to the shorter span. Hence, an additional term is added to Equation 9.30a or b in such cases so that

$$v_u = \frac{V_u}{A_c} + \frac{\gamma_v M_u c_{AB}}{J_c} + \frac{\gamma_v M'c}{J'_c} \tag{9.32}$$

where J'_c is the polar moment of inertia with moment areas taken in a direction perpendicular to that used for J_c.

9.9.3 Deflection Requirements for Minimum Thickness: An Indirect Approach

For a preliminary estimate of the two-way slab thickness, it is necessary to use some approximate guidelines in order to expeditiously select a trial depth. Span-to-depth ratios in prestressed slabs are expected to be higher than those in reinforced concrete slabs if the advantages of prestressing are not to be economically lost.

Service live loads rather than total dead plus live loads should be used to evaluate deflection. Load balancing from the transverse component of the prestressing force would have to be used to neutralize the dead-load deflection or even produce camber if the live load is excessively high. An approximate guideline of a span-to-depth ratio of 16 to 25 for solid cantilever slabs and 40 to 50 for two-way continuous slabs is not unreasonable to use. For waffle slabs, a lower value of 35 to 40 is recommended. For simply supported spans and for single- and double-T's, use 90 percent of these values for the first trial.

The ACI requires that the minimum span-to-deflection ratio be restricted depending on the type of loading and conditions of use. This limitation is obviated by the need to prevent cracking of plaster in the ceilings and damage to sensitive supported equipment as well as cracking of partitions and ponding of water in roofs. Table 9.1 gives recommended values of span-to-deflection ratios for deflection control.

Table 9.1 Minimum Permissible Ratios of Span *(l)* to Deflection *(a)* (*l* = longer span)

Type of member	Deflection *a* to be considered	$(l/a)_{min}$
Flat roofs not supporting and not attached to nonstructural elements likely to be damaged by large deflections	Immediate deflection due to live load *L*	180[a]
Floors not supporting and not attached to nonstructural elements likely to be damaged by large deflections	Immediate deflection due to live load *L*	360
Roof or floor construction supporting or attached to nonstructural elements likely to be damaged by large deflections	That part of total deflection occurring after attachment of nonstructural elements: sum of long-time deflection due to all sustained loads (dead load plus any sustained portion of live load) and immediate deflection due to any additional live load[b]	480[c]
Roof or floor construction supporting or attached to nonstructural elements not likely to be damaged by large deflections		240[c]

[a]Limit not intended to safeguard against ponding. Ponding should be checked by suitable computations of deflection, including added deflections due to ponded water, and considering long-term effects of all sustained loads, camber, construction tolerances, and reliability of provisions for drainage.

[b]Long-term deflection has to be determined, but may be reduced by the amount of deflection computed to occur before attachment of nonstructural elements. This reduction is made on the basis of accepted engineering data relating to time-deflection characteristics of members similar to those being considered.

[c]Ratio limit may be lower if adequate measures are taken to prevent damage to supported or attached elements, but should not be lower than tolerance of nonstructural elements.

A more accurate evaluation of the deflection/camber of reinforced concrete and prestressed concrete two-way-action slabs and plates is presented in Section 9.12. This approach utilizes the stiffnesses of the interconnecting members using the equivalent frame method in the deflection analysis. It is both logical and expedient to use, particularly since the stiffness factors of the various elements have already been computed in the flexural analysis of the equivalent continuous frames.

9.10 STEP-BY-STEP TRIAL-AND-ADJUSTMENT PROCEDURE FOR THE DESIGN OF A TWO-WAY PRESTRESSED SLAB AND PLATE SYSTEM

The following sequence of steps is suggested for the design or analysis of two-way action prestressed slabs:

1. Determine whether the slab geometry and loading require a two-way analysis by the equivalent frame method.
2. Select a trial slab thickness for the maximum of either the longitudinal $h = L/45$ or the transverse $h = L/45$. Compute the total service dead and live loads and the factored loads.
3. Assume a tendon profile across the continuous spans in both the E–W and N–S directions, and determine the prestressing force F, the concrete stress $f_c = F/A_c$, and the number of strands in a span. Compute the balancing load intensity $W_{bal} = 8Fa/L^2$ and the net load $W_{net}\downarrow = W_w\downarrow - W_{bal}\uparrow$.
4. Determine, by the equivalent frame method, the equivalent frame characteristics and the flexural and torsional stiffnesses of the slab given by

$$K_c \cong \frac{4EI}{L_n - 2h}$$

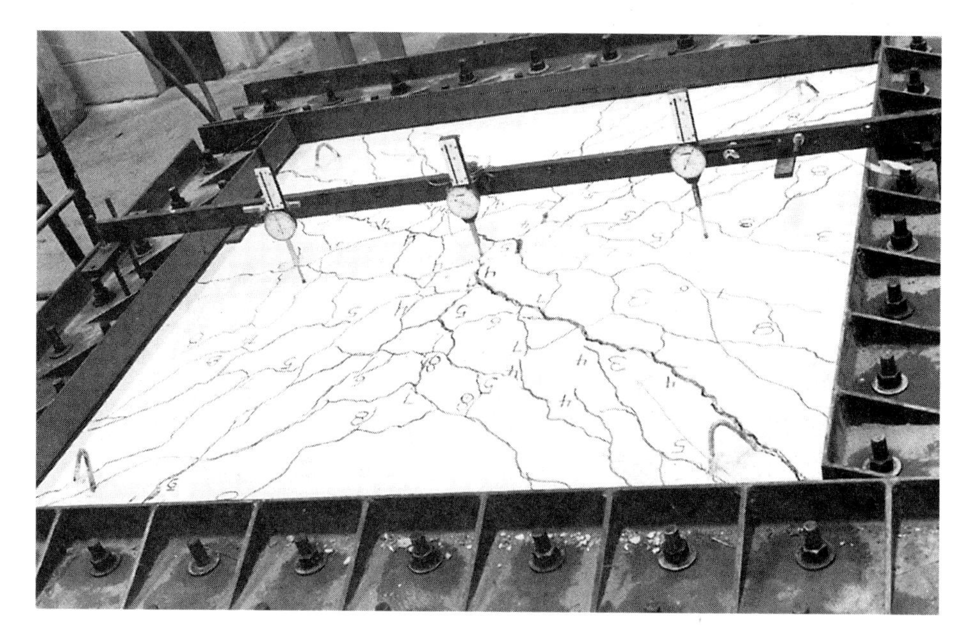

Photo 9.4 Flexural cracking in a restrained one-panel reinforced concrete slab. (Tests by Nawy et al.)

and

$$K_t = \Sigma \frac{9E_{cs}C}{L_2\left(1 - \dfrac{c_2}{L_2}\right)^3}$$

where $C = \Sigma(1 - 0.63x/y)x^3\, y/3$. Then determine

$$K_{ec} = \left(\frac{1}{K_c} + \frac{1}{K_t}\right)^{-1}$$

for the exterior and interior columns, and the slab stiffness

$$K_s \cong \frac{4EI}{L_1 - c_1/2}$$

where L_1 is the centerline span and c_1 is the column depth for each slab-column joint.

5. From the values of K_{ec} and K_s obtained at each joint, determine the moment distribution factors

$$DF = \frac{K_s}{\Sigma K}$$

for the slabs, where $\Sigma K = K_{ec} + K_{S(\text{left})} + K_{S(\text{right})}$. Then compute the fixed-end moments *FEM* at the joints for the net loads given by $FEM = WL^2/12$ for a distributed load.

6. Run a moment distribution for the net load moment M_{net}, and adjust the redistributed moments to obtain the net moment values at the face of the supports; the equation is M_n = centerline $M_n - Vc/3$. Then verify that the concrete stresses

$$f_t = -\frac{P}{A} + \frac{M_{\text{net}}}{S}$$

resulting from these moments are below the maximum allowable $f_t = 6\sqrt{f'_c}$ for support sections and $f_t = 2\sqrt{f'_c}$ for midspan sections.

7. Compute the balanced service-load fixed-end moments

$$FEM_{\text{bal}} = \frac{W_{\text{bal}}L^2}{12}$$

and run a moment distribution of the balanced load moments M_{bal}. Then find the primary moment $M_1 = P_e e$ and the secondary moment M_s = Distributed $(M_{\text{bal}} - M_1)$.

8. Compute the fixed-end factored load moments $FEM_u^- = (W_uL^2)/(12)$ and run a moment distribution of the factored moments. Then determine the required design moment M_u = Distributed $(M_u^- - M_s)$ for the slabs at all joints and at maximum positive M_u along the spans.

9. Determine the required nominal moment strengths $M_n = M_u/\phi$ for the negative support moments $-M_u$ and the positive span moments $+M_u$. Then check whether the available $-M_n$ and $+M_n$ for the slab and the prestressing steel are adequate. Next, determine the *inelastic* moment redistribution ΔM_R from

$$\rho_D = 20\left[1 - \frac{\omega_p + (d/d_p)(\omega - \omega')}{0.36\beta_1}\right] \text{ percent}$$

where $\Delta M_R = \rho_D$ (support M_u). Add mild steel to the support and midspan where necessary, recalling that the minimum nonprestressed steel $A_s = 0.00075hL$.

10. Check the nominal shear strength of the slab at the exterior and interior supports, and compute the shear-moment transfer and the flexure-moment transfer to the columns. The moment shear factor is

$$\gamma_v = 1 - \frac{1}{1 + \frac{2}{3}\sqrt{b_1/b_2}}$$

and the moment flexure factor is

$$\gamma_f = \frac{1}{1 + \frac{2}{3}\sqrt{b_1/b_2}}$$

where $b_1 = c_1 + d/2$ for an exterior column
$b_1 = c_1 + d$ for an interior column
$b_2 = c_2 + \mathrm{d}$

The value of γ_f can be increased by 25 percent at the interior supports and can be also increased to 1.0 at other supports as indicated in the discussion of Equation 9.29. Then compute c_{AB} and c_{CD} for exterior columns, as well as the total nominal unbalanced moment strength $M_n = M_{ue} + V_u g$.

11. Compute the shear ultimate stress due to perimeter shear and effect of $\gamma_v M_n$:

$$v_n = \frac{V_u}{\phi_v A_c} + \frac{\gamma_v c_{AB} M_n}{J_c} \leq \text{max. allowable } v_c \text{ where}$$

$$\text{max. allowable shear stress } v_c = \beta_p \sqrt{f_c'} + 0.3\bar{f}_c + \frac{V_p}{b_o d}$$

$$\beta_p = \text{the smaller of the two values of 3.5 or } (\alpha_s\, d/b_o + 1.5)$$

where α_s is 40 for interior columns, 30 for edge columns, and 20 for corner columns.
Column section shall be at least 4 in. from the face of the discontinuous edge, and f_c' shall not exceed 5000 psi and $\bar{f}$ shall be 125 psi min. and 500 psi max.; otherwise v_c shall be computed from the smaller of the values obtained from the following expressions.

$$v_c = (2 + 4/\beta_c)\sqrt{f_c'} \quad \text{or} \quad v_c = \left(\frac{\alpha_s\, d}{b_o} + 2\right)\sqrt{f_c'} \quad \text{or} \quad v_c = 4\sqrt{f_c'}$$

12. Compute the factored moment value $\gamma_f M_n$, and check the available moment strength M_n of the section, concentrating the steel in the column band $[c + 2(1.5h)]$.

13. Check the deflection and camber serviceability behavior of the critical slab panels.

14. Adopt the design if it satisfied all the preceeding criteria. Then perform the operations for the E–W and N–S directions of the floor system.

Figure 9.16 gives a flowchart for the design or analysis of two-way action prestressed concrete floor slabs and plates.

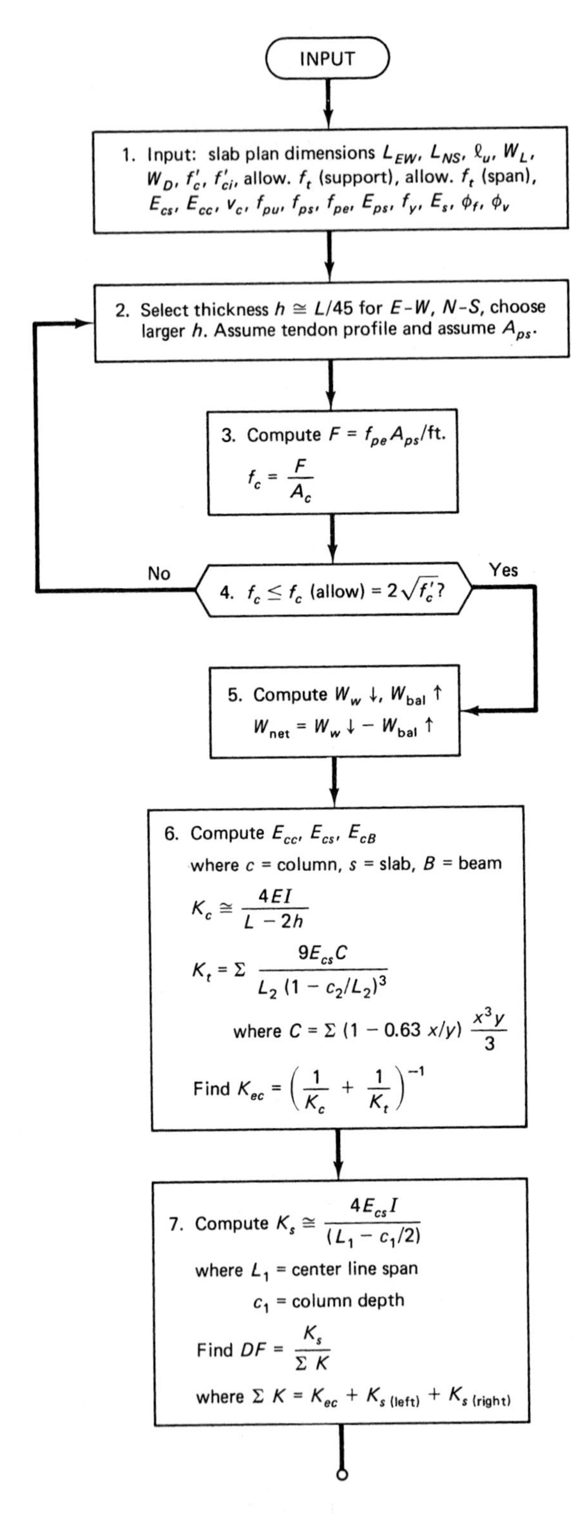

Figure 9.16 Flowchart for the design (analysis) of prestressed concrete two-way slabs and plates in flexure and shear.

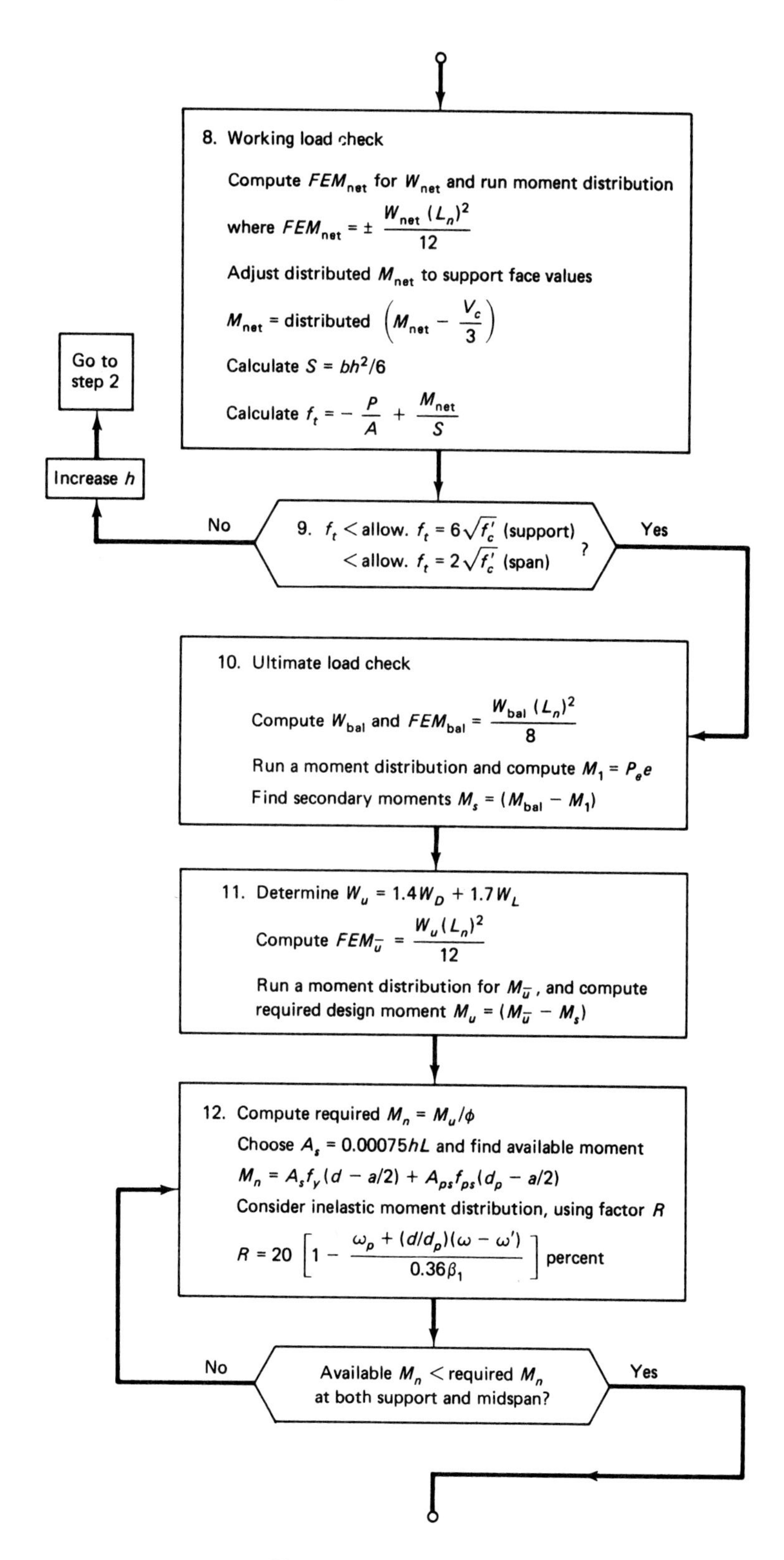

Figure 9.16 *Continued*

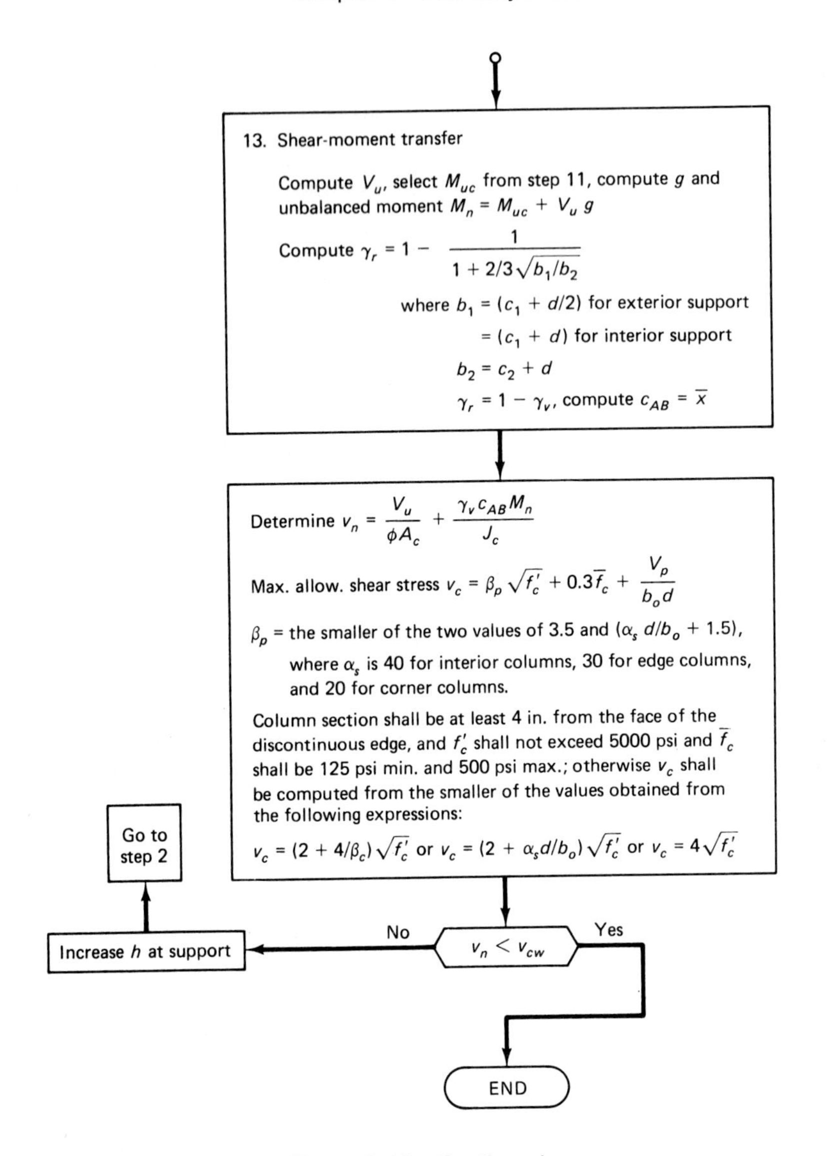

Figure 9.16 *Continued*

9.11 DESIGN OF PRESTRESSED POST-TENSIONED FLAT-PLATE FLOOR SYSTEM

Example 9.2

A post-tensioned prestressed nonbonded flat-plate floor system for an apartment complex is shown in Figure 9.17. The end-panel centerline dimensions are 17 ft, 6 in. × 20 ft, 0 in. (5.33 m × 6.10 m), and the interior panel dimensions are 24 ft, 0 in. × 20 ft, 0 in. (7.32 m × 6.10 m). The heights l_u of the intermediate floors are typically 8 ft, 9 in. (2.67 m). Design a typical floor panel to withstand a working live load $W_L = 40$ psf (1.92 kPa) and a superimposed dead load $W_D = 20$ psf (0.96 kPa) due to partitions and flooring. Assume in your solution that all panels are simultaneously loaded by the live load, and verify the shear-moment transfer capacity of the floor at the column supports. Use $\frac{1}{2}$-in. dia 7-wire 270-K prestressing strands and the equivalent frame method to arrive at your solution. Given data are as follows:

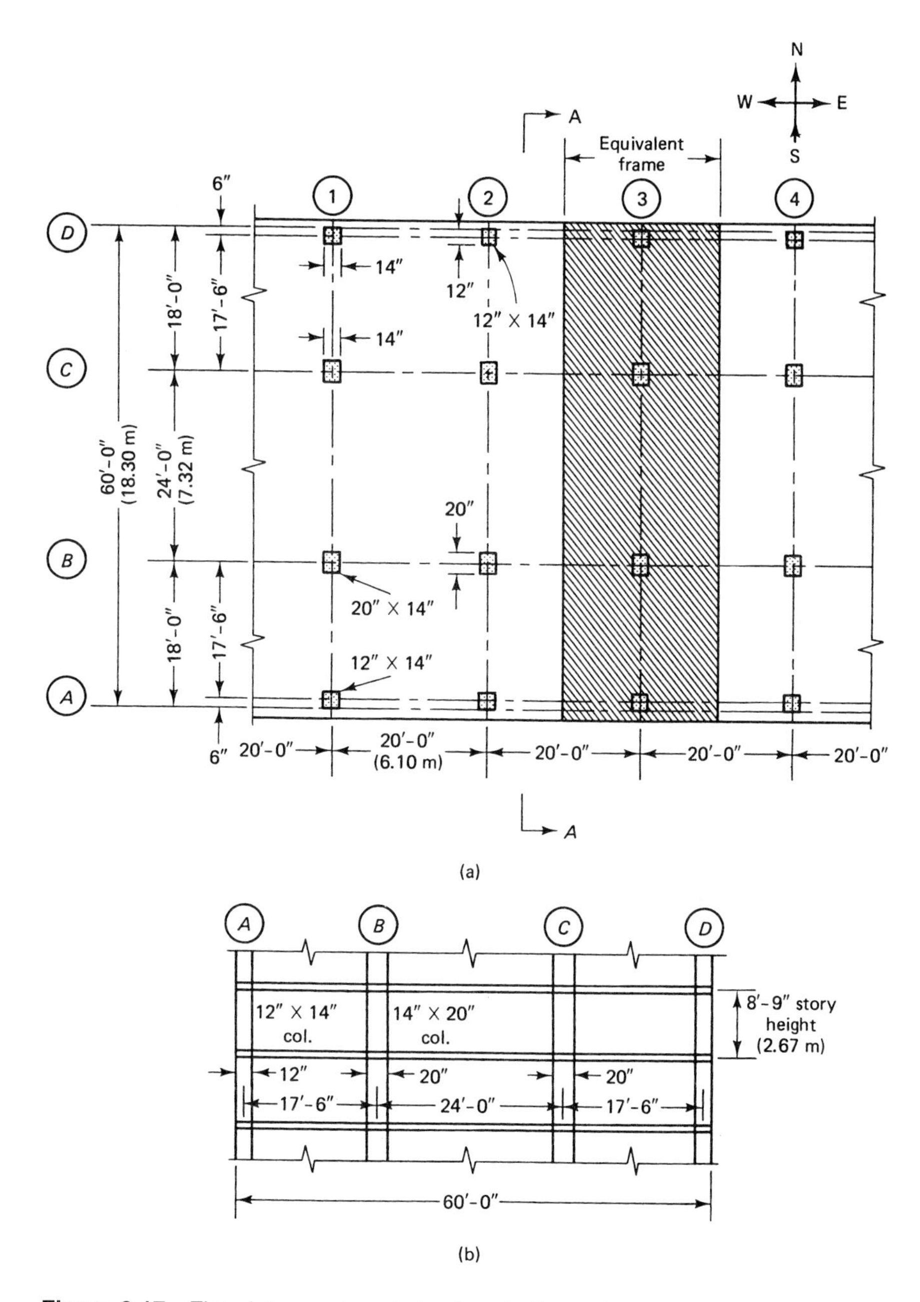

Figure 9.17 Flat-plate apartment structure in Example 9.1. (a) Plan. (b) Section A–A, N–S.

$f'_c = 4{,}000$ psi (27.6 MPa), normal weight

$f'_{ct} = 3{,}000$ psi (20.7 MPa)

Support $f_t = 6\sqrt{f'_c} = 380$ psi (2.62 MPa)

Midspan $f_t = 2\sqrt{f'_c} = 127$ psi (0.88 MPa)

Max. v_c required by the ACI Code

$f_{pu} = 270{,}000$ psi (1,862 MPa)

f_{ps} not to exceed 185,000 psi (1,276 MPa)

$f_{py} = 243{,}000$ psi (1,675 MPa)

$f_{pe} = 159{,}000$ psi (1,096 MPa)

$E_{ps} = 29 \times 10^6$ psi (200×10^3 MPa)

$f_y = 60{,}000$ psi (414 MPa)

Solution:

N–S Direction

I. *Service Load Analysis*

1. *Loads*

For deflection control, assume that the slab thickness $h \cong L/45$. Then the longitudinal direction $h = 20 \times 12/45 = 5.33$ in. and the transverse direction $h = 24 \times 12/45 = 6.40$ in. So try $h = 6\frac{1}{2}$ in. (165-mm) slabs = 81 psf. The superimposed dead load = 20 psf, and we have

$$\text{Total } W_D = 101 \text{ psf}$$

$$W_L = 40 \text{ psf}$$

$$\text{Total } W_w = W_{D+L} = 141 \text{ psf (6.75 kPa)}$$

$$W_u = 1.4W_D + 1.7W_L = 1.4 \times 101 + 1.7 \times 40 \cong 210 \text{ psf (10.05 kPa)}$$

$$L_n = \text{bay span (N–S for this part of the solution)}$$

$$L_2 = \text{band width (E–W direction)} = 20 \text{ ft (240 in.)}$$

2. *Load Balancing and Tendon Profile*

In order to make a preliminary estimate of the balanced load, assume an average intensity of compressive stress on the concrete due to load balancing of $f_c = 170$ psi (1.17 MPa). Then the unit $F = 170 \times 6.5 \times 12 = 13{,}260$ lb/ft (193.6 kN/m). So, trying $\frac{1}{2}$-in. dia 270-K seven-wire strands, we find that the effective force P_e per strand $= A_{ps} f_{pe} = 0.153 \times 159{,}000 = 24{,}327$ lb. For the L = 20-ft bay along the longitudinal direction of the structure, the total force is $F_e = FL = 13{,}260 \times 20 = 265{,}200$ lb (1,180 kN).

The number of strands per bay is $F_e/P_e = 265{,}200/24{,}327 \cong 11$ strands, and the total $P_e = F_e = 24{,}327 \times 11 = 267{,}597$ lb. The actual unit force $F = 267{,}597/20 = 13{,}380$ lb/ft (195.3 kN/m), and the actual $f_c = F/A = 13{,}380/(6.5 \times 12) \cong 172$ psi $\cong 170$ psi, which is satisfactory. Consequently, use $f_c = 172$ psi due to load balancing, and assume a parabolic tendon profile as shown in Figure 9.18.

Outside Spans AB or CD at Midspan

$$a_1 = a_3 = \frac{3.25 + 5.50}{2} - 1.75 = 2.625 \text{ in.}$$

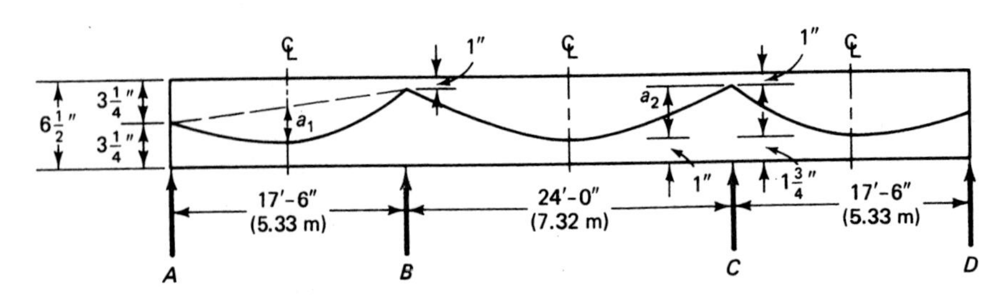

Figure 9.18 Tendon profile in N–S direction in Example 9.2.

From Equation 1.16 for a parabolic tendon,

$$W = \frac{8Fa}{L_n^2}$$

$$W_{\text{bal}} = \frac{8 \times 13{,}380 \times 2.625/12}{(18)^2} \cong 72 \text{ psf}$$

The net load intensity producing bending is

$$W_{\text{net}} = W_w - W_{\text{bal}} = 141 - 72 = 69 \text{ psf (3.30 kPa)}$$

Interior Span BC

$$a_2 = 6.5 - 1 - 1 = 4.5 \text{ in.}$$

$$W_{\text{bal}} = \frac{8Fa}{L_n^2} = \frac{8 \times 13{,}380 \times 4.5/12}{(24)^2} \cong 70 \text{ psf}$$

$$W_{\text{net}} = 141 - 70 = 71 \text{ psf (3.40 kPa)}$$

3. *Equivalent Frame Characteristics*

Take the equivalent frame in the N–S direction whose plan is shown in the shaded portion of Figure 9.17. The approximate flexural stiffness of the column above and below the floor joint (the moment per unit rotation), from Ref. 9.10 and Equation 9.9, is

$$K_c = \frac{4E_c I_c}{L_n - 2h}$$

where $L_n = l_u = 8$ ft, 9 in. = 105 in.

(a) *Exterior column (14 in. × 12 in.) stiffness*

For the exterior columns, $b = 14$ in., so $I_c = 14(12)^3/12 = 2{,}016$ in.^{4} Assume that $E_{col}/E_{slab} = E_{cc}/E_{cs} = 1.0$, and use $E_{cc} = E_{cs} = 1.0$ in the calculations as E_{cs} drops out in the equation for K_{ec}. Then we obtain

$$\text{Total } K_c = \frac{4 \times 1 \times 2{,}016}{105 - (2 \times 6.5)} \times 2 \text{ (for top and bottom columns)}$$

$$= 175.3 \text{ in.-lb/rad}/E_{cc}$$

From Equation 9.10b, the torsional constant is

$$C = \Sigma\left(1 - 0.63\frac{x}{y}\right)\frac{x^3 y}{3}$$

$$= \left(1 - 0.63 \times \frac{6.5}{12}\right)6.5^3 \times \frac{12}{3} = 724$$

The torsional stiffness of the slab at the column line is

$$K_t = \Sigma \frac{9E_{cs}C}{L_2\left(1 - \frac{c_2}{L_2}\right)^3}$$

$$= \frac{9 \times 1 \times 724}{20 \times 12(1 - 14/(12 \times 20))^3} + \frac{9 \times 1 \times 724}{20 \times 12(1 - 14/(12 \times 20))^3}$$

$$= 65.0 \text{ in-lb/rad}/E_{cc}$$

From Equation 9.7, the equivalent column stiffness is $K_{ec} = (1/K_c + 1/K_t)^{-1} = (1/175.3 + 1/65)^{-1} = 47$ in.-lb/rad/E_{cc}.

(b) *Interior column (14 in. × 20 in.) stiffness*
For the interior columns, $b = 14$ in., so $I = 14(20)^3/12 = 9{,}333$ in.^{4} Hence, we have

$$\text{Total } K_c = \frac{4 \times 1 \times 9{,}333}{105 - 2 \times 6.5} \times 2 = 812 \text{ in.-lb/rad}/E_{cc}$$

$$C = \left(1 - 0.63 \times \frac{6.5}{20}\right) \times (6.5)^3 \times \frac{20}{3} = 1{,}456$$

$$K_t = \frac{9 \times 1{,}456}{20 \times 12(1 - 14/(12 \times 20))^3} + \frac{9 \times 1{,}456}{20 \times 12(1 - 14/(12 \times 20))^3} = 131 \text{ in.-lb/rad}/E_{cs}$$

$$K_{ec} = (1/812 + 1/131)^{-1} = 113 \text{ in.-lb/rad}/E_{cc}$$

(c) *Slab stiffness*
From Equation 9.9 and Ref. 9.10,

$$K_s = \frac{4E_{cs} I_s}{L_n - \frac{c_1}{2}}$$

where L_n is the centerline span and c_1 the column depth. The slab band width in the E–W direction is $20/2 + 20/2 = 20$ ft. Thus, $I_s = 20 \times 12(6.5)^3/12 = 5{,}493$ in.^{4}, and for the slab at the right of exterior column A,

$$K_s = \frac{4 \times 1 \times 20(6.5)^3}{12 \times 17.5 - 12/2} = 108 \text{ in.-lb/rad}/E_{cs}$$

while for the slab at the left of interior column B,

$$K_s = \frac{4 \times 1 \times 20(6.5)^3}{12 \times 17.5 - 20/2} = 110 \text{ in.-lb/rad}/E_{cs}$$

and for the slab at the right of interior column B,

$$K_s = \frac{4 \times 1 \times 20(6.5)^3}{12 \times 24 - 20/2} = 79 \text{ in.-lb/rad}/E_{cs}$$

From Equation 9.12, the slab distribution factor at the joints is $DF = K_s/\Sigma K$, where $\Sigma K = K_{ec} + K_{s(left)} + K_{s(right)}$. So for the outer joint A slab, $DF = 108/(47 + 108) = 0.697$; for the left joint B slab, $DF = 110/(113 + 110 + 79) = 0.364$; and for the right joint B slab, $DF = 79/(113 + 110 + 79) = 0.262$.

4. *Design Service-load Moments and Stresses*
Design net load moments
For the exterior spans AB and CD, $W_{net} = 69$ psf. So the fixed-end moment is

$$FEM = \frac{WL_n^2}{12} = \frac{69 \times (17.5)^2}{12} \times 12 = 21.1 \times 10^3 \text{ in.-lb}$$

Similarly, for the interior span BC, $W_{net} = 71$ psf. So the fixed-end moment is

$$FEM = \frac{71(24)^2}{12} \times 12 = 40.9 \times 10^3 \text{ in.-lb}$$

By running a moment distribution analysis as shown in Table 9.2, a carryover factor $COF = \frac{1}{2}$ can be used for all spans. Such an assumption is justified, as the effect of nonprismatic sections would be negligible on the fixed-end moments and carryover factors. It can also be assumed in multispan frames that the frame at a joint two spans away from the left joint (joint C) can be considered fixed in the distribution of the moments.

Table 9.2 Moment Distribution of Net Load Moments M_{net}

	(A)		(B)	℄	(C)
DF	0.697	0.364	0.262	0.262	0.364
COF	0.5	0.5	0.5	0.5	0.5
FEM_{net} $\times 10^3$ in.-lb	−21.1	21.1	−40.9	40.9	−21.1
Dist.	+14.71	7.21	5.19	−5.19	
CO	3.61	7.36	−2.60		
Dist.	−2.52	−1.73	−1.25		
Final M_{net} $\times 10^3$ per ft	−5.30	33.94	−39.56		

Slab concrete tensile stress at support

The net moment at the interior face of column B is the difference of the centerline moment and $Vc/3$, i.e.,

$$M_{\text{net,max}} = 39.56 \times 10^3 - \frac{20}{3}\left(\frac{71 \times 24}{2}\right) = 33{,}880 \text{ in.-lb/ft}$$

The slab section modulus $S = bh^2/6 = 12(6.5)^2/6 = 84.5 \text{ in.}^3$, and we have, for the support concrete stress,

$$f_t = -\frac{P}{A} + \frac{M}{S} = -172 + \frac{33{,}880}{84.5} = +229 \text{ psi } (1.63 \text{ MPa})\ (T)$$

So the allowable $f_t = 6\sqrt{f_c'} = 380$ psi > 229 psi, which is satisfactory.

Slab concrete tensile stress at midspan

The net midspan maximum moment is $WL^2/8 - 39.56 \times 10^3$, or

$$M_{\text{net,max}} = \frac{71(24)^2}{8} \times 12 - 39.56 \times 10^3 = 21{,}784 \text{ in.-lb/ft } (7.85 \text{ kN/m})$$

Also,

$$\text{Midspan } f_t = -\frac{P}{A} + \frac{M}{S} = -172 + \frac{21{,}784}{84.5} = +86 \text{ psi } (0.545 \text{ MPa})\ (T)$$

So the allowable $f_t = 2\sqrt{f_c'} = 127$ psi > 86 psi, which is satisfactory.

If f_t were to exceed the allowable f_t, the entire tensile force would have to be taken by mild steel reinforcement at a stress $f_s = \frac{1}{2} f_y$.

Ultimate Flexural Strength Analysis

II. *Design Moments M_u.*

1. *Balanced moments M_{bal}*

The secondary moment is given by $M_s = M_{\text{bal}} - M_1$, where M_{bal} is the balanced moment and M_1 is the primary moment $= P_e e = Fe$. For the span AB or CD,

$$FEM_{\text{bal}} = \frac{72(17.5)^2}{12} \times 12 = 22{,}050 \text{ in.-lb/ft}$$

and for the span BC,

$$FEM_{\text{bal}} = \frac{70(24)^2}{12} \times 12 = 40{,}320 \text{ in.-lb/ft}$$

Table 9.3 Moment Distribution of Balanced Load Moments M_{bal}

	Ⓐ		Ⓑ	℄	Ⓒ
DF	0.697	0.364	0.262	0.262	0.364
COF	0.5	0.5	0.5	0.5	0.5
FEM_{bal} $\times 10^3$ in.-lb	−22.05	22.05	−40.32	40.3	−22.05
Dist.	+15.37	+6.65	+4.79	−4.79	−6.65
CO	3.33	7.69	−2.40		
Dist.	−2.32	−1.93	−1.39		
Final M_{bal} $\times 10^3$ per ft	−5.67	34.46	−39.32		

Running a moment distribution as in Table 9.3 will determine the maximum M_{bal} for the exterior column joints.

2. *Secondary moments M_s and factored load moment M_u*

Span AB
From the tendon profile of Figure 9.18, $e = 0$. So we have:

$$\text{Primary moment } M_1/\text{ft at } A = P_e e = 0$$

$$M_{bal} = 5{,}670 \text{ in.-lb/ft (from Table 9.3)}$$

$$M_s = M_{bal} - M_1 = 5{,}670 - 0 = 5.67 \times 10^3 = 5{,}670 \text{ in.-lb/ft}$$

$$\text{Factored load } FEM_u = \frac{W_u l^2}{12} = \frac{210(17.5)^2}{12} \times 12 = 64{,}313 \text{ in.-lb/ft}$$

Span BA
From the tendon profile in Figure 9.18, $e = 6.5/2 - 1 = 2.25$ in. So we have:

$$M_1 = 13{,}380 \times 2.25 = 30{,}105 \text{ in.-lb/ft (11.16 kN-m)}$$

$$M_{bal} = 34{,}460 \text{ in.-lb/ft (from Table 9.3)}$$

$$M_s = 34{,}460 - 30{,}105$$

$$= 4{,}355 \text{ in.-lb/ft (1.61 kN-m/m)}$$

Factored load $FEM_u = 64{,}313$ in.-lb/ft (23.84 kN-m/m)

Span BC

$$e = 2.25 \text{ in.}$$

$$M_1 = 30{,}105 \text{ in.-lb/ft}$$

$$M_{bal} = 39{,}320 \text{ in.-lb/ft (from Table 9.3)}$$

$$M_s = 39{,}320 - 30{,}105 = 9{,}215 \text{ in.-lb/ft (3.42 kN-m/m)}$$

$$\text{Factored load } FEM_u = \frac{210(24)^2}{12} \times 12 = 120{,}960 \text{ in.-lb/ft (44.84 kN-m/m)}$$

Run a moment distribution for the factored moments as in Table 9.4. Analysis of pattern loading of alternate spans should also be made to determine the worst conditions of service-load and factored-load moments.

Table 9.4 Moment Distribution of Factored Loads

	Ⓐ		Ⓑ	℄	Ⓒ
DF	0.697	0.364	0.262	0.262	0.364
COF	0.5	0.5	0.5	0.5	0.5
$FEM_u^- \times 10^3$ in.-lb per ft	−64.31	64.31	−120.96	120.96	−64.31
Dist.	+44.82	+20.62	+14.84	−14.84	−20.62
CO	10.31	22.41	−7.42		
Dist.	−7.19	−5.46	−3.93		
Final $M_u^- \times 10^3$ per ft	−16.37	101.88	−117.47		

3. *Design moment M_u*
The design moments M_u are the difference of the factored-load moments M_u^- and the secondary moments M_s, i.e., $M_u = M_u^- - M_s$ (from Equation 9.17).

Joint A (span AB) moment $-M_u$
For the joint A (span AB) moment, $M_s = 5{,}670$ in.-lb/ft (from before), and so the centerline $M_u = 16{,}370 - 5{,}670 = 10{,}700$ in.-lb/ft. The moment reduction to the column face of support A = $Vc/3$. Thus,

$$V_{AB} = \frac{W_u L}{2} - \frac{M_{u@B}^- - M_{u@A}^-}{L_n} = \frac{210 \times 17.5}{2} - \frac{10^3(101.88 - 16.37)}{17.5 \times 12}$$

$$= 1{,}837.5 - 407.2 = 1{,}430.3 \text{ lb/ft}$$

$$c = 12 \text{ in.}$$

$$\text{Centerline } M_u = 16{,}370 - 5{,}670 = 10{,}700 \text{ in.-lb/ft}$$

$$\text{Req column face } M_u = 10{,}700 - \frac{1{,}430.3 \times 12}{3}$$

$$= 10{,}700 - 5{,}720 - 4{,}980 \text{ in.-lb/ft } (1.85 \text{ kN-m/m})$$

$$\text{Req } - M_n = \frac{M_u}{\phi} = \frac{4{,}980}{0.9} = 5{,}533 \text{ in.-lb/ft } (2.05 \text{ kN-m/m})$$

Joint B (span BA) moment $-M_u$
For the joint B (span BA) moment, $M_s = 4{,}355$ in.-lb/ft (from before), and so the centerline $M_u = 101{,}880 - 4{,}355 = 97{,}525$ in.-lb/ft. Thus,

$$V_{BA} = 1{,}837.5 + 407.2 = 2{,}244.7 \text{ lb/ft.}$$

$$c = 20 \text{ in.}$$

$$\text{Req. column face } M_u = 97{,}525 - \frac{2{,}244.7 \times 20}{3}$$

$$= 97{,}525 - 14{,}965 = 82{,}560 \text{ in.-lb/ft } (30.61 \text{ kN-m/m})$$

$$\text{Req } - M_n = \frac{M_u}{\phi} = \frac{82{,}560}{0.9} = 91{,}733 \text{ in.-lb/ft } (34 \text{ kN-m/m})$$

Joint B (span BC) moment $-M_u$
For the joint B (span BC) moment, M_s = 9.215 in.-lb/ft, and so the centerline M_u = 117,470 − 9,215 = 108,255 in.-lb/ft. Thus,

$$V_{BC} = \frac{210 \times 24}{2} = 2{,}520 \text{ lb/ft}$$

$$\text{Req. column face } - M_u = 108{,}255 - \frac{2{,}520 \times 20}{3} = 108{,}255 - 16{,}800$$
$$= 91{,}455 \text{ in.-lb/ft } (37.32 \text{ kN-m/m})$$

$$\text{Req } - M_n = \frac{M_u}{\phi} = \frac{91{,}455}{0.9}$$
$$= 101{,}617 \text{ in.-lb/ft } (41.47 \text{ kN-m/m})$$

Span AB maximum positive moment $+M_u$
Assume that the point of zero shear and maximum moment is x ft from face A. Then $x = V_{AB}/W_u$ = 1,430/210 = 6.81 ft. Also, from Table 9.4, the end M_u^- at A = 16,370 in.-lb/ft., and from before, $M_s = \frac{1}{2}(5{,}670 + 4{,}355) = 5{,}013$ in.-lb/ft. So we have

$$\text{Max.} + M_u = V_{AB}x - \frac{W_u x^2}{2} - M_u^- + M_s$$
$$= 1{,}430.3 \times 6.81 \times 12 - \frac{210(6.81)^2}{2} \times 12 - 16{,}370 + 5{,}013$$
$$= 116{,}884 - 58{,}434 - 16{,}370 + 5{,}013$$
$$= 47{,}093 \text{ in.-lb/ft } (17.46 \text{ kN-m/m}) \text{ at 6.81 ft from A}$$

$$\text{Req} + M_n = \frac{M_u}{\phi} = \frac{47{,}093}{0.9} = 52{,}236 \text{ in.-lb/ft } (19.4 \text{ kN-m/m})$$

Span BC maximum positive moment $+M_u$
From before, V_{BC} = 2,520 lb/ft and $x = L_n/2$ = 24/2 = 12 ft. The simple span midspan moment is, then,

$$M = V_{BC} \times \frac{L_n}{2} - \left(W_u \times \frac{L}{2}\right)\frac{(L)}{4}$$
$$= 2{,}520 \times \frac{24}{2} - \frac{210(24)^2}{8} = 15{,}120 \text{ ft.-lb/ft} = 181{,}440 \text{ in.-lb/ft}$$

Alternatively, the simple span moment is

$$M = \frac{W_u L^2}{8} = \frac{210(24)^2}{8} \times 12$$
$$= 181{,}440 \text{ in.-lb/ft}$$

Now, $+M_u = M - M_u^- + M_s$. From Table 9.4, M_u^- = −117,470 in.-lb/ft, and M_s = 9,215 in.-lb/ft. So the required maximum $+M_u$ = 181,440 − 117,470 + 9,215 = 73,185 in.-lb/ft (27.13 kN-m/m) at midspan. And the required $+M_n = M_u/\phi$ = 73,185/0.9 = 81,317 in.-lb/ft (30.14 kN-m/m).

Figure 9.19 gives a plot of the required design moments M_u across the continuous spans and the peak values of the moments.

III. *Flexural Strength M_n (Nominal Moment Strength).* The ACI Code requires a minimum amount of nonprestressed reinforcement. From Equation 9.20,

$$A_s = 0.00075hL_n$$

1. *Interior support section at B*
For the interior support section at B, the controlling required M_n = 101,617 in.-lb/ft. We then have

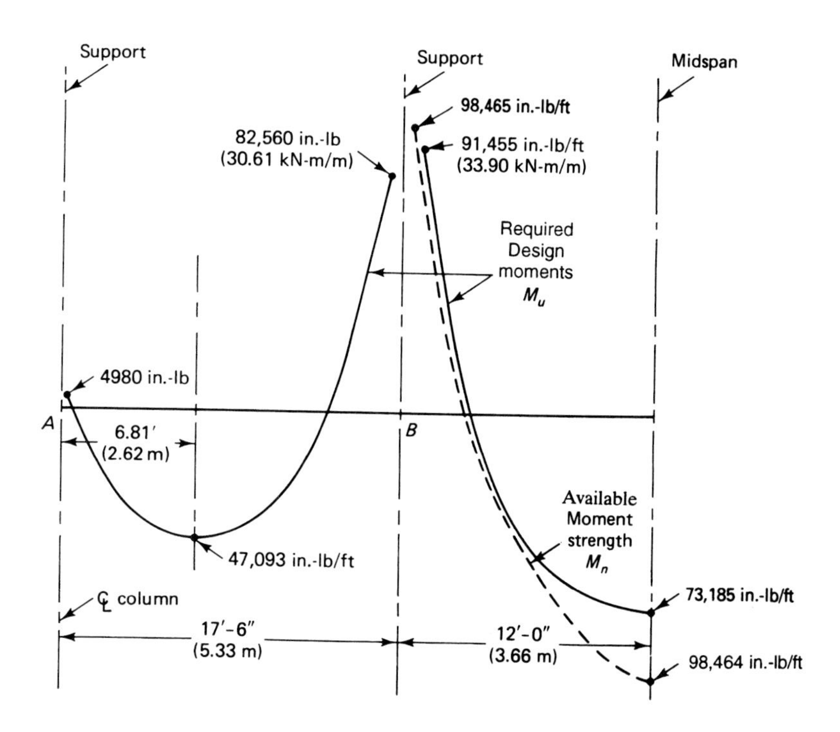

Figure 9.19 Maximum required design moments M_u and available nominal moment strengths M_n in Example 9.2, after redistribution.

$$A_s = 0.00075 \times 6.5\left(\frac{18 + 24}{2}\right) \times 12 = 1.23 \text{ in}^2 \text{ (7.93 cm}^2\text{)}$$

Hence, try six #4 bars of 11-ft length, and space the bars at a maximum of 6 in. (152 mm) center-to-center so that they are concentrated over the column on a band width equal to the column width plus $1\frac{1}{2}$-slab thicknesses on each side of the column. Then

$$A_s = 6 \times 0.20 = 1.20 \text{ in}^2 \cong \text{required } 1.23 \text{ in.}^2\text{, O.K.}$$

$$\text{Panel width} = 20 \text{ ft}$$

$$A_s/\text{ft} = \frac{1.2}{20} = 0.06 \text{ in.}^2$$

From Equation 9.23b, the design stress in the tendon is

$$f_{ps} = f_{pe} + \frac{f'_c}{300\rho_p} + 10{,}000 \text{ psi}$$

and

$$\rho_p = \frac{A_{ps}}{bd} = \frac{11 \times 0.153}{(20 \times 12)5.5} = 0.0013$$

$$f_{pe} = 159{,}000 \text{ psi}$$

$$f_{ps} = 159{,}000 + \frac{4{,}000}{300 \times 0.0013} + 10{,}000 = 179{,}256 \text{ psi (1,236 MPa)}$$

$$F_{ps} = \frac{179{,}256 \times 0.153 \times 11}{20} = 15{,}084 \text{ lb/ft}$$

$$F_s = 60{,}000 \times A_s/\text{ft} = 60{,}000 \times 0.06 = 3{,}600 \text{ lb/ft}$$

The the total force F/ft $= F_{ps} + F_s = 15{,}084 + 3{,}600 = 18{,}684$ lb/ft, and we also have

$$\text{Compression block depth } a = \frac{A_s f_y + A_{ps} f_{ps}}{0.85 f'_c b}$$

$$= \frac{18{,}684}{0.85 \times 4{,}000 \times 12} = 0.46 \text{ in. (11.7 mm)}$$

The bars and tendons are to be placed at the same level $d = 6.5 - 1 = 5.5$ in. Also, $M_n = (A_s f_y + A_{ps} f_{ps})\ (d - a/2)$, the available $-M_n = 18{,}684 \times (5.5 - 0.46/2) = 98{,}465$ in.-lb/ft (36.5 MPa), and the required $M_n = 101{,}617$ in.-lb/ft $> 98{,}465$, which is unsatisfactory. Thus, additional moment strength $\Delta M_n = 101{,}617 - 98{,}465 = 3{,}152$ in.-lb/ft is needed.

We next check the allowable inelastic moment redistribution to midspan.

2. *Midspan section at span BC*
From before, $F_{ps} = A_{ps} f_{ps} = 15{,}084$ lb/ft, and

$$a = \frac{A_{ps} f_{ps}}{0.85 f'_c b} = \frac{15{,}084}{0.85 \times 4{,}000 \times 12} = 0.37 \text{ in.}$$

So the available $-M_n = A_{ps} f_{ps}(d - a/2) = 15{,}084(5.5 - 0.37/2) = 80{,}171$ in.-/ft, and the required $M_n = 81{,}317$ in.-lb/ft $> 80{,}171$ in.-lb/ft, and hence is unsatisfactory. Accordingly, add two bars at midspan over a 20-ft width to get

$$A_s = 2 \times 0.20 = 0.40 \text{ in.}^2$$

$$A_s f_y = \frac{0.40 \times 60{,}000}{20} = 1{,}200 \text{ lb/ft}$$

$$a = \frac{(15{,}084 + 1{,}200)}{0.85 \times 4{,}000 \times 12} = 0.40 \text{ in.}$$

$$\text{Available } + M_n = (A_s f_y + A_{ps} f_{pf})\left(d - \frac{a}{2}\right)$$

$$= (15{,}084 + 1{,}200)\left(5.5 - \frac{0.4}{2}\right) = 86{,}305 \text{ in.-lb/ft}$$

$$> \text{req} + M_n = 81{,}317 \text{ in.-lb/ft, O.K.}$$

Thus, the available ΔM_R to accommodate moment redistribution from the support is $86{,}305 - 81{,}317 = 4{,}988$ in.-lb/ft.

3. *Allowable inelastic moment redistribution ΔM_R at support junction toward midspan*
From Equation 9.18,

$$\rho_D = 20\left[1 - \frac{\omega_p + (d/d_p)(\omega - \omega')}{0.36\beta_1}\right] \text{ percent}$$

$$\omega_p = \frac{\rho_p f_{ps}}{f'_c} = 0.0013 \times \frac{179{,}256}{4{,}000} = 0.0583$$

$$d = d_p = 5.5 \text{ in.}$$

$$\omega = \omega' = 0$$

$$\beta_1 = 0.85$$

$$\rho_D = 20\left(1 - \frac{0.0583 + 0}{0.36 \times 0.85}\right) = 16.2 \text{ percent}$$

So use a 12-percent redistribution factor.

The inelastic redistribution from the support to midspan is

$$\Delta M_R = 0.120 \times 98{,}465 = 11{,}816 \text{ in.-lb/ft}$$
$$> \text{available } \Delta M_R = 4.988 \text{ in.-lb/ft}$$

which is satisfactory. Thus, add four #4 bars to make a total of six #4 bars at midspan. Then

$$\Sigma A_s f_y = \frac{6 \times 0.2 \times 60{,}000}{20} = 3{,}600 \text{ lb/ft}$$

$$a = \frac{15{,}084 + 3{,}600}{0.85 \times 4{,}000 \times 12} = 0.46 \text{ in. (11.7 mm)}$$

$$\text{Available } + M_n = (15{,}084 + 3{,}600)\left(5.5 - \frac{0.46}{2}\right)$$
$$= 98{,}465 \text{ in.-lb/ft}$$

$$\text{Available } \Delta M_n \text{ at midspan} = 98{,}465 - 81{,}317 = 17{,}148 \text{ in.-lb/ft}$$
$$> \Delta M_R = 11{,}816$$

which can be elastically redistributed from the support. Hence, the design is satisfactory.

Summary

After redistribution, the required $-M_n = 101{,}617 - 17{,}148 = 84{,}469$ in.-lb/ft, which is less that the available support section nominal $-M_n = 98{,}465$ in.-lb/ft and, hence, satisfactory. The required $+M_n = 81{,}317 + 17{,}148 = 98{,}465$ in.-lb/ft, which is approximately equal to the available $+M_n = 98{,}465$ in.-lb/ft and, hence satisfactory. Accordingly, use six #4 (12.7-mm dia) nonprestressed mild steel bars at the bottom fibers at midspan in addition to the continuous prestressing tendon in the 20-ft segment. Also, use six #4 nonprestressed mild steel bars at the top fibers at the support, centered through the column at 6 in. center-to-center spacing (six 12.7-mm dia bars at 152 mm center-to-center).

The midspan sections of spans AB and CD would have more than adequate positive nominal moment strength to resist the positive factored moments. The nominal negative moment strength of the sections at the exterior supports A and D are governed by the moment-shear transfer stresses.

4. *Banding the reinforcement at the column region*

There are eleven $\frac{1}{2}$-in. dia strands, and the width of a column strip $= 2(\frac{1}{4} \times 20 \times 12) = 120$ in. Assume that 70 percent of the tendons are concentrated in the column strip. Then the number of strands $= 0.7 \times 11 = 7.7$. Accordingly, concentrate seven strands in the column strip, three of which are to pass through and be centered on the column section.

There are $11 - 7 = 4$ strands in the middle strip. On this basis, it can be reasonably assumed that the percentage distribution of moments between the column strip and the middle strip would be approximately as follows:

$$\text{Column strip moment factor} = 7/11 = 0.64$$
$$\text{Middle strip moment factor} = 0.36$$
$$\text{Max total } - M \text{ at } \textit{column face} \text{ B} = 33{,}880 \text{ in.-lb/ft (see Table 9.2)}$$
$$\text{Max total } + M \text{ at midspan} = 21{,}784 \text{ in.-lb/ft}$$

Consequently, distribute the prestressing tendons between the column strips and middle strips as shown subsequently.

IV. ***Nominal Shear Strength***

1. <u>*Exterior columns A and D*</u>

(a) *Geometry and external load*

From before, $V_{AB} = 1{,}430.3$ lb/ft, and the total shear is $V_B = 1{,}430.3 \times 20 = 28{,}606$ lb. Assume an exterior wall and glass averaging a load of 500 plf:

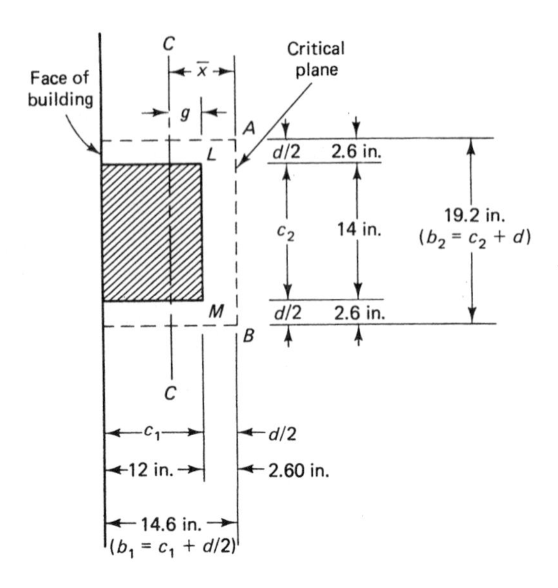

Figure 9.20 Critical planes for shear moment transfer in end column of Example 9.2 (line A, Figure 9.17).

$$\text{Wall } V_u = 1.4 \times 500 \times 20 = 14{,}000 \text{ lb}$$

$$\text{Slab } V_u = 28{,}606 \text{ lb}$$

$$\text{Total factored } V_{uA} = 42{,}606 \text{ lb } (189.5 \text{ kN})$$

The critical shear section is taken at $d/2$ from the face of the column, as shown in Figure 9.20. We have

$$d = 6.5 - 1.0 = 5.5 \text{ in.}$$

$$\text{Max } d_p = d_v = 0.8h = 0.8 \times 6.5 = 5.2 \text{ in. } (132 \text{ mm})$$

$$c_1 = 12 \text{ in.}$$

Photo 9.5 Placing concrete in a post-tensioned prestressed concrete slab.

$$c_2 = 14 \text{ in.}$$

$$b_1 = c_1 + \frac{d}{2} = 12 + \frac{5.2}{2} = 14.6 \text{ in.}$$

$$b_2 = c_2 + d = 14 + 5.2 = 19.2 \text{ in.}$$

$$A_c = b_0 d = 5.2(2 \times 14.6 + 19.2) = 252 \text{ in.}^2$$

From the figure,

$$d(2c_1 + c_2 + 2d)\bar{x} = d\left(c_1 + \frac{d}{2}\right)^2$$

or

$$5.2(2 \times 12 + 14 + 2 \times 5.2)\bar{x} = 5.2(14.6)^2$$

$$\bar{x} = c_{AB} = 4.40 \text{ in.}$$

$$g = \bar{x} - \frac{d}{2} = 4.4 - \frac{5.2}{2} = 1.8 \text{ in.}$$

Alternatively,

$$c_{AB} = \frac{b_1^2 d}{A_c} = \frac{(14.6)^2 \times 5.2}{252} = 4.4$$

$$c_{CD} = b_1 - c_{AB} = 14.6 - 4.4 = 10.2 \text{ in.}$$

From the geometrical properties of the exterior column shown in Figure 9.20, and from Equations 9.28 and 9.29,

$$\gamma_v = 1 - \frac{1}{1 + \frac{2}{3}\sqrt{b_1/b_2}} = 1 - \frac{1}{1 + \frac{2}{3}\sqrt{14.6/19.2}}$$
$$= 1 - 0.63 = 0.37$$

$$\gamma_f = \frac{1}{1 + \frac{2}{3}\sqrt{b_1/b_2}} = 0.63$$

Using d_v for d, the polar moment of inertia is

$$J_c = \frac{(c_1 + d/2)d^3}{6} + \frac{2d}{3}(c_{AB}^3 + c_{CD}^3) + (c_2 + d)(d)(c_{AB})^2$$

$$= \frac{14.6(5.2)^3}{6} + \frac{2 \times 5.2}{3}(4.4^3 + 10.2^3) + 19.2 \times 5.2(4.4)^2$$

$$= 342 + 3{,}974 + 1{,}933 = 6{,}249 \text{ in}^4$$

From before, the unit $-M_u = 10{,}700$ in.-lb/ft at the column centerline. So the total bay moment at the column centerline is $-M_c = 10{,}700 \times 20 = 214{,}000$ in.-lb. Now assume that the resultant V_u acts at the face of the column for shear-moment transfer. Then the shear moment transferred by eccentricity is $V_u g = -28{,}606 \times 1.8 = 51{,}491$ in.-lb, the total external factored moment $M_{ue} = -(214{,}000 + 51{,}491) = -265{,}491$ in.-lb, and the total required unbalanced moment strength $M_n = M_{ue}/\phi = 265{,}491/0.9 = 294{,}990$ in.-lb.

(b) *Shear-moment transfer*

The fraction of the nominal moment strength to be transferred by shear is $\gamma_v M_n = 0.37 \times 294{,}990 = 109{,}146$ in.-lb. From Equation 9.30a, the shearing stress due to perimeter shear, the effect of $\gamma_v M_n$ and the weight of the wall, is

$$
\begin{aligned}
v_n &= \frac{V_u}{\phi A_c} + \frac{\gamma_v c_{AB} M_n}{J_c} \\
&= \frac{42{,}606}{0.85 \times 252} + \frac{0.37 \times 4.4 \times 294{,}990}{6{,}249} \\
&= 198.9 + 76.9 \cong 276 \text{ psi}
\end{aligned}
$$

From the load balancing part of the solution, the average compressive stress in the concrete at the cross-section centroid due to externally applied load P_e is $\bar{f}_c = P_e / A_c = 172$ psi.

From Equations 9.24 and 9.25 and disregarding the effect of the vertical component V_p of the prestressing force, the maximum allowable shear strength becomes

$$v_c = \beta_p \sqrt{f'_c} + 0.3\bar{f}_c$$

where the factor β_p is the smaller of $(\alpha_s d/b_0 + 1.5)$ and 3.5, and $\alpha_s = 30$ for end column support. From Figure 9.20, $b_0 = 2 \times 14.6 + 19.2 = 48.2$ in., and

$$\frac{\alpha_s d}{b_0} + 1.5 = \frac{30 \times 5.5}{48.2} + 1.5 = 4.92 > 3.5.$$

Hence use $\beta_p = 3.5$.

$$
\begin{aligned}
\text{Max allowable } v_c &= 3.5\sqrt{4{,}000} + 0.3 \times 172 \\
&= 221 + 52 = 273 \text{ psi} \cong \text{actual } v_n = 276 \text{ psi, O.K.}
\end{aligned}
$$

If V_p were accounted for, the maximum allowable v_c would have been higher than 273 psi.

(c) *Flexure moment transfer*

The fraction of nominal moment strength to be transferred by flexure is $M_n = 0.63 \times 294{,}990 = 185{,}844$ in.-lb. From Equation 9.20, Min$A_s = 0.00075hL = 0.00075 \times 6.5 \times 17.5 \times 12 = 1.02$ in^2. So use six #4 bars × 6 ft, including the standard hook, yielding $A_s = 6 \times 0.2 = 1.2$ in^2. The stress in the tendon strands is computed from Equation 9.23 assuming that three strands pass the column at the exterior support at $e = 0$. We have $d = 6.5/2 = 3.25$ in., and the effective concrete width $b = c_2 + 2(1.5 + h) = 14 + 2(1.5 + 6.5) = 30$ in. Also,

$$\rho_p = \frac{A_{ps}}{bd_p} = \frac{3 \times 0.153}{30 \times 3.25} = 0.0047$$

$$
\begin{aligned}
f_{ps} &= f_{pe} + 10{,}000 + \frac{f'_c}{300\rho_p} \\
&= 159{,}000 + 10{,}000 + \frac{4{,}000}{300 \times 0.0047} \\
&= 171{,}837 \text{ psi}
\end{aligned}
$$

$$A_{ps} = 3 \times 0.153 = 0.459 \text{ in.}$$

$$
\begin{aligned}
a &= \frac{A_s f_y + A_{ps} f_{ps}}{0.85 f'_c b} = \frac{1.20 \times 60{,}000 + 0.459 \times 171{,}837}{0.85 \times 4{,}000 \times 30} \\
&= 1.48 \text{ in.}
\end{aligned}
$$

$$
\begin{aligned}
\text{Available } M_n &= A_s f_y \left(d - \frac{a}{2}\right) + A_{ps} f_{ps} \left(d_p - \frac{a}{2}\right) \\
&= 1.2 \times 60{,}000 \left(5.5 - \frac{1.48}{2}\right) + 0.459 \times 171{,}837 \left(3.25 - \frac{1.48}{2}\right) \\
&= 342{,}720 + 197{,}972 = 540{,}692 \text{ in.-lb} \\
&\gg \gamma_f M_n = 185{,}844 \text{ in.-lb}
\end{aligned}
$$

The available nominal moment strength is thus considerably larger than the moment being transferred by flexure. Figure 9.21 shows one scheme for banding both the prestressed and nonprestressed reinforcement to provide for shear-moment transfer at the exterior column zone.

2. *Interior columns B and C*

(a) *Geometry and external load*

From before, $V_{BA} + V_{BC} = 2{,}244.7 + 2{,}520 \cong 4{,}765$ plf. The total shear is $V_{uB} = 4{,}765 \times 20 = 95{,}300$ lb (423.9 kN), and also, $c_1 = 20$ in., $c_2 = 14$ in., and $d = 6.5 - 1 = 5.5$ in.

Assume that $d_v = 0.8h \cong 5.2$ in.; compute $g = \frac{1}{2}c_1 = 20/2 = 10$ in.

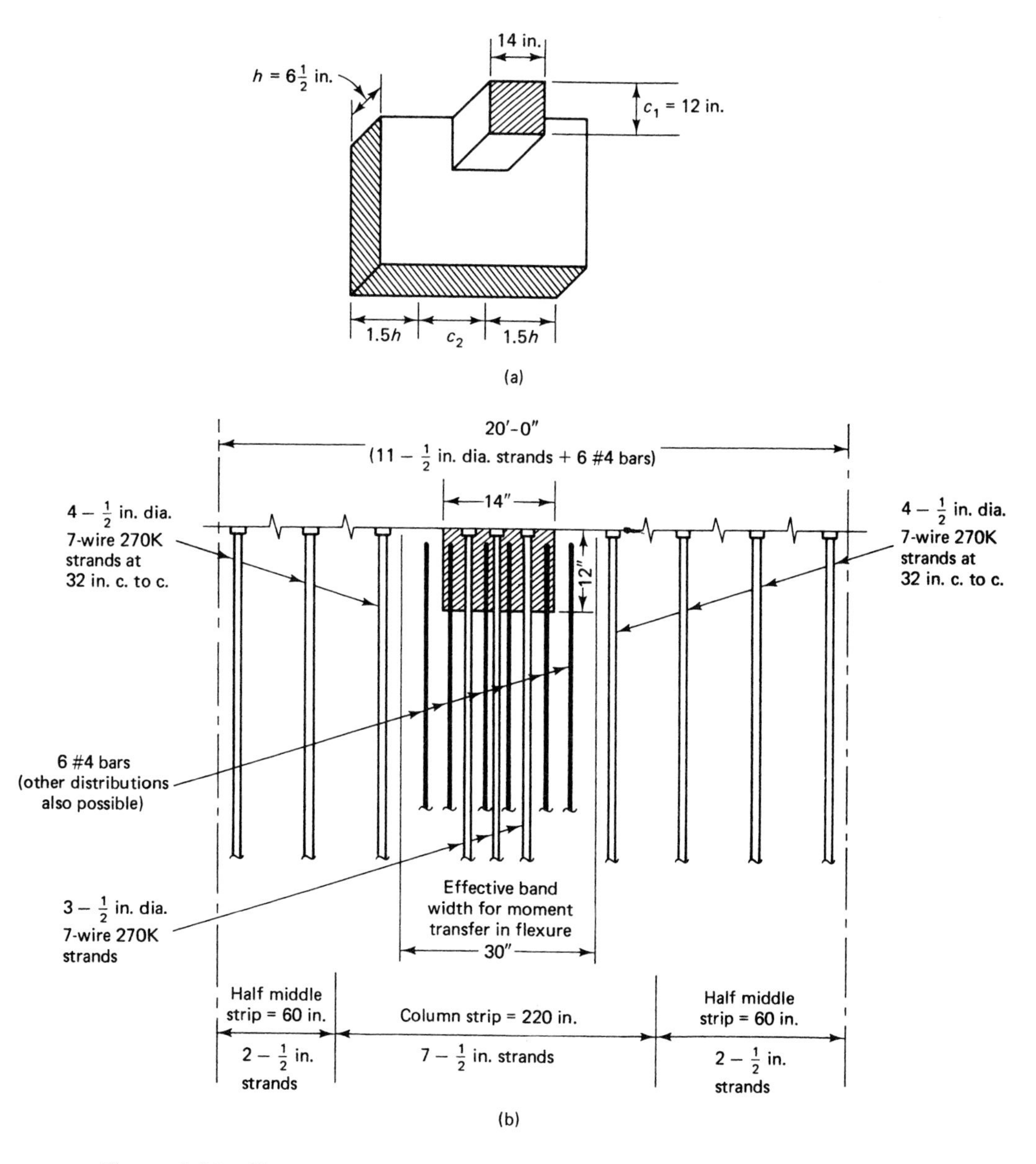

Figure 9.21 Shear-moment transfer zone and reinforcement distribution in Example 9.2. (a) Column zone band (33.5 in. wide). (b) Reinforcement distribution plan.

$$b_1 = c_1 + d = 20 + 5.2 = 25.2 \text{ in.}$$

$$b_2 = c_2 + d = 14 + 5.2 = 19.2 \text{ in.}$$

$$A_c = b_0 d = 2(25.2 \times 5.2 + 19.2 \times 5.2) = 462 \text{ in}^2$$

Using d_v for d, the polar moment of inertia is

$$J_c = \frac{d(c_1 + d)^3}{6} + \frac{d^3(c_1 + d)}{6} + \frac{d(c_2 + d)(c_1 + d)^2}{2}$$

$$= \frac{5.2(25.2)^3}{6} + \frac{(5.2)^3(25.2)}{6} + \frac{5.2(19.2)(25.2)^2}{2}$$

$$= 46{,}161 \text{ in}^4$$

Figure 9.22 shows the geometrical properties of the *interior columns:*

$$\gamma_v = 1 - \frac{1}{1 + \frac{2}{3}\sqrt{25.2/19.2}} = 0.433$$

$$\gamma_f = 1 - 0.433 = 0.567$$

The moment $M_{ue} = M_e$ for each interior column, and the net unit moment $+ M_e = 101{,}617 - 91{,}733 = 9{,}884$ in.-lb. The unbalanced shear moment is equal to the net $V_u \times g = 10(2{,}520 - 2{,}244.7) = 2{,}753$ in.-lb. Finally, the total moment $M_{ue} = 9{,}884 \times 20 + 2{,}753 = 200{,}433$ in.-lb, and the total required unbalanced moment strength is $M_n = M_{ue}/\phi = 200{,}433/0.9 = 222{,}703$ in.-lb.

(b) *Shear-moment transfer*

The fraction of nominal moment strength to be transferred by shear is $\gamma_v M_n = 0.433 \times 222{,}703 = 96{,}403$ in.-lb, and $c_{AB} = \frac{1}{2}(c_1 + d) = \frac{1}{2}b_1 = 25{,}2/2 = 12.6$ in.

From Equation 9.30a, the shear stress due to perimeter shear and the effect of M_n is

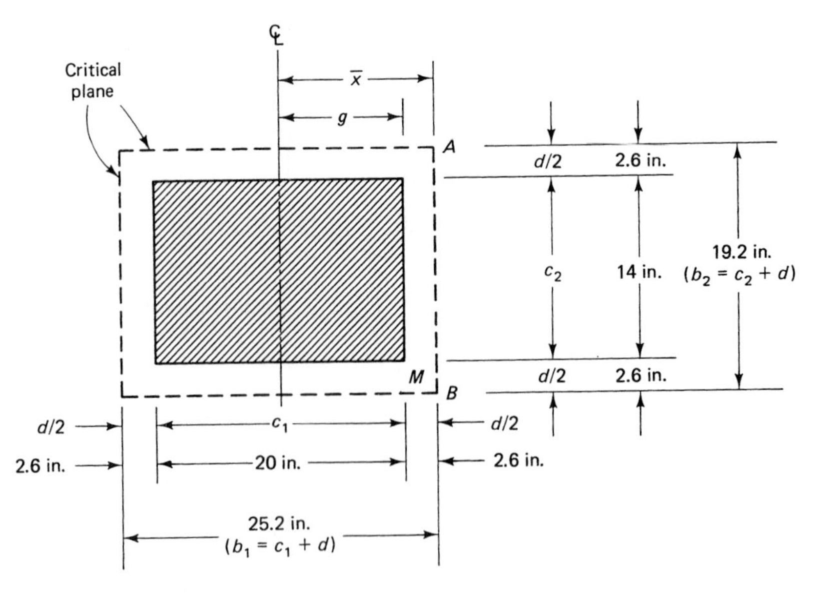

Figure 9.22 Critical plane for shear transfer in interior column of Example 9.2 (line B or C, Figure 9.17).

$$v_n = \frac{V_u}{\phi A_c} + \frac{\gamma_v c_{AB} M_n}{J_c}$$

$$= \frac{95{,}300}{0.85 \times 462} + \frac{96{,}403 \times 12.6}{46{,}161} = 242.7 + 26.3 = 269 \text{ psi (185 MPa)}$$

$$< \text{allowable } v_c = 273 \text{ psi, O.K.}$$

(c) *Flexure moment transfer*
The fraction of nominal moment strength to be transferred by flexure is $\gamma_f M_n = 0.567 \times 222{,}703 = 126{,}273$ in.-lb, and $b = c_2 + 2(1.5 + h) = 14 + 2(1.5 + 6.5) = 30$ in., the same as for exterior column A. Assume, as in the case of exterior columns, that three strands pass the interior columns B and C. We have

$$d_p = 6.5 - 1 = 5.5$$

$$\rho_p = \frac{A_{ps}}{bd_p} = \frac{3 \times 0.153}{30 \times 5.5} = 0.0028$$

$$f_{ps} = f_{pe} + 10{,}000 + \frac{f'_c}{300\rho_p}$$

$$= 159{,}000 + 10{,}000 + \frac{4{,}000}{300 \times 0.0028} = 173{,}762 \text{ psi}$$

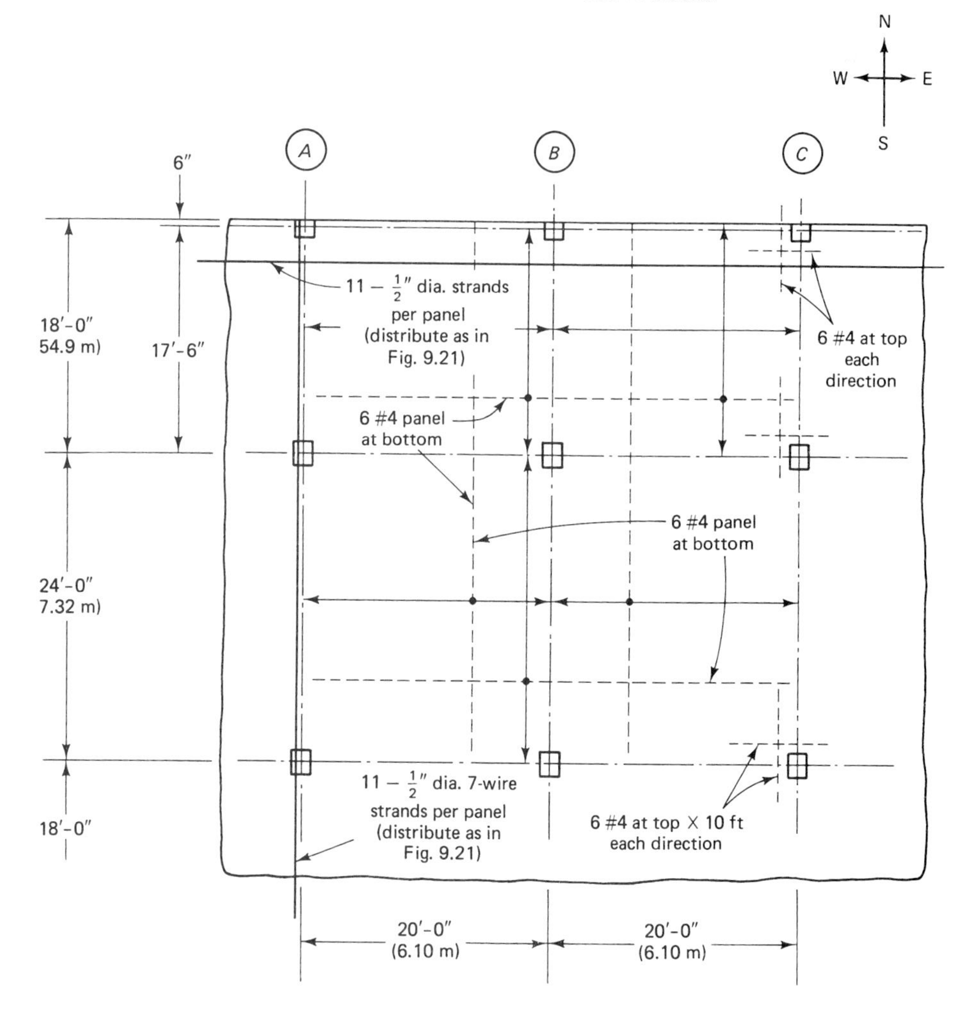

Figure 9.23 Schematic reinforcement distribution, partial floor plan for Example 9.2.

which is very close to f_{ps} for column A. Accordingly, using six #4 bars × 12 ft as minimum mild steel, as for the exterior columns, $a \cong 1.48$ in. and the available $M_n = 1.2 \times 60{,}000(5.5 - 1.48/2) + 0.459 \times 173{,}762(5.5 - 1.48/2) = 722{,}362$ in.-lb >> required $M_n = 126{,}273$ in.-lb, and hence satisfactory.

Figure 9.23 shows a schematic layout of the reinforcement in the continuous flat plate. The three $\frac{1}{2}$-in. dia strands in each direction should pass through the critical shear perimeter of the supporting columns. Of course, serviceability requirements for deflection should be checked, as in Section 9.13.

From the analysis made, we adopt the design and use the same pattern of reinforcement for both the N–S and E–W directions of the floor system, as the spans dimensions in both directions are very close in value.

9.12 DIRECT METHOD OF DEFLECTION EVALUATION

9.12.1. The Equivalent Frame Approach

As in the equivalent frame method for flexural analysis detailed in the preceding sections, the structure is divided into continuous frames centered on the column lines in each of the two perpendicular directions. Each frame is composed of a row of columns and a *broad* band of slab together with column line beams, if any, between panel centerlines.

By the requirement of statics, the applied load must be accounted for in each of the two perpendicular (orthogonal) directions. In order to account for the torsional deformations of the support beams, an *equivalent* column is used whose flexibility is the *sum* of the flexibilities of the actual column and the torsional flexibility of the transverse beam or slab strips (stiffness is the inverse of flexibility). In other words,

$$\frac{1}{K_{ec}} = \frac{1}{\Sigma K_c} + \frac{1}{K_t} \tag{9.33}$$

where K_{ec} = flexural stiffness of the equivalent column, bending moment per unit rotation

ΣK_c = sum of flexural stiffnesses of upper and lower columns, bending moment per unit rotation

K_t = torsional stiffness of the transverse beam or slab strip, torsional moment per unit rotation.

Photo 9.6 Rectangular concrete slab at rupture. (Tests by Nawy et al.)

The value of K_{ec} would thus have to be known in order to compute the deflection by this procedure.

The slab–beam strips are considered supported *not* on the columns, but on *transverse* slab–beam strips on the column centerlines. Figure 9.24(a) illustrates this point. Deformation of a typical panel is considered in *one direction at a time.* Thereafter, the contributions in each of the two directions, x and y, are added to obtain the total deflection at any point in the slab or plate.

First, the deflection due to bending in the x direction is computed [Figure 9.24(b)]. Then the deflection due to bending in the y direction is found. The midpanel deflection can now be obtained as the sum of the center-span deflections of the column strip in one direction and that of the middle strip in the orthogonal direction [Figure 9.24(c)].

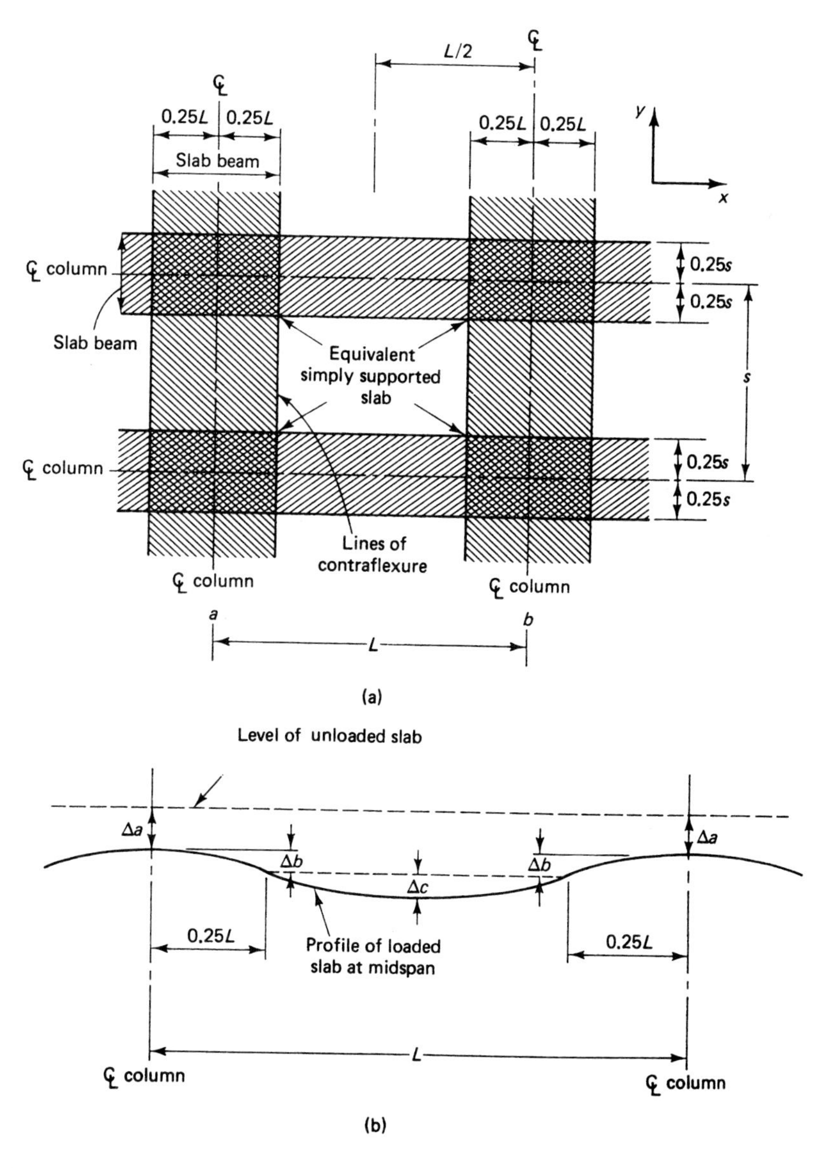

Figure 9.24 Equivalent frame method for deflection analysis. (a) Plate panel transferred into equivalent frames. (b) Profile of deflected shape at centerline. (c) Deflected shape of panel.

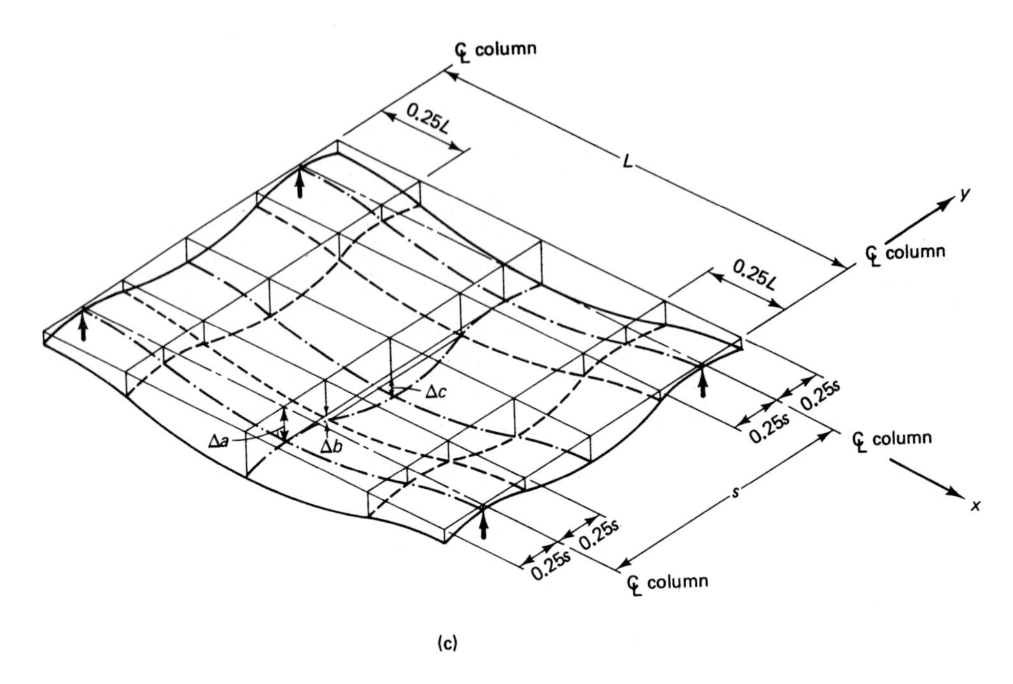

Figure 9.24 *Continued*

9.12.2 Column and Middle Strip Deflections

The deflection of each panel can be considered the sum of three components:

1. The basic midspan deflection of the panel, assumed fixed at both ends, given by

$$\delta' = \frac{wl^4}{384 E_c I_{\text{frame}}} \tag{9.34}$$

This has to be proportioned to separate deflections δ_c of the column strip and δ_s of the middle strip, such that

$$\delta_c = \delta' \frac{M_{\text{col strip}}}{M_{\text{frame}}} \frac{E_c I_{cs}}{E_c I_c} \tag{9.35a}$$

and

$$\delta_s = \delta' \frac{M_{\text{slap strip}}}{M_{\text{frame}}} \frac{E_c I_{cs}}{E_c I_s} \tag{9.35b}$$

where I_{cs} is the moment of inertia of the total frame, I_c the moment of inertia of the column strip, and I_s the moment of inertia of the middle slab strip.

2. The center deflection, $\delta''_{\theta L} = \frac{1}{8}\theta L$, due to rotation at the left end while the right end is considered fixed, where θL is the left M_{net}/K_{ec} and K_{ec} is the flexural stiffness of the equivalent column (moment per unit rotation).
3. The center deflection, $\delta''_{\theta R} = \frac{1}{8}\theta L$ due to rotation at the right end while the left end is considered fixed, where θL is the right M_{net}/K_{ec}. Hence,

$$\delta_{cx} \text{ or } \delta_{cy} = \delta_c + \delta''_{\theta L} + \delta''_{\theta R} \tag{9.36a}$$

$$\delta_{sx} \text{ or } \delta_{sy} = \delta_s + \delta''_{\theta L} + \delta''_{\theta R} \tag{9.36b}$$

In Equations 9.36a and 9.36b, use the values of δ_c, $\delta''_{\theta L}$, and $\delta''_{\theta R}$ which correspond to the applicable span directions. From Figures 9.24(b) and (c), the total deflection is

$$\Delta = \delta_{sx} + \delta_{cy} = \delta_{sy} + \delta_{cx} \tag{9.37}$$

9.13 DEFLECTION EVALUATION OF TWO-WAY PRESTRESSED CONCRETE FLOOR SLABS

Example 9.3

Compute the central deflection of the exterior panels of the two-way post-tensioned prestressed concrete floor designed in Example 9.2 for both short-term and long-term loading. Assume that the maximum allowable deflection is 1/480 of the span.

Solution:

Structural Data. From Example 9.2, we have the following data:

Plate thickness $h = 6\frac{1}{2}$ in. (165 mm)

Loads: $W_d = 101$ psf (4.84 kPa)

$W_L = 40$ psf (1.92 kPa)

Span AB $W_{\text{bal}} = 72$ psf (3.45 kPa)

$W_{\text{net}} = W_D + W_L - W_{\text{bal}} = 101 + 40 - 72 = 69$ psf (3.3 kPa)

Span BC $W_{\text{bal}} = 70$ psf (3.35 kPa)

$W_{\text{net}} = 141 - 70 = 71$ psf (3.4 kPa)

The floor plan is shown in Figure 9.25, and the overall details and vertical section of the building are presented in Figure 9.17. The distributed bending moments in the N–S direction taken from the flexural analysis for W_{net} in Table 9.2 are shown in Figure 9.26.

Stiffness Factors and Strip Moments *N–S Direction (Span 18 ft).* The column stiffness factor K_{ec} values were computed in Example 9.2, with the following results:

Exterior column A: $K_{ec} = 47E_c$ in.-lb/rad
Interior column B: $K_{ec} = 113E_c$ in.-lb/rad
Net Frame $M_A = 5.30 \times 10^3$ in.-lb/ft
Net Frame $M_B = (39.56 - 33.94)10^3 = 5.62 \times 10^3$ in.-lb/ft

As discussed in Example 9.2, the column strip takes 64 percent of the moment and the middle strip takes 36 percent of the moment. The frame total $I_{cs} = bh^3/12 = 20 \times 12(6.5)^3/12 = 5{,}493$ in.^{4}, while the column strip I_c = the middle strip $I_c = 5{,}493/2 = 2{,}747$ in.^{4}

From Equation 9.34, the basic midspan deflection in the N–S direction at central point O in Figure 9.27, assuming both ends of the panel fixed, is

$$\delta' = \frac{WL^4}{348E_cI_{cs}} = \frac{69 \times 20(18)^4(12)^3}{384 \times 4.03 \times 10^6 \times 5{,}493} = 0.029 \text{ in.}$$

This deflection has to be proportioned to separate deflections δ_c of the column strip and δ_s of the middle strip:

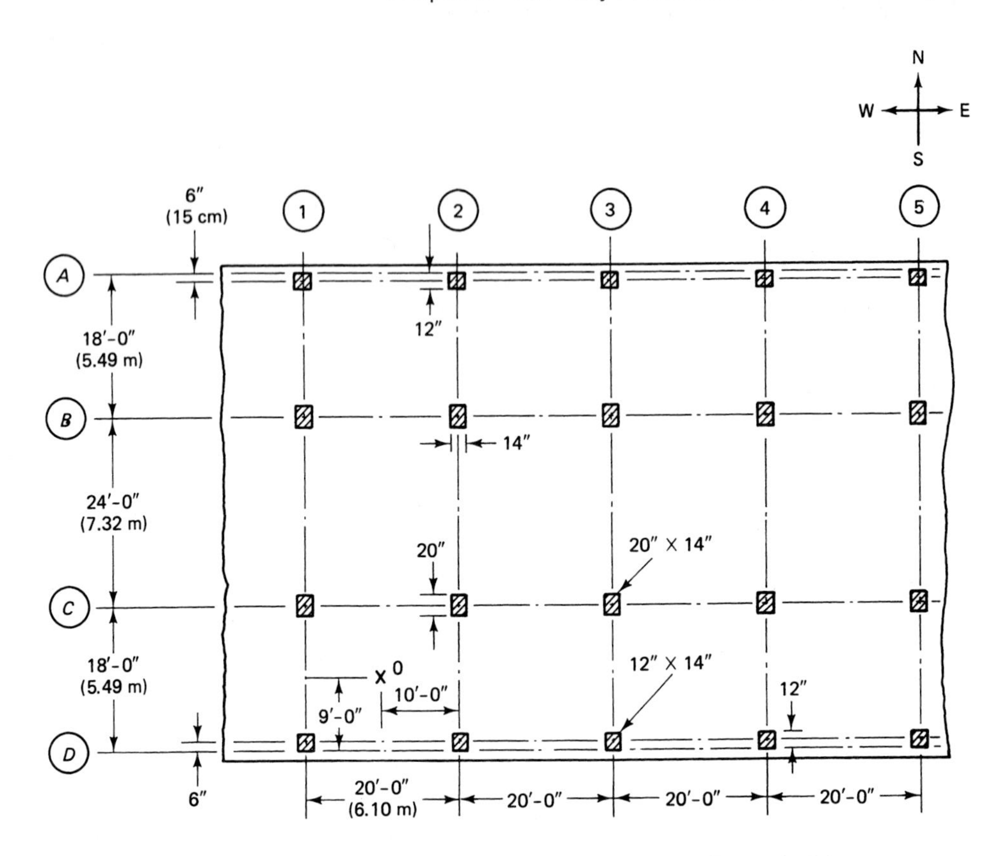

Figure 9.25 Two-way post-tensioned floor plan in Example 9.3

$$\delta_c = \delta' \frac{M_{\text{col. strip}}}{M_{\text{frame}}} \frac{E_c I_{cs}}{E_c I_c}$$

from Example 9.2, $M_{\text{col. strip}}/M_{\text{frame}} = 0.64$, so N–S $\delta_c = 0.029 \times 0.64 \times 2 = 0.037$ in., N–S $\delta_s = 0.029 \times 0.36 \times 2 = 0.021$ in., and the rotation at end A is

$$\theta_A = \frac{M_A}{K_{ec}} = \frac{5.30 \times 10^3 \times 20}{47 \times 4.03 \times 10^6} = 5.6 \times 10^{-4} \text{ rad}$$

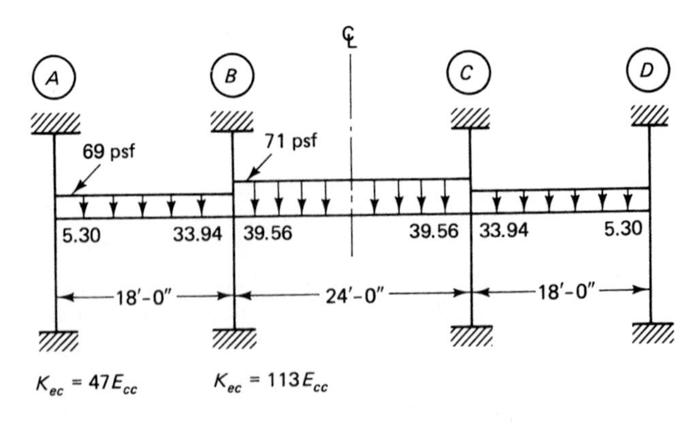

Figure 9.26 Service net load moments × 10^{-3} in.-lb in the N–S direction in Example 9.3.

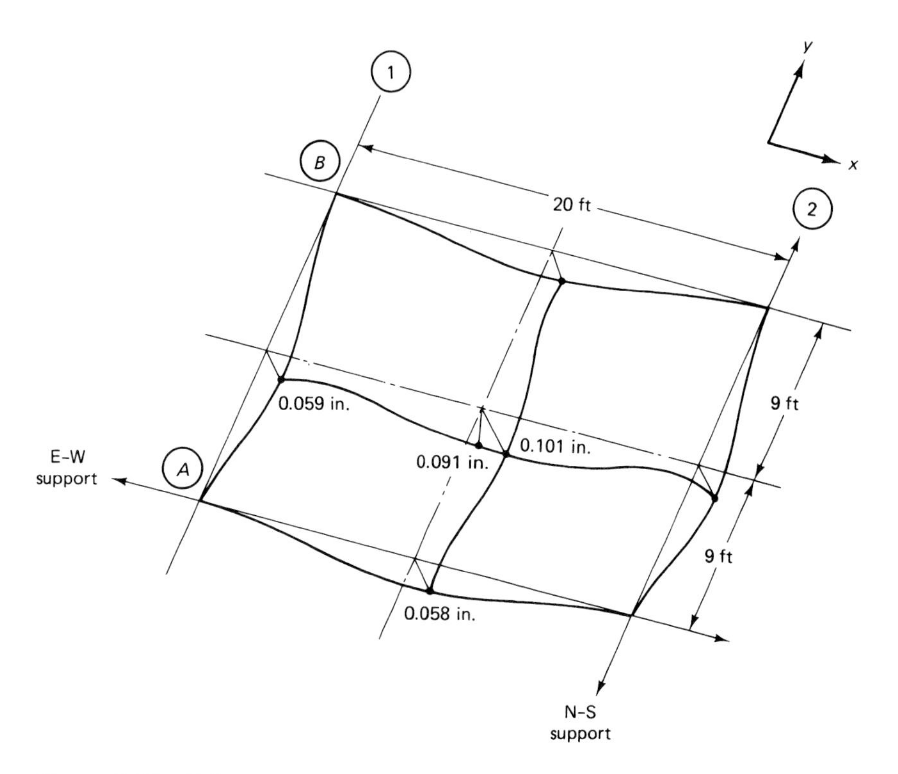

Figure 9.27 Column and middle strips immediate deflections in Example 9.3.

$$\theta_B = \frac{M_B}{K_{ec}} = \frac{5.62 \times 10^3 \times 20}{113 \times 4.03 \times 10^6} = 2.5 \times 10^{-4} \text{ rad}$$

$$\delta'' = \frac{\theta l}{8} = \frac{(5.6 + 2.5)10^{-4}(18 \times 12)}{8} = 0.022 \text{ in.}$$

Therefore, N–S net $\delta_{cy} = 0.037 + 0.022 = 0.059$ in. and N–S net $\delta_{sy} = 0.021 + 0.022 = 0.043$ in.

E–W Direction (Span 20 ft). For the E–W direction, the width b of an equivalent frame $= \frac{1}{2}(18 + 24) = 21.0$ ft. The frame total I_{cs} is

$$\frac{bh^3}{12} = \frac{21 \times 12(6.5)^3}{12} = 5{,}767 \text{ in}^4$$

and the column strip I_c = the middle strip $I_s = 5{,}767/2 = 2{,}884$ in.^{4}

From Equation 9.34, the fixed-end central deflection at O is

$$\delta' = \frac{WL^4}{384E_c I_{cs}} = \frac{69 \times 21(20)^4(12)^3}{384 \times 4.03 \times 10^6 \times 5{,}767} = 0.045 \text{ in.}$$

If the same distribution of moments is assumed to exist between the column and middle strips, then E–W $\delta_c = 0.045 \times 0.64 \times 2 = 0.058$ in. and E–W $\delta_s = 0.045 \times 0.36 \times 2 = 0.032$ in.

For the case of all panels loaded in this example, the net moments at each column due to the difference in negative moments from the spans to the left and to the right of the column are zero. Hence, consider the net rotation $\theta = 0$, and use E–W net $\delta_{cx} = 0.058$ in. and E–W net $\delta_{sx} = 0.032$ in.

Figure 9.27 gives the column and middle strip deflections in both the N–S and E–W directions.

Photo 9.7 Testing setup of four-panel prestressed concrete floor. (Tests by Nawy et al.)

Total Immediate Central Deflection. The total central deflection $\Delta = \delta_{xs} + \delta_{cy} = \delta_{sy} + \delta_{cx}$; therefore $\Delta_{N\text{-}S} = \delta_{sy} + \delta_{cx} = 0.043 + 0.058 = 0.101$ in. and $\Delta_{E\text{-}W} = \delta_{sx} + \delta_{cy} = 0.032 + 0.059 = 0.091$ in. Hence, the average immediate deflection due to the net load is $W_{net} = \frac{1}{2}(\Delta_{N\text{-}S} + \Delta_{E\text{-}W}) = \frac{1}{2}(0.101 + 0.091) = 0.096$ in. (2.44 mm).

Long-Term Deflection. For the long-term deflection, $W_{net} = 69$ psf and the live load $W_L = 40$ psf. Assuming that 65 percent of the live load is sustained, the total sustained load intensity is $W_{sust.} = (69 - 40) + 0.65 \times 40 = 55$ psf. Assuming further a total creep factor of 2, we have

$$\text{Long-term deflection} = \frac{55}{69} \times 0.096 \times 2 = 0.153 \text{ in. (4.09 mm)}$$

$$\text{Total deflection} = 0.096 + 0.153 = 0.249 \text{ in. (6.33 mm)}$$

The maximum allowable deflection in this structure is

$$\Delta_{allow.} = \frac{L}{480} = \frac{20 \times 12}{480} = 0.50 \text{ in. (12.7 mm)} > \text{actual } \Delta = 0.249 \text{ in., O.K.}$$

9.14 YIELD-LINE THEORY FOR TWO-WAY-ACTION PLATES

A study of the hinge-field mechanism in a slab or plate at loads close to failure aids the engineering student in developing a feel for the two-way-action behavior of plates. Hinge fields are successions of hinge bands which are idealized by lines; hence the name *yield-line theory* by K. W. Johansen.

To do justice to this subject, an extensive discussion over several chapters or a whole textbook is necessary. The intention of this chapter is only to introduce the reader to the fundamentals of the yield-line theory and its application.

The yield-line theory is an upper-bound solution to the plate problem. This means that the predicted moment capacity of the slab has the highest expected value in comparison with test results. Additionally, the theory assumes a totally rigid-plastic behavior, namely, that the plate stays plane at collapse. Consequently, deflection is not accounted for, nor are the compressive membrane forces that will act in the plane of the slab or plate considered. The plates are assumed to be considerably underreinforced, in such a manner that the maximum reinforcement percentage ρ does not exceed $\frac{1}{2}$ percent of the section *bd.*

Since the solutions are upper bound, the slab thickness obtained by this process is in many instances thinner than what is obtained by the lower bound solutions, such as the equivalent frame method. Consequently, it is important to apply rigorously the serviceability requirements for deflection control and for crack control in conjunction with the use of the yield-line theory.

One distinct advantage of this theory is that solutions are possible for any shape of plate, whereas most other approaches are applicable only to the rectangular shapes with rigorous computations for boundary effects. The engineer can, with ease, find the moment capacity for a triangular, trapezoidal, rectangular, circular, or any other conceivable shape, provided that the failure mechanism is known or predictable. Since most failure patterns are presently identifiable, solutions can be readily obtained.

9.14.1 Fundamental Concepts of Hinge-Field Failure Mechanisms in Flexure

Under action of a two-dimensional system of bending moments, yielding of a rigid-plastic plate occurs when the principal moments satisfy Johansen's square yield criterion as shown in Figure 9.28. According to this criterion, yielding is considered to have occurred when the numerically greater of the principal moments reaches the value of $\pm M$ at the yield-line cracks. The directions of the principal curvature rates are considered to coincide with the curvatures of the principal moments. The idealized moment–curvature relationship is shown as the solid line in Figure 9.29. Line *OA* is considered almost vertical at point *O,* and strain hardening is neglected.

If one considers the simplest case of a square slab with supports, with degree of fixity i varying from $i = 0$ for simply supported to $i = 1.0$ for fully restrained on all four sides, the failure mechanism would be as shown in Figure 9.30 when a uniformly distributed load is applied.

Consider the simply supported case (a). The yield-line moments along the yield lines are the principal moments. Hence, the twisting moments are zero in the yield lines and in most cases the shearing forces are also zero. Consequently, only the moment M per unit length of the yield line acts about the lines *AD* and *BE* in Figure 9.31. The total moments can be represented by a vector in the direction of the yield line whose value is

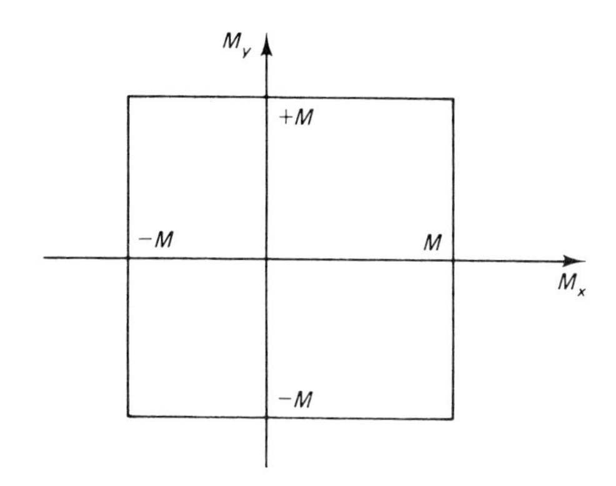

Figure 9.28 Johansen's square yield criterion.

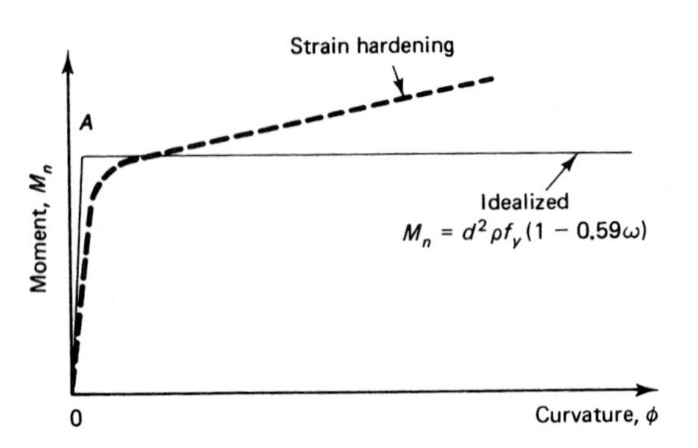

Figure 9.29 Moment-curvature relationship.

the product of M and the length of the yield line, that is, $M(a/2 \cos \theta)$ in Figure 9.31(c). The virtual work of the yield moments of the shaded triangular segment ABO is the scalar product of the two moment vectors $Ma/2 \cos \theta$ on fracture lines AO and BO and rotation θ. In other words, the internal work is

$$E_1 = \Sigma \overline{M}\,\overline{\theta}$$

If the displacement of the shaded segment at its center of gravity c is δ, the external work is

$$E_E = \text{force} \times \text{displacement} = \Sigma \int\int w_u dx\, dy\, \delta$$

where w_u is the intensity of external load per unit area. But $E_I = E_E$; hence,

$$\Sigma \overline{M}\,\overline{\theta} = \Sigma \int\int w_u dx\, dy\, \delta \tag{9.38}$$

Applying Equation 9.38 to the case under discussion gives us

$$\overline{M}\,\overline{\theta} = Ma \frac{\Delta}{a/2}$$

since angle θ in Figure 9.31(b) is small $[(\theta = \Delta/(a/2)]$.

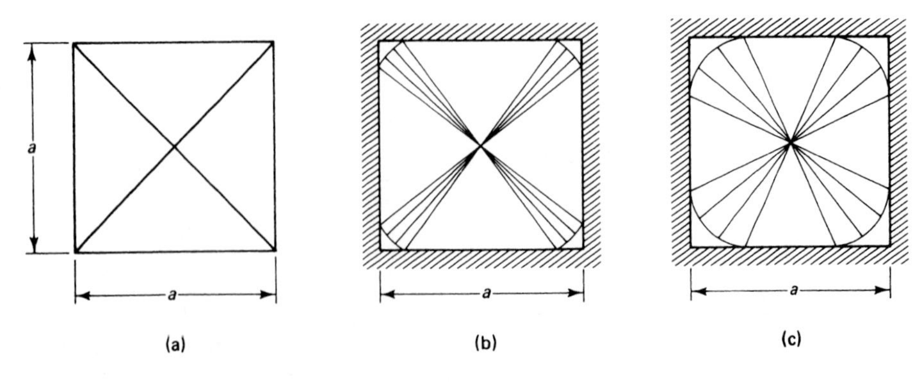

Figure 9.30 Failure mechanism of a square slab. (a) $i = 0$. (b) $i = 0.5$. (c) $i = 1.0$.

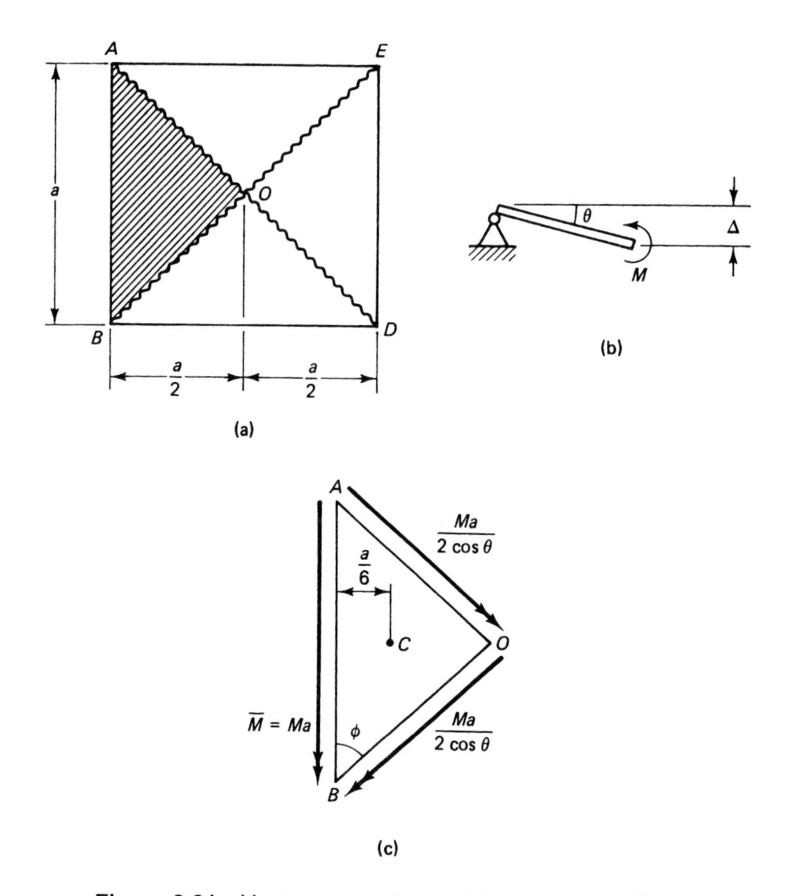

Figure 9.31 Vector moments on slab segment at failure.

The work per triangular segment is

$$E_I = \overline{M}\,\overline{\theta} = 2M\,\Delta$$

$$E_E = \frac{w_u a^2}{4} \times \frac{\Delta}{3}$$

where the deflection at the center of gravity of the triangle is $\Delta/3$. Therefore,

$$4(2M\,\Delta) = 4\left(\frac{w_u\,a^2}{12}\,\Delta\right)$$

and

$$\text{unit } M = \frac{w_u\,a^2}{24} \tag{9.39}$$

If the square slab is fully fixed on all four sides, $E_I = 4(4M\,\Delta)$ since fracture lines develop around not only the diagonals but also the four edges, as shown in Figure 9.30(c). Hence, for a fully fixed square slab,

$$\text{unit } M = \frac{w_u\,a^2}{48} \tag{9.40}$$

Observe that a lower-bound solution as proposed by Mansfield's failure pattern in Figure 9.30(c) gives a value $M = w_u a^2/42.88$. Hence, for a uniformly loaded square slab with load intensity w_u per unit area and degree of support fixity i on all sides,

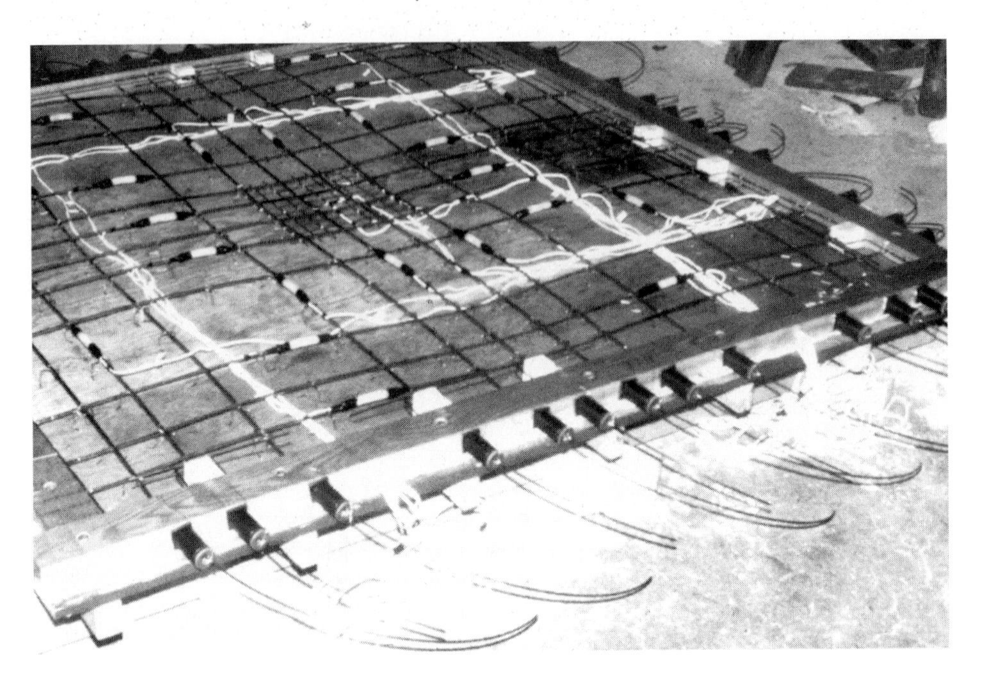

Photo 9.8 Preparing prestressing tendons in the forms for a four-panel continuous prestressed two-way-action plate (Nawy, Chakrabarti et al.).

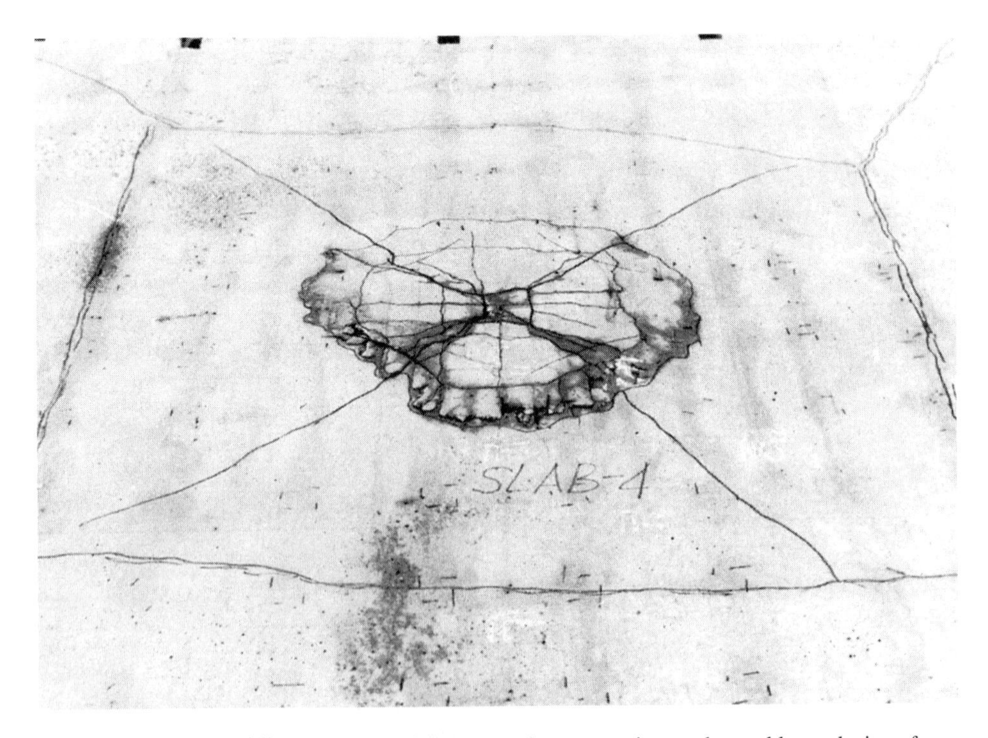

Photo 9.9 Yield-line pattern at failure at column reaction and panel boundaries of a two-way multipanel floor. (Tests by Nawy, Chakrabarti et al.)

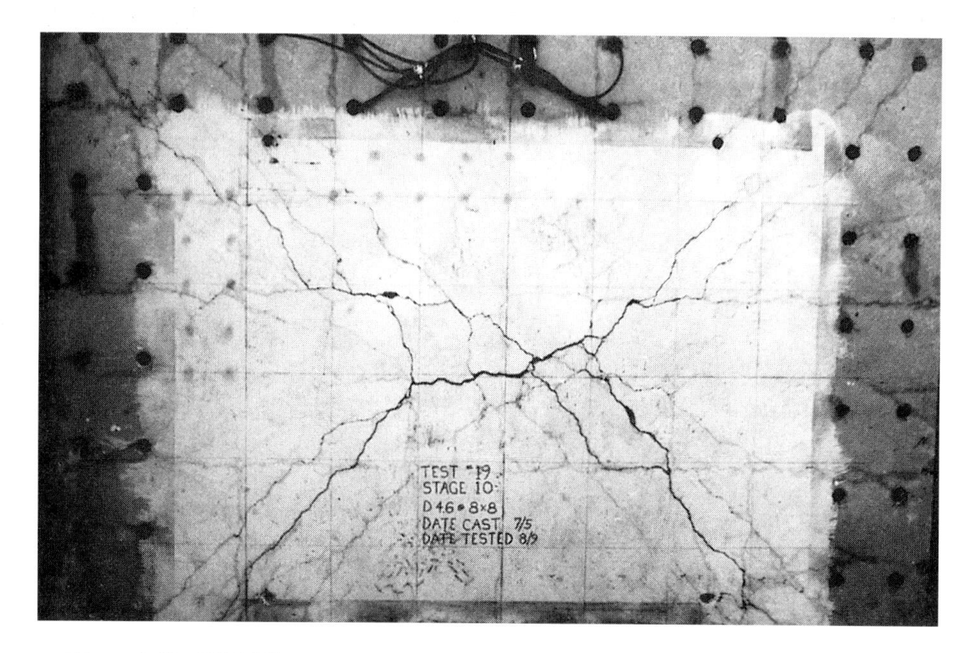

Photo 9.10 Yield-line patterns at failure at tensions face of rectangular restrained panel. (Tests by Nawy et al.)

$$w_u a^2 = M[24(1 + i)] \tag{9.41}$$

The general equation for the yield-line moment capacity of a rectangular isotropic slab on beams and having dimensions $a \times b$ as shown in Figure 9.32, with side a being the shorter dimension, is

$$\text{unit } M \frac{\text{ft-lb}}{\text{ft}} = \frac{w_u a_r^2}{24}\left[\sqrt{3 + \left(\frac{a_r}{b_r}\right)^2} - \frac{a_r}{b_r}\right]^2 \tag{9.42}$$

where $a_r = \dfrac{2a}{\sqrt{1 + i_2} + \sqrt{1 + i_4}}$

$b_r = \dfrac{2b}{\sqrt{1 + i_1} + \sqrt{1 + i_3}}$

i = degree of restraint, depending on stiffness ratios as discussed in Section 9.2.

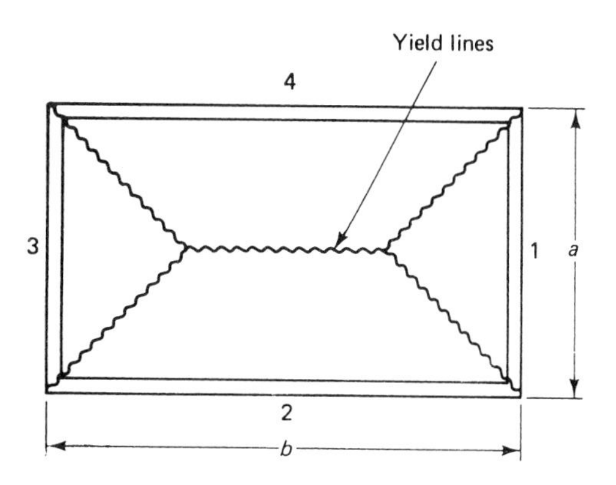

Figure 9.32 Rectangular slab. Note sequence of side numbers.

Note that Equation 9.42 reduces to the simplified form of Equation 9.40 or 9.41 for the case of a square slab restrained on all four sides ($i = 1.0$).

Affine Slabs. Slabs that are reinforced differently in the two perpendicular directions are called *orthotropic slabs* (or *plates*). The moment in the x direction equals M and in the y direction equals μM, where μ is a measure of the degree of orthotropy, or the ratio

$$\frac{M_y}{M_x} = \frac{(A_s)_y}{(A_s)_x}$$

To simplify the analysis, the slab should be converted to an *affine* (isotropic) slab, where the strength and reinforcement area in both the x and y directions are the same. Such conversion can be made as follows:

1. *Divide* the linear dimension in the M direction by $\sqrt{\mu}$ for a slab to be reinforced for a moment M in both directions using the same unit load intensity w_u per unit area.
2. In the case of concentrated loads or total loads, also divide such loads by $\sqrt{\mu}$.
3. In the case of line loads, the line load has to be divided by $\sqrt{\mu \cos^2 \theta + \sin^2 \theta}$, where θ is the angle between the line load and the M direction.

If the slab is to be analyzed as an affine slab with the moment μM in both directions, the dimension in the μM direction has to be *multiplied* by $\sqrt{\mu}$. In either case, the result is of course the same.

9.14.2 Failure Mechanisms and Moment Capacities of Slabs of Various Shapes Subjected to Distributed or Concentrated Loads

The preceding concise introduction to the virtual-work method of yield-line moment evaluation should facilitate a good understanding of the mathematical procedures of most standard rectangular shapes subjected to uniform loading. More complicated slab shapes and other types of symmetrical and nonsymmetrical loading require more advanced knowledge of the subject. Also, the assumed failure shape and minimization energy principles can give values for particular cases that differ slightly from one experimenter to another depending on the mathematical assumptions made with respect to the failure shape.

The following summary of failure patterns and the respective moment capacities in terms of load, many of them due to Mansfield (Ref. 9.14), should give the reader adequate coverage of solutions to most cases expected in today's and tomorrow's structures.

1. Point load to corner of rectangular cantilever plates:

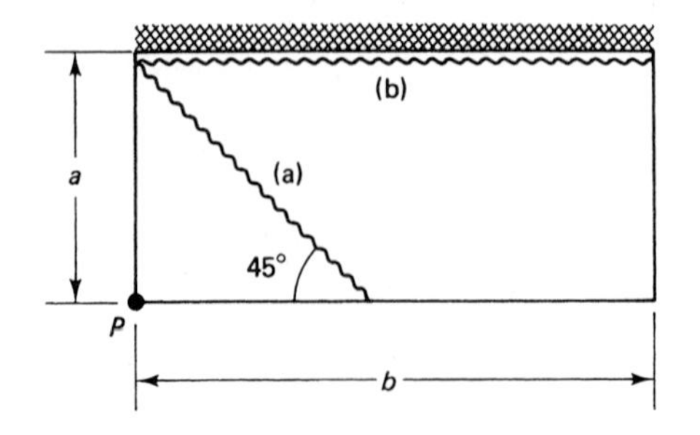

Case (a) $P = 2M$

Case (b) $P = \frac{b}{a}(M)$

2. Square plate centrally loaded and having boundaries simply supported against both downward and upward movements:

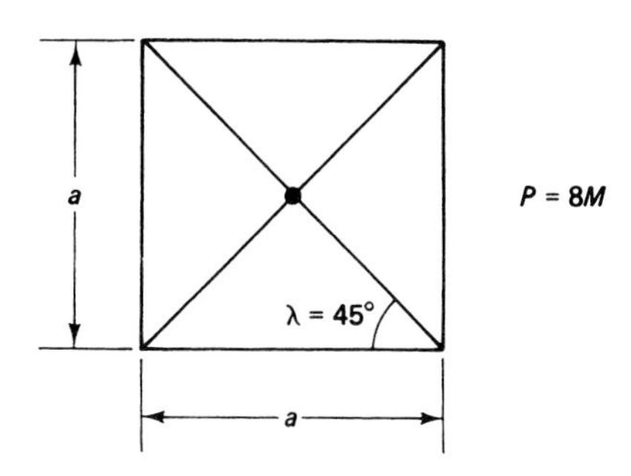

3. Regular n-sided plate with simply supported edges and centrally loaded ($n > 4$):

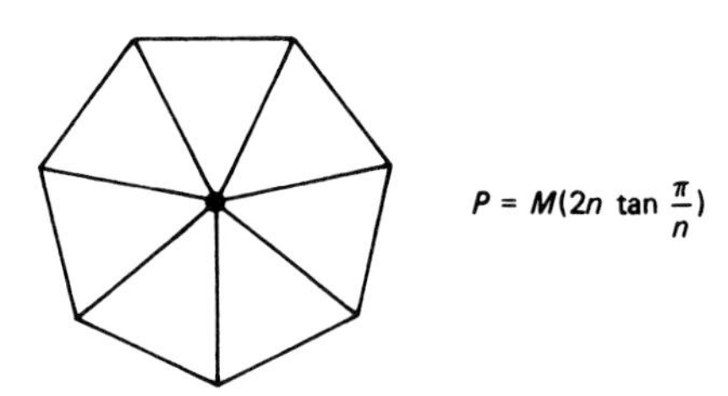

4. Square plates centrally loaded and having boundaries simply supported against downward movement, but free for upward movement:

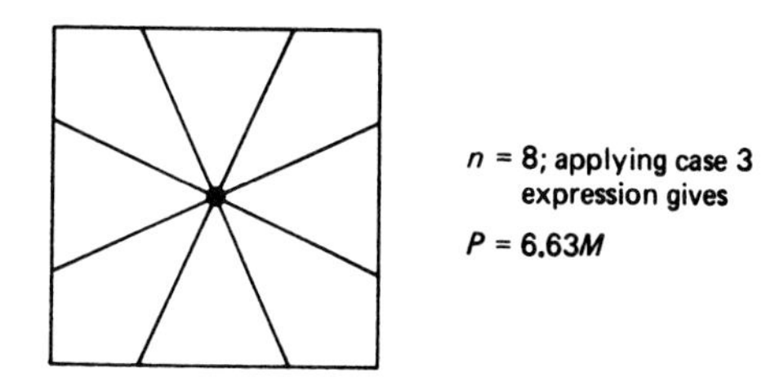

5. Circular centrally loaded plate simply supported along the edges:

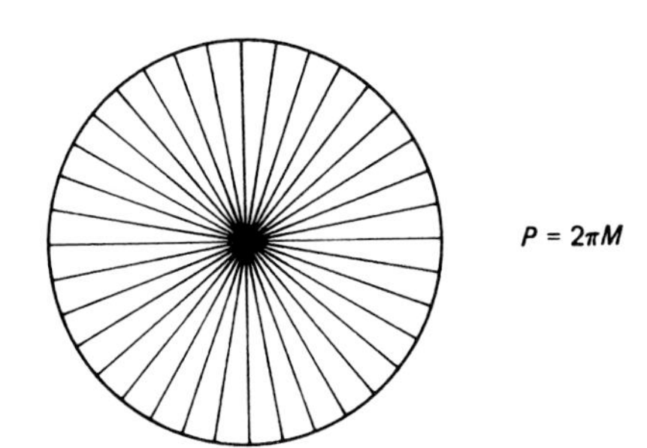

6. Circular plate with fully restricted edges and centrally loaded by point load P:

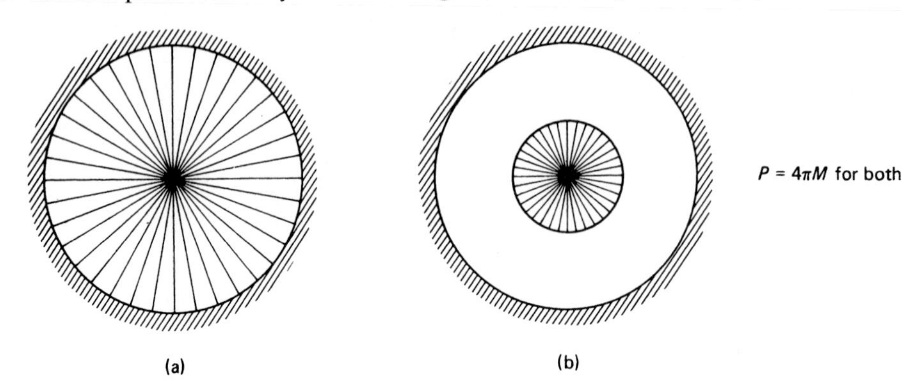

7. Point load P applied anywhere in arbitrarily shaped plate fully restrained on all boundaries:

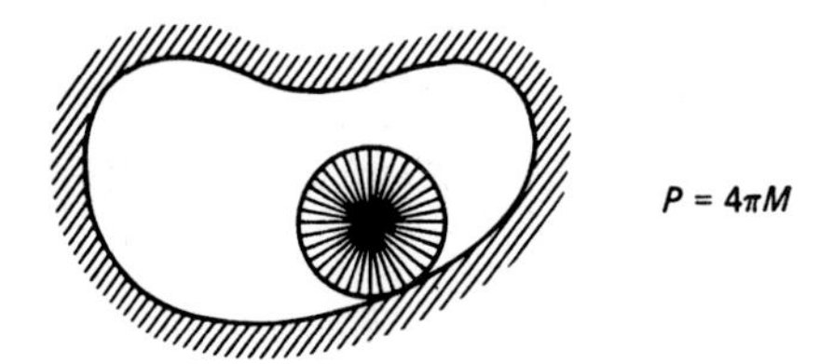

8. Equilateral triangular plate with simply supported edges and centrally loaded by point load P:

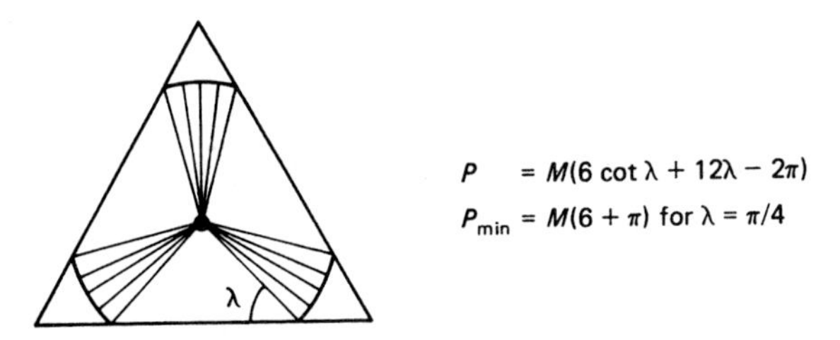

9. Acute-angled triangular plate on simply supported edges loaded with point load P at the center of the inscribed circle:

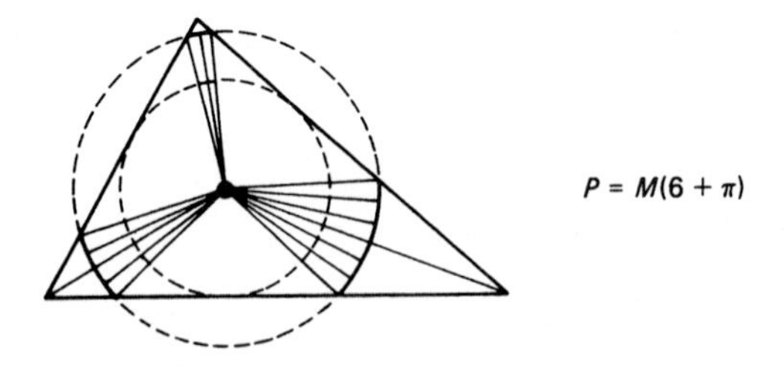

10. Obtuse-angled triangular plate with simply supported edges and load P at the center of the inscribed circle:

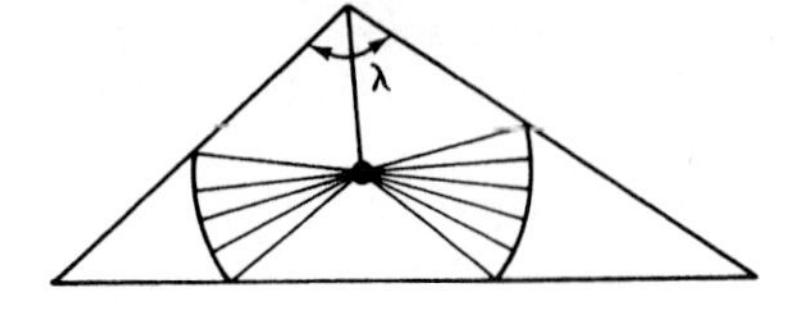

$P = M(4 + 2\lambda + 2 \cot 1/2\lambda)$, where ϕ is in radians

As λ approaches π, the plate degenerates into case 11

11. Long strip simply supported along the edges and loaded with point P midway between the edges:

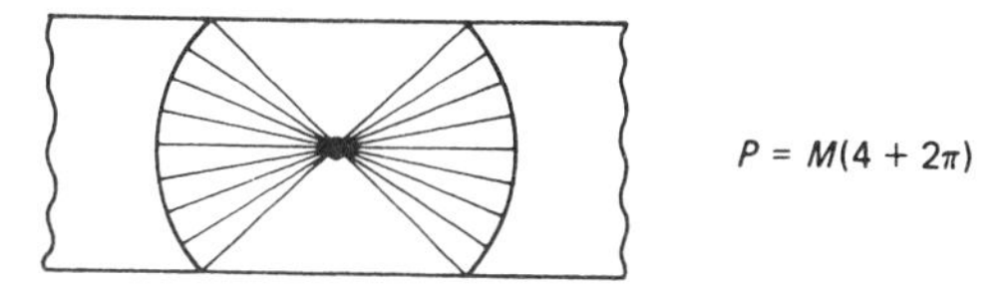

12. Simply supported strip with equal loads P between the edges:

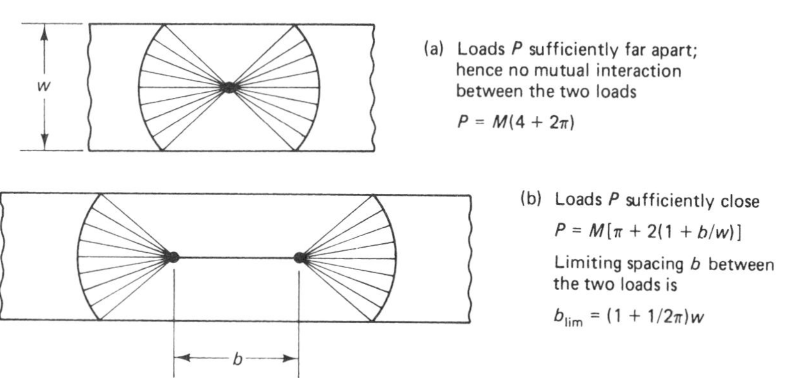

13. Simply supported strip with unequal loads P and kP midway between the edges, where $k < 1.0$ and the loads are sufficiently apart:

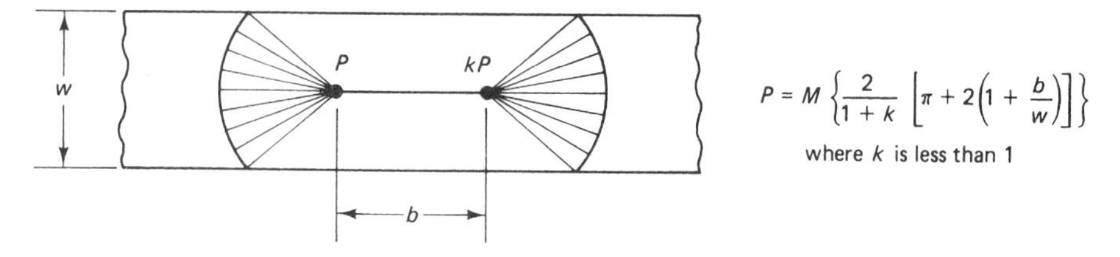

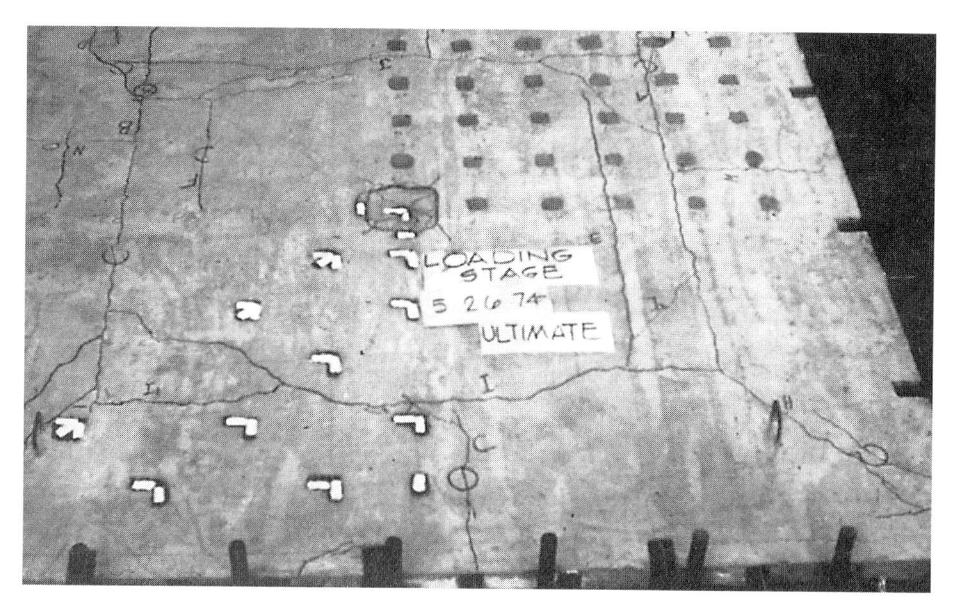

Photo 9.11 Four-panel slab at failure showing the yield-line patterns at the negative compression face of the supports. (Tests by Nawy and Chakrabarti.)

14. Uniformly loaded square slab with degree of fixity i varying between zero and 1.0:

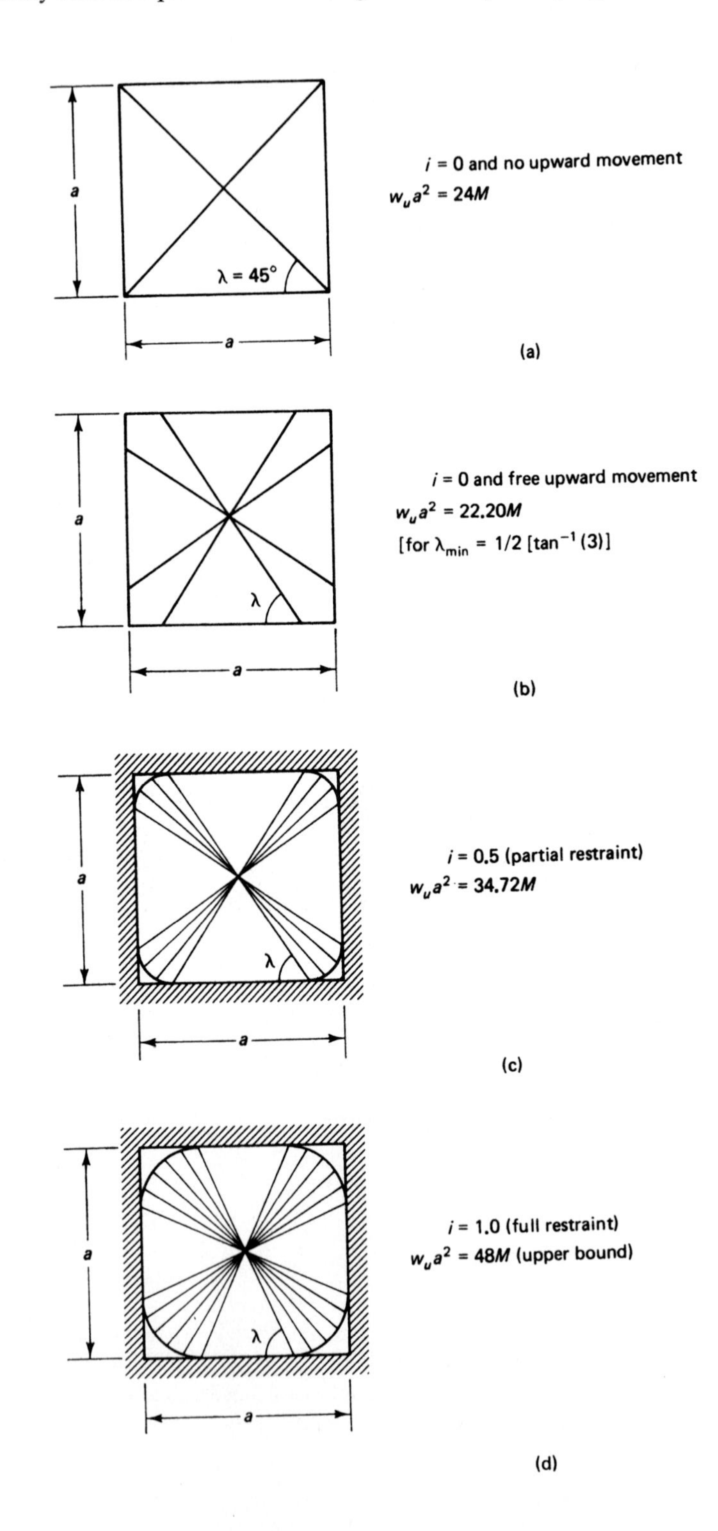

15. Equilateral triangular plate ($\lambda = 60°$) uniformly loaded:

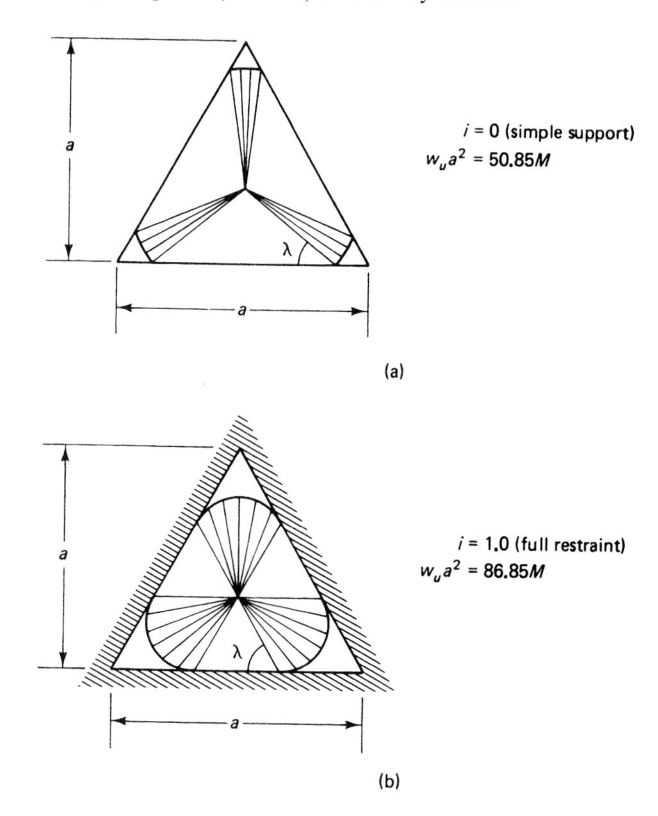

16. Rectangular slab uniformly loaded with unit load of intensity w_u supported on all four sides with degree of restraint i varying from zero to 1.0 (note the sequence of numbers assigned to the panel sides):

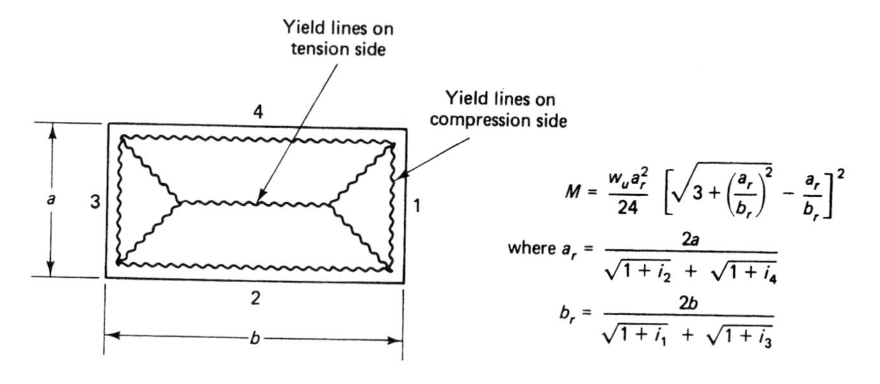

As a general note, in the foregoing equations relating the load P to the moment M, load P is assumed to act at a point. To adjust for the fact that P acts on a finite area, assume that it acts over a circular area of radius ρ. For a slab fully restrained on all boundaries, the hinge field would be bound by a circle touching the slab boundary (circle radius $= r$). In such a case,

$$M + M' = \frac{P}{2\pi}\left(1 - \frac{2\rho}{3r}\right) \tag{9.43}$$

where M is the positive unit moment and M' the negative unit moment.

The reaction of columns supporting flat plates can be similarly considered for analyzing the flexural local capacity of the plate in the column area. For rectangular supports, an approximation to an equivalent circular support can be made in the use of Equation 9.43.

9.15 YIELD-LINE MOMENT STRENGTH OF A TWO-WAY PRESTRESSED CONCRETE PLATE

Example 9.4

Find the nominal moment strength of the two-way prestressed concrete plate in Example 9.2, assuming that the prestressing strands are bonded.

Solution:

Loads. The total load intensity at the limit state of failure, from Example 9.2, is $W_u = 1.4W_D + 1.7W_L = 210$ psf. Assuming that the column reaction is an inverted concentrated load in the continuous plate field, the required moment strength M_n can be defined from case 7 of subsection 9.14.2 as follows:

$$P_A = 4\pi M_n$$

$$\text{Factored } P_u = 210 \times 20\left(\frac{24 + 18}{2}\right) = 88{,}200 \text{ lb (393 kN)}$$

(the column weight is negligible)

$$\text{Req } P_n = \frac{P_u}{\phi} = \frac{88{,}200}{0.9} = 98{,}000 \text{ lb (436 kN)}$$

$$\text{Req Unit } M_n \text{ for point load} = \frac{P_n}{4\pi} = \frac{98{,}000}{4 \times 3.14} = 7{,}799 \text{ lb (34.7 kN)}$$

or 7,779 in.-lb/in. (34.7 kN-m/m)

$$\text{Equivalent } \rho = \frac{20 \times 14}{\pi^2} = 28.4 \text{ in.}$$

Assume $r \cong 17.5$ ft = 210 in. and $M' = M$. Then

$$M'_n = \frac{P_n}{4\pi}\left(1 - \frac{2\rho}{3r}\right) = 7{,}799\left(1 - \frac{2 \times 28.4}{3 \times 210}\right) = 7{,}096 \text{ lb (31.6 kN)}$$

Available Moment Strength M_n. The available column area slab reinforcement is determined as follows.

Prestressing Steel

$$A_{ps} = \text{three } \tfrac{1}{2}\text{-in. dia 7-wire 270 K strands} = 3 \times 0.153 = 0.459 \text{ in.}^2$$

$$f_{py} = 243{,}000 \text{ psi (1,675 MPa)}$$

$$f_{ps} = 179{,}256 \text{ psi at interior column}$$

$$f'_c = 4{,}000 \text{ psi (27.58 MPa)}$$

Accordingly, use f_{py} at the limit state of failure.

Nonprestressed Steel

$$A_s = \text{six \#4 bars} = 6 \times 0.2 = 1.20 \text{ in.}^2$$

$$f_y = 60{,}000 \text{ psi}$$

Moment Strength M_n

$$d_p = d = 6.5 - 1 = 5.5 \text{ in.}$$

$$b = 33.5 \text{ in. (from Example 9.2)}$$

$$a = \frac{A_s f_y + A_{ps} f_{py}}{0.85 f'_c b} = \frac{1.2 \times 60{,}000 + 0.459 \times 243{,}000}{0.85 \times 4{,}000 \times 33.5} = 1.61 \text{ in. (40.9 mm)}$$

$$\text{Available } M_n = A_s f_y \left(d - \frac{a}{2}\right) + A_{ps} f_{py}\left(d_p - \frac{a}{2}\right)$$

$$= 1.2 \times 60{,}000\left(5.5 - \frac{1.61}{2}\right) + 0.459 \times 243{,}000\left(5.5 - \frac{1.61}{2}\right)$$

$$= 338{,}040 + 523{,}666 = 861{,}706 \text{ in.-lb}$$

$$\text{Unit } M_n = \frac{861{,}706 \text{ in.-lb}}{33.5 \text{ in.}} = 25{,}723 \text{ in.-lb/in.} = 25{,}723 \text{ lb}$$

Check M_n For the Entire Panel Width

$$\text{N–S slab band width} = \frac{18 + 24}{2} = 21 \text{ ft (6.4 m)}$$

$$\text{E–W slab band width} = 20 \text{ ft}$$

So use $b = 21$ ft $= 252$ in. Then the total A_{ps} = eleven $\frac{1}{2}$-in. (12.7 mm) dia 7-wire strands. Since the top prestressed steel is only in the column zone, disregarding it would be on the safe side. We then have

$$\text{Unit } A_{ps} = \frac{11 \times 0.153}{25.2} = 0.0067 \text{ in.}^2\text{/in.}$$

$$a = \frac{0.0067 \times 243{,}000}{0.86 \times 4{,}000 \times 1} = 0.48 \text{ in.}$$

$$\text{Unit } M_n = 0.0067 \times 243{,}000\left(5.5 - \frac{0.48}{2}\right) = 8{,}564 \text{ in.-lb/in.}$$

$$= 8{,}564 \text{ lb (38.09 kN), use}$$

$$\text{Req } M_n = 7{,}096 \text{ lb} < \text{available } M_n = 8{,}564 \text{ lb, O.K.}$$

Plainly, from this *limit theory* solution, quick analysis of a prestressed plate can be performed. Such an analysis, however, should also include an evaluation of yield-line shear strength at the support (Ref. 9.15) and serviceability checks for crack control and deflection control. The designer can easily choose the moment values for the applicable failure mechanism as presented in subsection 9.14.2. A serviceability check for crack control can be easily made using the criteria based on the extensive research reported in Refs. 9.19 to 9.21 and the discussion in Section 11.9 in this text on crack control in walls of large prestressed concrete tanks.

REFERENCES

9.1 Post-Tensioning Institute. *Post-Tensioning Manual.* 6th ed. Phoenix: Post-Tensioning Institute, 2000.

9.2 ACI Committee 318. *Building Code Requirements for Structural Concrete* (ACI 318–99) and *Commentary* (ACI 318R–99), American Concrete Institute, Farmington Hills, MI , 2000, pp. 392.

9.3 Nawy, E. G. *Reinforced Concrete—A Fundamental Approach.* 4th ed. Prentice Hall, Upper Saddle River, NJ. pp. 780.

9.4 Nawy, E. G., and Chakrabarti, P. "Deflection of Prestressed Concrete Flat Plates." *Journal of the Prestressed Concrete Institute* 21 (1976): 86–102.

9.5 Lin, T. Y. "Load-Balancing Method for Design and Analysis of Prestressed Concrete Structures." *Journal of the American Concrete Institute* 60 (1963): Farmington Hills, MI 719–742.

9.6 Burns, N. H., and Hemabom, R. "Test of Scale Model Post-Tensioned Flat Plate." *Journal of the Structural Division, American Society of Civil Engineers* 103 (1977): 1237–1255.

9.7 Nilson, A. H. *Design of Prestressed Concrete.* John Wiley & Sons, New York, 1987.

9.8 Scordelis, A. C., Lin, T. Y., and Itaya, R. "Behavior of a Continuous Slab Prestressed in Two Directions. *Journal of the American Concrete Institute* 56, Farmington Hills, MI (1959): 441–459.

9.9 Cross, H., and Morgan, N. *Continuous Frames of Reinforced Concrete.* John Wiley & Sons, New York, 1954.

9.10 Rice, P. F., Hoffman, E. S., Gustafson, D. P., and Gouwens, A. J. *Structural Design Guide to the ACI Building Code.* 3d ed. Van Nostrand Reinhold, New York, 1985.

9.11 Nawy, E. G. "Strength, Serviceability, and Ductility." In *Handbook of Structural Concrete,* pp. 12-1 to 12-88. McGraw Hill, New York, 1983.

9.12 Lin, T. Y., and Burns, N. H. *Design of Prestressed Concrete Structures.* 3d ed. John Wiley & Sons, New York, 1981.

9.13 Reynolds, C. E., and Steedman, J. C. *Reinforced Concrete Designer's Handbook.* 9th ed. Viewpoint Publications, London, 1981.

9.14 Mansfield, E. H. "Studies in Collapse Analysis of Rigid-Plastic Plates with a Square Yield Diagram." *Proceedings of the Royal Society* 241 (1957): 225–261.

9.15 Gesund, H., and Dikshit, O. P. "Yield Line Analysis of Punching Problem at Slab Column Intersections," *International Symposium on Cracking, Deflection, and Ultimate Load of Concrete Slab Systems,* pp. 177–203. American Concrete Institute, Farmington Hills, MI, 1971.

9.16 Hung, T. Y., and Nawy E. G. "Limit Strength and Serviceability Factors in Uniformly Loaded, Isotropically Reinforced Two-Way Slabs," *International Symposium on Cracking, Deflection, and Ultimate Load of Concrete Slab Systems,* pp. 1–41. American Concrete Institute, Farmington Hills, MI, 1971.

9.17 Nilson, A. H., and Walters, D. B. "Deflection of Two-Way Floor Systems by the Equivalent Frame Method." *Journal of the American Concrete Institute* 72, Farmington Hills, MI, (1975): 210–218.

9.18 Nawy, E. G., and Chakrabarti, P. "Serviceability Deflection Behavior of Two-Way Action Prestressed Concrete Plates." In *International Conference on Prestressed Concrete,* Concrete Institute of Australia, Sydney, Australia, 1976, pp. 1–10.

9.19 Nawy, E. G. "Crack Control Through Reinforcement Distribution in Two-Way Acting Slabs and Plates." *Journal of the American Concrete Institute* 69, Farmington Hills, MI, (1972): 217–219.

9.20 Nawy, E. G., and Blair, K. *Further Studies of Flexural Crack Control in Structural Slab Systems.* American Concrete Institute, Farmington Hills, MI, 1971.

9.21 Vessey, J. V., and Preston, R. L. *A Critical Review of Code Requirements for Prestressed Concrete Reservoirs.* Paris: F. I. P., 1978.

9.22 Cohn, M. Z. "Partial Prestressing, From Theory to Practice." In *NATO—ASI Applied Science Series,* Vol. 1, p. 405, and Vol. 2, p. 425. Martinus Nijhoff, in Cooperation with NATO Scientific Affairs Division, Dordrecht, Netherlands, 1986.

9.23 Nawy, E. G., editor-in-chief, *Concrete Construction Engineering Handbook,* CRC Press, Boca Raton, FL: 1998.

9.24 Nawy, E. G., *Fundamentals of High Performance Concrete,* 2nd Ed., John Wiley & Sons, 2000.

PROBLEMS

9.1 Design the two-way prestressed floor in Example 9.2 by the equivalent frame method if the spacing of the columns in the E-W direction is changed to 24 ft (7.32 m) center to center. Analyze the floor for the worst condition of pattern live loading, and find the maximum long-term deflection of both the central and the end floor panels and compare it with the maximum allowable deflection if the floor carries equipment that is sensitive to excessive deflection.

9.2 Design the flat plate in Problem 9.1 considering it as a lift slab supported on steel columns as shown in Figure P9.2. Assume that no negative moments are transferred from the slab panels to the supporting columns, and check for deflections accordingly.

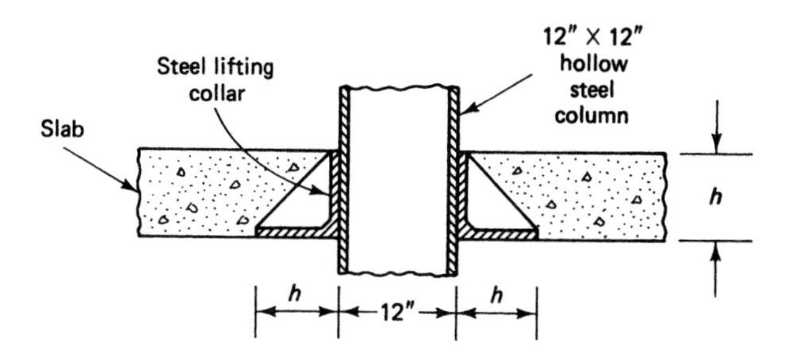

Figure P9.2 Lift slab at column support.

9.3 Analyze the flat plate in Problem 9.1 by the yield-line theory, and compare the design results with the equivalent frame design used in Problem 9.1.

10

CONNECTIONS FOR PRESTRESSED CONCRETE ELEMENTS

10.1 INTRODUCTION

The function of a connection is to economically transmit loads and stresses from one part of a structure to an adjoining part and provide stability to the structural system. The forces acting at the connection or joint are produced not only by gravity loads but by winds, seismic effects, volumetric changes due to long-term creep and shrinkage, differential movement of panels, and temperature effects.

Since a connection is the weakest link in the overall structural system, it has to be designed for nominal strength higher than the elements it connects. An additional load factor of at least 1.3 should be used in the design of connections, except in the case of insensitive connections, such as pads for column bases, where such an additional load factor is not necessary. All connections should be designed for a minimum horizontal tensile force of 0.2 times the vertical dead load, unless properly designed bearing pads are used.

The factors that have to be considered in design of a connection for strength are as follows:

1. The load transfer mechanism
2. Load factors

Gulf Life Center, Jacksonville, Florida. (*Courtesy,* Prestressed Concrete Institute.)

3. Volumetric changes
4. Ductility
5. Durability
6. Fire resistance
7. Required tolerances and clearances
8. Erection-related considerations
9. Considerations regarding hot weather and cold weather
10. The economics of the details of the connection

10.2 TOLERANCES

Clearances between elements must be realistically assessed. Where large tolerances are allowed in a supporting structure, or where no tolerances are specified, the clearances have to be increased to account for these factors. The following are recommended tolerances from Ref. 10.2 for deviations from idealized dimensions in beams, columns, and spandrel panels:

1. Variation in plan from specified location in plan: $\pm\frac{1}{2}$ in., any column or beam, any locations.
2. Deviation in plan from straight lines parallel to specified linear building lines: $\frac{1}{40}$ in. per ft, any beam less than 20 ft, or adjacent columns less than 20 ft apart; $\frac{1}{2}$ in., adjacent columns 20 ft or more apart.
3. Difference in relative position of adjacent columns from specified relative position: $\frac{1}{2}$in. at any deck level.
4. Deviation from plumb: $\frac{1}{4}$ in. for every 10 ft of height; 1 in. maximum for the entire height.
5. Variation in elevation of bearing surfaces from specified elevation: $\pm\frac{1}{2}$ in., any column or beam, any location.
6. Deviation of top of spandrel from specified elevation: $\frac{1}{2}$ in., any spandrel.
7. Deviation in elevation of bearing surfaces from lines parallel to specified grade lines: $\frac{1}{40}$ in. per ft, any beam less than 20 ft or adjacent columns less than 20 ft apart; $\frac{1}{2}$ in. maximum, any beam 20 ft or more in length or adjacent columns 20 ft or more apart.
8. Variation from specified bearing length on support: $\pm\frac{3}{4}$ in.
9. Variation from specified bearing width on support: $\pm\frac{1}{2}$ in.
10. Jog in alignment of matching edges: $\frac{1}{4}$ in.

Table 10.1 gives tolerances applicable to connections.

10.3 COMPOSITE MEMBERS

As discussed in detail in Chapter 5 Sections 5.7 through 5.11, full transfer of horizontal shear forces must be assured at the interface of the precast member and the situ-cast topping. Figure 5.14 presents the interacting forces, and the flowchart of Section 5.8.2 gives the operational step-by-step design procedure and the applicable design equations. Figure 5.18 of Example 5.3 and the accompanying design give the size and spacing of the

Table 10.1 Tolerances for Connections

Item	Recommended tolerances* in.
Field-placed anchor bolts (transit or template)	$\pm\frac{1}{2}$
Elevation of field cast footings and piers	± 1
Structural Precast Concrete	
Position of plates	± 1
Location of inserts	$\pm\frac{1}{2}$
Location of bearing plates	$\pm\frac{3}{4}$
Location of blockouts	$\pm\frac{1}{2}$
Length	$\pm\frac{3}{4}$
Overall depth	$\pm\frac{1}{4}$
Width of stem	$\pm\frac{1}{8}$
Overall width	$\pm\frac{1}{4}$
Horizontal deviation of ends from square	$\pm\frac{1}{2}$
Vertical deviation of ends from square	$\pm\frac{1}{8}$ per ft of height
Bearing deviation from plane	$\pm\frac{3}{16}$
Position of post-tensioning ducts in precast members	$\pm\frac{1}{2}$
Architectural Precast Concrete	
Length or width	$\pm\frac{1}{16}$ per 10 ft, but not less than $\pm\frac{1}{8}$
Thickness	$+\frac{1}{4}-\frac{1}{8}$
Location of blackouts	$\pm\frac{1}{2}$
Location of anchors and inserts	$\pm\frac{3}{8}$
Warpage or squareness	$\pm\frac{1}{8}$ in 6 ft
Joint widths	
—specified	$\frac{3}{8}-\frac{5}{8}$
—min. and max. dimensions	$\frac{1}{4}$ and $\frac{3}{4}$

*Other construction materials may control tolerances selected.

dowels necessary to effect the full transfer of the horizontal shear forces between the interconnected elements.

10.4 REINFORCED CONCRETE BEARING IN COMPOSITE MEMBERS

A typical composite-action dowel reinforcement is shown in Figure 10.1. In order to prevent the concrete that is in direct bearing contact in such reinforcements from crushing due to excess direct compressive load, the external load has to be applied to an adequate bearing area size such that the resulting limit-state stresses do not exceed the compressive strength of concrete. The nominal bearing strength of *plain* concrete can be defined as

$$V_n = C_r(0.85f'_c A_1)\sqrt{A_2/A_1} \leq 1.2f'_c A_1 \tag{10.1}$$

where C_r = 1.0 when reinforcement is provided in the direction of the horizontal frictional force N_u shown in Figure 10.2 or when N_u is taken to be zero. C_r can be defined as $(S \times W/200)^{N_u/V_u}$, where the area $S \times W$, shown in Figure 10.3, should not exceed 9.0 in.^{2}

A_1 = direct bearing area

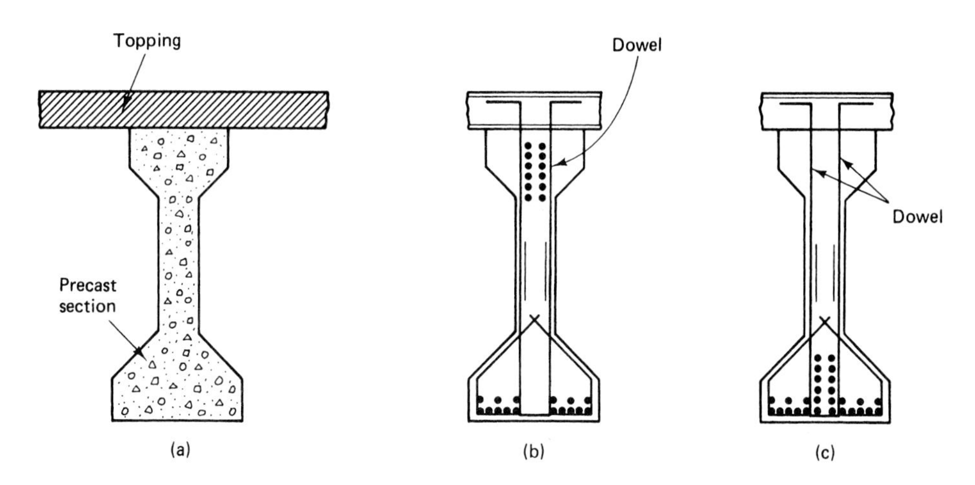

Figure 10.1 Typical dowel-action reinforcement. (a) Situ-cast topping on precast section. (b) Support section. (c) Midspan section.

A_2 = maximum area of the portion of the supporting surface that is geometrically similar to and concentric with the loaded area shown in Figure 10.3.

The design bearing strength is

$$V_u = \phi V_n$$

where ϕ = 0.70. In order to avoid accidental cracking or spalling at the ends of thin-stemmed members, a minimum reinforcement equal to $N_u/\phi f_y$, but not less than one #3 bar (9.52 mm dia), is recommended when the bearing area is less than 2 in.^{2} (12.9 cm^2).

If the applied factored load V_u exceeds the *design* bearing strength $V_u = \phi V_n$ as computed from Equation 10.1, reinforcement is required in the bearing area. This reinforcement can be designed by the shear-friction theory presented in Chapter 5. *All* precast members ought to be designed for reinforced bearing except solid and hollow-core slabs,

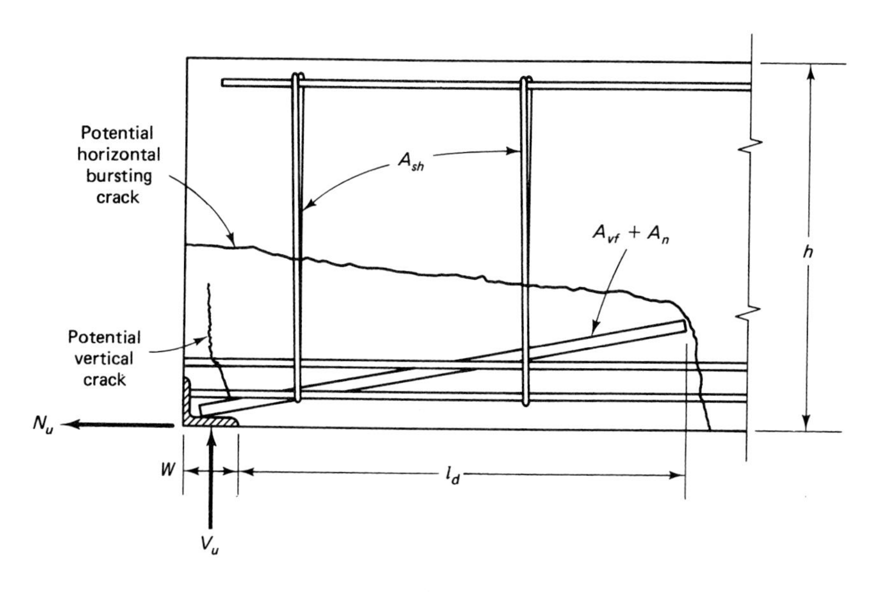

Figure 10.2 Reinforced bearing end in beam.

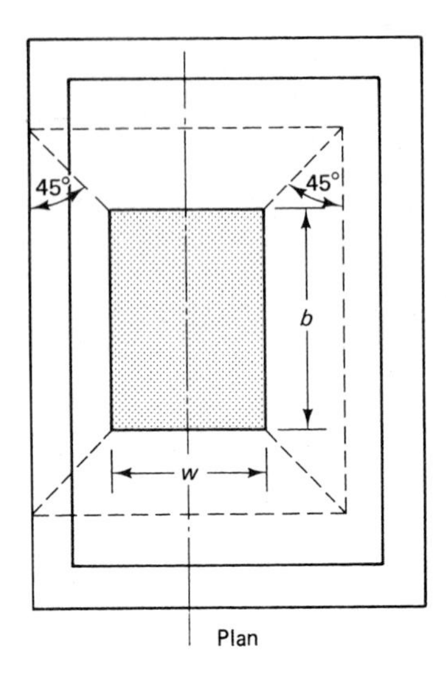

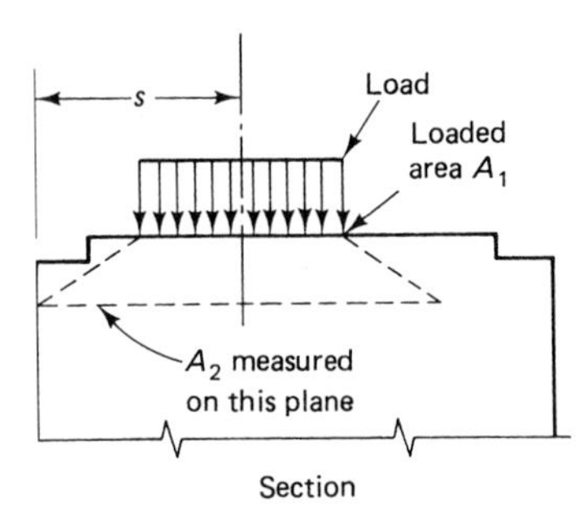

Figure 10.3 Bearing area on a concrete pad.

as recommended in Ref. 10.2, in order to prevent horizontal and vertical cracks from forming at the beam's extreme ties at the supports. The inclination of the end crack can be safely assumed to be approximately 20 degrees, as in Figure 10.2. Also, if V_u is equal to the applied factored shear force, in pounds, parallel to the assumed crack plane, it should be limited by the values given in Table 10.2 for the indicated maximum effective shear-friction coefficients μ_e.

The reinforcement area nominally perpendicular to the assumed crack plane can be found from

Table 10.2 Maximum Applied Factored Force V_u, *lb*

Crack Interface Condition	Recommended μ	Maximum μ_e	Maximum V_u, *lb*
1. Concrete to concrete, cast monolithically	1.4λ	3.4	$0.30\lambda^2 f'_c A_{cr} \leq 1{,}000\ \lambda^2 A_{cr}$
2. Concrete to hardened concrete with roughened surface	1.0λ	2.9	$0.25\lambda^2 f'_c A_{cr} \leq 1{,}000\ \lambda^2 A_{cr}$
3. Concrete to concrete	0.6λ	2.2	$0.20\lambda^2 f'_c A_{cr} \leq 800\ \lambda^2 A_{cr}$
4. Concrete to steel	0.7λ	2.4	$0.20\lambda^2 f'_c A_{cr} \leq 800\ \lambda^2 A_{cr}$

$$A_{vf} = \frac{V_{up}}{\phi \mu_e f_y} \tag{10.2}$$

where V_u/ϕ = nominal strength V_N

f_y = yield strength of A_{vf}, psi

V_{up} = applied factored shear force, limited by the values given in Table 10.2 and

$$\mu_e = \frac{1{,}000 \lambda A_{cr} \mu}{V_{up}} \tag{10.3}$$

in which λ = 1.0 for normal-weight, 0.85 for sand-lightweight, and 0.75 for all-lightweight concrete

A_{cr} = area of the crack plane interface (in.^{2}), which can be taken as $l_d b$, where l_d is the development length of the A_{vf} bars (in.) and b is the average member width (in.).

Table 10.3 gives the development length l_d for various bar sizes. The vertical reinforcement A_{sh} across potential horizontal cracks can be determined from

$$A_{sh} = \frac{(A_{vf} + A_n) f_y}{\mu_e' f_{ys}} \tag{10.4}$$

Table 10.3 Tension Reinforcement and Development Length (Inches) For $f_c' = 4{,}000$** psi concrete $f_y = 60{,}000$ psi Steel (α, β, $\lambda = 1.0$, $\gamma = 0.8$ for #6 Bars or Smaller and = 1.0 for #7 Bars and Larger)

Bar Size	Cross-Sectional Area (in.^{2})	Bar Diameter (in.)	Development length, l_d‡†† (in.): $s \geq 2d_b$ or *d_b — ≤ #6: $l_d = 38d_b$; ≥#7: $l_d = 48d_b$	Development length, l_d‡†† (in.): Other — ≥#6: $l_d = 57d_b$; ≥#7: $l_d = 72d_b$
(1)	(2)	(3)	(4)	(5)
3	0.11	0.375	15	21
4	0.20	0.500	19	29
5	0.31	0.625	24	36
6	0.44	0.750	29	43
7	0.60	0.875	42	63
8	0.79	1.000	48	72
9	1.00	1.128	54	81
10	1.27	1.270	61	92
11	1.56	1.410	68	102
14	2.25	1.693	82	122
18	4.00	2.257	108	163

*Confined

**For f_c' values different from 4000 psi, multiply table values by ($\sqrt{4{,}000/f_c'}$). For $f_y = 40{,}000$ psi, multiply by $\frac{2}{3}$; $\sqrt{f_c'}$ should not exceed 100.

‡For Compression development length, $l_d = \lambda l_{db}$ where $l_{db} = 0.02 d_b f_y / \sqrt{f_c'} \geq 0.0003 d_b f_y$ and λ_s = Required A_s/Provided A_s or $\lambda_{s1} = 0.75$ for spirally confined reinforcement.

††Multiply table values by:

α = 1.3 for top reinforcement

λ = 1.3 for lightweight aggregate

β = 1.5 for epoxy-coated bars with cover less than $3d_b$ or clear spacing less than $6d_b$ and β

where

$$\mu'_e = \frac{1{,}000\lambda A_{cr}\,\mu}{(A_{vf} + A_n)f_y} \tag{10.5}$$

and f_{ys} = yield strength of A_{sh}, psi
A_n = area of reinforcement to resist axial tension N_u in Figure 10.2, defined as

$$A_n = N_u/(\phi f_y) \tag{10.6}$$

in which N_u = factored applied horizontal tensile force nominally perpendicular to the assumed crack plane
ϕ = strength reduction factor = 0.85.

Note that all reinforcement on either side of the assumed crack plane should be properly anchored by development length or welding to angles, plates, or hooks in order to develop the computed resisting force.

10.4.1 Reinforced Bearing Design

Example 10.1

A PCI standard 16RB28 rectangular prestressed beam is subjected to a vertical factored end shear force V_u = 90,000 lb (400 kN) and a horizontal tensile force N_u = 21,000 lb (93.4 kN). The beam is supported on a teflon pad of size 4 in. × 4 in. (10 cm × 10 cm). Design the end reinforcement in the beam that can prevent the development of vertical or horizontal bearing cracks, given the following data:

f'_c = 5,000 psi (34.47 MPa), normal-weight concrete

f_y = 60,000 psi for all mild reinforcement (413.7 MPa)

θ = 20 degrees

Photo 10.1 Charlotte-Mecklenburg Government Center Parking Structure. (*Courtesy,* Prestressed Concrete Institute.)

Solution:

Horizontal Reinforcement **($A_{vf} + A_n$).** For the determination of the horizontal reinforcement, try No. 6 bars.

$$\text{Beam depth } h = 28 \text{ in.,} \qquad b = 16 \text{ in.}$$

From Table 10.3, $l_d = 29$ in.

$$A_{cr} = l_d b = 29 \times 16 = 464 \text{ in.}^2$$

From Table 10.2, $\mu = 1.4$, and from Equation 10.3,

$$\mu_e = \frac{1{,}000 \lambda A_{cr} \mu}{V_{up}} = \frac{1{,}000 \times 1.0 \times 464 \times 1.4}{90{,}000} = 10.61 > \text{allowable } \mu_e = 3.4$$

Thus, use $\mu_e = 3.4$.

From Equation 10.2,

$$A_{vf} = \frac{V_{up}}{\phi f_y \mu_e} = \frac{90{,}000}{0.85 \times 60{,}000 \times 3.4} = 0.52 \text{ in.}^2 \ (3.4 \text{ cm}^2)$$

$$N_u = 21{,}000 \text{ lb}$$

$$\frac{N_u}{V_u} = \frac{21{,}000}{90{,}000} = 0.23 > \text{minimum } 0.20$$

Hence, use $N_u = 21{,}000$ lb.

From Equation 10.6, $A_n = N_u/\phi f_y = 21{,}000/(0.85 \times 60{,}000) = 0.41 \text{ in}^2 \ (2.65 \text{ cm}^2)$

Total Steel

$$A_s = A_{vf} + A_n = 0.52 + 0.41 = 0.93 \text{ in.}^2 \ (6.0 \text{ cm}^2)$$

So use three #6 bars = 1.32 in.^{2} (8.52 cm^2).

Photo 10.2 Dallas Municipal Center, Dallas, Texas. (*Courtesy,* Post-Tensioning Institute.)

Vertical Reinforcement (A_{sh}). From Table 10.3, l_d = development length of #6 bars = 29 in. (74 cm) and $A_{cr} = l_d b = 29 \times 16 = 464$ in.^{2} (3,159 cm^2). From Equation 10.5,

$$\mu'_e = \frac{1{,}000\lambda A_{cr}\,\mu}{(A_{vf} + A_n)f_y} = \frac{1{,}000 \times 1.0 \times 464 \times 1.4}{0.93 \times 60{,}000} = 11.64 > \text{allowable } \mu_e = 3.4$$

So use $\mu'_e = 3.4$. Then, from Equation 10.4,

$$A_{sh} = \frac{(A_{vf} + A_n)f_y}{\mu'_e f_{ys}} = \frac{0.93 \times 60{,}000}{3.4 \times 60{,}000} = 0.27 \text{ in.}^2 \text{ (1.74 cm}^2\text{)}$$

Accordingly, use three #3 stirrups = 0.66 in.^{2} (4.26 cm^2).

10.5 DAPPED-END BEAM CONNECTIONS

A dapped-end beam is a structural element with abruptly reduced depth at its ends in order to provide the necessary seating or bearing on corbels or brackets without loss of clear height between floors. A typical dapped end in a prestressed beam is shown in Figure 10.4. Two types of cracks can develop: crack 2 is a direct shear crack, while cracks 3, 4, and 5 are diagonal tension cracks caused by flexure and axial tension in the extended reduced depth and the stress concentration at the reentrant corner. Therefore, the following types of reinforcement, as shown in the figure, have to be provided:

1. Flexural reinforcement A_f plus axial tension reinforcement A_n, where $A_s = A_f + A_n$, to resist the cantilever bending stresses.

Photo 10.3 Walt Disney World Monorail, Orlando, Florida, a series of hollow precast prestressed concrete 100-box girders that are individually post-tensioned to provide a six-span continuous structure, design by ABAM Engineers. (*Courtesy,* Walt Disney World Co.)

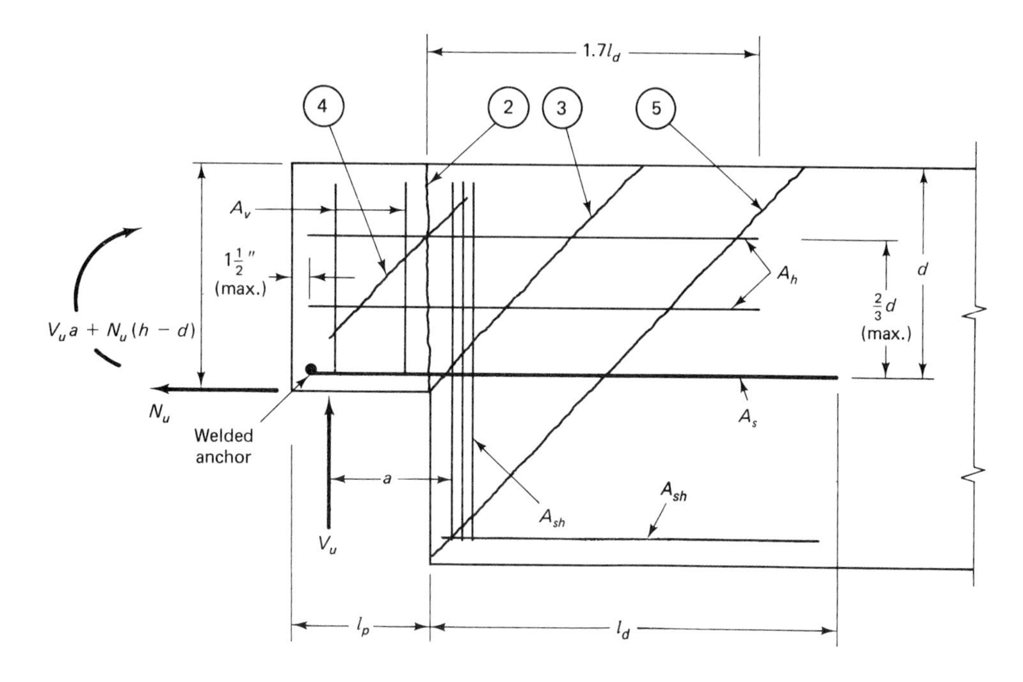

Figure 10.4 Cracking and reinforcement in dapped-end beam connections.

2. Shear-friction reinforcement $A_f + A_n$, plus axial tension reinforcement A_n, to resist the direct vertical shear force at the junction of the dapped and the undapped portion of the beam causing crack 2.
3. Shear reinforcement A_{sh}, to resist the diagonal tension generated at the reentrant corner causing crack 3.
4. Diagonal tension reinforcement $A_h + A_v$, to resist the potential diagonal tension crack 4 in the extended dapped portion of the beam.
5. Development length of $A_s = A_f + A_n$, to resist the potential diagonal tension crack 5 in the undapped portion of the beam.

10.5.1 Determination of Reinforcement to Resist Failure

10.5.1.1 Flexure and axial tension. For moment equilibrium in Figure 10.4, the total factored moment acting on the cantilever dapped portion at the plane of A_s is

$$M_u = V_u a + N_u(h - d) \tag{10.7a}$$

where h = depth of member above the dap
d = effective depth of the dap to center of reinforcement A_s
a = shear span.

M_u has to be resisted by a nominal moment strength $M_n = M_u/\phi$, or

$$M_n = \frac{V_u a + N_u(h - d)}{\phi} \tag{10.7b}$$

Assuming that the moment arm $jd \cong 0.9d$,

$$F_n = \frac{V_u a + N_u(h - d)}{0.9\phi d} \tag{10.8}$$

where $\phi = 0.90$ for flexure. Since $0.9\phi \cong 0.81$, for simplification use a value of $\phi = 0.85$ in Equation 10.8 to obtain

$$F_n = \frac{V_u a + N_u(h - d)}{\phi d} \tag{10.9a}$$

or

$$F_n = \frac{V_u}{\phi}\left(\frac{a}{d}\right) + \frac{N_u}{\phi}\left(\frac{h - d}{d}\right) \tag{10.9b}$$

The flexural reinforcement is then

$$A_f = \frac{F_n}{f_y} = \frac{V_u a + N_u(h - d)}{\phi f_y d} \tag{10.10}$$

and the direct tension reinforcement due to the tensile force N_u is

$$A_n = \frac{N_u}{\phi f_y} \tag{10.11}$$

The total area of the flexural and direct tension reinforcement then becomes, from Equations 10.10 and 10.11,

$$A_s = A_f + A_n = \frac{1}{\phi f_y}\left[V_u\left(\frac{a}{d}\right) + N_u\left(\frac{h}{d}\right)\right] \tag{10.12}$$

where, again, the adjusted $\phi = 0.85$.

10.5.1.2 Direct Vertical Shear. The potential direct shear crack 2 is resisted by the combination of reinforcements A_s and A_h in Figure 10.4. The horizontal reinforcement A_h needed to resist the direct shear can be evaluated as

$$A_h = 0.5(A_s - A_n) \tag{10.13}$$

where

$$A_s = \frac{2V_u}{3\phi f_y \mu_e} + A_n \tag{10.14a}$$

$$A_n = \frac{N_u}{\phi F_y} \tag{10.14b}$$

$$\mu_e = \frac{1{,}000\lambda b h \mu}{V_u} \tag{10.14c}$$

with $\phi = 0.85$ and μ_e the same as in Equation 10.3. Hence,

$$A_s = \frac{1}{\phi f_y}\left(\frac{2V_u}{3\mu_e} + N_u\right) \tag{10.15}$$

The value of A_s used in Equation 10.13 should be the greater of the two values obtained from Equations 10.12 and 10.15.

The reinforcement A_s should be extended a minimum of $1.7l_d$ past the end of the dap, or l_d past crack 5, and anchored at the end of the beam by welding to cross bars, angles, or plates. Horizontal bars A_h should be similarly extended, and vertical bars A_{sh} and vertical or inclined bars A_v should be well anchored by hooks as required by the ACI Code.

Photo 10.4 Jesse H. Jones Memorial Bridge, Houston, Texas. (*Courtesy,* Post-Tensioning Institute.)

The nominal shear strength of the dap end is limited to

$$V_n \leq 0.30f'_c\,bd \leq 1{,}000bd \tag{10.16a}$$

for normal-weight concrete, and

$$V_n \leq \left(0.20 - \frac{0.07a}{d}\right)f'_c\,bd \tag{10.16b}$$

or

$$V_n \leq \left(800 - \frac{280a}{d}\right)bd \tag{10.16c}$$

whichever is smaller, for sand-lightweight or all-lightweight concrete, where a is the shear span and d the effective depth of the beam.

10.5.1.3 Diagonal Tension at Reentrant Corner. The reinforcement needed to resist the inclined diagonal tension cracking propagating from the center of stress concentration at the reentrant corner towards the undapped portion can be obtained from the expression

$$A_{sh} = \frac{V_u}{\phi f_y} \tag{10.17}$$

where $\phi = 0.85$ and f_y is the yield strength of the A_{sh} reinforcement.

10.5.1.4 Diagonal Tension in the Dapped End. In order to resist the potential diagonal crack 4 in the dapped end, additional reinforcement A_v has to be provided such that the total nominal shear strength V_n satisfies the equation

$$V_n = \frac{V_u}{\phi} = A_v f_y + A_h f_y + 2\lambda bd\sqrt{f'_c} \tag{10.18}$$

At least half of the reinforcement in this area has to be placed vertically, so that Equation 10.18 gives

$$\text{Min } A_v = \frac{1}{2f_y}\left(\frac{V_u}{\phi} - 2\lambda b d\sqrt{f'_c}\right) \tag{10.19}$$

Note that performance considerations require the following:

1. The depth of the dapped end should be at least one-half the beam depth, unless the beam is significantly deeper than required by design for other than structural considerations.
2. If the flexural stress computed for the full depth of the section using factored loads and gross section properties exceeds $6\sqrt{f'_c}$ immediately beyond the dap, additional longitudinal reinforcement should be placed in the beam in order to develop the required flexural strength.
3. The diagonal tension reinforcement A_{sh} should be placed as closely as practicable to the reentrant corner. This reinforcement is in addition to the design shear reinforcement required for the full-depth beam section.

10.5.2 Dapped-End Beam Connection Design

Example 10.2

A PCI standard 16RB28 prestressed beam dapped at the end for bearing on a column corbel is subjected to a factored gravity end shear $V_u = 110{,}000$ lb (489 kN) and a horizontal axial tension $N_u = 20{,}000$ lb (97.9 kN). Design the flexural, direct shear, and diagonal tension reinforcements A_s, A_h, A_{sh}, and A_v required to prevent potential cracking due to dapping the beam ends. Given data are $f'_c = 5{,}000$ psi (34.5 MPa), normal weight, and $f_y = 60{,}000$ psi (414 MPa). Use bars for the reinforcement.

Solution: Assume that the shear span $a = 6$ in. (152 mm), the dapped-end effective $d = 16$ in. (406 mm), and $h = 18$ in. (457 mm).

Flexure and Axial Tension Reinforcement A_s

$$\frac{N_u}{V_u} = \frac{20{,}000}{110{,}000} = 0.18 < 0.20$$

Hence, $N_u = 0.20 \times 110{,}000 = 22{,}000$ lb (97.9 kN). Thus,

$$A_s = \frac{1}{\phi f_y}\left[V_u\left(\frac{a}{d}\right) + N_u\left(\frac{h}{d}\right)\right]$$

$$= \frac{1}{0.85 \times 60{,}000}\left[110{,}000 \times \frac{6}{16} + 22{,}000 \times \frac{18}{16}\right] = 1.29 \text{ in.}^2$$

Direct Shear Reinforcement A_s and A_h From Table 10.2 $\mu = 1.4\lambda$ where $\lambda = 1.0$. Then, from Equation 10.14c, where b for the 16RB28 section is 16 in.,

$$\mu_e = \frac{1{,}000\lambda b h \mu}{V_u} = \frac{1{,}000 \times 1.0 \times 16 \times 18 \times 1.4}{110{,}000} = 3.67$$

$$> \text{max allowable } \mu_e = 3.4$$

Thus, use $\mu_e = 3.4$. Then, from Equation 10.5,

$$A_s = \frac{1}{\phi f_y}\left(\frac{2V_u}{3\mu_e} + N_u\right) = \frac{1}{0.85 \times 60{,}000}\left(\frac{2 \times 110{,}000}{3 \times 3.4} + 22{,}000\right)$$

$$= 0.85 \text{ in.}^2 < A_s = 1.29 \text{ in.}^2 \text{ from before}$$

Hence, use $A_s = 1.29$ in.^{2} (8.32 cm^2). Then three #6 bars = 1.32 in.^{2}, which is satisfactory.

From Equation 10.4b,

$$A_n = \frac{N_u}{\phi f_y} = \frac{22{,}000}{0.85 \times 60{,}000} = 0.43 \text{ in}^2\ (2.77 \text{ cm}^2)$$

From Equation 10.13, the horizontal shear reinforcement across the depth of the beam is $A_h = 0.5\,(A_s - A_n) = 0.5(1.29 - 0.43) = 0.43$ in.^{2} (2.77 cm^2). So try two #3 U bars = $2(2 \times 0.11) = 0.44$ in.^{2} (2.84 cm^2), which will be verified subsequently. As a check, the nominal shear strength, from Equation 10.16a, is

$$\text{Available } V_n = 800bd = 800 \times 16 \times 16 = 204{,}800 \text{ lb}$$

$$\text{Required } V_n = \frac{V_u}{\phi} = \frac{110{,}000}{0.85} = 129{,}412 \text{ lb} < 204{,}800 \text{ lb, O.K.}$$

Diagonal Tension Vertical Reinforcement at Reentrant Corner. From Equation 10.17,

$$A_{sh} = \frac{V_u}{\phi f_y} = \frac{110{,}000}{0.85 \times 60{,}000} = 2.16 \text{ in.}^2\ (13.94 \text{ cm}^2)$$

So trying #4 closed ties, $A_s = 2 \times 0.20 = 0.40$ in.^{2} The number of ties = 2.16/0.4 = 5.4; hence, use six #4 ties, concentrated close to the reentrant corner.

Diagonal Tension Reinforcement A_v in the Dapped End. From Equation 10.19,

$$A_v = \frac{1}{2f_y}\left(\frac{V_u}{\phi} - 2\lambda bd\sqrt{f'_c}\right)$$

The nominal shear strength of the *plain concrete* is

$$2\lambda bd\sqrt{f'_c} = 2 \times 1.0 \times 16 \times 16\sqrt{5{,}000} = 36{,}204 \text{ lb}$$

Then

$$A_v = \frac{1}{2 \times 60{,}000}\left(\frac{110{,}000}{0.85} - 36{,}204\right) = 0.78 \text{ in.}^2$$

Try two #4 U stirrups = $2(2 \times 0.20) = 0.80$ in.^{2} From before, $A_h = 0.44$ in.^{2} So the total nominal shear strength of the section, from Equation 10.18, is

$$\text{Available } V_n = A_v f_y + A_h f_y + 2\lambda bd\sqrt{f'_c}$$

$$= 0.80 \times 60{,}000 + 0.44 \times 60{,}000 + 36{,}034$$

$$= 110{,}434 \text{ lb} < \frac{V_u}{\phi} = 129{,}412 \text{ lb}$$

which is unsatisfactory. Consequently, try increasing A_h by changing from #3 to #4 U stirrups:

$$A_h = 2(2 \times 0.2) = 0.80$$

$$\text{Revised available } V_n = 0.80 \times 60{,}000 + 0.80 \times 60{,}000 + 36{,}034$$

$$= 132{,}034 > \text{required } \frac{V_u}{\phi} = 129{,}412 \text{ lb, O.K.}$$

Check Development Length Requirements for Anchorage. The reinforcement A_s is three #6 bars. From Table 10.3 for #6 bars, $f'_c = 5{,}000$ psi and $l_d = 29$ in. Also, the undapped beam depth = 2 ft, 4 in. = 28 in., and the total development length = $28 - d + l_d = 28 - 16 + 29 = 41$ in. Since the minimum $l_d = 29$ in., use 42 in. = 3 ft, 6 in. (108 cm).

The reinforcement A_h is two #4 U bars. So from Table 10.3, $1.7l_d = 22$ in. (56 cm) beyond the beam dap. Figure 10.5 gives the reinforcement details for the dapped beam connection.

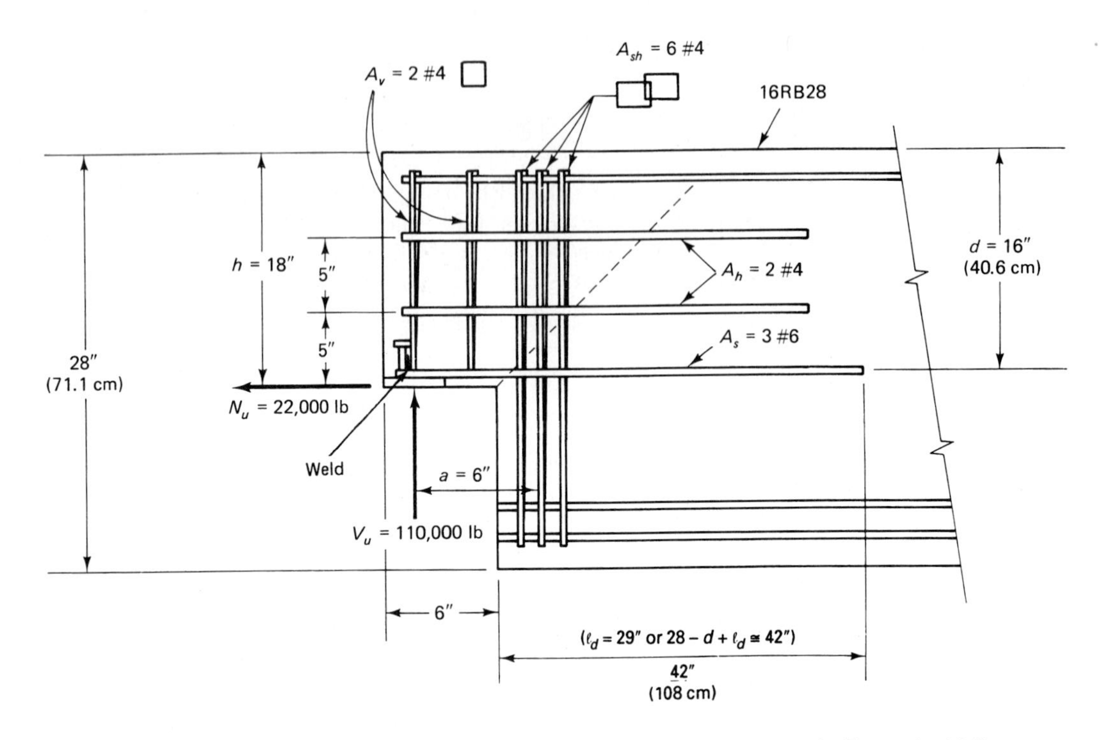

Figure 10.5 Reinforcing details for dapped-end beam connection in Example 10.2.

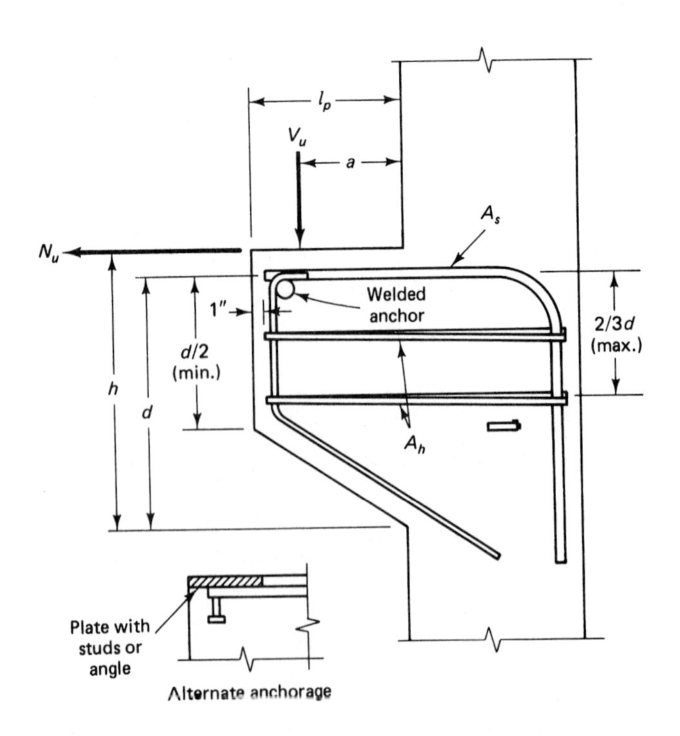

Figure 10.6 Typical corbel reinforcing details (see also Example 5.7.).

10.6 REINFORCED CONCRETE BRACKETS AND CORBELS

Corbels are short cantilevers whose shear span-to-depth radio a/d does not exceed a value of 1.0. They are subjected to a direct shear V_u and a horizontal tension N_u. Section 5.14 in Chapter 5, the design flowchart in Sec. 5.14.4 and the detailed design example 5.7 give a comprehensive discussion and application of the shear-friction theory in the design of corbels. Working out the details of reinforcement of the connection is of major significance in the success of the design of a corbel as regards its ability to resist applied loads. Typical corbel reinforcing details are shown in Figure 10.6.

10.7 CONCRETE BEAM LEDGES

Beam ledges are used to support transverse precast prestressed beam-end concentrated loads in a manner similar to the way corbels operate. Direct shear acting on the ledge can cause vertical cracks as shown in Figure 10.7. If the load is noncontinuous and comes from one side, the ledge beam in L-shaped form acts like a spandrel beam and is subjected to torsional moment in addition to direct shear. The design of the ledge beam itself follows the procedures and examples in Chapter 5. The discussion presented here covers the design of the shear reinforcement for the cantilevering ledge, which often has a shear span-to-depth ratio l_p/d of $\frac{1}{2}$ or less.

The nominal shear strength of the ledge at the reentrant corner can be determined by the lesser of the two values obtained from the following expressions under the given conditions:

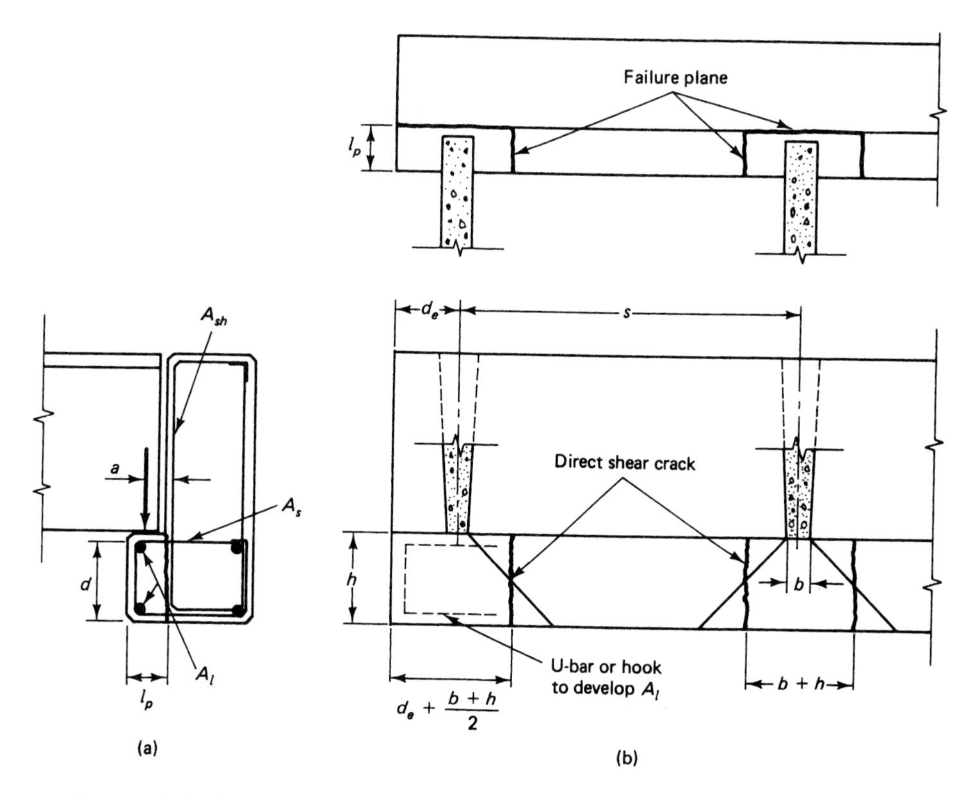

Figure 10.7 Connection design for beam ledges. (a) Ledge beam cross section. (b) Plan and elevation.

1. $s > b + h$

$$V_n = 3h\lambda\sqrt{f'_c}(2l_p + b + h) \tag{10.20a}$$

$$V_n = h\lambda\sqrt{f'_c}(2l_p + b + h + 2d_e) \tag{10.20b}$$

2. $s < b + h$, *and equal concentrated loads*

$$V_n = 1.5h\lambda\sqrt{f'_c}(2l_p + b + h + s) \tag{10.21a}$$

$$V_n = h\lambda\sqrt{f'_c}\left(l_p + \frac{b + h}{2} + d_e + s\right) \tag{10.21b}$$

where l_p = ledge projection, in.
b = width of bearing area, in.
h = depth of beam ledge, in.
s = spacing of concentrated loads, in.
d_e = distance from center of load to end of beam, in.

If the ledge supports a continuous load or closely spaced concentrated loads, the nominal shear strength of the ledge section has to be evaluated from

$$V_n = 24h\lambda\sqrt{f'_c} \tag{10.22}$$

where V_n is in lb per ft of length. The design strength V_u has to be at least equal to the factored force $V_u = \phi\, V_n$ for $\phi = 0.85$. If the applied factored load V_u exceeds the design strength as determined from Equation 10.20, 10.21, or 10.22, special reinforcement has to be provided in a design similar to the reinforcement required in a dapped beam end as discussed in Section 10.5. In such a case, the flexural reinforcement A_s is determined from Equation 10.12, the vertical diagonal tension "hanger" reinforcement A_{sh} from Equation

Photo 10.5 Hoisting double-T prestressed roof element at the Civil Enginerring Laboratory, Rutgers University.

10.17, and the longitudinal reinforcement A_l placed at the top and bottom fibers of the ledge from

$$A_l = \frac{200 l_p d}{f_y} \tag{10.23}$$

where A_l is the area of the longitudinal reinforcement in the ledge. The reinforcement A_{sh} can be uniformly spaced over a width $6h$ on either side of the bearing, but not to exceed half the distance to the next load. The bar spacing should not exceed the ledge depth h or 18 in., and the A_{sh} designed for the ledge need not be additive to the shear and torsional reinforcement of the total ledge beam.

10.7.1 Design of Ledge Beam Connection

Example 10.3

A garage floor structure is composed of 10-ft-wide double-T's supported at the exterior end by standard L-beam sections. The layout of the double-T's is such that a stem can be placed at any point on the ledge. The vertical factored end shear $V_u = 24{,}000$ lb (107 kN) per stem, and the horizontal tensile force $N_u = 5{,}000$ lb (22.4 kN) per stem. Compute the nominal shear strength of the ledge and design the reinforcement if necessary, given that

$$b = 4 \text{ in.}$$

$$h = 12 \text{ in.}$$

$$d = 10.5 \text{ in.}$$

$$l_p = 6 \text{ in. (15 cm)}$$

$$s = 48 \text{ in. (122 cm)}$$

$$f'_c = 5{,}000 \text{ psi (34.5 MPa), normal weight}$$

$$f_y = 60{,}000 \text{ psi (414 MPa)}$$

Photo 10.6 Mariners Island Office Building, San Mateo, California (*Courtesy,* Robert Englekirk Consulting Structural Engineers and W2MH Group Architects, Los Angeles, California. Photo by Dixie Carillo.).

Solution:

$$V_u = 24{,}000 \text{ lb}$$

$$N_u = 5{,}000 \text{ lb}$$

$$s = 48 \text{ in.}$$

$$b + h = 4 + 12 = 16 \text{ in.}$$

$$\text{Min. } d_e = \tfrac{1}{2} b = 2 \text{ in.}$$

$$2l_p + b + h = 2 \times 6 + 4 + 12 = 28 \text{ in.}$$

Since $s > b + h$, and $d_e < 2l_p + b + h$, Equation 10.20b applies, and the available $V_n = h\lambda\sqrt{f'_c}(2l_p + b + h + 2d_e) = 12 \times 1.0\sqrt{5{,}000}(2 \times 6 + 4 + 12 + 2 \times 2) = 27{,}153$ lb (120.8 kN). So the design $V_u = \phi V_n = 0.85 \times 27{,}153 = 23{,}080$ lb < Factored $V_u = 24{,}000$ lb, and we can use the dapped section reinforcement design.

Flexural Reinforcement A_s. The shear span $a \cong 3l_p/4 + 1.5 = 3 \times 6/4 + 1.5 = 6$ in. (15 cm). Since $N_u/V_u = 5{,}000/24{,}000 = 0.21 > 20\%$, use $N_u = 5{,}000$ lb. Then, from Equation 10.12,

$$A_s = \left[\frac{1}{\phi f_y}\left[V_u\left(\frac{a}{d}\right) + N_u\left(\frac{h}{d}\right)\right]\right]$$

$$= \frac{1}{0.85 \times 60{,}000}\left[24{,}000\frac{6}{10.5} + 5{,}000\frac{12}{10.5}\right] = 0.38 \text{ in}^2 \text{ (2.45 cm}^2\text{)}$$

Since $6h = 6 \times 12 > s/2 = 24$ in., distribute the reinforcement $s/2 = 24$ in. on each side of the load.

The width of the band for placement of flexural reinforcement $A_s = 2 \times 24 = 48$ in., and the maximum bar spacing $= h = 12$ in. So use four #3 bars in each 48-in. band width $= 0.44$ in.^{2} > required 0.38 in.^{2} Accordingly, place two additional bars at the beam end in order to provide equivalent reinforcement for the stem placed near the end.

Diagonal Tension Vertical Reinforcement A_{sh}. From Equation 10.17,

$$A_{sh} = \frac{V_u}{\phi f_y} = \frac{24{,}000}{0.85 \times 60{,}000} = 0.47 \text{ in.}^2 \text{ (3.03 cm}^2\text{)}$$

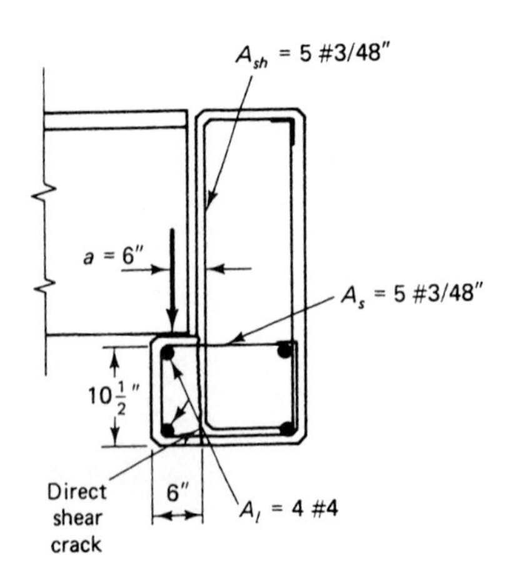

Figure 10.8 Ledge connection reinforcement in Example 10.3.

over a 48-in.-width band. Thus, A_{sh}/ft = 0.47/4 = 0.12 in.^{2}/ft, or #3 bars @ 11 in. Consequently, use five #3 ⊔ bars in each 48-in. band width = 0.55 in.^{2} > required 0.47 in.^{2} Then, for practical considerations, use the same number and spacing for both the A_s and A_{sh} steel, namely, five #3 closed hoops. Note that only one leg of the hanger steel hoop A_{sh} is accounted for in the selection of the five #3 bars in order to provide for the required concentration of the steel near the reentrant corner.

Longitudinal Reinforcement A_l. From Equation 10.23,

$$A_l = \frac{200 l_p d}{f_y} = \frac{200 \times 6 \times 10.5}{60{,}000} = 0.21 \text{ in}^2$$

For practical field considerations, use #4 bars, one on each corner of the ledge, giving four #4 bars = 0.80 in.^{2} (12.7 mm dia) > 0.21 in.^{2}, O.K.

The complete design of the ledge beam, of course, would require shear and torsional analysis of the total section to resist the total shear transmitted by all the double-T supported stems and the torsional moment caused by the application of the *eccentric* load from the supported stems. The designed ledge reinforcement area discussed in this example is in addition to the shear and torsional reinforcement required for the total beam.

Figure 10.8 gives the details of the ledge connection reinforcement, but does not include the shear and torsional reinforcement that has to be designed for the entire L-beam.

10.8 SELECTED CONNECTION DETAILS

As discussed in Section 10.1, connections are the major links in the overall structural system whose performance determines whether the structure will be safe and stable. Consequently, the design engineer has to be particularly cautious in the design and selection of the appropriate section for reasons of both safety and economy. Details of the design of numerous types of connections are given in Refs. 10.1 and 10.2, and Figures 10.9 through 10.16 show typical details of selected connections from these two references. The diagrams illustrate the application of the theories presented in this chapter.

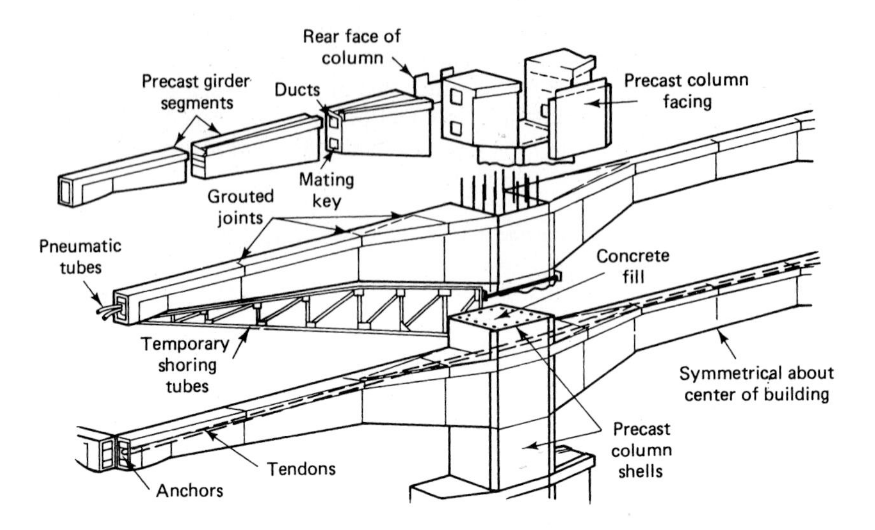

Beam assembly showing precast elements, temporary truss and prestressing

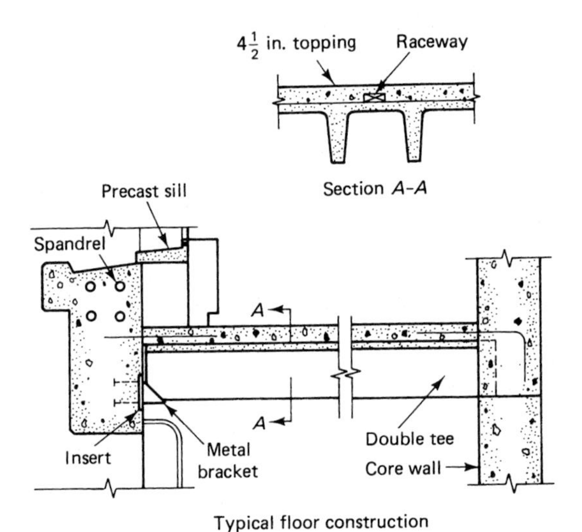

Typical floor construction

Figure 10.9 Construction method and typical joint details in Gulf Life Prestressed Building, Jacksonville, Florida.

Photo 10.7 Installation of prestressed double-T elements at the rod-suspended support, Civil Engineering Laboratory, Rutgers University.

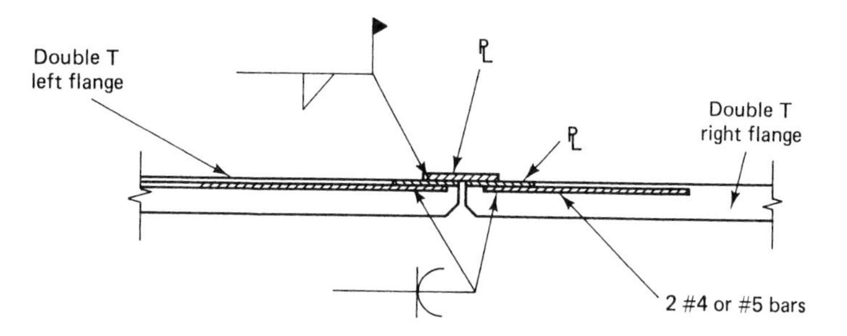

Figure 10.10 Connection of double-T flanges if no composite-action concrete topping is used.

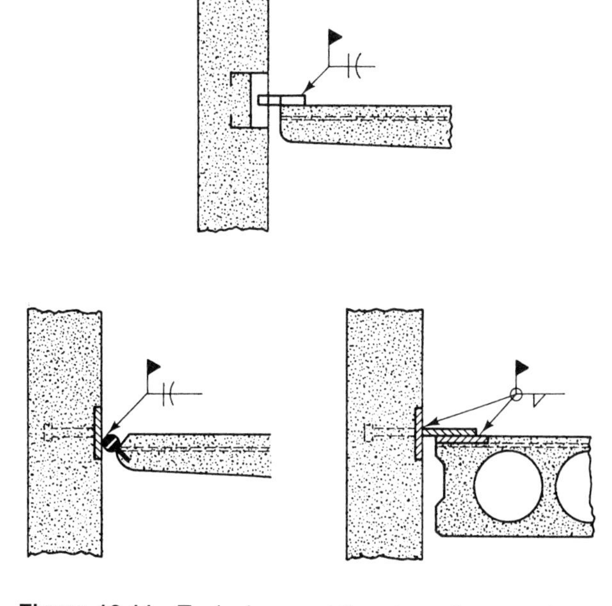

Figure 10.11 Typical precast floor-to-wall connections.

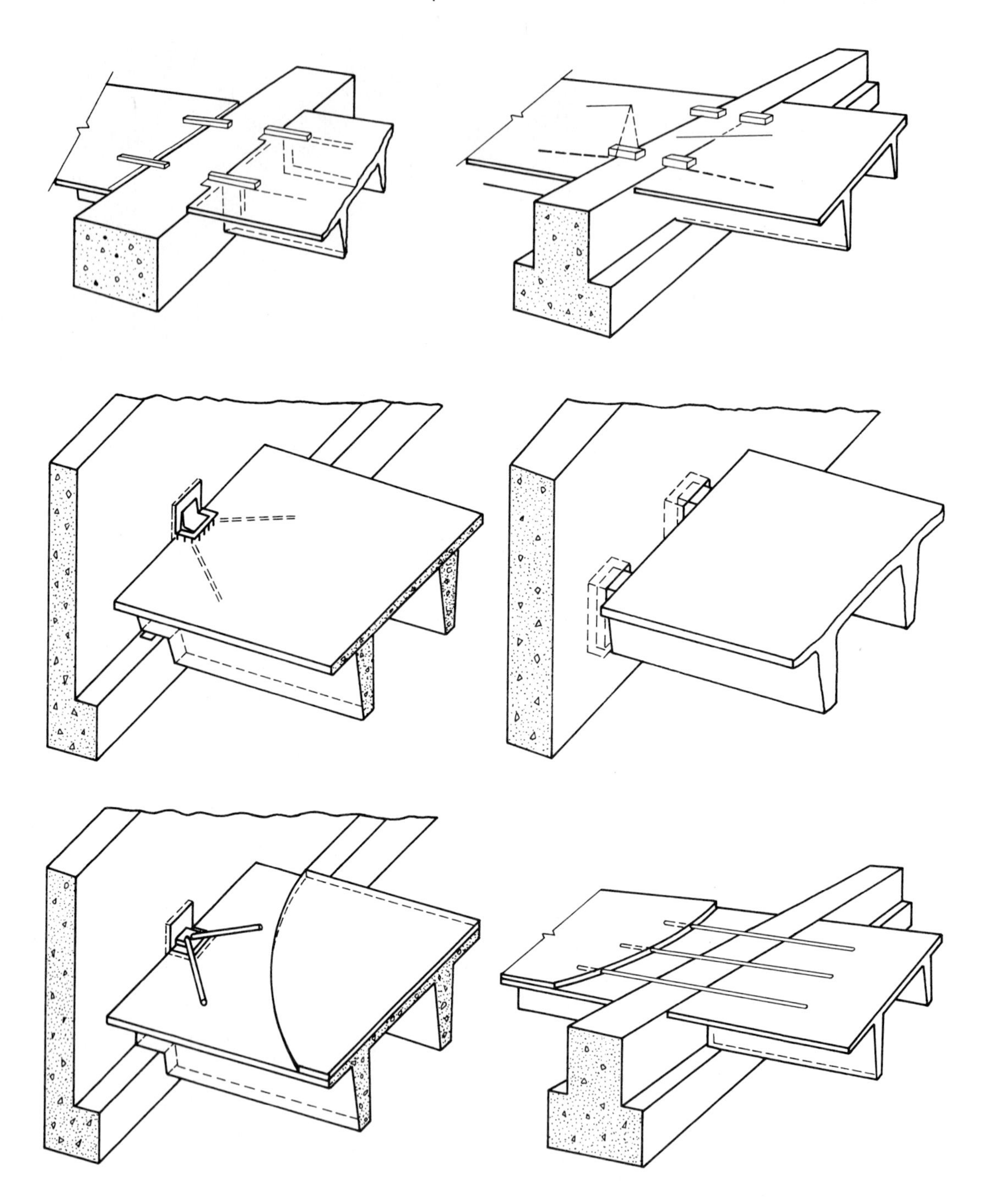

Figure 10.12 Typical double-T connections.

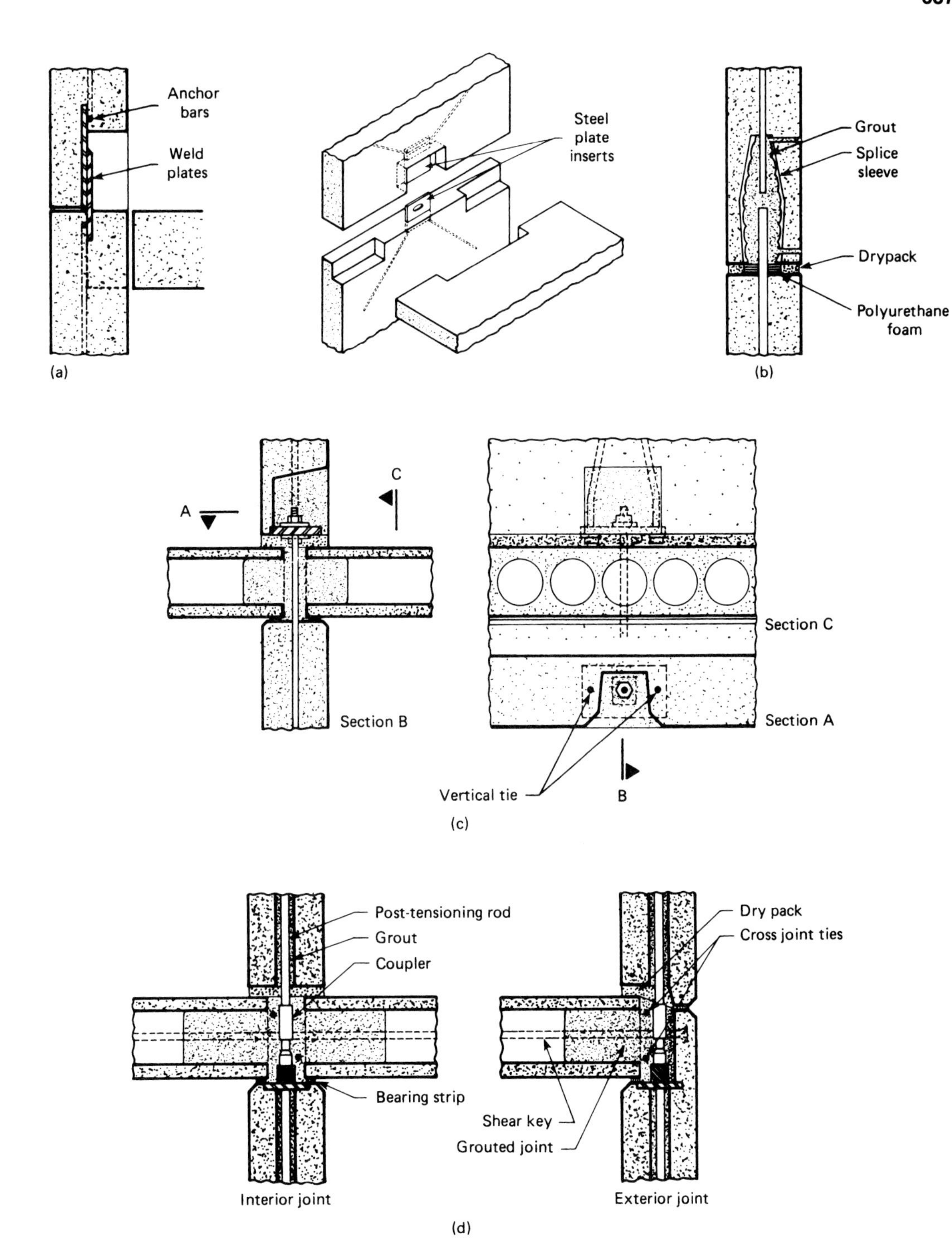

Figure 10.13 Connections through horizontal joints. (a) Weld plates. (b) Grouted splice sleeve. (c) Plate-bolt connections. (d) Post-tensioned connection.

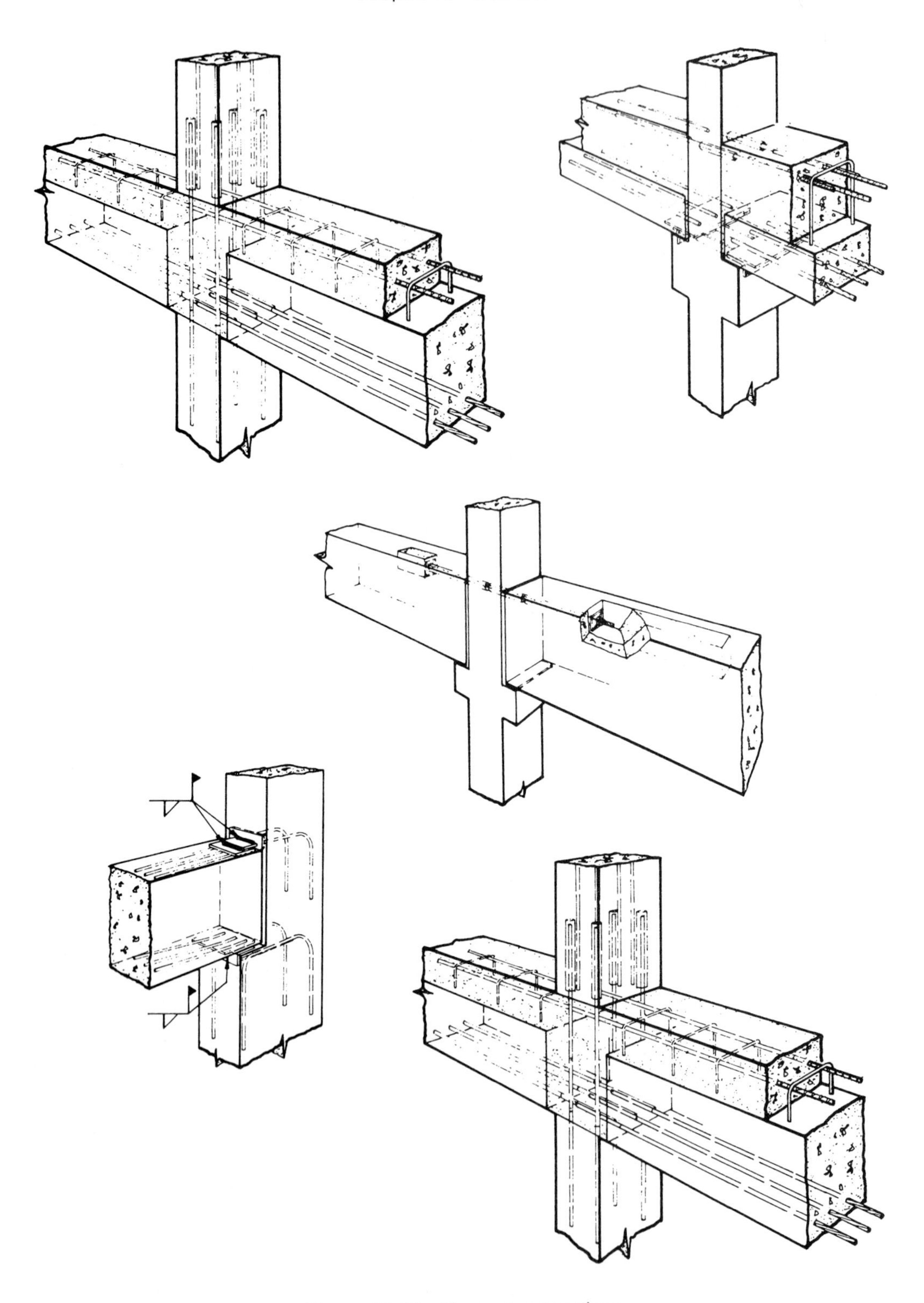

Figure 10.14 Moment connections.

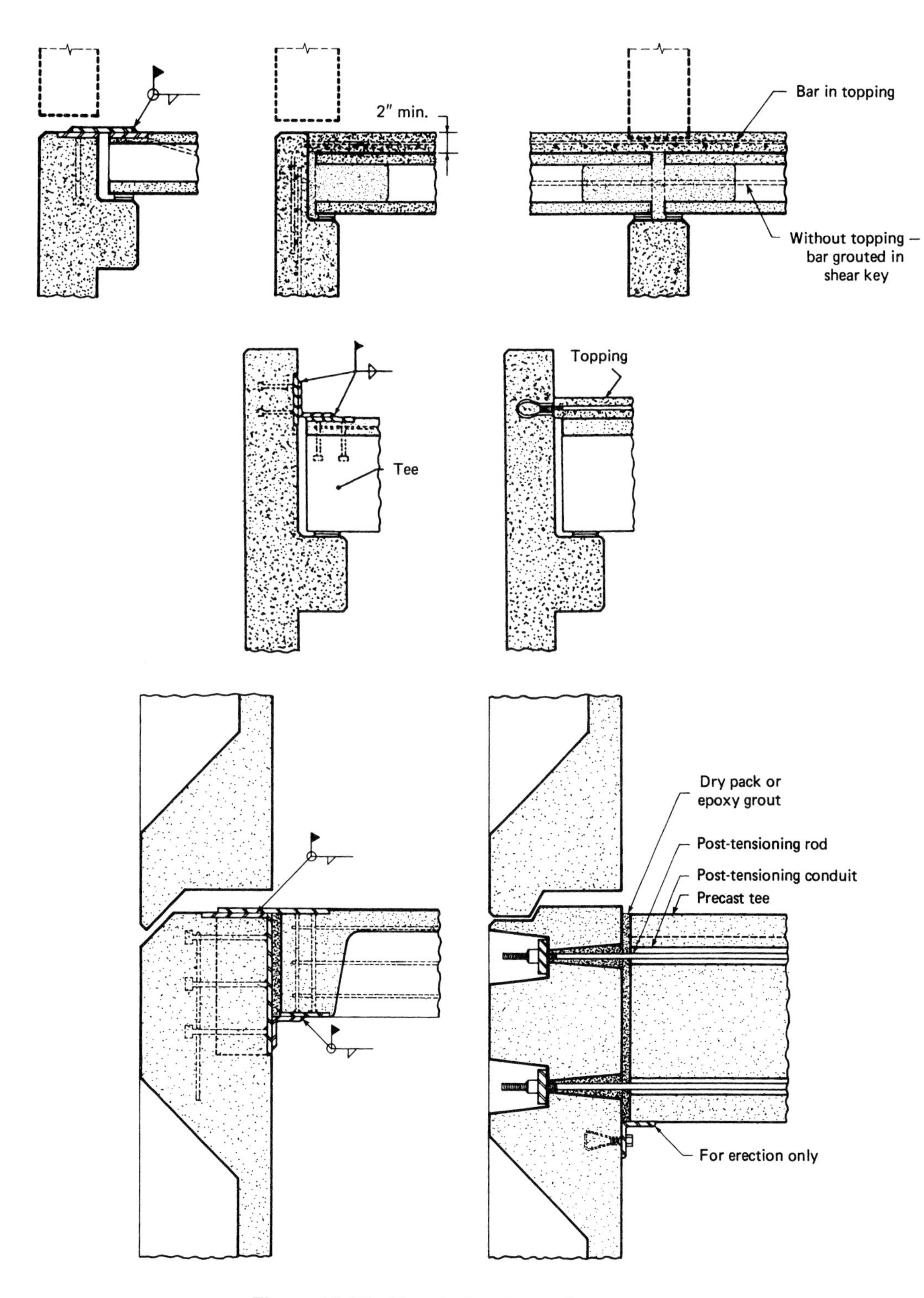

Figure 10.15 Floor-to-bearing wall connections.

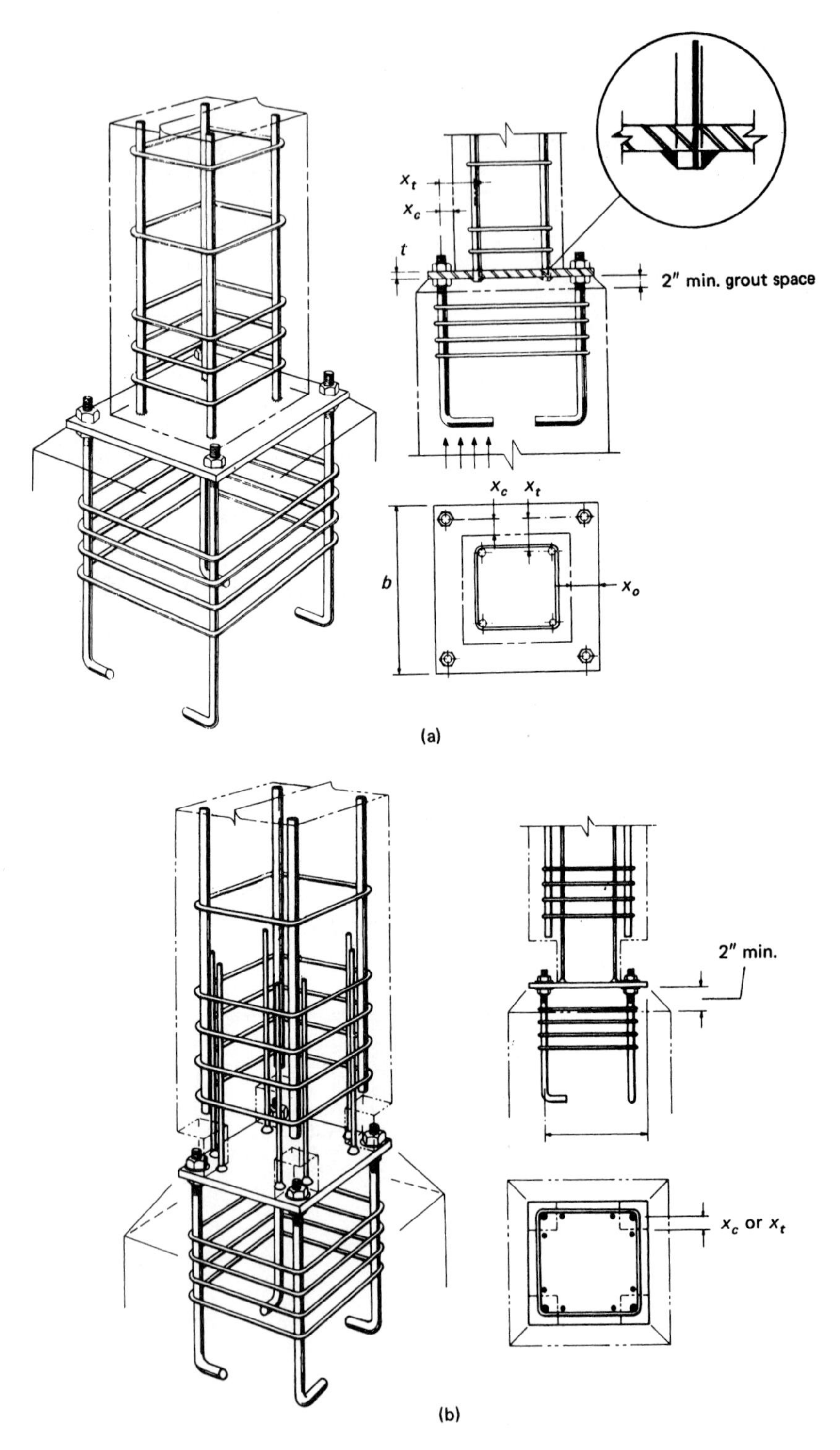

Figure 10.16 Column base connections. (a) Base plate larger than column. (b) Flush base plate.

REFERENCES

10.1 Prestressed Concrete Institute. *PCI Design Handbook—Precast and Prestressed Concrete.* 5th ed. Prestressed Concrete Institute, Chicago, 1999.

10.2 PCI Committee on Connection Details. *Manual on Design and Detailing of Connections for Precast and Prestressed Concrete.* Prestressed Concrete Institute, Chicago, 1988.

10.3 Post-Tensioning Institute. *Post-Tensioning Manual.* 5th ed. Post-Tensioning Institute, Phoenix, 2000.

10.4 Shaikh, A. F., and Yi, W. "In-Place Strength of Welded Headed Studs." *Journal of the Prestressed Concrete Institute* 30, 1985, 56–81.

10.5 Mattock, A. H., and Chan, T. C. "Design and Behavior of Dapped-End Beams." *Journal of the Prestressed Concrete Institute* 30, 1985, 28–45.

10.6 Mirza, S. A., and Furlong, R. W. "Serviceability Behavior of Concrete Inverted T-Beam Bridge Bent Caps." *Journal of the American Concrete Institute* 80, 1983, 294–304.

10.7 Clough, D. P. "Design of Connections for Precast Prestressed Concrete Buildings for the Effect of Earthquake." *PCI Technical Report No. 5.* Prestressed Concrete Institute, Chicago, 1985.

10.8 PCI Erectors Committee. *Recommended Practice for Erection of Precast Concrete.* Prestressed Concrete Institute, Chicago, 1985, pp. 1–198.

PROBLEMS

10.1 Design the end reinforcement required to resist the bearing stresses caused by an end vertical shear $V_u = 125{,}000$ lb (556 kN) and a horizontal tensile force $N_u = 24{,}000$ lb (107 kN) in a PCI 16RB28 rectangular beam supported at its ends on 5 in. × 5 in. pads. Take $f'_c = 5{,}000$ psi (34.47 MPa), normal weight, and $f_y = 60{,}000$ psi (413.7 MPa).

10.2 If the beam in Problem 10.1 is dapped at its ends and rests on concrete corbels, design the flexure and direct shear reinforcement for the dapped ends in order to prevent potential shear cracking and failure.

10.3 Solve Example 10.3 for factored loads $V_u = 29{,}000$ lb (128 kN) and factored $N_u = 2{,}500$ lb (11.1 kN).

PRESTRESSED CONCRETE CIRCULAR STORAGE TANKS AND SHELL ROOFS

11.1 INTRODUCTION

Prestressed concrete circular tanks are usually the best combination of structural form and material for the storage of liquids and solids. Their performance over the past half-century indicates that, when designed with reasonable skill and care, they can function for 50 years or more without significant maintenance problems.

The first effort to introduce circumferential prestressing into circular structures was that of W. S. Hewett, who applied the tie rod and turnbuckle principle in the early 1920s (Ref. 11.6). But the reinforcing steel available at that time had very low yield strength, limiting the applied tension to not more than 30,000 to 35,000 psi (206.9 to 241.3 MPa). Indeed, significant long-term losses due to concrete creep, shrinkage, and steel relaxation almost neutralized the prestressing force. As higher strength steel wires became available, J. M. Crom, Sr., in the 1940s, successfully developed the principle of winding high-tensile wires around the circular walls of prestressed tanks. Since that time, over 3,000 circular storage structures have been built of various dimensions up to diameters in excess of 300 feet (92 m).

Two 583,000-bbl (92,500-m^3) double-wall prestressed concrete tanks for liquefied natural gas storage, Philadelphia. (*Courtesy,* N.A. Legatos, Preload Technology, Inc., New York.)

The major advantage in performance and economy of using circular prestressing in concrete tanks over regular reinforcement is the requirement that no cracking be allowed. The circumferential "hugging" hoop stress in compression provided by external winding of the prestressing wires around the tank shell is the natural technique for eliminating cracking in the exterior walls due to the internal liquid, solid, or gaseous loads that the tank holds. Other techniques of circumferential prestressing using *individual* tendons which are anchored to buttresses have been more widely used in Europe than in North America for reasons of local economy and technological status.

Containment vessels utilizing circumferential prestressing, which can be either situ-cast or precast in segments, include water storage tanks, wastewater tanks and effluent clarifiers, silos, chemical and oil storage tanks, offshore oil platform structures, cryogenic vessels, and nuclear reactor pressure vessels. All these structures are considered thin shells because of the exceedingly small ratio of the container thickness to its diameter. Because no cracking at working-load levels is permitted, the shells are expected to behave elastically under working-load and overload conditions.

11.2 DESIGN PRINCIPLES AND PROCEDURES

11.2.1 Internal Loads

Considering the behavior of circular tanks involves examining both the interior pressure due to the material contained therein acting on a thin-walled cylindrical shell cross section and the exterior radial and sometimes vertical prestressing forces balancing the interior forces. The interior pressure is horizontally radial, but varies vertically depending on the type of material contained in the tank. If the material is water or a similar liquid, the vertical pressure distribution against the tank walls is *triangular,* with maximum intensity at the base of the wall. Other liquids which are accompanied by gas would give a *constant* horizontal pressure throughout the height of the wall. The vertical pressure distribution in tanks used for storage of granular material such as grain or coal would be essentially similar to the gas pressure distribution, with a constant value along most of the depth of the material contained. Figure 11.1 shows the pressure distributions for these three cases of loading.

The basic elastic theory of cylindrical shells applies to the analysis and design of the walls of prestressed tanks. A ring force causes ring tension in the thin cylindrical walls, assumed unrestrained at the ends at each horizontal section. The magnitude of the force is proportional to the internally applied pressure, and *no* vertical moment is produced along the height of the walls. If the wall ends are restrained, the magnitude of the ring force changes and a bending moment is induced in the vertical section of the tank wall. The magnitudes of the ring forces and vertical moments are thus a function of the degree of restraint of the cylindrical shell at its boundaries and are computed from the elastic shell theory and its simplifications and idealizations to be discussed subsequently.

Liquid Load and Freely Sliding Base. From basic mechanics, the ring force is

$$F = \frac{pd}{2} = pr \tag{11.1a}$$

and the ring stress is

$$f_R = \frac{pd}{2t} = \frac{pr}{t} \tag{11.1b}$$

where d = diameter of cylinder
r = radius of cylinder

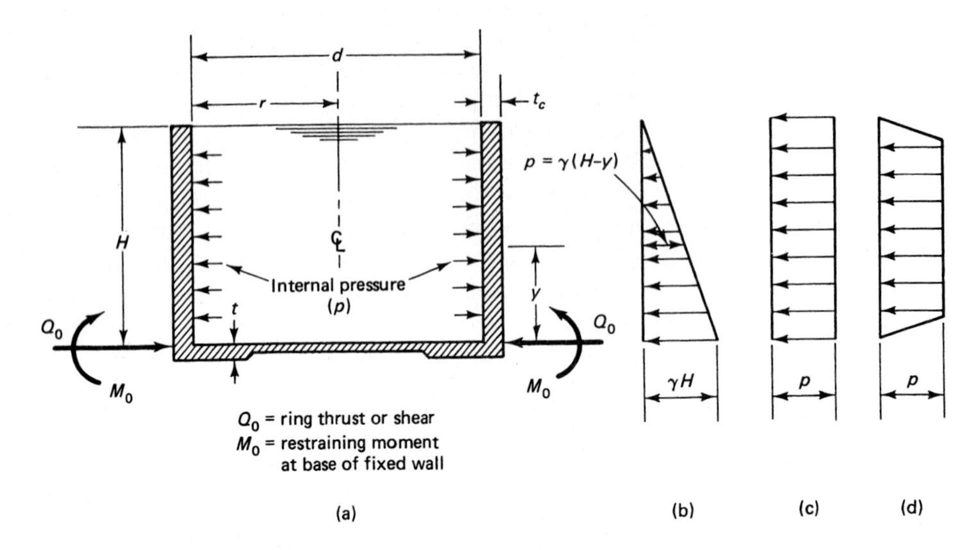

Figure 11.1 Tank internal pressure diagrams. (a) Tank cross section, showing radial shear Q_o and restraining moment M_o at base for fixed-base walls. (b) Liquid pressure, triangular load. (c) Gaseous pressure, rectangular load. (d) Granular pressure, trapezoidal load.

t = thickness of wall core
p = unit internal pressure at wall base = γH
γ = unit weight of material contained in vessel.

The tensile ring stress *at any point below the surface* of the material contained in the vessel becomes

$$f_R = \gamma(H - y)\frac{d}{2t} = \gamma(H - y)\frac{r}{t} \tag{11.2a}$$

where H is the height of the liquid contained and y is the distance *above* the base. The corresponding ring force is

$$F = \gamma(H - y)r \tag{11.2b}$$

The maximum tensile ring stress at the base of the freely sliding tank wall for $y = 0$ becomes, as in Equation 11.1b,

$$f_R(\max) = \frac{\gamma Hd}{2t} = \frac{\gamma hr}{t} \tag{11.2c}$$

Gaseous Load on Freely Sliding Base. Again from basic principles of mechanics, the constant tensile ring stress is

$$f_R = \frac{pd}{2t} = \frac{pr}{t} \tag{11.3}$$

Note that while theoretically the centerline diameter dimension is more accurate to use, the ratio t/d is so small that the use of the internal diameter d is appropriate.

Liquid and Gaseous Load on a Restrained Wall Base. If the base of the wall is fixed or pinned, the ring tension at the base vanishes. Because of the restraint imposed

Photo 11.1 4.0 Million Gallon Preload Tank, City of Troy, Ohio. (*Courtesy,* N.A. Legatos, Preload Inc., Garden City, New York.)

on the base, the simple membrane theory of shells is then no longer applicable, due to the imposed deformations of the restraining force at the wall base. Instead, bending modifications to the membrane stresses become necessary (see Refs. 11.2 and 11.6), and the deviation of the ring tension at intermediate planes along the wall height must be approximated as in Ref. 11.2 and the discussion in Sec. 11.3.

If the vertical bending moment in the horizontal plane of the wall at any height is M_y, the flexural stress in compression or tension in the concrete becomes

$$f_t = f_c = \frac{M_y}{S} = \frac{6M_y}{t^2} \text{ per unit height} \tag{11.4}$$

The distribution of the flexural stress across the thickness of the tank wall is shown in Figure 11.2.

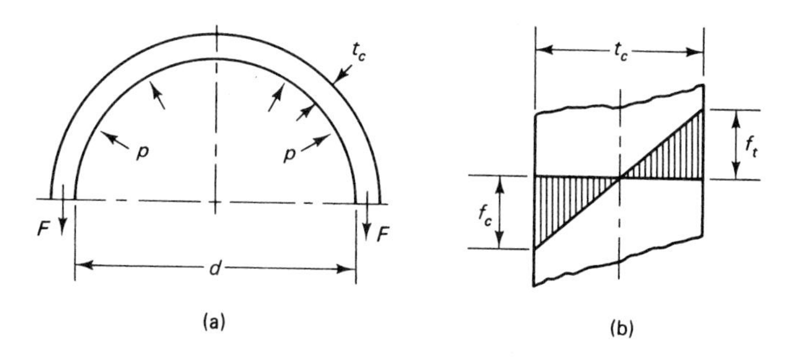

Figure 11.2 Ring tension and flexural stresses. (a) Ring tension internal force F in the horizontal section. (b) Flexural stress due to bending moment M in the wall thickness of the vertical section.

11.2.2 Restraining Moment M_0 and Radial Shear Force Q_0 at Freely Sliding Wall Base Due to Liquid Pressure

11.2.2.1 Membrane Theory. The study of forces and stresses in a circular uncracked tank wall is an elasticity problem in cylindrical shell analysis. If the shell is free to deform under the influence of the internal liquid pressure, the basic membrane equations of equilibrium apply. The longitudinal unit force N_y, the "hugging" circumferential unit force N_θ, and the central unit shears $N_{y\theta}$ and $N_{\theta y}$ are shown in the differential element of Figure 11.3(b). Note that these *four* unknowns all act in the plane of the shell.

The basic three equations of equilibrium for these four unknown unit forces are

$$\frac{\partial N_\theta}{\partial \theta} + r\frac{\partial N_{y\theta}}{\partial y} + p_\theta r = 0 \tag{11.5a}$$

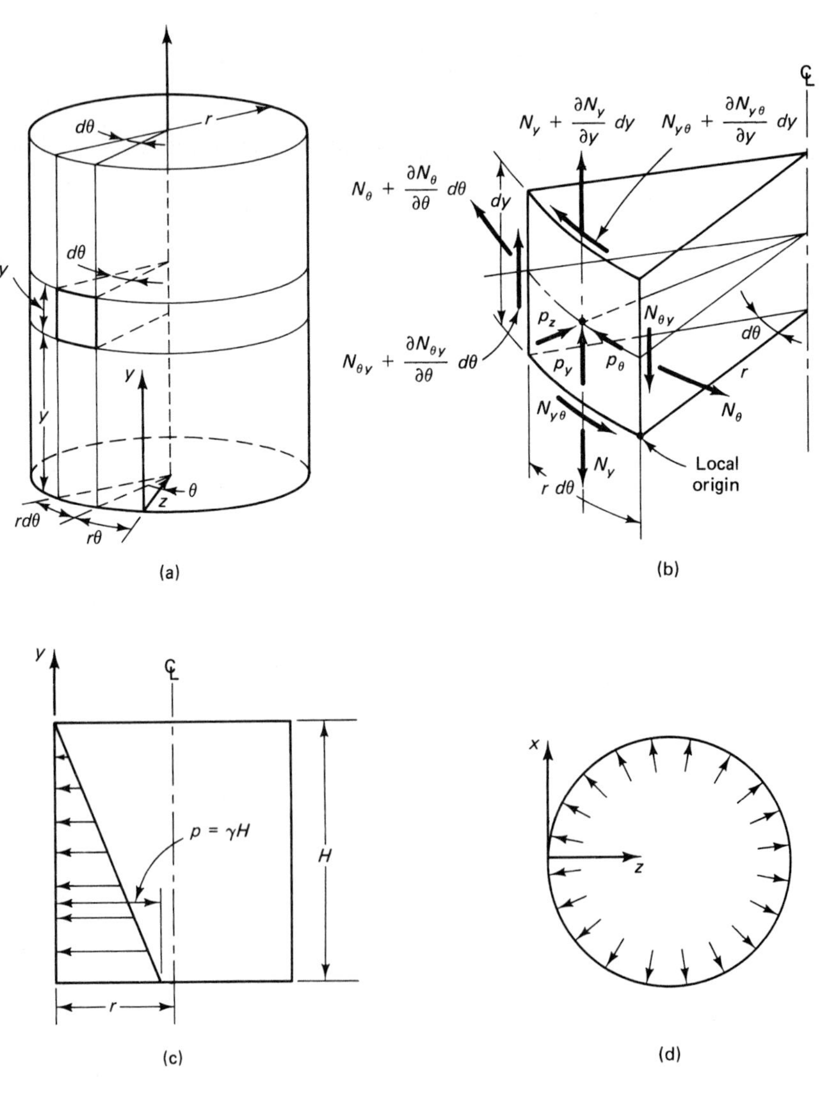

Figure 11.3 Membrane forces in cylindrical tank. (a) Tank shell geometry. (b) Shell membrane forces. (c) Liquid-filled tank elevation. (d) Axisymmetrical internal pressure at any horizontal plane.

Photo 11.2 Panel Being Lifted in a Preload Prestressed Tank (*Courtesy,* N.A. Legatos, Preload Inc., Garden City, New York.)

$$r\frac{\partial N_y}{\partial y} + \frac{\partial N_{\theta y}}{\partial \theta} + p_y r = 0 \tag{11.5b}$$

$$\frac{N_\theta}{r} = +p_z = 0 \tag{11.5c}$$

where $\partial N_{y\theta} = \partial N_{\theta y}$ due to loading symmetry. The unknowns are thus reduced to three, representing a statically *determinate* structure subjected to direct forces only.

For axisymmetrical loading as in Figure 11.3(c), $p_\theta = p_y = 0$ and $p_z = p \cdot f(y)$, independent of θ. Hence,

$$p_z = -\gamma(H - y) \tag{11.6}$$

and the solution to Equation 11.5 is

$$N_{y\theta} = N_y = 0$$

and

$$N_\theta = \gamma(H - y)r \tag{11.7}$$

11.2.2.2 Bending Theory. The introduction of restraint at the boundary of the vessel induces radial ring horizontal shear and vertical moments in the shell. Consequently, the membrane force equations presented in the previous section have to be modified by superimposing these additional moments and shears. The modified expressions are denoted the *bending theory of circular shells;* the theory accounts for strain compatibility requirements in the induced deformations caused by the induced shears and moments.

The bending moments and central shears in the axisymmetrically loaded cylindrical shell are shown by force and moment vectors in Figure 11.4. The infinitesimal element $ABCD$ shows the points of application and sense of the unit moments M_y about the x-axis and M_θ about the y-axis, the circumferential unit moments $M_{y\theta}$ and $M_{\theta y}$, the unit normal shear Q_y acting in the plane of the vertical shell generator and perpendicularly to the shell axis, and the unit radial shear Q_θ acting through the shell radius in the plane of the shell parallels.

Superposition of the moments and shears in Figure 11.4 on the forces in Figure 11.3(b) results in the following equilibrium equations:

$$\frac{\partial N_\theta}{\partial \theta} + \frac{\partial N_{y\theta}}{\partial y} - Q_\theta + p_\theta r = 0 \tag{11.8a}$$

$$\frac{\partial N_y}{\partial y} r + \frac{\partial N_{\theta y}}{\partial \theta} + p_y r = 0 \tag{11.8b}$$

$$\frac{\partial Q_\theta}{\partial \theta} + \frac{\partial Q_y}{\partial y} r + N_\theta + p_z r = 0 \tag{11.8c}$$

$$-\frac{\partial M_y}{\partial y} r + \frac{\partial M_{y\theta}}{\partial y} + Q_y r = 0 \tag{11.8d}$$

$$\frac{\partial M_\theta}{\partial \theta} + \frac{\partial M_{y\theta}}{\partial y} r - Q_\theta r = 0 \tag{11.8e}$$

Due to symmetry of loading, $N_{y\theta} = N_{\theta y} = M_{y\theta} = M_{\theta y} = 0$, and dQ_θ can be disregarded, reducing the partial differential equations 11.8 to the set of the ordinary differential equations

$$\frac{dN_y}{dy} r + p_y r = 0 \tag{11.9a}$$

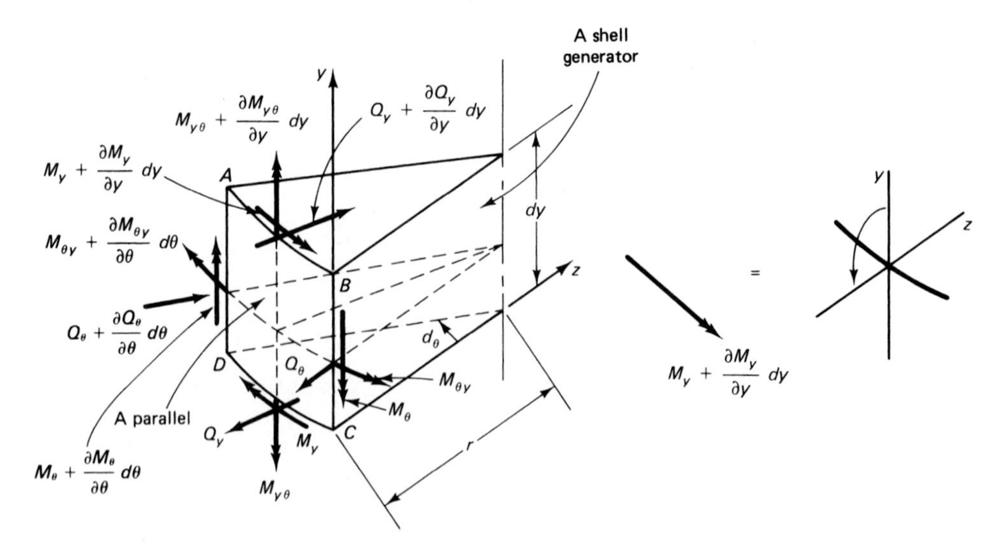

Figure 11.4 Bending moments and normal shears in a cylindrical shell wall.

$$\frac{dQ_y}{dy} r + N_\theta + p_z r = 0 \tag{11.9b}$$

$$-\frac{dM_y}{dy} r + Q_y r = 0 \tag{11.9c}$$

With the central membrane forces N_y constant and taken to be zero (see Refs. 11.1 and 11.3), the remaining equations 11.9b and 11.9c can be written in the following simplified form having the three unknowns N_θ, Q_y, and M_y:

$$\frac{dQ_y}{dy} + \frac{1}{r} N_\theta = -p_z \tag{11.10a}$$

$$\frac{dM_y}{dy} - Q_y = 0 \tag{11.10b}$$

In order to solve these equations, displacements have to be considered and equations of geometry developed.

Force Equations. If v and w are the displacements in the y and z directions, then the unit strains in these directions are, respectively,

$$\epsilon_y = \frac{dv}{dy}$$

and

$$\epsilon_\theta = -\frac{w}{r}$$

which give

$$N_y = \frac{Et}{1-\mu^2}(\epsilon_y + \mu\epsilon_\theta) = \frac{Et}{1-\mu^2}\left(\frac{dv}{dy} - \mu\frac{w}{r}\right) = 0 \tag{11.11a}$$

Photo 11.3 250,000-bbl (39,750-m^3) prestressed concrete propane gas storage container, Winnipeg, Manitoba, Canada. (*Courtesy,* N.A. Legatos, Preload Technology, Inc., New York.)

and

$$N_\theta = \frac{Et}{1-\mu^2}(\epsilon_\theta + \mu\epsilon_y) = \frac{Et}{1-\mu^2}\left(-\frac{w}{r} + \mu\frac{dv}{dy}\right) \tag{11.11b}$$

where μ = Poisson's ratio
t = thickness of the wall core.
From Equation 11.11a,

$$\frac{dv}{dy} = \mu\frac{w}{r} \tag{11.12a}$$

From Equation 11.11b,

$$N_\theta = -Et\frac{w}{r} \tag{11.12b}$$

Moment Equations. Due to symmetry, there is no change in curvature in the circumferential direction; hence, the curvature in the y direction has to be equal to $-d^2v/dy^2$. Using the same moment expressions for thin elastic plates results in

$$M_\theta = \mu M_y \tag{11.13a}$$

$$M_y = -D\frac{d^2w}{dy^2} \tag{11.13b}$$

where $D = Et^3/12(1-\mu^2)$ is the shell or plate flexural rigidity.

Introducing Equations 11.12 and 11.13 into Equations 11.10 results in

$$\frac{d^2}{dx^2}\left(D\frac{d^2w}{dy^2}\right) + \frac{Et}{r^2}w = p_z \tag{11.14}$$

If the wall thickness t is constant, Equation 11.14 becomes

$$D\frac{d^4w}{dy^2} + \frac{Et}{r^2}w = p_z \tag{11.15}$$

Letting

$$\beta^4 = \frac{Et}{4r^2D} = \frac{3(1-\mu^2)}{(rt)^2}$$

Equation 11.15 becomes

$$\frac{d^4w}{dy^4} + 4\beta^4 w = \frac{p_z}{D} \tag{11.16}$$

Equation 11.16 is the same as is obtained for a prismatic bar with flexural rigidity D supported by a continuous elastic foundation and subject to the action of a unit load intensity p_z. The general solution to this equation (Ref. 11.1) for the *radial* displacement in the z-direction is

$$\begin{aligned} w &= e^{\beta y}(C_1\cos\beta y + C_2\sin\beta y) \\ &\quad + e^{-\beta y}(C_3\cos\beta y + C_4\sin\beta y) + f(y) \end{aligned} \tag{11.17}$$

where $f(y)$ is the particular solution of Equation 11.16 as a membrane solution giving displacement

$$w = \frac{p_z r^2}{Et}$$

11.2.3 General Equations of Forces and Displacements

Solving Equation 11.17 and introducing the notation

$$\Phi(\beta y) = e^{-\beta y}(\cos \beta y + \sin \beta y)$$
$$\Psi(\beta y) = e^{-\beta y}(\cos \beta y - \sin \beta y)$$
$$\theta(\beta y) = e^{-\beta y} \cos \beta y$$
$$\zeta(\beta y) = e^{-\beta y} \sin \beta y$$

the expression for radial deformation in the z direction and its consecutive derivatives at any height y above the wall base can be evaluated from the following simplified expressions as a function of the wall base unit moments M_0 and unit radial shears Q_0:

$$\text{Deflection } w = -\frac{1}{2\beta^3 D}[\beta M_0 \psi(\beta y) + Q_0 \theta(\beta y)] \quad (11.18a)$$

$$\text{Rotation } \frac{dw}{dy} = \frac{1}{2\beta^2 D}[2\beta M_0\, \theta(\beta y) + Q_0 \Phi(\beta y)] \quad (11.18b)$$

$$\frac{d^2 w}{dy^2} = -\frac{1}{2\beta D}[2\beta M_0 \Phi(\beta y) + 2Q_0 \zeta(\beta y)] \quad (11.18c)$$

$$\frac{d^3 w}{dy^3} = \frac{1}{D}[2\beta M_0 \zeta(\beta y) - Q_0 \psi(\beta y)] \quad (11.18d)$$

The shell functions $\Phi(\beta y)$, $\psi(\beta y)$, $\theta(\beta y)$, and $\zeta(\beta y)$ are given in the standard influence coefficients of Table 11.1 (Ref. 11.1), for a range $0 \le \beta y \le 3.9$.

The maximum radial displacement or deflection at the restrained wall base, from Equation 11.18a, is

$$(w)_{y=0} = -\frac{1}{2\beta^3 D}(\beta M_0 + Q_0) \quad (11.19a)$$

and the maximum rotation of the wall at the base, from Equation 11.18b, becomes

$$\left(\frac{dw}{dy}\right)_{y=0} = \frac{1}{2\beta^2 D}(2\beta M_0 + Q_0) \quad (11.19b)$$

where M_0 and Q_0 are respectively the restraining moment and the ring shear at the base shown in Figure 11.1.

For tanks with constant wall thickness, the unit forces along the wall height are as follows:

$$N_\theta = -\frac{Etw}{r} \quad (11.20a)$$

$$Q_y = -D\frac{d^3 w}{dy^3} \quad (11.20b)$$

$$M_\theta = \mu M_y \quad (11.20c)$$

$$M_y = -D\frac{d^2 w}{dy^2} \quad (11.20d)$$

Table 11.1 Table of Functions Φ, ψ, θ, and ζ

βy	Φ	ψ	θ	ζ
0	1.0000	1.0000	1.0000	0
0.1	0.9907	0.8100	0.9003	0.0903
0.2	0.9651	0.6398	0.8024	0.1627
0.3	0.9267	0.4888	0.7077	0.2189
0.4	0.8784	0.3564	0.6174	0.2610
0.5	0.8231	0.2415	0.5323	0.2908
0.6	0.7628	0.1431	0.4530	0.3099
0.7	0.6997	0.0599	0.3798	0.3199
0.8	0.6354	−0.0093	0.3131	0.3223
0.9	0.5712	−0.0657	0.2527	0.3185
1.0	0.5083	−0.1108	0.1988	0.3096
1.1	0.4476	−0.1457	0.1510	0.2967
1.2	0.3899	−0.1716	0.1091	0.2807
1.3	0.3355	−0.1897	0.0729	0.2626
1.4	0.2849	−0.2011	0.0419	0.2430
1.5	0.2384	−0.2068	0.0158	0.2226
1.6	0.1959	−0.2077	−0.0059	0.2018
1.7	0.1576	−0.2047	−0.0235	0.1812
1.8	0.1234	−0.1985	−0.0376	0.1610
1.9	0.0932	−0.1899	−0.0484	0.1415
2.0	0.0667	−0.1794	−0.0563	0.1230
2.1	0.0439	−0.1675	−0.0618	0.1057
2.2	0.0244	−0.1548	−0.0652	0.0895
2.3	0.0080	−0.1416	−0.0668	0.0748
2.4	−0.0056	−0.1282	−0.0669	0.0613
2.5	−0.0166	−0.1149	−0.0658	0.0492
2.6	−0.0254	−0.1019	−0.0636	0.0383
2.7	−0.0320	−0.0895	−0.0608	0.0287
2.8	−0.0369	−0.0777	−0.0573	0.0204
2.9	−0.0403	−0.0666	−0.0534	0.0132
3.0	−0.0423	−0.0563	−0.0493	0.0071
3.1	−0.0431	−0.0469	−0.0450	0.0019
3.2	−0.0431	−0.0383	−0.0407	−0.0024
3.3	−0.0422	−0.0306	−0.0364	−0.0058
3.4	−0.0408	−0.0237	−0.0323	−0.0085
3.5	−0.0389	−0.0177	−0.0283	−0.0106
3.6	−0.0366	−0.0124	−0.0245	−0.0121
3.7	−0.0341	−0.0079	−0.0210	−0.0131
3.8	−0.0314	−0.0040	−0.0177	−0.0137
3.9	−0.0286	−0.0008	−0.0147	−0.0140

From Equations 11.18c, 11.18d, 11.20b, and 11.20d, the expressions for vertical moments and horizontal radial shears at the base of the wall, where y is zero, become (Ref. 11.1)

$$(M_y)_{y=0} = M_0 = \left(1 - \frac{1}{\beta H}\right)\frac{\gamma Hrt}{\sqrt{12(1-\mu^2)}} \tag{11.21a}$$

$$(Q_y)_{y=0} = Q_0 = -(2\beta H - 1)\frac{\gamma rt}{\sqrt{12(1-\mu^2)}} \tag{11.21b}$$

The expression for the vertical moment at any level y above the wall base can be obtained from

$$M_y = -\frac{1}{\beta}[\beta M_0 \Phi(\beta y) + Q_0 \zeta(\beta y)] \tag{11.22}$$

The *offset* ring shear force ΔQ_y corresponds to a radial displacement w_y of the wall at a height y above the base when the tank is *empty* and the values of Q_0 and M_0 due to a full liquid or full gas load are *induced,* as shown in Figure 11.5. This force can be expressed as either

$$\Delta Q_y = +\frac{Et}{r}(w_y)$$

or

$$\Delta Q_y = \frac{Et}{2r\beta^3 D}[\beta M_0 \psi(\beta y) + Q_0 \theta(\beta y)]$$

or

$$\Delta Q_y = +\frac{6(1-\mu^2)}{\beta^3 rt^2}[\beta M_0 \psi(\beta y) + Q_0 \theta(\beta y)] \tag{11.23}$$

The ring shear Q_y at a plane y above the base would be equal to the difference between the ring force for a freely sliding base and ΔQ_y:

$$Q_y = F - \Delta Q_y \tag{11.24}$$

It is important to be consistent in the sign convention used throughout a solution. The easiest approach is to draw the deflected shape of the wall and use a positive (+) notation for the following conditions:

1. Moment causing tension on the outside extreme fibers.
2. Ring tension radial forces.
3. Thrust inwards toward the vertical axis. Here, the same sense is used as for ring tension forces in order to draw the diagram for the balancing prestressing forces on the same side as the ring tension forces for comparison.
4. Lateral wall movement *inwards* toward the vertical axis.
5. Anticlockwise rotation.

Pinned Wall Base, Liquid Pressure. When the wall base is *pinned* and carrying a liquid load moment $M_0 = 0$ at the base,

$$Q_0 = +\frac{2\beta^3 \gamma H (rt)^2}{12(1-\mu^2)}$$

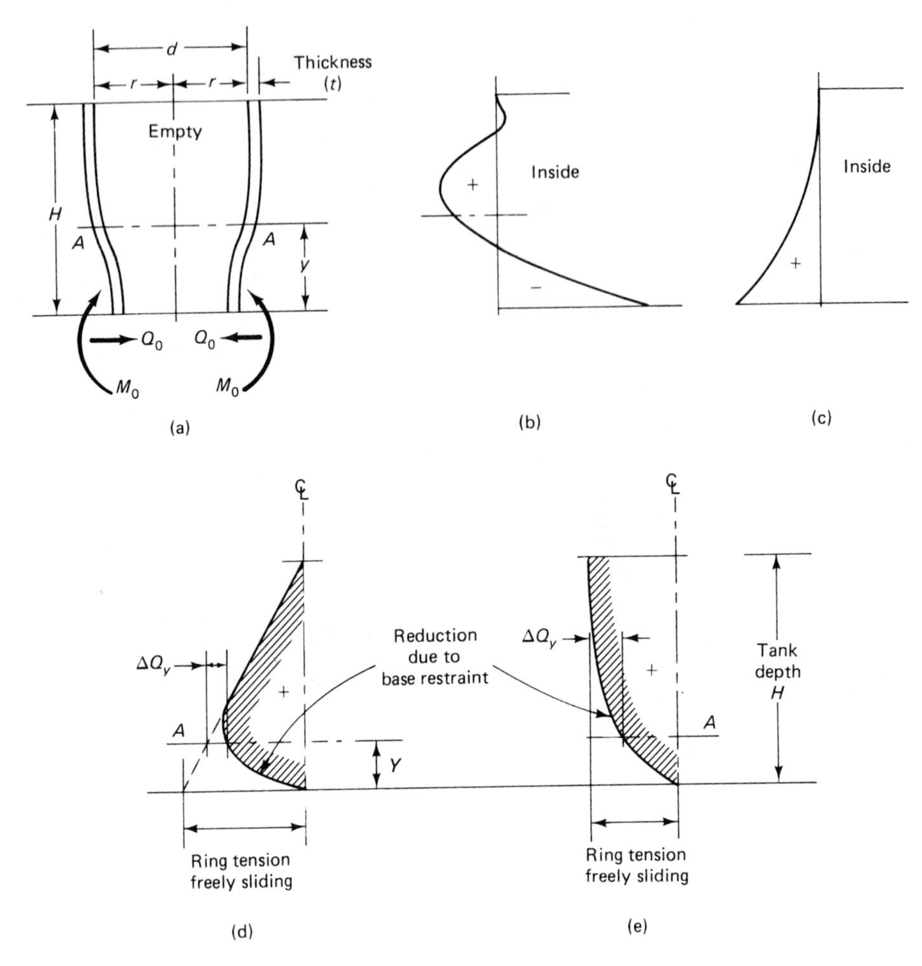

Figure 11.5 Wall base restraint in empty tank inducing M_o and Q_o for full liquid or gas pressure. (a) Deformed walls of empty tank. (b) Moment along vertical section (+ represents tension on outside). (c) Ring tension force F in horizontal section (always positive). (d) Offset ΔQ_v for liquid pressure. (e) Offset ΔQ_y for gas pressure.

or

$$Q_0 = +\frac{\gamma H}{[12(1-\mu^2)]^{1/4}}\left(\frac{rt}{2}\right)^{1/2} \tag{11.25}$$

The value of the shell constants β, β^2, and β^4 for use in the preceding equations can easily be computed from the expression for β^4 as follows:

$$\beta^4 = \frac{Et}{4r^2D} = \frac{3(1-\mu^2)}{(rt)^2} \tag{11.26a}$$

$$\beta^3 = \frac{[3(1-\mu^2)]^{3/4}}{(rt)^{3/2}} \tag{11.26b}$$

$$\beta^2 = \frac{[3(1-\mu^2)]^{1/2}}{(rt)} \tag{11.26c}$$

Photo 11.4 Wire Winding Operation (*Courtesy,* N.A. Legatos, Preload Inc., Garden City, New York.)

$$\beta = \frac{[3(1-\mu^2)]^{1/4}}{(rt)^{1/2}} \tag{11.26d}$$

11.2.4 Ring Shear Q_0 and Moment M_0 Gas Containment

If the edges of the shell are free at the wall base, the internal pressure produces only hoop stress $f_R = pr/t$ and the radius of the cylinder increases by the amount

$$w = \frac{rf_R}{E} = \frac{pr^2}{Et} \tag{11.27}$$

Also, for full restraint at the wall base,

$$(w)_{y=0} = \frac{1}{2\beta^3 D}(\beta M_0 + Q_0) \tag{11.28a}$$

and

$$\left(\frac{dw}{dy}\right)_{y=0} = \frac{1}{2\beta^2 D}(2\beta M_0 + Q_0) = 0 \tag{11.28b}$$

Solving for M_0 and Q_0 gives

$$M_0 = -2\beta^2 Dw = -\frac{p}{2\beta^2} = -\frac{prt}{\sqrt{12(1-\mu^2)}} \tag{11.29a}$$

and

$$Q_0 = +4B^3 Dw = +\frac{p}{\beta} = +\frac{p(2rt)^{1/2}}{[12(1-\mu^2)]^{1/4}} \tag{11.29b}$$

Table 11.2 Equations for Liquid-Retaining Tanks

Parameter	Equation	Number
Flexural rigidity, D	$Et^3/[12(1-\mu^2)]$	
Ring stress, f_R	$\gamma(H-Y)r/t$	11.2 a
Ring force, F	$\gamma(H-y)r$	11.2 b
Pressure, P_z	$\gamma(H-y)$	11.2 b
Radial deflection, w	$\frac{1}{2\beta^2 D}[\beta M_0\psi(\beta y)+Q_0\theta(\beta y)$	11.18 a
Rotation $\frac{dw}{dy}$	$\frac{1}{2\beta^3 D}[2\beta M_0\theta(\beta y)+Q_0\Phi(\beta y)]$	11.18 b
Maximum deflection, $(w)_{y=0}$	$\frac{1}{2\beta^3 D}(\beta M_0+Q_0)$	11.19 a
Maximum rotation $\left(\frac{dw}{dy}\right)_{y=0}$	$\frac{1}{2\beta^3 D}(2\beta M_0+Q_0)=0$	11.19 b
$M_0=(M_y)_{y=0}$	$-\left(1-\frac{1}{\beta H}\right)\frac{\gamma Hrt}{\sqrt{12(1-\mu^2)}}$	11.21 a
$Q_0=(Q_y)_{y=0}$	$+(2\beta H-1)\frac{\gamma rt}{\sqrt{12(1-\mu)^2}}$	11.21 b
M_y	$+\frac{1}{\beta}[\beta M_0\Phi(\beta y)+Q_0\zeta(\beta y)]$	11.22
Empty tank offset, ΔQ_y	$+\frac{6(1-\mu)^2}{\beta^3 rt^2}[\beta M_0\psi(\beta y)+(Q_0(\beta y)]$	11.23
Q_y	$+(F-\Delta Q_y)$	11.24
Q_0 when $M_0=0$ (Pinned base)	$+\frac{\gamma H\sqrt{rt/2}}{[12(1-\mu^2)]^{1/4}}$	11.25
Tank Constants: β^3	$[3(1-\mu^2)]^{3/4}/(rt)^{3/2}$	11.26 b
β^2	$[3(1-\mu^2]^{1/2}/rt$	11.26 c
β	$[3(1-\mu^2)]^{1/4}/(rt)^{1/2}$	11.26 d

Pinned Wall Base, Gas Pressure. If the wall base is *pinned* and carrying a gas load moment $M_0=0$ at the base,

$$Q_0 = 2\beta^3 D\left(\frac{pr^2}{Et}\right)$$

or

$$Q_0 = \frac{p}{[12(1-\mu^2)]^{1/4}}\left(\frac{rt}{2}\right)^{1/2} \tag{11.30}$$

Table 11.2 presents a summary of the design equations for liquid-retaining tanks, and Table 11.3 gives a similar summary for gas-retaining tanks.

11.3 MOMENT M_0 AND RING FORCE Q_0 IN LIQUID RETAINING TANK

Example 11.1

A prestressed concrete circular tank is fully restrained at the wall base. It has an interior diameter $d=125$ ft (38.1 m) and retains water having height $H=25$ ft (7.62 m). The wall thick-

Table 11.3 Equations for Gas-Retaining Tanks

Parameter	Equation	Number
Maximum deflection $(w)_{y=0}$	$\frac{1}{2\beta^3 D}(\beta M_0 + Q_0)$	11.28 a
Maximum rotation $\left(\frac{dw}{dy}\right)_{y=0}$	$\frac{1}{2\beta^2 D}(2\beta M_0 + Q_0) = 0$	11.28 b
$M_0 = (M_y)_{y=0}$	$-\frac{prt}{\sqrt{12(1-\mu^2)}}$	11.29 a
$Q_0 = (Q_y)_{y=0}$	$+\frac{p\sqrt{2rt}}{[12(1-\mu^2)]^{1/4}}$	11.29 b
Q_0 when $M_0 = 0$ (Pinned base)	$+\frac{p\sqrt{rt/2}}{[12(1-\mu^2)]^{1/4}}$	11.30

Note: Values of β, β^2, and β^3 constants as in Table 11.2.

ness $t = 10$ in. (25 cm). Compute (a) the unit vertical moment M_0 and the radial ring force Q_0 at the base of the wall, and (b) the unit vertical moment M_y at $7\frac{1}{2}$ ft (2.29 m) above the base. Use Poisson's ratio $\mu = 0.2$ and unit water weight $\gamma = 62.4$ lb/ft^3 (1,000 kg/m^3).

Solution:

(a) *At Wall Base*

$$r = \frac{1}{2} \times 125 = 62.5 \text{ ft (19 m)}$$

$$t = 10 \text{ in.} = 0.83 \text{ ft (.25 m)}$$

From Equation 11.26d,

$$\beta = \frac{[3(1-\mu^2)]^{1/4}}{(rt)^{1/2}} = \frac{[3(1-0.2\times 0.2)]^{1/4}}{(62.5\times 0.83)^{1/2}} = 0.181$$

From Equation 11.21a,

$$M_0 = -\left(1 - \frac{1}{\beta H}\right)\frac{\gamma H r t}{\sqrt{12(1-\mu^2)}}$$

$$= -\left(1 - \frac{1}{0.181\times 25}\right) \times \frac{62.4\times 25\times 62.5\times 0.83}{\sqrt{12(1-0.04)}}$$

$$= -18{,}574 \text{ ft-lb/ft (7.68 kN-m/m) of circumference}$$

From Equation 11.21b,

$$Q_0 = +(2\beta H - 1)\frac{\gamma r t}{\sqrt{12(1-\mu^2)}}$$

$$= +(2\times 0.181\times 25 - 1)\frac{62.4\times 62.5\times 0.83}{\sqrt{12(1-0.04)}}$$

$$= +7{,}677 \text{ lb/ft (112 kN/m) of circumference}$$

(b) *$7\frac{1}{2}$ ft above Wall Base*

$$y = 7.5 \text{ ft}$$

$$\text{Water height} = (H - y) = 25 - 7.5 = 17.5 \text{ ft (5.33 m)}$$

$$\text{Height ratio} = \left(1 - \frac{y}{H}\right) = 1 - \frac{7.5}{25} = 0.7$$

$$\beta y = 0.181 \times 7.5 = 1.36$$

From Equation 11.22,

$$M_y = +\frac{1}{\beta}[\beta M_0 \Phi(\beta y) + Q_0 \zeta(\beta y)]$$

From Table 11.1 for $\beta_y = 1.36$,

$$\Phi = 0.311$$

$$\zeta = 0.252$$

$$M_y = +\frac{1}{0.181}(-0.181 \times 18{,}574 \times 0.311 + 7{,}677 \times 0.252)$$

$$= +4{,}912 \text{ ft-lb/ft of circumference}$$

11.4 RING FORCE Q_Y AT INTERMEDIATE HEIGHTS OF WALL

Example 11.2

Compute the radial ring force Q_y in Example 11.1 at (a) $y = 7\frac{1}{2}$ ft (2.29 m) and (b) $y = 10$ ft (3.05 m) above the wall base.

Solution: The freely sliding base ring force $F = \gamma H r = 62.4 \times 25 \times 62.5 = 97{,}500$ lb/ft (1,423 kN/m). From Equation 11.23, the ring force offset is

$$\Delta Q_y = +\frac{6(1-\mu^2)}{\beta^3 r t^2}[\beta M_0 \psi(\beta y) + Q_0 \theta(\beta y)]$$

From Example 11.1, $\beta = 0.181$; hence, $\beta^3 = 0.0059$,

(a) *Q_y at 7.5 ft above Wall Base*

$$\beta y = 0.181 \times 7.5 = 1.36$$

From Table 11.1 for $\beta y = 1.36$,

$$\psi = -0.1965$$

$$\theta = +0.0543$$

$$\Delta Q_y = +\frac{6(1-0.04)}{0.0059 \times 62.5(0.83)^2}$$

$$\times [0.181(-18{,}574)(-0.1965) + 7{,}677(+0.0543)]$$

$$= 24{,}431 \text{ lb/ft (356 kN/m)}$$

From Equation 11.2b, the ring force $F = \gamma(H - y)r = 62.4 \times (25 \times 7.5) \times 62.5 = 68{,}250$ lb/ft. So $Q_{7.5} = F - \Delta Q_y = 68{,}250 - 24{,}431 = 43{,}819$ lb/ft (705 kN/m) of circumference, as shown in Figure 11.6(a): (a) At $7\frac{1}{2}$ ft above the base; (b) At 10 ft above the base.

(b) *Q_y at 10.0 ft above Wall Base*

$$\beta y = 0.181 \times 10 = 1.81$$

From Table 11.1 for $\beta_y = 1.81$,

$$\psi = -0.1984$$

$$\theta = -0.0387$$

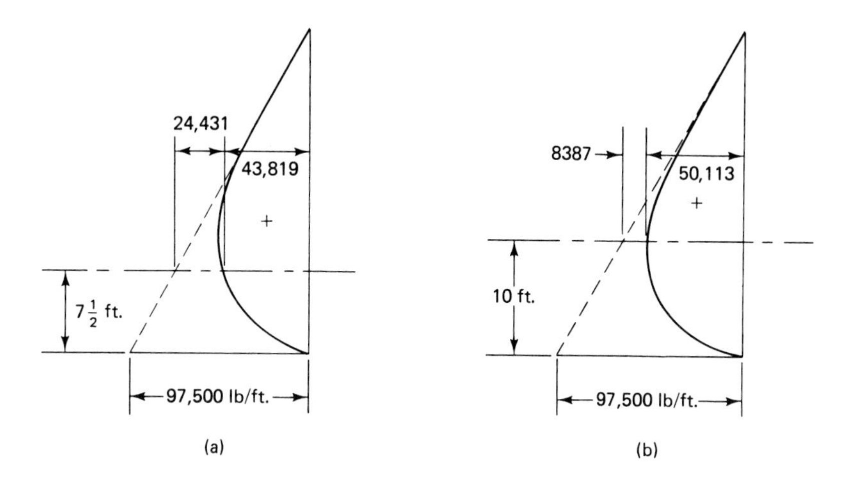

Figure 11.6 Radial ring force profile. (a) At $7\frac{1}{2}$ ft above the base. (b) At 10 ft above the base in Ex. 11.1.

$$\Delta Q_y = \frac{6(1 - 0.04)}{0.0059 \times 62.50(0.83)^2}$$

$$\times \left[0.181(-18{,}574)(-0.1984) + 7{,}677(-0.0387)\right] = 8.387 \text{ lb/ft}$$

The ring force $F = \gamma(H - y)r = 62.4(25 - 10)62.5 = 58{,}500$ lb/ft. So $Q_{10} = F - \Delta Q_y = 58{,}500 - 8{,}387 = 50{,}113$ lb/ft (731 kN/m) of circumference, as shown in Figure 11.6(b). Compare how close this value is to $Q = 50{,}115$ lb/ft obtained by using membrane coefficients in Example 11.3.

11.5 CYLINDRICAL SHELL MEMBRANE COEFFICIENTS

The bending moment at any level along the height above the base of a cylindrical tank can be computed from the bending moment expression for a cantilever beam. This is accomplished by multiplying the cantilever moment values by coefficients whose magnitudes are functions of the geometrical dimensions of the tank and which are termed *membrane coefficients.* The basic moment expressions developed in Section 11.2 for the circular container can be rearranged into a factor H^2/dt denoting *geometry* and a factor γH^3 or pH^2 denoting *cantilever effect,* for liquid and gaseous loading, respectively (Ref. 11.2).

The tank constant β in Equation 11.26d is a function of rt or dt, where d is the tank diameter. Using Poisson's ratio $\mu \cong 0.2$ for concrete, we have

$$\beta = \frac{[3(1 - \mu^2)]^{1/4}}{(rt)^{1/2}} = \frac{1.30}{(rt)^{1/2}} = \frac{1.84}{(dt)^{1/2}}$$

The factor $1/\beta H$ used in the basic bending expressions of Section 11.2 can be rewritten in terms of $(dt/H^2)^{1/2}$ since $\beta = 1.84/(dt)^{1/2}$. The product β_y can also be rewritten in terms of $\lambda(H^2/dt)^{1/2}$ using $y = \lambda H$, where y is the height above the base.

Consequently, the moment M_y of Equation 11.22 in a wall section a distance y above the base can be represented in terms of the form factor H^2/dt and the cantilever factor γH^3 or pH^2 as follows:

$$M_y = \text{numerical variant} \times \text{form factor} \times \text{cantilever factor}$$

Photo 11.5 Two-and-a-half-million-gallon tendon prestressed concrete tank with the horizontal and vertical tendons utilizing plastic sheathing to protect the prestressing steel from seepage through the wall. (*Courtesy,* Jorgenson, Hendrickson and Close, Denver, Colorado.)

or

$$M_y = \left[\text{variant} \times \frac{H^2}{dt} \right] \times [\gamma H^3 \quad \text{or} \quad pH^2] \tag{11.31}$$

The form factor H^2/dt is constant for the particular structure being designed. Hence, the product of the variant and the form factor produces the membrane coefficient C, so that Equation 11.31 becomes

$$M_y = C\gamma H^3 \tag{11.32a}$$

for a liquid load and

$$M_y = CpH^2 \tag{11.32b}$$

for a gaseous load.

Tables 11.4 to 11.16 from Ref. 11.5 give the membrane coefficients C for various form factors H^2/dt and most expected boundary and load conditions. They significantly reduce the computational efforts normally required in the design and analysis of shells, without loss of accuracy in the results. Using the membrane coefficients for the solution

Photo 11.6 Prestressing preload circular tank wall with wire winder. (*Courtesy,* N.A. Legatos, Preload Technology, Inc., New York.)

of the circular tank forces and moments should give results reasonably close to those obtained from the bending solutions presented in Section 11.2 and the sets of equations listed in Tables 11.2 and 11.3.

11.6 PRESTRESSING EFFECTS ON WALL STRESSES FOR FULLY HINGED, PARTIALLY SLIDING AND HINGED, FULLY FIXED, AND PARTIALLY FIXED BASES

The liquid or gas contained in a cylindrical tank exerts *outward* radial pressure γh or p on the tank walls, inducing ring tensions in each horizontal section of wall along its height. This ring tension in turn causes tensile stresses in the concrete at the *outside* extreme wall fibers, resulting in impermissible cracking. To eliminate this cracking that causes leaks and structural deterioration, external *horizontal* prestressing is applied which induces *inward* radial thrust that can balance the outward radial tension. Additionally, in order to prevent the development of cracks in the inside walls when the tank is empty, *vertical* prestressing is induced to reduce the residual tension within the range of the modulus of rupture of the concrete and with an adequate safety factor.

In order to ensure against the development of cracking at the outside face of the tank wall, it is good practice to apply somewhat larger horizontal prestressing forces than

(text continues on page 676)

Table 11.4 Moment Influence Coefficients, Triangular Load

Moments in Cylindrical Wall
Triangular Load
Fixed Base, Free Top
Mom. = coef. × γH^2 ft. lb. per ft.
Positive sign indicates tension in the outside

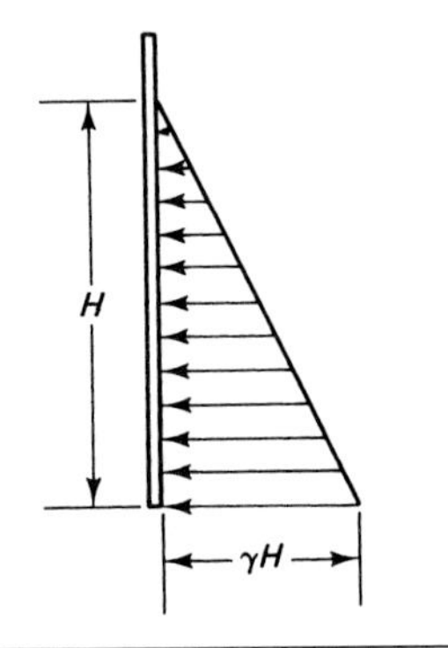

Liquid Load

$\frac{H^2}{dt}$	Coefficients at Point									
	0.1*H*	0.2*H*	0.3*H*	0.4*H*	0.5*H*	0.6*H*	0.7*H*	0.8*H*	0.9*H*	1.0*H*
0.4	+.0005	+.0014	+.0021	+.0007	−.0042	−.0150	−.0302	−.0529	−.0816	−.1205
0.8	+.0011	+.0037	+.0063	+.0080	+.0070	+.0023	−.0068	−.0224	−.0465	−.0795
1.2	+.0012	+.0042	+.0077	+.0103	+.0112	+.0090	+.0022	−.0108	−.0311	−.0602
1.6	+.0011	+.0041	+.0075	+.0107	+.0121	+.0111	+.0058	−.0051	−.0232	−.0505
2.0	+.0010	+.0035	+.0068	+.0099	+.0120	+.0115	+.0075	−.0021	−.0185	−.0436
3.0	+.0006	+.0024	+.0047	+.0071	+.0090	+.0097	+.0077	+.0012	−.0119	−.0333
4.0	+.0003	+.0015	+.0028	+.0047	+.0066	+.0077	+.0069	+.0023	−.0080	−.0268
5.0	+.0002	+.0008	+.0016	+.0029	+.0046	+.0059	+.0059	+.0028	−.0058	−.0222
6.0	+.0001	+.0003	+.0008	+.0019	+.0032	+.0046	+.0051	+.0029	−.0041	−.0187
8.0	.0000	+.0001	+.0002	+.0008	+.0016	+.0028	+.0038	+.0029	−.0022	−.0146
10.0	.0000	.0000	+.0001	+.0004	+.0007	+.0019	+.0029	+.0028	−.0012	−.0122
12.0	.0000	−.0001	+.0001	+.0002	+.0003	+.0013	+.0023	+.0026	−.0005	−.0104
14.0	.0000	.0000	.0000	.0000	+.0001	+.0008	+.0019	+.0023	−.0001	−.0090
16.0	.0000	.0000	−.0001	−.0002	−.0001	+.0004	+.0013	+.0019	+.0001	−.0079

Notes: 1-Tables 11.4 to 11.16 Adapted from Ref. 11.5.
2-0.0H is the top and 1.0H is the bottom of the wall, except if wall is fixed at top and with shear and moment at top.
3-Shear acting inwards is positive; moment applied at an edge is positive when outward rotation results at that edge.

Table 11.5 Moment Influence Coefficients, Rectangular Load

Moments in Cylindrical Wall
Rectangular Load
Fixed Base, Free Top
Mom. = coef. × pH^2 ft. lb. per ft.
Positive sign indicates tension in the outside

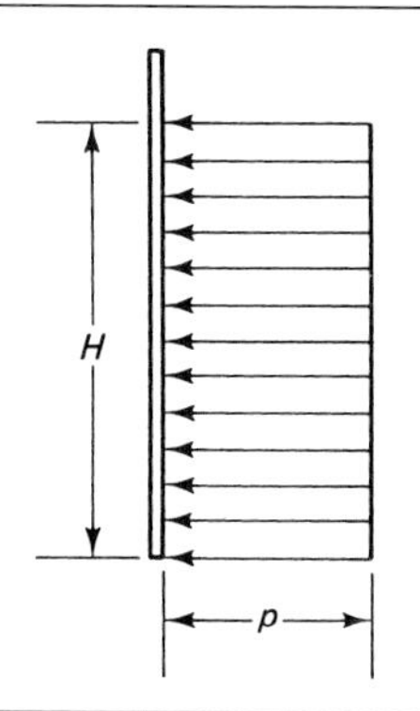

Gas Load

$\frac{H^2}{dt}$	Coefficients at Point									
	0.1*H*	0.2*H*	0.3*H*	0.4*H*	0.5*H*	0.6*H*	0.7*H*	0.8*H*	0.9*H*	1.0*H*
0.4	−.0023	−.0093	−.0227	−.0439	−.0710	−.1018	−.1455	−.2000	−.2593	−.3310
0.8	.0000	−.0006	−.0025	−.0083	−.0185	−.0362	−.0594	−.0917	−.1325	−.1835
1.2	+.0008	+.0026	+.0037	+.0029	−.0009	−.0089	−.0227	−.0468	−.0815	−.1178
1.6	+.0011	+.0036	+.0062	+.0077	+.0068	+.0011	−.0093	−.0670	−.0529	−.0876
2.0	+.0010	+.0036	+.0066	+.0088	+.0089	+.0059	−.0019	−.0167	−.0389	−.0719
3.0	+.0007	+.0026	+.0051	+.0074	+.0091	+.0083	+.0042	−.0053	+.0223	−.0483
4.0	+.0004	+.0015	+.0033	+.0052	+.0068	+.0075	+.0053	−.0013	−.0145	−.0365
5.0	+.0002	+.0008	+.0019	+.0035	+.0051	+.0061	+.0052	+.0007	−.0101	−.0293
6.0	+.0001	+.0004	+.0011	+.0022	+.0036	+.0049	+.0048	+.0017	−.0073	−.0242
8.0	+.0000	+.0001	+.0003	+.0008	+.0018	+.0031	+.0038	+.0024	−.0040	−.0184
10.0	.0000	−.0001	.0000	+.0002	+.0009	+.0021	+.0030	+.0026	−.0022	−.0147
12.0	.0000	.0000	−.0001	.0000	+.0004	+.0014	+.0024	+.0022	−.0012	−.0123
14.0	.0000	.0000	.0000	.0000	+.0002	+.0010	+.0018	+.0021	−.0007	−.0105
16.0	.0000	.0000	.0000	−.0001	+.0001	+.0006	+.0012	+.0020	−.0005	−.0091

Table 11.6 Moment Influence Coefficients, Trapezoidal Load

Moments in Cylindrical Wall
Trapezoidal Load
Hinged Base, Free Top
Mom. = coef. × ($\gamma H^2 + pH^2$) ft. lb. per ft.
Positive sign indicates tension in the outside

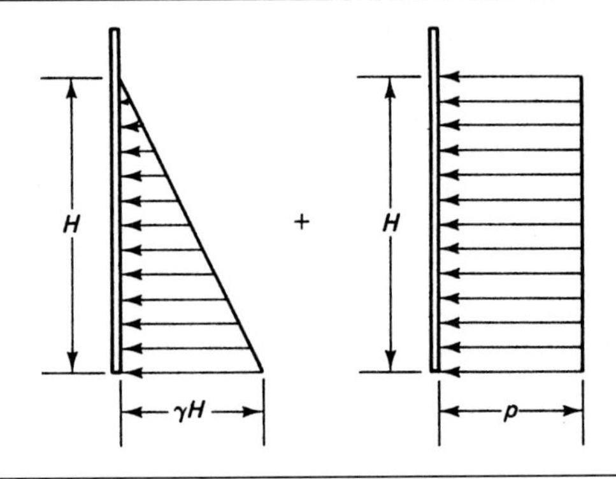

$\frac{H^2}{dt}$	Coefficients at Point									
	0.1*H*	0.2*H*	0.3*H*	0.4*H*	0.5*H*	0.6*H*	0.7*H*	0.8*H*	0.9*H*	1.0*H*
0.4	+.0020	+.0072	+.0151	+.0230	+.0301	+.0348	+.0357	+.0312	+.0197	0
0.3	+.0019	+.0064	+.0133	+.0207	+.0271	+.0319	+.0329	+.0292	+.0187	0
1.2	+.0016	+.0058	+.0111	+.0177	+.0237	+.0280	+.0296	+.0263	+.0171	0
1.6	+.0012	+.0044	+.0091	+.0145	+.0195	+.0236	+.0255	+.0232	+.0155	0
2.0	+.0009	+.0033	+.0073	+.0114	+.0158	+.0199	+.0219	+.0205	+.0145	0
3.0	+.0004	+.0015	+.0040	+.0063	+.0092	+.0127	+.0152	+.0153	+.0111	0
4.0	+.0001	+.0007	+.0016	+.0033	+.0057	+.0083	+.0109	+.0118	+.0092	0
5.0	.0000	+.0001	+.0006	+.0016	+.0034	+.0057	+.0080	+.0094	+.0078	0
6.0	.0000	.0000	+.0002	+.0008	+.0019	+.0039	+.0062	+.0078	+.0068	0
8.0	.0000	.0000	−.0002	.0000	+.0007	+.0020	+.0038	+.0057	+.0054	0
10.0	.0000	.0000	−.0002	−.0001	+.0002	+.0011	+.0025	+.0043	+.0045	0
12.0	.0000	.0000	−.0001	−.0002	.0000	+.0005	+.0017	+.0032	+.0039	0
14.0	.0000	.0000	−.0001	−.0001	−.0001	.0000	+.0012	+.0026	+.0033	0
16.0	.0000	.0000	.0000	−.0001	−.0002	−.0004	+.0008	+.0022	+.0029	0

Table 11.7 Moment Influence Coefficients, Empty Tank (Shear Applied at Top Base Fixed)

Moments in Cylindrical Wall
Shear Per Ft., Q, Applied at Top
Fixed Base, Free Top
Mom. = coef. × VH ft. lb. per ft.
Positive sign indicates tension in the outside

Empty tank

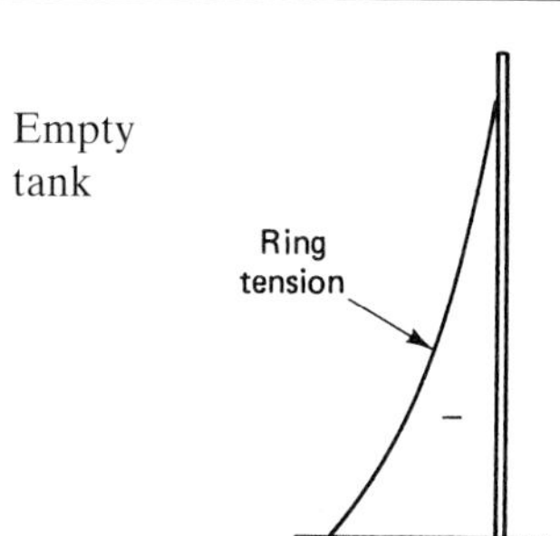

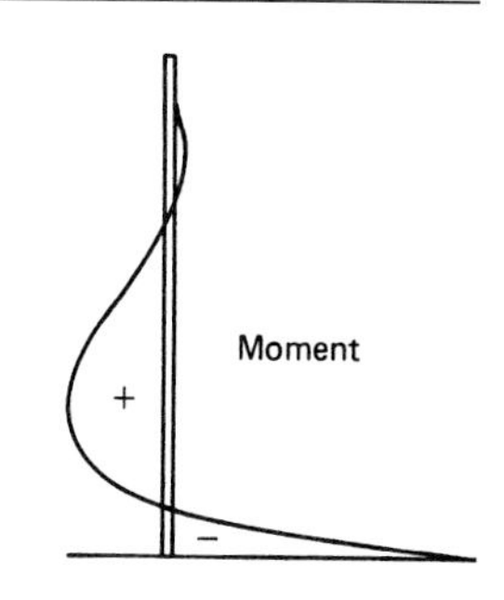

$\frac{H^2}{dt}$	Coefficients at Point									
	0.1H	0.2H	0.3H	0.4H	0.5H	0.6H	0.7H	0.8H	0.9H	1.0H
0.4	+0.093	+0.172	+0.240	+0.300	+0.354	+0.402	+0.448	+0.492	+0.535	+0.578
0.8	+0.085	+0.145	+0.185	+0.208	+0.220	+0.224	+0.223	+0.219	+0.214	+0.208
1.2	+0.082	+0.132	+0.157	+0.164	+0.159	+0.145	+0.127	+0.106	+0.084	+0.062
1.6	+0.079	+0.122	+0.139	+0.138	+0.125	+0.105	+0.081	+0.056	+0.030	+0.004
2.0	+0.077	+0.115	+0.126	+0.119	+0.103	+0.080	+0.056	+0.031	+0.006	+0.019
3.0	+0.072	+0.100	+0.100	+0.086	+0.066	+0.044	+0.025	+0.006	–0.010	–0.024
4.0	+0.068	+0.088	+0.081	+0.063	+0.043	+0.025	+0.010	–0.001	–0.010	–0.019
5.0	+0.064	+0.078	+0.067	+0.047	+0.028	+0.013	+0.003	–0.003	–0.007	–0.011
6.0	+0.062	+0.070	+0.056	+0.036	+0.018	+0.006	0.000	–0.003	–0.005	–0.006
8.0	+0.057	+0.058	+0.041	+0.021	+0.007	0.000	–0.002	–0.003	–0.002	–0.001
10.0	+0.053	+0.049	+0.029	+0.012	+0.002	–0.002	–0.002	–0.002	–0.001	–0.000
12.0	+0.049	+0.042	+0.022	+0.007	0.000	–0.002	–0.002	–0.001	0.000	0.000
14.0	+0.046	+0.036	+0.017	+0.004	–0.001	–0.002	–0.001	–0.001	0.000	0.000
16.0	+0.044	+0.031	+0.012	+0.001	–0.002	–0.002	–0.001	0.000	0.000	0.000

Table 11.8 Moment Influence Coefficients, Empty Tank (Shear Applied at Top Hinged Base)

Moments in Cylindrical Wall
Moment Per Ft., M, Applied at Base
Hinged Base, Free Top
Mom. = coef. × M ft. lb. per ft.
Positive sign indicates tension in the outside

Empty
Tank

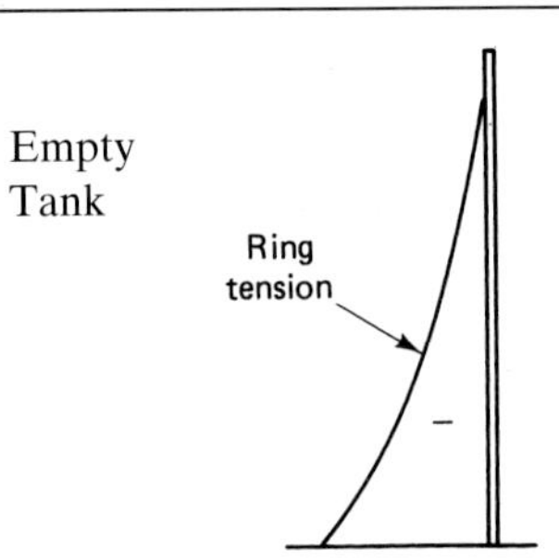

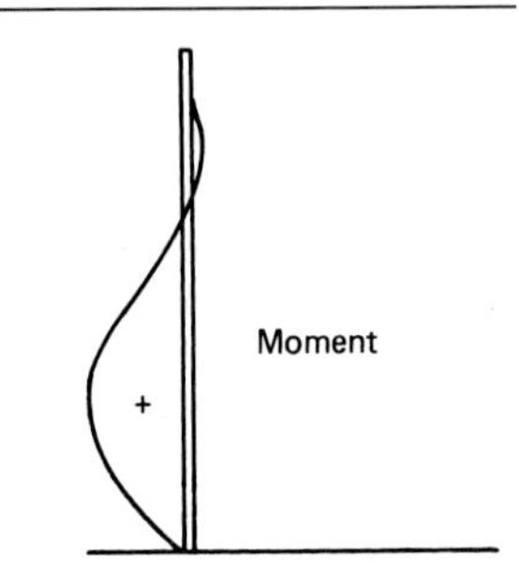

$\frac{H^2}{dt}$	Coefficients at Point									
	0.1*H*	0.2*H*	0.3*H*	0.4*H*	0.5*H*	0.6*H*	0.7*H*	0.8*H*	0.9*H*	1.0*H*
0.4	+0.013	+0.051	+0.109	+0.196	+0.296	+0.414	+0.547	+0.692	+0.843	+1.000
0.8	+0.009	+0.040	+0.090	+0.164	+0.253	+0.375	+0.503	+0.659	+0.824	+1.000
1.2	+0.006	+0.027	+0.063	+0.125	+0.206	+0.316	+0.454	+0.616	+0.802	+1.000
1.6	+0.003	+0.011	+0.035	+0.078	+0.152	+0.253	+0.393	+0.570	+0.775	+1.000
2.0	–0.002	–0.002	+0.012	+0.034	+0.096	+0.193	+0.340	+0.519	+0.748	+1.000
3.0	–0.007	–0.022	–0.030	–0.029	+0.010	+0.087	+0.227	+0.426	+0.692	+1.000
4.0	–0.008	–0.026	–0.044	–0.051	–0.034	+0.023	0.150	+0.354	+0.645	+1.000
5.0	–0.007	–0.024	–0.045	–0.061	–0.057	–0.015	+0.095	+0.296	0.606	+1.000
6.0	–0.005	–0.018	–0.040	–0.058	–0.065	–0.037	+0.057	+0.252	+0.572	+1.000
8.0	–0.001	–0.009	–0.022	–0.044	–0.068	–0.062	+0.002	+0.178	+0.515	+1.000
10.0	0.000	–0.002	–0.009	–0.028	–0.053	–0.067	–0.031	+0.123	+0.467	+1.000
12.0	0.000	0.000	–0.003	–0.016	–0.040	–0.064	–0.049	+0.081	+0.424	+1.000
14.0	0.000	0.000	0.000	–0.008	–0.029	–0.059	–0.060	+0.048	+0.387	+1.000
16.0	0.000	0.000	+0.002	–0.003	–0.021	–0.051	–0.066	+0.025	+0.354	+1.000

Table 11.9 Shear Q Influence Coefficients

Shear at Base of Cylindrical Wall

$$Q = \text{coef.} \times \begin{cases} \gamma H^2 \text{ lb. (triangular)} \\ pH \text{ lb. (rectangular)} \\ M/H \text{ lb. (mom. at base)} \end{cases}$$

Positive sign indicates shear acting inward

$\frac{H^2}{dt}$	Triangular load, fixed base	Rectangular load, fixed base	Triangular or rectangular load, hinged base
0.4	0.436	0.755	0.245
0.8	0.374	0.552	0.234
1.2	0.339	0.460	0.220
1.6	0.317	0.407	0.204
2.0	0.299	0.370	0.189
3.0	0.262	0.310	0.158
4.0	0.236	0.271	0.137
5.0	0.213	0.243	0.121
6.0	0.197	0.222	0.110
8.0	0.174	0.193	0.096
10.0	0.158	0.172	0.087
12.0	0.145	0.158	0.079
14.0	0.135	0.147	0.073
16.0	0.127	0.137	0.068

Table 11.10 Ring Tension Influence Coefficients, Triangular Load (Fixed Base)

Tension in Circular Rings
Triangular Load
Fixed base, Free Top
F = coef. × γHR lb. per ft.
Positive sign indicates tension

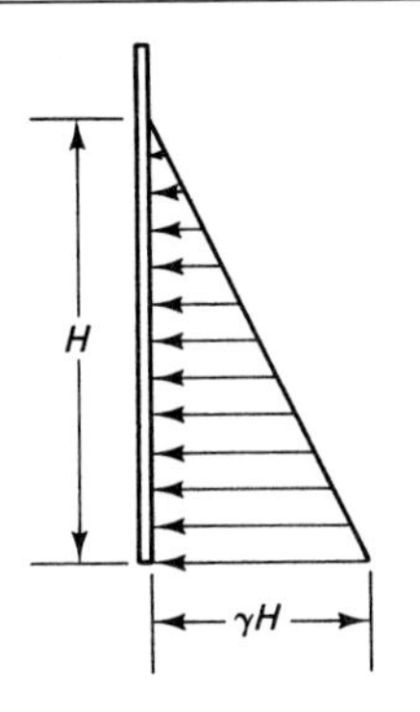

'Liquid Load'—Fixed

$\frac{H^2}{dt}$	Coefficients at Point									
	0.0*H*	0.1*H*	0.2*H*	0.3*H*	0.4*H*	0.5*H*	0.6*H*	0.7*H*	0.8*H*	0.9*H*
0.4	+0.149	+0.134	+0.120	+0.101	+0.082	+0.066	+0.049	+0.029	+0.014	+0.004
0.8	+0.263	+0.239	+0.215	+0.190	+0.160	+0.130	+0.096	+0.063	+0.034	+0.010
1.2	+0.283	+0.271	+0.254	+0.234	+0.209	+0.180	+0.142	+0.099	+0.045	+0.016
1.6	+0.265	+0.268	+0.268	+0.266	+0.250	+0.266	+0.185	+0.134	+0.075	+0.023
2.0	+0.234	+0.251	+0.273	+0.285	+0.285	+0.274	+0.232	+0.172	+0.104	+0.031
3.0	+0.134	+0.203	+0.267	+0.322	+0.357	+0.362	+0.330	+0.262	+0.157	+0.052
4.0	+0.067	+0.164	+0.256	+0.339	+0.403	+0.429	+0.409	+0.334	+0.210	+0.073
5.0	+0.025	+0.137	+0.245	+0.346	+0.428	+0.477	+0.469	+0.398	+0.259	+0.092
6.0	+0.018	+0.119	+0.234	+0.344	+0.441	+0.504	+0.514	+0.447	+0.301	+0.112
8.0	+0.011	+0.104	+0.218	+0.335	+0.443	+0.534	+0.575	+0.530	+0.381	+0.151
10.0	−0.011	+0.098	+0.208	+0.323	+0.437	+0.542	+0.608	+0.589	+0.440	+0.179
12.0	−0.005	+0.097	+0.202	+0.312	+0.429	+0.543	+0.628	+0.633	+0.494	+0.211
14.0	−0.002	+0.098	+0.200	+0.306	+0.420	+0.539	+0.639	+0.666	+0.541	+0.241
16.0	0.000	+0.099	+0.199	+0.304	+0.412	+0.531	+0.641	+0.687	+0.582	+0.265

Table 11.11 Ring Tension Influence Coefficients, Rectangular Load (Fixed Base)

Tension in Circular Rings
Rectangular Load
Fixed Base, Free Top
F = coef. × pR lb. per ft.
Positive sign indicates tension

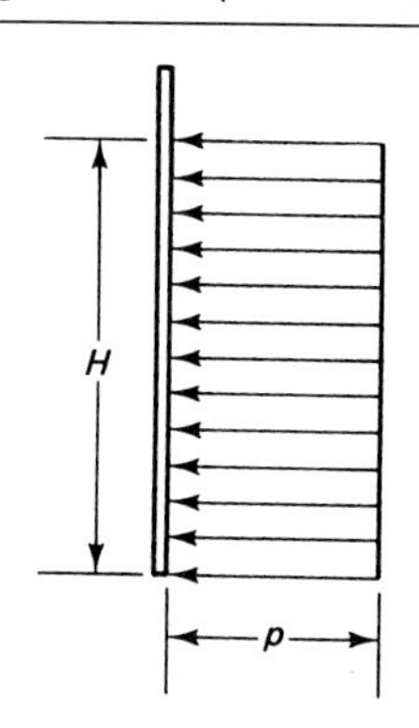

'Gas' Load—Fixed

$\frac{H^2}{dt}$	Coefficients at Point									
	0.0*H*	0.1*H*	0.2*H*	0.3*H*	0.4*H*	0.5*H*	0.6*H*	0.7*H*	0.8*H*	0.9*H*
0.4	+0.582	+0.505	+0.431	+0.353	+0.277	+0.206	+0.145	+0.092	+0.046	+0.013
0.8	+1.052	+0.921	+0.796	+0.669	+0.542	+0.415	+0.289	+0.179	+0.089	+0.024
1.2	+1.218	+1.078	+0.946	+0.808	+0.665	+0.519	+0.378	+0.246	+0.127	+0.034
1.6	+1.257	+1.141	+1.009	+0.881	+0.742	+0.600	+0.449	+0.294	+0.153	+0.045
2.0	+1.253	+1.144	+1.041	+0.929	+0.806	+0.667	+0.514	+0.345	+0.186	+0.055
3.0	+1.160	+1.112	+1.061	+0.998	+0.912	+0.796	+0.646	+0.459	+0.258	+0.081
4.0	+1.085	+1.073	+1.057	+1.029	+0.997	+0.887	+0.746	+0.553	+0.322	+0.105
5.0	+1.037	+1.044	+1.047	+1.042	+1.015	+0.949	+0.825	+0.629	+0.379	+0.128
6.0	+1.010	+1.024	+1.038	+1.045	+1.034	+0.986	+0.879	+0.694	+0.430	+0.149
8.0	+0.989	+1.005	+1.022	+1.036	+1.044	+1.026	+0.953	+0.788	+0.519	+0.189
10.0	+0.989	+0.998	+1.010	+1.023	+1.039	+1.040	+0.996	+0.859	+0.591	+0.226
12.0	+0.994	+0.997	+1.003	+1.014	+1.031	+1.043	+1.022	+0.911	+0.652	+0.262
14.0	+0.997	+0.998	+1.000	+1.007	+1.022	+1.040	+1.035	+0.949	+0.705	+0.294
16.0	+1.000	+0.999	+0.999	+1.003	+1.015	+1.032	+1.040	+0.975	+0.750	+0.321

Table 11.12 Ring Tension Influence Coefficients, Triangular Load (Pinned Base)

Tension in Circular Rings
Triangular Load
Hinged Base, Free Top
F = coef. × γHR lb. per ft.
Positive sign indicates tension

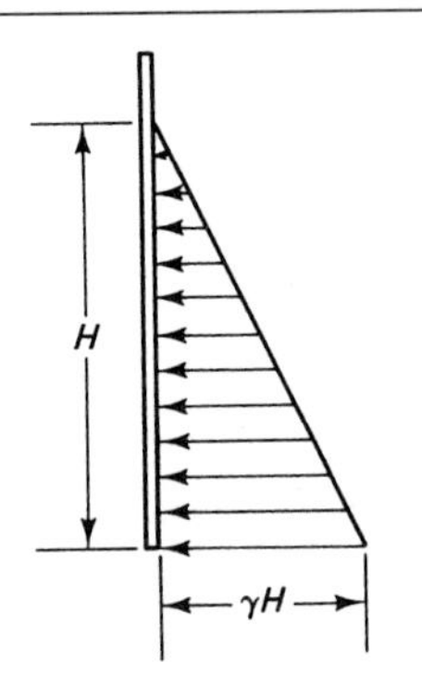

'Liquid Load'—Fixed

$\frac{H^2}{dt}$	Coefficients at Point									
	0.0*H*	0.1*H*	0.2*H*	0.3*H*	0.4*H*	0.5*H*	0.6*H*	0.7*H*	0.8*H*	0.9*H*
0.4	+0.474	+0.440	+0.395	+0.352	+0.308	+0.264	+0.215	+0.165	+0.111	+0.057
0.8	+0.423	+0.402	+0.381	+0.358	+0.330	+0.297	+0.249	+0.202	+0.145	+0.076
1.2	+0.350	+0.355	+0.361	+0.362	+0.358	+0.343	+0.309	+0.256	+0.186	+0.098
1.6	+0.271	+0.303	+0.341	+0.369	+0.385	+0.385	+0.362	+0.314	+0.233	+0.124
2.0	+0.205	+0.260	+0.321	+0.373	+0.411	+0.434	+0.419	+0.369	+0.280	+0.151
3.0	+0.074	+0.179	+0.281	+0.375	+0.449	+0.506	+0.519	+0.479	+0.375	+0.210
4.0	+0.017	+0.137	+0.253	+0.367	+0.469	+0.545	+0.579	+0.553	+0.447	+0.256
5.0	−0.008	+0.114	+0.235	+0.356	+0.469	+0.562	+0.617	+0.606	+0.503	+0.294
6.0	−0.011	+0.103	+0.223	+0.343	+0.463	+0.566	+0.639	+0.643	+0.547	+0.327
8.0	−0.015	+0.096	+0.208	+0.324	+0.443	+0.564	+0.661	+0.697	+0.621	+0.386
10.0	−0.008	+0.095	+0.200	+0.311	+0.428	+0.552	+0.666	+0.730	+0.678	+0.433
12.0	−0.002	+0.097	+0.197	+0.302	+0.417	+0.541	+0.664	+0.750	+0.720	+0.477
14.0	0.000	+0.098	+0.197	+0.299	+0.408	+0.531	+0.659	+0.761	+0.752	+0.513
16.0	+0.002	+0.100	+0.198	+0.299	+0.403	+0.521	+0.650	+0.764	+0.776	+0.543

Table 11.13 Ring Tension Influence Coefficients, Rectangular Load (Hinged Base)

Tension in Circular Rings
Rectangular Load
Hinged Base, Free Top
F = coef. × pR lb. per ft.
Positive sign indicates tension

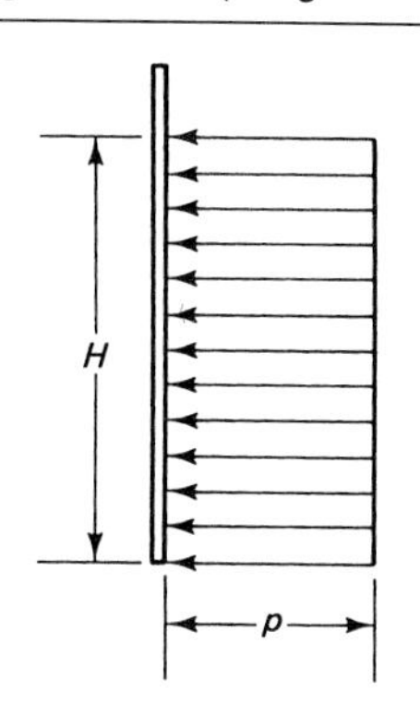

'Gas' Load—Pinned

$\frac{H^2}{dt}$	Coefficients at Point									
	0.0*H*	0.1*H*	0.2*H*	0.3*H*	0.4*H*	0.5*H*	0.6*H*	0.7*H*	0.8*H*	0.9*H*
0.4	+1.474	−1.340	+1.195	+1.052	+0.903	+0.764	+0.615	+0.465	+0.311	+0.154
0.8	+1.423	+1.302	+1.181	+1.058	+0.930	+0.797	+0.649	+0.502	+0.345	+0.166
1.2	+1.350	+1.255	+1.161	+1.062	+0.958	+0.843	+0.709	+0.556	+0.386	+0.198
1.6	+1.271	+1.203	+1.141	+1.069	+0.985	+0.885	+0.756	+0.614	+0.433	+0.224
2.0	+1.205	+1.160	+1.121	+1.173	+1.011	+0.934	+0.819	+0.669	+0.480	+0.251
3.0	+1.074	+1.079	+1.081	+1.075	+1.049	+1.006	+0.919	+0.779	+0.575	+0.310
4.0	+1.017	+1.037	+1.053	+1.067	+1.069	+1.045	+0.979	+0.853	+0.647	+0.356
5.0	+0.992	+1.014	+1.035	+1.056	+1.069	+1.062	+1.017	+1.906	+0.703	+0.394
6.0	+0.989	+1.003	+1.023	+1.043	+1.063	+1.066	+1.039	+0.943	+0.747	+0.427
8.0	+0.985	+0.996	+1.008	+1.024	+1.043	+1.064	+1.061	+0.997	+0.821	+0.486
10.0	+0.992	+0.995	+1.000	+1.011	+1.028	+1.052	+1.066	+1.030	+0.878	+0.523
12.0	+0.998	+0.997	+0.997	+1.002	+1.017	+1.041	+1.064	+1.050	+0.920	+0.577
14.0	+1.000	+0.998	+0.997	+0.999	+1.008	+1.031	+1.059	+1.061	+0.952	+0.613
16.0	+1.002	+1.000	+0.998	+0.999	+1.003	+1.021	+1.050	+1.064	+0.976	+0.543

Table 11.14 Empty Tank Ring Tension Influence Coefficients, Fixed Base

Tension in Circular Rings
Shear per Ft., Q, Applied at Top
Fixed Base, Free Top
F = coef. × VR/H lb. per ft.
Positive sign indicates tension

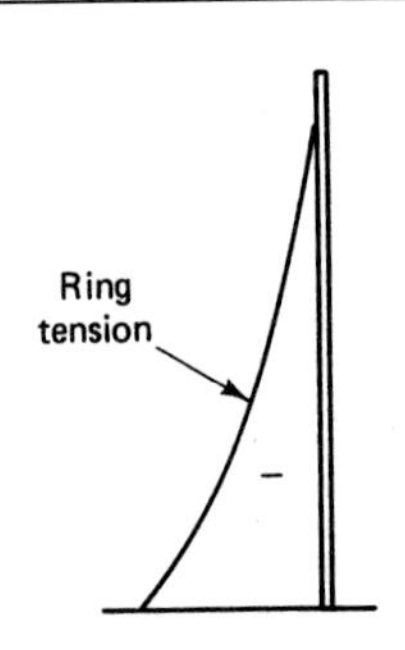

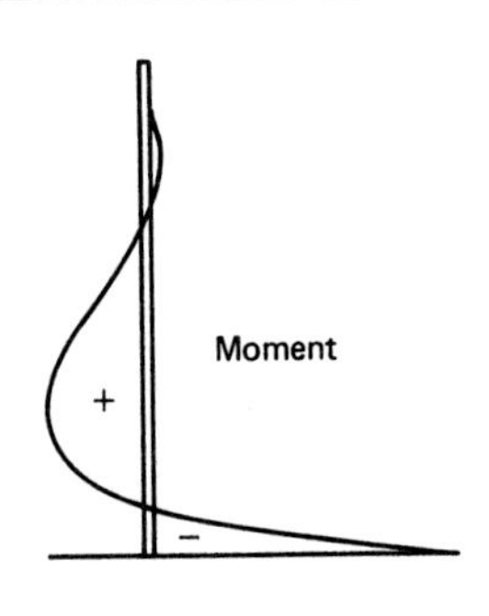

Empty Tank

$\frac{H^2}{dt}$	Coefficients at Point									
	0.0*H*	0.1*H*	0.2*H*	0.3*H*	0.4*H*	0.5*H*	0.6*H*	0.7*H*	0.8*H*	0.9*H*
0.4	– 1.57	–1.32	–1.08	–0.86	–0.65	–0.47	–0.31	–0.18	–0.08	–0.02
0.8	–3.09	–2.55	–2.04	–1.57	–1.15	–0.80	–0.51	–0.28	–0.13	–0.03
1.2	–3.95	–3.17	–2.44	–1.79	–1.25	–0.81	–0.48	–0.25	–0.10	–0.02
1.6	–4.57	–3.54	–2.60	–1.80	–1.17	–0.69	–0.36	–0.16	–0.05	–0.01
2.0	–5.12	–3.83	–2.68	–1.74	–1.02	–0.52	–0.21	–0.05	+0.01	+0.01
3.0	–6.32	–4.37	–2.70	–1.43	–0.58	–0.02	–0.15	+0.19	+0.13	+0.04
4.0	–7.34	–4.73	–2.60	–1.10	–0.19	+0.26	+0.38	+0.33	+0.19	+0.06
5.0	–8.22	–4.99	–2.45	–0.79	+0.11	+0.47	+0.50	+0.37	+0.20	+0.06
6.0	–9.02	–5.17	–2.27	–0.50	+0.34	+0.59	+0.53	+0.35	+0.17	+0.01
8.0	–10.42	–5.36	–1.85	–0.02	+0.63	+0.66	+0.46	+0.24	+0.09	+0.01
10.0	–11.67	–5.43	–1.43	+0.36	+0.78	+0.62	+0.33	+0.12	+0.02	0.00
12.0	–12.76	–5.41	–1.03	+0.63	+0.83	+0.52	+0.21	+0.04	–0.02	0.00
14.0	–13.77	–5.34	–0.68	+0.80	+0.81	+0.42	+0.13	0.00	–0.03	–0.01
16.0	–14.74	–5.22	–0.33	+0.96	+0.76	+0.32	+0.05	–0.04	–0.05	–0.02

Table 11.15 Empty Tank Ring Tension Influence Coefficients, Hinged Base

Tension in Circular Rings
Moment per Ft., M, Applied at Base
Hinged Base, Free Top
F = coef. × MR/H^2 lb. per ft.
Positive sign indicates tension

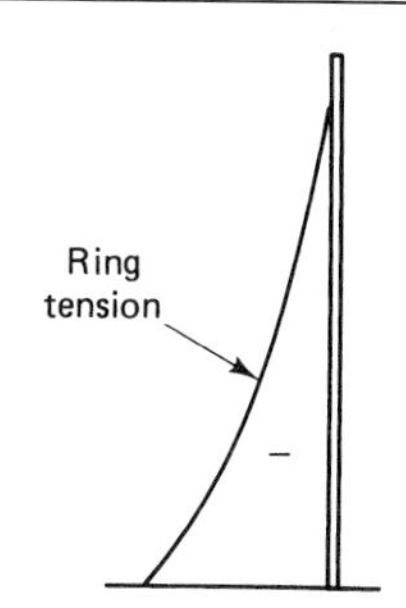

Empty Tank

$\frac{H^2}{dt}$	Coefficients at Point									
	0.0*H*	0.1*H*	0.2*H*	0.3*H*	0.4*H*	0.5*H*	0.6*H*	0.7*H*	0.8*H*	0.9*H*
0.4	+2.70	+2.50	+2.30	+2.12	+1.91	+1.69	+1.41	+1.13	+0.80	+0.44
0.8	+2.02	+2.06	+2.10	+2.14	+2.10	+2.02	+1.95	+1.75	+1.39	+0.80
1.2	+1.06	+1.42	+1.79	+2.03	+2.46	+2.65	+2.80	+2.60	+2.22	+1.37
1.6	+0.12	+0.79	+1.43	+2.04	+2.72	+3.25	+3.56	+3.59	+3.13	+2.01
2.0	–0.68	+0.22	+1.10	+2.02	+2.90	+3.69	+4.30	+4.54	+4.08	+2.75
3.0	–1.78	–0.71	+0.43	+1.60	+2.95	+4.29	+5.66	+6.58	+6.55	+4.73
4.0	–1.87	–1.00	–0.08	+1.04	+2.47	+4.31	+6.34	+8.19	+8.82	+6.81
5.0	–1.54	–1.03	–0.42	+0.45	+1.86	+3.93	+6.60	+9.41	+11.03	+9.02
6.0	–1.04	–0.86	–0.59	–0.05	+1.21	+3.34	+6.54	+10.28	+13.08	+11.41
8.0	–0.24	–0.53	–0.73	–0.67	–0.02	+2.05	+5.87	+11.32	+16.52	+16.06
10.0	+0.21	–0.23	–0.64	–0.94	–0.73	+0.82	+4.79	+11.63	+19.48	+20.87
12.0	+0.32	–0.05	–0.46	–0.96	–1.15	–0.18	+3.52	+11.27	+21.80	+25.73
14.0	+0.26	+0.04	–0.28	–0.76	–1.29	–0.87	+2.29	+10.55	+23.50	+30.34
16.0	+0.22	+0.07	–0.08	–0.64	–1.28	–1.30	+1.12	+9.67	+24.53	+34.65

Table 11.16 Supplementary Influence Coefficients for Values of H^2/dt Greater Than 16 for Tables 11.4–11.15

Table 11.4a

$\frac{H^2}{dt}$	Coefficients at Point				
	.80*H*	.85*H*	.90*H*	.95*H*	1.00*H*
20	+.0015	+.0014	+.0005	−.0018	−.0063
24	+.0012	+.0012	+.0007	−.0013	−.0053
32	+.0007	+.0009	+.0007	−.0008	−.0040
40	+.0002	+.0005	+.0006	−.0005	−.0032
48	.0000	+.0001	+.0006	−.0003	−.0026
56	.0000	.0000	+.0004	−.0001	−.0023

Table 11.5a

$\frac{H^2}{dt}$	Coefficients at Point				
	.80*H*	.85*H*	.90*H*	.95*H*	1.00*H*
20	+.0015	+.0013	+.0002	−.0024	−.0073
24	+.0012	+.0012	+.0004	−.0018	−.0061
32	+.0008	+.0009	+.0006	−.0010	−.0046
40	+.0005	+.0007	+.0007	−.0005	−.0037
48	+.0004	+.0006	+.0006	−.0003	−.0031
56	+.0002	+.0004	+.0005	−.0001	−.0026

Table 11.6a

$\frac{H^2}{dt}$	Coefficients at Point				
	.75*H*	.80*H*	.85*H*	.90*H*	.95*H*
20	+.0008	+.0014	+.0020	+.0024	+.0020
24	+.0005	+.0010	+.0015	+.0020	+.0017
32	.0000	+.0005	+.0009	+.0014	+.0013
40	.0000	+.0003	+.0006	+.0011	+.0011
48	.0000	+.0001	+.0004	+.0008	+.0010
56	.0000	.0000	+.0003	+.0007	+.0008

Table 11.7a

$\frac{H^2}{dt}$	Coefficients at Point				
	.05*H*	.10*H*	.15*H*	.20*H*	.25*H*
20	+0.032	+0.039	+0.033	+0.023	+0.014
24	+0.031	+0.035	+0.028	+0.018	+0.009
32	+0.028	+0.029	+0.020	+0.011	+0.004
40	+0.026	+0.025	+0.015	+0.006	+0.001
48	+0.024	+0.021	+0.011	+0.003	0.000
56	+0.023	+0.018	+0.008	+0.002	0.000

Table 11.8a

$\frac{H^2}{dt}$	Coefficients at Point				
	.80*H*	.85*H*	.90*H*	.95*H*	1.00*H*
20	−0.015	+0.095	+0.296	+0.606	+1.000
24	−0.037	+0.057	+0.250	+0.572	+1.000
32	−0.062	+0.002	+0.178	+0.515	+1.000
40	−0.067	−0.031	+0.123	+0.467	+1.000
48	−0.064	−0.049	+0.081	+0.424	+1.000
56	−0.059	−0.060	+0.048	+0.387	+1.000

Table 11.9a

$\frac{H^2}{dt}$	Coefficients at Point		
	Tri. Fixed	Rect. Fixed	T. or R. Hinged
20	+0.114	+0.122	+0.062
24	+0.102	+0.111	+0.055
32	+0.089	+0.096	+0.048
40	+0.080	+0.086	+0.043
48	+0.072	+0.079	+0.039
56	+0.067	+0.074	+0.036

Table 11.16 *Continued*

Table 11.10a

$\frac{H^2}{dt}$	Coefficients at Point				
	.75*H*	.80*H*	.85*H*	.90*H*	.95*H*
20	+0.716	+0.654	+0.520	+0.325	+0.115
24	+0.746	+0.702	+0.577	+0.372	+0.137
32	+0.782	+0.768	+0.663	+0.459	+0.182
40	+0.800	+0.805	+0.731	+0.530	+0.217
48	+0.791	+0.828	+0.785	+0.593	+0.254
56	+0.763	+0.838	+0.824	+0.636	+0.285

Table 11.11a

$\frac{H^2}{dt}$	Coefficients at Point				
	.75*H*	.80*H*	.85*H*	.90*H*	.95*H*
20	+0.949	+0.825	+0.629	+0.379	+0.128
24	+0.986	+0.879	+0.694	+0.430	+0.149
32	+1.026	+0.953	+0.788	+0.519	+0.189
40	+1.040	+0.996	+0.859	+0.591	+0226
48	+1.043	+1.022	+0.911	+0652	+0.262
56	+1.040	+1.035	+0.949	+0.705	+0.294

Table 11.12a

$\frac{H^2}{dt}$	Coefficients at Point				
	.75*H*	.80*H*	.85*H*	.90*H*	.95*H*
20	+0.812	+0.817	+0.756	+0.603	+0.344
24	+0.816	+0.839	+0.793	+0.647	+0.377
32	+0.814	+0.861	+0.847	+0.721	+0.436
40	+0.802	+0.866	+0.880	+0.778	+0.483
48	+0.791	+0.864	+0.900	+0.820	+0.527
56	+0.781	+0.859	+0.911	+0.852	+0.563

Table 11.13a

$\frac{H^2}{dt}$	Coefficients at Point				
	.75*H*	.80*H*	.85*H*	.90*H*	.95*H*
20	+1.062	+1.017	+0.906	+0.703	+0.394
24	+1.066	+1.039	+0.943	+0.747	+0.427
32	+1.064	+1.061	+0.997	+0.821	+0.486
40	+1.052	+1.066	+1.030	+0.878	+0.533
48	+1.041	+1.064	+1.050	+0.920	+0.577
56	+1.021	+1.059	+1.061	+0.952	+0.613

Table 11.14a

$\frac{H^2}{dt}$	Coefficients at Point				
	.00*H*	.05*H*	.10*H*	.15*H*	.20*H*
20	−16.44	− 9.98	−4.90	−1.59	+0.22
24	−18.04	−10.34	−4.54	−1.00	+0.68
32	−20.84	−10.72	−3.70	−0.04	+1.26
40	−23.34	−10.86	−2.86	+0.72	+1.56
48	−25.52	−10.82	−2.06	+0.26	+1.66
56	−27.54	−10.68	−1.36	+1.60	+1.62

Table 11.15a

$\frac{H^2}{dt}$	Coefficients at Point				
	.75*H*	.80*H*	.85*H*	.90*H*	.95*H*
20	+15.30	+25.9	+36.9	+43.3	+ 35.3
24	+13.20	+25.9	+40.7	+51.8	+ 45.3
32	+ 8.10	+23.2	+45.9	+65.4	+ 63.6
40	+ 3.28	+19.2	+46.5	+77.9	+ 83.5
48	− 0.70	+14.1	+45.1	+87.2	+103.0
56	− 3.40	+ 9.2	+42.2	+94.0	+121.0

Photo 11.7 Shotcrete Application Covering the Wire (*Courtesy,* N.A. Legatos, Preload Inc., Garden City, New York.)

are required to neutralize or balance the outward radial forces caused by the internal liquid or gas, thereby producing *residual* compression in the tank when it is full (Ref. 11.2). Such an increase in circumferential prestressing forces through the use of additional horizontal prestressing steel, and sometimes mild vertical steel, also counteracts the effects of temperature and moisture gradients across the wall thickness in an adverse environment.

11.6.1 Freely Sliding Wall Base

When the boundary condition is such that the wall at its base can freely slide when the tank is internally loaded, there is no moment in the vertical wall due either to liquid load or to prestressing when the tank is totally filled to height H. Only a small nominal moment develops when the tank is partially filled, partially prestressed, or empty, and *no* vertical prestressing is necessary. The deflected shape of the freely sliding tank is shown in Figure 11.7.

While free sliding is an ideal condition that renders the structure statically determinate and hence most economical, it is difficult to achieve in practice. Frictional forces produced at the wall base after the tank becomes operational and the difficulty of achieving liquid tightness render this alternative essentially unimplementable.

11.6.2 Hinged Wall Base

For walls with a hinged connection to the base, the maximum radial forces due to the liquid retained and the prestressing at the critical section a distance y above the base are almost equal to those in the freely sliding case at height y. But vertical moments are introduced, and vertical prestressing becomes necessary to reduce the tensile stresses in the concrete at the outer wall face.

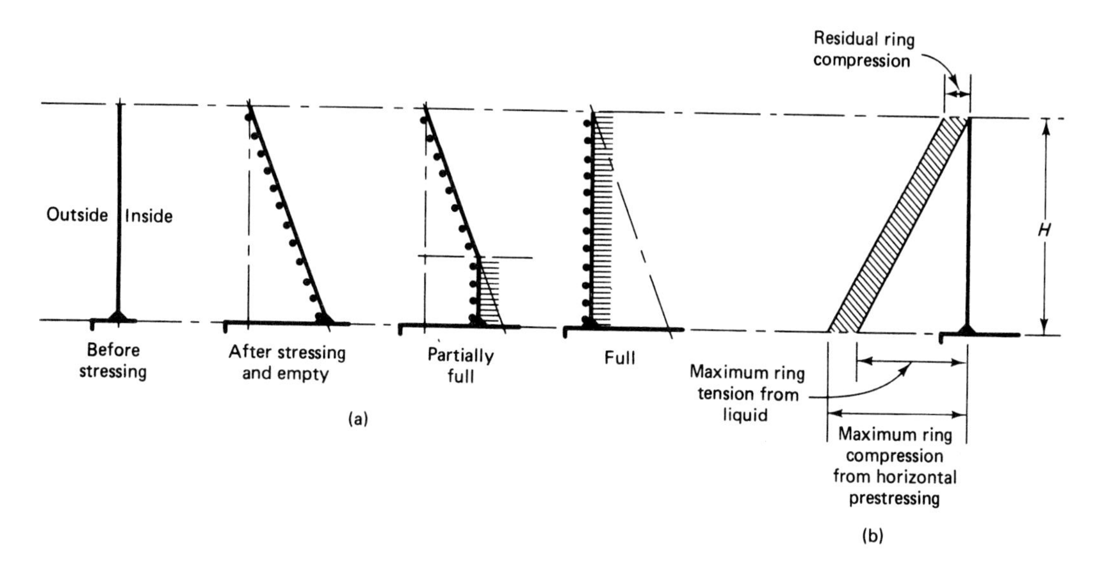

Figure 11.7 Freely sliding tank. (a) Deflected shape. (b) Residual ring compression.

The deflected shape of the hinged wall is shown in Figure 11.8. Note that the critical section for ring forces is not necessarily at the same height as the moment critical section.

In order to minimize the possibility of cracking, a residual ring compression of a minimum value of 200 psi (1.38 MPa) is necessary for wire-wrapped prestressed tanks without diaphragms, and 100 psi (0.7 MPa) for tanks with a continuous metal diaphragm. The maximum tension at the inside face of the wall should not exceed $3\sqrt{f'_c}$ at working-load level as given in Table 11.17 in a later section. The deflected shape of the tank walls and the stress variations in the concrete across the thickness of the section when the tank is empty and when it is full are shown in Figure 11.8. For tanks prestressed with pretensioned and post-tensioned tendons, the minimum residual compressive stress should be as stipulated in Section 11.10.

11.6.3 Partially Sliding and Hinged Wall Base

A partially sliding and hinged wall-base system is accomplished by providing a slot in the wall-base supporting slab such that the wall can slide within its base during the prestressing. After prestressing and all losses due to creep, shrinkage, and relaxation have taken place, the slot is sealed and the tank wall behaves as hinged under service-load conditions. The magnitude of sliding can be controlled such that either full or partial sliding is allowed before hinging is accomplished. A partial slide of about 50 percent of the full slide with hinging at the end of the wall movement has the structural advantages of both full sliding and hinging, and the sealing of the wall-base slab-pinned joint against leakage of liquids or gases is more dependable than if full sliding prior to anchorage is allowed. The deformed shape of the wall during the prestressing procedure, together with the ring forces, vertical moments, and concrete stress variations across the wall thickness, is shown in Figure 11.9. The vertical prestress needed for the partial slide-pinned case can be considerably smaller than the fully pinned case without sliding.

11.6.4 Fully Fixed Wall Base

Full fixity of the wall at its base means full restraint against rotation at the wall base. This condition can be accomplished if the lower segment of the wall is cast monolithically and is well anchored into a base slab of a similar stiffness. But such an indeterminate system

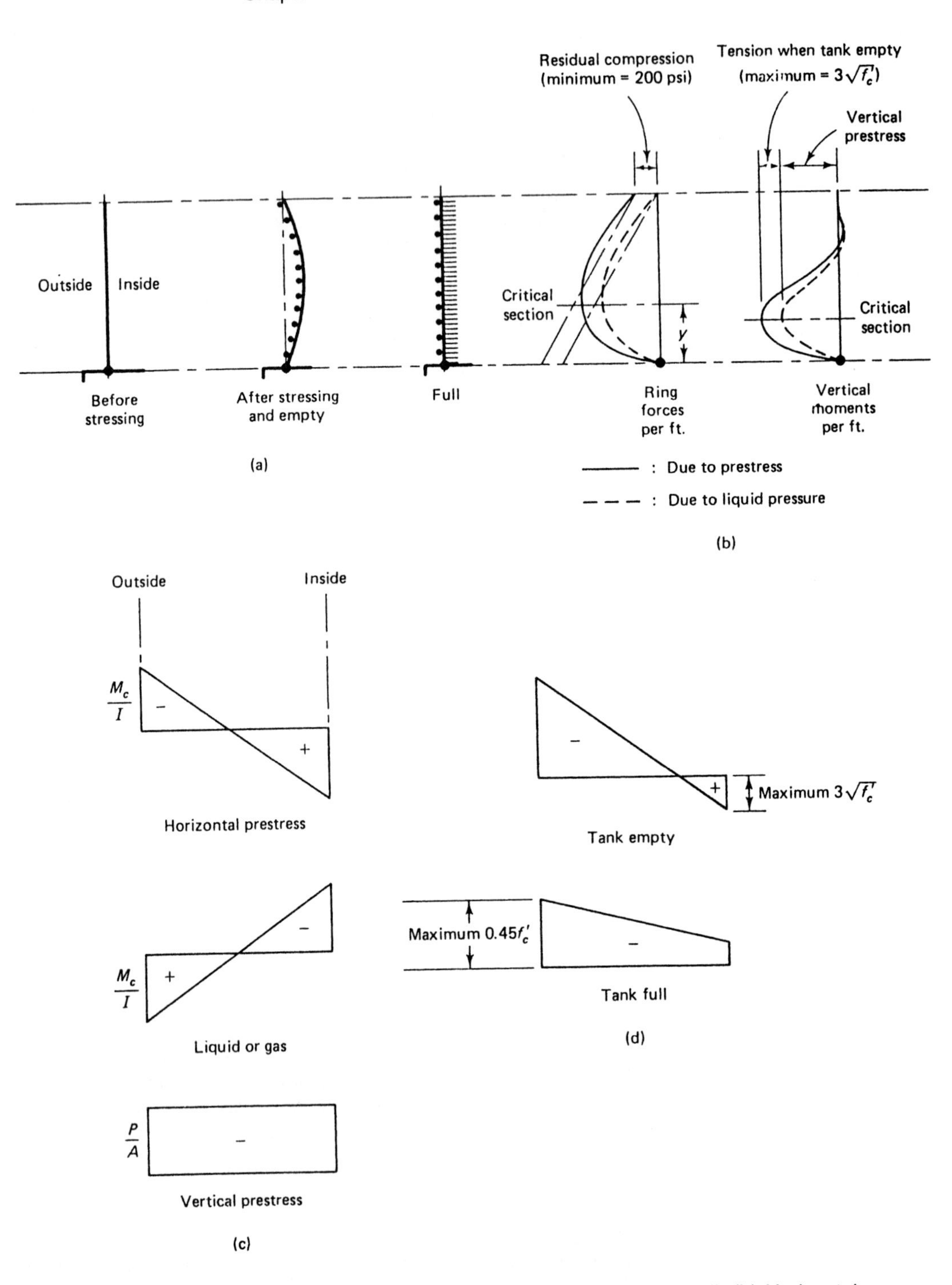

Figure 11.8 Hinged-base tank. (a) Deflected shape of tank wall. (b) Horizontal ring forces and vertical moments. (c) Concrete stresses across wall thickness. (d) Resultant wall stresses.

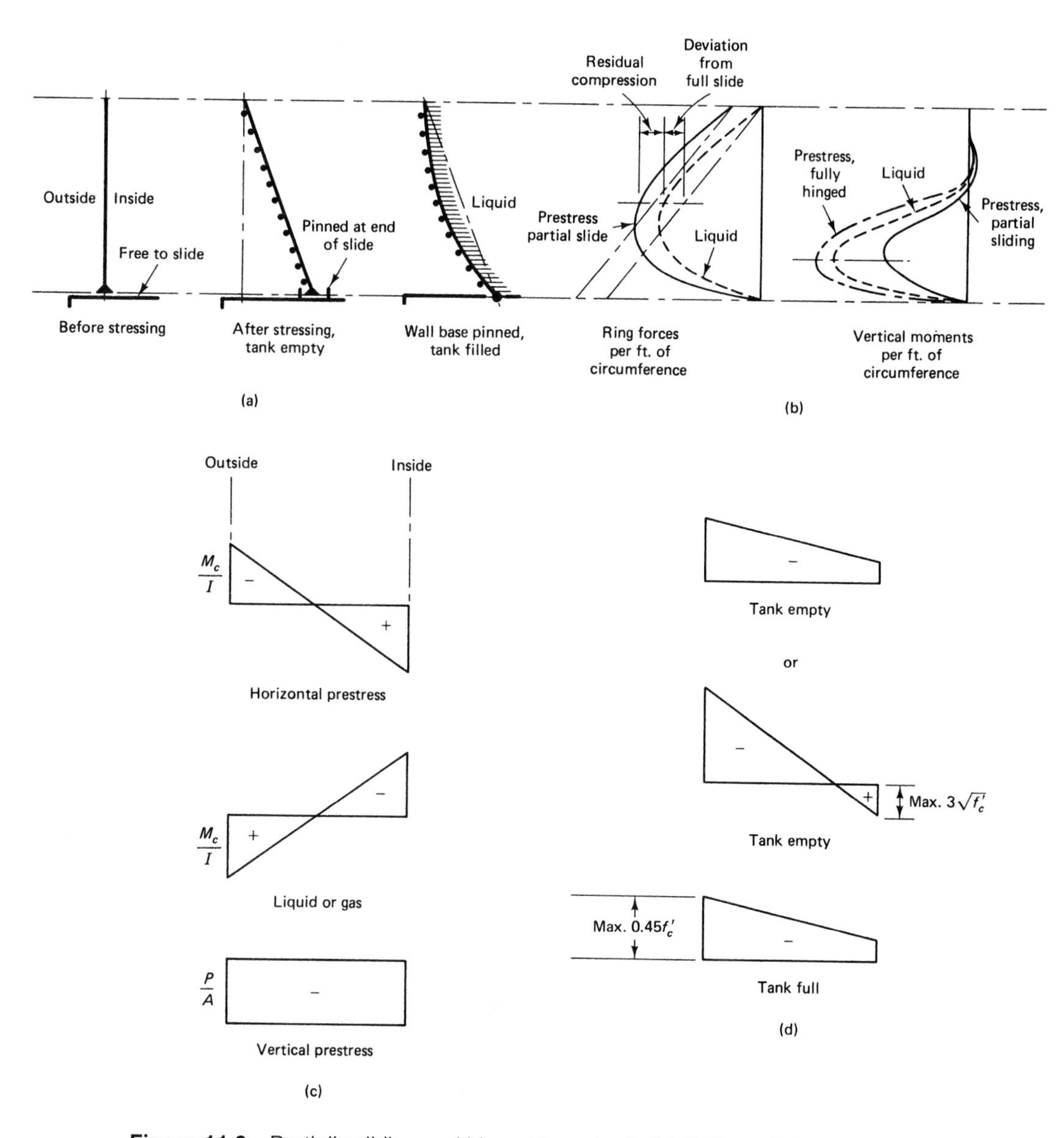

Figure 11.9 Partially sliding and hinged-base tank. (a) Deflected shape. (b) Horizontal ring forces and comparative vertical moments. (c) Concrete stresses across wall thickness. (d) Resultant wall stresses.

is difficult to fully achieve and is not economical as well, since a tank base area is very large and partial fixity becomes necessary (see shortly). The radial horizontal forces from both prestressing and the contained internal pressure are unchanged from the triangular form for liquid, rectangular for gas, and trapezoidal for granular contained material. The restraint imposed by the horizontal slab base, however, modifies the ring forces and introduces additional moment in the vertical section of the wall. Because of fixity at the base, no displacement takes place at either the bottom or the top of the wall, and a change in curvature along the height of the wall above the base takes place when the tank is empty, as is shown in Figure 11.10. Note that the wall should be designed to become essentially vertical, with a minimum residual compressive stress due to prestressing of 200

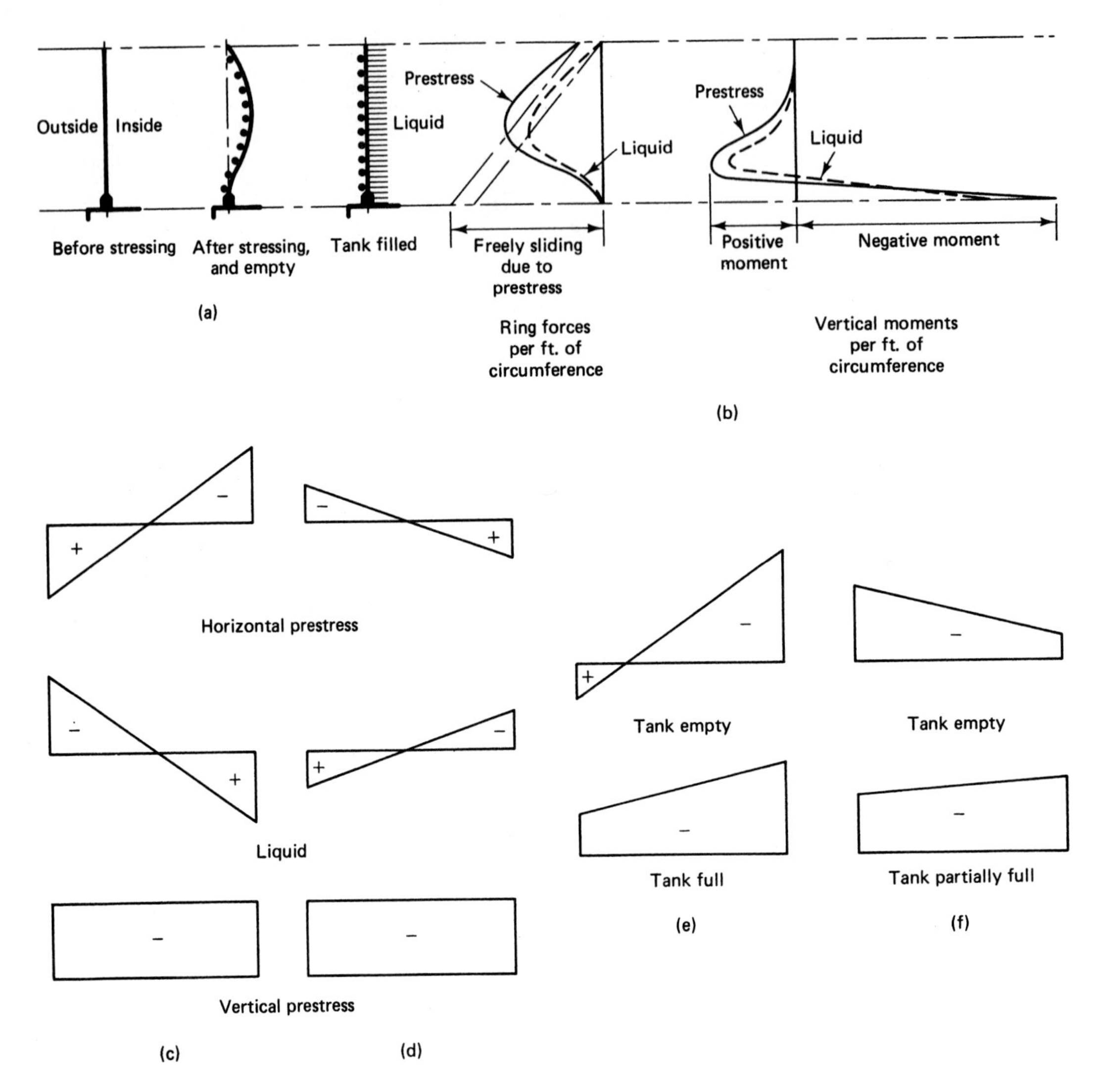

Figure 11.10 Fully fixed-base tank. (a) Deflected wall shape. (b) Horizontal ring forces and vertical moments. (c) Concrete stresses across wall for full tank. (d) Concrete stresses across wall for partially full tank. (e) Resultant stresses, full tank. (f) Resultant stresses, partially full tank.

psi as in the previous cases. The vertical prestress needed for tanks with fully fixed wall bases is considerably greater than the vertical prestress needed for the other boundary conditions. This is necessary in order to offset the high tensile stresses in the wall base at the *outside* face caused by the large negative movement at the base [see Figure 11.10(a) and (b)] and the reverse curvature near it. It is sometimes more economical to use mild steel reinforcement at the lower portion of the wall in addition to prestressing, in order to be able to use lesser vertical prestressing and assign the excess negative moment to the nonprestressed reinforcement. The tensile stresses in the concrete can also be reduced by using *eccentric* vertical prestressing with the appropriate eccentricity achieved by trial and adjustment, as well as by using additional mild steel. Vertical prestressing in tanks is expensive, however, due to the required anchorages at the top and bottom of the tank wall. Thus, reducing the level of vertical prestress needed in the design adds to the economy of the total design of the system.

11.6.5 Partially Fixed Wall Base

11.6.5.1 Rotational Restraint. As indicated previously, full restraint against rotation at the wall base is difficult to achieve. The reasons are essentially threefold: (1) one has to provide the necessary stiffness in the tank floor slab at the wall junction for total fixity; (2) subsoil movement under the wall can cause rotation of the wall base; and (3) a concentration of anchorages is required, for both the vertical prestressing of the wall and the horizontal circumferential prestressing of the wall-base segment since the wall and base rings are separately prestressed.

Because the floor slab area is large, its restraining or stiffening influence is limited to the narrow peripheral toe cantilevering from the wall bottom. The choice of the correct width of the toe or base ring determines whether or not the assumed degree of fixity of the wall base gives the correct stiffness values in the design. Figure 11.11 schematically demonstrates the effect of the base ring width on the rotation of the wall and the deformation of the ring. Part (c) of the figure gives an equilibrium state where the tip of the ring is at the same level as the bottom of the wall, whereas the conditions represented in parts (a) and (b) involve deformations below the bottom of the wall and are consequently unsatisfactory.

The theoretical formulation of the solution to the critical ring base width can be attained through the use of the principle of superposition by combining the case of a freely rotating wall with that of a totally fixed wall as shown in Figure 11.12. Let

M_0 = theoretical fully fixed moment at the wall base

M_p = partial moment at the wall base caused by the loaded cantilever toe

θ_1 = free rotation of wall base when pinned only, corresponding to deflection Δ_1 of a stiff unloaded toe

θ_2 = wall base rotation due to restraining moment M_p, corresponding to deflection Δ_2 of a straight unloaded toe

θ_3 = rotation of the tip of the stiffening toe as a cantilever under vertical load, corresponding to deflection Δ_3 of the toe tip due to the vertical load

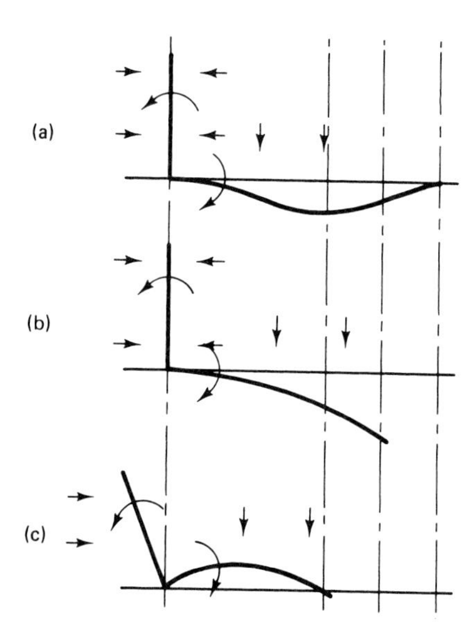

Figure 11.11 Base ring effective width. (a) Full base slab. (b) Large cantilever. (c) Equilibrium condition.

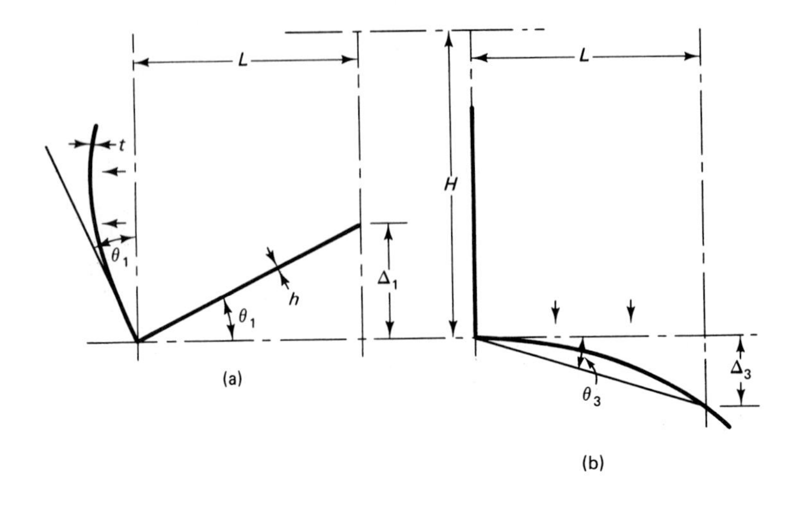

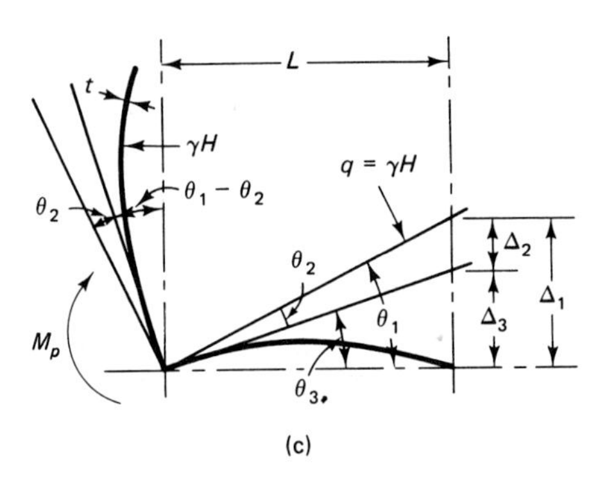

Figure 11.12 Deformation and rotation of wall base. (a) Fully free wall. (b) Fully fixed wall. (c) Superposition of (a) and (b).

L = width of stiffening toe

q = unit load applied to the stiffening toe = γH, where H is the height of a tank whose diameter is d, whose wall thickness is t, and whose base slab thickness is h.

Then the unit rotation θ of the wall at its base due to moment M_0, but without radial displacement, can be obtained from Equation 11.18a by setting $w = 0$ to get $Q = -\beta M$. Equation 11.18b for unit rotation then becomes

$$\theta_1 = \frac{M_o}{2\beta D}, \qquad \theta_2 = \frac{M_p}{2\beta D} \tag{11.33}$$

Hence, we have

$$\Delta_1 = \frac{LM_o}{2\beta D}, \qquad \Delta_2 = \frac{LM_p}{2\beta D} \tag{11.34}$$

If the stiffening wall toe is considered a cantilever subjected to a transverse load γH, the maximum cantilever moment M_p and the corresponding deflection Δ_3 are, respectively,

$$M_p = \frac{\gamma HL^2}{2}, \qquad \Delta_3 = \frac{3\gamma HL^4}{2Eh^3} \tag{11.35}$$

The moment at the fixed wall base can be obtained using the membrane coefficient C from Table 11.4 for the applicable form factor H^2/dt and type of load. For liquid load,

$$M_o = C\gamma H^3 \tag{11.36}$$

The deflected form due to full load, from Figure 11.12(c), is

$$\Delta_1 = \Delta_2 + \Delta_3$$

As a reasonable approximation, assume

$$\mu = 0.2 \quad \text{and} \quad \beta = 2/\sqrt{dt}.$$

Substituting for Δ_2 and Δ_3 from Equation 11.34 into Equations 11.35 and 11.36 and rearranging terms gives

$$L^2 = \frac{2CH^2}{1 + \dfrac{(t/h)^3}{(dt)^{1/2}}(L = 1)} \tag{11.37}$$

and

$$M_0 = \frac{\gamma H L^2}{2} \tag{11.38}$$

Now let the term

$$S = \frac{(t/h)^3}{(dt)^{1/2}} \tag{11.39}$$

in Equation 11.37 be designated a *modifying factor for partial fixity*. This factor is normally small and represents the difference between the total fixity moment M_0 and the partial restraint moment M_p. Hence,

$$M_p = M_o(1 - S) \tag{11.40}$$

The value of L in the *denominator* of Eq. 11.37 is conservatively assumed = 1 for simplification in modifying the factor S.

If the value of S is very small, as is the case in large-diameter tanks (diameter larger than 125 to 150 ft), the expressions for L and M_p become expressions for full fixity, namely,

$$L^2 = 2CH^2$$

and

$$M_p = C\gamma H^3$$

11.6.5.2 Base Radial Deformation. The radial deformation Δ_s of the base ring subjected to radial force in its plane can be obtained from the theory of circular plates with concentric holes. The expression for the deflection of the plate shown in Figure 11.13(a) is

$$\Delta_s = \frac{d_o Q}{2hE}\left(\frac{d_o^2 + d^2}{d_o^2 - d^2} - \mu\right) \tag{11.41}$$

where μ = Poisson's ratio ~ 0.2 for concrete and E is the modulus. The horizontal radial thrust per unit of circumference required to induce unit displacement in a solid circular slab is

$$Q_2 = \frac{2.5hE}{d_o} \tag{11.42}$$

and the corresponding value of the radiant thrust applied to the outer ring is

$$Q_3 = \frac{2hE}{d_o K} \tag{11.43}$$

where

$$K = \left(\frac{d_0^2 + d^2}{d_o^2 - d^2} - \mu \right)$$

and d = inside diameter of base ring = $(d_o - 2L)$.

The relative stiffness of the wall to the base is determined in terms of the force required to produce a *unit* deformation in the wall and the base slab from the principles of virtual work as shown in Figures 11.13(b) and (c). The distribution of the prestressing energy between the wall and base slab ring is a function of their relative radial stiffness; hence, determining the relative stiffness is necessary. In doing so, however, one must keep in mind that the stiffness response of the base ring in a prestressed tank to radial compression in its own plane is considerably larger than the response of the cylindrical wall of the tank under radial internal pressure. Thus, the loss of prestress from the differ-

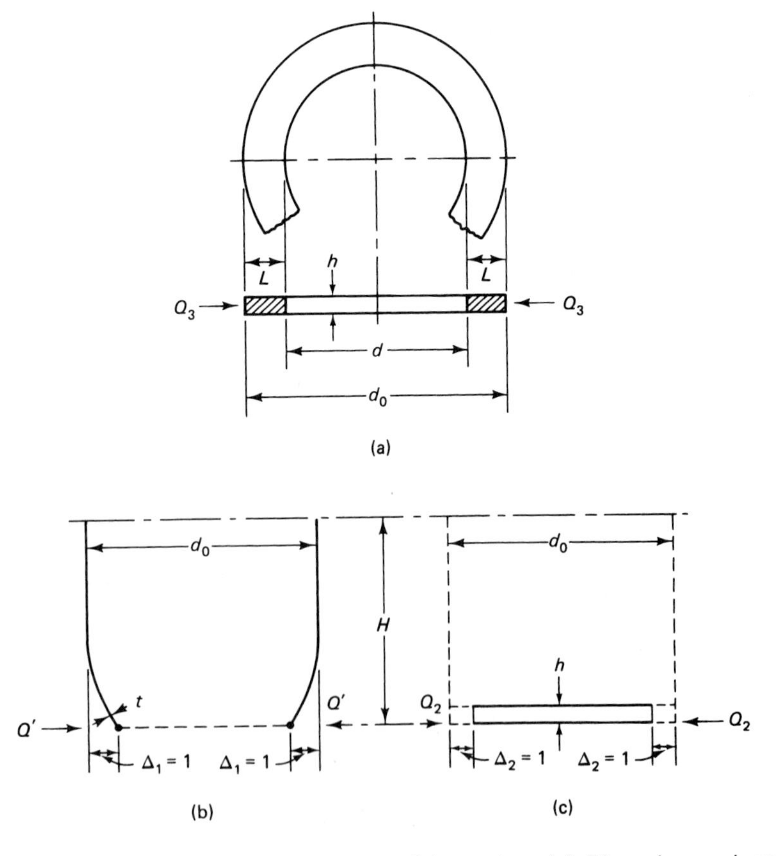

Figure 11.13 Deformation of circular wall base ring. (a) Ring plan and cross section. (b) Deflected wall bottom due to radial force Q'. (c) Deflected ring base due to radial force Q_2.

ence in stiffness is insignificant in large-diameter tanks (Ref. 11.2), but should be considered in small-diameter tanks.

The unit deformation Δ due to the radial force Q' per unit of circumference *without rotation* at the foot base can be obtained from Equation 11.18b using $2\beta M = -Q$ for rotation $dw/dy = 0$. The unit deflection Δ in Equation 11.18a becomes

$$\Delta = \frac{Q_3}{4\beta^3 D}$$

or

$$\Delta = \frac{Q'}{4\beta^3 D} \tag{11.44}$$

where

$$D = \frac{Et^3}{12(1 - \mu^2)}$$

Using $\mu \sim 0.2$, Equation 11.44 for unit radial displacement of the wall at the wall base without rotation becomes

$$Q' = 2.2E\left(\frac{t}{d}\right)^{3/2} \tag{11.45}$$

where E is the modulus of concrete. From Equation 11.42, the radial force per unit of circumference required to produce unit radial displacement in the solid circular slab is

$$Q_2 = 2.5E\left(\frac{h}{d_o}\right) \tag{11.46}$$

By superimposing Q' on Q_2, the total force exerted at the wall-slab base junction is distributed to the wall and the slab base in proportion to the relative energy required to produce unit deformation in each.

The proportion of the total force $Q' + Q_2$ to be carried by the wall is

$$R = \frac{Q'}{Q' + Q_2}$$

say

$$\frac{1}{1 + S_1}$$

Rearranging terms while combining Equations 11.45 and 11.46 results in

$$S_1 = \frac{2.5(h/d)}{2.2(t/d)^{3/2}}$$

assuming that $d \sim d_o$, or

$$S_1 = 1.1\left(\frac{h}{t}\right) \times \left(\frac{d}{t}\right)^{1/2} \tag{11.47}$$

If S_1 is small, the proportion of the horizontal force transferred from the slab base to the wall can be taken, with sufficient accuracy, to be

$$R = \frac{100}{S_1} \text{ percent} \tag{11.48}$$

Photo 11.8 Six-million-gallon tendon-prestressed circular tank seen from inside with situ-cast walls. (*Courtesy,* Jorgensen, Hendrickson and Close, Inc., Denver, Colorado.)

When only the outer ring of the slab is compressed by radial thrust at the rim, the value of Q_2 has to be modified from that obtained by Equation 11.42, and S_1 in Equation 11.48 becomes

$$S_1 = \frac{1}{K}\left(\frac{h}{t}\right) \times \left(\frac{d}{t}\right)^{1/2} \tag{11.49}$$

where, from before,

$$K = \left(\frac{d_0^2 + d^2}{d_0^2 - d^2} - \mu\right)$$

in which d is the inner slab ring diameter $= d_o = 2L$ and d_o is the outer diameter.

11.7 RECOMMENDED PRACTICE FOR SITU-CAST AND PRECAST PRESTRESSED CONCRETE CIRCULAR STORAGE TANKS

11.7.1 Stresses

General guidelines for situ-cast and precast prestressed concrete circular storage tanks are provided by the Prestressed Concrete Institute (Ref. 11.6), the American Concrete Institute (Refs. 11.7–11.9), and the Post-Tensioning Institute (Ref. 11.10) for choosing the applicable allowable stresses, dimensioning, minimum wall thickness, and construction and erection procedure. The allowable stresses in concrete and shotcrete are given in Table 11.17 (Ref. 11.7), with modifications to accommodate the recommended stresses in Ref. 11.6. Allowable stresses in the reinforcement are given in Table 11.18.

Table 11.17 Allowable Concrete Stresses in Circular Tanks

	Concrete situ-cast and precast		Shotcrete situ-cast	
Type and limit of stress	**Temporary[a] stresses f_{ci}, psi**	**Service load stresses f_c, psi**	**Temporary[a] stresses f_{gi}, psi**	**Service load stresses f_g, psi**
Axial compression, f_c	$0.55f'_{ci}$	$0.45f'_c$	$0.45f'_{gi}$ but not more than $1,600 + 40t_c$ psi	$0.38f'_g$
Axial tension	0	0	0	0
Flexural compression, f_c	$0.55f'_{ci}$	$0.4f'_c$	$0.45f'_{gi}$	$0.38f'_g$
Maximum flexural tension[b], f_t	$\approx 3\sqrt{f'_c}$	$3\sqrt{f'_c}$		
Minimum residual compression, f_{cv}	$200\left(\frac{f_{ci}}{f_c}\right)$	200 psi	$200\left(\frac{f_{ci}}{f_c}\right)$	200 psi

[a]Before creep and shrinkage losses.

[b]Fiber stress in precomposed tension zone.

Table 11.18 Stresses in Reinforcement

Type of Stress	Max allowable stress*
Tendon jacking force	$0.94f_{py} \le 0.85f_{pu}$
Immediately after prestress transfer	$0.82f_{py} \le 0.75f_{pu}$
Post-tensioning tendons at anchorage and couplers, immediately after tendon anchorage	$0.70f_{pu}$
Service load stress, f_{pe}	$0.55f_{pu}$
Nonprestressed mild steel at initial prestressing, f_{si}	$f_y/1.6$
Final service load stress, f_s (psi), potable water storage,	
60 grade steel	24,000
corrosive storage	18,000
dry storage	$f_y/1.8$

*1,000 psi = 6,895 Pa.

11.7.2 Required Strength Load Factors

The structure, together with its components and foundations, would have to be designed so that the design strength exceeds the effect of factored load combinations specified by ACI 318, ANSI/ASCE 7-95, or as justified by the engineer based on rational analysis, with the following exceptions:

Feature	Load factor
Initial liquid pressure	1.3
Internal lateral pressure from dry material	1.7
Prestressing forces:	
Final prestress after losses	1.7
Strength reduction factor for both reinforcement and concrete, ϕ	0.9

The nominal moment strength equation M_n is similar to the one used for linear prestressing, i.e.,

$$M_n = A_{ps} f_{ps}\left(d_p - \frac{a}{2}\right) \tag{11.50a}$$

or

$$M_n = A_{ps} f_{ps}\left(d_p - \frac{a}{2}\right) + A_s f_y\left(d = \frac{a}{2}\right) \tag{11.50b}$$

when mild vertical steel A_s is used and

where A_{ps} = vertical prestressing steel per unit width of circumference, in^2.
f_{ps} = stress in prestressed reinforcement at nominal strength, psi
f_y = yield strength of mild steel, psi

11.7.3 Minimum Wall-Design Requirements

11.7.3.1 Circumferential Forces

Liquid

$$\text{Initial } F_i = \gamma r(H - y)\frac{f_{pi}}{f_{ps}} \text{ per foot of wall} \tag{11.51a}$$

Backfill

$$\text{Initial } F_{bi} = p(r + t) \tag{11.51b}$$

where t is the total wall thickness.

11.7.3.2 Thickness and Stresses

Core Wall Thickness

$$t_{co} = \frac{F_i}{f_{ci}} \tag{11.52}$$

but not less than the minimum wall thickness to be set out in subsection 11.7.3.6.

Final Stress Due to Backfill and Initial Prestress

$$f = \frac{F_{bi}}{t} + \frac{F_i}{t_{co}}\frac{f_{pe}}{f_{pi}} \tag{11.53}$$

11.7.3.3 Deflections. The unrestrained initial elastic radial deflection of the wall due to initial prestressing is

$$\Delta_i = \frac{F_i r}{t_{co} E_c} \tag{11.54}$$

where r = tank inner radius
t_{co} = thickness of wall core at top or bottom of wall
$E_c = 57{,}000\sqrt{f'_c}$ psi for both normal-weight concrete and shotcrete.

The final radial deflection Δf may reach 1.5 to 3 times the initial unrestrained deflection. For normal conditions, the final permitted radial deflection can be taken as

$$\Delta f = 1.7\Delta_t \tag{11.55}$$

11.7.3.4 Restraint Effects

Maximum Vertical Wall Bending Due to Radial Shear

$$M_y = 0.24Q_0\sqrt{rt_{co}} \tag{11.56a}$$

This moment occurs at a distance

$$y = 0.68\sqrt{rt_{co}} \tag{11.56b}$$

from the base or top edge.

Radial Shear for Monolithic Base Details Which May be Assumed to Provide Hinged Connection

$$Q_o = 0.38\,F_i\sqrt{\frac{t_{co}}{r}} \tag{11.57}$$

This type of detail should be used only with situ-cast tanks which incorporate a diaphragm in their wall construction.

11.7.3.5 Mild Steel for Base Anchorage. If a diaphragm is used, extend the full area of the inside bars in a U-shape a distance

$$y_1 = 1.4\sqrt{rt_{co}} \tag{11.58a}$$

above the base. If no diaphragm is used, extend to

$$y_2 = 1.8\sqrt{rt_{co}} \tag{11.58b}$$

above the base. Note that anchorage length has to be added to y_1 or y_2. The minimum area of nominal vertical steel at the base region is

$$A_s = 0.005t_{co} \tag{11.59}$$

and should be extended above the base a distance of 3 ft or

$$y_3 = 0.75\sqrt{rt_{co}} \tag{11.60}$$

whichever is greater.

11.7.3.6 Minimum Wall Thickness

Situ-Cast Walls

Type of tank	Minimum wall thickness
Shotcrete-steel diaphragm tanks	$3\frac{1}{2}$ in.
Tanks without vertical prestressing	8 in.
Tanks with vertical prestressing	7 in.

Precast Walls

Type of tank	Minimum wall thickness
Tanks with vertical pretensioning and external circumferential prestress	5 in.
Tanks with vertical pretensioning and internal circumferential prestress	6 in.
Tanks with vertical post-tensioning and internal circumferential prestress	7 in.

It should be noted that for tanks prestressed with tendons, a thickness not less than 9 in. is advisable for practical considerations.

11.8 CRACK CONTROL IN WALLS OF CIRCULAR PRESTRESSED CONCRETE TANKS

Vessey and Preston in Ref. 11.14 recommend the following expression based on Nawy's work in Ref. 11.15 for the maximum crack width at the exterior surface of the prestressed tank wall:

$$w_{\max} = 4.1 \times 10^{-6}\, \epsilon_{ct}\, E_{ps} \sqrt{I_x} \tag{11.61}$$

where ϵ_{ct} = tensile surface strain in the concrete
I_x = grid index $= \dfrac{8}{\pi}\left(\dfrac{s_2\, s_1\, t_b}{\phi_1}\right)$
s_2 = reinforcement spacing in direction "2"
s_1 = reinforcement spacing in perpendicular direction "1" (horizontal)
t_b = concrete cover to center of steel
ϕ_1 = diameter of steel in main direction "1."

The tensile strain can be computed from

$$\epsilon_{ct} = \frac{\alpha_t f_{pi}}{E_{ps}} \tag{11.62}$$

where α_t = stress parameter $\cong f_p/f_{pi}$
f_p = actual stress in the prestressing steel
f_{pi} = initial prestress before losses.

For liquid-retaining tanks, the maximum allowable crack width is 0.004 in.

11.9 TANK ROOF DESIGN

Roofs for storage tanks are constructed in the form of a shell dome or as flat roofs supported internally on columns. The cost of the roof is generally about one-third of the overall cost of the structure. In the case of flat roofs, whether precast or situ cast, the design follows the normal design principles of floor systems for reinforced or prestressed concrete one-way- or two-way-action floors as stipulated in the ACI 318 Code. If the roof is made out of precast prestressed elements, and the tank diameter is not exceedingly large, no interior columns are necessary. Otherwise, the added cost of interior columns and the accompanying footings would increase the cost of the overall structure.

A shell roof in the form of a dome has distinct advantages for tanks not exceeding 150 ft. in diameter, namely, that the dome does not need supporting interior columns and can also be economical in underground storage tanks in withstanding backfill load. Hence, the shell form and the manner of its connection to the tank walls have a significant effect on cost. Preferably, the roof shell should be supported by tank walls with a completely *flexible* joint; otherwise the design of both the tank wall and the roof dome will have to be modified in relation to their degree of interrestraint and relative stiffness, with the concomitant added construction cost.

A spherical shell of low rise-to-diameter ratio h'/d of approximately $\frac{1}{8}$ is reasonable to use. Such a flat dome or axisymmetrical shell introduces outward horizontal thrust at the springing, which has to be resisted by a properly designed prestressed ring beam at the support level. The type of support of the ring beam determines the extent to which redundant reactions and moments due to end restraint impose additional direct and bending stresses in the shell near the springing. In other words, the membrane solution has to be adequately modified by superimposing on it the bending effects determined by the strain compatibility requirements of the bending theory.

11.9.1 Membrane Theory of Spherical Domes

11.9.1.1 Shell of Revolution. The basic membrane equations of equilibrium for the direct forces in a shell of revolution as shown in Figure 11.14 are used for defining the unit meridional forces N_ϕ, unit tangential forces N_θ, and unit central shears $N_{\phi\theta}$ and $N_{\theta\phi}$ in terms of the gravity loads p_ϕ, p_θ, and p_z. These equations are as follows:

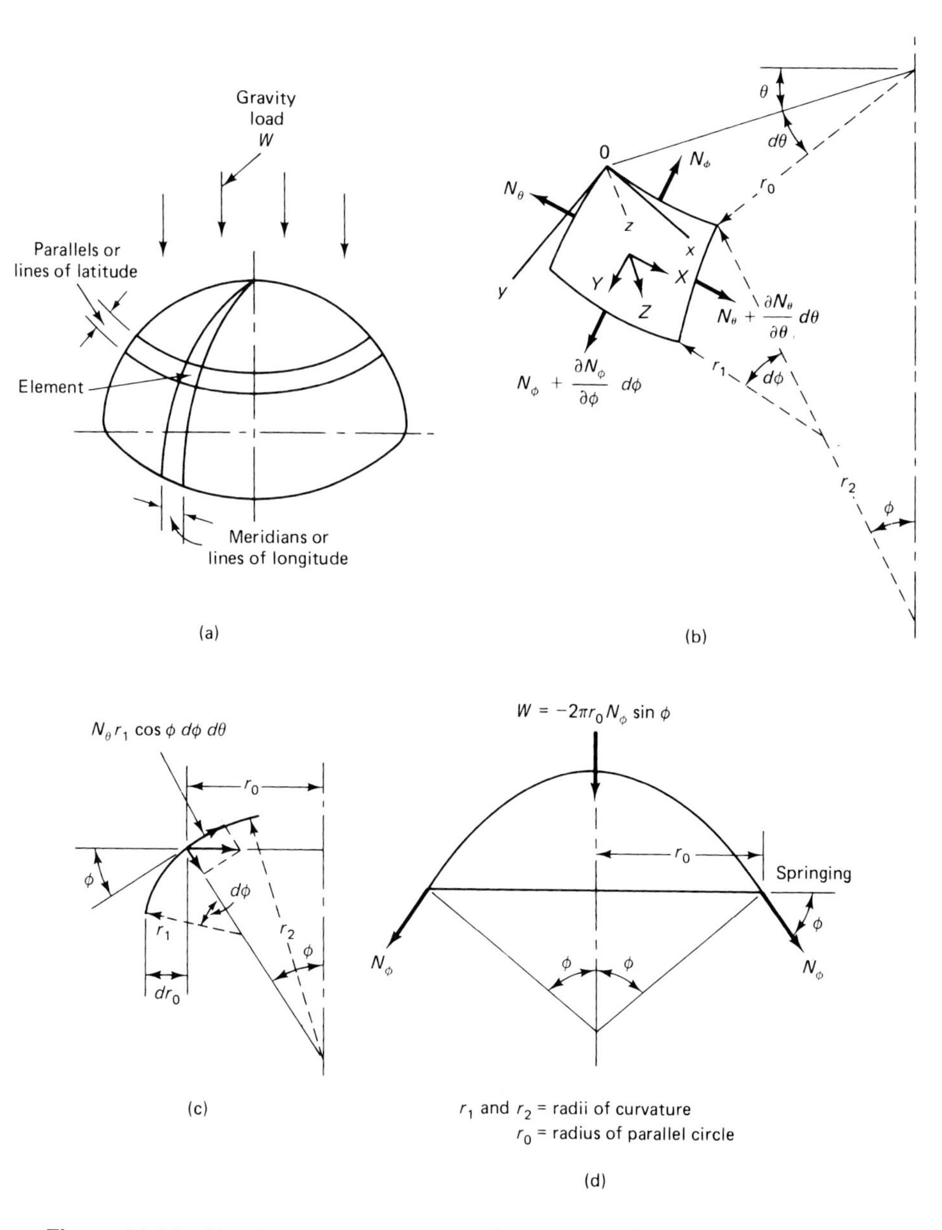

Figure 11.14 Membrane forces in a shell of revolution. (a) Meridian and parallel lines. (b) Membrane forces on infinitesimal surface element. (c) Component of force $N_\theta r_1 d\phi$ in the y direction needed to simplify the basic equation 11.63a. (d) Dome cross section with total gravity load W.

$$\text{Meridional:} \quad \frac{\partial(N_\phi r_o)}{\partial \phi} - N_\theta \frac{\partial r}{\partial \phi} + \frac{\partial N_{\theta\phi}}{\partial \theta} r_1 + p_\phi r_o r_1 = 0 \tag{11.63a}$$

$$\text{Tangential:} \quad \frac{\partial N_\theta}{\partial \theta} r_1 + N_{\theta\phi} \frac{\partial r_0}{\partial \phi} + \frac{\partial N_{\theta\phi}}{\partial \phi} r_1 + p_\theta r_o r_1 = 0 \tag{11.63b}$$

$$z\text{-direction:} \quad \frac{N_\phi}{r_1} + \frac{N_\theta}{r_2} + p_z = 0 \tag{11.63c}$$

Because of loading symmetry, all terms involving $\partial\theta$ vanish, and those involving $\partial\theta$ can be rewritten as total differentials $d\phi$ since nothing varies with respect to θ. Also, the circumferential load component $p_\theta = 0$, as the shear resultants vanish along the meridional and parallel circles. Hence, Equations 11.63 can be rewritten as

$$\frac{d}{d\phi}(N_\phi r_o) - N_\theta r_1 \cos\phi + p_y r_1 r_o = 0 \tag{11.64a}$$

$$\frac{N_\phi}{r_1} + \frac{N_\theta}{r_2} + p_z = 0 \tag{11.64b}$$

11.9.1.2 Spherical Dome

Membrane Analysis of the Equilibrium Forces. The spherical dome has a uniform curvature. Consequently, $r_1 = r_2 = r_o$. Assuming that the radius of the sphere $= a$, then $r_o = a \sin\phi$ in Figure 11.14(c), and, setting $p_z = w_D$ for self-weight, the general equilibrium equations 11.64 become

$$N_\theta = a w_D\left(\frac{1}{1+\cos\phi} - \cos\phi\right) \tag{11.65a}$$

and

$$N_\phi = -\frac{a w_D}{1+\cos\phi} \tag{11.65b}$$

where w_D is the intensity of self-weight per unit area. It is plain from Equation 11.65b that the meridional force N_ϕ is always negative. Therefore, *compression* develops along the meridians and increases as the angle ϕ increases: when $\phi = 0$, $N_\phi = -\frac{1}{2} a w_D$; and when $\phi = \pi/2$, $N_\phi = -a w_D$.

The tangential force N_θ is negative, i.e., compressive, only for limited values of the angle ϕ. Setting $N_\theta = 0$ in Equation 11.65a, $1/(1+\cos\phi) - \cos\phi = 0$ gives $\phi = 51°49'$. This determination indicates that for ϕ greater than 51°49′, tensile stresses develop in the direction perpendicular to the meridians. The distribution of the meridional stresses N_ϕ and the tangential stresses N_θ for both the self-weight w_D and the external live load w_L is shown in Figure 11.15.

If the external load is uniform, such as snow, giving a projection intensity w_L, the meridional force N_ϕ is obtained from free-body equilibrium by equating the external load to the internal meridional force, i.e., $-\pi(d/2)^2 w_L = 2\pi(a \sin\phi)N_\phi$. Since $d/2 = a \sin\phi$, we obtain

$$N_\phi = -\frac{w_L a}{2} \tag{11.66a}$$

Hence, N_ϕ is constant throughout the shell depth, as is plain in Figure 11.15.

N_θ due to the live load w_L is

$$N_\theta = -a w_L \cos^2\phi + \frac{a w_L}{2} = a w_L\left(\frac{1}{2} - \cos^2\phi\right) = \frac{a w_L}{2}\cos 2\phi \tag{11.66b}$$

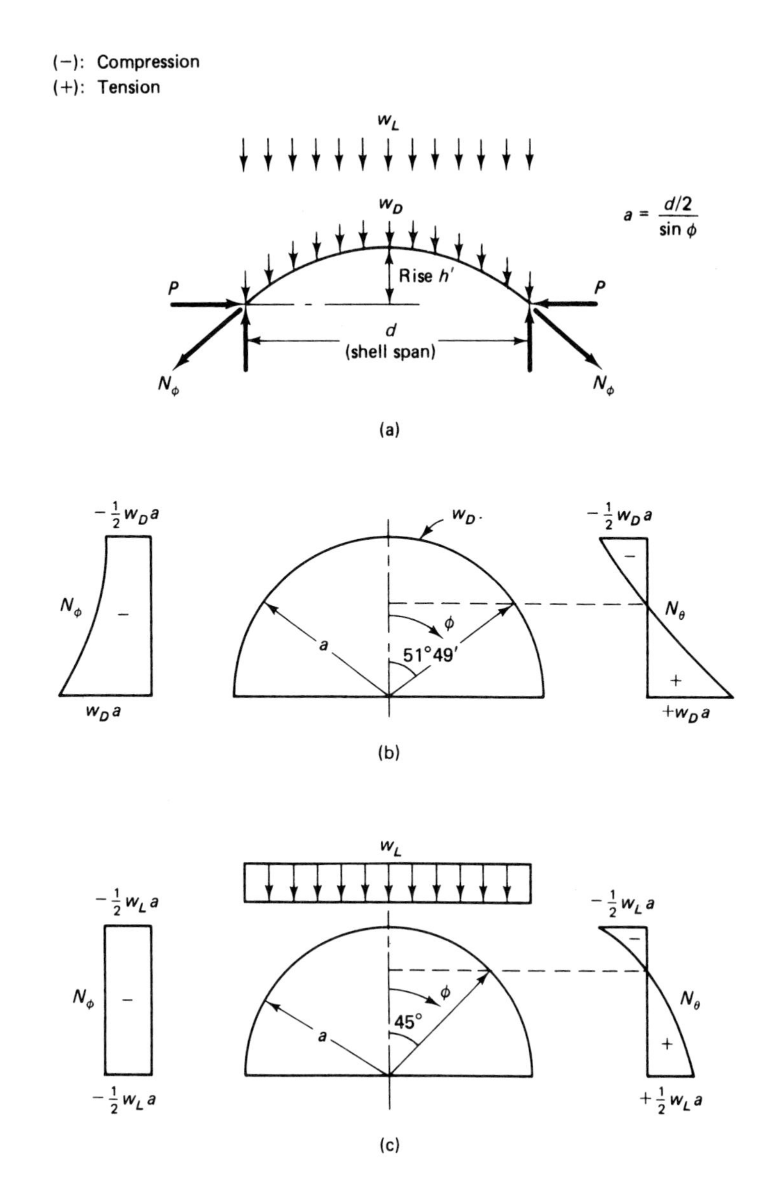

Figure 11.15 Gravity membrane force distribution in a spherical dome. (a) Flat dome segment of rise h'. (b) Membrane stresses due to self-weight w_D ($N_\theta = 0$ for $\phi = 51°, 49'$). (c) Membrane stresses due to snow load w_L ($N_\theta = 0$ for $\phi = 45°$).

For the case of $N_\theta = 0$, the shell angle $\phi = 45°$. Consequently, shell stresses due to tangential forces N_θ for ϕ less than 45 degrees are compressive, eliminating cracking. From the distribution of the tangential forces N_θ, it can be concluded that roofs of storage tanks should be *flat,* i.e., the ratio h'/d in Figure 11.15(b) should not exceed $\frac{1}{8}$, so that the concrete will be totally in compression due to both N_ϕ and N_θ, as angle ϕ is less than 51°49′ for meridional forces and 45° for tangential forces.

As discussed at the outset, the support type at the springing level, if restrained, introduces indeterminate reactions that result in direct and bending stresses in the shell near the springing level. Accordingly, the bending theory, a rigorous procedure beyond the scope of this text, has to be applied. Refs. 11.1 and 11.3, on the subject of plates and shells, can be used for determining the resulting bending stresses. The following covers the design of the prestressed ring beam at the springing level to counter the horizontal component of the meridional compressive thrust N_ϕ which causes the edge of the dome to move inwards.

From Equations 11.65b and 11.66a, the meridional thrust, N_ϕ, for self-weight w_D per unit surface area and uniform live load w_L per unit projected area can be written as

$$N_\phi = -a\left(\frac{w_D}{1+\cos\phi} + \frac{w_L}{2}\right) \tag{11.67}$$

where $a = d/2 \sin\phi$ is the radius of the shell.

Note that the thrust, N_ϕ, becomes vertical at the springing ($\phi = \pi/2$) of a hemispherical dome and is equal to $W = a/2(2w_D + w_L)$ per unit width. At other values of ϕ, N_ϕ, it is inclined and the value of its horizontal component is needed for the design of the prestressed ring beam at the springing level, namely, the shell rim. This horizontal component is $p = N_\phi \cos\phi$. If P is the prestressing force per beam height in the ring beam, then $P = pd/2$ from Equation 11.1a, and

$$P = \frac{d}{2}(N_\phi \cos\phi) \tag{11.68}$$

Evidently, if the force P could be applied directly to the dome rim, the stresses in the dome would be those defined by Equation 11.67. This is usually not feasible, since the large amount of prestressing steel needed due to P cannot be accommodated in the small thickness of the shell, and the stress in the concrete in the rim zone would be very high indeed. Thus, an edge beam has to be provided, transforming the shell into a statically determinate structure.

Prestressing the Statically Indeterminate Flat Dome. The simplest boundary condition is obtained when the edge beam reaction is vertical and without any support restraint, as shown in Figure 11.16, where the dome thrust N_ϕ passes through the beam centroid. If an imaginary cut along line A–A is made, the horizontal thrust $N_\phi \cos\phi$ causes the dome edge to move *inwards* a distance (Ref. 11.16)

$$\Delta_s = \frac{d}{2Et}(N_\theta - \mu N_\phi) \tag{11.69}$$

where μ = Poisson's ratio ≈ 0.2 for concrete
d = shell span

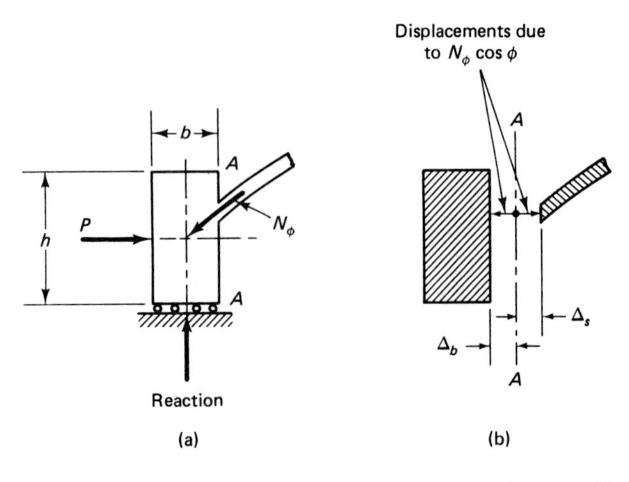

Figure 11.16 Ring beam effects. (a) Simply supported beam with thrust line passing through ring beam centroid. (b) Shell displacements at rim; rotations disregarded.

and the tangential unit force is obtained from Equation 11.65a as

$$N_\theta = \frac{w_D d}{2 \sin \phi}\left(\frac{1}{1 + \cos \phi} - \cos \phi\right) - \frac{w_L d}{4 \sin \phi}(\cos 2\phi) \tag{11.70}$$

Conversely, the meridional thrust N_ϕ causes the ring beam to move *outwards* a distance

$$\Delta_b = \frac{N_\phi(\cos \phi)d^2}{4Ebh} \tag{11.71}$$

The prestressing force must therefore be sufficient to move the ring beam *inwards* a total distance

$$\Delta_T = \Delta_s + \Delta_b$$

so that the total force acting on the ring beam cross section is

$$P = \frac{bh}{t}(N_\theta - \mu N_\phi) + \frac{d(N_\phi \cos \phi)}{2} \tag{11.72}$$

where h is the total ring beam depth. A comparison of Equations 11.72 and 11.68 shows that the effective prestressing force needed in the former is greater than that required in the latter. The magnitude of this increase is about 5 to 10 percent. The same conditions also hold true for domes in which the line of thrust from the dome does not pass through the centroid of the ring beam and the beam itself is rigidly attached to the wall as in Figure 11.17(a). The required prestressing force P can be obtained approximately by increasing the value of P in Equation 11.68 by 10 percent (Ref. 11.16). In such a case, the stresses in the shell itself at the springing level zone can significantly differ from those obtained in the membrane solution, and the bending solution modifications have to be made as in Ref. 11.1 or 11.3.

If the horizontal radial prestressing force in the ring beam is larger than required, excessive bending deformation develops in the shell rim, as is shown in Figure 11.17(b), with a significant increase in the value of the tangential force N_θ as compared to the increase in the meridional force N_ϕ. As a result, the bending stresses in the concrete in the affected zone could exceed the maximum allowable at service load. If the initial prestress before losses is P_i, the area of the beam cross section is

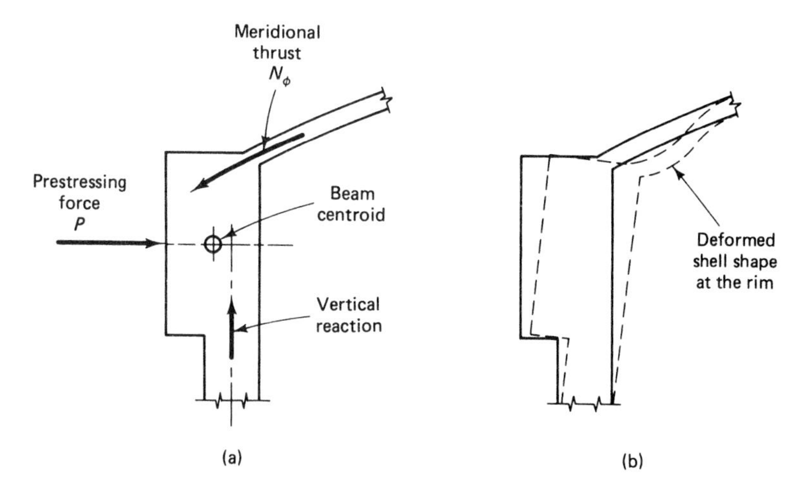

Figure 11.17 Edge ring beam monolithic with tank wall. (a) Thrust N_ϕ not passing through ring beam centroid—general case. (b) Shell deformed shape due to excessive prestressing.

Photo 11.9 1.55 Million Gallon Reactor Tank, Bishop Texas. (*Courtesy,* N.A. Legatos, Preload Inc., Garden City, New York.)

$$A_c = \frac{P_i}{f_c} \tag{11.73}$$

where P_i = initial prestressing force $P/\bar{\gamma}$
f_c = allowable compressive stress in the concrete
$\bar{\gamma}$ = residual stress percentage.

It is desirable to maintain a low value of f_c, about $0.2f'_c$ and not exceeding 800 to 900 psi, in order to minimize any excessive strain that develops in the edge ring beam, which in turn could produce high stresses in the shell at the springing zone.

The area of the prestressing steel in the dome ring is

$$\text{Unit } A_{ps} = \frac{P_i}{f_{pi}} \tag{11.74a}$$

where f_{pi} is the allowable stress, in psi, in the prestressing reinforcement before losses. If accurate analysis to determine A_{ps} is not required, the steel area can be taken as

$$A_{ps} = \frac{W \cot \phi}{2\pi f_{pe}} \tag{11.74b}$$

where W = total dead and live load on the dome due to $w_D + w_L$
f_{pe} = effective steel prestress after losses, psi.

The minimum thickness of the dome required to withstand buckling (Ref. 11.7) may be taken to be

$$\text{Min } h_d = a\sqrt{\frac{1.5p_u}{\phi\beta_i\,\beta_c\,E_c}} \tag{11.75}$$

where a = radius of dome shell

p_u = ultimate uniformly distributed design unit pressure due to dead load and live load = $(1.4D + 1.7L)/144$

ϕ = strength reduction factor for material variability = 0.7

β_i = buckling reduction factor for deviations from true spherical surface due to imperfections

$\beta_i = (a/r_i)^2$, where $r_i \leq 1.4a$

β_c = buckling reduction factor for creep, material nonlinearity, and cracking = $0.44 + 0.003\ W_L$, but not to exceed 0.53

E_c = initial modulus of concrete = $57{,}000\sqrt{f'_c}$ psi.

11.10 PRESTRESSED CONCRETE TANKS WITH CIRCUMFERENTIAL TENDONS

Instead of wrapping the prestressing wires or strands, as is done in the Preload System, internal or external horizontal tendons are used. These tendons are stressed after they are placed within or on the wall. Vertical post-tensioning is incorporated in the walls as part of the vertical reinforcement. The concrete walls are either cast in place or precast, and the core wall is considered to be the portion of the concrete wall that is circumferentially prestressed. No steel diaphragms are used in this type of construction as compared with wrapped-wire prestressing, where the tank walls can be either with or without steel diaphragms.

The internal prestressed reinforcement is protected by the concrete cover as required in ACI 318, and the ducts or sheathing have to be filled with corrosion-inhibiting materials or grouted. The bonded post-tensioned tendon reinforcement has to be protected by portland cement grout as required in the ACI 318 code, and external tendons should be protected by a shotcrete cover of 1-in. (25-mm) minimum thickness.

The wall design procedures are similar to those of circular tanks prestressed by wire or strand wrapping, and the same requirements for crack control and water or liquid tightness apply. A minimum residual compressive stress of 200 psi (1.4 MPa) in the concrete wall after all prestress losses has to be provided in the design when the tank is filled to the design level. If the tank is not covered, a residual compressive stress of 400 psi (2.8 MPa) has to be provided at the wall top, reducing linearly to not less than 200 psi at $0.6\sqrt{Rh}$ from the top of the liquid level.

Typical Wall Base and Dome Roof Connections. From the foregoing discussions, it is clear that the boundary conditions at the base of the circular prestressed tank and at the ring beam support for the roof dome determine the practicality, economy, and success of the entire design. Consequently, accumulated experience in developing the connections at these boundary conditions is invaluable. A selection of connection details taken from Refs. 11.6 to 11.9 is given in Figures 11.18 through 11.22.

11.11 STEP-BY-STEP PROCEDURE FOR THE DESIGN OF CIRCULAR PRESTRESSED CONCRETE TANKS AND DOME ROOFS

The following trial-and-adjustment procedure is recommended for designing a prestressed concrete circular tank and its roof shell:

1. Select the prestressing system, the type of prestressing wire, the concrete strength, and the type of restraint that can be accomplished under local conditions.

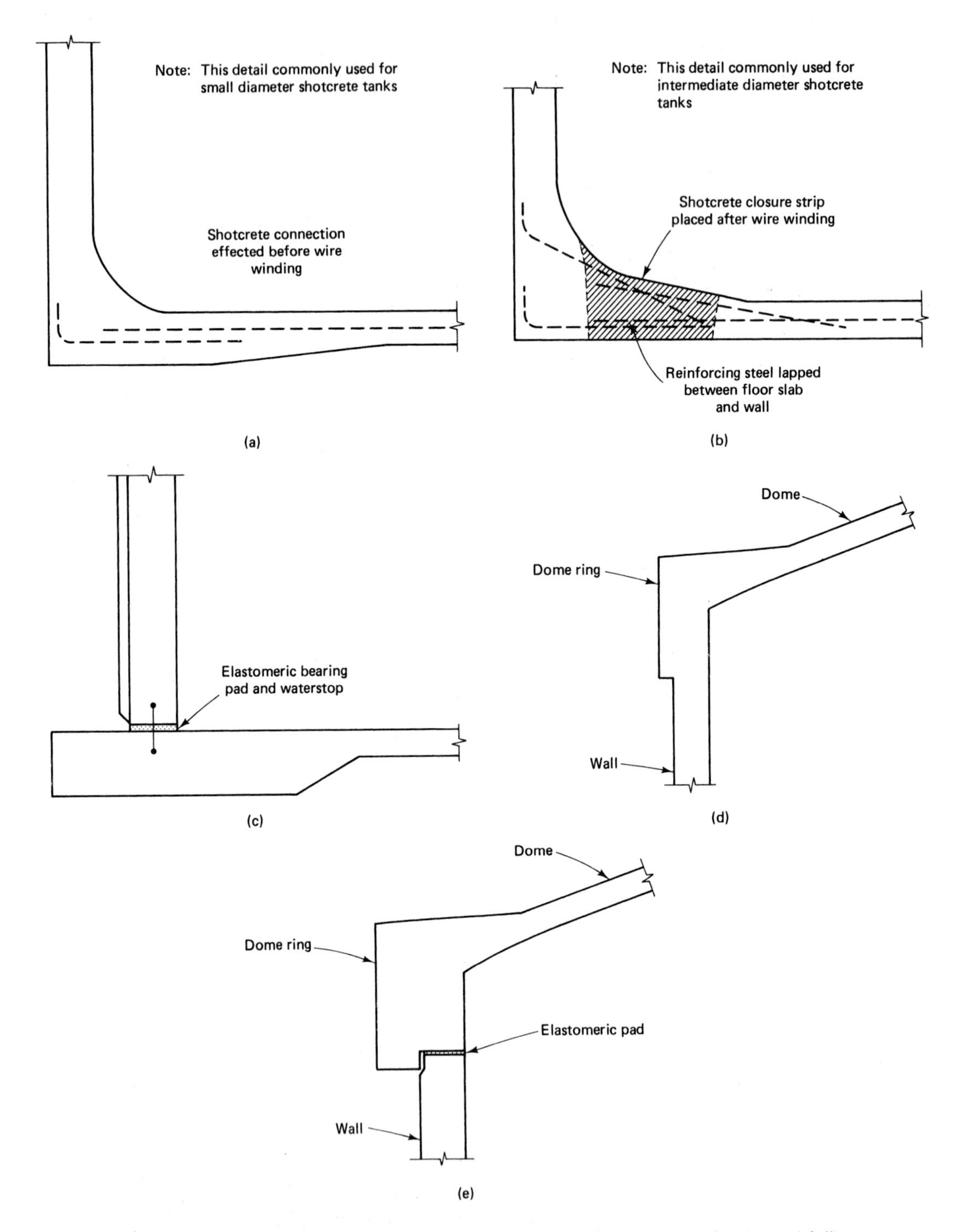

Figure 11.18 Cast-in-place tanks. (a) Monolithic base joint; monolithic and fully restrained against translation before and after wire winding. (b) Monolithic base joint; hinged with limited restraint against translation during wire winding, and monolithic and fully restrained against translation after wire winding. (c) Separated base joint, allows translation, rotation, or both (d) Monolithic dome-wall connection. (e) Separated dome-wall connection.

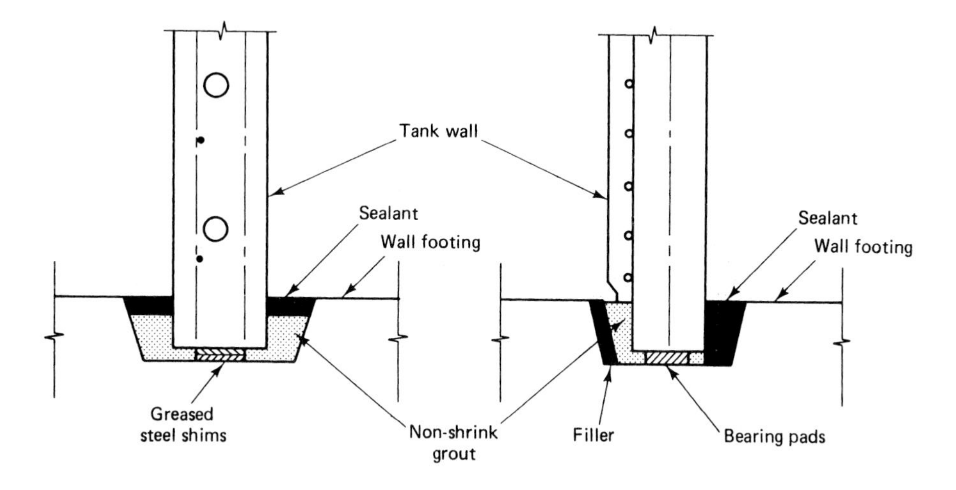

Figure 11.19 Wall base joints for precast tanks.

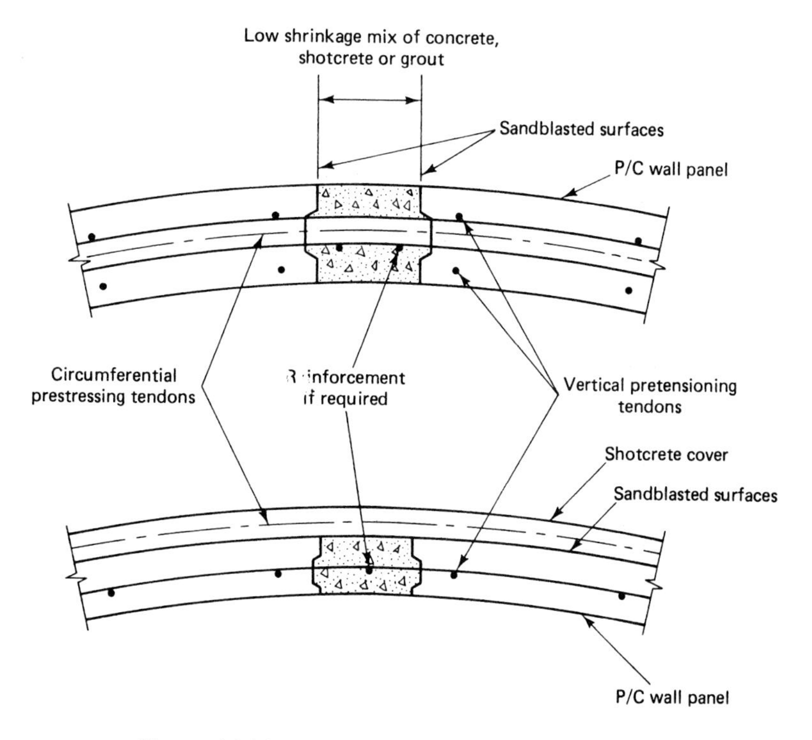

Figure 11.20 Vertical wall joints for precast tanks.

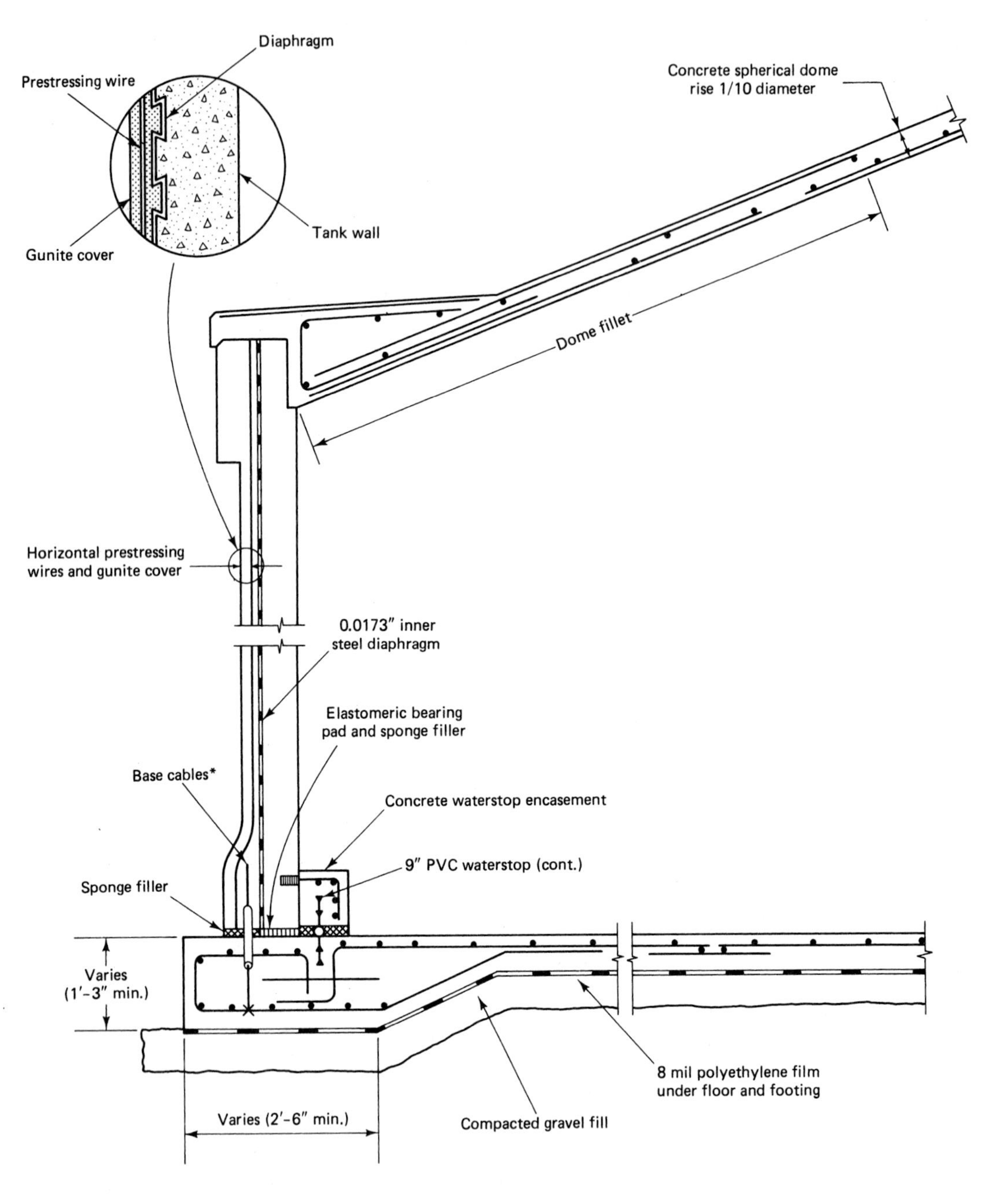

Figure 11.21 Typical tank section of a domed preload prestressed concrete tank with an inner steel diaphragm. (*Courtesy,* Preload Technology, Inc., New York.)

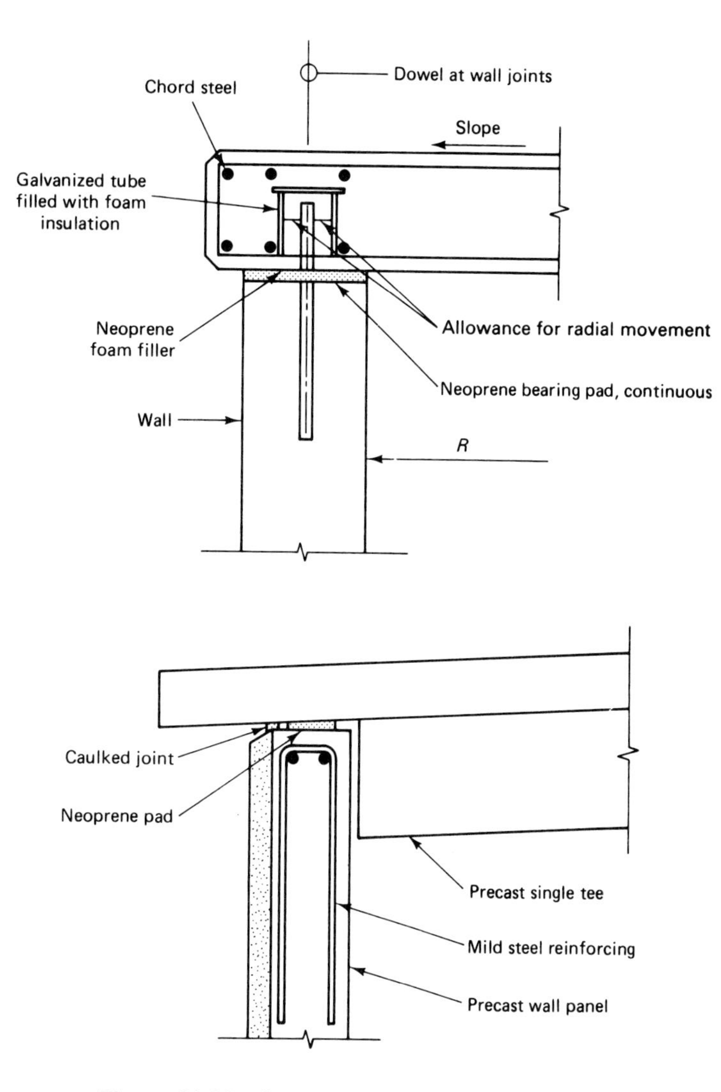

Figure 11.22 Connections for precast tank roofs.

2. Determine the contained material pressure on the wall: γH for liquid and p for gas. Use the trapezoidal distribution for granular or solid containment.

   Find the unit ring force $F = \gamma(H - y)r$ for a completely sliding base, where r is the radius of the tank and y is the distance above the base.

3. Choose, from Tables 11.4 through 11.16, the applicable vertical moment coefficients for the particular load type and wall base restraint condition caused by liquid pressure

$$M_y = +\frac{1}{\beta}[\beta M_o \phi(\beta y) + Q_o \zeta(\beta y)]$$

and determine the corresponding horizontal radial ring tensions

$$Q_o = +(2\beta H - 1)\frac{\gamma r t}{\sqrt{12(1 - \mu^2)}}$$

and $Q_y = (F - \Delta Q_y)$, where the offset

$$\Delta Q_y = +\frac{6(1-\mu^2)}{\beta^3 rt^2}(\beta M_o \psi(\beta y) + Q_o\,\theta(\beta y))$$

and

$$\beta = \frac{[3(1-\mu^2)]^{1/4}}{(rt)^{1/2}}$$

where $\mu \approx .20$ for concrete.

4. Find the applicable membrane coefficients C from Tables 11.4 through 11.16. Compute the applicable ring force $F = C\gamma Hr$.
5. Compute the critical vertical moments in the wall using the applicable membrane coefficient C. The equation for moment due to liquid load is

$$M_y = C(\gamma H^3 + pH^2)$$

or

$$M_y = CpH^2$$

due to gas load if applicable. Compute the moment at the base, where applicable, and at the critical y plane above the base.

6. Choose the level of vertical prestressing force.
7. Compute the concrete stresses across the thickness of the wall both for the condition when the tank is empty and for when it is totally full. Allow maximum residual axial compressive stress f_{cv} = 200 psi at service and a maximum tensile stress $f_t = 3\sqrt{f'_c}$ as shown in Table 11.17.
8. Design both the horizontal and the vertical prestressing steel limiting stresses to those given in Table 11.18.
9. Compute the factored moment M_u using the applicable load factors given in subsection 11.7.2. The required $M_n = M_u/\phi$, where $\phi = 0.9$. Compute the available nominal moment strength $M_n = A_{ps}f_{ps}(d_p - a/2)$, or $M_n = A_{ps}f_{ps}(d_p - a/2) + A_s f_y(d - a/2)$. The available M_n has to be greater than or equal to the required M_n.
10. Design the length L of the annular ring at the base of the wall from the equation

$$L^2 = \frac{2CH^2}{1 + \dfrac{(t/h)^3}{(dt)^2}}$$

where t is the thickness of the wall and h the thickness of the base slab.

11. Compute the percentage of prestress in the base to be transferred to the wall from the formula

$$\text{Percentage } R = \frac{1}{1+S}$$

where $S = 1.1(h/t) \times (d/t)^{1/2}$.

When only the outer rim of the slab ring is compressed by radial thrust at the rim, the value of S is modified to

$$S_1 = \frac{1}{K}\left(\frac{h}{t}\right)\left(\frac{d}{t}\right)^{1/2}$$

where

$$K = \left(\frac{d_o^2 + d^2}{d_0^2 - d^2} - \mu \right)$$

in which d_o = outer diameter
d = inner slab ring diameter = $d_o - 2L$.

12. Check the minimum wall thickness requirements, and evaluate the unrestrained initial elastic radial deflection

$$\Delta_i = \frac{F_i r}{t_{co} E_c}$$

where $E_c = 57{,}000\sqrt{f'_c}$
t_{co} = thickness of wall core at top or bottom of wall
$r = \frac{1}{2}d$.

The final radial deflection $\Delta_f = 1.7\Delta_i$.

13. Anchor the steel from the base to the wall such that the steel extends into the wall a distance $y_2 = 1.8\sqrt{rt_{co}}$ or 3 ft, whichever is greater. Also, ensure that the minimum nominal vertical steel at the base region is

$$A_s = 0.005t_{co}$$

14. Verify the maximum crack width $w_{max} = 4.1 \times 10^{-6}\epsilon_{ct}E_{ps}\sqrt{I_x}$,

where ϵ_{ct} = tensile surface strain in the concrete = $(\lambda_t f_p)/(E_{ps})$
f_p = actual stress in the steel
f_{pi} = initial prestress before losses
$\lambda_t \sim f_p/f_{pi}$
I_x = grid index = $\frac{8}{\pi}\left(\frac{s_2 s_1 t_b}{\phi_1}\right)$
s_1 = spacing of reinforcement in direction "1"
ϕ_1 = diameter of steel in direction "1"
s_2 = spacing of reinforcement in direction "2"
t_b = concrete cover to center of steel, in.

Note that maximum allowable $w_{max} = 0.004$ in. for liquid-retaining tanks.

15. Design the roof cover dome after selecting the type of connection at the top of the tank wall. Limit the ratio of the rise h' of the dome to its base d such that h'/d does not exceed $\frac{1}{8}$.

Compute the required horizontal radial prestressing force P for the edge beam from the equation

$$P = \frac{bh}{t}(N_\theta - \mu N_\theta) + \frac{d(N_\phi \cos \phi)}{2}$$

where

$$N_\theta = \frac{w_D d}{2 \sin \phi}\left[\frac{1}{1 + \cos \phi} - \cos \phi\right] - \frac{w_L d}{4 \sin \phi}(\cos 2\phi)$$

$$N_\phi = -a\left(\frac{w_D}{1 + \cos \phi} + \frac{w_L}{2}\right)$$

and

h = total depth of rim beam
b = ring beam width

Photo 11.10 Two prestressed concrete anaerobic digester tanks during construction. (*Courtesy,* N.A. Legatos, Preload Technology, Inc., New York.)

w_D = intensity of self-weight of shell per unit area (dead load)
w_L = intensity of live-load projection.

16. Compute the ring-edge beam cross section

$$A_c = \frac{P_i}{f_c}$$

where P_i = initial prestressing force = $P/\bar{\gamma}$
$\bar{\gamma}$ = residual stress percentage
f_c = allowable compressive stress in the concrete, not to exceed $0.2f'_c$, but not more than 800–900 psi, in the edge beam.

17. Compute the area of the edge beam prestressing tendon

$$A_{ps} = \frac{P_i}{f_{si}}$$

where f_{si} is the allowable stress in the prestressing steel before losses, or

$$A_{ps} = \frac{W \cot \phi}{2\pi f_{pe}}$$

if accurate analysis is not performed. In the latter, W is the total dead and live load on the dome due to $w_D + w_L$ and f_{pe} is the effective prestress after losses.

18. Check the minimum dome thickness required to withstand buckling, i.e.,

$$\text{Min. } h_d = a\sqrt{\frac{1.5p_u}{\phi\beta_i\,\beta_c\,E_c}}$$

where a = radius of dome shell

P_u = ultimate uniformly distributed design unit pressure due to dead load and live load = $(1.4D + 1.7L)/144$

ϕ = strength reduction factor for material variability = 0.7

β_i = buckling reduction factor for deviations from true spherical surface due to imperfections

$\beta_i = (a/r_i)^2$, where $r_i \leq 1.4a$

β_c = buckling reduction factor for creep, material nonlinearity, and cracking = $0.44 + 0.003W_L$, but not to exceed 0.53

E_c = initial modulus of concrete = $57{,}000\sqrt{f'_c}$ psi.

Figure 11.23 gives a step-by-step flowchart for a recommended sequence of operations to be performed in the design of circular prestressed concrete tanks and their shell roofs.

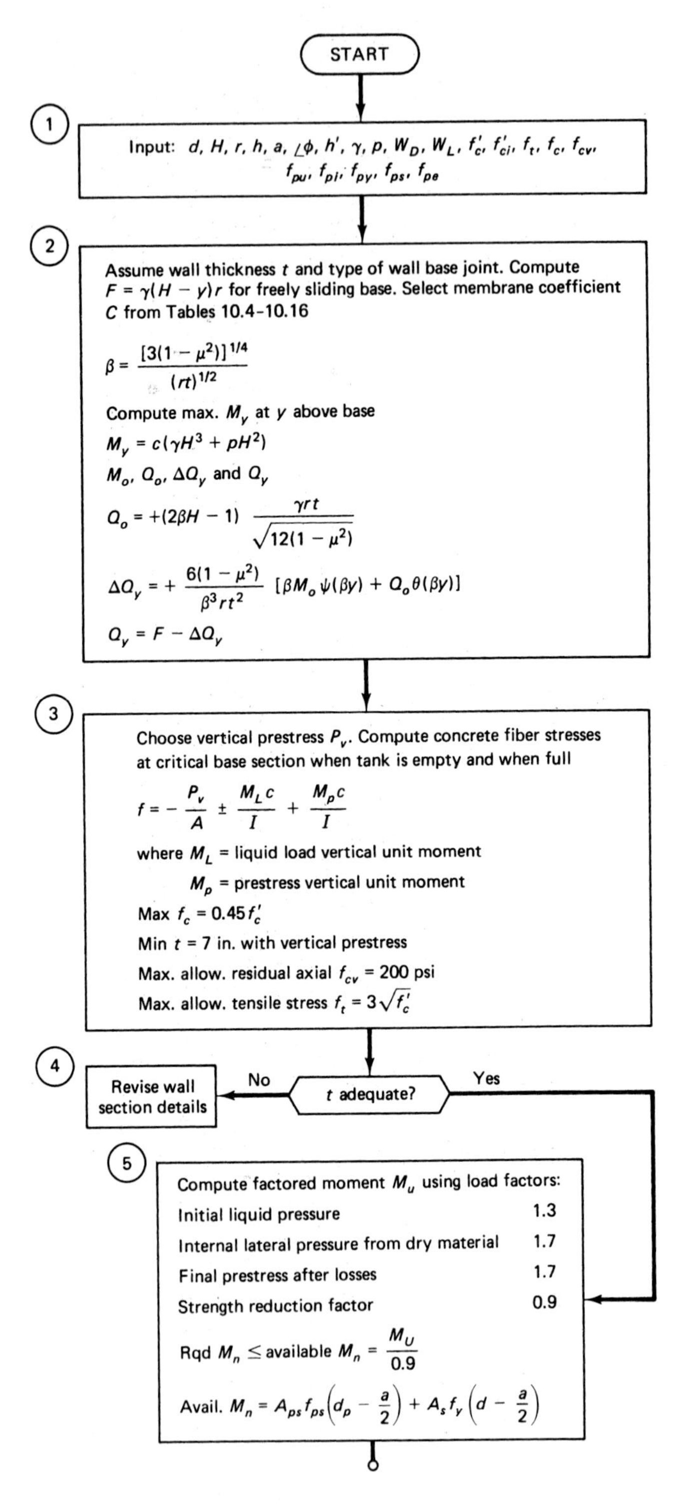

Figure 11.23 Flowchart for the design of circular prestressed tanks and their flat dome roofs.

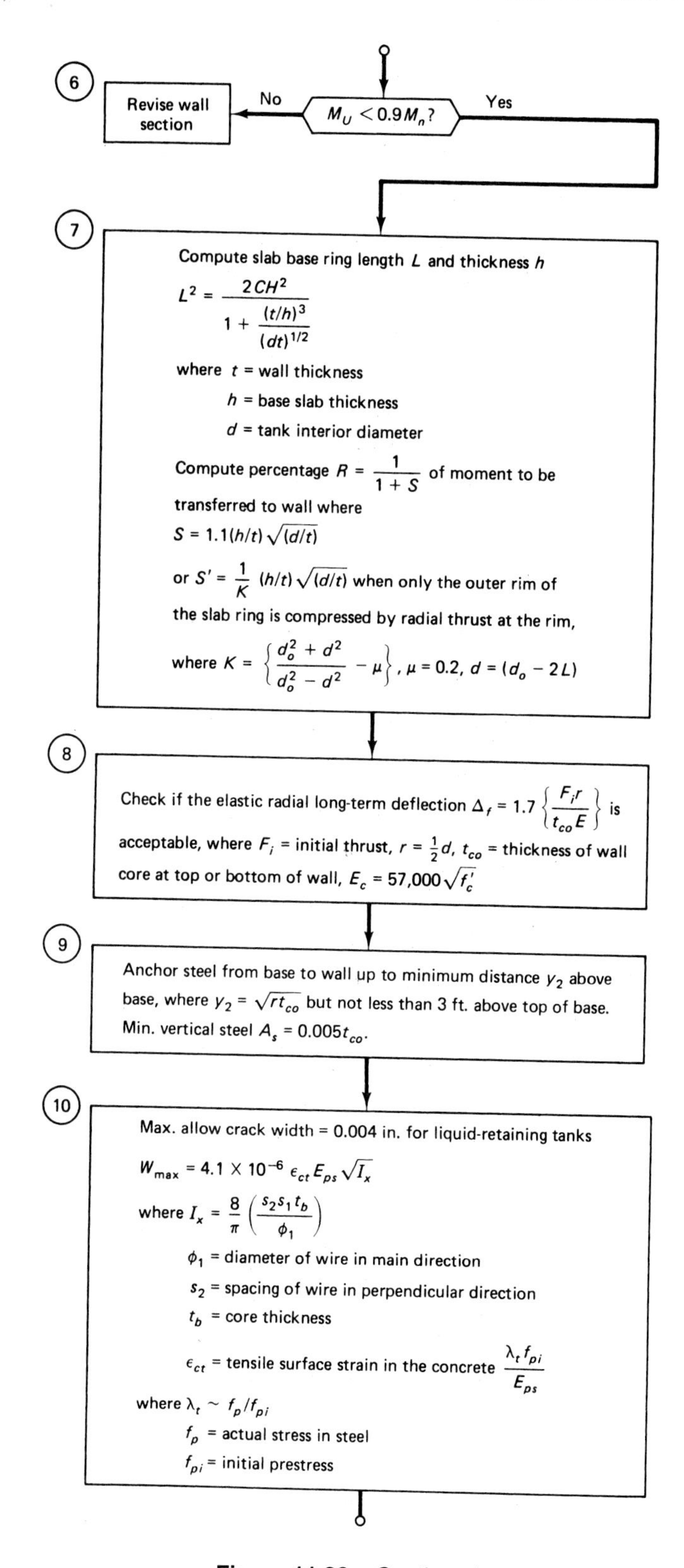

Figure 11.23 *Continued*

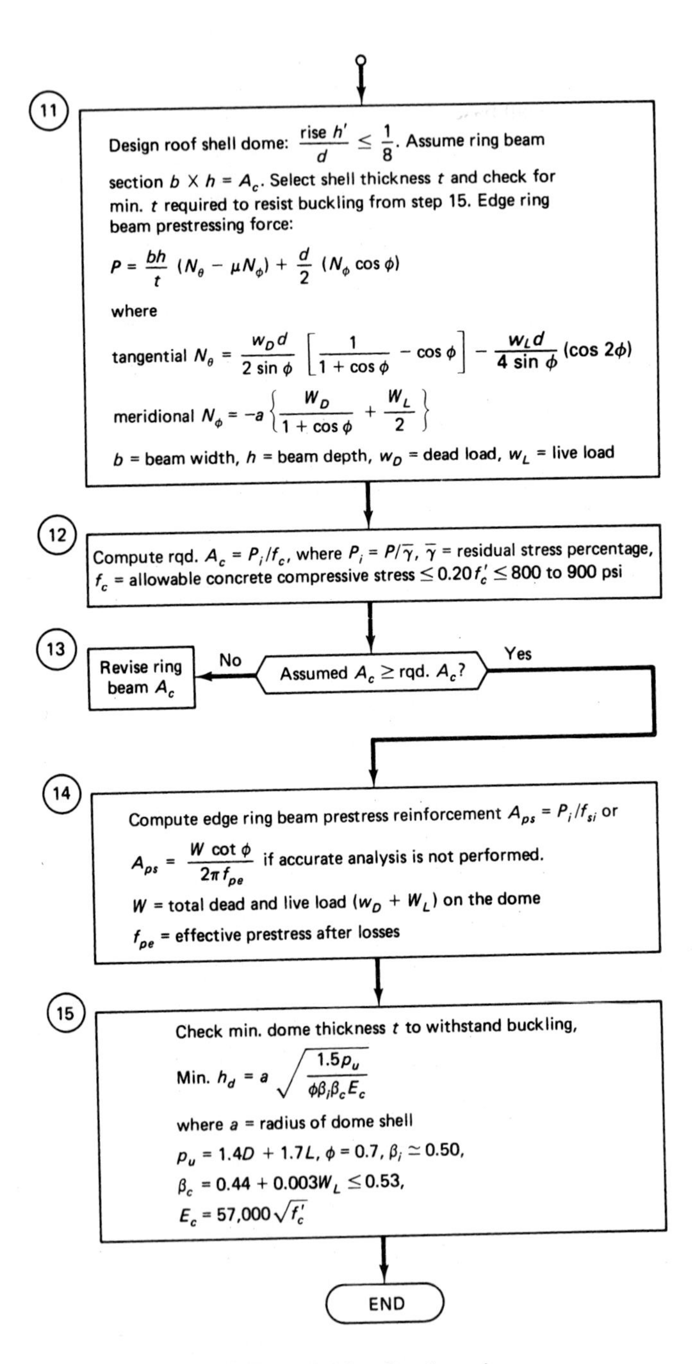

Figure 11.23 *Continued*

11.12 DESIGN OF CIRCULAR PRESTRESSED CONCRETE WATER-RETAINING TANK AND ITS DOMED ROOF

Example 11.3

Determine the maximum horizontal ring forces and vertical moments, and design the wall prestressing reinforcement, for a circular prestressed concrete tank whose diameter $d = 125$ ft (38.1 m) and which retains a water height $H = 25$ ft (7.62 m) for the following conditions of wall base support: (a) hinged, (b) fully fixed, (c) semisliding, and (d) partially fixed. Also, design the prestressed concrete ring edge beam for the domed roof shell assuming that the shell rise-span ratio $h'/d = \frac{1}{8}$. Use a flat shell roof having shell angle $\phi = 36°$, and find the area of prestressing reinforcement for both wire-wrapped and tendon reinforced conditions. Given data are as follows:

$$f'_c = 5{,}000 \text{ psi (34.5 MPa), normal-weight concrete}$$
$$f'_{ci} = 3{,}750 \text{ psi (25.9 MPa)}$$
$$f_t = 212 \text{ psi (0.86 MPa)} \leq 3\sqrt{f'_c}$$
$$f_c = 0.45f'_c = 2{,}250 \text{ psi (15.5 MPa)}$$
$$\text{residual } f_{cv} = 225 \text{ psi (1.55 MPa)}$$
$$f_{pu} \text{ (wire)} = 250{,}000 \text{ psi (1,724 MPa)}$$
$$f_{pu} \text{ (strands and tendons)} = 250{,}000 \text{ psi (1,724 MPa)}$$
$$f_{pi} = 0.7f_{pu} = 175{,}000 \text{ psi (1,207 MPa)}$$
$$f_{ps} = 220{,}000 \text{ psi (1,517 MPa)}$$
$$w_L = 15 \text{ psf (718 Pa) for snow load on dome}$$

Assume 26 percent total loss in prestress for all long-term effects.

Solution: Disregard the weight of the wall and the roof dome effect as insignificant on the stresses as compared to the effect of the vertical prestress forces. Consider the water pressure distribution shown in Figure 11.24 on the tank wall giving

$$\gamma = 62.4 \text{ lb/ft}^3 \text{ (1,000 kg/m}^3\text{)}$$

$$r = \frac{d}{2} = \frac{125}{2} = 62.5 \text{ ft (19.1 m)}$$

Assume the wall thickness $t = 10$ in. = 0.83 ft (25.4 cm). Then the form factor

$$\frac{H^2}{dt} = \frac{25 \times 25}{125 \times 0.83} = 6$$

and $\gamma Hr = 62.4 \times 25 \times 62.5 = 97{,}500$ lb/ft of circumference.

Basic Forces and Moments. Tables 11.19 through 11.21 give the basic forces and moments in the tank wall.

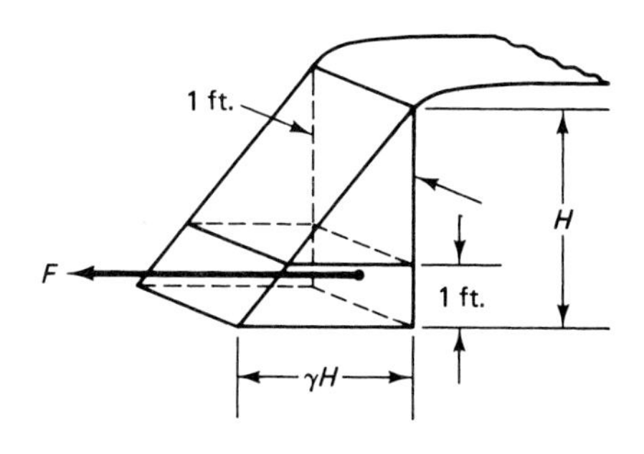

Figure 11.24 Liquid ring tension F, wall base freely sliding.

Table 11.19 Maximum Ring Tension $F = C(\gamma Hr)$ lb/ft Circumference, Example 11.3

Freely Sliding Wall Base	Fixed Base	Hinged Base
	Table 11.10 for $\frac{H^2}{dt} = 6$	Table 11.12 for $\frac{H^2}{dt} = 6$
$C = 1$	$C = 0.514$	$C = 0.643$
$F = 97{,}500$	$F = 0.514 \times 97{,}500 = 50{,}115$	$F = 0.643 \times 97{,}500 = 62{,}693$
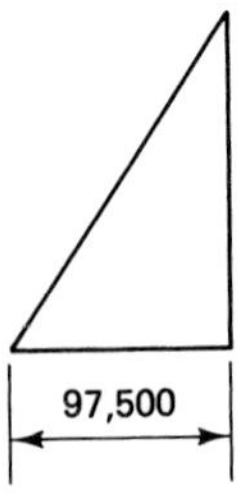 97,500	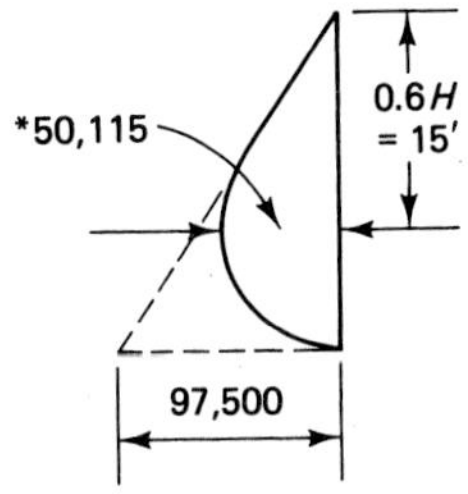 *50,115; 0.6H = 15′; 97,500	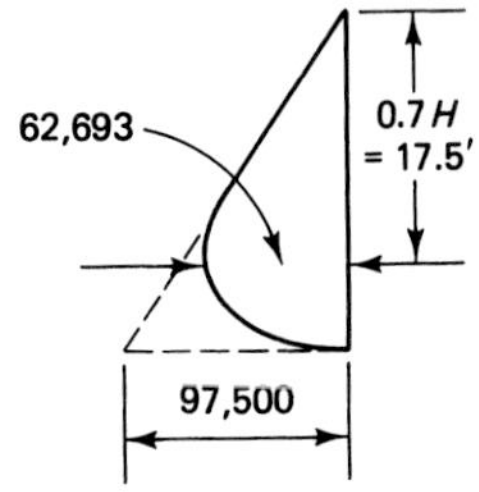 62,693; 0.7H = 17.5′; 97,500

*Compare with 50,113 lb/ft in the detailed method of Example 11.2.

Table 11.20 Vertical Moments $M = C(\gamma H^3)$ ft-lb/ft, Example 11.3. Positive (+) = Tension in Outside Face

Freely Sliding Base	Fixed Wall Base	Hinged Base
	Table 11.4	Table 11.6
	$C = +0.0051$ for $0.7H = 17.5$ ft	$C = +0.0078$ for $0.8H = 20$ ft
	$C = -0.0187$ for $1.0H = 25$ ft	$C = 0$ for $1.0H$, or full height
$M_y = M_o = 0$	$M_y = +0.0051 \times 62.4(25)^3 = +4{,}973$	$M_y = +0.0078 \times 62.4(25)^3 = +7{,}605$
	$M_o = -0.0187 \times 62.4(25)^3 = -18{,}233$	$M_o = 0$
	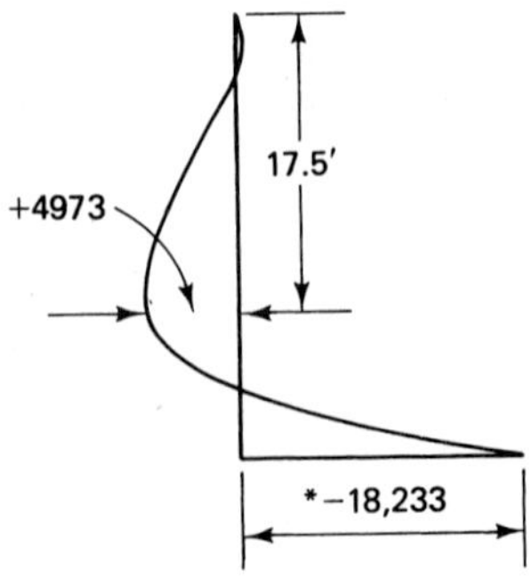 +4973; 17.5′; *−18,233	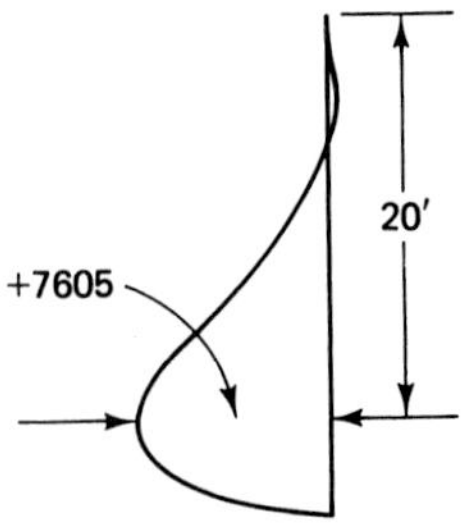 +7605; 20′

*This moment value is very close to the value obtained by using the detailed method and the moment functions of Table 11.1 and Example 11.1 ($M_o = -18{,}574$).

Table 11.21 Prestressing Effects Using 225-psi Residual Radial Compression, Example 11.3. Ring Forces Q lb/ft, Vertical Moments M_y ft-lb/ft

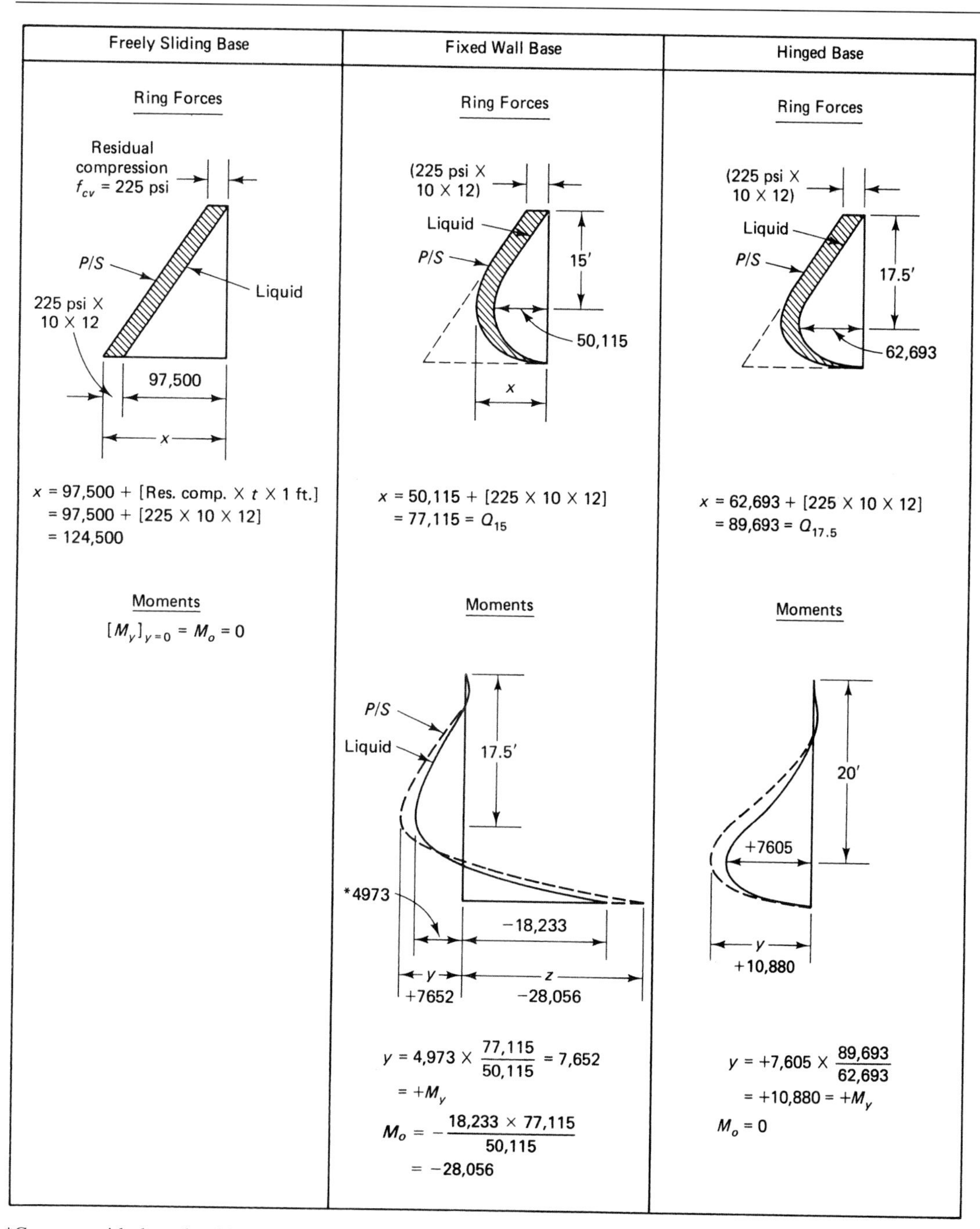

*Compare with the value $M = +4{,}912$ ft-lb/ft obtained by the detailed method of Example 11.1.

Wall Maximum Concrete Stresses at 20 ft from Top: Hinged Base. By trial and adjustment, provide vertical concentric prestress $P_v = 50{,}000$ lb/ft (730 kN/m) of circumference. Then for a wall thickness t = 10 in. compute the resulting stresses as shown in Figure 11.25.

$$f_+ = \frac{M}{S} = \frac{10{,}880 \times 12}{\dfrac{12(10)^2}{6}} = \mp 653 \text{ psi}$$

$$f_+ = \frac{M}{S} = \frac{7{,}605 \times 12}{\dfrac{12(10)^2}{6}} = \pm 456 \text{ psi}$$

$$f_v = \frac{P_v}{A_c} = \frac{50{,}000}{12 \times 10} = -417 \text{ psi}$$

f_4 = ① + ③

Max. $f_t = 236 \text{ psi} \cong 3\sqrt{5{,}000}$

$\cong 212$ psi, O.K.

Max. $f_c = -1{,}070 \text{ psi} < 0.45 f'_c$, O.K.

f_5 = ① + ② + ③

Max. $f_c = -614 \text{ psi} < 0.45 f'_c$, O.K.

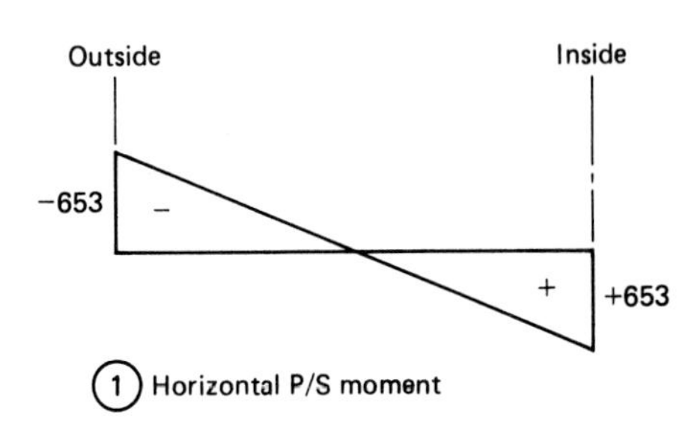

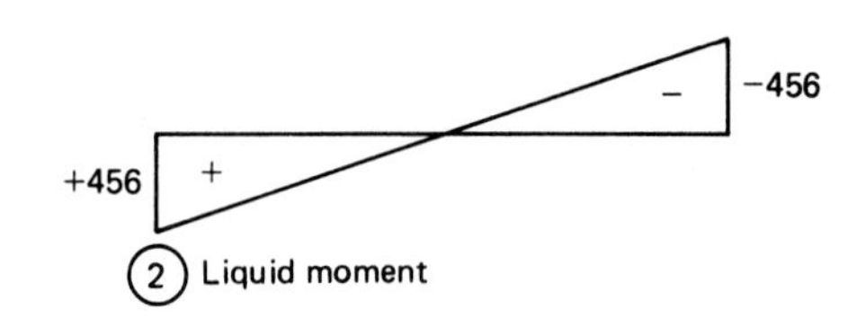

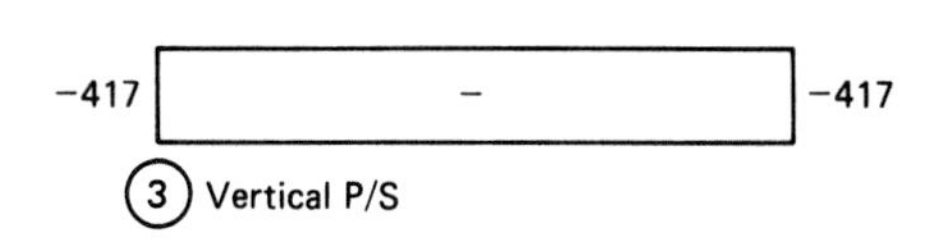

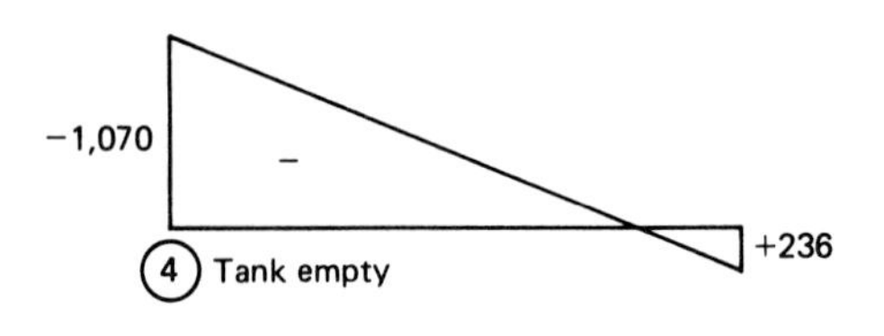

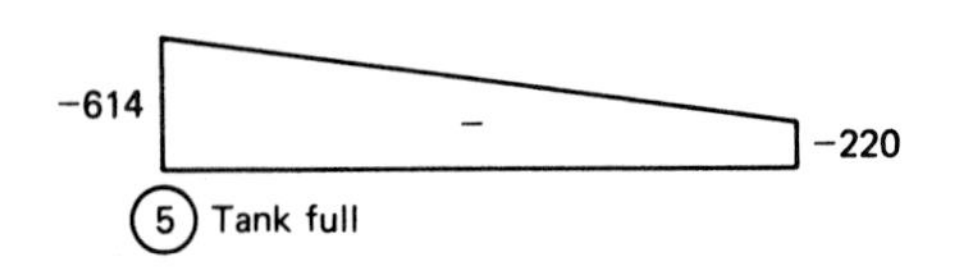

Figure 11.25 Stress at maximum moment, 20 ft from top, psi. Negative (−) = compression, positive (+) = tension.

Wall Maximum Concrete Stress at 17 ft 6 in. from Top: Fully Fixed Base. The maximum positive moment M_y is at 17 ft 6 in. from the top of the wall. By trial and adjustment, use *eccentric* vertical prestressing P_v = 100,000 lb/ft closer to the outer face [e = 1.05 in. (26.7 m)]. Then compute the resulting stresses in the wall as shown in Figure 11.26.

$$f_+ = \frac{M}{S} = \frac{7{,}652 \times 12}{\dfrac{12(10)^2}{6}} = \mp 459 \text{ psi}$$

$$f_+ = \frac{M}{S} = \frac{4{,}973 \times 12}{\dfrac{12(10)^2}{6}} = \pm 298 \text{ psi}$$

$$f_v = \frac{P_v}{A_c} = \frac{100{,}000}{12 \times 10} = -833 \text{ psi}$$

$$f_v = \frac{P_v e(c)}{I} = \frac{100{,}000 \times 1.05}{\dfrac{12(10)^2}{6}} = \mp 525 \text{ psi}$$

$$f_5 = (1) + (3) + (4)$$

$$\text{Max. } f_t = +151 \text{ psi} < 3\sqrt{f'_c} = 212, \text{ O.K.}$$

$$\text{Max. } f_c = -1{,}817 \text{ psi} < 0.45 f'_c = -2{,}250 \text{ psi, O.K.}$$

$$f_6 = (1) + (2) + (3) + (4)$$

$$\text{Max. } f_c = -1{,}519 \text{ psi} < 0.45 f'_c, \text{ O.K.}$$

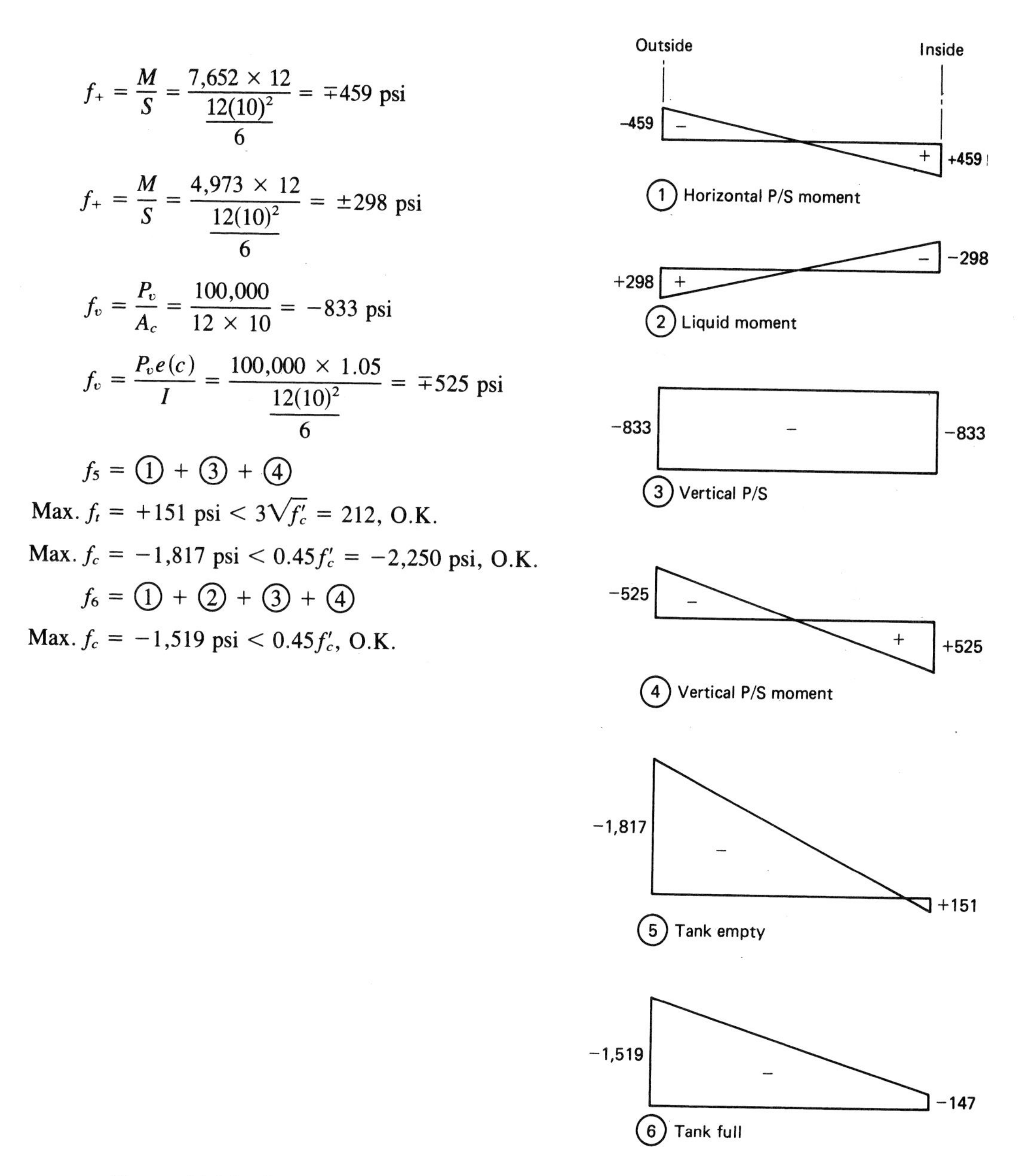

Figure 11.26 Stresses at maximum positive (+) moment, 17 ft, 6 in. from top, psi. Negative (–) = compression, positive (+) = tension.

Wall Maximum Concrete Stress at Base: Fully Fixed Base. Use eccentric vertical prestress $P_v = 100{,}000$ lb closer to the outer face ($e = 1.05$ in.). Then compute the resulting stresses in the wall as shown in Figure 11.27.

$$f_+ = \frac{M}{S} = \frac{-28{,}056 \times 12}{\dfrac{12(10)^2}{6}} = \pm 1{,}683 \text{ psi}$$

$$f_+ = \frac{M}{S} = \frac{-18{,}233 \times 12}{\dfrac{12(10)^2}{6}} = \mp 1{,}094 \text{ psi}$$

$$f_v = \frac{P_v}{A_c} = \frac{-100{,}000}{12 \times 10} = -833 \text{ psi}$$

$$f_v = \frac{P_v e(c)}{I} = \frac{100{,}000 \times 1.05}{\dfrac{12(10)^2}{6}} = \mp 525 \text{ psi}$$

f_5 = ① + ③ + ④

Max. $f_c = -1{,}991 \text{ psi} < 0.45 f'_c$, O.K.

Max. $f_t = +325$ psi at base when tank is empty. This stress will rapidly decrease well below $3\sqrt{f'_{ci}}$ within one foot above base, hence O.K.

f_6 = ① + ② + ③ + ④

Max. $f_c = -897 \text{ psi} < 0.45 f'_c$, O.K.

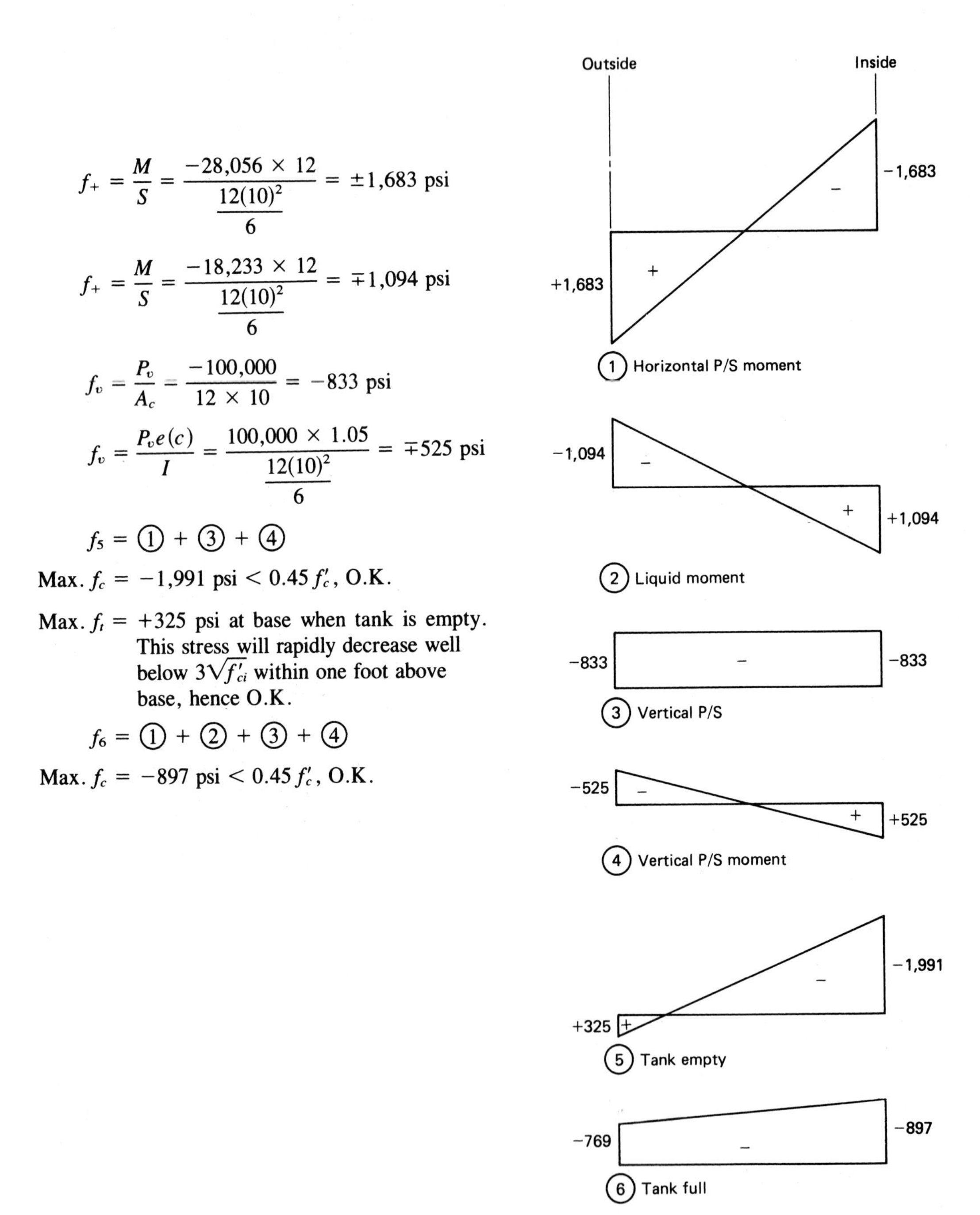

Figure 11.27 Stresses at maximum negative (−) moment at wall base, psi. Negative (−) = compression, positive (+) = tension.

Wall Maximum Concrete Stress: Semisliding Base. By trial and adjustment, use concentric vertical prestress $P_v = 20{,}400$ lb/ft (297 kN/m). Then semislide $M = \frac{1}{2}(+10{,}880) = 5{,}440$ ft-lb/ft, and compute the resulting stresses in the wall as shown in Figure 11.28.

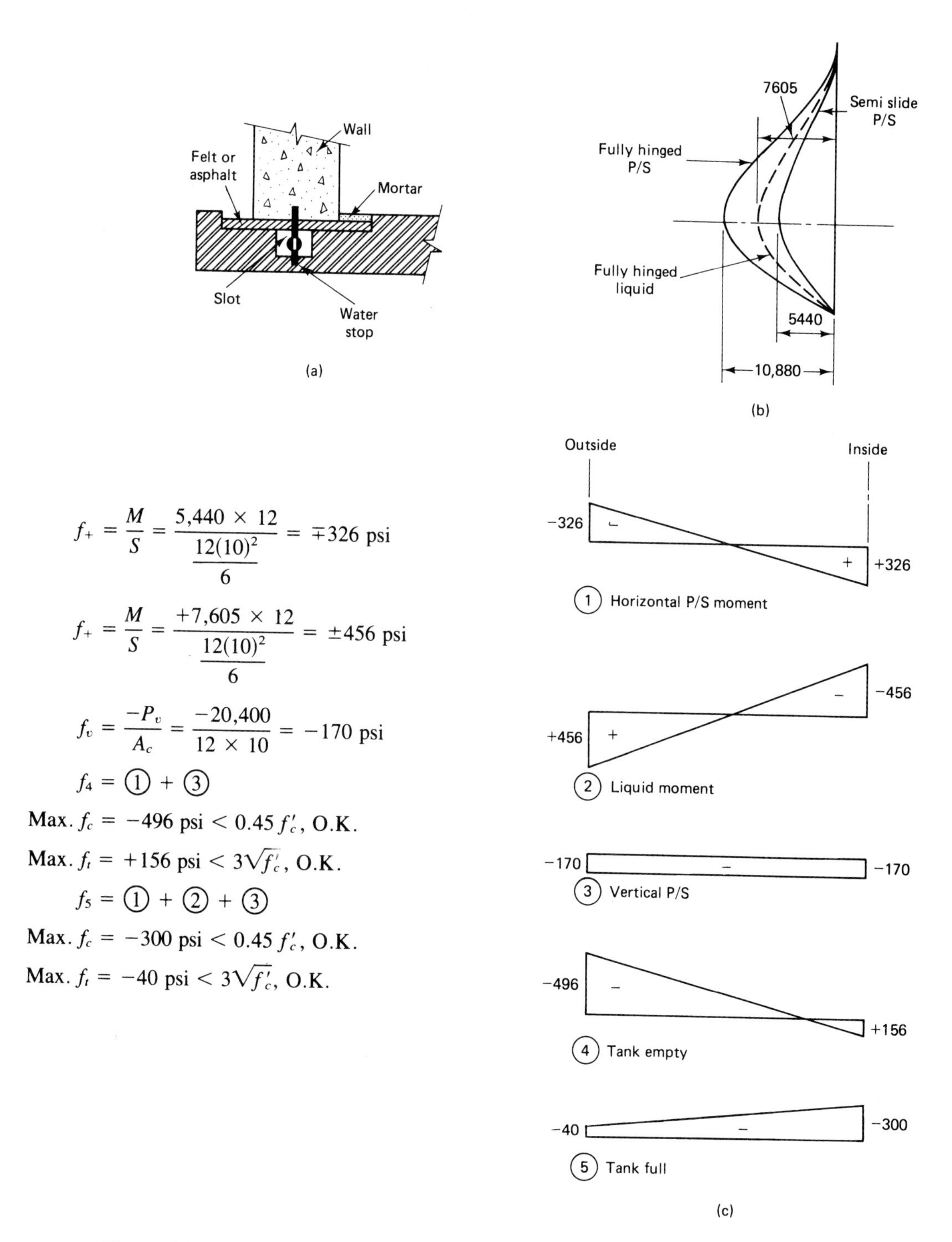

$$f_+ = \frac{M}{S} = \frac{5{,}440 \times 12}{\dfrac{12(10)^2}{6}} = \mp 326 \text{ psi}$$

$$f_+ = \frac{M}{S} = \frac{+7{,}605 \times 12}{\dfrac{12(10)^2}{6}} = \pm 456 \text{ psi}$$

$$f_v = \frac{-P_v}{A_c} = \frac{-20{,}400}{12 \times 10} = -170 \text{ psi}$$

$f_4 = (1) + (3)$

Max. $f_c = -496$ psi $< 0.45 f'_c$, O.K.

Max. $f_t = +156$ psi $< 3\sqrt{f'_c}$, O.K.

$f_5 = (1) + (2) + (3)$

Max. $f_c = -300$ psi $< 0.45 f'_c$, O.K.

Max. $f_t = -40$ psi $< 3\sqrt{f'_c}$, O.K.

Figure 11.28 Stresses at maximum positive (+) moment, psi. (a) Wall base details. (b) Semislide moment, ft-lb/ft. (c) Concrete stresses, psi.

Photo 11.11 Arco Floating LPG Barge: ABAM-designed largest floating prestressed hull in the world. (*Courtesy,* ABAM Engineers, Tacoma, Washington.)

Partial Fixity at the Wall Base. The restraint moment is $M_p = M_o\,(1 - S)$, where the full fixity moment M_o = 18,233 ft-lb/ft. The modifying factor for partial fixity $S = (t/h)^3/(dt)^{1/2}$.

Figure 11.29 shows the deformed shape of the base slab. If the base slab thickness $h = 10$ in., then, from Equations 11.39 and 11.40,

$$S = \frac{(10/10)^3}{(125 \times 0.83)^{1/2}} = 0.10$$

and

$$M_p = M_o(1 - S) = 18{,}233(1 - 0.1) = 16{,}410 \text{ ft-lb/ft.}$$

The moment loss due to partial fixity = 18,233 − 16,410 = 1,823 ft-lb/ft. From Equation 11.37 for the base ring width L,

$$L^2 = \frac{2CH^2}{1 + S}$$

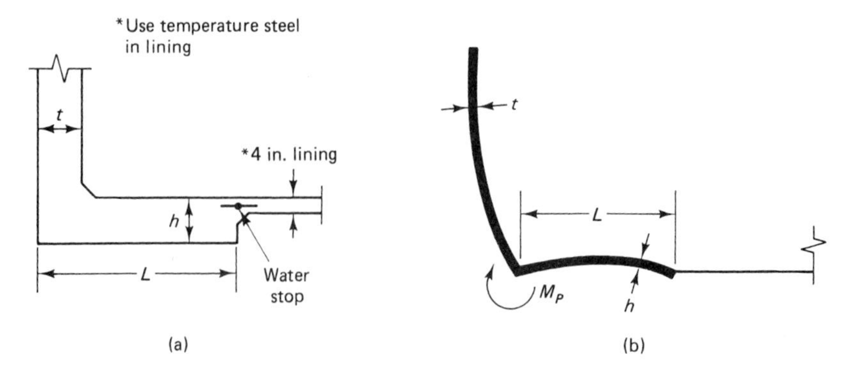

Figure 11.29 Deformed shape of base slab. (a) Wall base. (b) Deformed section.

Also, from Table 11.4, the membrane coefficient at the base for form factor $(H^2)/(dt) = 6$ is $C = -0.0187$. Thus, we have

$$L^2 = \frac{2 \times 0.0187(25)^2}{1 + 0.1} = 21.25$$

and it follows that

$$L = 4.61 \text{ ft} = 4 \text{ ft } 7\tfrac{1}{2} \text{ in.}$$

Accordingly, use a ring slab base width $L = 4$ ft 9 in. (145 cm). Since for large-diameter tanks S has a very small value, the degree of fixity, as the solution shows, is almost the same for both fully fixed and partially fixed wall bases.

From Equations 11.47 and 11.48, the percent R of prestress in the base that is transferred to wall $= 100/S_1$, where

$$S_1 = 1.1\left(\frac{h}{t}\right)\left(\frac{d}{t}\right)^{1/2} = 1.1\left(\frac{10}{10}\right)\left(\frac{125}{0.83}\right)^{1/2} = 13.50\%$$

Consequently,

$$R = \frac{100}{13.50} = 7.4\%$$

which means that the required design prestress for the wall can be slightly reduced, as some compression is available from the base ring.

Design of Prestressing Reinforcement

Horizontal Prestressing. Use the same size wire to wrap the circular wall, varying the spacing of the wire hoops in 5-ft bands along the tank height. In the case of the freely sliding tank wall, the minimum spacing is in the lowest band at the base, as presented graphically in Figure 11.30.

In order to determine the variation of wire pitch throughout the height of the wall, additional computations of the horizontal ring thrust Q_y have to be made at the bottom of each band. Consequently, only one typical calculation of size and wire distribution will be made for purposes of illustration.

Taking the case of the fixed wall base from Table 11.21, the maximum $Q_{15} = 77{,}115$ lb/ft of circumference per foot height of wall. So trying 0.192-in. dia (4.88 mm) prestressing 250-K wire, we obtain $A_{ps} = 0.0289$ in.^{2} per wire and $f_{pi} = 0.7 f_{pu} = 0.7 \times 250{,}000 =$ 175,000 psi (1,207 MPa).

Now assume 26-percent prestress loss for elastic shortening, seating, creep, shrinkage, and steel relaxation. Then

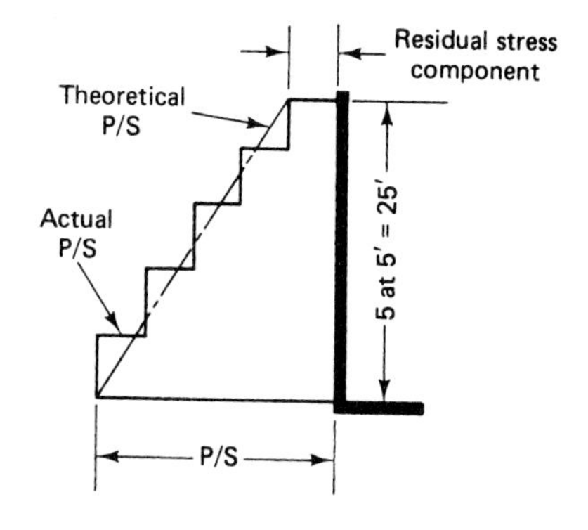

Figure 11.30 Horizontal-prestress wire distribution bands.

$$f_{pe} = 0.74 \times 175{,}000 = 129{,}500 \text{ psi (893 MPa)}$$

$$A_{ps} = \frac{77{,}115}{129{,}500} = 0.60 \text{ in.}^2 \text{ per 1 ft of wall height}$$

$$\text{No. of wire loops in 5-ft band} = \frac{0.60 \times 5}{0.0289} = 104$$

Hence, use 104 wire loops in the 5-ft wall band whose base is 15 ft below the top of the water level. Also, use 2-in. shotcrete to cover the wrapped horizontal 0.192-in. dia wires.

If the tank were prestressed with $\frac{1}{2}$-in. dia 250-K 7-wire strand tendons, A_{ps} would be 0.144 in.^{2}/strand and the required number of strands in a 5-ft-height band would be $0.60 \times 5/0.144 \cong 20$ tendons.

Vertical Prestressing. For proportioning the vertical prestressing reinforcement, $P_v = 100{,}000$ lb/ft at $e = 1.05$ in. (1,459 N/m at $e = 26.7$ mm) on the outer force side. Hence, try $\frac{1}{2}$-in. dia (17.7-mm dia) 7-wire 250-K strands. We obtain

$$A_{ps} = 0.144$$

$$f_{pu} = 250{,}000 \text{ psi (1,724 MPa)}$$

$$f_{pi} = 0.7 f_{pu} = 0.7 \times 250{,}000 = 175{,}000 \text{ psi (1,207 MPa)}$$

Assume 26-percent total prestress loss. Then $f_{pe} = 0.74 \times 175{,}000 = 129{,}500$ psi (889 MPa), the required A_{ps} per foot of circumference = 100,000/129,500 = 0.772 in.^{2} (4.98 cm^2), and the number of vertical strands per foot of circumference = 0.772/0.144 = 5.36. Thus, use $\frac{1}{2}$-in. dia 7-wire 250-K strands for vertical prestressing at $2\frac{1}{4}$ in. center-to-center spacing = 0.769 in.$^2 \cong 0.772$ in.^{2}, O.K.

Nominal Moment Strength Check of Tank Wall. The maximum wall vertical moment for a fixed-base wall, from Table 11.21, is $M = 28{,}056$ ft-lb/ft or in.-lb/in. of circumference. We thus have:

$$\text{S.F.} = 1.3 \text{ (step 5 of flowchart)}$$

$$M_u = 1.3 \times 28{,}056 = 36{,}473 \text{ in.-lb/in.}$$

$$\text{Rqd } M_n = \frac{M_u}{0.9} = \frac{36.473}{0.9} = 40{,}525 \text{ in.-lb/in.}$$

$$d = \frac{10}{2} + 1.05 = 6.05 \text{ in. (15.37 cm)}$$

$$A_{ps} = \frac{0.144}{2.25} = 0.064 \text{ in.}^2\text{/in. width}$$

$$a = \frac{A_{ps} f_{ps}}{0.85 f'_{cb}} = \frac{0.064 \times 220{,}000}{0.85 \times 5{,}000 \times 1} = 3.31 \text{ in.}$$

$$\text{Available } M_n = A_{ps} f_{ps}\left(d - \frac{a}{2}\right) = 0.064 \times 220{,}000\left(6.05 - \frac{3.31}{2}\right)$$

$$= 61{,}882 \text{ in.-lb/in.} \gg \text{Rqd. } M_n = 40{,}525 \text{ in.-lb/in., O.K.}$$

The wall design should include a check of the deflection as described in step 8 of the flowchart. Also, a determination should be made of the anchor steel at the base of the wall as well as the crack width w_{max} in step 9 of the flowchart. Finally, a check of temper-

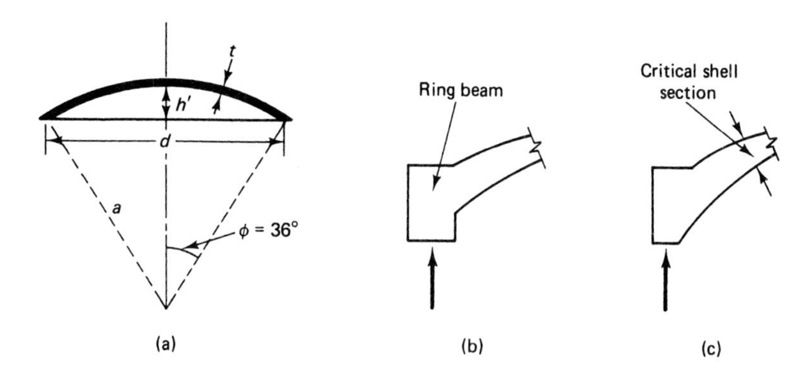

Figure 11.31 Tank dome shell roof. (a) Geometry of dome. (b) Edge ring beam. (c) Equivalent ring beam.

ature and creep effects has to be made to ascertain whether any additional nonprestressed mild steel has to be added to the prestressed wall reinforcement.

Design of Roof Dome Prestressed Edge Ring Beam. Use a rise-span ratio $h'/d = \frac{1}{8}$. Also, choose a freely supporting reaction at the top of the tank wall, using a neoprene pad under the edge ring beam. The shell would then have the form shown in Figures 11.31 and 11.32.

Since $d = 125$ ft., $h' = 125/8 = 15.63$ ft (4.76 m). Also, since $\phi = 36°$ is less than $51°49'$, the entire shell would be in compression, and only temperature reinforcement is needed.

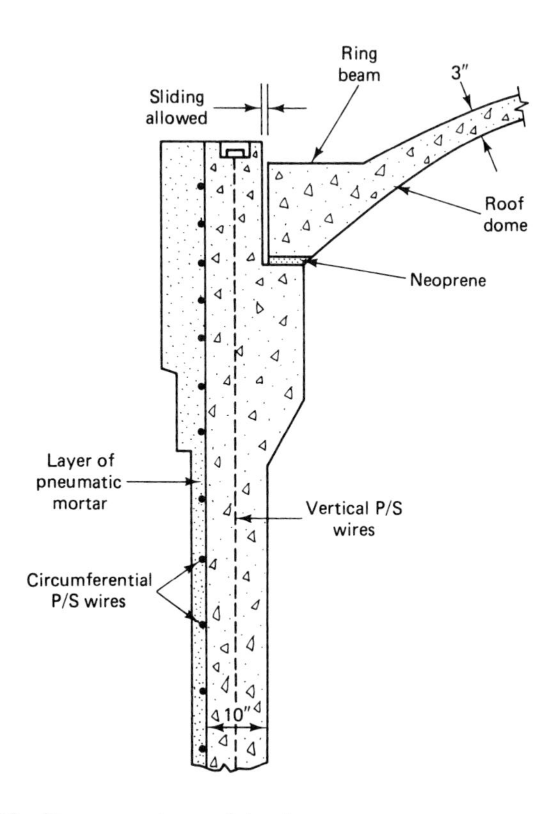

Figure 11.32 Dome prestressed ring beam support detail in Example 11.3.

Photo 11.12 Olympic oval at the University of Calgary, Calgary, Canada. Structural engineers: Simpson, Lester, Goodrich; Calgary, Alberta, Canada. (*Courtesy,* Prestressed Concrete Institute.)

The shell radius is

$$a = \frac{d/2}{\sin \phi} = \frac{62.5}{0.588} = 106 \text{ ft (32.3 m)}$$

From Equation 11.75, the minimum shell thickness to withstand buckling is

$$h_d = a\sqrt{\frac{1.5P_u}{\phi\beta_i\,\beta_c\,E_c}}$$

Hence, assuming that $t = 3.0$ in., we have

$$P_u = 1.4D + 1.7L = 1.4\left(\frac{3}{12} \times 150\right) + 1.7 \times 15 = 78 \text{ lb/ft}^2$$

$$\phi = 0.7$$

$$\beta_i = (a/r_i)^2 = \left(\frac{106}{1.4 \times 106}\right)^2 = 0.51$$

$$\beta_c = 0.44 + 0.003 \times 15 = 0.49 < 0.53, \text{ use } \beta_c = 0.49$$

$$E_c = 57{,}000\sqrt{5{,}000} = 4.03 \times 10^6 \text{ psi}$$

$$\text{Min } h = a\sqrt{\frac{1.5p_u}{\phi\beta_i\,\beta_c\,E_c}} = 106\sqrt{\frac{1.5 \times 78}{0.7 \times 0.51 \times 0.49 \times 4.03 \times 10^6}}$$

$$= 1.36 \text{ in. (3.5 cm)} < 3 \text{ in., O.K.}$$

So use a shell $t = 3$ in. (7.6 cm). Then $\sin\phi = \sin 36° = 0.59$, $\cos\phi = \cos 36° = 0.81$, and $a =$ sphere radius $= 106$ ft.

From Equation 11.70, the tangential force per unit length of circumference is

$$N_\theta = \frac{W_D d}{2 \sin \phi}\left[\frac{1}{1+\cos\phi} - \cos\phi\right] - \frac{W_L d}{4 \sin\phi}(\cos 2\phi)$$

$$= \frac{37.5 \times 125}{2 \times 0.59}\left[\frac{1}{1+0.81} - 0.81\right] - \frac{15 \times 125}{4 \times 0.59}(0.31)$$

$$= -1{,}269 \text{ lb/ft}$$

From Equation 11.67, the meridional force per unit length of circumference, with $a = 106$ ft, is

$$N_\phi = -a\left(\frac{w_D}{1+\cos\phi} + \frac{w_L}{2}\right)$$

$$= -106\left(\frac{37.5}{1.81} + \frac{15}{2}\right) = -2{,}991 \text{ lb/ft } (43.6 \text{ kN/m})$$

From Equation 11.72, the radial prestressing force in the ring beam required to produce compatibility of deformation with the shell rim is

$$P = \frac{bh}{t}(N_\theta - \mu N_\phi) + \frac{d}{2}(N_\phi \cos\phi)$$

To determine the cross-sectional area bh of the ring beam, use $P = (d/2)(N_\phi \cos\phi)$ for the first trial, since the first term of the equation has less than 10 percent of the total value of P (see the discussion accompanying Equation 11.62). We obtain

$$P = \frac{d}{2}(N_\phi \cos\phi) = \frac{125}{2}(-2{,}991 \times 0.81) = -151{,}149 \text{ lb per ft}$$

Given that the total prestress loss is 26 percent, it follows that

$$\bar{\gamma} = 1 - 0.26 = 0.74$$

and

$$P_i = \frac{151{,}419}{0.74} = 204{,}620 \text{ lb/ft}$$

Use a maximum concrete compressive stress $f_c = 800$ psi (5.52 MPa) in order to minimize excess strain in the edge beam, which could produce high stresses in the shell rim. The required cross-sectional area of the prestressed ring beam is

$$A_c = bh = \frac{P_i}{f_c} = \frac{204{,}620}{800} = 256 \text{ in.}^2$$

Try $b = 14$ in. and $h = 20$ in. Then $A_c = 280$ in.^{2} Substituting into Equation 11.72, we get

$$P = \frac{280}{3.0}\left[-\frac{1}{12} \times 1{,}269 - 0.2(-2{,}991)\right] + \frac{125}{2}(-2{,}991 \times 0.81)$$

$$= -5{,}217 - 151{,}419 = -156{,}636 \text{ lb/ft}$$

Use

$$P_i = \frac{156{,}636}{0.74} = 211{,}671 \text{ lb } (717 \text{ kN})$$

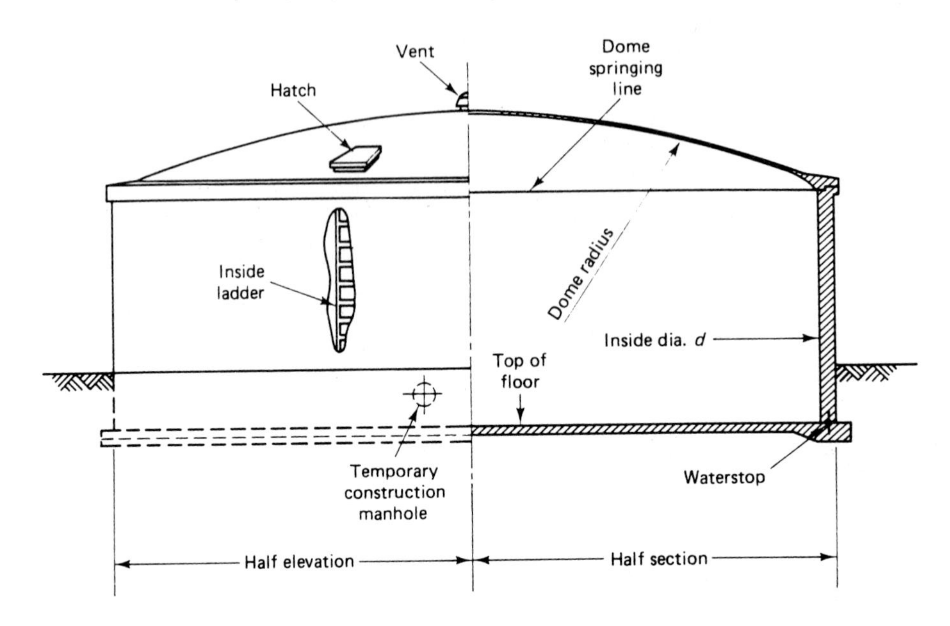

Figure 11.33 Typical elevation and section of a domed prestressed concrete circular tank.

From before,

$$f_{pi} = 0.7f_{pu} = 175{,}000 \text{ psi}$$

so

$$A_{ps} = \frac{P_i}{f_{pi}} = \frac{211{,}671}{175{,}000} = 1.21 \text{ in}^2 \text{ (7.56 cm}^2\text{)}$$

Trying $\frac{1}{2}$-in. dia (12.7-mm) 7-wire 250-K strands, we obtain

$$A_{ps}/\text{strand} = 0.144 \text{ in.}^2$$

and

$$\text{No. of strands} = \frac{1.21}{0.144} = 8.4$$

If the prestress loss is slightly more than 26 percent, the number of strands should be approximately 9. Hence, use nine $\frac{1}{2}$-in. dia 7-wire strands to prestress the edge ring beam.

Check the Concrete Stress in the Critical Section $t = 3$ in. of the Shell Rim. The meridional compression $N_\phi = -2{,}991$ lb/ft of circumference, and the compressive stress $f_c = 2{,}991/(12 \times 3) = 83$ psi only, which is satisfactory. The support details of the edge ring beam and the roof are shown in Figure 11.32. Note that the ring beam is supported vertically on a neoprene pad, which enables sliding. A typical elevation and section of a domed prestressed circular tank is shown in Figure 11.33.

REFERENCES

11.1 Timoshenko, S., and Woinowsky-Krieger, S. *Theory of Plates and Shells.* 2d ed. McGraw Hill, New York, 1959.

11.2 Creasy, L. R. *Prestressed Concrete Cylindrical Tanks.* John Wiley & Sons, New York, 1961.

11.3 Billington, D. P. *Thin Shell Concrete Structures.* 2d ed. McGraw Hill, New York, 1982.

11.4 Ghali, A. *Circular Storage Tanks and Silos.* E. & F. N. Spon Ltd., London, 1979.

11.5 PCA, "Circular Concrete Tanks without Prestressing," Concrete Information Series ST-57, Portland Cement Association, Skokie, Ill., 1957, 32 pp.

11.6 PCI Committee on Precast Prestressed Concrete Storage Tanks. "Recommended Practice for Precast Prestressed Concrete Circular Storage Tanks." Prestressed Concrete Institute, Chicago, 1987.

11.7 ACI Committee 344. *Design and Construction of Circular Prestressed Concrete Structures, ACI 344R.* American Concrete Institute, Farmington Hills, MI, 1970.

11.8 ACI Committee 344. *Design and Construction of Circular Wire and Strand Wrapped Prestressed Concrete Structures, ACI 344-R,* American Concrete Institute, Farmington Hills, MI, 1989.

11.9 ACI Committee 344. *Design and Construction of Circular Prestressed Concrete Structures with Circumferential Tendons, ACI 344.2R,* American Concrete Institute, Farmington Hills, MI, 1989.

11.10 Post-Tensioning Institute. *Post-Tensioning Manual.* 6th ed. Post-Tensioning Institute, Phoenix, 2000.

11.11 Prestressed Concrete Institute. *PCI Design Handbook.* 5th ed. Prestressed Concrete Institute, Chicago, 1999.

11.12 Tadros, M. K. "Expedient Service Load Analysis of Cracked Prestressed Concrete Sections." *Journal of the Prestressed Concrete Institute,* Vol. 27, No. 6, Nov–Dec, Chicago, 1983, 137–158.

11.13 Brondum-Nielsen, T. "Prestressed Tanks." *Journal of the American Concrete Institute,* Detroit, July–August 1985, pp. 500–509.

11.14 Vessey J.V., and Preston, R. L. *A Critical Review of Code Requirements for Circular Prestressed Concrete Reservoirs.* F.I.P., Paris, 1978.

11.15 Nawy, E. G., and Blair, H., *Further Studies of Flexural Crack Control in Structural Slab Systems.* American Concrete Institute, Farmington Hills, MI, SP-30, 1971.

11.16 Abeles, P. W., and Bardhan-Roy, B. K. *Prestressed Concrete Designer's Handbook.* 3d ed. Viewpoint Publications, London, 1981.

PROBLEMS

11.1 Solve Example 11.3 if the tank diameter is 120 ft (36.6 m) and the water height is 30 ft (9.1 m). Assume that the total prestress loss is 20 percent, and use a rise-span ratio $h'/d = \frac{1}{10}$ for the roof dome, assuming that half the shell angle is $\phi = 45°$.

11.2 A circular prestressed concrete tank has an internal diameter $d = 85$ ft (26 m) and retains water to a height $H = 22$ ft (6.7 m). Determine the maximum horizontal ring forces and vertical moment, and design the prestressing reinforcement using both horizontal and vertical prestressing. Also, design a roof dome shell for the tank assuming a rise-span ratio $h'/d = \frac{1}{8}$ and half shell angle $\phi = 30°$. Solve for (a) hinged, (b) partially fixed, and (c) sliding wall base fixity, and design the prestressing reinforcement for both wire-wrapped and tendon prestressing conditions. Given data are:

$f'_c = 6{,}000$ psi (41.4 MPa), normal weight

$f'_{ci} = 4{,}250$ psi (29.3 MPa)

$f_t \leq 3\sqrt{f'_c} = 230$ psi (1.59 MPa)

$f_c = 0.45f'_c = 2{,}700$ psi (18.6 MPa)

$f_{cv} = 250$ psi (1.72 MPa)-residual compressive stress

f_{pu} for both wire and strand or tendon $= 250{,}000$ psi (1,724 Pa)

$f_{pi} = 0.7f_{pu} = 175{,}000$ psi (1,207 MPa)

Snow load intensity $w_L = 20$ lb/ft^2 (985 Pa)

Assume 20-percent total loss in prestress.

12

LRFD AND STANDARD AASHTO DESIGN OF CONCRETE BRIDGES

12.1 INTRODUCTION: SAFTEY AND RELIABILITY

As discussed in Section 4.10.1, a load-resistance factor design method (LRFD) is a reliability-based approach for evaluating probability-based factored design criteria (Ref. 12.1). It is intended for proportioning structural members based on the load types such that the resisting strength levels are *greater* than the factored load or moment distributions.

Figures 12.1(a) and (b), as in Figure 4.36 of Chapter 4, show a plot of separate frequency distributions of the actual load W and the resistance R with mean value R. Figure 12.1(c) gives the two distributions superimposed and intersecting at point C in the diagram.

The safety and reliable integrity of the structure can be expected to exist if the load effect W falls at a point to the left of intersection C on the resistance curve. Failure, on the other hand, would be expected to occur if the load effect on the resistance curve falls within the shaded area in Fig. 12.1(c). If $\overline{\beta}$ is a safety index, then:

$$\overline{\beta} = \frac{\overline{R} - \overline{W}}{\sqrt{\sigma_R^2 + \sigma_W^2}} \tag{12.1}$$

West Kowloon Expressway Viaduct, Hong Kong, 1997. A 4.2-Km dual three-lane causway connecting Western Harbor Crossing to new airport (*Courtesy* Institution of Civil engineers, London)

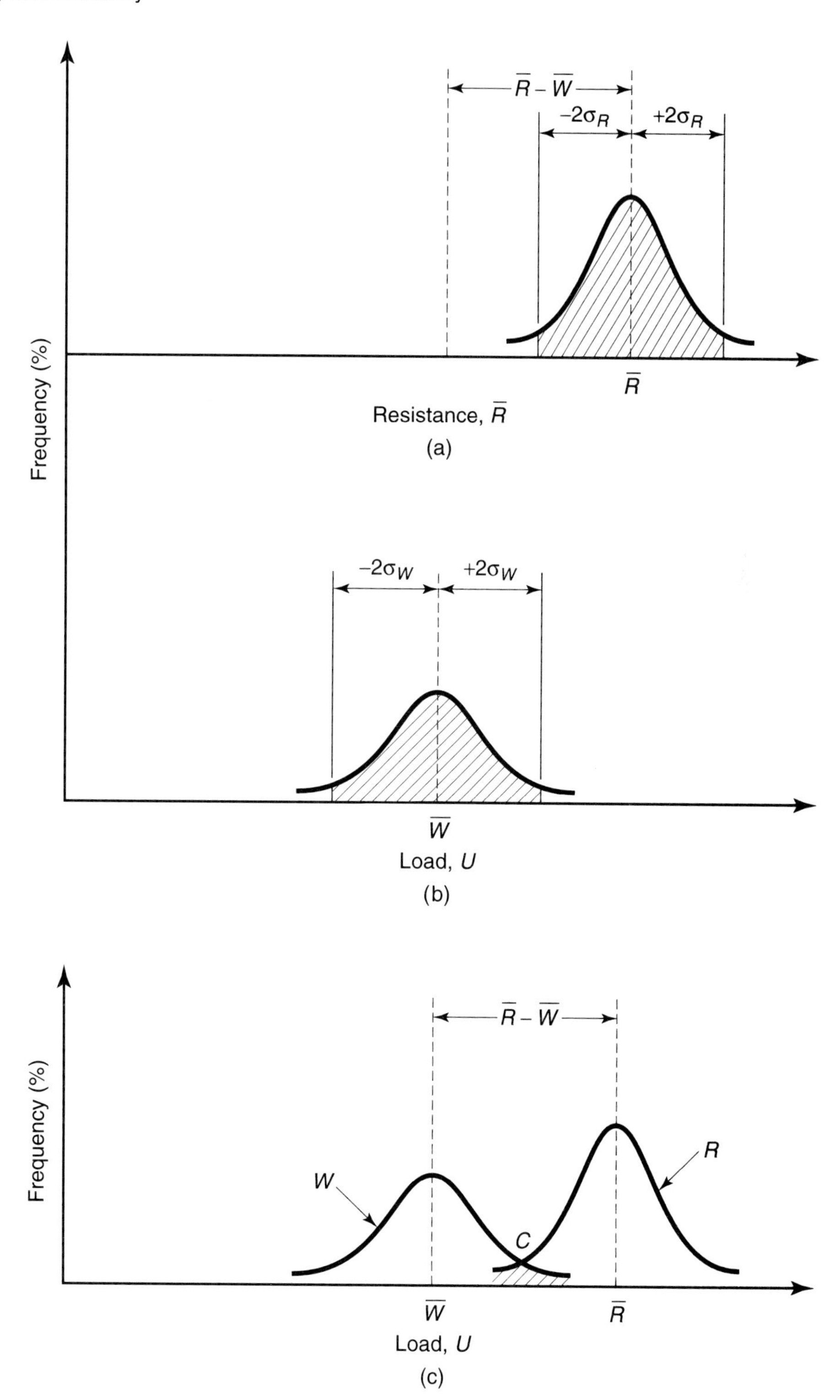

Figure 12.1 Frequency Distribution of Load vs. Resistance

Table 12.1(a) LRFD Resistance Factors ϕ

	ϕ
Flexure and tension of reinforced concrete	0.90
Flexure and tension in prestressed concrete	1.00
Shear and torsion:	
normal density concrete	0.90
low-density concrete	0.70
Axial compression with spirals of ties	0.75
Bearing on concrete	0.70
Compression in strut and tie	0.70
Compression in anchorage zones:	
normal density concrete	0.80
low-density concrete	0.65
Tension in steel in anchorage zones	1.00
For partially prestressed components in flexure with or without tension, where PPR = $A_{ps}f_{py}/(A_{ps}f_{py} + A_s f_y)$	0.90 + 0.10(PPR)

where, σ_R and σ_W are the standard deviations of the resistance and the load, respectively.

The different load combinations in Eq. 4.29, Chapter 4, are based on giving a reasonable difference between $\overline{R}$ and $\overline{W}$ as dictated by economical considerations.

The reliability of safe performance of the structure is, hence, controlled by the load-resistance considerations in the load factors used in the design.

AASHTO's LRFD approach (Ref. 12.2–12.3) is intended to extend the load-resistance considerations to the expressions for deformations and forces and modified load resistance factors ϕ from those used by ACI 318 (Ref. 12.4) where necessary. Those LRFD ϕ factors are listed in Table 12.1(a).

This chapter presents, and uses in design examples, the LRFD expressions where they differ from the standard AASHTO and ACI-318 expressions. Otherwise, the expressions used in the previous Chapters 3, 4, and 5 and the principles enunciated would apply. The student and the design engineer will easily recognize these expressions. Hence the need for redefining them becomes unnecessary.

12.2 AASHTO STANDARD AND LRFD TRUCK LOAD SPECIFICATIONS

The design of prestressed concrete elements of a bridge is governed by requirements of the American Association of Highway and Transportation Officials (AASHTO). The traffic lanes and the loads they contain for the design of the bridge superstructure have to be chosen and placed in such numbers and positions on the roadway that they produce the maximum stress in the constituent members.

The bridge live loadings should consist of standard truck or lane loads that are equivalent to truck trains. For railway bridges, the requirements are set by the American

Railway Engineering Association (AREA). Requirements for the structural proportioning of the supporting members usually follow the ACI and PCI standards.

12.2.1 Loads

There are four standard classes of highway loading: H 20, H 15, HS 20, and HS 15. Loading HS 15 is 75 percent of HS 20. If loadings other than these are to be considered, they should be obtained by proportionally adjusting the weights for the standard trucks and the corresponding lane loads. Bridges supporting interstate highways should be redesigned for HS 20–44 loading or an alternate military loading of two axles 4 ft apart, with each axle weighing 24,000 lb, whichever loading produces the larger stress value.

Figure 12.2 shows the standard H truck loading, while Figure 12.3 shows the standard HS truck loading giving wheel spacing and load distribution. Figure 12.4 gives the equivalent lane loading for both the H and HS 20–44 and the H and HS 15–44 categories (Ref. 12.1). Figure 12.5 gives an overview of the different bridge deck systems in common use.

Figure 12.5 gives typical dech bridge structures.

(i) Impact. Movable loads require impact allowance as a fraction of the live load stress. It can be expressed by

$$I = \frac{50}{L + 125} \leq 30\% \tag{12.2}$$

where I = Impact fraction
L = Length in feet of the portion of the span that is loaded resulting in maximum stress in that member.

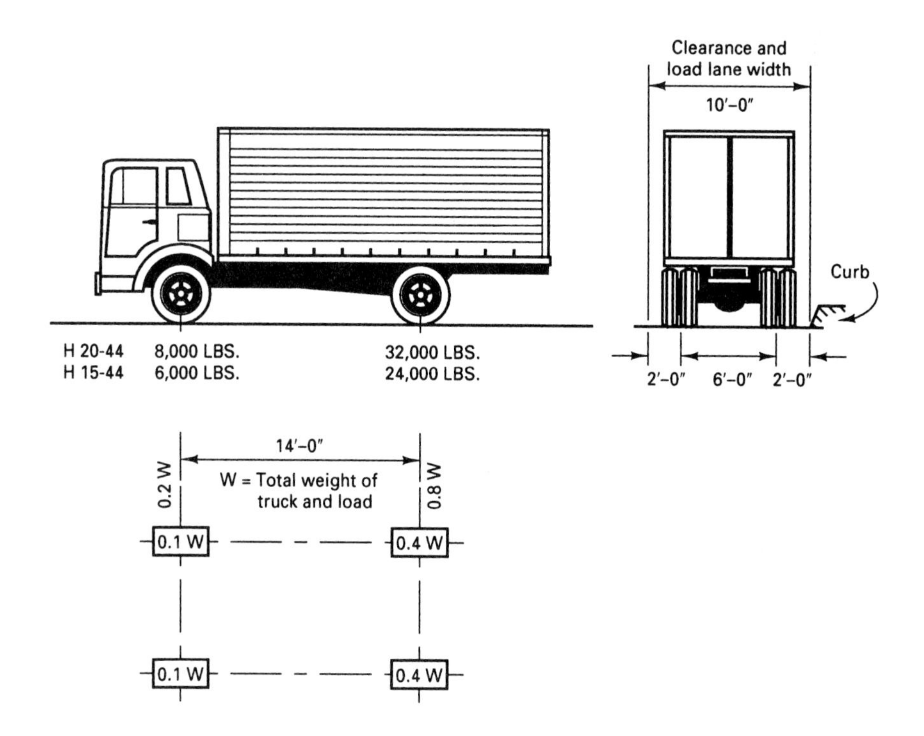

Figure 12.2 Wheel loads and geometry for H trucks

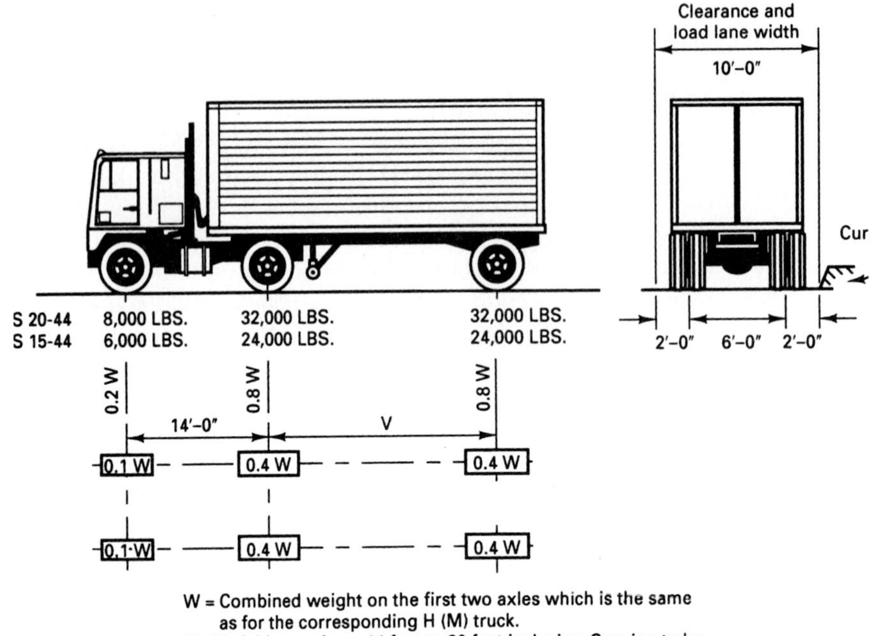

Figure 12.3 Wheel loads and geometry for HS trucks

The loaded length L for transverse members, such as floor beams, is the span length of the member center to center of the supports.

(ii) Longitudinal Forces. Provision should be made for the effect of a longitudinal force of 5 percent of the live load in all lanes carrying traffic headed in the same direction. All lanes should be loaded in the case of bridges which could likely become one-directional in the life of the structure. The load area, without impact, should be as follows:

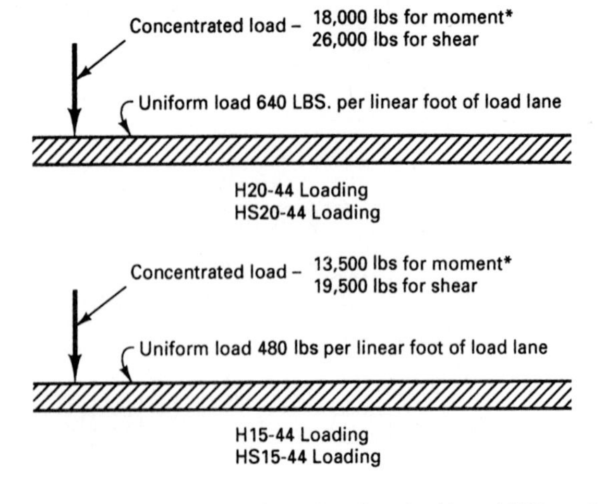

Figure 12.4 Equivalent lane loading for H and HS trucks

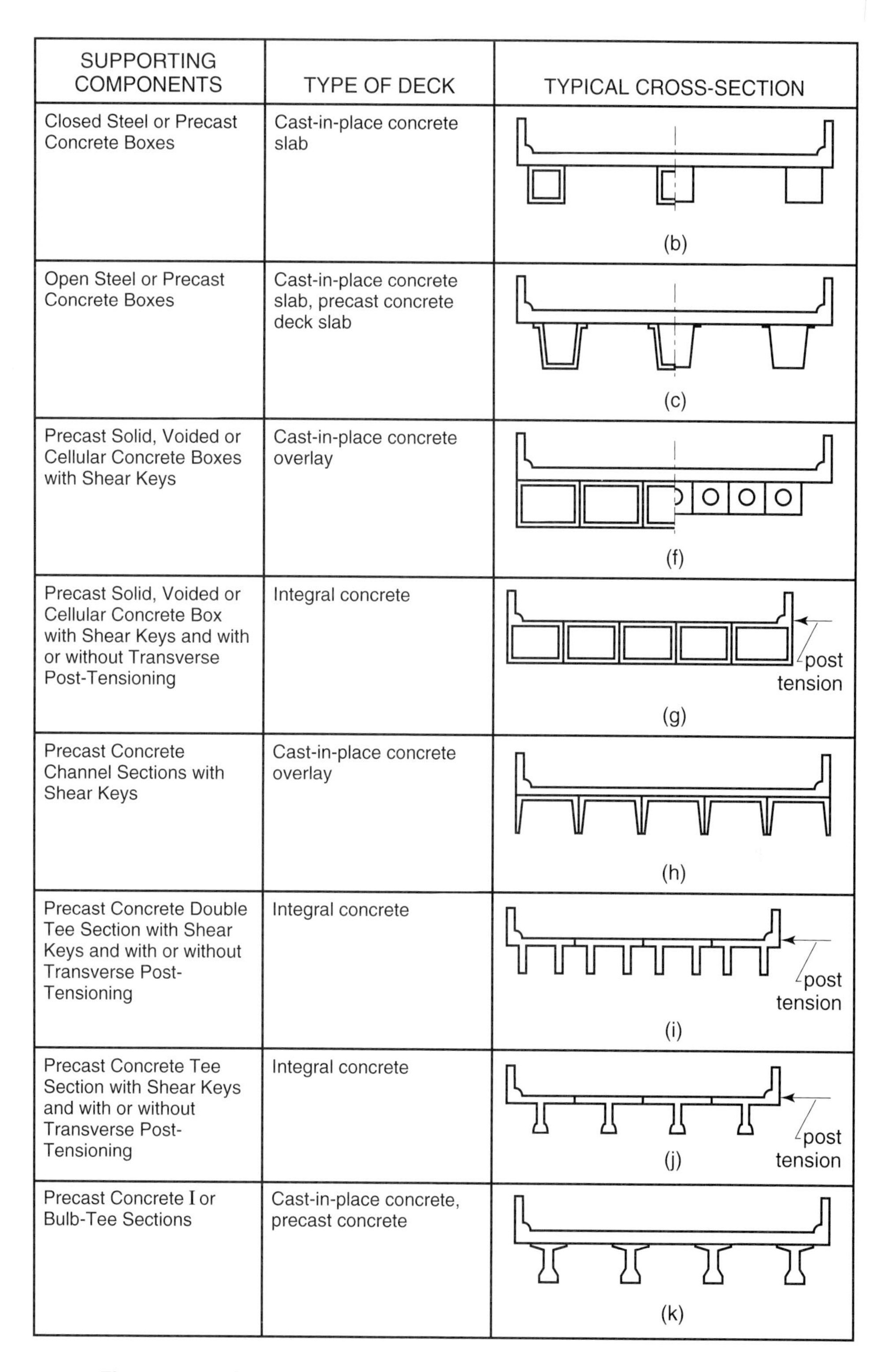

SUPPORTING COMPONENTS	TYPE OF DECK	TYPICAL CROSS-SECTION
Closed Steel or Precast Concrete Boxes	Cast-in-place concrete slab	(b)
Open Steel or Precast Concrete Boxes	Cast-in-place concrete slab, precast concrete deck slab	(c)
Precast Solid, Voided or Cellular Concrete Boxes with Shear Keys	Cast-in-place concrete overlay	(f)
Precast Solid, Voided or Cellular Concrete Box with Shear Keys and with or without Transverse Post-Tensioning	Integral concrete	post tension (g)
Precast Concrete Channel Sections with Shear Keys	Cast-in-place concrete overlay	(h)
Precast Concrete Double Tee Section with Shear Keys and with or without Transverse Post-Tensioning	Integral concrete	post tension (i)
Precast Concrete Tee Section with Shear Keys and with or without Transverse Post-Tensioning	Integral concrete	post tension (j)
Precast Concrete I or Bulb-Tee Sections	Cast-in-place concrete, precast concrete	(k)

Figure 12.5 Cross sections of Typical Bridge Deck Structures (Ref. 12.11)

Lane load + concentrated load so placed on the span as to produce maximum stress. The concentrated load and uniform load should be considered as uniformly distributed over a 10 foot width on a line normal to the centerline of the lane. The center of gravity of the longitudinal force is to be assumed located 6 feet above the floor slab.

A reduction factor should be applied when a number of traffic lanes are simultaneously loaded, as in Section (iv) to follow.

(iii) Centrifugal Horizontal Force. This force is produced by vehicle motion on curves. It is a percentage of the live load, without impact, as follows:

$$C = 0.00117S^2D = \frac{6.68S^2}{R} \tag{12.3}$$

where C = centrifugal force in percent of the live load without impact
S = design speed in miles per hour
D = degree of curve
R = radius of curve in feet.

(iv) Reduction in Load Intensity. When maximum stresses are produced in any member by loading a number of traffic lanes simultaneously, a reduction in the live load intensity can be made as follows:

	Percent
One or two lanes	100
Three lanes	90
Four lanes or more	75

12.2.2 Wheel Load Distribution on Bridge Decks: Standard AASHTO Specifications (LFD)

(i) Shear. No longitudinal distribution of wheel loads can be made for wheel or axle load adjacent to the end when computing end shears and reactions in transverse or longitudinal beams.

(ii) Bending Moments: Longitudinal Beams. In computing bending moments in longitudinal beams or stringers, no longitudinal distribution of the wheel loads is permitted. In the case of interior stringers, the live load bending moment for each stringer should be determined by applying to the stringer a fraction of the wheel load as follows for prestressed concrete elements

	Bridge designed for one traffic lane	**Bridge designed for two or more traffic lanes**
Prestressed concrete girders	$S/7.0$ if $S > 6$ ft.*	$S/5.5$ if $S > 10$.*
Non-attached Concrete Box girders	$S/8.0$ if $S > 12$ ft.*	$S/7.0$ if $S > 16$ ft.*

*If S exceeds denominator, the load on the beam should be the reaction of the wheel loads assuming the flooring between beams to act as a simple beam.

S = spacing of floor beams in feet.

(iii) Side by Side Precast Beams in Multi-Beam Decks.[12.2] A multi-beam bridge is constructed with precast reinforced or prestressed concrete beams that are placed *side by side* on the supports. The interaction between the beams is developed by continuous longitudinal shear keys used in combination with transverse tie assemblies which may, or may not, be prestressed, such as bolts, rods, or prestressing strands, or other mechanical means. Full-depth rigid end diaphragms are needed to ensure proper load distribution for channel, single- and multi-stemmed tee beams.

In computing bending moments in multi-beam precast concrete bridges, conventional or prestressed, no longitudinal distribution of wheel load shall be assumed. The live load bending moment for each section is determined by applying to the beam the fraction of a wheel load (both front and rear) determined by the following equation:

$$\text{Load Fraction} = \frac{S}{D}$$

where,

S = width of precast member;
$D = (5.75 - 0.5N_L) + 0.7N_L(1 - 0.2C)^2$ when $C \leq 5$
$D = (5.75 - 0.5N_L)$ when $C > 5$
N_L = number of traffic lanes
$C = K(W/L)$

where,

W = overall width of bridge measured perpendicular to the longitudinal girders in feet;
L = span length measured parallel to longitudinal girders in feet; for girders with cast-in-place end diaphragms, use the length between end diaphragms;
$K = \{(1 + \mu)\, I/J\}^{\frac{1}{2}}$

If the value of $\sqrt{I/J}$ exceeds 5.0, the live load distribution should be determined using a more precise method, such as the Articulated Plate Theory or Grillage Analysis.

where,

I = moment of inertia;
J = Saint-Venant torsion constant;
μ = Poisson's ratio for girders.

In lieu of more exact methods, "J" may be estimated using the following equations:

For Non-voided Rectangular Beams, Channels, Tee Beams:

$$J = \Sigma\{(1/3)bt^3(1 - 0.630t/b)\}$$

where,

b = the length of each rectangular component within the section,
t = the thickness of each rectangular component within the section.

The flanges and stems of stemmed or channel sections are considered as separate rectangular components whose values are summed together to compute "J". Note that for "Rectangular Beams with Circular Voids" the value of "J" can usually be approximated by using the equation above for rectangular sections and neglecting the voids.

For Box-Section Beams:

$$J = \frac{2tt_f(b - t)^2(d - t_f)^2}{bt + dt_f - t^2 - t_f^2}$$

where

b = the overall width of the box,
d = the overall depth of the box,
t = the thickness of either web,
t_f = the thickness of either flange.

The formula assumes that both flanges are the same thickness and uses the thickness of only one flange. The same is true of the webs.

For preliminary design, the following values of K may be used:

Bridge type	Beam Type	K
Multi-beam	Non-voided rectangular beams	0.7
	Rectangular beams with circular voids	0.8
	Box section beams	1.0
	Channel, single- and multi-stemmed tee beams	2.2

(iv) Stresses in Concrete

Case I: *All Loads including Prestress (D + L + P/S)*

$$f_c = 0.60\, f'_c$$
$$f_t = 6\sqrt{f'_c}$$

Case II: *Prestress + All Dead Loads (D + P/S)*

$$f_c = 0.40\, f'_c$$
$$f_t = 6\sqrt{f'_c}$$

Case III: $\frac{1}{2}$*Prestress + Dead) + Live Load [0.5 (D + P/S) + L]*

$$f_c = 0.40\, f'_c$$
$$f_t = 6\sqrt{f'_c}$$

12.2.3 Bending Moments in Bridge Deck Slabs: Standard AASHTO Specifications

There are two categories for bending moment calculations: category A and category B for reinforcement perpendicular and parallel respectively to the traffic.

S = effective span length in feet
E = width of slab in feet over which a wheel load is distributed
P = load on one rear wheel of truck (P_{15} or P_{20})
P_{15} = 12,000 lbs. for H 15 loading
P_{20} = 16,000 lbs. for H 20 loading

(a) *Case A—Main Reinforcement Perpendicular to Traffic (spans 2 to 24 feet)*

The live load moments for simple spans are to be determined in accordance with the following expressions:

H 20 Loading,

$$M_L = \left(\frac{S+2}{32}\right) P_{20} \tag{12.4a}$$

H 15 Loading,

$$M_L = \left(\frac{S + 2}{32}\right) P_{15} \qquad (12.4b)$$

where M_L is in ft-lb/ft of slab width

In slabs continuous over three or more supports, a continuity factor of 0.8 should be applied to Equations 12.4(a) and 12.4(b)

(b) *Case B—Main Reinforcement Parallel to Traffic*

For wheel loads, the distribution width, E, should be $= 4 + 0.06S \leq 7.0$ ft. Lane loads are distributed over a width $2E$ as follows:

H 20 Loading

$$S \leq 50 \text{ ft:} \qquad M_L = 900S \qquad (12.4c)$$

$$S = 50 - 100 \text{ ft: } M_L = 1000S \qquad (12.4d)$$

where M_L is in ft-lb

For H 15 loading, reduce the values in Equations 12.4(c), and 12.4(d) by 25 percent.

12.2.4 Wind Loads

In accounting for the wind loads, the exposed area is equal to the sum of the areas of all members including floor system and railings as seen in an elevation 90 degrees to the longitudinal axis of the structure. Design should be based on a wind velocity $V = 100$ miles per hour. The area may be reduced as stipulated in Ref. 12.2.

12.2.5 Seismic Forces

Both the equivalent static force method and the response spectrum method can be used for the design of structures with supporting members of approximately equal stiffnesses. Details are given in Ref. 12.2. Additional basic discussion of earthquake response, the fundamental period of vibration and the International Building Code (IBC 2000) are given in Ref. 12.5.

12.2.6 Load Combinations

The design should consider such a group of load combinations that results in the maximum stress condition in the member under consideration. There are ten groups of loadings under service load conditions:

Group I: $D + (L + I) + CF + E + B + SF$
Group II: $D + E + B + SF + W$
Group III: $D + (L + I) + CF + E + B + SF + W + WL + LF$
Group IV: $D + (L + I) + CF + E + SF + (R + S + T)$
Group V: $D + E + B + SF + W + (R + S + T)$
Group VI: $D + (L + I) + CF + E + B + SF + W + WL + LF + (R + S + T)$
Group VII: $D + E + B + SF + EQ$
Group VIII: $D + (L + I) + CF + E + B + SF + ICE$
Group IX: $D + E + B + SF + W + ICE$
Group X: $D + (L + I) + E$

where D = dead load

L = live load

I = live load impact

E = earth pressure

B = buoyancy

W = wind load on the structure

WL = wind load on live load – 100 lb./linear foot

LF = longitudinal force from live load

CF = centrifugal force

R = rib shortening

S = shrinkage

T = temperature

EQ = earthquake

SF = stream flow pressure

ICE = ice pressure

For load factor design, the proceeding parameters are multiplied by the load factors in Table 12.1(b)

For factor loads, the group value is

$$\begin{aligned}\text{Group number } (N) = \; & \gamma[\beta_D D + \beta_L(L + I) + \beta_C CF + \beta_E E \\ & + \beta_B B + \beta_S SF + \beta_W W + \beta_{WL} WL + \beta_L LF \\ & + \beta_R(R + S + T) + \beta_{EQ} EQ + \beta_{ICE} ICE]\end{aligned} \tag{12.5}$$

The load factors to be applied to any particular load combination are as follows:

β_E = 0.7 for vertical loads on reinforced concrete boxes.

= 1.00 for lateral loads on reinforced concrete boxes.

= 1.00 for vertical and lateral loads on all other culverts.

= 1.0 and 0.5 for lateral loads on rigid frames (check which loading governs for the particular group).

β_E = 1.3 for lateral earth pressure when checking positive moment in rigid frames, culverts or reinforced box culverts.

β_D = 0.75 when checking member for minimum axial load and maximum moment for maximum eccentricity for column design.

= 1.0 when checking for maximum axial load and minimum moment.

= 1.0 for flexural and tension members.

Table 12.1(b) gives the values of the β coefficients for the various load parameters in Equation 12.5 for Standard AASHTO Specifications.

Table 12.1(b) β Coefficients for LAOD Group Parameters: Standard AASHTO Specifications (Ref. 12.2)

Col. No.		1	2	3	3A	4	5	6	7	8	9	10	11	12	13	14
			β FACTORS													
GROUP		γ	D	$(L+I)_n$	$(L+I)_p$	CF	E	B	SF	W	WL	LF	R+S+T	EQ	ICE	%
SERVICE LOAD	I	1.0	1	1	0	1	β_E	1	1	0	0	0	0	0	0	100
	IA	1.0	1	2	0	0	0	0	0	0	0	0	0	0	0	150
	IB	1.0	1	0	1	1	β_E	1	1	0	0	0	0	0	0	**
	II	1.0	1	0	0	0	1	1	1	1	0	0	0	0	0	125
	III	1.0	1	1	0	1	β_E	1	1	0.3	1	1	0	0	0	125
	IV	1.0	1	1	0	1	β_E	1	1	0	0	0	1	0	0	125
	V	1.0	1	0	0	0	1	1	1	1	0	0	1	0	0	140
	VI	1.0	1	1	0	1	β_E	1	1	0.3	1	1	1	0	0	140
	VII	1.0	1	0	0	0	1	1	1	0	0	0	0	1	0	133
	VIII	1.0	1	1	0	1	1	1	1	0	0	0	0	0	1	140
	IX	1.0	1	0	0	0	1	1	1	1	0	0	0	0	1	150
	X	1.0	1	1	0	0	β_E	0	0	0	0	0	0	0	0	100
LOAD FACTOR DESIGN	I	1.3	β_D	1.67*	0	1.0	β_E	1	1	0	0	0	0	0	0	Not Applicable
	IA	1.3	β_D	2.20	0	0	0	0	0	0	0	0	0	0	0	
	IB	1.3	β_D	0	1	1.0	β_E	1	1	0	0	0	0	0	0	
	II	1.3	β_D	0	0	0	β_E	1	1	1	0	0	0	0	0	
	III	1.3	β_D	1	0	1	β_E	1	1	0.3	1	1	0	0	0	
	IV	1.3	β_D	1	0	1	β_E	1	1	0	0	0	1	0	0	
	V	1.25	β_D	0	0	0	β_E	1	1	1	0	0	1	0	0	
	VI	1.25	β_D	1	0	1	β_E	1	1	0.3	1	1	1	0	0	
	VII	1.3	β_D	0	0	0	β_E	1	1	0	0	0	0	1	0	
	VIII	1.3	β_D	1	0	1	β_E	1	1	0	0	0	0	0	1	
	IX	1.20	β_D	0	0	0	β_E	1	1	1	0	0	0	0	1	
	X	1.30	1	1.67	0	0	β_E	0	0	0	0	0	0	0	0	

$(L+I)_n$ - Live load plus impact for AASHTO Standard Highway H or HS loading

$(L+I)_p$ - Live load plus impact consistent with the overload criteria of the operation agency.

12.2.7 LRFD Load Combinations

The load combinations using the LRFD specifications differ from the standard specifications. The following tables: 12.2 to 12.3, give the required load combinations, and Tables 12.4 to 12.7 the shear and moment expressions to be used in design. Section 12.1.1 gives the LRFD resistance factors, ϕ, which differ from the standard reduction factor ϕ. It should be noted that in the standard specifications, either the lane load or the truck load is used in the live-load calculations. The LRFD specifications require that the *combined* lane and truck loads be used in the live-load computations.

Table 12.2(a) LRFD Load Combinations and Load Factors

Load Combination / Limit State	DC DD DW EH EV ES	LL IM CE BR PL LS	WA	WS	WL	FR	TU CR SH	TG	SE	Use One of These at a Time			
										EQ	IC	CT	CV
STRENGTH-I	γ_P	1.75	1.00	-	-	1.00	0.50/1.20	γ_{TG}	γ_{SE}	-	-	-	-
STRENGTH-II	γ_P	1.35	1.00	-	-	1.00	0.50/1.20	γ_{TG}	γ_{SE}	-	-	-	-
STRENGTH-III	γ_P	-	1.00	1.40	-	1.00	0.50/1.20	γ_{TG}	γ_{SE}	-	-	-	-
STRENGTH-IV EH, EV, ES, DW DC ONLY	γ_P	-	1.00	-	-	1.00	0.50/1.20	-	-	-	-	-	-
STRENGTH-V	γ_P	1.35	1.00	0.40	0.40	1.00	0.50/1.20	γ_{TG}	γ_{SE}	-	-	-	-
EXTREME EVENT-I	γ_P	γ_{EQ}	1.00	-	-	1.00	-	-	-	1.00	-	-	-
EXTREME EVENT-II	γ_P	0.50	1.00	-	-	1.00	-	-	-	-	1.00	1.00	1.00
SERVICE-I	1.00	1.00	1.00	0.30	0.30	1.00	1.00/1.20	γ_{TG}	γ_{SE}	-	-	-	-
SERVICE-II	1.00	1.30	1.00	-	-	1.00	1.00/1.20	-	-	-	-	-	-
SERVICE-III	1.00	0.80	1.00	-	-	1.00	1.00/1.20	γ_{TG}	γ_{SE}	-	-	-	-
FATIGUE-LL, IM & CE ONLY	-	0.75	-	-	-	-	-	-	-	-	-	-	-

where: γ_p = Load factor for permanent loads

Permanent Loads
- DD = downdrag
- DC = dead load of structural components and nonstructural attachments
- DW = dead load of wearing surfaces and utilities
- EH = horizontal earth pressure load
- ES = earth surcharge load
- EV = vertical pressure from dead load of earth fill

Transient Loads
- BR = vehicular braking force
- CE = vehicular centrifugal force
- CR = creep
- CT = vehicular collision force
- CV = vessel collision force
- EQ = earthquake
- FR = friction
- IC = ice load
- IM = vehicular dynamic load allowance
- LL = vehicular live load
- LS = live load surcharge
- PL = pedestrian live load
- SE = settlement
- SH = shrinkage
- TG = temperature gradient
- TU = uniform temperature
- WA = water load and stream pressure
- WL = wind on live load

Strength I: Basic load combination, no wind
Strength II: Load on bridge with owner-specified design, no wind
Strength III: Load includes wind
Strength IV: Very high ratio of dead to live load
Service I: Normal operational use load combinations with deflection and crack control
Service II: Load combinations with control of yielding of steel structures
Service III: Load combinations relating only to tension in prestressed concrete

Photo 12.1 Stoney Trail Bow River segmental bridge, Calgary, Alberta, utilizing the incremental launch method—span 1562 ft, deck width 69 ft, and the deck rises 89 to 118 ft above the river valley (*Courtesy* James Skeet–Reid Crowther Engineering, Calgary)

The following expressions in Table 12.4 and 12.5 (Ref. 12.2) may be used to compute the maximum bending moments and the maximum shear force per *lane* of any point in a span for HS20 truck, with the limitations indicated in the table. The computed values have to be multiplied by a factor of $\frac{1}{2}$ in order to obtain the shear force and moment per *line* of wheels.

The expressions in the tables are limited to simply supported spans and do not include the impact factors.

The maximum bending moments and maximum shear forces per lane at any point on a span for a lane load of 0.64 kip/ft may be computed from the following simplified expressions:

Table 12.2(b) LRFD Permanent Loads

Type of Load	Load Factor	
	Maximum	**Minimum**
DC: Component and Attachments	1.25	0.90
DD: Downdrag	1.80	0.45
DQ: Wearing surface and utilities	1.50	0.65
EH: Horizontal Earth Pressure		
Active	1.50	0.90
At-Rest	1.35	0.90
EV: Vertical Earth Pressure	1.35	N/A
Overall Stability	1.35	1.00
Retaining Structure	1.30	0.90
Rigid Buried Structure	1.35	0.90
Rigid Frame	1.95	0.90
Flexible Buried Structure other than Metal Box Culvert	1.50	0.90
ES: Earth Surcharge	1.50	0.75

$$\text{Maximum } V_{LL} = \frac{0.64}{2L}(L - x)^2 \text{ kip} \tag{12.6a}$$

$$\text{Maximum } M_{LL} = \frac{0.64(x)(L - X)}{2} \text{ ft-kip} \tag{12.6b}$$

where, x = distance from left support, ft
L = beam span, ft
LL = lane load

The LRFD specifications require a higher impact factor than the standard specifications. They also require consideration of the fatigue state limits. For fatigue, a special truck load is considered. It consists of a single design truck which has the same axle weight used in all other limit states, but with a constant spacing of 30 ft between the 32-kip axles. Table 12.6 gives the impact factor IM for the various types of limit states:

Table 12.3(a) Distribution of Live Load Per Lane for Shear in Interior Beams

Section	One Design Lane	Two or More Design Lanes
Concrete Box Beams in Multi-beam Decks	$\left(\frac{b}{130L}\right)^{0.15}\left(\frac{I}{J}\right)^{0.05}$	$\left(\frac{b}{156}\right)^{0.4}\left(\frac{b}{12L}\right)^{0.1}\left(\frac{I}{J}\right)^{0.05}$
Concrete Deck, I-, T- and Double-T Sections	$0.36 + \left(\frac{s}{25.0}\right)$	$0.20 + \left(\frac{s}{12}\right) - \left(\frac{s}{36}\right)^2$

1. Ranges for b, d, L, s, S, t_s, Kg are given in Ref. 12.3.
2. For exterior beams, see Ref. 12.3, Section 4.6.2

Table 12.3(b) Distribution of Live Load Per Lane For Moment in Interior Beams

Section	One Design Lane	Two or More Design Lanes
Concrete Box Beams in Multi-beam Decks	$k\left(\frac{b}{33.5L}\right)^{0.05}\left(\frac{I}{J}\right)^{0.25}$	$k\left(\frac{b}{305}\right)^{0.6}\left(\frac{b}{12L}\right)^{0.2}\left(\frac{I}{J}\right)^{0.6}$
Concrete Deck, I-, T- and Double-T Sections	$0.06+\left(\frac{s}{14}\right)^{0.4}\left(\frac{S}{L}\right)^{0.3}\left(\frac{K_g}{12t_s^3L}\right)^{0.1}$	$0.075+\left(\frac{s}{9.5}\right)^{0.6}\left(\frac{S}{L}\right)^{0.2}\left(\frac{K_g}{12t_s^3L}\right)^{0.1}$

1. Ranges for b, d, L, s, S, t_s, K_g are given in Ref. 12.3.
2. For exterior beams, see Ref. 12.3, Section 4.6.2
3. Notation:

b = Beam width, in.

J = St. Vincent's torsional constant, in 4 = $4\,A_0^2/\Sigma\, s/t$

Kg = Longitudinal Stiffness parameter distribution factor for multi-beam bridges, where

$Kg = n(I + A_c e_g^2)$

e_g = distance between centers of gravity of members

$k = 2.5(N_b)^{-0.2}$ where N_b = number of beams

A_c = cross-sectional area

L = span, ft

A_o = area enclosed by centerlines of the beam elements

s = length of an element of box beam

S = beam spacing

t = thickness of the beam elements Bending Moments and Shear Forces due to Vehicular Live Loads

Table 12.4 Maximum Shear Force per Lane for HS20 Truck Load (V_{LT})

Load Type	x/L	Formula for maximum bending moment, ft-kips	Minimum		Maximum
			x,* ft	L, ft	L, ft
HS20 Truck	0–0.500	$\frac{72[(L-x)-4.67]}{L}-8$	14	28	42
	0–0.500	$\frac{72[(L-x)-9.33]}{L}$	0	42	-

*x is the distance from left support to the section being considered, ft; LT = truck load

Table 12.5 Maximum Bending Moment per Lane for HS20 Truck Load (M_{LT})

Load Type	x/L	Formula for maximum bending moment, ft-kips	Minimum	
			x,*ft	L,ft
HS20 Truck	0–0.333	$\frac{72(x)[(L-x)-9.33]}{L}$	14	28
	0.333–0.500	$\frac{72(x)[(L-x)-4.67]}{L}-112$	0	42

*x is the distance from left support to the section being considered, ft; LT = truck load

Table 12.6 Impact Factors

Component	*IM*
Deck Joints—All Limit States	**15%**
All other Components	
Fatigue and Fracture Limit States	15%
All Other Limit States	33%

Table 12.7 Fatigue Bending Moment per Lane

Load Type	*x/L*	*Formula for maximum bending moment, ft-kips*	*Minimum*	
			x, * ft	**L, ft**
Fatigue Truck Loading (LRFD)	0–0.241	$\dfrac{72(x)[(L-x)-18.22]}{L}$	0	44
	0.241–0.500	$\dfrac{72(x)[L-x)-11.78]}{L} - 112$	14	28

*x is the distance from left support to the section being considered, ft; LT = truck load

Table 12.7 (Ref 12.3) gives expressions for computing the maximum bending moments per land due to HL-93 fatigue truck loading. The values obtained from the table have to be multiplied by a factor of $\frac{1}{2}$ in order to obtain the values per line of wheels.

The LRFD design live load is an HL-93 truck configuration which consists of a *combination* of:

(a) design truck or design tandem with dynamic allowance. The design truck is the same as the HS20 design truck specified in the Standard AASHTO specifications. The design tandem consists of a pair of 25 kip axles spaced at 4-ft apart.

(b) design lane load of 0.64 kip/ft without dynamic allowance.

12.3 FLEXURAL DESIGN CONSIDERATIONS

12.3.1 Strain ϵ and Factor ϕ Variations: The Strain Limits Approach

For ductile behavior of sections, the reinforcement percentage has to be considerably smaller than the balanced percentage in flexure. Net strain ϵ_T and a reduction factor range of $0.70 \le \phi \le 0.90$ are used. No upper limits on the amount of reinforcement needs to be used in a beam provided that the strain limit is not exceeded and the appropriate ϕ factor is used. An approach that can be used both in reinforced and in prestressed beams using strain limits rather than stress limits through reinforcement percentage limits is given in Appendix B of ACI 318-99 (Ref. 12.4). An upper limit tensile strain $\epsilon_T = 0.005$ in./in. as the limiting strain is comparable to the 75% of the balanced reinforcement percentage and is the basis of this approach (Figure 12.6). This limiting strain is considered at the extreme tensile steel reinforcement level, namely, at the centroid of the layer closest to the tensile face of the section. More precisely, $\epsilon_t = 0.0041$ corresponds to $f_y \simeq 230{,}000$ psi in the prestressing steel.

In the AASTHO LRFD procedure, a limiting value of the ratio of the neutral axis depth, c, to the effective beam depth, d_e, to the centroid of the reinforcement is taken as

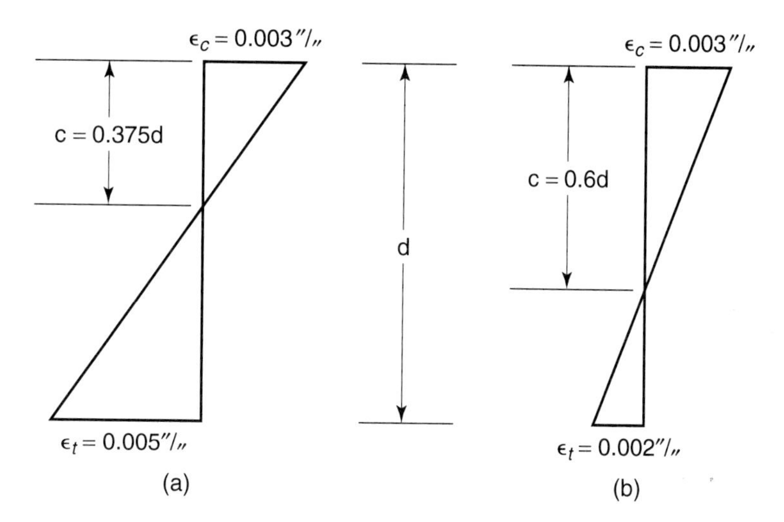

Figure 12.6 Strain Limits (a) Tension-Controlled, (b) Compression-Controlled

0.42 in this strain limits approach, invariably called as a unified approach (Refs. 12.14–12.16). The approach is a rational way of assuring strain-compatibility using common strain and stress expressions regardless of whether the member is reinforced, prestressed or partially prestressed. The depth d_e in the ratio c/d_e becomes d_p if no mild steel reinforcement is used. Table 12.11 of section 12.8 presents a general comparison between the ACI and the LRFD procedures for determining the required reinforcement for flexure (Ref. 12.14).

A strain value ϵ_t considerably higher than 0.005 in./in. has to be used, such as 0.007 to 0.009 in./in. The lower limit for beam-column sections is $\epsilon_t = 0.002$ in./in. The $\epsilon_t = 0.002$ is used as a basis for first yield strain. $\epsilon_y - f_y/E_s = 0.002$, although this value can vary depending on the type of reinforcement used. Fig. 12.7 gives on this basis the limits of strain for tension-controlled and compression-controlled concrete sections for all cases, reinforced and prestressed.

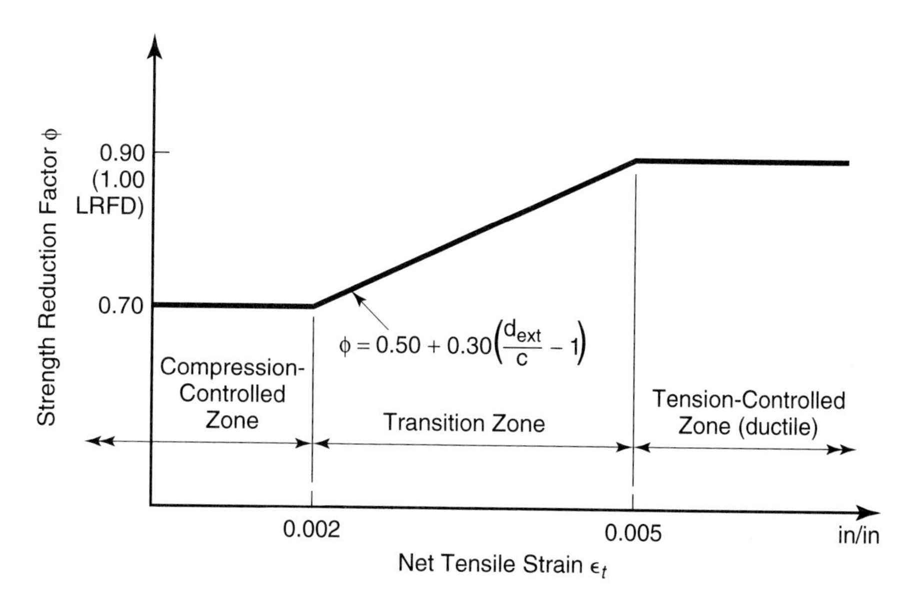

Figure 12.7 Variation of Strength Reduction Factor ϕ with the Net Tensile Strain ϵ_t.

When the net tensile strain in the extreme tension reinforcement is sufficiently large (equal to or greater than 0.005), the section is defined as tension-controlled where ample warning of failure with extensive deflection and cracking can occur. When the net tensile strain in the extreme tension reinforcement is small (less than or equal to the compression-controlled strain limit), a brittle failure condition is expected to develop, with little warning of impending failure.

A balanced strain condition develops at a section when the maximum strain at the extreme compression fibers just reaches 0.003 in./in. simultaneously with the first yield strain $\epsilon_y = f_y/E_s$ in the tension reinforcement corresponding to a net tensile strain in the tension reinforcement set in this method at a value $\epsilon_t = 0.002$ in./in.

This condition cannot be used in the flexural design of beams not subjected to compression. In such members, a strain ϵ_t in the extreme tensile reinforcement should not exceed 0.0075 in./in. for practical purposes.

12.3.2 Factored Flexural Resistance

The factored flexural resisting moment,

$$M_r = \phi M_n \tag{12.7}$$

where the resistance factor $\phi = 1.0$.

It is recommended in strain compatibility analysis that ϕ be reduced from a value $\phi = 1.0$ for net tensile strain of 0.005 in./in. to $\phi = 0.7$ for net tensile strain of 0.002 in./in. in the extreme tension steel, namely,

$$0.7 \le \phi = 0.50 + 0.30\left[\frac{d_{ext}}{c} - 1\right] \le 1.0 \tag{12.8}$$

where d_{ext} is d_e of the extreme layer of reinforcement, namely the one closest to the extreme tension fibers of the prestressed concrete section.

12.3.3 Flexural Design Parameters

The expression for computing the nominal moment strength of the prestressed sections by the LRFD method are similar to the standard AASHTO and ACI 318 strength design procedures given in Section 4.11 of Chapter 4. The ultimate design strength, f_{ps}, of the reinforcement can be computed either by strain-compatibility procedures such as in Example 4.19 or by an approximate method using the following expression:

$$f_{ps} = f_{pu}\left(1 - k\frac{c}{d_p}\right) \tag{12.9a}$$

where,

$$k = 2\left(1.04 - \frac{f_{py}}{f_{pu}}\right) \tag{12.9b}$$

$$= 0.28 \text{ for low relaxation steel}$$

In the Standard AASHTO specifications:

$$f_{ps} = f_{pu}\left(1 - \frac{\gamma}{\beta_1}\rho\frac{f_{pu}}{f'_c}\right) \tag{12.9c}$$

The depth, c, of the neutral axis is obtained from the following expressions:

(a) ***Doubly reinforced sections:***

$$c = \frac{A_{ps}f_{pu} + A_s f_y - A'_s f'_y}{0.85f'_c\beta_1 + kA_{ps}\dfrac{f_{pu}}{d_p}} \tag{12.10}$$

where f'_y = yield strength of the compression reinforcement

(b) ***Flanged Sections:***

$$c = \frac{A_{ps}f_{pu} + A_s f_y - 0.85f'_c\beta_1(b - b_w)h_f}{0.85f'_c\beta_1 b_w + kA_{ps}\dfrac{f_{pu}}{d_p}} \tag{12.11}$$

where b_w = web width

d_p = distance from the extreme compression fiber to the centroid of the prestressing tendons.

12.3.4 Reinforcement Limits

(a) ***Maximum reinforcement limit***

The maximum amount of prestressed and non-prestressed reinforcement should be such that,

$$\frac{c}{d_e} \leq 0.42 \tag{12.12a}$$

$$\text{where } d_e = \frac{A_{ps}f_{ps}d_p + A_s f_y d_s}{A_{ps}f_{ps} + A_s f_y} \tag{12.12b}$$

(b) ***Minimum reinforcement***

At any section, the amount of prestressed and non-prestressed reinforcement should be adequate to develop a factored flexural resistance, M_r, at least equal to the *lesser* of 1.2 M_{cr} determined on the basis of elastic analysis or 1.33 times the factored moment required by the applicable strength load combinations.

$$M_{cr} = (f_r + f_{ce})S_b - M_{dnc}\left[\frac{S_{bc}}{S_b} - 1\right] \tag{12.13}$$

where,

M_{dnc} = moment due to non-composite dead loads

S_b = non-composite section modulus

S_{bc} = composite section modulus

f_r = modulus of rupture = $7.5\sqrt{f'_c}$ psi = $0.24\sqrt{f'_c}$ ksi

f_{ce} = compressive stress in the concrete due to effective prestress *only*, after losses, at the extreme tensile fibers of the section where tensile stresses are caused by external loads.

12.4 SHEAR DESIGN CONSIDERATIONS

12.4.1 The Modified Compression Field Theory

The compression field theory for both shear and shear combined with torsion is discussed in Section 5.17.3 of Chapter 5. When torsion exists, it assumes that concrete carries no tension after cracking and the field of diagonal compressive struts carries the torsional shear. The inclination angle θ of these struts varies depending on the longitudinal, transverse, and principal strains (Ref. 12.6) in the web such that:

$$\tan^2 \theta = \frac{\varepsilon_x - \varepsilon_2}{\varepsilon_t - \varepsilon_2} \tag{12.14}$$

where ϵ_x = longitudinal strain of web, tension positive
ϵ_t = transverse strain, tension positive
ϵ_2 = principal compressive strain, negative

Figure 12.8 shows the stress field in the web of a non-prestressed beam before and after cracking. Before the beam cracks, the shear is equally carried by the diagonal tensile and diagonal compressive stresses acting at a 45° angle (Figure 12.8a). After cracking, the diagonal cracks from the tensile stresses in the concrete are considerably reduced (Ref. 12.6, 12.7; also the shear and torsion equilibrium theory by Hsu in Ref. 12.8, 12.9).

In the compression field theory, the assumption is made that the principal tensile stress, f_1, equals zero as in Figure 12–8(b) after the concrete has cracked. The modified compression field theory takes into account the contribution of the tensile stresses in the concrete between the cracks as in Figure 12.8(c). From Mohr's stress circle in Figure 5.2(b), in Chapter 5, in conjunction with Figure 12.8(c), the following expression can be obtained:

$$f_2 = (\tan\theta + \cot\theta)v - f_1 \tag{12.15a}$$

where the applied shear stress is:

$$v = \frac{V}{b_w jd} = \frac{(V_u - \theta V_p)}{\phi\, b_w d_v} \tag{12.15b}$$

$d_v = (d_p - a/2)$and b_w = effective web width. The tension web reinforcement, A_v, required to balance the compressive stresses would have to be expressed as:

$$A_v f_v = (f_2 \sin^2 \theta - f_1 \cos^2 \theta) b_w s \tag{12.16}$$

where $A_v f_v$ is the vertical component of the balancing tensile force to close the diagonal crack inclined at angle θ and f_v is the average stress in the vertical stirrups. Substituting for f_2 in equation 12.15(a) into equation 12.16 gives:

$$V = f_1 b_w d_v \cot\theta + \frac{A_v f_v}{s} d_v \cot\theta \tag{12.17}$$

where V represents V_n and is equal to $(V_c + V_s)$, V_s being the shear force taken by the vertical stirrups.

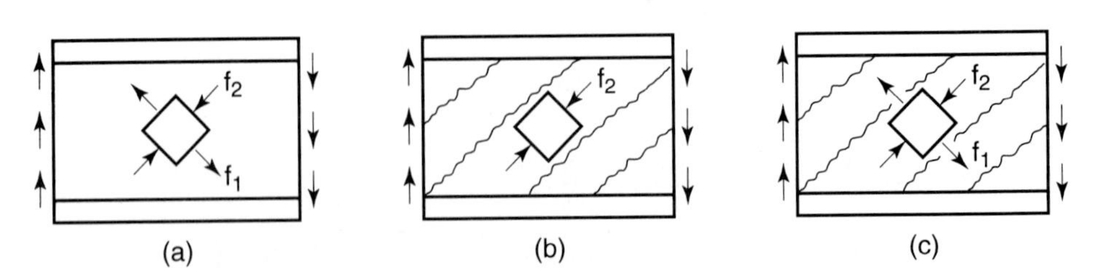

Figure 12.8 Stress fields in web of reinforced concrete beam (Ref. 12.6) (a) before cracking $f_1 = f_2$, $\theta = 45°$, (b) compression field theory, $f_1 = 0$, (c) modified compression field theory, $f_1 \neq 0$.

12.4.2 Design Expressions

As discussed in detail in Ref. 12.6, by making simplifying assumptions, the basic equations of the modified compression field theory can be rearranged so that the nominal shear resistance, V_n, in a prestressed beam can be evaluated, where:

$$V_n = V_c + V_s + V_p \tag{12.18}$$

where, V_c = nominal shear strength provided by the tensile stresses in the concrete
V_s = nominal shear strength provided by the tensile stresses in the web reinforcement.
V_p = nominal shear strength provided by the vertical component of the harped or draped longitudinal tendons.

12.4.2.1 AASHTO Standard Specifications. The AASHTO standard provisions, similar to the ACI-318 provision (Ref. 12.1–12.4) provide that V_c would be the smaller of the following two expressions presented and discussed in detail in Sections 5.5.1 and 5.5.2 of chapter 5:

(a) ***Flexural shear:***

$$V_{ci} = 0.6\sqrt{f'_c}\, b_w d + \frac{V_i M_{cr}}{M_{\max}} \tag{12.19}$$

(b) ***Web shear:***

$$V_{cw} + [3.5\sqrt{f'_c} + 0.3\bar{f}_c] b_w d + V_p \tag{12.20}$$

where, in AASHTO, the cracking moment is expressed as:

$$M_{cr} = S_t(6\sqrt{f'_c} + f_{pe} - f_d)$$

12.4.2.2 LRFD Specifications. The LRFD AASSHTO provisions recognize two methods:

(a) Strut-and-tie model applicable to any section geometry with regular or discontinuity features

(b) Modified compression field model (Ref. 12.3, 12.6). This model is based on variable angle truss model in which the inclination of the diagonal compression field is allowed to vary. It differs from the standard method where the angle θ is always assumed as 45°, in that the plain concrete contribution, V_c is attributed to the tension carried across the compression diagonals as discussed in section 12.4.1.

The nominal resistance is taken as the lesser of:

$$V_n = V_c + V_s + V_p \tag{12.21}$$

or,

$$V_n = 0.25 f'_c b_v d_v \tag{12.22}$$

where, b_v = effective web width
d_v = effective shear depth $\approx (d_p - a/2)$
a = depth of the compressive block

This critical section for shear is located at distance d_v or ($0.5 d_v \cot\theta$), whichever is larger. The value of d_v is taken from midspan flexural capacity computations.

The nominal shear resistance of the plain concrete, V_c, in psi is:

$$V_c = \beta\sqrt{f'_c}\, b_v d_v \tag{12.23}$$

and in ksi,

$$V_c = 0.0316\beta\sqrt{f'_c}\, b_v d_v \tag{12.24}$$

The factor 0.0316 is $1/\sqrt{1000}$, which converts the expression from psi to ksi.

The contribution of the vertical web reinforcement is taken as:

$$V_s + \frac{A_v f_y d_v Cot\theta}{s} \tag{12.25}$$

Transverse shear reinforcement should always be provided when the factored shear, V_u, exceeds the plain concrete shear capacity, namely when

$$V_u > 0.5\phi(V_c + V_p) \tag{12.26}$$

Additionally, when the beam reaction induces compression into the ends of the members as occurs in the majority of cases, the critical section for shear is taken as the larger of: 0.5 d_v cot θ or d_v, measured from the face of the support.

In order to determine the nominal shear resistance of the prestressed member, the design engineer has to determine the values of β and θ needed for computing V_c and V_s in equations 12.21 and 12.22. For non-prestressed concrete sections use $\beta = 2.0$ and $\theta = 45°$. For prestressed concrete sections, lower variable β values are to be used by trail and adjustment. Figure 12.9 (Ref. 12.3) graphically enables choosing the appropriate value.

The strain, ϵ_x, in the tensile reinforcement is obtained from the following expression:

$$\epsilon_x = \frac{\dfrac{M_u}{d_u} + 0.5N_u + 0.5V_u Cot\theta - A_{ps}f_{po}}{E_s A_s + E_{ps}A_{ps}} \leq 0.002 \tag{12.27}$$

The strain ϵ_x is shown in Figure 12.10.

$$f_{po} = f_{pe} + \frac{\bar{f}_{ce}E_{ps}}{E_c} \tag{12.28a}$$

The stress f_{po} represents the stress in the prestressing strands when the stress in the surrounding concrete is zero, namely, the stress in the reinforcement at decompression.

$\bar{f}_{ce}$ = concrete compressive stress at the centroid of the composite section resisting live load or at the junction of the web and the flange if it lies within the flange due to both prestress and the bending moments resisted by precast section acting alone, namely, prior to composite action.

f_{pe} = effective stress in the prestressing steel after losses. f_{po} can conservatively taken as the effective prestress f_{pe}.

If the strain in the tensile reinforcement is negative, ϵ_x should be multiplied by the factor F_ϵ, in the following expression:

$$F_\varepsilon = \frac{E_s A_s + E_{ps}A_{ps}}{E_c A_c + E_s A_s + E_{ps}A_{ps}} \tag{12.28b}$$

where A_c = area of the concrete of the flexural tension side of the member as shown in the shaded portion of Figure 12.10.

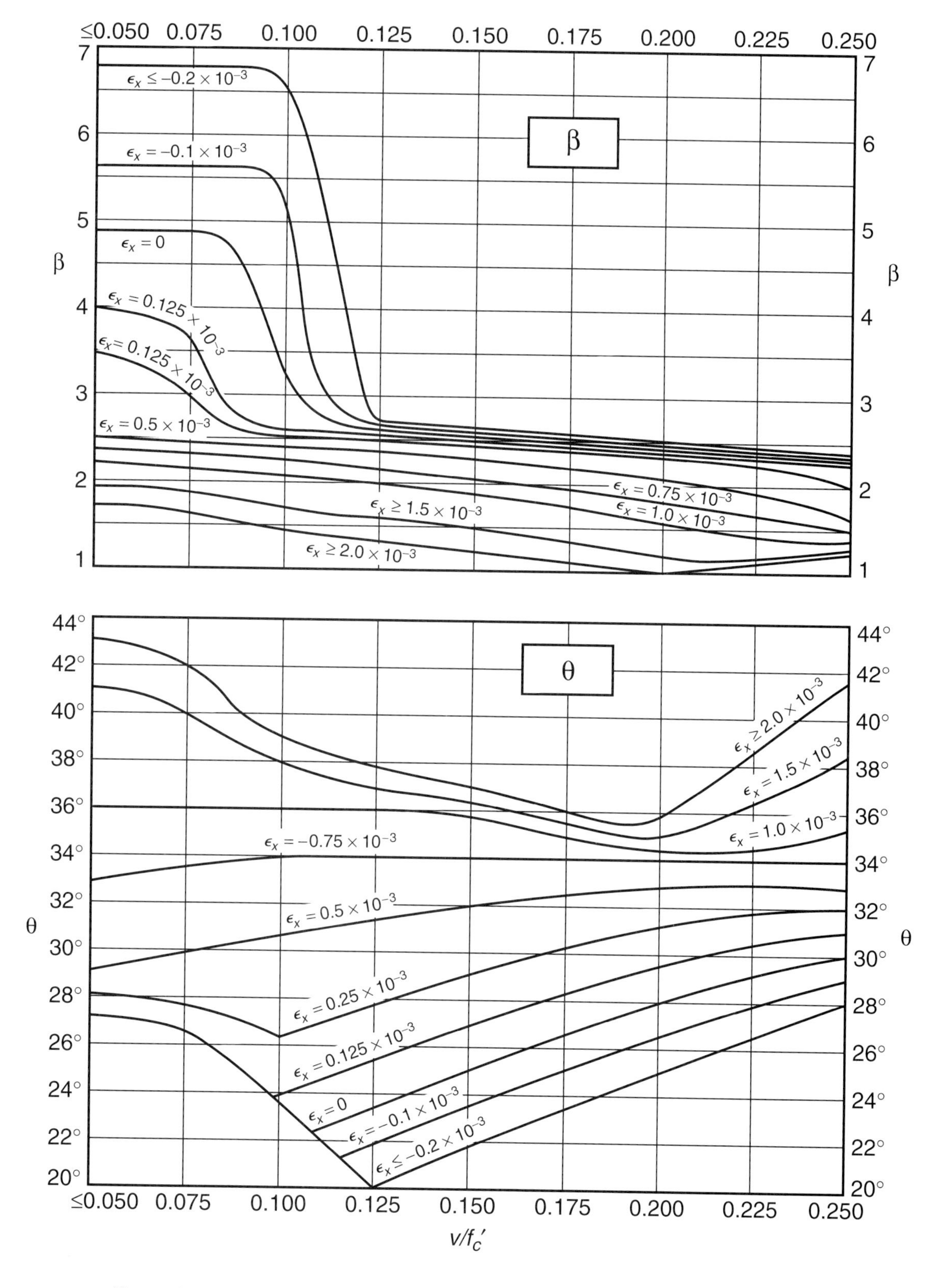

Figure 12.9 Values of θ and β for Sections with Transverse Reinforcement (Ref. '12.3)

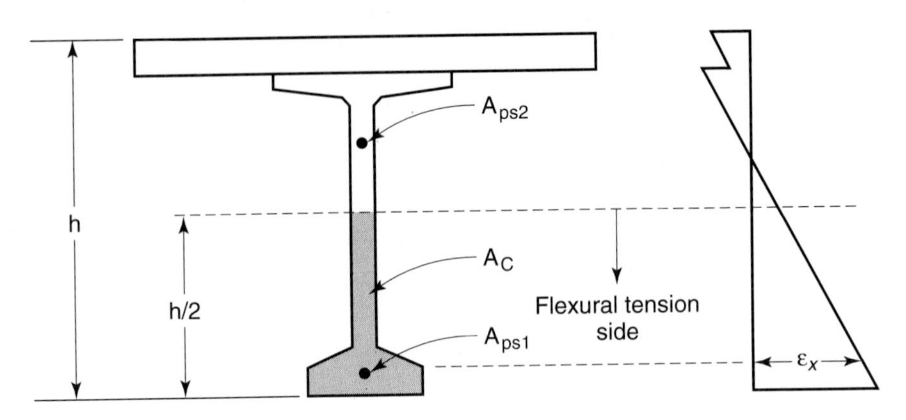

Figure 12.10 Strain distribution in prestressed flanged section

The longitudinal reinforcement should be so proportioned that each beam section has to satisfy the following expression:

$$A_s f_s + A_{ps} f_{ps} \geq \frac{M_u}{d_v \phi} + 0.5 \frac{N_u}{\phi} + \left(\frac{V_u}{\phi} + 0.5 V_s + V_p\right) \cot\theta \tag{12.29}$$

From the forgoing AASHTO expressions, the variable β is an essential determinant for evaluating the nominal shear resistance V_c, as in Equation 12.21. The chart for selecting β from Figure 12.9 based on the Modified Compression Field Theory seems to be insensitive for ratios (v/f'_c) in excess of 0.125 when the strain is less than 0.005 in./in. Hsu's discussion in Ref. 12.21 points to this difficulty, partly arising from assigning a numerical value to the crack shear stress, v_{ci}, namely, the ability of the crack interface to transmit a shear stress value dependant on the crack width, w, in the following expression:

$$v_{ci} \leq \frac{2.16\sqrt{f'_c}}{0.3 + \dfrac{24w}{a + 0.63}} \text{ psi, } w(\text{in.}) \qquad v_{ci} \leq \frac{0.18\sqrt{f'_c}}{0.3 + \dfrac{24w}{a + 16}} \, MPa, \text{ w(mm)}$$

Hsu's work (Ref. 12.21, 12.22) proposes using $v_{ci} = 0$ in order to maintain equilibrium and compatibility. Also, the crack angle θ in the V_s term of Equation 12.25 is the angle between the longitudinal steel and the principal compression stress (strain) of concrete. As such, the shear stress along the principal axis is zero. This discussion also applies to the LRFD provision for the case of combined shear and torsion. Future AASHTO modifications might become necessary in order to rectify the discrepancy.

12.2.4.3 Maximum Spacing of Web Reinforcement. The maximum allowable spacing, s, of the web reinforcement is the smaller of

$$s \leq 0.75 \text{ h} \quad \text{or} \quad 24 \text{ in.}$$

If $V_s > 4\sqrt{f'_c}\, b_w d$, the maximum allowable spacing is reduced by 50 percent.

12.5 HORIZONTAL INTERFACE SHEAR

The principle of horizontal interface shear both at service and ultimate load levels are fully discussed in Chapter 5 Section 5.7, including illustrative examples in accordance with ACI 318 and PCI requirements. AASHTO Standard specifications requirements for the nominal horizontal shear strength, V_{nh}, are similar to those of ACI when no dowel reinforcement is used, namely, the maximum allowable stress is 80 psi. They differ when minimum dowel reinforcement is used in that the maximum allowable horizontal shear stress is 350 psi instead of the 500 psi allowed by the ACI.

Extensive investigations and tests by the author (Ref. 12.13) have shown that these are indeed very low allowable stresses. These tests demonstrate that even in early strength under sub-freezing temperature conditions, it is possible to obtain a strength at ultimate load in excess of 1200 psi (8.3 MPa) using vertical dowel reinforcement.

The standard AASHTO requirements are as follows (Ref. 12.2):

(a) When no vertical ties are provided:

$$V_{nh} = 80 b_v d \quad (12.30a)$$

(b) When minimum vertical ties are provided:

$$V_{nh} = 350 b_v d \quad (12.30b)$$

(c) Required area of ties, A_{vh}, exceeds the minimum area:

$$V_{nh} = 330 b_v d + 0.40 A_{vh} f_y \frac{d_p}{s} \quad (12.30c)$$

where, factored vertical shear $V_u = \phi\, V_{nh}$

V_{nh} = nominal horizontal shear strength

ϕ = 0.90

minimum $A_{vh} = 50 b_v s / f_y$

b_v = width of cross-section at the contact surface being analyzed for horizontal shear

b_p = distance from extreme compression fibers to centroid of prestressing steel, but not to be taken less than $0.80h$

s = maximum spacing of the dowels, but not to exceed four times the least-web width of the support element, nor 24 in.

The LRFD specifications do not give guidance for computing the horizontal shear V_{nh}. The following expression can be used:

$$v_{uh} = \frac{V_u}{b_v d_v} \quad (12.31)$$

where

u_{nh} = horizontal factored shear stress

V_u = factored vertical shear

d_v = distance between resultants of tensile and compressive forces = $(d - a/2)$

b_v = interface width

LRFD specifies that the nominal shear resistance of the interface surface, V_n, be computed using the following expression:

$$V_n = c A_{cv} + \mu [A_v f_y + P_c] \quad (12.32)$$

and that

$$V_{uh} A_{cv} \leq \phi V_n \quad (12.33)$$

where

c = cohesion factor
μ = friction factor
A_{cv} = interface area of concrete engaged in shear transfer
A_{vf} = area of shear reinforcement crossing the shear plane within area A_c
P_c = permanent net compressive force normal to the shear plane (may be conservatively neglected)
f_y = yield strength of dowel reinforcement.

Typically, the top surface of the precast element is intentionally roughened to an amplitude of $\frac{1}{4}$ in. as discussed in section 5.7. Hence, for normal weight concrete, LRFD recommends simplifying equations 12.32 and 12.33 as follows:

$$v_{uh} \leq \phi\left(0.1 + \frac{A_{vf}}{A_{cv}}\right) \tag{12.34}$$

where the minimum

$$A_{vf} = \frac{(0.05b_v s)}{f_y} \tag{12.35}$$

and the nominal shear resistance is to be taken as the lesser of

$$V_n \leq 0.20f'_c A_{cv} \tag{12.36a}$$

or

$$V_n = 0.80A_{cv} \tag{12.36b}$$

The cohesion factor c and the friction factor μ in equation 12.32 have the following values for the particular conditions of the interacting surfaces:

(a) Monolithically placed concrete:

$$c = 145 \text{ psi} \qquad \mu = 1.4\,\lambda$$

(b) Concrete placed against clean, hardened concrete with surface intentionally roughened

$$c = 100 \text{ psi} \qquad \mu = 1.0\,\lambda$$

(c) Concrete placed against hardened concrete clean and free of laitance but not intentionally roughened

$$c = 75 \text{ psi} \qquad \mu = 0.6\,\lambda$$

(d) Concrete anchored to as-rolled structural steel by headed studs or by reinforcing bars, where all steel in contact with the concrete is clean and free of paint

$$c = 25 \text{ psi} \qquad \mu = 0.7\,\lambda$$

where λ = 1.0 for normal-density concrete
= 0.85 for sand-low-density concrete
= 0.75 all other low-density concrete.

While the LRFD AASHTO specifications require that minimum reinforcement is to be provided regardless of the stress level at the interface, designers may choose to limit this reinforcement to cases in which V_{uh}/ϕ is greater than 100 psi (0.7 MPa). Doing so would be consistent with the ACI 318 code and the standard AASHTO specifications.

12.5.1 Maximum Spacing of Dowel Reinforcement

The maximum allowable spacing of the dowels is:

(i) If $V_u < 0.1 f'_c b_v d_v$, maximum $s \leq 0.8 d_v \leq 24$ in.
(ii) If $V_u > 0.1 f'_c b_v d_v$, maximum $s \leq 0.4 d_v \leq 12$ in.

12.6 COMBINED SHEAR AND TORSION

The discussion in Section 12.4.1 on the modified compression field theory in conjunction with section 5.17.3 give an ample treatment of the strains, shear forces and the resisting diagonal compression struts. Figures 5.38, 5.39 and 5.40 illustrate the deformed shape of the critical section when subjected to torsional moments. The shear stresses due to torsion and shear are assumed in this hypothesis to add on one side of the section and counteract on the opposite side. The transverse closed tie reinforcement is designed for the side in which the combined shear and torsional effects are additive.

The external loading which causes the highest torsional moment is not the same as the loading that causes the highest shear at the critical section. The tendency by the designer is to combine the highest value of torsion and the highest value of shear in the design of the web reinforcement. This is, naturally, conservative. It is possible to utilize the fact that the two loads are different and thus design the transverse reinforcement for the highest torsion and its concurrent shear or the highest shear and its concurrent torsion, whichever leads to a higher resistance capacity. The LRFD uses the same nominal torsional resisting moment as the ACI:

$$T_n = \frac{2A_0 A_t f_y \cot\theta}{s} \tag{12.37}$$

where

A_o = cross-section area enclosed by the shear flow path, including are of holes
A_t = area of one leg of the enclosed transverse tension reinforcement
θ = variable angle of crack chosen by trial and adjustment using the chart in Figure 12.9 or from the AASHTO LRFD tables (Ref. 12.3)

In order to determine the value of θ, the strain, ϵ_x, in the tensile reinforcement is obtained from equation 12.27, except that V_u should be replaced

$$V_u = \sqrt{V_u^2 + \left(\frac{P_h T_u}{2A_0}\right)^2} \tag{12.38}$$

The required amount of transverse reinforcement for shear is obtained from equations 12.21(a) in conjunction with equations 12.23(a) and 12.25, namely,

$$V_n = \beta\sqrt{f'_c}\, b_v d_v + \frac{A_v f_y d_v \cot\theta}{s} + V_p \tag{12.39}$$

so that for shear in lb units and stress in psi,

$$\frac{A_v}{s} = \frac{V_n - (\beta\sqrt{f'_c}\, b_v d_v + V_p)}{f_y d_v \cot\theta} \tag{12.40a}$$

Photo 12.2 PCI bulb tee precast prestressed bridge deck beams (Courtesy Ms. Monica Schultese, Mid Atlantic Precast Concrete Association).

If ksi units are used, multiply β by 0.0316 and for torsion, from equation 12.31,

$$\frac{A_t}{s} = \frac{T_n}{2A_0 f_y \cot\theta} \tag{12.40b}$$

the total area of web reinforcement would be:

$$\frac{A_{vt}}{s} = \frac{A_v}{s} + 2\frac{A_t}{s} \tag{12.40c}$$

The angle θ is obtained from Figure 12.9 using shear, v, as follows:

(a) Box sections:

$$V = \frac{V_u - \phi V_p}{\phi b_v d_v} + \frac{T_p p_h}{\phi A_{oh}^2} \tag{12.41}$$

(b) Other Sections:

$$V = \sqrt{\left(\frac{V_u - \phi V_p}{\phi b_v d_v}\right)^2 + \left(\frac{T_u P_h}{\phi A_{oh}^2}\right)^2} \tag{12.42}$$

where

p_h = perimeter of the center line of the enclosed transverse torsion reinforcement

A_{oh} = area enclosed by the center line of the outermost closed torsional reinforcement

A_o = gross area enclosed by the shear flow path (see Figure 5.45 for graphical representation of A_o and A_{oh} where $A_o \approx 0.85 A_{oh}$)

T_u = factored torsional moment

ϕ = resistance factor

The value of β in equation 12.39 for determining the shear capacity, V_c, of the plain concrete in the web is obtained from the chart in Figure 12.9. In order to avoid yielding of the longitudinal reinforcement, a check has to be made that the flexural reinforcement on the tension face is so proportioned as to satisfy the following condition:

$$\phi(A_s f_y + A_{ps} f_{ps}) \geq \frac{M_u}{d_v} + 0.5N_u + \cot\theta\sqrt{(V_u + 0.5V_s - V_p)^2 + \left(\frac{0.45T_u p_o}{2A_0}\right)^2} \qquad (12.43)$$

where p_0 = Perimeter of the shear flow path
N_u = Applied axial force, taken as positive if compressive

12.7 AASHTO-LRFD FLEXURAL-STRENGTH DESIGN SPECIFICATIONS VS. ACI CODE PROVISIONS

There are fundamental differences in the approaches of the AASHTO-LRFD flexural-strength design specifications and the ACI-318 code provisions. The LRFD approach is based on strain limit values as discussed in Section 12.3 and controlled by the ratio of the neutral axis depth, c, to the effective depth, d_e (Ref. 12.14–12.16). This approach is variably termed as a unified approach since it is applicable to reinforced, prestressed, and partially prestressed concrete ultimate limit state design. The ACI-318 code strength provisions have been applied for determining the ultimate design strength f_{ps} in several examples in other sections of the book such as sections 4.9 and 4.10. They have to be applied in the design of fully and partially prestressed concrete members in building structures. The current AASHTO standard specifications (Ref. 12.2) for proportioning

Photo 12.3 Segmental bridge launching, (Courtesy Monica Schultese, MidAtlantic Precast Concrete Association)

Table 12.8(a) LRFD and ACI Provisions for Ultimate Strength Flexural Design

ACI Code	AASHTO-LRFD
Notation	
d to nonprestressed reinforcement d_p to prestressed reinforcement d' to compression steel $\rho_p, \rho, \rho' \ldots\ldots \rho = \dfrac{A_s}{bd}$ $\omega = \rho \dfrac{f_y}{f'_c} = \dfrac{A_s f_y}{bdf'_c}$ $\omega_p = \rho_p \dfrac{f_{ps}}{f'_c} = \dfrac{A_{ps} f_{ps}}{bd_p f'_c}$ $\omega_p + \dfrac{d}{d_p}(\omega - \omega')$	d_s to nonprestressed reinforcement $d_e = \dfrac{A_{ps} f_{ps} d_p + A_s f_y d_s}{A_{ps} f_{ps} + A_s f_y}$
Maximum Flexural Reinforcement	
Reinforced Concrete $20\left(1 - \dfrac{\rho - \rho'}{\rho_b}\right)$ In % provided: $(\rho - \rho') \leq 0.50\, \rho_b$ Prestressed and Partially Prestressed Concrete $20\left[1 - \dfrac{\omega_p + \dfrac{d}{d_p}(\omega - \omega')}{0.36\beta_1}\right]$ provided: $\omega_p + \dfrac{d}{d_p}(\omega - \omega') \leq 0.24\beta_1$ for rectangular sections and: $\omega_{pw} + \dfrac{d}{d_p}(\omega_w - \omega_w') < 0.24\beta_1$ for T-section behavior	All Cases - RC, PC, PPC Rectangular or T Section: $20\left(1 - 2.36\dfrac{c}{d_e}\right)$ in % provided: $\dfrac{c}{d_e} \leq 0.28$ 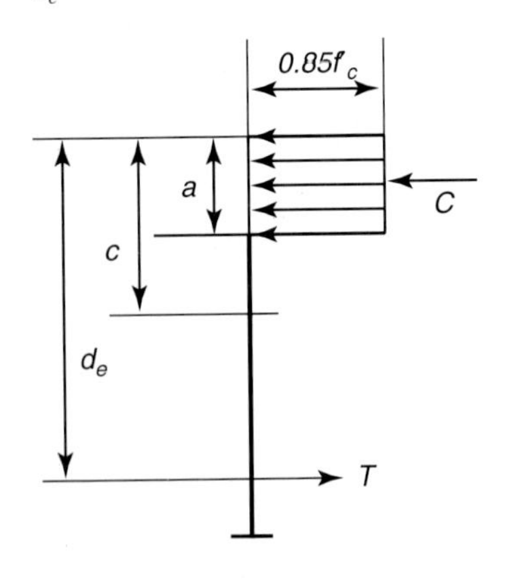

Table 12.8 *Continued from (a)*

ACI Code	AASHTO-LRFD
Minimum Reinforcement	
Reinforced Concrete $\rho \geq \rho_{\min} \geq \dfrac{200}{f_y} \geq \dfrac{3.5\sqrt{f'_c}}{f_y}$ (For T-sections, ρ is based on web only) Prestressed and Partially Prestressed Concrete: $\phi P_n \geq 1.2 P_{cr}$	All Cases: $\phi M_n \geq 1.2 M_{cr}$ $\phi M_n \geq 1.33 M_u$ Particular result for reinforced concrete: $\rho \geq \rho_{\min} = \dfrac{0.03 f'_c}{fy}$ (For T-sections, ρ is based on web only)
Stress in Bonded Prestressing Steel at Ultimate Resistance in Bending	
Prestressed and Partially Prestressed Concrete-Bonded Tendons $f_{ps} = f_{pu}\left[1 - \dfrac{\gamma_p}{\beta_1}\left\{\rho_p \dfrac{f_{pu}}{f'_c} + \dfrac{d}{d_p}(\omega - \omega')\right\}\right]$ where: $\left[\rho_p \dfrac{f_{pu}}{f'_c} + \dfrac{d}{d_p}(\omega - \omega')\right] \geq 0.17$ $d' < 0.15\, d_p$ $\gamma_p = 0.28$ for $f_{py} \geq 0.90 f_{pu}$ (Low Lax) 0.40 for $f_{py} \geq 0.85 f_{pu}$ [normal] 0.55 for $f_{py} \geq 0.80 f_{pu}$ (bars) $\beta_1 = 0.85$ for $f'_c \leq 4$ ksi 0.65 for $f'_c \geq 8$ ksi $0.85 - 0.05\,(f'_c - 4)$ for $4 \leq f'_c \leq 8$ ksi	PC and PPC Bonded Tendons: $f_{ps} = f_{pu}\left(1 - k\dfrac{c}{d_p}\right)$ $k = 2\left(1.04 - \dfrac{f_{py}}{f_{pu}}\right)$

prestressed concrete members in flexure generally follow the ACI code provisions. The LRFD alternative, which is a rational design approach, requires applying the strain limits unified procedure. It is useful to give a comparison summary showing the differences between the expressions specified in these two approaches as shown in Tables 12.8 (a), (b) adapted from Ref. 12.15.

12.8 STEP-BY-STEP DESIGN PROCEDURE (LRFD)

The following is a summary of a recommended sequence of design steps:

1. Determine whether or not partial prestressing is to be chosen.

2. Select the bending moments and shear forces from Table 12.2, Section 12.7.

3. Follow the step sequence for flexural design of the member outlined in steps 2 through 10 of Section 4.13 in Chapter 4 and the flowchart of Figure 12.11 when using the LRFD method for flexure. Generally, $d_v = (d_e - a/2)$.

4. Determine the factored shear force V_u due to all applied loads at the critical section located at a distance d_v or $0.5\ d_v \cot\theta$ from the face of the support, whichever is larger, where

$$d_e = \text{effective depth as shown in Table 12.8(a) and (b)}$$
$$= d_p \text{ if no mild steel is used.}$$

5. Compute the tendon shear component V_p. The factored shear stress is:

$$v = \frac{V_u - \phi V_p}{\phi b_v d_v}$$

The nominal available shear stress $v_c = v/h$.

6. Compute the quantity v/f'_c and assume a value of θ. A good initial assumption for prestressed beams is $\theta = 25°$

7. Compute the strain in the tensile reinforcement in order to enter Figure 12.9 to obtain a trial value of θ and β.

$$\epsilon_x = \frac{\dfrac{M_u}{d_v} + 0.5N_u + 0.5V_u \cot\theta - A_{ps}f_{po}}{E_s A_s + E_{ps}A_{ps}} \leq 0.002$$

where $f_{po} = f_{pe} + f_{ce}E_{ps}/E_c$. It represents the stress in the prestressing strands when the stress in the surrounding concrete is zero, namely, the stress in the reinforcement at decompression.

f_{ce} = compressive stress in the concrete at the centroid of the tension reinforcement considering both rpestress after losses and all permanent loads.

If the strain in the tensile reinforcement is negative, ϵ_x should be multiplied by the factor F_ϵ:

$$F_\epsilon = \frac{E_s A_s + E_{ps} A_{ps}}{E_c A_c + E_s A_s + E_p A_{ps}}$$

A_c = area of the concrete in the flexural tension side of the member.

8. Enter LRFD Figure 12.9 again, with the value of v/f'_c and strain ϵ_x if the strut angle θ is not close to the one assumed in the first trial, in order to obtain an adjusted value of β. Otherwise, compute V_c from Equation 12.23, namely $V_c = \beta\sqrt{f'_c}\, b_v d_d$ (lb) or $V_c = 0.0316\beta\sqrt{f'_c}\, b_v d_v$ (kip) using the β value obtained from the chart in Figure 12.9.

9. Compute V_s for the web reinforcement after the value of V_c has been determined. Find the corresponding shear reinforcement spacing from:

$$A_v = 0.036\sqrt{f'_c}\,\frac{b_v s}{f_y}$$

10. In regions of high shear stresses, ensure that the amount and development of the longitudinal reinforcement A_s and A_{ps} should satisfy the following expression:

$$A_s f_y + A_{ps} f_{ps} \geq \left[\frac{M_u}{d_v \phi} + 0.5\frac{N_u}{\phi} + \left(\frac{V_u}{\phi} - 0.5V_s - V_p\right)\cot\theta\right]$$

It is recommended that this check be made at the face of the bearing which lies within the transfer length of the strands where the effective prestressing force is not fully developed.

11. When torsion exists combined with shear and flexure, the following steps need to be followed:

$$\text{nominal torsion } T_n = \frac{2A_0 A_t f_y \cot\theta}{s}$$

Strain in tensile reinforcement:

$$\epsilon_x = \frac{\dfrac{M_u}{d_v} + 0.5N_u + 0.5\cot\theta\sqrt{V_u^2 + \left(\dfrac{P_h T_u}{2A_0}\right)^2} - A_{ps} f_{po}}{E_s A_s + E_{ps} A_{ps}} \leq .002$$

where $f_{po} = f_{pe} + \dfrac{f_{ce} E_{ps}}{E_c}$, or conservatively take $f_{po} = f_{pe}$

Nominal shear resistance:

$$V_n = V_c + V_s + V_p = \beta\sqrt{f'_c}\, b_v d_v + \frac{A_v f_y d_v \cot\theta}{s} + V_p$$

where $d_v = (d_p - a/2)$

Shear reinforcement:

$$\frac{A_v}{s} = \frac{V_n - 0.0316\beta\sqrt{f'_c}\, b_v d_v + V_p}{f_y d_v \cot\theta}$$

Forces are in kips and the stresses in ksi. For using lb and psi units, remove factor 0.0316.

Torsion reinforcement:

$$\frac{A_t}{s} = \frac{T_n}{2A_0 f_y \cot\theta}$$

Total web closed ties reinforcement:

$$\frac{A_{vt}}{s} = \frac{A_v}{s} + 2\frac{A_t}{s}$$

Shear stress v for obtaining angle θ:

(a) Box sections:

$$v = \frac{V_u - \phi V_p}{\phi\, b_v d_v} + \frac{TP_h}{\phi\, A_0 h^2}$$

(b) Other sections

$$v = \sqrt{\left(\frac{V_u - \phi V_p}{\phi b_v d_v}\right)^2 + \left(\frac{TP_h}{\phi\, A_{oh}^2}\right)^2}$$

For avoiding yield of the longitudinal tensile reinforcement:

$$\phi(A_s f_s + A_{ps} f_{ps}) \geq \frac{M_u}{d_v} + 0.5N_u + \cot\theta\sqrt{(V_u + 0.5V_s - V_p)^2 + \left(\frac{0.45T_u P_h}{2A_o}\right)^2}$$

12. Check the horizontal interface shear:

$$v_n A_{cv} \leq \phi\, V_n$$

where

$$V_n = cA_{cv} + \mu\,(A_{vf} f_y)$$

$$v_{uh} \leq \phi\left(0.1 + \frac{A_{vf}}{A_{cv}}\right)$$

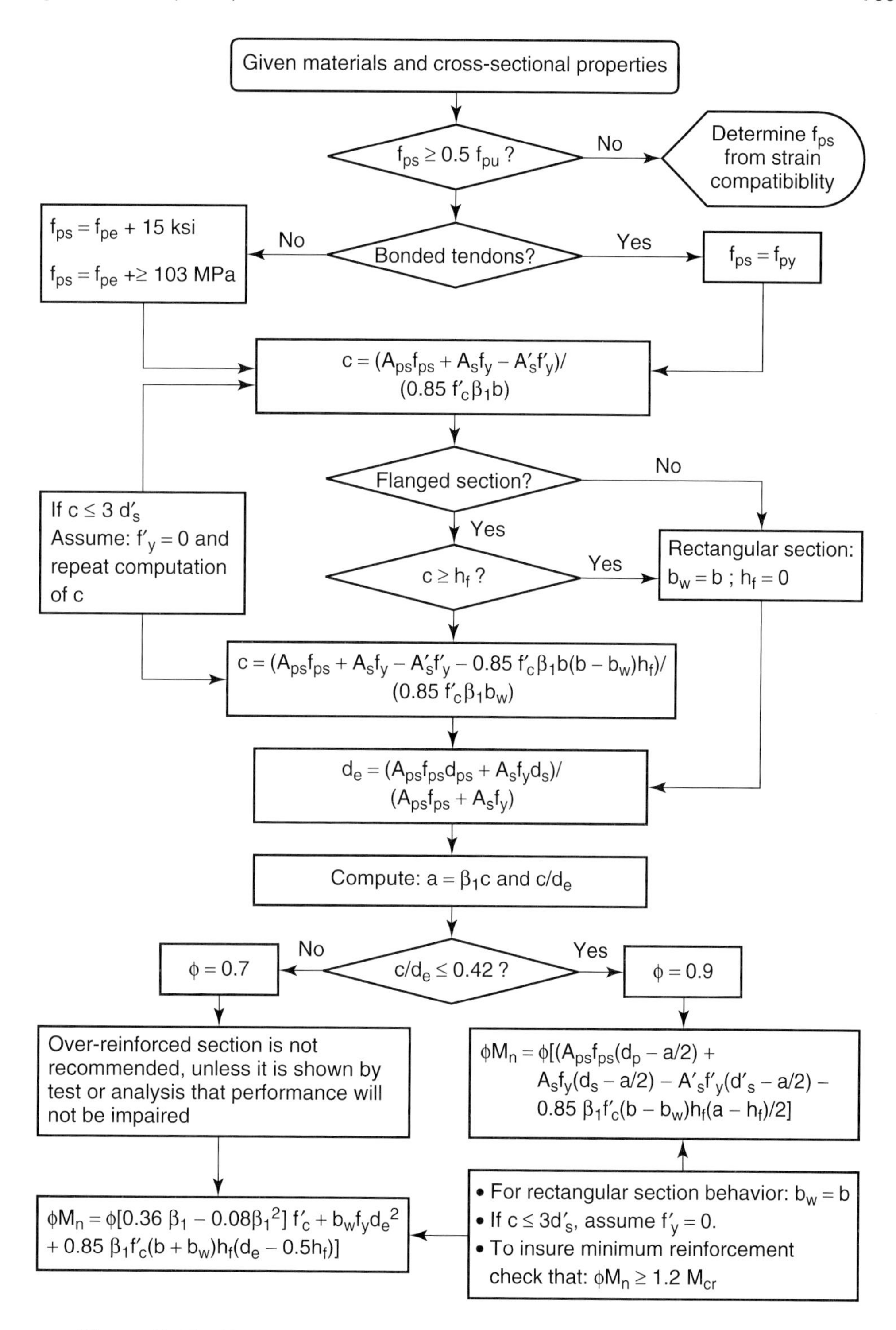

Figure 12.11 Nominal moment strength of prestressed section with bonded and unbonded reinforcement (Ref. 12.15).

where $$A_{vf} = \frac{0.05b_v s}{f_y}$$

Take the nominal shear resistance as the lesser of

$$V_n \leq 0.20f'_c A_{cv}$$

or

$$V_n \leq 0.80A_{cv}$$

c = cohesion factor
μ = friction factor
A_{cr} = concrete interface area = $b_v l_v$
A_{vf} = area of shear reinforcement crossing the shear plane within area A_{cv}
ϕ = strength reduction factor = 0.90.

Limit A_{vf} to cases in which v_{uh}/ϕ is greater than 100 psi.
Figure 12.11 gives flowchart (Ref. 12.15) for steps to be followed in determining the nominal moment strength, namely, the nominal bending resistance for bonded and unbonded tendons.

13. Maximum allowable spacing of web shear reinforcement:

$$s \leq 0.75 \text{ h} \quad \leq 24 \text{ in.}$$

If $V_s > 4\sqrt{f'_c}\, b_w d$, reduce spacing by 50%
For Dowel reinforcement spacing:
If $V_u < 0.1\, f'_c\, b_v d_v$, $s \leq 0.8\, d_v \leq 24$ in.
If $V_u > 0.1\, f'_c\, b_v d_v$, $s \leq 0.4\, d_v \leq 12$ in.,
where b_v = width of contact for horizontal shear

12.9 LRFD DESIGN OF BULB-TEE BRIDGE DECK

Example 12.1

Design for flexure a 120 ft (36.6 m) simply supported AASHTO-PCI bulb-tee composite bridge deck with no skews (adapted from Ref. 12.11). The superstructure is composed of six pretensioned beams at 9′–0″ (2.74 m) on centers as shown in Figure 12.12. The bridge has an 8-in. (203-mm) situ-cast concrete deck with the top ½-in to be considered as wearing surface. The design live load is the HL-93 AASHTO-LRFD fatigue loading.

Assume the bridge is to be located in a low seismicity zone.

Given:

Maximum allowable stresses:

Deck	f'_c = 4000 psi, normal weight
	$f_c = 0.60\, f'_c - 2400$ psi
Bulb-tee	f'_c = 6500 psi
	f'_{ci} = 5500 psi

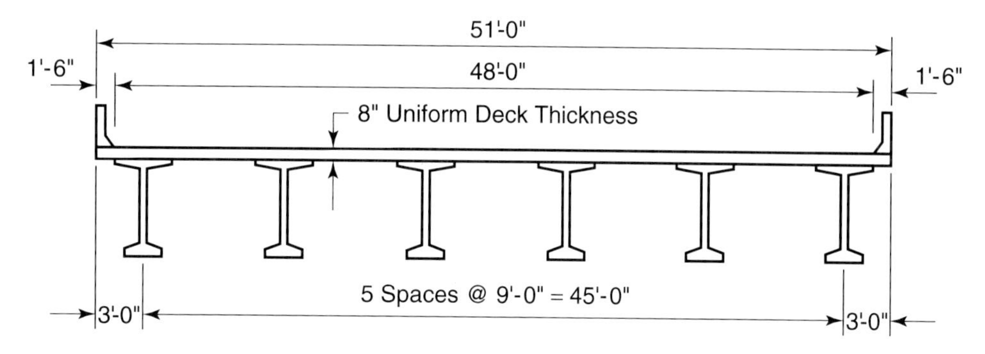

Figure 12.12 Bulb-tee Bridge Deck Cross Section in Example 12.1 (Ref. 12.11)

$$f_c = 0.60\, f'_c = 3900 \text{ psi, Service III}$$
$$f_c = 0.45\, f'_c = 2925 \text{ psi, Service I}$$
$$f_{ci} = 0.60\, f'_c = 3480 \text{ psi}$$
$$f_t = 6\sqrt{f'_c} = 484 \text{ psi}$$
$$f_{pu} = 270{,}000 \text{ psi}$$
$$f_{py} = 0.90\, f_{pu} = 243{,}000 \text{ psi}$$
$$f_{pi} = 0.75\, f_{pu} = 202{,}500 \text{ psi}$$
$$f_y = 60{,}000 \text{ psi}$$
$$E_{ps} = 28.5 \times 10^6 \text{ psi}$$
$$E_s = 29.0 \times 10^6 \text{ psi.}$$

Section Properties

$$A_c = 767 \text{ in.}^2$$
$$h = 72 \text{ in.}$$
$$I_c = 545{,}894 \text{ in.}^4$$
$$c_b = 36.60 \text{ in.}$$
$$c_t = 35.40 \text{ in.}$$
$$S_b = 14{,}915 \text{ in.}^3$$
$$S^t = 15{,}421 \text{ in.}^3$$
$$r^2 = \frac{I_c}{A_c} = \frac{545{,}894}{767} = 712 \text{ in.}^2$$
$$W_D = 779 \text{ plf.}$$

Solution:

1. ***Transformed Deck slab controlling width***
   Compute the transformed flange width:

$$E_{cs} = 33w^{1.5}\sqrt{f'_c} = 33 \times (1.5)^{1.5}\sqrt{4000} = 3830 \text{ psi}$$

$$E_{ci} \text{ at transfer} = 33(1.5)^{1.5}\sqrt{5500} = 4500 \text{ psi}$$

E_{ce} at service $= 33(1.5)^{1.5}\sqrt{6500} = 4890$ *psi*

Effective flange width is the lesser of

(i) $(\frac{1}{4})$ span $= \dfrac{120 \times 12}{4} = 360$ in.

(ii) $12\, h_f +$ greater of web thickness or $\frac{1}{2}$-beam top flange width, $b = 12 \times 7.5 + 0.5 \times 42 = 111$ in.

Photo 12.4 Chesapeake and Delaware Canal cable-stayed bridge on S.R. 1. Length 4650 ft. with two 12-ft. deep precast girders carrying a 112-ft. wide roadway; typical spans are 150 ft. with precast box piers (*Courtesy* Figg Engineering Group, Tallahassee, Florida)

(iii) average spacing between beams = 9 × 12 = 108 in.
hence, controlling flange width = 108 in.

$$\text{Modular ratio } n_s = \frac{E_{cs}}{E_c} = \frac{3870}{4890} = 0.78$$

Transformed width $b_m = n_s b = 0.78 \times 108 = 84$ in.

2. *Properties of composite section*
Disregard as insignificant the contribution of the deck concrete haunch to I'_c which is needed because of the precast element camber.

$A'_c = 1402$ in.^{2}

$h = 80$ in.

$I_{cc} = 1{,}095{,}290$ in.^{4}

$c_{bc} = 54.6$ in. to the bottom fibers

$c_{tc} = 72 - 54.6 = 17.4$ in. – precast

$c_{tsc} = 80 - 54.6 = 25.4$ in. – deck top

$$S_{bc} = \frac{1{,}095{,}290}{54.6} = 20{,}060 \text{ in.}^3$$

$$S_c^t = \frac{1{,}095{,}290}{17.4} = 62{,}950 \text{ in.}^3$$

$$S_c^{ts} = \frac{1,095,290}{25.4 \times 0.78} = 55,284 \text{ in.}^3$$

3. *Bending moments and shear forces*

$$\text{Slab: } W_{SD1} = \frac{8}{12} \times 9 \times 150 = 900 \text{ lb/ft}$$

$$\text{Barrier weight: } W_{SD2} = \frac{2 \text{ barriers (300 lb/ft)}}{6 \text{ beams}} = 100 \text{ lb/ft}$$

$$\text{2 in. future-wearing surface: } W_{SD3} = \frac{2}{12} \times \frac{48 \text{ ft}}{6 \text{ beams}} \times 150 = 200 \text{ lb/ft}$$

Live load (truck load) in LRFD would be based on HL-93 truck fatigue loading.
Clear width from figure 12.12 = 48 ft (14.6 cm)

$$\text{Number of lanes} = \frac{48}{12} = 4 \text{ lanes}$$

(a) *Distribution factor for moment*

For two or more lanes loaded (Ref. 12.3), the distribution factor for bending moment (Table 12.3b)

$$\text{DFM} = 0.075 + \left(\frac{S}{9.5}\right)^{0.6}\left(\frac{S}{L}\right)^{0.2}\left(\frac{K_g}{12t_s^3L}\right)^{0.1}$$

provided that

beam spacing:	$3.5 \le S \le 16$	Actual S = 9.0 ft	O.K.
deck slab:	$4.5 \le T_s \le 12$	Actual T_s = 7.5 in.	O.K.
span:	$20 \le L \le 240$	Actual L = 120 ft	O.K.
no. of beams:	$N_b > 4$	Actual $N_b = 6$	O.K.

e_g = distance between the center of gravity of the beam and the slab

$$= \frac{7.5}{2} + 0.5 + 35.4 = 39.65 \text{ in.}$$

$$n = \frac{E_c}{E_{sc}} = \frac{4890}{3830} = 1.28$$

$$K_g = n(I_c + A_c e_g^2)$$

$$= 1.28\left[545,894 + 767\,(39.65)^2\right] = 2,242,191 \text{ in.}^4$$

hence,

$$\text{DFM} = 0.075 + \left(\frac{9}{9.5}\right)^{0.6}\left(\frac{9}{120}\right)^{0.2}\left[\frac{2,242,191}{12(7.5)^3\,(120)}\right]^{0.1}$$

$$= 0.732 \text{ lanes/beam}$$

For one design lane loaded, from Table 12.3b,

$$\text{DFM} = 0.06 + \left(\frac{S}{14}\right)^{0.4}\left(\frac{S}{L}\right)^{0.3}\left(\frac{K_g}{12t_s^3L}\right)^{0.1}$$

$$= 0.06 + \left(\frac{9}{14}\right)^{0.4}\left(\frac{9}{120}\right)^{0.3}\left[\frac{2,242,191}{12(7.5)^3(120)}\right]^{0.1} = 0.499 \text{ lanes/beam;}$$

consequently, the case of two or more lanes loaded controls so that the DFM = 0.732 lanes per beam.

Fatigue Moments:

The moment is taken for a single design truck having the same axle weight as in all other limit states, but with a constant spacing of 30 ft between the 32-kip axles.

A multiple lane factor of 1.2 for fatigue is used to reduce the controlling DFM factor. From table 12.2a, the load factor is 0.75 and the impact factor (IM) for fatigue = 15%.

Hence, the fatigue truckload bending moment becomes:

$$M_f = \text{(bending moment per lane)}\,(\text{DFM}/1.2)(1 + \text{IM})$$

$$\text{or}\quad M_f = \text{(bending moment per lane)}\left(\frac{0.499}{1.2}\right)(1 + 0.15)$$

$$= \text{(bending moment per lane)}\quad (0.415)(1 + 0.15)$$

$$= (0.478)\,\text{(bending moment per lane)}$$

(b) *Distribution factor for shear*

From Table 12.3(a),

For two or more lanes loaded

$$\text{DFV} = 0.2 + \left(\frac{S}{12}\right) - \left(\frac{S}{36}\right)^2$$

provided that:

beam spacing:	$3.5 \le S \le 16$	Actual $S = 9.0$ ft	O.K.
deck slab:	$4.5 \le T_s \le 12$	Actual $T_s = 7.5$ in.	O.K.
span:	$20 \le L \le 240$	Actual $L = 120$ ft	O.K.
$10{,}000 \le K_g \le 7{,}000{,}000$		Actual $K_g = 2{,}242{,}191$ in.^{4}	O.K.

$$\text{hence,}\quad \text{DFV} = 0.2 + \left(\frac{9}{12}\right) - \left(\frac{9}{36}\right)^2 = 0.887 \text{ lanes/beam.}$$

For one design lane loaded (Table 12.3a)

$$\text{DFV} = 0.36 + \left(\frac{S}{25.0}\right) = 0.36\left(\frac{9.0}{25.0}\right) = 0.720 \text{ lanes/beam;}$$

consequently, the case of two or more lanes loaded controls and DFV = 0.887 lanes per beam.

4. *Load Combinations*

Total factored load, $Q = \eta\Sigma\gamma_i q_i$

where η = a factor relating to ductility, redundancy and operational importance.
γ_i = load factors
q_i = special loads;

use $\eta = 1.0$ for all practical purposes in this example.

Investigate all the load combinations from table 12.2(a) and (b). The cases that control are as follows:

(a) Service I for compressive stresses in the prestressed concrete components: $Q = 1.0\,(\text{DC} + \text{DW}) + 1.0\,(\text{LL} + \text{IM})$

(b) Service III for tensile stresses in the prestressed concrete components: $Q = 1.0\,(\text{DC} + \text{DW}) + 0.8\,(\text{LL} + \text{IM})$

(c) Strength I for ultimate strength:

$$\text{Maximum}\quad Q = 1.25\,\text{DC} + 1.50\,\text{DW} + 1.75\,(\text{LL} + \text{IM})$$

$$\text{Minimum}\quad Q = 0.90\,\text{DC} + 0.65\,\text{DW} + 1.75\,(\text{LL} + \text{IM})$$

(d) Fatigue for checking stress range in the strands

$$Q = 0.75\,(\text{LL} + \text{IM})$$

(The fatigue Q is a special load combination for checking the tensile stress range in the strands due to live load and dynamic allowance.)

5. ***Unfactored shear forces and bending moments***

(a) *Truck Loads*

Truck load shear force:

$$V_{LT} = \text{(shear force per lane)(DFV)(I + IM)}$$
$$= \text{(shear force per lane)}(0.887)(1 + 0.33)$$
$$= 1.180 \text{ (shear force per lane) kips.}$$

Truck load bending moment:

$$M_{LT} = \text{(moment per lane)(DFM)(I + IM)}$$
$$= \text{(moment per lane)}(0.732)(1 + 0.33)$$
$$= 0.974 \text{ (moment per lane) ft-kips.}$$
$$LT = \text{Truck live load}$$

(b) *Lane Loads*

For lane loads, no dynamic allowance is applied, hence,

$$V_{LL} = \text{(shear force per lane)(DFV)}$$
$$= \text{(shear force per lane)}(0.884) \text{ kips}$$
$$M_{LL} = \text{(moment per lane)(DFM)}$$
$$= \text{(moment per lane)}(0.732) \text{ ft-kips.}$$

The lane loads from Figure 12.4, the load on this bridge is as follows in Figure 12.13.

6. ***Computation of moments and shears***

(a) *Lane live load ($DFV = 0.884$, $DFM = 0.732$)*

(i) *Support section:*

shear at the left support ($x = 0$) from equation 12.6(a) and Figure 12.12:

$$V_{LL} = \frac{0.64}{2L}(L - x)^2 \text{ (DFV)}$$

$$= \frac{0.64}{2 \times 120}(120)^2 (0.887) = 34.1 \text{ kips}$$

From equation 12.6(b), and DFM = 0.732

$$M_{LL} = \frac{0.64(x)(L - x)}{2} \text{(DFM)} = 0 \text{ ft-kip}$$

(ii) *Section at 24 ft from support:*

As an example, find V_{LL} and M_{LL} at $x = 24$ ft from the left support.

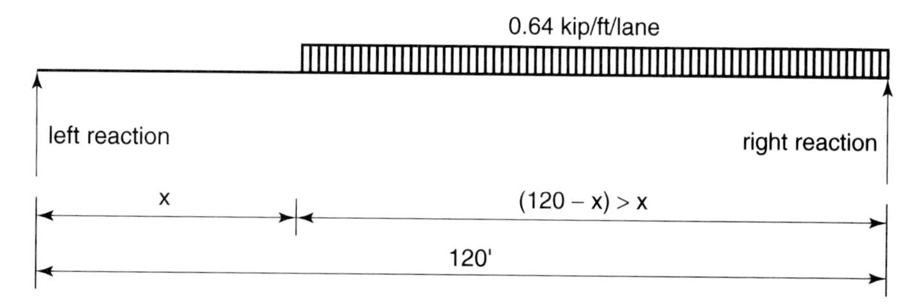

Figure 12.13 Truck load per lane

$$V_{LL} = \frac{0.64}{2 \times 120}(120 - 24)^2(0.887) = 21.8 \text{ kips}$$

$$M_{LL} = \frac{0.64(24)(120 - 24)}{2}(0.732) = 539.7 \text{ ft-kip}$$

(b) *Truck live load (DFV = 1.180, DFM = 0.974)*
Here, the impact factor IM = 33% has to be included, hence, larger DFV and DFM values.

(i) *Support sections:*
From Table 12.4,

$$V_{LT} = \frac{72[(L - x) - 9.33]}{L}(\text{DFV})$$

$$= \frac{72[(120 - 0.0) - 9.33]}{120}(1.180) = 78.1 \text{ kips}$$

From Table 12.5,

$$M_{LT} = \frac{72(x)[(L - x) - 9.33]}{L}(\text{DFM})$$

$$= 0 \text{ ft-kip for the support moment.}$$

(ii) *Section at 24 ft from support:*

$$V_{LT} = \frac{72[(120 - 24) - 9.33]}{120}(1.180) = 61.4 \text{ kips}$$

$$M_{LT} = \frac{72(24)[(120 - 24) - 9.33]}{120}(0.974) = 1215.0 \text{ ft-kip}$$

(c) *Fatigue moment at 24 ft (DFF = 0.478)*
From Table 12.7,

$$M_f = \frac{72(x)[(L - x) - 18.22]}{L}(\text{DFF})$$

From before, DFF = 0.478
hence,

$$M_f = \frac{72(24)[(120 - 24) - 18.22]}{120}(0.478) = 535.8 \text{ ft-kip}$$

(d) *Shears and moments due to dead loads:*
The loads to be considered are beam weight *(W_D)* plus deck slab and haunches *(W_{SD1})*, and future wearing surface *(W_{SD3})*.
The beam is simply supported, hence, the shear and moment at any cross section along the span are:

$$V_x = W_D(0.5L - x)$$

$$M_x = 0.5W_D x(L - x)$$

As an example, consider a section at 24 ft from the left support and compute the shear and moment due to self-weight $W_D = 0.799$ Kip/ft:

$$V_x = 0.799(0.5 \times 120 - 24) = 28.8 \text{ kips}$$

$$M_x = 0.5 \times 0.799 \times 24(120 - 24) = 920.4 \text{ ft-kip.}$$

Tables 12.9 and 12.10 (Ref. 12.11) list the forces and moments required for the design of the interior beam elements. It should be noted that long-hand computa-

Table 12.9 LRFD Service Shear and Moment Due to Dead Load

Distance X	Section X/L	Beam Weight W_D		(Slab + Haunch) Weight W_{SD1}		Barrier Weight W_{DS2}		Wearing Surface W_{SD3}	
		Shear	Moment M_g	Shear	Moment M_s	Shear	Moment M_b	Shear	Moment M_{ws}
ft		kips	ft-kips	kips	ft-kips	kips	ft-kips	kips	ft-kips
0	0.0	47.9	0.0	55.3	0.0	6.0	0.0	12.0	0.0
6.00*	0.05	43.1	274.3	49.8	315.3	5.4	34.2	10.8	68.4
12	0.1	38.4	517.8	44.3	597.5	4.8	64.8	9.6	129.6
24	0.2	28.8	920.4	33.2	1,062.1	3.6	115.2	7.2	230.4
36	0.3	19.2	1,208.1	22.1	1,394.1	2.4	151.2	4.8	302.4
48+	0.4	9.6	1,380.7	11.1	1,593.2	1.2	172.8	2.4	345.6
60	0.5	0.0	1,438.2	0.0	1,659.6	0.0	180.0	0.0	360.0

*Critical section for shear
+ Harp point

tions to develop such a table are time consuming. Computer programs developed by several state DOTs are available, some on the internet, such as the Washington State DOT Program.

7. *Design of the Bulb-tee prestressed interior beam*

(1) *Selection of Prestressing Strands*

For Service-III load combination, bottom fiber stress f_b is:

$$f_b = \frac{M_D + M_S}{S_b} + \frac{M_b + M_{WS} + 0.8(M_{LT} + M_{LL})}{S_{be}}$$

where M_D = unfactored self-weight moment, ft-kip
M_S = unfactored moment due to slab and haunch weight, ft-kip

Table 12.10 LRFD Service Shear and Moment Due to Truck and Lane Loads

Distance X	Section X/L	Truck Load with Impact W_{LT+I}		Lane Load W_{LL}		Fatigue Truck with Impact W_f
		Shear V_{LT}	Moment M_{LT}	Shear V_{LL}	Moment M_{LL}	Moment M_f
ft		kips	ft-kips	kips	ft-kips	ft-kips
0	0.0	78.1	0.0	33.9	0.0	0.0
6.00*	0.05	73.8	367.8	30.6	160.2	165.0
12	0.1	69.6	691.6	27.5	303.6	309.2
24	0.2	61.1	1,215.0	21.7	539.7	535.8
36	0.3	52.7	1,570.2	16.6	708.3	692.7
48+	0.4	44.2	1,778.9	12.2	809.5	776.2
60	0.5	35.7	1,830.2	8.5	843.3	776.9

*Critical section for shear
+ Harp point

M_b = unfactored barrier moment, ft-kip
M_{WS} = unfactored future wearing surface moment, ft-kip
M_{LT} = unfactored truck load moment, ft-kip
M_{LL} = unfactored lane load moment, ft-kip

From before, $S_b = 14{,}915$ in.^{3}

$S_{bc} = 20{,}090$ in.^{3}

From Tables 12.9 and 12.10,
Midspan stresses at bottom fibers at service:

$$f_{bc} = \frac{(1438.2 + 1659.6)}{14{,}915}(12) + \frac{(180 + 360) + 0.8(1830.3 + 843.2)}{20{,}090}(12)$$

$$= 2.50 + 1.60 \cong 4.10 \text{ ksi (T)}.$$

The 4.10 ksi (T) will be neutralized by prestressing the beam. Maximum allowable tensile stress:

$$f_t = 6.0\sqrt{f'_c} \text{ psi} = 6\sqrt{6500} = 484 \text{ psi} = 0.484 \text{ ksi}$$

Required prestress compressive stress at the bottom fibers:

$$f_{cb} = (4.1 - 0.48) = 3.62 \text{ ksi}$$

Assume that the distance from the centroid of the prestressing reinforcement and the section bottom fibers = $0.05h$

$$= (0.05)(72) = 3.6 \text{ in; use 4.0 in., hence } e_c = 36.6 - 4.0 = 32.6 \text{ in.}$$

As presented in the examples in Chapter 4,

$$f_{bp} \text{ due to prestress} = \frac{P_e}{A_c} + \frac{P_e \times e_c}{S_b}$$

$$\text{or} \quad f_{bp} = \frac{P_e}{767} + \frac{P_e \times 32.6}{14{,}915} = 3.62 \text{ ksi}$$

to give prestressing force $P_e = 1037$ kips
Assume total prestress loss = 25%

$$P_i = \frac{1037}{1 - 0.25} = 1383 \text{ kips}$$

assume using $\frac{1}{2}$ in.-dia 7-wire 270-K low-relaxation strands ($A_{ps} = 0.153$ in.^{2})

$$\text{Required number of strands} = \frac{1383}{0.153 \times 202.5} = 44.6 \text{ strands.}$$

After two trials and adjustments, 48 strands with the configuration shown in Figure 12.13 are tried. Less than 48 strands result in tensile stresses at the bottom fibers at service which exceed the maximum allowable $f_t = 484$ psi. Twelve strands are harped at 0.4 L. Accordingly, 36 strands remain straight at the beam (see Figure 12.14).

From data, $c_b = 36.60$ in. and $c_t = 72 - 36.60 = 35.40$ in.

$$e_e = c_b - [2 \times 70 + 2 \times 68 + 2 \times 66 + 2 \times 64 + 2 \times 62 + 2 \times 60 + 4 \times 8 + 8 \times 6 + 12 \times 4 + 12 \times 2]/48$$

$$= 36.60 - 19.42 = 17.28 \text{ in.}$$

$$e_c = c_b - [2 \times 12 + 12 \times 4 + 8 \times 6 + 8 \times 4 + 2 \times 10 + 2 \times 12 + 2 \times 14 + 2 \times 16 + 2 \times 18 + 2 \times 20]/48$$

$$= 36.6 - 6.92 = 29.68 \text{ in.}$$

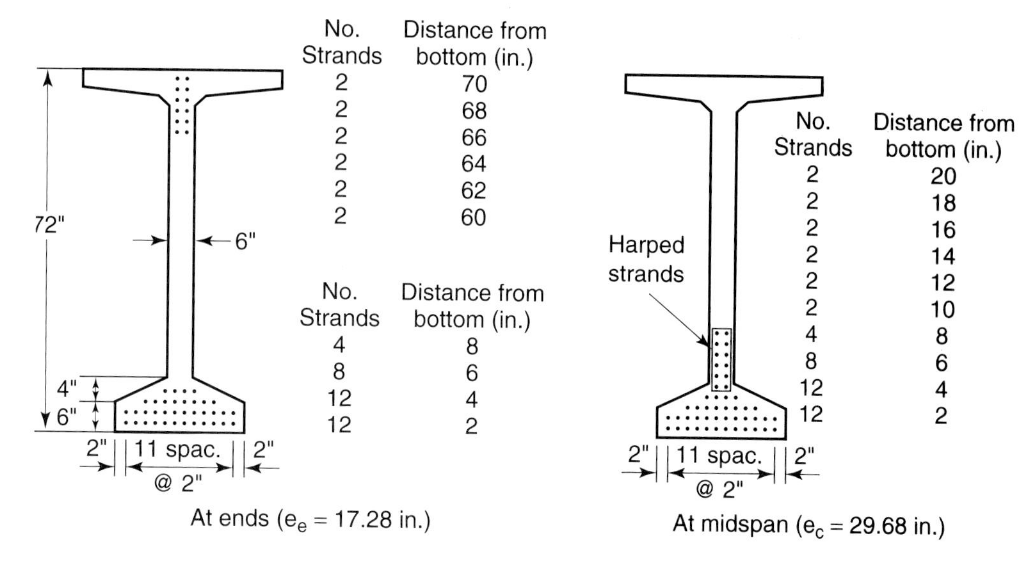

Figure 12.14 Bulb-Tee Prestressing Strand Pattern

Given $f_{pi} = 0.75\,f_{pu} = 202{,}500$ psi,

$$P_i = (48)(0.153)(202.5) = 1488 \text{ kips.}$$

After running a detailed step-by-step analysis of prestress losses as in chapter 3, Section 3.9, the total prestress loss was determined to be 26.4%.

$$f_{pe} = 202.5(1 - 0.264) = 149.0 \text{ ksi}$$

hence, $P_e = 1488(1 - 0.264) = 1095.0$ kips

(2) ***Check of Concrete Unfactored Stresses***

(a) *Stresses at Transfer*

Initial $f_{pi} = 0.7\,f_{pu} = 0.7 \times 270 = 202.5$ ksi. Common practice assumes that initial relaxation losses at prestressing amount to 9 to 10%. Use 10% reduction in f_{pi}.

$$P_i = 0.90 \times 1488 = 1339 \text{ kips}$$

Hence, $P_i = (0.9)(202.5)(0.153 \times 48) = 1338$ kips

(i) *Support Section*

From Chapter 4, Equation 4.1 (a),

$$f^t = -\frac{P_i}{A_c}\left(1 - \frac{e_e c_t}{r^2}\right) - \frac{M_D}{S^t}$$

$$= -\frac{1338}{767}\left(1 - \frac{17.28 \times 35.4}{712}\right) - 0 = -0.25 \text{ ksi (C), no tension, O.K.}$$

$$f_b = -\frac{P_i}{A_c}\left(1 + \frac{e_e c_b}{r^2}\right) + \frac{M_D}{S_b}$$

$$= -\frac{1339}{767}\left(1 + \frac{17.28 \times 36.3}{712}\right) + 0$$

$$= 3.29 \text{ ksi (C)} < \text{allowable } f_c = 3.48 \text{ ksi} \quad \text{O.K.}$$

(ii) *Midspan Section*

$$f^t = -\frac{1338}{767}\left(1 - \frac{29.68 \times 36.60}{712}\right) - \frac{1438 \times 12}{15{,}421}$$

$= 0.917 - 1.119 = -0.202$ ksi (C), no tension allowed, hence, O.K.

$$f_b = -\frac{1339}{767}\left(1 + \frac{29.68 \times 36.6}{712}\right) + \frac{1438 \times 12}{14{,}915}$$

$= -4.513 + 1.157 = -3.356$ ksi (C) < than allowable $f'_{ci} = 5.50$ ksi, O.K.

(b) *Stresses at Service:*

(i) *Midspan Section:*

From chapter 4, Equations 4.3(a) and 4.3(b):

$$f^t = -\frac{P_e}{A_c}\left(1 - \frac{e_c c_t}{r^2}\right) - \frac{M_T}{S_c^t} \leq f_c$$

$$f_b = -\frac{P_e}{A_c}\left(1 + \frac{e_c c_b}{r^2}\right) + \frac{M_T}{S_{cb}} \leq f_t$$

Since the loads are placed at different stages of construction, for Service-I precast sections,

$$f^t = -\frac{P_e}{A_c}\left(1 - \frac{e_c c_t}{r^2}\right) - \frac{M_D + M_S}{r^2} - \frac{M_{WS} + M_b}{S_c^t}$$

$$= -\frac{1095}{767}\left(1 - \frac{29.68 \times 35.40}{712}\right) - \frac{(1438 + 1660)12}{15{,}421} - \frac{(360 + 180)12}{62{,}950}$$

$= 0.679 - 2.411 - 0.103 = -1.835$ ksi (C)

< Service allowable $f_c = 2925$ psi O.K.

$$f_b = -\frac{1095}{767}\left(1 + \frac{29.68 \times 36.60}{712}\right) + \frac{(1438 + 1660)12}{14{,}915} + \frac{(360 + 180)12}{20{,}060}$$

$= -3.605 + 2.493 + 0.323 = -0.789$ ksi (C) O.K.

(3) ***Including stresses due to the transient lane and truck loads***

$$f^t = -1.835 - \frac{(1830 + 843)12}{62{,}950}$$

$= -1.835 - 0.510 = -2.345$ ksi (C) < Service-III $f_c = 3900$ psi O.K.

$$f_b = -0.789 + \frac{(1830 + 843)12}{20{,}060}$$

$= -0.789 + 1.110 = 0.321$ ksi (T) < Allowable $f_t = 0.484$, O.K.

(4) ***Concrete stresses at top deck fibers***

(i) *Under permanent Service I loads*

$$f_c^t = \frac{M_{ws} + M_b}{S_c^t} = \frac{(360 + 180)12}{55{,}284}$$

$= -0.117$ ksi (c) < allowable $f_c = 2.4$ ksi O.K.

(ii) *Under permanent and transient lane and truck loads, Service I:*

$$f_c^t = \frac{M_{ws} + M_b}{S_c^t} + \frac{M_{LT} + M_{LL}}{S_c^t}$$

$$= -0.117 - \frac{(1830 + 843)12}{55{,}284}$$

$= -0.697$ ksi (C) < allowable $f_c = 2.4$ ksi, O.K.

(5) ***Concrete Stresses at beam bottom fibers, Service III***

$$f_b = -\frac{P_e}{A_c}\left(1 + \frac{e_c c_b}{r^2}\right) + \frac{M_D + M_S}{S_b} + \frac{(M_{ws} + M_B) + 0.8(M_{LT} + M_{LL})}{S_{bc}}$$

$$= -\frac{1095}{767}\left(1 + \frac{29.68 \times 36.60}{712}\right) + \frac{(1438 + 1660)12}{14{,}915}$$

$$+ \frac{[(360 + 180) + 0.8(1830 + 843)]12}{20{,}060}$$

$$= -3.606 + 2.492 + 1.602 = 0.488 \text{ ksi (T)} \cong \text{allowable } f_t = 0.484 \text{ ksi, O.K.}$$

(6) ***Fatigue stresses***

LRFD specifies that in regions of compressive stress due to permanent loads and prestress, fatigue is only considered if the compressive stress is less than twice the maximum tensile live load stress resulting from the fatigue:

Thus, for permanent loads only, the term $(M_{LT} + M_{LL})/S_c^t$ is taken out to give:

$$f_b = -\frac{P_e}{A_c}\left(1 + \frac{e_c c_b}{r^2}\right) + \frac{M_D M_S}{S_b} + \frac{(M_{ws} + M_B)}{S_{bc}}$$

$$= -3.606 + 2.492 + \frac{(360 + 180)12}{20{,}060} = -790 \text{ ksi (c) O.K.}$$

From table 12.10, fatigue moment $M_f = 777$ ft-kips,
Tensile fatigue stress at the bottom fibers,

$$f_b = \frac{0.75 M_f}{S_{bc}} = \frac{0.75 \times 777 \times 12}{20{,}060} = 0.348 \text{ ksi (T)}$$

Since twice $0.348 = 0.696 < 0.790$ ksi (which is a compressive stress), a fatigue check is unnecessary.

From the forgoing computations, the flexural design is O.K. at the initial and service load conditions. To be complete and also determine the reserve strength available for overload conditions, the limit state at failure design is necessary as in the following section. The total design has to include shear, torsion, if any, and serviceability as in Example 12.2.

8. ***Ultimate strength (Limit state of failure)***

(a) *Nominal flexural resistance moment*

From Tables 12.2 (a) and (b) total factored moment for Strength I Load:
$M_u = 1.25\text{DC} + 1.5\text{DW} + 1.75(\text{LL} + \text{IM})$
From Table 12.9

$$M_u = 1.25(1438 + 1660) + 1.5(360 + 180) + 1.75(1830 + 843) = 9316 \text{ ft-kip}$$

$$\text{Required } M_r = \frac{M_U}{\phi} = \frac{9316}{1.0} = 9316 \text{ ft-kip}$$

Average stress in the prestressing reinforcement when $f_{pe} \geq 0.5 f_{pu}$, from

equation 12.9: $f_{ps} = f_{pu}\left(1 - k\frac{c}{d_p}\right)$ where $k = 2\left(1.04 - \frac{f_{py}}{f_{pu}}\right)$.

For the depth of the compressive block, use slab $f'_c = 4.0$ ksi.

k = 0.28 for low-relaxation steel

d_p = (h-cover to c.g.s.) = [(72 + 8) − 6.92] = 73.08 in.

b = effective compression flange width = 9′ - 0″ = 108 in.

$A_{ps} = 48 \times 0.153 = 7.344$ in.^{2}

From equation 12.10,

$$c = \frac{A_{ps}f_{pu} + A_sf_y - A'_sf'_y}{0.85f'_c\beta_1 b + kA_{ps}\dfrac{f_{pu}}{d_p}}$$

$$= \frac{7.344 + 0 - 0}{0.85 \times 4.0 \times 0.85 \times 108 + 0.28 \times 7.344\left(\dfrac{270}{73.08}\right)}$$

$$= 6.20 \text{ in.} < t_s = 7.5 \text{ in.}$$

$$a = \beta_1 c = 0.85 \times 6.20 = 5.27 \text{ in.},$$

hence, neutral axis is within the flange and the section is considered rectangular.

Average design reinforcement strength f_{ps}:

$$f_{ps} = 270\left(1 - 0.28\frac{6.20}{73.08}\right) = 263.6 \text{ ksi}$$

nominal flexural resistance $M_r = A_{ps}f_{ps}\left(d - \dfrac{a}{2}\right)$

$$M_n = M_r = 7.344 \times 263.6\left(73.08 - \frac{5.27}{2}\right)\left(\frac{1}{12}\right)$$

$$= 11{,}364 \text{ ft-kips} > \text{required } M_n = 9316 \text{ ft-kip} \quad \text{O.K.}$$

$\dfrac{c}{d_e} \le 0.42$ for ductile behavior discussed in section 12.3 and table 12.11 (a).

$$\text{Actual } \frac{c}{d_e} = \frac{6.20}{73.08} = 0.085 < 0.42 \quad \text{O.K.}$$

(b) *Minimum reinforcement*

As discussed in section 12.3.4, the minimum reinforcement has to be the lesser of 1.2 M_{cr} or 1.33 M_u required by the applicable load combinations. See also flowchart Figure 12.11 and Tables 12.11.

$$f_r = 7.5\sqrt{f'_c} = 7.5\sqrt{6500} = 605 \text{ psi} = 0.6 \text{ ksi}$$

f_{ce} = compressive stress due to effective prestress only at the bottom fibers as defined in section 12.3.4,

$$= -\frac{P_e}{A_c}\left(1 + \frac{e_c c_b}{r^2}\right) = -3.606 \text{ ksi from before.}$$

Non-composite $M_{dnc} = M_D + M_S = 1438 + 1660 = 3098$ ft-kip

$$S_{bc} = 20{,}060 \text{ in.}^4$$

$$S_b = 14{,}915 \text{ in.}^4$$

From equation 12.13,

$$M_{cr} = (f_r + f_{ce})S_b - M_{dnc}\left(\frac{S_{bc}}{S_b} - 1\right)$$

$$= (0.6 + 3.6)\frac{14{,}915}{12} - 3098\left(\frac{20{,}060}{14{,}915} - 1\right)$$

$$= 5220 - 1069 = 4151 \text{ ft-kip}$$

$$1.2\,M_{cr} = 1.2 \times 4151 = 4981 \text{ ft-kip}$$

$$1.33\,M_u = 1.33 \times 9316 = 12{,}390 \text{ ft-kip} > 4981 \text{ ft-kip}$$

hence, the lesser of the two moments controls, namely, $1.2\ M_{cr} = 4981$.
M_n or $M_r = 11{,}364 > 4981$ O.K.

9. ***Pretensioned Anchorage Zone***

The zone reinforcement is designed using the force in the strands just prior to release transfer. The LRFD specifications require that the bursting resistance, P_r should not be less than 4.0% of the force in the strands, F_{pi}, before release, namely:

$f_{pr} = f_s A_s \geq 0.4 F_{pi}$

$F_{pi} = 48 \times 0.153 \times 202.5 = 1488$ kips

$P_r = 0.04 \times 1488 = 59.5$ kips

Use a stress, f_s, in the anchorage reinforcement not exceeding 20 ksi.

Required area $= 59.5/20 = 2.98$ in.^{2}

Try No. 5 closed ties; $A_s = 2 \times 0.31 = 0.62$ in.^{2}

Number of ties $= 2.98/0.62 = 4.8$

Distance within which anchorage reinforcement has to be provided from beam end $= h/5 = 72/5 = 14.4$ in.

Use No. 5 closed ties at 3 in. center-to-center, with the first tie starting at 2 in. from the beam end.

Conclusion:

Accept the design of the bulb-tee bridge for flexure. For the design to be complete, design for shear, interface shear transfer and deflection/camber checks have to be performed as in Example 12.2.

Photo 12.5 Hanging Lake Viaduct, Glenwood Canyon, Colorado, total length 1297 ft., consisting of 34 spans, primarily 200-ft. lengths (*Courtesy* Figg Engineering Group, Tallahassee, Florida)

12.10 LRFD SHEAR AND DEFLECTION DESIGN

Example 12.2

Design the web shear reinforcement for the bulb-tee beam in Example 12.1 at the critical section near the supports and the interface shear transfer reinforcement at the interface plane between the precast section and the deck situ-cast concrete. Also, verify if the span deflection is within the allowable limits.

Solution:

1. ***Web Shear Design***

   (a) ***Strain at centroid level of reinforcement***

$$\phi = 0.90$$

$$e_e = 17.7 \text{ in.}$$

Provide web steel when $V_u > 0.5\phi\,(V_c + V_p)$
Critical section is the greater of $0.5\,d_v \cot\theta$ or d_v

$$d_e = d_c = h - e_e = 80.0 - 17.28 = 62.72 \text{ in.}$$

$$d_v = \left(d_e - \frac{a}{2}\right) = 62.72 - \frac{5.27}{2} = 60.08 \text{ in.}$$

$$\geq 0.9d_c = 0.9 \times 60.08 = 54.07 \text{ in.}$$

$$\geq 0.72h = 0.72 \times 80 = 57.6 \text{ in.}$$

$d_v = 60.08$ in. controls as the largest of the three values

Assume $\theta = 22°$ for a first trial

$$0.5d_v\cot\theta = 0.5 \times 60.08 \cot 22°$$

$$= 74.35 > 59.67 \text{ in.}$$

As the support bearing width is not yet determined, assume it conservatively = 0. Consequently, the critical section for shear is 74.35 in. ≅ 6.2 ft from the support, being larger than the dimension $d_v = 60.08$ in. as stipulated by the LRFD AASHTO requirement; hence distance 74.35 in. controls for the critical shear section.

$$\frac{x}{L} = \frac{6.2}{120} \cong 0.05L \text{ from the support face.}$$

From equation 12.23,

$$V_c = \beta\sqrt{f'_c}\,b_v d_v$$

In order to determine the value of β several computations have to be performed. Reinforcement strain ϵ_x from equation 12.27 is

$$\epsilon_x = \frac{\dfrac{M_u}{d_v} + 0.5N_u + 0.5V_u Cot\theta - A_{ps}f_{po}}{E_s A_s + E_{ps}A_{ps}} \leq 0.002$$

At plane 0.05L, from Table 12.9

$$M_u = 1.25(275 + 315 + 34) + 1.5\,(68) + 1.75(368 + 160) = 1802 \text{ ft-kip}$$

Corresponding shear:

$$V_u = 1.25(43 + 50 + 5) + 1.50(11) + 1.75(74 + 31) = 323 \text{ kips}$$

$$N_u = \text{applied normal force at } 0.05L \text{ plane} = 0$$

f_{po} = effective stress in the prestressing reinforcement at the decompression state in the concrete

$= f_{pe} + \dfrac{\bar{f}_{ce} E_{ps}}{E_c}$. It can however, be conservatively taken as the effective prestress f_{pe}

$\bar{f}_{ce}$ = concrete compressive stress at the centriod of the composite section due to both prestressing and the bending moments resisted by the precast section acting alone.

Distance from the c.g.c. of the composite section to the c.g.c. of the precast section.

$$c_1 = c_{bc} - c_b = 54.6 - 36.6 = 18.0 \text{ in.}$$

At the critical section $e_{el} = 18.9$ in.

$$\text{Section modulus} \quad S_1 = \frac{I_c}{c_1} = \frac{545{,}894}{18.0} = 30{,}327 \text{ in.}^4$$

$$e_e = 17.28 \text{ in.}$$

$$r^2 = 712 \text{ in.}^2$$

$$\bar{f}_{ce} = -\frac{P_e}{A_c}\left(1 - \frac{e_e c_1}{r^2}\right) - \frac{(M_D + M_{SD})}{S_1}$$

$$= -\frac{1095}{767}\left(1 - \frac{18.9 \times 18.0}{712}\right) - \frac{(275 + 315)(12)}{30{,}327}$$

$$= -0.746 - 0.233 = -0.979 \text{ ksi (c)}$$

$$f_{pe} = \frac{1095}{48 \times 0.153} = 149.0 \text{ ksi}$$

$$E_c = 4890 \text{ ksi}$$

$$V_u = 323 \text{ kips}$$

$$M_u = 1802 \text{ kip-ft}$$

$$A_{ps} = 48 \times 0.153 = 7.344$$

$$f_{po} = 149.0 + \frac{0.979 + 28{,}500}{4890} = 154.7 \text{ ksi}$$

(f_{po} could be conservatively taken at $f_{pe} = 149.0$ ksi)

$$\epsilon_x = \frac{\dfrac{1802 \times 12}{59.67} + 0 + 0.5 \times 232 \times \cot 22° - 7.344 \times 154.7}{0 + 28{,}500 \times 7.344}$$

$$= -1.787 \times 10^{-3} \text{ in./in.}$$

Since the value of strain ϵ_x at the level of the reinforcement centroid is negative, its value has to be reduced by the factor F_ϵ

$$F_\epsilon = \frac{E_S A_S + E_{PS} A_{ps}}{E_c A_c + E_S A_S + E_{PS} A_{PS}}$$

From Figures 12.10 and 12.12, $h = 80$ in.

$$A_c = 26 \times 6 + 2 \times 0.5 \times 10 \times 4.5 + 6\left[\frac{80}{2} - (6 + 4.5)\right] = 378 \text{ in.}$$

$$F_\epsilon = \frac{0 + 28{,}000 \times 7.344}{4890 \times 378 + 0 + 28{,}000 \times 7.344} = 0.101$$

Adjusted $\epsilon_x = (-1.787 \times 10^{-3})(0.101) = -0.180 \times 10^{-3}$

(b) ***Wed shear strength, V_c, from $\theta - \beta$ analysis.***

$$V_u = 323 \text{ kips}$$

From equation 12.15(b)

$$\text{Shear stress } v = \frac{(V_u - \phi V_P)}{\phi b_v d_v}, \qquad \text{where } \phi = 0.90 \text{ for shear}$$

$$f_{pe} \text{ from before} = 149.0 \text{ ksi}$$

Figure 12.15 shows the inclination angle, ψ, of the 12 harped strands,

$$\sin \psi = \frac{(65 - 15)}{48.5 \times 12} = 0.086$$

Harped tendon force $= 12 \times 0.153 \times 149.0 = 273.6$ kips

$$V_p = 273.6 \sin \psi = 273.6 \times 0.086 = 23.5 \text{ kips}$$

$$\text{Required } v = \frac{323 - 0.9 \times 23.5}{0.9 \times 6 \times 59.67} = 0.94 \text{ ksi}$$

$$\text{Ratio } \frac{v}{f'_c} = \frac{0.94}{6.5} = 0.145$$

$$\epsilon_x \text{ from before} = (-1.787 \times 10^{-3})(0.101) = -0.180 \times 10^{-3} \text{ in./in.}$$

Entering the values of ϵ_x and v/f'_c in the chart of Figure 12.9,

$\theta = 22.3°$ (assumed $\theta = 22°$) accept

$\beta = 2.6$ (factor indicating the ability of the compression strut, namely, the diagonal cracked concrete, to transmit tension)

Hence, $V_c = 0.0316\beta\sqrt{f'_c}\, b_v d_v$

$$= 0.0316 \times 2.6\sqrt{6.5} \times 6 \times 59.67 = 75.0 \text{ kips}$$

(c) ***Selection of web reinforcement***

From equation 12.26, check whether web reinforcement is needed, namely, if $V_u > 0.5\phi\,(V_c + V_p)$

$$0.5\phi\,(V_c + V_p) = 0.5 \times 0.9(75.0 + 23.5) = 44.3 \text{ kips}, < V_u = 323 \text{ kips}$$

Use web steel.

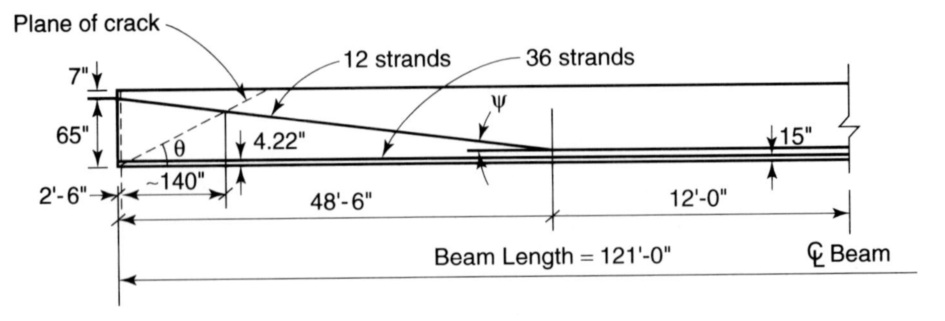

Figure 12.15 Beam tendon geometry

$$\text{Required } V_s = \frac{V_u}{\phi} - V_c - V_p = \left(\frac{323}{0.9}\right) - 75.0 - 23.5 = 260.4 \text{ kips}$$

From Equation 12.25,

$$\text{Available } V_s = \frac{A_v f_y d_v \cot\theta}{s}$$

Trying No. 4 stirrups, $A_v = 2 \times 0.20 = 0.40$ in.^{2}

$$\cot\theta = \cot 22.3° = 2.438$$

$$\text{hence, } 260.4 = \frac{0.4 \times 60.0 \times 59.67 \times 2.438}{s}, \qquad \text{giving } s = 13.4 \text{ in.}$$

(i) *Maximum allowable web stirrup spacing:*

$$0.10 f'_c d_v = 0.1 \times 6.5 \times 6 \times 60.08 = 234 \text{ kips} < V_u = 323 \text{ kips}$$

hence, maximum spacing $s = 12$ in.
If $0.10\, f_c\, b_v\, d_v > V_u$, maximum $s = 24$ in.

(ii) *Minimum area of transverse reinforcement:*

$$A_v/ft = 0.0316\sqrt{f'_c}\,\frac{b_v s}{f_y}$$

$$= 0.0316\sqrt{6.5}\left(\frac{6 \times 12}{60}\right) - 0.10 \text{ in.}^2/\text{ft}$$

Use No. 4 stirrups at 12 in. center-to-center with the spacing to be increased along the span.

(iii) *Maximum shear resistance:*
To ensure that the concrete in the web does not crush prior to yielding of the stirrups,

$$(V_n - V_p) \le 0.25 f'_c b_v d_v$$

$$(V_n - V_p) = V_c + V_S = 75.0 + 260.4 = 335.4 \text{ kips}$$

$$0.25 f'_c b_v d_v = 0.25 \times 6.5 \times 6 \times 60.08 = 585.7 \text{ kips} > 335.4, \quad \text{O.K.}$$

2. *Interface shear transfer*

(a) *Dowel reinforcement design*
Assume that the critical section for shear transfer is the same as the vertical shear at plane 0.05 L from the support face.
From load combination Strength I:

$$V_u = 1.25(5.4) + 1.5(10.8) + 1.75(73.8 + 30.5)$$

$$= 205.5 \text{ kips}$$

$$d_v = 60.08 \text{ in.}$$

$$V_{uh} = \frac{205.5}{60.08} = 3.42 \text{ kip/in.}$$

$$\text{Required } V_n = \frac{V_{uh}}{\phi} = \frac{3.42}{0.9} = 3.80 \text{ kip/in.}$$

From Equation 12.32,

$$\text{available } V_n = cA_{cv} + \mu[A_{vf} f_y + P_c]$$

For concrete placed clean, hardened concrete with interface contact not intentionally roughened,

$c = 0.075$ ksi $\quad \mu = 0.6$

b_v = contact width between slab and precast flange top = 42 in.

$$\frac{A_{cv}}{\text{in. depth}} = 42.0 \times 1.0 = 42.0 \text{ in.}^2$$

hence, $3.80 = 0.075 \times 42.0 + 0.6(A_{vf} \times 60 + 0)$ to give $Avf = 0.0181$ in.^{2}/in. = 0.217 in.^{2}/ft, $< A_v = 0.40$ in.^{2} at 12 in. c/c vertical stirrups.

On this basis, no special additional dowel reinforcement is needed. LRFD, however, also requires that if the width b_v exceeds 36 in., a minimum of four bars are required as dowel reinforcement. Thus, use also two No. 3 dowels at 12 in. c/c in addition to the No. 4 vertical stirrups at 12 in. c/c, to give total $A_{vf} = 0.62$ in.^{2}/ft.

(b) ***Maximum and minimum dowel reinforcement***

$$f'_c = 4.0 \text{ ksi for the deck concrete}$$

$$\text{Actual provided } V_n = 0.075 \times 42 + 0.6\left(\frac{0.62}{12} \times 60\right) = 5.01 \text{ kips/in.}$$

From Equations 12.36 (a) and (b), the maximum allowable:

$$0.2f'_c A_{cv} = 0.2 \times 4.0 \times 42.0 + 33.6 \text{ kip/in.}$$

In both cases, more than provided V_n, O.K.

3. ***LRFD Minimum Longitudinal Reinforcement***

The longitudinal reinforcement at *each* beam section along the span has to satisfy equation 12.29:

$$A_s f_s + A_{ps} f_{ps} \geq \frac{M_u}{d_v \phi} + 0.5\frac{N_u}{\phi} + \left(\frac{V_u}{\phi} - 0.5V_s - V_p\right)\cot\theta$$

From Tables 12.8 and 12.9 at $x = 0$ from support, $V_u = 1.25(47.9 + 55.3 + 6.0) + 1.50(12.0) + 1.75(78.1 + 33.9) = 350$ kip

V_s based on only the No. 4 stirrups = 260.4 kips

$$M_u = 0$$

$$N_u = 0$$

$$\cot\theta = \cot 22.3° = 2.438$$

hence,
$$\frac{M_u}{d_v \phi} + 0.5\frac{N_u}{\phi} + \left(\frac{V_u}{\phi} - 0.5V_s - V_P\right)\cot\theta$$

$$= 0 + 0 + \left(\frac{350}{0.9} - 0.5 \times 260.4 - 23.5\right)(2.438) = 573.4 \text{ kips}$$

Number of straight strands at the support = 36

Number of draped strands at the support = 12

From the assumed crack plane intersection with the strands in Figure 12.15, the distance of the intersection from the support = 6 + 4.22 cot 22.3° = 16.9 in. where the strand stops at 6 in. from face of the support.

Transfer length = 60 × strand diameter = 30 in.

The available prestress of the 36 straight strands at the support face is a portion of the effective prestress, f_{pe}.

$$\text{Hence, use } f_{pe} = 149.0 \times \frac{16.29}{30.0} = 80.9 \text{ ksi}$$

For the top draped strands, the crack in Figure 12.15 intersects the strands at a distance ≅ 140 in. from the support face (compute from geometry of the dimensions in

Figure 12.14). Consequently the effective prestress can be approximated at $f_{pc} = 149.0$ ksi.

$$A_s f_s + A_{ps} f_{ps} = 0 + 36 \times 0.153(80.9) + 12 \times 0.153(149.0)$$
$$= 445.6 + 273.6 = 719.2 \text{ kips}$$
$$> 573.4 \text{ kips, hence, no additional longitudinal reinforcement is needed.}$$

4. ***Deflection and camber***

(i) ***Immediate deflection due to permanent loads***

Compute the camber and deflection of the beam as detailed in the discussions and numerous examples of chapter 7.

From the deflection table in Figure 7.6,

$$\text{Midspan } \delta = \frac{PL^2}{8E_c I_g}\left[e_c + (e_e - e_c)\frac{4}{3}\frac{a^2}{L^2}\right]$$

$$\text{here, } a = \frac{L}{2}$$

$$\delta = \frac{PL^2}{E_c I_g}\left[\frac{e_c}{8} + \frac{(e_e - e_c)}{24}\right]$$

From Example 12.1:

$$P_i = 1488 \text{ kips}$$
$$e_c = 29.68 \text{ in.}$$
$$e_e = 17.7 \text{ in.}$$
$$E_{ci} = 4620 \text{ ksi}$$
$$E_{ce} = 4890 \text{ ksi}$$
$$w_D = 0.779 \text{ kip/ft}$$
$$I_{cc} = 1{,}095{,}290 \text{ in.}^4$$
$$w_{s+h} = 0.922 \text{ kip/ft}$$
$$w_{barrier} = 0.300 \text{ kip/ft}$$
$$A_c = 767 \text{ in.}^2$$
$$S_b = 14{,}915 \text{ in.}^3$$
$$S^t = 15{,}421 \text{ in.}^3$$
$$I_c = 545{,}897 \text{ in.}^4$$

$$\delta_i = \frac{1488(120 \times 12)^2}{4620 \times 545{,}897}\left[\frac{29.68}{8} + \frac{(17.7 - 29.68)}{24}\right]$$
$$= 1.22\,(3.71 - 0.50) = 3.92 \text{ in.} \uparrow \text{(camber)}$$

w_D per inch = 0.799/12 = 0.065 kip/in.

$$\delta_D = \frac{5wL^4}{384E_{ci}I_g} = \frac{5(0.065)(120 \times 12)^4}{384 \times 4620 \times 545{,}894}$$
$$= 1.44 \text{ in.} \downarrow$$

$$w_{s+h} = 0.922 \text{ kip/ft} = 0.077 \text{ kip/in.}$$

$$\delta_D = \frac{5(0.077)(120 \times 12)^4}{384 \times 4888 \times 545{,}894} = 1.61 \text{ in.} \downarrow$$

$$w_{\text{barrier}} = 0.300 \text{ kip/ft} = 0.025 \text{ kip/in.}$$

$$\delta_D = \frac{5(0.025)(120 \times 12)^4}{384 \times 4888 \times 1,095,290} = 0.26 \text{ in.} \downarrow$$

(ii) ***Immediate deflection due to transient loads***

Live load deflection limit = L/800.

LRFD specifications require that all the bridge deck beams be assumed to deflect equally under applied live load and impact. They also stipulate that the long-term deflection may be taken as four times the immediate deflection. This stipulation is too general and the designer is well-advised to use other more refined methods. The larger is the span the more is the needed accuracy. It should be emphasized that computed deflection values can differ from actual deflections by as much as 30 to 40% depending on the concrete modulus and stress-strain relationship assumed and the degree of accuracy of the method used in the computation.

The following deflection computation methods from chapter 7 can give reasonable step-by-step values during the loading history

- PCI multipliers method (Sec. 7.7.1)
- Incremental time-step method (Sec. 7.7.2)
- Approximate time-step method (Sec. 7.7.3)

From Figure 12.12 in Example 12.1, the number of bridge beams = 7 and the number of lanes = 4

$$\begin{aligned} \text{DFD} &= \text{distribution factor for deflection} \\ &= \text{number of lanes divided by number of beams} \\ &= \frac{4}{7} = 0.571 \text{ lanes/beam} \end{aligned}$$

It is more conservative to use moment distribution factor DFM = 0.732
Design lane load, $W = 0.64$ DFM

$$\begin{aligned} &= 0.64 \text{ kip/ft } (0.732) = 0.468 \text{ kip/ft/beam} \\ &= 0.039 \text{ kip/in./beam} \end{aligned}$$

$$\delta_{LL} = \frac{5}{384 E_c I_{cc}} = \frac{5(0.039)(120 \times 12)^4}{384 \times 4888 \times 1,095,290} = 0.41 \text{ in.} \downarrow$$

Table 12.11 Long-Term Camber and Deflection

	Transfer δ_p (in.)	Non-composite PCI Multipliers	Composite PCI Multipliers	δ_{final} (in.)
Prestress	3.92↑	1.80	2.20	8.62↑
W_d	−1.44↓	1.85	2.40	−3.47↓
	Net 2.48↑			Net 5.15↑
w_{S+h}	−1.61↓	1.85	2.40	−3.84↓
$w_{barrier}$	−0.26↓	1.85	2.30	−0.60↓
δ_{LL}	−0.41↓			−0.41↓
δ_{Lt}	−0.78↓			−0.78↓
Final δ	2.48↑			−0.48↓

Photo 12.6 West Kowloon Expressway Viaduct, Hong Kong, 1997. 4.2-Km dual three-lane causway connecting Western Harbor Crossing to new airport (*Courtesy* Institution of Civil Engineers, London)

The transient truck load and impact deflection is determined from influence lines of wheel position for maximum moment. For a 120-ft span, the 72 kip resultant of the axial loads falls at 2.33 ft from the midspan. The deflection at midspan = 0.8 in. ↓

$$\delta_{LT} = 0.8(IM)(DFM) = 0.8(1.33)(0.732) = 0.78 \text{ in.} \downarrow$$

Using the PCI multipliers from Table 7.1, a summary of the long-term cambers and deflections are given in Table 12.11.

$$\text{Allowable deflection } \delta = \frac{L}{800} = \frac{120 \times 12}{800} = 1.80 \text{ in. (down)}$$

$$> \text{actual} = 0.49 \text{ in.} \quad \text{O.K.}$$

Adopt the bridge deck design of the interior beam in Example 12.1 and 12.2.

12.11 STANDARD AASHTO FLEXURAL DESIGN OF PRESTRESSED BRIDGE DECK BEAMS

Example 12.3

Design for flexure, an interior beam of the bridge deck in Example 12.1 (adopted from Ref. 12.11) using the standard AASHTO Design Specifications for HS-20 lane and truck loads. Use the same data and allowable stresses of the materials as in the indicated example except where they differ from the LRFD allowable stresses.

Solution:

1. ***Transformed deck slab controlling width***
   From example 12.2 Step 1,

$$E_{cs} = 3830 \text{ psi}$$

$$E_{ci} = 4620 \text{ psi at transfer}$$

$$E_{ce} = 4890 \text{ psi at service}$$

Average spacing between beams = 108 in.
Transformed flange width b_m = 84 in.

2. ***Properties of Section***

Non-Composite	*Composite*
$A_c = 767 \text{ in.}^2$	$A_c = 1042 \text{ in.}^3$
$h = 72 \text{ in.}$	$h = 80 \text{ in.}$
$I_c = 545{,}894 \text{ in.}^4$	$I_{cc} = 1{,}095{,}290 \text{ in.}^4$
$c_b = 36.60 \text{ in.}$	$c_{bc} = 54.6 \text{ in.}$
$c_t = 35.40 \text{ in.}$	$c_{tc} = 17.4 \text{ in.}$
	$c_{tsc} = 25.4 \text{ in.}$
$r^2 = 1051 \text{ in.}^2$	$r^2 = 712 \text{ in.}^2$
$S_b = 14.915 \text{ in.}^3$	$S_{bc} = 20{,}060 \text{ in.}^3$
$S^t = 15{,}421 \text{ in.}^3$	$S_c^t = 62{,}950 \text{ in.}^3$
	$S_c^{ts} = 55{,}284 \text{ in.}^3$

3. ***Bending moment and shear forces***

self-weight	W_D	= 799 lb/ft
slab	W_{SD1}	= 900 lb/ft
haunch		= 22 lb/ft
Barrier weight	W_{SD2}	= 100 lb/ft
2-in. future wearing surface	W_{SD3}	= 200 lb/ft

Live load (truck load) in AASHTO standard specifications would be based on HS-20 trucks.
Number of lanes = 48/12 = 4 lanes

(a) ***Distribution factor for moment***
Live load in the standard specifications is *either* the standard truck or lane loading corresponding to HS-20. In LRFD, both have to be used in the design. From Section 12.2.2, the live load distribution factor for moment for a precast beam is $DF_m = S/5.5 = 9.0/5.5 = 1.636$ wheels per beam, where S = average spacing between beams in feet.

$$\frac{1}{2} DF_m = 0.818 \text{ lanes per beam}$$

$$\text{the live load impact factor } I = \frac{50}{L + 125} \leq 30\% \quad \text{or,}$$

$$= \frac{50}{120 + 150} = 0.204$$

In LRFD, This factor has a maximum 33% value hence,

$$V_{LL+I} = (\text{shear force per lane})\,(\text{DFM})\,(1 + I)$$

$$= (\text{shear force per lane})\,(0.818)\,(1 + 0.204) \text{ kips}$$

$$= \text{(shear force per lane) (0.985) kips}$$

$$M_{LL+I} = \text{(moment per lane) (DFM) } (1 + I)$$

$$= \text{(moment per lane) (0.818) } (1 + 0.204) \text{ kips}$$

$$= \text{(moment per lane) (0.985) kips.}$$

Load contributions from Equation 12.5 and Table 12.1 show that load combination Group I controls.

$$\text{Group I service load design} = 1.00\,D + 1.00\,(L + I)$$

$$\text{Group I factored load design} = 1.3\,[1.00D + 1.67(L + I)]$$

(b) ***Shear and bending moments***

$$V_x = (w)\,(0.5\,L - X)$$

$$M_x = 0.5\,(w)\,(X)\,(L - X)$$

As an example, the following are computations for the shear and moment at midspan, at the support and at the critical shear section:
At midspan, $V_x = 0$

$$M_D = 0.5(0.799)\,(60)\,(60) = 1438 \text{ ft-kip}$$

At support, $V = 0.799(60) = 47.9$ kip
At critical shear section,
The cgs of the prestressing steel is $e = 17.1$ in. near the support section (see subsequent computations)

$$d = 80 - 17.1 = 62.9 \text{ in.} > 0.80\,h = 64 \text{ in.}$$

Use $d = 64.0$ in.
Critical shear section at $h/2 = 8\text{-}0/2 = 40$ in. $= 3.33$ ft

$$V_{3.33} = 0.799\,[(0.5)\,(120) - 3.33] = 45.3 \text{ kips}$$

$$M_{3.33} = 0.5(0.799)\,(3.33)\,(120 - 3.33) = 155.4 \text{ ft-kip}$$

The values for shear and moment for all permanent and transient loads are tabulated in Table 12.12 (Ref. 12.11). Compare the tabulated values with those computed by the LRFD method in Table 12.9 and 12.10.

Table 12.12 Standard AASHTO (LFD) Service Shear and Moment Due to Dead Load

Distance X	Section X/L	Beam Weight W_D Shear V_g	Beam Weight W_D Moment M_g	(Slab + Haunch) Weight W_{SD1} Shear V_S	(Slab + Haunch) Weight W_{SD1} Moment M_S	Barrier Weight W_{SD2} Shear V_b	Barrier Weight W_{SD2} Moment M_b	Wearing Surface W_{SD3} Shear V_{WS}	Wearing Surface W_{SD3} Moment M_{WS}	Live Load Plus Impact Shear V_{LL+I}	Live Load Plus Impact Moment M_{LL+I}
ft		kips	ft-kips	kips	ft-kips	kips	ft-kips	kips	ft-kips	kips	ft-kips
0	0.0	47.9	0.0	55.3	0.0	6.0	0.0	12.0	0.0	65.4	0.0
3.33*	0.028	45.3	155.4	52.2	179.3	5.7	19.4	11.3	38.9	63.6	211.5
12	0.1	38.4	517.8	44.3	597.5	4.8	64.8	9.6	129.6	58.3	699.7
24	0.2	28.8	920.4	33.2	1,062.1	3.6	115.2	7.2	230.4	51.2	1,229.1
36+	0.3	19.2	1,208.1	22.1	1,394.1	2.4	151.2	4.8	302.4	44.1	1,588.4
48	0.4	9.6	1,380.7	11.1	1,593.2	1.2	172.8	2.4	345.6	37.0	1,799.6
60	0.5	0.0	1,438.2	0.0	1,659.6	0.0	180.0	0.0	360.0	29.9	1,851.6

*Critical section for shear
+Harp point

4. *Design of Bulb-Tee Prestressed Interior Beam*

1. *Selection of prestressing strands*

Due to applied gravity loads, the unfactored stress at bottom fibers:

$$f_b = \frac{M_D + M_{SD1}}{S_b} + \frac{M_{SD2} + M_{SD3} + M_{LL+I}}{S_{bc}}$$

$$= \frac{(1438 + 1660)12}{14{,}915} + \frac{(180 + 360 + 1852)12}{20{,}060} = 3.923 \text{ ksi}$$

Allowable tensile stress $f_t = 6\sqrt{f'_c} = 6\sqrt{6500} = 484$ psi $= 0.484$ ksi

Required precompressive stress at the bottom fibers after losses = $(f_b - f_t)$

$f_{bp} = 3.923 - 0.484 = 3.439$ ksi

Assume that the tendon c.g.s. is at a distance $y_b = 4$ in. from the bottom fibers.

$$e_c = c_c - y_b = 36.60 - 4.00 = 32.60 \text{ in.}$$

$$f_{bp} \text{ due to prestress} = \frac{P_e}{A_c} + \frac{P_e e_c}{S_b} \text{ or } 3.439 = \frac{P_e}{767} + \frac{P_e \times 32.6}{14{,}915}$$

$$P_e = 986 \text{ kips}$$

Assume total prestress loss = 25%

$$P_i = \frac{986}{1 - 0.25} = 1315 \text{ kips}$$

Assume using 1/2-in. diameter 7-wire 270-K low-relaxation strands (A_{ps} = 0.153 in.^{2}),

$$\text{Required number of strands} = \frac{1315}{0.153 \times 202.5} = 42.44 \text{ strands}$$

Try 44 strands.

After trials and adjustments, assume that the strands pattern is as shown in Figure 12.16 with 10 strands harped at 48 ft from the support.

From data, $c_b = 36.60$ in. and $c_t = 72 - 36.60 = 35.40$ in.

$$e_e = c_b - [2 \times 70 + 2 \times 68 + 2 \times 66 + 2 \times 64 + 2 \times 62 + 2 \times 8 + 8 \times 6 + 12 \times 4 + 12 \times 2]/44 = 36.60 - 18.09$$

$$= 18.51 \text{ in.}$$

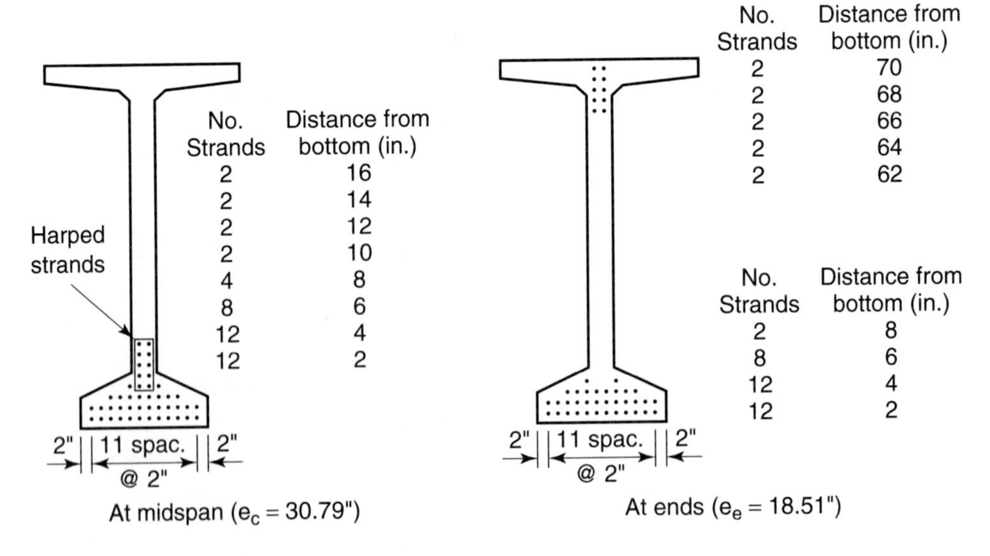

Figure 12.16 Bulb-tee prestressing strand pattern

$e_{3.33} = 17.1$ in. at the critical shear section.

$$e_c = c_b - [2 \times 16 + 2 \times 14 + 2 \times 12 + 2 \times 10 + 4 \times 8 + 8 \times 6 + 12 \times 4 + 12 \times 2]/48$$

$$= 36.6 - 5.81 = 30.79 \text{ in.}$$

Computing the total losses in prestress by the detailed method and the examples of Chapter 3, the total loss of prestress was 24.9%.

$$f_{pe} \text{ before losses} = 0.75(270) = 202.5 \text{ ksi}$$

hence, adjusted $f_{pe} = 202.5(1 - 0.249) = 152.1$ ksi

$$P_e = 48(0.153)(15.2) = 1024 \text{ kips}$$

Common practice assumes that a prestress relaxation and other losses at prestressing amount to 9 to 10%
Use 9% here to get $f_{pi} = 202.5(1 - 0.09) = 184.3$ ksi

$$P_i = 44(0.153)(184.3) = 1240 \text{ kips}$$

From Example 12.2,

$$f_{ci} = 0.6 f'_c = 0.6(5500) = 3300 \text{ psi (c)}$$

$$f_{ti} = 7.5\sqrt{f'_{ci}} = 7.5\sqrt{5500} = 556 \text{ psi (T)}$$

If the computed tensile stress at transfer exceeds 200 psi or $3\sqrt{f'_{ci}} = 220$ psi, whichever is small, bonded reinforcement has to be provided to resist the total tensile force in the concrete, computed on the basis of uncracked section.

2. *Check of concrete unfactored stresses*

The standard AASHTO allowable stresses are as follows, Case I for all load combinations:

Precast beam $f_c = 0.60 f'_c = 0.6(6500) = 3900$ psi

Deck slab $f_c = 0.60 f'_c = 0.6(4000) = 2400$ psi

Case (II) for effective pretension force + permanent dead loads:

Precast beam $f_c = 0.40 f'_c = 0.4(6000) = 2400$ psi

Deck slab $f_c = 0.40 f'_c = 0.4(4000) = 1600$ psi

Case (III) for live load + $\frac{1}{2}$(pretensioning force + dead load)

Precast beam $f_c = 0.40 f'_c = 0.4(6500) = 2600$ psi

Deck slab $f_c = 0.40 f'_c = 0.4(4000) = 1600$ psi

Allowable tension $f_t = 6\sqrt{f'_{ci}} = 6\sqrt{6500} = 484$ psi

Allowable f_t at transfer $= 3\sqrt{f'_c} = 242$ psi

a. *Stresses at Transfer*

Initial $f_{pi} = 0.7 f_{pu} = 202.5$ ksi. Common practice assumes that initial relaxation losses at prestressing amount to 9 to 10%. Use 9% reduction in f_{pi}.
hence, $P_i = (0.9)(202.5)(0.153 \text{ x } 48) = 1338$ kips

i. *Support Section*

From Chapter 4, Equation 4.1 (a),

$$f^t = -\frac{P_i}{A_c}\left(1 - \frac{e_e c_t}{r^2}\right) - \frac{M_D}{S^t}$$

Photo 12.7 Natchez Parkway Arches, Nashville Tennessee, America's first segmental arch bridge; principal arch span is 582-ft. long and has a vertical clearance of 137 ft (*Courtesy* Figg Engineering Group, Tallahassee, Florida)

$$= -\frac{1240}{767}\left(1 - \frac{18.51 \times 35.40}{712}\right) - 0 = 0.129 \text{ ksi (T)} < 242 \text{ psi}, \quad \text{O.K.}$$

$$f_b = -\frac{P_i}{A_c}\left(1 + \frac{e_e c_b}{r^2}\right) + \frac{M_D}{S^t}$$

$$= -\frac{1240}{767}\left(1 + \frac{18.51 \times 36.30}{712}\right) + 0$$

$$= -3.14 \text{ ksi (C)} \cong f_{ci} 3.3 \text{ ksi} \quad \text{O.K.}$$

(ii) *Midspan Section.*

$$f^t = -\frac{1240}{767}\left(1 - \frac{30.79 \times 35.40}{712}\right) - \frac{1438 \times 12}{15{,}421}$$

$$= 0.858 - 1.119 = -0.261 \text{ ksi (C), no tension allowed, hence,} \quad \text{O.K.}$$

$$f_b = -\frac{1240}{767}\left(1 + \frac{30.79 \times 36.60}{712}\right) + \frac{1438 \times 12}{14{,}915}$$

$$= -4.175 + 1.157 = -3.018 \text{ ksi (C)} < 3.300 \text{ ksi allowed} \quad \text{O.K.}$$

b. ***Stresses at Service load:***

(i) *Midspan Section precast section fiber stresses:*
concrete stress at top fibers at midspan due to all loads:

$$f^t = -\frac{P_e}{A_c}\left(1 - \frac{e_c c_t}{r^2}\right) - \frac{M_D + M_{DS1}}{S^t} - \frac{M_{DS2} + M_{DS3}}{S_c^t} - \frac{M_{LL+I}}{S_c^t}$$

$$e_c = 30.79 \text{ in.} \qquad e_e = 18.51 \text{ in.}$$

From the moment values at midspan tabulated in table 12.10 for load combinations:

Case(I):

$$f^t = -\frac{1024}{767}\left(1 - \frac{30.79 \times 35.40}{712}\right) - \frac{(1438 + 1660)12}{15{,}421} - \frac{(360 + 180)12}{62{,}950} - \frac{(1852)12}{62{,}950}$$

$$= 0.708 - 2.411 - 0.103 - 0.353 = -2.159 \text{ ksi (C)}$$

Case (II):

$$f^t = 0.708 - 2.411 - 0.103 = -1.806 \text{ ksi (C)}$$

Case (III):

$$f^t = 0.5(0.708 - 2.411 - 0.103) = -1.256 \text{ ksi (C)}$$

All compressive stress are less than the allowable $f_c = 3900$ psi O.K.

(ii) *Midspan section bottom fiber stresses*

$$f_b = -\frac{P_e}{A_c}\left(1 + \frac{e_c c_b}{r^2}\right) + \frac{M_D + M_{DS1}}{S_b} + \frac{M_{DS2} + M_{DS3}}{S_{bc}} + \frac{M_{LL+I}}{S_{bc}}$$

$$f_b = -\frac{1024}{767}\left(1 + \frac{30.79 \times 36.60}{712}\right) + \frac{(1438 + 1660)12}{14{,}915} + \frac{(360 + 180)12}{20{,}060} + \frac{(1852)12}{20{,}060}$$

$$= -3.437 + 2.493 + 0.323 + 1.10 = 0.486 \text{ ksi (T)}$$

$$\cong \text{allowable } f_t = 0.484 \text{ ksi} \quad \text{O.K.}$$

(iii) *Midspan slab top-fiber stresses*

Case (I):

$$f^t = -\frac{M_D + M_{DS1}}{S_b} - \frac{M_{LL+I}}{S_{bc}}$$

$$f^t = -\frac{(180 + 360)}{55{,}284} - \frac{1852(12)}{55{,}284} = -0.117 - 0.402$$

$$= -0.519 \text{ (C)} < \text{allowable } f_c = 2400 \text{ ksi} \quad \text{O.K.}$$

Case (II):

$$f^t = -0.117 \text{ ksi (C)} \qquad \text{O.K.}$$

Case (III):

$$f^t = 0.5(-0.117) - 0.402 = -0.461 \text{ ksi (C)} \qquad \text{O.K.}$$

3. *Ultimate Strength (Limit State at Failure)*

a. *Normal flexural resistance moment*

$$M_U = 1.3\left[M_D + M_{SD1} + M_{SD2} + M_{SD3} + 1.67\,(M_{LL+I})\right]$$

$$= 1.3(1438 + 1660 + 180 + 360 + 1.67 \times 1852) = 8750 \text{ ft-kip}$$

From equation 12.9(c)

$$f_{ps} = f_{pu}\left[1 - \frac{\gamma}{\beta_1}\rho\frac{f_{pu}}{f'_c}\right]$$

For the depth of the compressive block use the slab $f'_c = 4.0$ ksi, $\beta_1 = 0.85$

$$\gamma = 0.28 \text{ for low-relaxation strands}$$

$$b = \text{flange wdith} = 108 \text{ in.}$$

$$e_c = 30.79 \text{ in.}, \qquad c_b = 36.60,$$

hence, $y_b = 36.60 - 30.79 = 5.81$ in.

$$d_p = \text{distance from the top of the deck to the centroid of the prestressing strands.}$$

$$= \text{beam depth (h)} + \text{haunch} + \text{slab thickness } (h_f - y_b)$$

$$= 72 + (0.5 + 7.5) - 5.81 = 74.19 \text{ in.}$$

$$A_{ps} = 44(0.153) = 6.732 \text{ in.}^2$$

$$\rho = \frac{A_{ps}}{bd_p} = \frac{6.732}{108 \times 74.19} = 0.00084$$

Consequently,

$$f_{ps} = 270\left[1 - \frac{0.28}{0.85}0.00084\frac{270}{4.0}\right] = 265.0 \text{ ksi}$$

$$a = \frac{A_{ps}f_{ps}}{0.85f'_c b} = \frac{6.732 \times 265.0}{0.85 \times 4.0 \times 108} = 4.86 \text{ in.} < 7.5 \text{ in.}$$

hence design as rectangular section.

$$\text{Available } M_U = \phi M_n = A_{ps}f_{ps}\left(d - \frac{a}{2}\right)$$

$$= 1.0\left[6.732\,(265.0)\left(74.19 - \frac{4.86}{2}\right)\right]$$

$$= 128{,}018 \text{ in.-kip} > \text{required } M_U = 8750 \text{ ft-kip}, \quad \text{O.K.}$$

b. *Maximum Reinforcement*

$$\rho\frac{f_{ps}}{f'_c} \leq 0.36\beta_1 \leq 0.36 \times 0.85 = 0.306$$

$$\rho\frac{f_{ps}}{f'_c} = 0.00084 \times \frac{265}{4.0} = 0.0557 < 0.306 \quad \text{O.K.}$$

c. *Minimum Reinforcement*

The total amount of pretensioned and post-tensioned reinforcement should be adequate to develop an ultimate moment such that

$$\phi\, M_n \geq 1.2\, M_{cr}$$

From equation 12.13,

$$M_{cr} = (f_r + f_{ce})S_b - M_{dnc}\left[\frac{S_{bc}}{S_b} - 1\right]$$

$$f_r = 7.5\sqrt{f'_c} = 7.5\sqrt{6500} = 605 \text{ psi} = 0.605 \text{ ksi}$$

Concrete stress due to prestressing only, after all losses is $f_b = 3.437$ ksi (C)

$$M_{dnc} = \text{non-composite dead load moment due to beam self-weight and slab weight}$$

$$= 1438 + 1660 = 3098 \text{ ft-kip}$$

$$1.2\, M_{cr} = \frac{1}{12}(0.605 + 3.437)20{,}060 - 3098\left(\frac{20{,}060}{14{,}915} - 1\right) = 5688 \text{ ft-kip}$$

It should be noted that contrary to the LRFD specifications, the standard specification stipulates that this requirement has to be satisfied only at the critical section.

d. ***Pretensioned anchorage zone***

$$\text{Before initial losses, } f_{pi} = A_{ps}(0.75\, f_{pu})$$
$$= 6.732(0.75 \times 270) = 1363 \text{ kips}$$
$$4\%\ f_{pi} = 0.04 \times 1363 = 54.5 \text{ kips}$$
$$\text{Allowable } f_s = 20 \text{ ksi}$$
$$\text{Hence, required } A_v = \frac{54.5}{20} = 2.73 \text{ in.}^2$$

Try No. 5 vertical stirrups in the rectangular anchorage zone region

$$A_v = 2 \times 0.305 = 0.61 \text{ in.}^2$$
$$\text{No. of stirrups} = \frac{2.73}{0.61} = 4.5$$
$$\text{Precast Section } d_p = (h - y_b) = 72 - 5.81 = 66.19 \text{ in.}$$
$$d_v = \left(d_p - \frac{a}{2}\right) = 66.19 - \frac{4.86}{2} = 63.76 \text{ in.}$$

Distance within which anchorage reinforcement has to be provided

$$= \frac{d_v}{4} = \frac{63.76}{4} = 15.94 \text{ in.}$$

Use five # 5 closed U-stirrups at 3 in. center-to-center at each beam end.

12.12 STANDARD AASHTO SHEAR-REINFORCEMENT DESIGN OF BRIDGE DECK BEAMS

Example 12.4

Design for shear, an interior beam of the bridge deck in Example 12.3 using the standard AASHTO design specifications for HS-20 Lane and Truck loads. Use the same data and allowable stresses of the materials as in the indicated example. Use the refined flexural and web shear approach for determining the nominal strength of the plain concrete in the web. Also design the interface shear transfer reinforcement and check the deflection and camber of the beam.

Solution:

1. ***Shear Reinforcement***
The AASHTO standard specification follows the ACI-318 code for shear and torsion which are detailed in Chapter 5, Sections 5.5 and 5.6 as well Section 5.18 for torsion.

$$V_s \leq \phi\,(V_c + V_S), \text{ where } \phi = 0.90 \text{ vs. } \phi = 0.85 \text{ in ACI.}$$

Other strength reduction factors, ϕ, also differ from the ACI factors. The computations have to be based on a factored shear value at a distance 1/2 h from the face of the support. The nominal shear strength, V_c, of the plain concrete in the web has to be the lesser of the flexural shear, V_{ci}, and the web shear, V_{cw}.

a. ***Flexural shear, V_{ci}***
From Equation 5.11,

$$V_{ci} = 0.6\lambda\sqrt{f'_c}\, b_w d_p + V_d + \frac{V_i}{M_{\max}}(M_{cr}) \geq 1.7\lambda\sqrt{f'_c}\, b_w d_p$$

where $b_w = 6$ in.
From table 12.12 for standard AASHTO loads in Example 12.3

$$V_d = \text{total unfactored dead load at the critical section}$$
$$= 45.3 + 52.2 + 5.7 + 11.3 = 114.5 \text{ kips}$$
$$V_{LL+I}\,(\text{unfactored}) = 63.6 \text{ kips}$$
$$V_U = \text{factored shear force at the critical section}$$
$$= 1.3(V_d + 1.67V_{LL+I})$$
$$= 1.3(114.5 + 1.67 \times 63.6) = 286.9 \text{ kips}$$
$$M_d = 155.4 + 179.3 + 19.4 + 38.9 = 393.0 \text{ kips}$$
$$M_{LL+I} = 211.5 \text{ kips}$$
$$M_U = 1.3\,(M_d + 1.67\,M_{LL+I})$$
$$= 1.3(393.0 + 1.67 \times 211.5) = 970.1 \text{ ft-kips}$$

V_i = factored shear force at the section due to externally applied loads occurring simultaneously with M_{max}.

$= (V_U - V_D) = 286.9 - 114.5 = 172.4$ kips. This is on the conservative side since the factored V_U is reduced by the unfactored V_d.

$$M_{max} = (M_U + M_d) = 970.1 - 393.0 = 577.1 \text{ ft-kip}$$

From equation 5.12,

$M_{cr} = S_{bc}\,(6\sqrt{f'_c} + f_{ce} - f_d)$. Note that the factor "6" in the term $6\sqrt{f'_c}$ is conservative and maybe unjustified since the modulus of rupture, f_r, is taken as $7.5\sqrt{f'_c}$ and tests indicate even higher values.

f_{ce} = compressive stress due to prestress after losses at the extreme fibers of the section where tensile stress is caused by externally applied load.

$$= -\frac{P_e}{A_c}\left(1 + \frac{e_{e1}c_b}{r^2}\right), \text{ where } e_{e1} = 19.5 \text{ in. at the critical section}$$

$$= -\frac{1024}{767}\left(1 + \frac{19.5 \times 36.60}{712}\right) = -2.673 \text{ ksi (C)}$$

f_d = stress due to unfactored dead load at the extreme fibers of the section where tensile stress is caused by externally applied load

$$= \left[\frac{(155.4 + 179.3)(12)}{14{,}915} + \frac{(19.4 + 38.9)12}{20{,}060}\right] = 0.304 \text{ ksi}$$

$$\text{Hence, } M_{cr} = \frac{20{,}060}{12}\left(\frac{6\sqrt{6500}}{1000} + 2.673 - 0.304\right) = 4776 \text{ ft-kip}$$

y_b = distance of cgs of the prestressing strands at the critical section from the bottom fibers = 17.12 in.

$$d_{pc} = h_c - y_b = 80 - 17.12 = 62.88 \text{ in.}$$
$$0.8\,h_c = 64 \text{ in., controls}$$
$$\text{used } d_p = 64 \text{ in.}$$

Hence,

$$V_{ci} = \frac{0.6\sqrt{6500}(6.0 \times 64)}{1000} + 114.5 - \frac{172.4 \times 4776}{577.1} = 1{,}559.8 \text{ kips}$$

$$\text{Minimum } V_{ci} = 1.7\sqrt{f'_c}\, b_w d$$

$$= \frac{0.6\sqrt{6500}(6.0 \times 64)}{1000} = 52.6 \text{ kips} \ll 1{,}559.8 \text{ kips} \quad \text{O.K.}$$

b. *Web shear V_{cw}*
From Equation 5.15,

$$V_{cw} = (3.5\lambda\sqrt{f'_c} + 0.3\bar{f}_c)b_w d_p + V_P$$

where $\bar{f}_c$ is termed as F_{pc} in AASHTO notation. It is the concrete stress at the centroid of the section resisting all externally applied load.

$$\bar{f}_c = -\frac{P_e}{A_c}\left(1 - \frac{e_{e1}(c_{bc} - c_b)}{r^2}\right) + \frac{M_D(c_{bc} - c_b)}{I_c}$$

From section properties, $c_{bc} = 54.6$ in. and $c_b = 36.6$ in.

$$\bar{f}_c = -\frac{1024}{767}\left(1 - \frac{19.5(54.60 - 36.60)}{712}\right) + \frac{334.7 \times 12(54.60 - 36.60)}{545{,}894}$$

$$= -0.676 - 0.147 = -0.823 \text{ ksi (C)}$$

V_p = vertical component of prestressing force.

From Figure 12.15, the tangent of angle ψ subtended by the 10 harped

$$\text{tendons} \cong \frac{65 - 15}{48.5 \times 12} \cong \sin\psi \cong 0.086$$

$$V_p = A_{ps} f_{pe} \sin\psi = (10 \times 0.153)\, 149.0 \times 0.086 = 19.61 \text{ kips}$$

Hence,

$$V_{cw} = \left(\frac{3.5 \times 1.0\sqrt{6500}}{1000} + 0.3 \times 0.823\right) 6.0 \times 64 + 19.61$$

$$= 203.17 + 19.61 = 222.8 \text{ kips, controls since it is less than } V_{ci}$$

c. *Selection of web steel*

$$V_c = 222.8 \text{ kips}$$

$$V_U < \phi(V_c + V_s) \quad \text{or,} \quad V_s = \left(\frac{V_U}{\phi} - V_C\right)$$

$$\text{Required } V_s = \frac{286.9}{0.90} - 222.8 = 95.0 \text{ kips}$$

$$\text{Maximum allowable } V_s = 8\sqrt{f'_c}\, b_w d_v = 9\sqrt{6500}\,\frac{6.0 \times 64}{1000} = 247.7 \text{ kips}$$

> 95.0 kips O.K. (the section depth adequate for shear).

$$V_s = \frac{A_v f_y d_v}{s}$$

$$\text{Required } \frac{A_v}{s} = \frac{V_s}{f_y d_v} = \frac{95.0}{60.0 \times 64.0} = 0.0247 \text{ in.}^2/\text{ft} = 0.0021 \text{ in.}^2/\text{in.}$$

$$\text{Minimum } \frac{A_v}{s} = \frac{50 b_w}{f_y} = \frac{(50 \times 6.0)}{60{,}000} = 0.005 \text{ in.}^2/\text{in., controls.}$$

Using No. 4 two-legged U-stirrups in the rectangular end section, $A_v = 2 \times 0.20 = 0.40$ in.^{2}

$$\text{Spacing } s = \frac{A_v}{\text{unit } A_v} = \frac{0.40}{0.005} = 80 \text{ in.}$$

Maximum allowable spacing $s = 0.75\, h_c$ or 24 in.

$$= 0.75(72 + 7.5 + 0.5) = 60 \text{ in. or } 24 \text{ in.}$$

Use No. 4 U-stirrups at 12 in. center-to-center in the rectangular end block section over a width $= h = 80$ in. Beyond the end of the anchorage block, stirrups would no longer be needed. However, it is useful to use minimum vertical mesh reinforcement in the web along the span. The 12-in. spacing is necessitated by the interface horizontal shear requirement.

2. *Interface shear reinforcement*

Determine the interface shear force for the critical section at $\frac{1}{2}\, h_c$ from the support.

a. *Contact surface roughened or minimum ties used*

$$V_U = 286.9 \text{ kips} < V_{nh}$$

$$V_{nh} = \text{nominal horizontal shear strength}$$

$$\geq \frac{V_U}{\phi} = \frac{286.9}{0.9} = 318.8 \text{ kips}$$

Allowable $V_{nh} = b_v d_{pc}$ where

b_c = width of cross-section at the contact surface being investigated for horizontal shear = 42.0 in.

d_{pc} = 62.88 in. from before

$$\text{Available } V_{nh} = \frac{80(42.0 \times 62.88)}{1000} = 211.3 \text{ kips} < 318.9 \text{ kips}$$

hence, dowel reinforcement is needed.

b. *Minimum ties provided and contact surface roughened*

$$V_{nh} = 350 b_v d_{pc} = \frac{350(42.0 \times 62.88)}{1000} = 924.3 \text{ kips} > 318.8 \text{ kips O.K.}$$

Use $V_{nh} = 318.8$ kips
From equation 5.33,

$$\text{Minimum } A_{vh} = \frac{50 b_v l_{vh}}{f_y} = \frac{50 b_v s}{f_y} \text{ per dowel}$$

Assume 12-in. dowel spacing,

$$\text{Minimum } A_{vh} \text{ per dowel} = \frac{(50)(42 \times 12)}{60{,}000} = 0.42 \text{ in.}^2/\text{ft}$$

Available dowels form shear reinforcement:

$$= \text{No. 4 U-stirrups at 12 in. c./c.} = 0.40 \text{ in.}^2/\text{ft} \quad \text{O.K.}$$

Beyond the rectangular end block zone of 80 in., add in the web additional #4 dowels at 12 in. c./c. to compliment the single #4 bars available in the 6-in. thick web.

$$\text{Maximum allowable spacing} = 4\, b_w = 4 \times 6 = 24 \text{ in.} \quad \text{O.K.}$$

Photo 12.8 Pier support for Stoney Trail Bow River segmental Bridge, Calgary, Alberta, span 1562 ft, deck width 69 ft, the deck rises 89 to 118 ft above the river valley (*Courtesy* James Skeet–Reid Crowther Engineering, Calgary)

Note that the vertical web shear reinforcement is utilized here to provide for the required dowel reinforcement.

3. *Deflection Computations*

The deflection computations are similar to those given in Example 12.2 except that in the standard AASHTO specifications, fatigue live load for deflection is disregarded. from Table 12.11, Example 12.2, the final long-term deflection becomes:

$$\delta = -8.62 + 3.46 + 3.86 + 0.60 + 0.41 = -0.29 \text{ in. (camber)}$$

$$< \frac{L}{800} = \frac{120 \times 12}{800} = 1.80 \text{ in.} \quad \text{O.K.}$$

Adopt the bridge-deck design of the interior prestressed beam in Examples 12.3 and 12.4.

12.13 SHEAR AND TORSION REINFORCEMENT DESIGN OF A BOX-GIRDER BRIDGE

Example 12.5

A single span composite two-lane box girder bridge has a span of 90'–0" (27.5m) The deck is composed of seven AASHTO BIII-48 box beams at 4'-0" on centers to form a 28'–0" bridge deck with a traffic pathway width = 25'–0" as shown in Figure 12.17. Each beam is subjected

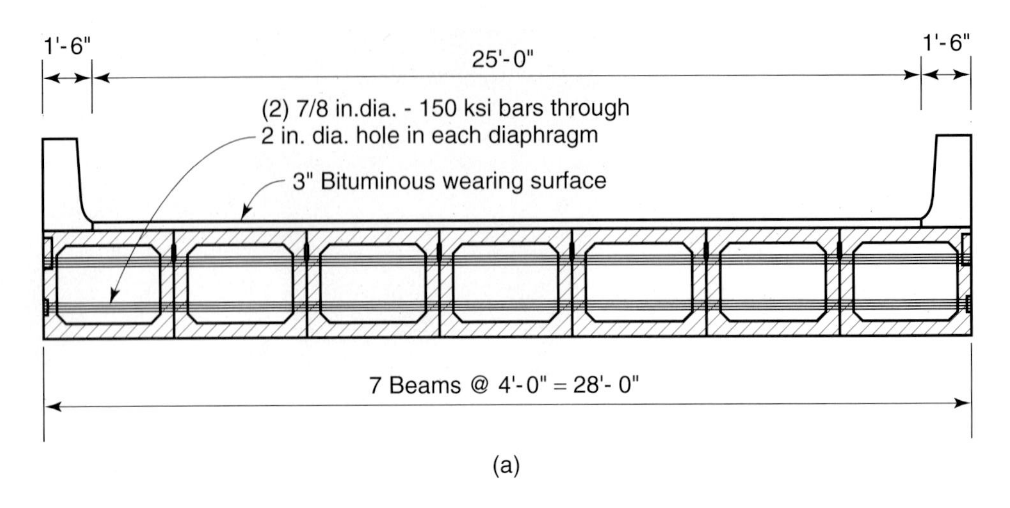

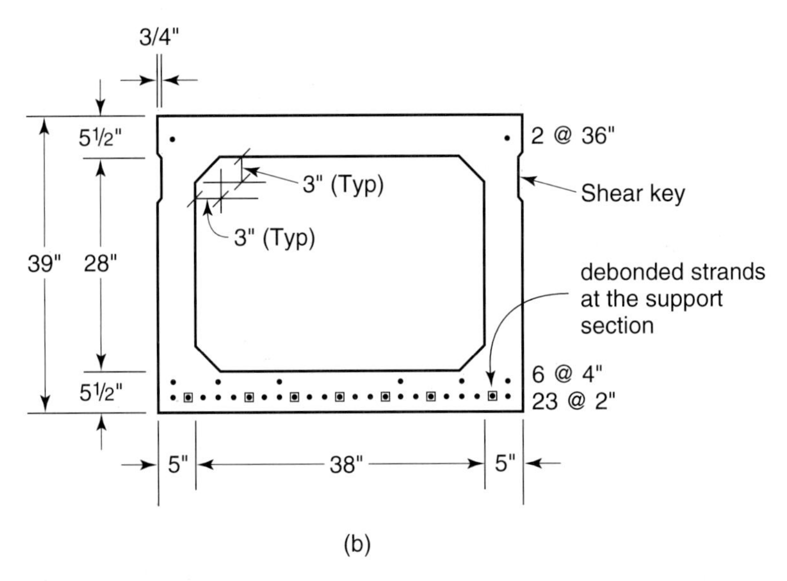

Figure 12.17 Two-Lane Box-Girder Bridge. (a) Roadway cross section, (b) Cross section of a component beam unit at midspan and at end sections. The end section has seven strands de-bonded.

to a factored shear $V_U = 140$ kips at the critical support section, a corresponding moment at that section = 320 ft-kip and a torsional moment $T_U = 165$ ft-kip.

Design the shear and torsional reinforcement for this bridge section, using the LRFD expressions, given:

$$f'_c = 5.0 \text{ ksi}$$

$$f'_{ci} = 4.0 \text{ ksi}$$

$$f_{pu} = 270.0 \text{ ksi}$$

$$f_{py} = 0.90\, f_{pu} = 243.0 \text{ ksi}$$

$$f_{pi} < 0.75\, f_{pu} = 202.5 \text{ ksi}$$

$$\text{Total prestress loss} = 22\%$$

$$f_y = 60.0 \text{ ksi}$$

$$E_{ps} = 28{,}500 \text{ ksi}$$
$$E_s = 29{,}000 \text{ ksi}$$
$$A_c = 813 \text{ in.}^2$$
$$h = 39 \text{ in.}$$
$$I_g = 168{,}367 \text{ in.}^4$$
$$c_b = 19.29 \text{ in.}$$
$$c_t = 19.71 \text{ in.}$$
$$S_b = 8{,}721 \text{ in.}^3$$
$$S^t = 8{,}542 \text{ in.}^3$$
$$E_{ci} = 3840 \text{ ksi}$$
$$E_c = 4290 \text{ ksi}$$

Solution:

1. ***Effective shear depth, d_v***
The strands are horizontal, $V_p = 0$
From Figure 12.17,

$$\text{At midspan, } A_{ps} = 29\ \tfrac{1}{2}\text{-in. dia., 7-wire, low-relaxation 270-K strands}$$
$$= 4.437 \text{ in.}^2 \text{ at the bottom fibers.}$$
$$\text{At support, } A_{ps} = 22 \text{ stands, since 7 are de-bonded} = 3.366 \text{ in.}^2$$

Midspan c.g.s. of strands from bottom:

$$y_{bc} = \frac{23 \times 2 + 6 \times 4 + 2 \times 36}{31} = 4.58 \text{ in.}$$

Midspan $d_p = h - y_{bc} = 39.0 - 4.58 = 34.42$ in.

Support cgs of strands from bottom:

$$y_{be} = \frac{(23 - 7)2 + 6 \times 4 + 2 \times 36}{31 - 7} = 5.33 \text{ in.}$$

support $d_p = h - y_{be} = 39.0 - 5.33 = 33.67$ in.

$$\text{where, } d_c = \frac{A_{ps}f_{ps}d_p + A_s f_y d}{A_{ps}f_{ps} + A_s f_y}$$

d_v = effective shear depth
= distance between resultants of tensile and compressive forces, but not less than $0.90d_c$ or $0.72\,h$.

To determine the neutral axis depth, c, and the equivalent rectangular block depth, a, use the midspan section for the computations as the section of maximum moment ($M_U = 0$ at support).

Assume the neutral axis within the $5\frac{1}{2}$ in. "flange."

$$c = \frac{A_{ps}f_{pu} + A_s f_y - A'_s f'_y}{0.85 f'_c \beta_1 b + kA_{ps}\left(\dfrac{f_{pu}}{d_p}\right)} = \frac{4.437(270) + 0 + 0}{0.85 \times 5.0 \times 0.80 \times 48 + 0.28 \times 4.437\left(\dfrac{270}{34.42}\right)}$$
$$= 6.95 \text{ in.}$$

$$a = \beta_1 c = 0.80 \times 6.95 = 5.56 \text{ in.} > 5.5 \text{ in., hence treat as a flanged}$$
$$\text{section with width } b = \text{web width } b_w$$

$$b_w = 2 \times 5 = 10 \text{ in.}$$

$$c = \frac{A_{ps}f_{pu} + A_sf_y - A'_sf'_y - 0.85f'_c\beta_1(b - b_w)h_f}{0.85f'_c\beta_1 b_w + kA_{ps}\left(\dfrac{f_{pu}}{d_p}\right)}$$

$$= \frac{4.437 \times 270 + 0 - 0 - 0.85 \times 0.80 \times 5.0(48 - 10)5.5}{0.85 \times 5.0 \times 0.80 \times 10 + 0.28 \times 4.437\left(\dfrac{270}{34.42}\right)}$$

$$= 11.14 \text{ in.} > 5.5 \text{ in.} \qquad \text{O.K.}$$

$$a = \beta_1 c = 0.80 \times 11.14 = 8.91 \text{ in.}$$

$$f_{ps} = f_{pu}\left(1 - k\frac{c}{d_p}\right) = 270\left(1 - 0.28\left\{\frac{11.14}{34.4}\right\}\right) = 245.4 \text{ ksi}$$

$$d_v = \left(d_p - \frac{a}{2}\right) = 33.67 - \frac{8.91}{2} = 29.22 \text{ in.}$$

$0.9\,d_c = 0.9 \times 33.67 \quad = 30.30$ in. (controls)

$0.72h = 0.72 \times 39 \quad = 28.08$ in.

2. *Angle of inclination θ of the diagonal compression struts*

Critical section near the support is the larger of $0.5d_v \cot\theta$ or d_v from the face of the support. θ is obtained from Figure 12.9 using the values of v/f'_c and ϵ_x.

From equation 12.41,

$$v = \frac{V_u - \phi V_p}{\phi\, b_v d_v} + \frac{T_p P_h}{\phi\, A_{oh}^2}$$

$$\phi = 0.9$$

$$A_{oh} = (48 - 2 \times 1.5 \text{ for clear cover} - 2 \times 0.25 \text{ for stirrups}) \times (39 - 2 \times 1.5 - 2 \times 0.25) = 44.5 \times 35.5 = 1580 \text{ in.}^2$$

$$P_h = 2(44.5 + 35.5) = 160 \text{ in.}$$

$$b_v = 2 \times 5 = 10 \text{ in.}$$

$$v = \frac{140 - 0}{0.9 \times 10 \times 30.3} + \frac{165 \times 12 \times 160}{0.9(1580)^2}$$

$$= 0.515 + 0.141 = 0.656 \text{ ksi}$$

$$\frac{v}{f'_c} = \frac{0.656}{5.0} = 0.131$$

From Equations 12.27 for torsion adjustment and
Eq. 12.28(a) for stress f_{po},

$$\epsilon_x = \frac{\dfrac{M_u}{d_u} + 0.5N_u + 0.5V_u \cot\theta\sqrt{V_u^2 + \left(\dfrac{p_h T_u}{2A_o}\right)^2} - A_{ps}f_{po}}{E_sA_s + E_{ps}A_{ps}}$$

Use $f_{po} \cong f_{pe} = f_{pi}(1 - 0.22) = 202.5 \times 0.78 = 157.9$ ksi

$N_u = 0$

$A_o \cong 0.85A_{oh} = 0.85 \times 1580 = 1343 \text{ in.}^2$

Try $\theta = 22.5°$ for a first trial

Photo 12.9 Launching the segmental bridge segments for Stoney Trail Bow River segmental Bridge, Calgary, Alberta, span 1562 ft, deck width 69 ft, the deck rises 89 to 118 ft above the river valley (Courtesy James Skeet–Reid Crowther Engineering, Calgary)

$$\epsilon_x = \frac{\dfrac{320 \times 12}{30.3} + 0 + 0.5 \cot 22.5° \sqrt{(140)^2 + \left(\dfrac{160 \times 165 \times 12}{2 \times 1343}\right)^2} - (3.366 \times 157.9)}{0 + 28{,}500 \times 3.366}$$

$$= -1.84 \times 10^{-3} \text{ in./in.}$$

Since the value of the tensile strain ϵ_x at the level of the reinforcement centroid is negative, its value has to be reduced by a factor

$$F_\epsilon = \frac{E_s A_s + E_{ps} A_{ps}}{E_c A_c + E_s A_s + E_{ps} A_{ps}} = \frac{0 + 28{,}500 \times 4.437}{4290 \times 813 + 0 + 28{,}500 \times 4.437} = 0.035$$

Adjusted $\epsilon_x = 0.035\,(-1.84 \times 10^{-3}) = -0.064 \times 10^{-3}$ in./in.

Entering the chart in Figure 12.9 for $v/f'_c = 0.131$ and $\epsilon_x = -0.064 \times 10^{-3}$ give $\theta = 22.6°$ which is $\approx$ the θ assumed in the first trial, hence accept.

Corresponding $\beta = 2.7$

3. *Design of transverse closed stirrups*

$$V_c = 0.0316\beta\sqrt{f'_c}\, b_v d_v \text{ using ksi units}$$

$$= 0.0316 \times 2.7\sqrt{5.0}\,(10 \times 30.3) = 57.0 \text{ kips} < 140 \text{ kips}$$

hence web shear reinforcement is necessary.

From equation 12.40(a)

$$\frac{A_v}{s} = \frac{V_n - (0.0361\beta\sqrt{f'_c}\, b_v d_v + V_p)}{f_y d_v \cot\theta}$$

$$= \frac{\dfrac{140}{0.9} - (0.0316 \times 2.7\sqrt{5.0} \times 10 \times 30.3 + 0)}{60.0 \times 30.3 \times 1.376}$$

$$= 0.039 \text{ in.}^2/\text{in./two legs.}$$

From equation 12.40(b),

$$\frac{A_t}{s} = \frac{T_n}{2A_0 f_y \cot\theta} = \frac{\dfrac{165 \times 12}{0.9}}{2 \times 1343 \times 60 \times 1.376} = 0.010 \text{ in.}^2/\text{in./one leg}$$

$$\frac{A_{vt}}{s} = \frac{A_v}{s} + 2\frac{A_t}{s} = 0.039 + 2(0.01) = 0.059 \text{ in.}^2/\text{in./two legs}$$

Trying No. 4 closed stirrups with each of the two legs of a stirrup in each vertical wall.

$$\text{spacing } s = \frac{2 \times 0.20}{0.059} = 6.78 \text{ in.}$$

Use No. 4 closed stirrups at $6\frac{3}{4}$ in. center to center throughout the span. Note that the spacing of the transverse reinforcement can be increased along the span if the shear and torsion envelopes warrant it.

4. ***Longitudinal reinforcement check***
From Equation 12.40,

$$\phi(A_s f_y + A_{ps} f_{ps}) \geq \frac{M_u}{\phi d_v} + 0.5\frac{N_u}{\phi} + \cot\theta\sqrt{\left(\frac{V_U}{\phi} + 0.5V_s - V_p\right)^2 + \left(\frac{0.45T_u p_o}{\phi 2A_o}\right)^2}$$

$$A_{ps} f_{ps} = 3.366 \times 245.5 = 826 \text{ kips}$$

$$A_o \approx 0.85\, A_{oh} = 0.85 \times 1580 = 1343 \text{ in.}^2$$

Photo 12.10 State Route 509 Elevated Single-Point Urban Interchange, Tacoma, Washington: a situ-cast post-tensioned box girder bridge featuring tight radius curved exterior webs; the footprint of the interchange is approximately two football fields in size (Designed by BERGER/ABAM Engineers, Federal Way, Washington, courtesy Robert Mast, Senior Principal)

Photo 12.11 Valley Ave. Bridge, Fife, Washington: 4-span bridge consisting of two structural types, situ-cast post-tensioned concrete box girder with precast prestressed end girders (Designed by BERGER/ABAM Engineers, Federal Way, Washington, courtesy Robert Mast, Senior Principal)

$$p_o \approx 0.85\, p_h = 0.85 \times 160 = 136 \text{ in.}$$

$$V_s = \frac{A_v f_y d_v}{s} = \frac{0.40 \times 60 \times 30.3}{6.75} = 108 \text{ kips.}$$

Hence,

$$\frac{320 \times 12}{0.9 \times 30.2} + 0 + 1.376\sqrt{\left(\frac{140}{0.9} + 0.5 \times 102\right)^2 + \left(\frac{0.45 \times 165 \times 12 \times 136}{0.9 \times 2 \times 1343}\right)^2}$$

$$= 140.8 + 1.376(212.6) = 431 \text{ kips} < 826 \text{ kips.}$$

Hence no additional longitudinal reinforcement is needed. Adopt the No. 4 vertical ties at 6-3/4 in. on centers in each of the two beam box walls. Each vertical transverse tie, if not in a single piece, has to be fully developed to satisfy the development length requirements of the specifications.

12.14 LRFD MAJOR DESIGN EXPRESSIONS IN SI FORMAT

Eq. 12.9(a):

$$f_{ps} = f_{pu}\left(1 - k\frac{c}{d_p}\right)$$

$$k = 2\left(1.04 - \frac{f_{py}}{f_{pu}}\right)$$

For non-bonded tendons,

$$f_{ps} = f_{pe} + 1.5\frac{L}{d_p}E_{ps}\epsilon_{cu}\left(\frac{d_p}{c} - 1.0\right)\frac{L_1}{L_2}$$

$< 0.94\, f_{py}$ where L_1 = span and tendon length

L_2 = stresses in MPa.

Eq. 12.12(a):

$$\frac{c}{d_e} \leq 0.42$$

$$\text{Moment distribution factor } p_d = 20\left(1 - 2.36\frac{c}{d_e}\right)$$

Eq. 12.24(a)

$$V_c = 0.083\beta\sqrt{f'_c}\, b_v d_v$$

where b and d_y (mm), f'_c (MPa)

Eq. 12.29, when torsion is present,

$$\phi(A_s f_y + A_{ps} f_{ps}) \geq \frac{M_u}{d_v} + 0.5N_u + \cot\theta\sqrt{(V_U - 0.5V_p)^2 + \left(\frac{0.45T_u P_h}{2A_0}\right)^2}$$

Eq. 12.35:

$$A_{vf} = 0.35b_v\frac{s}{f_y}, \text{ where } b_v, s \text{ (mm).}$$

Eq. 12.38:

$$V_u = \sqrt{V_u^2 + \left(\frac{0.9p_h T_u}{2A_o}\right)^2}$$

where V_u (Newton), T_u (N-mm), A_o (mm^2).

Eq. 12.40(b):

$$\frac{A_t}{s} = \frac{T_n}{2A_0 f_y \cot\theta} \text{ where } s \text{ (mm).}$$

SELECTED REFERENCES

12.1 ASCE, "Minimum Design Loads for Buildings and Other Structures," ANSI-ASCE 7-95 Standard, American Society of Civil Engineers, Reston, VA, 1995, pp. 214.

12.2 AASHTO, "Standard Specifications for Highway Bridges," 16th Ed. and 1997, 1998 Supplements, American Association of State Highway and Transportation Officials, Washington, D.C., 1996.

12.3 AASHTO, "LRFD Bridge Design Specifications," American Association of State Highway and Transportation Officials, Washington, D.C., 1998.

12.4 ACI, "Building Code Requirements for Structural Concrete (ACI 318-99) and Commentary (ACI 318R-99), American Concrete Institute, Farmington Hills, MI.

12.5 Nawy, E.G., *Reinforced Concrete—A Fundamental Approach,* 4th Ed., Prentice Hall, Upper Saddle River, NJ., pp. 786.

12.6 Collins, M. P., and Mitchel, D., *Prestressed Concrete Structures,* Prentice Hall, Upper Saddle River, NJ, 1991.

12.7 Collins, M. P., and Mitchel, D, "Shear and Torsion Design of Prestressed and Non-Prestressed Concrete Beams," *PCI Journal,* Precast/Prestressed Concrete Institute, Chicago, 1980, pp. 12–100.

12.8 Hsu, T. T. C., *Torsion in Reinforced Concrete,* Van Nostrand Reinhold, New York, 1983.

12.9 Hsu, T. T. C., "Torsion in Structural Concrete—Uniformly Prestressed Members Without Web Reinforcement," *PCI Journal,* V. 13, Precast/Prestressed Concrete Institute, Chicago, 1968, pp. 34–44.

12.10 Hsu, T. T. C., *Unified Theory of Reinforced Concrete,* CRC Press, Boca Raton, FL, 1993.

12.11 PCI, *Bridge Design Manual,* Precast/Prestressed Concrete Institute, Chicago, 1999.

12.12 PTI, *Post-Tensioning Manual,* 6th Ed., Post-Tensioning Institute, Phoenix, AX, 2000.

12.13 Kudlapur, S. T., and Nawy, E. G., "Early Age Shear Friction Behavior of High Strength Concrete Layered Systems at Subfreezing Temperatures," Proceedings, *Symposium on Designing Concrete Structures for Serviceability and Safety,* ACI SP-133.9, E. G. Nawy and A. Scanlon, ed., American Concrete Institute, Farmington Hills, MI, 1992, pp. 159–185.

12.14 Naaman, A. E., "Unified Design Recommendations for Reinforced, Prestressed and Partially Prestressed Concrete Bending and Compression Members," *ACI Structural Journal,* American Concrete Institute, Farmington Hills, MI, March–April 1992, pp. 200–210.

12.15 Naaman, A. E., "Unified Bending Strength Design of Concrete Members: AASHTO-LRFD Code," *Journal of Structural Engineering,* American Society of Civil Engineers, Reston, VA, June 1995, pp. 964–970.

12.16 Mast, R. F., "Unified Design Provisions for Reinforced and Prestressed Concrete Flexural and Compression Members," *ACI Structural Journal,* American Concrete Institute, Farmington Hills, MI, April 1992, pp. 185–199.

12.17 Badie, S. S., Baishya, M. C., and Tadros, M. K., "NUDECK—An Efficient and Economical Precast Prestressed Bridge Deck System," *PCI Journal,* Vol. 43 No. 5, Precast/Prestressed Concrete Institute, Chicago, September–October 1998, pp. 56–71.

12.18 Ma, Z., Huo, X., and Tadros, M. K., "Restraint Moment in Precast/Prestressed Concrete Continuous Bridges," *PCI Journal,* Vol. 43, No. 6, Precast/Prestressed Concrete Institute, Chicago, November–December 1998, pp. 40–57.

12.19 Nawy, E. G., editor-in-chief, *Concrete Construction Engineering Handbook,* CRC Press, Boca Raton, FL, 1998, pp. 1–1250.

12.20 Nawy, E. G., *Fundamentals of Performance Concrete,* 2nd Ed., John Wiley & Sons, New York, 2000.

12.21 Hsu, T. T. C., Zhu, R. H. and Lee, J. Y., "A Critique on the Modified Compression Field Theory," Presented at the 78th Annual Meeting of the Transportation Research Board, Washington, D.C., January 10–14., 23 pp.

PROBLEMS FOR SOLUTION

12.1 Design for flexure a 100 ft (30.5m) simply supported AASHTO-PCI bulb-tee composite bridge deck with no skews using the LRFD AASHTO specifications. The superstructure is composed of six pretensioned beams at 9'-0" (2.74 m) on centers. The bridge has an 8 in. (203 mm) situ-cast concrete deck with the top one half inch to be considered as wearing surface. The design live load is the HL-93 AASHTO-LRFD fatigue loading. Assume the bridge is to be located in a low seismicity zone. Given, the following maximum allowable stresses:

Deck $f'_c = 4000$ psi, normal weight

$f_c = 0.60\,f'_c = 2400$ psi

Bulb-tee $f'_c = 6500$ psi

$f'_{ci} = 5500$ psi

$f_c = 0.60\,f'_c = 3900$ psi, Service III

$f_c = 0.45\,f'_c = 2925$ psi, Service I

$f_{ci} = 0.60\,f'_c = 3480$ psi

$f_t = 6\sqrt{f'_c} = 484$ psi

$f_{pu} = 270{,}000$ psi

$f_{py} = 0.90\,f_{pu} = 243{,}000$ psi

$f_{pi} = 0.75\,f_{pu} = 202{,}500$ psi

$f_y = 60{,}000$ psi

$E_{ps} = 28.5 \times 10^6$ psi

$E_s = 29.0 \times 10^6$ psi.

12.2 Design the web shear reinforcement for the bulb-tee beam in Problem 12.1 at the critical section near the supports and the interface shear transfer reinforcement at the interface plane between the precast section and the deck situ-cast concrete. Also, verify if the span deflection is within the allowable limits.

12.3 A single-span two-lane unskewed AASHTO Type BIII-48 bridge has an overall span of 96 ft and the cross-section shown in Figure P12.1 (Adapted from the PCI Manual - Ref. 12.11). The total deck width is 28 ft and the clear roadway is 25 ft wide. The deck has a 3-in. bituminous wearing surface. Design for flexure and shear an interior box element using the AASHTO LRFD specifications in the design. Given:

Effective span = 95 ft.

$f'_c = 5000$ psi, normal weight

$f_c = 0.60\,f'_c = 3000$ psi, Service III

$f_c = 0.45\,f'_c = 2250$ psi, Service I

$f_{ci} = 0.60\,f'_c = 4000$ psi

$f_t = 6\sqrt{f'_c} = 424$ psi

$f_{pu} = 270{,}000$ psi

$f_{py} = 0.90\,f_{pu} = 243{,}000$ psi

$f_{pi} = 0.75\,f_{pu} = 202{,}500$ psi

$f_y = 60{,}000$ psi

$E_{ps} = 28.5 \times 10^6$ psi

$E_s = 29.0 \times 10^6$ psi.

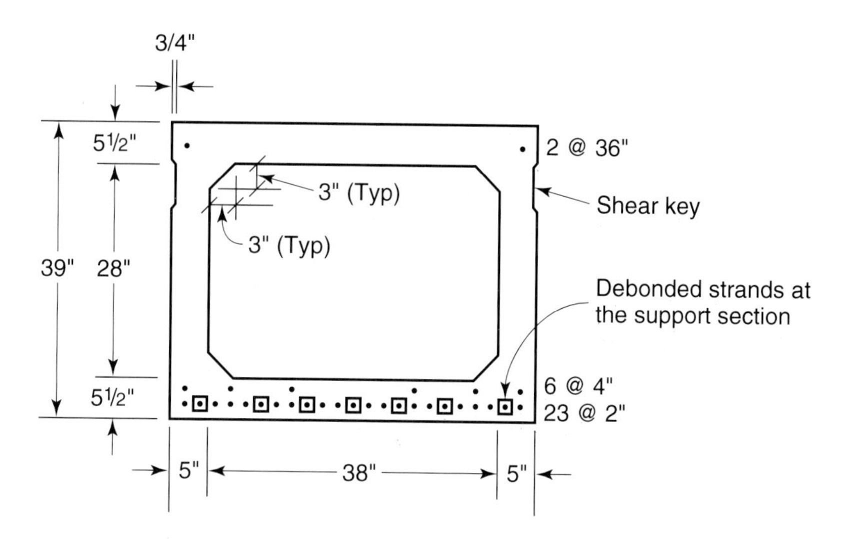

Figure P12.1 Box Beam Geometry

Section Properties:

$$A_c = 813 \text{ in.}^2$$

$$h = 39 \text{ in.}$$

$$I_c = 168{,}367 \text{ in.}^4$$

$$c_b = 19.29 \text{ in.}$$

$$c_t = 19.71 \text{ in.}$$

$$S_b = 8728 \text{ in.}^3$$

$$S^t = 8542 \text{ in.}^3$$

$$W_D = 847 \text{ lb/ft.}$$

12.4 Solve Problem 12.3 using the AASHTO Standard specifications for both flexure, shear and deflection.

13

Northridge, California, 1994 earthquake structural failure. (*Courtesy,* Dr. Murat Saatcioglu.)

SEISMIC DESIGN OF PRESTRESSED CONCRETE STRUCTURES

13.1 INTRODUCTION: MECHANISM OF EARTHQUAKES

The earth crust is composed of several layers of hard "tectonic" plates, called *lithospheres,* which float on the softer, underpinning, fluid medium called *mantle.* These plates or rock masses, when fractured, *form fault lines.* The adjoining plates or rock masses are prevented by the interacting frictional forces from moving past one another most of the time. However, when this frictional ultimate resistance is reached because of the continuous motion of the underlying fluid, any two plates can impact on one another, generating seismic waves that can cause large horizontal and vertical ground motions. These ground motions translate into inertia forces in structures.

The length and width of a fault are interrelated to the magnitude of the earthquake. The fault is the cause rather than the result of the earthquake. A fault can cause an earthquake due to the following reasons (Ref. 13.5):

1. Cumulative strain in the fault over a long period of time reaches the rupture level.
2. Slip of the tectonic plates at the fault zones causes a rebound, as in Fig. 13.1(a),

Photo 13.1 311 S. Wacker Street, Chicago, 12,000 concrete (*Courtesy* Portland Cement Association.)

3. Sudden push and pull forces at the fault lead to reverse moment couples, as in Fig. 13.1(b). The moment caused by these couples as a measure of earthquake size can be termed the *seismic moments*. The magnitude is equal to rock rigidity × fault area × amount of slip. The range of slip velocity in such faults as the San Andreas Fault in California is 30 to 100 mm per year. On this basis, a slippage or horizontal motion of 3 m at such faults in one single earthquake is expected to occur at intervals of 30 to 100 years.

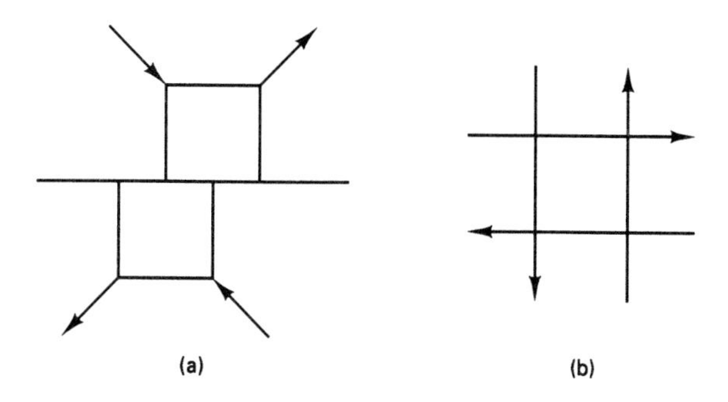

Figure 13.1 Mechanism of earthquakes: (a) slip of tectonic plates; (b) reverse moment couples.

Photo 13.2 Bridge girder collapse in the San Francisco 1989 earthquake. (*Courtesy* Portland Cement Association.)

Earthquakes may be characterized by three categories: low, moderate, and high intensity. The intensity is governed by ground motion accelerations, represented by response spectra and coefficients derived from such spectra. A structure is expected to respond essentially elastically to low-intensity earthquakes. In such a case, the stresses are expected to remain within the elastic range, with a slight possibility of developing limited inelasticity with no appreciable structural or non-structural damage. Structural response is expected to be inelastic under high-intensity earthquakes having an intensity of 5 or higher on the Richter scale and in regions close to the epicenter. For the design of structures in seismic zones, two methods are presented in the IBC 2000 code: the spectral response method and the equivalent lateral force method. The latter has certain limitations that will be discussed later.

A detailed discussion of the subject of earthquakes is beyond the scope of this book since the primary aim of this chapter is the proportioning of seismic resistant components of concrete structures. However, some of the basic underlying characteristics are important to cover. They are intended to help define the magnitude of the lateral seismic base shear forces that determine the geometry and form of the earthquake resisting components of a structure, namely, the lateral force resisting system (LFRS).

Such a system has two components: horizontal and vertical. The horizontal elements are the components that resist the seismic forces. They can be diaphragms, coupling beams, and shear walls. The vertical component comprises the walls and vertical frames of the structure.

13.1.1 Earthquake Ground Motion Characteristics

Ground motion, caused by seismic tremors, involves acceleration, velocity, and displacement. These are in the majority amplified, thereby producing forces and displacements, which can exceed those which the structure is able to sustain (Ref. 13.13). The maximum value of the ground motion magnitude, namely, the peak ground velocity, peak ground acceleration and peak ground displacement become the principal parameters in the seismic design of structures.

Additional factors also affect the response of a structure. They include frequency, amplitude of motion, shaking duration, and site soil characteristics. These can all be represented by a response spectrum which idealizes a structure into a dampened, single degree of freedom system (SDF) oscillating at various periods and frequencies. The maximum vibration magnitude reached during any time duration after the base ground motion is its spectral value.

13.1.2 Fundamental Period of Vibration

The basic natural period T of a simple one-degree-of-freedom system is the time required to complete one whole cycle during dynamic loading. In other words, it is the time required for a phase angle ωt to travel from 0 to 2π, where ω is the angular frequency of the system. Hence $\omega t = 2\pi$, leading to the expression

$$T = \frac{2\pi}{\omega} = 2\pi\left(\frac{m}{k}\right)^{1/2} \tag{13.1}$$

where m = mass of system
k = spring constant and damping is not considered

Most reinforced concrete structures are multidegrees-of-freedom systems, as in Fig. 13.2. In this case the structural mass can be assumed to be concentrated in the vertical spring element at the floor level, resulting in multiple modes with frequencies (periods) for each mode. The compound natural period T is then evaluated with due consideration given to the distribution of mass and stiffness. Codes require that T be established using the structural properties and deformation characteristics of the resisting elements in a properly substantiated analysis using expressions such as those given by the International Building Code—IBC 2000 (Ref. 13.2), or The Uniform Building Code (Ref. 13.3) integrated into the IBC provisions.

Since a structure is composed of a series of single degrees of freedom components subjected to the *same* base motion, a series of maximum values related to the SDF system's fundamental periods, T, would ensue. These, in turn, form a spectral curve for that base ground motion. By knowing the base motion, the SDF fundamental period and the percent critical damping, one can obtain from the applicable curve the maximum acceleration, velocity, and displacement relative to the base (Ref. 13.14). Evidently, computer

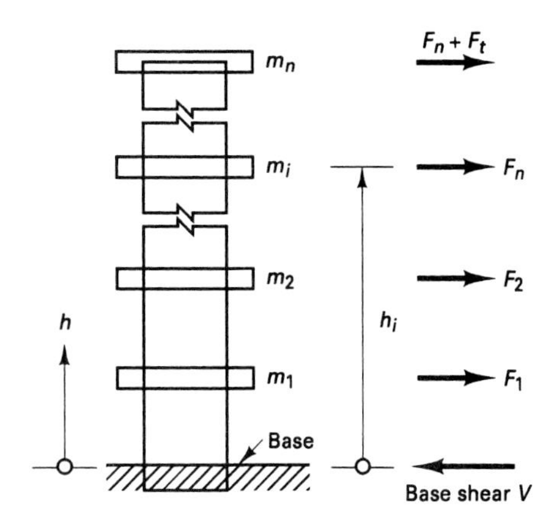

Figure 13.2 Modeling multistory structures.

Photo 13.3 Northridge, California, 1994 earthquake structural failure. (Courtesy Dr. Murat Saatcioglu.)

use is needed to obtain a complete spectral response of the multi-degree state of a structure.

It should be recognized that a structure is designed to resist earthquake motion such that it is able to sustain and survive the earthquake through large inelastic deformations and energy dissipation through cracking and limited local material failure, but without loss of stability. It would be highly uneconomical to design the lateral force resisting system to the earthquake forces such that the structure deforms only elastically as a result of these forces. The codes have this as a basic philosophy particularly for major earthquakes in which some structural damage can result.

13.1.3 Design Philosophy

The International Building Code (IBC 2000) on seismic design consolidates the three major existing regional codes:

1. Building Officials Code Administration International (BOCA)
2. International Conference of Building Officials (ICBO): Uniform Building Code (UBC)
3. Southern Building Code Congress International (SBCCI)

Underlying its seismology design provisions are:

1. Recommended design levels related to effective peak accelerations that can resist minor earthquakes without damage, moderate earthquakes without structural damage, and major earthquakes in which some structural damage can result.
2. Minimum design criteria for all types of buildings, low and high rise, with and without shear walls.
3. Spectral response values for various ground motion intensities, mainly within the elastic range.
4. Provide design criteria for lateral ground motion, unidirectional and bi-directional, addressing them one at a time.
5. Limit the story drift and displacement magnitudes of the building structures within acceptable ranges, through control of stiffness of components and shear walls, diaphragms, and coupling beams.

13.2 SPECTRAL RESPONSE METHOD

13.2.1 Spectral Response Acceleration Maps

As discussed in Ref. 13.15, prior to the Northridge and Kobe earthquakes, the Uniform Building Code (UBC) provisions performed satisfactorily in the United States in past earthquakes. The failures in these two cases were determined to be due to "related configurations of the structural systems, inadequate connection detailing, incompatibility of deformations and design or construction deficiencies. They were not due to deficiency in strength (Structural Engineers Association of California, 1995).

The UBC provisions incorporated in the International Building Code (IBC) are based on consideration of the site conditions of the structure and the application of maximum considered earthquake ground motion maps for site class B, prepared by the United States Geological Survey (USGS). The equivalent maximum considered earthquake ground motion values for the ceiling were determined to be 1.50 g for the short period and 0.60 g for the long period (Ref. 13.15).

The high seismicity regions, where the maximum considered earthquake ground motion values are greater than 0.75 g for the 1.0 sec, peak acceleration additional requirements are imposed on irregular structures exceeding five stories in height and a period T in excess of 0.5 sec, such as increasing the ground motion spectral acceleration values by 50 percent. The USGS large-scale maps for the 1.0 sec and the 0.2 sec levels of spectral response acceleration, site-B class, and 5 percent critical damping are condensed and abridged in Figs. 13.3(a) and (b) for general guidance. They show the relative values of the peak spectral response accelerations at the two ground motion levels of 0.2 and 1.0 sec. Values have to be extrapolated linearly from the USGS large-scale maps for use in the seismic design of structures.

13.2.2 Design Parameters

Both the spectral response method and the equivalent lateral force method are based on the same code principles and formulations presented in this chapter. Sites are classified into six categories A, B, C, D, E, and F as shown in Table 13.1 on site properties.

Ground motion accelerations and the maximum considered earthquake spectral response acceleration are considered at 1.0 sec period (S_1) and at short periods (S_s) such as 0.2 sec obtained from seismic contour maps discussed in Section 13.2.1.

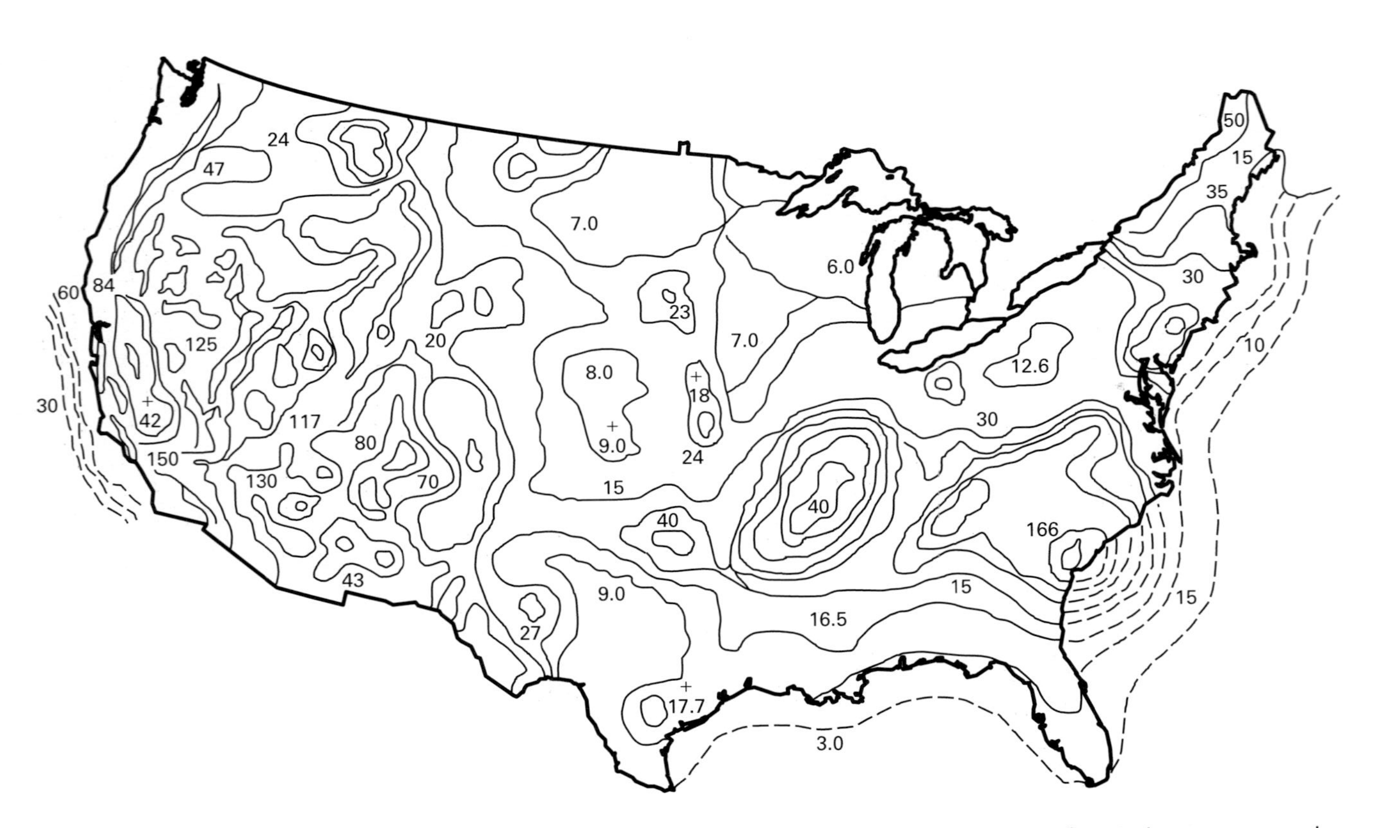

Figure 13.3(a) Maximum considered earthquake ground motion for the United States, 0.2 sec. Spectral response acceleration S_s as a percent of gravity, site-class B with 5 percent critical damping.

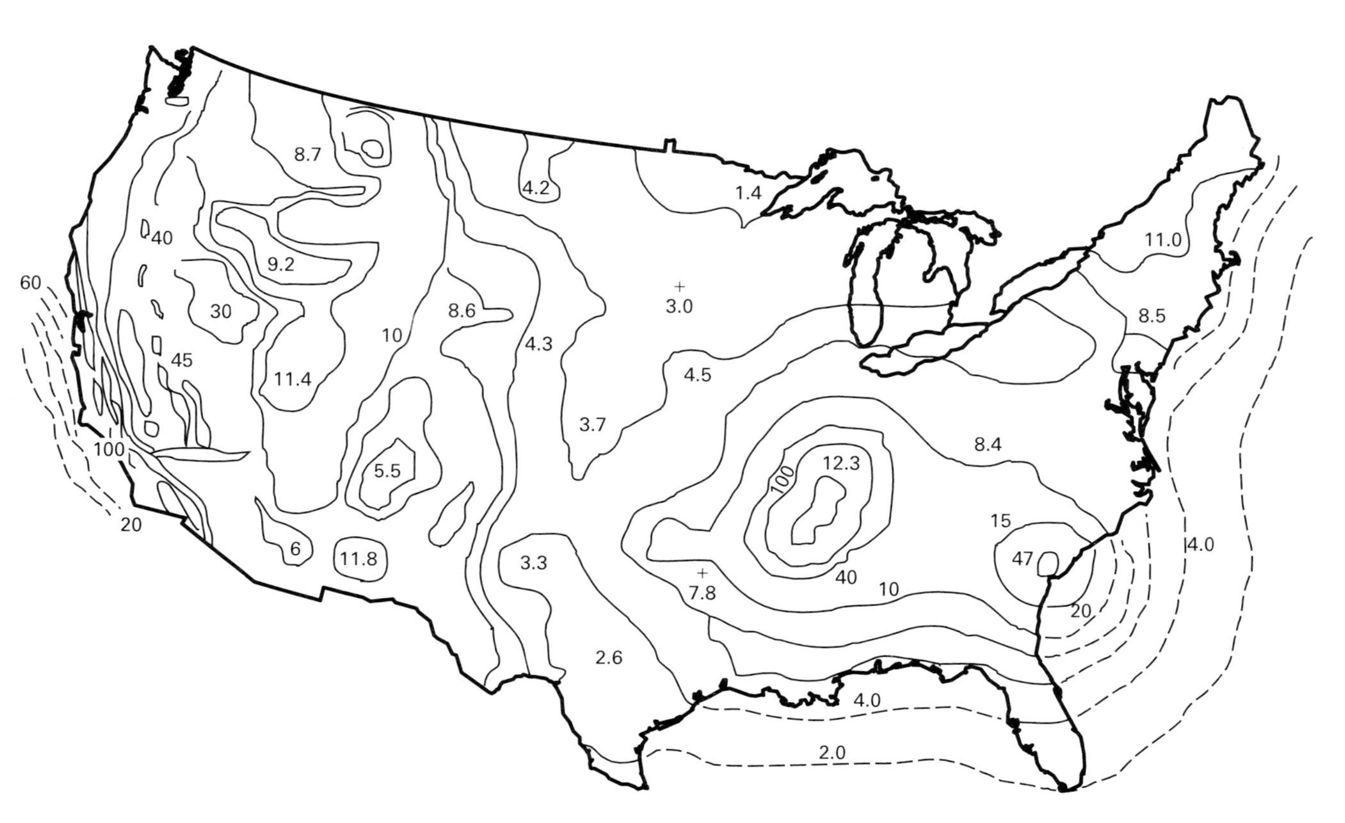

Figure 13.3(b) Maximum considered earthquake ground motion for the United States, 1.0 sec. Spectral response acceleration S_1 as a percent of gravity, site-class B with 5 percent critical damping.

Table 13.1 Site Classifications

Site Class	Soil Profile Name	Average Properties in Top 100Ft (30 m), As in Section 1615.1.5		
		Soil Shear Wave Velocity, $\overline{V}_{s1}$ (ft/S)	**Standard Penetration Resistance, $\overline{N}$**	**Soil Unconfined Shear Strength $\overline{S}_U$ (PSF)**
A	Hard rock	$\overline{V}_s > 5{,}000$	not applicable	not applicable
B	rock	$2{,}500 \leq \overline{V}_s \leq 5{,}000$	not applicable	not applicable
C	Very dense soil and soft Rock	$1{,}200 \leq \overline{V}_s \leq 2{,}500$	$\overline{N} > 50$	$\overline{S}_U > 2{,}000$
D	Stiff soil profile	$600 \leq \overline{V}_s \leq 2{,}500$	$15 \leq \overline{N} \leq 50$	$1{,}000 \leq \overline{S}_U \leq 2{,}000$
E	Soft soil profile	$\overline{V}_s < 600$	$\overline{N} < 15$	$\overline{S}_U < 1{,}000$
E		Any profile with more than 10 ft of soil having the following characteristics: -plasticity index PI > 20; -moisture content w > 40% and -unconfined shear strength $\overline{S}_U < 500$ psf		
F		Any profile containing soil having one or more of the following characteristics: 1. Soils vulnerable to potential failure or collapse under seismic loading such as liquefiable soils, quick and highly sensitive clays, collapsible weakly cemented soils. 2. Peats and/or highly organic clays ($H > 10$ ft of peat and/or highly organic clay where H = thickness of soil) 3. Very high plasticity clays ($H > 25$ ft with plasticity index PI > 75) 4. Very thick soft/medium stiff clays ($H > 120$ ft)		

For S1: 1 ft/sec = 305 mm/sec; 1 psf = 0.0479 kP; 1 ft. = 305 mm.

The design spectral response accelerations at short periods (S_S) and at 1 sec second (S_1) are to be adjusted for site class effect (S_{MS}) at short periods and (S_{M1}) for 1 sec based on Table 13.1 in conjunction with Tables 13.2(a) and 13.2(b) for site coefficients.

The maximum considered earthquake spectral response for short and one second periods are respectively defined by the following expressions:

$$S_{MS} = F_a S_S \tag{13.2a}$$

$$S_{M1} = F_v S_1 \tag{13.2b}$$

Table 13.2(a) Values of Site Coefficient F_a as a Function of Site Class and Mapped Spectral Response Acceleration at Short Periods (S_s)

Site Class	Mapped Spectral Response Acceleration at Short Periods				
	$S_s \leq 0.25$	$S_s = 0.50$	$S_s = 0.75$	$S_s = 1.00$	$S_s \leq 1.25$
A	0.8	0.8	0.8	0.8	0.8
B	1.0	1.0	1.0	1.0	1.0
C	1.2	1.2	1.1	1.0	1.0
D	1.6	1.4	1.2	1.1	1.0
E	2.5	1.7	1.2	0.9	Note a
F	Note a	Note a	Note a	Note a	Note a

Table 13.2(b) Values of Site Coefficient F_V as a Function of Site Class and Mapped Spectral Response Acceleration at 1.0 Sec Periods (S_1)

Site Class	Mapped Spectral Response Acceleration at 1.0-Sec Periods				
	$S_1 \leq 0.1$	$S_1 = 0.2$	$S_1 = 0.3$	$S_1 = 4$	$S_1 \leq 5$
A	0.8	0.8	0.8	0.8	0.8
B	1.0	1.0	1.0	1.0	1.0
C	1.7	1.6	1.5	1.4	1.3
D	2.4	2.0	1.8	1.6	1.5
E	3.5	3.2	2.8	2.4	Note a
F	Note b	Note b	Note b	Note b	Note b

NOTES: a—Straight line interpolation for intermediate values are to be made.
b—Site geotechnical investigation and dynamic site response analyses are to be performed

where,

F_a = Site coefficient from Table 13.2a
F_v = Site coefficient from Table 13.2b
S_s = Mapped spectral acceleration for short periods (See Ref. 13.2 for map contour values)
S_1 = Mapped spectral acceleration for 1.0-sec periods (See Ref. 13.2 for map contour values)

For 5 percent damped design, the spectral response acceleration becomes:

$$S_{DS} = \frac{2}{3} S_{MS} \tag{13.3a}$$

$$S_{D1} = \frac{2}{3} S_{M1} \tag{13.3b}$$

13.2.3 Earthquake Design Load Classifications

The International Building Code (IBC-2000) classifies the seismic design categories into three seismic use groups in lieu of the former zones 0 to 4 of the UBC Code. The three groups for short period and 1.0-sec period response acceleration are given in Tables 13.3(a) and 13.3(b) respectively. These seismic use groups can be defined as follows (Ref. 13.2):

Table 13.3(a) Seismic Design Category Based on Short Period Response Accelerations

Value of S_{DS}	Seismic Use Group		
	I	II	III
$S_{DS} < 0.167$g	A	A	A
$0.167\text{g} \leq S_{DS} < 0.33$g	B	B	C
$0.33\text{ g} \leq S_{DS} < 0.50$g	C	C	D
$0.50\text{ g} \leq S_{DS}$	D[a]	D[a]	D[a]

Table 13.3(b) Seismic Design Category
Based on 1 Second Period Response Accelerations

Value of S_{D1}	Seismic Use Group		
	I	II	III
$S_{D1} < 0.067g$	A	A	A
$0.067g \leq S_{D1} < 0.133g$	B	B	C
$0.133\text{ g} \leq S_{D1} < 0.20g$	C	C	D
$0.20\text{ g} \leq S_{D1}$	D[a]	D[a]	D[a]

NOTE: [a]- Seismic Use Groups I and II structures located on sites with mapped maximum considered earthquake spectral response acceleration at 1-sec period, S_1, equal to or greater than 0.75 g shall be assigned to seismic design category E and seismic use group III structures located on such sites shall be assigned to seismic design category F.

i. *Seismic Use Group I:* These are structures that are not assigned to either seismic use group II or III

ii. *Seismic Use Group II:* The structures in this group are those the failure of which would result in substantial public hazard due to occupancy or use described in Table 13.5

iii. *Seismic Use Group III:* The structures in these groups are those the failure of which would result in having essential facilities for post-earthquake recovery and those containing substantial quantities of hazardous substances in jeopardy.

In Tables 13.3(a) and (b), categories B and C range from low- to moderate-risk regions where categories D and E are designated as high-risk seismic regions.

These tables enable the designer to choose from the spectral maps the S_S and S_1 values pertinent to the structure site location. They are based on classifying the map regions into three as follows (Ref. 13.15, Part 2):

Region 1—Regions of Negligible Seismicity with Very Low Probability of Collapse of the Structure (No Spectral Values)

Region definition: Regions for which $S_S < 0.25$ g and $S_1 < 0.10$ g.

Design values: No spectral ground motion values required. Use a minimum *lateral* force level of 1 percent of the dead load for seismic *design* Category A.

Region 2—Regions of Low and Moderate to High Seismicity (Probabilistic Map Values)

Region definition: Regions for which 0.25 g $< S_S <$ 1.5 g and 0.25 g $< S_1 <$ 0.60 g.

Maximum considered earthquake map values: Use S_S and S_1 map values.

Transition between Regions 2 and 3—Use values of S_S = 1.5 g and $S_1 = 0.60$ g.

Region 3—Regions of High Seismicity Near Known Faults (Deterministic Values)

Regional definition: Regions for which 1.5 g < S_S and 0.60 g < S_1.

The structural analysis based on the worst load combinations should be the basis for determining the seismic forces E for combined gravity and seismic load effects when they are additive and the maximum seismic load effect E_m. The value of E and E_m are determined from the following expressions detailed in Ref. 13.2 for additive seismic force and dead load:

$$E = \rho Q_E + 0.2\, S_{DS} D \tag{13.4a}$$

$$E = \Omega_o Q_E + 0.2\, S_{DS} D \tag{13.4b}$$

For counteracting seismic forces and dead load:

$$E = \rho Q_E - 0.2\, S_{DS} D \tag{13.5a}$$

$$E = \Omega_o Q_E - 0.2\, S_{DS} D \tag{13.5b}$$

where, E = combined effect of horizontal and vertical earthquake-induced forces
ρ = reliability factor based on system redundancy = 1.0 for categories A, B, and C
Q_E = effect of horizontal seismic forces
S_{DS} = spectral response acceleration at short periods obtained from IBC Sec. 1615.1.3 or 1615.2.2.5
Ω_o = system over-strength factor given in Table 13.4
D = effect of dead load

13.2.4 Redundancy

A redundancy coefficient ρ has to be assigned to all structures based on the extent of structural redundancy inherent in the lateral force resisting system. For structures in seismic design categories A, B, and C, the value of the redundancy coefficient ρ is to be taken as 1.0. For structures in seismic design categories D, E, and F, the redundancy coefficient ρ has to be taken as the largest of the values ρ_1 computed at each story level "i" of the structure in accordance with the expression

$$\rho_1 = 2 - \frac{20}{r_{maxi}\sqrt{A_i}} \tag{13.6a}$$

In SI Units, the expression becomes

$$\rho_1 = 2 - \frac{6.1}{r_{maxi}\sqrt{A_i}} \tag{13.6b}$$

where

$r_{max\,i}$ = ratio of the design story shear resisted by the most heavily loaded single element in the story to the total story shear for a given loading condition,

A_I = Floor area in square feet (m^2) of the diaphragm level immediately above the story

The value of ρ cannot be less than 1.0 and need not exceed 1.5.

13.2.5 General Procedure Response Sepectrum

The design response can be idealized by the fundamental period-response acceleration relationship shown in Fig. 13.4 for three fundamental period levels.

Table 13.4 Design Coefficients and Factors for Basic Seismic-Force-Resisting Systems (abridged from table 1617.6, Ref. 13.2)

BASIC SEISMIC-FORCE-RESISTING SYSTEM	RESPONSE MODIFICATION COEFFICIENT R[a]	SYSTEM OVER-STRENGTH FACTOR Ω_o	DEFLECTION AMPLIFICATION FACTOR, c_D[b]	SYTEM LIMITATIONS AND BUILDING HEIGHT LIMITATIONS (FT) BY SEISMIC DESIGN CATEGORY[c] AS DETERMINED IN IBC SECTION 1616.				
				A & B	C	D[d]	E[e]	F[f]
Bearing Wall System								
Special reinforced concrete shear walls	5.5	2.5	5	NL	NL	160	160	100
Ordinary reinforced concrete shear walls	4.5	2.5	4	NL	NL	NP	NP	NP
Detailed plain concrete shear walls	2.5	2.5	2	NL	NL	NP	NP	NP
Ordinary plan concrete shear walls	1.5	2.5	1.5	NL	NP	NP	NP	NP
Building Frame System								
Ordinary reinforced concrete shear walls	5	2.5	4.5	NL	NL	NP	NP	NP
Detailed plain concrete shear walls	3	2.5	2.5	NL	NL	NP	NP	NP
Ordinary plain concrete shear walls	2	2.5	2	NL	NP	NP	NP	NP
Moment Resistant Frames								
Special reinforced concrete moment frames	8	3	5.5	NL	NL	NL	NL	NL
Intermediate reinforced concrete moment frames	5	3	4.5	NL	NL	NP	NP	NP
Ordinary reinforced concrete moment frames	3	3	2.5	NL[h]	NP	NP	NP	NP
Dual System with Special Moment Frames								
Special reinforced concrete shear wall	8	2.5	6.5	NL	NL	NL	NL	NL
Ordinary reinforced concrete shear wall	7	2.5	6	NL	NL	NP	NP	NP
Dual System with Intermediate Moment Frames								
Special reinforced concrete shear wall	6	2.5	5	NL	NL	160	100	100
Ordinary reinforced concrete shear wall	5.5	2.5	4.5	NL	NL	NP	NP	NP
Shear Wall-Frame interactive system with ordinary reinforced concrete moment frames and ordinary reinforced concrete shear walls	5.5	2.5	5	NL	NP	NP	NP	NP

For SI 1 ft = 305 mm

[a]Response modification coefficient R, for use throughout

[b]Deflection amplification factor, C_d

[c]NL = not limited and NP = not permitted

[d]limited to buildings with a height of 240 ft or less.

[e]limited to buildings with a height of 160 ft or less.

[f]Ordinary moment frame is permitted to be used in lieu of Intermediate moment frame in Seismic Design Categories B, and C.

[g]The tabulated value of the overstrength factor, Ω_o may be reduced by subtracting $\frac{1}{2}$ for structures with flexible diaphragms but shall not be taken as less than 2.0 for any structure.

[h]Ordinary moment frames of reinforced concrete are not permitted as a part of the seismic-force-resisting system in Seismic Design Category B structures founded on Site-Class E or F soils

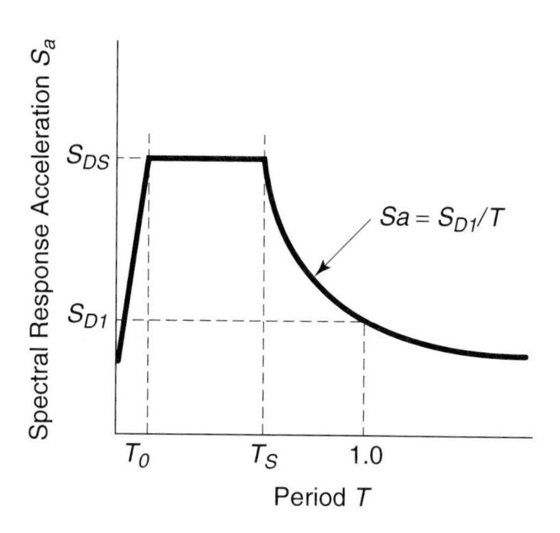

Figure 13.4 Design response spectrum.

1. For periods in seconds less than or equal to T_o, the design spectral response acceleration S_a is determined from the following equation:

$$S_a = 0.6 \frac{S_{DS}}{T_o} T + 0.4\, S_{DS} \tag{13.7a}$$

2. For periods greater than or equal to T_o, and less than or equal to T_s, the design spectral response acceleration S_a, is taken equal to S_D.
3. For periods greater than T_s, the design spectral response acceleration, S_a, is determined from the expression:

$$S_a = \frac{S_{D1}}{T} \tag{13.7b}$$

where,

S_{DS} = the design spectral response acceleration at short periods
S_{D1} = the design spectral response acceleration at 1-sec periods
T = Fundamental period (in seconds) of the structure
$T_o = 0.2\, S_{D1}/S_{DS}$
$T = S_{D1}/S_{DS}$

The sites have to be classified for determining the shear wave velocity and the maximum considered earthquake ground motion. Details are given in the IBC (Ref. 13.2) section 1615.

13.3 EQUIVALENT LATERAL FORCE METHOD

13.3.1 Horizontal Base Shear

In this method, a building is considered fixed at the base. The seismic base shear, V, in a given direction is determined from the expression(Ref. 13.2):

$$V = C_S W \tag{13.8}$$

where,

C_S = seismic response coefficient
W = The effective seismic weight of the structure, including the total dead loads and other loads listed herein:

1. In areas used for storage, a minimum of 25 percent of the reduced floor live load (floor live load in public garages and open parking structures need not be included).
2. Where an allowance for partition load is included in the floor load design, the actual partition weight or a minimum weight of 10 psf (500 Pa/m^2) of floor area, whichever is greater.
3. Total operating weight of permanent equipment.
4. 20 percent of flat roof snow load where the flat roof snow load exceeds 30 psf.

$$C_S = \frac{S_{DS}}{(R/I)} \tag{13.9}$$

But C_S cannot exceed the value:

$$C_S = \frac{S_{D1}}{\left(\frac{R}{I}\right)T} \tag{13.10}$$

nor can it be taken less than:

$$C_S = 0.044\, S_{DS} \tag{13.11}$$

where,

S_{DS} = Design spectral response acceleration at short period as determined in Section 13.2.2
R = Response modification factor from Table 13.4
I = Occupancy importance factor from Table 13.5
T = fundamental period of building (seconds)

For buildings and structures in seismic design categories E or F and in buildings and structures for which the 1-sec spectral response, S_1 is equal to or greater than 0.6 g, the value of the seismic coefficient C_S should not be taken less than:

$$C_S = \frac{0.5S_1}{R/I} \tag{13.12}$$

The fundamental period T in the direction under consideration has to be determined by analysis basis on the structural and deformational characteristics of the resisting element. In lieu of an analysis, an approximate fundamental period T_a, in seconds, can be used from the following expression:

$$T_a = C_T h^{3/4} \tag{13.13}$$

where,

C_T = Building Period Coefficient

- 0.035 for moment resisting frame systems of steel in which the frames resist 100 percent of the required seismic force and are not enclosed or adjoined by more rigid components that will prevent the frames from deflecting when subjected to seismic forces (the metric coefficient is 0.085)
- 0.030 for moment resisting frame systems of reinforced concrete in which the frames resist 100 percent of the required seismic force and are not enclosed or adjoined by more rigid components that will prevent the frames from deflecting when subjected to seismic forces (the metric coefficient is 0.073)
- 0.030 for eccentrically braced steel frames (the metric coefficient is 0.073)
- 0.020 for all other building systems (the metric coefficient is 0.049)

h_n = the height (ft or m) above the base to the highest level of the building.

Table 13.5 Occupancy Importance Factor
Classification of Buildings and Other Structures for Importance Factors

Category[a]	Nature of Occupancy	Siesmic Factor I_E	Snow Factor I_S	Wind Factor I_W
I	Building and other structures except those listed in Categories II, III, and IV	1.00	1.0	1.00
II	Buildings and other structures that represents a substantial hazard to human life in the event of failure including, but not limited to: • Buildings and other structures where more than 300 people congregate in one area • Buildings and other structures for elementary school, secondary school or day-care facilities with capacity greater than 250 • Buildings and other structures with a capacity greater than 500 for colleges or adult education facilities • Health care facilities with a capacity of 50 or more resident patients but not having surgery or emergency treatment facilities • Jail or detention facilities • Any other occupancy with an occupant load greater than 5,000 • Power generating stations, water treatment for potable water, waste water treatment facilities and other public utility facilities not included in category IV • Buildings and other structures not included in category IV containing sufficient quantities of toxic or explosive substances to be dangerous to the public if released.	1.25	1.1	1.15
III	Building and other structures designated as essential facilities including, but not limited to: • Hospitals and other health care facilities having surgery or emergency treatment facilities • Fire, rescue and police stations and emergency vehicle garages • Designated earthquake, hurricane, or other emergency shelters • Designated emergency preparedness, communication, and operation centers and other facilities required for emergency response • Power generating stations and other public utility facilities required as emergency back-up facilities for category IV structures • Structures containing highly toxic material • Aviation control towers, air traffic control centers and emergency aircraft hangers • Buildings and other structures having critical national defense functions • Water treatment facilities required to maintain water pressure for fire suppression	1.50	1.2	1.15
IV	Buildings representing low hazard to human life at failure such as agricultural and storage facilities	1.00	0.8	0.87

[a] "Category" is equivalent to "Seismic Use Group" for the purpose of Section 13.2.3.

Table 13.6 Coefficient for Upper Limit On Computed Fundamental Period

Design Spectral Response Acceleration at 1-sec period, S_{D1}	Coefficient C_u
> 0.4	1.2
0.3	1.3
0.2	1.4
0.15	1.5
< 0.1	1.7

In cases where moment resisting frames do not exceed 12 stories in height and having a minimum story height of 10 ft (3 m), an approximate period T_a in seconds in the following form can be used:

$$T_a = 0.1\,N \tag{13.14}$$

where

$$N = \text{number of stories}$$

The computed fundamental period, T, cannot exceed the product of the coefficient, C_n, in Table 13.6 for the upper limit on the computed period and the approximate fundamental period, T_a.

13.3.2 Vertical Distribution of Forces

The lateral force F_x (kips or kN) induced at any level can be determined from the following expressions:

$$F_x = C_{vx}\,V \tag{13.15a}$$

$$C_{vx} = \frac{W_x h_x^k}{\sum_{i=1}^{n} W_i h_i^k} \tag{13.15b}$$

where

C_{xv} = vertical distribution factor
V = total design lateral force or shear at the base of the building (kips or kN),
W_i and W_x = the portion of the total gravity load of the building, W, located or assigned to level i or x
h_i and h_x = the height (ft or m) from the base to level i or x
k = a distribution exponent related to the building period as follows:

- For buildings having a period of 0.5 sec or less, $k = 1$
- For buildings having a period of 2.5 sec or more, $k = 2$
- for buildings having a period between 0.5 and 2.5 seconds, k shall be 2 or shall be determined by linear interpolation between 1 and 2

13.3.3 Horizontal Distribution of Story Shear V_x

The seismic design story horizontal shear in any story, V_x (kips or kN) should be determined from the following expression:

$$V_x = \sum_{i=1}^{x} F_i \tag{13.16}$$

where

F_i = the portion of the seismic base shear, V (kips or kN) introduced at level i.

13.3.4 Rigid and Flexible Diaphragms

(a) *Rigid diaphragms:* The seismic design story shear, V_x, has to be distributed to the various vertical elements of the system in the story under consideration. This distribution is to be based on the relative stiffness of the vertical resisting elements and the diaphragms.

(b) *Flexible Diaphragms:* The seismic design story shear, V_x, in this case has to be distributed to the various vertical elements based on the tributary area of the diaphragms to each line of resistance. The vertical elements of the lateral force resisting system can be considered to be in the same line of resistance, if the maximum out of plane offset between such elements in less than 5 percent of the building's dimension *perpendicular* to the direction of the lateral load.

13.3.5 Torsion

If the diaphragms are not flexible, the design has to include the torsional moment M_t (Kip-ft or kN-m) resulting from the difference in location between the center of mass and the center of stiffness. Dynamic amplification of torsion for structures in seismic design category C, D, E or F has to be accounted for by multiplying the torsional moments by a torsional amplification factor presented in Ref. 13.2, Sec. 16.17.4.

13.3.6 Story Drift and the P-Delta Effect

(a) *Drift:* The design story drift, Δ, is computed as the difference between the deflections of the center of mass at the top and bottom of the story being considered. If allowable stress design is used Δ is computed using earthquake forces without dividing by 1.4.

The deflection of level X is to be determined from the following expression,

$$\delta_x = \frac{C_d \delta_{xe}}{I} \tag{13.17}$$

where,

C_d = Deflection amplification factor (Table 13.4)

δ_x = Deflections (in. or mm) determined by an elastic analysis of the seismic forces resisting system.

I = Occupancy importance factor (Table 13.5)

The design story drift, Δ, has to be increased by an incremental factor relating to the P-delta effects. The redundancy coefficient, ρ, in the case of drift should be taken as 1.0.

(b) *P-delta effects:* The P-delta effects can be disregarded if the stability coefficient, θ, from the following expression is equal or less than 0.10,

$$\theta = \frac{P_x \Delta}{V_x h_{sx} C} \tag{13.18}$$

where,

P_x = The total unfactored vertical design load at and above Level x (kip or kN); when computing the vertical design load for purposes of determining P-delta, the individual load factors need not exceed 1.0

Δ = The design story drift (in. or mm) occurring simultaneously with V_x

V_x = The seismic shear force (kip of kN) acting between level x and $x - 1$

h_{sx} = The story height (ft or m) below level x

C_d = The deflection amplification factor in Table 13.4.

The stability coefficient, θ, shall not exceed θ_{max} determined as follows:

$$\theta_{max} = \frac{0.5}{C_d \beta} \leq 0.25$$

where:

β = The radio of shear demand to shear capacity for the story between level x and $x - 1$. Where the ratio β is not calculated, a value of $\beta = 1.0$ shall be used.

When the stability coefficient, θ, is greater than 0.10 but less than or equal to θ_{max}, inter-story drifts and element forces shall be computed including P-delta effects. To obtain the story drift for including the P-delta effect, the design story drift shall be multiplied by $1.0/(1 - \theta)$.

When θ is greater than θ_{max}, the structure is potentially unstable and has to be redesigned.

The allowable story drifts are given in Table 13.7 as follows:

Table 13.7 Allowable Story Drift, Δ (in. or mm)[a]

Building	Seismic Use Group		
	I	II	III
Buildings, other than masonry shear wall or masonry wall frame buildings, four stories or less in height with interior walls, partitions, ceilings, and exterior wall systems that have been designed to accommodate the story drifts.	$0.025h_{sx}$[b]	$0.020h_{sx}$	$0.015h_{sx}$
Masonry cantilever shear wall buildings[c]	$0.010h_{sx}$	$0.010h_{sx}$	$0.010h_{sx}$
Other masonry shear wall buildings	$0.007h_{sx}$	$0.007h_{sx}$	$0.007h_{sx}$
Masonry wall frame buildings	$0.013h_{sx}$	$0.013h_{sx}$	$0.010h_{sx}$
All other buildings	$0.020h_{sx}$	$0.015h_{sx}$	$0.010h_{sx}$

[a] There shall be no drift limit for single-story buildings with interior walls, partitions, ceilings, and exterior wall systems that have been designed to accommodate the story drifts.

[b] h_{sx} is the story height below level x.

[c] Buildings in which the basic structural system consists of masonry shear walls designed as vertical elements cantilevered from their base or foundation support which are so constructed that moment transfer between shear walls (coupling) is negligible.

13.3.7 Overturning

Ground motion can result in overturning of a structure. At any story, the increment of overturning moment in the story under consideration would have to be distributed to the various vertical force-resisting elements, in the same proportion as the distribution of the horizontal shear forces to these elements. The overturning moment at level *x*, M_x (kip-ft or kN-m), is determined from the following expression:

$$M_x = \tau \sum_{i=x}^{n} F_i(h_i - h_x) \tag{13.19}$$

where

F_i = Portion of h_i and h_x = height (ft or m) from the base to the level *i* or *x*.
τ = Overturning moment reduction factor
= 1.0 for the top 10 stories
= 0.8 for the 20th story from the top and below
= values between 1.0 and 0.8 determined by a straight line interpolation for stories between the 20th and 10th stories below the top. The seismic base shear, *V*, is induced at level *i*.

13.3.8 Simplified Analysis Procedure for Seismic Design of Buildings

This procedure can be used for structures in seismic use group I, subject to the following limitations, otherwise either the method in Section 13.2 or this section has to be used.

1. Buildings of light-framed construction not exceeding *three* stories in height, excluding basement.
2. Buildings of any construction other than light framed, not exceeding two stories in height, excluding basement.

   The seismic base shear, *V*, can be computed from the following expression,

$$V = \frac{1.2 S_{ds} W}{R} W \tag{13.20}$$

where

S_{DS} = Design elastic response acceleration at short periods as determined from Section 13.2
R = Response modification factor from Table 13.4
W = The effective seismic weight of the structure, including the total dead load and other loads listed below.

In areas used for storage, a minimum of 25 percent of the reduced floor live load (floor live load in public garages and open parking structures need not be included.)

1. Where an allowance for partition load is included in the floor load design, the actual partition weight of 10 psf of floor area, whichever is greater.
2. Total weight of permanent operating equipment
3. 20 percent of flat roof snow load where flat snow load exceeds 30 psf (1.44 kN/m^2).

The vertical distribution of forces at each level would be computed from the following expression:

$$F_x = \frac{1.2 S_{DS}}{R} W_x \tag{13.21}$$

where,

W_x = The portion of the effective seismic weight of the total structure, W, at story level x.

For structures satisfying this section, the design story drift, Δ, is taken as 1 percent of the story height unless a more exact analysis is made.

13.3.9 Other Aspects in Seismic Design

The discussion presented in the previous sections is intended only to highlight the most important basic considerations for establishment of the seismic basic shear force values and their distribution over the height of a structure, at all story levels. The scope of this book does not permit more coverage of other essential topics such as modeling, model forces, deflections and drifts, diaphragms, coupling beams, interconnecting shear walls, connections, irregularity of structures, out-of-plane loading, torsion, and foundations.

Through a careful review of the details presented, the numerical examples and solving the assignments, the reader becomes well equipped to handle the design requirement aspects of the topics listed. The International Building Code—IBC 2000(Ref. 13.2) detailed provisions give all the additional provisions and guidance needed for safe complete designs of concrete structures that can successfully resist severe earthquakes. The ensuing sections will present ACI 318-99 code provisions for proportioning and detailing of reinforced concrete elements that can withstand Seismic loading through conformity with the IBC 2000 requirements.

13.4 SEISMIC SHEAR FORCES IN BEAMS AND COLUMNS OF A FRAME: STRONG COLUMN–WEAK BEAM CONCEPT

13.4.1 Probable Shears and Moments

Shear failure in reinforced concrete members is regarded as brittle failure. Therefore, in designing earthquake-resistant structures, it is important to provide excess shear capacity over and above that corresponding to flexural failure. The ACI 318-99 requirements are based on the strong column-weak beam concept subsequently discussed. Hence, plastification of the critical regions at the ends of the beams will have to be considered as a possible loading condition.

The shear force is then computed based on the moment resistances in the developed plastic hinges, labeled as probable moment resistance, M_{pr}, developed when the longitudinal flexural steel enters into the hardening stage. Consequently, in the computation of the probable moment resistance, $1.25 f_y$ is used as the stress in the longitudinal reinforcement. This is because the development of inelastic rotation at the faces of the joints is associated with strains in the flexural reinforcement well in excess of the yield strain. As a result, the joint shear force generated by the flexural reinforcement is computed for an increased stress $\lambda_0 f_y$ where $\lambda_o = 1.25$, namely, an increase in stress of 25 percent. In order to absorb the energy that can cause plastic hinging, the earthquake resistant frame has to be ductile in part through confinement of the longitudinal reinforcement of the columns and the beam-column joints and in part through the provision of the excess shear capacity previously discussed.

Fig. 13.5 shows the deformed geometry of and the moment and shear forces for a beam subjected to gravity loading and reversible side-sway. If the intensity of gravity load is W_u then, ACI 318-99 stipulates:

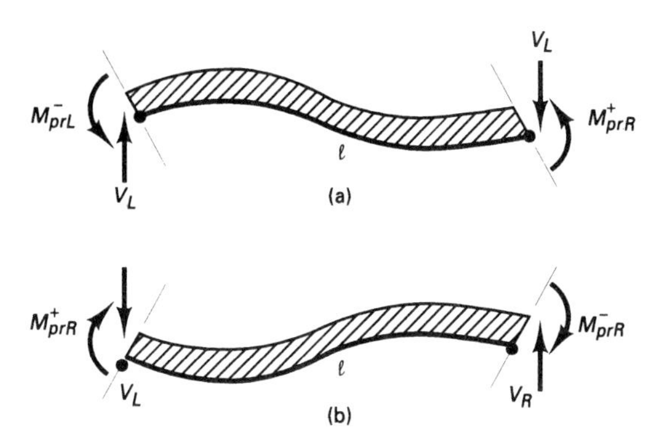

Figure 13.5 Seismic moments and shears at beam ends: (a) sidesway to the left; (b) sidesway to the right

$$W_u = 0.75(1.4D + 1.7L + 1.87E) \tag{13.22}$$

The IBC (Sec. 1605.2) stipulates the following load combinations:

$1.4D$
$1.2D + 1.6L + 0.5(L_r \text{ or } S \text{ or } R)$
$1.2D + 1.6L(L_r \text{ or } S) + (f_1L \text{ or } 0.8W)$
$1.2D + 1.3W + f_1L + 0.5(Lr \text{ or } S \text{ or } R)$
$1.2D + 1.0E + (f_1L \text{ or } f_2S)$
$0.9 +/- (1.0E \text{ or } 1.3W)$

Photo 13.4 Skybridge, Vancouver, Canada, a 2020-ft long cable-stayed bridge and the world's longest transit bridge. (Courtesy Portland Cement Association.)

where,

f_1 = 1.0 for floors in places of public assembly, for live loads in excess of 100 lb/ft^2 (4,79 kN/m^2), and for parking garage live load
= 0.5 for other live loads

f_2 = 0.7 For roof configurations (such as saw tooth) that do not shed snow off the structure
= 0.2 for other roof configurations

L = Live load except roof load

L_r = Roof live load including any live load reduction

R = Rain load

S = Snow load

W = Wind load

The seismic shear forces are:

$$V_L = \frac{M_{prL}^- + M_{prR}^+}{l} + 0.75\frac{1.4D + 1.7L}{2} \tag{13.23}$$

$$V_R = \frac{M_{prL}^+ + M_{prR}^-}{l} - 0.75\frac{1.4D + 1.7L}{2} \tag{13.24}$$

where l = span, L and R subscripts = left and right ends and M_{pr} = probable moment strength at the end of the beam based on steel reinforcement tensile strength of 1.25 f_y and strength reduction factor $\phi = 1.0$. These instantaneous moments, M_{pr}, should be computed on the basis of equilibrium of moments at the joint where the beam moments are equal to the probable moments of resistance.

The shear forces in the columns are computed in a similar manner so that the horizontal shear force, V_e at top and bottom of the column is

$$V_e = \frac{M_{prl} + M_{pr2}}{h} \tag{13.25}$$

except that end moments for columns M_{pr1} and M_{pr2} need not be greater than the moments generated by the M_{pr} of beams framing into the beam-column joint. h = column height and the subscripts 1 and 2 indicate the top and bottom column end moments respectively as seen in Figure 13.6. The sense of moments at the joints is shown in Figure 13.7.

13.4.2 Strong Column Weak Beam Concept

As previously stated, U.S. seismic codes require that earthquake induced energy be dissipated by plastic hinging of the beams rather than the columns. This hypothesis is due to the fact that compression members such as columns have lower ductility than flexure-dominant beams. If the columns are not stronger than the beams framing into a joint, inelastic action can develop in the column, and if large enough, can cause the column to collapse. Furthermore, the consequence of a column failure is far more severe than a local beam failure. Therefore, the ACI 318-99 Code as well as the IBC stipulates "strong columns and weak beams". This is ensured by the following inequality

$$\sum M_{col} \geq \left(\frac{6}{5}\right)\sum M_{bm} \tag{13.26}$$

Where ΣM_{col} = sum of moments, at the face of the joint, corresponding to the nominal flexural strength of the columns framing into that joint.

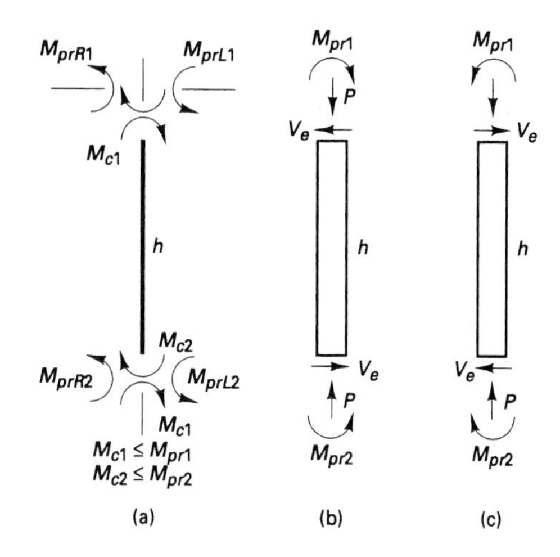

Figure 13.6 Seismic moments and shears at column ends: (a) joint moments (b) sway to right; (c) sway to left.

ΣM_{bm} = sum of moments, at the face of the joint, corresponding to the nominal flexural strengths of the beams framing into that joint.

For a joint subjected to reversible base shear forces, as shown in Fig. 13.7, Eq. 13.26 becomes

$$(\phi\, M_n^+ + \phi\, M_n^-)_{col} \geq \frac{6}{5}(\phi\, M_n^+ + \phi\, M_n^-)_b \tag{13.27}$$

where ϕ = 0.90 for beams
= 0.70 for tied and 0.75 for spiral columns.
= 0.90 to 0.7 for beam-columns.

13.5 ACI CONFINING REQUIREMENTS FOR STRUCTURAL CONCRETE MEMBERS

13.5.1 Longitudinal Reinforcement in Compression Members

1. In seismic design, when the factored axial load P_u is negligible or significantly less than $A_g\, f'_c/10$, the member is considered a flexural member (beam). If $P_u >$

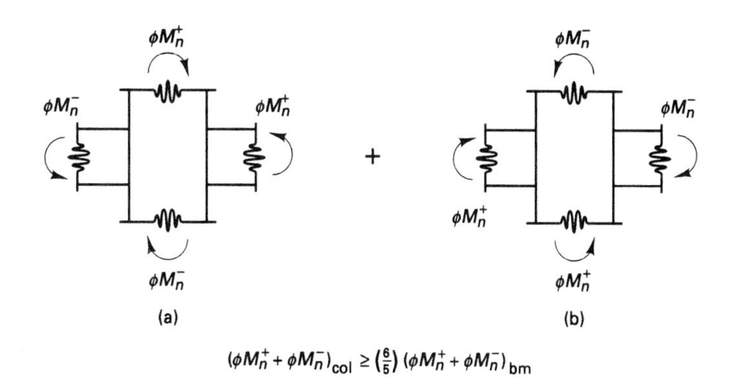

Figure 13.7 Seismic moment summation at beam-column joint: (a) sidesway to left; (b) sidesway to right.

$A_g f'_c/10$, the member is considered beam-column, because it is subjected to both axial and flexural loads as columns and shear walls are.

2. The shortest cross-sectional dimension ≥ 12 in. (300 mm).
3. The limitation on the longitudinal reinforcement ratio in the beam-column element is $0.01 \leq \rho_g = A_s/A_g \leq 0.06$. For practical considerations, an upper limitation of 6 percent is too excessive, because it results in impractical congestion of longitudinal reinforcement. A practical maximum total percentage ρ_g of 3.5 percent to 4.0 percent should be a reasonable limit.
4. A minimum percentage of longitudinal reinforcement in flexural members (beams) for sections requiring tensile reinforcement.

$$\rho \geq \frac{3\sqrt{f'_c}}{f_y} \geq \frac{200}{f_y} \tag{13.28}$$

But under no condition should the value of ρ exceed 0.025. The stresses f'_c and f_y in these expressions are in psi units. All reinforcement has to be continued through the joint. At least two bars have to be continuously provided *both* at top and bottom.

5. Main reinforcement should be chosen on the basis of the strong column-weak beam concept of the ACI Code, namely, $\Sigma M_{col} \geq 6/5\ \Sigma\ M_{bm}$.
6. The nominal moment strength requirements are:
   (a) M_n^+ at joint face $\geq 1/2\ M_n^-$ at that face.
   (b) Neither the negative nor the positive moment strength at *any* section along the span can be less than *one quarter* the maximum moment strength provided at the face of either joint. Hence,

Photo 13.5 Column localized damage in a high-rise frame building, Los Angeles 1994 Earthquake. (Courtesy Portland Cement Association.)

at joint face:

$$M_n^+ \geq \frac{1}{2} M_a \tag{13.29a}$$

at any section:

$$M_a^+ \geq \frac{1}{4} (M_a^-)_{\max} \tag{13.29b}$$

$$M_a^- \geq \frac{1}{4} (M_a^-)_{\max} \tag{13.29c}$$

13.5.2 Transverse Confining Reinforcement

Transverse reinforcement in the form of closely spaced hoops (ties) or spirals has to be adequately provided. The aim is to produce adequate rotational capacity within the elastic hinges that may develop as a result of the seismic forces.

1. For column spirals, the minimum volumetric ratio of the spiral hoops needed for the concrete core confinement cannot be less than the larger of:

$$\rho_s \geq \frac{0.12 f'_c}{f_{yh}} \tag{13.30a}$$

or

$$\rho_s \geq 0.45 \left(\frac{A_g}{A_{ch}} - 1 \right) \frac{f'_c}{f_{yh}} \tag{13.30b}$$

whichever is greater, where

ρ_s = ratio of volume of spiral reinforcement to the core volume measured out to out.

A_g = gross area of the column section.

A_{ch} = core area of section measured to the outside of the transverse reinforcement (sq. in.).

f_{yh} = specified yield of transverse reinforcement, psi.

2. For column rectangular hoops, the total cross-sectional area within spacing s, cannot be less than the larger of:

$$A_{sh} \geq 0.09\, s h_c \frac{f'_c}{f_{yh}} \tag{13.31a}$$

or

$$A_{sh} \geq 0.3\, s h_c \left(\frac{A_g}{A_{ch}} - 1 \right) \frac{f'_c}{f_{yh}} \tag{13.31b}$$

where

A_{sh} = total cross-sectional area of transverse reinforcement (including cross ties) within spacing s and perpendicular to dimension h_c.

h_c = cross-sectional dimension of column core measured c.–c. of confining reinforcement, in.

h_x = maximum horizontal spacing of hoops or ties on all faces of the column, in.

A_{ch} = cross-sectional area of structural member, measured out-to-out of transverse reinforcement

s = spacing of transverse reinforcement measured along the longitudinal axis of the member, in.

s_{max} = one-quarter of the smallest cross-sectional dimension of the member or 6 times the diameter of longitudinal reinforcement,

Also $$s_x = 4 + \left(\frac{14 - h_x}{3}\right)$$

s_x = longitudinal spacing of the transverse reinforcement within length l_o. Its value should not exceed 6 in. and need not be taken less than 4 in.

Additionally, if the thickness of the concrete outside the confining transverse reinforcement exceeds 4 in., additional transverse reinforcement has to be provided at a spacing not to exceed 12 in. The concrete cover on the additional reinforcement should not exceed 4 in.

3. The confining transverse reinforcement in *columns* should be placed on *both* sides of a potential hinge over a distance l_o. The largest of the following three conditions governs l_o:
   (a) depth of member at joint face
   (b) one-sixth of the clear span
   (c) 18 in.

   Increase the distance l_o by 50% or more in locations of high axial loading and flexural demand such as at the base of a building. When transverse reinforcement is not provided throughout the column length, the remainder of the column length has to contain spiral of hoop reinforcement with spacing not exceeding the smaller of six times the diameter of the longitudinal bars of 6 in.

4. For beam confinement, the confining transverse reinforcement at *beam* ends should be placed over a length equal to *twice* the member depth h from the face of the joint on either side or of any other location where plastic hinges can develop. The maximum hoop spacing should be the smallest of the following four conditions:
   (a) One-fourth effective depth d.
   (b) 8 × diameter of longitudinal bars.
   (c) 24 × diameter of the hoop.
   (d) 12 in. (300 mm).

   Figure 13.7(a) from Ref. 13.14 summarizes typical detailing requirements for a confined column in a monolithic ductile connection and Figure 13.7(b) from Ref. 13.24 for a hybrid precast prestressed assembly.

5. Reduction in confinement at joints: a 50 percent reduction in confinement and an increase in the minimum tie spacing to 6 in. is allowed by the ACI Code, if a monolithic joint is confined on all *four* faces by adjoining beams with each beam wide enough to cover three quarters of the adjoining face.

6. The yield strength of reinforcement in seismic zones should not exceed 60,000 psi.

13.5.3 Horizontal Shear at the Joint of Beam-Column Connections

Test of joints and deep beams have shown that shear strength is not as sensitive to joint shear reinforcement as for that along the span. On this basis, the ACI Code has assumed the joint strength as a function of only the compressive strength of the concrete and requires a minimum amount of transverse reinforcement in the joint. The effective area A_j within the joint should in no case be greater than the column cross-sectional area.

The minimal shear strength of the joint should not be taken greater than the forces V_n specified below for normal weight concrete,

1. Confined on all faces by beams framing into the joint,

$$V_n \leq 20\sqrt{f'_c}A_j \tag{13.32a}$$

2. Confined on three faces or on two opposite faces,

$$V_n \leq 15\sqrt{f'_c}A_j \tag{13.32b}$$

3. All other cases,

$$V_n \leq 12\sqrt{f'_c}A_j \tag{13.32c}$$

A framing beam in a monolithic joint is considered to provide confinement to the joint only if at least three-quarters of the joint is covered by the beam.

The value of allowable V_n should be reduced by 25 percent if lightweight concrete is used. Also, test data indicates that the value in Eq. 13.32c is unconservative when applied to corner joints. A_j = effective cross-sectional area within a joint as in Figure 13.8, in a plane parallel to the plane of reinforcement generating shear at the joint. The reversible seismic forces at the joint are shown in Figure 13.9. The ACI Code assumes that the horizontal shear in the joint is determined on the basis that the stress in the flexural tensile steel = 1.25 f_y. Figure 13.9 shows the forces acting on a beam-column connection at the joint.

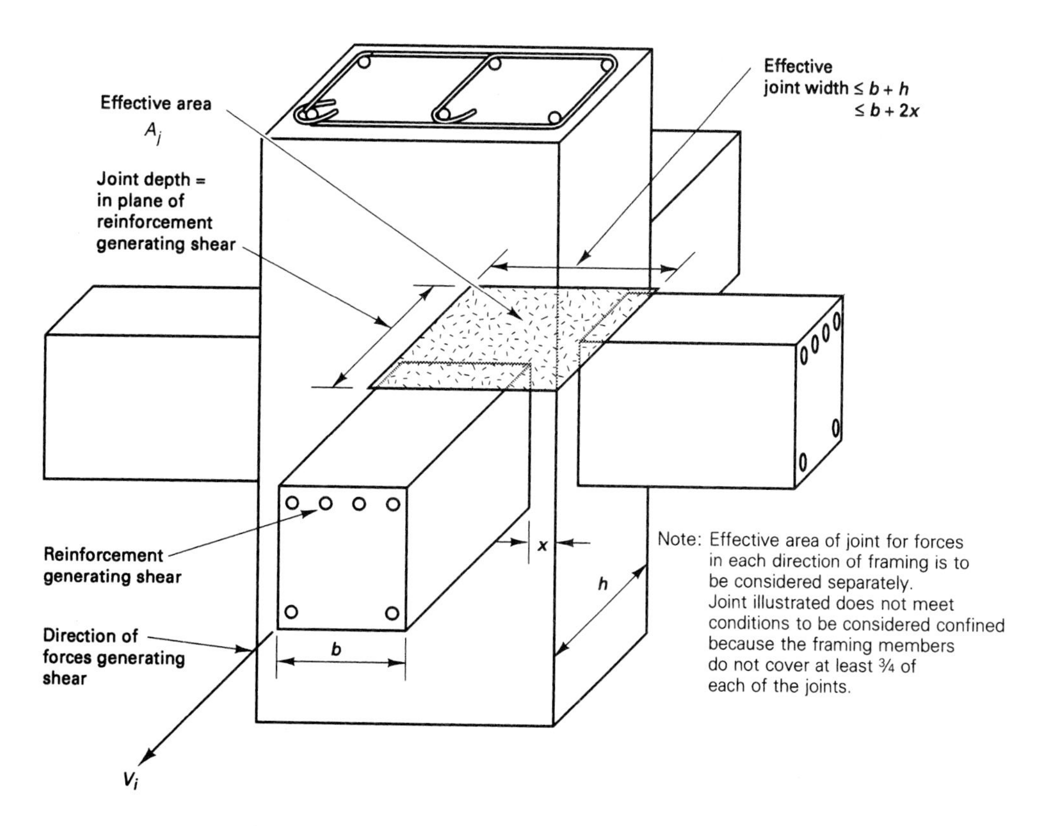

Figure 13.8 Seismic effective area of joint (Ref. 13.1).

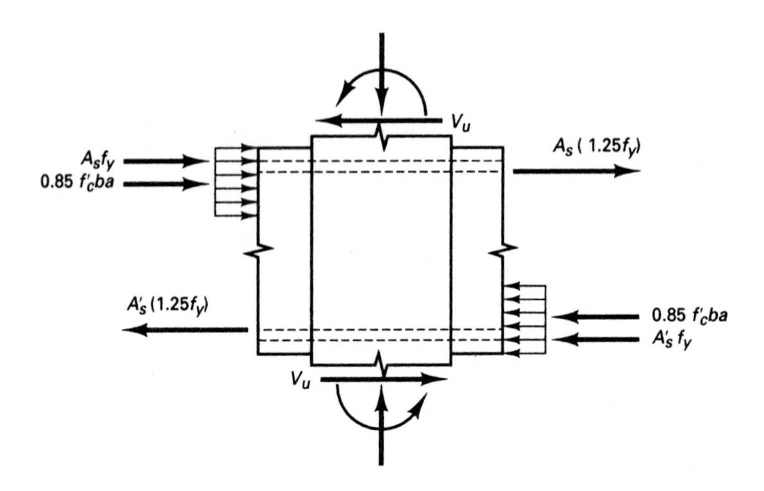

Figure 13.9 Reversible forces at beam-column joint connection. (V_u = horizontal shear at joint).

13.5.4 Development of Reinforcement

For bars of sizes No. 3 through 11 terminating at an exterior joint with standard 90° hooks in normal concrete, the development length ℓ_{dh} beyond the column face, as required by the ACI 318 Code, should not be less than the largest of following:

$$\ell_{dh} \geq f_y d_b / 65 \sqrt{f'_c} \tag{13.33a}$$

$$\ell_{dh} \geq 8\, d_b \tag{13.33b}$$

where d_b = bar diameter

$$\ell_{dh} \geq 6 \text{ in.} \tag{13.33c}$$

The development length provided beyond the column face must be no less than $\ell_d = 2.5\, \ell_{dh}$ when the depth of concrete cast in a monolithic joint in one lift beneath the bar ≤ 12 in., or $\ell_d = 3.5\, \ell_{dh}$ when the depth of concrete cast in one lift beneath the bar exceeds 12 in.

All straight bars terminated at a joint are required to pass through the confined core of the column or shear wall boundary member. Any portion of the straight embedment length not within the confined core should be increased by a factor of 1.6.

13.5.5 Allowable Shear Stresses in Structural Walls, Diaphragms, and Coupling Beams

1. Structural Walls and Diaphragms

High shear walls, that is, structural walls, with height-to-depth ratio in excess of 2.0 essentially act as vertical cantilever beams. As a result, their strength is principally determined by flexure rather than by shear.

Flexural considerations:

(a) *Displacement-based Approach:* For walls or piers continuous in cross section from the base of the structure to the top of the wall and designed to have a single critical section for flexure and axial loads, the compressive zones have to be reinforced with boundary elements with a geometry defined as follows:

$$c \geq \frac{l_w}{600\,(\delta_u/h_w)} \tag{13.34a}$$

but that δ_u/h_w is taken not less than 0.007. The reinforcement has to extend vertically along the wall a distance not less than the larger of l_w or $M_u/4V_u$ from the critical section.

c = distance from the extreme compression fibers to the neutral axis computed from the factored axial force and nominal moment strength.
h_w = height of entire wall.
δ_u = design displacement

(b) *Stress-based Approach:* This alternative design procedure requires that boundary elements in structural walls have to be provided whenever the extreme fiber compressive stresses exceed 0.20 f'_c. The boundary elements have to extend along the vertical boundaries of the entire wall and around the edges of openings. They can be discontinued where the computed compressive stress is less than 0.15 f'_c. The stresses are computed for factored forces using a linearly elastic model and cross-section properties.

It should be noted that when boundary elements are required, the wall is essentially detailed in a similar manner in both approaches.

Shear considerations:

If the shear wall is subjected to factored in-plane seismic shear forces $V_{uh} > A_{cv}\sqrt{f'_c}$, then it should be reinforced with a reinforcement percentage $\rho_v \geq 0.0025$. Spacing of the reinforcement each way should not exceed 18 in. center to center. If $V_{uh} < A_{cv}\sqrt{f'_c}$, the reinforcement percentage can be reduced to 0.0012 for No. 5 bars or less in diameter and 0.0015 for larger deformed bar sizes. Reinforcement provided for shear strength has to be continuous and distributed across the shear plane.

At least two curtains of reinforcement are need in such a wall if the in-plane factored shear forces exceed a value of $2A_{cr}\sqrt{f'_c}$.
where

$\rho_v = A_{sv}/A_{cv}$
A_{cv} = net area of concrete cross section = thickness × length of section in direction of shear considered.
A_{sv} = projection on A_{cv} of area of distributed shear reinforcement crossing the plane A_{cv}.

The nominal shear strength V_n of structural walls and diaphragms of high-rise buildings with aspect ratio greater than 2 should not exceed the shear force computed from:

$$V_n = A_{cv}(2\sqrt{f'_c} + \rho_n f_y) \tag{13.34b}$$

where

ρ_n = ratio of distributed shear reinforcement of a plane perpendicular to the plane of A_{cv}.

For low-rise walls with aspect ratio h_w/l_w less than 2, the ACI Code requires that the coefficient in Eq. 13.34b be increased linearly up to a value of 3 when the h_w/l_w ratio reaches 1.5 in order to account for the higher shear capacity of low-rise walls. In other words,

$$V_n = A_{cv}\,(\alpha_c\sqrt{f'_c} + \rho_n f_y) \tag{13.34c}$$

Photo 13.6 Northridge, California, 1994 earthquake structural failure. (Courtesy Dr. Murat Saatcioglu.)

where

$\alpha_c = 2$ when $h_w/l_w \geq 2$ and $\alpha_c = 3$ when $h_w/l_w = 1.5$; $V_u = \phi\, V_n$

$\phi = 0.6$ for designing the joint, if nominal shear is less than the shear corresponding to the development of the nominal flexural strength of that corresponding to the development of the nominal flexural strength of the member.

The nominal flexural strength is determined considering the most critical factored axial loads including earthquake effects. The maximum allowable nominal unit shear strength in structural walls is $8A_{cv}\sqrt{f'_c}$ where A_{cv} is the total cross-sectional area (in.^{2}) previously defined and f'_c is in psi. However, the nominal shear strength of any one of the individual wall piers can be permitted to have a maximum value of $10\, A_{cp}\sqrt{f'_c}$, where A_{cp} is the cross-sectional area of the individual pier.

2. Coupling Beams:

The provisions for allowable shear stresses in coupling beams are as follows. Coupling beams are structural elements connecting structural walls to provide additional stiffness and energy dissipation. In many cases, geometrical limits result in coupling beams whose depth to clear span ratio is high (Ref. 13.1, 13.2). Hence, they can be controlled by shear and subjected to strength and stiffness deterioration in earthquakes. To reduce the extent of the deterioration, the span to depth ratio l_n/d is limited to a value of 4.0. Coupling beams should only be used in locations where damage to them would not impair the vertical load carrying capacity of the structure or the integrity of the nonstructural components and their connection to the structure (Ref. 13.1).

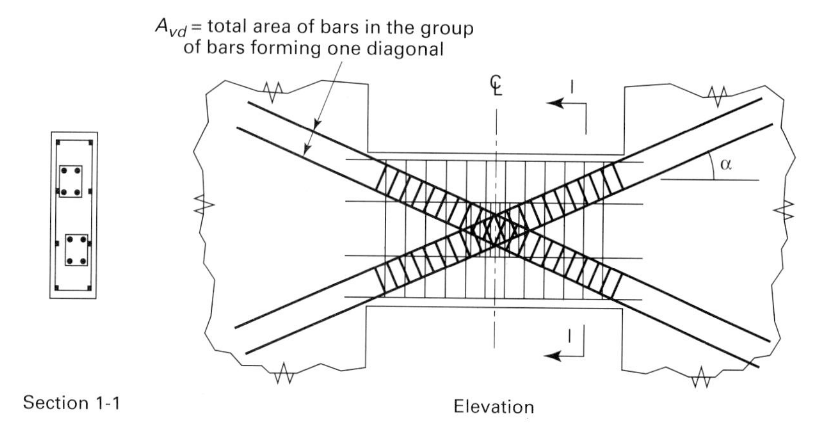

Figure 13.10 Coupling Beam with Diagonally Oriented Reinforcement

If the factored shear force V_u exceeds $4\sqrt{f'_c}\, b_w d$, two intersecting groups of diagonally-placed bars symmetrical about the midspan have to be used. This requirement can be waived if it can be demonstrated that their stiffness loss does not impair the vertical load carrying capacity of the structure. The nominal shear strength, V_n, is determined from the following expression.

$$V_n = 2A_{Vd}\, f_y \sin\alpha \leq 10\sqrt{f'_c}\, b_w d \tag{13.35}$$

A typical illustration of a diagonally reinforced coupling beam is shown in Fig. 13.10. The diagonally-placed bars have to be developed in tension within the wall and also considered to contribute to the nominal flexural strength of the coupling beam.

13.6 SEISMIC DESIGN CONCEPTS IN HIGH-RISE BUILDINGS AND OTHER STRUCTURES

13.6.1 General Concepts

The design of concrete structures in seismic regions has to take into consideration the impact of the large reversible seismic horizontal forces that act on a structure during an earthquake. For the main elements of a structure to service such high-intensity forces, the structure must have adequate ductility in the joints of the principal components or in the response of solid vertical elements such as structural shear walls to ground motion. Excessive strength is not necessarily desirable or essential in earthquake-resistant design. Inelastic response can overcome service damage if adequate ductility is available through proper design and confinement. Shear strength has to exceed the flexural strength of the components and joints in order that shear deformations do not occur as a result of significant loss of stiffness and strength (Ref 13.17).

The failure due to severe ground motion is accentuated in stories with sudden stiffness changes. The dynamic response of the total structure is determined by the flexible stories. Since loss of stiffness results in large inelastic deformations, such deformations, if of sufficient magnitude, would lead to the collapse of the total structure. Therefore, the design has to proportion the detailing of the members to such a degree that the components can tolerate the expected large inelastic deformations without rupture. Such detailing will be discussed in subsequent sections.

In the design of high-rise buildings, a number of analytical tools are usually used to identify the required strength and probable deformations demand (Refs. 13.14, 13.16, 13.17). The required strength is the factored load or required ultimate strength of the component along the lateral load path ("ductile-link") that is expected to absorb the anticipated post-yield deformation. The strength of this "ductile-link" is usually developed by combining load effects (D, L, E, etc.), although moment redistribution may be used to attain a more rational development of the system (Ref. 13.17). The *design* earthquake load (E) must exceed that required by the IBC or other controlling codes. Typically, the strength of this "ductile-link" is developed from site-specific ground motion studies.

The probable deformation demand is the level of deformability likely to be imposed on a structure and most importantly on the component that is expected to deform in the post-yield range during a catastrophic earthquake. Deformation objectives will be many times greater than those associated with the objective strength level. The designer should endeavor to have those post-yield deformations occur where they are likely to create the least potential for collapse of the structure and minimize component damage. It is for this reason that the weak beam/strong column philosophy is adopted in the design of special moment frames, be they constructed of concrete or steel. The brief design examples that are in subsequent sections will presume that the level of strength and deformation required of the "ductile-link" has been determined. The development of the ductile-link is essential to the success of the adopted seismic bracing system, but it is not sufficient. The other members along the lateral load path must be protected so that they do not fail as the ductile-link deforms. This member protection hypothesis is generically referred to as *capacity-based design* (Ref. 13.16, 13.17).

13.6.2 Ductility of Elements and Plastic Hinging

Ductility is an essential property in structures which have to respond to inelasticity in severe earthquakes. It is measured in terms of strain, displacement, and rotation. High ductility enables a member or a joint to sustain plastic strains without a significant reduction of stress. Hence, large rotations are essential as a measure of curvature if discontinuity, unsustainable displacements, or rupture are to be avoided. Three measures of ductility are identified:

(a) Strain ductility defined by

$$\mu_e = \frac{\epsilon}{\epsilon_y} \tag{13.36a}$$

where ϵ = maximum sustainable strain
ϵ_y = yield strain ductility

(b) Curvature ductility defined by:

$$\mu_\phi = \frac{\phi_m}{\phi_y} \tag{13.36b}$$

where ϕ_m = maximum sustainable curvature
ϕ_y = yield curvature

(c) Displacement ductility defined by:

$$\mu_\Delta = \frac{\Delta}{\Delta_y} \tag{13.36c}$$

where Δ = maximum sustainable displacement = $\Delta_y + \Delta_p$
Δ_y = yield displacement
Δ_p = plastic displacement

The values of all these ductility factors have to be considerably greater than 1.0 for inelastic behavior to be sustainable. Ductility can effectively be achieved through adequate confinement as stipulated in the *ACI 318–99 code* (Ref. 13.1) and the *International Building Code, IBC 2000* (Ref. 13.2)

Due to large rotations, the structure at imposed locations reaches the limit ultimate state through the development of plastic hinges. The plastic hinges generated by seismic action would generally develop close to the side of the column since weak beam–strong column design is generally used, as stipulated in ACI 318 (Ref. 13.1). For the plastic hinge to develop in the beams rather than the columns of a multistory frame, special confinements have to be provided over a beam's length ahead of the columns face, equal to twice the beam depth. Figures 13.11 (a) and (b) schematically demonstrate the imposed locations of the plastic hinges in monolithic construction.

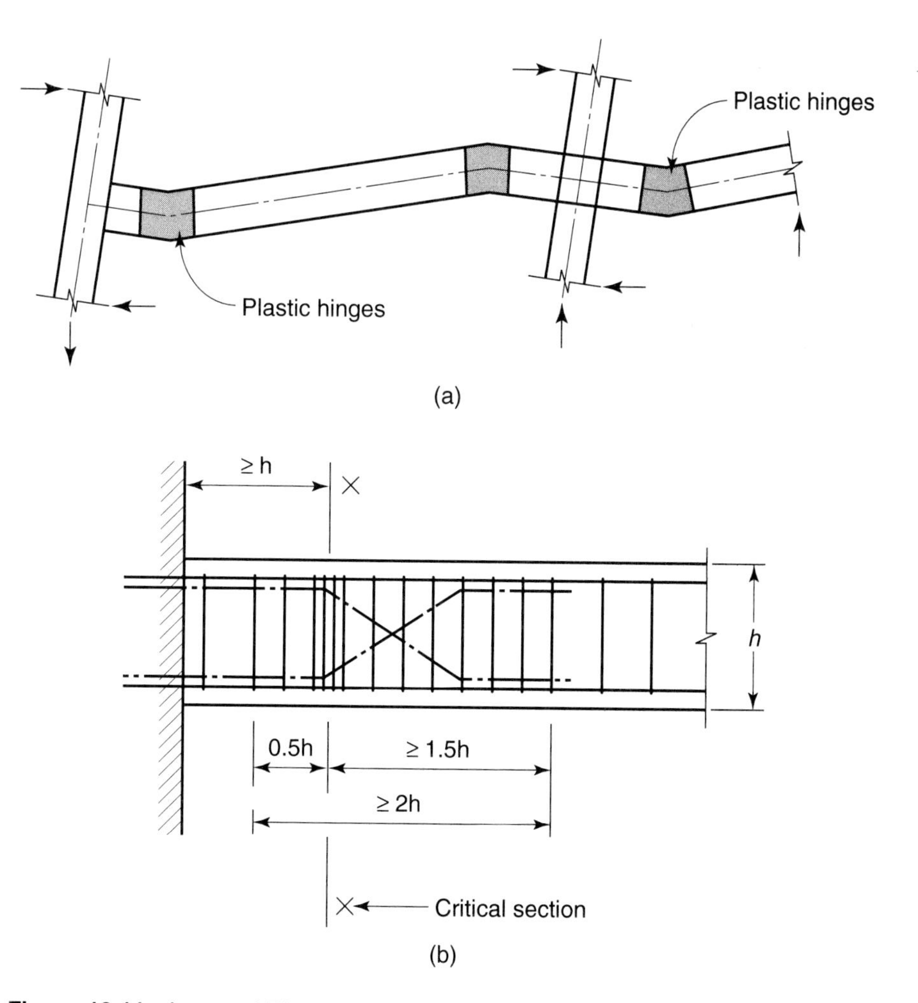

Figure 13.11 Imposed Plastic Hinge Locations: (a) Transformed hinge location in monolithic construction (b) critical hinge section.

Hence, the columns would be large enough to resist the design seismic forces while the beams possess the required ductility to respond to the seismic strains imposed by the earthquake. In the case of using precast ductile moment resisting frames, a hybrid connection can be used and proportioned by a capacity-based design. An example is the Dywidag Ductile Assembly described in Section 13.7.2 or a dual system as in Section 13.7.5, providing a large level of energy dissipation.

13.6.3 Ductility Demand Due to Drift Effect

As the multistory floors drift in response to the horizontal seismic force, the drift increases in the lower levels due to the P-Δ effect. The plastic rotation demand increases. If ignoring the P-Δ effect results in inelastic drift significantly larger than 1.5 percent of the story height, the drift, with the P-Δ influence, would be considerably magnified. In such a case, the plastic rotation demands in both beams and first-story columns would exceed the levels achieved with normal detailing in seismic design (Ref. 13.17).

It must be emphasized that design joint deformations associated with shear and bond mechanisms should *not* result in excessive drift. This is because large shear forces can develop in the beam-column joints under seismic action regardless whether plastic hinges develop close to the column face or ahead in the beam span. In order to prevent shear failure at the joint, both vertical and horizontal shear reinforcement is necessary, with the horizontal reinforcement significantly more than is normally provided by ties or hoops. Also, full anchorage development lengths or bond mechanisms have to be ensured in the reinforcement embedded within the beam-column joint.

13.7 STRUCTURAL SYSTEMS IN SEISMIC ZONES

In general, three systems are applicable in medium- and high-seismicity zones

1. Structural ductile frames
2. Shear wall systems
3. Dual systems, which are a combination of the two

13.7.1 Structural Ductile Frames

Present building codes when used in high-seismicity zones have generally been limited to situ-cast special moment resisting ductile frame and shear walls. From the discussion in Sections 13.6.2 and 13.6.3 it is clear that the beam-column connection is the major part of the frame that has to sustain large seismically imposed deformations. Both reinforced and monolithic prestressed concrete frames have been designed and built for some time (Ref. 13.13, 13.14, 13.17).

Precast concrete, on the other hand, has traditionally been viewed as an assembly of components that attempts to emulate a situ-cast structure. This approach disregards the advantages presented by the discrete elements that make up a total precast structure. If, by design, a post-yield deformation can be imposed to occur where precast elements are joined, damage to the structure can be significantly reduced. This is because a weakened plane already exists at the point where a post-yield rotation has to be accommodated. A monolithically cast element, on the other hand, must crack, usually along *several* planes, in order to accommodate the required rotation. Given this advantage, precast concrete structures can be created capable of surviving earthquakes with lower levels of damage than those created from other materials or by other processes.

The use of precast prestressed concrete elements in ductile frame construction is coming of age. Extensive research is available to justify use of precast elements safely in ductile beam-column frames in high seismicity zones (Ref. 13.18–13.27). Figure 13.12 from Ref. 13.18 shows a hybrid connection. The connection would have well-bonded mild (ductile) steel reinforcing bars at top and bottom of the beam and high-strength prestressing tendons at mid-depth of the beam. The mild steel is intended to dissipate the seismic energy by yielding. The prestressing steel provides the shear resistance from the friction developed by the prestressing force. The system is defined as hybrid because of using two types of reinforcement.

The hybrid system is the evolution of an assemblage of precast concrete components by post-tensioning that was first proposed in New Zealand in the early 1970s. The hybrid system was developed largely through an interactive test program (Ref. 13.17, 13.22, 13.23). The objective of the tests performed was to improve upon the energy dissipation characteristics of assemblies connected exclusively by post-tensioning (Ref. 13.24).

The basic objectives of the hybrid system are to mainly accomplish the following results:

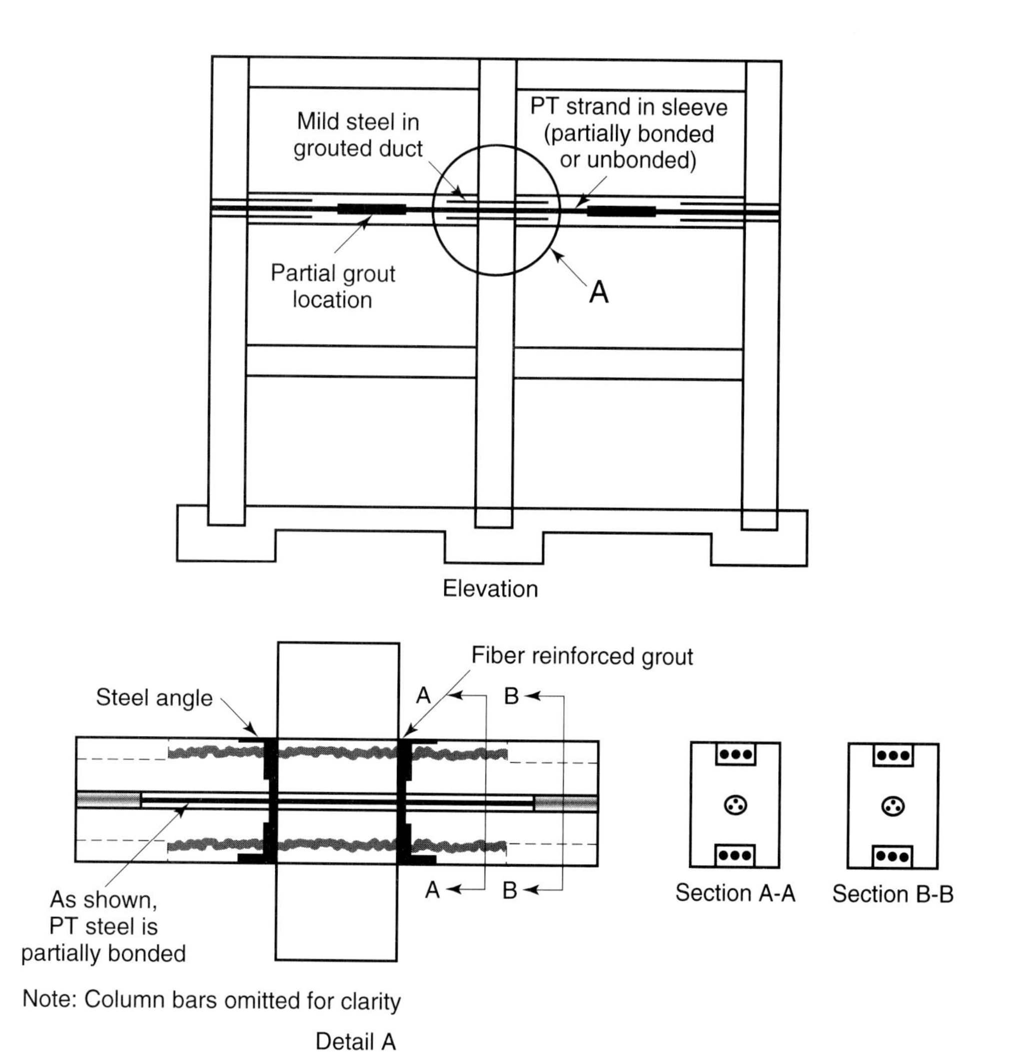

Figure 13.12 Precast Hybrid Moment Connection (Ref. 13.18)

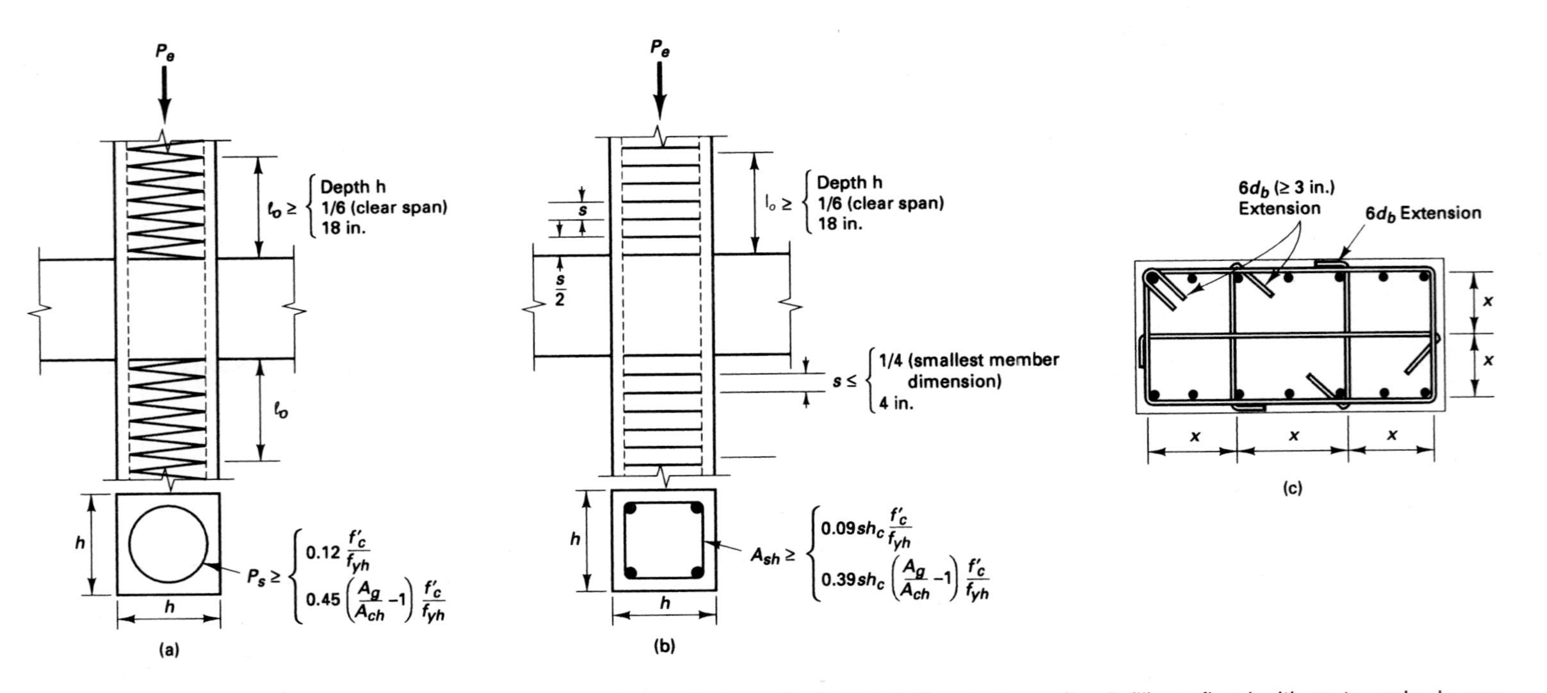

Figure 13.13 (a) Typical Detailing of Seismically Reinforced Column (Ref. 13.14) (i) spirally confined, (ii) confined with rectangular hoops, (iii) cross-sectional detailing of ties.

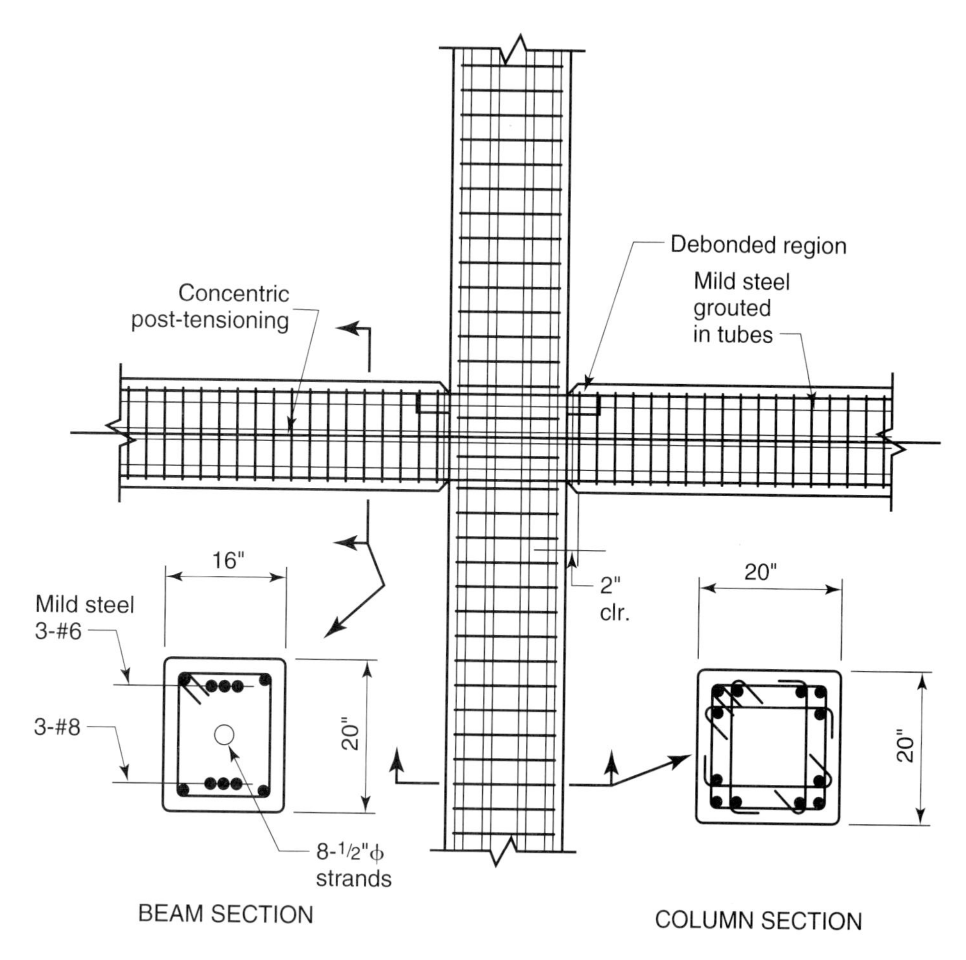

Figure 13.13 (b) Hybrid Frame Assembly (Ref. 13.24)

- Balance the restoring force provided by the concentric post-tensioning with the strength developed by the mild steel so that a restorative or self-centering force exists after the earthquake. This should reduce the potential for permanent deformation.
- Maintain a strain state in the post-tensioning reinforcement at the deformation limit state that is within the elastic range ($f_{ps} < 0.9\, f_{pu}$).
- Localize the post-yield deformation so as to cause the post-yield rotation to occur *along the interface* between the beam and the column. This reduces the potential for nonstructural damage to the beam.

Figure 13.13 (a) gives typical detailing of monolithic situ-cast reinforced concrete ductile connection. Figure 13.13 (b) demonstrates typical details of the reinforcement in a hybrid precast frame assembly.

The performance of the hybrid moment-resisting beam-column connection has been thoroughly verified through tests in several centers of research as listed in the selected references. The crack widths in all the specimens in both beams and columns were very small, in the 0.04 in. range (Ref. 13.26). Research test results have demonstrated that hybrid precast systems have the following performance capabilities:

(a) Can be designed to have the same flexural strength as conventionally reinforced systems.
(b) Have large drift capacity.
(c) Dissipate more energy than conventional systems up to 1.5 percent drift.
(d) Have concrete in the hybrid system that suffers negligible damage even if the drift is in the range of 6%.

Figure 13.14 (a) demonstrates the narrow cracking pattern and negligible damage at 3.5 percent drift while Figure 13.14 (b) shows the contrasting behavior of the monolithically cast assembly.

Additionally, studies on large-scale prototype tests have been conducted by Pessiki et al. on precast beam-column non-bonded post-tensioned connections in ductile frames under high-seismic loading (Ref. 13.31). They demonstrate that such assemblages can perform satisfactorily for frames on hard soil conditions. Their tests also indicate that displacement of the frames on medium or soft soil conditions in high-seismicity regions are difficult to reasonably estimate using elastic analysis under the equivalent lateral base force code approach.

13.7.2 Dywidag Ductile Beam-Column Connection: DDC Assembly

The DDC assembly was developed by Dr. R. E. Englekirk (Ref. 13.19) and produced by Dywidag Systems International (DSI). It allows the precast concrete beams to be bolted to the column, simplifying construction while at the same time improving seismic behavior. The system seems to satisfy the ductility requirements for both the shear forces at the column joint and the deformation and rotational ductility requirements in high seismicity zones. The design is also simple and easy.

The assembly consists of ductile rods embedded in the concrete column. The precast beam contains high-strength (F_{ymin} = 120 ksi) Dywidag Bars® connected to a transfer block. The beam is connected to the column by high-strength steel bolts ($1\frac{1}{2}$ in. ϕ-A490). The flexural strength of the beam is limited by the capacity of the ductile rod (F_y = 141 kips). The other components along the load path are designed to the probable strength of the ductile rod (1.25 F_y), a capacity-based approach. Shear is transferred by steel-to-steel friction at the interface (transfer block to ductile rod) and bearing of the head of the ductile rod on the confined concrete of the column.

Figure 13.15 illustrates a typical single DDC assembly unit with the high-strength bolts connecting the precast prestressed beams to the assembly embedded in the columns at the joint. Example 13.2 gives the design computational steps for a typical ductile connection in a high-rise frame building. Figure 13.16 gives an example of this application to a parking garage in high seismicity zones showing beams-to-column DDC details in a parking garage (Ref. 13.19). Photo 13.7 shows the ductile garage frame structure at completion.

13.7.3 Structural Walls in High-Seismicity Zones (Shear Walls)

Shear walls from efficient and reliable lateral force resisting systems. They are designed to account for the total lateral base shear force generated by an earthquake. This condition assumes that the wall has an adequate foundation, which can transmit deformational actions from the structure to the ground without rocking to any measurable extent. They also provide tortional stability to the multi-story system. Figure 13.17 shows a typical torsional stability arrangement of walls both in the E-W and N-S directions, with Figure 13.17 (b) the torsional stability provided by an interior core.

(a)

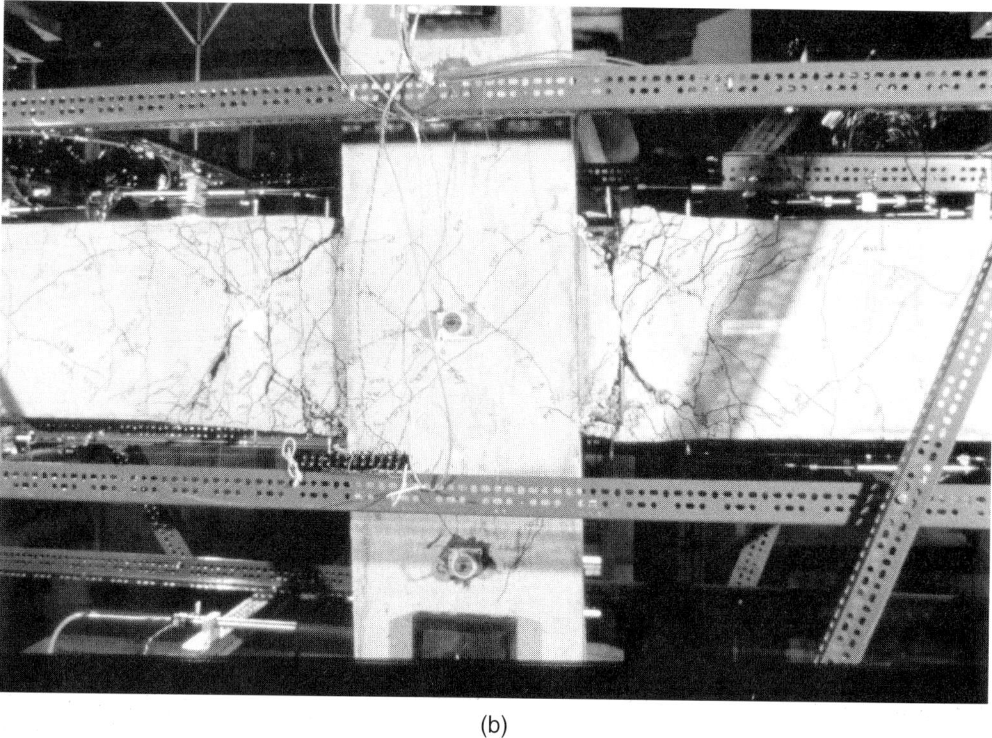

(b)

Figure 13.14 Beam-Column Assembly at 3.5 percent Drift (Courtesy Dr. R. E. Englekirk) (a) Precast Connection Assembly (b) Monolithically-cast Assembly

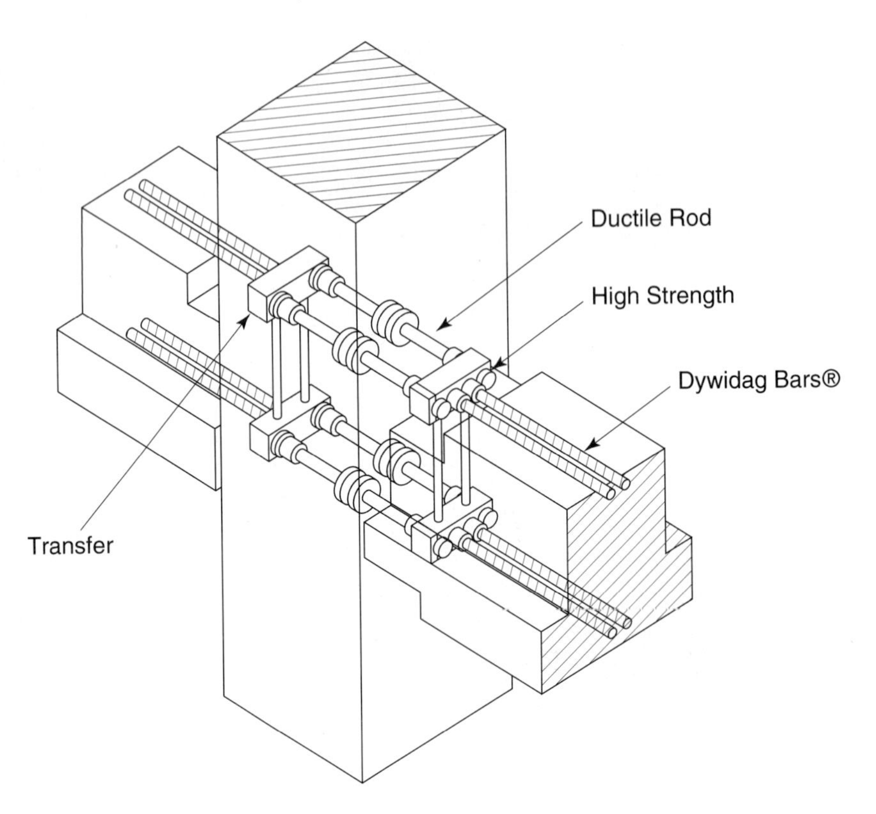

Figure 13.15 Single DDC Assembly, Tensile Strength at Yield = 282 Kips (Courtesy Dywidag-Systems International and Dr. R. E. Englekirk)

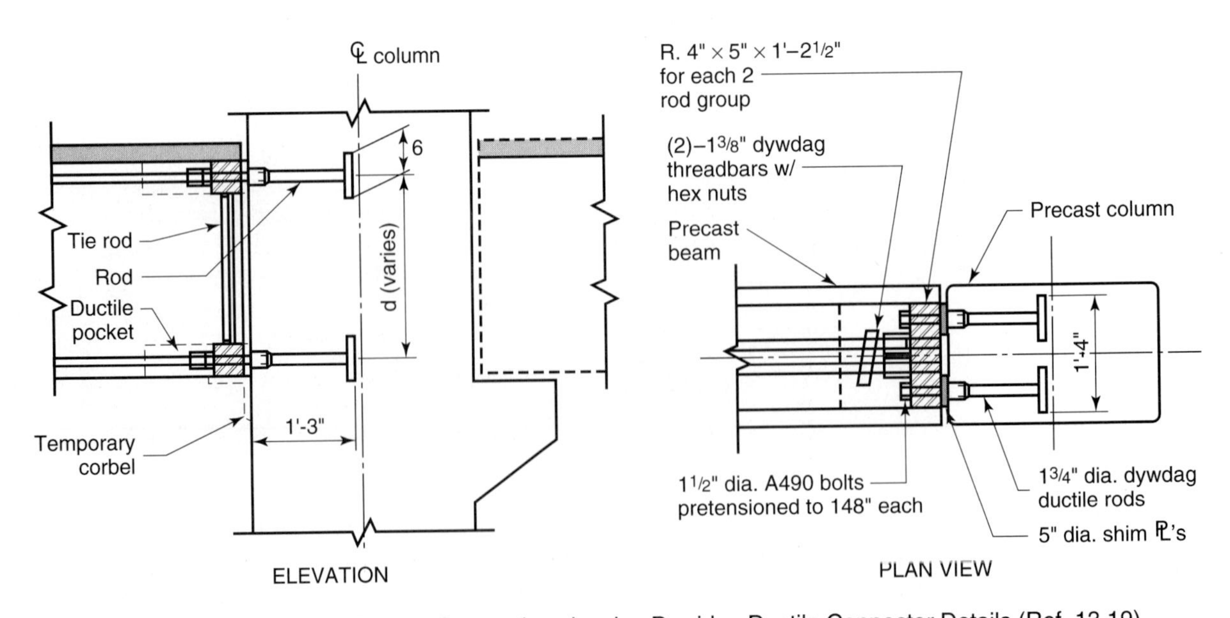

Figure 13.16 Beam-to-Column Connection showing Dywidag Ductile Connector Details (Ref. 13.19)

Photo 13.7 Wiltern Center Parking Structure, Los Angeles, California (Courtesy Dr. Robert E. Englekirk).

Structural (shear) walls have been successfully used for more than 35 years. They are essentially vertical cantilevers designed to receive lateral forces from diaphragms or coupling beams and then transmit the forces to the ground. The forces in these walls may often be predominantly shear forces for low-rise buildings. Slender walls will also undergo significant bending, mainly flexural stresses.

One of the main objectives of the structural analysis is to determine in what proportion the applied wind or seismic forces are distributed among the various shear walls. For the case where no ductile moment frames are present, one can assume that each floor diaphragm displaces in its plane as a rigid body. In such an analysis, the magnitude of lateral displacement becomes the dominant factor in determining the proportion of loads resisted by each wall.

If the wall is treated as a *deep* vertical beam cantilevering from the foundation, shear deformations become a major component of the displacement and have to be

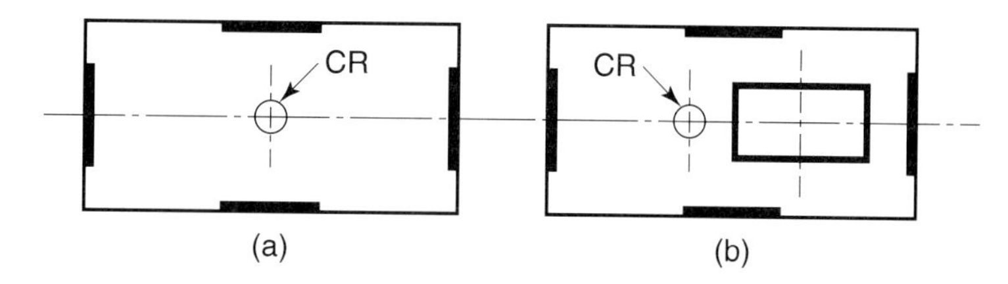

Figure 13.17 Torsionally Stable Shear Wall Systems (a) Boundary walls arrangement with concentric resistance center; (b) Core wall system with eccentric resistance center.

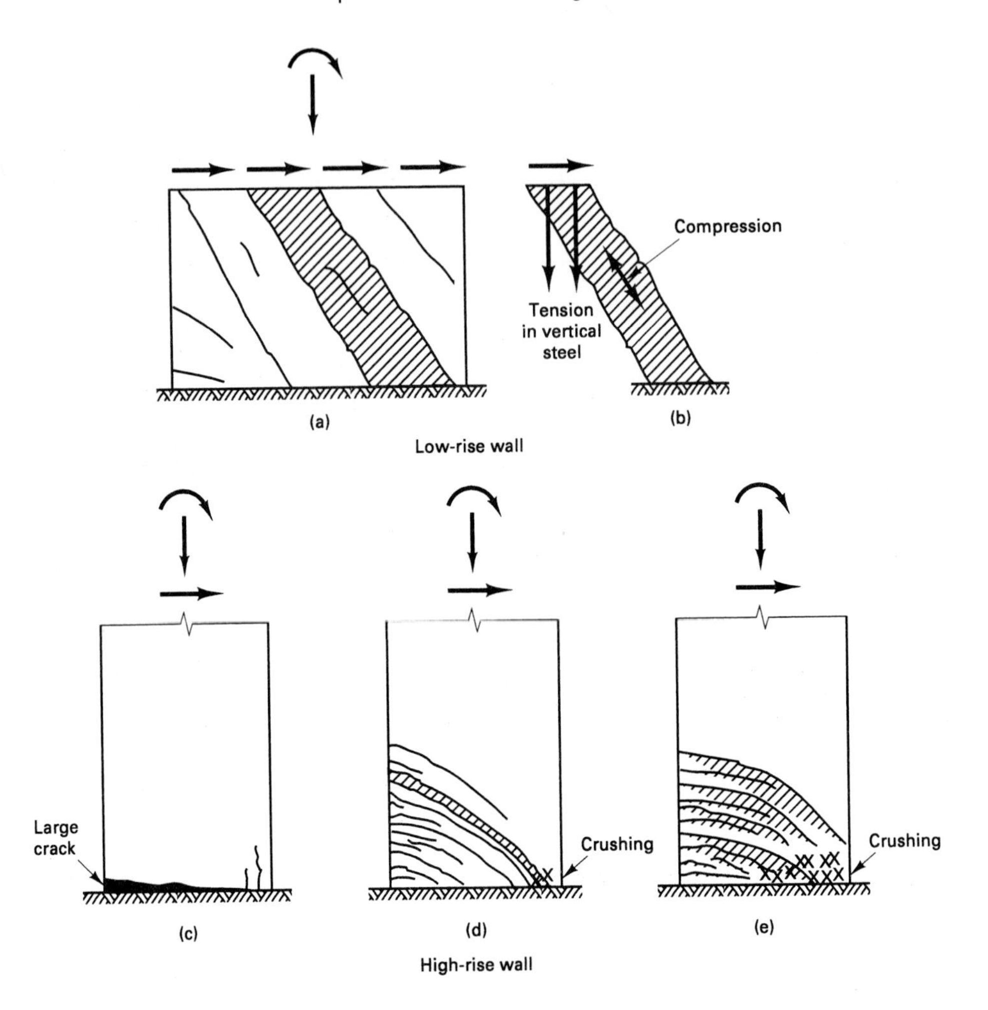

Figure 13.18 Typical failure modes of structural walls: (a) Shear cracking pattern; (b) compression strut between cracks; (c) fracture of the reinforcement; (d) flexure-shear failure pattern; (e) failure by crushing of concrete.

taken into account. Based on the analysis by Aswad et al. in Ref. 13.28, it has been shown that the "beam element" method including the shear deformations is quite accurate for evaluating the shear and overturning moments in plan layouts with shear walls. Figure 13.18 shows the modes of failure of structural walls subjected to seismic lateral loading. Figure 13.19 schematically illustrates the drift due to both bending and shear, and Figure 13.20 shows a precast shear wall connection to the foundation using a Dywidag threaded bar connector. For small uplift forces, Figure 13.21 gives a typical welded angle connector to the foundation.

13.7.4 Unbonded Precast Post-Tensioned Walls

Unbonded precast post-tensioned walls are constructed by vertically joining precast wall panels along horizontal connections using post-tensioning reinforcement not bonded to the concrete. Precast concrete walls with substantial initial lateral stiffness can be designed to soften and satisfy estimated nonlinear displacement demands under code-

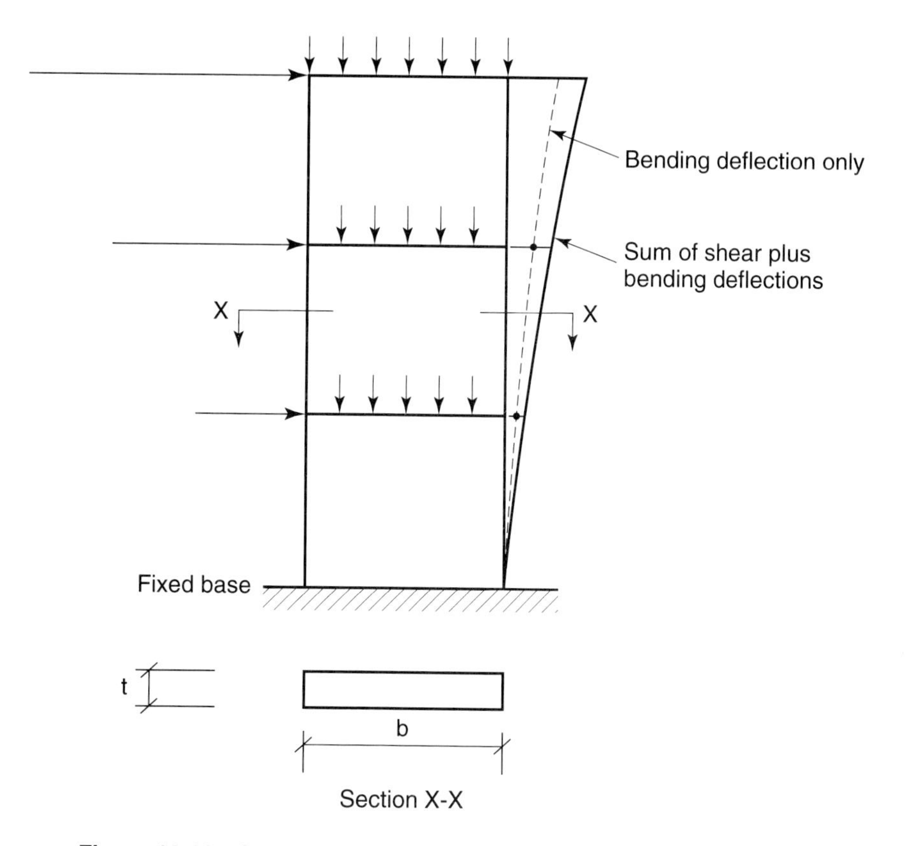

Figure 13.19 Schematic of shear wall drift due to bending and shear

specified design level motion, *without yielding* in the post-tensioning reinforcement or significant damage in the wall panel (Ref. 13.32–13.33). Figure 13.22 shows a prototype wall from Ref. 13.33 with unbounded tendons prestressing the six structural wall panels. The tests showed that the nonlinear elastic behavior resulted in small inelastic energy dissipation per hysterisis cycle. Because of the small inelastic energy dissipation, larger lateral displacements of the unbounded post-tensioned precast concrete walls are larger than the displacements of convensional reinforced concrete systems. More research is obviously needed in this area particularly for application to designs in high-seismicity regions.

13.8 DUAL SYSTEMS

Ductile frames interacting with shear walls can provide a large level of energy dissipation in a major earthquake. They would also significantly reduce the story drift and the development of pronounced hinges. Since the precast frame primarily deforms in shear due to lateral loading and the wall deforms primarily in flexure with some shear deformations, the combination of both types in a dual system can result in a more efficient structure.

Part of the lateral forces in such a system is allocated to the ductile frame. The balance is assigned to the shear wall. In such dual systems, the walls can be either free-standing or connected to the frames by the floor diaphragms or by coupling beams which are continuous beams in their planes connected to the abutting frames.

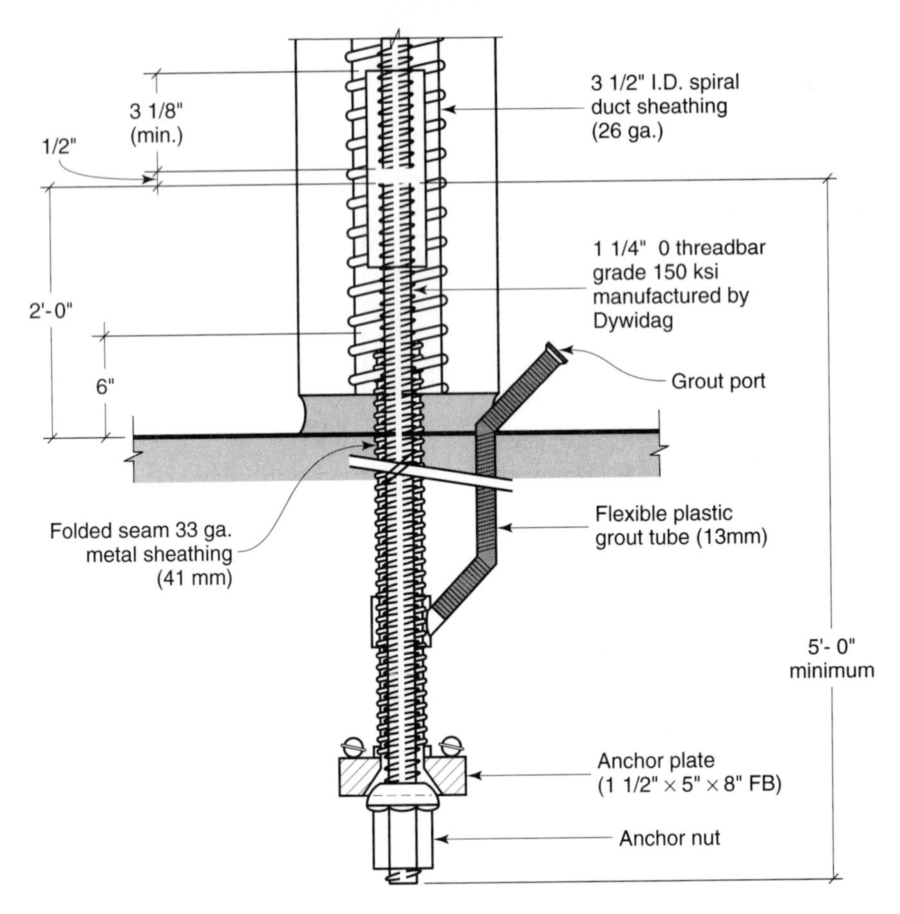

Figure 13.20 Precast shear wall connection to continuous foundation using Dywidag threaded bar connector

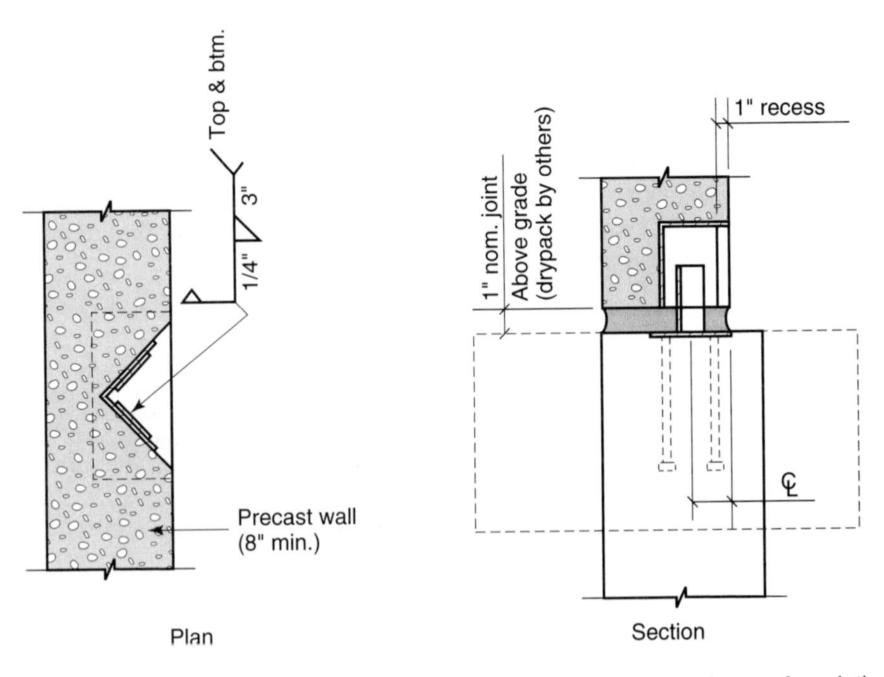

Figure 13.21 Welded angle connection of precast shear wall to continuous foundation

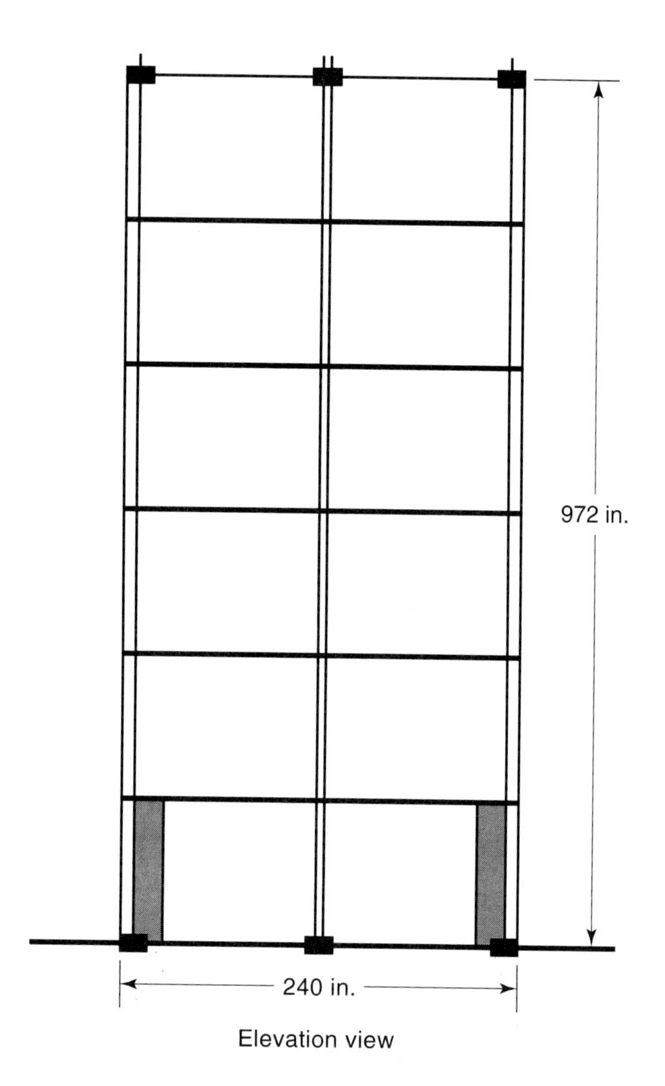

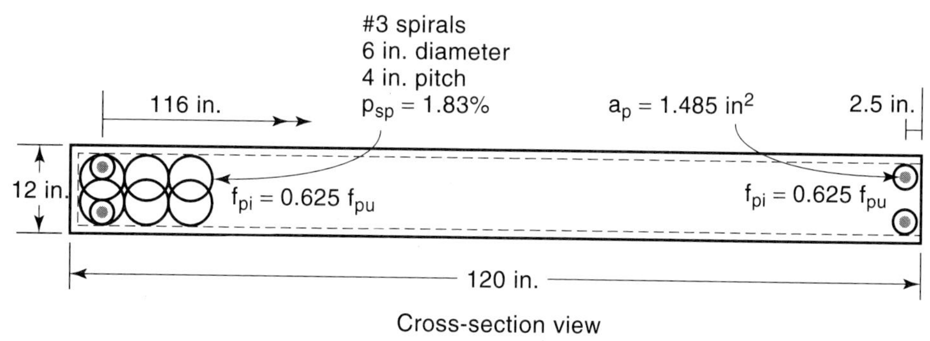

Figure 13.22 Post-tensioned unbounded precast shear wall (Ref. 13.33)

In all systems where nonbonded prestressing is used in high-seismicity regions, it is important that the actual stress in the prestressing reinforcement can achieve and sustain the design ultimate stress level and beyond the yield strength level of 1.25 f_{py}.

13.9 DESIGN PROCEDURE FOR EARTHQUAKE-RESISTANT STRUCTURES

1. Determine the earthquake seismicity region, namely whether it is in a low, moderate, or high seismicity region and the site classification (A, B, C, D, E, and F) from Table 13.1
2. Determine from the maximum considered earthquake ground motion maps, the maximum spectral response S_s for 0.2 sec and S1 for 1 sec, site-class B, Figure 13.3a and b respectively using the large scale FEMA maps of USGS (Ref. 13.15)
3. Compute for the particular seismic use group (Table 13.3), the design spectral response S_{DS} and S_{DI} from Equations 13.2 and 13.3:

$$S_{DS} = \frac{2}{3} S_{MS} \quad \text{where,} \quad S_{MS} = F_a S_s$$

$$S_{D1} = \frac{2}{3} S_{M1} \quad \text{where,} \quad S_{M1} = F_v S_1$$

There are three seismic use groups I, II, and III with groups II and III structures that require full seismic design consideration.

4. Compute the seismic base shear $V = C_S W$

$$C_S = \frac{S_{DS}}{(R/I)}$$

But C_S cannot exceed $CS = S_{D1}/(R/I)T$ or less than $C_S = 0.044\ C_{DS}$

R = Response modification factor from Table 13.4

I = Occupancy importance factor from Table 13.5

T = Fundamental period of vibration of a structure, Sec. 13.3.1, $T_a = C_t h^{\frac{3}{4}}$

where, C_T = building period coefficient ranging between 0.035 – 0.020 as given in the text.

In cases where moment resisting frames do not exceed twelve stories in height, an approximate period $T_a = 0.1\ N$ can be used where N = number of stories.

For structures in seismic design categories E or F and for other structures having a spectral response $S_1 \geq 0.6$ g, the value of $C_S \geq (0.5S_1)R/I$

5. Vertically distribute the base shear force, V, to forces F_x to the floors above the base level:

$$F_x = C_{vx} V$$

$$C_{vx} = \frac{W_x h_x^k}{\sum_{i=1}^{n} W_i h_i^k}$$

6. Horizontally distribute the shear $V_x = \sum_{i=1}^{x} F_i$

   where F_i = the portion of the seismic base shear, V, introduced at level i.

7. Tabulate these forces at all story levels.

8. Evaluate the torsional moments, story drift, the P-Δ effect and the overturning moment to ensure they are within permissible limits.

9. Execute a structural frame analysis to determine all shears and moments in the frame beams, columns, shear walls diaphragms and/or coupling beams if these are used to connect shear walls.

10. Proportion members of the ductile moment-resistant frame, that is, all beams, columns, and beam-columns. If the frame is not a ductile moment-resisting frame, the designer has the uneconomical and inefficient alternative of choosing a brittle system using a low R_w factor.

11. Using the strong column-weak beam concept, plastic hinges are assumed to form in the beams.

Seismic beam shear forces

$$V_L = \frac{M^-_{prL} + M^+_{prR}}{\ell} + 0.75\frac{1.4D + 1.7L}{2}$$

$$V_R = \frac{M^+_{prL} + M^-_{prR}}{\ell} - 0.75\frac{1.4D + 1.7L}{2}$$

ℓ = beam span, M_{pr} = probable moment of resistance, and L, R = left and right.

Seismic column shear force

$$V_e = \frac{M_{pr1} + M_{pr2}}{h}$$

where h = column height.

$$\Sigma M_{\text{col}} \geq \frac{6}{5}\Sigma M_{\text{bm}}$$

at joint to ensure hinges form in the beams; hence

$$(\phi M^+_n + \phi M^-_n)_{\text{col}} \geq \frac{6}{5}(\phi M^+_n + \phi M^-_n)_{\text{bm}}$$

The nominal moment strengths M_n have to be evaluated and the member proportioned prior to evaluating the seismic beam shear forces.

Beam: flexural design, P_u insignificant
Column: combined bending and axial load P_u
Beam–column: $P_u > A_g f'_c/10$
Shortest cross-sectional dimension $\geq$ 12 in.

12. *Longitudinal reinforcement*

Beam–column or columns

$$0.01 \leq \rho_g = \frac{A_s}{A_g} \leq 0.06$$

For practical considerations, $\rho_g \leq 0.035$.

Beam (positive reinforcement):

$$\rho_{\min} \geq \frac{200}{f_y} \geq \frac{3\sqrt{f'_c}}{f_y}$$

Beam (flange in tension):

$$\rho_{\min} \geq \frac{200}{f_y} \geq \frac{6\sqrt{f'_c}}{f_y}$$

The factor value, 6, in the numerator instead of 3 is because a flange width twice the web width or more is used.

where f_y is in psi units. ρ should never exceed 0.025.

For proportioning reinforcement in beams, the nominal moment strength requirements are

(a) M_n^+ at face of joint $\geq \frac{1}{2} M_n^-$ at the face.
(b) M_n^+ or M_n^- at any section $\geq \frac{1}{4} M_{a,\max}$ at the face.

13. *Transverse confining reinforcement*

(a) *Spirals*

$$\rho_s \geq \frac{0.12 f'_c}{f_{yh}} \quad \text{or} \quad \rho_s \geq 0.45\left(\frac{A_g}{A_{ch}} - 1\right)\frac{f'_c}{f_{yh}}$$

whichever is greater.

A_g = gross area
A_{ch} = core area to outside of spirals
f_{yh} = specified yield strength

(b) *Rectangular hoops in columns:* Total cross-sectional area within spacing s:

$$A_{sh} \geq 0.09\, s h_c \frac{f'_c}{f_{yh}}$$

$$\geq 0.3\, s h_c \left(\frac{A_g}{A_{ch}} - 1\right)\frac{f'_c}{f_{yh}}$$

whichever is greater.

A_{sh} = total cross-sectional area of transverse reinforcement (including cross ties) within spacing s and perpendicular to dimension h_c
h_c = cross-sectional dimension of column core, in.
s = spacing of transverse hoops
$s_{\max}$ = one-quarter of the smallest cross-sectional dimension or 4 in., whichever is smaller

Placement of confining reinforcement: Place confining reinforcement on either side of potential hinge over a distance the largest of

(i) Depth of member at joint face
(ii) One-sixth clear span
(iii) 18 in.

The spacing of the ties in the balance of column height follows normal column tie requirements.

(c) *Confining reinforcement in beam ends:* Should be placed on a length = $2h$ on both sides of the joint if it is internal; otherwise, maximum hoop spacing, smallest of

(i) One-quarter effective depth d
(ii) $8 \times$ diameter of longitudinal bar
(iii) $24 \times$ diameter of hoop
(iv) 12 in.

The ties in the balance of the beam span follow the standard shear web reinforcement requirements. If the joint is confined on all four sides, 50 percent reduction in confinement and increase in minimum tie spacing to 6 in. in the columns are allowed. No smooth bar reinforcement is allowed in seismic structures.

14. ***Beam–column connections (joints):*** Normal concrete nominal shear strength V_n at a joint:

Photo 13.8 NCNB Tower, Charlotte, North Carolina, 9000-psi concrete. (Courtesy Portland Cement Association.)

(a) Confined on all faces: $V_n \leq 20\sqrt{f'_c}\,A_j$
(b) Confined on three faces or two opposite faces: $V_n \leq 15\sqrt{f'_c}\,A_j$
(c) All other cases: $V_n \leq 12\sqrt{f'_c}\,A_j$

where A_j is effective area at joint (Fig. 13.8). The value of allowable V_n should be reduced by 25% for lightweight concrete. Note from Fig. 13.9 that the horizontal shear in the joint is determined by assuming a stress = $1.25f_y$ in the tensile reinforcement.

15. ***Development length of reinforcing bars:*** For bar sizes Nos. 3 to 11 without hooks, the largest of

$$\ell_d = 2.5\ell_{dh} \text{ when concrete below bars} \leq 12 \text{ in.}$$
$$\ell_d = 3.5\ell_{dh} \text{ when concrete below bars} \geq 12 \text{ in.}$$

where for normal-weight concrete

$$\ell_{dh} \geq f_y d_b/65\sqrt{f'_c}$$
$$\geq 8\,d_b$$
$$\geq 6 \text{ in.}$$

When standard 90° hooks are used, $\ell_d = \ell_{dh}$. Any portion of straight embedment length not within the confined core should be increased by a factor of 1.6.

16. ***Shear walls: height/depth > 2.0***

(i) Minimum $\rho_v = 0.0025$ if $V_{uh} > A_{cv}\sqrt{f'_c}$. At least two curtains of reinforcement needed if in-plane factored shear force $V_{uh} > 2A_{cv}\sqrt{f'_c}$, where A_{cv} = net area of concrete cross section = thickness × length of section in direction of the considered shear.

(ii) If extreme fiber compressive stresses exceed $0.2f'_c$, shear walls have to be provided with boundary elements along their vertical boundaries and around the edges of openings.

(iii) Available $V_n = A_{cv}(2\sqrt{f'_c} + \rho_n f_y)$ for $h_w/\ell_w \geq 2.0$. For $h_w/\ell_w < 2$, the factor of 2 inside the parenthesis varies linearly from 3.0 for $h_w/\ell_w = 1.5$ to 2.0 for $h_w/\ell_w = 2.0$; $V_u = \phi V_n$, where $\phi = 0.60$.

(iv) Maximum allowable nominal unit shear $V_n = 8A_{cv}\sqrt{f'_c}$ for total wall, but can be increased to $V_n = 10A_{cp}\sqrt{f'_c}$ for an individual pier, where A_{cp} is the cross-sectional area of the individual pier.

Figure 13.23 gives a logic flowchart for the preceeding sixteen steps.

13.10 SI SEISMIC DESIGN EXPRESSIONS

compressive strength $f'_c \geq 20$ MPa

$$E_c = w_c^{1.5}\,0.043\sqrt{f'_c} \text{ MPa}$$
$$E_s = 200{,}000 \text{ MPa}$$

Equation 13.22
$$w_u = 0.75\,[1.4D + 1.7L + 1.87E]$$

Equation 13.23
$$V_L = \frac{M^-_{prL} + M^+_{prR}}{\ell} + 0.75\left(\frac{1.4D + 1.7L}{2}\right)$$

Equation 13.24
$$V_R = \frac{M^+_{prL} + M^-_{prR}}{\ell} - 0.75\left(\frac{1.4D + 1.7L}{2}\right)$$

Equation 13.25
$$V_e = \frac{M_{pr1} + M_{pr2}}{h}$$

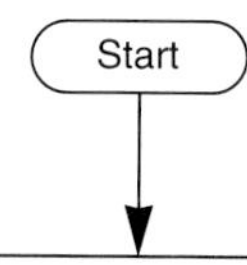

Determine earthquake seismic region, select IBC seismic coefficients S_S, S_1, S_{DS}, S_{D1}, R, I, C_s. Determine period T by IBC Eqs. 15.12, 15.13, or 15.14 and the n, W_s, W values.

Compute $V = C_S W$ and $V = F_t + \sum_{i=1}^{n} F_i$ $F_t = 0$ when $T = 0.7$ s $F_t = 0.07\ TV \leq 0.25\ V$.

Tabulate base lateral force and each story force $F_x = C_{vx}\ V$ using the summation

$$C_{vx} = \frac{w_x h_x^k}{\sum_{i=1}^{n} w_i h_i^k} V.$$ Find each story shear and moment where $V_x = \sum_{i=1}^{x} F_i$

V = Seismic base shear

Execute a structural frame analysis to determine all shears and moments in the frame beams, columns, and shear walls.

Proportion for flexure and revise where necessary the size and main reinforcement of the moment-resistant frame members: beams, and beam-columns (beam – column when $P_u > A_g f'_c/10$).

Use strong column–weak beam concept, plastic hinges in beams and not columns.

$\Sigma M_{col} \geq 6/5\ M_{bm}$ at joint.

$$\text{Beams: } V_L = \frac{M_{prL}^- + M_{prR}^+}{\ell} + 0.75\,\frac{1.4D + 1.7L}{2}$$

$$V_R = \frac{M_{prL}^+ + M_{prR}^-}{\ell} - 0.75\,\frac{1.4D + 1.7L}{2}$$

$$\text{Columns: } V_e = \frac{M_{prL} + M_{pr2}}{h}$$

Design longitudinal reinforcement.

(a) Beam-columns or columns: $0.01 \leq \rho_g \leq \dfrac{A_s}{A_g} \leq 0.06$

For practical considerations $\rho_g \leq 0.035$:

$$\rho_{min} \geq \frac{200}{f_y} \geq \frac{3\sqrt{f'_c}}{f_y} \text{ (for } +M) \geq \frac{6\sqrt{f'_c}}{f_y} \text{ (for negative region T-beam)}$$

(b) Beams: M_n^+ at joint face $\geq 1/2\ M_a^-$ at that face

M_n^+ or M_n^- at any section $\geq 1/4\ M_{a,max}$ at face

Figure 13.23 Flowchart for seismic design of ductile monolithic (strong column-weak beam concept) structures.

Transverse confining reinforcement.

(a) Spirals for columns: $\rho_s \geq \dfrac{0.12 f'_c}{f_{yh}}$ or $\geq 0.45\left(\dfrac{A_g}{A_{ch}} - 1\right)\dfrac{f'_c}{f_{yh}}$

Whichever is greater.

(b) hoops for columns: $A_s \geq 0.09\, s\, h_c \dfrac{f'_c}{f_{yh}}$

$$\geq 0.3\, s\, h_c\left(\frac{A_g}{A_{ch}} - 1\right)\frac{f'_c}{f_{yh}}$$

s = 1/4 of smallest cross-sectional dimension or 6 times diameter of longitudinal reinforcement or $S_x \leq 4 + \left(\dfrac{14 - h_x}{3}\right)$ and need not exceed 4 in.

Use standard tie spacing for the balance of the length.

(c) Beams: Place hoops over a length = $2h$ from face of columns. Maximum spacing: smaller of $s = 1/4d$, 8 d_b main bar, $24d_b$ hoop, or 12 in. If joint confined on all four sides, 50% reduction in confining steel and increase in minimum spacing of ties to 6 in. in columns is allowed. Use the standard size and spacing of stirrups for the balance of the span as needed for shear.

Beam–column connection (joint)

Available nominal shear strength ≥ applied V_u

Confined on all faces: $V_n = 20\sqrt{f'_c}\; A_j$

Confined on three faces or two opposite faces: $V_n \leq 15\sqrt{f'_c}\; A_j$

All other cases: $V_n \leq 12\sqrt{f'_c}\; A_j$

Check development length, normal-weight concrete,

$\ell_{dh} \geq f_y d_b / 65\sqrt{f'_c} \geq 8d_b \geq 6$ in.

$\ell_d = 2.5\,\ell_{dh}$ for 12 in. or less concrete below straight bar

$\ell_d = 3.5\,\ell_{dh}$ for > 12 in. in one pour

If bars have 90° hooks, $\ell_d = \ell_{dh}$. For lightweight concrete, adjust as in the ACI Code.

Design shear wall.

$V_{uh} > 2A_{cv}\sqrt{f'_c}$; use two reinforcement curtains in wall.

If wall $f_c > 0.2\sqrt{f'_c}$, provide boundary elements.

Available $V_n = A_{cv}(\alpha_s\sqrt{f'_c} + \rho_n f_y)$

For $h_w/\ell_w \geq 2.0$, $\alpha_s = 2.0$

For $h_w/\ell_w = 1.5$, $\alpha_s = 3.0$

Interpolate intermediate values of h_w/ℓ_w.

Maximum allowance: $v_n = 8A_{cv}\sqrt{f'_c}$ for total wall

$v_n = 10A_{cp}\sqrt{f'_c}$ for individual pier

Design diaphragms and coupling beams when used as indicated in the text and as detailed in the IBC Code

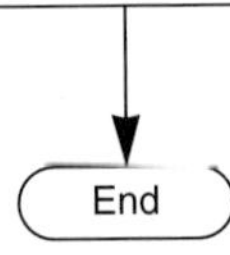

Figure 13.23 *Continued*

Photo 13.9 Masonry collapse in Los Angeles earthquake, 1994. (Courtesy Portland Cement Association.)

Equation 13.27
$$(\phi M_n^+ + \phi M_n^-)_{col} \geq \frac{6}{5}(\phi M_n^+ + \phi M_n^-)_{bm}$$

$\phi = 0.9$ for beams and 0.7 or 0.75 for columns.

Equation 13.28

$$\text{For positive moment: } \rho \geq \frac{\sqrt{f'_c}}{4f_y} \geq \frac{1.4}{f_y}$$

where f_c, f_y are in MPa

Equation 13.29(a)

At joint face: $M_n^+ \geq \frac{1}{2} M_a^-$

At any section:

Equation 13.29(b)
$$M_a^+ \geq \frac{1}{4}(M_a^-)_{max}$$

Equation 13.29(c)
$$M_a^- \geq \frac{1}{4}(M_a^-)_{max}$$

13.11 SEISMIC BASE SHEAR AND LATERAL FORCES AND MOMENTS BY THE IBC APPROACH

Example 13.1:

A moment-resisting, five-story building with shear walls is idealized as in Figure 13.2. Each floor has a weight W_s and a height $h = 9'$–$6''$ (2.9 m). Compute the seismic base shear, V, and the overturning moment, M, at each story level in terms of single floor weight W_s, assuming the idealized mass of each floor is W_s. Consider the structure a building category II is site-class B and seismic use group II.

Given: Response modification factor $R = 3.0$
Occupancy importance factor $I = 1.25$

Use the equivalent lateral force method in the solution.

Solution:

(a) *Spectral response period and base shear*
Total building height $= 5 \times 9.5 = 47.5$ ft
From the FEMA ground motion maps (Figure 13.3) spectral response accelerations

$S_1 = 0.42$ sec and $S_S = 0.85$ sec, with a site-B class and 5 percent damping.

Adjusted spectral response accelerations for site class effects: from Tables 13.2(a) and (b),

$$\text{for } S_1 = 0.42 \text{ sec}, F_v = 1.0 \text{ and for } S_S = 0.85 \text{ sec}, F_a = 1.0$$

From Equations 13.2(a) and (b),

$$S_{MS} = F_a S_S = 1.0 \times 0.85 = 0.85$$

$$S_{M1} = F_V S_1 = 1.0 \times 0.42 = 0.42$$

For 5 percent damped design spectral response acceleration using Eqs. 13.3(a) and (b),

$$S_{DS} = \frac{2}{3} S_{MS} = \frac{2}{3} 0.85 = 0.567$$

$$S_{D1} = \frac{2}{3} S_{M1} = \frac{2}{3} 0.42 = 0.278$$

The seismic base shear V from Eq. 13.8 is $V = C_S W = C_S(5W_S)$ for the five stories where W_S is the idealized weight of each story.

From Table 13.4, the response modification coefficient for ordinary reinforced concrete moment frame is given as $R = 3$.

The occupancy importance factor for building category II from Table 13.5 is: $I = 1.25$.

From Eq. 13.9, $C_S = \dfrac{S_{DS}}{(R/I)} = \dfrac{0.567}{3/1.25} = 0.236,$

But C_S cannot exceed the value: $C_S = \dfrac{S_{D1}}{\left(\dfrac{R}{I}\right)T}$ from Eq. 13.10.

For $S_{DI} = 0.278$ and from Table 13.6, $C_u \cong 1.37$.

For moment resistant concrete frame systems, a building period coefficient $C_T = 0.035$ will be used in this example.

From Eq. 13.13, the approximate fundamental period,

$$T_a = C_T h^{3/4} = 0.035\,(47.5)^{3/4} = 0.63 \text{ sec}$$

Maximum allowable $T_a = C_u T_a = 1.37 \times 0.63 = 0.86$ sec
Hence $T_a = 0.63$ sec controls.

$$C_S = \frac{S_{D1}}{\left(\frac{R}{I}\right)T} = \frac{0.278}{\left(\frac{3}{1.25}\right)0.63} = 0.184 \text{ sec}$$

From Eq. 13.11, C_S cannot be less than $C_S = 0.044\, S_{DS} - 0.044 \times 0.567 = 0.025$
Hence, $C_s = 0.184$ sec controls.
$\therefore$ base shear $V = C_S W = C_S(5W_S) = 0.184 \times 5\, W_S = 0.92\, W_S$

(b) ***Vertical Distribution of Forces and Overturning Moments:***

From Eqs. 13.15 (a) and (b), the lateral force induced at any story level is:

$$F_x = C_{vx}V \quad \text{where,} \quad C_{vx} = \frac{W_x h_x^k}{\sum_{i=1}^{n} W_i h_i^k}$$

$$k = \frac{0.63 - 0.50}{0.50} \times 1.0 + 1.0 = 1.26 \text{ (by linear interpolation)}$$

Since h is constant for all the floors, C_{VX} becomes $\dfrac{W_x}{\sum_{i=1}^{n}}$ where $i = 5$ at the top floor.

$$\sum_{i=1}^{n} = 1W_s + 2W_s + 3W_s + 4W_s + 5W_s = 15W_s$$

Lateral force $F_x = C_{vx}V = 0.92\ C_i W_s$

Overturning moment from Eq. 13.19 is $M_x = \tau \sum_{i=x}^{n} F_i(h_i - h_x)$ for the top ten stories, overturning moment reduction factor $\tau = 1.0$.

$$\text{Hence, } M_x = \sum_{i=x}^{n} F_i\,(h_i - h_x)$$

Computing and tabulating the story forces F_i and the overturning moment M_i for all stories,

Photo 13.10 Overpass collapse in 1971 Los Angeles earthquake. (Courtesy Portland Cement Association.)

Floor	C_i	Lateral force F_i = $0.92W_sC_i$	Story Shear	Story Moment
(1)	(2)	(3)	(4)	(5)
5	$C_5 = \frac{5W_s}{15W_s} = 0.333$	$0.3064W_s$	0	0
4	$C_4 = \frac{4}{15} = 0.267$	$0.2456W_s$	$0.3064W_s$	$0.3064W_sh$
3	$C_3 = \frac{3}{15} = 0.200$	$0.1840W_s$	$0.5520W_s$	$0.8584W_sh$
2	$C_2 = \frac{2}{15} = 0.133$	$0.1224W_s$	$0.7360W_s$	$1.5944W_sh$
1	$C_1 = \frac{1}{15} = 0.067$	$0.0616W_s$	$0.8584W_s$	$2.4528W_sh$
Wall base	$C_0 = 0$	0	$0.9200W_s$	$3.3728W_sh$

hence seismic base shear $V = 0.9200W_s$. The moments at each story level are tabulated in column (5).

13.12 SEISMIC SHEAR WALL DESIGN AND DETAILING

Example 13.2

Design by the ACI 318 Code the reinforcement for a shear wall in a multibay, ductile frame, twelve-story structure (adapted from Ref. 13.9) having a total height $h_w = 148$ ft (45 m) and having equal spans of 22 ft (6.7 m). Except for the ground story, which is 16 ft (4.88 m) high, all other stories have 12 ft (3.67 m) heights. The total gravity factored load on the shear wall is $W_u = 4{,}800{,}000$ lb (21.4 *MN*). The factored moment at the base of the wall due to seismic loads from the lateral load analysis of the transverse frames is $M_u = 554 \times 10^6$ in.-lb (62.6 *MN*-m). The maximum axial force on the boundary element is $P_u = 4{,}500{,}000$ lb (20 *MN*). The horizontal shear force at the base is 885,000 lb (3940 kN).

Given:

$$\text{wall length (horizontally)} = 26' - 2'' = 26.17 \text{ ft} = 314 \text{ in. (7980 mm)}$$

$$\text{thickness } t = 20 \text{ in.} = 1.67 \text{ ft (508 mm)}$$

$$\text{boundary element width} = 32 \text{ in. (813 mm)}$$

$$\text{depth} = 50 \text{ in. (1270 mm)}$$

$$A_s = 30 \text{ No. 11 bars (30 bars of 35-mm diameter) in each boundary element}$$

$$f'_c = 4000 \text{ psi (27.6 MPa), normal weight}$$

$$f_y = 60{,}000 \text{ psi (414 MPa)}$$

Use $\phi = 0.60$ as the strength reduction factor for shear in this example.

Solution:

1. ***Wall geometry and forces:*** $\ell_n = 22$ ft (6.7 m), ℓ_w (horizontal dimension) = 26.17 ft, $b_{\text{web}} = 20$ in. = 1.67 ft, and $b_{\text{bound}} = 32$ in. = 2.67 ft.

$$\text{factored } W_u = 4{,}800{,}000 \text{ lb (21.4 MN)}$$

$$M_u = 554 \times 10^6 \text{ in. lb (62.6 MN-m)}$$

$$P_u = 4{,}500{,}000 \text{ lb (20 MN)}$$

Photo 13.11 Northridge, California, 1994 earthquake structural failure. (Courtesy Dr. Murat Saatcioglu.)

2. ***Boundary element check:*** $\ell_w = 26.17$ ft, $b = 1.67$ ft, $P_u = 4{,}500{,}000$ lb, and $M_u = 550 \times 10^6$ in.-lb. Assume that the wall will not be provided with confinement over its entire section.

$$\text{gross } I_g = \frac{bh^3}{12} = \frac{1.67(26.17)^3}{12} = 2495 \text{ ft}^4$$

$$A_g = 1.67 \times 26.17 = 43.7 \text{ ft}^2$$

$$f_c = \frac{P}{A} \pm \frac{M_C}{I}, \quad c = \frac{26.17}{2} \times 12 = 157 \text{ in. (3990 mm)}$$

Concrete compressive stress in the wall is

$$f_c = \frac{4{,}500{,}000}{43.7\,(12)^2} - \frac{554 \times 10^6 \times 157}{2494(12)^4}$$

$$= -715 - 1682 = -2400 \text{ psi (C) (16.5 MPa)}$$

Maximum allowable $f_c = 0.2 f'_c = 0.2 \times 4000 = 800$ psi (5.52 MPa) in compression if a boundary element is not required. Hence boundary elements are needed subject to the confinement and loading requirements of Section 13.5.

3. ***Longitudinal and transverse reinforcement:*** Check if two curtains of reinforcement are needed, that is, if in-plane factored shear > $2A_{cv}\sqrt{f'_c}$ (Section 13.5.5).

$$V_u = 885{,}000 \text{ lb}$$

$$A_{cv} = \text{area bound by web thickness and length of section in direction of shear force}$$

$$= 20 \times 314 = 6280 \text{ in.}^2$$

$$2A_{cv}\sqrt{f'_c} = 2 \times 6280\sqrt{4000} = 799{,}400 \text{ lb (353 kN)} < V_u = 885{,}000 \text{ lb}$$

Hence two curtains of reinforcement are required.

$$\min \rho_v = \frac{A_{sv}}{A_{cv}} = \rho_n = 0.0025 \quad \text{and} \quad \max s = 18 \text{ in.}$$

$$A_{cv} \text{ per ft of wall} = 20 \times 12 = 240 \text{ in.}^2$$

$$\text{required } A_s \text{ in each direction} = 0.0025 \times 240 = 0.60 \text{ in.}^2/\text{ft}$$

Trying No. 5 bars (15.8-mm diameter), $A_s = 2(0.31) = 0.62$ in.^{2} in two curtains.

$$s = \frac{\text{one bar area}}{\text{required } A_s/12 \text{ in.}} = \frac{0.62}{0.60/12}$$

$$= 12.4 \text{ in. (315 mm)} < 18 \text{ in. limit} \quad \text{O.K.}$$

Use $s = 12$ in.

Check for shear reinforcement capacity

A check is needed in order to determine that the No. 5 bars in two curtains at 12 in. c-c both ways are adequate for the wall section to sustain the applied shear force at the base. The shear wall aspect ratio is

$$\frac{h_w}{\ell_w} = \frac{148}{26.17} = 5.66 > 2$$

Hence from Eq. 13.34 b

$$\phi V_n = \phi A_{cv}(2\sqrt{f'_c} + \rho_n f_y)$$

where $\phi = 0.60$ in this example; otherwise, refer to the ACI 318-99 Code for other conditions.

$$A_{cv} = 20(26.17 \times 12) = 6280 \text{ in.}^2$$

$$\rho_n = \frac{2(0.31)}{20 \times 12} = 0.0026$$

$$\text{available } \phi V_n = 0.60 \times 6280\,(2\sqrt{4000} + 0.0026 \times 60{,}000)$$

$$= 1{,}065{,}000 \text{ lb} > V_u = 885{,}000 \text{ lb (4.7 MN} > \text{required 3.9 MN)}$$

Hence the wall section is adequate. Therefore, use two curtains of No. 5 bars spaced at 12 in. c-c in both horizontal and vertical directions.

4. ***Boundary element check if acting as a short column under factored vertical forces due to gravity and lateral loads:*** P_u acting on wall = 4,500,000 lb. From before, $b = 32$ in., $h = 50$ in., $A_s = 30$ No. 11 bars = $30 \times 1.56 = 46.8$ in.^{2} (30,190 mm^2) in each boundary element.

$$\rho_{st} = \frac{A_s}{A_g} = \frac{46.8}{32 \times 50 = 1600} = 0.0293$$

$$\rho_{min} = 0.01 < \rho_{st} < \rho_{max} = 0.06 \text{ O.K.}$$

The axial load capacity of the boundary element acting as a short column is

$$\phi P_{n(max)} = 0.80\,\phi\,[0.85 f'_c(A_g - A_{st}) + A_{st} f_y]$$

$$= 0.80 \times 0.70\,[0.85 \times 4000\,(1600 - 46.8) + 46.8 \times 60{,}000]$$

$$= 4{,}530{,}000 \text{ lb} > P_u = 4{,}500{,}000 \text{ lb} \quad \text{O.K.}$$

5. ***Boundary element transverse confining reinforcement:*** $b_w = 20$ in., $b_b = 32$ in., h or $\ell_w = 314$ in., and $A_g = 1600$ in.^{2}. From Eqs. 13.31(a) and (b)

$$\rho_s \geq \frac{0.12 f'_c}{f_{yh}}$$

and

$$A_{sh} \geq 0.3\, s\, h_c \left(\frac{A_g}{A_{ch}} - 1\right)\frac{f'_c}{f_{yh}}$$

Assume No. 5 hoops and crossties spaced at 4 in. c-c.

(a) *Short direction*

$$h_c = 50 - 2\left(1.5 + \frac{5}{16}\right) = 46.37 \text{ in.}$$

$$b_c = 32 - 2\left(1.5 + \frac{5}{16}\right) = 28.37 \text{ in.}$$

$$A_{ch} = 46.33 \times 28.37 = 1314 \text{ in.}^2 \quad \text{(core area)}$$

$$A_{sh} = \frac{0.09 f'_c s h_c}{f_{yh}} = \frac{0.09 \times 4000 \times 4 \times 46.37}{60{,}000} = 1.08 \text{ in.}^2$$

$$A_{sh} = 0.3 \times 4 \times 46.37\left(\frac{1600}{1314} - 1\right)\frac{4000}{60{,}000} = 0.80 \text{ in.}^2$$

$A_{sh} = 1.08$ in.^{2} governs.

Use three No. 5 crossties, for a total of five legs being provided including the hoop, every 4 in. along the boundary length (wall length ℓ_w). A_{sh} provided $= 5 \times 0.31 = 1.55$ in.^{2}, O.K., on the conservative side

(b) *Longitudinal direction*

$$h_c = 28.37 \text{ in.}, \quad A_{ch} = 1314 \text{ in.}^2$$

or

$$A_{sh} = \frac{0.12 f'_c h_c s}{f_{yh}} = \frac{0.12 \times 4000 \times 4 \times 28.37}{60{,}000} = 0.91 \text{ in.}^2$$

$$A_{sh} = 0.3 \times 4 \times 28.37\left(\frac{1600}{1314} - 1\right)\frac{4000}{60{,}000} = 0.49 \text{ in.}^2$$

$A_{sh} = 0.91$ in.^{2} (587 mm^2) controls. With one No. 5 crosstie, a total of three legs is provided every 4 in. c-c. A_{sh} provided $= 3 \times 0.31 = 0.93$ in.^{2} (600 mm^2).

6. ***Check for maximum hoop spacing:***

$$s \leq \frac{1}{4} \times 32 = 8 \text{ in.}$$

$$s \leq 6 \text{ times dia. of longitudinal bar} = 6 \times \frac{11}{8} = 8.25 \text{ in.}$$

$$s_x \leq 4 + \left(\frac{14 - h_x}{3}\right) = 4 + \left(\frac{14 - 2.6}{3}\right) = 7.8 \text{ in.} < 6.0 \text{ in. within length } l_o$$

Maximum spacing of cross-ties or hoops = 4 in. (100 mm)

7. *Development of reinforcement:* Development length of No. 5 horizontal bars assuming no hooks are used within the boundary element: From Eqs. 13.33(a), b, and c,

$$\ell_{dh} \geq \frac{f_y d_b}{65\sqrt{f'_c}} = \frac{60{,}000 \times 0.625}{65\sqrt{4000}} = 9 \text{ in.}$$

$$\geq 8 d_b = 8 \times 0.625 = 5 \text{ in.}$$

$$\geq 6 \text{ in.}$$

Photo 13.12 Interfirst Plaza, Dallas, Texas, 10,000-psi concrete. (Courtesy Portland Cement Association.)

$$\ell_{dh} = 9 \text{ in. (229 mm)} \quad \text{governs}$$

$$\ell_d = 3.5\ell_{dh} = 3.5 \times 9 \cong 32 \text{ in. (815 mm)}$$

If bars are straight as in this example, ensure that development length is provided. If 90° hooks are used, $\ell_d = \ell_{dh} = 9$ in. Note that no lap splices should be allowed for the No. 5 horizontal bars.

8. ***Verify adequacy of shear wall section at its base under combined axial load and bending in its plane:*** From before,

$P_u = 4{,}800{,}000$ lb (total gravity factored load)

$M_u = 554 \times 10^6$ in.-lb, $e = \dfrac{M_u}{P_u} = 115$ in., $b_{\text{web}} = 20$ in., $b_{\text{bound}} = 32$ in.

$\ell_w = 26.17$ ft $= 314$ in., wall height $h_w = 148$ ft $= 1776$ in. (45 m)

column action $\phi = 0.70$, beam action $\phi = 0.90$

no. of longitudinal bars in wall plane = 110 composed of two (30 No. 11 bars) for *both* boundary elements and two curtains of No. 5 bars at 12 in. center-to-center over $\ell_w = 314$ in.

total A_{st} in the lateral cross section $= 2 \times 46.8 + 2\,(25 \times 0.31) = 109.1$ in.^{2}

$A_g = 2(32 \times 50) + 20(314 - 2 \times 50) = 7480$ in.^{2} (4,830,000 mm^2)

$$\rho = \frac{109.1}{7480} = 0.015 > 0.01 \text{ and } < 0.06 \quad \text{O.K.}$$

$$\frac{e}{\ell_w} = \frac{115}{314} = 0.366$$

$$\frac{\phi M_n}{A_g \ell_w} = \frac{554 \times 10^6}{7480 \times 314} = 236 \text{ psi}$$

From a column interaction curve, such as Fig. 5.20 in Ref. 13.14, with $h_e/h = 1.0$, enter the plot with $\phi M_n / A_g \ell_w = 236$ and $e/\ell_w = 0.366$ coordinates. This gives a value of $\phi P_n/A_g = 0.85$ ksi or available $P_u = 850 A_g = 850 \times 7480 = 6{,}360{,}000$ lb > actual $P_u = 4{,}800{,}000$ lb (available 28.2 M_n > required 21.4 MN). Figure 13.24 gives the detailing of the shear wall longitudinal and boundary element confining reinforcement.

13.13 EXAMPLE 13.3 STRUCTURAL PRECAST WALL BASE CONNECTION DESIGN

A precast structural (shear) wall for a five-story building in a moderate seismicty zone is $b = 8$ in. (203 mm) thick (Ref. 13.29). The length of the interior wall is 24 ft (7.32 m) and the height of each story is 13′–0″ (3.96 m). Structural analysis showed that the wall is subjected to an unfactored seismic base shear force $V_e = 124.7$ kips (555 kN) and an unfactored overturning base moment $M_e = 4854$ ft-kip (6581 kN-m). The total weight of each floor including attached masses is 2,400 kips (10,675 kN). Design the connection at the base of the wall assuming that it is so reinforced that the neutral axis obtained by trial and adjustment and strain compatibility analysis is $c = 17.62$ in. (447 mm). Use either welded connection as in Figure 13.21 having a rated capacity of 25 kips per connection (111 kN) or Dywidag rods grade 150 ksi (1034 MPa) with thread bar couplers as in Fig. 13.20.

Given:

Sliding shear friction coefficient $\mu = 0.60$
Maximum allowable horizontal concrete shear interaction stress (sliding friction):

$$f_{cv} = 1200 \text{ psi (8.2 MPa)}$$

$$f'_c = 5{,}000 \text{ psi (34.5 MPa) for shear wall and for grout (dry pack)}$$

Solution: The system forces acting on the structural wall are shown in Figure 13.25. The ACI load factors governing the design are:

- Load Case I: = 0.75 (1.40D + 1.7L + 1.1 × 1.7E) = 1.05D + 1.275L + 1.40E
- Load Case II: = 0.90D + 1.1 × 1.3E = 0.90D + 1.43E

Usually, Load Case II controls the majority of design cases for gravity walls. Use case II for seismic effects. As given in the problem statement, computer analysis using strain compatibility and trial and adjustment for the reinforcement used in the wall (Ref. 13.29) gave a neutral axis depth $c = 17.62$ in. for load case II. The analysis gave the following factored overturning moments:

Load case I, $M_u = 9{,}596$ ft-kip

Load case II, $M_u = 7{,}114$ ft-kip

Required seismic overturning moment for case II = Load factor × (M_e)
or

$$M_u = 1.43 \times 4{,}854 = 6{,}941 \text{ ft-Kip} < 7114 \text{ ft-kip, hence O.K.}$$

Actual Seismic Shear:

$$V_u = 1.43\, V_e = 1.43 \times 124.7 = 178.3 \text{ kips}$$

$$V_n = \frac{V_u}{\phi} = \frac{178.3}{0.85} = 209.8 \text{ kips (930 kN)}$$

$$C_c = 0.85 f'_c b \beta_1 c = 0.85 \times 5.0 \times 8 \times 0.80 \times 17.62 = 476.5 \text{ kips.}$$

From Figure 13.25, the sliding friction contribution,

$$V_1 = \mu\, C_c = 0.60 \times 476.5 = 285.9 \text{ kips.}$$

Upper bound of V_1 = [(wall thickness h) (N.A. depth c) (allowable horizontal frictional stress f_{cv})]

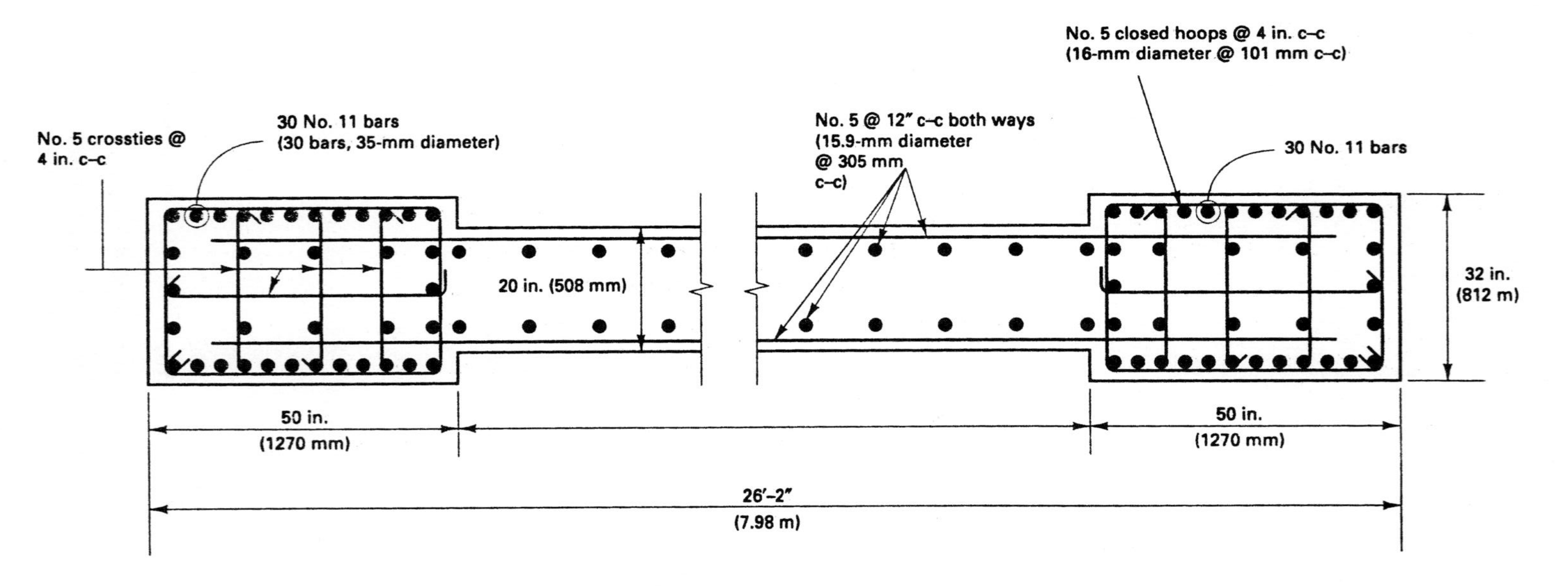

Figure 13.24 Plan and detailing of shear wall in Example 13.2.

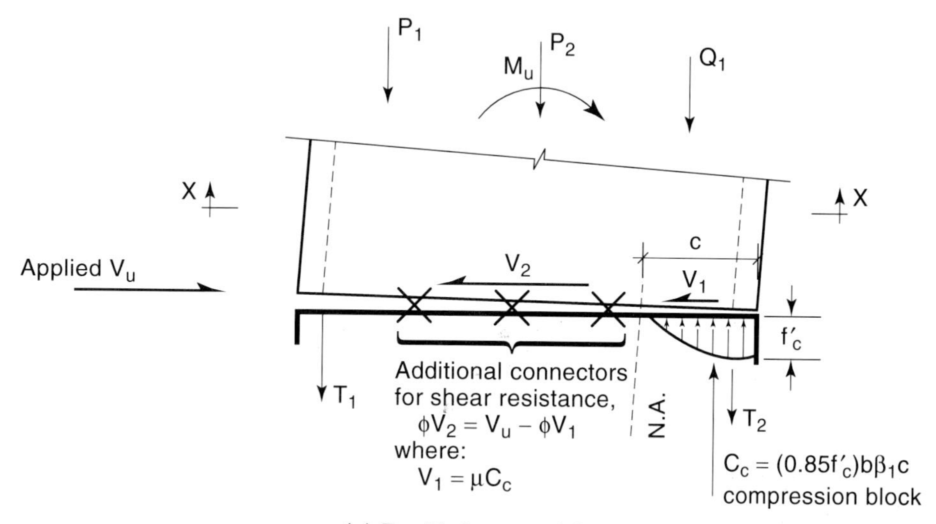

(a) Equilibrium at a joint

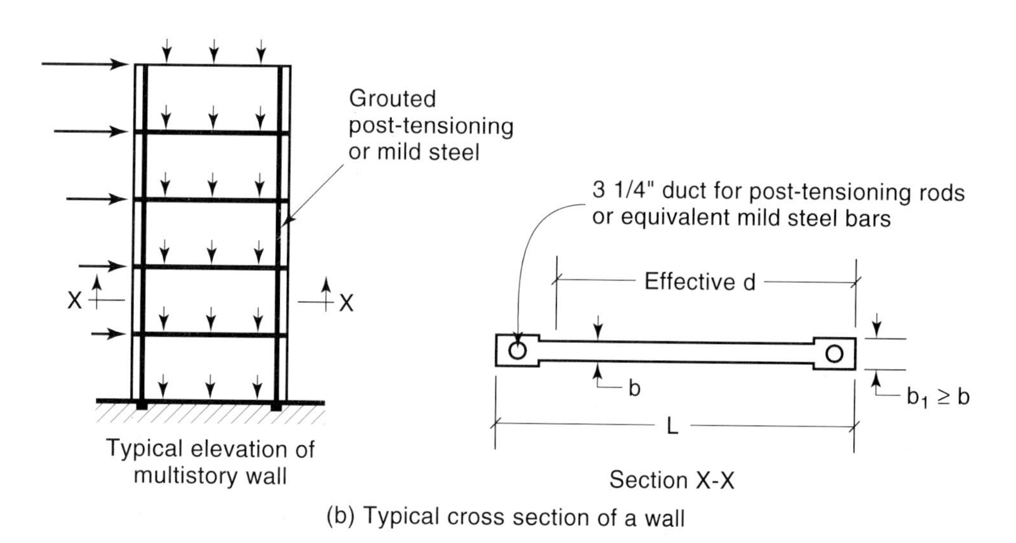

(b) Typical cross section of a wall

Figure 13.25 Equilibrium forces and stresses at base of structural wall

$$V_1 = 8.0 \times 17.62 \times 1.2 = 169.2 \text{ kips} \leftarrow \text{controls.}$$

$$\text{Net } V_2 = \text{Actual } V_n - V_1 = 209.8 - 169.2 = 40.6 \text{ kips.}$$

Hence, required horizontal force contribution for designing the connection at the wall base is 40.6 kips.

If a welded connection is used with the given rated shear capacity of 25 kips, two welded connections have to be used per wall, which is the minimum per panel (see Figure 13.20).

If Dywidag connector and *grouted* post-tensioning is used throughout the wall height, use Dywidag type A722, Grade 150 ksi connectors with minimum yield strength of 120 ksi (f_{pu} = 150 ksi and effective pull after seating losses = 0.75 in. minimum). Grout the horizontal joint as in Figure 13.20.

The vertical flexural reinforcement in the precast 8-in. wall panel elements is a typical 6×6 – W5 $\times$ W5 welded wire fabric reinforcement for such standard flexural wall design.

It should be noted that the design engineer has to consult with the local precasters for the appropriate connection configuration and its rated capacity.

13.14 DESIGN OF PRECAST PRESTRESSED DUCTILE FRAME CONNECTION IN A HIGH RISE BUILDING IN HIGH-SEISMICITY ZONE USING DYWIDAG DUCTILE CONNECTION ASSEMBLY (DDC)

Example 13.4

Design a typical ductile precast prestressed concrete moment-resisting frame connection in a forty-one-story high rise building in a high-intensity seismic zone with design data from Ref. 3.21. Use the Dywidag ductile connector assembly (DDC) described in Section 13.7.2 and Figure 13.15. The frame analysis output for this connection gave a factored moment $M_u = 1{,}150$ ft-kip (1559 kN-m) and a post-yield rotation $\theta_p = 3.0$ percent. The floors are post-tensioned with 200 psi stress limit in the slab concrete at service.

Given:

Precast beam span:	18 feet (5.49 cm)
Clear Span:	15 feet (4.57 m)
Story Height:	9ft 8 in. (2.95 m)
Column Size:	36 in. × 36 in. (914 mm × 914 mm)
Beam Size:	30 in. × 36 in. (762 mm × 914 mm) : [b × h]
Center to Center of Ductile Rods:	2.33 ft (710 mm) : [d-d′]
Objective Strength	1150 ft-kips (1559 kN-m) : [Mu]
Objective Post-Yield Rotation:	3% (θ_p)

$f'_c = 5000$ psi (34.5 MPa), normal weight
$f_y = 60{,}000$ psi (414 MPa)
$f_{pu} = 270{,}000$ psi (1861 MPa)
$f_{ps} \leq 0.90\, f_{pu}$
$f_{pe} = 162{,}000$ psi (1117 MPa)
$E_s = 29{,}000{,}000$ psi (200,000 MPa)
$E_{ps} = 28{,}000{,}000$ psi (193,000 MPa)
Maximum concrete stress, f_c, at post-tensioning = 1000 psi (6.9 MPa)

Note that $\lambda_o\, f_y = 1.25\, f_y$ due to probable seismic increase in the longitudinal reinforcement strain at the joint well beyond the yield strain, as stipulated in the ACI 318 and IBC 2000 codes (see Sec. 13.4.1).

Solution:

1. ***Determine nominal capacity of a double DDC connection.***

$$T_y = 2(282) = 564 \text{ kips}$$

$$M_n = T_y(d - d') = 564(2.33) = 1314 \text{ ft-kips}$$

$$M_u = \phi M_n = 0.9(1314) = 1183 \text{ ft-kips} > 1150 \text{ ft-kips}$$

As discussed in Sec. 13.6 and 13.7, it is important to note that a *capacity-based approach* is being used in order to develop the strength required along the seismic load path.

2. ***Determine the shear imposed on the connector at probable demand of the DDC assembly.***

$$W_D = 0.6 \text{ k/ft. (beam)} + 0.7 \text{ k/ft. (slab)} = 1.3 \text{ k/ft.}$$

$$W_L = 0.16 \text{ k/ft.}$$

The subscripts *pr* denote the probable seismic force, shear, or moment.

$$V_{bpr} = \left(\frac{M_A + M_B}{L_{cl}}\right) + [1.2(1.3) + 0.5(0.16)]\frac{L_{cl}}{2}$$

$$= \frac{2.5(1314)}{15} + 12.3$$

$$= 231 \text{ kips}$$

where M_A and M_B are the probable flexural strengths that can be developed in the beams at the column face (1.25 M_n). Dead and live loads are factored.

Required friction factor between transfer block and face of ductile rod,

$$f_v = \frac{V_b}{1.25T_y} = \frac{231}{(1.25)564} = 0.33$$

Note that class-A slip critical connectors develop a friction factor of 0.33 (AISC Specifications) but this includes a safety factor of about 40 percent (Ref. 13.16). Observe also that the available friction increases in direct proportion to the tensile load developed in the ductile rod.

3. ***Check the induced bearing pressure (ρ) under a ductile rod at the probable strength of the ductile rod.***

$$\text{Shear/bolt} = \frac{V_{bpr}}{4} = \frac{231}{4} = 58.6 \text{ kips}$$

$$\rho = \frac{58.6}{2.95(2.95)} = 6.73 \text{ ksi}$$

Note that the shear transfer mechanism is assumed conservatively to flow through the compression node (Figure 13.26). The bearing area under the ductile rod is confined on all sides. On the open face where it meets the beam, an oversized washer is provided to accomplish this objective.

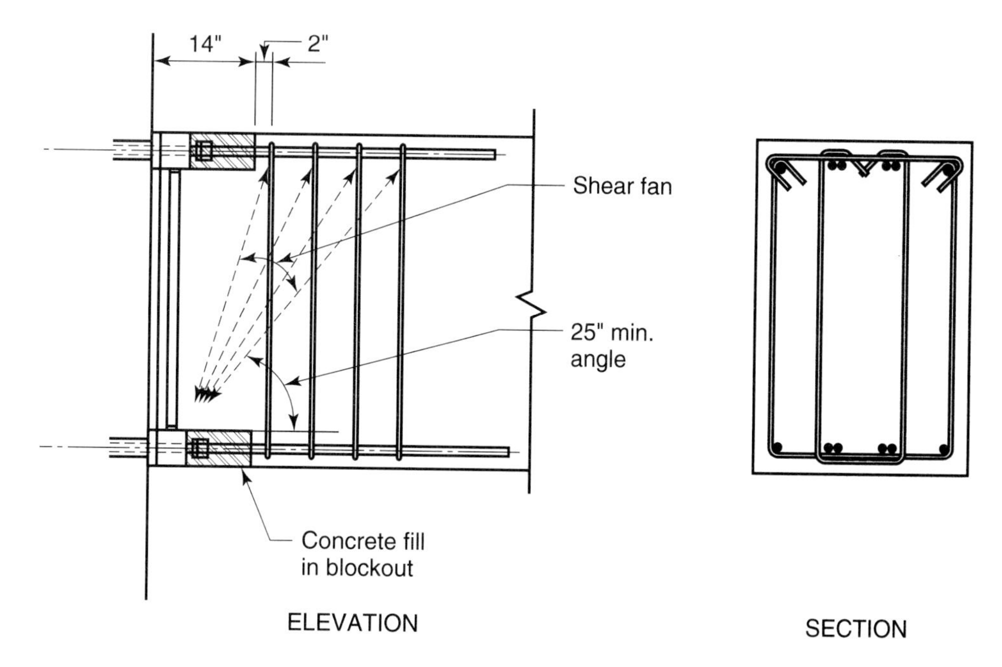

Figure 13.26 Shear transfer mechanism in the discontinuous region of a DDC® frame beam (Ref. 13.21)

$$\rho_{allow} = 1.7 f'_c$$
$$= 1.7(5) = 8.5 \text{ ksi}$$
$$\phi\rho_{allow} = 0.7(8.5) = 6.0 \text{ ksi}$$

4. ***Design of shear reinforcement for the beam.***

$$V_{bpr} = 231 \text{ kips} \qquad \text{(see step 2)}$$

$$V_u = \frac{V_{bpr}}{bd} = \frac{231}{30(33)} = 0.23 \text{ ksi}$$

It should be noted that since inelastic behavior will not occur in the beam, the ability of the concrete to carry shear is not diminished (Ref. 13.17). Hence, $v_c = 2\sqrt{f'_c} = 2\sqrt{5000} = 141$ psi $= 0.141$ ksi.

$$v_s = \frac{v_{upr}}{\phi} - v_c$$
$$= \frac{0.23}{0.85} - 0.141 = 0.130 \text{ ksi}$$
$$s = \frac{0.4(60)}{0.130(30)} = 6.15 \text{ in. c./c.}$$

Provide #4 closed U-stirrups (ties) at 6 in. c./.c.

It is suggested that the first two stirrups should be hoop sets and include an inner hoop set to provide lateral support for the flexural bars (see Figure 13.26). The first hoop set should be placed at the edge of the blockout. A shear fan describes the shear transfer mechanism in this discontinuous region.

Shear transfer in the Discontinuous Region
A shear fan describes the shear transfer mechanism in this discontinuous region. Capacity of one #4 tie set:

$$V_s = A_v f_y = 0.4(60) = 24 \text{ kips}$$

$$\text{Number Required} = \frac{231}{24} \cong 10$$

Provide five double-#4 closed U-stirrup sets within the shear fan region.

5. ***Load transfer mechanism within the joint.***

The bearing plate on the interior end of the ductile rod develops the tensile strength of the rod in bearing. Joint behavior is significantly improved because this load transfer mechanism does not slip, as a conventional bar will as it debonds when subjected to load reversals.

Bearing on the end plate:

$$\lambda_o R = 1.25(141) = 176.5 \text{ kips}$$

$$\rho = \frac{176.5}{28} = 6.3 \text{ ksi} \cong 6.0 \text{ ksi, O.K.} \qquad \text{(see step 3)}$$

Load transfer within a beam-column joint
It is accomplished through the activation of a strut mechanism and a truss mechanism (Ref. 13.17, 13.22). As seen from Figure 13.27; The strut mechanism is activated by bearing on the column face in the case of the compression side at the peak load. On the tension side the load is delivered to the truss mechanism. Since the stiffer load path is, at least initially, through the strut mechanism, it is advisable to provide a sufficient number of ties in the vicinity of the ductile rod to transfer the developed rod force to the node region. The suggested tie force (T_T) is:

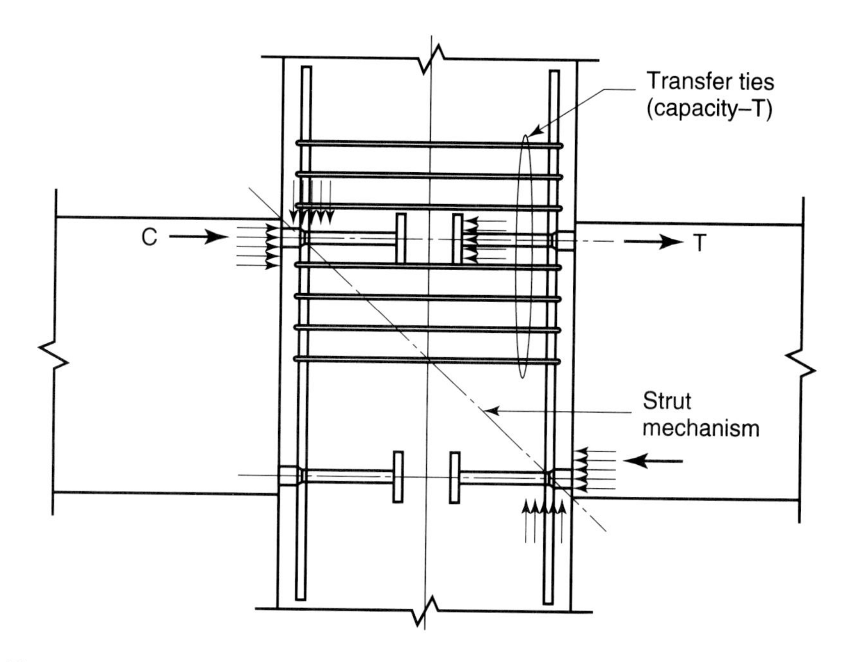

Figure 13.27 Load path within the beam-column joint of a DDC assembly (Ref. 13.21)

$$T_T = 4\,(\lambda_o R) = 4\,(176.4) = 706 \text{ kips}$$

$$A_T = \frac{706}{60} = 11.7 \text{ in.}^2$$

Use seven triple-#5 tie sets ($A_S = 1.86$ in.^{2})—three above and four below the ductile rods. Tie sets should be located within a 65° angle of the ductile rod bearing area.

Joint Shear Stress—ACI 318 Code basis

$$V_{bpr} = \frac{\lambda_o M_n}{L_c/2} = \frac{1.25(1314)}{7.5} = 219 \text{ kips}$$

$$V_{cpr} = \frac{V_{bp}L}{h} = \frac{219(18)}{9.67} = 408 \text{ kips}$$

$$V_{jh} = 2\lambda_o T_y - V_{cpr} = 2(1.25)(564) - 408 = 1002 \text{ kips}$$

$$V_{jh} = \frac{V_{jh}}{A_j} = \frac{1002}{36(36)} = 0.773 \text{ ksi}$$

$$v_{jh\,allow} = 15\phi\sqrt{f_c'} = 15(0.85)\sqrt{5000} = 0.90 \text{ ksi}$$

6. *Column Design Criterion*

Design of the column follows standard situ-cast concrete design procedures using the strong column–weak beam requirement in the ACI Code:

$$\Sigma M_{col} > \frac{6}{5}\Sigma M_{bm}$$

Alternatively, and consistent with the capacity design approach, the moment demand can be developed directly from probable column shears.

$$M_{col} \geq \frac{h_c}{2}(V_{cpr})$$

Photo 13.13 Failure of piers at Hanshin Highway Bridge, Kobe Earthquake, Japan, 1995 (Courtesy Professor Megumi Tominaga, Kyoto University, Japan)

where h_c is the clear height of the column.

$$M_{col} = \frac{6.67}{2}(408)$$

$$= 1359 \text{ ft-kips}$$

6. ***Post-yield deformability***

The portion of the ductile rod, which was designed to absorb post-yield deformations, is 9 in. long. The elongation required of the ductile rod (Δ_ℓ) is:

$$\Delta_\ell = \theta_p \frac{(d - d')}{2} = 0.03(14) = 0.42 \text{ in.}$$

$$\epsilon_p = \frac{0.42}{9} = 0.047$$

The strain associated with the fracture of the ductile rod is in excess of 50 percent or 0.50.

13.15 DESIGN OF PRECAST PRESTRESSED DUCTILE FRAME CONNECTION IN A HIGH-RISE BUILDING IN HIGH-SEISMICITY ZONE USING A HYBRID CONNECTOR SYSTEM

Example 13.5:

Design a typical moment-resisting connection for the ductile frame building in Example 13.4 using a hybrid system as described in Section 13.7.1. Use both well-bonded mild reinforcing bars at top and bottom of the frame beams and concentric post-tensioning steel reinforcement at their mid-depth. Given:

Beam size: 24 in. × 36 in. (610 mm × 914 mm)

$f'_c = 5000$ psi (34.5 MPa), normal weight

$f_y = 60{,}000$ psi (414 MPa)
$f_{pu} = 270{,}000$ psi (1861 MPa)
$f_{ps} \leq 0.90\, f_{pu}$
$f_{pe} = 162{,}000$ psi (1117 MPa)
$E_s = 29{,}000{,}000$ psi (200,000 MPa)
$E_{ps} = 28{,}000{,}000$ psi (193,000 MPa)
Maximum concrete stress, $\bar{f}_c$, at post-tensioning = 1000 psi (6.9 MPa)

Solution: The design procedure in this example differs in that the beam width here does not need to be a function of the hardware as is the case for the DDC in the previous example.

1. ***Determine the amount of post-tensioning required.***
The post-tensioning should be capable of resisting about 60 percent of the moment demand.

$$M_u = 1150 \text{ ft-kips}$$

$$M_n = \frac{M_u}{\phi} = \frac{1150}{0.9} = 1278 \text{ ft-kips}$$

$$M_{nps} = 0.6(1278) = 767 \text{ ft-kips}$$

Assume that the effective lever arm is about 16 in. $(h/2 - a/2) = 1.33$ ft.

$$T_{nps} = \frac{M_{nps}}{\left(\frac{h}{2} - \frac{a}{2}\right)} = \frac{767}{1.33} = 576 \text{ kips}$$

$$A_{ps} = \frac{T_{nps}}{f_{pse}} = \frac{576}{162} = 3.56 \text{ in.}^2$$

Use 0.6 in. diameter strands, $A_{ps} = 0.213$ in.^{2} / strand.

$$\text{Required number of strands, } N = \frac{3.56}{0.213} = 16.7$$

It should be noted that the selection of the amount of post-tensioning is fairly arbitrary. The objective is to provide a restoring force (see design objectives) and satisfy objective strength requirements. The designer may select a 17-strand tendon or opt to use a 19-strand tendon. The latter will be used in this example since a 19-strand tendon anchorage is available and would be used for either choice. Also observe that post-tensioning strands are more cost effective than Grade 60 reinforcing when developed as described in Figure 13.13(b). The consequence associated with the use of proportionately higher amounts of post-tensioning is the loss of some energy dissipation.

$$A_{ps} = 19(0.213) = 4.05 \text{ in.}^2$$

$$T_{ps} = A_{ps} f_{pse} = 4.05(162) = 656 \text{ kips}$$

$$a = \frac{T_{ps}}{0.85 f'_c b} = \frac{656}{0.85(5)24} = 6.43 \text{ in.} \qquad \text{(from data, } b = 24 \text{ in.)}$$

$$M_{nps} = T_{ps}\left(\frac{h}{2} - \frac{a}{2}\right) = \frac{656(18 - 3.21)}{12} = 813 \text{ ft-kips}$$

Check compressive stress on the concrete = P_c/A_c

$$\frac{P_e}{A_c} = \frac{T_{ps}}{A_g} = \frac{656}{24(36)} = 0.76 \text{ ksi}$$

Note that the post-tensioning stress level in the concrete should not exceed 1000 psi. This is because stress compatibility with the other components of the structure will become more of a problem and system shortening is likely to be excessive. In the building being designed and used in this example (Ref. 13.21), the floor slab is post-tensioned and the level of effective post-tensioning in the floor will be on the order of 200 psi (or less).

2. ***Determine the amount of mild steel required to attain the objective strength.***

$$M_{ns} = M_n - M_{nps} = 1278 - 813 = 465 \text{ ft.kips}$$

$$T_{ns} = \frac{M_{ns}}{d - d'} = \frac{465}{(33 - 3)/12} = 186 \text{ kips} \qquad (d' = 3 \text{ in.})$$

$$A_s = \frac{T_{ns}}{f_y} = \frac{186}{60} = 3.1 \text{ in.}^2$$

Hence, provide four No. 8 bars ($A_s = 3.16$ in.^{2}/$M_{ns} = 474$ ft.-kips).

3. ***Limit state at ultimate strength.***

The original design guidelines for this construction system (Ref. 13.18) require that the flexural overstrength (probable strength) provided by the mild steel be less than that provided by the post-tensioning steel. This provision is intended to produce the objective of a self-restoring bracing system. The mild steel reinforcing bars are debonded at the beam-column interface as in Figure 13.28. This is because the precast beam will tend to separate or "lift-off" the column, introducing large strains in the mild steel in this region. The extent of the debonding determines the strain/stress induced in the mild steel when the deformation limit state is reached. It should be noted that as the strain limit in the mild steel reinforcing bars is approached, their stress level becomes on the order of approximately 105 ksi. The probable strain in the mild steel would be comfortably less than the strain limit state at the deformation limit state.

Determine the mild steel stress at a post-yield rotation of 3% (see stress-strain diagram in Figurer 2.18 and Ref. 13.19). The unbonded mild steel length includes a region of probable debonding of 2.75 d_b on either side of the intentionally debonded region. If the intentionally debonded length is 6 in., the total debonded length becomes:

$$L_u + 5.5d_b = 6 + 5.5(1) = 11.5 \text{ in.}$$

The strain states in the mild and post-tensioned steel are developed from the post-yield deformation state of the connection in Fig. 13.28 and the stress-strain diagram of the mild steel reinforcement in Figure 2.16 or 2.18b of Chapter 2. The process is an iterative trial-and-adjustment procedure since the location of the neutral axis and steel strain states are mutually dependent. A reasonable estimate of the neutral axis is possible. T_{ps} will increase by about 25% to about 200 ksi. T_{ns} will increase by about 67 percent (100/60). The compressive force that must be resisted by the concrete is:

$$\begin{aligned} C &= 1.25T_{ps} + 0.67T_{ns} \\ &= 1.25(656) + 0.67(190) = 820 + 127 = 947 \text{ kips} \end{aligned}$$

$$a = \frac{947}{0.85(5)(24)} = 9.3 \text{ in.}$$

$$c = \frac{a}{\beta_1} = \frac{9.3}{0.80} = 11.6 \text{ in.}$$

Assume $c = 11.6$ in. for this trial cycle

$$\Delta_{su} = \theta(d - c) = 0.03(33 - 11.5) = 0.645 \text{ in.} \quad \text{(elongation of the mild steel)}$$

$$\Delta\epsilon_s = \frac{\Delta_s}{L_u + 5.5d_b} = \frac{0.645}{11.5} = 0.056$$

$$\epsilon_{su} = 0.056 + \epsilon_y = 0.058$$

$$f_{su} \cong 100 \text{ ksi} \qquad \text{(Refer to Figure 2.16 or Figure 2.18b)}$$

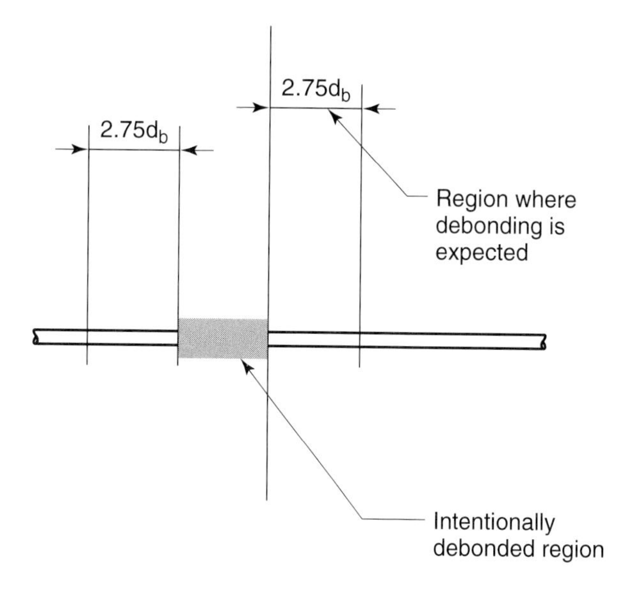

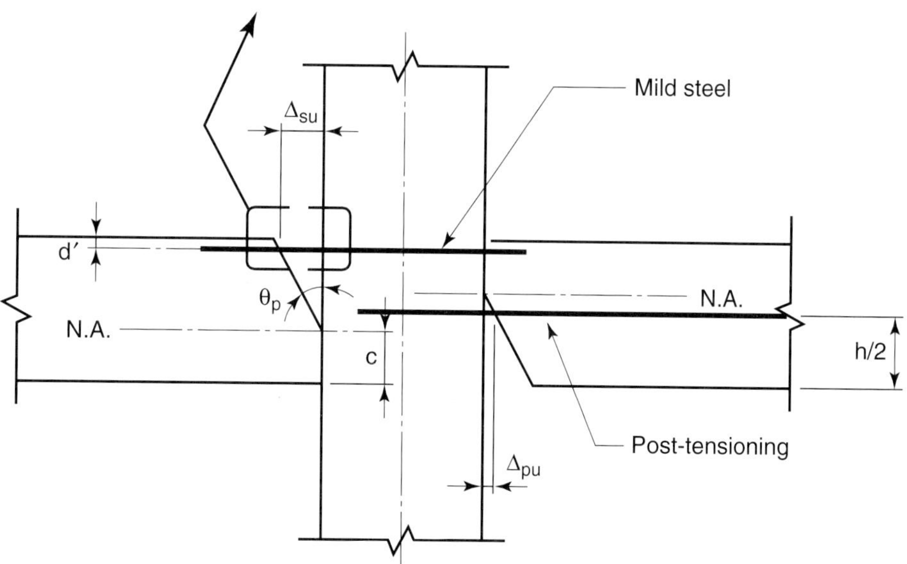

Figure 13.28 Reinforcing elongation at post-yield rotation limit state hybrid beam subassembly (Ref. 13.21)

$$\Delta_{pu} = \theta\left(\frac{h}{2} - c\right) = 0.03(18 - 11.6) = 0.195$$ (elongation required of the post-tensioning tendon)

$$\Delta\epsilon_{pu} = \frac{\Delta_{pu}}{L/2} = \frac{0.195}{18(12)(0.5)} = 0.0018$$

$$\Delta f_{ps} = \Delta\epsilon_{pu} E_{ps} = 0.0018(28{,}000) = 50 \text{ ksi}$$

$$f_{psu} = f_{pse} + \Delta f_{ps} = 162 + 50 = 212 \text{ ksi} < 0.9\,f_{pu}$$ (Refer to Figure 2.16 or Figure 2.18b)

$$T_{su} = A_s f_{su} = 3.16(100) = 316 \text{ kips}$$

$$T_{psu} = A_{ps} f_{psu} = 4.05(212) = 859 \text{ kips}$$

$$C = T_{psu} + T_{su} - C'_s = 859 + 316 - 190 = 985 \text{ kips}$$

$$a = \frac{985}{0.85(5)(24)} = 9.65 \text{ in.}$$

$$c = \frac{9.65}{0.8} = 12 \text{ in. (which is close to the } c = 11.6 \text{ in., hence O.K.)}$$

$$M_{pr} = T_{psu}\left(\frac{h}{2} - \frac{a}{2}\right) + (T_{su} - T_{ns})\left(d - \frac{a}{2}\right) + T_{ns}(d - d')$$

$$= 212(4.05)(18 - 4.83) + 40(3.16)(33 - 4.83) + 60(3.16)(30)$$

$$= 11{,}308 + 3561 + 5688$$

$$= 20{,}560 \text{ in.-kips}$$

$$= 1713 \text{ ft-kips}$$

$$M_{su} = 3561 + 5688 = 9249 \text{ in.-kips} < M_{psu} = 11{,}308 \text{ in.-kips}$$

$$\text{Overstrength factor achieved} = \frac{M_{pr}}{M_n} = \frac{1713}{1278} = 1.34, \quad \text{O.K.}$$

4. ***Verify that the capacity of the joint is sufficient.***
Comparing with the ACI 318 Code requirements, the ratio of probable strength (M_{pr}) to nominal strength (M_n) is 1.33 for the hybrid beam as opposed to the 1.25 stipulated by the ACI for conventional cases. This would probably be the conclusion resulting from the analysis of a conventionally reinforced frame beam at a post-yield rotation demand of 3 percent.

$$T_{ps-pr} = 1.25(162)(4.05) = 820 \text{ kips}$$

$$T_{s-pr} = 1.25(60)(3.16) = 237 \text{ kips}$$

$$T_s = A_s f_y = 60 \times 3.16 = 190 \text{ kips}$$

$$a = \frac{T_{ps-pr} + 0.25T_s}{0.85 f'_c b} = \frac{867}{0.85(5)(24)} = 8.5 \text{ in.}$$

$$M_{pr} = T_{ps-pr}\left(\frac{h}{2} - \frac{a}{2}\right) + T_s(d - d') + 0.25T_s\left(d - \frac{a}{2}\right)$$

$$= 820\,(18 - 4.25) + 190\,(33 - 3) + 47\,(33 - 4.25)$$

$$= 18{,}326 \text{ in.-kips}$$

$$V_{b-pr} = \frac{M_{pr}}{L_c/2} = \frac{18{,}326}{9(12)} = 170 \text{ kips}$$

$$V_{c-pr} = \frac{V_{b-pr} L}{h} = \frac{170(18)}{9.67} = 316 \text{ kips}$$

$$V_j = T_{ps-pr} + 2T_{s-pr} - V_{c-pr}$$

$$= 820 + 2(237) - 316 = 978 \text{ kips}$$

$$v_j = \frac{V_j}{A_j} = \frac{978}{36(36)} = 0.75 \text{ ksi}$$

$$v_{j-allow} = \phi 15\sqrt{f'_c} = 0.85(15)\sqrt{5000} = 892 \text{ psi} = 0.89 \text{ ksi} > 0.75 \text{ ksi, O.K.}$$

5. ***Design the shear reinforcement for the beam.***
Since post-yield behavior is anticipated in the end of the beam, the procedures used for the design of a monolithically cast "special" concrete frame beam are required. From step 4,

$$V_{b-pr} = 170 \text{ kips}$$

$$v_{b-pr} = \frac{V_{b-pr}}{bd} = \frac{170}{24(33)} = 0.215 \text{ ksi}$$

(a) *Hinge region:*
Disregard in the hinge region the shear strength of the plain concrete, namely, $v_c = 0$.

$$v_s = 0.215 \text{ ksi}$$

Trying two No. 4 stirrup sets, $A_v = 0.20$ in.^{2}/one leg of a stirrup.

$$s = \frac{\phi f_y A_v}{v_s b} = \frac{0.85(60)(4)(0.20)}{0.215(24)} = 7.91 \text{ in.}$$

where s is the maximum spacing of stirrups and A_v is the stirrup area.
Provide two No. 4 closed U-stirrups (hoops) at 7 in. center-to-center in the hinge region.

(b) *Outside the hinge region:*
Assume No. 4 U-stirrups, $A_v = 0.40$ in.^{2} per stirrup.

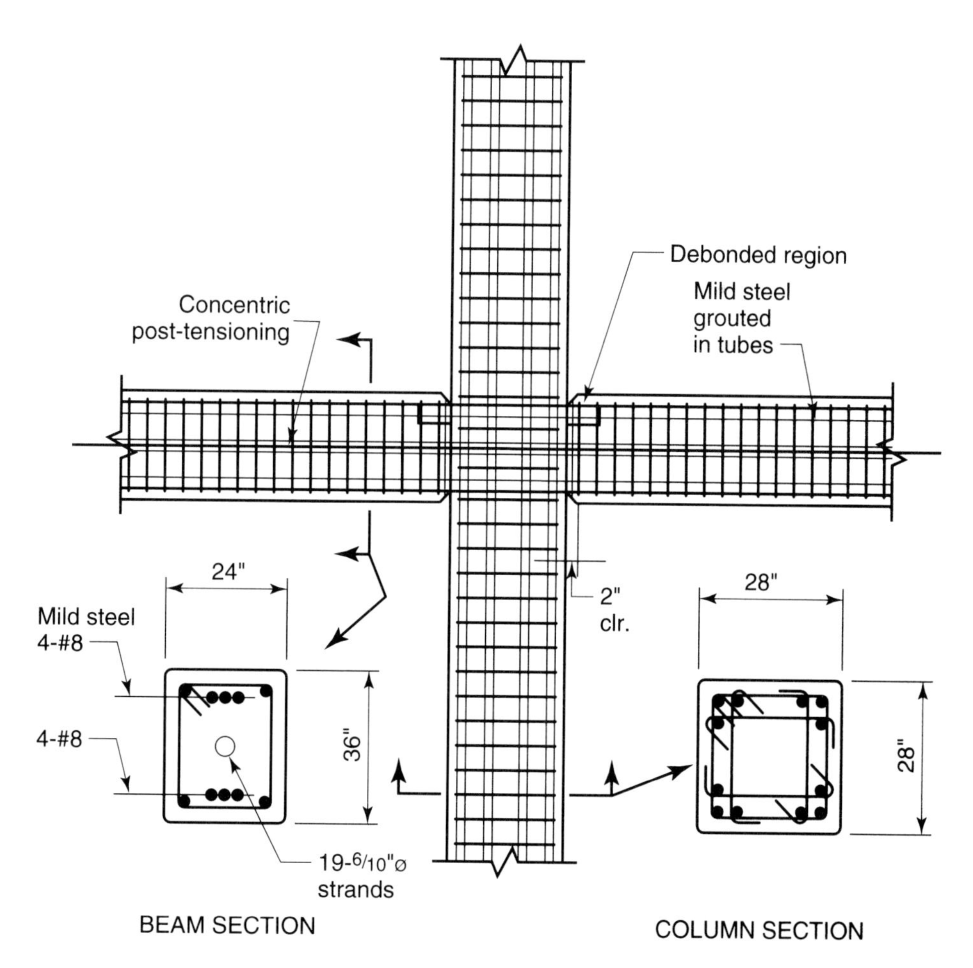

Figure 13.29 Schematic of the hybrid ductile beam–column connection in Example 13.5

$$v_c = 2\sqrt{f_c'} = 141 \text{ psi} = 0.141 \text{ ksi}$$

$$v_s = \frac{v_{b-pr}}{\phi} - v_c = \frac{0.215}{0.85} - 0.141 = 0.111 \text{ ksi}$$

$$s = \frac{f_y A_v}{v_s b} = \frac{60(0.40)}{0.111(24)} = 9.0 \text{ in. c./c.}$$

Provide single No. 4 U-stirrups at 9 in. center-to-center outside the hinge region which do not necessarily have to be hooped. Figure 13.29 schematically shows details of the ductile beam–column connection designed in this example.

SELECTED REFERENCES

13.1 ACI Committee 318, *Building Code Requirements for Structural Concrete (ACI 318–99) and Commentary (ACI 318R-99).* American Concrete Institute, Farmington Hills, MI, 2000, pp. 392.

13.2 International Code Council, *International Building Code 2000 (IBC),* Joint UBC, BOCA, SBCCI, Whittier, CA, 2000.

13.3 International Conference of Building Officials, *Uniform Building Code (UBD),* Vol. 2, ICBO, Whittier, CA, 1997.

13.4 Norris, H. C., Hansen, R. J., Holley, M. J., Biggs, J. M., Namyet, S., and Minami, K., *Structural Design for Dynamic Loads,* McGraw-Hill, New York, 1959.

13.5 Wakabayashi, M., *Design of Earthquake-resistant Buildings,* McGraw-Hill, New York, 1986.

13.6 Englekirk, R. E., and Hart, G. C., *Earthquake Design of Concrete Masonry Buildings,* Vols. I & II, Prentice Hall, Upper Saddle River, N.J., 1982.

13.7 Schneider, R. R., and Dickey, W. L., *Reinforced Masonry Design,* Prentice Hall, Upper Saddle River, N.J.,1994.

13.8 Clough R. W., "Dynamic Effects of Earthquakes," *Proc. ASCE,* Vol. 86, ST4, New York, April 1960, pp. 49–65.

13.9 Ghosh, S. K., "Special Provisions for Seismic Design," PCA Publication, *Notes on ACI318-89 Code,* Chapter 31, Portland Cement Association, Skokie, IL, 1990, pp. 31–1 to 31–81.

13.10 Derecho, A. T., Fintel, M., and Ghosh, S. K., "Earthquake-resistant Structures," Ch. 12 in *Handbook of Concrete Engineering,* 2nd ed., Van Nostrand Reinhold, New York, 1985, pp. 411–513.

13.11 Borg, S. E., *Earthquake Engineering-Damage Assessment and Structural Design,* John Wiley & Sons, New York, 1983.

13.12 ASCE Standard 7-95, *Minimum Design Loads for Buildings and other Structures,* American Society of Civil Engineers, Reston, VA, 1995.

13.13 Naja, W. M., and Barth, F. G., "Seismic Resisting Construction" Chapter 26, in E. G. Nawy, editor-in-chief, *Concrete Construction Engineering Handbook,* CRC Press, Boca Raton, FL, 1998, pp. 26-1, 26-8.

13.14 Nawy, E. G., *Reinforced Concrete—A Fundamental Approach,* 4th Ed., Prentice Hall, Upper Saddle River, N.J., 2000, pp. 786.

13.15 Federal Emergency Management Agency (FEMA), "NEHRP Recommended Provisions for Seismic Regulations for New buildings and Other Structures," FEMA 302, Part I & II, Building Seismic Safety Council, Washington, D.C., 1998.

13.16 Englekirk, R. E., *Steel Structures—Controlling Behavior through Design,* John Wiley & Sons, New York, 1994.

13.17 Pauley, T., and Priestely, M. J. N., *Seismic Design of Reinforced Concrete and Masonry Walls,* John Wiley & Sons, New York, 1992, 744 pp.

13.18 Cheok, G. S., Stone, W. C. and Nakaki, S. D., "Simplified Design Procedure for Hybrid Precast Concrete Connection," NISTIR Report No. 5765, National Institute of Standards and Technology, Gaithersburg, MD, February 1996, 82 pp.

13.19 Englekirk, R. E., "An Innovative Design Solution for Precast Prestressed Concrete Buildings in High Seismic Zones," Vol. 41 No. 4, *PCI Journal,* Precast/Prestressed Concrete Institute, Chicago, IL, July/August 1996, pp. 44–53.

13.20 Ghosh, S. K., Nakaki, S. W., and Krishnan, K., "Precast Structures in Regions of High Seismicity: 1997 UBC Design Provisions," Vol. 42, No. 6, *PCI Journal,* Precast/Prestressed Concrete Institute, Chicago, IL, November–December 1997, pp. 76–93.

13.21 Englekirk, R. E., Design of a Forty-one Story Ductile Frame Precast Prestressed Concrete Building in San Francisco. Private Communication, 1999.

13.22 Englekirk, R. E., and Llovet, D., "Cyclic Test of Cast-in-Place High Strength Beam-Column Joints," Concrete International, American Concrete Institute, Farmington Hills, MI, 1999.

13.23 Choek, G. S., and Lew, H. S., "Model Precast Beam-to-Column Connections Subject to Cyclic Loading," *PCI Journal,* Vol. 38, No 4, July–August 1993, Chicago, IL: Precast/Prestressed Concrete Institute, Chicago, IL, July/August, pp. 80–100.

13.24 Choek, G. S., and Stone, W. C., and Kunnath, S. K., "Seismic Response of Precast Prestressed Concrete Frames with Hybrid Connections," *ACI Structural Journal,* American Concrete Institute, Farmington Hills, MI, September/October 1998, pp. 527–546.

13.25 Priestly, M. J. N. and Marcie, G. A., "Seismic Tests of Precast Beam-to-Column Joint Subassemblages with Unbonded Tendons," *PCI Journal,* Vol. 41 No. 1, Precast/Prestressed Concrete Institute, Chicago, IL, January–February, 1996, pp. 64–81.

13.26 Stone, W. C., Choek, G. S., and Stanton, J. F., "Performance of Hybrid Moment-Resisting Precast Beam-Column Concrete Connections Subjected to Cyclic Loading," Title 92-S22, *ACI Structural Journal,* American Concrete Institute, Farmington Hills, MI, March–April, 1995, pp. 229–249.

13.27 Stanton, J., Stone, W. C., and Choek, G. S., "A Hybrid Reinforced Precast Frame for seismic Regions," *PCI Journal,* Precast/Prestressed Concrete Institute, Chicago, IL, March–April, 1997, pp. 20–32.

13.28 Aswad, G. S., Djazmati and Aswad, A., "Comparison of Shear Wall Deformations and forces Using Two Approaches," *PCI Journal,* Vol. 44, No. 1, Precast/Prestressed Concrete Institute, Chicago, IL, January–February, 1999, pp. 34–46.

13.29 Aswad, A., Private Communication and "Analysis of Reinforced or Post-Tensioned Shear Walls," SHEARWAL Computer Program, Version 2.1, Jacques and Aswad Inc., Denver, CO, 1990.

13.30 Cleland, N. M., "Design for Lateral Resistance with Precast Concrete Shear Walls," Col. 42, No. 5, *PCI Journal,* Precast/Prestressed Concrete Institute, Chicago, IL, September–October 1997, pp. 44–63.

13.31 Pessiki, S. et al. "Seismic Analysis, Behavior and Design of Unbonded Post-Tensioned Precast Concrete Frames." Report No. EQ-97-02, Dept. of Civil and Environmental Engineering, Lehigh University, Bethlehem, PA, November 1997, pp. 1–315.

13.32 Pessiki, S. et al. "Analytical Modeling and Lateral Load Behavior of Unbonded Post-Tensioned Precast concrete walls," Reports No. EQ-96-02, Dept. of Civil and Environmental Engineering, Lehigh University, Bethlehem, PA, November 1996, pp. 1–192.

13.33 Pessiki, S. et al. "Seismic Design and Response evaluation of Unbonded Post-Tensioned Precast Concrete Walls" Report No. EQ-97-01, Dept. of Civil and Environmental Engineering, Lehigh University, Bethlehem, PA, November 1997, pp. 185.

PROBLEMS FOR SOLUTION

13.1 A 3 x 18 panel ductile, moment-resistant frame category-II site-class B frame building has a ground story 15 ft. high (4.6 m) and ten upper stories of equal height of 11'–6" (3.5 m). Compute the seismic base shear V and the overturning moment at each story level in terms of the weight W_s of each floor. Use the equivalent lateral force method in the solution. Given:

$$S_1 = 0.34 \text{ sec}, S_s = 0.90 \text{ sec } R = 5,$$

$$W_s \text{ per floor} = 2400 \text{ kips (9560 kN)}$$

13.2 A moment-resisting ductile frame building is located in a high-seismic-intensity zone. The earthquake forces are resisted equally as a dual system by the ductile frame and a monolithic reinforced concrete shear wall over the total height of the building. The geometry of the structure is given below. Design the shear wall assuming that the magnitude of the loads, forces and moments applied to the wall are 110 percent of the values used in Ex. 13.2. Given:

floors have slabs of thickness $h_f = 7$ in. (178 mm)

clear beam spans in both longitudinal and transverse directions = 20′–0″ (6.1 m)

shear wall base length $l_w = 25$ ft (39.6 m)

shear wall height $h_w = 130$ ft (39.6 m)

$f'_c = 5000$ psi, normal weight (34.5 MPa)

$f_{yv} = f_{yh} = 60{,}000$ psi (414 MPa)$_v$

Sketch the wall reinforcement.

13.3 A precast shear wall in a moderate seismicity zone for a six story frame building has a wall thickness of 10 in. (254 mm). The height of each story is 11′–6″ (3.5 m) and the wall segments extend the height of the building and prestressed vertically. The wall is subjected to an unfactored seismic base shear $V_e = 165$ kips (734 kN) and an unfactored overturning moment $M_e = 5600$ ft-kip (7594 kN-m). The total weight of each floor including any attached masses is 2800 kips (12,454 kN). Design the connection at the base of the wall assuming that the neutral axis obtained by strain-compatibility analysis is $c = 21.5$ in. (546 mm). Given:

sliding coefficient $\mu = 0.60$

beams size: 24 in. × 28 in. (610 mm × 711 mm)

effective beam spans: 22 ft 6 in. (6.9 m)

allowable horiz. shear stress $f_{cv} = 1200$ psi (8.3 MPa)

$f'_c = 5000$ psi, normal weight (34.5 MPa)

$f_y = 60{,}000$ psi (414 MPa)

Assume using either a welded connection or a Dywidag connector.
Also assume that analysis of the wall gave the following factored overturning moments:

Load Case I: $M_u = 10{,}500$ ft-kip
Load Case II: $M_u = 8600$ ft-kip

13.4 Design a ductile precast prestressed concrete moment-resistant connection of a ductile frame in a high intensity seismic zone subjected to a factored seismic moment $M_u = 1350$ ft-kip (1831 kN-m) and a post-yield rotation $\theta_p = 2.75$ percent. The frame precast beams have spans of 20′–0″ and the clear spans are 17′–4″ (5.3 m). Each story height is 9′–0″ (2.74 m). Use the Dywidag Ductile Connection assembly (DDC) in your solution. Given:

column sizes: 38 in. × 38 in. (965 mm × 965 mm)

center to center Ductile Rods, $(d - d')$: 27 in. (686 mm)

$f'_c = 5000$ psi, normal weight concrete (34.5 MPa)

13.5 Design the moment resisting connection in Problem 13.4 as a hybrid connection using both mild steel and prestressing post-tensioned reinforcement. Given:

$$f'_c = 5000 \text{ psi, normal weight concrete (34.5 MPa)}$$
$$f_y = 60{,}000 \text{ psi (414 MPa)}$$
$$f_{pu} = 270{,}000 \text{ psi (1862 MPa)}$$
$$f_{ps} = <0.90\, f_{pu} \text{ (determine from compatibility analysis)}$$
$$f_{pe} = 160{,}000 \text{ psi (1103 MPa)}$$
$$E_s = 29{,}000 \text{ ksi (200,000 MPa)}$$
$$E_{ps} = 28{,}000 \text{ ksi (193,000 MPa)}$$
$$f_c = 1000 \text{ psi (6.9 MPa) maximum concrete stress at post-tensioning}$$

COMPUTER PROGRAMS IN Q-BASIC

The programs presented in this appendix are furnished as a guide to the development of a user's own programs for the design of prestressed concrete members. While every effort has been made to utilize the existing state of the art and to assure accuracy of the analytical solution and design techniques, neither the author nor the publisher make any warranty, either expressed or implied, regarding the use of these programs for other than informational purposes. The user of the program is responsible for the final evaluation as to the validity, accuracy, and applicability of any results obtained. The listed programs can be purchased on $3\frac{1}{2}$-in. diskette from NC SOFTWARE, Box 161, East Brunswick, New Jersey 08816.

A-1 COMPUTER PROGRAM EGNAWY10 FOR ESTIMATION OF TIME-DEPENDENT LOSSES IN PRESTRESSED CONCRETE BEAMS

A computer program in Q-Basic for personal computers attempts to analyze the partial loss of prestress in pretensioned and post-tensioned beams due to time-dependent effects. It is based on the step-by-step flowchart Figure 3.9 and the discussions and examples in Chapter 3.

"Reflections"—High strength polymer concrete sculpture at Rutgers University. Work by R. H. Karol, the civil engineering class of 1982, and the author.

The program uses equations for incremental time-dependent steps that assess steel relaxation, shrinkage and creep. It also differentiates between stress-relieved and low relaxation strands. Input values are A_c, T_c, f_{pu}, E_{ps}, f'_c, f'_{ci}, span L, A_{ps}, W_D, W_{SD}, W_L and anchorage seating loss Δ_A.

Equations 3.7 and 3.8 give the loss due to relaxation as a function of time range $(t_2 - t_1)$.

A basic maximum creep coefficient $C_u = 2.35$ is used in the analysis that is reduced by a time function such that

$$C_t = \left(\frac{t^{0.60}}{1 + t^{0.60}} \right) C_u$$

as discussed in Equations 3.9a, 3.9b, and 3.10. The user can input other maximum creep values into the programs which are different than $C_u = 2.35$ and the computer run would base the computations on the user's value of C_u.

Shrinkage loss is computed using time-dependent Equation 3.15a for moist-cured concrete with maximum $\epsilon_{SH} = 800 \times 10^{-6}$ in./in. and Equation 3.15b for steam-cured concrete with maximum $\epsilon_{SH} = 730 \times 10^{-6}$ in./in.

$$\text{Moist-cured:} \quad \epsilon_{SH,t} = \left(\frac{t}{t + 35} \right) \epsilon_{SH}$$

$$\text{Steam-cured:} \quad \epsilon_{SH,t} = \left(\frac{t}{t + 55} \right) \epsilon_{SH}$$

The user can input other shrinkage strain values directly into the program that are normally lower than $\epsilon_{SH} = 0.0008$ in./in.

For post-tensioned beams, the program computes frictional losses due to six different types of sheathing for tendons or strands using the governing ACI Code coefficients μ for curvature friction effect, and K for wobble effect. Then, it sums up all the losses in prestress and tabulates them as seen in one typical computer run of a double-T beam presented in this appendix.

```
llist
1 REM -- EGNAWY 10 "TIME DEPENDANT LOSSES IN PRESTRESSED CONCRETE BEAMS"
2 REM -- ****************************************************************
3 REM -- COPYRIGHT 1988 BY DR. EDWARD G. NAWY
4 REM -- ALL RIGHTS RESERVED
6 CLS: LOCATE 5,33: PRINT "PROGRAM EGNAWY 10"
7 LOCATE 7,16: PRINT "TIME DEPENDANT LOSSES IN PRESTRESSED CONCRETE BEAMS"
8 LOCATE 11,33: PRINT "COPYRIGHT 1988"
9 LOCATE 13,32: PRINT "DR. EDWARD G. NAWY"
10 LOCATE 15,32: PRINT "ALL RIGHTS RESERVED"
14 J=0: FOR I = 1 TO 1500: 7=J + 1: NEXT I
20 SCREEN 1
30 PAINT ( 15, 15) ,2
35 SCREEN 2
40 PRINT
41 PRINT "****************************************************************"
42 PRINT "* PROGRAM TO CALCULATE TIME DEPENDANT LOSSES OF                 *"
43 PRINT "*                                                               *"
44 PRINT "*   PRESTRESSED CONCRETE SIMPLY SUPPORTED BEAMS                 *"
45 PRINT "*                                                               *"
46 PRINT "****************************************************************"
47 PRINT :PRINT :PRINT
```

```
48 INPUT "PRESS (RETURN) KEY TO CONTINUE ";PPPPPP
49 SCREEN 1
50 PAINT (15, 15) ,2
51 PRINT :PRINT :PRINT
60 SCREEN 2
100 DIM K(6) ,Mu(6) ,PP(10) ,TIME(50) ,LOSS(50) ,DC(50) ,DS(50) ,DR(50)
299 PRINT"---------------------------------------------"
300 PRINT "READ SECTION DIMENSIONS AND SPAN"
305 PRINT"---------------------------------------------" :PRINT :PRINT
306 INPUT "THE BEAM SPAN IS (FEET) = ";L:PRINT :PRINT
307 INPUT "DO YOU WANT TO INPUT SECTION PROPERTIES (Y/N) " ;XX$:IF XX$="N" THEN
310S SAVE "B:EGNAWY 10"
308 INPUT "AC =" ,AC: INPUT"IC =",IC: INPUT"CT =" ,CT: INPUT "CB =" ,CB: INPUT
"WD =", WD: GOTO 635
310 PRINT TAB(5) ;"*** THE SECTIONS MENU ***"
320 PRINT TAB(5) ;"*************************" :PRINT : PRINT
330 PRINT TAB(2) ;" 1-RECTANGULAR SECTION" :PRINT TAB(2) ;"2-t-SECTION"
340 PRINT TAB(2) ;" 3-I-SECTION" :PRINT
350 INPUT "WHAT IS YOUR CHOICE ? ";WW : PRINT
360 PRINT :PRINT" INPUT DATA ":PRINT:-------------------------" :PRINT
370 ON WW GOTO 510, 450, 380
380 INPUT "THE TOP FLANGE WIDTH        =";B1:PRINT
390 INPUT "THE TOP FLANGE DEPTH        =";H1:PRINT
400 INPUT "THE BOTTOM FLANGE WIDTH     =";B2:PRINT
410 INPUT "THE BOTTOM FLANGE DEPTH     =";H2:PRINT
420 INPUT "THE TOTAL BEAM DEPTH        =";H:PRINT
430 INPUT "THE WEB WIDTH               =";BW:PRINT
435 INPUT "DO YOU WANT TO CORRECT DIMENSIONS (Y/N) ";A$:IF A$="Y" THEN 380
440 GOTO 570
450 INPUT "THE FLANGE WIDTH            =";B1:PRINT
460 INPUT "THE FLANGE DEPTH            =";H1:PRINT
470 INPUT "THE TOTAL DEPTH             =";H:PRINT
480 INPUT "THE WEB WIDTH               =";BW:PRINT
490 B2=BW
495 INPUT "DO YOU WANT TO CORRECT DIMENSIONS (Y/N) ";A$:IF A$="Y" THEN 450
500 GOTO 570
510 INPUT "THE WIDTH                   =";B1:PRINT
520 INPUT "THE TOTAL DEPTH             =";H:PRINT
530 BW=B1
540 B2=B1
550 H1=H
560 INPUT"DO YOU WANT TO CORRECT DIMENSIONS (Y/N) ";A$:IF A$="Y" THEN 510
569 REM ---------------------------------------------------------------
570 REM CALCULATION OF SECTION GEOMETRIC PROPERTIES AND ITS WEIGHT
580 REM ---------------------------------------------------------------
590 AC= (H*BW) + ((B1-BW) *H1-) + ((B2-BW) *H2)
620 CB=H-CT
630 WD=AC*150/144
635 RR = IC/AC
640 ST=IC/CT
650 SB=1C/CB
764 PRINT"-----------------------"
765 PRINT" CONCRETE PROPERTIES  "
770 PRINT"-----------------------"
775 PRINT : INPUT "SPECIFIED CONCRETE COMPRESSIVE STRENGTH f'c (psi) ? =";FPC
```

```
780 PRINT : INPUT "CONCRETE STRENGTH AT INITIAL PRESTRESS f ' ci (psi) ? ="
;FPCI
782 PRINT : INPUT "DO YOU WANT TO CORRECT THE PREVIOUS DATA (Y/N)?" ;B$:IF
B$="Y"TH EN 775
899 PRINT"---------------------------------"
900 PRINT "DESIGN LOAD, ECCENTRICITIES  "
910 PRINT"---------------------------------" :PRINT:PRINT
920 PRINT:INPUT"THE IMPOSED DEAD LOAD (LB/FT)         =";WSD
930       INPUT"THE LIVE LOAD (LB/FT) =" ;WL
940       INPUT"TENDONS ECCENTRICITY AT MIDSPAN     =";ECEN
950       INPUT"TENDONS ECCENTRICITY AT SUPPORT     =";ESUP
980 PRINT"INPUT"DO YOU WANT TO CORRECT THE PREVIOUS DATA (Y/N)?";B$:IF B$="Y" TH
EN 920
1049 REM -------------------------------------
1050 REM CONCRETE ALLOWABLE STRESS CALCULATIONS
1060 REM -------------------------------------
1110 EC=57000!*SQR(FPC)
1120 ECI-57000!*SQR(FPCI)
1190 PRINT"----------------"
1200 PRINT"STEEL PROPERTIES"
1210 PRINT"----------------"
1220 PRINT
1230 INPUT "ULTIMATE STRENGTH OF PRESTRESSING STEEL  fpu (psi)?   ="; FPU
1240 INPUT "INITIAL PRESTRESSING STRESS             fpi (psi)  ?   ="; FPI
1250 INPUT "YIELD STRENGTH OF PRESTRESSING STEEL    fpy (psi)  ?   ="; FPY
1260      INPUT "YOUNG'S MODULUS OF PRESTRESSED STEEL (psi)       =";EPS
1262 PRINT:INPUT "AREA OF PRESTRESSED STEEL              =";APS
1267 INPUT "NUMBER OF PRESTRESSED TENDONS            =";N
1269 PRINT:INPUT"DO YOU WANT TO CORRECT THE PREVIOUS DATA (Y/N) ";B$:IF B$="Y"
TH EN 1220
1270 PRINT:PRINT"----------------------"
1280 PRINT      " ADDITIONAL DATA      "
1290 PRINT      "------------------------":PRINT
1300 INPUT "PRETENSIONED TYPE (1), POST-TENSIONED TYPE (2)     =";S
1310 PRINT "MOIST CURED FOR 7 DAYS TYPE 1"
1312 PRINT "STEAM CURED FOR 3 DAYS TYPE 2"
1315 INPUT SSS
1320 IF S=1 THEN 1340
1330 INPUT "NUMBER OF TENDONS JACKED AT THE TIME          ="; NN
1340 INPUT " ANCHORAGE SLIP                  (in)          =";SL
1350 PRINT:PRINT
1365      INPUT "DEPTH OF PRESTRESSED STEEL       =" ;DPS
1368 PI= APS*FPI
1380 NEXT I
1390 IF S= 1 THEN 1550
1400 REM --------------------------
1410 REM FRICTION LOSS
1420 REM --------------------------
1430 PRINT TAB (10) "TENDONS TYPE MENU "
1440 PRINT "1-TENDONS IN FLEXIBLE METAL SHEATING (WIRE TENDONS )       "
1450 PRINT "2-TENDONS IN FLEXIBLE METAL SHEATING ( 7 WIRE STRANDS)     "
1460 PRINT "3-TENDONS IN FLEXIBLE METAL SHEATING (HIGH STRENGTH BAR)   "
1470 PRINT "4-TENDONS IN RIGID METAL DUCT, 7 WIRE STRAND               "
1480 PRINT "5-PREGREASED TENDONS , WIRE TENDONS , 7 WIRE STRANDS       "
```

```
1490 PRINT "6-MASTIC COATED TENDONS , WIRE TENDONS AND 7 WIRE STRANDS "
1500 PRINT :INPUT"TYPE YOUR CHOICE "; 1 :PRINT:PRINT
1510 *=8*ECCEN/(L*12)
1520 DF = FPI*((MU(I)*X) + (K(I)*L))
1530 PP(1) =  (DF/FPI*100)
1540 PRINT "LOSSES DUE TO FRICTION" TAB(50) "= " DF " psi " TAB(67) PP(1) " % "
1550 REM ----------------------
1560 REM ANCHORAGE SLIP LOSS
1570 REM ----------------------
1580 DA = SL*EPS/(L*12)
1590 PP(S)= (DA/FPI*100)
1600 PRINT "LOSSES DUE TO ANCHORAGE SLIP"TAB(50) " = " DA " psi " TAB(67) PR(2)"
%"
1610 FFI = FPI-DF-DA
1620 IF S=2 THEN 1690
1630 PRINT:PRINT
1640 INPUT "NUMBER OF DAYS BETWEEN JACKING AND TRANSFER = " ;TTRANSFER
1650 TTRANSFER= TTRANSFER*24
1660 DR=(LOG(TTRANSFER)/LOG(10))
1670 DR=DR*FFI*((FFI/FPY)-.55)/10
1675 PRINT"LOSSES DUE TO RELAXATION BETWEEN JACKING AND TRANSFER =" :DR " psi "
:PR INT:PRINT:PRINT
1680 FFI = FFI-DR
1690 REM -----------------------
1700 REM ELASTIC SHORTENING LOSS
1710 REM -----------------------
1820 IF S=1 THEN 1920
1830 IF NN=N THEN 1910
1840 NN=N/NN
1850 SUM=0
1860 FOR 1=1 TO (NN-1)
1870 SUM=SUM + I/(NN-1)
1880 NEXT I
1890 DE=SUM *DE/NN
1900 GOTO 1920
1910 DE =0
1920 PP(3)= (DE/FPI*100)
1930 PRINT "LOSSES DUE TO ELASTIC SHORTENING"TBA(50) " = "DE" psi "TAB(67)
PP(3)" %"
1931 PRINT "DO YOU WANT TO INPUT SPECIAL SHRINKAGE STRAIN VALUE(Y/N)"
1932 INPUT 000$:IF 000$="N" THEN 1935
1933 INPUT "SHRINKDAGE STRAIN VALUE" :SHRINKS
1935 INPUT"ARE STRANDS STRESS RELIEVED" :QQQ$
1940 REM -----------------------
1950 REM CREEP LOSS
1960 REM -----------------------
1961 INPUT"DO YOU WANT TO INPUT CREEP FACTOR";YYY$:IF YYY$="N" THEN 1965
1962 INPUT "ULTIMATE CREEP FACTOR IS ";CU
1963 GOTO 1970
1965 CO=2.35
1970 I=0
1975 I=I + 1
1980 INPUT "TIME AFTER JACKING THAT LOSSES ARE NEEDED (DAYS)":TIME(I)
2050 MOMENT = MD
```

```
2080 X=MOMENT*ECEN /IC
2090 AA=FFI*APS*(1+(ECEN^2/RR))
2100 AA=AA/AC
2110 X=X-AA
2120 CTIME=(TIME(1)^,6)/(10+TIME(I)^.6) *CU
2130 DC(I)=CTIME*ABS(X) *EPS/EC
2140 PP(4)=INT(DC(I)/FPI*100)
2150 PRINT"LOSSES DUE TO CREEP"TAB(50)" = " DC(I) " psi " TAB(67) PP(4) " %"
2160 REM ------------------
2170 REM SHRINKAGE LOSS
2180 REM ------------------
2181 IF SHRINKS=0! THEN 2190
2184 X=TIME(I)/(TIME(I)+35)*SHRINKS
2185 GOTO 2230
2190 ON SSS GOTO 2200,2220
2200 X=TIME(I)/(TIME(I)+35)*8.000001E-04
2210 GOTO 2230
2220 X=TIME(I)/(TIME(I)+35)*7.300001E-04
2230 DS(I)=X*EPS
2240 PP(5)-(DS(I)/FPI*100)
2250 PRINT"LOSSES DUE SHRINKAGE"TAB(50) " = " DS(I) " psi " TAB(67)PP(5)" %"
2260 REM ------------------
2270 REM RELAXATION STEEL LOSS
2280 REM ------------------
2290 FSTEEL=FFI-DE
2300 TIME(I)=TIME(I)*24
2310 IF S=2 THEN 2340
2345 Y=FSTEEL/FPY
2347 IF Y>.55 THEN 2350
2348 Y=.6
2350 DR(I)=X*FSTEEL*((Y)-.55)/10
2351 IF QQQ$="Y" THEN 2360
2355 DR(I)=XXFSTEEL*((Y)-.55)/45
2360 PP(6)=(DR(I)/FPI*100)
2370 PRINT "LOSSES     TO STEEL RELAXATION"TAB(50) " = "DR(I) " psi
"TAB(67)00(6)" %"
2380 LOSS(I) = DF+DA+DE+DC(I) +DS(I) +DR(I)
2450 INPUT "DO YOU NEED LOSSES AFTER ANOTHER TIME INTERVAL" ;YY$
2460 IF YY$="Y" THEN 1975
2500 PRINT TAB(10) "-------------------------------"
2510 PRINT TAB(10) "    TABLE OF FINAL RESULTS      "
2520 PRINT TAB(10) "-------------------------------"
2530 PRINT"TIME(DAYS) "TAB(15) "TOTAL"TAB(25) "FRICTI."TAB(35) "ANCHOR."TAB(45)
"ELAS T.) "TAB(55) "CREEP"TAB(65) "SHRI."TAB(75) "RELAX."
2535 PRINT TAB (15) "LOSSES"TAB(25) "LOSS"TAB(35) "LOSS"TAB(45) "LOSS"TAB(55)
"LOSS"TAB(65) "LOSS"TAB(75) "LOSS"
2540 FOR K=1 TO I
2550 PRINT TIME(K)/24 TAB(15) LOSS(K) TAB(25) DF TAB(35) DA TAB(45)DE TAB(55) I
NT(DC(K)) TAB(65) INT(DS(K)) TAB(75) INT(DR(K))
2560 PRINT
2570 NEXT K
8000 DATA 0.0015,0.25,0.002,0.25,0.0006,0.3
8010 DATA 0.0002,0.25,0.002,0.15,0.002,0.15
9999 END
```

Example A-1 Time Dependent Partial Losses in Pretensioned Prestressed Concrete Beam

Compute the long-term partial losses in prestress for the pretensioned T beam in example 3.8 using the computer program EGNAWY 10.

Input

Beam span $L = 70$ ft.

Flange width $b = 120$ in.

Flange depth $h_1 = 2$ in.

Total depth $h = 32$ in.

Web width $b_w = 12.5$ in.

$f'_c = 5{,}000$ psi

$f'_{ci} = 3{,}500$ psi

$W_{SD} = 250$ plf

$W_L = 400$ plf

Midspan tendon eccentricity $\epsilon_c = 18.73$ in.

Support tendon eccentricity $\epsilon_e = 12.98$ in.

$f_{pu} = 270{,}000$ psi

$f_{pi} = 189{,}000$ psi

$f_{py} = 229{,}500$ psi

$E_{ps} = 28{,}000{,}000$ psi

$A_{ps} = 1.836$ in.^{2}

No. of tendons = 12

Depth of prestressing steel $d_p = 28.75$ in.

Number of days between jacking and transfer = 0.75

Output

```
************************************************************
* PROGRAM TO CALCULATE THE TIME DEPENDANT LOSSES OF        *
*                                                          *
*   PRESTRESSED CONCRETE SIMPLY SUPPORTED BEAMS            *
*                                                          *
************************************************************

PRES (RETURN) KEY TO CONTINUE ?

-------------------------------------
READ SECTION DIMENSIONS AND SPAN
-------------------------------------
```

```
THE BEAM SPAN IS (FEET) = ? 70
              --------------------------------
                TABLE OF FINAL RESULTS
              --------------------------------
TIME(DAYS)   TOTAL      FRICTI.   ANCHOR.   ELAST.)     CREEP   SHRI.    RELAX.
             LOSSES     LOSS      LOSS      LOSS        LOSS    LOSS     LOSS
.75          9787.352   0         0         8236.696    1256    293      0
1            10553.68   0         0         8236.696    1472    388      455
7            18048.41   0         0         8236.696    3939    2333     3539
14           22179.41   0         0         8236.696    5305    4000     4637
21           24973.09   0         0         8236.696    6206    5250     5280
30           27586.71   0         0         8236.696    7043    6461     5845
45           30621.6    0         0         8236.696    8022    7875     6487
90           35587.91   0         0         8236.696    9684    10080    7586
365          43368.96   0         0         8236.696    12552   12775    9804
730          46091.9    0         0         8236.696    13592   13359    10903
1825         48911.89   0         0         8236.696    14583   13736    12355
OK
```

A-2 COMPUTER PROGRAM EGNAWY12 FOR SERVICE LOAD ANALYSIS AND DESIGN IN FLEXURE OF PRESTRESSED CONCRETE BEAMS

This computer program in Q-BASIC for personal computers is intended to proportion prestressed concrete T and I beams data pretensioned and post-tensioned. The flow chart Figure 4.34 and the discussions and example solutions of Chapter 4 are the background of the program.

The input data comprise f_{pu}, f'_c, f_{ci}, W_D, W_{SD}, W_L, span L and effective prestress coefficient γ. It computes the service load level flexural moments M_D, M_{SD}, M_L, and M_I and selects the cross-sectional dimensions of the prestressed beam on the basis of the maximum allowable service load stresses in the concrete and the prestressing steel as set by the ACI-318 Code.

The basis for the section selection is the section moduli values S_b and S^t obtained from the computed moment values and the allowable concrete stresses at service load. By trial and adjustment, the program iterates to the closest T- or I-beam section that can sustain the service level load within the maximum allowable service load stresses at effective prestress both for tension and compression.

Standard PCI sections are contained in the program when standard sections are to be chosen by the user. The computer run of one typical example in this appendix illustrates the output resulting from use of this program.

Example A-2 Service Load Proportioning of Prestressed Concrete Beams

Select the appropriate section of Example 4.2 of a prestressed pretensioned beam having a span of 65 ft. (19.8 in.) using the computer program EGNAWY12.

Input

$$f_{pu} = 270{,}000 \text{ psi}$$

$$f'_c = 6{,}000 \text{ psi}$$

$$f'_{ci} = 4{,}500 \text{ psi}$$

$$\text{Span } L = 65 \text{ ft.}$$

$$W_{SD} = 100 \text{ plf}$$

W_L = 1,100 plf

Effectiveness ratio γ = 0.82

Trial I section chosen

Top flange width b_1 = 17 in.

Top flange thickness h_1 = 4.85 in.

Bottom flange width b_2 = 18 in.

Bottom flange thickness h_2 = 7 in.

Total depth h = 40 in.

Web width b_w = 6 in.

Initial prestressing force = 376,110 lb.

Depth of prestressing steel = 36.15 in.

Output

```
***************************************
*     SERVICE LOAD DESIGN OF          *
*  SIMPLY SUPPORTED PRESTRESSED BEAMS *
***************************************

PRESS (RETURN) KEY TO CONTINUE ?

RANGE OF INIT.PRES.FORCE     PRES.STEEL DEPTH
  474155  TO  299175          38
  492498  TO  310749          37
  512316  TO  323253          36
  533796  TO  336806          35
  557156  TO  351546          34
  582654  TO  432208          33

INPUT YOUR CHOSEN INITIAL PRESTRESSED FORCE THEN THE CORRESPONDING DEPTH OF PRE-
STRESSED STEEL
? 376110, 36. 15
PERMISSIBLE LINEAR STRESSES
---------------------------

F.in.top= 402.4923
F.in.bot=-2700
F.top=-2700
F.bot= 929.5161

ACTUAL LINEAR STRESSES
----------------------

F.in.top= 67.87824
F.in.bot=-1821.373
F.top=-2450.798
F.bot= 632.9773

DO YOU WISH TO REENTER OTHER DIMENSIONS(Y/N)    ? Y

RANGE OF INIT.PRES.FORCE     PRES.STEEL DEPTH
  171381  TO -159926          38
```

```
196223  TO -166113          37
229488  TO -172797          36
276334  TO -180042          35
347210  TO -187921          34
466987  TO -196522          33
490540  TO -205947          32
515251  TO -216321          31

INPUT YOUR CHOSEN INITIAL PRESTRESSED FORCE THEN THE CORRESPONDING DEPTH OF
PRESSTRESSED STEEL
? 376110, 33.675
PERMISSIBLE LINEAR STRESSES
---------------------------

F.in.top= 402.4923
F.in.bot=-2700
F.top=-2700
F.bot= 929.5161

ACTUAL LINEAR STRESSES
----------------------

F.in.top= 396.8533
F.in.bot=-2233.981
F.top= 325.4197
F.bot=-1831.865

ACTUAL STRESSES CALCULATED AT END SECTION

DO YOU WISH TO REENTER OTHER DIMENSIONS(Y/N)
? N
OK
```

A-3 COMPUTER PROGRAM EGNAWY14 FOR THE STRENGTH ANALYSIS AND DESIGN IN FLEXURE OF PRESTRESSED CONCRETE BEAMS

This computer program in Q-BASIC for personal computers analyzes the sections already proportioned by the service load level requirements in flexure. It follows the step-by-step flowchart Figure 4.50 and the discussions and example solutions of Chapter 4.

The input data formats the section type, mainly, whether it is a T, I or rectangular section, the dimensions b, d, d_p, the maximum stresses f'_c, f_{pu}, f_{py}, f_{ps}, and reinforcement moduli E_s and E_{ps}. The strain-compatibility solution gives the moment strength M_n for flanged sections where the neutral axis falls within or outside the flange for typical T-beams sections. It also verifies that the reinforcement, both mild and prestressed, is within the maximum allowable by the ACI-318 Code for both underreinforced and over-reinforced sections.

For under-reinforced rectangular sections, it computes the flexural moment strength from

$$M_n = A_{ps}f_{ps}\left(d_p - \frac{a}{2}\right) + A_sf_y\left(d - \frac{a}{2}\right) + A'_sf_y\left(\frac{a}{2} - d\right)$$

For under-reinforced flanged sections, it computes the flexural moment strength from

$$M_n = A_{pw}f_{ps}\left(d_p - \frac{a}{2}\right) + A_s f_y (d - d_p) + 0.85 f'_c (b - b_w) h_f \left(d_p - \frac{h_f}{2}\right)$$

where

$$A_{pw}f_{ps} = A_{ps}f_{ps} - 0.85 f'_c (b - b_w) h_f$$

A strength reduction factor $\phi = 0.9$ is used in the program. The computer run gives a typical output for the evaluation of the nominal moment strength of prestressed sections.

A-4 COMPUTER PROGRAM EGNAWY16 FOR THE SHEAR STRENGTH DESIGN AND SHEAR REINFORCEMENT SELECTION IN PRESTRESSED CONCRETE BEAMS

This is a computer program in Q-BASIC for personal computers and is intended to determine the shear strength and shear reinforcement required in prestressed concrete beams. It follows the step-by-step flowchart Figure 5.16 and the discussions of flexure shear V_{ci} and web shear W_{cw} in Chapter 5 and the detailed example solutions and diagrams.

The input data compares the load data W_D, W_{SD}, and W_L and the stresses f'_c, f_{pu}, f_y, f_{pe}, effective prestressing force P_e and the section dimensions of the top flange, b_1 and h_1, total depth h and the web width b_w.

It computes the ACI approximate shear strength:

$$V_c = b_w d_p \left(0.6\lambda\sqrt{f'_c} + 700\frac{V_u d}{M_u}\right)$$

and the other alternate more refined solution:
flexure shear:

$$V_{ci} = 0.6\lambda\sqrt{f'_c}\, b_w d + V_d + \frac{V_i M_{cr}}{M_{\max}} - 1.7\lambda\sqrt{f'_c}\, b_w d_p$$

and the web shear:

$$V_{cw} = (3.5\lambda\sqrt{f'_c} + 0.3\bar{f}_c) b_w d_p + V_p$$

where λ is a function of type of concrete used, namely, normalweight, sand-lightweight, or all-lightweight.

Based on choosing the controlling shear strength, the program proceeds to compute the required area of the web stirrups for shear, checking for the minimum area A_v as required by the ACI Code. Then it selects the required spacing of the stirrups based on the stirrup size being used in the design.

A-5 COMPUTER PROGRAM EGNAWY18 FOR THE DESIGN OF CONCRETE BRACKETS AND CORBELS

This computer program in Q-BASIC for personal computers is intended to proportion brackets or corbels that are essential components in precast prestressed concrete prestressing systems. The step-by-step procedure given in the flowchart Figure 5.27 forms the basis of the program as well as the discussion and example calculations in Chapter 5.

The input data comprise the allowable stress values f'_c, f_y, the vertical shear load V_u, the horizontal frictional force N_{uc}, and the section dimensions, namely, the depth h, the effective depth d, the moment arm a, and the width b_w of the section.

By trial and adjustment the program computes areas of the main steel A_s and the horizontal steel closed steel stirrups A_h. The program uses the ACI 318 Code requirements for determining the areas of the steel reinforcement, and assumes that the minimum horizontal frictional force N_{uc} is 20 percent of the vertical shear V_u acting on the corbel.

A-6 COMPUTER PROGRAM EGNAWY20 FOR MOMENT CURVATURE ANALYSIS OF BONDED PARTIALLY PRESTRESSED BEAMS

A computer program in Q-BASIC for personal computers is intended to evaluate the moment-curvature relationships in a bonded, prestressed concrete beam. It computes these values for three loading stages:

1. Linear Uncracked Stage
2. Linear Cracked Stage
3. Non-linear Cracked Stage

The user should use the "interactive format" response to a prompt in the program for both T and I beam sections.

Input data comprise section dimensions b_1 and h_1 for the top flange; b_2 and h_2 for the bottom flange; total depth h and web width b_w. It also includes input of the effective prestressing force P_e after losses, prestressing a steel area A_{ps}, prestressing steel depth d_p, mild steel area A_s, mild steel depth d, the yield strengths f_{py} and f_y and the ultimate prestressing strength f_{pu}.

The program internally generates the co-ordinates of the moment-curvature relationships using strain increments of $\epsilon = 0.0005$ in./in. beyond zero strain, starting at a strain $\epsilon_1 = 0.001$ in./in. for the linear post-cracking stage through the nonlinear cracking stages up to strain of $\epsilon_c = 0.003$ in./in. It is intended to evaluate the curvature of individual sections of prestressed concrete elements. The moment-curvature plot $(M - \phi)$ between the stages $\epsilon_1 = 0.001$ and $\epsilon_c = 0.003$ in./in., if assumed a straight line, permits the user to interpolate intermediate $M - \phi$ values when necessary.

The program prompts input of data on the number of layers of prestressing steel reinforcement and the computer run checks the stress level in the concrete section at the extreme fibers at each loading and cracking stage.

A-7 COMPUTER PROGRAM EGNAWY22 FOR TIME DEPENDENT DEFLECTION EVALUATION OF PRESTRESSED CONCRETE SIMPLY SUPPORTED BEAMS

This is a computer program in Q-BASIC for personal computers intended to evaluate the time-dependent deflection and camber in simply supported bonded prestressed concrete beams by the approximate time step method. It follows the steps shown in the flowchart Fig. 7.18 and requires the input of the value of the effective prestressing force P_e and the prestress loss ΔP obtained from Program EGNAWY10 on time-dependent losses.

Input data comprise section properties A_c, I_c, c_t, c_b, self-weight W_D, superimposed dead load W_D and live load W_L. Also, the tendon eccentricities e_c and e_e at midspan and support sections are input into the program.

The program prompts the user to input any time intervals for which deflection is to be computed starting from the prestress transfer load application, then throughout the time-history of the prestressed member. A table of final results is given in the output.

Example A-7 Time Dependent Deflection Computation for Prestressed Concrete Beams

Compute by the approximate incremental time-step method the long-term deflections of the beam in Example 7.9 if it were a single T for the time intervals of 7, 30, 90, 365 and 1825 days (5 years) using the computer program EGNAWY22. Assume a maximum shrinkage coefficient $\epsilon_{SH} = 0.0005$ in./in. and a maximum creep coefficient $C_u = 2.35$.

Input

Beam span L = 65 ft.

Beam section = T

Flange width = 120 in.

Flange depth h_1 = 2.25 in.

Total depth h = 48 in.

Web width b_w = 8 in.

f'_c = 5,000 psi

f'_{ci} = 3,750 psi

W_{SD} = 100 plf

W_L = 1,100 plf

Midspan tendon eccentricity ϵ_c = 33.14 in.

Support tendon eccentricity ϵ_c = 20.0 in.

f_{pi} = 189,000 psi

f_{pe} = 154,900 psi

E_{ps} = 27,500,000 psi

E_s = 29,000,000 psi

A_{ps} = 2.142 in.

d_p = 45.95 in.

A_s = 0

d = 0

Prestress force after transfer = 331,967 lb.

Time for casting to curing = 3 days

Time for casting to prestressing = 4 days

Time for casting to load application = 30 days

Tendon profile = harped

Distance from support to harped point = 32.5 ft.

Curing process = steam

Output

```
*********************************************************
* PROGRAM TO CALCULATE TIME DEPENDANT DEFLECTION OF    *
*                                                       *
* PRESTRESSED CONCRETE SIMPLY SUPPORTED BEAMS           *
*********************************************************

PRESS ANY KEY TO CONTINUE ?
-------------------------------------
READ OF SECTION DIMENSION AND SPAN
-------------------------------------

THE BEAM'S SPAN IS (FEET)  =? 65

    DATA OBTAINED FROM LOSSES PROGRAM
    *********************************

PRESTRESSING FORCE AFTER TRANSFER (after elastic loss) IN LBS=? 331967
    TIME SCHEDULE FOR CONSTRUCTION
    ***************************

TIME FROM SECTION CASTING TO END OF CURING (DAYS)        =? 3
TIME FROM SECTION CASTING TO PRESTRESSING (posttensioned) OR
        OR TRANSFER (pretensioned) (DAYS)                =? 4
TIME FROM SECTION CASTING TO APPLICATION OF SUPERIMPOSED DEAD
       LOAD AND LIVE LOAD (DAYS) =? 30

SELECT THE PRESTRESSED CABLES LAYOUT

1-PARABOLIC CABLES
2-HARPED CABLES
3-STRAIGHT CABLES

? 2
DISTANCE FROM SUPPORT TO HARPED POINT (FT.)    = ? 32.5

ENTER 1 FOR MOIST CURED CONCRETE AND 2 FOR STEAM-CURED CONCRETE ? 2

DO YOU WISH TO CALCULATE DEFLECTION AFTER ANOTHER TIME INTERVAL? N

   *********************************************
   *         TABLE OF FINAL RESULTS            *
   *********************************************

TIME (days)                    DEFLECTION
  7                            -1.160123
  30                           -.744374
  90                           -.9480114
  365                          -1.081456
  1825                         -1.079121
OK
```

B

UNIT CONVERSIONS, DESIGN INFORMATION, PROPERTIES OF REINFORCEMENT

Table B–1 Conversion to International System of Units (SI)

To convert from	to	Multiply by
Length		
inch (in.)	millimeter (mm)	25.4
inch (in.)	meter (m)	0.0254
foot (ft)	meter (m)	0.3048
yard (yd)	meter (m)	0.9144
Area		
square foot (sq ft)	square meter (sq m)	0.09290
square inch (sq in.)	square millimeter (sq mm)	645.2
square inch (sq in.)	square meter (sq m)	0.0006452
square yard (sq yd)	square meter (sq m)	0.8361
Volume		
cubic inch (cu in.)	cubic meter (cu m)	0.00001639
cubic foot (cu ft)	cubic meter (cu m)	0.02832
cubic yard (cu yd)	cubic meter (cu m)	0.7646
gallon (gal) Can. liquid*	liter	4.546
gallon (gal) Can. liquid*	cubic meter (cu m)	0.004546

Table B–1 *Continued*

To convert from	to	Multiply by
gallon (gal) U.S. liquid*	liter	3.785
gallon (gal) U.S. liquid*	cubic meter (cu m)	0.003785
Force		
kip	kilogram (kgf)	453.6
kip	newton (N)	4448.0
pound (lb)	kilogram (kgf)	0.4536
pound (lb)	newton (N)	4.448
Pressure or Stress		
kips/square inch (ksi)	megapascal (MPa)**	6.895
pound/square foot (psf)	kilopascal (kPa)**	0.04788
pound/square inch (psi)	kilopascal (kPa)**	6.895
pound/square inch (psi)	megapascal (MPa)**	0.006895
pound/square foot (psf)	kilogram/square meter (kgf/sq m)	4.882
Mass		
pound (avdp)	kilogram (kg)	0.4536
ton (short, 2000 lb)	kilogram (kg)	907.2
ton (short, 2000 lb)	tonne (t)	0.9072
grain	kilogram (kg)	0.00006480
tonne (t)	kilogram (kg)	1000
Mass (weight) per Length		
kip/linear foot (klf)	kilogram/meter (kg/m)	0.001488
pound/linear foot (plf)	kilogram/meter (kg/m)	1.488
pound/linear foot (plf)	newton/meter (N/m)	14.593
Mass per volume (density)		
pound/cubic foot (pcf)	kilogram/cubic meter (kg/cu m)	16.02
pound/cubic yard (pcy)	kilogram/cubic meter (kg/cu m)	0.5933
Bending Moment or Torque		
inch-pound (in.-lb)	newton-meter	0.1130
foot-pound (ft-lb)	newton-meter	1.356
foot-kip (ft-k)	newton-meter	1356
Temperature		
degree Fahrenheit (deg F)	degree Celsius (C)	$t_C = (t_F - 32)/1.8$
degree Fahrenheit (deg F)	degree Kelvin (K)	$t_K = (t_F + 459.7)/1.8$
Energy		
British thermal unit (Btu)	joule (j)	1056
kilowatt-hour (kwh)	joule (j)	3,600,000
Power		
horsepower (hp) (550 ft lb/sec)	watt (W)	745.7
Velocity		
mile/hour (mph)	kilometer/hour	1.609
mile/hour (mph)	meter/second (m/s)	.04470
Other		
Section modulus (in.^{3})	mm^3	16.387
Moment of inertia (in.^{4})	mm^4	416.231
Coefficient of heat transfer (Btu/ft^2/h/°F)	W/m^2/°C	5.678
Modulus of elasticity (psi)	MPa	0.006895
Thermal conductivity (Btu-in./ft^2/h/°F)	Wm/m^2/°C	0.1442
Thermal expansion (in./in./°F)	mm/mm/°C	1.800
Area/length (in.^{2}/ft)	mm^2/m	2116.80

*One U.S. gallon equals 0.8321 Canadian gallon.

**A pascal equals one newton/square meter.

Table B–2 Recommended Minimum Floor Live Loads*

Uniformly Distributed Loads	
Occupancy or Use	**Live Load (psf)**
Apartments (*see* Residential)	
Armories and drill rooms	150
Assembly halls and other places of assembly:	
Fixed seats	60
Movable seats	100
Platforms (assembly)	100
Balcony (exterior)	100
On one- and two-family residences only and not exceeding 100 sq ft	60
Bowling alleys, poolrooms, and similar recreational areas	75
Corridors:	
First floor	100
Other floors, same as occupancy served except as indicated	
Dance halls and ballrooms	100
Dining rooms and restaurants	100
Dwellings (*see* Residential)	
Fire escapes	100
On multi- or single-family residential buildings only	40
Garages (passenger cars only)	50
For trucks and buses use AASHTO lane loads (1)	
Grandstands (*see* Stadium and arena bleachers)	
Gymnasiums, main floors and balconies	100
Hospitals:	
Operating rooms, laboratories	60
Private rooms	40
Wards	40
Corridors, above first floor	80

Uniformly Distributed Loads	
Occupancy or Use	**Live Load (psf)**
Hotels (*see* Residential)	
Libraries:	
Reading rooms	60
Stack rooms (books & shelving at 65 pcf) but not less than	150
Corridors, above first floor	80
Manufacturing:	
Light	125
Heavy	250
Marquees and canopies	75
Office buildings:	
Offices	50
Lobbies	100
File and computer rooms require heavier loads based upon anticipated occupancy	
Penal institutions:	
Cell blocks	40
Corridors	100
Residential:	
Dwellings (one- and two-family)	
Uninhabitable attics without storage	10
Uninhabitable attics with storage	20
Habitable attics and sleeping areas	30
All other areas	40
Hotels and multifamily houses:	
Private rooms and corridors serving them	40
Public rooms and corridors serving them	100

Table B–2 *Continued*

Uniformly Distributed Loads

Occupancy or Use	Live Load (psf)
Schools:	
Classrooms	40
Corridors above first floor	80
Sidewalks, vehicular driveways, and yards, subject to trucking (2)	250
Stadiums and arena bleachers (3)	100
Stairs and exitways	100
Storage warehouse:	
Light	125
Heavy	250
Stores:	
Retail:	
First floor	100
Upper floors	75
Wholesale, all floors	125
Walkways and elevated platforms (other than exitways)	60

Concentrated Loads

Location	Load (lb)
Elevator machine room grating (on area of 4 sq in)	300
Finish light floor plate construction (on area of 1 sq in)	200
Garages	(4)
Office floors	2000
Scuttles, skylight ribs, and accessible ceilings	200
Sidewalks	8000
Stair treads (on area of 4 sq in at center of tread)	300

(1) American Association of State Highway and Transportation Officials.

(2) AASHTO lane loads should also be considered where appropriate.

(3) For detailed recommendations, see Assembly Seating, Tents and Air Supported Structures, ANSI/NFPA 102-1978 [Z20.3].

(4) Floors in garages or portions of buildings used for storage of motor vehicles shall be designed for the uniformly distributed live loads shown or the following concentrated loads: (1) for passenger cars accommodating not more than nine passengers, 2000 pounds acting on an area of 20 in.^{2} (2) mechanical parking structures without slab or deck, passenger cars only, 1500 pounds per wheel; (3) for trucks or buses, maximum axle load on an area of 20 in.^{2}

*Source: American National Standard ANSI A58.1-1982.

Local building codes take precedence.

Table B–3 Dead Weights of Floors, Ceilings, Roofs, and Walls

Floorings	Weight (psf)
Normal weight concrete topping, per inch of thickness	12
Sand-lightweight (120 pcf) concrete topping, per inch	10
Lightweight (90–100 pcf) concrete topping, per inch	8
$\frac{7}{8}$ in. hardwood floor on sleepers clipped to concrete without fill	5
$1\frac{1}{2}$ in. terrazzo floor finish directly on slab	19
$1\frac{1}{2}$ in. terrazzo floor finish on 1 in. mortar bed	30
1 in. terrazzo finish on 2 in. concrete bed	38
$\frac{3}{4}$ in. ceramic or quarry tile on $\frac{1}{2}$ in. mortar bed	16
$\frac{3}{4}$ in. ceramic or quarry tile on 1 in. mortar bed	22
$\frac{1}{4}$ in. linoleum or asphalt tile directly on concrete	1
$\frac{1}{4}$ in. linoleum or asphalt tile on 1 in. mortar bed	12
$\frac{3}{4}$ in. mastic floor	9
Hardwood flooring, $\frac{7}{8}$ in. thick	4
Subflooring (soft wood), $\frac{3}{4}$ in. thick	$2\frac{1}{2}$
Asphaltic concrete, $1\frac{1}{2}$ in. thick	18
Ceilings	
$\frac{1}{2}$ in. gypsum board	2
$\frac{5}{8}$ in. gypsum board	$2\frac{1}{2}$
$\frac{3}{4}$ in. plaster directly on concrete	5
$\frac{3}{4}$ in. plaster on metal lath furring	8
Suspended ceilings	2
Acoustical tile	1
Acoustical tile on wood furring strips	3
Roofs	
Ballasted inverted membrane	16
Five-ply felt and gravel (or slag)	$6\frac{1}{2}$
Three-ply felt and gravel (or slag)	$5\frac{1}{2}$
Five-ply felt composition roof, no gravel	4
Three-ply felt composition roof, no gravel	3
Asphalt strip shingles	3
Rigid insulation, per inch	$\frac{1}{2}$
Gypsum, per inch of thickness	4
Insulating concrete, per inch	3

Walls	Un-plastered	One side plastered	Both sides plastered
4 in. brick wall	40	45	50
8 in. brick wall	80	85	90
12 in. brick wall	120	125	130
4 in. hollow normal weight concrete block	28	33	38
6 in. hollow normal weight concrete block	36	41	46
8 in. hollow normal weight concrete block	51	56	61
12 in. hollow normal weight concrete block	59	64	69

Table B–3 *Continued*

Walls	**Un-plastered**	**One side plastered**	**Both sides plastered**
4 in. hollow lightweight block or tile	19	24	29
6 in. hollow lightweight block or tile	22	27	32
8 in. hollow lightweight block or tile	33	38	43
12 in. hollow lightweight block or tile	44	49	54
4 in. brick 4 in. hollow normal weight block baking	68	73	78
4 in. brick 8 in. hollow normal weight block backing	91	96	101
4 in. brick 12 in. hollow normal weight block backing	119	124	129
4 in. brick 4 in. hollow lightweight block or tile backing	59	64	69
4 in. brick 8 in. hollow lightweight block or tile backing	73	78	83
4 in. brick 12 in. hollow lightweight block or tile backing	84	89	94
4 in. brick, steel or wood studs, $\frac{5}{8}$ in. gypsum board	43		
Windows, glass, frame and sash	8		
4 in. stone	55		
Steel or wood studs, lath, $\frac{3}{4}$ in. plaster	18		
Steel or wood studs, $\frac{5}{8}$ in. gypsum board each side	6		
Steel or wood studs, 2 layers $\frac{1}{2}$ in. gypsum board each side	9		

Table B–4 Area of Bars in a 1-Foot-wide Slab Strip

Spacing, in.	Cross Section Area of Bar, A_s (or A'_s), in.^{2}											Spacing, in.
	Bar Size											
	#3	#4	#5	#6	#7	#8	#9	#10	#11	#14	#18	
4	0.33	0.60	0.93	1.32	1.80	2.37	3.00	3.81	4.68			4
$4\frac{1}{2}$	0.29	0.53	0.83	1.17	1.60	2.11	2.67	3.39	4.16	6.00		$4\frac{1}{2}$
5	0.26	0.48	0.74	1.06	1.44	1.90	2.40	3.05	3.74	5.40	9.60	5
$5\frac{1}{2}$	0.24	0.44	0.68	0.96	1.31	1.72	2.18	2.77	3.40	4.91	8.73	$5\frac{1}{2}$
6	0.22	0.40	0.62	0.88	1.20	1.58	2.00	2.54	3.12	4.50	8.00	6
$6\frac{1}{2}$	0.20	0.37	0.57	0.81	1.11	1.46	1.85	2.34	2.88	4.15	7.38	$6\frac{1}{2}$
7	0.19	0.34	0.53	0.75	1.03	1.35	1.71	2.18	2.67	3.86	6.86	7
$7\frac{1}{2}$	0.18	0.32	0.50	0.70	0.96	1.26	1.60	2.03	2.50	3.60	6.40	$7\frac{1}{2}$
8	0.17	0.30	0.47	0.66	0.90	1.19	1.50	1.91	2.34	3.38	6.00	8
$8\frac{1}{2}$	0.16	0.28	0.44	0.62	0.85	1.12	1.41	1.79	2.20	3.18	5.65	$8\frac{1}{2}$
9	0.15	0.27	0.41	0.59	0.80	1.05	1.33	1.69	2.08	3.00	5.33	9
$9\frac{1}{2}$	0.14	0.25	0.39	0.56	0.76	1.00	1.26	1.60	1.97	2.84	5.05	$9\frac{1}{2}$
10	0.13	0.24	0.37	0.53	0.72	0.95	1.20	1.52	1.87	2.70	4.80	10
$10\frac{1}{2}$	0.13	0.23	0.35	0.50	0.69	0.90	1.14	1.45	1.78	2.57	4.57	$10\frac{1}{2}$
11	0.12	0.22	0.34	0.48	0.65	0.86	1.09	1.39	1.70	2.45	4.36	11
$11\frac{1}{2}$	0.11	0.21	0.32	0.46	0.63	0.82	1.04	1.33	1.63	2.35	4.17	$11\frac{1}{2}$
12	0.11	0.20	0.31	0.44	0.60	0.79	1.00	1.27	1.56	2.25	4.00	12
13	0.10	0.18	0.29	0.41	0.55	0.73	0.92	1.17	1.44	2.08	3.69	13
14	0.09	0.17	0.27	0.38	0.51	0.68	0.86	1.09	1.34	1.93	3.43	14
15	0.09	0.16	0.25	0.35	0.48	0.63	0.80	1.02	1.25	1.80	3.20	15
16	0.08	0.15	0.23	0.33	0.45	0.59	0.75	0.95	1.17	1.69	3.00	16
17	0.08	0.14	0.22	0.31	0.42	0.56	0.71	0.90	1.10	1.59	2.82	17
18	0.07	0.13	0.21	0.29	0.40	0.53	0.67	0.85	1.04	1.50	2.67	18

Table B–5 Properties and Design Strengths of Prestressing Strand and Wire

Seven-Wire Strand, f_{pu} = 270 ksi					
Nominal Diameter, in.	3/8	7/16	1/2	9/16	0.600
Area, sq in.	0.085	0.115	0.153	0.192	0.215
Weight, plf	0.29	0.40	0.53	0.65	0.74
0.7 f_{pu} A_{ps}, kips	16.1	21.7	28.9	36.3	40.7
0.75 f_{pu} A_{ps}, kips	17.2	23.3	31.0	38.9	43.5
0.8 f_{pu} A_{ps}, kips	18.4	24.8	33.0	41.4	46.5
f_{pu} A_{ps}, kips	23.0	31.0	41.3	51.8	58.1

Seven-Wire Strand, f_{pu} = 250 ksi						
Nominal Diameter, in.	1/4	5/16	3/8	7/16	1/2	0.600
Area, sq in.	0.036	0.058	0.080	0.108	0.144	0.215
Weight, plf	0.12	0.20	0.27	0.37	0.49	0.74
0.7 f_{pu} A_{ps}, kips	6.3	10.2	14.0	18.9	25.2	37.6
0.8 f_{pu} A_{ps}, kips	7.2	11.6	16.0	21.6	28.8	43.0
f_{pu} A_{ps}, kips	9.0	14.5	20.0	27.0	36.0	53.8

Three- and Four-Wire Strand, f_{pu} = 250 ksi				
Nominal Diameter, in.	1/4	5/16	3/8	7/16
No. of wires	3	3	3	4
Area, sq in.	0.036	0.058	0.075	0.106
Weight, plf	0.13	0.20	0.26	0.36
0.7 f_{pu} A_{ps}, kips	6.3	10.2	13.2	18.6
0.8 f_{pu} A_{ps}, kips	7.2	11.6	15.0	21.2
f_{pu} A_{ps}, kips	9.0	14.5	18.8	26.5

Prestressing Wire										
Diameter, in.	0.105	0.120	0.135	0.148	0.162	0.177	0.192	0.196	0.250	0.276
Area, sq in.	0.0087	0.0114	0.0143	0.0173	0.0206	0.0246	0.0289	0.0302	0.0491	0.0598
Weight, plf	0.030	0.039	0.049	0.059	0.070	0.083	0.098	0.10	0.17	0.20
Ult. strength, f_{pu}, ksi	279	273	268	263	259	255	250	250	240	235
0.7 f_{pu} A_{ps}, kips	1.70	2.18	2.68	3.18	3.73	4.39	5.05	5.28	8.25	9.84
0.8 f_{pu} A_{ps}, kips	1.94	2.49	3.06	3.64	4.26	5.02	5.78	6.04	9.42	11.24
f_{pu} A_{ps}, kips	2.43	3.11	3.83	4.55	5.33	6.27	7.22	7.55	11.78	14.05

Table B–6 Properties and Design Strengths of Prestressing Bars

Smooth Prestressing Bars, f_{pu} = 145 ksi*

Nominal Diameter, in.	3/4	7/8	1	1 1/8	1 1/4	1 3/8
Area, sq in.	0.442	0.601	0.785	0.994	1.227	1.485
Weight, plf	1.50	2.04	2.67	3.38	4.17	5.05
0.7 f_{pu} A_{ps}, kips	44.9	61.0	79.7	100.9	124.5	150.7
0.8 f_{pu} A_{ps}, kips	51.3	69.7	91.0	115.3	142.3	172.2
f_{pu} A_{ps}, kips	64.1	87.1	113.8	144.1	177.9	215.3

Smooth Prestressing Bars, f_{pu} = 160 ksi*

Nominal Diameter, in.	3/4	7/8	1	1 1/8	1 1/4	1 3/8
Area, sq in.	0.442	0.601	0.785	0.994	1.227	1.485
Weight, plf	1.50	2.04	2.67	3.38	4.17	5.05
0.7 f_{pu} A_{ps}, kips	49.5	67.3	87.9	111.3	137.4	166.3
0.8 f_{pu} A_{ps}, kips	56.6	77.0	100.5	127.2	157.0	190.1
f_{pu} A_{ps}, kips	70.7	96.2	125.6	159.0	196.3	237.6

Deformed Prestressing Bars

Nominal Diameter, in.	5/8	1	1	1 1/4	1 1/4	1 3/8
Area, sq. in.	0.28	0.85	0.85	1.25	1.25	1.58
Weight, plf	0.98	3.01	3.01	4.39	4.39	5.56
Ult. strength, f_{pu}, ksi	157	150	160*	150	160*	150
0.7 $f_{pu}A_{ps}$, kips	30.5	89.3	95.2	131.3	140.0	165.9
0 8 $f_{pu}A_{ps}$, kips	34.8	102.0	108.8	150.0	160.0	189.6
$f_{pu}A_{ps}$,kips	43.5	127.5	136.0	187.5	200.0	237.0

Stress-strain characteristics (all prestressing bars):

For design purposes, following assumptions are satisfactory:

E_s = 29,000 ksi

f_y = 0.95 f_{pu}

*Verify availability before specifying

Table B–7 Moments in Beams with Fixed Ends

Loading	Moment at A	Moment at center	Moment at B
(1) 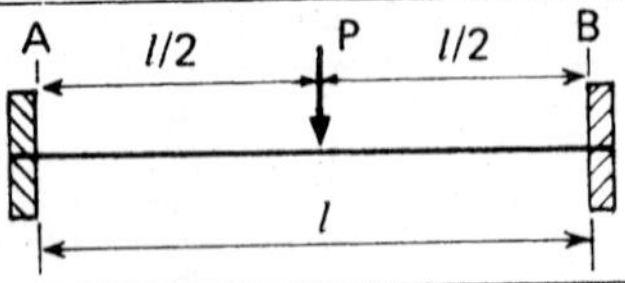 A, B; $l/2$, P, $l/2$; l	$-\frac{Pl}{8}$	$+\frac{Pl}{8}$	$-\frac{Pl}{8}$
(2) 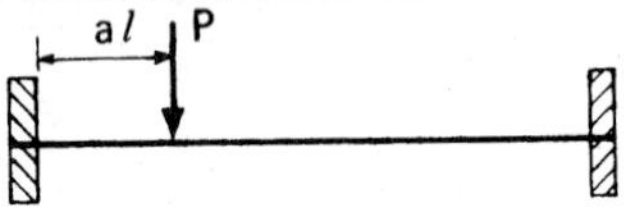 al, P	$-Pla(1-a)^2$		$-Pla^2(1-a)$
(3) 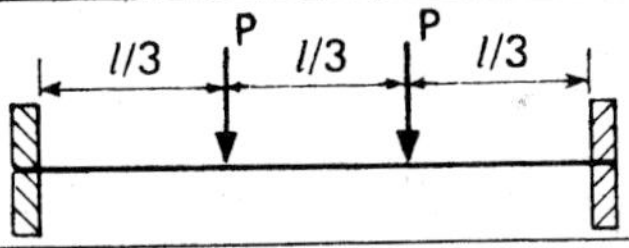 $l/3$, P, $l/3$, P, $l/3$	$-\frac{2Pl}{9}$	$+\frac{Pl}{9}$	$-\frac{2Pl}{9}$
(4) $l/4$, P, $l/4$, P, $l/4$, P, $l/4$ 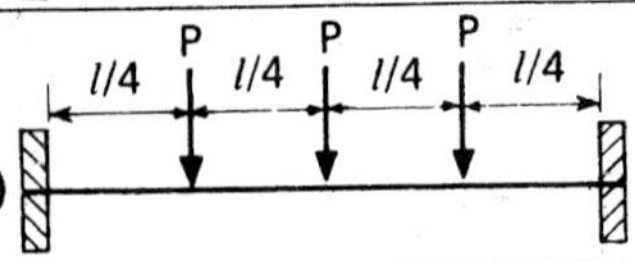	$-\frac{5Pl}{16}$	$+\frac{3Pl}{16}$	$-\frac{5Pl}{16}$
(5) 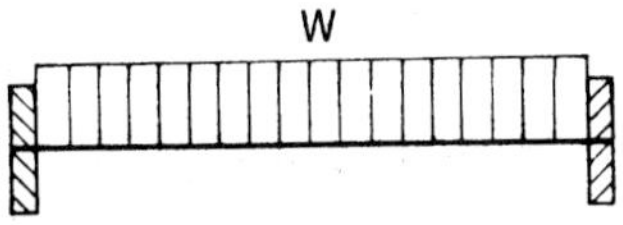 W	$-\frac{Wl}{12}$	$+\frac{Wl}{24}$	$-\frac{Wl}{12}$
(6) 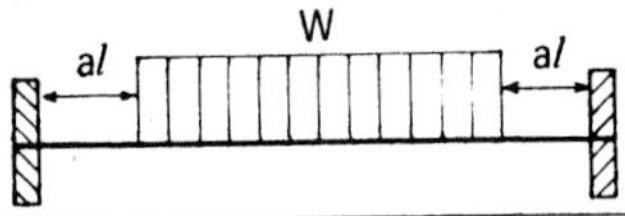 al, W, al	$-\frac{Wl(1+2a-2a^2)}{12}$	$+\frac{Wl(1+2a+4a^2)}{24}$	$-\frac{Wl(1+2a-2a^2)}{12}$
(7) 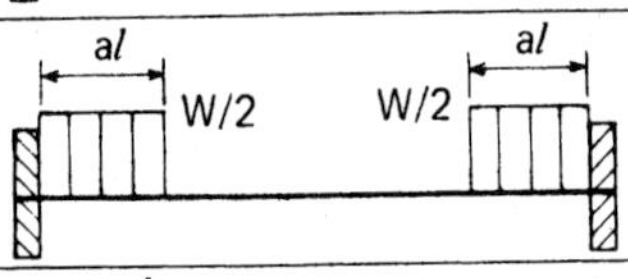 al, W/2, W/2, al	$-\frac{Wl(3a-2a^2)}{12}$	$+\frac{Wla^2}{6}$	$-\frac{Wl(3a-2a^2)}{12}$
(8) 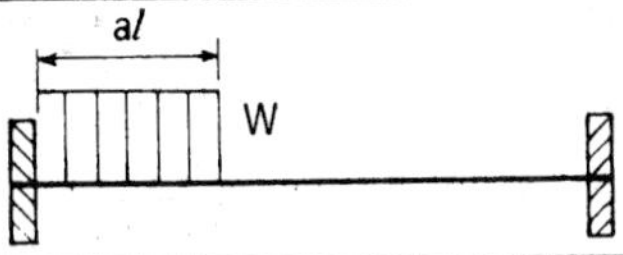 al, W	$-\frac{Wla(6-8a+3a^2)}{12}$		$-\frac{Wla^2(4-3a)}{12}$
(9) 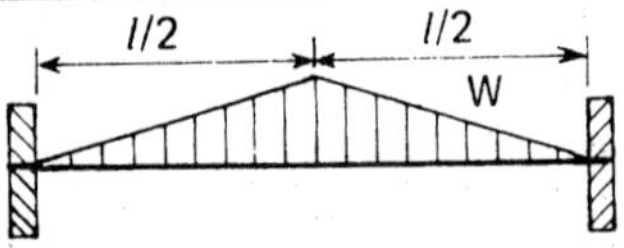 $l/2$, $l/2$, W	$-\frac{5Wl}{48}$	$+\frac{3Wl}{48}$	$-\frac{5Wl}{48}$
(10) 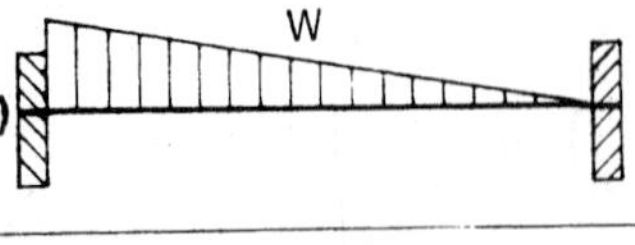 W	$-\frac{Wl}{10}$		$-\frac{Wl}{15}$
W = Total load on beam			

Table B–8 Camber (Deflection) and Rotation Coefficients for Prestress Force and Loads*

Prestress Pattern	Equivalent Moment or Load	Equivalent Loading	Camber	End Rotation	
			+ ↑	+ (left end)	+ (right end)
(1) 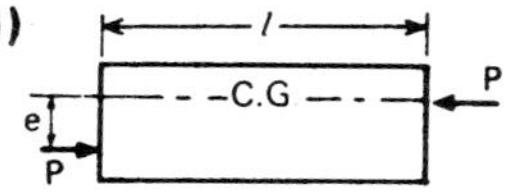	$M = Pe$	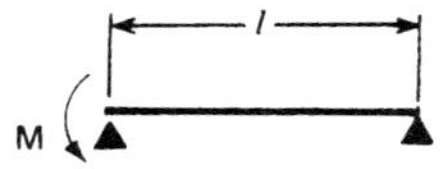	$+\frac{Ml^2}{16\,EI}$	$+\frac{Ml}{3\,EI}$	$-\frac{Ml}{6\,EI}$
(2) 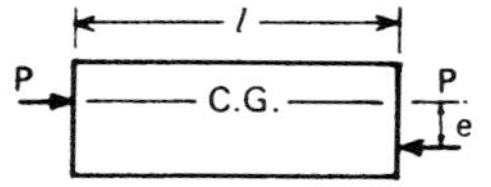	$M = Pe$	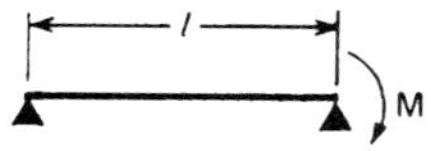	$+\frac{Ml^2}{16\,EI}$	$+\frac{Ml}{6\,EI}$	$-\frac{Ml}{3\,EI}$
(3) 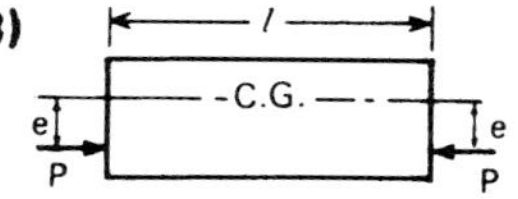	$M = Pe$	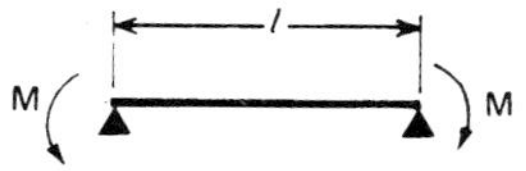	$+\frac{Ml^2}{8\,EI}$	$+\frac{Ml}{2\,EI}$	$-\frac{Ml}{2\,EI}$
(4) 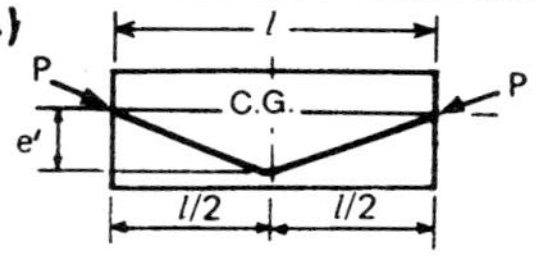	$N = \frac{4\,Pe'}{l}$	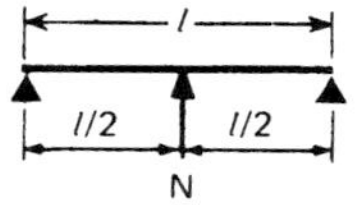	$+\frac{Nl^3}{48\,EI}$	$+\frac{Nl^2}{16\,EI}$	$-\frac{Nl^2}{16\,EI}$
(5) 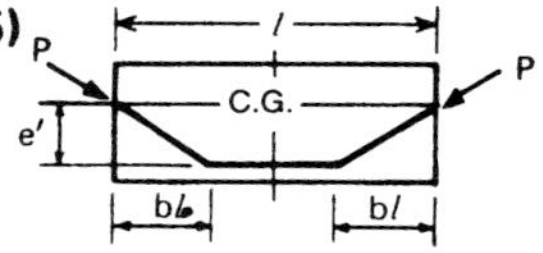	$N = \frac{Pe'}{bl}$	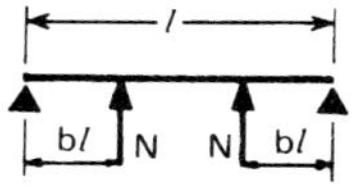	$+\frac{b\,(3-4b^2)\,Nl^3}{24\,EI}$	$+\frac{b\,(1-b)\,Nl^2}{2\,EI}$	$-\frac{b\,(1-b)\,Nl^2}{2\,EI}$

Table B–8 *Continued*

Pattern	Equivalent load	Loading	Deflection	Rotation (left)	Rotation (right)
(6) 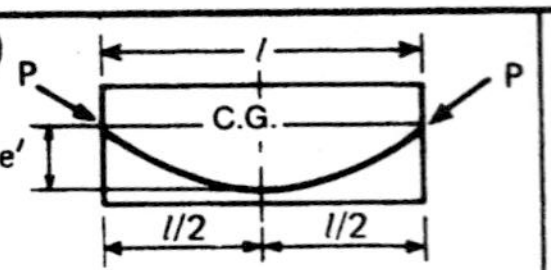	$w = \frac{8\,Pe'}{l^2}$	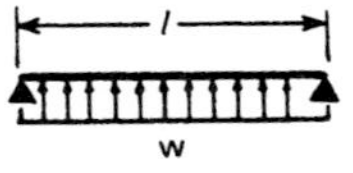	$+\frac{5wl^4}{384\,FI}$	$+\frac{wl^3}{24\,EI}$	$-\frac{wl^3}{24\,EI}$
(7) 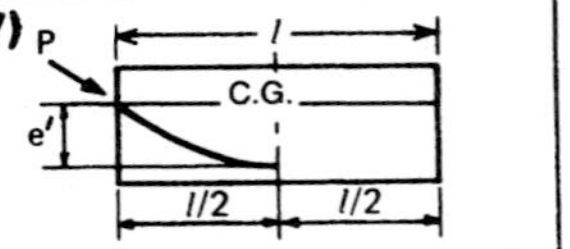	$w = \frac{8\,Pe'}{l^2}$	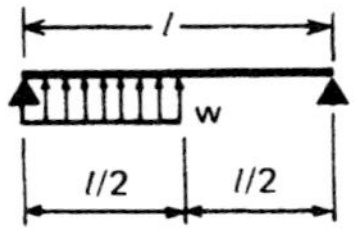	$+\frac{5\,wl^4}{768\,EI}$	$+\frac{9\,wl^3}{384\,EI}$	$-\frac{7\,wl^3}{384\,EI}$
(8) 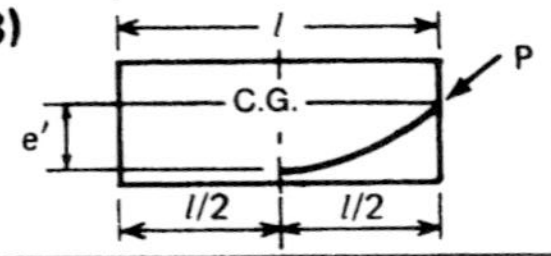	$w = \frac{8\,Pe'}{l^2}$	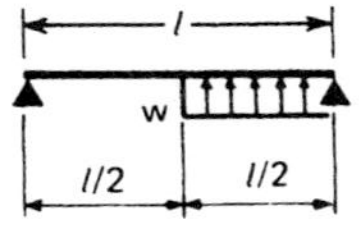	$+\frac{5\,wl^4}{768\,EI}$	$+\frac{7\,wl^3}{384\,EI}$	$-\frac{9\,wl^3}{384\,EI}$
(9) 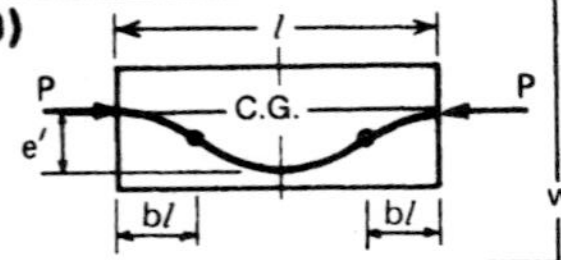	$w = \frac{4\,Pe'}{(0.5 \cdot b)\,l^2}$ $w_1 = \frac{w}{b}\,(0.5 \cdot b)$	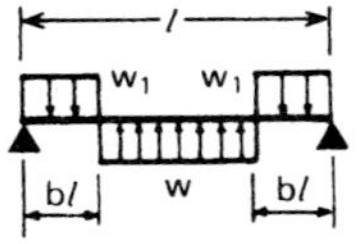	$\left[\frac{5}{8} - \frac{b}{2}\,(3-2b^2)\right]\frac{wl^4}{48\,EI}$	$+\frac{(1-b)\,(1-2b)\,wl^3}{24\,EI}$	$-\frac{(1-b)\,(1-2b)\,wl^3}{24\,EI}$
(10) 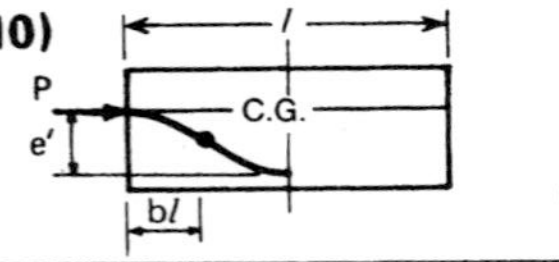	$w = \frac{4\,Pe'}{(0.5 \cdot b)\,l^2}$ $w_1 = \frac{w}{b}\,(0.5 \cdot b)$	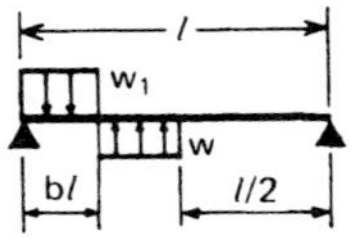	$\left[\frac{5}{16} - \frac{b}{4}\,(3-2b^2)\right]\frac{wl^4}{48\,EI}$	$\left[\frac{9}{8} - b\,(2-b)^2\right]\frac{wl^3}{48\,EI}$	$\left[-\frac{7}{8} + b\,(2-b^2)\right]\frac{wl^3}{48\,EI}$
(11) 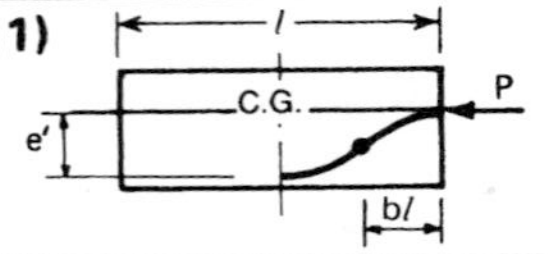	$w = \frac{4\,Pe'}{(0.5 \cdot b)\,l^2}$ $w_1 = \frac{w}{b}\,(0.5 \cdot b)$	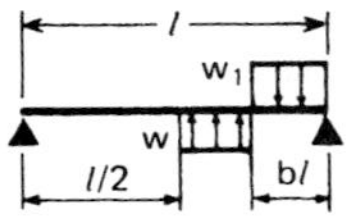	$\left[\frac{5}{16} - \frac{b}{4}\,(3-2b^2)\right]\frac{wl^4}{48\,EI}$	$\left[\frac{7}{8} - b\,(2-b^2)\right]\frac{wl^3}{48\,EI}$	$\left[-\frac{9}{8} + b\,(2-b)^2\right]\frac{wl^3}{48\,EI}$

* The tabulated values apply to the effects of prestressing. By adjusting the directional notation, they may also be used for the effects of loads.
For patterns 4–11, superimpose on 1, 2, or 3 for other C.G. locations.

Table B–9 Presumptive Bearing Capacity (tons/ft^2)

Type of Soil	Bearing Capacity
Massive crystalline bedrock, such as granite, diorite, gneiss, and trap rock	100
Foliated rocks, such as schist or slate	40
Sedimentary rocks, such as hard shales, sandstones, limestones, and siltstones	15
Gravel and gravel–sand mixtures (GW and GP soils)	
Densely compacted	5
Medium compacted	4
Loose, not compacted	3
Sands and gravely sands, well graded (SW soil)	
Densely compacted	$3\frac{3}{4}$
Medium compacted	3
Loose, not compacted	$2\frac{1}{4}$
Sands and gravely sands, poorly graded (SP soil)	
Densely compacted	3
Medium compacted	$2\frac{1}{2}$
Loose, not compacted	$1\frac{3}{4}$
Silty gravels and gravel–sand–silt mixtures (GM soil)	
Densely compacted	$2\frac{1}{2}$
Medium compacted	2
Loose, not compacted	$1\frac{1}{2}$
Silty sand and silt-sand mixtures (SM soil)	2
Clayey gravels, gravel–sand–clay mixtures, clayey sands, sand–clay mixtures (GC and SC soils)	2
Inorganic silts, and fine sands; silty or clayey fine sands and clayey silts, with slight plasticity; inorganic clays of low to medium plasticity; gravely clays; sandy clays; silty clays; lean clays (ML and CL soils)	1
Inorganic clays of high plasticity, fat clays; micaceous or diatomaceous fine sand or silty soils, elastic silts (CH and MH soils)	1

Table B–10 Standard Wire Reinforcement

W&D Size		U.S. Customary			Area (in.^{2}/ft of width for various spacings)						
					Center-to-center spacing (in.)						
Smooth	Deformed	Nominal diameter (in.)	Nominal area (in.^{2})	Nominal weight (lb/ft)	2	3	4	6	8	10	12
W31	D31	0.628	0.310	1.054	1.86	1.24	0.93	0.62	0.465	0.372	0.31
W30	D30	0.618	0.300	1.020	1.80	1.20	0.90	0.60	0.45	0.366	0.30
W28	D28	0.597	0.280	0.952	1.68	1.12	0.84	0.56	0.42	0.336	0.28
W26	D26	0.575	0.260	0.934	1.56	1.04	0.78	0.52	0.39	0.312	0.26
W24	D24	0.553	0.240	0.816	1.44	0.96	0.72	0.48	0.36	0.288	0.24
W22	D22	0.529	0.220	0.748	1.32	0.88	0.66	0.44	0.33	0.264	0.22
W20	D20	0.504	0.200	0.680	1.20	0.80	0.60	0.40	0.30	0.24	0.20
W18	D18	0.478	0.180	0.612	1.08	0.72	0.54	0.36	0.27	0.216	0.18
W16	D16	0.451	0.160	0.544	0.96	0.64	0.48	0.32	0.24	0.192	0.16
W14	D14	0.422	0.140	0.476	0.84	0.56	0.42	0.28	0.21	0.168	0.14
W12	D12	0.390	0.120	0.408	0.72	0.48	0.36	0.24	0.18	0.144	0.12
W11	D11	0.374	0.110	0.374	0.66	0.44	0.33	0.22	0.165	0.132	0.11
W10.5		0.366	0.105	0.357	0.63	0.42	0.315	0.21	0.157	0.126	0.105
W10	D10	0.356	0.100	0.340	0.60	0.40	0.30	0.20	0.15	0.12	0.10
W9.5		0.348	0.095	0.323	0.57	0.38	0.285	0.19	0.142	0.114	0.095
W9	D9	0.338	0.090	0.306	0.54	0.36	0.27	0.18	0.135	0.108	0.09
W8.5		0.329	0.085	0.289	0.51	0.34	0.255	0.17	0.127	0.102	0.085
W8	D8	0.319	0.080	0.272	0.48	0.32	0.24	0.16	0.12	0.096	0.08
W7.5		0.309	0.075	0.255	0.45	0.30	0.225	0.15	0.112	0.09	0.075
W7	D7	0.298	0.070	0.238	0.42	0.28	0.21	0.14	0.105	0.084	0.07
W6.5		0.288	0.065	0.221	0.39	0.26	0.195	0.13	0.097	0.078	0.065
W6	D6	0.276	0.060	0.204	0.36	0.24	0.18	0.12	0.09	0.072	0.06
W5.5		0.264	0.055	0.187	0.33	0.22	0.165	0.11	0.082	0.066	0.055
W5	D5	0.252	0.050	0.170	0.30	0.20	0.15	0.10	0.075	0.06	0.05
W4.5		0.240	0.045	0.153	0.27	0.18	0.135	0.09	0.067	0.054	0.045
W4	D4	0.225	0.040	0.136	0.24	0.16	0.12	0.08	0.06	0.048	0.04
W3.5		0.211	0.035	0.119	0.21	0.14	0.105	0.07	0.052	0.042	0.035
W3		0.195	0.030	0.102	0.18	0.12	0.09	0.06	0.045	0.036	0.03
W2.9		0.192	0.029	0.098	0.174	0.116	0.087	0.058	0.043	0.035	0.029
W2.5		0.178	0.025	0.085	0.15	0.10	0.075	0.05	0.037	0.03	0.025
W2		0.159	0.020	0.068	0.12	0.08	0.06	0.04	0.03	0.024	0.02
W1.4		0.135	0.014	0.049	0.084	0.056	0.042	0.028	0.021	0.107	0.014

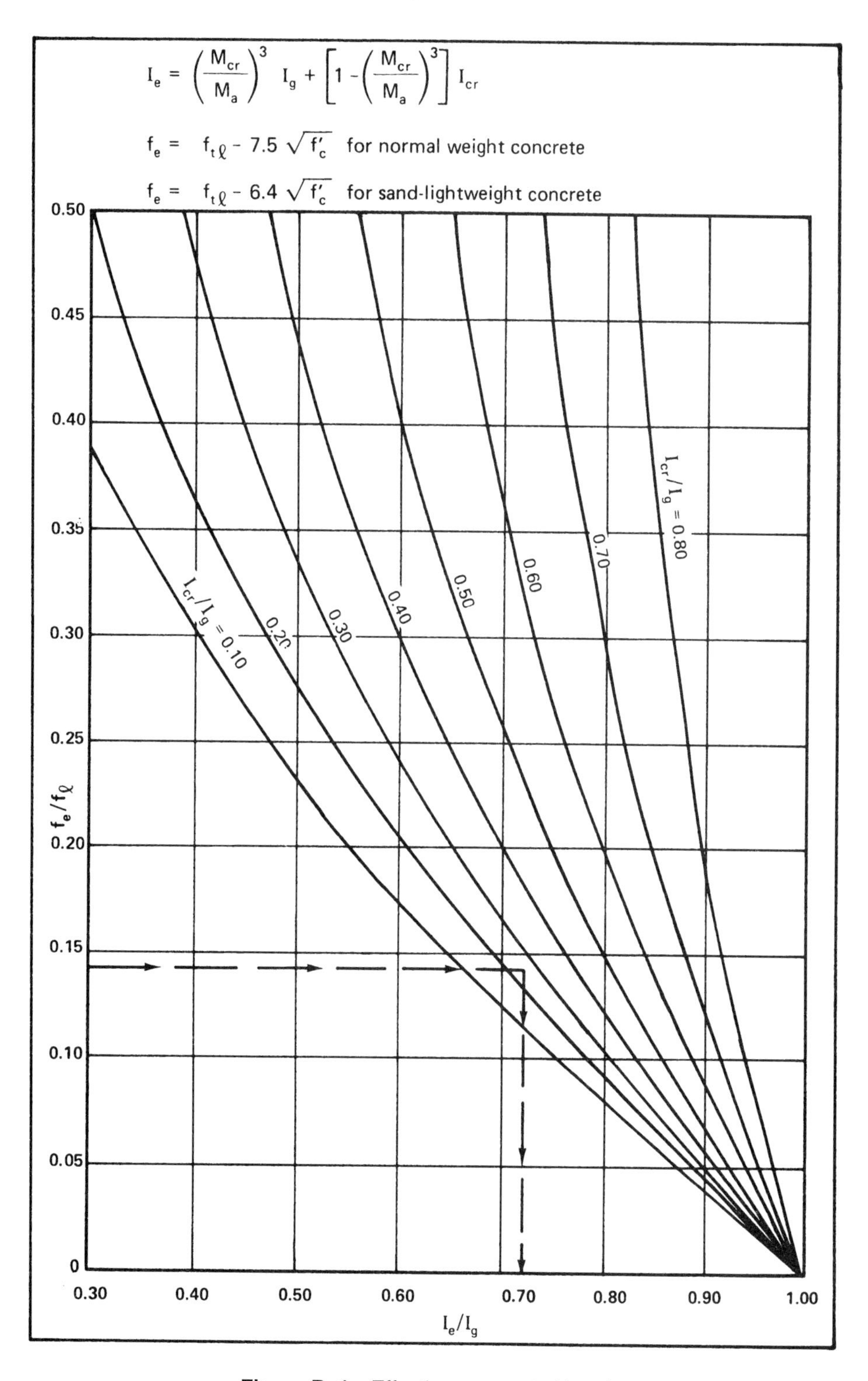

Figure B–1 Effective moment of inertia

Section	Geometric section modulus, Z_s, in.^{3}	Shape factor
Rectangle (b, h)	$\dfrac{bh^2}{4}$	1.5
I-section (b, h, t, w)	x-x axis: $bt\,(h - t) + \dfrac{w}{4}\,(h - 2t)^2$	1.12 (approx)
	y-y axis: $\dfrac{b^2 t}{2} + \dfrac{(h - 2t)w^2}{4}$	1.55 (approx)
Channel (t(ave.), w, h, b)	$bt\,(h - t) + \dfrac{w(h - 2t)^2}{4}$	1.12 (approx)
Circle (h)	$\dfrac{h^3}{6}$	1.70
Tube (t, h)	$\dfrac{h^3}{6}\left[1 - \left(1 - \dfrac{2t}{h}\right)^3\right]$ th^2 for $t \ll h$	$\dfrac{16}{3\pi}\left[\dfrac{1 - \left(1 - \dfrac{2t}{h}\right)^3}{1 - \left(1 - \dfrac{2t}{h}\right)^4}\right]$ 1.27 for $t \ll h$
Box (t, w, h, b)	$\dfrac{bh^2}{4}\left[1 - \left(1 - \dfrac{2w}{b}\right)\left(1 - \dfrac{2t}{h}\right)^2\right]$	1.12 (approx) for thin walls
Diamond (b, h)	$\dfrac{bh^2}{12}$	2

Figure B–2 Geometric section moduli and shape factors

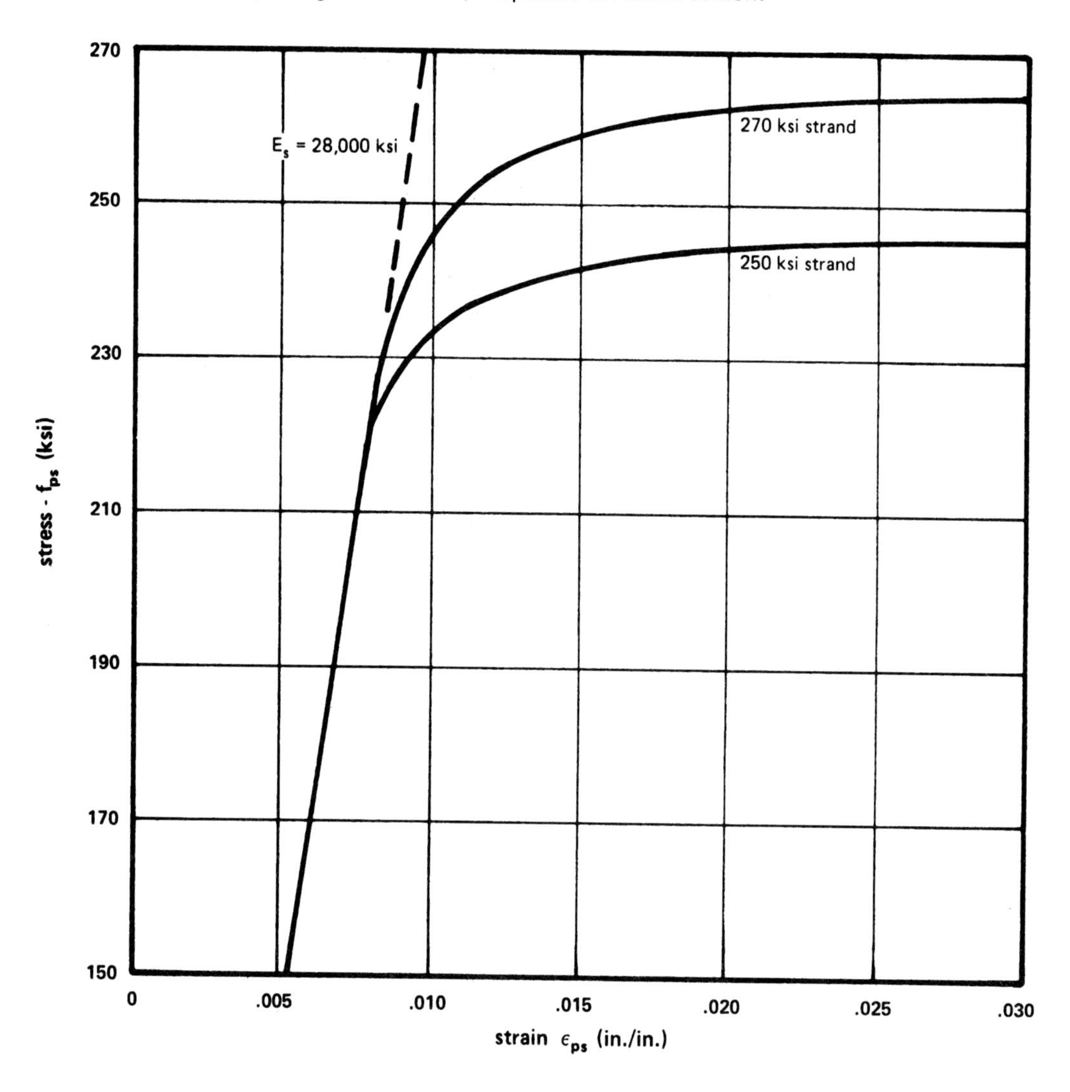

These curves can be approximated by the following equations:

$\epsilon_{ps} \leqslant 0.008$: $f_{ps} = 28{,}000\,\epsilon_{ps}$ (ksi)

$\epsilon_{ps} > 0.008$:

250 ksi strand: $$f_{ps} = 248 - \frac{0.058}{\epsilon_{ps} - 0.006} < 0.98\,f_{pu} \text{ (ksi)}$$

270 ksi strand: $$f_{ps} = 268 - \frac{0.075}{\epsilon_{ps} - 0.0065} < 0.98\,f_{pu} \text{ (ksi)}$$

Figure B–3 Typical stress-strain curve, 7-wire stress-relieved and low-relaxation prestressing strand

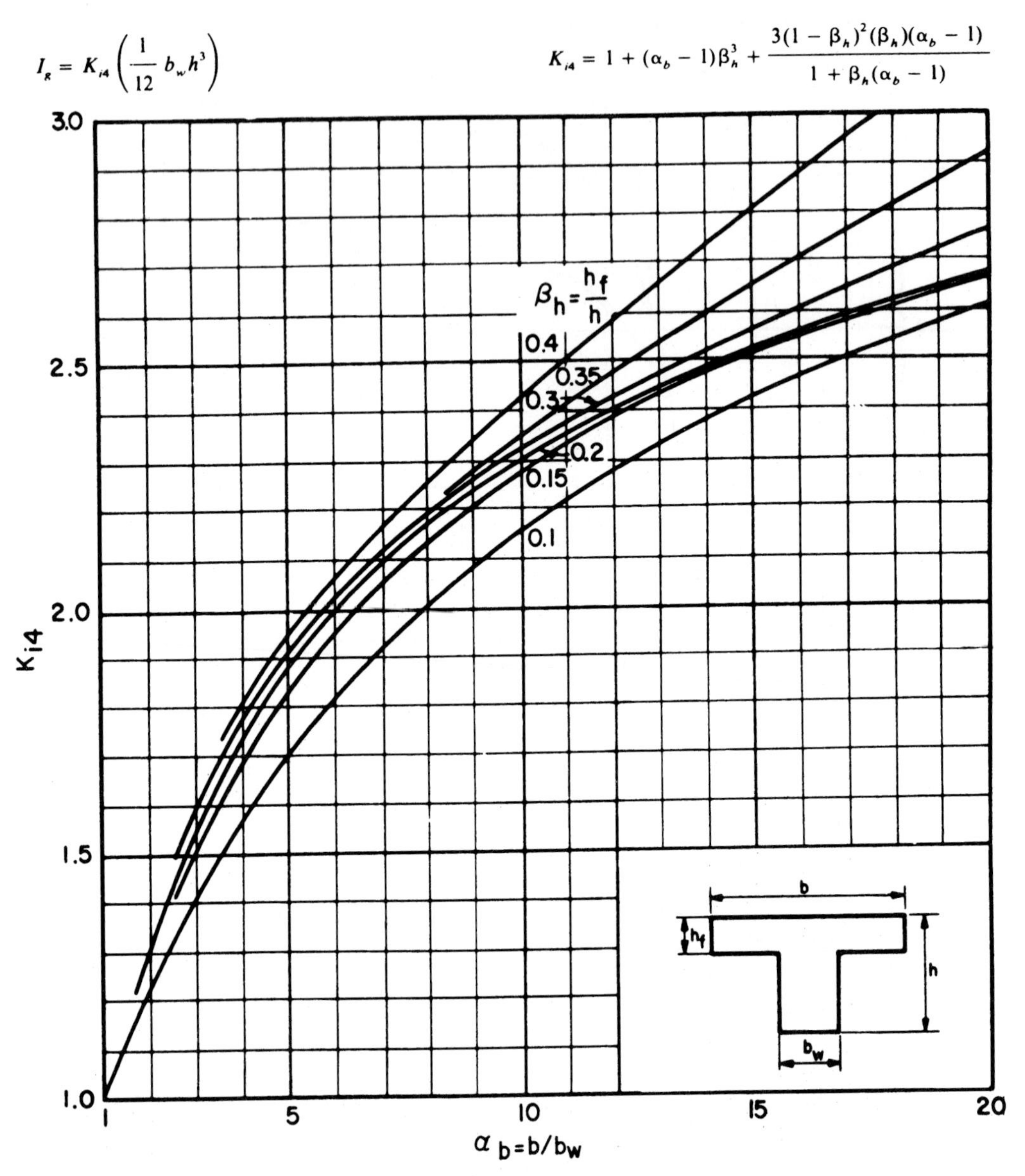

Example: For the T-beam shown, find the moment of inertia I_g:

$$\alpha_b = b/b_w = 143/15 = 9.53$$

$$\beta_h = h_f/h = 8/36 = 0.22$$

Interpolating between the curves for $\beta_h = 0.2$ and 0.3, read $K_{i4} = 2.28$

$$I_g = K_{i4}\frac{b_w h^3}{12} = 2.28\,\frac{15(36)^3}{12} = 133{,}000 \text{ in.}^4$$

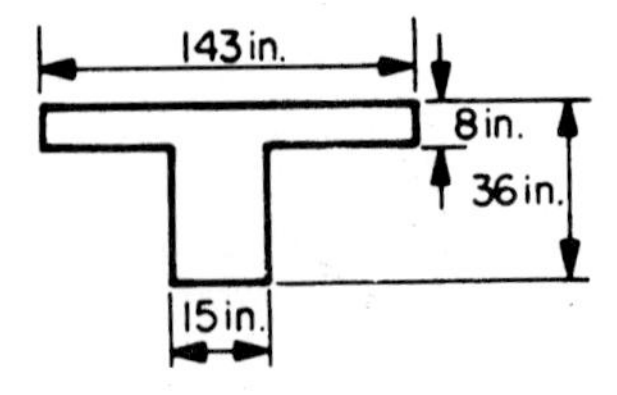

Figure B–4 Gross moment of inertia of T sections

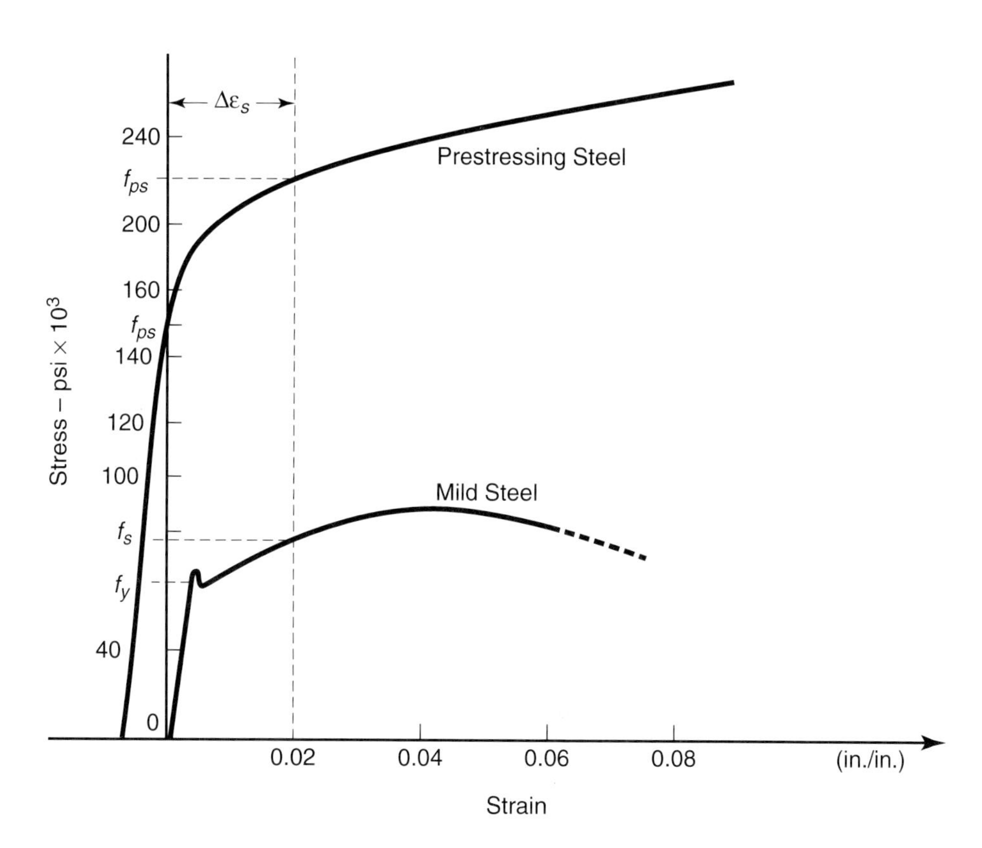

Figure B–5 Stress-strain diagram for prestressing steel strands in comparison with mild steel bar reinforcement

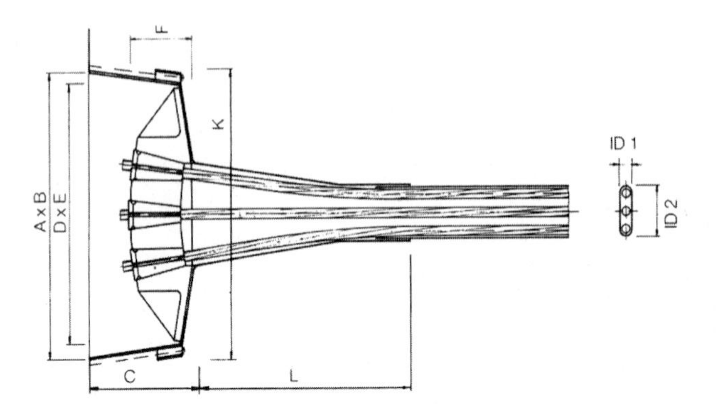

Flat Anchorage FA

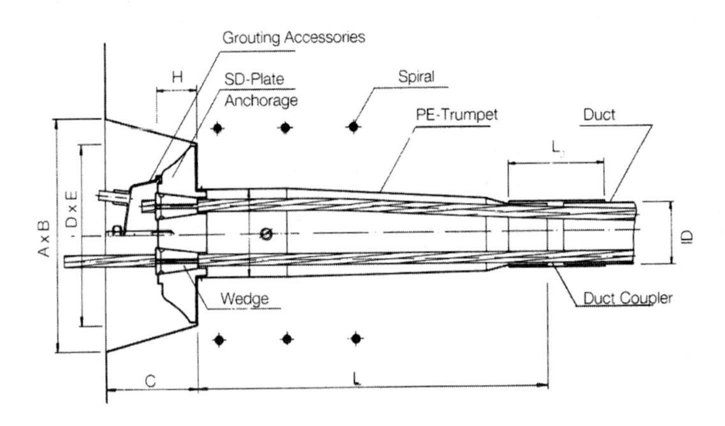

Plate Anchorage SD

Flat Anchorage FA

Tendon Size		3-0.6" or 4-0,5"	4-0.6" or 5-0,5"
Flat Anchor	D	10 \ 255	13 \ 330
	E	4 \ 100	4 \ 100
	F	2-1/4 \ 57	2-1/4 \ 57
Tran-sition	K	12-1/4 \ 310	--
	L	4-1/2 \ 115	8-5/8 \ 220
Pocket Former	A	10-3/4 \ 275	13-3/4 \ 350
	B	4-1/2 \ 115	4-7/8 \ 124
	C	5-1/2 \ 140	5-7/8 \ 148
Duct	ID1	1 \ 25	1 \ 25
	ID2	3 \ 75	3 \ 75

Combination Plate Anchorage SD

Tendon Size		3-0.6" or 4-0,5"	4-0.6" or 5-0,5"	4-0.6"	5-0.6" or 7-0,5"	6-0.6" or 8-0,5"	7-0.6" or 9-0,5"
Combin. Plate	D	4-15/16 \ 125	5-5/16 \ 135	5 \ 127	5-7/8 \ 150	6-1/2 \ 165	6-11/16 \ 170
	E	5-1/2 \ 140	6-5/16 \ 160	9 \ 229	7-1/16 \ 180	8-1/16 \ 205	8-1/2 \ 215
	H	1-5/8 \ 41	1-5/8 \ 41	2 \ 51	1-9/16 \ 40	1-3/4 \ 44	1-3/4 \ 44
Tran-sition	Ø	2-9/16 \ 65	2-15/16 \ 75	2-13/16 \ 72	3-3/8 \ 85	3-3/4 \ 95	3-3/4 \ 95
	L	11-3/8 \ 290	10-7/16 \ 265	11 \ 280	14 \ 355	15-15/16 \ 405	15-15/16 \ 405
Pocket Former	A	6-1/2 \ 165	6-1/2 \ 165	7-1/16 \ 179	7-1/16 \ 180	7-7/8 \ 200	7-7/8 \ 200
	B	7-5/16 \ 185	7-5/16 \ 185	9-7/8 \ 251	8-1/4 \ 210	9-7/16 \ 240	9-7/16 \ 240
	C	3-15/16 \ 100	3-15/16 \ 100	5-1/8 \ 130	3-15/16 \ 100	4-5/16 \ 110	4-5/16 \ 110
Duct	ID	1-9/16 \ 40	1-13/16 \ 46	2 \ 51	2-1/16 \ 52	2-7/16 \ 62	2-7/16 \ 62
	L2	8 \ 200	8 \ 200	4 \ 100	8 \ 200	8 \ 200	8 \ 200

All dimensions are nominal and are expressed in inch \ mm. Technical data subject to change.

Figure B–6 Dywidag flat anchorage for prestressing strands (*Courtesy* Dywidag Systems International).

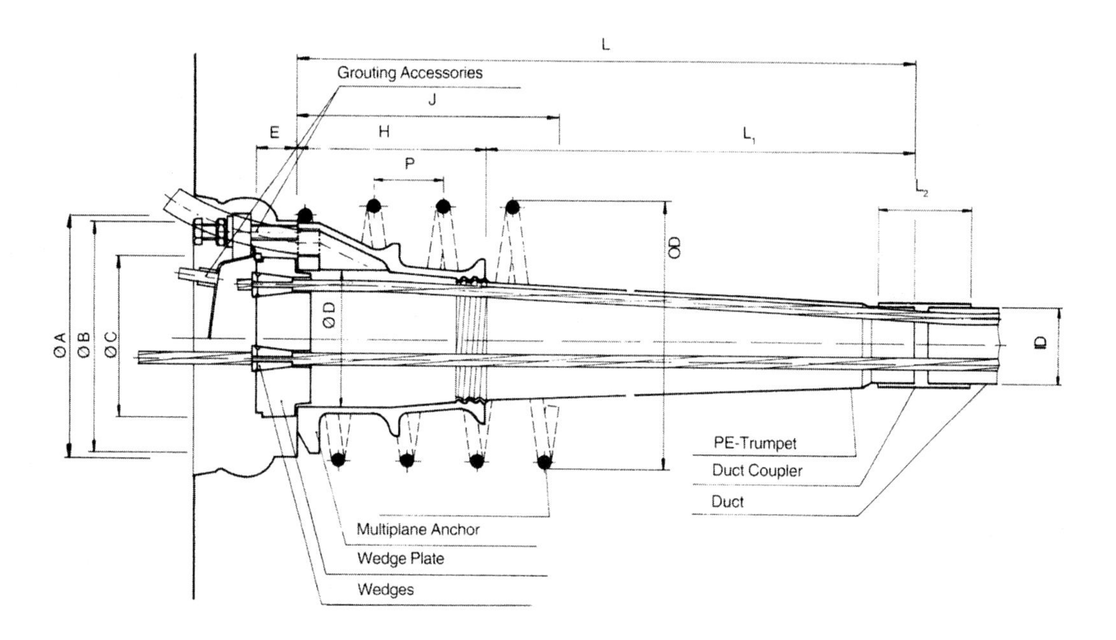

Anchorage Size		5-0.6" or 7-0.5"	7-0.6" or 9-0.5"	9-0.6" or 12-0.5"	12-0.6" or 15-0.5"	15-0.6" or 20-0.5"	19-0.6" or 27-0.5"	27-0.6" or 37-0.5"	37-0.6" or 59-0.5"
Min. Block-out Dia.	A	7 \ 179	8 \ 203	9 \ 229	10 \ 254	11 \ 279	12 \ 305	13-1/2 \ 343	16 \ 407
Transition Length		12-3/8 \ 314	13-7/16 \ 341	15-3/4 \ 400	20 \ 508	22-5/8 \ 575	25-3/16 \ 640	27-5/8 \ 702	35 \890
Anchor Dia.	B	5-15/16 \ 150	6-11/16 \ 170	7-1/2 \ 190	8-5/8 \ 220	9-7/8 \ 250	11 \ 280	12-3/8 \ 315	14-1/8 \ 360
	D	3-9/16 \ 90	3-7/8 \ 98	4-7/16 \ 113	5-1/16 \ 128	5-13/16 \ 148	6-3/8 \ 162	7-1/2 \ 190	8-1/2 \ 220
	H	3-9/16 \ 90	3-15/16 \ 100	4-15/16 \ 125	7-1/16 \ 180	7-7/8 \ 200	8-5/8 \ 220	9-7/16 \ 240	12-1/2 \ 320
Wedge Plate	C	5-1/8 \ 130	5-1/8 \ 130	5-1/2 \ 140	6-5/16 \ 160	7-1/16 \ 180	7-7/8 \ 200	9-7/16 \ 240	10-2/3 \ 270
	E	2 \ 50	1-9/16 \ 40	1-11/16 \ 43	1-11/16 \ 43	2 \ 50	2-3/16 \ 55	2-15/16 \ 75	3-1/2 \ 90
Trumpet	L1	8-7/8 \ 225	9-1/2 \ 241	10-13/16 \ 275	12-7/8 \ 327	14-3/4 \ 375	16-1/2 \ 419	18-1/8 \ 460	22-1/2 \ 600
Rebar Spiral *	Size	# 4 \ 15M	# 4 \ 15M	# 4 \ 15M	# 5 \ 15M	# 5 \ 15M	# 5 \ 15M	# 6 \ 20M	# 7 \ 22M
	Grade 400 MPa	60 KSI \ 400 MPa	60 KSI \ 400 MPa	60 KSI \ 400 MPa	60 KSI \ 400 MPa	60 KSI \ 400 MPa	60 KSI \ 400 MPa	60 KSI \ 400 MPa	60 KSI \ 400 MPa
	Pitch	1-7/8 \ 50	1-7/8 \ 50	1-7/8 \ 50	2-1/4 \ 55	1-7/8 \ 50	1-7/8 \ 50	2-1/4 \ 55	2-3/8 \ 60
	J	10 \ 255	10-1/2 \ 265	10-5/8 \ 270	14 \ 355	14-3/4 \ 365	15 \ 380	16-5/8 \ 420	18 \ 460
	OD	7-3/4 \ 190	9 \ 230	9-1/2 \ 240	11-1/4 \ 285	12-1/2 \ 315	14-1/2 \ 365	17 \ 430	22 \ 560
Duct	ID	2 \ 50	2-3/8 \ 60	3 \ 75	3-3/8 \ 85	3-3/4 \ 95	4 \ 100	4-1/2 \ 115	5-1/8 \ 130
Duct Coupler	L2	8 \ 200	8 \ 200	8 \ 200	8 \ 200	8 \ 200	8 \ 200	8 \ 200	12 \ 300
Grout Requirements	gal/ft \ l/m	0.12 \ 1.5	0.17 \ 2.1	0.28 \ 3.46	0.35 \ 4.39	0.44 \ 5.48	0.47 \ 5.80	0.58 \ 7.25	0.72 \ 8.90

Spiral required in "local anchor" zone.
\ll dimensions are nominal and are expressed in inch \ mm. Technical data subject to change.

Figure B–7 Dywidag multiple anchorage for prestressing strands (*Courtesy* Dywidag Systems International)

SELECTED TYPICAL STANDARD PRECAST DOUBLE TEES, INVERTED TEES, HOLLOW CORE SECTIONS, AND AASHTO BRIDGE SECTIONS

Strand Pattern Designation

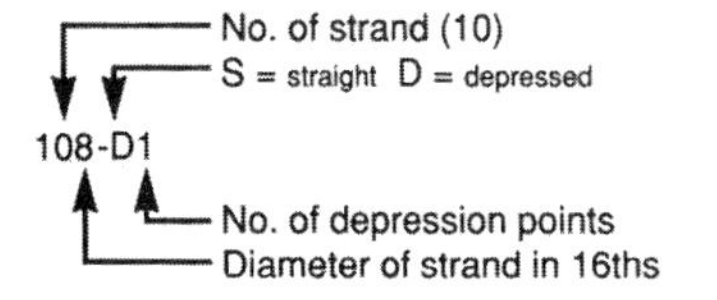

Safe loads shown include dead load of 10 psf for untopped members and 15 psf for topped members. Remainder is live load. Long-time cambers include superimposed dead load but do not include live load.

Key

173 — Safe superimposed service load, psf
0.5 — Estimated camber at erection, in.
0.7 — Estimated long-time camber, in.

DOUBLE TEE

8'-0" x 24"
Normal Weight Concrete

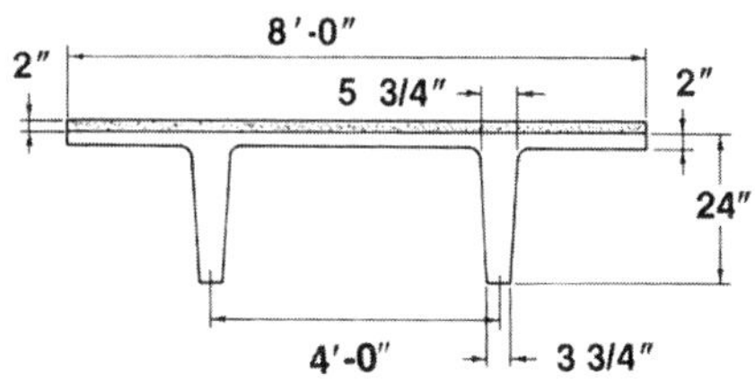

f'_c = 5,000 psi

f_{pu} = 270,000 psi

Section Properties

		Untopped		Topped	
A	=	401	in.^{2}	—	
I	=	20.985	in.^{4}	27,720	in.^{4}
y_b	=	17.15	in.	19.27	in.
y_t	=	6.85	in.	6.73	in.
Z_b	=	1,224	in.^{3}	1,438	in.^{3}
Z_t	=	3,063	in.^{3}	4,119	in.^{3}
wt	=	418	plf	618	plf
		52	psf	77	psf
V/S	=	1.41	in.		

8DT24

Table of safe superimposed service load (psf) and cambers — **No Topping**

Strand Pattern	e_e e_c	30	32	34	36	38	40	42	44	46	48	50	52	54	56	58	60	62	64	66	68	70	72	74
		Span, ft.																						
68-S	11.15 11.15	173 0.5 0.7	147 0.6 0.8	126 0.6 0.8	108 0.7 0.8	92 0.7 0.8	79 0.7 0.8	68 0.7 0.8	58 0.7 0.8	50 0.7 0.7	43 0.6 0.6	36 0.6 0.4	30 0.5 0.2											
88-S	9.15 9.15		180 0.7 0.9	155 0.7 0.9	134 0.8 1.0	116 0.8 1.0	100 0.8 1.0	87 0.8 1.0	76 0.9 1.0	66 0.8 0.9	57 0.8 0.8	49 0.8 0.7	43 0.7 0.5	36 0.6 0.3	31 0.5 0.1									
88-D1	9.15 14.40				190 1.1 1.5	166 1.2 1.5	146 1.3 1.6	129 1.4 1.7	114 1.5 1.8	100 1.5 1.8	89 1.6 1.8	79 1.6 1.8	70 1.6 1.7	62 1.6 1.6	54 1.6 1.5	48 1.5 1.4	42 1.4 1.2	37 1.3 0.9	32 1.2 0.5					
108-D1	7.15 14.15								145 1.7 2.2	129 1.8 2.2	116 1.9 2.3	103 2.0 2.3	92 2.0 2.3	83 2.1 2.3	74 2.1 2.2	66 2.1 2.1	59 2.1 2.0	53 2.0 1.8	47 2.0 1.6	42 1.8 1.3	37 1.7 1.0	32 1.5 0.5		
128-D1	5.48 13.90															83 2.5 2.7	75 2.5 2.7	68 2.5 2.5	61 2.5 2.3	55 2.5 2.1	49 2.4 1.8	44 2.3 1.5	40 2.1 1.1	35 1.9 0.6
148-D1	4.29 13.65																				61 2.9 2.6	55 2.9 2.3	50 2.8 1.9	45 2.6 1.5

8DT24 + 2

Table of safe superimposed service load (psf) and cambers — **2" Normal Weight Topping**

Strand Pattern	e_e e_c	26	28	30	32	34	36	38	40	42	44	46	48	50	52	54	56	58	60	62	64
		Span, ft.																			
48-S	14.15 14.15	183 0.4 0.4	149 0.4 0.4	122 0.4 0.4	100 0.5 0.4	82 0.5 0.4	66 0.5 0.3	53 0.5 0.3	42 0.5 0.2	33 0.5 0.0											
68-S	11.15 11.15			175 0.5 0.5	147 0.6 0.6	123 0.6 0.6	103 0.7 0.5	86 0.7 0.5	72 0.7 0.4	60 0.7 0.3	49 0.7 0.2	39 0.7 0.0									
68-D1	11.15 14.65					184 0.7 0.8	156 0.8 0.8	133 0.9 0.8	113 0.9 0.8	96 1.0 0.7	81 1.0 0.7	69 1.0 0.6	58 1.0 0.5	48 1.0 0.3	39 1.0 0.1						
88-D1	9.15 14.40							190 1.1 1.1	165 1.2 1.1	143 1.3 1.2	124 1.4 1.2	107 1.5 1.1	93 1.5 1.1	80 1.6 0.9	69 1.6 0.8	59 1.6 0.6	51 1.6 0.4	43 1.6 0.1			
108-D1	7.15 14.15										142 1.7 1.5	124 1.8 1.5	109 1.9 1.4	96 2.0 1.3	84 2.0 1.2	74 2.1 1.0	64 2.1 0.7	56 2.1 0.5	48 2.1 0.1		
128-D1	5.48 13.90																	74 2.5 1.0	65 2.5 0.7	57 2.5 0.3	49 2.5 0.1

Strength based on strain compatibility; bottom tension limited to $12\sqrt{f'_c}$*;*
Shaded values require release strengths higher than 3500 psi.

Figure C–1 8″–0″ × 24″ Double Tee (Courtesy PCI, Ref. 4.9)

Strand Pattern Designation

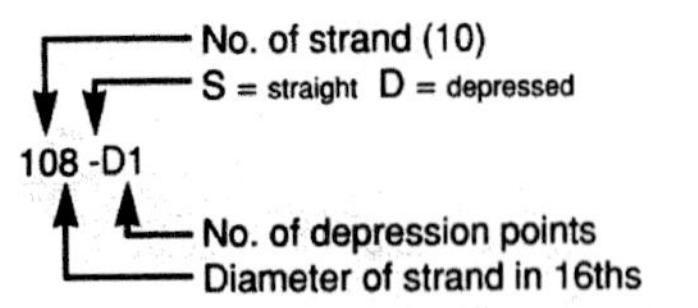

Because these units are pretopped and are typically used in parking structures, safe loads shown do not include any superimposed dead loads. Loads shown are live load. Long-time cambers do not include live load.

Key

196 — Safe superimposed service load, psf
0.4 — Estimated camber at erection, in.
0.5 — Estimated long-time camber, in.

PRETOPPED DOUBLE TEE

10'-0" x 26"

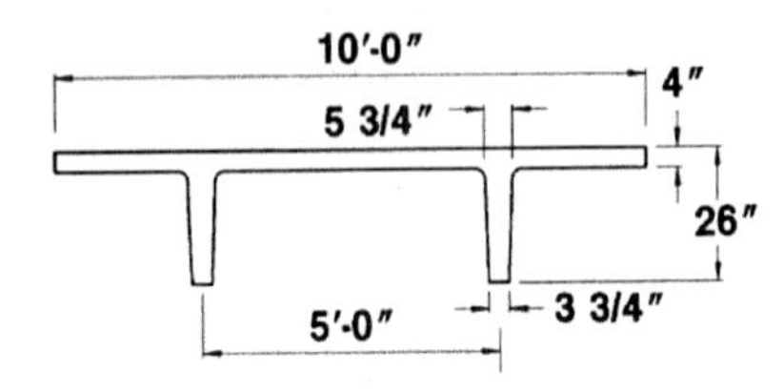

f'_c = 5,000 psi

f_{pu} = 270,000 psi

Section Properties

		Normal Weight		Lightweight	
A	=	689	in.^{2}	689	in.^{2}
I	=	30,716	in.^{4}	30,716	in.^{4}
y_b	=	20.29	in.	20.29	in.
y_t	=	5.71	in.	5.71	in.
Z_b	=	1,514	in.^{3}	1,514	in.^{3}
Z_t	=	5,379	in.^{3}	5,379	in.^{3}
wt	=	718	plf	550	plf
		72	psf	55	psf
V/S	=	2.05	in.	2.05	in.

10DT26

No Topping

Table of safe superimposed service load (psf) and cambers

Strand Pattern	e_e / e_c	26	28	30	32	34	36	38	40	42	44	46	48	50	52	54	56	58	60	62	64	66	68
68-S	14.29 14.29	196 0.4 0.5	161 0.4 0.5	133 0.4 0.6	109 0.4 0.6	90 0.5 0.6	74 0.5 0.6	60 0.5 0.6	49 0.4 0.6	39 0.4 0.6	30 0.4 0.5												
88-S	12.29 12.29			169 0.5 0.7	142 0.5 0.8	119 0.6 0.8	100 0.6 0.8	83 0.6 0.9	69 0.6 0.8	57 0.6 0.8	47 0.6 0.8	38 0.5 0.7	30 0.4 0.6										
88-D1	12.29 17.54				200 0.7 1.0	170 0.8 1.1	146 0.9 1.2	125 0.9 1.3	107 1.0 1.3	91 1.0 1.3	78 1.0 1.4	66 1.0 1.4	56 1.0 1.3	47 0.9 1.3	39 0.9 1.2	32 0.8 1.0							
108-D1	10.29 17.29						189 1.1 1.5	163 1.1 1.6	142 1.2 1.7	123 1.3 1.7	107 1.3 1.8	93 1.4 1.8	80 1.4 1.8	69 1.4 1.8	60 1.3 1.8	51 1.3 1.7	43 1.2 1.6	36 1.1 1.5					
128-D1	8.62 17.04										134 1.6 2.1	118 1.6 2.2	103 1.7 2.3	90 1.7 2.3	79 1.7 2.3	69 1.7 2.3	60 1.7 2.2	52 1.6 2.1	45 1.5 2.0	38 1.4 1.9	32 1.3 1.7		
148-D1	7.43 16.79														98 2.0 2.8	87 2.1 2.8	76 2.1 2.8	67 2.1 2.8	59 2.1 2.7	51 2.0 2.5	45 1.9 2.4	38 1.7 2.2	33 1.5 1.9

Span, ft.

10LDT26

No Topping

Table of safe superimposed service load (psf) and cambers

Strand Pattern	e_e / e_c	28	30	32	34	36	38	40	42	44	46	48	50	52	54	56	58	60	62	64	66	68	70	72
68-S	14.29 14.29	175 0.6 0.9	146 0.7 1.0	123 0.7 1.0	104 0.8 1.1	88 0.8 1.2	74 0.9 1.2	63 0.9 1.3	53 0.9 1.3	44 0.9 1.2	36 0.9 1.2	30 0.8 1.1												
88-S	12.29 12.29		183 0.8 1.1	156 0.9 1.2	133 1.0 1.3	113 1.0 1.4	97 1.1 1.5	83 1.1 1.6	71 1.2 1.6	61 1.2 1.6	52 1.2 1.6	44 1.2 1.6	37 1.1 1.5	31 1.0 1.4										
88-D1	12.29 17.54				184 1.3 1.8	159 1.4 1.9	138 1.5 2.1	120 1.7 2.2	105 1.7 2.3	92 1.8 2.4	80 1.9 2.4	70 1.9 2.5	61 1.9 2.5	53 1.9 2.5	46 1.9 2.5	39 1.8 2.4	34 1.7 2.3							
108-D1	10.29 17.29						177 1.8 2.5	155 2.0 2.7	137 2.1 2.9	121 2.2 3.0	106 2.3 3.2	94 2.5 3.3	83 2.6 3.4	73 2.6 3.4	65 2.6 3.4	57 2.6 3.4	50 2.6 3.3	44 2.5 3.2	38 2.4 3.1	33 2.3 3.0				
128-D1	8.62 17.04												104 3.0 4.0	93 3.1 4.1	83 3.2 4.2	74 3.3 4.3	66 3.3 4.3	59 3.4 4.3	52 3.3 4.2	46 3.2 4.0	41 3.1 3.8	36 3.0 3.6	31 2.8 3.4	
148-D1	7.43 16.79																	73 4.0 5.3	65 4.0 5.3	58 4.0 5.2	52 4.0 5.1	47 4.0 5.0	41 3.8 4.7	37 3.6 4.3

Span, ft.

Strength based on strain compatibility; bottom tension limited to $12\sqrt{f'_c}$;
Shaded values require release strengths higher than 3500 psi.

Figure C–2 10′–0″ × 26″ Double Tee (Courtesy PCI, Ref. 4.9)

Strand Pattern Designation

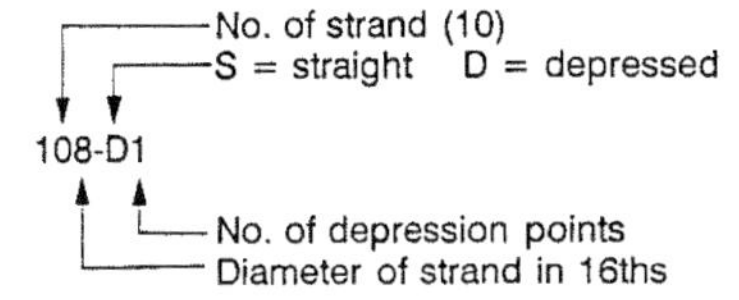

Because these units are pretopped and are typically used in parking structures, safe loads shown do not include any superimposed dead loads. Loads shown are live load. Long-time cambers do not include live load.

Key
176 — Safe superimposed service load, psf
0.8 — Estimated camber at erection, in.
1.1 — Estimated long-time camber, in.

PRETOPPED DOUBLE TEE

12′-0″ x 34″

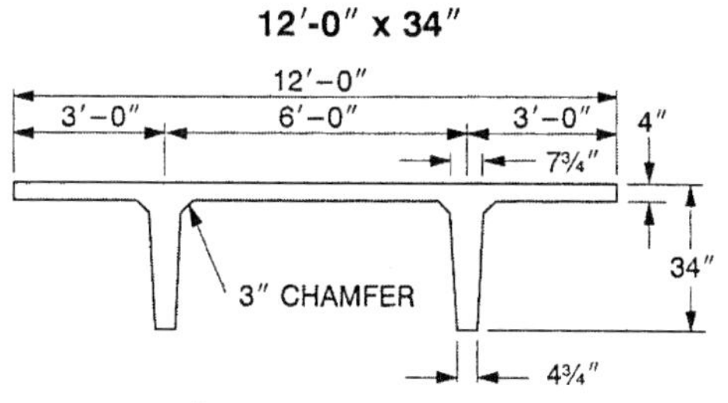

f'_c = 5,000 psi
f_{pu} = 270,000 psi

Section Properties

		Normal Weight		Lightweight	
A	=	978	in²	978	in²
I	=	86,072	in⁴	86,072	in⁴
y_b	=	25.77	in.	25.77	in.
y_t	=	8.23	in.	8.23	in.
S_b	=	3,340	in³	3,340	in³
S_t	=	10,458	in³	10,458	in³
wt	=	1,019	plf	781	plf
		85	psf	65	psf
V/S	=	2.39	in.	2.39	in.

12DT34

Table of safe superimposed service load (psf) and cambers (in.) — **No Topping**

Strand Pattern	e_e, in. / e_c, in.	42	44	46	48	50	52	54	56	58	60	62	64	66	68	70	72	74	76	78	80	82	84	86
128-D1	14.10 / 22.52	176	155	135	119	104	91	79	69	59	51	43	36	30										
		0.8	0.8	0.9	0.9	0.9	0.9	0.9	0.8	0.8	0.7	0.7	0.6	0.5										
		1.1	1.1	1.2	1.2	1.2	1.2	1.2	1.2	1.1	1.0	0.9	0.7	0.6										
148-D1	12.91 / 22.27		187	165	146	129	114	101	89	78	68	60	52	44	38	32	26							
			1.0	1.0	1.1	1.1	1.1	1.1	1.1	1.1	1.1	1.0	1.0	0.9	0.7	0.6	0.4							
			1.4	1.4	1.4	1.5	1.5	1.5	1.5	1.5	1.5	1.4	1.3	1.1	1.0	0.8	0.5							
168-D1	12.77 / 22.02			196	174	155	138	123	110	97	86	76	67	59	52	45	39	33	28					
				1.2	1.3	1.3	1.4	1.4	1.4	1.4	1.4	1.4	1.4	1.3	1.2	1.1	1.0	0.8	0.6					
				1.7	1.7	1.8	1.9	1.9	1.9	1.9	1.9	1.9	1.8	1.7	1.6	1.5	1.3	1.0	0.7					
188-D1	11.38 / 21.77					178	160	143	128	115	102	91	82	73	64	57	50	43	38	32	27			
						1.5	1.5	1.6	1.6	1.7	1.7	1.7	1.7	1.6	1.5	1.5	1.4	1.2	1.1	0.9	0.6			
						2.0	2.1	2.1	2.2	2.2	2.2	2.2	2.2	2.1	2.0	1.9	1.8	1.6	1.4	1.1	0.7			
208-D1	10.27 / 21.52									131	118	106	95	85	76	68	61	54	47	41	36	31	26	
										1.8	1.9	1.9	1.9	1.9	1.9	1.8	1.7	1.6	1.5	1.3	1.1	0.9	0.6	
										2.5	2.5	2.5	2.5	2.5	2.4	2.3	2.2	2.1	1.9	1.7	1.4	1.1	0.7	
228-D1	9.36 / 21.27												109	98	88	79	71	64	57	50	44	39	34	29
													2.1	2.1	2.1	2.1	2.1	2.0	1.9	1.7	1.6	1.4	1.1	0.8
													2.8	2.8	2.8	2.7	2.6	2.5	2.4	2.2	2.0	1.7	1.4	1.0

(Span, ft: 42–86)

12LDT34

Table of safe superimposed service load (psf) and cambers (in.) — **No Topping**

Strand Pattern	e_e, in. / e_c, in.	42	44	46	48	50	52	54	56	58	60	62	64	66	68	70	72	74	76	78	80	82	84	86
128-D1	14.10 / 22.52	193	171	152	135	120	107	95	85	76	67	59	52	46	40	35	30	26						
		1.3	1.4	1.5	1.5	1.6	1.6	1.7	1.7	1.7	1.7	1.7	1.6	1.6	1.5	1.4	1.2	1.0						
		1.8	1.9	2.0	2.0	2.1	2.1	2.2	2.2	2.2	2.2	2.2	2.1	2.1	2.0	1.8	1.6	1.3						
148-D1	12.91 / 22.27			182	162	146	130	117	105	94	85	76	68	61	54	48	42	37	33	28				
				1.7	1.8	1.9	2.0	2.0	2.1	2.1	2.1	2.1	2.1	2.1	2.1	2.0	1.9	1.8	1.6	1.4				
				2.3	2.4	2.5	2.6	2.7	2.8	2.8	2.8	2.8	2.8	2.7	2.7	2.6	2.4	2.3	2.1	1.8				
168-D1	12.77 / 22.02				191	172	155	139	126	114	103	93	84	76	68	61	55	49	44	39	34	30	26	
					2.1	2.2	2.3	2.4	2.5	2.6	2.7	2.7	2.7	2.7	2.7	2.7	2.6	2.5	2.4	2.3	2.1	1.9	1.6	
					2.8	3.0	3.1	3.2	3.3	3.4	3.5	3.6	3.5	3.5	3.5	3.4	3.3	3.2	3.1	2.9	2.7	2.4	2.1	
188-D1	11.38 / 21.77								144	131	119	108	98	89	81	73	66	60	54	48	43	39	34	30
									2.7	2.8	2.9	3.0	3.1	3.2	3.2	3.2	3.2	3.1	3.0	2.9	2.8	2.6	2.4	2.1
									3.7	3.8	3.9	4.0	4.1	4.1	4.2	4.1	4.0	3.9	3.7	3.6	3.4	3.2	3.0	2.7
208-D1	10.27 / 21.52													102	93	85	77	70	64	58	52	47	42	38
														3.5	3.6	3.6	3.6	3.6	3.6	3.5	3.4	3.3	3.1	2.9
														4.6	4.7	4.7	4.7	4.7	4.6	4.4	4.2	3.9	3.7	3.5
228-D1	9.36 / 21.27																	80	73	67	61	82	50	45
																		4.1	4.1	4.1	4.0	3.9	3.8	3.6
																		5.3	5.2	5.2	5.0	4.8	4.6	4.2

(Span, ft: 42–86)

Strength based on strain compatibility; bottom tension limited to $12\sqrt{f'_c}$; see pages 2-2–2-6 for explanation.
Shaded values require release strengths higher than 3500 psi.

Figure C–3 Pretopped 12′–0″ × 34″ Double Tee (Courtesy PCI, Ref. 4.9)

INVERTED TEE BEAMS

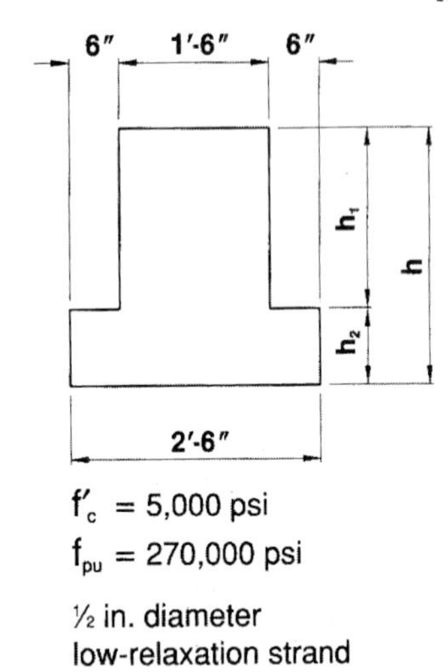

f'_c = 5,000 psi

f_{pu} = 270,000 psi

½ in. diameter low-relaxation strand

Normal Weight Concrete

Section Properties								
Designation	h (in.)	h_1/h_2 (in.)	A (in.^{2})	I (in.^{4})	y_b (in.)	Z_b (in.^{3})	Z_t (in.^{3})	wt (plf)
30IT20	20	12/8	456	15,240	8.74	1,744	1,354	475
30IT24	24	12/12	576	26,352	10.50	2,510	1,952	600
30IT28	28	16/12	648	41,824	12.22	3,423	2,650	675
30IT32	32	20/12	720	62,400	14.00	4,457	3,467	750
30IT36	36	24/12	792	88,678	15.82	5,605	4,394	825
30IT40	40	24/16	912	121,923	17.47	6,979	5,412	950
30IT44	44	28/16	984	162,161	19.27	8,415	6,557	1,025
30IT48	48	32/16	1,056	210,199	21.09	9,967	7,811	1,100
30IT52	52	36/16	1,128	266,627	22.94	11,623	9,175	1,175
30IT56	56	40/16	1,200	332,032	24.80	13,388	10,642	1,250
30IT60	60	44/16	1,272	406,997	26.68	15,255	12,215	1,325

1. Check local area for availability of other sizes.

2. Safe loads shown include 50% dead load and 50% live load. 800 psi top tension has been allowed, therefore additional top reinforcement is required.

3. Safe loads can be significantly increased by use of structural composite topping.

Key

8,428 — Safe superimposed service load, plf
0.4 — Estimated camber at erection, in.
0.2 — Estimated long-time camber, in.

Table of safe superimposed service load (plf) and cambers

Designation	No. Strand	e	Span, ft. 18	20	22	24	26	28	30	32	34	36	38	40	42	44	46	48	50
30IT20	14	6.65	8,428 0.4 0.2	6,736 0.5 0.2	5,485 0.6 0.2	4,533 0.7 0.3	3,792 0.9 0.3	3,204 1.0 0.3	2,730 1.1 0.3	2,342 1.2 0.3	2,020 1.3 0.3	1,751 1.4 0.3	1,523 1.4 0.3	1,332 1.5 0.3	1,167 1.6 0.3	1,024 1.6 0.2			
30IT24	17	7.67		9,736 0.4 0.2	7,942 0.5 0.2	6,578 0.6 0.2	5,516 0.7 0.2	4,673 0.8 0.3	3,994 0.9 0.3	3,437 1.0 0.3	2,976 1.1 0.3	2,592 1.2 0.3	2,269 1.2 0.3	1,993 1.3 0.3	1,755 1.4 0.2	1,550 1.4 0.2	1,370 1.5 0.2	1,212 1.5 0.1	1,073 1.5 0.0
30IT28	20	9.06				9,087 0.6 0.2	7,643 0.6 0.2	6,497 0.7 0.3	5,573 0.8 0.3	4,816 0.9 0.3	4,189 1.0 0.3	3,664 1.1 0.3	3,219 1.2 0.3	2,839 1.2 0.3	2,513 1.3 0.3	2,334 1.4 0.3	1,990 1.4 0.3	1,776 1.5 0.3	1,588 1.5 0.2
30IT32	23	10.50						8,647 0.7 0.2	7,436 0.7 0.3	6,445 0.8 0.3	5,623 0.9 0.3	4,935 1.0 0.3	4,352 1.1 0.4	3,855 1.2 0.4	3,426 1.2 0.4	3,055 1.3 0.4	2,732 1.4 0.4	2,448 1.5 0.4	2,201 1.5 0.3
30IT36	24	12.32							9,492 0.7 0.2	8,243 0.7 0.2	7,207 0.8 0.3	6,340 0.9 0.3	5,605 1.0 0.3	4,978 1.0 0.3	4,439 1.1 0.3	3,971 1.2 0.3	3,563 1.3 0.3	3,205 1.3 0.3	2,892 1.4 0.3
30IT40	30	12.92									9,077 0.8 0.3	7,994 0.8 0.3	7,077 0.9 0.3	6,295 1.0 0.4	5,621 1.1 0.4	5,037 1.2 0.4	4,528 1.2 0.4	4,081 1.3 0.4	3,687 1.4 0.4
30IT44	30	14.73										9,659 0.7 0.3	8,564 0.8 0.3	7,629 0.9 0.3	6,825 1.0 0.3	6,127 1.0 0.3	5,519 1.1 0.3	4,985 1.2 0.3	4,514 1.2 0.3
30IT48	33	16.17												9,222 0.8 0.3	8,262 0.9 0.3	7,431 1.0 0.3	6,705 1.0 0.3	6,068 1.1 0.3	5,506 1.2 0.3
30IT52	36	17.62													9,836 0.9 0.3	8,858 0.9 0.3	8,004 1.0 0.3	7,255 1.1 0.3	6,594 1.1 0.4
30IT56	39	19.06															9,407 1.0 0.3	8,538 1.0 0.4	7,770 1.1 0.4
30IT60	42	20.49																9,917 1.0 0.3	9,036 1.0 0.4

Figure C–4 Inverted Tee Beam Sections (Courtesy PCI, Ref. 4.9)

HOLLOW-CORE SLABS

Section Properties — normal weight concrete **Dy-Core**

Trade name: Dy-Core®
Licensing Organization: Dy-Core Systems, Inc., Vancouver, British Columbia

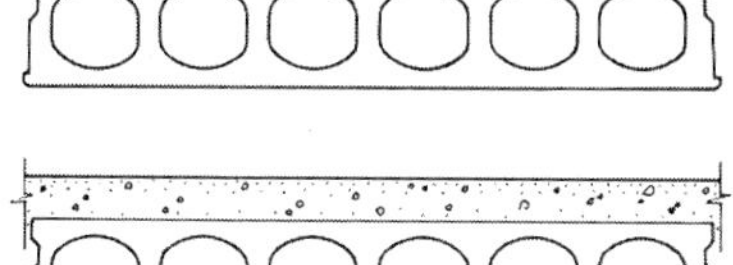
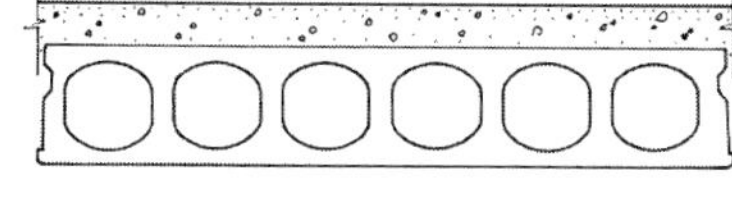
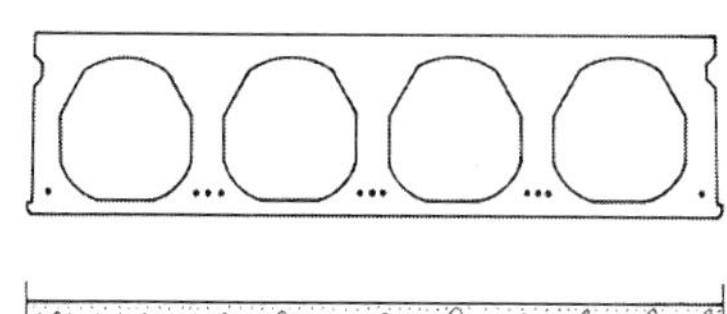

Section	Untopped				With 2" topping		
width x depth	A in.²	y_b in.	I in.⁴	wt psf	y_b in.	I in.⁴	wt psf
4'-0" x 6"	151	3.11	683	40	4.54	1,552	65
4'-0" x 8"	190	3.95	1,568	51	5.54	3,130	76
4'-0" x 10"	216	5.10	2,892	58	6.80	5,097	83
4'-0" x 12"	262	6.34	4,875	71	8.01	7,823	96
4'-0" x 15"	289	7.34	8,701	78	9.36	13,776	103

Note: All sections not available from all producers. Check availability with local manufacturers.

Section Properties — normal weight concrete **Dynaspan**

Trade name: Dynaspan®
Equipment Manufacturers: Dynamold Corporation, Salina, Kansas

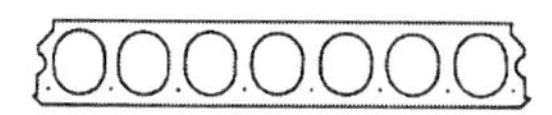
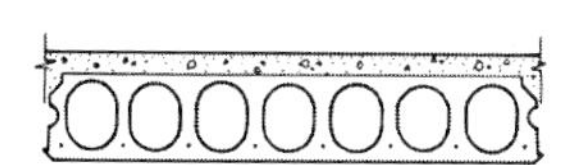
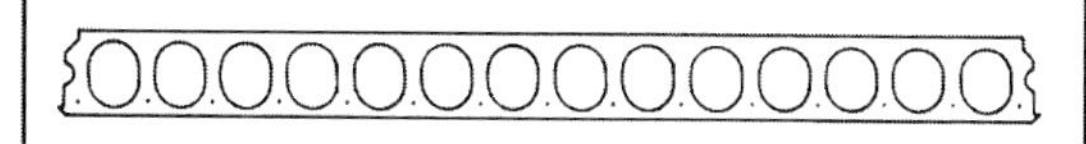
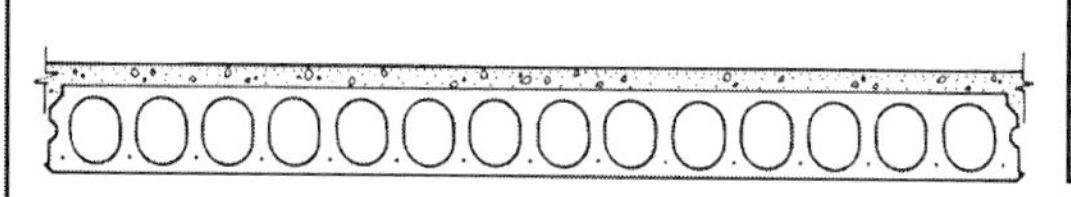

Section	Untopped				With 2" topping		
width x depth	A in.²	y_b in.	I in.⁴	wt psf	y_b in.	I in.⁴	wt psf
4'-0" x 4"	133	2.00	235	35	3.08	689	60
4'-0" x 6"	165	3.02	706	43	4.25	1,543	68
4'-0" x 8"	233	3.93	1,731	61	5.16	3,205	86
4'-0" x 10"	260	4.91	3,145	68	6.26	5,314	93
8'-0" x 6"	338	3.05	1,445	44	4.26	3,106	69
8'-0" x 8"	470	3.96	3,525	61	5.17	6,444	86
8'-0" x 10"	532	4.96	6,422	69	6.28	10,712	94
8'-0" x 12"	615	5.95	10,505	80	7.32	16,507	105

Note: All sections not available from all producers. Check availability with local manufacturers.

Figure C–5 Hollow Core Slab Sections (Courtesy PCI, Ref. 4.9)

AASHTO I-Beams

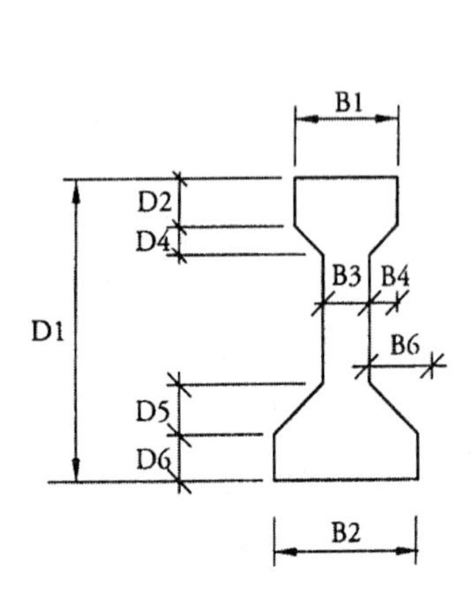

Type I-IV

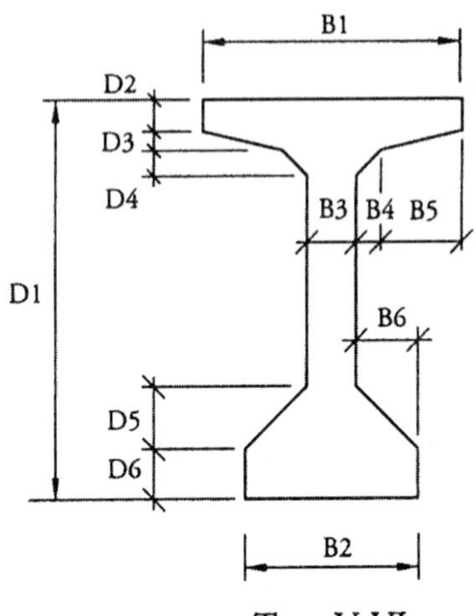

Type V-VI

Dimensions (inches)

Type	D1	D2	D3	D4	D5	D6	B1	B2	B3	B4	B5	B6
I	28.0	4.0	0.0	3.0	5.0	5.0	12.0	16.0	6.0	3.0	0.0	5.0
II	36.0	6.0	0.0	3.0	6.0	6.0	12.0	18.0	6.0	3.0	0.0	6.0
III	45.0	7.0	0.0	4.5	7.5	7.0	16.0	22.0	7.0	4.5	0.0	7.5
IV	54.0	8.0	0.0	6.0	9.0	8.0	20.0	26.0	8.0	6.0	0.0	9.0
V	63.0	5.0	3.0	4.0	10.0	8.0	42.0	28.0	8.0	4.0	13.0	10.0
VI	72.0	5.0	3.0	4.0	10.0	8.0	42.0	28.0	8.0	4.0	13.0	10.0

Properties

Type	Area in.^{2}	y_{bottom} in.	Inertia in.^{4}	Weight kip/ft	Maximum Span,* ft
I	276	12.59	22,750	0.287	48
II	369	15.83	50,980	0.384	70
III	560	20.27	125,390	0.583	100
IV	789	24.73	260,730	0.822	120
V	1,013	31.96	521,180	1.055	145
VI	1,085	36.38	733,320	1.130	167

*Based on simple span, HS-25 loading and f'_c = 7,000 psi.

Figure C–6 (a) AASHTO/PCI Standard Bridge Sections (Courtesy PCI, Ref. 12.11)

AASHTO I-Beams

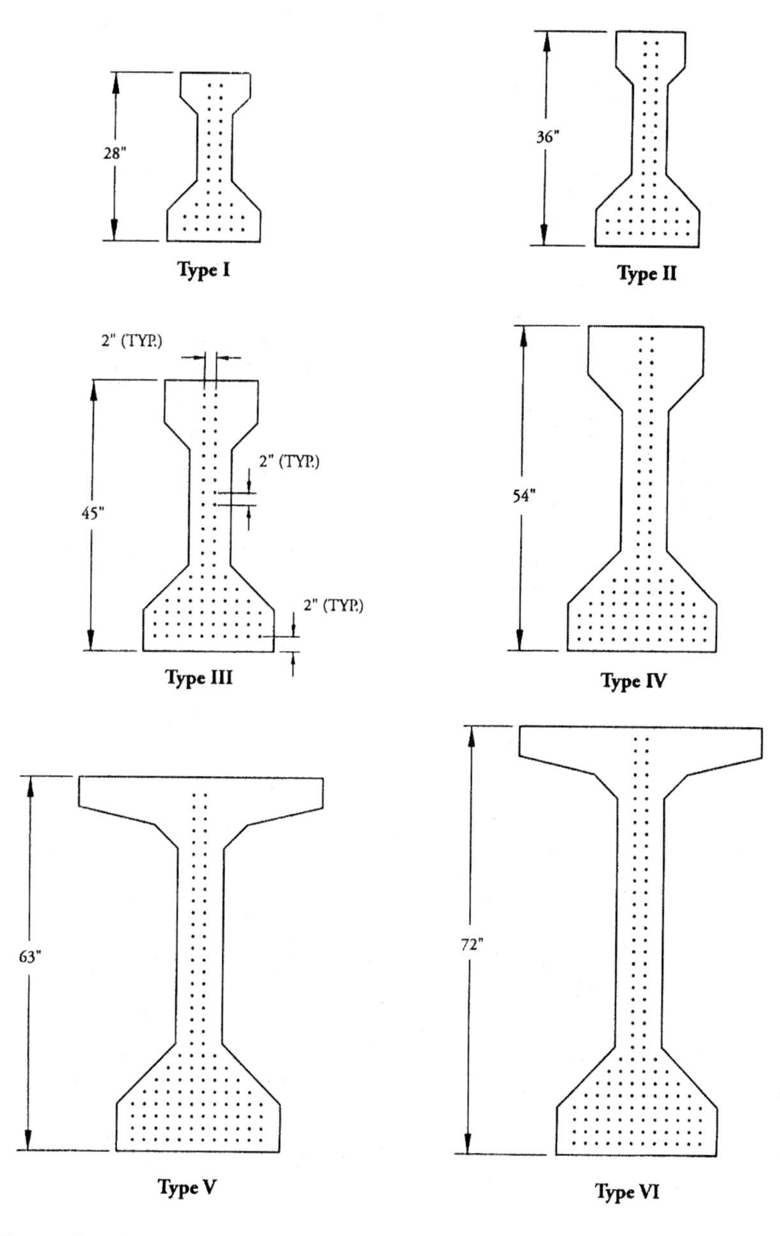

Figure C–6 (b) Possible Strand Arrangement for Sections in Figure C-6 (a)

AASHTO-PCI Bulb-Tees

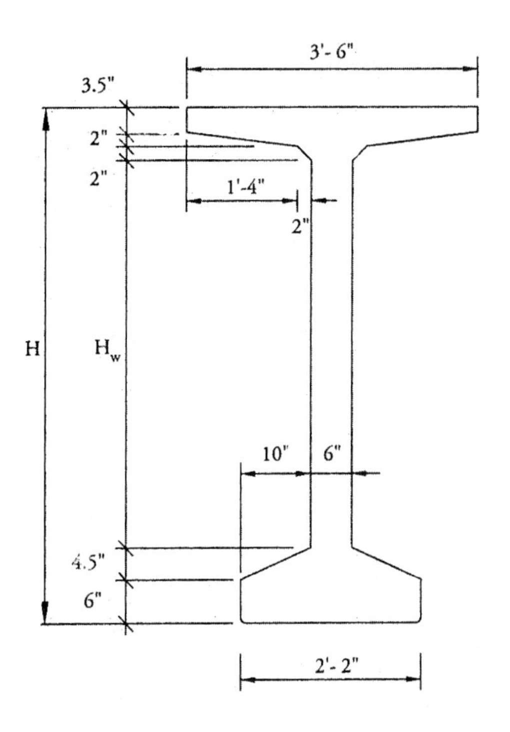

Properties

Type	H in.	H_w in.	Area in.^{2}	Inertia in.^{4}	y_{bottom} in.	Weight kip/ft	Maximum Span,* ft
BT-54	54	36	659	268,077	27.63	0.686	114
BT-63	63	45	713	392,638	32.12	0.743	130
BT-72	72	54	767	545,894	36.60	0.799	146

*Based on simple span, HS-25 loading and f'_c = 7,000 psi.

Figure C–7 (a) AASHTO/PCI Bulb Tees (Courtesy PCI, Ref. 12.11)

AASHTO-PCI Bulb-Tees

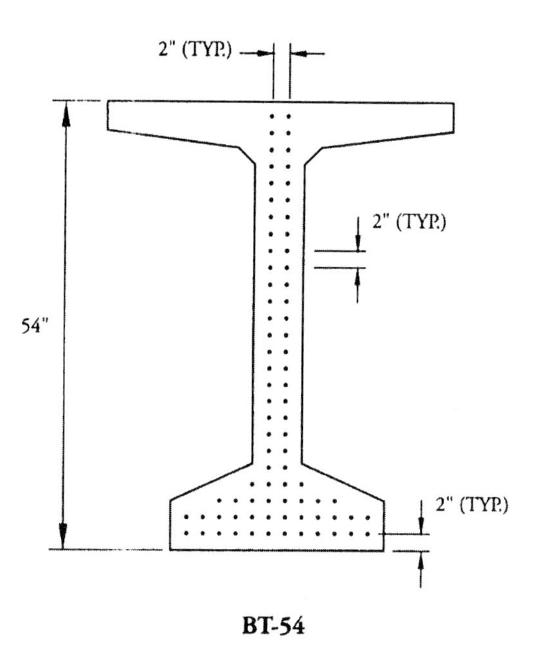

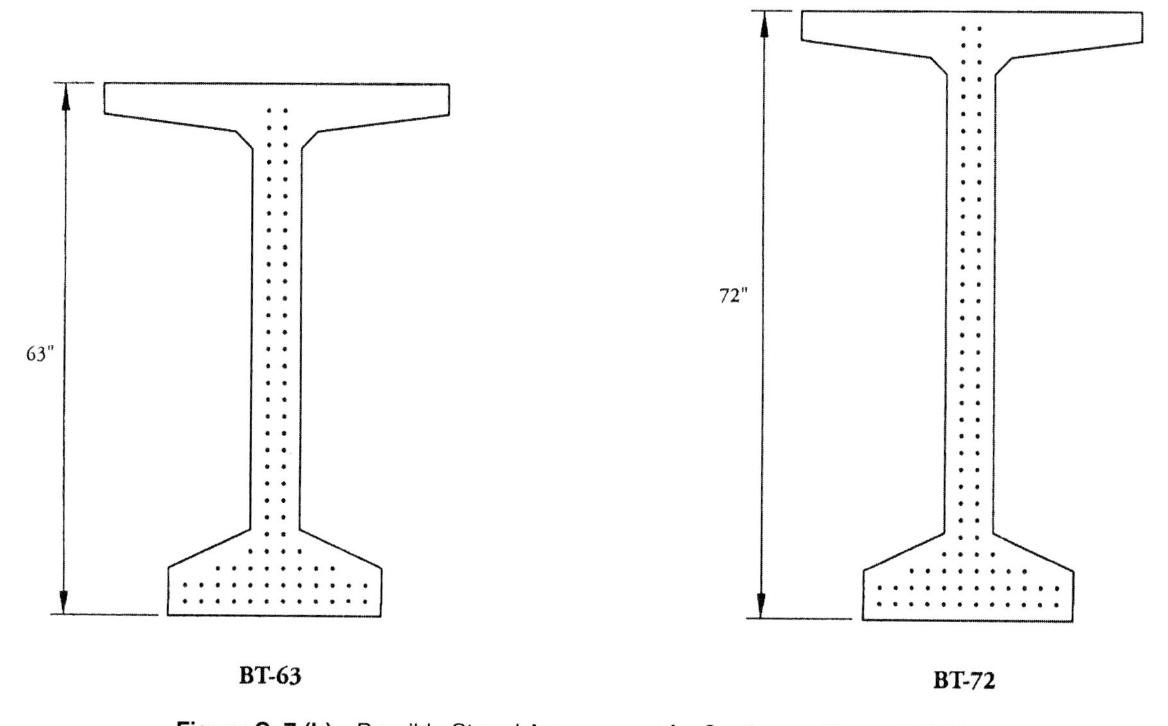

Figure C–7 (b) Possible Strand Arrangement for Sections in Figure C–7 (a)

Deck Bulb-Tees

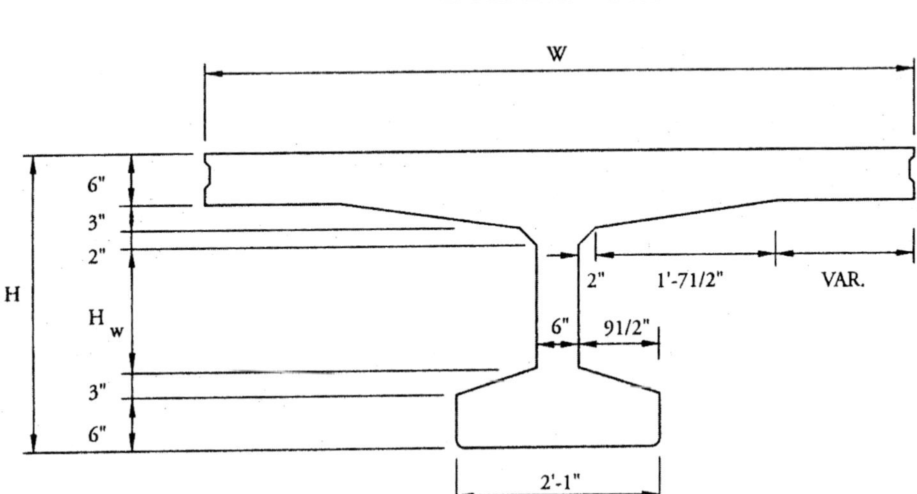

Dimensions and Properties

H in.	H_w in.	W in.	Area in.^{2}	Inertia in.^{4}	y_{bottom} in.	Weight kip/ft	Maximum Span* ft
35	15	48	677	101,540	21.12	0.75	100
		72	823	116,071	23.04	0.91	78
		96	967	126,353	24.37	1.07	65
53	33	48	785	294,350	31.71	0.87	145
		72	931	335,679	34.56	1.03	121
		96	1,075	365,827	36.63	1.19	105
65	45	48	857	490,755	38.55	0.95	168
		72	1,003	559,367	41.95	1.11	148
		96	1,147	610,435	44.46	1.27	130

*Based on simple span, HS-25 loading and f'_c = 7,000 psi.

Figure C–8 (a) AASHTO/PCI Shallow Bridge Deck Bulb Tees (Courtesy PCI, Ref. 12.11)

Deck Bulb-Tees

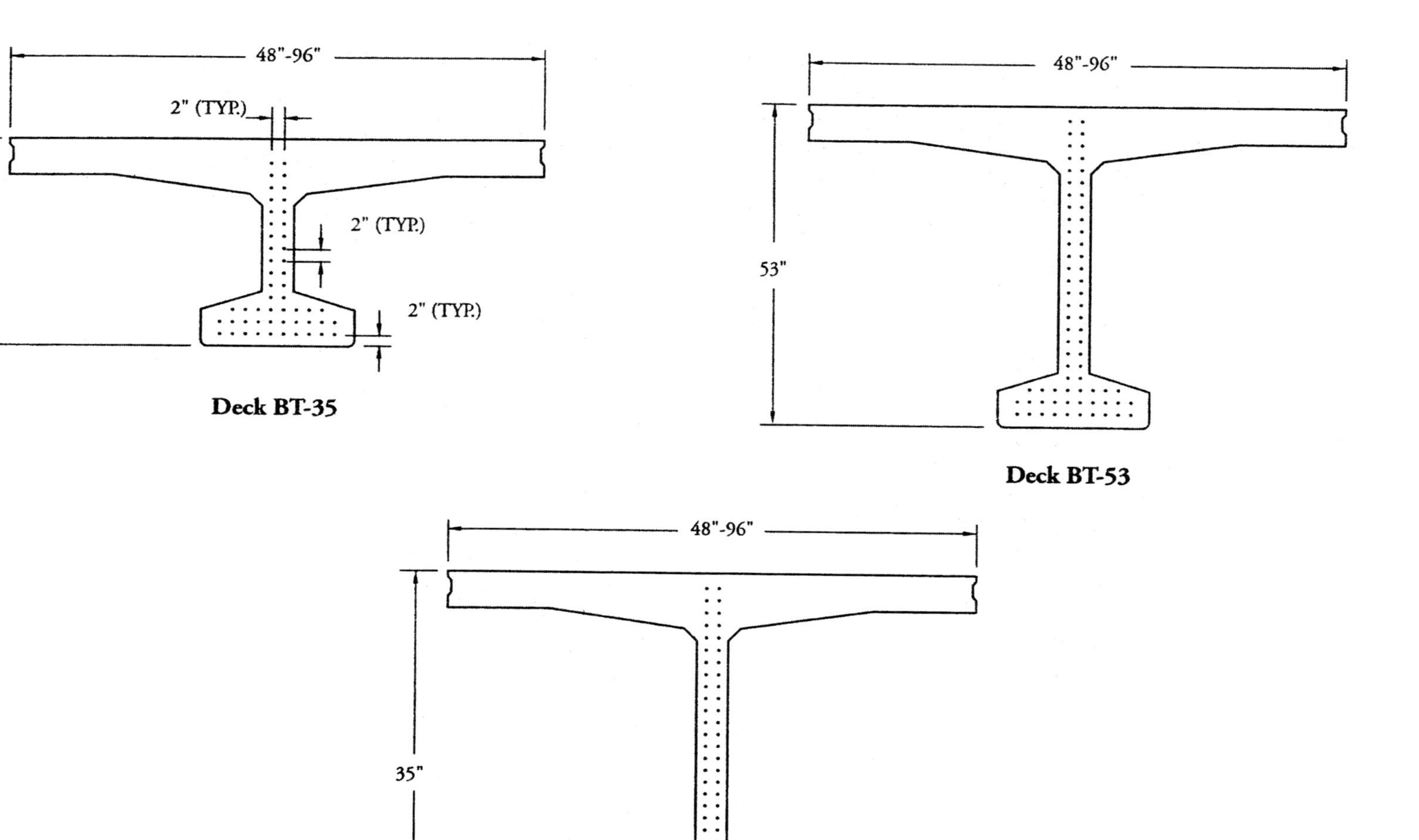

Figure C–8 (b) Possible Strand Arrangement for Sections in Figure C–8 (a)

AASHTO Box Beams

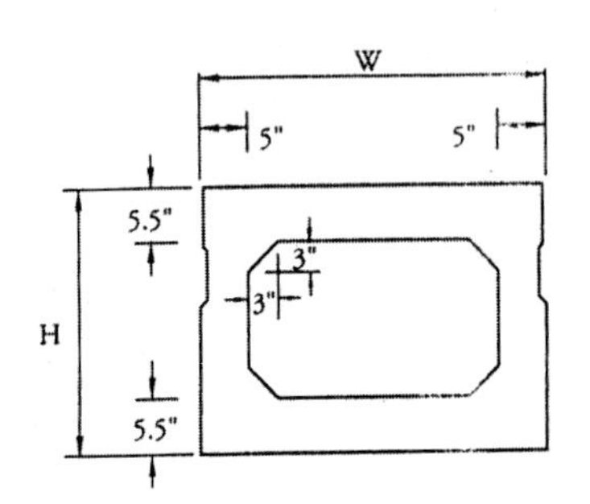

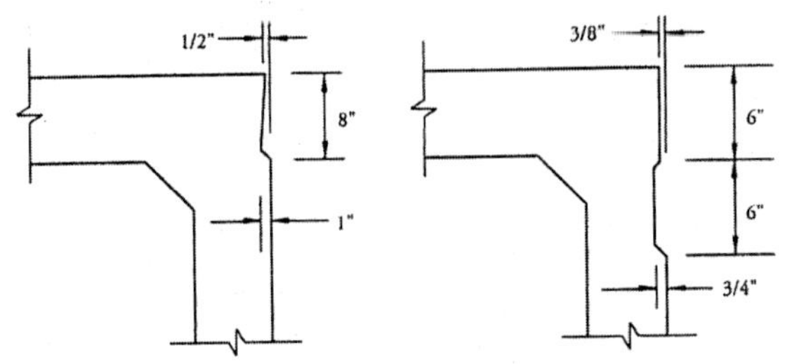

Typical Keyway Details

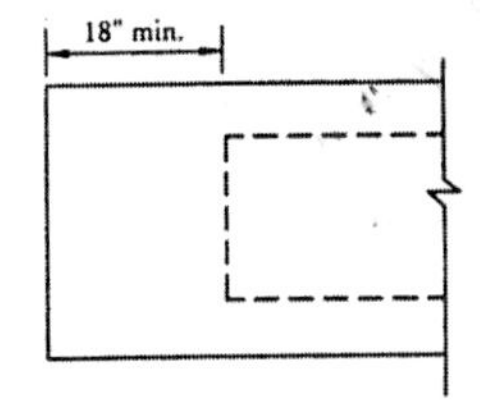

Typical Longitudinal Section

Dimensions (inches)

Type	W	H
BI-36	36	27
BI-48	48	27
BII-36	36	33
BII-48	48	33
BIII-36	36	39
BIII-48	48	39
BIV-36	36	42
BIV-48	48	42

Properties

Type	Area in.^{2}	y_{bottom} in.	Inertia in.^{4}	Weight kip/ft	Max. Span* ft
BI-36	560.5	13.35	50,334	0.584	92
BI-48	692.5	13.37	65,941	0.721	92
BII-36	620.5	16.29	85,153	0.646	107
BII-48	752.5	16.33	110,499	0.784	108
BIII-36	680.5	19.25	131,145	0.709	120
BIII-48	812.5	19.29	168,367	0.846	125
BIV-36	710.5	20.73	158,644	0.740	124
BIV-48	842.5	20.78	203,088	0.878	127

*Based on simple span, HS-25 loading and f'_c = 7,000 psi.

Figure C–9 (a) AASHTO/PCI Bridge Box Girders (Courtesy PCI, Ref. 12.11)

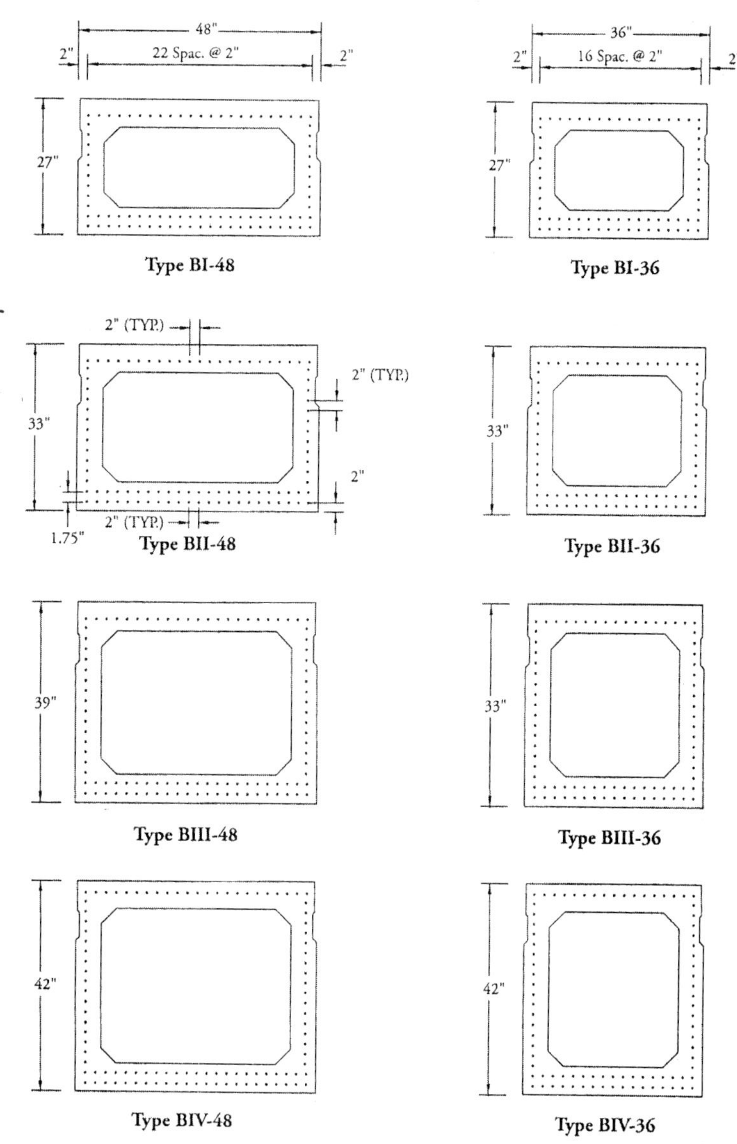

Figure C–9 (b) Possible Strand Arrangement for Sections in Figure C–9 (a)

INDEX

V_i = factored shear force at section due to externally applied loads occurring simultaneously with M_{max}.

V_n = nominal shear strength.

w_u = factored load per unit length of beam or per unit area of slab.

x = shorter overall dimension of rectangular part of cross section.

x_1 = shorter center-to-center dimension of closed rectangular stirrup.

y = longer overall dimension of rectangular part of cross section.

y_t = distance from centroidal axis of gross section, neglecting reinforcement, to extreme fiber in tension.

y_1 = longer center-to-center dimension of closed rectangular stirrup.

α = total angular change of prestressing tendon profile in radians from tendon jacking end to any point x.

α = ratio of flexural stiffness of beam section to flexural stiffness of a width of slab bounded laterally by centerlines of adjacent panels (if any) on each side of the beam.

$= \dfrac{E_{cb}I_b}{E_{ca}I_s}$

α_m = average value of α for all beams on edges of a panel.

β_a = ratio of dead load per unit area to live load per unit area (in each case without load factors).

β_d = ratio of maximum factored dead load moment to maximum factored total load moment, always positive.

β = a ratio of clear spans in long to short direction of two-way slabs.

γ_f = fraction of unbalanced moment transferred by flexure at slab-column connections.

γ_p = factor for type of prestressing tendon.

= 0.55 for f_{py}/f_{pu} not less than 0.80

= 0.40 for f_{py}/f_{pu} not less than 0.85

= 0.28 for f_{py}/f_{pu} not less than 0.90

γ_v = fraction of unbalanced moment transferred by eccentricity of shear at slab-column connections.

$= 1 - \gamma_f$

δ_{ns} = moment magnification factor for frames braced against sidesway, to reflect effects of member curvature between ends of compression member.

δ_s = moment magnification factor for frames not braced against sidesway, to reflect lateral drift resulting from lateral and gravity loads.

μ = curvature friction coefficient.

$\xi_{(xi)}$ = time-dependent factor for sustained load.

$\rho_{(rho)}$ = ratio of nonprestressed tension reinforcement.

$= A_s/bd$

ρ' = ratio of nonprestressed compression reinforcement

$= A'_s/bd$

ρ_b = reinforcement ratio producing balanced strain conditions.

ρ_p = ratio of prestressed reinforcement.

$= A_{ps}/bd_p$

ρ_v = A_{sv}/A_{cv}; where A_{sv} is the projection on A_{cv} of area of distributed shear reinforcement crossing the plane of A_{cv}.

θ = angle of compression diagonals in truss analogy for torsion.

ϕ = strength reduction factor.

ω = pf_y/f'_c.

ω' = $p'f_y/f'_c$.

ω_p = $\rho_p f_{pe}/f'_c$.

$\omega_{pw}, \omega_w, \omega'_w$

= reinforcement indices for flanged sections computed as for ω, ω_p, and ω' except that b shall be the web width, and reinforcement area shall be that required to develop compressive strength of web only.